CONCEPTS AND APPLICATIONS OF FINITE ELEMENT ANALYSIS

CONCEPTS AND APPLICATIONS OF FINITE ELEMENT ANALYSIS

THIRD EDITION

ROBERT D. COOK
DAVID S. MALKUS
MICHAEL E. PLESHA

University of Wisconsin—Madison

WILEY

JOHN WILEY & SONS

New York Chichester Brisbane Toronto Singapore

Library of Congress Cataloging in Publication Data:

Cook, Robert Davis.
Concepts and applications of finite element analysis.

Bibliography: p.
Includes index.
1. Structural analysis (Engineering) 2. Finite
element method. I. Malkus, David S. II. Plesha,
Michael E. III. Title.
TA646.C66 1989 624.1'71 88-27929
ISBN 0-471-84788-7

Printed in the United States of America

Printed and bound by the Hamilton Printing Company.

10

About the Authors

Robert D. Cook received his Ph.D. degree from the University of Illinois in 1963. He then went to the University of Wisconsin—Madison, where he is Professor of Engineering Mechanics. His research interests include stress analysis and finite element methods. He is a member of the American Society of Mechanical Engineers. With Warren C. Young, he is coauthor of *Advanced Mechanics of Materials* (Macmillan, 1985).

The first edition of *Concepts and Applications of Finite Element Analysis* was published in 1974 and the second in 1981, both with Dr. Cook as sole author.

David S. Malkus received his Ph.D. from Boston University in 1976. He spent two years at the National Bureau of Standards and seven years in the Mathematics Department of Illinois Institute of Technology. He is now Professor of Engineering Mechanics and a professor in the Center for Mathematical Sciences at the University of Wisconsin—Madison. His research interests concern the application of the finite element method to problems of structural and continuum mechanics, in particular the flow of non-Newtonian fluids. He is a member of the Rheology Research Center (University of Wisconsin—Madison), the American Academy of Mechanics, the Society for Industrial and Applied Mathematics, and the Society of Rheology.

Michael E. Plesha received his B.S. degree from the University of Illinois at Chicago, and his M.S. and Ph.D. degrees from Northwestern University, the Ph.D. degree in 1983. After a short stay at Michigan Technological University, he joined the Engineering Mechanics Department at the University of Wisconsin—Madison, where he is an associate professor. His research interests include constitutive modeling and finite element analysis of contact-friction problems, transient finite element analysis, and geomechanics.

PREFACE

The finite element method is firmly established as a powerful and popular analysis tool. It is applied to many different problems of continua but is most widely used for structural mechanics. Accordingly, structural mechanics is emphasized in this book, with lesser excursions into other areas such as heat conduction.

The finite element literature is very large. In a book this size it would scarcely be possible even to list all publications, let alone discuss all useful procedures. This text is introductory and is oriented more toward the eventual practitioner than toward the theoretician. The book contains enough material for a two-semester course.

We assume that the reader has the following background. Undergraduate courses in calculus, statics, dynamics, and mechanics of materials must be mastered. Matrix operations (summarized in Appendix A) must be understood. More advanced studies—theory of elasticity, energy methods, numerical analysis, and so on—are not essential. Occasionally these studies must be called upon, but only for their elementary concepts.

The specific elements discussed are often quite good, but we do not claim that they are the best available. Rather, these elements illustrate useful concepts and procedures. Similarly, blocks of Fortran code in the book illustrate the steps of an element formulation, of an algorithm for equation solving, or of finite element bookkeeping, but they may not be the most efficient coding available. These blocks of code can form the basis of various semester projects if so desired. However, the principal purpose of most of these blocks of code is to state precisely the content of certain procedures, and they thereby serve as aids to understanding.

Software entitled FEMCOD is intended for use with the book. FEMCOD is a "framework" program for time-independent finite element analysis: it provides the machinery for input of data, assembly of elements, assignment of loads and boundary conditions, and solution of equations. The user may supply coding for a particular element and for postprocessing (such as stress calculation). To institutions that adopt this textbook, FEMCOD, with instructions for use and examples, is available on diskette from the publisher (John Wiley & Sons, Inc., 605 Third Avenue, New York, N.Y., 10158).

Our presentation of structural dynamics is based partially on the finite element course notes of Ted Belytschko. We gratefully acknowledge his advice and assistance. The inspiration for the discussion of optimal lumping came originally from Isaac Fried. We are also grateful to T. J. R. Hughes, W. K. Liu, and V. Snyder for their insights. Not the least of our thanks is to Beth Brown, who typed and retyped with her usual intelligence and dependability, despite substantial other commitments, and without ever suggesting that the task might be tiresome.

Madison, Wisconsin
October 1988

R. D. Cook
D. S. Malkus
M. E. Plesha

CONTENTS

NOTATION

What follows is a list of principal symbols. Less frequently used symbols, and symbols that have different meanings in different contexts, are defined where they are used. Matrices and vectors are denoted by boldface type.

MATHEMATICAL SYMBOLS

$[\ ]$	Rectangular or square matrix.
$\{\ \}$, $\lfloor\ \rfloor$, $\lceil\ \rfloor$	Column, row, and diagonal matrices.
$[\ ]^T$	Matrix transpose.
$[\ ]^{-1}$, $[\ ]^{-T}$	Matrix inverse and inverse transpose; that is, $([\ ]^{-1})^T \equiv ([\ ]^T)^{-1}$.
$\Vert\ \Vert$	Norm of a matrix or a vector.
$\cdot$	Time differentiation; for example, $\dot{u} = du/dt$, $\ddot{u} = d^2u/dt^2$.
,	Partial differentiation if the following subscript(s) is literal; for example, $w_{,x} = \partial w/\partial x$, $w_{,xy} = \partial^2 w/\partial x\, \partial y$.
$\left\{\dfrac{\partial \Pi}{\partial \mathbf{a}}\right\}$	Represents $\left\lfloor \dfrac{\partial \Pi}{\partial a_1}\ \dfrac{\partial \Pi}{\partial a_2}\ \cdots\ \dfrac{\partial \Pi}{\partial a_n} \right\rfloor^T$, where Π is a scalar function of $a_1, a_2, \ldots, a_n$.

LATIN SYMBOLS

A	Area or cross-sectional area.
$[\mathbf{A}]$	Relates $\{\mathbf{d}\}$ to $\{\mathbf{a}\}$; $\{\mathbf{d}\} = [\mathbf{A}]\{\mathbf{a}\}$.
$\{\mathbf{a}\}$	Generalized coordinates.
B	Bulk modulus, $B = E/(3 - 6\nu)$.
$[\mathbf{B}]$	Spatial derivative(s) of the field variable(s) are $[\mathbf{B}]\{\mathbf{d}\}$.
C^m	Field continuity of degree m (Section 3.11).
$[\mathbf{C}]$	Damping matrix. Constraint matrix.
d.o.f.	Degree(s) of freedom.
D	Displacement. Flexural rigidity of a plate or shell.
$\{\mathbf{D}\}$, $\{\mathbf{d}\}$	Nodal d.o.f. of structure and element, respectively.
E	Modulus of elasticity.
$[\mathbf{E}]$	Matrix of elastic stiffnesses (Section 1.7).
$\{\mathbf{F}\}$	Body forces per unit volume.
G	Shear modulus.
I	Moment of inertia of cross-sectional area.
$[\mathbf{I}]$	Unit matrix (also called identity matrix).
J	Determinant of $[\mathbf{J}]$ (called the Jacobian).
$[\mathbf{J}]$	The Jacobian matrix.
k	Spring stiffness. Thermal conductivity.
$[\mathbf{K}]$, $[\mathbf{k}]$	Structure and element conventional stiffness matrices.
$[\mathbf{K}_\sigma]$, $[\mathbf{k}_\sigma]$	Structure and element stress stiffness matrices.
L, L_T	Length of element, length of structure.
ℓ, m, n	Direction cosines.
n_{eq}	Number of equations.
$[\mathbf{M}]$, $[\mathbf{m}]$	Structure and element mass matrices.

$[\mathbf{N}]$, $\lfloor \mathbf{N} \rfloor$	Shape (or basis, or interpolation) functions.
O	Order; for example, $O(h^2)$ = a term of order h^2.
$[\mathbf{0}]$, $\{\mathbf{0}\}$	Null matrix, null vector.
$\{\mathbf{P}\}$	Externally applied concentrated loads on structure nodes.
q	Distributed load (surface or line).
$\{\mathbf{R}\}$	Total load on structure nodes; $\{\mathbf{R}\} = \{\mathbf{P}\} + \Sigma \{\mathbf{r}_e\}$.
$\{\mathbf{r}_e\}$	Loads applied to nodes by element, for example, by temperature change or distributed load (Eq. 4.1-6).
S, S_e	Surface, element surface.
T	Temperature.
t	Thickness. Time.
$[\mathbf{T}]$	Transformation matrix.
U, U_0	Strain energy, strain energy per unit volume.
u, v, w	Displacements, for example, in directions x, y, z.
$\{\mathbf{u}\}$	Vector of displacements; $\{\mathbf{u}\} = [u \quad v \quad w]^T$.
V, V_e	Volume, element volume.
x, y, z	Cartesian coordinates.

GREEK SYMBOLS

α	Coefficient of thermal expansion, penalty number.
$[\mathbf{\Gamma}]$	Jacobian inverse; $[\mathbf{\Gamma}] = [\mathbf{J}]^{-1}$.
$\{\boldsymbol{\epsilon}\}$, $\{\boldsymbol{\epsilon}_0\}$	Strains, initial strains.
$[\boldsymbol{\kappa}]$, $\{\boldsymbol{\kappa}\}$	Matrix of thermal conductivities, vector of curvatures.
λ	Eigenvalue. Lagrange multiplier.
ν	Poisson's ratio of an isotropic material.
ξ, η, ζ	Isoparametric coordinates.
ξ_1, ξ_2, ξ_3	Area coordinates.
Π	A functional; for example, Π_p = potential energy.
ρ	Mass density.
$\{\boldsymbol{\sigma}\}$, $\{\boldsymbol{\sigma}_0\}$	Stresses, initial stresses.
ϕ	A dependent variable. Meridian angle of a shell.
$\{\mathbf{\Phi}\}$	Surface tractions.
ω	Circular frequency in radians per second.

CHAPTER 1

INTRODUCTION

A brief overview of the finite element method and its concepts is presented. Background information used for finite element applications in structural mechanics is discussed.

1.1 THE FINITE ELEMENT METHOD

The finite element method is a numerical procedure for analyzing structures and continua. Usually the problem addressed is too complicated to be solved satisfactorily by classical analytical methods. The problem may concern stress analysis, heat conduction, or any of several other areas. The finite element procedure produces many simultaneous algebraic equations, which are generated and solved on a digital computer. Finite element calculations are performed on personal computers, mainframes, and all sizes in between. Results are rarely exact. However, errors are decreased by processing more equations, and results accurate enough for engineering purposes are obtainable at reasonable cost.

The finite element method originated as a method of stress analysis. Today finite elements are also used to analyze problems of heat transfer, fluid flow, lubrication, electric and magnetic fields, and many others. Problems that previously were utterly intractable are now solved routinely. Finite element procedures are used in the design of buildings, electric motors, heat engines, ships, airframes, and spacecraft. Manufacturing companies and large design offices typically have one or more large finite element programs in-house. Smaller companies usually have access to a large program through a commercial computing center or use a smaller program on a personal computer.

Figure 1.1-1 shows a very simple problem that illustrates *discretization,* a basic finite element concept. Imagine that the displacement of the right end of the bar is required. The classical approach is to write the differential equation of the continuously tapered bar, solve this equation for axial displacement u as a function of x, and finally substitute $x = L_T$ to find the required end displacement. The finite element approach to this problem does not begin with a differential equation. Instead, the bar is *discretized* by modeling it as a series of *finite elements,* each uniform but of a different cross-sectional area A (Fig. 1.1-1*b*). In each element, u varies linearly with x; therefore, for $0 < x < L_T$, u is a piecewise-smooth function of x. The elongation of each element can be determined from the elementary formula PL/AE. The end displacement, at $x = L_T$, is the sum of the element elongations. Accuracy improves as more elements are used.

In the foregoing example, and in general, the finite element method models a structure as an assemblage of small parts (elements). Each element is of simple geometry and therefore is much easier to analyze than the actual structure. In

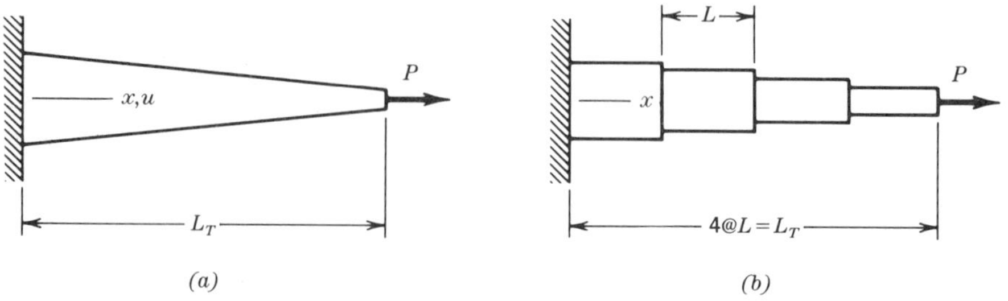

Figure 1.1-1. (*a*) A tapered bar under end load *P*. (*b*) A model built of four uniform (nontapered) elements of equal length.

essence, we approximate a complicated solution by a model that consists of piecewise-continuous simple solutions. Elements are called "finite" to distinguish them from differential elements used in calculus.

In a heat transfer context, Fig. 1.1-1 might represent a bar with insulated sides, prescribed temperature at the left end, and prescribed heat flow at the right end. One might ask for the temperature in the bar as a function of x and time.

Figure 1.1-2*a* shows a plane structure. Displacements and stresses caused by pressure p are required. The finite element model, Fig. 1.1-2*b*, consists of plane areas, some triangular and some quadrilateral (if done properly, there is no difficulty in combining the different element types). Black dots, called *nodes* or *node points,* indicate where elements are connected to one another. In this model each node has two degrees of freedom (d.o.f.): that is, each node can displace in both the x direction and the y direction. Thus, if there are n nodes in Fig. 1.1-2*b*, there are $2n$ d.o.f. in the model. (In the real structure there are infinitely many d.o.f. because the structure has infinitely many particles.) Algebraic equations that describe the finite element model are solved to determine the d.o.f. Use of only $2n$ d.o.f. in analysis is similar to use of the first $2n$ terms of a convergent infinite series. (In heat transfer, each node has only one d.o.f.—namely, the temperature of the node. Thus a finite element model of n nodes has n d.o.f.)

We see that in going from Fig. 1.1-2*a* to 1.1-2*b* the distributed pressure p has been converted to concentrated forces at nodes. The analysis procedure gives a prescription for making conversion, as will be shown subsequently.

From Fig. 1.1-2 it may appear that discretization is accomplished simply by

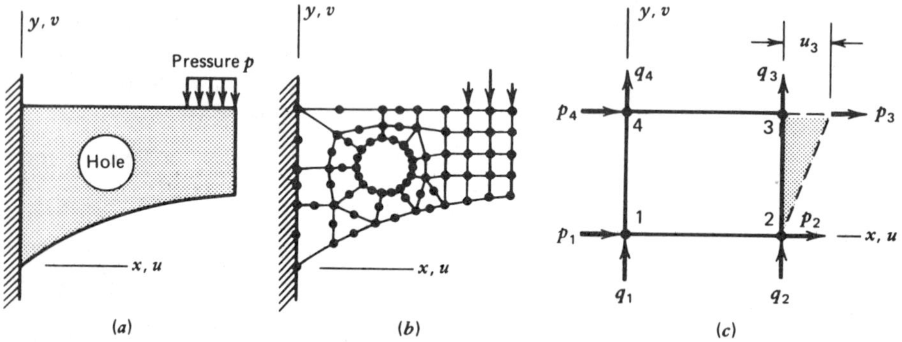

Figure 1.1-2. (*a*) A plane structure of arbitrary shape. (*b*) A possible finite element model of the structure. (*c*) A plane rectangular element showing nodal forces p_i and q_i. The dashed line shows the deformation mode associated with x-direction displacement of node 3.

sawing the continuum into pieces and then pinning the pieces together again at node points. But such a model would not deform like the continuum. Under load, strain concentrations would appear at the nodes, and the elements would tend to overlap or separate along the saw cuts. Clearly, the actual structure does not behave in this way, so the elements must be restricted in their deformation patterns. For example, if elements are allowed to have only such deformation modes as will keep edges straight (Fig. 1.1-2*c*), then adjacent elements will neither overlap nor separate. In this way we satisfy the basic requirement that deformations of a continuous medium must be *compatible*.

An important ingredient in a finite element analysis is the behavior of the individual elements. A few good elements may produce better results than many poorer elements. We can see that several element types are possible by considering Fig. 1.1-3. Function ϕ, which might represent any of several physical quantities, varies smoothly in the actual structure. A finite element model typically yields a *piecewise*-smooth representation of ϕ. Between elements there may be jumps in the x and y derivatives of ϕ. Within each element ϕ is a smooth function that is usually represented by a simple polynomial. What shall the polynomial be? For the triangular element, the linear polynomial

$$\phi = a_1 + a_2x + a_3y \tag{1.1-1}$$

is appropriate, where the a_i are constants. These constants can be expressed in terms of ϕ_1, ϕ_2, and ϕ_3, which are the values of ϕ at the three nodes. Triangles model the actual ϕ by a surface of flat triangular facets. For the four-node quadrilateral, the "bilinear" function

$$\phi = a_1 + a_2x + a_3y + a_4xy \tag{1.1-2}$$

is appropriate. The eight-node quadrilateral in Fig. 1.1-3 has eight a_i in its polynomial expansion and can represent a parabolic surface.

Equations 1.1-1 and 1.1-2 are *interpolations* of function ϕ in terms of the position (x,y) within an element. That is, when the a_i have been determined in terms of nodal values ϕ_i, Eqs. 1.1-1 and 1.1-2 define ϕ within an element in terms of the ϕ_i and the coordinates. Clearly, if the mesh of elements is not too coarse and if the ϕ_i happened to be exact, then ϕ away from nodes would be a good approx-

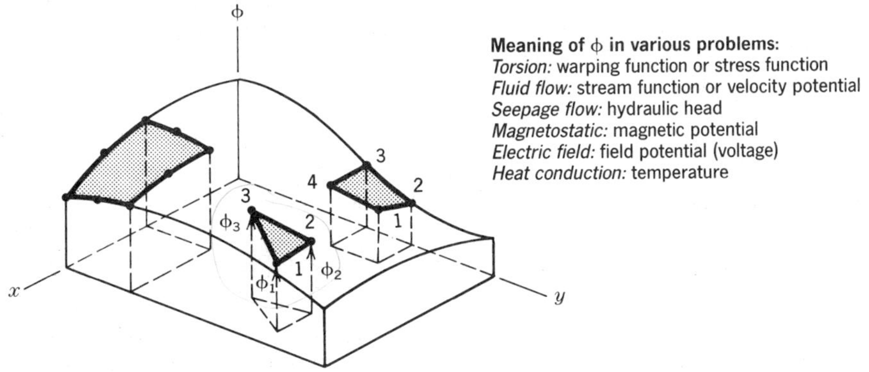
φ
x
y
3
4
2
1
3
2
1
φ3
φ1
φ2
Meaning of φ in various problems:
Torsion: warping function or stress function
Fluid flow: stream function or velocity potential
Seepage flow: hydraulic head
Magnetostatic: magnetic potential
Electric field: field potential (voltage)
Heat conduction: temperature

Figure 1.1-3. A function $\phi = \phi(x,y)$ that varies smoothly over a rectangular region in the xy plane, and typical elements that might be used to approximate it.

imation. Nodal values ϕ_i are *close* to exact if the mesh is not too coarse and if element properties are properly formulated.

How can the user decide which element to use? Unfortunately, the answer is not simple. An element that is good in one problem area (such as magnetic fields) may be poor in another (such as stress analysis). Even in a specific problem area an element may behave well or badly, depending on the particular geometry, loading, and boundary conditions. A competent user of finite elements must be familiar with how various elements behave under various conditions.

We may now venture some definitions. The *finite element method* is a method of piecewise approximation in which the approximating function ϕ is formed by connecting simple functions, each defined over a small region (element). A *finite element* is a region in space in which a function ϕ is interpolated from nodal values of ϕ on the boundary of the region in such a way that interelement continuity of ϕ tends to be maintained in the assemblage.

A finite element analysis typically involves the following steps. Again we will cite stress anaylsis and heat transfer as typical applications. Steps 1, 4, and 5 require decisions by the analyst and provide input data for the computer program. Steps 2, 3, 6, and 7 are carried out automatically by the computer program.

1. Divide the structure or continuum into finite elements. Mesh generation programs, called preprocessors, help the user in doing this work.
2. Formulate the properties of each element. In stress analysis, this means determining nodal loads associated with all element deformation states that are allowed. In heat transfer, it means determining nodal heat fluxes associated with all element temperature fields that are allowed.
3. Assemble elements to obtain the finite element model of the structure.
4. Apply the known loads: nodal forces and/or moments in stress analysis, nodal heat fluxes in heat transfer.
5. In stress analysis, specify how the structure is supported. This step involves setting several nodal displacements to known values (which often are zero). In heat transfer, where typically certain temperatures are known, impose all known values of nodal temperature.
6. Solve simultaneous linear algebraic equations to determine nodal d.o.f. (nodal displacements in stress analysis, nodal temperatures in heat transfer).
7. In stress analysis, calculate element strains from the nodal d.o.f. and the element displacement field interpolation, and finally calculate stresses from strains. In heat transfer, calculate element heat fluxes from the nodal temperatures and the element temperature field interpolation. Output interpretation programs, called postprocessors, help the user sort the output and display it in graphical form.

The power of the finite element method resides principally in its versatility. The method can be applied to various physical problems. The body analyzed can have arbitrary shape, loads, and support conditions. The mesh can mix elements of different types, shapes, and physical properties. This great versatility is contained within a single computer program. User-prepared input data controls the selection of problem type, geometry, boundary conditions, element selection, and so on.

Another attractive feature of finite elements is the close physical resemblance between the actual structure and its finite element model. The model is not simply an abstraction. This seems especially true in structural mechanics, and may account for the finite element method having its origins there.

The finite element method also has disadvantages. A specific numerical result is found for a specific problem: a finite element analysis provides no closed-form solution that permits analytical study of the effects of changing various parameters. A computer, a reliable program, and intelligent use are essential. A general-purpose program has extensive documentation, which cannot be ignored. Experience and good engineering judgment are needed in order to define a good model. Many input data are required and voluminous output must be sorted and understood.

Example Applications. Figure 1.1-4 shows a finite element model of an axially symmetric rocket nozzle [10.1].[1] The axis, not shown, is horizontal and lies above the cross section in Fig. 1.1-4. Each element is a toroidal ring of triangular cross section. Each element has a node (actually a nodal circle) at each vertex. Each nodal circle has axial and radial displacements as d.o.f. Stresses caused by temperature gradient and internal pressure are desired.

Figure 1.1-5 shows three ways of modeling an arch dam using "isoparametric" solid elements (discussed in Chapter 6). One might ask for the stresses produced by hydrostatic and gravity loads. Or, the response to earthquake motion might be required, in which fluid-structure interaction is taken into account.

Figures 1.1-6 and 1.1-7 show typical problems in structural mechanics. The structure in Fig. 1.1-6 consists primarily of plate-bending elements. The structure in Fig. 1.1-7 consists of three-dimensional solid elements. The postprocessor has removed hidden lines. The deformation and stress plots display the results of analysis.

Figure 1.1-8 shows a nonstructural problem. Lines of magnetic flux are crowded

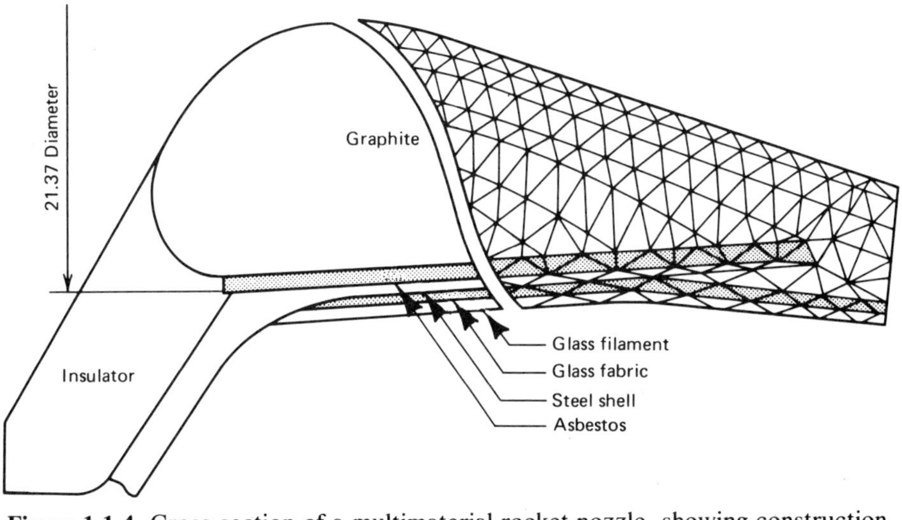

Figure 1.1-4. Cross section of a multimaterial rocket nozzle, showing construction (left portion) and possible finite element mesh (right portion). This problem was solved in the early days of finite element technology [10.1].

[1]Numbers within brackets indicate references listed at the back of the book.

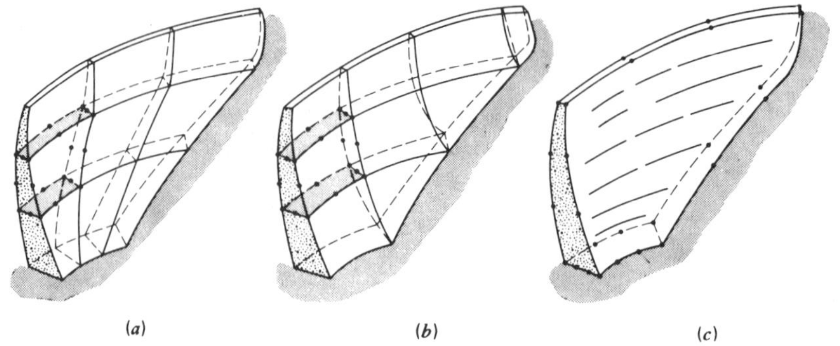

Figure 1.1-5. Half of an arch dam, modeled by (a,b) quadrilateral and triangular "quadratic" elements, and (c) a single "cubic" element [1.1]. Nodes of a typical element are shown by dots.

near the gap between rotor and stator, which means that gradients are large in this region. Areas of large gradient are areas of particular interest. The analyst places more elements there in order to calculate the magnetic field in greater detail. The mesh shown is adequate for the analysis of magnetic flux, but is probably too crude to be used for stress analysis.

Clearly some problems use a great many d.o.f. How many d.o.f. must a problem have to be considered "large"? In 1960, perhaps 1000; in 1980, over 10,000. Improvements in hardware and software have made this increase possible.

Why Study the Theory of Finite Elements? Many satisfactory elements have already been formulated and reside in popular computer programs. The practitioner desires to understand how various elements behave. Clearly, engineers who understand analysis tools will be able to use them to better advantage and will be less likely to *mis*use them. Such an understanding cannot be achieved if theory is ignored. In this book we intend to aid the eventual practitioner and to present theory to an adequate but not excessive degree. We recognize that for engineers the study of finite elements is more than a theoretical study of mathematical foundations and formulation procedures for various types of finite elements.

Complete computer codes need not be studied in detail, but concepts and assumptions behind the coding should be mastered. Otherwise, the treatment of loads and boundary conditions may be confusing, the variety of program options and element types may be baffling, and error messages may provide no clue as to the source of difficulty or how to correct it.

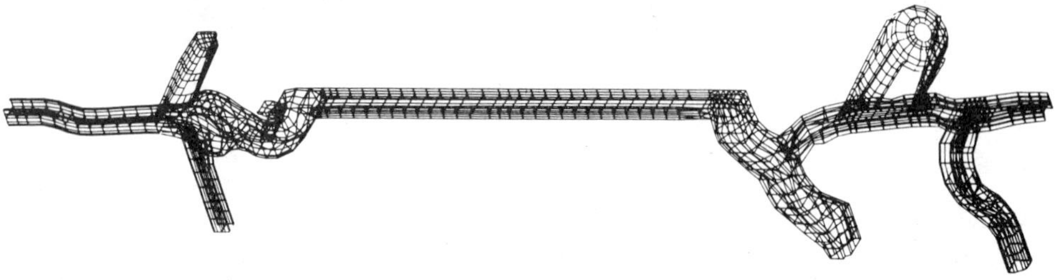

Figure 1.1-6. A detailed model of half of an automobile frame, used to find deformations, stresses, natural frequencies, and mode shapes. (*Courtesy of A. O. Smith Corp., Data Systems Division, Milwaukee, Wisconsin.*)

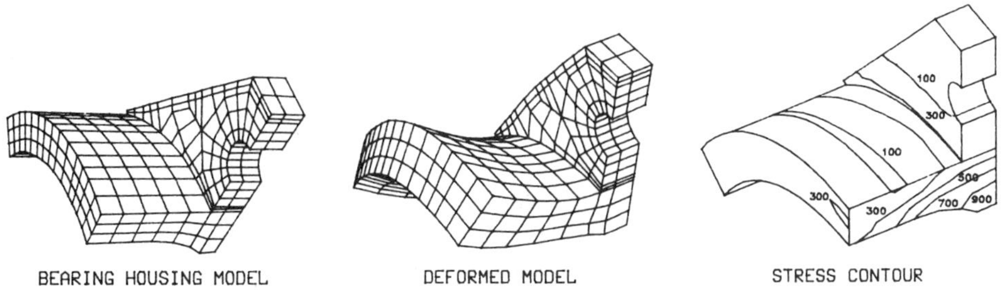

Figure 1.1-7. Finite element mesh and computed deformations and stresses in a portion of a bearing housing. (*Courtesy of Algor Interactive Systems Inc., Pittsburgh, Pennsylvania.*)

1.2 THE ELEMENT CHARACTERISTIC MATRIX

The element characteristic matrix has different names in different problem areas. In structural mechanics it is called a *stiffness matrix:* it relates nodal displacements to nodal forces. In heat conduction it is called a *conductivity matrix:* it relates nodal temperatures to nodal fluxes. There are three important ways to derive an element characteristic matrix.

1. *The direct method* is based on physical reasoning. It is limited to very simple elements, but is worth studying because it enhances our physical understanding of the finite element method.

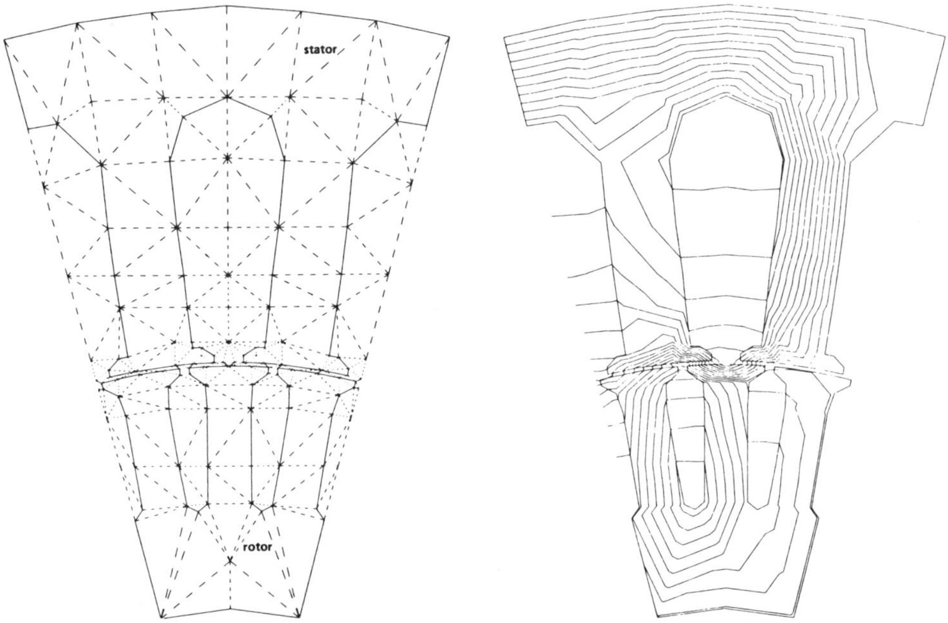

Figure 1.1-8. Part of an induction motor. Elements model the solid parts as well as the spaces between them. Symmetry is exploited by modeling only a half-pole. The computed magnetic flux contours for zero rotor speed are shown by the right-hand figure. (*Courtesy of A. O. Smith Corp., Data Systems Division, Milwaukee, Wisconsin.*)

2. *The variational method* is applicable to problems that can be stated by certain integral expressions such as the expression for potential energy. This method is discussed in Chapters 3 and 4.
3. *Weighted residual methods* are particularly suited to problems for which differential equations are known but no variational statement is available. For stress analysis and some other problem areas, the variational method and the most popular weighted residual method (the Galerkin method) yield identical finite element formulations. Weighted residual methods are discussed in Chapter 15.

In the present section we consider applications of the direct method.

The Elastic Bar: Direct Method. Consider a weightless straight bar of length L, elastic modulus E, and cross-sectional area A. We regard the bar as a finite element and place a node at each end. If only axial loads and axial displacements are allowed, nodal d.o.f. are displacements u_i and u_j (Fig. 1.2-1). The element stiffness matrix is formulated by determining the nodal forces that must be applied in order to produce nodal displacements u_i and u_j. Our sign convention is that both force and displacement are positive when directed toward the right. Accordingly, when $u_i > 0$ but $u_j = 0$ (Fig. 1.2-1*a*), nodal forces consistent with static equilibrium and a linearly elastic material are

$$F_i = \frac{AE}{L} u_i \qquad \text{and} \qquad F_j = -\frac{AE}{L} u_i \tag{1.2-1}$$

Similarly, when $u_i = 0$ but $u_j > 0$ (Fig. 1.2-1*b*),

$$F_i = -\frac{AE}{L} u_j \qquad \text{and} \qquad F_j = \frac{AE}{L} u_j \tag{1.2-2}$$

If both u_i and u_j can be simultaneously nonzero, then nodal forces are $F_i = (AE/L)(u_i - u_j)$ and $F_j = (AE/L)(-u_i + u_j)$. In matrix format these two equations are

$$\begin{bmatrix} AE/L & -AE/L \\ -AE/L & AE/L \end{bmatrix} \begin{Bmatrix} u_i \\ u_j \end{Bmatrix} = \begin{Bmatrix} F_i \\ F_j \end{Bmatrix} \tag{1.2-3}$$

or

$$[\mathbf{k}]\{\mathbf{d}\} = \{\bar{\mathbf{r}}\} \tag{1.2-4}$$

where [**k**] is the *element stiffness matrix*. By considering first $u_i = 1$ and $u_j = 0$, then $u_i = 0$ and $u_j = 1$, each time computing nodal forces F_i and F_j by the matrix-

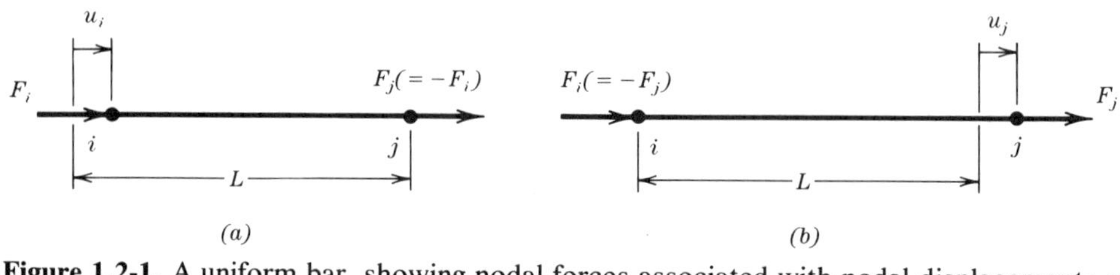

Figure 1.2-1. A uniform bar, showing nodal forces associated with nodal displacements u_i and u_j. Displacements u_i and u_j are greatly exaggerated in the drawing.

times-vector multiplication indicated in Eq. 1.2-3, we reach the following conclusion: a column of [**k**] lists nodal loads that must be applied to nodal d.o.f. in order to create the deformation state associated with unit value of the corresponding element d.o.f. while all other element d.o.f. are zero. Later in this section we will use this procedure in calculating [**k**] for a beam element, Fig. 1.2-2.

Remarks. No approximation was used in deriving the foregoing stiffness matrix. It is exact. Of course, if used to construct a stepwise model of a continuously tapered bar, as in Fig. 1.1-1, the stepwise model is inexact and computed displacements will differ from those of the actual structure.

The stiffness matrix is symmetric. This is always true when displacements are directly proportional to applied loads.

By trying to apply the direct method of element derivation to Fig. 1.1-2*c*, we can see that the direct method is limited to simple elements: physical reasoning, based on elementary mechanics of materials, does not quantify the eight nodal forces associated with a nodal displacement such as u_3.

Heat or Current Conduction. A uniform bar element for heat conduction analysis has an element characteristic matrix that resembles stiffness matrix [**k**] of the elastic bar. With s a coordinate along the bar, the Fourier heat conduction equation becomes

$$q = -k\frac{dT}{ds} = -k\frac{T_j - T_i}{L} \tag{1.2-5}$$

where q = heat flux per unit area, k = thermal conductivity, and T = temperature. If A is the cross-sectional area and nodal heat flux is considered to be positive when directed into the bar at either end, then the equation analogous to Eq. 1.2-3 is

$$\frac{Ak}{L}\begin{bmatrix} 1 & -1 \\ -1 & 1 \end{bmatrix}\begin{Bmatrix} T_i \\ T_j \end{Bmatrix} = \begin{Bmatrix} Aq_i \\ Aq_j \end{Bmatrix} \tag{1.2-6}$$

Similarly, if the bar is regarded as an electrical resistor, Ohm's law becomes $I = (V_i - V_j)/r$, where r is the resistance of the bar. In matrix form,

$$\begin{bmatrix} 1/r & -1/r \\ -1/r & 1/r \end{bmatrix}\begin{Bmatrix} V_i \\ V_j \end{Bmatrix} = \begin{Bmatrix} I_i \\ I_j \end{Bmatrix} \tag{1.2-7}$$

where V_i and V_j are nodal voltages. Nodal currents I_i and I_j are considered positive when flowing into the element.

One can envision networks of the foregoing heat conduction or current flow elements, in which a single node may be shared by several elements. If no external source supplies heat or current to such a node, the net flow into the node is zero.

The Elastic Beam: Direct Method. Consider a uniform beam that deforms in the plane of the paper and undergoes no axial deformation. This element has four d.o.f.: a lateral displacement w and a rotation θ at each end (Fig. 1.2-2). Nodal forces F_i and F_j correspond to nodal displacements w_i and w_j. Nodal moments M_i

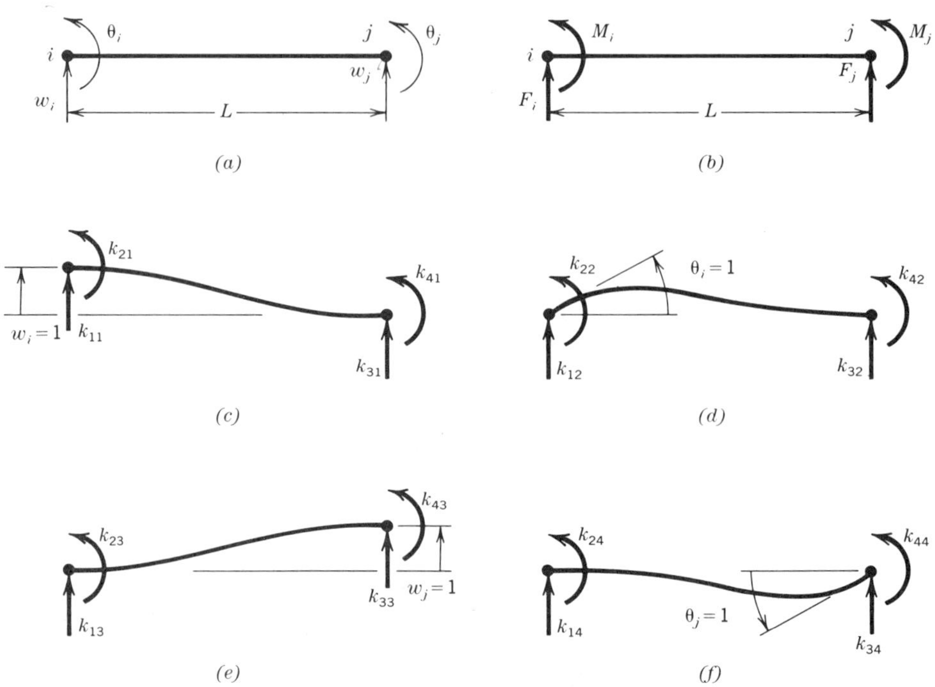

Figure 1.2-2. (*a*) A uniform beam element and its four nodal d.o.f. (*b*) Associated nodal forces and moments. (*c–e*) Deformation states associated with activation of each d.o.f. in turn, showing the required nodal forces and moments labeled according to their position in [**k**].

and M_j correspond to nodal rotations θ_i and θ_j. Positive directions of these quantities are shown in Figs. 1.2-2*a* and 1.2-2*b*.

The element equation is $[\mathbf{k}]\{\mathbf{d}\} = \{\bar{\mathbf{r}}\}$, where [**k**] is a 4 by 4 stiffness matrix for the present beam element. The element nodal displacement vector is

$$\{\mathbf{d}\} = \lfloor w_i \quad \theta_i \quad w_j \quad \theta_j \rfloor^T \tag{1.2-8}$$

(A different ordering of d.o.f. in {**d**} would change the ordering of coefficients in [**k**] but would not change their numerical values.) Vector $\{\bar{\mathbf{r}}\}$ contains nodal forces and moments applied to the element to maintain the deformation state {**d**}. Written out, the element stiffness equation is

$$\begin{bmatrix} k_{11} & k_{12} & k_{13} & k_{14} \\ k_{21} & k_{22} & k_{23} & k_{24} \\ k_{31} & k_{32} & k_{33} & k_{34} \\ k_{41} & k_{42} & k_{43} & k_{44} \end{bmatrix} \begin{Bmatrix} w_i \\ \theta_i \\ w_j \\ \theta_j \end{Bmatrix} = \begin{Bmatrix} F_i \\ M_i \\ F_j \\ M_j \end{Bmatrix} \tag{1.2-9}$$

We must express each stiffness coefficient in [**k**] in terms of element geometry and elastic modulus. To thus determine the stiffness coefficients in a single column of [**k**], we set the corresponding d.o.f. to unity while keeping all other d.o.f. zero, and calculate the values of F_i, M_i, F_j, and M_j needed in order to produce this deformation state. Thus, the first column of [**k**] is determined by activating only the first d.o.f. Specifically, for the case $w_i = 1$ and $\theta_i = w_j = \theta_j = 0$, we see from Eq. 1.2-9 that $k_{11} = F_i$, $k_{21} = M_i$, $k_{31} = F_j$, and $k_{41} = M_j$. These quantities are shown in Fig. 1.2-2*c*, all directed in the assumed positive sense. Let E = elastic modulus and I = moment of inertia of the cross-sectional area. We regard Fig. 1.2-2*c* as a beam cantilevered from its right end and loaded at its left end, and apply equations of beam theory and statics. Thus

$$w = 1 \text{ at node } i \qquad 1 = \frac{k_{11}L^3}{3EI} - \frac{k_{21}L^2}{2EI} \tag{1.2-10}$$

$$\theta = 0 \text{ at node } i \qquad 0 = \frac{k_{11}L^2}{2EI} - \frac{k_{21}L}{EI} \tag{1.2-11}$$

$$\Sigma \text{ (forces)} = 0 \qquad 0 = k_{11} + k_{31} \tag{1.2-12}$$

$$\Sigma \text{ (moments)} = 0 \qquad 0 = k_{21} + k_{41} - k_{11}L \tag{1.2-13}$$

Solution of these equations yields

$$k_{11} = -k_{31} = \frac{12EI}{L^3} \quad \text{and} \quad k_{21} = k_{41} = \frac{6EI}{L^2} \tag{1.2-14}$$

Stiffness coefficients in columns 2, 3, and 4 of [**k**] are determined by applying similar arguments to Figs. 1.2-2*d*, 1.2-2*e*, and 1.2-2*f* in turn. The resulting stiffness matrix is exact, not approximate (provided that transverse shear deformation is ignored and deflections are small, as is commonly the case).

Having defined [**k**] in terms of E, I, and L, one is prepared to solve many problems of plane beams, such as that in Fig. 1.2-3. Generating the finite element model involves converting the distributed load into concentrated nodal loads. The conversion recipe for the loading shown in Fig. 1.2-3 is discussed in Section 4.3.

One can combine the stiffness matrices of Eqs. 1.2-3 and 1.2-9 to produce a 6 by 6 matrix [**k**] for an element that has two translational and one rotational d.o.f. at each end. Such elements can be used to analyze a plane frame. These elements are discussed in Sections 4.2 and 7.5.

1.3 ELEMENT ASSEMBLY AND SOLUTION FOR UNKNOWNS

We consider a very simple example, which illustrates briefly how elements are put together to form a finite element structure and how a solution for displacements and stresses is obtained. These matters are discussed in detail in Chapter 2. In

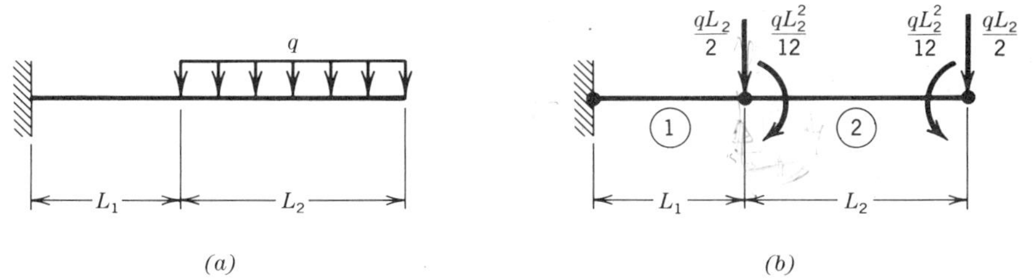

(*a*) (*b*)

Figure 1.2-3. (*a*) Cantilever beam carrying a uniformly distributed load q (force per unit length). (*b*) A two-element model, showing nodal loads produced by q.

nonstructural problems the matrices have other names, but manipulations are the same.

Assembly. An axially loaded bar structure is shown in Fig. 1.3-1*a*; a two-element model of it is shown in Fig. 1.3-1*b*. The stiffness matrix of a typical element is given by Eq. 1.2-3. Stiffness coefficients associated with elements 1 and 2 in Fig. 1.3-1 are abbreviated as

$$k_1 = \frac{A_1E_1}{L_1} \quad \text{and} \quad k_2 = \frac{A_2E_2}{L_2} \tag{1.3-1}$$

The d.o.f. are axial displacements u_1, u_2, and u_3, where 1, 2, and 3 are arbitrary labels assigned to identify the structure nodes.[2] Now imagine two hypothetical states: in the first, only element 1 is present; in the second, only element 2 is present. Thus the respective structure stiffness matrices would be

$$\underset{\text{element 1 only}}{\overset{\begin{matrix} u_1 & u_2 & u_3 \end{matrix}}{\begin{bmatrix} k_1 & -k_1 & 0 \\ -k_1 & k_1 & 0 \\ 0 & 0 & 0 \end{bmatrix}}} \quad \text{and} \quad \underset{\text{element 2 only}}{\overset{\begin{matrix} u_1 & u_2 & u_3 \end{matrix}}{\begin{bmatrix} 0 & 0 & 0 \\ 0 & k_2 & -k_2 \\ 0 & -k_2 & k_2 \end{bmatrix}}} \tag{1.3-2}$$

where column headings indicate the d.o.f. associated with the matrix coefficients. As elements are put together to form a finite element model, element matrices are put together to form the structure matrix. By direct addition of the preceding matrices, the structure stiffness matrix is

$$[\mathbf{K}] = \begin{bmatrix} k_1 & -k_1 & 0 \\ -k_1 & k_1 + k_2 & -k_2 \\ 0 & -k_2 & k_2 \end{bmatrix} \tag{1.3-3}$$

One can easily check that each column of $[\mathbf{K}]$ represents an equilibrium set of nodal forces associated with activation of the corresponding d.o.f. This should be no surprise, as the method of activating each d.o.f. in turn can be used to generate either an element matrix $[\mathbf{k}]$ or a structure matrix $[\mathbf{K}]$.

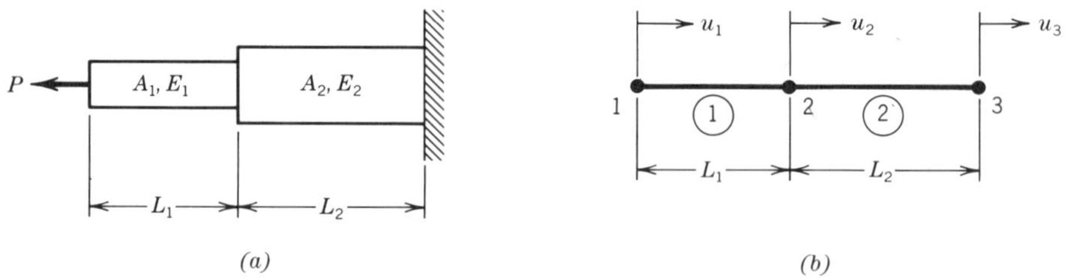

Figure 1.3-1. (*a*) An axially loaded bar. (*b*) Finite elements used to model the bar.

[2]Node labels *i* and *j* in Figs. 1.2-1 and 1.2-2 are called *element* or *local* node labels. Numbers would serve as well. They are used only temporarily, when element properties are defined. When elements are assembled, local labels are replaced by *structure* or *global* node labels (such as 1, 2, and 3 in Fig. 1.3-1). These matters are discussed in Section 2.7.

Matrix [**K**] of Eq. 1.3-3 is singular. Physically, this means that the structure in Fig. 1.3-1*b* is unsupported and can undergo a rigid-body translation. For solving a particular problem, at least one of the d.o.f. must be prescribed.

Solution for Unknowns. To impose the displacement boundary condition appropriate to Fig. 1.3-1, we must enforce the constraint $u_3 = 0$. This can be done by discarding row 3 and column 3 from [**K**] of Eq. 1.3-3. What remains is a 2 by 2 system to be solved for u_1 and u_2. (This system can also be obtained by activating u_1 and u_2 in turn, while u_3 is kept at zero, and calculating the loads that must be applied to d.o.f. u_1 and u_2.) Thus the problem of Fig. 1.3-1*a* is described by the matrix equation

$$\begin{bmatrix} k_1 & -k_1 \\ -k_1 & k_1 + k_2 \end{bmatrix} \begin{Bmatrix} u_1 \\ u_2 \end{Bmatrix} = \begin{Bmatrix} -P \\ 0 \end{Bmatrix} \tag{1.3-4}$$

The right-hand side indicates that node 1 carries a load P in the negative direction and that node 2 carries no externally applied load. The stiffness matrix in Eq. 1.3-4 is nonsingular. Therefore, Eqs. 1.3-4 can be solved for u_1 and u_2. These results are

$$u_1 = -\frac{P}{k_1} - \frac{P}{k_2} \qquad \text{and} \qquad u_2 = -\frac{P}{k_2} \tag{1.3-5}$$

Finally we convert displacements to strains and strains to stresses as follows. In elements 1 and 2, respectively, with $u_3 = 0$,

$$\sigma_1 = E_1\epsilon_1 = E_1\frac{u_2 - u_1}{L_1} = \frac{E_1}{L_1}\frac{P}{A_1E_1/L_1} = \frac{P}{A_1} \tag{1.3-6a}$$

$$\sigma_2 = E_2\epsilon_2 = E_2\frac{u_3 - u_2}{L_2} = \frac{E_2}{L_2}\frac{P}{A_2E_2/L_2} = \frac{P}{A_2} \tag{1.3-6b}$$

The preceding displacements and stresses are exact for this particular problem.

Notation. Our symbols for element and structure stiffness equations are, respectively,

$$[\mathbf{k}]\{\mathbf{d}\} = \{\bar{\mathbf{r}}\} \qquad \text{and} \qquad [\mathbf{K}]\{\mathbf{D}\} = \{\mathbf{R}\} \tag{1.3-7}$$

where [**k**] and [**K**] are stiffness matrices, {**d**} and {**D**} are vectors of nodal d.o.f., and $\{\bar{\mathbf{r}}\}$ and {**R**} are vectors of nodal loads. Subsequently (in Section 2.6) it will be worthwhile to distinguish between loads applied *to* an element and loads applied *by* an element. Displacements {**d**} produce loads $\{\bar{\mathbf{r}}\}$ consistent with static equilibrium and applied *to* an element. If we define {**r**} as loads equal and opposite to $\{\bar{\mathbf{r}}\}$, that is,

$$\{\mathbf{r}\} = -\{\bar{\mathbf{r}}\} \tag{1.3-8}$$

then {**r**} represents loads applied *to* nodes *by* deformed elements. Similarly, in the assembled structure, {**R**} represents loads applied *to* nodes, now by external sources rather than by deformed elements.

1.4 SUMMARY OF FINITE ELEMENT HISTORY

Beginning in 1906, researchers suggested a "lattice analogy" for stress analysis [1.2–1.4]. The continuum was replaced by a regular pattern of elastic bars. Properties of the bars were chosen in a way that caused displacements of the joints to approximate displacements of points in the continuum. The method sought to capitalize on well-known methods of structural analysis.

Courant appears to have been the first to propose the finite element method as we know it today. In a 1941 mathematics lecture, published in 1943, he used the principle of stationary potential energy and piecewise polynomial interpolation over triangular subregions to study the Saint-Venant torsion problem [1.5]. Courant's work was ignored until engineers had independently developed it.

None of the foregoing work was of much practical value at the time because there were no computers available to generate and solve large sets of simultaneous algebraic equations. It is no accident that the development of finite elements coincided with major advances in digital computers and programming languages.

By 1953 engineers had written stiffness equations in matrix format and solved the equations with digital computers [1.6]. Most of this work took place in the aerospace industry. At the time, a large problem was one with 100 d.o.f. In 1953, at the Boeing Airplane Company, Turner suggested that triangular plane stress elements be used to model the skin of a delta wing [1.7]. This work, published almost simultaneously with similar work done in England [1.8,1.9], marks the beginning of widespread use of finite elements. Much of this early work went unrecognized because of company policies against publication [1.10].

The name "finite element method" was coined by Clough in 1960. The practical value of the method was soon obvious. New elements for stress analysis applications were developed, largely by intuition and physical argument. In 1963 the finite element method gained respectability when it was recognized as having a sound mathematical foundation: it can be regarded as the solution of a variational problem by minimization of a functional. Thus the method was seen as applicable to all field problems that can be cast in a variational form. Papers about the application of finite elements to problems of heat conduction and seepage flow appeared in 1965.

Large general-purpose finite element computer programs emerged during the late 1960s and early 1970s. Examples include ANSYS, ASKA, and NASTRAN. Each of these programs included several kinds of elements and could perform static, dynamic, and heat transfer analysis. Additional capabilities were soon added. Also added were preprocessors (for data input) and postprocessors (for results evaluation). These processors rely on graphics and make it easier, faster, and cheaper to do finite element analysis. Graphics development became intensive in the early 1980s as hardware and software for interactive graphics became available and affordable.

A general-purpose finite element program typically contains over 100,000 lines of code and usually resides on a mainframe or a superminicomputer. However, in the mid-1980s, adaptations of general-purpose programs began to appear on personal computers. Hundreds of analysis and analysis-related programs are now available, large and small, general and narrow, cheap and expensive, for lease or for purchase.

Ten papers about finite elements were published in 1961, 134 in 1966, and 844 in 1971. By 1976, two decades after engineering applications began, the cumulative total of publications about finite elements exceeded 7000. By 1986, the total was about 20,000.

1.5 STRAIN–DISPLACEMENT RELATIONS

The relation between strain and displacement is a key ingredient in the formulation of finite elements for stress analysis problems. In the present section we consider general strain-displacement relations in Cartesian coordinates. Alternative special forms of the relations, such as forms used for solids of revolution and for plate bending, are stated where they are used.

In Fig. 1.5-1, perpendicular lines 01 and 02 are drawn on a plane sheet of material before the sheet is loaded. As a result of loading the lines become 0′1′ and 0′2′. Displacements u and v are functions of the coordinates: $u = u(x,y)$ and $v = v(x,y)$. We assume that displacement increments[3] such as $u_{,x}\,dx$ are small in comparison with u and v. By definition, normal strain is the ratio of change in length to original length. Therefore,

$$\epsilon_x = \frac{L_{0'2'} - L_{02}}{L_{02}} = \frac{[dx + (u + u_{,x}\,dx) - u] - dx}{dx} = u_{,x} \qquad (1.5\text{-}1)$$

A similar analysis yields the y-direction normal strain as

$$\epsilon_y = v_{,y} \qquad (1.5\text{-}2)$$

Shear strain in the "engineering definition" is defined as the amount of change in a right angle. Because displacement increments are small, $\beta_1 \approx \tan \beta_1$ and $\beta_2 \approx \tan \beta_2$. Therefore, the engineering shear strain is

$$\gamma_{xy} = \beta_1 + \beta_2 = \frac{(u + u_{,y}\,dy) - u}{dy} + \frac{(v + v_{,x}\,dx) - v}{dx} = u_{,y} + v_{,x} \qquad (1.5\text{-}3)$$

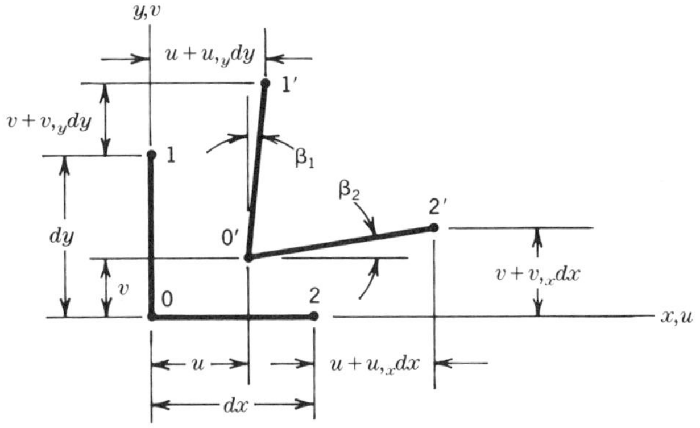

Figure 1.5-1. Displacement and distortion of differential lengths dx and dy.

[3]The notation $u_{,x}$ means $\partial u/\partial x$. In general, a comma followed by a literal subscript indicates partial differentiation with respect to that subscript. Thus, for example, $\phi_{,xy} = \partial^2\phi/\partial x\,\partial y$.

Collecting results, we have the *two-dimensional strain–displacement relations:*

$$\epsilon_x = u_{,x} \qquad \epsilon_y = v_{,y} \qquad \gamma_{xy} = u_{,y} + v_{,x} \tag{1.5-4}$$

In *three dimensions,* with w the z-direction displacement, a similar analysis yields the strains of Eqs. 1.5-4 and also

$$\epsilon_z = w_{,z} \qquad \gamma_{yz} = v_{,z} + w_{,y} \qquad \gamma_{zx} = u_{,z} + w_{,x} \tag{1.5-5}$$

where now u, v, and w are all functions of x, y, and z. The foregoing relations can be stated in matrix operator form as follows. In two and three dimensions, respectively,

$$\begin{Bmatrix} \epsilon_x \\ \epsilon_y \\ \gamma_{xy} \end{Bmatrix} = \begin{bmatrix} \dfrac{\partial}{\partial x} & 0 \\ 0 & \dfrac{\partial}{\partial y} \\ \dfrac{\partial}{\partial y} & \dfrac{\partial}{\partial x} \end{bmatrix} \begin{Bmatrix} u \\ v \end{Bmatrix} \quad \text{and} \quad \begin{Bmatrix} \epsilon_x \\ \epsilon_y \\ \epsilon_z \\ \gamma_{xy} \\ \gamma_{yz} \\ \gamma_{zx} \end{Bmatrix} = \begin{bmatrix} \dfrac{\partial}{\partial x} & 0 & 0 \\ 0 & \dfrac{\partial}{\partial y} & 0 \\ 0 & 0 & \dfrac{\partial}{\partial z} \\ \dfrac{\partial}{\partial y} & \dfrac{\partial}{\partial x} & 0 \\ 0 & \dfrac{\partial}{\partial z} & \dfrac{\partial}{\partial y} \\ \dfrac{\partial}{\partial z} & 0 & \dfrac{\partial}{\partial x} \end{bmatrix} \begin{Bmatrix} u \\ v \\ w \end{Bmatrix} \tag{1.5-6}$$

Plane Beams. We adopt standard beam theory in which transverse shear deformation is ignored and all deformations and strains are expressed in terms of the lateral displacement w. In Fig. 1.5-2, rotation $w_{,x}$ is assumed to be small, and plane cross sections are assumed to remain plane and normal to the deformed

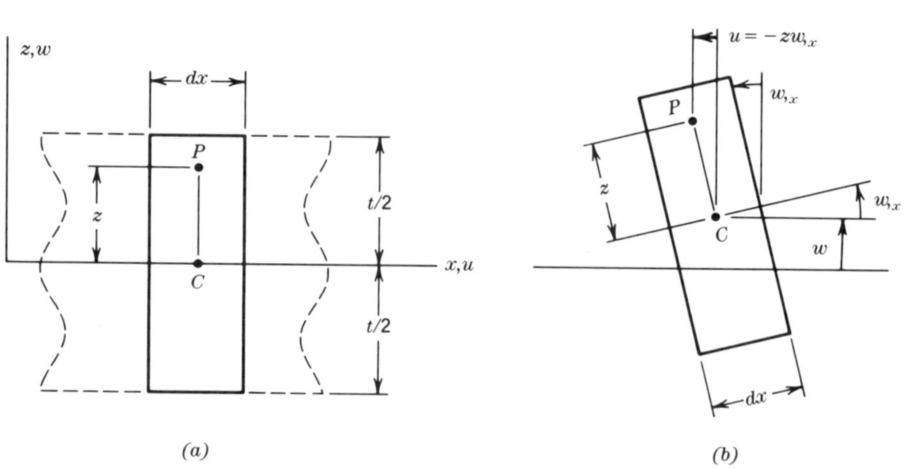

Figure 1.5-2. (*a*) Differential slice of a beam that lies along the x axis, before loading. (*b*) The same slice after loading. Transverse shear deformation is assumed to be negligible.

axis of the beam after it is bent. Therefore, axial displacement u is seen to be $u = -zw_{,x}$. The negative sign means that with z and $w_{,x}$ both positive, displacement u is in the negative x direction. Accordingly, since $\epsilon_x = u_{,x}$,

$$\epsilon_x = -zw_{,xx} \tag{1.5-7}$$

where $w_{,xx}$ is called the curvature of the beam.

Notation. A compact notation serves to symbolize any strain–displacement relation. Let $\{\boldsymbol{\epsilon}\}$ represent the strains and $\{\mathbf{u}\}$ represent the displacements. Then the equation

$$\{\boldsymbol{\epsilon}\} = [\partial]\{\mathbf{u}\} \tag{1.5-8}$$

represents *either* of Eqs. 1.5-6. It is only necessary to infer the proper size and proper contents of the three matrices $\{\boldsymbol{\epsilon}\}$, $[\partial]$, and $\{\mathbf{u}\}$. For beam bending, the matrices are all 1 by 1 (scalars):

$$\text{In beams:}\quad \epsilon_x = [\partial]w, \qquad \text{where}\quad [\partial] = -z\frac{d^2}{dx^2} \tag{1.5-9}$$

1.6 THEORY OF STRESS AND DEFORMATION

Fundamental concepts, definitions, and equations used in the analysis of stress and deformation are discussed in the discipline called theory of elasticity. These fundamentals are used in solving problems by both classical and finite element methods. Elasticity theory states conditions that must be met by an exact solution, and therefore helps us in judging the shortcomings or range of applicability of approximate solutions. For simplicity, the following summary is presented in Cartesian coordinates only, and the two-dimensional case is emphasized. More general arguments may be found in texts on theory of elasticity.

Equilibrium. Figure 1.6-1*a* shows a plane differential element (not a finite element!). We will develop equations stating that the differential element is in equilibrium under forces applied to it. Forces come from stresses on the edges and from body forces.

Body forces F_x and F_y are applied to all material points and have dimensions of force per unit volume. They can arise from gravity, acceleration, a magnetic field, and so on. They are considered positive when acting in positive coordinate directions. On each differential element of volume ($dV = t\,dx\,dy$, where t = thickness), F_x and F_y produce differential forces $F_x\,dV$ and $F_y\,dV$.

In general, stresses and body forces are functions of the coordinates. Thus, for example, $\sigma_{x,x}$ is the *rate* of change of σ_x with respect to x, and $\sigma_{x,x}\,dx$ is the *amount* of change of σ_x over distance dx. For constant thickness t, static equilibrium of forces in the x direction, $\Sigma f_x = 0$, requires that

$$\begin{aligned} -\sigma_x t\,dy - \tau_{xy} t\,dx &+ (\sigma_x + \sigma_{x,x}\,dx)t\,dy \\ &+ (\tau_{xy} + \tau_{xy,y}\,dy)t\,dx + F_x t\,dx\,dy = 0 \end{aligned} \tag{1.6-1}$$

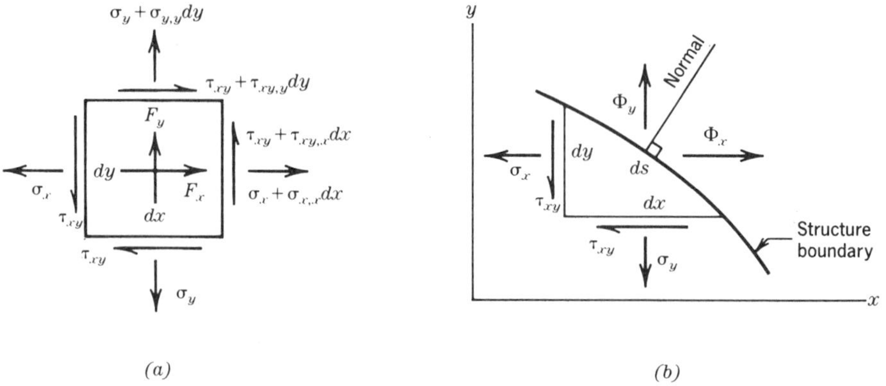

Figure 1.6-1. (*a*) Stresses and body forces that act on a plane differential element of constant thickness. (*b*) Surface tractions Φ_x and Φ_y on an arbitrarily oriented edge in the *xy* plane.

There is a corresponding *y*-direction equilibrium equation, $\Sigma f_y = 0$. After simplification, the two equilibrium equations are

$$x \text{ direction:} \quad \sigma_{x,x} + \tau_{xy,y} + F_x = 0 \tag{1.6-2a}$$

$$y \text{ direction:} \quad \tau_{xy,x} + \sigma_{y,y} + F_y = 0 \tag{1.6-2b}$$

In many problems the effects of body forces are far less important than the effects of loads applied to the surface of the structure. Then F_x and F_y are set to zero. Note that Eqs. 1.6-2 are derived from equilibrium considerations alone: as no material properties are invoked, Eqs. 1.6-2 are applicable whether or not the body is linearly elastic.

Although derived for the case of static equilibrium, Eqs. 1.6-2 can also be used if acceleration is present, provided that F_x *and* F_y include D'Alembert or "effective" body forces. For example, *x*-direction acceleration a_x implies that F_x includes the term $-\rho a_x$, where ρ is the mass density.

Compatibility. When a body is deformed without breaking, no cracks appear in stretching, no kinks appear in bending, and no part overlaps another. Stated more elegantly, this is the *compatibility condition:* the displacement field is continuous and single valued.

The *compatibility equation* $\epsilon_{x,yy} + \epsilon_{y,xx} = \gamma_{xy,xy}$ states the relation that exists among the strains if a plane displacement field is compatible. Most finite element methods are based on displacements rather than stresses. Thus, each element invokes a displacement field that is continuous and single valued. Therefore the compatibility equation is automatically satisfied, as one may see by substituting Eqs. 1.5-4 into it. (However, the displacement field may not satisfy the equilibrium equations. This point is discussed further in the following.)

Boundary Conditions. Boundary conditions consist of prescriptions of displacement and of stress. For example, in Fig. 1.1-2*a* the left edge does not move, so the displacement boundary condition along $x = 0$ is $u = v = 0$. Along the top edge, the stress boundary condition is $\sigma_y = \tau_{xy} = 0$ to the left of the zone that carries pressure *p*, and $\sigma_y = -p$ and $\tau_{xy} = 0$ beneath pressure *p*. On the right edge, $\sigma_x = \tau_{xy} = 0$. Along the curved lower edge, surface tractions Φ_x and Φ_y are zero. Surface tractions are related to stresses in the manner now described.

Surface tractions Φ_x and Φ_y in Fig. 1.6-1*b* have units of stress. They are applied on a boundary (in contrast to body forces, which act throughout a volume). Tractions are *x*- and *y*-direction force increments divided by the boundary area increment dA on which they act. In Fig. 1.6-1*b*, $dA = t\,ds$, where t = thickness. Note that dA is not either of the projected areas $t\,dx$ or $t\,dy$. Equilibrium of *x*- and *y*-direction forces in Fig. 1.6-1*b* requires that $\Phi_x t\,ds = \sigma_x t\,dy + \tau_{xy} t\,dx$ and $\Phi_y t\,ds = \tau_{xy} t\,dy + \sigma_y t\,dx$. But $dy/ds = \ell$ and $dx/ds = m$, where ℓ and m are direction cosines of the outward normal to the boundary. Thus

$$x \text{ direction:} \quad \Phi_x = \ell\sigma_x + m\tau_{xy} \qquad (1.6\text{-}3a)$$

$$y \text{ direction:} \quad \Phi_y = \ell\tau_{xy} + m\sigma_y \qquad (1.6\text{-}3b)$$

Equations 1.6-3 define a relation among stresses at an arbitrarily curved edge when tractions Φ_x and Φ_y are prescribed along that edge. Like Eqs. 1.6-2, Eqs. 1.6-3 do not require that the body be linearly elastic. One can check that Eqs. 1.6-3 yield the correct stress boundary conditions along the top and right edges of Fig. 1.1-2*a*.

Three Dimensions. In solids, arguments analogous to those preceding lead to analogous results. The equilibrium and surface traction equations are, respectively,

$$\begin{aligned} \sigma_{x,x} + \tau_{xy,y} + \tau_{zx,z} + F_x &= 0 \\ \tau_{xy,x} + \sigma_{y,y} + \tau_{yz,z} + F_y &= 0 \\ \tau_{zx,x} + \tau_{yz,y} + \sigma_{z,z} + F_z &= 0 \end{aligned} \quad \text{and} \quad \begin{aligned} \Phi_x &= \ell\sigma_x + m\tau_{xy} + n\tau_{zx} \\ \Phi_y &= \ell\tau_{xy} + m\sigma_y + n\tau_{yz} \\ \Phi_z &= \ell\tau_{zx} + m\tau_{yz} + n\sigma_z \end{aligned} \qquad (1.6\text{-}4)$$

where ℓ, m, and n are direction cosines of an outward normal to the surface. There are six compatibility equations that relate the six strains used in solids. They are not presented here.

We observe that if stresses $\{\boldsymbol{\sigma}\}$ are arrayed in the order used in Section 1.7, then the notation of Eq. 1.5-8 permits us to write the equilibrium equations in either two or three dimensions as

$$[\partial]^T\{\boldsymbol{\sigma}\} + \{\mathbf{F}\} = \{\mathbf{0}\} \qquad (1.6\text{-}5)$$

where $[\partial]$ is given by Eqs. 1.5-6 and $\{\mathbf{F}\}$ is $\lfloor F_x \quad F_y \rfloor^T$ or $\lfloor F_x \quad F_y \quad F_z \rfloor^T$.

Remarks. Consider a linearly elastic body. If one finds a stress field or a displacement field that simultaneously satisfies equilibrium, compatibility, and boundary conditions, then one has found a solution to the problem posed. The solution is both unique and exact within the assumptions made (such as linearity and homogeneity).

How do these observations relate to finite element analysis? If elements are based on polynomial displacement fields, then compatibility prevails within elements. Suitably chosen displacement fields also provide compatibility *between* elements and satisfy displacement boundary conditions. The differential equations of equilibrium and boundary conditions on stress are satisfied only approximately. The approximation improves as more elements are used and, barring computational difficulties, the exact solution is achieved in the limit of an infinitely refined mesh.

Some finite elements are incompatible: adjacent elements can overlap or gap apart between nodes. Compatibility is then satisfied only within elements and at node points. Thus there is only approximate satisfaction of equilibrium, stress boundary conditions, and compatibility. If interelement compatibility tends to be restored as more elements are used, it is still possible to converge toward the exact solution as the mesh is refined. (Incompatible elements are subsequently treated in more detail.)

Other finite element methods are based on fields other than displacement. A stress field model may satisfy equilibrium equations *a priori*. Mesh refinement would then yield a better approximation of compatibility conditions.

1.7 STRESS–STRAIN–TEMPERATURE RELATIONS

The finite element method deals easily with rather general material properties and with both thermal and mechanical loading. We therefore devote this entire section to elastic properties and thermal relations.

General. The stress vector $\{\boldsymbol{\sigma}\}$ and the strain vector $\{\boldsymbol{\epsilon}\}$ are, respectively,

$$\{\boldsymbol{\sigma}\} = \lfloor \sigma_x \quad \sigma_y \quad \sigma_z \quad \tau_{xy} \quad \tau_{yz} \quad \tau_{zx} \rfloor^T \tag{1.7-1a}$$

and

$$\{\boldsymbol{\epsilon}\} = \lfloor \epsilon_x \quad \epsilon_y \quad \epsilon_z \quad \gamma_{xy} \quad \gamma_{yz} \quad \gamma_{zx} \rfloor^T \tag{1.7-1b}$$

Ignoring the effect of temperature change, which will be treated subsequently, we symbolize the isothermal stress–strain relation as

$$\{\boldsymbol{\epsilon}\} = [\mathbf{C}]\{\boldsymbol{\sigma}\} \qquad \text{or as} \qquad \{\boldsymbol{\sigma}\} = [\mathbf{E}]\{\boldsymbol{\epsilon}\} \tag{1.7-2}$$

where [**C**] is a symmetric matrix of material compliances, [**E**] is a symmetric matrix of material stiffnesses, and $[\mathbf{E}] = [\mathbf{C}]^{-1}$ (examples follow). In the most general case of anisotropy, [**C**] and [**E**] each contain 21 independent coefficients.

An *orthotropic* material is an anisotropic material that displays extreme values of stiffness in mutually perpendicular directions. These directions are called *principal directions* of the material. An example is wood cut from a log, which is stiffest in the axial direction, least stiff in the circumferential direction, and of intermediate stiffness in the radial direction. If x, y, and z are principal directions, then normal stresses σ_x, σ_y, and σ_z are independent of shear strains γ_{xy}, γ_{yz}, and γ_{zx}. [**E**] for an orthotropic material contains only nine independent coefficients.

Equations 1.7-2 express Hooke's law: stress is directly proportional to strain. This rule is an approximation limited to small strains and certain materials.

Isotropy. An isotropic material has no preferred directions. Material properties are commonly expressed as a combination of two of the following: elastic modulus E, Poisson's ratio ν, and shear modulus G. In its upper triangle, **[E]** contains 12 zero coefficients and the 9 nonzero coefficients

$$\begin{aligned} E_{11} &= E_{22} = E_{33} = (1 - \nu)c \\ E_{12} &= E_{13} = E_{23} = \nu c \\ E_{44} &= E_{55} = E_{66} = G \end{aligned} \tag{1.7-3}$$

where $$c = \frac{E}{(1 + \nu)(1 - 2\nu)} \quad \text{and} \quad G = \frac{E}{2(1 + \nu)}$$

Because of the relation $E = 2(1 + \nu)G$ and the symmetry of **[E]**, we see that **[E]** for an isotropic material contains only two independent coefficients.

Plane strain is defined as a deformation state in which $w = 0$ everywhere and u and v are functions of x and y but not of z. Thus, $\epsilon_z = \gamma_{yz} = \gamma_{zx} = 0$. A typical slice of an underground tunnel that lies along the z axis might deform in essentially plane strain conditions. The stress–strain relation $\{\boldsymbol{\sigma}\} = [\mathbf{E}]\{\boldsymbol{\epsilon}\}$ for isotropic and isothermal plane strain is

$$\begin{Bmatrix} \sigma_x \\ \sigma_y \\ \tau_{xy} \end{Bmatrix} = \frac{E}{(1 + \nu)\,(1 - 2\nu)} \begin{bmatrix} 1 - \nu & \nu & 0 \\ \nu & 1 - \nu & 0 \\ 0 & 0 & \dfrac{1 - 2\nu}{2} \end{bmatrix} \begin{Bmatrix} \epsilon_x \\ \epsilon_y \\ \gamma_{xy} \end{Bmatrix} \tag{1.7-4}$$

In Eq. 1.7-4, **[E]** is obtained by discarding rows and columns 3, 5, and 6 from the 6 by 6 matrix **[E]** of an isotropic solid.

Stress σ_z does not appear in Eq. 1.7-4, even though it is usually not zero. If needed, σ_z can be obtained from the relation $\epsilon_z = 0 = (\sigma_z - \nu\sigma_y - \nu\sigma_x)/E$ after σ_y and σ_x are known.

Plane stress is a condition that prevails in a flat plate in the xy plane, loaded only in its own plane and without z-direction restraint, so that $\sigma_z = \tau_{yz} = \tau_{zx} = 0$. Then, for isotropic and isothermal conditions,

$$\begin{Bmatrix} \epsilon_x \\ \epsilon_y \\ \gamma_{xy} \end{Bmatrix} = \frac{1}{E} \begin{bmatrix} 1 & -\nu & 0 \\ -\nu & 1 & 0 \\ 0 & 0 & E/G \end{bmatrix} \begin{Bmatrix} \sigma_x \\ \sigma_y \\ \tau_{xy} \end{Bmatrix}$$

or

$$\begin{Bmatrix} \sigma_x \\ \sigma_y \\ \tau_{xy} \end{Bmatrix} = \frac{E}{1 - \nu^2} \begin{bmatrix} 1 & \nu & 0 \\ \nu & 1 & 0 \\ 0 & 0 & \dfrac{1 - \nu}{2} \end{bmatrix} \begin{Bmatrix} \epsilon_x \\ \epsilon_y \\ \gamma_{xy} \end{Bmatrix} \tag{1.7-5}$$

where $E/G = 2(1 + \nu)$. The square matrices, including their scalar multipliers $1/E$ and $E/(1 - \nu^2)$, are, respectively, **[C]** and **[E]**.

Axially symmetric solids require a 4 by 4 matrix **[E]**. This problem is discussed in Chapter 10.

Beam Bending. Consider again the beam of Fig. 1.5-2. Uniaxial stress prevails, so $\sigma_x = E\epsilon_x$. Combining this equation with Eq. 1.5-7, we obtain $\sigma_x = -Ezw_{,xx}$. Let the beam have cross-sectional area A and a constant modulus E. Multiplying by $z\,dA$ and integrating over a cross section of depth t, we obtain

$$\int_{-t/2}^{t/2} \sigma_x z\, dA = -Ew_{,xx} \int_{-t/2}^{t/2} z^2\, dA \tag{1.7-6}$$

The former integral is identified as bending moment M, positive when it bends the beam of Fig. 1.5-2 concave down. The latter integral is identified as the moment of inertia I of the cross-sectional area. Thus Eq. 1.7-6 becomes

$$M = -EIw_{,xx} \tag{1.7-7}$$

This is a familiar expression from elementary mechanics of materials. It is the form of $\{\boldsymbol{\sigma}\} = [\mathbf{E}]\{\boldsymbol{\epsilon}\}$ that applies to beam bending and is called a load–displacement relation. A similar expression, expanded to two dimensions, is used for plate bending (Chapter 11).

Initial Stress and Strain. Thermal Effects. The term "initial stress" signifies a stress present before deformations are allowed. Effectively, it is a residual stress to be superposed on stress caused by deformation. With the addition of initial stresses $\{\boldsymbol{\sigma}_0\}$ and initial strains $\{\boldsymbol{\epsilon}_0\}$, Eq. 1.7-2 becomes

$$\{\boldsymbol{\sigma}\} = [\mathbf{E}](\{\boldsymbol{\epsilon}\} - \{\boldsymbol{\epsilon}_0\}) + \{\boldsymbol{\sigma}_0\} \tag{1.7-8}$$

As examples, $\{\boldsymbol{\epsilon}_0\}$ might describe moisture-induced swelling and $\{\boldsymbol{\sigma}_0\}$ might describe stresses produced by heating. Alternatively, both effects can be placed in $\{\boldsymbol{\epsilon}_0\}$, or $\{\boldsymbol{\epsilon}_0\}$ and $\{\boldsymbol{\sigma}_0\}$ can be viewed as alternative ways to express the same thing. For example, free expansion of an orthotropic material with principal axes xyz produces the initial strains

$$\{\boldsymbol{\epsilon}_0\} = \lfloor \alpha_x T \quad \alpha_y T \quad \alpha_z T \quad 0 \quad 0 \quad 0 \rfloor^T \tag{1.7-9}$$

where T is the temperature relative to an arbitrary reference temperature at which the body is free of stress, and the α's are coefficients of thermal expansion in the principal material directions. Thus, to account for the effects of temperature change, we can substitute Eq. 1.7-9 into Eq. 1.7-8 and set $\{\boldsymbol{\sigma}_0\} = \{\mathbf{0}\}$. Alternatively, we can substitute $\{\boldsymbol{\sigma}_0\} = -[\mathbf{E}] \lfloor \alpha_x T \quad \alpha_y T \quad \alpha_z T \quad 0 \quad 0 \quad 0 \rfloor^T$ into Eq. 1.7-8 and set $\{\boldsymbol{\epsilon}_0\} = \{\mathbf{0}\}$. Stresses $\{\boldsymbol{\sigma}\} = \{\boldsymbol{\sigma}_0\}$ prevail when mechanical strains $\{\boldsymbol{\epsilon}\}$ are prohibited.

In the special case of isotropy we have, in three dimensions,

$$\{\boldsymbol{\sigma}_0\} = -\frac{E\alpha T}{1 - 2\nu} \lfloor 1 \quad 1 \quad 1 \quad 0 \quad 0 \quad 0 \rfloor^T \tag{1.7-10}$$

and, in plane stress,

$$\{\boldsymbol{\epsilon}_0\} = \lfloor \alpha T \quad \alpha T \quad 0 \rfloor^T \qquad \text{and} \qquad \{\boldsymbol{\sigma}_0\} = -\frac{E\alpha T}{1 - \nu} \lfloor 1 \quad 1 \quad 0 \rfloor^T \tag{1.7-11}$$

and, finally, in plane strain,

$$\{\boldsymbol{\epsilon}_0\} = (1 + \nu) \lfloor \alpha T \quad \alpha T \quad 0 \rfloor^T \qquad \text{and} \qquad \{\boldsymbol{\sigma}_0\} = -\frac{E\alpha T}{1 - 2\nu} \lfloor 1 \quad 1 \quad 0 \rfloor^T \tag{1.7-12}$$

Temperature-dependent moduli can be accommodated by using the [**E**] appropriate to the temperature that prevails. A temperature-dependent expansion coefficient is more troublesome. By definition, $\alpha = \partial\epsilon/\partial T$ when deformation is unrestrained, so $\epsilon = \alpha T$ only if α is independent of T. Otherwise,

$$\epsilon = \int_0^T \alpha \, dT = \overline{\alpha} T, \qquad \text{where} \quad \overline{\alpha} = \frac{1}{T} \int_0^T \alpha \, dT \tag{1.7-13}$$

Here $\overline{\alpha}$ is an average expansion coefficient, valid only over the temperature range from 0 to T.

Remarks. For plane stress or plane strain conditions to prevail in the xy plane, the xy plane must be a plane of elastic symmetry. Thus, if the material is orthotropic, the z axis must be a principal material direction. If, in addition, the x and y axes are principal material directions, then $E_{13} = E_{31} = E_{23} = E_{32} = 0$ in the 3 by 3 matrix [**E**].

Poisson's ratio ν is little affected by temperature. Modulus E is affected more: for stainless steel E decreases about 20% when the temperature rises from 0° to 450°C. Barring plastic flow, elastic properties are almost independent of stress. For example, an increase in hydrostatic pressure from 0 to 350 MPA increases the moduli of steel and aluminum 0.8% and 2.6%, respectively. Like E, thermal expansion coefficient α is relatively insensitive to stress but may vary appreciably with temperature.

When strain rates are high, as in wave propagation, modulus E is higher than its static value. The difference is appreciable for rubber-like materials but negligible for common metals unless strain rates are extreme, as in explosive forming processes.

If a body is isotropic and linearly elastic, and its supports do not inhibit thermal expansion or contraction, then the body deforms but remains free of stress when T is a harmonic function; that is, when $\nabla^2 T = 0$. A special case of this is when T is a linear function of the coordinates, for which both isotropic and rectilinearly orthotropic bodies remain free of stress (but not curvilinearly orthotropic bodies, such as tree trunks).

Capabilities of the finite element method far exceed the knowledge of material behavior on which an analysis must be based. If test data are lacking, as is often the case with anisotropic materials, we can only estimate the elastic constants. Even when the constants are known, anisotropy has an adverse effect on the accuracy of finite element solutions [8.34].

1.8 WARNING: THE COMPUTED ANSWER MAY BE WRONG

Users of finite element programs may be so impressed by the power of the method that its limitations are ignored. Whether computer-based or not, analytical methods rely on assumptions and on theory that is not universally applicable. The analyst may overlook or misjudge important aspects of physical behavior. There may be an error in the computer program. A large program has many options and many computational paths. Perhaps some paths have never before been exercised, were not anticipated by the program designers, and have never been checked. *Far more likely* causes of incorrect results are user errors, such as using an inappropriate program or supplying an appropriate program with the wrong data. A poor mesh may be used, an inappropriate element type may be chosen, yielding or buckling may be overlooked, support conditions may be misrepresented, and so on. Users must remember that a structure is not obliged to behave as a computer says it should, regardless of how expensive the program, how many digits are printed in the results, or how elegant the graphic display. Computer graphics has achieved such a level of polish and versatility as to inspire great trust in the underlying analysis, a trust that may be unwarranted. (One can now make mistakes with more confidence than ever before.)

The finite element method is a most versatile tool, but not the best analytical tool for every problem. It is foolish, but not unheard of, to use three-dimensional finite elements to compute stresses obtainable by the flexure formula $\sigma = Mc/I$. In other cases experiment may be the most appropriate method, especially if experiment is needed anyway to obtain data needed for analysis (data such as material properties, the effective stiffness of joints, damping properties, the time history of loads, etc.). If an analysis is to be done by numerical methods, finite elements are not the only choice. For example, finite difference methods are effective for shells of revolution, and boundary elements are effective for some problems with boundaries at infinity.

Powerful computer programs cannot be used without training. Their results cannot be trusted if users have no knowledge of their internal workings and little understanding of the physical theories on which they are based. An error caused by misunderstanding or oversight is not correctible by mesh refinement or by use of a more powerful computer. Some authorities have suggested that users be "qualified," somewhat in the manner of practitioners having to be licensed before engaging in a profession in which the potential for damage to the public is substantial. Although the finite element method can make a good engineer better, it can make a poor engineer more dangerous.

In years past, when analysis was done by hand, the analyst was required to invent a mathematical model before undertaking its analysis. Invention of a good model required sound physical understanding of the problem. Understanding can now be replaced by activation of a computer program. Having had little need to sharpen intuitions by devising simple models, the computer user may lack the physical understanding needed to prepare a good model and to check computed results. Or, what the user perceives as understanding may instead be familiarity with previous computer output.

Computed results must in some way be judged or compared with expectations. Alternative results, useful for comparison, might be obtained from a different computer program that relies on a different analytical basis, from a simplified model amenable to hand calculation, from the behavior of similar structures already built, and from experiment. Experiment may be expensive and has its own pitfalls, but is desirable if the analytical process is pushed beyond previous experience and established practice.

The overall message of this discussion is that a competent analyst must have sound engineering judgment and experience, and that doubts raised in the course of an analysis should be taken seriously.

PROBLEMS

Section 1.1

1.1 In Fig. 1.1-1*a*, let cross-sectional area A vary linearly from $3A_0$ at $x = 0$ to A_0 at $x = L_T$. Model the bar by uniform elements. Let the cross-sectional area of each element be that of the actual bar at the x coordinate of the element midpoint. Assume that elastic modulus E is constant. Solve for the displacement of load P in the following ways, and compute the percentage error of each result. The exact answer is $PL_T \ln 3/2EA_0$.
(a) Use a single uniform element. Let $L_1 = L_T$ (and $A = 2A_0$).
(b) Use two uniform elements. Let $L_1 = L_2 = L_T/2$.
(c) Use three uniform elements. Let $L_1 = L_2 = L_3 = L_T/3$.
(d) Use four uniform elements, each of length $L_T/4$.

1.2 Derive the "exact answer" stated in Problem 1.1.

1.3 The bar shown has a uniform circular cross section of radius a. It carries only torsional loads. Develop the 2 by 2 element stiffness matrix in terms of a, L, and shear modulus G.

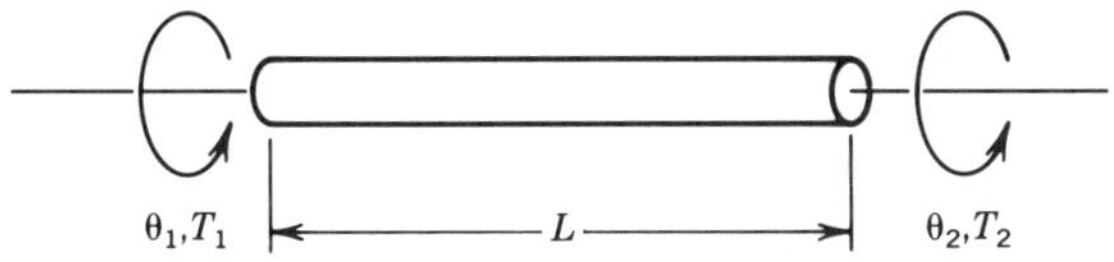

Problem 1.3

1.4 Let ϕ be a function that is interpolated over an element, where ϕ is a function of x, y, element dimensions, and nodal values of ϕ. For each element shown, write this expression for ϕ.
(a) Consider an element of length L. Let ϕ vary linearly with axial coordinate x.
(b) Consider the triangular element (see Eq. 1.1-1).
(c) Consider the rectangular element (see Eq. 1.1-2).

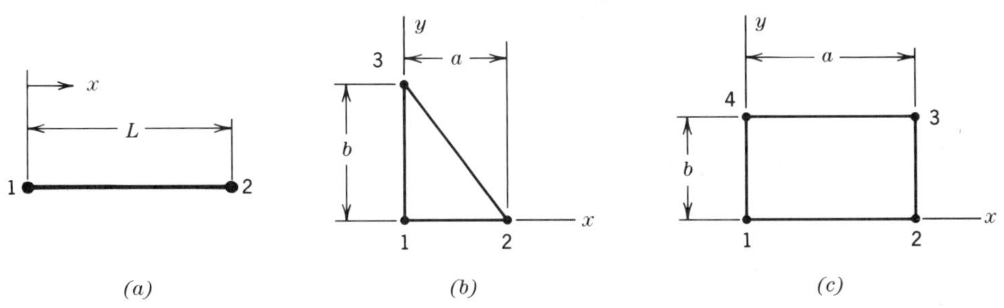

Problem 1.4

1.5 The quadrilateral element shown might be used in the finite element model of Fig. 1.1-2*b*. Imagine that its *x*-direction displacement field is $u = a_1 + a_2x + a_3y + a_4xy$, where the a_i are constants. How does u vary with x or y along each side? Do you think this element will be compatible with its neighbors?

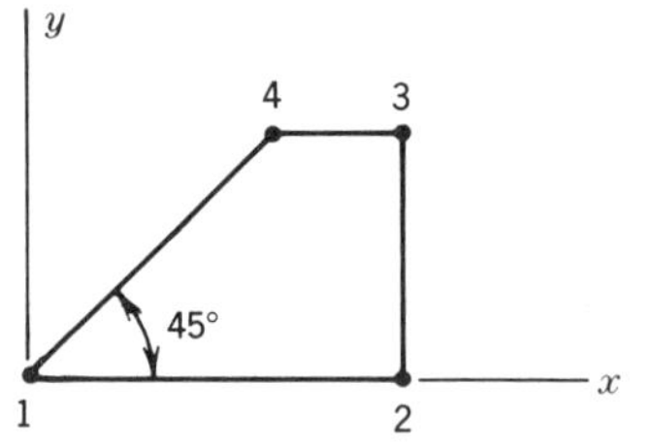

Problem 1.5

Section 1.2

1.6 Following the procedure used in Eqs. 1.2-10 to Eq. 1.2-14, derive the stiffness coefficients in columns 2, 3, and 4 of the stiffness matrix of a uniform beam element.

1.7 The bar shown can have both axial and bending deformations. Regard the bar as a single element with the four translational and two rotational d.o.f. shown. Without calculation, determine the *algebraic sign* of each coefficient in the 6 by 6 element stiffness matrix [**k**] (or enter zero for a null coefficient). Assume that displacements and rotations are small. *Suggestion:* Sketch each of six separate deformation states. Show the forces and moments needed to produce these states while satisfying static equilibrium. Finally, compare the directions of these loads with the assumed positive directions (which are those of the six nodal d.o.f.).

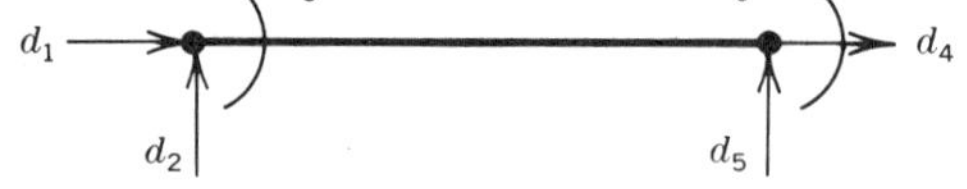

Problem 1.7

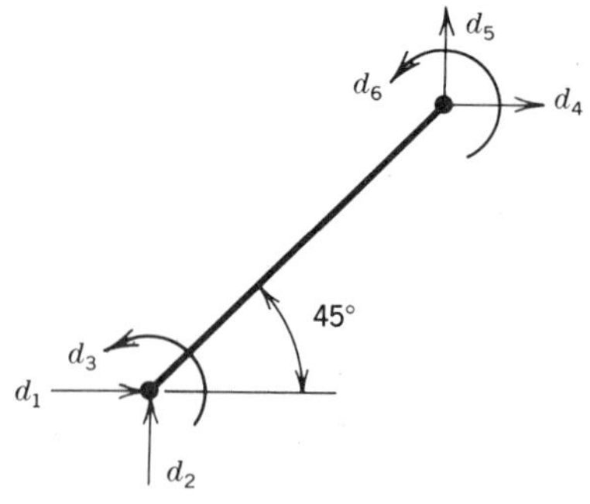

Problem 1.8

1.8 Repeat the instructions of Problem 1.7, but with reference to the inclined bar shown. Assume that the bar has an axial stiffness much greater than its bending stiffness. Also assume that displacements and rotations are small.

1.9 Repeat the instructions of Problem 1.7, but with reference to the bent bar shown. *Suggestion:* Assume that [**k**] is symmetric. Also, start with column 2 of [**k**], then proceed to columns 1, 3, 4, 5, and 6.

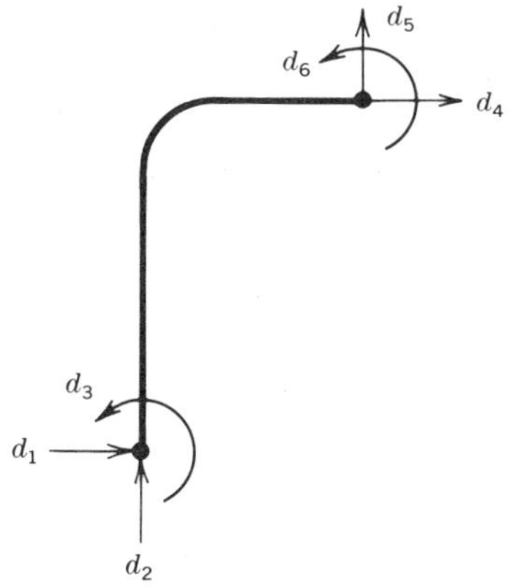

Problem 1.9

Section 1.3

1.10 The structure shown consists of a rigid, weightless bar and two linear springs of stiffnesses k_1 and k_2. Only small vertical displacements are permitted. The stiffness matrix [**K**] of this structure is 2 by 2 but can have various forms depending on the choice of d.o.f. Write [**K**] for each of the following choices of d.o.f.

(a) Displacements v_1 at $x = 0$ and v_2 at $x = L$ (shown in the second sketch).
(b) Displacements v_1 at $x = 0$ and v_A at $x = L/2$.
(c) Displacements v_2 at $x = L$ and v_B at $x = 2L$.
(d) Displacement v_1 at $x = 0$ and a small rotation θ about $x = 0$.
(e) Displacement v_B at $x = 2L$ and a small rotation θ about $x = 2L$.

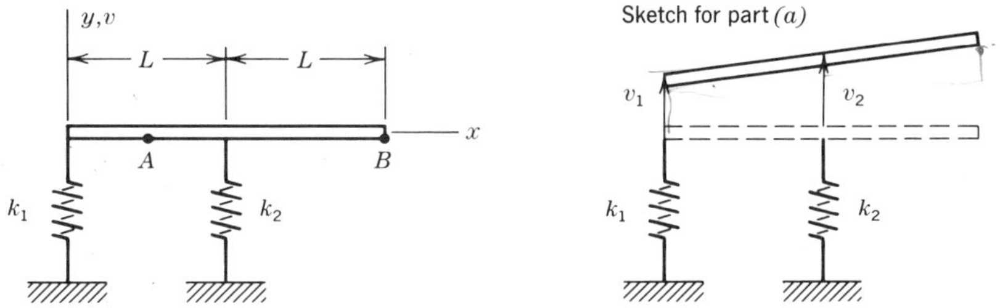

Problem 1.10

1.11 The angled bar is rigid and weightless. With its two linear springs it forms a structure whose stiffness matrix [**K**] is 2 by 2. Various forms of [**K**] are

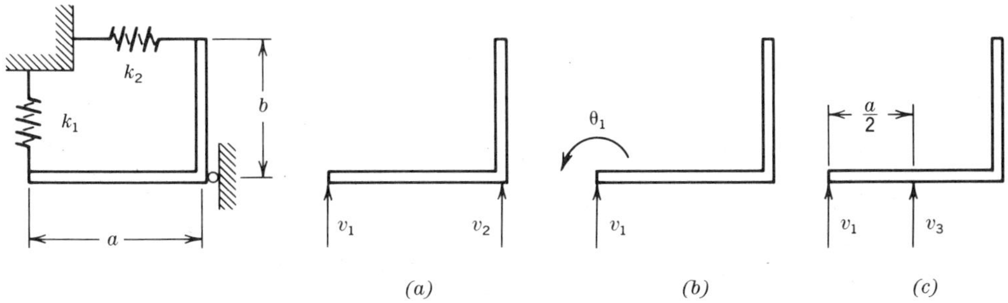

Problem 1.11

possible for various choices of d.o.f. Write [**K**] for each of the choices (*a*), (*b*), and (*c*) shown in the sketch. Displacements are small in each case.

1.12 The structure shown consists of linear springs whose stiffnesses are k_1, k_2, k_3, and k_4. Only horizontal displacements are allowed.
(a) In matrix form, write the three equilibrium equations of the structure. The d.o.f. are u_1, u_2, and u_3.
(b) Let $k_1 = k_2 = k_3 = k_4 = k$ and $F_1 = F_2 = 0$. Determine u_1, u_2, and u_3 in terms of k and F_3.

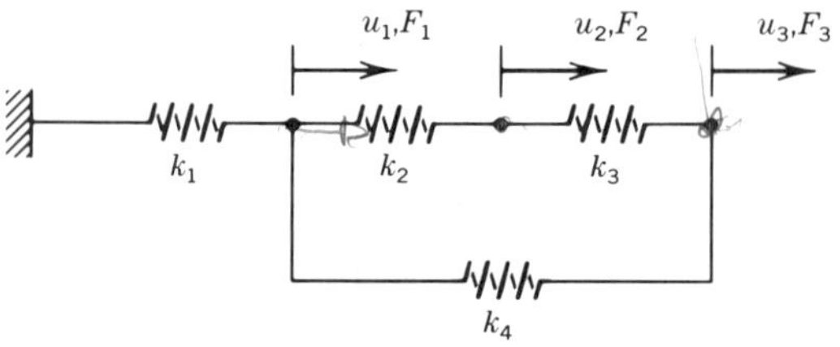

Problem 1.12

1.13 The structure shown consists of rigid bars *AB* and *CD* and linear springs that connect *A* to *C* and *B* to *D*. Only horizontal motions and small rotations of the bars are permitted. In terms of k_1, k_2, and a, determine the 4 by 4 stiffness matrix that operates on $\lfloor u_1 \quad \theta_1 \quad u_2 \quad \theta_2 \rfloor^T$ to yield $\lfloor F_1 \quad M_1 \quad F_2 \quad M_2 \rfloor^T$. *Suggestion:* Since rotations are small, $u_A = u_1 - a\theta_1$, with similar relations for u_B, u_C, and u_D.

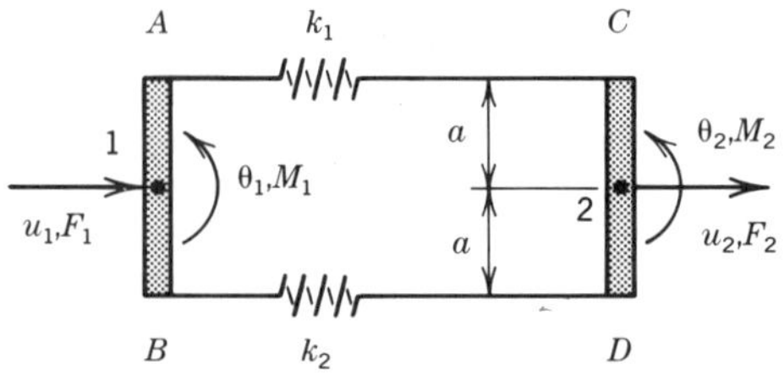

Problem 1.13

1.14 The uniform, linearly elastic bar shown is fixed at both ends. Force *P* is applied at node 2. Use the finite element method to compute nodal displacements and stresses in each element in terms of *P*, *L*, *A*, and *E*. Compare these results with exact values.

1.15 The uniform, linearly elastic bar shown carries a uniformly distributed load q_0. Assume that q_0 produces the respective nodal loads $q_0L/2$, q_0L, and $q_0L/2$.
(a) Compute nodal displacements by the finite element method and compare them with the exact values, which are $u_2 = 3q_0L^2/2AE$ and $u_3 = 2q_0L^2/AE$.
(b) Using the procedure of Eqs. 1.3-6, compute the axial stress in each element. On a single set of axes, plot these stresses as well as the actual stress distribution along the bar. Do the results suggest a rule regarding where stresses should be calculated in a finite element?

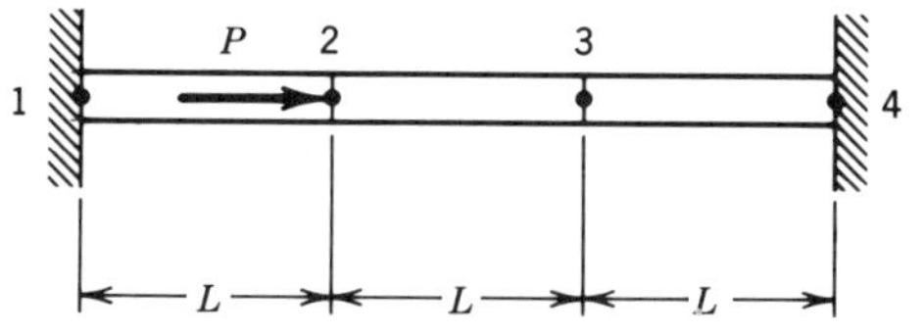

Problem 1.14

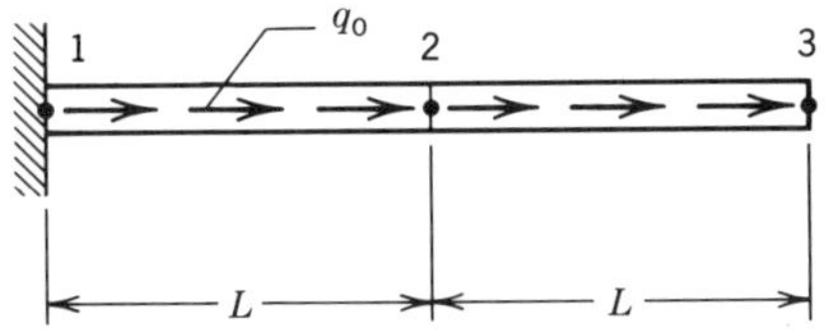

Problem 1.15

Section 1.6

1.16 (a) The sketch shows a plane differential element similar to that in Fig. 1.6-1*a* but in polar coordinates. The stresses are σ_r (radial), σ_θ (circumferential), and $\tau_{r\theta}$ (shear). Derive the differential equations of equilibrium in polar coordinates.

(b) Similarly, use cylindrical coordinates to derive equilibrium equations analogous to the first set of Eqs. 1.6-4.

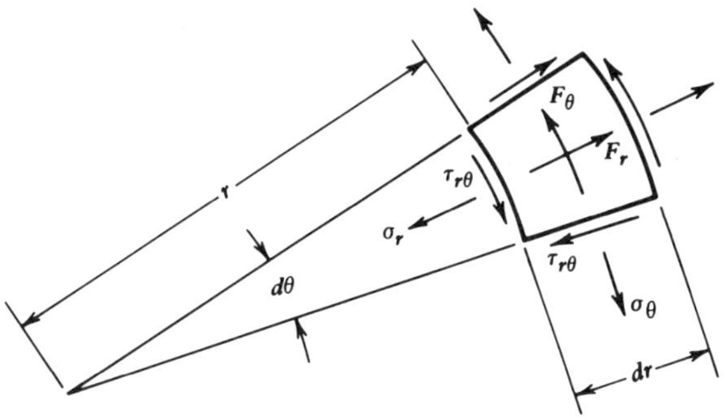

Problem 1.16

1.17 Imagine that stresses in the *xy* plane are given as $\sigma_x = -6a_1x^2$, $\sigma_y = 12a_1x^2$, and $\tau_{xy} = 12a_1y^2$, where a_1 is a constant.

(a) Consider the square region $0 \leq x \leq b$, $0 \leq y \leq b$. Write expressions for tractions Φ_x and Φ_y on each side of this square, in terms of x, y, b, and a_1.

(b) If body forces are zero, is the given stress field in fact possible? Explain.

1.18 In a certain approximation method, stresses in a plane region are assumed to have the forms

$$\sigma_x = a_1 + a_2x + a_3y, \qquad \sigma_y = a_4 + a_5x + a_6y, \qquad \tau_{xy} = a_7 + a_8x + a_9y$$

where each a_i is a constant. Let all body forces vanish. What must be the relation among the a_i if equilibrium is to be satisfied?

1.19 Determine whether or not the following stress field is a valid solution of a plane elasticity problem: $\sigma_x = 3a_1x^2y$, $\sigma_y = a_1y^3$, and $\tau_{xy} = -3a_1xy^2$, where a_1 is a constant. The body is isotropic and linearly elastic and body forces are zero.

Section 1.7

1.20 Judging by Eq. 1.7-4, what property does a material display if $\nu = 0.5$?

1.21 Combine the latter form of Eqs. 1.7-5, the strain–displacement relations, and the equilibrium equations with $F_x = F_y = 0$, and show that

$$u_{,xx} + u_{,yy} = (1 + \nu)(u_{,yy} - v_{,xy})/2$$

This equation, and its companion (obtained by interchange of u with v and x with y), are known as the equilibrium equations expressed in terms of displacements.

1.22 Let displacements in a plane stress problem be given by

$$u = a_1 + a_2x + a_3y + a_4x^2 + a_5xy + a_6y^2$$

$$v = a_7 + a_8x + a_9y + a_{10}x^2 + a_{11}xy + a_{12}y^2$$

where each a_i is a constant. Let all body forces vanish. What relation among the a_i is needed to satisfy equilibrium? The medium is isotropic.

1.23 Start with Eqs. 1.7-8 and 1.7-9, and derive the plane strain form of $\{\epsilon_0\}$ (Eq. 1.7-12).

CHAPTER 2

THE STIFFNESS METHOD AND THE PLANE TRUSS

Matrix procedures used in the analysis of framed structures and other finite element structures are described. The plane truss is used as the principal vehicle for the discussion.

2.1 INTRODUCTION

Certain matrix procedures of structural mechanics are described in this chapter. These methods are also used in finite element analysis of many other physical problems. These procedures include assembly of elements to form a structure, imposition of boundary or support conditions, solution of simultaneous equations to obtain nodal quantities, and processing of elements to obtain quantities such as stresses or flows. The plane truss is a very simple structure that serves to explain these concepts and procedures.

The truss can be called a "discrete element" structure. Its elements are the individual bars, already present as separate pieces. Thus we bypass the important finite element processes of dividing a continuum into appropriate elements and idealizing the behavior of each element. In the present chapter we are primarily concerned with manipulation procedures that apply to a previously discretized structure.

Each bar of a truss is assumed to be uniform, linearly elastic, pin-connected to nodes at its ends, and axially loaded. Displacements shown in sketches are *greatly exaggerated.* Actual displacements are assumed to be small, so that if θ is the angle of rotation under load of any bar, then $\sin \theta \approx \theta$ and $\cos \theta \approx 1$. We consider only statically loaded structures. Within these restrictions, the analysis is exact, not approximate.

Degrees of Freedom (d.o.f.). A structure has n d.o.f. if n independent quantities are needed to uniquely define the deformed configuration of the structure. The structure stiffness matrix will have n rows and n columns. In a plane truss, n is equal to two times the number of nodes allowed to displace. The individual d.o.f. are the x- and y-direction displacement components of each structure node. In nonstructural problems, d.o.f. are analogously defined: they are the independent quantities needed to define a field, such as the temperature field in a heat conduction problem. In heat conduction analysis, there is a single d.o.f. per node—namely, the nodal temperature.

2.2 STRUCTURE STIFFNESS EQUATIONS

We begin by generating the structure stiffness matrix [**K**] of a plane truss by a direct attack on the structure as a whole. The result will be used to illustrate certain concepts. Later we will show how [**K**] can be built by assembly of element matrices, which is the process actually used in computer programs.

Consider, for example, the three-bar truss of Fig. 2.2-1. Nodes and elements (bars) are numbered arbitrarily. For element i, where $i = 1, 2, 3$ in this example, let A_i = cross-sectional area, E_i = elastic modulus, and L_i = length. From elementary mechanics of materials, axial force F_i and change in length e_i have the relation

$$e_i = \frac{F_i L_i}{A_i E_i} \tag{2.2-1}$$

Stiffness is defined as the ratio of force to displacement and is by custom given the symbol k. Thus, the axial stiffness of any uniform bar of a truss is

$$k_i = \frac{F_i}{e_i} = \frac{A_i E_i}{L_i} \tag{2.2-2}$$

Supports at nodes 2 and 3 in Fig. 2.2-1 are temporarily removed, so that nonzero values can be assigned to all nodal displacements. Now let a node be displaced a small amount, first in the x direction and then in the y direction, while all other nodes are held at zero displacement. Thus there are six possible deformation states for a three-node plane truss. In each of these six states we calculate forces that must be applied to the nodes to maintain the deformation state. These forces, acting on the truss cut free of its supports, place the truss in static equilibrium. The first two free-body diagrams are shown in Fig. 2.2-2.

As an example of the force computation, consider the forces in Fig. 2.2-2a. Because displacement u_1 is small, its component along bar 2, which is the change in length of bar 2, is $e_2 = 0.6u_1$. This deformation produces the axial force $F_2 = k_2 e_2$, whose horizontal and vertical components have the respective magnitudes $0.6F_2 = 0.36k_2u_1$ and $0.8F_2 = 0.48k_2u_1$. Bar 3 has elongation $e_3 = u_1$ and contributes forces $F_3 = k_3e_3 = k_3u_1$.

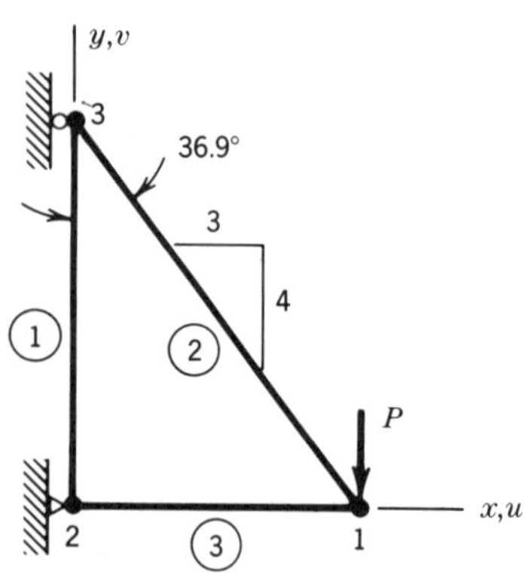

Figure 2.2-1. A three-bar plane truss. D.o.f. u_2, v_2, and u_3 are restrained. D.o.f. u_1, v_1, and v_3 are active (allowed to displace). Externally applied loading consists of force P.

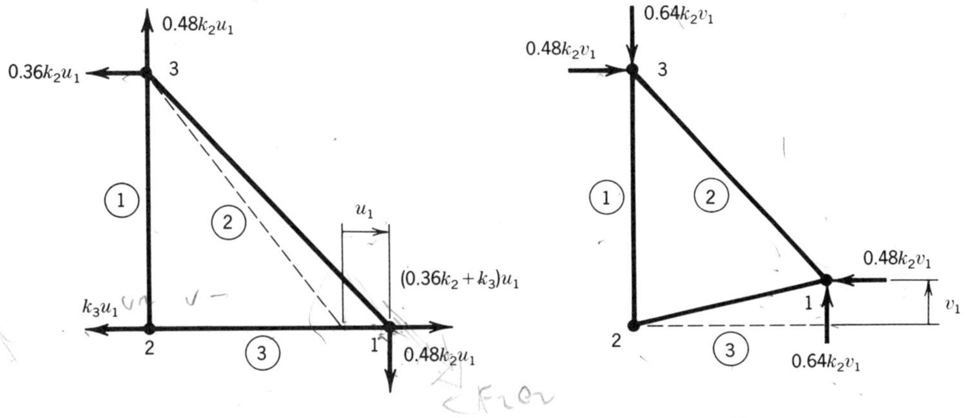

(*a*) (*b*)

Figure 2.2-2. Nodal loads consistent with the respective displacement states $\{\mathbf{D}\} = \lfloor u_1 \quad 0 \quad 0 \quad 0 \quad 0 \quad 0 \rfloor^T$ and $\{\mathbf{D}\} = \lfloor 0 \quad v_1 \quad 0 \quad 0 \quad 0 \quad 0 \rfloor^T$.

Let $\{\mathbf{Q}_1\}$ represent the vector of forces in Fig. 2.2-2*a* associated with *unit* displacement, $u_1 = 1$. Thus forces that appear in Fig. 2.2-2*a* are $\{\mathbf{Q}_1\}u_1$:

$$\{\mathbf{Q}_1\}u_1 = \lfloor k_3 + 0.36k_2 \quad -0.48k_2 \quad -k_3 \quad 0 \quad -0.36k_2 \quad 0.48k_2 \rfloor^T u_1 \tag{2.2-3}$$

Similarly, forces that appear in Fig. 2.2-2*b* are $\{\mathbf{Q}_2\}v_1$, where $\{\mathbf{Q}_2\}$ is the force vector associated with the unit displacement $v_1 = 1$:

$$\{\mathbf{Q}_2\}v_1 = \lfloor -0.48k_2 \quad 0.64k_2 \quad 0 \quad 0 \quad 0.48k_2 \quad -0.64k_2 \rfloor^T v_1 \tag{2.2-4}$$

Let $\{\mathbf{Q}_3\}$, $\{\mathbf{Q}_4\}$, $\{\mathbf{Q}_5\}$, and $\{\mathbf{Q}_6\}$ represent the equilibrium nodal force vectors associated with the remaining four unit displacement states $u_2 = 1$, $v_2 = 1$, $u_3 = 1$, and $v_3 = 1$. Then, if all six nodal d.o.f. may be nonzero *simultaneously*, the associated nodal loads are obtained by adding the six separate cases,

$$[\mathbf{Q}_1 \quad \mathbf{Q}_2 \quad \mathbf{Q}_3 \quad \mathbf{Q}_4 \quad \mathbf{Q}_5 \quad \mathbf{Q}_6] \begin{Bmatrix} u_1 \\ v_1 \\ u_2 \\ v_2 \\ u_3 \\ v_3 \end{Bmatrix} = \begin{Bmatrix} p_1 \\ q_1 \\ p_2 \\ q_2 \\ p_3 \\ q_3 \end{Bmatrix} \tag{2.2-5}$$

where forces applied to node i are called p_i and q_i, positive in $+x$ and $+y$ directions, respectively. Written out, Eq. 2.2-5 is

$$\begin{bmatrix} k_3 + 0.36k_2 & -0.48k_2 & -k_3 & 0 & -0.36k_2 & 0.48k_2 \\ -0.48k_2 & 0.64k_2 & 0 & 0 & 0.48k_2 & -0.64k_2 \\ -k_3 & 0 & k_3 & 0 & 0 & 0 \\ 0 & 0 & 0 & k_1 & 0 & -k_1 \\ -0.36k_2 & 0.48k_2 & 0 & 0 & 0.36k_2 & -0.48k_2 \\ 0.48k_2 & -0.64k_2 & 0 & -k_1 & -0.48k_2 & k_1 + 0.64k_2 \end{bmatrix} \begin{Bmatrix} u_1 \\ v_1 \\ u_2 \\ v_2 \\ u_3 \\ v_3 \end{Bmatrix} = \begin{Bmatrix} p_1 \\ q_1 \\ p_2 \\ q_2 \\ p_3 \\ q_3 \end{Bmatrix} \tag{2.2-6}$$

In our standard abbreviation, these structure stiffness equations are

$$[\mathbf{K}]\{\mathbf{D}\} = \{\mathbf{R}\} \tag{2.2-7}$$

where $[\mathbf{K}]$ is the *structure stiffness matrix*. As shown by the development, Eqs. 2.2-7 are equilibrium equations.

The physical meaning of $[\mathbf{K}]$, as well as a procedure for formulating $[\mathbf{K}]$, are contained in the following statement. *The jth column of* $[\mathbf{K}]$ *is the vector of loads that must be applied to nodal d.o.f. in order to maintain the deformation state associated with unit value of d.o.f. j while all other nodal d.o.f. are zero.* For a frame, "loads" include moments as well as forces. By this procedure—activating one d.o.f. at a time—we can generate the stiffness matrix of any truss or frame, regardless of the number of bars or the degree of static indeterminacy. The stiffness matrix is square; that is, there are as many equations as there are d.o.f.

2.3 PROPERTIES OF [K]. SOLUTION FOR UNKNOWNS

Properties of [K]. Each diagonal stiffness coefficient K_{ii} in Eq. 2.2-6 is positive. This is physically reasonable, as it means that a force R_i directed toward (say) the right will not produce a displacement directed toward the left (here it helps to imagine that all d.o.f. in $\{\mathbf{D}\}$ except D_i are constrained to be zero). In general, for any structure, no diagonal coefficient K_{ii} is negative or zero unless the structure is unstable.

$[\mathbf{K}]$ is symmetric. This is true of *any* structure that displays a linear relationship between applied loads and the resulting displacements. The symmetry of $[\mathbf{K}]$ may be proved by use of the procedure suggested in Problem 2.7.

If the structure is not attached to supports, its $[\mathbf{K}]$ does not resist rigid-body motion of the structure. An infinite number of vectors $\{\mathbf{D}\}$ that represent rigid-body motion can be written. For the plane truss of Fig. 2.2-1, with all supports removed, four of them are

$$\begin{aligned} \{\mathbf{D}\}_1 &= \lfloor \delta \quad 0 \quad \delta \quad 0 \quad \delta \quad 0 \rfloor^T & \{\mathbf{D}\}_2 &= \lfloor 0 \quad \delta \quad 0 \quad \delta \quad 0 \quad \delta \rfloor^T \\ \{\mathbf{D}\}_3 &= \lfloor \delta \quad \delta \quad \delta \quad \delta \quad \delta \quad \delta \rfloor^T & \{\mathbf{D}\}_4 &= \lfloor 4\theta \quad 3\theta \quad 4\theta \quad 0 \quad 0 \quad 0 \rfloor^T \end{aligned} \tag{2.3-1}$$

where δ is a small displacement and θ is a small angle of rotation. Respectively, the foregoing vectors represent translation along the x axis, translation along the y axis, translation along the line $x = y$, and rotation about node 3. For any plane structure only three of the infinitely many rigid-body $\{\mathbf{D}\}_i$ are linearly independent. The choice of three is not unique. For example, from Eqs. 2.3-1 we could choose $\{\mathbf{D}\}_4$ and any two of $\{\mathbf{D}\}_1$, $\{\mathbf{D}\}_2$, and $\{\mathbf{D}\}_3$. The first three $\{\mathbf{D}\}_i$ in Eqs. 2.3-1 are linearly dependent because $\{\mathbf{D}\}_3 = \{\mathbf{D}\}_1 + \{\mathbf{D}\}_2$.

A rigid-body motion does not deform a structure. Therefore, $[\mathbf{K}]\{\mathbf{D}\}_i = \{\mathbf{0}\}$ for any rigid-body motion $\{\mathbf{D}\}_i$. With reference to our three-bar truss example, and for $\delta = 1$ in Eqs. 2.3-1, the equations $[\mathbf{K}]\{\mathbf{D}\}_3 = \{\mathbf{0}\}$ state that coefficients in each row of $[\mathbf{K}]$ sum to zero. Note, however, that row sums of $[\mathbf{K}]$ will not vanish for any and all structures, since setting each entry in $\{\mathbf{D}\}$ to unity does not in general constitute rigid-body motion. Cases in point include structures that contain beam or plate elements, for which rotational d.o.f. are present. To summarize: for an unsupported structure, (a) $[\mathbf{K}]\{\mathbf{D}\} = \{\mathbf{0}\}$ when $\{\mathbf{D}\}$ represents rigid-body motion, and (b) each column of $[\mathbf{K}]$ represents a set of nodal forces and/or moments in static equilibrium.

Solution for Unknowns. The stiffness matrix of Eq. 2.2-6 is singular. Its order is 6 but its rank is 3. $[\mathbf{K}]$ cannot be inverted, nor can a unique $\{\mathbf{D}\}$ be obtained by solving equations. The physical reason for this is that rigid-body motion is still possible. Without supports, the structure will float away if the slightest external load is applied. Before continuing with the truss example we state a more general argument about solution for unknowns, as follows.

One must remove the singularity of $[\mathbf{K}]$ in order to solve for the unknown d.o.f. in $\{\mathbf{D}\}$. We now show a formal procedure by which this may be done. Let $\{\mathbf{D}_c\}$ and $\{\mathbf{R}_c\}$ be known d.o.f. and known loads, and $\{\mathbf{D}_x\}$ and $\{\mathbf{R}_x\}$ be as yet unknown d.o.f. and loads. By partitioning, accompanied by such rearrangement of matrix coefficients as may be necessary, the structural equations $[\mathbf{K}]\{\mathbf{D}\} = \{\mathbf{R}\}$ can be written in the form

$$\begin{bmatrix} \mathbf{K}_{11} & \mathbf{K}_{12} \\ \mathbf{K}_{21} & \mathbf{K}_{22} \end{bmatrix} \begin{Bmatrix} \mathbf{D}_x \\ \mathbf{D}_c \end{Bmatrix} = \begin{Bmatrix} \mathbf{R}_c \\ \mathbf{R}_x \end{Bmatrix} \tag{2.3-2}$$

KNOWN

KNOWN

or, in a more expanded form,

$$[\mathbf{K}_{11}]\{\mathbf{D}_x\} + [\mathbf{K}_{12}]\{\mathbf{D}_c\} = \{\mathbf{R}_c\} \tag{2.3-3}$$

$$[\mathbf{K}_{21}]\{\mathbf{D}_x\} + [\mathbf{K}_{22}]\{\mathbf{D}_c\} = \{\mathbf{R}_x\} \tag{2.3-4}$$

(Note that at this stage we know either a d.o.f. or its corresponding load, but not both.) $[\mathbf{K}_{11}]$ is nonsingular if the prescribed d.o.f. $\{\mathbf{D}_c\}$ are sufficient in arrangement and number to prevent rigid-body motion. Therefore, the unknown d.o.f. $\{\mathbf{D}_x\}$ can be found from Eq. 2.3-3:

$$\{\mathbf{D}_x\} = [\mathbf{K}_{11}]^{-1}(\{\mathbf{R}_c\} - [\mathbf{K}_{12}]\{\mathbf{D}_c\}) \tag{2.3-5}$$

Finally, unknown loads $\{\mathbf{R}_x\}$ can be found from Eq. 2.3-4 after substitution of d.o.f. $\{\mathbf{D}_x\}$, which are now known. In structural mechanics, $\{\mathbf{R}_x\}$ usually represents support reactions.

(*) In practice, the foregoing rearrangement of coefficients, partitioning, and matrix inversion are avoided by use of other operations that implicitly accomplish the same ends. These operations are discussed in Section 2.10.

We now apply the foregoing solution procedure to the truss of Fig. 2.2-1. Support conditions $\{\mathbf{D}_c\}$ are

$$u_2 = v_2 = u_3 = 0 \tag{2.3-6}$$

which means that $\{\mathbf{D}_c\} = \{\mathbf{0}\}$. Known loads $\{\mathbf{R}_c\}$, which correspond to as yet unknown d.o.f., are

$$p_1 = 0 \qquad q_1 = -P \qquad q_3 = 0 \tag{2.3-7}$$

Equation 2.3-3 becomes $[\mathbf{K}_{11}]\{\mathbf{D}_x\} = \{\mathbf{R}_c\}$, or

$$\begin{bmatrix} k_3 + 0.36k_2 & -0.48k_2 & 0.48k_2 \\ -0.48k_2 & 0.64k_2 & -0.64k_2 \\ 0.48k_2 & -0.64k_2 & k_1 + 0.64k_2 \end{bmatrix} \begin{Bmatrix} u_1 \\ v_1 \\ v_3 \end{Bmatrix} = \begin{Bmatrix} 0 \\ -P \\ 0 \end{Bmatrix} \tag{2.3-8}$$

Since $[\mathbf{K}_{11}]$ is nonsingular, Eq. 2.3-8 has a unique solution and can be solved for u_1, v_1, and v_3. Then, since $\{\mathbf{D}_c\} = \{\mathbf{0}\}$, Eq. 2.3-4 becomes $\{\mathbf{R}_x\} = [\mathbf{K}_{21}]\{\mathbf{D}_x\}$, or

$$\begin{Bmatrix} p_2 \\ q_2 \\ p_3 \end{Bmatrix} = \begin{bmatrix} -k_3 & 0 & 0 \\ 0 & 0 & -k_1 \\ -0.36k_2 & 0.48k_2 & -0.48k_2 \end{bmatrix} \begin{Bmatrix} u_1 \\ v_1 \\ v_3 \end{Bmatrix} \tag{2.3-9}$$

which can be solved for the support reactions p_2, q_2, and p_3 by substituting the known values of u_1, v_1, and v_3.

The case $\{\mathbf{D}_c\} = \{\mathbf{0}\}$, which says that all prescribed d.o.f. are zero, is common. When $\{\mathbf{D}_c\} = \{\mathbf{0}\}$ one can obtain $[\mathbf{K}_{11}]$ from $[\mathbf{K}]$ by discarding each row i and column i for which $D_i = 0$. In the present example problem, we obtain Eq. 2.3-8 by discarding rows and columns 3, 4, and 5 from Eq. 2.2-6.

Note also that $[\mathbf{K}_{11}]$ can be obtained by applying the procedure described in Section 2.2, using only active d.o.f. and calculating loads associated with these d.o.f. Thus, for the truss of Fig. 2.2-1, one obtains the first column of $[\mathbf{K}_{11}]$ by setting $u_1 = 1$ and $u_2 = v_2 = u_3 = 0$, then computing the nodal loads p_1, q_1, and q_3.

2.4 ELEMENT STIFFNESS EQUATIONS

In Section 2.2, $[\mathbf{K}]$ is generated directly by considering the structure as a whole. This approach clarifies the physical meaning of $[\mathbf{K}]$ but does not lend itself to computer implementation. In practice, $[\mathbf{K}]$ is built by summation of coefficients from element stiffness matrices $[\mathbf{k}]$. The summation process is easily computerized. In the present section we formulate the necessary element $[\mathbf{k}]$ matrix for a uniform plane truss member.

Let the element of Fig. 2.4-1 have constant cross-sectional area A and elastic modulus E. Everything needed to generate $[\mathbf{k}]$ can be found from A, E, and the four nodal coordinates x_i, x_j, y_i, and y_j. First, we compute

$$L = [(x_j - x_i)^2 + (y_j - y_i)^2]^{1/2} \tag{2.4-1a}$$

$$s = \sin\beta = \frac{y_j - y_i}{L} \tag{2.4-1b}$$

$$c = \cos\beta = \frac{x_j - x_i}{L} \tag{2.4-1c}$$

Next, as in Section 2.2, we generate columns of $[\mathbf{k}]$ by activating each d.o.f. in turn while keeping the others zero. The first of these four cases is shown in Fig. 2.4-2. Axial shortening cu_i produces an axial compressive force $F = (AE/L)cu_i$,

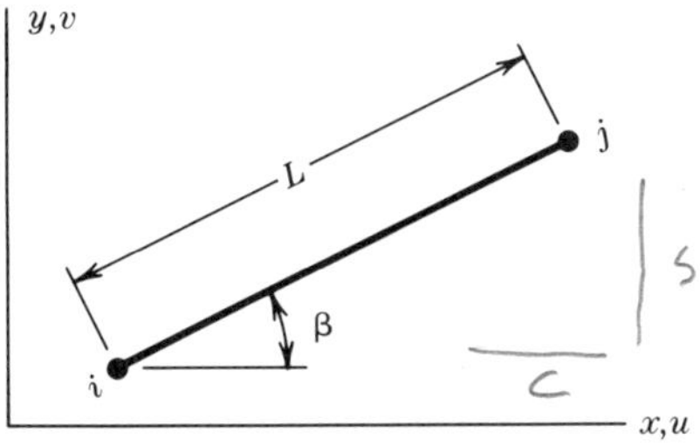

Figure 2.4-1. A uniform truss element, arbitrarily oriented in the *xy* plane.

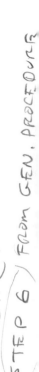

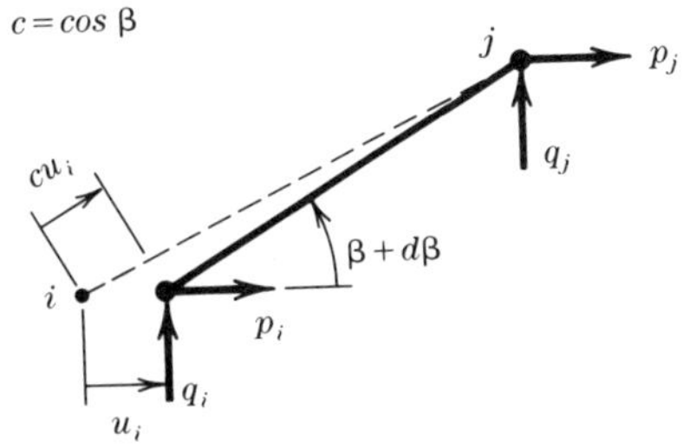

Figure 2.4-2. The truss element after nodal displacements $u_i > 0$, $v_i = u_j = v_j = 0$ have been imposed.

whose x and y components are $p_i = -p_j = Fc$ and $q_i = -q_j = Fs$. These components provide static equilibrium. Thus

$$\frac{AE}{L}\begin{Bmatrix} c^2 \\ cs \\ -c^2 \\ -cs \end{Bmatrix} u_i = \begin{Bmatrix} p_i \\ q_i \\ p_j \\ q_j \end{Bmatrix} \qquad (2.4\text{-}2)$$

Similar results are given by the remaining displacements, v_i, u_j, and v_j, when each acts alone. If all four d.o.f. may be nonzero simultaneously, we superpose results, just as in Eq. 2.2-5, and obtain

$$\frac{AE}{L}\begin{bmatrix} c^2 & cs & -c^2 & -cs \\ cs & s^2 & -cs & -s^2 \\ -c^2 & -cs & c^2 & cs \\ -cs & -s^2 & cs & s^2 \end{bmatrix}\begin{Bmatrix} u_i \\ v_i \\ u_j \\ v_j \end{Bmatrix} = \begin{Bmatrix} p_i \\ q_i \\ p_j \\ q_j \end{Bmatrix} \qquad (2.4\text{-}3)$$

where $c = \cos\beta$ and $s = \sin\beta$. The square matrix, including the factor AE/L, is the element stiffness matrix **[k]**. We abbreviate Eq. 2.4-3 as

$$[\mathbf{k}]\{\mathbf{d}\} = \{\bar{\mathbf{r}}\} \qquad (2.4\text{-}4)$$

The jth column of **[k]** is the vector of loads that must be applied to element nodes to maintain the deformation state when $d_j = 1$ and all other element nodal d.o.f. are zero.

Subsequently it will be desirable to distinguish between loads applied *by* the element and loads applied *to* the element. Loads $\{\bar{\mathbf{r}}\} = [\mathbf{k}]\{\mathbf{d}\}$ are applied *to* the element in order to sustain nodal d.o.f. $\{\mathbf{d}\}$.

Why are rotational d.o.f. not present in $\{\mathbf{d}\}$? Such d.o.f. would be present if bending stiffness of the bar were taken into account. Then the structure would be a beam or a plane frame. By definition, truss bars do not resist bending. In a truss, rotational d.o.f. are not introduced because there is no resistance to them. Their presence would make **[K]** a singular matrix.

Special Cases. If $\beta = 0$, as for bar 3 in Fig. 2.2-1, **[k]** remains 4 by 4 but contains only four nonzero coefficients, $k_{11} = k_{33} = -k_{13} = -k_{31} = AE/L$. Therefore, displacements v_i and v_j produce no nodal loads $\{\bar{\mathbf{r}}\}$. This is correct: in accordance with our assumptions, *small* lateral displacements v_i and v_j do not strain the bar and therefore generate no force. Similar remarks apply if $\beta = \pi/2$, π, and so on. That some diagonal coefficients in [**k**] are null does not necessarily mean that any diagonal coefficient in the assembled [**K**] will be null.

Imagine that $\beta = 0$ and that v_i and v_j are suppressed by striking out rows and columns 2 and 4 from Eq. 2.4-3. What remains is

$$\frac{AE}{L}\begin{bmatrix} 1 & -1 \\ -1 & 1 \end{bmatrix}\begin{Bmatrix} u_i \\ u_j \end{Bmatrix} = \begin{Bmatrix} p_i \\ p_j \end{Bmatrix} \tag{2.4-5}$$

This is the stiffness equation of a bar allowed only the axial d.o.f. u_i and u_j. It was previously derived (Eq. 1.2-3). The orientation of the bar in space does not matter if u_i, u_j, p_i, and p_j remain axially directed. In Section 7.5 we will see that Eq. 2.4-3 can be obtained from Eq. 2.4-5 by a simple coordinate transformation procedure.

The 2 by 2 stiffness matrix in Eq. 2.4-5 is used in various examples throughout this book.

2.5 ASSEMBLY OF ELEMENTS. PLANE TRUSS EXAMPLE

The process of assembling elements to form a structure can be symbolized as $[\mathbf{K}] = \Sigma\,[\mathbf{k}]$. In this section we consider arguments that apply to the three-bar truss of Fig. 2.2-1. In subsequent sections we present more general arguments and provide computer algorithms. Fortunately, the concepts and procedures of assembly depend very little on whether the structure is a truss, a frame, or a discretized continuum.

Physically, construction of the truss of Fig. 2.2-1 can be visualized as follows. Structure nodes are positioned in space and are assigned labels, such as 1, 2, and 3 in Fig. 2.2-1. Bars are at first unassembled, but each bar is tagged with a node label at each end to show where it is to be placed. One by one, the bars are attached to the appropriate structure nodes. The structure gains stiffness as each bar is added.

Symbolically, the foregoing process is that of starting with a null structure stiffness matrix [**K**], then adding to it the [**k**] of each element. When the last element has been added, the structure is complete and [**K**] is complete.

The summation $[\mathbf{K}] = \Sigma\,[\mathbf{k}]$ can be performed if each [**k**] is made to operate on {**D**}, the vector of structure d.o.f. If the structure has n d.o.f., this means that each [**k**] must be expanded to become an n by n matrix. Such expansion to "structure size" is a helpful conceptual device. Computationally it would be cumbersome. In Section 2.7 we will show how to perform the summation without expansion.

To begin, we write [**k**] for each bar in Fig. 2.2-1 as a 4 by 4 matrix. Element node labels i and j can be interchanged: in other words, adding π to angle β does not change Eq. 2.4-3. To this extent, node labels on the bars are arbitrary. Let k_1, k_2, and k_3 represent the AE/L factors of the respective bars, and apply Eq. 2.4-3.

Bar 1: Let $i = 2$ and $j = 3$. Hence $\beta = 90°$, $c = 0$, and $s = 1$.

$$[\mathbf{k}]_1\{\mathbf{d}\}_1 = k_1 \begin{bmatrix} 0 & 0 & 0 & 0 \\ 0 & 1 & 0 & -1 \\ 0 & 0 & 0 & 0 \\ 0 & -1 & 0 & 1 \end{bmatrix} \begin{Bmatrix} u_2 \\ v_2 \\ u_3 \\ v_3 \end{Bmatrix} \tag{2.5-1}$$

Bar 2: Let $i = 1$ and $j = 3$. Hence $\beta = 126.9°$, $c = -0.6$, and $s = 0.8$.

$$[\mathbf{k}]_2\{\mathbf{d}\}_2 = k_2 \begin{bmatrix} 0.36 & -0.48 & -0.36 & 0.48 \\ -0.48 & 0.64 & 0.48 & -0.64 \\ -0.36 & 0.48 & 0.36 & -0.48 \\ 0.48 & -0.64 & -0.48 & 0.64 \end{bmatrix} \begin{Bmatrix} u_1 \\ v_1 \\ u_3 \\ v_3 \end{Bmatrix} \tag{2.5-2}$$

Bar 3: Let $i = 1$ and $j = 2$. Hence $\beta = 180°$, $c = -1$, and $s = 0$.

$$[\mathbf{k}]_3\{\mathbf{d}\}_3 = k_3 \begin{bmatrix} 1 & 0 & -1 & 0 \\ 0 & 0 & 0 & 0 \\ -1 & 0 & 1 & 0 \\ 0 & 0 & 0 & 0 \end{bmatrix} \begin{Bmatrix} u_1 \\ v_1 \\ u_2 \\ v_2 \end{Bmatrix} \tag{2.5-3}$$

Each of the foregoing three [**k**] matrices must be expanded from 4 by 4 to 6 by 6. This is done by adding two rows of zeros and two columns of zeros—at the start in Eq. 2.5-1, in the middle in Eq. 2.5-2, and at the end in Eq. 2.5-3. Thus each element displacement vector becomes identical to the structure displacement vector {**D**}:

$$\{\mathbf{d}\}_1 = \{\mathbf{d}\}_2 = \{\mathbf{d}\}_3 = \{\mathbf{D}\} = \lfloor u_1 \quad v_1 \quad u_2 \quad v_2 \quad u_3 \quad v_3 \rfloor^T \tag{2.5-4}$$

Equation 2.5-4 enforces compatibility; that is, it makes end points of the pin-connected bars coincident under any displacement {**D**}.

The addition of rows and columns of zeros can be physically justified as follows. Consider, for example, the expanded [**k**] of element 2, in which rows 3 and 4 and columns 3 and 4 contain only zeros. Element 2 and node 2 are not connected. Therefore, no displacement of node 2 can strain element 2. Thus displacements u_2 and v_2 are associated with zero force, which accounts for the two columns of zeros. In addition, none of the six d.o.f. can produce forces at node 2 because there is no connecting material to resist strain or transmit load. This accounts for the two rows of zeros.

The reader can easily check that the expanded [**k**]'s from Eqs. 2.5-1, 2.5-2, and 2.5-3 do indeed add up to the structure [**K**] in Eq. 2.2-6. When one regards a column of [**K**] as a set of resisting forces, with each contributory force coming from elements connected to a common node, it becomes clear that one obtains [**K**] by adding element stiffness matrices.

Stiffness Coefficients That Remain Zero. When is $K_{ij} = 0$ in the assembled structure? A column of [**K**] represents nodal loads associated with activation of one and only one d.o.f. Activation of a d.o.f. creates nodal loads in only the element or elements that contain the d.o.f. in question. Other elements are not strained and produce no nodal loads. Therefore, in a column j of [**K**], coefficient K_{ij} is zero unless structure d.o.f. i and j are *both* present in at least one element. (K_{ij} *may* be zero even if d.o.f. i and j *are* shared by an element. For example, in Eq. 2.2-6, $K_{41} = K_{14} = 0$ because bar 3 of the truss happens to be horizontal.)

2.6 ASSEMBLY REGARDED AS SATISFYING EQUILIBRIUM

In stress analysis, assembly of elements can be regarded as a process of writing equations stating that each node of the structure is in static equilibrium under all loads applied to it. Nodal loads come from elements because of self-weight, temperature change, and lack of fit, from deformations associated with nodal displacements, and from external sources. The equilibrium argument is now explained with particular reference to the plane truss.

Loads applied to nodes because of gravity in the negative y direction are shown in Fig. 2.6-1*a*. Written formally for a four d.o.f. element, these loads are

$$\{\mathbf{r}_W\} = \frac{W}{2} \lfloor 0 \quad -1 \quad 0 \quad -1 \rfloor^T \tag{2.6-1}$$

where the total element weight W is equally apportioned to the two nodes. (We will assume that bending of an individual bar under its own weight can be neglected.) If a fully restrained bar is initially stress-free and then is uniformly heated T degrees, it sustains an axial compressive force $F = \alpha EAT$, where α is the coefficient of thermal expansion (Fig. 2.6-1*b*). The resulting nodal load vector is

$$\{\mathbf{r}_T\} = \alpha EAT \lfloor -c \quad -s \quad c \quad s \rfloor^T \tag{2.6-2}$$

where $c = \cos\beta$ and $s = \sin\beta$. The same forces $\{\mathbf{r}_T\}$ would arise from the force-fitting of a bar that is initally αLT units too long. We will use $\{\mathbf{r}_e\}$ to symbolize element loads. For the loads mentioned here,

$$\{\mathbf{r}_e\} = \{\mathbf{r}_W\} + \{\mathbf{r}_T\} \tag{2.6-3}$$

Loads $\{\bar{\mathbf{r}}\} = [\mathbf{k}]\{\mathbf{d}\}$ are loads applied *to* an element to sustain its deformation state $\{\mathbf{d}\}$. Therefore, equal and opposite loads $\{\mathbf{r}\} = -\{\bar{\mathbf{r}}\}$ are applied *by* the element

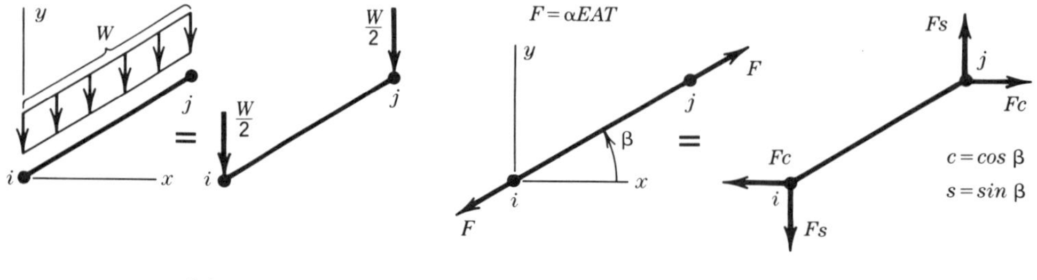

Figure 2.6-1. (*a*) Allocation of the weight W of a truss bar to its nodes. (*b*) Nodal loads associated with uniform heating of T degrees above the unstressed temperature.

to structure nodes, in accord with Newton's third law. That is, nodal loads associated with element deformation are

$$\{\mathbf{r}\} = -[\mathbf{k}]\{\mathbf{d}\} \tag{2.6-4}$$

Finally, loads applied to structure nodes by external sources are called $\{\mathbf{P}\}$. For example, $\{\mathbf{P}\} = \lfloor 0 \quad -P \quad 0 \quad 0 \quad 0 \quad 0 \rfloor^T$ for the truss of Fig. 2.2-1. Loads applied by fixed supports are usually not included in $\{\mathbf{P}\}$ because they would immediately be discarded by standard methods of imposing support conditions (see Section 2.10).

The set of equations that places each node in static equilibrium is

$$\{\mathbf{P}\} + \sum_{n=1}^{\text{numel}} \{\mathbf{r}\}_n + \sum_{n=1}^{\text{numel}} \{\mathbf{r}_e\}_n = \{\mathbf{0}\} \tag{2.6-5}$$

where *numel* is the number of elements in the structure. Summations are written because a typical node is connected to more than one element. However, a node receives $\{\mathbf{r}\}$ and $\{\mathbf{r}_e\}$ contributions only from the elements to which it is connected; thus, Eq. 2.6-5 implies the expansion of element vectors to "structure size" by addition of many zeros.

Substitution of Eq. 2.6-4 into 2.6-5 yields

$$[\mathbf{K}]\{\mathbf{D}\} = \{\mathbf{R}\} \tag{2.6-6}$$

where

$$[\mathbf{K}] = \sum_{n=1}^{\text{numel}} [\mathbf{k}]_n \qquad \text{and} \qquad \{\mathbf{R}\} = \{\mathbf{P}\} + \sum_{n=1}^{\text{numel}} \{\mathbf{r}_e\}_n \tag{2.6-7}$$

Summations imply the expansion of element arrays $[\mathbf{k}]$ and $\{\mathbf{r}_e\}$ to "structure size" so that $\{\mathbf{d}\}_n$ of each element n becomes identical to the structure displacement vector $\{\mathbf{D}\}$.

For a plane truss, Eq. 2.6-6 contains two equations per node. For a space truss there would be three equations per node.

2.7 ASSEMBLY AS DICTATED BY NODE NUMBERS

Element node labels, such as i and j in Fig. 2.4-1, serve only as convenient tags during the generation of element matrices. In the assembly process it is the *structure* node labels, such as 1, 2, and 3 in Fig. 2.2-1, that determine the locations in $[\mathbf{K}]$ and $\{\mathbf{R}\}$ to which coefficients in element arrays $[\mathbf{k}]$ and $\{\mathbf{r}_e\}$ are assigned. This is true of any finite element, regardless of its type, size, shape, or number of nodes.

As a simple example, consider a hypothetical structure that has two triangular elements and one d.o.f. per node (Fig. 2.7-1). This structure is *not* a truss. We need not know what physical problem is being modeled. We need say only that the characteristic matrix $[\mathbf{k}]$ of each element is 3 by 3, that structure nodes of

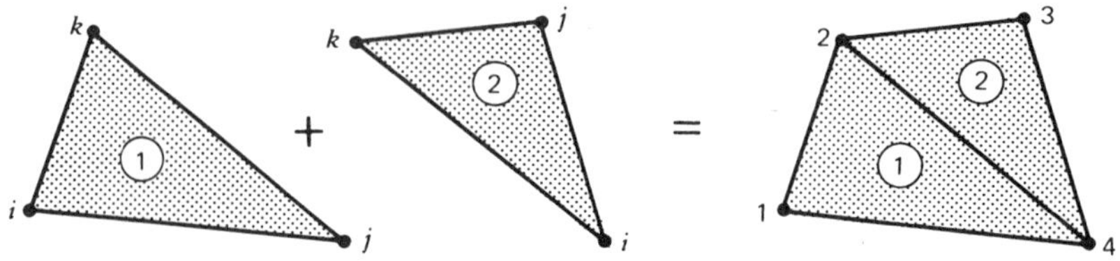

Figure 2.7-1. A hypothetical four-node structure built of two triangular elements. This structure is not a truss. Each node has a single d.o.f. Node labels are assigned arbitrarily.

element 1 are numbered 1, 4, and 2, and that structure nodes of element 2 are numbered 4, 3, and 2. For elements 1 and 2 in Fig. 2.7-1, we write

$$[\mathbf{k}]_1\{\mathbf{d}\}_1 = \begin{bmatrix} a_1 & a_2 & a_3 \\ a_4 & a_5 & a_6 \\ a_7 & a_8 & a_9 \end{bmatrix} \begin{Bmatrix} d_i \\ d_j \\ d_k \end{Bmatrix} \quad \text{and} \quad [\mathbf{k}]_2\{\mathbf{d}\}_2 = \begin{bmatrix} b_1 & b_2 & b_3 \\ b_4 & b_5 & b_6 \\ b_7 & b_8 & b_9 \end{bmatrix} \begin{Bmatrix} d_i \\ d_j \\ d_k \end{Bmatrix} \tag{2.7-1}$$

where d's are nodal d.o.f. Letter subscripts indicate *element* node labels. It does not matter how the a's and b's are calculated; it matters only that they exist. In the following explanation we ignore symmetry of the $[\mathbf{k}]$'s to show more clearly what happens to the a's and b's. Let nodal "loads" be called $\{\bar{\mathbf{r}}\}$, where $\{\bar{\mathbf{r}}\} = [\mathbf{k}]\{\mathbf{d}\}$. Now consider element 1. Its nodal loads, first in element labeling and then in structure labeling, are

$$\begin{aligned} \bar{r}_i &= a_1 d_i + a_2 d_j + a_3 d_k \\ \bar{r}_j &= a_4 d_i + a_5 d_j + a_6 d_k \\ \bar{r}_k &= a_7 d_i + a_8 d_j + a_9 d_k \end{aligned} \quad \text{and} \quad \begin{aligned} \bar{r}_1 &= a_1 D_1 + a_2 D_4 + a_3 D_2 \\ \bar{r}_4 &= a_4 D_1 + a_5 D_4 + a_6 D_2 \\ \bar{r}_2 &= a_7 D_1 + a_8 D_4 + a_9 D_2 \end{aligned} \tag{2.7-2}$$

To the latter group of equations we can add the equation $\bar{r}_3 = 0$ because node 3 is not attached to element 1. After this addition, and after rearrangement to place the D's in numerical order, we have for element 1

$$\begin{Bmatrix} \bar{r}_1 \\ \bar{r}_2 \\ \bar{r}_3 \\ \bar{r}_4 \end{Bmatrix} = \begin{bmatrix} a_1 & a_3 & 0 & a_2 \\ a_7 & a_9 & 0 & a_8 \\ 0 & 0 & 0 & 0 \\ a_4 & a_6 & 0 & a_5 \end{bmatrix} \begin{Bmatrix} D_1 \\ D_2 \\ D_3 \\ D_4 \end{Bmatrix} \tag{2.7-3}$$

in which the square matrix is $[\mathbf{k}]_1$. Element 2 can be treated similarly. Then, because the two matrices $[\mathbf{k}]_1$ and $[\mathbf{k}]_2$ have the same size and operate on the same vector of d.o.f. $\{\mathbf{D}\}$, we can write $[\mathbf{K}]\{\mathbf{D}\} = (\Sigma\, [\mathbf{k}])\{\mathbf{D}\}$, where

$$[\mathbf{K}] = [\mathbf{k}]_1 + [\mathbf{k}]_2 = \begin{bmatrix} a_1 & a_3 & 0 & a_2 \\ a_7 & a_9 & 0 & a_8 \\ 0 & 0 & 0 & 0 \\ a_4 & a_6 & 0 & a_5 \end{bmatrix} + \begin{bmatrix} 0 & 0 & 0 & 0 \\ 0 & b_9 & b_8 & b_7 \\ 0 & b_6 & b_5 & b_4 \\ 0 & b_3 & b_2 & b_1 \end{bmatrix} \tag{2.7-4}$$

We see that coefficients below the diagonal of an element [**k**] matrix (before reordering) may appear above the diagonal in [**K**]. This happens in the present example (but not in the truss example of Section 2.5) because expansion *and*

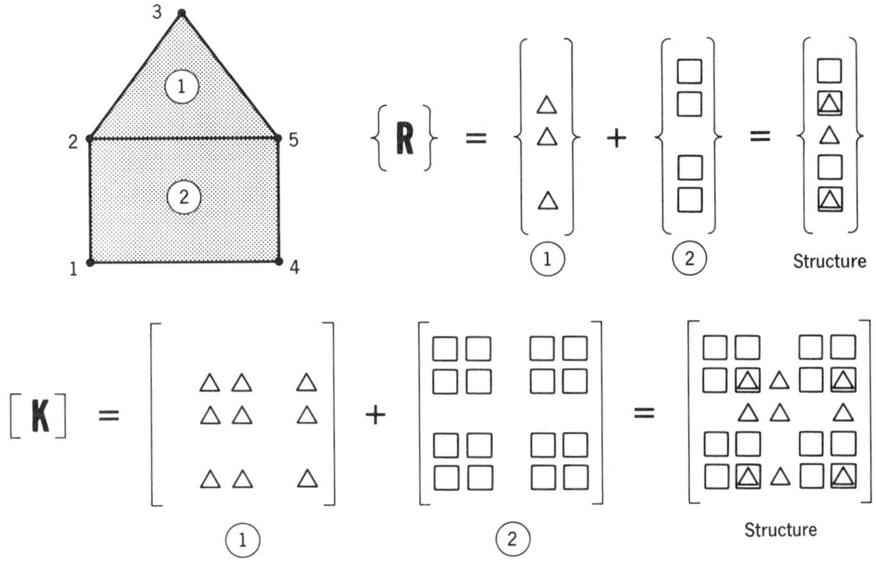

Figure 2.7-2. A hypothetical structure having one d.o.f. per node and built from a three-node element and a four-node element.

rearrangement of coefficients are needed to make element d.o.f. vectors {**d**} identical to the structure d.o.f. vector {**D**}.

If element d.o.f. labels are interchanged, then coefficients a_i and b_i in Eqs. 2.7-1 will be rearranged in the element [**k**] matrices. However, if structure node labels are preserved, the a_i and b_i are assigned to the same locations in the structure matrix [**K**] as before. For example, if in Fig. 2.7-1 the labels *ijk* are permuted to *jki*, then the first of Eqs. 2.7-2 becomes $\bar{r}_j = a_1 d_j + a_2 d_k + a_3 d_i$, but $i = 2$, $j = 1$, and $k = 4$, so that $\bar{r}_1 = a_1 D_1 + a_2 D_4 + a_3 D_2$ as before.

Another example of assembly appears in Fig. 2.7-2. Again the structure is hypothetical and is not a truss. Element matrices are shown already expanded to "structure size." Because element node labels are not shown and specific k_{ij} are not identified, Fig. 2.7-2 shows only the matrix topology of assembly. Note that {**R**} and [**K**] have the same row topology.

```
    DO 500 N=1,NUMEL
    CALL ELEMNT
    KK(1) = NOD(1,N)
    KK(2) = NOD(2,N)
    KK(3) = NOD(3,N)
    DO 400 I=1,3
    K = KK(I)
    R(K) = R(K)+RE(I)
    DO 300 J=1,3
    L = KK(J)
    S(K,L) = S(K,L)+SE(I,J)
300 CONTINUE
400 CONTINUE
500 CONTINUE
```

(*a*)

Terms:

S = [**K**]

SE = [**k**]

R = {**R**}

RE = {$\mathbf{r}_e$}

(*b*)

Example (Fig. 2.7-1):

NOD(1,1) = 1

NOD(2,1) = 4

NOD(3,1) = 2

NOD(1,2) = 4

NOD(2,2) = 3

NOD(3,2) = 2

(*c*)

Figure 2.7-3. (*a*) Fortran coding for assembly of element matrices. Each element has three nodes and one d.o.f. per node. NUMEL = number of elements in the structure. (*b*) Typical structural notation (as in Eqs. 2.6-7). (*c*) Example of array NOD.

```
      DO 500 N=1,NUMEL
      CALL ELEMNT
      KK(2) = 2*NOD(1,N)
      KK(1) = KK(2) - 1
      KK(4) = 2*NOD(2,N)
      KK(3) = KK(4) - 1
      DO 400 I=1,4
      K = KK(I)
      R(K) = R(K) + RE(I)
      DO 300 J=1,4
      L = KK(J)
      S(K,L)=S(K,L)+SE(I,J)
300   CONTINUE
400   CONTINUE
500   CONTINUE
```

(*a*)

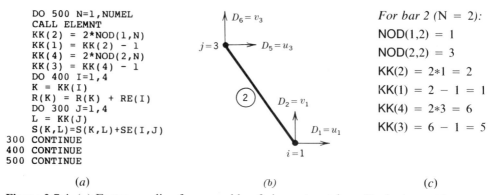

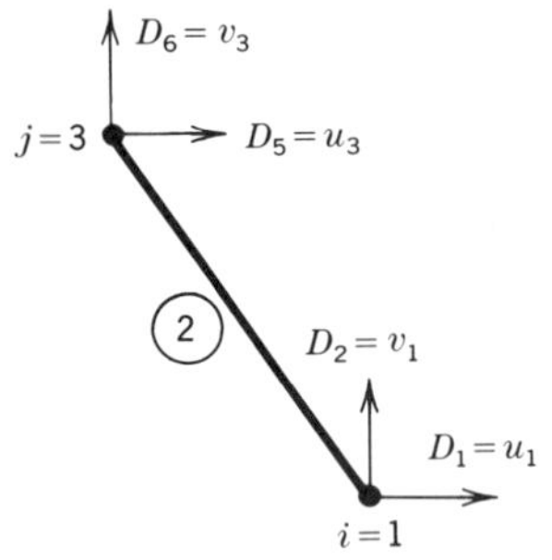

(*b*)

For bar 2 (N = 2):

NOD(1,2) = 1

NOD(2,2) = 3

KK(2) = 2*1 = 2

KK(1) = 2 − 1 = 1

KK(4) = 2*3 = 6

KK(3) = 6 − 1 = 5

(*c*)

Figure 2.7-4. (*a*) Fortran coding for assembly of element matrices. Each element has two nodes and two d.o.f. per node. NUMEL = number of elements in the structure. (*b*) Bar 2 of the truss of Fig. 2.2-1, showing the numbering of nodes and d.o.f. (*c*) Contents of arrays for bar 2.

Expansion to "structure size" is a conceptual device that need not actually be carried out. Assembly can be stated as an addition algorithm that assigns element coefficients to positions dictated by structure node numbers associated with the element. Such an algorithm appears in Fig. 2.7-3. Here it is assumed that each element has three nodes, each with a single d.o.f., as in Fig. 2.7-1. Structure node numbers that correspond to element node labels i, j, and k are assumed to have been previously stored in rows 1, 2, and 3 of arrary NOD, which has as many columns as there are elements in the structure. Subroutine ELEMNT (not shown) is assumed to return element matrices $[\mathbf{k}]$ and $\{\mathbf{r}_e\}$ in arrays SE and RE, respectively, via a COMMON block (not shown). Information in SE and RE is repeatedly created and destroyed as subroutine ELEMNT is called for each element in turn. Information in SE and RE is added into structural arrays S and R as part of the assembly process $[\mathbf{K}] = \Sigma\ [\mathbf{k}]$ and $\{\mathbf{R}\} = \Sigma\ \{\mathbf{r}_e\} + \{\mathbf{P}\}$. Externally applied loads $\{\mathbf{P}\}$ must be separately added to $\{\mathbf{R}\}$ after completing the algorithm in Fig. 2.7-3. It is assumed that arrays S and R are null before executing this assembly algorithm.

A similar assembly algorithm, applicable to a plane truss, is shown in Fig. 2.7-4. Each element has two nodes and each node has two d.o.f. Otherwise, this algorithm is like that of Fig. 2.7-3.

The algorithm of Fig. 2.7-4 can easily be altered to deal with a space truss, where each node has three d.o.f. and each $[\mathbf{k}]$ is 6 by 6. Array KK must contain six entries. The first three are KK(3) = 3*NOD(1,N), KK(2) = KK(3)-1, and KK(1) = KK(3)-2. Loop indices I and J must run from 1 to 6.

2.8 NODE NUMBERING THAT EXPLOITS MATRIX SPARSITY

A finite element structure with many d.o.f. has a sparse coefficient matrix $[\mathbf{K}]$. That is, most of the individual coefficients K_{ij} are zero. Sparsity should be exploited in order to economize on computer storage space and running time. Sparsity may be exploited by various schemes. In the present section we emphasize *bandedness,* which is among the simpler schemes.

The number of nonzero coefficients in [**K**], and their numerical values, are independent of how structure nodes are numbered. A change in structure node numbers changes only the *arrangement* of nonzero K_{ij}. Figure 2.8-1 is a case in point. The topology of nonzero coefficients in Fig. 2.8-1 can be understood by recalling that for any structure, a structure stiffness coefficient K_{ij} can be nonzero only if d.o.f. i and j are *both* present in at least one element.

Consider next the plane truss of Fig. 2.8-2. [**K**] is 12 by 12. For the first numbering, the topology of [**K**] is shown in Fig. 2.8-3*a*. The *semibandwidth* (also called the *half-bandwidth*) is given the symbol b. Here $b = 6$. Matrix [**K**] is symmetric and has a *total* bandwidth $2b - 1$. Bandwidth $2b - 1$ indicates the horizontal span of the zone in which all nonzero K_{ij} reside. This zone lies along the principal diagonal of [**K**]. Some zeros may appear within the band, but *only* zeros appear outside it.

A small semibandwidth is usually achieved by placing consecutive node numbers along the shorter dimension of a structure. The reader may check that the alternative numbering, in Fig. 2.8-2*b*, achieves the maximum possible semibandwidth for this problem ($b = 12$). A *small* value of b is desired.

The entire information content of a symmetric banded matrix resides in coefficients within the semiband. In practice, matrix order n_{eq} may greatly exceed semibandwidth b. If we store and process only the semiband rather than *all* coefficients in the upper triangle of [**K**], we decrease storage requirements by a factor of about $n_{eq}b/(n_{eq}^2/2) = 2b/n_{eq}$. In addition, as compared with processing a full but symmetric matrix, we reduce equation-solving expense by a factor of about $3b^2/n_{eq}^2$. For example, if $n_{eq} = 10b$, then the time needed to solve for d.o.f. {**D**} is reduced by a factor of about 30.

A simple storage format for the semiband is shown in Fig. 2.8-3*b*. Each row is shifted left: 1 space for row 2, 2 spaces for row 3, and in general $i - 1$ spaces for row i. Thus all diagonal coefficients K_{ii} of the matrix are stored in column 1 of the semiband array. To program the assembly of [**K**] in this form we need change only the innermost loop of the algorithm in Fig. 2.7-4. The required form of this innermost loop is shown in Fig. 2.8-4. The IF statement avoids coefficients K_{ij} below the main diagonal of [**K**], which would fall outside the stored semiband.

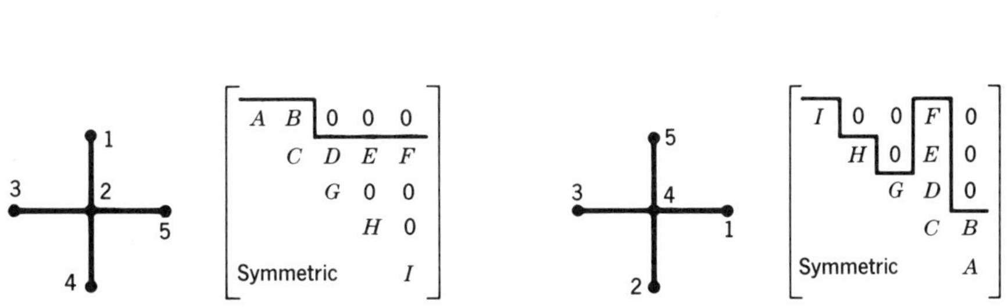

(*a*) (*b*)

Figure 2.8-1. A hypothetical four-element structure. Each element has two nodes. Each node has one d.o.f. Two different numberings (one the reverse of the other) and their associated stiffness matrices are shown. Capital letters indicate nonzero stiffness coefficients.

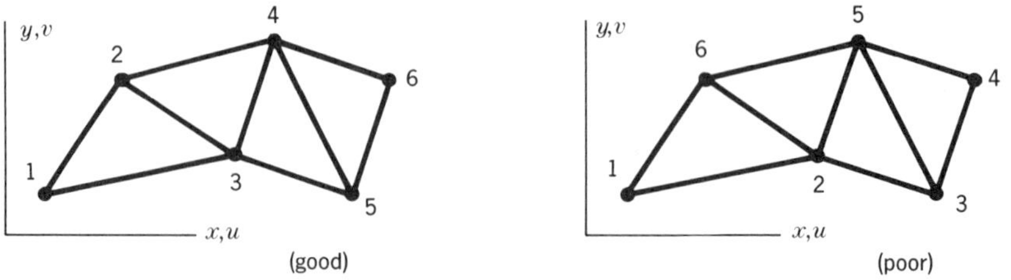

(a) (b)

Figure 2.8-2. A plane truss, showing node numberings that are (*a*) favorable and (*b*) unfavorable for achieving a banded stiffness matrix [**K**].

Some Computational Details. From Fig. 2.8-1 we see that the rightmost nonzero coefficient K_{ij} in any row may be close to the diagonal or far away. To compute b we must look for the K_{ij} farthest away. Specifically, b is the maximum of all n_{eq} values of b_i, where b_i is the number of columns from *and including* the diagonal to the rightmost nonzero K_{ij} in row i. Or, we can compute b by adding one to the magnitude of the maximum difference in active global d.o.f. in an element, using the element that displays the largest difference. For example, in Fig. 2.8-1*a* we find from element 2–5 that $b = (5 - 2) + 1 = 4$. In Fig. 2.8-2*a* we obtain $b = 6$ from elements 1–3, 2–4, 3–5, and 4–6; respectively, they give the d.o.f. differences $6 - 1 = 5$, $8 - 3 = 5$, $10 - 5 = 5$, and $12 - 7 = 5$. (D.o.f. that are suppressed, as by a fixed support, may not be listed in {**D**}. Then the numbering of active d.o.f. will not correspond to node numbers in such a convenient way. Information needed to determine b can still be found in columns of array ID. See Section 2.10.)

Both parts of Fig. 2.8-1 display a solid line that bounds the uppermost nonzero coefficient in each column of the matrix. This line is called the *skyline* (or *envelope*, or *profile*). We see that the matrices in Fig. 2.8-1 have the same semibandwidth but different skylines. As an alternative to the band storage scheme of Fig. 2.8-3, one could elect to store only the portion of each matrix column between

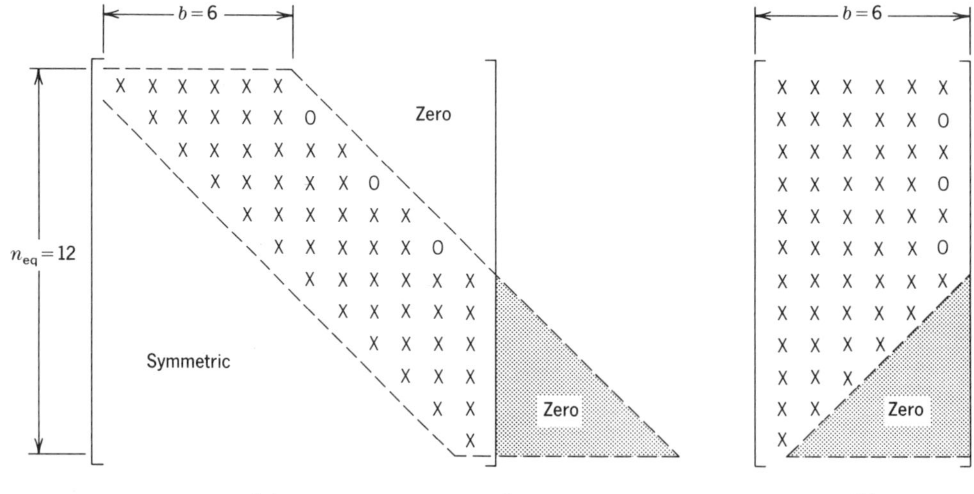

(a) (b)

Figure 2.8-3. (*a*) [**K**] for the plane truss of Fig. 2.8-2*a* (the "good" numbering). X = nonzero coefficient. (*b*) Band form storage of the same matrix.

```
      DO 300 J=1,4
      IF (KK(J) .LT. K)  GO TO 300
      L = KK(J) - K + 1
      S(K,L) = S(K,L) + SE(I,J)
300 CONTINUE
```

Figure 2.8-4. Altered form of the innermost loop of Fig. 2.7-4 to achieve the semiband storage format of Fig. 2.8-3*b*.

the skyline and the diagonal. Then Fig. 2.8-1*b* would be preferable to Fig. 2.8-1*a*. Another reason to prefer Fig. 2.8-1*b* is that equation solving creates *fills*; that is, a zero beneath the skyline is often changed to nonzero by the equation-solving process. There are three such zeros in Fig. 2.8-1*a* but none in Fig. 2.8-1*b*. Storage space must be reserved for fills, and fills must be processed after they are created.

The attention given to exploiting sparsity is worthwhile in reducing computation cost. This is clear from Fig. 2.8-5: for every nonzero coefficient in [**K**] there are over 100 zero coefficients. Many different storage schemes and equation-solving algorithms are available. Extensive discussion appears in Ref. 2.1.

2.9 AUTOMATIC ASSIGNMENT OF NODE NUMBERS

Imagine that a node must be added to the left of the truss of Fig. 2.8-2*a*. Two new bars will connect the new node to existing nodes 1 and 2. The entire structure must be renumbered if low bandwidth is to be preserved. It would be far more convenient if the new node could be given the next available number, 7, and the computer program could do the rest. That is, the computer program should accept arbitrary node numbers, adopt new numbers for efficient internal operations, and produce results in the user's original numbering system.

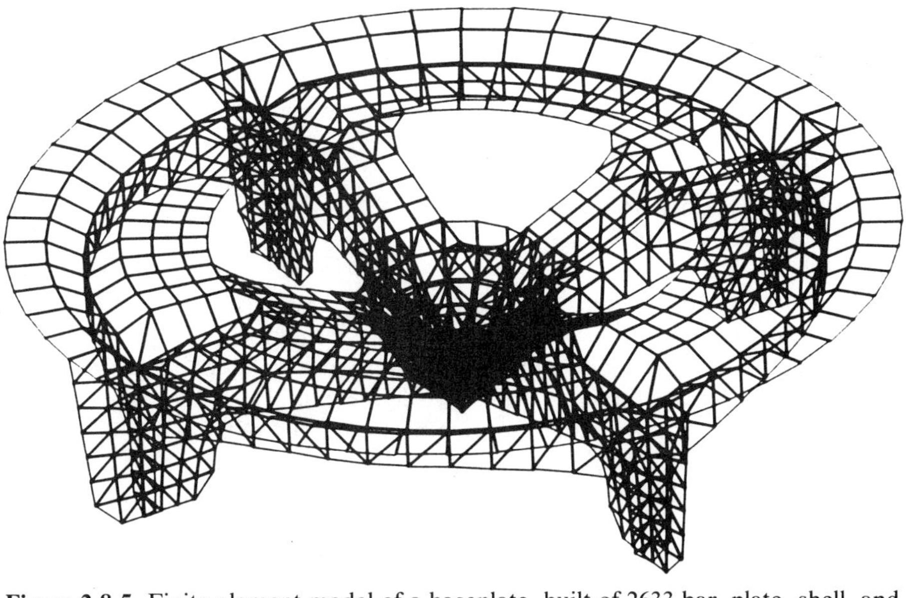

Figure 2.8-5. Finite element model of a baseplate, built of 2633 bar, plate, shell, and solid elements. There are 1005 nodes and from 3 to 6 d.o.f. per node. The semi-bandwidth is 852 and the density of [**K**] is 0.85% [2.2] (*Courtesy of G.C. Everstine, David W. Taylor Naval Ship R&D Center, Bethesda, Maryland.*)

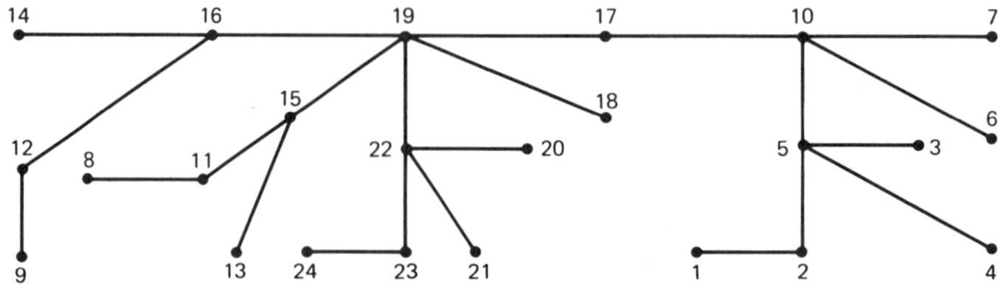

Figure 2.9-1. A graph that might represent a piping system. Node numbers have been chosen by a formal algorithm rather than by inspection.

Figure 2.9-1 shows the numbering achieved by a node-renumbering algorithm. For a single d.o.f. per node the semibandwidth is $b = 8$. Numbering by inspection, one would probably not achieve so low a value of b without considerable thought and several trials.

Renumbering algorithms exploit graph theory and its terminology [2.2,2.3]. Details are not presented here. Broadly speaking, a typical reordering algorithm examines a few promising numbering patterns (of the $N!$ possibilities for a N by N matrix) and selects the best one. The algorithm does not guarantee an optimum numbering, or even guarantee improvement over the numbering supplied to it. But the goal of reducing bandwidith—or skyline, or fills—is usually achieved. No single strategy is best for all goals or for all finite element meshes. Interestingly, simple reversal of node numbers (as in Fig. 2.8-1) may reduce the skyline while leaving bandwidth unchanged.

A "frontal" or "wave front" equation solver processes equations in element order rather than in node order. One then uses an algorithm that produces good *element* numbering.

A table of nodal connectivity is needed for node numbering. There is some computational expense in generating this table and doing the renumbering. But automatic renumbering is cost-effective, especially if the same numbering is used in repeated solutions, as is the case in nonlinear problems. Automatic renumbering is always worthwhile from the viewpoint of user convenience.

2.10 DISPLACEMENT BOUNDARY CONDITIONS

In the structure stiffness equations $[\mathbf{K}]\{\mathbf{D}\} = \{\mathbf{R}\}$, matrix $[\mathbf{K}]$ is singular and no unique solution for d.o.f. $\{\mathbf{D}\}$ is possible if the structure is unsupported. Some d.o.f. in $\{\mathbf{D}\}$ must be prescribed to enable a solution. Similarly, in a nonstructural problem where the matrices have other physical meanings, one or more d.o.f. D_i must be prescribed. The method described by Eqs. 2.3-2 to 2.3-4 usually requires row and column interchanges. Therefore, it is not well suited to computer programming. In this section we consider alternative procedures for imposing prescribed values of one or more d.o.f. D_i. Initially we assume that all prescribed D_i are prescribed as *zero*. Prescribed *non*zero D_i are considered subsequently.

A general-purpose program for structural analysis typically allows six d.o.f. per node (displacement in each coordinate direction and rotation about each coordinate axis). Often, not all of these d.o.f. are needed in the analysis of a particular structure. Indeed, for a plane structure, some *must* be eliminated: if no element resists a displacement D_i, then $K_{ii} = 0$ and [**K**] is singular. A plane structure, by definition, resists only in-plane distortions. Therefore, nodal d.o.f. that represent z-direction motion and rotations about x and y axes must be eliminated (i.e., prohibited) at all nodes. For a plane truss (but not a plane frame) we must also prohibit rotation θ_z about the z axis at all nodes. If θ_z is prohibited at all nodes, then truss elements, and the truss itself, can still have rigid body rotation in the xy plane because the prohibited rotations are not among their nodal d.o.f. (One can imagine that nodes of a plane truss are frictionless pins that connect bars together. A rotation θ_z of a pin does not deform the truss. Accordingly, θ_z is not resisted, and the associated rotational stiffness is zero.)

ID Array. We introduce a "destination array" ID, which is to be filled with numbers that indicate the locations in [**K**] to which element coefficients k_{ij} are to be assigned. Array ID has as many columns as there are nodes in the structure and as many rows as the maximum number d.o.f. allowed per node (typically six rows, for three displacement d.o.f. and three rotation d.o.f.). By use of array ID we will directly assemble matrix [$\mathbf{K}_{11}$] of Eq. 2.3-3, although we will call it simply [**K**] in what follows. Coefficients in [$\mathbf{K}_{12}$] of Eq. 2.3-3 will be discarded, which is acceptable if $\{\mathbf{D}_c\} = \{\mathbf{0}\}$ as is currently assumed.

Consider, for example, Fig. 2.10-1. If this structure is to be analyzed by use of a program that allows six d.o.f. per node, array ID has 6 rows and 8 columns. We start with ID null and, by means of input data, insert a 1 for each d.o.f. to be eliminated because it has a prescribed zero displacement. Support conditions in Fig. 2.10-1 dictate that $v_1 = u_5 = v_5 = u_7 = 0$. In addition, at each node i we must suppress z-direction displacement w_i (normal to the xy plane) and rotations θ_{xi}, θ_{yi}, and θ_{zi} about x, y, and z axes, respectively. The resulting ID array is shown in Fig. 2.10-2.

The next step is to convert array ID to a list of equation numbers by counting zeros in successive columns and converting each 1 to a zero. This counting, accomplished by the algorithm of Fig. 2.10-4, produces the result shown in Fig. 2.10-3. Zeros now indicate d.o.f. that are *not* to appear in vector {**D**} of active d.o.f. Nonzeros indicate equation numbers associated with active d.o.f. For example, to locate the D_i associated with node 7, we go to column 7 in Fig. 2.10-3, and find that all d.o.f. are suppressed except v_7, which appears as D_{10} in {**D**}. Matrix [**K**] for the supported structure is 12 by 12, where NEQ = 12 is computed in Fig. 2.10-4.

To assemble structural equations in the band format of Fig. 2.8-3*b*, while allowing only active d.o.f. to be present in {**D**}, we can make use of array ID. An assembly algorithm, obtained by combining and modifying Figs. 2.7-4 and 2.8-4, is shown in Fig. 2.10-5. Array KK is filled with structural equation numbers for

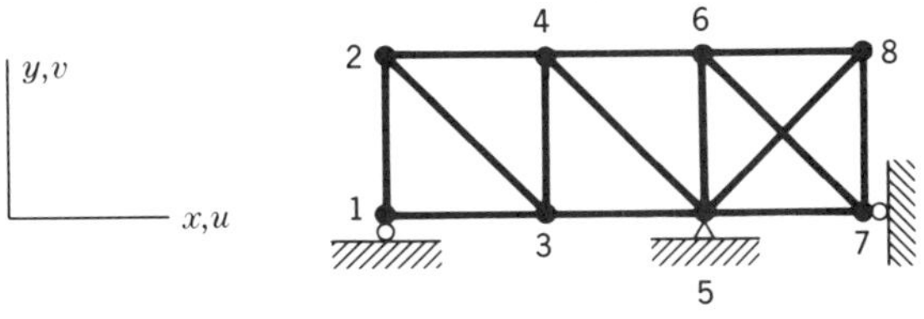

Figure 2.10-1. A plane truss, showing support conditions.

$$
\underset{\text{(input)}}{[\text{ID}]} = \begin{bmatrix} 0 & 0 & 0 & 0 & 1 & 0 & 1 & 0 \\ 1 & 0 & 0 & 0 & 1 & 0 & 0 & 0 \\ 1 & 1 & 1 & 1 & 1 & 1 & 1 & 1 \\ 1 & 1 & 1 & 1 & 1 & 1 & 1 & 1 \\ 1 & 1 & 1 & 1 & 1 & 1 & 1 & 1 \\ 1 & 1 & 1 & 1 & 1 & 1 & 1 & 1 \end{bmatrix} \begin{matrix} u \\ v \\ w \\ \theta_x \\ \theta_y \\ \theta_z \end{matrix}
$$

Figure 2.10-2. Array ID for the truss of Fig. 2.10-1, after input data has supplied 1's for d.o.f. to be suppressed. Types of nodal d.o.f. associated with each row are shown at the right.

$$
\underset{\text{(converted)}}{[\text{ID}]} = \begin{bmatrix} 1 & 2 & 4 & 6 & 0 & 8 & 0 & 11 \\ 0 & 3 & 5 & 7 & 0 & 9 & 10 & 12 \\ 0 & 0 & 0 & 0 & 0 & 0 & 0 & 0 \\ 0 & 0 & 0 & 0 & 0 & 0 & 0 & 0 \\ 0 & 0 & 0 & 0 & 0 & 0 & 0 & 0 \\ 0 & 0 & 0 & 0 & 0 & 0 & 0 & 0 \end{bmatrix}
$$

Figure 2.10-3. Array ID for the truss of Fig. 2.10-1, after conversion to a table of equation numbers (by means of Fig. 2.10-4 with NUMNP = 8 and NDOF = 6). It is now a "destination array."

each bar in turn. For example, bar 1–3 of Fig. 2.10-1 yields KK entries of 1, 0, 4, and 5, from columns 1 and 3 of array ID. The first IF statement in Fig. 2.10-5 discards K_{ij} and R_i in *rows* associated with suppressed d.o.f., such as the preceding KK(2) = 0, which indicates that $v_1 = 0$. The second IF statement discards K_{ij} in *columns* associated with suppressed d.o.f. and also discards k_{ij} that would fall below the diagonal of $[\mathbf{K}]$. Any externally applied loads $\{\mathbf{P}\}$ must be separately added to $\{\mathbf{R}\}$ after completing the assembly algorithm.

Figure 2.10-5 requires element $[\mathbf{k}]$'s to be 4 by 4. For generality, an arbitrary size should be allowed. Such generalizations are discussed in Ref. 2.1. As an exercise, one may imagine that the structure in Fig. 2.10-1 is a plane *frame,* whose nodal d.o.f. are u, v, and θ_z. How is the preceding discussion altered? If θ_z is not suppressed at any node, Fig. 2.10-2 is altered only by setting all entries in row 6 to zero. Figure 2.10-5 must be altered to allow for larger element arrays and one more d.o.f. per node.

Penalty Method. We now describe a method of imposing boundary conditions that allows prescribed d.o.f. to be either zero *or nonzero*. Consider again the truss of Fig. 2.2-1. Let nodal loads be R_1, R_2, and R_3, as shown in Fig. 2.10-6. Imagine that vertical displacement v_1 is to be forced to have a value $\bar{v}_1$. This condition can be treated as follows. Let k_s be a large positive stiffness, say 10^6 times K_{22}. Add a spring of stiffness k_s as shown in Fig. 2.10-6, and apply the large force $k_s\bar{v}_1$ in the direction of $\bar{v}_1$. Load R_2 is discarded. Now solve for all three d.o.f. (u_1, v_1, and v_3) in the usual way. If force $k_s\bar{v}_1$ were applied to *only* the added spring, its displacement would be precisely $\bar{v}_1$. In our model, the added spring is only slightly

```
      NEQ = 0
      DO 62 N=1,NUMNP
      DO 60 J=1,NDOF
C --- Transfer if D.O.F. is fixed. Otherwise increment NEQ.
      IF (ID(J,N) .GT. 0)  GO TO 58
      NEQ = NEQ + 1
      ID(J,N) = NEQ
      GO TO 60
   58 ID(J,N) = 0
   60 CONTINUE
   62 CONTINUE
```

Figure 2.10-4. Fortran statements that generate a table of equation numbers. This algorithm converts Fig. 2.10-2 to Fig. 2.10-3. Here NUMNP = total number of structure nodes, NDOF = number of d.o.f. per structure node allowed by the program, and NEQ = number of active d.o.f. = order of $[\mathbf{K}]$ for the constrained structure.

```
      DO 500 N=1,NUMEL
      CALL ELEMNT
      I = NOD(1,N)
      J = NOD(2,N)
      KK(1) = ID(1,I)
      KK(2) = ID(2,I)
      KK(3) = ID(1,J)
      KK(4) = ID(2,J)
      DO 400 I=1,4
      IF (KK(I) .LE. 0)  GO TO 400
      K = KK(I)
      R(K) = R(K) + RE(I)
      DO 300 J=1,4
      IF (KK(J) .LT. K)  GO TO 300
      L = KK(J) - K + 1
      S(K,L) = S(K,L) + SE(I,J)
  300 CONTINUE
  400 CONTINUE
  500 CONTINUE
```

Figure 2.10-5. Assembly of active stiffness equations in the banded format described by Fig. 2.8-3*b*. Arrays S and R must initially be null.

restrained by the comparatively flimsy truss, and we compute v_1 to be only slightly less than $\bar{v}_1$. Computed values of u_1 and v_3 are those appropriate to the remaining loads R_1 and R_3 and the computed value of v_1. The value $\bar{v}_1 = 0$ is permissible, in which case the force $k_s\bar{v}_1$ is zero.

Any or all structural d.o.f. can be prescribed in the foregoing way. Each prescription adds a large diagonal stiffness to [**K**], and also adds a large load to {**R**} if the prescribed d.o.f. is nonzero. Mathematically, this procedure is called a *penalty method* and k_s is called a *penalty number*. As k_s approaches infinity, the constraint $v_1 = \bar{v}_1$ is exactly enforced. Of course, for computational purposes, k_s is given a finite value. A large k_s greatly increases the maximum eigenvalues of [**K**] and may therefore cause trouble in a dynamic analysis.

Each prescription of a d.o.f. decreases the number of unknowns in {**D**} by one. Accordingly, the size of the system [**K**]{**D**} = {**R**} should contract. This is true if formal procedures such as Eqs. 2.3-2 through 2.3-5 are used, but rearrangement of coefficients in [**K**] is then required, which we wish to avoid in computation. Instead, by using the penalty method, we have elected to keep {**D**} the same size, and populated only with unknowns, by changing the *structure:* each prescription of a d.o.f. adds a stiff element, creating a new structure but leaving the number of active d.o.f. unchanged.

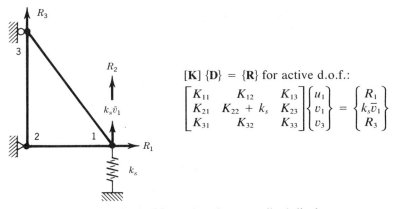

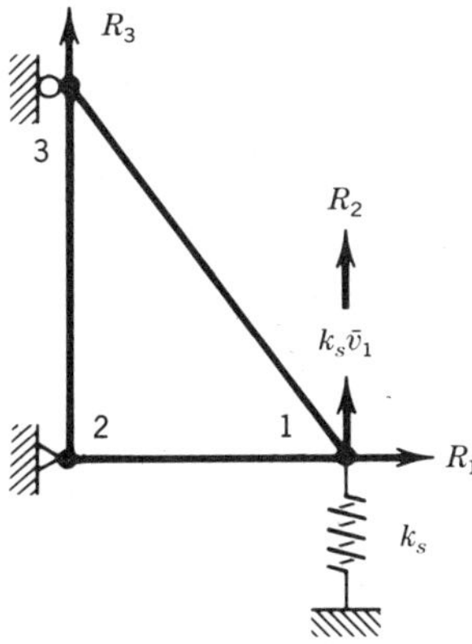

[**K**] {**D**} = {**R**} for active d.o.f.:

$$\begin{bmatrix} K_{11} & K_{12} & K_{13} \\ K_{21} & K_{22} + k_s & K_{23} \\ K_{31} & K_{32} & K_{33} \end{bmatrix} \begin{Bmatrix} u_1 \\ v_1 \\ v_3 \end{Bmatrix} = \begin{Bmatrix} R_1 \\ k_s\bar{v}_1 \\ R_3 \end{Bmatrix}$$

Figure 2.10-6. A method of imposing the prescribed displacement $v_1 = \bar{v}_1$ by adding a large stiffness k_s. External loads R_1 and R_3 may continue to act.

The procedure illustrated in Fig. 2.10-6, while couched in structural terminology, is generally applicable. For example, in heat conduction analysis [**K**] would be a conductance matrix, {**D**} a vector of temperatures, and {**R**} a vector of heat fluxes. The matrix operations in Fig. 2.10-6 then represent the prescription of a nodal temperature by adding a large conductance to [**K**] and placing a large flux in {**R**}.

Caution. A very stiff element should be parallel to a d.o.f. as is the case for k_s in Fig. 2.10-6. If k_s were inclined, or were placed within a structure, it would contribute to both diagonal and off-diagonal coefficients in [**K**]. This circumstance can lead to numerical difficulties (see Section 18.2).

More About Prescribed Nonzero D.O.F. Another method of imposing a prescribed displacement, either zero or nonzero, is illustrated with reference to the truss of Fig. 2.10-6. Again imagine that [**K**] is 3 by 3 for the supported structure and that displacement $v_1 = \bar{v}_1$ is to be imposed. As a first step we take known forces to the right side (Fig. 2.10-7*a*). But now the square matrix is unsymmetric and singular. We can restore symmetry and nonsingularity by replacing the second equation by the trivial equation $v_1 = \bar{v}_1$ (Fig. 2.10-7*b*). Solution of the latter set of equations gives $v_1 = \bar{v}_1$ and values of u_1 and v_3 appropriate to the system

$$\begin{bmatrix} K_{11} & K_{13} \\ K_{31} & K_{33} \end{bmatrix} \begin{Bmatrix} u_1 \\ v_3 \end{Bmatrix} = \begin{Bmatrix} R_1 - K_{12}\bar{v}_1 \\ R_3 - K_{32}\bar{v}_1 \end{Bmatrix} \tag{2.10-1}$$

The effect of the treatment in Fig. 2.10-7 is to obtain Eqs. 2.10-1 but without changing the size of [**K**].

When several d.o.f. are prescribed, one merely applies the foregoing treatment to each d.o.f. in turn. The result is a [**K**] with several rows and columns that are null except for 1's on the diagonal. [**K**] remains symmetric and banded. Results are exact, not approximate. Prescribed d.o.f. may be zero or nonzero. The method is not limited to structural problems.

Except for the load terms $K_{12}\bar{v}_1$ and $K_{32}\bar{v}_1$, Eq. 2.10-1 could be obtained by use of array ID. This observation suggests that we use array ID as before, so as to retain only d.o.f. not prescribed, but augment the procedure so as to obtain the extra load terms. More specifically, as each element is assembled, calculate loads $\{\bar{\mathbf{r}}\} = [\mathbf{k}]\{\mathbf{d}\}$ produced by prescribed d.o.f. in {**d**}, subtract $\{\bar{\mathbf{r}}\}$ from element loads $\{\mathbf{r}_e\}$, then assemble the net loads as before. If no d.o.f. are prescribed for the element at hand, or if the prescribed d.o.f. are zero, then $\{\bar{\mathbf{r}}\} = \{\mathbf{0}\}$, and {**R**} is not changed.

$$\begin{bmatrix} K_{11} & 0 & K_{13} \\ K_{21} & 0 & K_{23} \\ K_{31} & 0 & K_{33} \end{bmatrix} \begin{Bmatrix} u_1 \\ v_1 \\ v_3 \end{Bmatrix} = \begin{Bmatrix} R_1 - K_{12}\bar{v}_1 \\ R_2 - K_{22}\bar{v}_1 \\ R_3 - K_{32}\bar{v}_1 \end{Bmatrix}, \qquad \begin{bmatrix} K_{11} & 0 & K_{13} \\ 0 & 1 & 0 \\ K_{31} & 0 & K_{33} \end{bmatrix} \begin{Bmatrix} u_1 \\ v_1 \\ v_3 \end{Bmatrix} = \begin{Bmatrix} R_1 - K_{12}\bar{v}_1 \\ \bar{v}_1 \\ R_3 - K_{32}\bar{v}_1 \end{Bmatrix}$$

(*a*) (*b*)

Figure 2.10-7. Use of the "zero-one" treatment to impose displacement $v_1 = \bar{v}_1$ on the truss of Fig. 2.10-6. (*a*) Intermediate form. (*b*) Final form.

Bandwidth Calculation. A disadvantage of the treatment in Fig. 2.10-7 is that if many d.o.f. are prescribed, many useless zeros are stored and processed.[1] Moreover, calculation of semibandwidth b should recognize that whole groups of d.o.f. may be suppressed (e.g., rows 3, 4, 5, and 6 in the ID arrays of Figs. 2.10-2 and 2.10-3). As explained in Section 2.8, semibandwidth b can be calculated from knowledge of which structural d.o.f. are associated with nodes of each element. This information resides in columns of ID. Consider, for example, bar 2-4 in Fig. 2.10-1. We consult columns 2 and 4 in Fig. 2.10-3, ignore the zeros, and find that the largest difference among the d.o.f. numbers 2, 3, 6, and 7 is $7 - 2 = 5$. No other element yields a larger difference. Therefore, adding 1 to include the diagonal of $[\mathbf{K}]$, we conclude that $b = 5 + 1 = 6$ for this problem.

2.11 GAUSS ELIMINATION SOLUTION OF EQUATIONS

Structural equations $[\mathbf{K}]\{\mathbf{D}\} = \{\mathbf{R}\}$ can be solved by a direct method or an indirect (iterative) method. In either case there are many algorithms to choose from. Direct algorithms are favored in practice. Computational aspects of equation solving are discussed in Appendix B. In the present section we summarize Gauss elimination, which is a direct method, and illustrate its physical meaning.

Consider the application of Gauss elimination to the n_{eq} by n_{eq} system of stiffness equations $[\mathbf{K}]\{\mathbf{D}\} = \{\mathbf{R}\}$. The first equation is solved for D_1, then substituted into the subsequent equations. Thus, D_1 is said to be "eliminated." Then the second equation is solved for D_2 and substituted into subsequent equations, and so on. This forward-reduction process alters $\{\mathbf{R}\}$ and changes $[\mathbf{K}]$ to upper triangular form with 1's on the diagonal. Finally, numerical values of unknowns are computed by back-substitution, so that $D_{n_{eq}}$ is found first and D_1 is found last.

An example appears in Fig. 2.11-1. All d.o.f. are restrained except u_2, u_3, and u_4. Starting with the original matrix equation, Fig. 2.11-1*b*, we divide the first row by 2 and add it to the second row. This completes the substitution of u_2 into the remaining equations (Fig. 2.11-1*c*). Since u_2 does not appear in the third equation, the third row is unaffected. A similar substitution, now of row 2 into row 3 by multiplication of row 2 by 2/3 and addition, is shown in Fig. 2.11-1*d*. Figure 2.11-1*e* shows the result of dividing each equation by its diagonal coefficient (in an actual algorithm, this step may not be postponed until last). Solution for the d.o.f. by back-substitution is shown in Fig. 2.11-1*f*.

The foregoing process admits a physical interpretation: that each elimination releases the corresponding d.o.f., freeing it to move as dictated by applied loads and elastic properties of the structure. Consider the result of eliminating u_2, Fig. 2.11-1*c*. In the original structure, the diagonal coefficient is $K_{22} = 12$. That is, a force of 12 is needed to produce $u_3 = 1$ while $u_2 = u_4 = 0$; or, $K_{22} = 6 + 6$ is the sum of the adjacent bar stiffnesses seen by d.o.f. u_3. Elimination of u_2 effectively eliminates the constraint $u_2 = 0$ and places bars 1 and 2 in series, forming

[1] The disadvantage disappears if an "active column" equation solver is used. Each column of useless zeros may be discarded, leaving only the 1 on the diagonal (that is, a column of height one).

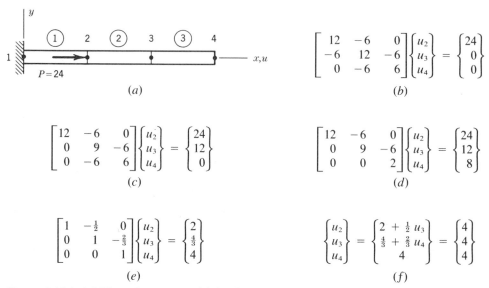

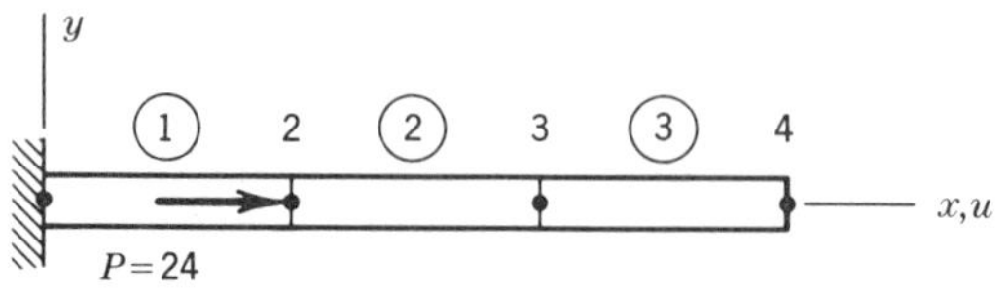

(*a*)

$$\begin{bmatrix} 12 & -6 & 0 \\ -6 & 12 & -6 \\ 0 & -6 & 6 \end{bmatrix} \begin{Bmatrix} u_2 \\ u_3 \\ u_4 \end{Bmatrix} = \begin{Bmatrix} 24 \\ 0 \\ 0 \end{Bmatrix}$$

(*b*)

$$\begin{bmatrix} 12 & -6 & 0 \\ 0 & 9 & -6 \\ 0 & -6 & 6 \end{bmatrix} \begin{Bmatrix} u_2 \\ u_3 \\ u_4 \end{Bmatrix} = \begin{Bmatrix} 24 \\ 12 \\ 0 \end{Bmatrix}$$

(*c*)

$$\begin{bmatrix} 12 & -6 & 0 \\ 0 & 9 & -6 \\ 0 & 0 & 2 \end{bmatrix} \begin{Bmatrix} u_2 \\ u_3 \\ u_4 \end{Bmatrix} = \begin{Bmatrix} 24 \\ 12 \\ 8 \end{Bmatrix}$$

(*d*)

$$\begin{bmatrix} 1 & -\frac{1}{2} & 0 \\ 0 & 1 & -\frac{2}{3} \\ 0 & 0 & 1 \end{bmatrix} \begin{Bmatrix} u_2 \\ u_3 \\ u_4 \end{Bmatrix} = \begin{Bmatrix} 2 \\ \frac{4}{3} \\ 4 \end{Bmatrix}$$

(*e*)

$$\begin{Bmatrix} u_2 \\ u_3 \\ u_4 \end{Bmatrix} = \begin{Bmatrix} 2 + \frac{1}{2} u_3 \\ \frac{4}{3} + \frac{2}{3} u_4 \\ 4 \end{Bmatrix} = \begin{Bmatrix} 4 \\ 4 \\ 4 \end{Bmatrix}$$

(*f*)

Figure 2.11-1. (*a*) Three-bar truss, with load $P = 24$ at node 2. (*b*) The structure equations $[\mathbf{K}]\{\mathbf{D}\} = \{\mathbf{R}\}$, if $AE/L = 6$ for each bar and only active d.o.f. are retained. (*c–f*) Stages in a Gauss elimination solution for nodal d.o.f. u_2, u_3, and u_4 (which are respectively D_1, D_2, and D_3).

a bar with nodes 1 and 3 whose axial stiffness is 3 rather than 6. Adjacent bar stiffnesses seen by d.o.f. u_3 are now 3 (from bars 1 and 2 in series) plus 6 (from bar 3), for a total of 9. Similarly, after elimination of u_3 and u_4 (Fig. 2.11-1*d*), the stiffness coefficient $K_{33} = 2$ represents the stiffness seen by d.o.f. u_4 when all three elements are connected in series with u_2 and u_3 free to move.

From the foregoing physical argument we conclude that each elimination reduces the stiffnesses seen by d.o.f. not yet eliminated, but does not reduce these stiffnesses to zero unless the structure is badly modeled or is without adequate support (Fig. 2.11-2). Accordingly, if a structure is properly modeled and adequately supported, we can proceed as in Fig. 2.11-1: use the *i*th equation to eliminate the *i*th d.o.f., without rearranging coefficients and without special coding to avoid zeros on the diagonal. If a zero diagonal coefficient *is* encountered, the user should check for an error in modeling or support conditions. (However, if [**K**] is not a true stifffness matrix, as for a "mixed" structural model or a nonstructural problem, zero and/or negative diagonal coefficients do not necessarily signal an error.)

After elimination of d.o.f. D_1 through D_i, where $1 \leq i < n_{eq}$, the lower right portion of [**K**] below row i remains symmetric if [**K**] was originally symmetric. Semibandwidth b is also preserved. Each elimination affects only b rows and up to b coefficients per row. These attributes are exploited in programming (see Appendix B and Ref. 2.1).

Often it is necessary to compute the response of a structure to several different sets of loads. Then one must process a single [**K**] but several vectors {**R**}. The processing of [**K**] need not be repeated in order to treat another {**R**}. This is fortunate, as the processing of [**K**] is by far the more expensive operation.

The notation $\{\mathbf{D}\} = [\mathbf{K}]^{-1}\{\mathbf{R}\}$ does not necessarily mean that matrix inversion is used. Often it means only that the equations $[\mathbf{K}]\{\mathbf{D}\} = \{\mathbf{R}\}$ are to be solved for

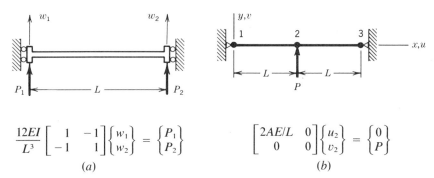

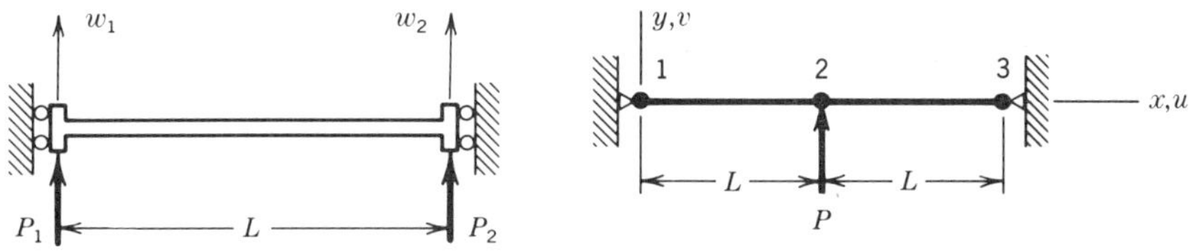

$$\frac{12EI}{L^3}\begin{bmatrix} 1 & -1 \\ -1 & 1 \end{bmatrix}\begin{Bmatrix} w_1 \\ w_2 \end{Bmatrix} = \begin{Bmatrix} P_1 \\ P_2 \end{Bmatrix} \qquad \begin{bmatrix} 2AE/L & 0 \\ 0 & 0 \end{bmatrix}\begin{Bmatrix} u_2 \\ v_2 \end{Bmatrix} = \begin{Bmatrix} 0 \\ P \end{Bmatrix}$$

(*a*) (*b*)

Figure 2.11-2. Structures for which Gauss elimination fails because **[K]** is singular. (*a*) One-element beam, with $\theta_1 = \theta_2 = 0$ the only d.o.f. prescribed. (*b*) Two-element truss with bars collinear at node 2.

{D} by any convenient or efficient method. Solving equations is faster than inverting a matrix. In addition, the inverse of a banded matrix is a *full* matrix. For these reasons matrix inversion is usually avoided.

2.12 STRESS COMPUTATION. SUPPORT REACTIONS

Stress Computation. After solving the global equations **[K]{D}** = **{R}** for **{D}**, all nodal d.o.f. of the structure are known. To compute stress in a given element we extract nodal d.o.f. **{d}** of that element from **{D}**, compute element strains from **{d}**, and finally compute stresses from strains.

Extraction of **{d}** from **{D}** is straightforward if *all* d.o.f. of the structure reside in **{D}** in a regular pattern. For a plane truss of n nodes, we may have **{D}** = $\lfloor u_1 \quad v_1 \quad u_2 \quad v_2 \cdots u_n \quad v_n \rfloor^T$. This is the case if the boundary condition treatment of Fig. 2.10-7 is applied to all prescribed d.o.f. of the structure. Thus, if i and j represent structure node numbers of a particular bar, then d.o.f. in **{d}** are $u_i = D_{2i-1}$, $v_i = D_{2i}$, $u_j = D_{2j-1}$, and $v_j = D_{2j}$. It does not matter that some of these d.o.f. were initially prescribed (usually as zero) rather than calculated by solving equations.

Things are not as simple if assembly makes use of array ID and the algorithm of Fig. 2.10-5. Now d.o.f. that are initially prescribed do not appear in **{D}**, thus destroying the regular pattern. However, columns of array ID associated with element nodes still contain the location in **{D}** of each nodal d.o.f. calculated by solving equations. An algorithm for extracting **{d}** from **{D}** appears in Fig. 2.12-1, in which it is assumed that all prescribed d.o.f. are zero. Prescribed *non*zero d.o.f. require that the statement DE(M) = 0 be altered.

Elongation e of a plane truss bar is computed from components of nodal d.o.f. parallel to the bar (see Fig. 2.4-2):

$$e = (u_j - u_i)\cos\beta + (v_j - v_i)\sin\beta \tag{2.12-1}$$

Axial strain is $\epsilon = e/L$. The bar is in uniaxial stress. Therefore, the axial stress caused by strain is

```
      M = 0
      DO 220 K=1,NNEL
C --- N is the structure number of node K of the Nth element.
      N = NOD(K,NTH)
      DO 200 L=1,NDOF
      M = M + 1
      DE(M) = 0.
      J = ID(L,N)
C --- J is zero only if the D.O.F. is fixed.
      IF (J .GT. 0)  DE(M) = D(J)
  200 CONTINUE
  220 CONTINUE
```

Figure 2.12-1. Fortran code to extract the displacement vector DE of the Nth element from the structure solution vector D when prescribed d.o.f. are zero and D lists only nonzero d.o.f. Here NNEL = number of nodes per element and NDOF = number of degrees of freedom per node allowed by the program (e.g., NDOF = 6 in Fig. 2.10-3).

$$\sigma = E\epsilon = E\frac{e}{L} \tag{2.12-2}$$

If the bar has thermal expansion coefficient α and is uniformly heated T degrees from its stress free state, then initial stress must be superposed on stress owing to mechanical strain. Thus, instead of Eq. 2.12-2, we have

$$\sigma = E\epsilon - E\alpha T = E\left(\frac{e}{L} - \alpha T\right) \tag{2.12-3}$$

The physical argument associated with thermal stress analysis is as follows. The argument applies to finite element structures in general, not only to a truss. With all d.o.f. fixed, compute loads that each element applies to its nodes because of heating or cooling (e.g., as in Fig. 2.6-1*b*). Add mechanical loads (if any). Release the d.o.f.; that is, find nodal displacements by solving $[\mathbf{K}]\{\mathbf{D}\} = \{\mathbf{R}\}$ for $\{\mathbf{D}\}$. Compute stress caused by nodal displacements (e.g., Ee/L in Eq. 2.12-3). Algebraically add stress associated with heating or cooling of the restrained element (e.g., $-E\alpha T$ in Eq. 2.12-3). Note that stress will be zero if thermal strain is uninhibited (e.g., if $e = \alpha TL$ in Eq. 2.12-3).

In stress analysis, displacements $\{\mathbf{d}\}$ yield strains $\{\boldsymbol{\epsilon}\}$, and strains yield stresses $\{\boldsymbol{\sigma}\}$ when multiplied by elastic constants. In more general terms, $\{\boldsymbol{\epsilon}\}$ is the gradient of the element displacement field produced by element d.o.f. $\{\mathbf{d}\}$. A similar remark applies to nonstructural problems, where typically a flow quantity is analogous to $\{\boldsymbol{\sigma}\}$. As examples, if nodal d.o.f. $\{\mathbf{d}\}$ are voltages or temperatures, the gradient of voltage or temperature produced by $\{\mathbf{d}\}$ yields a flow of current when multiplied by electrical conductivity, or of heat when multiplied by thermal conductivity.

Support Reactions. When all d.o.f. are known, support reactions $\{\mathbf{R}_x\}$ can be obtained from Eq. 2.3-4. Unfortunately, while imposing displacement boundary conditions by methods discussed in Section 2.10, the K_{ij} belonging to arrays $[\mathbf{K}_{21}]$ and $[\mathbf{K}_{22}]$ of Eq. 2.3-4 have been discarded. One can either save the necessary coefficients in a file before imposing displacements, or regenerate the coefficients later.

A particular reaction R_i in the list $\{\mathbf{R}_x\}$ can be computed as

$$\sum_j K_{ij}D_j = R_i \qquad \text{or} \qquad \sum_m \left(\sum_j k_{ij}d_j\right) = R_i \tag{2.12-4}$$

The first equation uses the j nonzero entries in row i of [**K**]. The second equation is a similar sum, taken over the m elements joined to the node that R_i acts upon.

What is the meaning of {**R**} if large stiffnesses have been added in the process of imposing nonzero values of certain d.o.f. (as in Fig. 2.10-6)? If [**K**] pertains to the original structure, before the addition of large stiffnesses, then loads {**R**} = [**K**]{**D**} include loads originally applied to d.o.f. that are free to move and loads that must be applied to the original structure in order to produce the prescribed d.o.f.

2.13 SUMMARY OF PROCEDURE

The principal computational steps of linear static stress analysis by the finite element method are now listed. Analogous steps are used for linear time-independent analysis of a nonstructural problem.

1. *Input and initialization.* Input the number of nodes and elements, nodal coordinates, structure node numbers of each element, material properties, temperature changes, mechanical loads, and boundary conditions. Reserve storage space for structure arrays [**K**] and {**R**}. Initialize [**K**] and {**R**} to null arrays. If array ID is used to manage boundary conditions, initialize ID and then convert it to a table of equation numbers.
2. *Compute element properties.* For each element: compute element property matrix [**k**] and element load vector {$\mathbf{r}_e$}.
3. *Assemble the structure.* Add [**k**] into [**K**] and {$\mathbf{r}_e$} into {**R**}. Go back to step 2. Repeat steps 2 and 3 until all elements are assembled. Add external loads {**P**} to {**R**}. Impose displacement boundary conditions (if not imposed implicitly during assembly by use of array ID).
4. *Solve the equations* [**K**]{**D**} = {**R**} *for* {**D**}.
5. *Stress calculation.* For each element, extract {**d**} from {**D**}. Compute mechanical strains produced by {**d**}. Include initial strains, if any, and convert resultant strains to stresses.

The foregoing steps outline an austere computer program, without preprocessors or postprocessors, automatic node renumbering, and other conveniences for the user. Modifications of the procedure are possible, such as computing properties of all elements before assembling any, and alternating steps of assembly with steps of equation solving.

Example. We illustrate the foregoing steps by applying them to the three-bar truss of Fig. 2.13-1. Only axial displacements and axial loads are present. For clarity we will use symbols as well as numbers. In actual computation only numbers would be present.

1. *Input and initialization.* Read the number of nodes and the number of elements: NUMNP = 4 and NUMEL = 3. Nodal coordinates are

$$x_1 = 0 \qquad x_2 = L \qquad x_3 = 2L \qquad x_4 = 3L$$

$$y_1 = z_1 = 0 \qquad y_2 = z_2 = 0 \qquad y_3 = z_3 = 0 \qquad y_4 = z_4 = 0$$

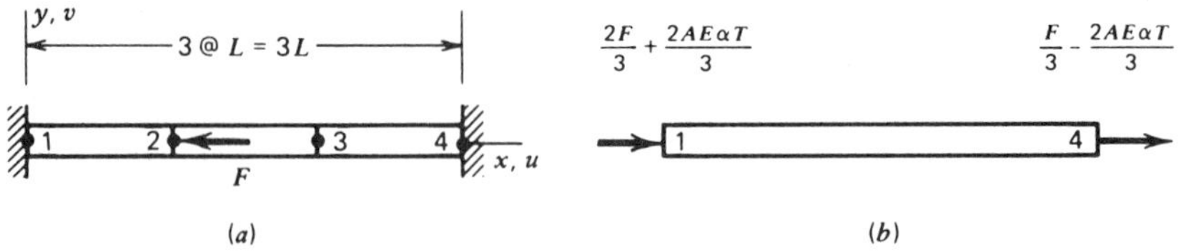

Figure 2.13-1. (*a*) Example problem. The bar is divided into three identical elements and has four nodes. Elements 1 and 2 *only* are uniformly heated T degrees. (*b*) Support reactions predicted by elementary mechanics of materials theory.

Node numbers associated with the three elements are

$$\text{NOD}(1,1) = 1 \qquad \text{NOD}(1,2) = 2 \qquad \text{NOD}(1,3) = 3$$
$$\text{NOD}(2,1) = 2 \qquad \text{NOD}(2,2) = 3 \qquad \text{NOD}(2,3) = 4$$

Read cross-sectional area A, elastic modulus E, and coefficient of thermal expansion α (the same for each element in this example). The left two elements *only* are uniformly heated T degrees above the stress-free temperature of the structure. External force F is applied in the negative direction at node 2. Boundary conditions prohibit all nodal motions except u_2 and u_3. For the sake of explanation we presume that the computer program allows only three d.o.f. per node (u, v, and w). Thus for array ID we have

$$\underset{\text{(input)}}{[\text{ID}]} = \begin{bmatrix} 1 & 0 & 0 & 1 \\ 1 & 1 & 1 & 1 \\ 1 & 1 & 1 & 1 \end{bmatrix} \quad \text{and} \quad \underset{\text{(converted)}}{[\text{ID}]} = \begin{bmatrix} 0 & 1 & 2 & 0 \\ 0 & 0 & 0 & 0 \\ 0 & 0 & 0 & 0 \end{bmatrix}$$

2. *Compute element properties*. The stiffness matrix of each element is

$$[\mathbf{k}] = \frac{AE}{L}\begin{bmatrix} 1 & 0 & 0 & -1 & 0 & 0 \\ 0 & 0 & 0 & 0 & 0 & 0 \\ 0 & 0 & 0 & 0 & 0 & 0 \\ -1 & 0 & 0 & 1 & 0 & 0 \\ 0 & 0 & 0 & 0 & 0 & 0 \\ 0 & 0 & 0 & 0 & 0 & 0 \end{bmatrix}$$

where, in element 1 for example, the associated structural nodal d.o.f. are $\{\mathbf{d}\} = \lfloor u_1 \quad v_1 \quad w_1 \quad u_2 \quad v_2 \quad w_2 \rfloor^T$. From Eq. 2.6-2, nodal loads of the three elments are:

$$\{\mathbf{r}_e\}_1 = \{\mathbf{r}_e\}_2 = \alpha EAT \lfloor -1 \quad 0 \quad 0 \quad 1 \quad 0 \quad 0 \rfloor^T$$
$$\{\mathbf{r}_e\}_3 = \alpha EAT \lfloor 0 \quad 0 \quad 0 \quad 0 \quad 0 \quad 0 \rfloor^T$$

3. *Assemble the structure*. In the assembly algorithm of Fig. 2.10-5, the contents of array KK for the successive elements are $\lfloor 0 \quad 0 \quad 0 \quad 1 \quad 0 \quad 0 \rfloor$, $\lfloor 1 \quad 0 \quad 0 \quad 2 \quad 0 \quad 0 \rfloor$, and $\lfloor 2 \quad 0 \quad 0 \quad 0 \quad 0 \quad 0 \rfloor$. Information associated with nodes 1 and 4 is discarded, and the "active" structure stiffness matrix $[\mathbf{K}]$ is 2 by 2. For illustration, consider where this 2 by 2 matrix would appear in a 4 by 4 stiffness matrix that operates on d.o.f. u_1 through u_4. (The entire 4 by 4 matrix need not actually be formed.) The contribution of the leftmost element to structure arrays $[\mathbf{K}]$ and $\{\mathbf{R}\}$ is

$$\frac{AE}{L}\begin{bmatrix} \overset{u_1}{1} & \overset{u_2}{-1} & \overset{u_3}{\cdot} & \overset{u_4}{\cdot} \\ -1 & 1 & \cdot & \cdot \\ \cdot & \cdot & \cdot & \cdot \\ \cdot & \cdot & \cdot & \cdot \end{bmatrix} \quad \text{and} \quad \alpha EAT\begin{Bmatrix} -1 \\ 1 \\ \cdot \\ \cdot \end{Bmatrix}$$

where dashed lines enclose [**K**] and {**R**}. Dots indicate locations in the structure arrays that receive no contribution from the element. In similar notation, the assembly of all three elements is written

$$[\mathbf{K}] = \frac{AE}{L}\begin{bmatrix} 1 & \cdot \\ \cdot & \cdot \end{bmatrix} + \frac{AE}{L}\begin{bmatrix} 1 & -1 \\ -1 & 1 \end{bmatrix} + \frac{AE}{L}\begin{bmatrix} \cdot & \cdot \\ \cdot & 1 \end{bmatrix} = \frac{AE}{L}\begin{bmatrix} 2 & -1 \\ -1 & 2 \end{bmatrix}$$

$$\{\mathbf{R}\} = \alpha EAT\begin{Bmatrix} 1 \\ \cdot \end{Bmatrix} + \alpha EAT\begin{Bmatrix} -1 \\ 1 \end{Bmatrix} + \alpha EAT\begin{Bmatrix} \cdot \\ 0 \end{Bmatrix} + \begin{Bmatrix} -F \\ 0 \end{Bmatrix} = \begin{Bmatrix} -F \\ \alpha EAT \end{Bmatrix}$$

4. *Solve the equations* $[\mathbf{K}]\{\mathbf{D}\} = \{\mathbf{R}\}$ *for* $\{\mathbf{D}\}$.

$$D_1 = u_2 = -\frac{2FL}{3AE} + \frac{\alpha LT}{3} \quad \text{and} \quad D_2 = u_3 = -\frac{FL}{3AE} + \frac{2\alpha LT}{3}$$

5. *Stress calculation.* With $u_1 = u_4 = 0$ and the rightmost element not heated,

$$\{\mathbf{d}\}_1 = \begin{Bmatrix} 0 \\ u_2 \end{Bmatrix} \quad \text{and} \quad \sigma_1 = E\left(\frac{u_2 - 0}{L} - \alpha T\right) = -\frac{2F}{3A} - \frac{2\alpha ET}{3}$$

$$\{\mathbf{d}\}_2 = \begin{Bmatrix} u_2 \\ u_3 \end{Bmatrix} \quad \text{and} \quad \sigma_2 = E\left(\frac{u_3 - u_2}{L} - \alpha T\right) = \frac{F}{3A} - \frac{2\alpha ET}{3}$$

$$\{\mathbf{d}\}_3 = \begin{Bmatrix} u_3 \\ 0 \end{Bmatrix} \quad \text{and} \quad \sigma_3 = E\left(\frac{0 - u_3}{L} - 0\right) = \frac{F}{3A} - \frac{2\alpha ET}{3}$$

These results agree with results given by elementary mechanics of materials. Thermal stress is constant over the entire length of the structure, as should be expected.

PROBLEMS

Section 2.2

2.1 For the plane truss of Fig. 2.2-1, sketch the remaining four free-body diagrams not shown in Fig. 2.2-2 and write equations analogous to Eqs. 2.2-3 and 2.2-4.

2.2 Four springs, each of stiffness k, are constrained to slide in a circular frictionless track as shown. Nodes 1, 2, 3, and 4 are allowed only small displacements u, tangent to the circular track and positive counterclockwise. Write the structure stiffness matrix [**K**]. What is its rank?

2.3 A plane truss is shown in the sketch. Set up an initially null matrix [**K**] having 12 rows and 12 columns. Indentify locations of nonzero coefficients K_{ij} by inserting at proper positions in [**K**] a "+" sign if $K_{ij} > 0$ or a "−" sign if $K_{ij} < 0$.

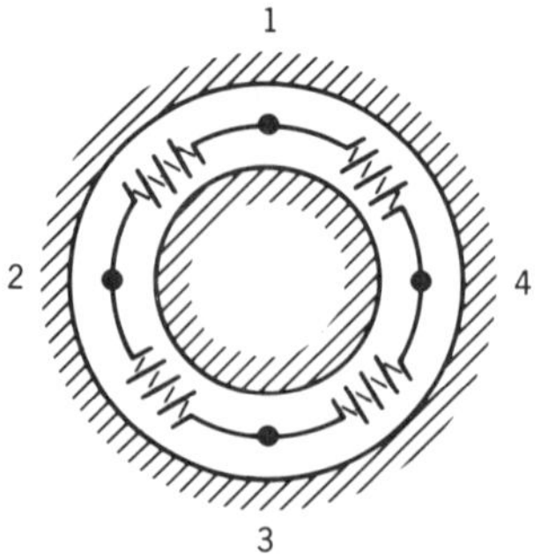

Problem 2.2

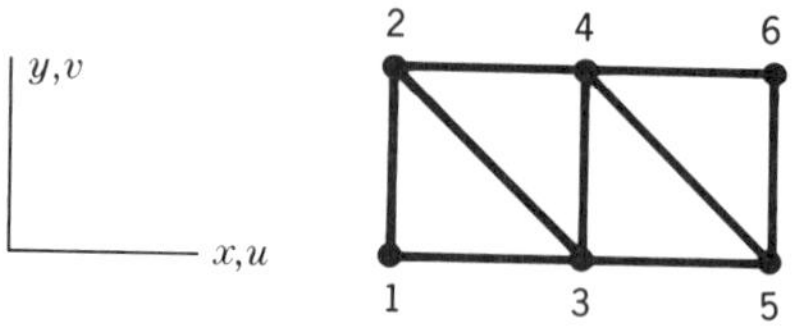

Problem 2.3

2.4 (a) Follow the instructions of Problem 2.3, but with reference to the six-d.o.f. beam shown in Problem 2.12. Let the beam be uniform and let $L_1 = L_2$.

(b) Repeat part (a), but let $L_1 > L_2$.

(c) Repeat part (a), but let $L_1 < L_2$.

2.5 Follow the instructions of Problem 2.3 but with reference to the eight-d.o.f. truss shown. All bars have the same A and the same E. Each of the nodal d.o.f. (D_1 through D_8) is parallel or perpendicular to one or more bars.

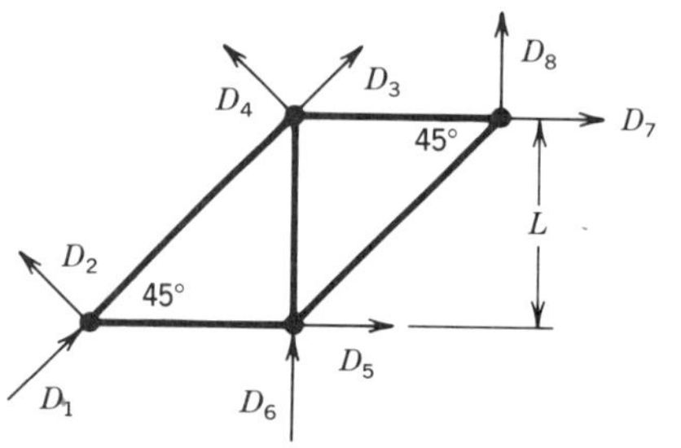

Problem 2.5

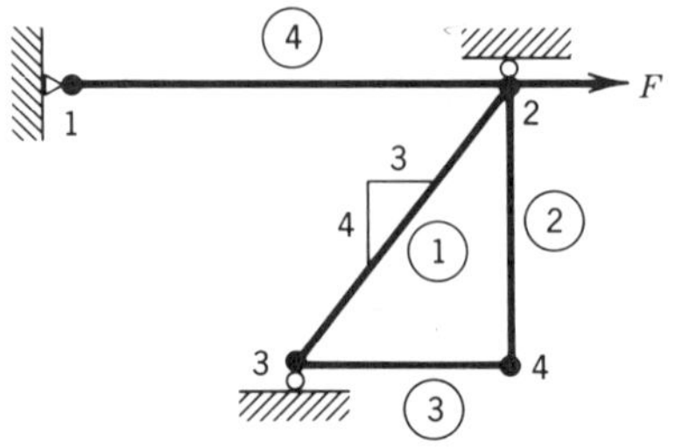

Problem 2.6

2.6 The plane truss shown has bars of stiffness k_1, k_2, k_3, and k_4, where $k_i = A_iE_i/L_i$. Free all d.o.f., and write the structure stiffness matrix in terms of the k_i. Let nodal d.o.f. have the order $\{\mathbf{D}\} = \lfloor u_1 \quad v_1 \quad u_2 \quad v_2 \quad u_3 \quad v_3 \quad u_4 \quad v_4 \rfloor^T$.

Section 2.3

2.7 The Betti–Maxwell reciprocal theorem states that if two sets of loads $\{\mathbf{R}\}_1$ and $\{\mathbf{R}\}_2$ act on a structure, work done by the first set in acting through displacements caused by the second set is equal to work done by the second set in acting through displacements caused by the first set. Symbolically, $\{\mathbf{D}\}_1^T\{\mathbf{R}\}_2 = \{\mathbf{D}\}_2^T\{\mathbf{R}\}_1$. Substitute $\{\mathbf{D}\}_1 = [\mathbf{K}]^{-1}\{\mathbf{R}\}_1$ and $\{\mathbf{D}\}_2 = [\mathbf{K}]^{-1}\{\mathbf{R}\}_2$ and show that $[\mathbf{K}]$ is symmetric.

2.8 Consider the truss of Fig. 2.2-1. Write rigid-body motion vectors for the following cases and show that each produces zero forces $\{\mathbf{R}\}$. Are the three cases linearly independent?

(a) Translation in the direction of bar 2.
(b) Rotation through a small angle about node 1 (the point $x = L_3$, $y = 0$).
(c) Rotation through a small angle about the point $x = L_3$, $y = L_1$.

2.9 For the truss of Fig. 2.2-1, let bar lengths be $L_1 = 4$, $L_2 = 5$, and $L_3 = 3$. Now consider the rigid-body motion $\{\mathbf{D}\} = \lfloor 1 \quad 3 \quad 4 \quad 0 \quad 0 \quad -4 \rfloor^T$. Sketch the displaced truss. Explain why the product $[\mathbf{K}]\{\mathbf{D}\}$ is not zero.

2.10 Consider the circular four-spring structure of Problem 2.2. Write the displacement vector $\{\mathbf{D}\}$ for each possible rigid-body motion, and show that $[\mathbf{K}]\{\mathbf{D}\} = \{\mathbf{0}\}$.

2.11 Imagine that a 90° curved-beam element is formulated using the six d.o.f. shown.
(a) Does each row (or each column) of $[\mathbf{k}]$ sum to zero? Why or why not?
(b) Sketch the approximate displaced shape if the d.o.f. are $\{\mathbf{d}\} = c\lfloor 1 \quad 1 \quad 1 \quad 1 \quad 1 \quad 1 \rfloor^T$, where c is a small number.
(c) Write $\{\mathbf{d}\}$ (all six terms) such that $[\mathbf{k}]\{\mathbf{d}\} = \{\mathbf{0}\}$. There are infinitely many possibilities. Can you write three $\{\mathbf{d}\}$'s that are linearly independent?

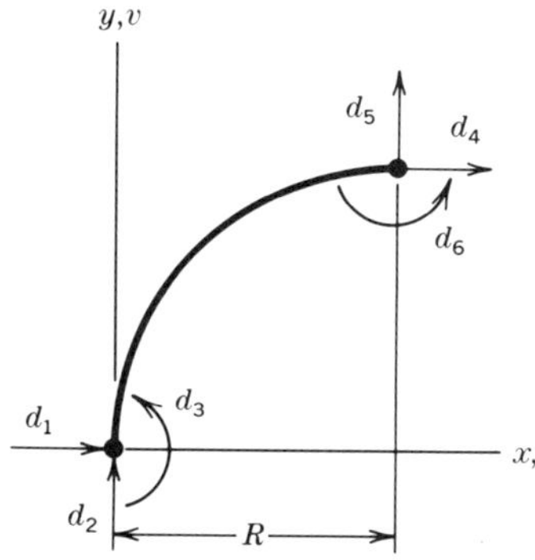

Problem 2.11

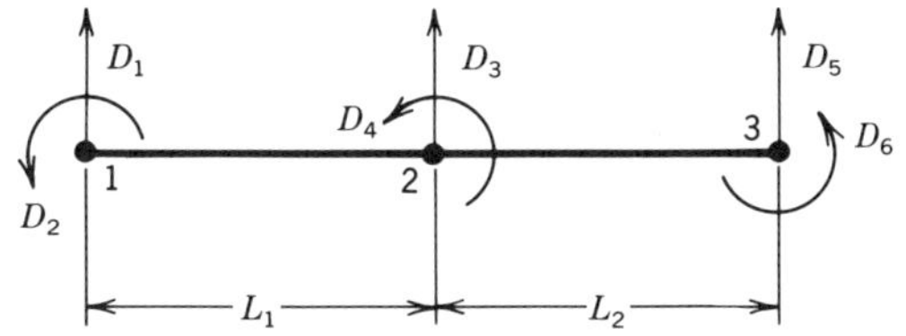

Problem 2.12

2.12 The beam shown contains two elements. Each node has two d.o.f., one in translation and one in rotation.
(a) How many rigid-body motions are possible? Write a suitable d.o.f. vector $\{\mathbf{D}\}$ for each.
(b) Let $\{\mathbf{D}\} = c\lfloor 1 \quad 1 \quad 1 \quad 1 \quad 1 \quad 1 \rfloor^T$, where c is a small number. Sketch the deformed structure. Is $[\mathbf{K}]\{\mathbf{D}\} = \{\mathbf{0}\}$? (Do not derive $[\mathbf{K}]$.)
(c) Let $\{\mathbf{D}\} = c\lfloor 1 \quad 0 \quad 0 \quad 0 \quad 0 \quad 0 \rfloor^T$. Sketch the deformed structure and show qualitatively by properly directed arrows the nodal loads required.

2.13 Consider the plane truss of Problem 2.6.
(a) Impose support conditions implied by the sketch. That is, by discarding the appropriate rows and columns, obtain a smaller $[\mathbf{K}]$ that operates on only the active d.o.f.
(b) For the loading by force F shown, write the 4 by 1 vector $\{\mathbf{D}\}$ by inspection (not by solving simultaneous equations). Hence, find the nodal loads $\{\mathbf{R}\} = [\mathbf{K}]\{\mathbf{D}\}$. Are these loads physically reasonable?

2.14 The two-element structure shown is built of standard beam elements with two d.o.f. per node (see Fig. 1.2-2). By an error, the boundary condtions

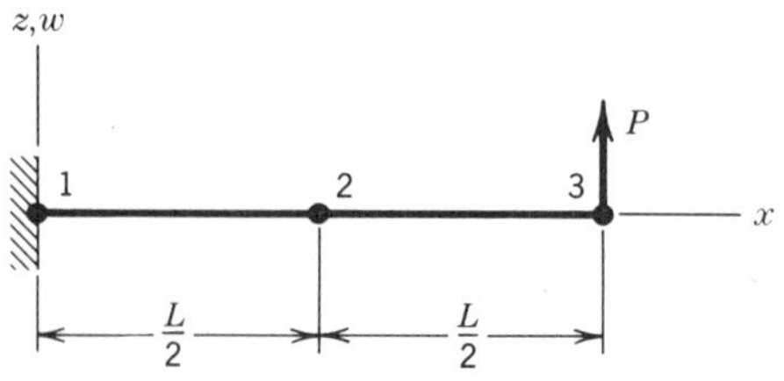

Problem 2.14

given to a computer program are $w_1 = \theta_1 = \theta_2 = \theta_3 = 0$. The expected result, $w_3 = PL^3/3EI$, is not computed. What value of w_3 is in fact computed by the program? (The question can be answered by sketching the deformed structure and applying elementary beam theory.)

2.15 (a) Let $k_1 = k_2 = k_3 = k$ in Eq. 2.3-8. Solve for u_1, v_1, and v_3 in terms of P and k.
(b) Using the results of part (a), compute p_2, q_2, and p_3 (Eq. 2.3-9). Show these forces applied to a free-body diagram of the truss, and check that static equilibrium conditions are satisfied.

Section 2.4

2.16 (a) Derive a 4 by 4 element stiffness matrix for a uniform plane truss member, using d.o.f. shown in the sketch.
(b) Check that nodal loads [**k**]{**d**} are zero for the following rigid-body motions: x-direction translation, y-direction translation, and a small counterclockwise rotation about node i.

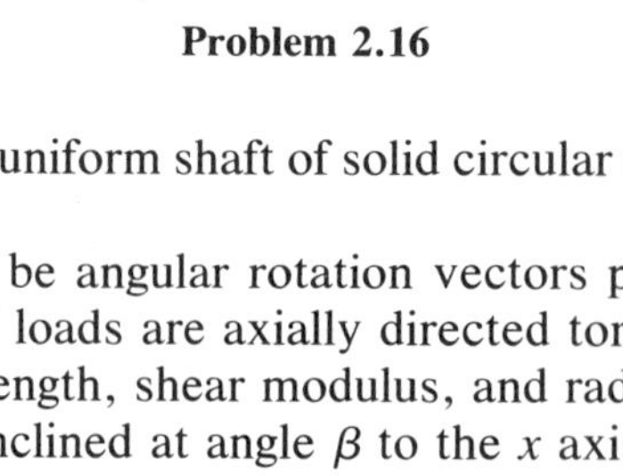

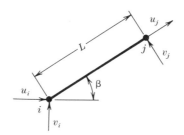

Problem 2.16

2.17 Consider a straight, uniform shaft of solid circular cross section, with a node at each end.
(a) Let nodal d.o.f. be angular rotation vectors parallel to the bar, one at each end. Nodal loads are axially directed torque vectors. What is [**k**], in terms of the length, shear modulus, and radius of the cross section?
(b) Let the bar be inclined at angle β to the x axis, with θ_x and θ_y as d.o.f. at each node (rotations about the x and y coordinate axes). What is [**k**]? As in part (a), consider torsional stiffness only.

2.18 A uniform bar of axial stiffness $k = AE/L$ is arbitrarily oriented in space. Cosines of angles between the bar and the x, y, and z coordinate axes are ℓ, m, and n. Nodal d.o.f. are translations u, v, and w at each end. Derive the 6 by 6 element stiffness matrix.

Section 2.5

2.19 For each of the following structures, generate the structure stiffness matrix by writing element matrices "structure size" and assembling them.
(a) The four-spring circular structure of Problem 2.2.
(b) The four-bar truss of Problem 2.6 (without supports).

Section 2.6

2.20 The uniform bar shown hangs under its own weight W. Compute the deflection of the lower end in terms of W, L, A, and E. (Obtain $[\mathbf{K}_{11}]$ of Eq. 2.3-3 by retaining only active d.o.f. The uppermost node is fixed.)
(a) Use one element of length L.
(b) Use two elements, each of length $L/2$.

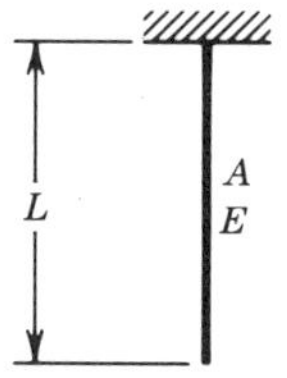

Problem 2.20

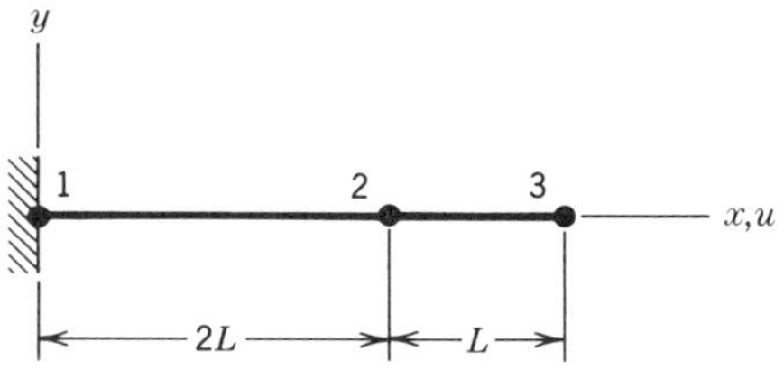

Problem 2.21

2.21 The uniform bar shown is built of two elements. Both elements are uniformly heated T degrees. Obtain $[\mathbf{K}_{11}]$ of Eq. 2.3-3 by retaining only the active d.o.f. (u_2 and u_3). Solve for u_2 and u_3 in terms of α, L, and T.

Section 2.7

2.22 As suggested in Section 2.7, permute node labels of both elements in Fig. 2.7-1 so that j replaces i, k replaces j, and i replaces k. Maintain structure node labels 1, 2, 3, and 4 where they are shown in Fig. 2.7-1. Show that $[\mathbf{K}]$ of Eq. 2.7-4 is again produced.

2.23 Change structure node labels in Fig. 2.7-1 from 1, 2, 3, and 4 to 1, 3, 4, and 7, respectively. Thus, the two elements shown are regarded as a fragment of a larger structure. To what row and column location in a 7 by 7 array $[\mathbf{K}]$ is each of the a's and b's in Eqs. 2.7-1 assigned?

2.24 Add the following elements to Fig. 2.7-2. Show the locations of nonzero element coefficients in $\{\mathbf{R}\}$ and in $[\mathbf{K}]$, as in Fig. 2.7-2.
(a) Attach a triangular element 1–4–6 to nodes 1 and 4 of existing element 2.
(b) Attach a rectangular element 3–5–7–8 to nodes 3 and 5 of existing element 1.

2.25 Manually apply the assembly algorithm of Fig. 2.7-3 to matrix $[\mathbf{k}]_1$ of Eq. 2.7-1. Specifically, by supplying numerical indexes for arrays, discover where the a's are placed in array S for
(a) I = 1 in the DO 400 loop.
(b) I = 2 in the DO 400 loop.
(c) I = 3 in the DO 400 loop.

2.26 Imagine that coefficients in the plane truss element stiffness matrix are arranged to suit the order of d.o.f. $\{\mathbf{d}\} = \lfloor u_i \quad u_j \quad v_i \quad v_j \rfloor^T$. If structure d.o.f. still have the order $\{\mathbf{D}\} = \lfloor u_1 \quad v_1 \quad u_2 \ldots u_N \quad v_N \rfloor^T$, revise the assembly algorithm of Fig. 2.7-4 as required.

2.27 Revise the assembly algorithm of Fig. 2.7-4 to deal with the following elements:

(a) a plane frame element (three d.o.f. per node).

(b) a space frame element (six d.o.f. per node).

Section 2.8

2.28 (a) For each of the plane trusses shown, show the topology of the structure stiffness matrix, in the manner of Fig. 2.8-3*a*. However, let each X represent a 2 by 2 submatrix: thus, the sketch of [**K**] will have eight rows and eight columns of submatrices X.

(b) How would your answer to part (a) change if the structure were a plane frame? Or a network of electrical resistors?

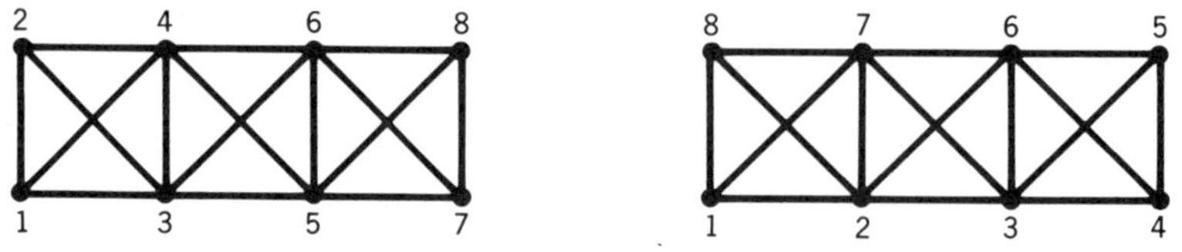

Problem 2.28

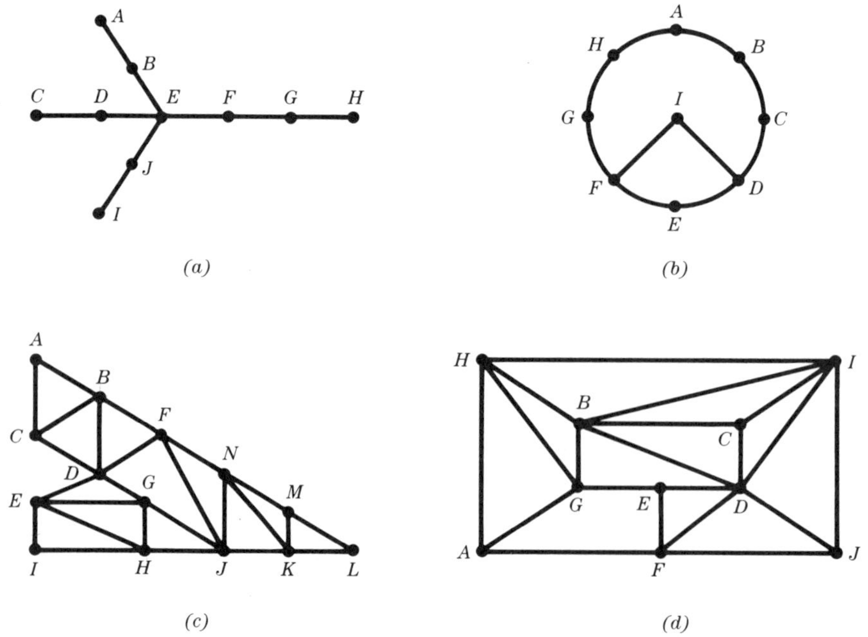

Problem 2.29

2.29 Each of the structures shown is hypothetical and has one d.o.f. per node. Solid lines indicate connectivity between nodes. (Letters are for use in another problem.) Number the nodes so as to achieve minimum bandwidth of the coefficient matrix [**K**]. Sketch the topology of the assembled matrix [**K**] (as in Fig. 2.8-3*a*).

2.30 In an element having many nodes, let i and j represent respectively the highest and lowest structure node numbers connected to that element. If $i - j$ happens to be the largest difference for any element of the structure, and if n is the number of d.o.f. per node, what is semibandwidth b in terms of i, j, and n?

2.31 Apply the formula for b devised in Problem 2.30 to the following structures:
(a) the plane truss in Fig. 2.2-1.
(b) the first structure shown in Problem 2.28, regarded as a plane frame (three d.o.f. per node).

2.32 Consider the structures of Problem 2.29. Assign node numbers by the following system. Pick a starting node (say A) and call it 1. Number as 2, 3, and so on, nodes that share an element with node 1 (thus, in (*b*), node numbers become $H = 2$ and $B = 3$). Next, number nodes that share an element with nodes 2, 3, and so on. (Figure 2.9-1 shows the results of such a scheme, but starting with the highest number and counting down.) For one d.o.f. per node, what semibandwidth b do you obtain?

2.33 *Reverse* the node numberings found in Problem 2.32. For each structure, how many fills are there during equation solving, both in the original numbering of Problem 2.32 and in the reversed numbering?

Section 2.10

2.34 For each of the plane trusses shown in Problem 2.28, write the "input" and "converted" forms of array ID (see Figs. 2.10-2 and 2.10-3). As support conditions, assume that all d.o.f. are set to zero at the upper left node and at the lower right node.

2.35 Repeat Problem 2.34, but regard each structure as a plane frame (nodal d.o.f. u, v, and θ_z).

2.36 (a) The unsupported plane truss shown has eight d.o.f. Set up an 8 by 8 stiffness array [**K**]. Write a bar number in those positions of [**K**] that

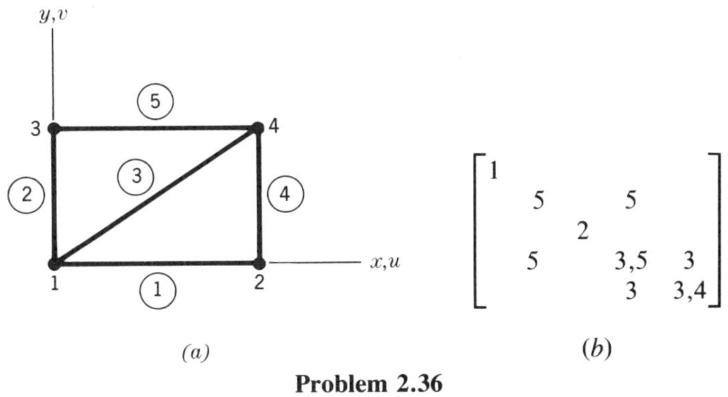

Problem 2.36

receive nonzero stiffness contributions from that bar (e.g., write $K_{22} = K_{26} = K_{62} = K_{66} = 2$, from bar 2).

(b) After support conditions are imposed by use of array ID, a similarly diagrammed matrix [**K**] appears as shown. Sketch the truss, showing the supports implied.

2.37 Imagine that the structure in Fig. 2.10-1 is a plane frame, for which nodal d.o.f. are u, v, and θ_z. Supports apply no nodal moments.

(a) Modify the ID arrays in Figs. 2.10-2 and 2.10-3 as required.

(b) Modify the assembly algorithm of Fig. 2.10-5 as required.

2.38 Consider the 2 by 2 system of equations

$$\begin{bmatrix} K_1 & K_2 \\ K_3 & K_4 \end{bmatrix} \begin{Bmatrix} x \\ y \end{Bmatrix} = \begin{Bmatrix} 0 \\ b \end{Bmatrix}$$

Use the penalty method to impose the result $x = c$. Show that exact values of x and y are approached as the added stiffness approaches infinity.

2.39 How would you impose a prescribed *relative* displacement by the penalty method? Consider, for example, imposing $u_4 - u_2 = c$ in Fig. 2.11-1*a*, where c is a constant. Give a physical explanation, then state exactly which coefficients in the equations [**K**]{**D**} = {**R**} must be changed and how you would change them. (This procedure is *not recommended;* see the *Caution* in Section 2.10.)

2.40 Consider the axially loaded structure in Fig. 2.11-1*a* and the [**K**]{**D**} = {**R**} equation in Fig. 2.11-1*b*. Impose the displacement $u_3 = 6$ and solve for u_2 and u_4.

(a) Use the penalty method of Fig. 2.10-6.

(b) Use the "zero-one" procedure of Fig. 2.10-7.

(c) Use Eq. 2.10-1, and determine the supplementary terms on the right-hand side by summing element contributions, as suggested below Eq. 2.10-1.

2.41 Consider again the four-spring circular structure of Problem 2.2. No forces are applied, but displacements $u_2 = u_4 = c$ are prescribed, where c is a constant. Impose these displacements and solve for u_1 and u_3. Use the "zero-one" procedure of Fig. 2.10-7.

2.42 From the "converted" ID arrays found in Problem 2.34, compute semi-bandwidth b by applying the method described at the end of Section 2.10.

2.43 Write a Fortran algorithm that calculates semibandwidth b according to the procedure outlined at the end of Section 2.10.

Section 2.11

2.44 Let $k_1 = k_2 = k_3 = k$ in Eq. 2.3-8. Calculate u_1, v_1, and v_3 in terms of P and k by applying the Gauss elimination method.

2.45 Consider the circular four-spring structure of Problem 2.2. Without imposing any support condition, show the four modified [**K**]'s produced by successive steps of Gauss elimination (as in Fig. 2.11-1).

2.46 A one-element cantilever beam is shown. Also shown is the stiffness matrix that operates on the unrestrained d.o.f. w_2 and θ_2. In parts (a) and (b) carry

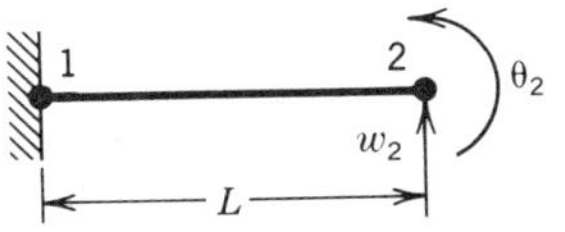

$$[\mathbf{K}] = \begin{bmatrix} 12EI/L^3 & -6EI/L^2 \\ -6EI/L^2 & 4EI/L \end{bmatrix} \begin{matrix} w_2 \\ \theta_2 \end{matrix}$$

Problem 2.46

out one step of Gauss elimination, and explain the physical meaning of the diagonal coefficient that remains.
(a) Eliminate w_2 (reduction of [**K**] to upper triangular form).
(b) Eliminate θ_2 (reduction of [**K**] to *lower* triangular form).

2.47 Consider reduction of [**K**] to upper triangular form by Gauss elimination. Coefficients that are initially zero may become nonzero in this process. In the following matrices, which zeros above the diagonal become nonzero?
(a) [**K**] of Eq. 2.2-6.
(b) [**K**] of Fig. 2.7-2.

2.48 Imagine that no boundary conditions are imposed, so that too many structure d.o.f. remain active. A solution for the d.o.f. by Gauss elimination is started, but fails during the attempt to eliminate the nth d.o.f. For the following structures, what is n, and why?
(a) The plane truss of Fig. 2.8-2*a* (allow two d.o.f. per node).
(b) Imagine that Fig. 2.8-2*a* represents a plane frame (allow three d.o.f. per node).
(c) The network of Fig. 2.9-1 (allow one d.o.f. per node).
(d) The beam of Problem 2.12 (allow the six d.o.f. shown).

Section 2.12

2.49 Consider the uniform hanging bar of Problem 2.20. Assume that finite element analysis yields nodal displacements that are exact. Plot the correct distribution of axial stress (from W/A at the top to zero at the bottom). On the same plot show the stress distribution predicted by finite elements using
(a) one element.
(b) two identical elements.
(c) four identical elements.

2.50 The uniform bar shown is built of two identical bar elements and is loaded by axially directed forces P_2 and P_3 at nodes 2 and 3, respectively. Impose

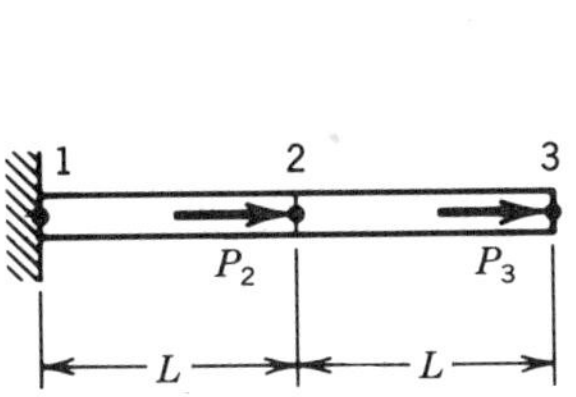

Problem 2.50

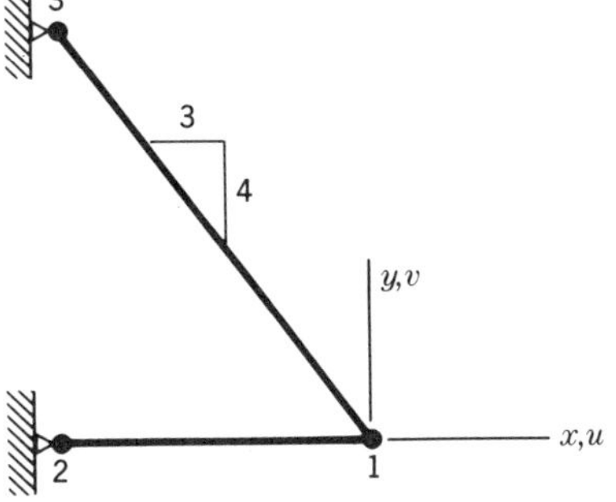

Problem 2.51

the displacement $u_2 = \bar{u}_2$ by the penalty method described in Section 2.10, using $k_s = 1000AE/L$. Solve for u_2 and u_3, then compute loads $\{\mathbf{R}\} = [\mathbf{K}]\{\mathbf{D}\}$, where $[\mathbf{K}]$ pertains to the original structure. Interpret the result for the special cases $\bar{u}_2 = 0$ and $P_3 = 0$.

2.51 Let $AE/L = 2(10)^6$ N/m for each bar of the two-bar truss shown.
(a) Set up the 2 by 2 structure stiffness matrix that operates on $\{\mathbf{D}\} = \lfloor u_1 \quad v_1 \rfloor^T$.
(b) Let there be a prescribed downward displacement of 0.0001 m at node 1. No horizontal load is applied and no horizontal displacement is prescribed at node 1. Modify the structure equations using the penalty method procedure described in Section 2.10, using $k_s = 1000\ AE/L$.
(c) Solve for u_1 by Gauss elimination.
(d) Solve for the vertical force applied to node 1.

Section 2.13

2.52 Repeat the example given in Section 2.13, but allow node 1 to move axially, so that the active d.o.f. are u_1, u_2, and u_3.

2.53 Analyze the truss shown for nodal displacements and element stresses. Follow the steps used in the example problem of Section 2.13. Let $E = 200$ GPa for each bar.

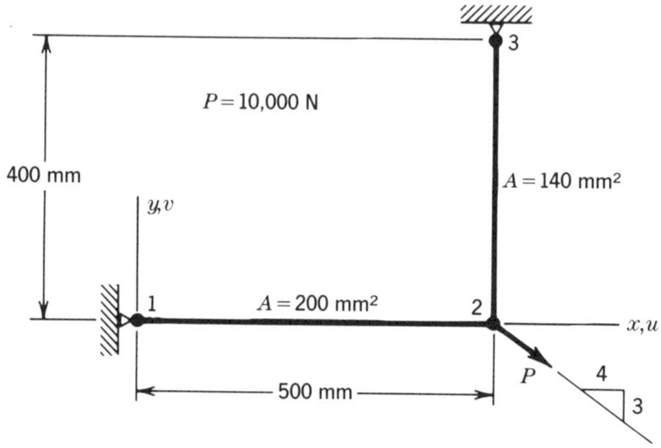

Problem 2.53

2.54 Using the steps listed in Section 2.13 as a guide, write a computer program for the analysis of plane trusses. An algorithm for equation solving is given in Appendix B.

CHAPTER 3

STATIONARY PRINCIPLES, THE RAYLEIGH–RITZ METHOD, AND INTERPOLATION

The equilibrium configuration of a system is found by analysis of its potential energy. Expressions for potential energy are presented. These and other integral expressions, called *functionals,* are introduced as a starting point for an approximation technique—namely, the Rayleigh–Ritz method—whose modern form is the finite element method. Interpolation, necessary to the method, is described.

3.1 INTRODUCTION

In preceding chapters, element stiffness matrices [**k**] have been formulated by direct physical argument. This is easily done for truss and beam elements by activating d.o.f. in turn and computing the nodal loads required to maintain the deformation state. Finite elements obtained by discretization of a continuum are not as easily formulated. (For example, is there an easy way to find nodal forces that appear in response to displacement u_3 in Fig. 1.1-2*c*?) A systematic and general way of obtaining [**k**] is needed. One of the best ways is the Rayleigh–Ritz method. An alternative, the method of weighted residuals, is discussed in Chapter 15.

The Rayleigh–Ritz method has a classical form and a finite element form. In the classical form, an approximating field is defined over the entire region of interest. In the finite element form, the approximating field is defined in piecewise fashion. As degrees of freedom, finite elements use nodal values of the field (and perhaps nodal values of one or more spatial derivatives of the field as well). By degrees of freedom (d.o.f.) we mean independent quantities used to define a configuration of a system that violates neither compatibility conditions nor constraints such as support conditions. Using more general terms, one can say that d.o.f. are quantities used to define the spatial variation of an approximating field.

In order to analyze a continuum by use of the Rayleigh–Ritz method, one must have a *functional.* A functional is an integral expression that implicitly contains differential equations that describe the problem. In structural mechanics the most widely used functional is the expression for potential energy. Functionals are also available for problems of heat conduction, acoustic modes in cavities, certain types of fluid flow, and other problems. We will present some of these functionals and will show how they are used to produce finite element formulations.

Differential equations are said to state a problem in the *strong form*. An integral expression such as a functional that implicity contains the differential equations is called the *weak form*. The strong form states conditions that must be met at every material point, whereas the weak form states conditions that must be met only in an average sense.

A functional, such as that for potential energy Π_p, contains integrals that span the line, area, or volume of interest. After applying the Rayleigh–Ritz method, the Π_p expression contains no integrals and is no longer called a functional. Rather, Π_p is then a function of a finite number of d.o.f. Indeed, for an initially discrete structure such as a truss, no integrals need be invoked in writing the Π_p expression. We will consider these "initially discrete" forms first, then return to integral forms later in this chapter.

Physical insight was responsible for the early rapid development of the finite element method and for its ready appeal to stress analysts. A more mathematical approach augments physical understanding by placing the finite element method on a sound foundation, thus allowing statements to be made regarding bounds and convergence, and suggesting solution tactics that are not apparent from physical reasoning alone.

3.2 PRINCIPLE OF STATIONARY POTENTIAL ENERGY

In the present section we consider time-independent problems of structural mechanics. We define a *system* as the physical structure and the loads applied to it. The *configuration* of a system is the set of positions of all particles of the structure. Let the system have a reference configuration C_R and a displaced configuration C_D. A system is called *conservative* if work done by internal forces and work done by external loads are each independent of the path taken between C_R and C_D. In an elastic structure, work done by internal forces is equal in magnitude to the change in strain energy.

The loaded spring of Fig. 3.2-1 is a case in point. Let C_R and C_D refer to unstretched and stretched configurations, respectively. If the spring dissipates no energy, then the work of internal forces (i.e., strain energy in the spring) depends only on stretch D, not on whether the passage from C_R to C_D is via path A or path B. Similarly, if external load P has constant magnitude and constant direction, it does negative work of magnitude PD regardless of the path taken from C_R to C_D. We conclude that because internal forces and external loads are both conservative, so is the system.

Boundary conditions are of two types: *essential* (or *principal*) and *nonessential* (often called *natural*). In the finite element method, essential boundary conditions are prescribed values of nodal d.o.f., and nonessential boundary conditions are prescribed values of higher derivatives of the field quantity than are usually used as nodal d.o.f. For example, if standard beam elements are used, nodal d.o.f. are lateral deflection w and its first derivative, $w_{,x}$. When these elements are used to analyze the beam of Fig. 3.2-2*a*, essential boundary conditions (which can also be called *geometric* or *kinematic* in this problem) are that $w = 0$ and $w_{,x} = 0$

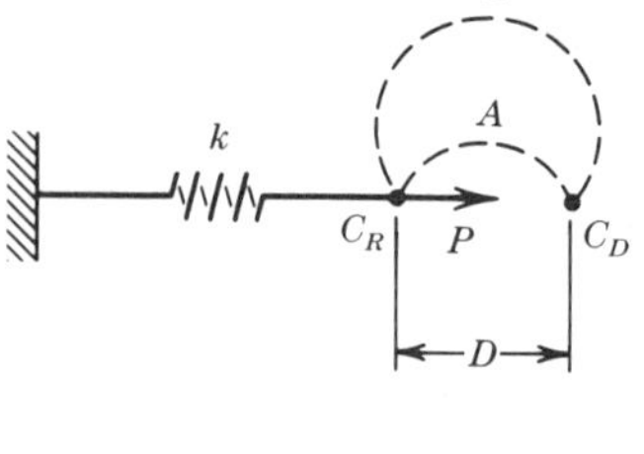

Figure 3.2-1. A linear spring of stiffness k loaded by a constant force P that acts parallel to the x axis. Hypothetical displacement paths A and B of the loaded point are shown by dashed lines.

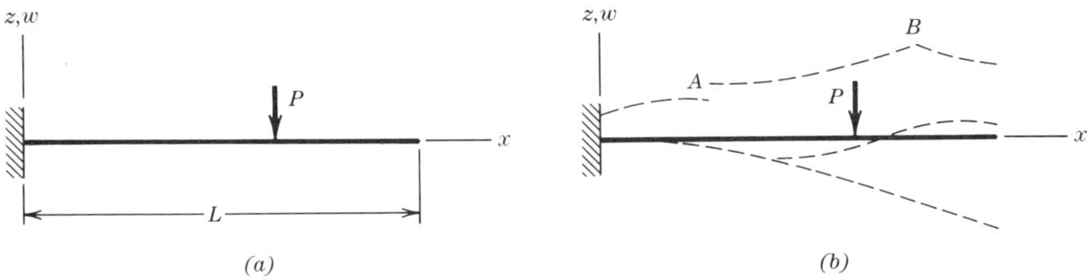

Figure 3.2-2. (*a*) A cantilever beam. (*b*) An inadmissible configuration (upper dashed line) and two admissible configurations (lower dashed lines).

at $x = 0$. Nonessential boundary conditions are that $w_{,xx} = 0$ and $w_{,xxx} = 0$ at $x = L$, since bending moment $M = EIw_{,xx}$ and transverse shear force $V = EIw_{,xxx}$ are both zero at $x = L$.

An *admissible configuration* is any configuration that satisfies internal compatibility and essential boundary conditions. Examples appear in Fig. 3.2-2*b*. The uppermost curve, which is inadmissible, has four faults: it violates the two essential boundary conditions $w = 0$ and $w_{,x} = 0$ at $x = 0$, and it violates compatibility because of the jump at A and the cusp at B. The lower two curves are *both* admissible, even though only the lower one seems physically reasonable. An admissible configuration need not satisfy nonessential boundary conditions. Thus, at $x = L$, neither of the two lower curves need display $w_{,xx} = 0$ or $w_{,xxx} = 0$.

A conservative mechanical system has a potential energy. That is, one can express the energy content of the system in terms of its configuration, without reference to whatever deformation history or path may have led to that configuration [3.1]. Potential energy, also called total potential energy, includes (a) the strain energy of elastic distortion, and (b) the potential possessed by applied loads, by virtue of their having the capacity to do work if displaced through a distance. The *principle of stationary potential energy* states that

> *Among all admissible configurations of a conservative system, those that satisfy the equations of equilibrium make the potential energy stationary with respect to small admissible variations of displacement.*

This principle is applicable whether or not the load versus deformation relation is linear. If the stationary condition is a relative minimum, the equilibrium state is stable. Note that loads are kept constant while displacements are varied.

Example. Linear Spring with Axial Load. A very simple system is shown in Fig. 3.2-3. Its (total) potential energy Π_p has two parts;

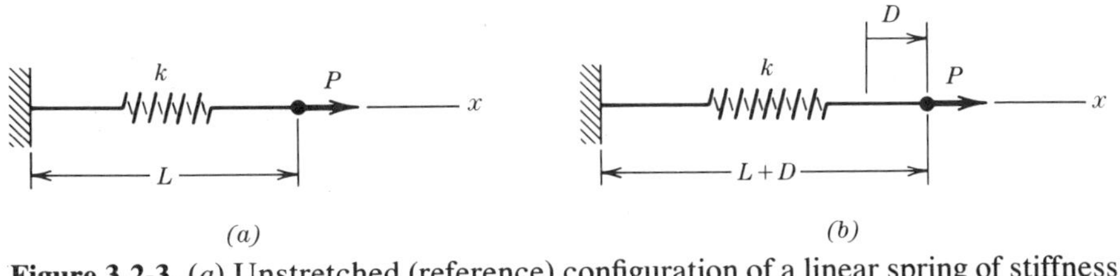

Figure 3.2-3. (*a*) Unstretched (reference) configuration of a linear spring of stiffness k. (*b*) Configuration after force P is applied, stretching the spring D units.

$$\Pi_p = U + \Omega \tag{3.2-1}$$

where U is the strain energy of the system, which in the present example is given by

$$U = \tfrac{1}{2} kD^2 \tag{3.2-2}$$

The potential of loads, called Ω, is here given by

$$\Omega = -PD \tag{3.2-3}$$

The load is regarded as always acting at its full value P. In moving through displacement D it does work in the amount PD, thereby *losing* potential of equal amount; hence the negative sign in the expression $\Omega = -PD$. The potential energy

$$\Pi_p = \tfrac{1}{2} kD^2 - PD \tag{3.2-4}$$

can be regarded as the total internal and external work done in changing the configuration from the reference state $D = 0$ to the displaced state $D \neq 0$. Note that if P were directed toward the left, while D remains positive toward the right, then Ω would become $+PD$. This is in essence the same as increasing potential energy by increasing the elevation of a weight.

If only displacements along the x axis are allowed, then the single d.o.f. D defines all admissible configurations. The equilibrium configuration D_{eq} is found from the stationary value of Π_p:

$$d\Pi_p = (kD_{eq} - P)\, dD = 0, \qquad \text{hence} \quad D_{eq} = \frac{P}{k} \tag{3.2-5}$$

The equation $(kD_{eq} - P)\, dD = 0$ is an instance of the virtual work principle: zero net work is done by all forces during a small admissible displacement dD from the equilibrium configuration. This is graphically apparent in Fig. 3.2-4. We see also that Π_p is a relative minimum, which means that the equilibrium state is stable.

The reference datum for Ω can be arbitrarily changed by a constant. For example, if we say that Ω is zero at the equilibrium configuration, then $\Omega = P(D_{eq} - D)$. The added constant PD_{eq} disappears in the process of writing $d\Pi_p = 0$, and the same value of D_{eq} is again obtained.

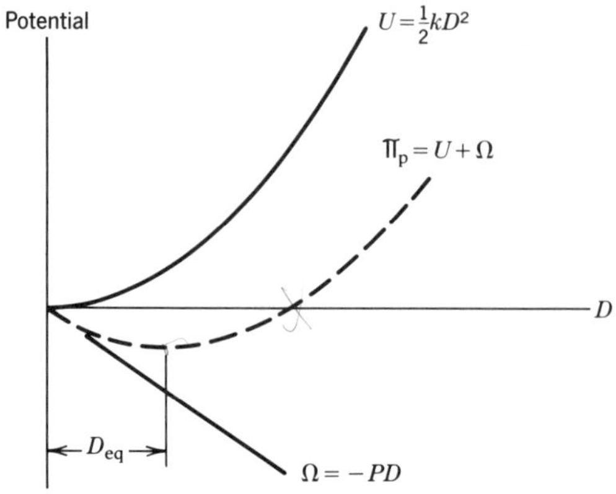

Figure 3.2-4. Graphical interpretation of the potential energy relations for the problem of Fig. 3.2-3.

In Eq. 3.2-2 we write $\Omega = -PD$ rather than $\Omega = -PD/2$. This is because P is regarded as always acting at full intensity. True, we could bypass the stationary potential energy principle and imagine that D is produced by a gradually increasing load whose final value is P. Thus, equating work done against a linear spring to strain energy stored, we have

$$\tfrac{1}{2}PD_{\text{eq}} = \tfrac{1}{2}kD_{\text{eq}}^2 \qquad \text{from which} \qquad D_{\text{eq}} = \frac{P}{k} \tag{3.2-6}$$

This energy balance argument is valid but rarely helpful. It yields but one equation, even if there are a great many d.o.f. that must be determined.

3.3 PROBLEMS HAVING MANY D.O.F.

A finite element analysis typically uses hundreds of d.o.f. They may be the x and y displacements of nodes (as in a plane stress problem), or lateral displacement w and its first derivative $w_{,x}$ at nodes (as in a beam problem), and so on. Let n be the number of d.o.f. that must be calculated, and let them be collected in the structure displacement vector $\{\mathbf{D}\} = \lfloor D_1 \quad D_2 \ldots D_n \rfloor^T$. We assume here that support conditions are already imposed, so that arbitrary values of the D_i always create admissible configurations.

Potential Π_p is a function of the D_i. Symbolically, $\Pi_p = \Pi_p(D_1, D_2, \ldots, D_n)$. Applying the principle of stationary potential energy, we write

$$d\Pi_p = \frac{\partial \Pi_p}{\partial D_1} dD_1 + \frac{\partial \Pi_p}{\partial D_2} dD_2 + \cdots + \frac{\partial \Pi_p}{\partial D_n} dD_n = 0 \tag{3.3-1}$$

The stationary principle states that equilibrium prevails when $d\Pi_p = 0$ for *any* small admissible variation of the configuration. We can imagine that only dD_1 is nonzero, or that only dD_2 and dD_3 are nonzero, and so on. For any and all such choices, $d\Pi_p$ must vanish. This is possible only if coefficients of the dD_i vanish separately. Thus, for $i = 1, 2, 3, \ldots, n$,

$$\frac{\partial \Pi_p}{\partial D_i} = 0 \qquad \text{or, in alternative notation,} \qquad \left\{\frac{\partial \Pi_p}{\partial \mathbf{D}}\right\} = \{\mathbf{0}\} \tag{3.3-2}$$

These are n equations to be solved for the n values of d.o.f. D_i that define the static equilibrium configuration.

Example. Springs in Series. The structure shown in Fig. 3.3-1 has the potential

$$\Pi_p = \tfrac{1}{2}k_1D_1^2 + \tfrac{1}{2}k_2(D_2 - D_1)^2 + \tfrac{1}{2}k_3(D_3 - D_2)^2 - P_1D_1 - P_2D_2 - P_3D_3 \tag{3.3-3}$$

Equations 3.3-2 and 3.3-3 yield, for $i = 1, 2, 3$,

$$\begin{aligned} k_1D_1 - k_2(D_2 - D_1) - P_1 &= 0 \\ k_2(D_2 - D_1) - k_3(D_3 - D_2) - P_2 &= 0 \\ k_3(D_3 - D_2) - P_3 &= 0 \end{aligned} \tag{3.3-4}$$

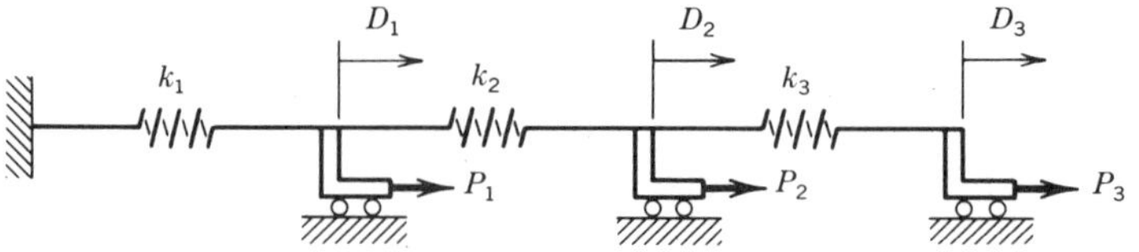

Figure 3.3-1. A three d.o.f. system of three linear springs and three axial loads P_1, P_2, and P_3. D.o.f. D_i are axial displacements relative to a fixed point, such as the left support. The springs are unstretched when $D_1 = D_2 = D_3 = 0$.

In the matrix form $[\mathbf{K}]\{\mathbf{D}\} = \{\mathbf{R}\}$, Eqs. 3.3-4 are

$$\begin{bmatrix} k_1 + k_2 & -k_2 & 0 \\ -k_2 & k_2 + k_3 & -k_3 \\ 0 & -k_3 & k_3 \end{bmatrix} \begin{Bmatrix} D_1 \\ D_2 \\ D_3 \end{Bmatrix} = \begin{Bmatrix} P_1 \\ P_2 \\ P_3 \end{Bmatrix} \tag{3.3-5}$$

The correctness of stiffness matrix $[\mathbf{K}]$ in Eq. 3.3-5 can be checked by the procedure of activating one d.o.f. at a time, as described in Section 2.2.

Example. Plane Truss Problems. Consider the plane truss element of Figs. 2.4-1 and 2.4-2. Its potential expression is

$$\Pi_p = \tfrac{1}{2} ke^2 - p_i u_i - q_i v_i - p_j u_j - q_j v_j \tag{3.3-6}$$

where $k = AE/L$ and elongation e is given by

$$e = (u_j - u_i)\cos\beta + (v_j - v_i)\sin\beta \tag{3.3-7}$$

Setting to zero the four derivatives of Π_p with respect to d.o.f. u_i, v_i, u_j, and v_j, we obtain the four rows of Eq. 2.4-3.

The same procedure can be applied to an entire structure. Consider the three-bar truss of Fig. 2.2-1, with support conditions $u_2 = v_2 = u_3 = 0$ already imposed. Its potential energy is

$$\Pi_p = \tfrac{1}{2} k_3 u_1^2 + \tfrac{1}{2} k_1 v_3^2 + \tfrac{1}{2} k_2 \left[(0 - u_1)(-0.6) + (v_3 - v_1)(0.8)\right]^2 + Pv_1 \tag{3.3-8}$$

The Pv_1 term bears a positive sign because the potential of the (downward) load P is increased by a positive (upward) displacement v_1. The derivatives of Π_p with respect to u_1, v_1 and v_3, when equated to zero, are found to yield Eqs. 2.3-8.

From the foregoing examples we draw the following conclusions, which are true in general.

1. A system that has linear load versus displacement characteristics has a symmetric stiffness matrix; that is, $K_{ij} = K_{ji}$. This happens because each symmetrically located pair of off-diagonal coefficients comes from a single term in Π_p whose form is a constant times $D_i D_j$. Thus $K_{ij} = \partial^2\Pi_p/\partial D_i\,\partial D_j = \partial^2\Pi_p/\partial D_j\,\partial D_i = K_{ji}$.
2. If D_i is a nodal displacement (or rotation), the equation $\partial\Pi_p/\partial D_i = 0$ is a nodal equilibrium equation stating that forces (or moments) applied to the

node sum to zero in the direction of D_i. Included in the sum are (a) loads applied externally, and (b) loads applied internally, owing to deformation of structural components (and perhaps also to thermal load, body force, etc.).

3. Static indeterminacy does not alter the procedure or make a problem more difficult. For example, in Fig. 3.3-1 we could connect a fourth spring between the fixed support and node 3. Then Π_p is augmented by $k_4 D_3^2/2$, and the last stiffness coefficient in Eq. 3.3-5 is changed from k_3 to $k_3 + k_4$, but the same three d.o.f. still suffice.

4. The potential energy of a structure can be written in the form

$$\Pi_p = U + \Omega, \quad \text{where} \quad U = \tfrac{1}{2}\{D\}^T[K]\{D\} \quad \text{and} \quad \Omega = -\{D\}^T\{R\} \quad (3.3\text{-}9)$$

If $U = 0$, then either $\{D\} = \{0\}$ or $\{D\}$ expresses a rigid-body motion. If the structure is stable and is supported so that rigid-body motion is not possible (as in Eqs. 3.3-3 and 3.3-5), then $\tfrac{1}{2}\{D\}^T[K]\{D\} > 0$ for any nonzero $\{D\}$; that is, $[K]$ is said to be *positive definite*.

3.4 POTENTIAL ENERGY OF AN ELASTIC BODY

The potential energy of an elastic body consists of the strain energy contained in elastic distortions and the potential of loads that act within the body or on its surface. The potential energy expression can be used to formulate element stiffness matrices and element load vectors. Simple finite element formulations appear later in this chapter. Additional formulations appear in subsequent chapters.

In this section we present formulas, argue their validity, show that special cases yield correct results, and consider examples. Derivations and detailed arguments may be found elsewhere [3.1, 3.2].

Consider a linearly elastic body that carries conservative loads. Let its volume be V and its surface area be S. The expression for its potential energy is

FOR UNIAXIAL
SEE P. 77

$$\Pi_p = \int_V \overbrace{(\tfrac{1}{2}\{\epsilon\}^T[E]\{\epsilon\}}^{U_0} - \underset{\substack{\text{INITIAL} \\ \text{STRAIN}}}{\{\epsilon\}^T[E]\{\epsilon_0\}} + \underset{\substack{\text{INITIAL} \\ \text{STRESS}}}{\{\epsilon\}^T\{\sigma_0\}})\, dV$$

$$\underset{\substack{\text{FINAL} \\ \text{STRAIN}}}{}$$

$$- \int_V \underset{\substack{\text{BODY} \\ \text{FORCES}}}{\{u\}^T\{F\}}\, dV - \int_S \underset{\substack{\text{SURFACE} \\ \text{TRACTIONS}}}{\{u\}^T\{\Phi\}}\, dS - \underset{\substack{\text{CONCENTRATED} \\ \text{FORCES APPLIED} \\ \text{TO BODY}}}{\{D\}^T\{P\}} \quad (3.4\text{-}1)$$

The notation is explained in Section 1.6 and in the following.

Explanation and Justification. In the first integral of Eq. 3.4-1, the expression within parentheses represents U_0, the strain energy per unit volume. The expression is derived as follows. Consider a unit cube (i.e., a cube of unit length along each edge). Stresses that act on faces of the cube do work during infinitesimal straining of the cube. This work is stored as an increment of strain energy dU_0. On a unit cube, stress and force have the same magnitude on each face, and strain and elongation have the same magnitude along each edge. Therefore, since work is equal to force times displacement,

$$dU_0 = \sigma_x\, d\epsilon_x + \sigma_y\, d\epsilon_y + \sigma_z\, d\epsilon_z + \tau_{xy}\, d\gamma_{xy} + \tau_{yz}\, d\gamma_{yz} + \tau_{zx}\, d\gamma_{zx} \quad (3.4\text{-}2)$$

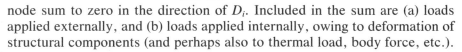

Changes in stress produced by infinitesimal strain increments have been discarded from dU_0 because they produce terms of higher order. For example, $(\sigma_x + d\sigma_x)\,d\epsilon_x \approx \sigma_x\,d\epsilon_x$. From Eq. 3.4-2 we conclude that $\partial U_0/\partial\epsilon_x = \sigma_x$, $\partial U_0/\partial\epsilon_y = \sigma_y, \ldots, \partial U_0/\partial\gamma_{zx} = \tau_{zx}$. Expressing these six derivatives in matrix format and using the stress–strain relation (Eq. 1.7-8), we obtain

$$\left\{\frac{\partial U_0}{\partial \boldsymbol{\epsilon}}\right\} = \{\boldsymbol{\sigma}\} \qquad \text{or} \qquad \left\{\frac{\partial U_0}{\partial \boldsymbol{\epsilon}}\right\} = [\mathbf{E}]\{\boldsymbol{\epsilon}\} - [\mathbf{E}]\{\boldsymbol{\epsilon}_0\} + \{\boldsymbol{\sigma}_0\} \tag{3.4-3}$$

Integration of the latter equation with respect to the strains yields the parenthetic expression in Eq. 3.4-1. That the integration is correct may be shown by applying differentiation rules given in Appendix A. A constant of integration has been discarded. It is superfluous because it disappears during the differentiation process that makes Π_p stationary.

Integrals in Eq. 3.4-1 that contain body forces $\{\mathbf{F}\}$ and surface tractions $\{\boldsymbol{\Phi}\}$ represent work done (hence potential lost) by $\{\mathbf{F}\}$ and $\{\boldsymbol{\Phi}\}$ as the body is deformed. Displacements in the x, y, and z coordinate directions are

$$\{\mathbf{u}\}^T = \lfloor u \quad v \quad w \rfloor \tag{3.4-4}$$

Thus potential changes associated with $\{\mathbf{F}\}$ and $\{\boldsymbol{\Phi}\}$, per unit volume and per unit area, respectively, are

$$-F_x u - F_y v - F_z w \qquad \text{and} \qquad -\Phi_x u - \Phi_y v - \Phi_z w \tag{3.4-5}$$

These are the products $-\{\mathbf{u}\}^T\{\mathbf{F}\}$ and $-\{\mathbf{u}\}^T\{\boldsymbol{\Phi}\}$. It is assumed that positive senses correspond—for example, that F_x, Φ_x, and u are all considered positive when acting in the $+x$ direction. Integrals that contain $\{\mathbf{F}\}$ and $\{\boldsymbol{\Phi}\}$ are evaluated only over the portions of V and S where $\{\mathbf{F}\}$ and $\{\boldsymbol{\Phi}\}$ are prescribed.

The final term in Eq. 3.4-1, $-\{\mathbf{D}\}^T\{\mathbf{P}\} = -D_1P_1 - D_2P_2 - \cdots - D_nP_n$, accounts for work done (hence potential lost) by concentrated forces and/or moments applied to the body. The potential of these loads could be included in the surface integral by imagining large tractions to be applied over small areas, but is easier to regard the potential of such loads as $-\{\mathbf{D}\}^T\{\mathbf{P}\}$. As usual, the same sense is considered positive for a displacement or rotation D_i and for its corresponding force or moment P_i.

In a typical problem, many of the load terms $\{\boldsymbol{\epsilon}_0\}$, $\{\boldsymbol{\sigma}_0\}$, $\{\mathbf{F}\}$, $\{\boldsymbol{\Phi}\}$, and $\{\mathbf{P}\}$ are zero. For example, if heating is localized $\{\boldsymbol{\epsilon}_0\}$ is zero over most of the body. If gravity or acceleration loads are considered unimportant, $\{\mathbf{F}\} = \{\mathbf{0}\}$. Surface tractions $\{\boldsymbol{\Phi}\}$ usually act on only a portion of surface S and may be absent altogether. Indeed, *all* these load terms might be zero if nonzero values of one or more d.o.f. are prescribed instead.

Equation 3.4-1 is not restricted to rectangular coordinates. It requires only that x, y, and z refer to three mutually perpendicular directions at each material point.

Particular Cases. In its most general form, Eq. 3.4-1 includes all six strains in $\{\boldsymbol{\epsilon}\}$. Material property matrix $[\mathbf{E}]$ is 6 by 6. Its coefficients are stated in Eq. 1.7-3 for the case of isotropy.

If the problem is one of plane stress or plane strain, then $\{\boldsymbol{\epsilon}\}^T = \lfloor \epsilon_x \quad \epsilon_y \quad \gamma_{xy} \rfloor$ and $[\mathbf{E}]$ is a 3 by 3 array whose coefficients are given by Eq. 1.7-4 or Eq. 1.7-5 for the case of isotropy.

The simplest special case is that of uniaxial stress. For this particular case, and perhaps for others as well, it may be easiest to derive the strain energy expression afresh, as follows, rather than specialize Eq. 3.4-1. Equations 3.4-2 and 3.4-3 become $dU_0 = \sigma_x \, d\epsilon_x$ and $dU_0/d\epsilon_x = \sigma_x = E\epsilon_x - E\epsilon_{x0} + \sigma_{x0}$. After integration and inclusion of the load terms, we obtain in place of Eq. 3.4-1

FOR UNIAXIAL STRESS

$$\Pi_p = \int_0^L \left(\tfrac{1}{2} E\epsilon_x^2 - \epsilon_x E\epsilon_0 + \epsilon_x \sigma_0 \right) A \, dx - \int_0^L uF_x A \, dx - \{D\}^T\{P\} \quad (3.4\text{-}6)$$

∅ SEE P. 76

NOTES 10-26
P. 4

where E = elastic modulus, $dV = A \, dx$, A = cross-sectional area, and L = length. In the second integral, the integrand could be regarded as $uF_x \, dV$ or as $uq \, dx$, where $q = F_x A$: axial body force F_x and axial line load q have the same effect when the body is mathematically one-dimensional.

In beam bending, Fig. 3.4-1, each z = constant layer of the beam is regarded as being in a state of uniaxial stress if, as is common, transverse shear deformation is neglected. Accordingly, the expression for strain energy in a beam can be derived from Eq. 3.4-6. Let b represent the width of the beam. Since $\epsilon_x = u_{,x}$ and $u = -zw_{,x}$, the first term in Eq. 3.4-6 yields

FOR BEAM BENDING

$$\int \tfrac{1}{2} E\epsilon_x^2 \, dV = \iint \tfrac{1}{2} E(-zw_{,xx})^2 \, b \, dz \, dx = \int \tfrac{1}{2} EIw_{,xx}^2 \, dx \quad (3.4\text{-}7)$$

BEST

where, if the cross section is rectangular, $I = bt^3/12$ is the moment of inertia of cross-sectional area A. If terms analogous to ϵ_0 and σ_0 are omitted (but which the reader may add as an exercise), the expression for potential of a straight beam without transverse shear deformation is

FOR BEAM

$$\Pi_p = \int_0^L \tfrac{1}{2} EIw_{,xx}^2 \, dx - \int_0^L wq \, dx - \{w\}^T\{F\} - \{\theta\}^T\{M\} \quad (3.4\text{-}8)$$

where $\{w\}^T = \lfloor w_1 \quad w_2 \ldots \rfloor$ and $\{\theta\}^T = \lfloor \theta_1 \quad \theta_2 \ldots \rfloor$ are lateral deflections and rotations ($\theta = w_{,x}$) at locations where lateral forces $\{F\}$ and moments $\{M\}$ are applied as loads. The second integral in Eq. 3.4-8 accounts for work done by lateral force increments $q \, dx$ during lateral displacement w. Axial stress in the beam is $\sigma_x = E\epsilon_x = Eu_{,x} = -Ezw_{,xx}$.

(a) (b)

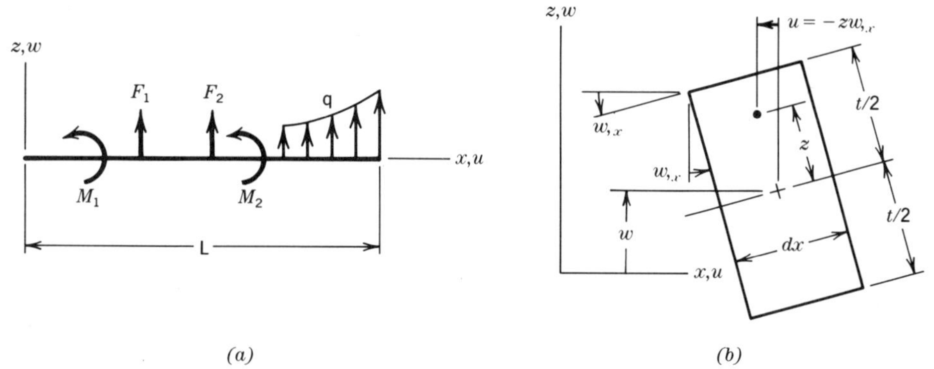

Figure 3.4-1. (*a*) A beam loaded by lateral forces F_i, moments M_i, and distributed lateral load q (force per unit length). (*b*) A slice cut from the beam, shown after it has undergone lateral deflection w and small rotation $w_{,x}$. Transverse shear deformation is ignored.

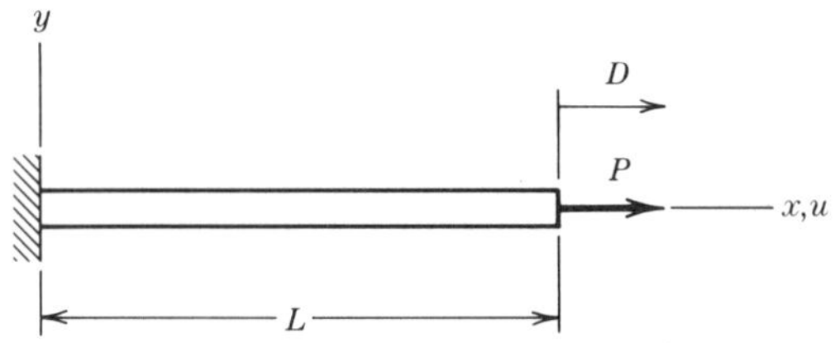

Figure 3.4-2. Uniform bar under axial load P.

Plates in bending are analogous to beams in that strain energy is conveniently expressed in terms of curvatures rather than strains. Specifically, in a thin plate having lateral deflection w, strain energy can be expressed in terms of $w_{,xx}$, $w_{,xy}$, and $w_{,yy}$ instead of ϵ_x, ϵ_y, and γ_{xy}. Potential expressions for problems of plates, shells, and other special cases are introduced in subsequent chapters where they are used.

Example. Bar Under Axial Load. The correctness of Eq. 3.4-6 can be established by solving simple test problems. For example, let the uniform bar of Fig. 3.4-2 carry an end load P and be uniformly heated T degrees. D designates the end displacement produced by P and T. With $\epsilon_0 = \alpha T$, $\sigma_0 = 0$, and $\epsilon_x = D/L$, Eq. 3.4-6 becomes

$$\Pi_p = \int_0^L \left(\tfrac{1}{2} E \frac{D^2}{L^2} - \frac{D}{L} E\alpha T \right) A\, dx - DP = \frac{EAD^2}{2L} - DEA\alpha T - DP \qquad (3.4\text{-}9)$$

End displacement D is found from the equation $d\Pi_p/dD = 0$:

$$D = \frac{PL}{AE} + \alpha TL \qquad (3.4\text{-}10)$$

Finally, the axial stress σ_x is, with $\epsilon_x = D/L$,

$$\sigma_x = E\epsilon_x - E\epsilon_0 = E\left(\frac{P}{AE} + \alpha T\right) - E\alpha T = \frac{P}{A} \qquad (3.4\text{-}11)$$

which is the result expected.

3.5 THE RAYLEIGH–RITZ METHOD

A structure composed of discrete members, such as a truss or a frame, can be represented exactly by a finite number of d.o.f.; these d.o.f. are motions of the joints. A continuum, such as an elastic solid, has infinitely many d.o.f.; these d.o.f. are the displacements of every material point. The behavior of a continuum is described by partial differential equations. For all but the simplest problems there is little hope of discovering a stress field or a displacement field that solves the differential equations and satisfies boundary conditions. The need to solve differential equations can be avoided by applying the Rayleigh–Ritz method to a functional such as Π_p that describes the problem. The result is a substitute problem that has a finite number of d.o.f. and is described by algebraic equations rather than by differential equations. A Rayleigh–Ritz solution is rarely exact but becomes more accurate as more d.o.f. are used.

The Rayleigh–Ritz method began in 1870 with studies of vibration problems by Lord Rayleigh. He used an approximating field that contained a single d.o.f. In 1909, Ritz generalized the method by building an approximating field from several functions, each satisfying essential (i.e., kinematic) boundary conditions, and each associated with a separate d.o.f. Ritz applied the method to equilibrium problems and to eigenvalue problems. The procedure for an equilibrium (static) problem is as follows.

Consider an elastic solid. Displacements and stresses produced by applied loads are required. The displacement of a point is described by the displacement components u, v, and w. A Rayleigh–Ritz solution begins with approximating fields for u, v, and w. Each field is a series, whose typical term is a function of the coordinates, $f_i = f_i(x,y,z)$, times an amplitude a_i whose value is yet to be determined. The a_i may be called *generalized coordinates*. We write

$$u = \sum_{i=1}^{\ell} a_i f_i \qquad v = \sum_{i=\ell+1}^{m} a_i f_i \qquad w = \sum_{i=m+1}^{n} a_i f_i \tag{3.5-1}$$

Each of the functions $f_i = f_i(x,y,z)$ must be *admissible;* that is, each must satisfy compatibility conditions and essential boundary conditions. It is not required that any of the f_i satisfy nonessential boundary conditions (but doing so yields a more accurate approximation for a given number of d.o.f.). Usually, but not necessarily, the f_i are polynomials. The analyst must estimate how many terms are needed in each series in order to achieve the accuracy required. Thus the series are truncated rather than infinite, having, respectively, ℓ, $m - \ell$, and $n - m$ terms, for a total of n terms.

The d.o.f. of the problem are the n amplitudes a_i. They are determined as follows. Substitute Eqs. 3.5-1 into the strain–displacement relations (Eqs. 1.5-6) to find strains $\{\epsilon\}$, then use Eq. 3.4-1 to evaluate Π_p. Thus Π_p becomes a function of d.o.f. a_i, just as Π_p is a function of d.o.f. D_i in Eq. 3.3-1. According to the principle of stationary potential energy, the equilibrium configuration is defined by the n algebraic equations

$$\frac{\partial \Pi_p}{\partial a_i} = 0 \qquad \text{for} \quad i = 1, 2, \ldots, n \tag{3.5-2}$$

After Eqs. 3.5-2 are solved for numerical values of the a_i, the displacement fields of Eqs. 3.5-1 are completely defined. Differentiation of the displacement fields yields strains, which enter the stress–strain relations to produce stresses.

The foregoing procedure has two principal steps. First, establish a trial family of admissible solutions. Second, apply a criterion to select the best form of the family. Here the criterion is that Π_p be stationary. Alternative criteria are available, such as methods of weighted residuals (Chapter 15).

Equations 3.5-1 create a substitute problem because the infinitely many d.o.f. of the real structure are replaced by the finite number of d.o.f. in the mathematical model. A Rayleigh–Ritz solution is usually approximate because the functions f_i are usually incapable of exactly representing the actual displacements. The solution process selects amplitudes a_i so as to combine the functions f_i to best advantage. When Π_p is the functional, "best" means tending to satisfy differential equations of equilibrium and stress boundary conditions more and more closely as more and more terms $a_i f_i$ are added to the series.

Equations 3.5-2 are found to be stiffness equations. They can be written in the usual form $[\mathbf{K}]\{\mathbf{D}\} = \{\mathbf{R}\}$, where $\{\mathbf{D}\} = \lfloor a_1 \quad a_2 \ . \ . \ . \ a_n \rfloor^T$. Not all D_i have units of displacement and not all R_i have units of force, but each product R_iD_i has units of work or energy.

Example. Bar Under Axial Load. Consider the uniform bar of Fig. 3.5-1*a*. The load is distributed along the length of the bar in linearly varying fashion: $q = cx$, where c is a constant that has units of force divided by the square of length. Axial displacement u and axial stress σ_x are to be computed by the Rayleigh–Ritz method.

Axial strain is $\epsilon_x = u_{,x}$. Thus eq. 3.4-6 becomes

$$\Pi_p = \int_0^{L_T} \tfrac{1}{2} E u_{,x}^2 A \, dx - \int_0^{L_T} u(cx) \, dx \tag{3.5-3}$$

Equation 3.5-1 becomes, with $f_i = f_i(x)$ and only polynomial functions considered,

$$u = \sum_{i=1}^{n} a_i f_i = a_1 x + a_2 x^2 + a_3 x^3 + \cdots + a_n x^n \tag{3.5-4}$$

Note that there is no initial term a_0: the displacement mode $u = a_0$ is inadmissible because it violates the essential boundary condition—namely $u = 0$ at $x = 0$.

The simplest approximation results from using only the first term of the series, $u = a_1 x$. From Eqs. 3.5-2, 3.5-3, and 3.5-4,

$$\Pi_p = \frac{AEL_T}{2} a_1^2 - \frac{cL_T^3}{3} a_1 \tag{3.5-5a}$$

$$\frac{d\Pi_p}{da_1} = 0 \qquad \text{yields} \qquad a_1 = \frac{cL_T^2}{3AE} \tag{3.5-5b}$$

$$\text{hence} \qquad u = \frac{cL_T^2}{3AE} x \qquad \text{and} \qquad \sigma_x = Eu_{,x} = \frac{cL_T^2}{3A} \tag{3.5-5c}$$

Before commenting on these results we consider a two-term solution, using the field

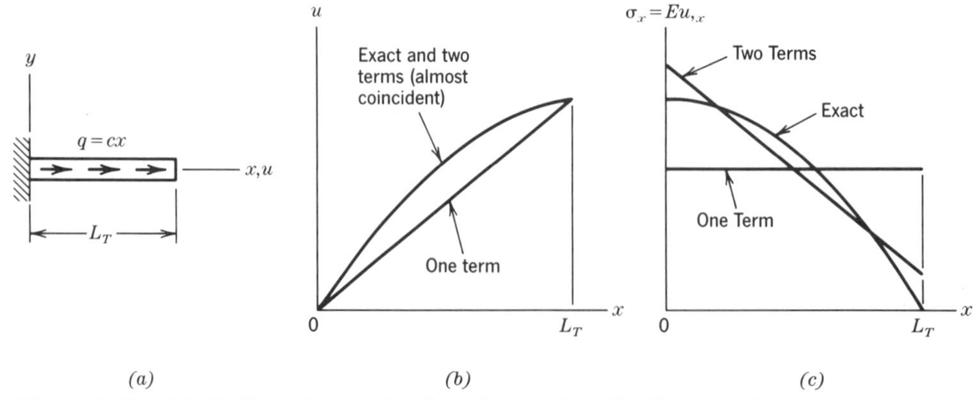

Figure 3.5-1. (*a*) Uniform bar under linearly varying distributed axial load of intensity $q = cx$, where c is a constant. (*b*) Exact and approximate axial displacements. (*c*) Exact and approximate axial stresses.

$u = a_1 x + a_2 x^2$. After substituting into Eq. 3.5-3 and writing $\dfrac{\partial \Pi_p}{\partial a_1} = 0$ and $\dfrac{\partial \Pi_p}{\partial a_2} = 0$, we obtain

$$AEL_T \begin{bmatrix} 1 & L_T \\ L_T & 4L_T^2/3 \end{bmatrix} \begin{Bmatrix} a_1 \\ a_2 \end{Bmatrix} = \frac{cL_T^3}{12} \begin{Bmatrix} 4 \\ 3L_T \end{Bmatrix} \quad \text{or} \quad \begin{Bmatrix} a_1 \\ a_2 \end{Bmatrix} = \frac{cL_T}{12AE} \begin{Bmatrix} 7L_T \\ -3 \end{Bmatrix} \tag{3.5-6a}$$

hence $$u = \frac{cL_T}{12AE}(7L_T x - 3x^2) \quad \text{and} \quad \sigma_x = Eu_{,x} = \frac{cL_T}{12A}(7L_T - 6x) \tag{3.5-6b}$$

Figures 3.5-1*b* and 3.5-1*c* show the comparison between exact and approximate results. As might be expected, the two-term results (Eqs. 3.5-6) are better than the one-term results (Eqs. 3.5-5). With the body force term $F_x = cx/A$, the differential equation of equilibrium (Eq. 1.6-2a) becomes

$$\sigma_{x,x} + \frac{cx}{A} = 0 \quad \text{or} \quad AEu_{,xx} + cx = 0 \tag{3.5-7}$$

The latter equation results from the substitution $\sigma_x = Eu_{,x}$. Neither Eq. 3.5-7, nor the natural boundary condition $\sigma_x = 0$ at $x = L_T$, is satisfied by the foregoing two approximate solutions.

Notice that approximate displacements are more accurate than approximate stresses. This is to be expected, because stresses are calculated from derivatives of the approximating field. (To see that differentiation emphasizes discrepancies, consider the functions $f_1 = 4x(1 - x)$ and $f_2 = \sin \pi x$. In the range $0 < x < 1$, functions f_1 and f_2 look much alike, but successive derivatives of f_1 and f_2 have less and less resemblance.)

The exact solution of the problem of Fig. 3.5-1*a* is

$$u = \frac{c}{6AE}(3L_T^2 x - x^3) \tag{3.5-8}$$

Use of the series $u = a_1 x + a_2 x^2 + a_3 x^3$ in the Rayleigh–Ritz method produces

$$a_1 = \frac{cL_T^2}{2AE} \qquad a_2 = 0 \qquad a_3 = -\frac{c}{6AE} \tag{3.5-9}$$

which is the exact solution. Use of still more terms leads again to the exact solution: one finds a_1, a_2, and a_3 to be the values given in Eq. 3.5-9 and $a_4 = a_5 = a_6 = \cdots = a_n = 0$. In general, the Rayleigh–Ritz method yields the exact solution if the approximating field is capable of representing the exact field by appropriate choice of d.o.f. a_i. In practice this circumstance is rare.

3.6 COMMENTS ON THE RAYLEIGH–RITZ METHOD BASED ON ASSUMED DISPLACEMENT FIELDS

Approximating fields must be admissible and should be easy to use. Only polynomials, and occasionally sine and cosine functions, are simple enough to be practicable. Beyond this there are no easy answers to important questions: What sort of assumption for the field is best? What particular terms and how many of them? How good are the computed results? These difficulties and uncertainties are increased for multidimensional problems.

Let a problem be solved repeatedly, each time with another term added to the assumed field (as in Eqs. 3.5-5 and 3.5-6, for example). Thus we generate a sequence of trial solutions. We expect the sequence to converge: to the exact Π_p, to the exact displacements, and to the exact stresses. A necessary condition for convergence is that trial field be *complete*.

Completeness is achieved if the exact displacements, *and* their derivatives that appear in Π_p, can be matched arbitrarily closely if enough terms appear in the trial field. A polynomial series is complete if it is of high enough degree and if no terms are omitted. Fourier series are also complete.

Completeness demands that the lowest-order admissible terms be included. For example, consider the end-loaded bar of Fig. 3.4-2. If we omit the term $a_1 x$ from Eq. 3.5-4, we omit the very term that contains the exact answer—namely, $u = (P/AE)x$. Thus completeness is destroyed, and the sequence of approximate solution does not produce the exact answer even if the number of terms approaches infinity. This is easy to see: if Eq. 3.5-4 begins with $a_2 x^2$, then $\sigma_x = Eu_{,x} = 0$ at $x = 0$, which is incorrect. The term $a_1 x$ represents the essential *constant-strain capability*. (In a finite element context, this requirement means that each element must be capable of represeting a state of constant strain.)

Completeness also requires that no series terms be omitted. Referring again to Eq. 3.5-4, a two-term approximation should be $a_1 x + a_2 x^2$ but not $a_1 x + a_3 x^3$, and not $a_1 x + a_4 x^4$, and so on.

In two dimensions, a polynomial is of degree n if it contains a term of the form $x^\ell y^m$, where ℓ and m are nonnegative integers and $\ell + m = n$. The polynomial is complete if it contains all combinations of ℓ and m for which $\ell + m = n$ and if no lower-order terms are omitted. For example, a complete quadratic, $n = 2$, has the form $u = a_1 + a_2 x + a_3 y + a_4 x^2 + a_5 xy + a_6 y^2$. A complete polynomial of degree n in two dimensions contains $(n + 1)(n + 2)/2$ terms (see Fig. 3.6-1).

In three dimensions, similar remarks apply, so that a complete quadratic contains 10 terms, which include a constant term, the linear terms x, y, and z, and the quadratic terms x^2, xy, y^2, yz, z^2, and zx.

A Rayleigh–Ritz solution is either exact or it is too stiff. This happens because the mathematical structure is permitted to displace only into shapes that can be described by superposing the finite number of functions f_i present in the assumed displacement field that the analyst selects. Therefore, the correct shape is excluded, unless the assumed field happens to contain it. Effectively, the assumed field imposes constraints that prevent the structure from deforming the way it wants to. Constraints stiffen a structure. In effect, the solution method creates a substitute structure that is stiffer than the real one.

The work W done by loads that gradually increase from zero to $\{\mathbf{R}\}$ is $W = \{\mathbf{D}\}^T \{\mathbf{R}\}/2$ if the structure is linearly elastic. An approximate solution yields d.o.f.

Pascal triangle	*Degree and number of terms*
1	↓ 0 (constant), 1 term
$x \quad y$	↓ 1 (linear), three terms
$x^2 \quad xy \quad y^2$	↓ 2 (quadratic), six terms
$x^3 \quad x^2y \quad xy^2 \quad y^3$	↓ 3 (cubic), ten terms

Figure 3.6-1. Pascal triangle, showing the number of terms in complete polynomials in two independent variables x and y.

{**D**} such that work W is less than the exact value. This does not necessarily mean that *every* d.o.f. in {**D**} is underestimated. But if the structure carries a single load P, we can say that its computed displacement D is a lower bound to the correct magnitude.

Strain energy U is numerically equal to W, so the approximate solution underestimates U when loads are prescribed. If displacements are prescribed instead, U is overestimated because extra force is needed to deform an overly stiff structure. When loads *and* displacements are prescribed, U may be high or low.

Stresses are calculated from displacements, so we expect that a too-stiff structure will underestimate stress magnitudes. However, as seen in Fig. 3.5-1*c*, approximate stresses may be too low in one place but too high in another, even when the stress is derived from an approximate displacement field that is everywhere too low. Accordingly, a rule about stress magnitudes would be either so crude or so equivocal as to be of little value.

3.7 STATIONARY PRINCIPLES AND GOVERNING EQUATIONS

The principle of stationary potential energy is but one of many stationary principles of mathematical physics. Central to each is a functional, of which Π_p is but one. Rayleigh–Ritz approximations and finite element formulations can be derived from functionals. For this reason the following brief remarks are offered.

Consider the functional for Π_p, Eq. 3.4-1. It depends on displacements {**u**} and on strains {$\boldsymbol{\epsilon}$}, which are derivates of {**u**}. The term "functional" indicates that Π_p depends not on {**u**} and its derivatives at a point but upon their integrated effect over a region of interest. The stationary condition $d\Pi_p = 0$ may be applied directly to Eq. 3.4-1, without first expressing Π_p in terms of a finite number of d.o.f. This is accomplished by using the *calculus of variations*, the procedures of which are beyond the scope of this book [3.1]. However, the end results of setting $d\Pi_p$ to zero are found to be the differential equations of equilibrium (Eqs. 1.6-2) and the nonessential boundary conditions (the stress boundary conditions, Eqs. 1.6-4). Thus, if the field {**u**} is admissible, the statement $d\Pi_p = 0$ implies all components of a valid solution: satisfaction of equilibrium, compatibility, and boundary conditions. In an approximate solution, equilibrium conditions and stress boundary conditions are satisfied only in an average or integral sense, not at every point.

In other physical problems there exist other functionals Π. Instead of displacements {**u**}, the primary field may be temperature, or pressure, or voltage, and so on. In each case the functional Π can be tested for correctness by applying the calculus of variations to see if the condition $d\Pi = 0$ yields the appropriate governing differential equation and nonessential boundary conditions.

Boundary Conditions. In order to use variational methods, one must be able to distinguish between essential and nonessential boundary conditions. For a problem having one dependent field variable, the rule is as follows. Let $2m$ be the highest-order derivative of the dependent field variable in the governing differential equation. (Derivatives of order m then appear in the functional.) Essential boundary conditions involve derivatives of order zero through $m - 1$, the zeroth derivative being the dependent variable itself. Nonessential boundary conditions

FROM P. 70
ESSENTIAL B.C. – VALUES OF NODAL D.O.F.
NON ESSENTIAL – HIGHER DERIVATIVES

involve derivatives of order m and higher, up to and including $2m - 1$. The following are examples.

Problem	Bar (Fig. 3.5-1*a*)	Beam bending	Two-dimensional heat conduction
Differential equation	$AEu_{,xx} + q = 0$	$EIw_{,xxxx} - q = 0$	$k\nabla^2 T + Q = c\rho\dot{T}$
$2m$, $m - 1$, $2m - 1$	2, 0, 1	4, 1, 3	2, 0, 1
Essential B.C.	On u only	On w and $w_{,x}$	On T only
Nonessential B.C.	On $\sigma_x = Eu_{,x}$	On $M = EIw_{,xx}$ and $V = EIw_{,xxx}$	On $q = -k(T_{,x}\,\ell_B + T_{,y}m_B)$

In these examples the dependent field variables are axial displacement u, lateral displacement w, and temperature T. Nonessential boundary conditions concern axial stress σ_x, bending moment M, transverse shear force V, and heat flow q. In the heat conduction example, k is thermal conductivity, $\dot{T}$ means the time derivative of T, and ℓ_B and m_B are direction cosines of a normal to the boundary (see Chapter 16).

The foregoing remarks are little changed if there is more than one field. Imagine, for example, that there are dependent field variables u and v, with second derivatives $u_{,xx}$, $u_{,xy}$, $u_{,yy}$, $v_{,xx}$, $v_{,xy}$, and $v_{,yy}$ in the governing differential equations and first derivatives $u_{,x}$, $u_{,y}$, $v_{,x}$ and $v_{,y}$ in the functional. Then $2m = 2$ and $m = 1$ for both u and v. Essential boundary conditions are prescriptions of u and v at particular locations. Nonessential boundary conditions involve first derivatives of u and v, either singly or in combination.

Functionals and Governing Differential Equations. Imagine that a functional Π depends on two dependent field variables, $u = u(x,y)$ and $v = v(x,y)$, in which independent variables x and y are Cartesian coordinates:

$$\Pi = \iint F(x,y,u,v,u_{,x},u_{,y},v_{,x},v_{,y}, \ldots ,v_{,yy})\,dx\,dy \tag{3.7-1}$$

Here we will assume that F contains no derivatives of order higher than second. There are as many "Euler equations" as there are dependent field variables. An Euler equation is a governing differential equation of the physical problem. Methods of calculus of variations extract from Eq. 3.7-1 the Euler equations

$$\frac{\partial F}{\partial u} - \frac{\partial}{\partial x}\frac{\partial F}{\partial u_{,x}} - \frac{\partial}{\partial y}\frac{\partial F}{\partial u_{,y}} + \frac{\partial^2}{\partial x^2}\frac{\partial F}{\partial u_{,xx}} + \frac{\partial^2}{\partial x\,\partial y}\frac{\partial F}{\partial u_{,xy}} + \frac{\partial^2}{\partial y^2}\frac{\partial F}{\partial u_{,yy}} = 0 \tag{3.7-2a}$$

$$\frac{\partial F}{\partial v} - \frac{\partial}{\partial x}\frac{\partial F}{\partial v_{,x}} - \frac{\partial}{\partial y}\frac{\partial F}{\partial v_{,y}} + \frac{\partial^2}{\partial x^2}\frac{\partial F}{\partial v_{,xx}} + \frac{\partial^2}{\partial x\,\partial y}\frac{\partial F}{\partial v_{,xy}} + \frac{\partial^2}{\partial y^2}\frac{\partial F}{\partial v_{,yy}} = 0 \tag{3.7-2b}$$

Equations 3.7-1 and 3.7-2 both describe the same problem, Eq. 3.7-1 being called the "weak form" and Eqs. 3.7-2 the "strong form."

As a specific example of Eq. 3.7-2, consider the axially loaded uniform bar described by Fig. 3.5-1*a*. Here there is one independent variable, one dependent variable, and no second derivative. Equation 3.7-1 reduces to Eq. 3.5-3. There is but one Euler equation, Eq. 3.7-2*a*, which reduces to

$$\frac{\partial F}{\partial u} - \frac{d}{dx}\frac{\partial F}{\partial u_{,x}} = 0 \qquad \text{in which} \qquad F = \tfrac{1}{2}AEu_{,x}^2 - u(cx) \tag{3.7-3}$$

Derivatives in the Euler equation are

$$\frac{\partial F}{\partial u} = -cx \qquad \text{and} \qquad \frac{d}{dx}\frac{\partial F}{\partial u_{,x}} = \frac{d}{dx}(AEu_{,x}) = AEu_{,xx} \tag{3.7-4}$$

from which Eq. 3.5-7 is obtained, as expected.

As another example, consider plane heat conduction in an isotropic material. A suitable functional is

$$\Pi = \iint (\tfrac{1}{2}kT_{,x}^2 + \tfrac{1}{2}kT_{,y}^2 - QT + \rho cT\dot{T})\,dx\,dy \qquad \text{or} \qquad \Pi = \iint F\,dx\,dy \tag{3.7-5}$$

in which T = temperature, k = thermal conductivity, Q = internally generated heat flow, ρ = mass density, c = specific heat, and $\dot{T}$ is the time derivative $\partial T/\partial t$. Unit thickness is assumed. (Equation 3.7-5 omits certain boundary terms of practical interest. See Section 16.3 for a more detailed treatment.) Again there is one dependent field variable, T, and one Euler equation—namely,

$$\frac{\partial F}{\partial T} - \frac{\partial}{\partial x}\frac{\partial F}{\partial T_{,x}} - \frac{\partial}{\partial y}\frac{\partial F}{\partial T_{,y}} = 0 \tag{3.7-6}$$

where F is the integrand of Eq. 3.7-5. Equations 3.7-5 and 3.7-6 yield

$$k(T_{,xx} + T_{,yy}) + Q - \rho c\dot{T} = 0 \tag{3.7-7}$$

as the differential equation that describes the temperature distribution in the region of interest.

Potential energy functionals Π_p for problems of beam bending and plate bending contain second derivatives of lateral displacement w. These and other examples are left as exercises.

The calculus of variations also produces natural boundary conditions. An explanation of their derivation and interpretation takes more space than we can allot to it. The reader is referred to other texts [3.1,3.2].

Finally, we remark that although there is always a differential equation associated with a functional, the reverse is not necessarily true. For example, a differential equation that contains an odd-numbered derivative does not have an associated functional of the form of Eq. 3.7-1.

Variational Methods: A Brief Example. Figure 3.7-1 shows a uniform bar loaded by distributed axial load $q = q(x)$ and prescribed stress σ_L at $x = L$. We will use this problem to illustrate that the calculus of variations produces the governing differential equation and the nonessential boundary condition. In this way we will discover the origin of terms seen in Eqs. 3.7-3 and 3.7-4. The development will

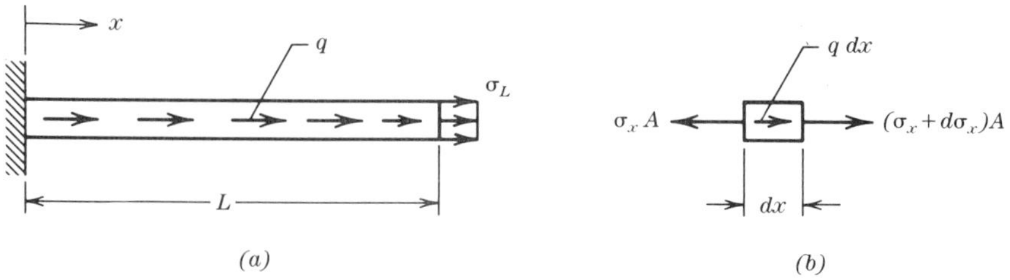

Figure 3.7-1. (*a*) Uniform elastic bar loaded by distributed axial load q and end stress σ_L. (*b*) Forces that act on a differential element of the bar.

also be seen to produce the virtual work equation and to suggest an alternative formulation method (the *method of weighted residuals*).

In Eq. 3.4-6, let $\epsilon_x = u_{,x}$, $F_x = q/A$, $D = u_L$, and $P = A\sigma_L$. Thus the potential energy functional for the bar in Fig. 3.7-1*a* is

$$\Pi_p = \frac{AE}{2}\int_0^L u_{,x}^2\, dx - \int_0^L qu\, dx - (A\sigma_L)u_L \tag{3.7-8}$$

We presume that $u = u(x)$ is an admissible displacement field for this problem—that is, a field that is continuous and satisfies the essential boundary condition $u = 0$ at $x = 0$. Let u be perturbed by an amount δu, which we elect to write as $\delta u = e\eta$, where e is a small number and $\eta = \eta(x)$ is an admissible field. Thus the perturbed field $u + e\eta$ is also admissible and satisfies the same *essential* boundary condition as u. Hence, $u_{,x}$ becomes $u_{,x} + e\eta_{,x}$, u_L becomes $u_L + e\eta_L$, and Π_p becomes $\Pi_p + \delta\Pi_p$. The change in energy $(\Pi_p + \delta\Pi_p) - \Pi_p$ is

$$\delta\Pi_p = e\left[AE\int_0^L u_{,x}\,\eta_{,x}\, dx - \int_0^L q\eta\, dx - (A\sigma_L)\eta_L\right] + e^2\frac{AE}{2}\int_0^L \eta_{,x}^2\, dx \tag{3.7-9}$$

According to the potential energy principle, stable equilibrium occurs when Π_p is a relative minimum. This implies that $\delta\Pi_p > 0$ for any admissible η. Now $e^2\eta_{,x}^2$ is never negative, and the remaining term $e[\text{---}]$ changes sign when e changes sign. We conclude that if $\delta\Pi_p$ is to be positive for all small values of e, the bracketed expression in Eq. 3.7-9 must vanish. Setting this expression to zero, and integrating its first term by parts according to the standard formula $\int u\, dv = -\int v\, du + uv$, we obtain

$$0 = -AE\int_0^L u_{,xx}\,\eta\, dx + \Big[AEu_{,x}\,\eta\Big]_0^L - \int_0^L q\eta\, dx - (A\sigma_L)\eta_L \tag{3.7-10}$$

But $\eta = 0$ at $x = 0$, so Eq. 3.7-10 becomes

$$0 = -\int_0^L (AEu_{,xx} + q)\eta\, dx + A(Eu_{,x} - \sigma_L)\eta_L \tag{3.7-11}$$

Since $\eta = \eta(x)$ is admissible but otherwise arbitrary, an arbitrary value of η_L can be assigned while infinitely many functions η are yet possible in the range $0 <$

$x < L$. Accordingly, Eq. 3.7-11 can be satisfied only if the coefficients of η and η_L vanish separately. Thus we obtain

$$AEu_{,xx} + q = 0 \qquad \text{for} \quad 0 < x < L \tag{3.7-12a}$$

$$Eu_{,x} - \sigma_L = 0 \qquad \text{at} \quad x = L \tag{3.7-12b}$$

Equation 3.7-12a is the governing differential equation. It can be written in the alternative form $A\sigma_{x,x} + q = 0$, and can also be derived by considering the equilibrium of axial forces in Fig. 3.7-1*b*. Equation 3.7-12*b* is a nonessential (or *natural*) boundary condition, which says that $\epsilon_x = \sigma_L/E$ at $x = L$.

The vanishing of the bracketed expression in Eq. 3.7-9 can be regarded as an expression of the virtual work principle, which states that the total work of internal and external forces must vanish for any admissible infinitesimal displacement from an equilibrium configuration. In Eq. 3.7-9 internal forces $AEu_{,x}\,dx = A\sigma_x\,dx$ do work (and store strain energy) when strains $\eta_{,x}$ occur, and external forces $q\,dx$ and $A\sigma_L$ do negative work (and lose potential energy) when positive displacements η and η_L occur.

Matrices used in finite element analysis can be generated from either Eq. 3.7-8 or the bracketed expression in Eq. 3.7-9. (In Eq. 3.7-9, if we identify $u_{,x}$ as ϵ_x and $e\eta_{,x}$ as $\delta\epsilon_x$, the first integrand becomes $\delta\epsilon_x AE\epsilon_x$, which will subsequently be recognized as a familiar form.)

The vanishing of the bracketed expression in Eq. 3.7-9 can also be obtained by "working backward," as follows. Imagine that we seek an approximate solution $\tilde{u} = \tilde{u}(x)$—for example, the admissible polynomial $\tilde{u} = a_1x + a_2x^2 + a_3x^3 + \cdots$, where the a_i are constants that must be determined. Now $\tilde{u}$ does not satisfy Eq. 3.7-12a for all x: a "residual," $R = R(x) = AE\tilde{u}_{,xx} + q \neq 0$, is left over. Nevertheless, we can select the a_i so as to satisfy Eq. 3.7-12a in an average or integral sense by writing

$$\int_0^L (AE\tilde{u}_{,xx} + q)\eta\,dx = 0 \tag{3.7-13}$$

where $\eta = \eta(x)$ may now be called a "weight function." Applying integration by parts to Eq. 3.7-13, we obtain

$$-AE\int_0^L \tilde{u}_{,x}\,\eta_{,x}\,dx + \Big[AE\tilde{u}_{,x}\eta\Big]_0^L + \int_0^L q\eta\,dx = 0 \tag{3.7-14}$$

But $\eta = 0$ at $x = 0$. In addition, at $x = L$, we may replace $E\tilde{u}_{,x}$ by σ_L, thus introducing the nonessential boundary condition. Equation 3.7-14 becomes

$$-AE\int_0^L \tilde{u}_{,x}\,\eta_{,x}\,dx + \int_0^L q\eta\,dx + (A\sigma_L)\eta_L = 0 \tag{3.7-15}$$

which agrees with the vanishing of the bracketed expression in Eq. 3.7-9. This method of formulating a problem for approximate solution is called a *weighted residual method*. It can be applied to problems for which one knows the differential equation but not the functional or the variational principle. If $\eta_i = \partial\tilde{u}/\partial a_i$ the method is known as the *Galerkin method* (for which Eq. 3.7-15 yields as many equations as there are a_i to be determined). A more detailed discussion appears in Chapter 15.

3.8 A PIECEWISE POLYNOMIAL FIELD

In this section we use a one-dimensional example to illustrate how a displacement field can be written in terms of physical displacements $\{\mathbf{d}\}$ rather than parameters a_i (as in Eq. 3.5-4). The use of $\{\mathbf{d}\}$, in combination with a piecewise polynomial field, leads to the finite element method in a form that is easy to program for computer solution. We begin by using the a_i, then show how to replace them by functions of nodal d.o.f. d_i.

Consider the bar of Fig. 3.8-1*a*. It is to be loaded axially. Axial displacement u over the length from $x = 0$ to $x = L_T$ is to be approximated as three separate linear fields,

$$u = a_1 + a_2 x \quad \text{for} \quad 0 \leq x \leq x_2 \tag{3.8-1a}$$

$$u = a_3 + a_4 x \quad \text{for} \quad x_2 \leq x \leq x_3 \tag{3.8-1b}$$

$$u = a_5 + a_6 x \quad \text{for} \quad x_3 \leq x \leq x_4 \tag{3.8-1c}$$

where the a_i are d.o.f. to be determined in a subsequent Rayleigh–Ritz solution. For the field of Eqs. 3.8-1 to be admissible we must have $u = 0$ at $x = 0$. In addition, the first and second expressions must yield the same u at $x = x_2$, and the second and third expressions must yield the same u at $x = x_3$. These three conditions yield $a_1 = 0$, $a_3 = (a_2 - a_4)x_2$, and $a_5 = (a_2 - a_4)x_2 + (a_4 - a_6)x_3$. Thus Eqs. 3.8-1 assume the form

$$u = a_2 x \quad \text{for} \quad 0 \leq x \leq x_2 \tag{3.8-2a}$$

$$u = a_2 x_2 + a_4(x - x_2) \quad \text{for} \quad x_2 \leq x \leq x_3 \tag{3.8-2b}$$

$$u = a_2 x_2 + a_4(x_3 - x_2) + a_6(x - x_3) \quad \text{for} \quad x_3 \leq x \leq x_4 \tag{3.8-2c}$$

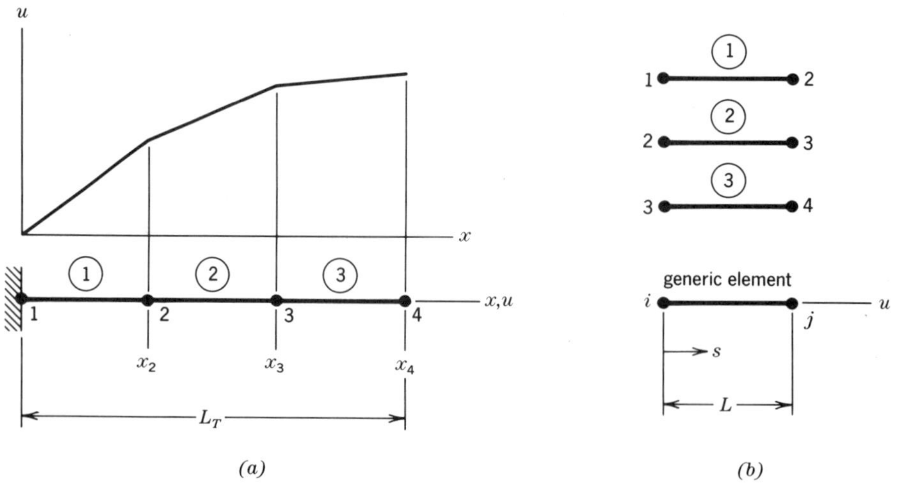

(*a*) (*b*)

Figure 3.8-1. (*a*) Bar whose axial displacement field $u = u(x)$ is approximated by a piecewise linear fit. (*b*) Separate regions (elements) of the bar.

The foregoing procedure has drawbacks. First, the a_i do not have an obvious physical meaning. Second, neither the passage from Eqs. 3.8-1 to 3.8-2 nor the subsequent determination of d.o.f. a_i from the equations $\partial\Pi_p/\partial a_i = 0$ is readily coded, especially if the user of the program is to be allowed to use various loadings, other boundary conditions, and more a_i than six. These drawbacks are neatly avoided by expressing the displacement field in terms of nodal d.o.f. rather than the a_i. The procedure is as follows.

Shape Function Matrix. Consider the generic element, Fig. 3.8-1*b*. Its axial displacement u is to be linear in axial coordinate s. The displacement field must yield $u = u_i$ at one end and $u = u_j$ at the other. *By inspection,* we write

$$u = \frac{L - s}{L} u_i + \frac{s}{L} u_j \qquad \text{or} \qquad u = \lfloor \mathbf{N} \rfloor \{\mathbf{d}\} \tag{3.8-3}$$

where

$$\lfloor \mathbf{N} \rfloor = \left\lfloor \frac{L - s}{L} \quad \frac{s}{L} \right\rfloor \qquad \text{and} \qquad \{\mathbf{d}\} = \begin{Bmatrix} u_i \\ u_j \end{Bmatrix} \tag{3.8-4}$$

Checking, we see the u is indeed linear in s and assumes the values $u = u_i$ at $s = 0$ and $u = u_j$ at $s = L$. Therefore, the expression written is the one desired.

Matrix $\lfloor \mathbf{N} \rfloor$ is usually called a *shape function matrix*. Its terms N_i may be called *shape*, *basis*, or *interpolation functions*. Each of the two terms N_1 and N_2 defines how displacement u varies with s when the corresponding d.o.f. has unit value while the other d.o.f. is zero. Matrix $\lfloor \mathbf{N} \rfloor$ describes how u is to be interpolated from nodal values u_i and u_j over the elements—that is, over the range $0 \leqslant s \leqslant L$. Here the shape function matrix happens to be a row matrix. For many other elements, treated in subsequent chapters, it is a rectangular matrix. Often, as in Eq. 3.8-3, it is possible to write $\lfloor \mathbf{N} \rfloor$ by inspection and a bit of trial. Alternatively, $\lfloor \mathbf{N} \rfloor$ can be formally derived, as we now illustrate.

Consider again the generic element in Fig. 3.8-1*b*. We begin with the linear displacement field

$$u = a_1 + a_2 s \qquad \text{or} \qquad u = \lfloor 1 \quad s \rfloor \begin{Bmatrix} a_1 \\ a_2 \end{Bmatrix} \tag{3.8-5}$$

This field must take on the values $u = u_i$ at $s = 0$ and $u = u_j$ at $s = L$:

$$\begin{Bmatrix} u_i \\ u_j \end{Bmatrix} = \begin{bmatrix} 1 & 0 \\ 1 & L \end{bmatrix} \begin{Bmatrix} a_1 \\ a_2 \end{Bmatrix} \qquad \text{or} \qquad \{\mathbf{d}\} = [\mathbf{A}]\{\mathbf{a}\} \tag{3.8-6}$$

Solving for $\{\mathbf{a}\}$ and substituting into Eq. 3.8-5, we obtain

$$\{\mathbf{a}\} = [\mathbf{A}]^{-1}\{\mathbf{d}\} \qquad \text{and} \qquad u = \lfloor 1 \quad s \rfloor [\mathbf{A}]^{-1} \begin{Bmatrix} u_i \\ u_j \end{Bmatrix} \tag{3.8-7}$$

The shape function matrix is

$$\lfloor \mathbf{N} \rfloor = \lfloor 1 \quad s \rfloor [\mathbf{A}]^{-1} = \lfloor 1 \quad s \rfloor \begin{bmatrix} 1 & 0 \\ -1/L & 1/L \end{bmatrix} = \left\lfloor \frac{L-s}{L} \quad \frac{s}{L} \right\rfloor \tag{3.8-8}$$

which agrees with Eq. 3.8-4.

In all elements of Fig. 3.8-1, displacement u has the same *form*—a linear variation—but not the same *value* because lengths L and nodal d.o.f. $\{\mathbf{d}\}$ are in general different for different elements. Compatibility between elements is assured because elements share a common d.o.f. where they meet; for example, at $x = x_2$, $u = u_2$ in element 1 *and* in element 2.

Additional examples of displacement fields and shape function matrices appear in Sections 3.12 and 3.13 and in subsequent chapters.

3.9 FINITE ELEMENT FORM OF THE RAYLEIGH–RITZ METHOD

The finite element method can be defined as a Rayleigh–Ritz method in which the approximating field is interpolated in piecewise fashion from d.o.f. that are nodal values of the field. This viewpoint is illustrated by the following treatment of the axially loaded bar depicted in Fig. 3.8-1. As in preceding sections of the present chapter, we will form potential $\Pi_p = U + \Omega$, make Π_p stationary with respect to the d.o.f., then solve the resulting equations $[\mathbf{K}]\{\mathbf{D}\} = \{\mathbf{R}\}$ for d.o.f. $\{\mathbf{D}\}$. In an actual computer program equations $[\mathbf{K}]\{\mathbf{D}\} = \{\mathbf{R}\}$ would be written directly, without using Π_p at all.

Element Stiffness Matrix. Consider a typical bar element of length L that lies along the x axis (Fig. 3.8-1*b*). Axial strain is $\epsilon_x = du/dx = du/ds$. Hence, from Eq. 3.8-3,

$$\epsilon_x = \lfloor \mathbf{B} \rfloor \{\mathbf{d}\}, \qquad \text{where} \quad \lfloor \mathbf{B} \rfloor = \frac{d}{ds} \lfloor \mathbf{N} \rfloor = \left\lfloor -\frac{1}{L} \quad \frac{1}{L} \right\rfloor \tag{3.9-1}$$

Matrix $\lfloor \mathbf{B} \rfloor$ is called the *strain–displacement matrix*. From the first integral in Eq. 3.4-6, with $L_T = L$ and $dx = ds$, strain energy in an element is

$$U = \int_0^L \tfrac{1}{2} E \epsilon_x^2 A \, ds = \tfrac{1}{2} \int_0^L \epsilon_x^T AE \epsilon_x \, ds \tag{3.9-2}$$

The purpose of writing $\epsilon_x^T \epsilon_x$ instead of ϵ_x^2 is to simplify subsequent differentiation of matrix forms. From Eqs. 3.9-1 and 3.9-2,

$$U = \tfrac{1}{2} \{\mathbf{d}\}^T [\mathbf{k}] \{\mathbf{d}\}, \qquad \text{where} \quad [\mathbf{k}] = \int_0^L \lfloor \mathbf{B} \rfloor^T AE \lfloor \mathbf{B} \rfloor \, ds \tag{3.9-3}$$

If AE is constant, the 2 by 2 element stiffness matrix $[\mathbf{k}]$ is found to be the familiar result for a bar element, seen previously in Eqs. 1.2-3 and 2.4-5.

Loads. From the second integral in Eq. 3.4-6, with the substitution of $u = u^T$ for axial displacement and $F_x = q/A$ for axial body force, the potential of the load is

$$\Omega = -\int_0^L uF_xA\,ds = -\int_0^L u^T q\,ds \tag{3.9-4}$$

From Eqs. 3.8-3 and 3.9-4,

$$\Omega = -\{\mathbf{d}\}^T\{\mathbf{r}_e\}, \qquad \text{where} \quad \{\mathbf{r}_e\} = \int_0^L \lfloor \mathbf{N} \rfloor^T q\,ds \tag{3.9-5}$$

Vector $\{\mathbf{r}_e\}$ is called a *consistent load vector*. It tells how a distributed load should be allocated to nodes in a way that is consistent with the displacement field assumed. Further explanation appears in Section 4.3.

For the present illustration we will take $q = cx$, which is the linearly varying distributed axial load used in Fig. 3.5-1. In elements 1, 2, and 3, respectively, $q = cs$, $q = c(x_2 + s)$, and $q = c(x_3 + s)$. For convenience we now give all elements the same length, $L = L_T/3$. Thus, for elements 1, 2, and 3,

$$\{\mathbf{r}_e\}_1 = \frac{cL^2}{6}\begin{Bmatrix}1\\2\end{Bmatrix} \qquad \{\mathbf{r}_e\}_2 = \frac{cL^2}{6}\begin{Bmatrix}4\\5\end{Bmatrix} \qquad \{\mathbf{r}_e\}_3 = \frac{cL^2}{6}\begin{Bmatrix}7\\8\end{Bmatrix} \tag{3.9-6}$$

Global (Structural) Equations. The total potential of the three-element structure is the sum of the three element contributions:

$$\Pi_p = U + \Omega = U_1 + U_2 + U_3 + \Omega_1 + \Omega_2 + \Omega_3 \tag{3.9-7}$$

Let element matrices be expanded to "structure size" as explained in Section 2.5, so that nodal d.o.f. vector $\{\mathbf{d}\}$ of each element is replaced by the "global" vector $\{\mathbf{D}\} = \lfloor u_1 \quad u_2 \quad u_3 \quad u_4 \rfloor^T$, which contains all d.o.f. of the structure. Thus, from Eqs. 3.9-3 and 3.9-5, with AE taken as constant over the entire length $L_T = 3L$,

$$\begin{aligned}\Pi_p = \frac{1}{2}\{\mathbf{D}\}^T \left(\frac{AE}{L}\begin{bmatrix}1 & -1 & 0 & 0\\ -1 & 1 & 0 & 0\\ 0 & 0 & 0 & 0\\ 0 & 0 & 0 & 0\end{bmatrix} + \frac{AE}{L}\begin{bmatrix}0 & 0 & 0 & 0\\ 0 & 1 & -1 & 0\\ 0 & -1 & 1 & 0\\ 0 & 0 & 0 & 0\end{bmatrix} \right. \\ \left. + \frac{AE}{L}\begin{bmatrix}0 & 0 & 0 & 0\\ 0 & 0 & 0 & 0\\ 0 & 0 & 1 & -1\\ 0 & 0 & -1 & 1\end{bmatrix}\right)\{\mathbf{D}\} - \{\mathbf{D}\}^T\left(\frac{cL^2}{6}\begin{Bmatrix}1\\2\\0\\0\end{Bmatrix} + \frac{cL^2}{6}\begin{Bmatrix}0\\4\\5\\0\end{Bmatrix} + \frac{cL^2}{6}\begin{Bmatrix}0\\0\\7\\8\end{Bmatrix} + \begin{Bmatrix}R_1\\0\\0\\0\end{Bmatrix}\right)\end{aligned} \tag{3.9-8}$$

in which R_1 represents the support reaction at $x = 0$. Four equilibrium equations $[\mathbf{K}]\{\mathbf{D}\} = \{\mathbf{R}\}$ are provided by the stationary condition $\{\partial\Pi_p/\partial\mathbf{D}\} = \{\mathbf{0}\}$. Differentiation rules given in Appendix A make this process easy. The resulting global equations $[\mathbf{K}]\{\mathbf{D}\} = \{\mathbf{R}\}$ are

$$\frac{AE}{L}\begin{bmatrix} 1 & -1 & 0 & 0 \\ -1 & 2 & -1 & 0 \\ 0 & -1 & 2 & -1 \\ 0 & 0 & -1 & 1 \end{bmatrix}\begin{Bmatrix} u_1 \\ u_2 \\ u_3 \\ u_4 \end{Bmatrix} = \frac{cL^2}{6}\begin{Bmatrix} 1 \\ 6 \\ 12 \\ 8 \end{Bmatrix} + \begin{Bmatrix} R_1 \\ 0 \\ 0 \\ 0 \end{Bmatrix} \tag{3.9-9}$$

The essential boundary condition $u_1 = 0$ is imposed by striking out the first equation and the first column of the matrix, as explained in Section 2.3. Solution of the remaining three equations yields nodal d.o.f. u_2, u_3, and u_4. The complete solution vector is

$$\{\mathbf{D}\} = \begin{Bmatrix} u_1 \\ u_2 \\ u_3 \\ u_4 \end{Bmatrix} = \frac{cL^3}{3AE}\begin{Bmatrix} 0 \\ 13 \\ 23 \\ 27 \end{Bmatrix} \tag{3.9-10}$$

These are the axial displacements at $x = 0$, $x = L$, $x = 2L$, and $x = 3L$.

Inspection of the Solution. Axial displacements are plotted in Fig. 3.9-1*a*. Upon comparing Eq. 3.9-10 with the exact solution, Eq. 3.5-8, we find that nodal d.o.f. u_1, u_2, and u_3 are exact. This happens only because the present problem is of a special mathematical type [3.3]. *In most problems, nodal d.o.f.* $\{\mathbf{D}\}$ *are not exact.* Between nodes, the linear finite element field cannot match the exact cubic field. For example, at the midpoint $x = 3L_T/2$, for which $s = L/2$ in element 2, Eq. 3.8-3 yields

$$u = \left\lfloor \frac{L - (L/2)}{L} \quad \frac{L/2}{L} \right\rfloor \begin{Bmatrix} u_2 \\ u_3 \end{Bmatrix} = \frac{u_2 + u_3}{2} = \frac{6cL^3}{AE} \tag{3.9-11}$$

The exact result at $x = 3L_T/2$, from Eq. 3.5-8, is $u = 6.1875cL^3/AE$.

The state of stress is uniaxial. Therefore, in a typical element, using Eq. 3.9-1,

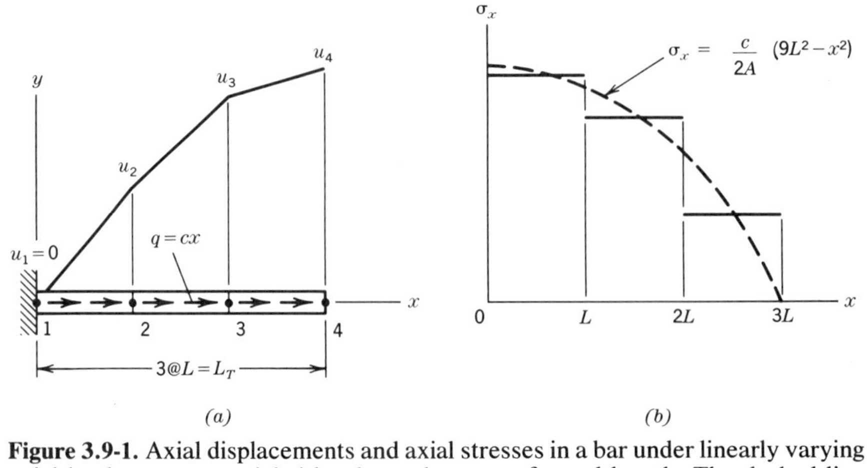

Figure 3.9-1. Axial displacements and axial stresses in a bar under linearly varying axial load $q = cx$, modeled by three elements of equal length. The dashed line represents the exact stress field.

$$\sigma_x = E\epsilon_x = E\lfloor \mathbf{B} \rfloor \{\mathbf{d}\} = \frac{E}{L}\lfloor -1 \quad 1 \rfloor \{\mathbf{d}\} \tag{3.9-12}$$

Applying Eq. 3.9-12 to the three elements in turn, we obtain the stresses plotted in Fig. 3.9-1*b*. As is typical in finite element problems, stresses change abruptly at nodes and are less accurate than displacements. A bar element based on a linear axial displacement field will always approximate the exact stress curve in stairstep fashion. Nevertheless, the exact curve can be approached arbitrarily closely by using more and more elements.

That stresses are most accurate near element centers is a consequence of the mean value theorem for derivatives. As applied to Fig. 3.5-1*b*, the theorem says that slopes of the two curves agree near the middle of the interval even though slopes disagree at $x = 0$ and at $x = L_T$. The same argument applies to Fig. 3.9-1*a*: slopes of the piecewise linear curve match the exact $u_{,x}$ from Eq. 3.5-8 somewhere near element centers but not at nodes.

3.10 FINITE ELEMENT FORMULATIONS DERIVED FROM A FUNCTIONAL

In Section 3.9, a finite element solution based on a two-d.o.f. bar element is derived from the potential energy functional Π_p. In the present section we consider another example, this time in two-dimensional heat conduction, but without requiring that the element have a particular shape or a particular number of nodes. Similar formulations for stress analysis are considered in Chapter 4.

Let temperature T within a plane element be interpolated from n nodal temperatures $\{\mathbf{T}_e\}$,

$$T = \underset{1\times n}{\lfloor \mathbf{N} \rfloor} \underset{n\times 1}{\{\mathbf{T}_e\}} \tag{3.10-1}$$

where each of the n shape functions in $\lfloor \mathbf{N} \rfloor$ is a function of x and y. If elements have corner nodes only, then $n = 3$ for a triangle, $n = 4$ for a rectangle, and so on.

A functional for plane heat conduction is given in Eq. 3.7-5. With $T = T^T$, $T_{,x}^2 = T_{,x}^T T_{,x}$, and $T_{,y}^2 = T_{,y}^T T_{,y}$, Eq. 3.7-5 has the form

$$\Pi = \iint \tfrac{1}{2}(T_{,x}^T T_{,x} + T_{,y}^T T_{,y})k \, dx \, dy - \iint T^T Q \, dx \, dy + \iint T^T \dot{T} \rho c \, dx \, dy \tag{3.10-2}$$

As in Eq. 3.9-2, the transposition symbol is placed on the scalars to make it easier to differentiate subsequent matrix expressions. Differentiation of Eq. 3.10-1 yields

$$T_{,x} = \lfloor \mathbf{N}_{,x} \rfloor \{\mathbf{T}_e\} \qquad T_{,y} = \lfloor \mathbf{N}_{,y} \rfloor \{\mathbf{T}_e\} \qquad \dot{T} = \lfloor \mathbf{N} \rfloor \{\dot{\mathbf{T}}_e\} \tag{3.10-3}$$

in which, for example,

$$\lfloor \mathbf{N}_{,x} \rfloor = \lfloor N_{1,x} \quad N_{2,x} \quad \dots \quad N_{n,x} \rfloor \tag{3.10-4}$$

The expression for $\dot{T}$ in Eqs. 3.10-3 indicates that nodal temperatures are regarded as functions of time, but the N_i are independent of time. Thus T and $\dot{T}$ are interpolated from nodal values $\{\mathbf{T}_e\}$ and $\{\dot{\mathbf{T}}_e\}$ by means of the same shape function matrix $\lfloor \mathbf{N} \rfloor$.

Substitution of Eqs. 3.10-3 into 3.10-2 yields, for a single element,

$$\Pi = \tfrac{1}{2}\{\mathbf{T}_e\}^T[\mathbf{k}]\{\mathbf{T}_e\} + \{\mathbf{T}_e\}^T[\mathbf{c}]\{\dot{\mathbf{T}}_e\} - \{\mathbf{T}_e\}^T\{\mathbf{r}_Q\} \tag{3.10-5}$$

in which we have defined terms as follows:

$$[\mathbf{k}] = \iint (\lfloor \mathbf{N}_{,x} \rfloor^T \lfloor \mathbf{N}_{,x} \rfloor + \lfloor \mathbf{N}_{,y} \rfloor^T \lfloor \mathbf{N}_{,y} \rfloor) k \, dx \, dy \tag{3.10-6a}$$

$$[\mathbf{c}] = \iint \lfloor \mathbf{N} \rfloor^T \lfloor \mathbf{N} \rfloor \rho c \, dx \, dy \tag{3.10-6b}$$

$$\{\mathbf{r}_Q\} = \iint \lfloor \mathbf{N} \rfloor^T Q \, dx \, dy \tag{3.10-6c}$$

Finite element equations are obtained by making Π stationary with respect to variations of nodal temperature:

$$\left\{\frac{\partial \Pi}{\partial \mathbf{T}_e}\right\} = \{\mathbf{0}\} \qquad \text{yields} \qquad [\mathbf{k}]\{\mathbf{T}_e\} + [\mathbf{c}]\{\dot{\mathbf{T}}_e\} = \{\mathbf{r}_Q\} \tag{3.10-7}$$

Upon assembly of elements, $\{\mathbf{T}_e\}$ is replaced by the global vector $\{\mathbf{T}\}$, which contains all nodal temperatures of the structure. The global equations are therefore

$$(\textstyle\sum [\mathbf{k}])\{\mathbf{T}\} + (\sum [\mathbf{c}])\{\dot{\mathbf{T}}\} = \sum \{\mathbf{r}_Q\} \tag{3.10-8}$$

in which summation signs imply the usual assembly process of summing overlapping terms of element matrices. (Equation 3.10-8 is again obtained if assembly is indicated earlier, i.e., by summing element contributions from Eq. 3.10-5 to a global Π.)

For the sake of having notation like that used in structural mechanics, we can write $[\mathbf{k}]$ of Eq. 3.10-6a in the form

$$[\mathbf{k}] = \iint [\mathbf{B}]^T k [\mathbf{B}] t \, dx \, dy, \qquad \text{where} \quad [\mathbf{B}] = \begin{bmatrix} \mathbf{N}_{,x} \\ \mathbf{N}_{,y} \end{bmatrix} \tag{3.10-9}$$

Thickness t is taken as unity in the preceding development.

Remarks. The foregoing derivation has significance that goes beyond the problem of plane heat conduction. The derivation shows that a finite element formulation of a physical problem is available from only two basic ingredients—namely, a functional that describes the physical problem and a shape function matrix $\lfloor \mathbf{N} \rfloor$ that describes the element. From these we obtain definitions of element properties (e.g., Eqs. 3.10-6) and algebraic equations of the structure (e.g., Eqs. 3.10-8).

To obtain numerical results one must next attend to specifics by choosing the element shape, number of d.o.f., distribution of d.o.f. over the element, and the shape function matrix $\lfloor \mathbf{N} \rfloor$. These choices have great influence on the efficiency of calculation and the accuracy of results.

3.11 INTERPOLATION

To interpolate is to approximate the value of a function between known values by operating on the known values with a formula different from the function itself. This is done in each of the three spans of length L in Fig. 3.9-1*a*, where a linear operator $\lfloor \mathbf{N} \rfloor$ is applied to known values u_1, u_2, u_3, and u_4 that happen to lie on a cubic curve. In a finite element context, the "known values" are d.o.f. to be found by solving algebraic equations, and they are usually approximate rather than exact. Operator $\lfloor \mathbf{N} \rfloor$, the shape function matrix, serves as a basis from which a finite element can be formulated.

One can regard interpolation as the basic motivation of the finite element method, in that a sufficiently small portion of even a complicated field can be modeled well enough by a simpler interpolating field. A linear interpolating field, as used in each element of Fig. 3.9-1*a*, may be adequate if many elements are used. Elements based on a quadratic field or a cubic field would provide a better fit of the actual field: fewer elements would be needed, but each element would be more complicated. The limiting case of a single element with many d.o.f. yields the classical Rayleigh–Ritz method.

Should one use many simple elements or a few complicated elements? There is no easy answer. A good analyst is familiar with how various elements behave in various circumstances. In solving a transient or nonlinear problem, for example, many analysts prefer simpler (and therefore cheaper) elements because of the need to seek low cost in every computational step.

Degree of Continuity. For future use, we introduce the following symbolism to define the degree of continuity of a function or a field. A field is said to have C^m continuity if derivatives of the field through order m are continuous. Thus $\phi = \phi(x)$ is C^0 continuous if ϕ is continuous but $\phi_{,x}$ is not. An example of C^0 continuity appears in Fig. 3.11-1*a*. Another example of C^0 continuity is axial displacement u in Fig. 3.9-1. Figure 3.11-1*b* shows an example of C^1 continuity: both ϕ and $\phi_{,x}$ are continuous but $\phi_{,xx}$ is discontinuous at $x = x_c$. In general, it is necessary that derivatives of ϕ of order m be used as nodal d.o.f. if the field ϕ produced by a mesh of finite elements is to be C^m continuous.

Element Node Identification. Heretofore we have labeled element nodes with letters and structure nodes with numbers. We will subsequently encounter ele-

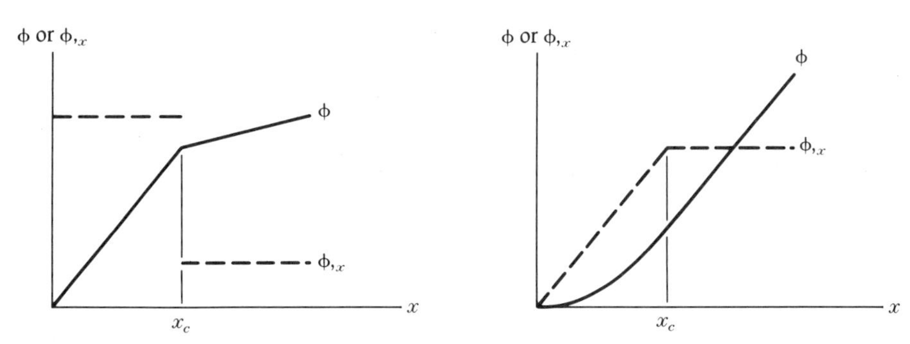

(*a*) (*b*)

Figure 3.11-1. Function $\phi = \phi(x)$ is (*a*) C^0 continuous and (*b*) C^1 continuous.

ments that have many nodes, for which letters would be awkward as node labels. Therefore, we will henceforth use numbers as element node labels. At this stage in our study, the reader should be able to distinguish between an element and a structure without the artifice of separate labeling systems.

3.12 SHAPE FUNCTIONS FOR C^0 ELEMENTS

A C^0 element provides interelement continuity of the field quantity ϕ but not interelement continuity of all first derivatives of ϕ. Thus, in a mesh of C^0 elements, $\phi_{,x}$, $\phi_{,y}$, and/or $\phi_{,z}$ exhibit a jump as one passes from one element into another.

Interpolation formulas lead to shape functions $\lfloor \mathbf{N} \rfloor$, from which finite elements can be formulated. In this section and the next we consider shape functions for some simple elements. These and other elements are considered in more detail in subsequent chapters.

A field ϕ is interpolated over an element from n element nodal values $\{\boldsymbol{\phi}_e\} = \lfloor \phi_1 \quad \phi_2 \; . \; . \; . \; \phi_n \rfloor^T$ according to the formula

$$\phi = \lfloor \mathbf{N} \rfloor \{\boldsymbol{\phi}_e\} \qquad \text{that is,} \quad \phi = \sum_{i=1}^{n} N_i \phi_i \tag{3.12-1}$$

where the N_i are functions of the coordinates. A shape function N_i defines the distribution of ϕ within the element when the ith nodal d.o.f. ϕ_i has unit value and all other nodal ϕ's are zero.

One Dimension. Linear interpolation in one dimension is depicted in Fig. 3.12-1a. The interpolated function $\phi = \phi(x)$ is to have value ϕ_1 at $x = x_1$ and

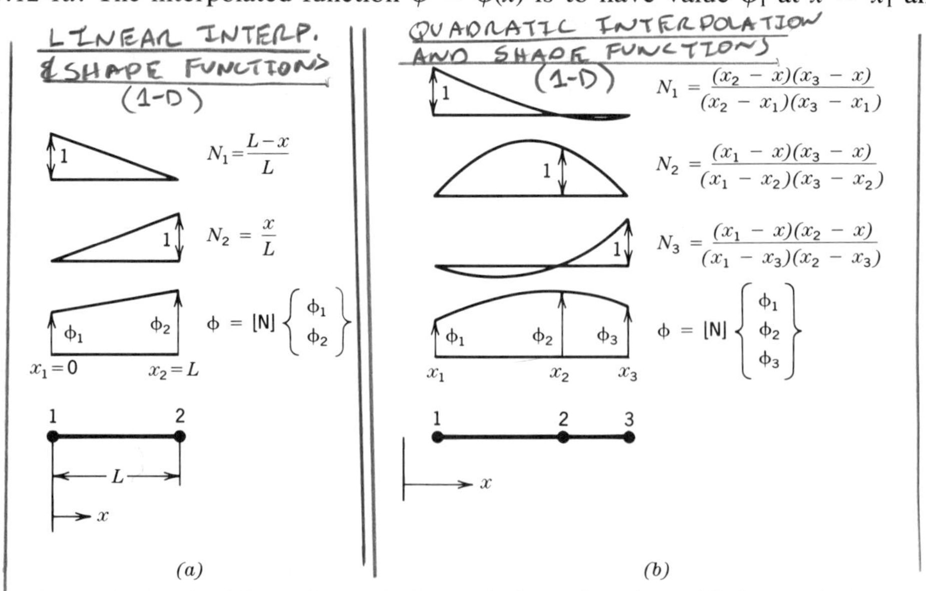

Figure 3.12-1. (a) Linear interpolation and shape functions. (b) Quadratic interpolation and shape functions.

value ϕ_2 at $x = x_2$. Two data points define a linear polynomial $\phi = a_1 + a_2 x$. The procedure for obtaining shape functions N_i from this polynomial is discussed in Section 3.8. In the notation of Fig. 3.12-1*a*, with $x_1 = 0$ and $x_2 = L$, the result is

$$\lfloor \mathbf{N} \rfloor = \left\lfloor \frac{L - x}{L} \quad \frac{x}{L} \right\rfloor \tag{3.12-2}$$

Quadratic interpolation in one dimension is depicted in Fig. 3.12-1*b*. Here there are three data points. They define a parabola $\phi = a_1 + a_2 x + a_3 x^2$, which must display the values $\phi = \phi_1$, $\phi = \phi_2$, and $\phi = \phi_3$ at $x = x_1$, $x = x_2$, and $x = x_3$, respectively. The x_i values need not be uniformly spaced. Quadratic shape functions can be determined by the procedure used for linear shape functions in Eqs. 3.8-6 to 3.8-8, but now there are three a_i and the algebra is more tedious. The three shape functions are most easily obtained from Lagrange's interpolation formula, which is discussed subsequently. The result is

$$\lfloor \mathbf{N} \rfloor = \left\lfloor \frac{(x_2 - x)(x_3 - x)}{(x_2 - x_1)(x_3 - x_1)} \quad \frac{(x_1 - x)(x_3 - x)}{(x_1 - x_2)(x_3 - x_2)} \quad \frac{(x_1 - x)(x_2 - x)}{(x_1 - x_3)(x_2 - x_3)} \right\rfloor \tag{3.12-3}$$

In Eqs. 3.12-2 and 3.12-3 we note the following characteristics, which are true of all C^0 polynomial shape functions (in one dimension).

1. All shape functions N_i, and function ϕ itself, are polynomials of the same degree.
2. For any shape function N_i, $N_i = 1$ when $x = x_i$ and $N_i = 0$ when $x = x_j$ where $i \neq j$.
3. C^0 shape functions sum to unity. This is not obvious in Eq. 3.12-3 but can be shown as follows. If $\phi_i = 1$ at all n data points, then $\phi = 1$ everywhere in the interpolated function ϕ. Equation 3.12-1 becomes

$$1 = \sum_{i=1}^{n} N_i \tag{3.12-4}$$

For C^1 elements, in which derivatives of ϕ are also used as nodal d.o.f., Eq. 3.12-4 is valid if the N_i are those associated with translational d.o.f. only.

Lagrange's Interpolation Formula. A function $\phi = \phi(x)$, of degree $n - 1$ and defined by n values ϕ_i at corresponding abscissae x_i, has the form

$$\phi = \sum_{i=1}^{n} N_i \phi_i \qquad \text{or} \qquad \phi = N_1\phi_1 + N_2\phi_2 + \cdots + N_n\phi_n \tag{3.12-5}$$

in which shape functions N_i have been devised by Lagrange as follows:

$$N_1 = \frac{(x_2 - x)(x_3 - x)(x_4 - x) \cdots (x_n - x)}{(x_2 - x_1)(x_3 - x_1)(x_4 - x_1) \cdots (x_n - x_1)}$$

$$N_2 = \frac{(x_1 - x)(x_3 - x)(x_4 - x) \cdots (x_n - x)}{(x_1 - x_2)(x_3 - x_2)(x_4 - x_2) \cdots (x_n - x_2)} \tag{3.12-6}$$

$$\vdots$$

$$N_n = \frac{(x_1 - x)(x_2 - x)(x_3 - x) \cdots (x_{n-1} - x)}{(x_1 - x_n)(x_2 - x_n)(x_3 - x_n) \cdots (x_{n-1} - x_n)}$$

Note that the N_i have characteristics 1 and 2 just cited. Characteristic 3 is present but is not obvious. Note also that the N_i in Eqs. 3.12-2 and 3.12-3 are special cases of Eq. 3.12-6, for which $n = 2$ and $n = 3$, respectively.

If there is a "true curve" for which the interpolated curve $\phi = \Sigma N_i\phi_i$ is but an approximation, the two curves are coincident only at the n values of x_i that provide the ϕ_i used for interpolation. Moreover, the interpolated curve yields only exact ordinates ϕ_i, not exact slopes $\phi_{,xi}$ as well. An example appears in Fig. 3.12-2.

Two Dimensions. Imagine that a dependent variable $\phi = \phi(x,y)$ is to be interpolated from four nodal ϕ_i at corners of a rectangle (Fig. 3.12-3). Here ϕ has the form

$$\phi = a_1 + a_2x + a_3y + a_4xy \tag{3.12-7}$$

Shape functions are products of the N_i of Lagrange's formula. We argue as follows.

In Fig. 3.12-3, one can linearly interpolate ϕ along the left edge between nodal values ϕ_1 and ϕ_4, and along the right edge between nodal values ϕ_2 and ϕ_3. Thus, in Eqs. 3.12-6, y replaces x and $n = 2$. Calling the edge values ϕ_{14} and ϕ_{23}, we have

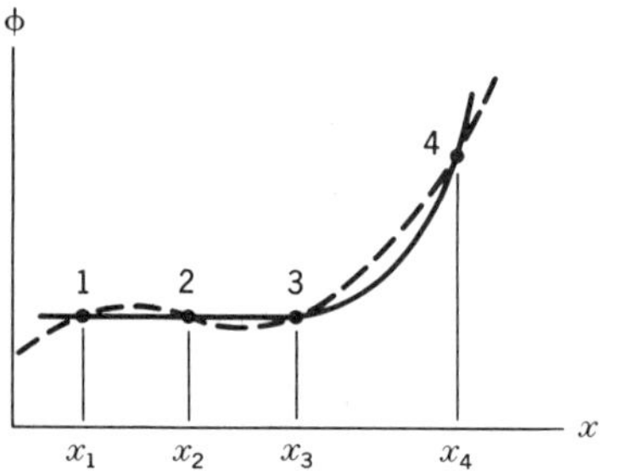

Figure 3.12-2. Possible discrepancies between a "true curve" (solid line) and the fit produced by Lagrange's formula (dashed line).

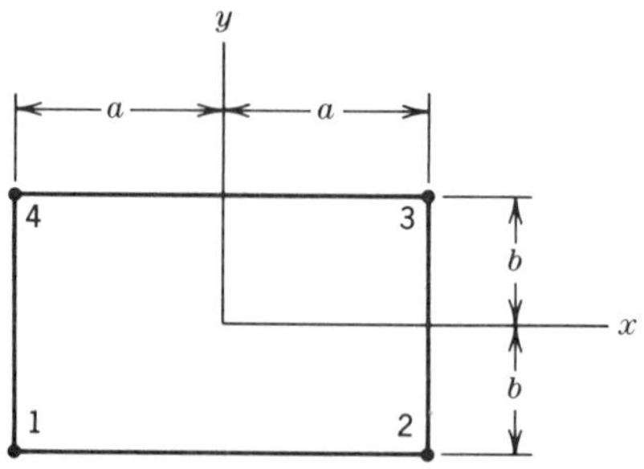

Figure 3.12-3. Four-node "bilinear" element.

$$\phi_{14} = \frac{b-y}{2b}\phi_1 + \frac{b+y}{2b}\phi_4 \quad \text{and} \quad \phi_{23} = \frac{b-y}{2b}\phi_2 + \frac{b+y}{2b}\phi_3 \tag{3.12-8}$$

Next we linearly interpolate in the x direction between ϕ_{14} and ϕ_{23}:

$$\phi = \frac{a-x}{2a}\phi_{14} + \frac{a+x}{2a}\phi_{23} \tag{3.12-9}$$

Substitution of Eq. 3.12-8 into Eq. 3.12-9 yields $\phi = \Sigma N_i\phi_i$, where

$$N_1 = \frac{(a-x)(b-y)}{4ab} \qquad N_2 = \frac{(a+x)(b-y)}{4ab}$$
$$N_3 = \frac{(a+x)(b+y)}{4ab} \qquad N_4 = \frac{(a-x)(b+y)}{4ab} \tag{3.12-10}$$

One can easily check that each $N_i = 1$ at the coordinates of node i, is zero at other nodes, and that $N_1 + N_2 + N_3 + N_4 = 1$.

The element associated with Eqs. 3.12-10 is called "bilinear," as each of its shape functions is a product of two linear polynomials. Similarly, a nine-node element (nodes at corners, midsides, and the center) is called "biquadratic," a 16-node element in which four of the nodes are internal is called "bicubic," and so on. Shape functions for all these elements are products of one-dimensional Lagrange interpolation shape functions [3.4]. These elements, and analogous elements in three dimensions, are called *Lagrange elements*.

Additional Dependent Variables. The element of Fig. 3.12-3 can be used to solve problems of plane stress and plane strain. For such problems there are *two* dependent field variables—namely, $u = u(x,y)$ and $v = v(x,y)$. The four-node element then has eight d.o.f. Displacements u and v are each interpolated from four nodal values, that is,

$$u = \sum_{i=1}^{4} N_i u_i \quad \text{and} \quad v = \sum_{i=1}^{4} N_i v_i \tag{3.12-11}$$

in which the N_i are defined by Eqs. 3.12-10.

3.13 SHAPE FUNCTIONS FOR C^1 ELEMENTS

A C^1 element provides interelement continuity of the field quantity ϕ and its first derivatives at nodes, but not interelement continuity of all second derivatives of ϕ. An example appears in the analysis of a thin plate in bending, where the field quantity is displacement w in the z direction, where $w = w(x,y)$, and x and y are coordinates in the plane of the plate. Typical thin-plate elements use w, $w_{,x}$, and $w_{,y}$ as nodal d.o.f. Second derivatives $w_{,xx}$, $w_{,yy}$, and $w_{,xy}$ are not all continuous across interelement boundaries. (Even the first derivative $w_{,n}$, where n is a direction normal to an edge, is typically discontinuous except at nodes.)

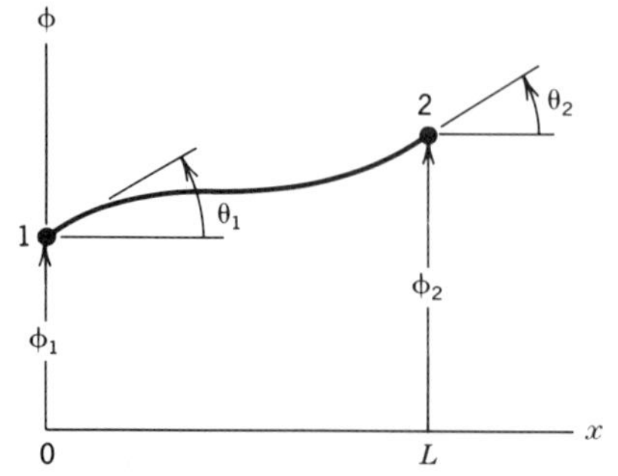

Figure 3.13-1. A curve interpolated between two points at which ordinates ϕ_1 and ϕ_2 and inclinations θ_1 and θ_2 are known.

One Dimension. Fitting a curve to both ordinate and slope information at data points is known as *Hermitian interpolation*. The simplest and most common Hermitian interpolation is between two points at which both ordinate and slope are known (Fig. 3.13-1). We will assume that slope $\phi_{,x}$ is small, so that rotation θ is practically the same as $\phi_{,x}$. Four data items define a cubic curve,

$$\phi = a_1 + a_2x + a_3x^2 + a_4x^3 \qquad \text{or} \qquad \phi = \lfloor \mathbf{X} \rfloor \{\mathbf{a}\} \tag{3.13-1a}$$

where

$$\lfloor \mathbf{X} \rfloor = \lfloor 1 \quad x \quad x^2 \quad x^3 \rfloor \qquad \text{and} \qquad \{\mathbf{a}\} = \lfloor a_1 \quad a_2 \quad a_3 \quad a_4 \rfloor^T \tag{3.13-1b}$$

To express the a_i in terms of ordinates and slopes at $x = 0$ and at $x = L$, we make the substitutions

$$\begin{aligned} \phi &= \phi_1 \quad \text{and} \quad \phi_{,x} = \theta_1 \quad \text{at } x = 0 \\ \phi &= \phi_2 \quad \text{and} \quad \phi_{,x} = \theta_2 \quad \text{at } x = L \end{aligned} \tag{3.13-2}$$

Thus Eq. 3.13-1 yields

$$\begin{Bmatrix} \phi_1 \\ \theta_1 \\ \phi_2 \\ \theta_2 \end{Bmatrix} = \begin{bmatrix} 1 & 0 & 0 & 0 \\ 0 & 1 & 0 & 0 \\ 1 & L & L^2 & L^3 \\ 0 & 1 & 2L & 3L^2 \end{bmatrix} \begin{Bmatrix} a_1 \\ a_2 \\ a_3 \\ a_4 \end{Bmatrix} \qquad \text{or} \qquad \{\mathbf{d}\} = [\mathbf{A}]\{\mathbf{a}\} \tag{3.13-3}$$

Therefore $\{\mathbf{a}\} = [\mathbf{A}]^{-1}\{\mathbf{d}\}$, and Eq. 3.13-1 becomes

$$\phi = \lfloor \mathbf{N} \rfloor \{\mathbf{d}\}, \qquad \text{where} \qquad \lfloor \mathbf{N} \rfloor = \lfloor \mathbf{X} \rfloor [\mathbf{A}]^{-1} \tag{3.13-4}$$

Shape function matrix $\lfloor \mathbf{N} \rfloor$ is 1 by 4. The four N_i are shown in Fig. 3.13-2. As expected, three of the N_i and three of the dN_i/dx are zero at end $x = 0$, whereas the remaining N_i and the remaining dN_i/dx have unit value. The same is true at end $x = L$. This behavior is required if Eqs. 3.13-2 are to be satisfied by the interpolation $\phi = \Sigma N_i d_i$. These shape functions may be used to generate the stiffness matrix of a beam element (Section 4.2).

Two Dimensions. Hermitian interpolation of a function $w = w(x,y)$ can be used to generate elements for the analysis of thin plates in bending. A great many

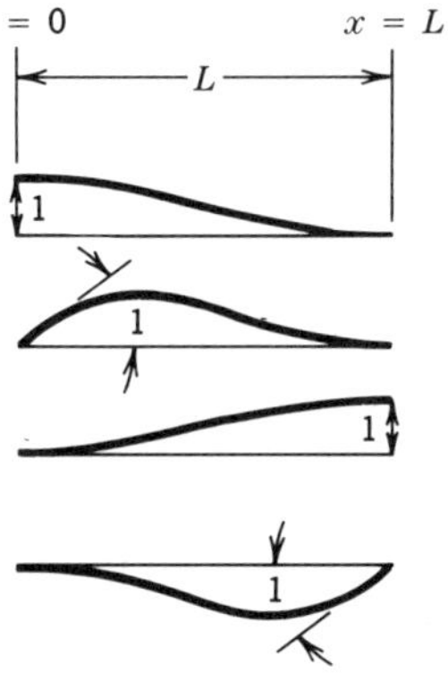

	At $x = 0$		At $x = L$	
	N_i	$N_{i,x}$	N_i	$N_{i,x}$
$N_1 = 1 - \dfrac{3x^2}{L^2} + \dfrac{2x^3}{L^3}$	1	0	0	0
$N_2 = x - \dfrac{2x^2}{L} + \dfrac{x^3}{L^2}$	0	1	0	0
$N_3 = \dfrac{3x^2}{L^2} - \dfrac{2x^3}{L^3}$	0	0	1	0
$N_4 = -\dfrac{x^2}{L} + \dfrac{x^3}{L^2}$	0	0	0	1

Figure 3.13-2. Shape functions of a cubic curve fitted to ordinates and slopes at $x = 0$ and at $x = L$.

different interpolation schemes are possible. All are lengthy to write out, and their comparative merits are intimately connected with plate theory (Chapter 11).

PROBLEMS

Section 3.2

3.1 The system shown consists of a rigid half-cylinder that can roll without friction on a horizontal surface, a linear nondissipative spring, and a force P of constant magnitude that can move around the cylinder, but is always directed toward point C on the cylinder. Show that this system is not conservative.

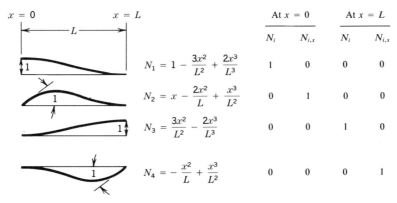

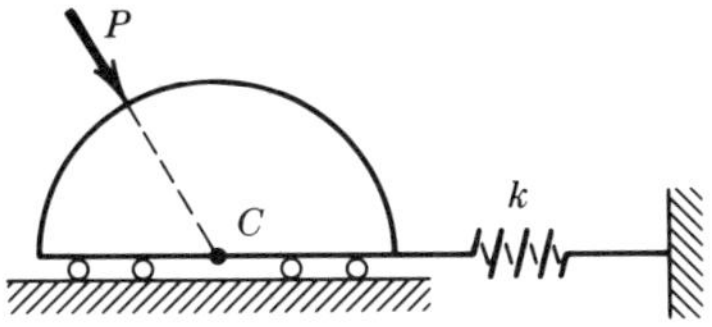

Problem 3.1

3.2 Redefine Ω to be $\Omega = P(D_{eq} - D)$, as suggested near the end of Section 3.2. Redraw Fig. 3.2-4 as required.

3.3 Reverse the direction of load P in Fig. 3.2-3. Solve for the equilibrium value of D by use of Π_p. Revise Fig. 3.2-4 as required.

3.4 Imagine that the spring in Fig. 3.2-3 is not linear but exerts a force proportional to the square of its stretch. Write an expression for Π_p, and from it determine the equilibrium value of D.

Section 3.3

3.5 Redefine D_2 and D_3 in Fig. 3.3-1 so that D_2 is an axial displacement *relative* to D_1 and D_3 is an axial displacement *relative* to D_2. Write an expression for Π_p, analogous to Eq. 3.3-3. For the special case $k_1 = k_2 = k_3 = k$ and $P_1 = P_2 = P_3 = P$, solve for the D_i and show that they give the same *absolute* axial displacements as Eq. 3.3-3.

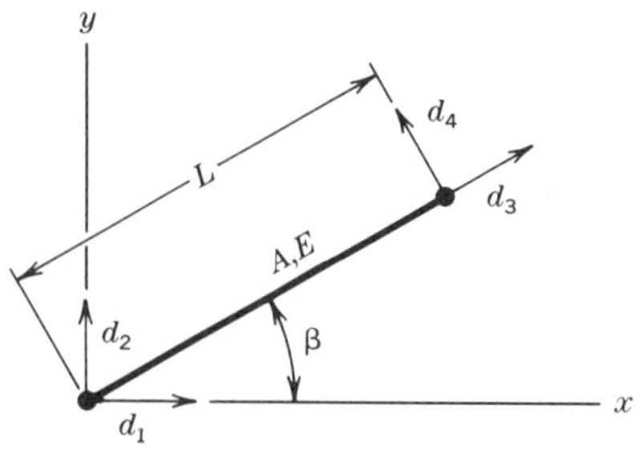

Problem 3.7

3.6 Show that Eqs. 3.3-6 and 3.3-7 yield the rows of Eq. 2.4-3 from the stationary condition $d\Pi_p = 0$.

3.7 Displacement d.o.f. d_i at ends of a uniform bar element have the directions shown. Write an expression for Π_p in terms of these d.o.f. From the stationary condition $d\Pi_p = 0$, obtain the element stiffness matrix that operates on these four d.o.f.

3.8 Verify that Eqs. 2.3-8 are produced when the stationary condition $d\Pi_p = 0$ is applied to Eq. 3.3-8.

3.9 Use the method of stationary potential energy to derive the 4 by 4 stiffness matrix for the structure described in (a) Problem 1.13, and (b) Problem 2.2.

3.10 Verify that Eqs. 3.3-5 and 3.3-9 yield Eq. 3.3-3.

Section 3.4

3.11 Write the term $\frac{1}{2}\{\boldsymbol{\epsilon}\}^T[\mathbf{E}]\{\boldsymbol{\epsilon}\}$ in Eq. 3.4-1 for an isotropic material in a condition of plane stress in the xy plane (for which $[\mathbf{E}]$ is 3 by 3). Specialize this expression for the case of uniaxial stress σ_x. Check your result against Eq. 3.4-6.

3.12 In the beam of Fig. 3.4-1*b*, let initial strain and initial stress be given by $\epsilon_0 = -z\kappa_0$ and $\sigma_0 = -m_0 z/I$, respectively. Here κ_0 and m_0 are regarded as initial curvature and initial moment, both considered positive when associated with a concave-up condition of the beam. Determine the contributions of κ_0 and m_0 to Eq. 3.4-8.

3.13 Repeat the example of Eqs. 3.4-9 to 3.4-11, but account for heating by use of σ_0 rather than ϵ_0.

Section 3.5

3.14 Verify that the first of Eqs. 3.5-6a is indeed given by the conditions $\partial\Pi_p/\partial a_1 = \partial\Pi_p/\partial a_2 = 0$.

3.15 Verify that the a_i of Eq. 3.5-9 result from the use of the three-term polynomial $u = a_1x + a_2x^2 + a_3x^3$ in a Rayleigh–Ritz solution.

3.16 Consider the two approximate solutions and the one exact solution in Section 3.5 (Eqs. 3.5-5, 3.5-6, and 3.5-8). At what point or points is the differential equation of equilibrium satisified by each solution?

3.17 Obtain one-term and two-term solutions for a uniform axially loaded bar, analogous to Eqs. 3.5-5 and 3.5-6, if load q is replaced by concentrated forces at $x = L_T/3$, $x = 2L_T/3$, and $x = L_T$. Each of the three forces is directed to the right and is of magnitude P.

3.18 Consider a cantilever beam of length L, fixed at end $x = 0$ and carrying a moment load M_L at $x = L$. Write an admissible series for lateral displacement w based on either sine or cosine functions.

3.19 Consider a uniform cantilever beam of length L, fixed at end $x = 0$ and carrying a transverse force F at $x = L$.
(a) Let the lateral displacement field be $w = a_1x^3$, where a_1 is a constant. Is this field admissible? Explain.
(b) Write a polynomial field for w that is better than that of part (a). Let the field contain three terms, each of the form a_ix^j, where $i = 1,2,3$ and j is an integer such that the term is admissible.
(c) Without calculation, can you predict the quality of the answers obtainable from the field of part (b) and the numerical value of any of the a_i?
(d) Use the field of part (a) to find the deflection of force F.

3.20 A uniformly loaded beam of constant flexural stiffness EI is simply supported at its ends $x = 0$ and $x = L$. In parts (a) and (b), determine the deflection and bending moment predicted at $x = L/2$ by a Rayleigh–Ritz solution that has a single d.o.f. Compare exact and approximate results.
(a) Use a single-d.o.f. algebraic expression—that is, a d.o.f. a_1 times a function that contains x and x^2.
(b) Use one term of a sine series.
(c) Why should you anticipate that part (b) will be better than part (a) if part (a) is the simplest admissible function?

3.21 The uniform cantilever beam shown carries uniformly distributed load of intensity q, tip force P_L, and tip moment M_L. In parts (a) and (b), compute Rayleigh–Ritz approximations for displacement and rotation at the tip. Compare these results with formulas from beam theory, and explain why the Rayleigh–Ritz result is or is not exact.
(a) Use one term of a polynomial series.
(b) Use two terms of a polynomial series.
(c) In part (a), for which of the given loadings is $w = w(x)$ exact *away* from the tip—that is, for $0 < x < L$? Why?

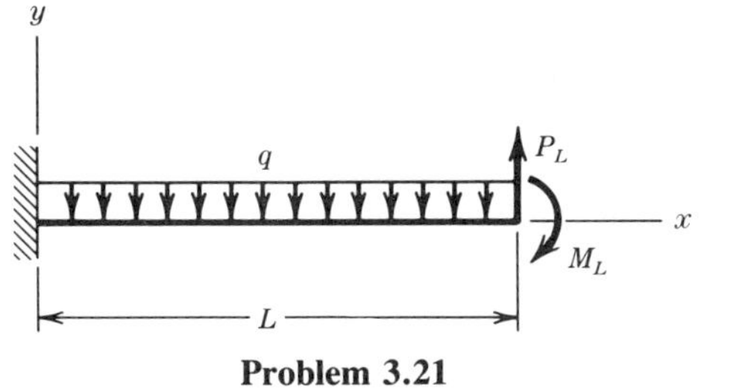

Problem 3.21

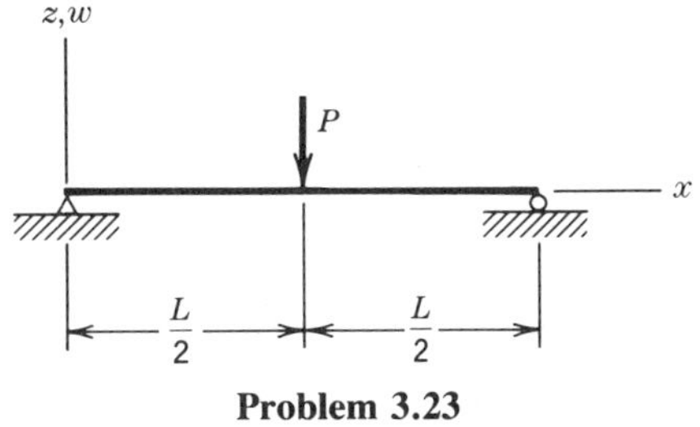

Problem 3.23

3.22 If loads P_L and M_L in Problem 3.21 were applied at $x = L/2$ rather than at the tip, how many series terms would be needed in order to obtain the exact displacement at $x = L$? Explain.

3.23 The uniform beam shown is simply supported and carries a force P at its center. Use the infinite series

$$w = \sum_{i=1}^{n} a_i \sin \frac{i\pi x}{L}$$

in a Rayleigh–Ritz solution for deflection and bending moment at the center. Compare exact and approximate results for $n = 1, 2, 3,$ and 4.

3.24 Repeat Problem 3.23 with load P replaced by a uniformly distributed downward load of intensity q.

Section 3.6

3.25 A complete polynomial of degree n in x, y, and z contains $(n + 3)!/6n!$ terms. With this in mind, construct a "Pascal tetrahedron" analogous to the Pascal triangle in Fig. 3.6-1.

3.26 (a) Compute the work done by applied load $q = cx$ in going through the exact displacement u of Eq. 3.5-8.

(b) Similarly, compute the work done by load q in each of the two approximate solutions (Eqs. 3.5-5 and 3.5-6). What conclusion can you draw?

3.27 (a) Compute the strain energy associated with the exact solution given in Eq. 3.5-8. How is this energy related to the work computed in Problem 3.26a, and why?

(b) Similarly, compute the strain energy associated with the two approximate solutions (Eqs. 3.5-5 and 3.5-6), and compare answers with work values computed in Problem 3.26b.

Section 3.7

3.28 A certain functional is $\Pi = \int F\,dx$, in which $F = c_1\phi_{,xx}^2 + c_2\phi_{,x}^2 + c_3\phi^2 + c_4\phi + c_5$ and the five c_i are constants. What is the Euler equation?

3.29 A certain physical problem has the functional

$$\Pi = \int_0^L \left(\tfrac{1}{2}\phi_{,x}^2 - 50\phi\right) dx$$

Essential boundary conditions are $\phi = 0$ at $x = 0$ and $\phi = 20$ at $x = L$. What is ϕ as a function of x and L?

3.30 Discard terms that contain $\{\mathbf{F}\}$ and $\{\mathbf{M}\}$ from Eq. 3.4-8, and find the Euler equation of a uniform beam under lateral load q. What is the Euler equation if EI is not constant?

3.31 The potential energy of an isotropic plate that carries lateral pressure q is

$$\Pi_p = \frac{D}{2}\int\int\left\{(w_{,xx} + w_{,yy})^2 - 2(1 - \nu)[w_{,xx}w_{,yy} - w_{,xy}^2] - \frac{2q}{D}\right\} dx\,dy$$

where D is a constant called *flexural rigidity*. Show that the Euler equation is $\nabla^4 w = q/D$, where ∇^4 is the biharmonic operator.

3.32 Consider the functional

$$\Pi = \int\left(\frac{M^2}{2EI} + M_{,x}w_{,x} + qw\right) dx$$

for a uniform beam that carries distributed lateral load q. Bending moment M and lateral deflection w are each regarded as dependent variables. Derive the two Euler equations. Do they have the form expected from beam theory?

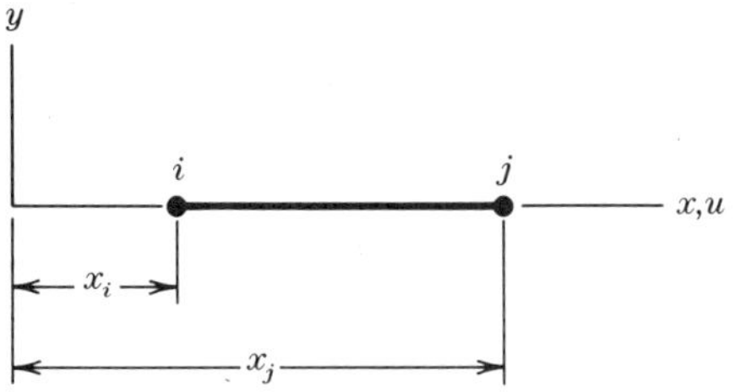

Problem 3.33

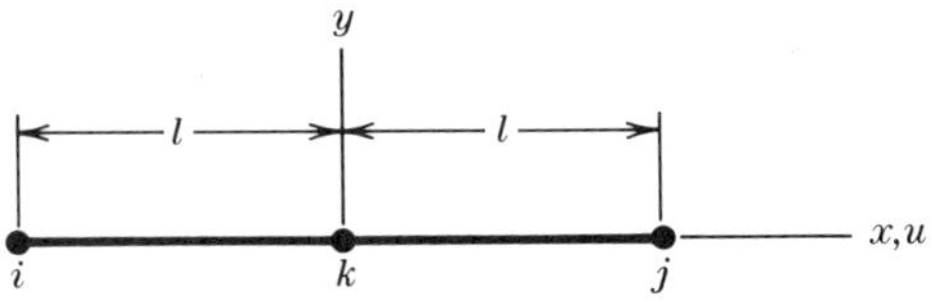

Problem 3.34

Section 3.8

3.33 The bar element shown is to have an axial displacement field u that is linear in x and depends on nodal d.o.f. u_i and u_j. The shape function matrix $\lfloor \mathbf{N} \rfloor$ is a function of x_i, x_j, and x. Write $\lfloor \mathbf{N} \rfloor$ by inspection if you can. Then derive $\lfloor \mathbf{N} \rfloor$ by use of the [**A**]-matrix method described in Section 3.8.

3.34 The three-node bar element shown is to have an axial displacement field u that is quadratic in x and depends on nodal d.o.f. u_i, u_j, and u_k. Determine the shape function matrix $\lfloor \mathbf{N} \rfloor$ in terms of x and ℓ.

Section 3.9

3.35 Use {**D**} as given by Eq. 3.9-10, and [**K**] and {**R**} as given in Eq. 3.9-9, to compute $U = \frac{1}{2} \{\mathbf{D}\}^T[\mathbf{K}]\{\mathbf{D}\}$ and $\Omega = -\{\mathbf{D}\}^T\{\mathbf{R}\}$. How is U related to Ω? What are the precentage errors of U and Ω? (Exact values of U and Ω are computed in Problems 3.26a and 3.27a.)

3.36 Consider a uniform bar under axial load. Displacements and stresses may be obtained by either the classical or the finite element form of the Rayleigh–Ritz method. Which form (if either) gives exact results, and how many d.o.f. are required for exactness, if the axial loading is (a) distributed in the form $q = c(L_T - x)$, and (b) concentrated, with equal forces P at $x = L$, $x = 2L$, and $x = 3L$? The total length of the bar is $L_T = 3L$.

3.37 Consider a bar of length L_T that is fixed at $x = 0$ and free at $x = L_T$. The bar carries a distributed axial loading of intensity $q = c(L_T - x)$, where c is a constant. Generate a finite element solution like that in Section 3.9, using
(a) a single element of length L_T.
(b) two elements, each of length $L = L_T/2$.
(c) three elements, each of length $L = L_T/3$.

3.38 Repeat the finite element solution of Section 3.9 but let the respective elements have lengths $L_T/6$, $L_T/3$, and $L_T/2$ (reading left to right).

3.39 (a) Apply Eq. 3.9-3, with $\lfloor \mathbf{B} \rfloor$ taken from Eq. 3.9-1, to determine the stiffness matrix of a bar element that is tapered: its cross-sectional area is $A = A_0(3L - 2x)/L$, where A_0 is the cross-sectional area at the right end $(x = L)$.
(b) What is the *exact* stiffness matrix of this tapered bar? Find out by displacing d.o.f. one at a time and using an elementary mechanics of materials analysis to compute the axial force required.

3.40 The *rigid* bar shown rests on an elastic foundation. When displaced laterally an amount w, the foundation applies a force $kw\ dx$ to a length dx of the bar, where k is a constant. Determine the 2 by 2 stiffness matrix that operates

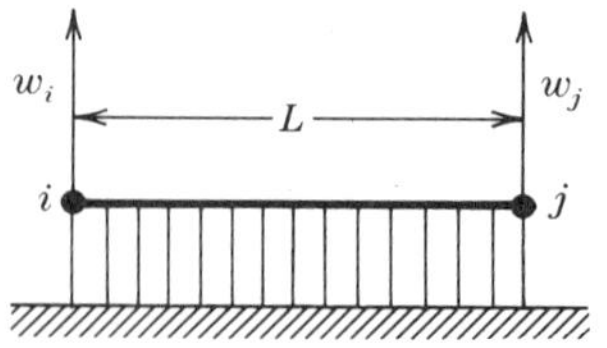

Problem 3.40

on w_i and w_j. *Suggestion*: Express strain energy U in terms of k, L, w_i, and w_j, then write U in the form $\{\mathbf{d}\}^T[\mathbf{k}]\{\mathbf{d}\}/2$, and identify $[\mathbf{k}]$.

Section 3.10

3.41 As suggested below Eq. 3.10-8, indicate assembly of elements when Π is written, and see whether Eq. 3.10-8 again results.

3.42 For each of the following problems, obtain finite element formulations in a form analogous to Eqs. 3.10-6 and 3.10-7. Details such as specific element shape functions are not required.

(a) Plane beam (use the integrals in Eq. 3.4-8).

(b) Plane beam (see Problem 3.32). Use two interpolating fields, one for M that depends on nodal moments $\{\mathbf{M}_e\}$ and one for w that depends on nodal lateral deflections $\{\mathbf{w}_e\}$.

(c) Acoustic modes in a cavity with rigid walls. The functional is

$$\Pi = \int \left(p_{,x}^2 + p_{,y}^2 + p_{,z}^2 - \frac{\omega^2}{c^2} p^2 \right) dV$$

where $p = p(x,y,z)$ is the amplitude of gas pressure that varies with time, ω is the circular frequency, and c is the speed of sound. Let $p = \lfloor \mathbf{N} \rfloor \{\mathbf{p}_e\}$, where $\{\mathbf{p}_e\}$ represents nodal pressures.

3.43 The uniform bar shown is to act as a heat conduction element, with T = temperature, k = thermal conductivity, q = axial heat flux in the bar per unit of cross-sectional area A, and H = lateral heat flux per unit length. From the functional

$$\Pi = Aq_jT_j - Aq_iT_i + \frac{1}{2}\int_0^L kT_{,x}^2 A\, dx - \int_0^L HT\, dx$$

determine expressions for $[\mathbf{k}]$ and $\{\mathbf{r}_Q\}$ in the element equations $[\mathbf{k}]\{\mathbf{T}_e\} = \{\mathbf{r}_Q\}$, where $\{\mathbf{T}_e\} = \lfloor T_i \quad T_j \rfloor^T$. Let $T = \lfloor \mathbf{N} \rfloor \{\mathbf{T}_e\}$ and $H = \lfloor \mathbf{N} \rfloor \lfloor H_i \quad H_j \rfloor^T$, where $\lfloor \mathbf{N} \rfloor = \left\lfloor \dfrac{L-x}{L} \quad \dfrac{x}{L} \right\rfloor$.

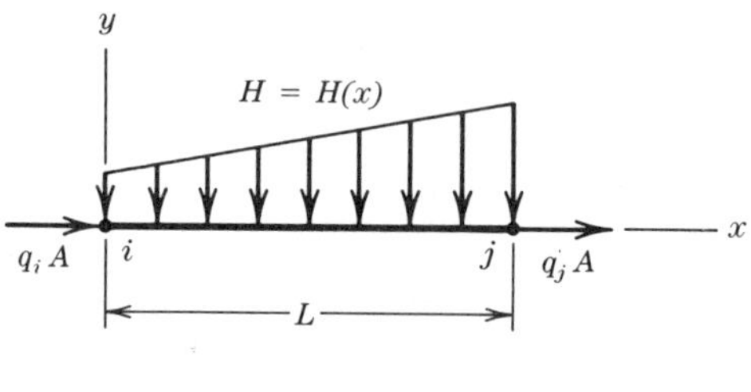

Problem 3.43

Section 3.12

3.44 In Fig. 3.12-1*b*, let $x_1 = 0$, $x_2 = 2$, and $x_3 = 3$. Then do as follows.
 (a) Verify numerically that $\Sigma\, N_i = 1$.
 (b) According to Eq. 3.12-4, $\Sigma\, N_{i,x} = 0$. Verify this property numerically.

3.45 For the four points in Fig. 3.12-2, let the respective x_i be 1, 3, 5, and 8, and let the respective ϕ_i be 2, 2, 2, and 5.
 (a) Use Lagrange's formula to obtain the interpolating curve.
 (b) What values of ϕ does Lagrange's formula predict at $x = 2$, $x = 4$, and $x = 7$?

3.46 Sketch the four N_i of Eqs. 3.12-10 in a fashion analogous to Fig. 3.12-1. That is, sketch the element in isometric view, with $\phi = N_i\phi_i$ shown normal to the xy plane, as in Fig. 1.1-3.

3.47 (a) Determine shape functions N_i for the nine-node Lagrange element shown.
 (b) What is $\phi = \phi(x)$ for this element when written in the form of Eq. 3.12-7?
 (c) Show how N_1 varies over the element by making an isometric sketch and showing $\phi = N_1\phi_1$ normal to the xy plane (similar to Fig. 1.1-3). Do the same for N_8 and N_9.

THIS IS IMPORTANT

SEE NOTES P.4 11-11-96

SEE EQ 3.12-3

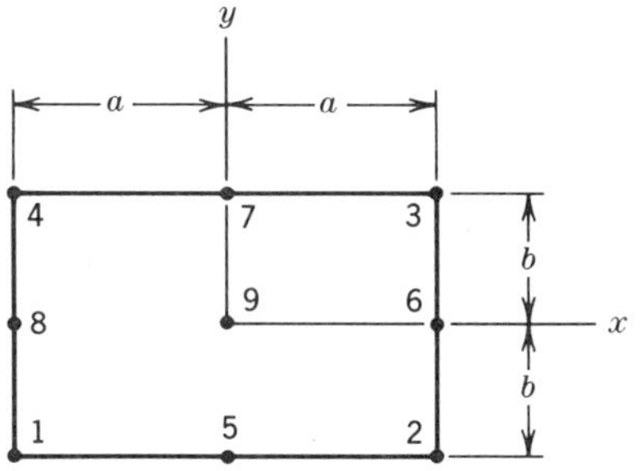

Problem 3.47

3.48 The element shown has three linear edges and one quadratic edge. Determine the five shape functions N_i. *Suggestion*: after interpolating along edges $y = -b$ and $y = +b$, interpolate linearly in the y direction.

3.49 Determine shape functions N_i for the eight-node rectangular parallelepiped shown. Overall side lengths are $2a$, $2b$, and $2c$. *Suggestion*: Infer answers from the pattern seen in Eqs. 3.12-10, then test the answers.

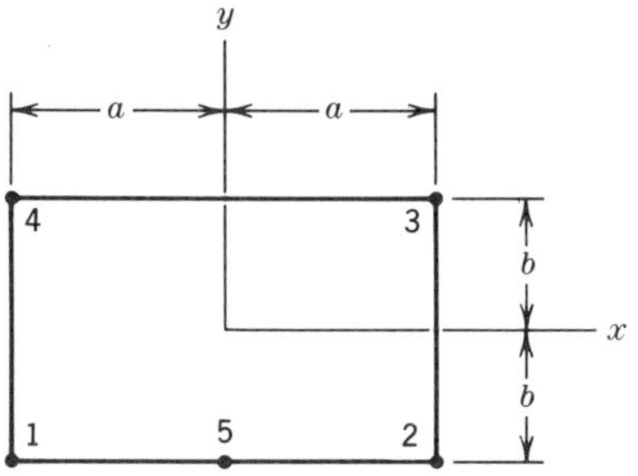

Problem 3.48

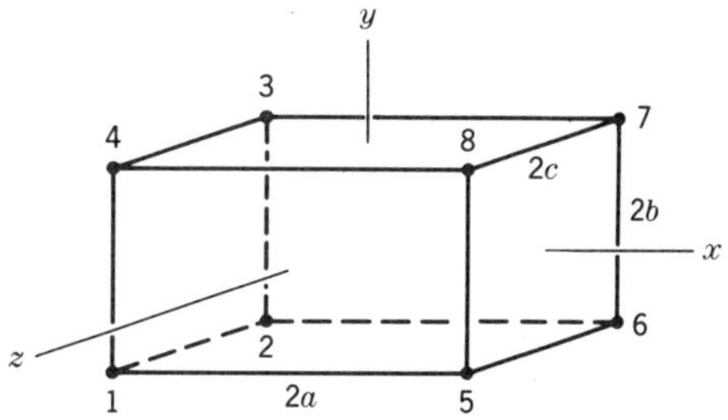

Problem 3.49

Section 3.13

3.50 Determine matrix $[\mathbf{A}]^{-1}$ of Eq. 3.13-4 and verify the shape functions given in Fig. 3.13-2. *Suggestion*: Regard Eqs. 3.13-3 as four equations to be solved for the four a_i. Arrange the results in matrix format and identify a 4 by 4 matrix as $[\mathbf{A}]^{-1}$.

3.51 Imagine that a curve $\phi = \phi(x)$ is to be fitted to three data values: ϕ_1 and θ_1 at $x = 0$ and ϕ_2 at $x = L$ (analogous to Fig. 3.13-1, but with θ_2 unspecified). Determine the shape functions. Also sketch them and check their behavior at $x = 0$ and at $x = L$ (in the fashion of Fig. 3.13-2).

3.52 Show that cubic shape functions (Fig. 3.13-2) do not provide C^2 continuity. *Suggestion*: Examine the node shared by two adjacent elements, only one of which has nonzero d.o.f.

3.53 Imagine that at points A, B, and C in the sketch one knows both ordinate and slope data. Slope is indicated by a short line through a data point. Without calculation, sketch

(a) a Lagrange interpolation curve through all three points.
(b) piecewise interpolation of C^0 continuity.
(c) piecewise interpolation of C^1 continuity.

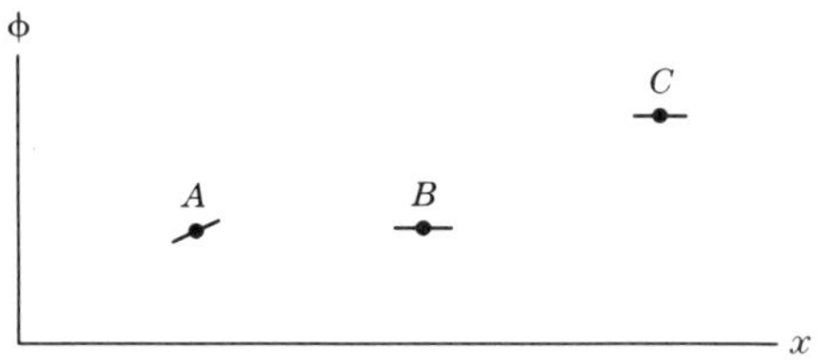

Problem 3.53

3.54 Shape functions N_i of C^0 elements satisfy the relation $\Sigma N_i = 1$. Such is not the case for the N_i of Fig. 3.13-2. Why?

CHAPTER 4

DISPLACEMENT-BASED ELEMENTS FOR STRUCTURAL MECHANICS

General expressions for the element stiffness matrix **[k]** and the element load vector $\{\mathbf{r}_e\}$ are derived, then used to formulate simple elements. Also discussed are how equilibrium and compatibility are approximated by the solution, requirements for convergence with mesh refinement, and procedures for stress computation.

4.1 FORMULAS FOR ELEMENT MATRICES [k] AND $\{\mathbf{r}_e\}$

Our discussion is restricted to elements based on displacement fields. Other elements, often equally good but less popular, are based on stress fields. Stress field elements are not discussed in this chapter.

The derivation of finite element formulas is a straightforward procedure, which can be verbally summarized as follows [4.1]. Displacements are taken as the dependent variables. Therefore, the appropriate functional for a Rayleigh–Ritz solution is Π_p, the expression for potential energy. We select an admissible displacement field, defined in piecewise fashion so that displacements within any element are interpolated from nodal d.o.f. of that element, then evaluate Π_p in terms of nodal d.o.f. Using the principle of stationary potential energy, we write $d\Pi_p = 0$, from which we obtain algebraic equations to be solved for the nodal d.o.f. In the course of this argument we identify certain expressions as the element stiffness matrix **[k]** and the element load vector $\{\mathbf{r}_e\}$. Details of this derivation are now described.

The starting point is the expression for potential energy in a linearly elastic body, Eq. 3.4-1, which is repeated here as Eq. 4.1-1,

$$\begin{aligned}\Pi_p = &\int_V \left(\tfrac{1}{2}\{\boldsymbol{\epsilon}\}^T[\mathbf{E}]\{\boldsymbol{\epsilon}\} - \{\boldsymbol{\epsilon}\}^T[\mathbf{E}]\{\boldsymbol{\epsilon}_0\} + \{\boldsymbol{\epsilon}\}^T\{\boldsymbol{\sigma}_0\}\right) dV \\ &- \int_V \{\mathbf{u}\}^T\{\mathbf{F}\}\, dV - \int_S \{\mathbf{u}\}^T\{\boldsymbol{\Phi}\}\, dS - \{\mathbf{D}\}^T\{\mathbf{P}\}\end{aligned} \tag{4.1-1}$$

in which $\{\mathbf{u}\} = \lfloor u \quad v \quad w \rfloor^T$, the displacement field
$\{\boldsymbol{\epsilon}\} = \lfloor \epsilon_x \quad \epsilon_y \quad \epsilon_z \quad \gamma_{xy} \quad \gamma_{yz} \quad \gamma_{zx} \rfloor^T$, the strain field
$[\mathbf{E}]$ = the material property matrix (e.g., Eq. 1.7-3)
$\{\boldsymbol{\epsilon}_0\}, \{\boldsymbol{\sigma}_0\}$ = initial strains and initial stresses (Eq. 1.7-8)
$\{\mathbf{F}\} = \lfloor F_x \quad F_y \quad F_z \rfloor^T$, body forces (Eq. 1.6-4)
$\{\boldsymbol{\Phi}\} = \lfloor \Phi_x \quad \Phi_y \quad \Phi_z \rfloor^T$, surface tractions (Eq. 1.6-4)

$\{\mathbf{D}\}$ = nodal d.o.f. of the structure
$\{\mathbf{P}\}$ = loads applied to d.o.f. by external agencies
S, V = surface area and volume of the structure

Shorter expressions may be used for problems that are two- or one-dimensional, as will be seen subsequently.

Displacements within an element are interpolated from element nodal d.o.f. $\{\mathbf{d}\}$,

$$\{\mathbf{u}\} = [\mathbf{N}]\{\mathbf{d}\} \qquad (4.1\text{-}2)$$

WE USE $[N]^T$ (WRITTEN AS A COLUMN). SEE NOTES

where $[\mathbf{N}]$ is the shape function matrix. The particular form of $[\mathbf{N}]$ need not be specified yet, but a form must eventually be written. The form selected has much to do with the quality of the approximate solution.

From here onward, the derivation requires only straightforward manipulation. Strains are obtained from displacements by differentiation. Thus

$$\{\boldsymbol{\epsilon}\} = [\partial]\{\mathbf{u}\} \qquad \text{yields} \qquad \{\boldsymbol{\epsilon}\} = [\mathbf{B}]\{\mathbf{d}\}, \qquad \text{where} \quad [\mathbf{B}] = [\partial][\mathbf{N}] \quad (4.1\text{-}3)$$

The differential operator matrix $[\partial]$ is given by Eq. 1.5-6; its size is 6 by 3 for three-dimensional problems and 3 by 2 for two-dimensional problems. Substitution of the expressions for $\{\mathbf{u}\}$ and $\{\boldsymbol{\epsilon}\}$ into Eq. 4.1-1 yields

$$\Pi_p = \frac{1}{2} \sum_{n=1}^{\text{numel}} \{\mathbf{d}\}_n^T [\mathbf{k}]_n \{\mathbf{d}\}_n - \sum_{n=1}^{\text{numel}} \{\mathbf{d}\}_n^T \{\mathbf{r}_e\}_n - \{\mathbf{D}\}^T \{\mathbf{P}\} \qquad (4.1\text{-}4)$$

where summation symbols indicate that we include contributions from all *numel* elements of the structure, and we have defined

element stiffness matrix $\boxed{[\mathbf{k}] = \int_{V_e} [\mathbf{B}]^T [\mathbf{E}][\mathbf{B}] \, dV}$ $(4.1\text{-}5)$

EACH OF THESE IN 11-25 NOTES $\{P\} + \{r_e\} = [\mathcal{R}]$ *NOTES*

$[\sigma]$ IN NOTES

$\{F\} = BODY \ FORCES = \{P\}$

element load vector $\boxed{\begin{aligned} \{\mathbf{r}_e\} &= \int_{V_e} [\mathbf{B}]^T [\mathbf{E}]\{\boldsymbol{\epsilon}_0\} \, dV - \int_{V_e} [\mathbf{B}]^T \{\boldsymbol{\sigma}_0\} \, dV \\ &+ \int_{V_e} [\mathbf{N}]^T \{\mathbf{F}\} \, dV + \int_{S_e} [\mathbf{N}]^T \{\boldsymbol{\Phi}\} \, dS \end{aligned}}$ $(4.1\text{-}6)$

SURFACE FORCES $\{\Phi\}$ *SEE NOTES 11-25-96 P. 3*

$[\mathcal{RS}]$

where V_e denotes the volume of an element and S_e its surface. In the surface integral, $[\mathbf{N}]$ is evaluated on S_e.

Despite its formidable appearance, the expression for $\{\mathbf{r}_e\}$ is less important than the expression for $[\mathbf{k}]$. In essence, the $\{\mathbf{r}_e\}$ expression states how certain loads can be dealt with to best advantage. Use of an ad hoc method instead may provide acceptable accuracy. Examples follow (see below Eq. 4.3-14).

To complete the derivation we must determine the algebraic equations to be solved for nodal d.o.f., as follows. Every d.o.f. in an element vector $\{\mathbf{d}\}$ also appears in the vector of global (i.e., structural) d.o.f. $\{\mathbf{D}\}$. Therefore, as argued in Sections 2.6 and 2.7, $\{\mathbf{D}\}$ can replace $\{\mathbf{d}\}$ in Eq. 4.1-4 if $[\mathbf{k}]$ and $\{\mathbf{r}_e\}$ of every element are conceptually expanded to structure size. Thus Eq. 4.1-4 becomes

$$\Pi_p = \tfrac{1}{2} \{\mathbf{D}\}^T [\mathbf{K}]\{\mathbf{D}\} - \{\mathbf{D}\}^T \{\mathbf{R}\} \qquad (4.1\text{-}7)$$

$\{P\} + \{r_e\}$

SEE P. 6 11-25

where

NUMBER OF ELEMENTS

$$[\mathbf{K}] = \sum_{n=1}^{\text{numel}} [\mathbf{k}]_n \qquad \text{and} \qquad \{\mathbf{R}\} = \{\mathbf{P}\} + \sum_{n=1}^{\text{numel}} \{\mathbf{r}_e\}_n \tag{4.1-8}$$

Summations indicate assembly of element matrices by addition of overlapping terms, as in Section 2.7. Now Π_p is a function of d.o.f. {**D**}. Making Π_p stationary with respect to small changes in the D_i by use of convenient differentiation rules given in Appendix A, we write

$$\left\{\frac{\partial \Pi_p}{\partial \mathbf{D}}\right\} = \{\mathbf{0}\} \qquad \text{yields} \qquad [\mathbf{K}]\{\mathbf{D}\} = \{\mathbf{R}\} \tag{4.1-9}$$

The latter matrix equation is a set of simultaneous algebraic equations to be solved for d.o.f. {**D**}. As noted in the latter part of Section 3.3, $\partial\Pi_p/\partial D_i = 0$ is a nodal equilibrium equation, [**K**] is a symmetric matrix, and $K_{ij} = \partial^2\Pi_p/\partial D_i\, \partial D_j$.

Some Particular Cases. The preceding derivation may have suggested that {$\boldsymbol{\epsilon}$} must always be a 6 by 1 vector, as for a three-dimensional problem. However, no change in Eqs. 4.1-5 and 4.1-6 need be made if the problem is two-dimensional. For plane stress or plane strain, {$\boldsymbol{\epsilon}$} is 3 by 1 and contains only ϵ_x, ϵ_y, and γ_{xy}, and matrix [**E**] is 3 by 3 (see Eq. 1.7-5, for example). For a symmetrically loaded solid of revolution, {$\boldsymbol{\epsilon}$} is 4 by 1 and contains ϵ_r, ϵ_z, ϵ_θ, and γ_{zr}, and [**E**] is 4 by 4.

Formulas for [**k**] and {$\mathbf{r}_e$} applicable to a bar under axial load can be obtained by specialization of Eqs. 4.1-5 and 4.1-6. However, it is easier to rederive the formulas, using at the outset special forms applicable to a bar. Such a derivation has already been given in Section 3.9. In that development, note Eqs. 3.9-2 and 3.9-3 in particular. They state that the strain energy of a one-element structure is

$$U_e = \int_{V_e} (\text{strain energy density})\, dV = \tfrac{1}{2}\{\mathbf{d}\}^T[\mathbf{k}]\{\mathbf{d}\} \tag{4.1-10}$$

Equation 4.1-10 summarizes the argument that leads to Eq. 4.1-5: namely, that an element stiffness matrix is obtained by substitution of an interpolation scheme into a strain energy expression. Equation 4.1-6 can be explained similarly. Element nodal loads {$\mathbf{r}_e$} are obtained by substitution of an interpolation scheme into an expression for work done by distributed loads that act on an element:

$$\Omega_e = -\int_{V_e} (\text{work per unit volume})\, dV = -\{\mathbf{d}\}^T\{\mathbf{r}_e\} \tag{4.1-11}$$

(Work done by externally applied nodal loads {**P**} is excluded from {$\mathbf{r}_e$}.)

In flexural problems, such as beam and plate bending, it is convenient to define [**B**] in such a way that the product [**B**]{**d**} represents curvatures rather than strains. Consider beam bending. Expressions for U_e and Ω_e come from the first two integrals in Eq. 3.4-8. Noting that a scalar is its own transpose, we write $w_{,xx}^2 = w_{,xx}^T\, w_{,xx}$ and $w = w^T$. Hence

$$U_e = \frac{1}{2}\int_0^L w_{,xx}^T\, EIw_{,xx}\, dx \qquad \text{and} \qquad \Omega_e = -\int_0^L w^T q\, dx \qquad (4.1\text{-}12)$$

where w is lateral displacement and L is the element length. Interpolating w from element nodal d.o.f. $\{\mathbf{d}\}$, we have

$$w = \lfloor \mathbf{N} \rfloor \{\mathbf{d}\} \qquad \text{and} \qquad w_{,xx} = \lfloor \mathbf{B} \rfloor \{\mathbf{d}\}, \qquad \text{where } \lfloor \mathbf{B} \rfloor = \frac{d^2}{dx^2}\lfloor \mathbf{N} \rfloor \qquad (4.1\text{-}13)$$

Therefore, in view of Eqs. 4.1-10 and 4.1-11, $[\mathbf{k}]$ and $\{\mathbf{r}_e\}$ for a straight beam element are

$$[\mathbf{k}] = \int_0^L \lfloor \mathbf{B} \rfloor^T EI \lfloor \mathbf{B} \rfloor\, dx \qquad \text{and} \qquad \{\mathbf{r}_e\} = \int_0^L \lfloor \mathbf{N} \rfloor^T q\, dx \qquad (4.1\text{-}14)$$

Analogous expressions are written for flat plates in bending, where EI becomes a matrix of flexural rigidities and integration is over the area of the plate midsurface.

***a*-Basis Formulation.** If displacements are initially expressed in terms of d.o.f. $\{\mathbf{a}\}$, one usually replaces $\{\mathbf{a}\}$ by element nodal d.o.f. $\{\mathbf{d}\}$ before generating a stiffness matrix. The process of replacing $\{\mathbf{a}\}$ by $\{\mathbf{d}\}$ is explained in Sections 3.8 and 3.13. Sometimes it is convenient to delay the replacement of $\{\mathbf{a}\}$ by $\{\mathbf{d}\}$. The procedure is as follows.

Displacements and strains, expressed in terms of $\{\mathbf{a}\}$, are

$$\{\mathbf{u}\} = [\mathbf{N}_a]\{\mathbf{a}\} \qquad \text{and} \qquad \{\boldsymbol{\epsilon}\} = [\mathbf{B}_a]\{\mathbf{a}\}, \qquad \text{where } [\mathbf{B}_a] = [\partial][\mathbf{N}_a] \qquad (4.1\text{-}15)$$

(For example, $[\mathbf{N}_a] = \lfloor 1 \quad s \rfloor$ in Eq. 3.8-5.) But $\{\mathbf{a}\} = [\mathbf{A}]^{-1}\{\mathbf{d}\}$, so

$$[\mathbf{N}] = [\mathbf{N}_a][\mathbf{A}]^{-1} \qquad \text{and} \qquad [\mathbf{B}] = [\mathbf{B}_a][\mathbf{A}]^{-1} \qquad (4.1\text{-}16)$$

Substitution of Eqs. 4.1-16 into Eqs. 4.1-5 and 4.1-6 yields the "*d*-basis" matrices $[\mathbf{k}]$ and $\{\mathbf{r}_e\}$,

$$[\mathbf{k}] = [\mathbf{A}]^{-T}[\mathbf{k}_a][\mathbf{A}]^{-1} \qquad \text{and} \qquad \{\mathbf{r}_e\} = [\mathbf{A}]^{-T}\{\mathbf{r}_{ea}\} \qquad (4.1\text{-}17)$$

where the "*a*-basis" matrices are defined as

$$[\mathbf{k}_a] = \int_{V_e} [\mathbf{B}_a]^T[\mathbf{E}][\mathbf{B}_a]\, dV \qquad \text{and} \qquad \{\mathbf{r}_{ea}\} = \int_{V_e} [\mathbf{B}_a]^T[\mathbf{E}]\{\boldsymbol{\epsilon}_0\}\, dV - \cdots \qquad (4.1\text{-}18)$$

The *a*-basis matrices $[\mathbf{k}_a]$ and $\{\mathbf{r}_{ea}\}$ are converted to *d*-basis matrices $[\mathbf{k}]$ and $\{\mathbf{r}_e\}$ before global equations are assembled, as the relationship between d.o.f. $\{\mathbf{a}\}$ in neighboring elements is complicated and unwieldy.

An example application appears in Eqs. 4.2-6 and 4.2-7.

4.2 OVERVIEW OF ELEMENT STIFFNESS MATRICES

In this section we outline the formulation of selected element stiffness matrices, with the intent of showing the conceptual simplicity of the process. Details of manipulations may be found in the text sections cited.

Bar. Figure 4.2-1 shows a straight bar whose nodal d.o.f. are axial displacements u_1 and u_2. A linear axial displacement field, as used in Section 3.9, is appropriate,

$$u = \lfloor \mathbf{N} \rfloor \begin{Bmatrix} u_1 \\ u_2 \end{Bmatrix}, \qquad \text{where} \quad \lfloor \mathbf{N} \rfloor = \left\lfloor \frac{L-x}{L} \quad \frac{x}{L} \right\rfloor \tag{4.2-1}$$

Using Eq. 3.9-3, with $\lfloor \mathbf{B} \rfloor = d\lfloor \mathbf{N} \rfloor / dx = \lfloor -1 \quad 1 \rfloor / L$, we obtain

$$[\mathbf{k}] = \int_0^L \lfloor \mathbf{B} \rfloor^T AE \lfloor \mathbf{B} \rfloor \, dx = \frac{AE}{L} \begin{bmatrix} 1 & -1 \\ -1 & 1 \end{bmatrix} \tag{4.2-2}$$

where the latter expression is for a *uniform* bar, AE = constant. The integral expression for $[\mathbf{k}]$ does not demand that AE be independent of x.

Beam. Figure 4.2-2 shows a four-d.o.f. straight beam element. Rotation θ is assumed to be small, so that $\theta \approx dw/dx$. Four d.o.f. define a cubic lateral displacement field,

$$w = \lfloor \mathbf{N} \rfloor \lfloor w_1 \quad \theta_1 \quad w_2 \quad \theta_2 \rfloor^T \tag{4.2-3}$$

where the four N_i are given in Fig. 3.13-2. The curvature field is $w_{,xx} = \lfloor \mathbf{B} \rfloor \{\mathbf{d}\}$, where

$$\lfloor \mathbf{B} \rfloor = \frac{d^2}{dx^2} \lfloor \mathbf{N} \rfloor = \left\lfloor -\frac{6}{L^2} + \frac{12x}{L^3} \qquad -\frac{4}{L} + \frac{6x}{L^2} \qquad \frac{6}{L^2} - \frac{12x}{L^3} \qquad -\frac{2}{L} + \frac{6x}{L^2} \right\rfloor \tag{4.2-4}$$

For constant EI, the element stiffness matrix given by Eq. 4.1-14 is

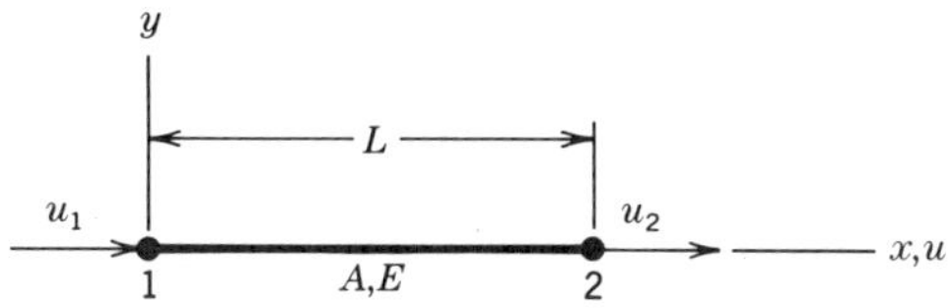

Figure 4.2-1. Bar element with two d.o.f. (u_1 and u_2).

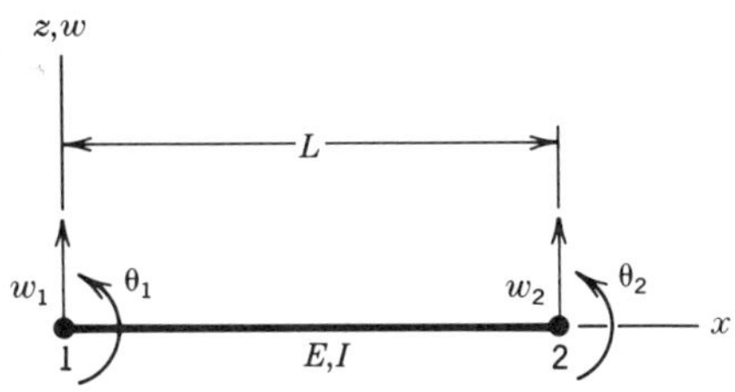

Figure 4.2-2. Standard four-d.o.f. beam element.

$$[\mathbf{k}] = \int_0^L \lfloor \mathbf{B} \rfloor^T EI \lfloor \mathbf{B} \rfloor \, dx = \frac{EI}{L^3} \begin{bmatrix} 12 & 6L & -12 & 6L \\ 6L & 4L^2 & -6L & 2L^2 \\ -12 & -6L & 12 & -6L \\ 6L & 2L^2 & -6L & 4L^2 \end{bmatrix} \tag{4.2-5}$$

This [**k**] operates on d.o.f. {**d**} having the order shown in Eq. 4.2-3. A slightly modified form of Eq. 4.2-5 is able to account for transverse shear deformation [4.2,4.11].

The manipulations needed to obtain [**k**] are shortened by using the "*a*-basis" of Eqs. 4.1-17 and 4.1-18. With $\lfloor \mathbf{X} \rfloor$ and [**A**] given by Eqs. 3.13-1 and 3.13-3, we write

$$\lfloor \mathbf{B} \rfloor = \frac{d^2}{dx^2} \lfloor \mathbf{X} \rfloor [\mathbf{A}]^{-1} = \lfloor 0 \quad 0 \quad 2 \quad 6x \rfloor [\mathbf{A}]^{-1} \tag{4.2-6}$$

$$[\mathbf{k}] = [\mathbf{A}]^{-T} \int_0^L \lfloor 0 \quad 0 \quad 2 \quad 6x \rfloor^T EI \lfloor 0 \quad 0 \quad 2 \quad 6x \rfloor \, dx \, [\mathbf{A}]^{-1} \tag{4.2-7}$$

and obtain the same [**k**] as given in Eq. 4.2-5.

The reader should understand the sign conventions for nodal moments and bending moment. Nodal moments M_1 and M_2 are positive when acting in the directions of θ_1 and θ_2 in Fig. 4.2-2. Bending moment $M = EIw_{,xx}$ is positive when it creates tension on the bottom of the beam. Therefore, $M = -M_1$ at the left end and $M = +M_2$ at the right end.

Plane Frame. A plane frame member can deform both axially and in bending. Effectively, to obtain a plane frame element we superpose the bar and beam elements of Figs. 4.2-1 and 4.2-2 (and of Eqs. 4.2-2 and 4.2-5). Nodal d.o.f. are $\{\mathbf{d}\} = \lfloor u_1 \quad w_1 \quad \theta_1 \quad u_2 \quad w_2 \quad \theta_2 \rfloor^T$. If the element is uniform and lies along the x axis, its stiffness matrix is

$$[\mathbf{k}] = \frac{AE}{L} \begin{bmatrix} 1 & 0 & 0 & -1 & 0 & 0 \\ 0 & 0 & 0 & 0 & 0 & 0 \\ 0 & 0 & 0 & 0 & 0 & 0 \\ -1 & 0 & 0 & 1 & 0 & 0 \\ 0 & 0 & 0 & 0 & 0 & 0 \\ 0 & 0 & 0 & 0 & 0 & 0 \end{bmatrix} + \frac{EI}{L^3} \begin{bmatrix} 0 & 0 & 0 & 0 & 0 & 0 \\ 0 & 12 & 6L & 0 & -12 & 6L \\ 0 & 6L & 4L^2 & 0 & -6L & 2L^2 \\ 0 & 0 & 0 & 0 & 0 & 0 \\ 0 & -12 & -6L & 0 & 12 & -6L \\ 0 & 6L & 2L^2 & 0 & -6L & 4L^2 \end{bmatrix} \tag{4.2-8}$$

If a frame element is arbitrarily oriented in the plane, its stiffness matrix is easily determined from [**k**] of Eq. 4.2-8 by a coordinate transformation (see Section 7.5). Similarly, a plane *truss* element of arbitrary orientation can be obtained by coordinate transformation of Eq. 4.2-2 (again, see Section 7.5).

Constant-Strain Triangle. This element, shown in Fig. 4.2-3, is one of the earliest finite elements [1.8]. It can be used to solve problems of plane stress and plane strain. However, it is not a very good element for this purpose. We introduce it here primarily because it serves as a good example in subsequent discussions of why elements behave as they do.

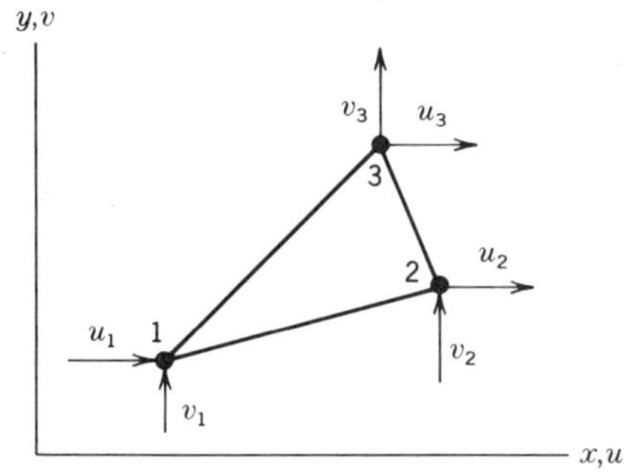

Figure 4.2-3. Constant-strain triangle (six d.o.f.).

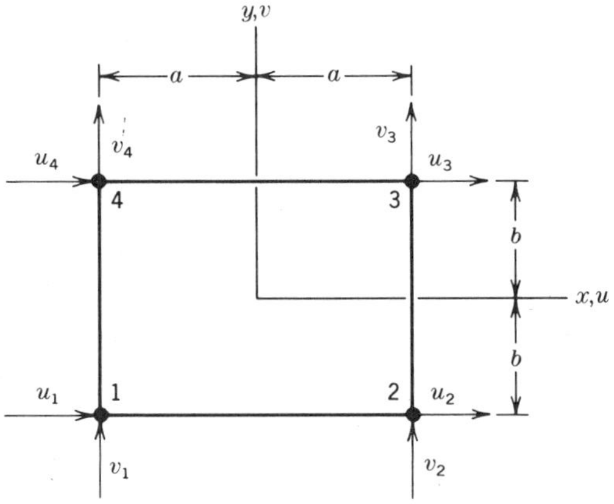

Figure 4.2-4. Plane rectangular bilinear element (eight d.o.f.).

The constant-strain triangle is based on a complete linear polynomial for both x- and y-direction displacements:

$$u = a_1 + a_2 x + a_3 y \tag{4.2-9a}$$

$$v = a_4 + a_5 x + a_6 y \tag{4.2-9b}$$

Strains are $\epsilon_x = u_{,x}$, $\epsilon_y = v_{,y}$, and $\gamma_{xy} = u_{,y} + v_{,x}$. Thus

$$\epsilon_x = a_2 \qquad \epsilon_y = a_6 \qquad \gamma_{xy} = a_3 + a_5 \tag{4.2-10}$$

We see that strains are independent of x and y within the element; hence the name "constant-strain triangle."

The algebra of generating the element stiffness matrix is most easily carried out in area coordinates, as described in Chapter 5. Strain–displacement matrix $[\mathbf{B}]$ is 3 by 6 and contains only constants, which depend on the x and y coordinates of the three nodes. Hence, if material property matrix $[\mathbf{E}]$ and element thickness t are constant over the element,

$$\underset{6\times 6}{[\mathbf{k}]} = \iint \underset{6\times 3}{[\mathbf{B}]^T} \underset{3\times 3}{[\mathbf{E}]} \underset{3\times 6}{[\mathbf{B}]}\, t\, dx\, dy = At[\mathbf{B}]^T[\mathbf{E}][\mathbf{B}] \tag{4.2-11}$$

where A is the area of the triangle.

The behavior of the constant-strain triangle is illustrated by numerical examples in Fig. 5.4-2.

Plane Rectangular Bilinear Element. This element, shown in Fig. 4.2-4, is based on the bilinear displacement field

$$u = a_1 + a_2 x + a_3 y + a_4 xy \tag{4.2-12a}$$

$$v = a_5 + a_6 x + a_7 y + a_8 xy \tag{4.2-12b}$$

In terms of nodal d.o.f., the displacement field $\{\mathbf{u}\} = [\mathbf{N}]\{\mathbf{d}\}$ is

SAME AS MY NOTES, WRITTEN DIFFERENTLY

$$\begin{Bmatrix} u \\ v \end{Bmatrix} = \begin{bmatrix} N_1 & 0 & N_2 & 0 & N_3 & 0 & N_4 & 0 \\ 0 & N_1 & 0 & N_2 & 0 & N_3 & 0 & N_4 \end{bmatrix} \begin{Bmatrix} u_1 \\ v_1 \\ u_2 \\ \vdots \\ v_4 \end{Bmatrix} \tag{4.2-13}$$

Shape functions N_i of this element are presented as Eqs. 3.12-10. Each N_i has the form $(a \pm x)(b \pm y)/4ab$. (With an adequate understanding of shape functions, the reader should have no difficulty in choosing the proper algebraic signs for each of the N_i by inspection.) The strain–displacement matrix is

$$\underset{3\times 8}{[\mathbf{B}]} = \begin{bmatrix} \partial/\partial x & 0 \\ 0 & \partial/\partial y \\ \partial/\partial y & \partial/\partial x \end{bmatrix} [\mathbf{N}] = \frac{1}{4ab} \begin{bmatrix} -(b-y) & 0 & (b-y) & \cdots \\ 0 & -(a-x) & 0 & \cdots \\ -(a-x) & -(b-y) & -(a+x) & \cdots \end{bmatrix} \tag{4.2-14}$$

We see that ϵ_x depends on y, ϵ_y depends on x, and γ_{xy} depends on both x and y. The element stiffness matrix is

$$\underset{8\times 8}{[\mathbf{k}]} = \int_{-b}^{b} \int_{-a}^{a} \underset{8\times 3}{[\mathbf{B}]^T} \underset{3\times 3}{[\mathbf{E}]} \underset{3\times 8}{[\mathbf{B}]} \, t \, dx \, dy \tag{4.2-15}$$

where t is the element thickness. The integrand involves polynomials in x and y and is easily evaluated.

If elements are rectangular, the bilinear element and the four-node plane isoparametric element of Section 6.3 are identical. Numerical examples that use isoparametric elements appear in Table 6.14-1.

Solid Rectangular Trilinear Element. This element, shown in Fig. 4.2-5, is a simple generalization of the plane bilinear element. Its x-direction displacement is

$$u = a_1 + a_2x + a_3y + a_4z + a_5xy + a_6yz + a_7zx + a_8xyz \tag{4.2-16}$$

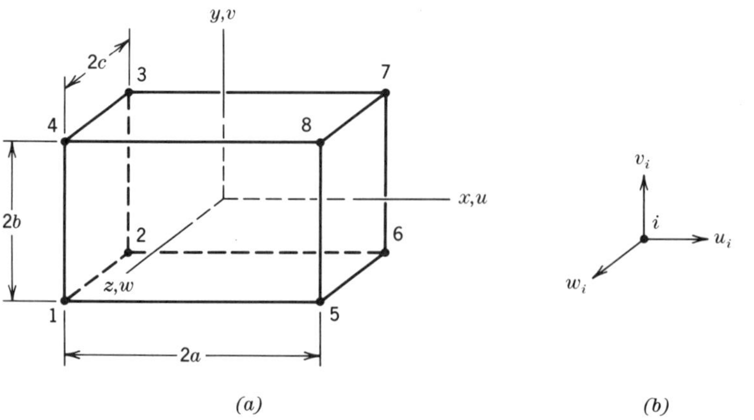

Figure 4.2-5. (*a*) Solid rectangular trilinear element (24 d.o.f.). (*b*) D.o.f. at a typical node *i*.

and similarly for v and w, for a total of 24 d.o.f. In terms of nodal d.o.f., the displacement field $\{\mathbf{u}\} = [\mathbf{N}]\{\mathbf{d}\}$ is

$$\begin{Bmatrix} u \\ v \\ w \end{Bmatrix} = \begin{bmatrix} N_1 & 0 & 0 & N_2 & 0 & 0 & \cdots \\ 0 & N_1 & 0 & 0 & N_2 & 0 & \cdots \\ 0 & 0 & N_1 & 0 & 0 & N_2 & \cdots \end{bmatrix} \begin{Bmatrix} u_1 \\ v_1 \\ w_1 \\ u_2 \\ \vdots \\ w_8 \end{Bmatrix} \tag{4.2-17}$$

where each N_i has the form

$$\frac{(a \pm x)(b \pm y)(c \pm z)}{8abc} \tag{4.2-18}$$

The signs are all negative for N_2, all positive for N_8, and so on. The strain–displacement matrix is $[\mathbf{B}] = [\partial][\mathbf{N}]$, where $[\partial]$ is given in Eqs. 1.5-6. The element stiffness matrix is

$$\underset{24\times 24}{[\mathbf{k}]} = \int_{-c}^{c}\int_{-b}^{b}\int_{-a}^{a} \underset{24\times 6}{[\mathbf{B}]^T}\ \underset{6\times 6}{[\mathbf{E}]}\ \underset{6\times 24}{[\mathbf{B}]}\ dx\, dy\, dz \tag{4.2-19}$$

Again the integrations are straightforward.

Note that on any face (e.g., $z = c$), Eqs. 4.2-16 and 4.2-17 yield forms used for the bilinear element, Eqs. 4.2-12 and 4.2-13, respectively.

A possible application of solid elements is shown in Fig. 4.2-6.

Remark. According to terminology discussed in Section 3.11, the bar, triangular, bilinear, and trilinear elements are all C^0 elements, and the beam is a C^1 element. The frame element is C^0 in axial deformation and C^1 in bending.

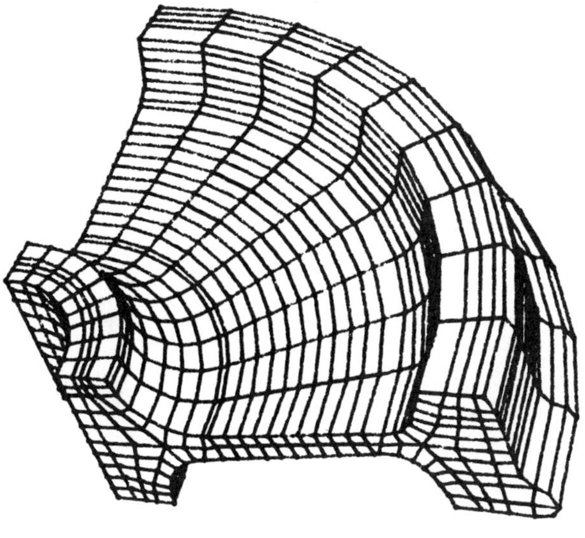

Figure 4.2-6. Quarter of a railway wheel, modeled by solid elements of a type not restricted to rectangular shape. Hidden lines are removed. (*Courtesy of Algor Interactive Systems, Inc., Pittsburgh, Pennsylvania.*)

4.3 CONSISTENT ELEMENT NODAL LOADS $\{\mathbf{r}_e\}$

In this section we consider the element nodal load vector $\{\mathbf{r}_e\}$, which is stated in Eq. 4.1-6. Equation 4.1-6 converts loads distributed throughout an element or over its surface to discrete loads at element nodes.

Any of the four integrals in Eq. 4.1-6 may vanish. For example, the surface integral is zero unless the element has an edge on the structure boundary *and* traction is applied to that edge. Then we integrate over only that edge of the element. All four integrals vanish if externally applied nodal loads $\{\mathbf{P}\}$ make up the entire load vector $\{\mathbf{R}\}$.

Initial Strain and Initial Stress. Terms in Eq. 4.1-6 that contain $\{\boldsymbol{\epsilon}_0\}$ and $\{\boldsymbol{\sigma}_0\}$ produce nodal loads that account for heating or cooling of the element, swelling (due perhaps to irradiation), and initial lack of fit. These nodal loads are self-equilibrating; that is, $\{\mathbf{r}_e\}$ produces zero resultant force and zero resultant moment.

The familiar bar element provides a convenient example (Fig. 4.3-1). Here

$$\lfloor \mathbf{N} \rfloor = \left\lfloor \frac{L-x}{L} \quad \frac{x}{L} \right\rfloor \qquad \text{and} \qquad \lfloor \mathbf{B} \rfloor = \frac{1}{L} \lfloor -1 \quad 1 \rfloor \tag{4.3-1}$$

Imagine that the bar is ΔL units too long, so that the initial strain is $\epsilon_0 = \Delta L/L$. Also, let the bar be heated T° so that (with expansion prohibited) the initial stress caused by heating is $\sigma_0 = -E\alpha T$. The first two integrals in Eq. 4.1-6 yield

$$\underset{2\times 1}{\{\mathbf{r}_e\}} = \int_0^L \frac{1}{L} \begin{Bmatrix} -1 \\ 1 \end{Bmatrix} E \frac{\Delta L}{L} A\, dx - \int_0^L \frac{1}{L} \begin{Bmatrix} -1 \\ 1 \end{Bmatrix} (-E\alpha T) A\, dx \tag{4.3-2a}$$

hence

$$\underset{2\times 1}{\{\mathbf{r}_e\}} = \begin{Bmatrix} -F \\ F \end{Bmatrix}, \qquad \text{where} \quad F = EA\left(\frac{\Delta L}{L} + \alpha T\right) \tag{4.3-2b}$$

Forces F are shown in Fig. 4.3-1*b*. It is a worthwhile exercise for the reader to show that the same forces F are produced by regarding the lack of fit as an initial stress and the temperature effect as an initial strain.

Although forces F sum to zero, they act to deform the bar an amount FL/AE if axial deformation of the bar is uninhibited. This deformation creates a stress $\sigma = F/A$, which is exactly canceled by the initial stress $\sigma = -E\epsilon_0 + \sigma_0$. Thus we obtain axial strain without axial stress, which is entirely correct for an unrestrained bar.

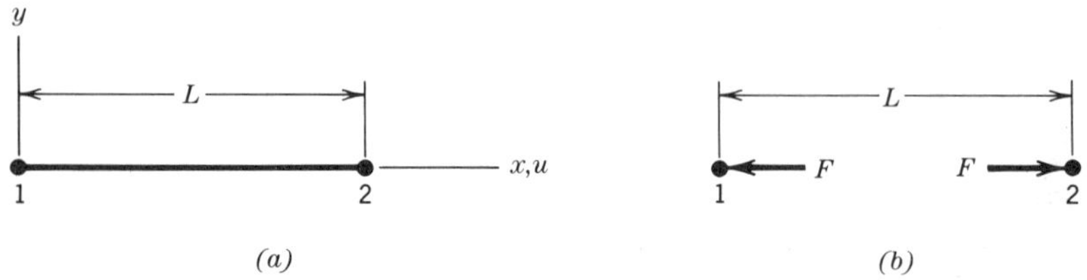

Figure 4.3-1. (*a*) Bar element. (*b*) Nodal forces caused by heating and initial lack of fit (from Eq. 4.3-2).

In this example, nodal forces could easily be deduced by direct physical argument. Elements more complicated than bars and beams often require formal use of Eq. 4.1-6.

Mechanical Loads. Loads $\{\mathbf{r}_e\}$ produced by body forces $\{\mathbf{F}\}$ and surface tractions $\{\mathbf{\Phi}\}$ are given by the latter two integrals in Eq. 4.1-6. These loads are called *work-equivalent loads* for the following reason: work done by nodal loads $\{\mathbf{r}_e\}$ in going through nodal displacements $\{\mathbf{d}\}$ is equal to work done by distributed loads $\{\mathbf{F}\}$ and $\{\mathbf{\Phi}\}$ in going through the displacement field associated with the element shape function. To show this we argue as follows. Work W done by loads $\{\mathbf{r}_e\}$ during small nodal displacements $\{\mathbf{d}\}$ is $W = \{\mathbf{d}\}^T\{\mathbf{r}_e\}$. Taking for example the surface integral in Eq. 4.1-6, and substituting the displacement field $\{\mathbf{u}\}^T = \{\mathbf{d}\}^T[\mathbf{N}]^T$, we have

$$W = \{\mathbf{d}\}^T\{\mathbf{r}_e\} = \int_{S_e} \{\mathbf{d}\}^T[\mathbf{N}]^T\{\mathbf{\Phi}\}\, dS = \int_{S_e} \{\mathbf{u}\}^T\{\mathbf{\Phi}\}\, dS \qquad (4.3\text{-}3)$$

The latter integral sums the work of force increments $\{\mathbf{\Phi}\}\, dS$ in going through displacements $\{\mathbf{u}\}$, where $\{\mathbf{u}\}$ are field displacements created by $\{\mathbf{d}\}$ via shape functions $[\mathbf{N}]$. (See Eqs. 4.3-8 and 4.3-9 for an illustrative example.)

Loads $\{\mathbf{r}_e\}$ calculated by Eq. 4.1-6 are also called *consistent* because they are based on the same shape functions as used to calculate the element stiffness matrix. Finally, loads $\{\mathbf{r}_e\}$ calculated by Eq. 4.1-6 are *statically equivalent* to the original distributed loading; that is, both $\{\mathbf{r}_e\}$ and the original loading have the same resultant force and the same moment about an arbitrarily chosen point. That this is true may be seen by considering the work equivalence of the two loadings during a rigid-body translation and a small rigid-body rotation about an arbitrarily chosen point.

We define *inconsistent loading* or *lumping* as the conversion of a distributed load to nodal loads that are inconsistent with Eq. 4.1-6, but are statically equivalent to the distributed load in that they provide the same resultant force. Typically, lumping is achieved by (a) computing the total force on an element caused by distributed loading, then assigning the same fraction of the total force to each element node, and (b) ignoring any nodal moments that would be present in the consistent vector $\{\mathbf{r}_e\}$. As will be seen in subsequent examples, the consistent method generally does *not* allot the total force on an element equally to element nodes. Lumping can yield poor answers in a coarse mesh, produce locally poor answers in a fine mesh, and lead to failure of the patch test (the patch test is discussed in Section 4.6).

Concentrated Loads Not at Nodes. We denote by $\{\mathbf{p}\}$ a concentrated force that has components p_x, p_y, and p_z. The contribution of $\{\mathbf{p}\}$ to $\{\mathbf{r}_e\}$ can be evaluated from the surface integral in Eq. 4.1-6 by regarding a concentrated force as a large traction $\{\mathbf{\Phi}\}$ acting on a small area dS. Thus $\{\mathbf{p}\} = \{\mathbf{\Phi}\}\, dS$. The integral of $[\mathbf{N}]^T\{\mathbf{\Phi}\}\, dS$ is simply $[\mathbf{N}]^T\{\mathbf{p}\}$ where the concentrated force acts and is zero elsewhere. Thus, if there are n concentrated forces applied to an element, Eq. 4.1-6 yields

$$\text{Owing to concentrated forces } \{\mathbf{p}\}_i, \quad \{\mathbf{r}_e\} = \sum_{i=1}^{n} [\mathbf{N}]_i^T\{\mathbf{p}\}_i \qquad (4.3\text{-}4)$$

where $[\mathbf{N}]_i$ is the value of $[\mathbf{N}]$ at the location of $\{\mathbf{p}\}_i$.

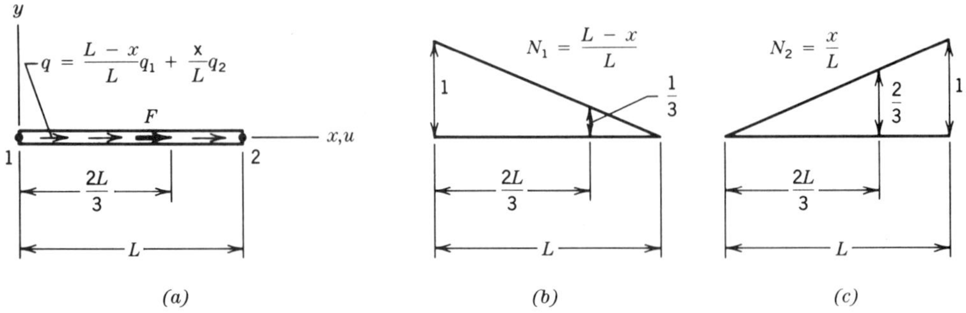

Figure 4.3-2. (*a*) Linearly varying distributed load q and concentrated force F on a bar. (*b,c*) Shape functions for axial displacement u.

Similarly, if n concentrated moments $\{\mathbf{m}\} = \lfloor m_x \quad m_y \quad m_z \rfloor^T$ act at n nonnodal locations on an element,

$$\text{Owing to concentrated moments} \quad \{\mathbf{m}\}_i, \quad \{\mathbf{r}_e\} = \sum_{i=1}^{n} [\mathbf{N}']_i^T \{\mathbf{m}\}_i \tag{4.3-5}$$

where $[\mathbf{N}']$ contains *derivatives* of $[\mathbf{N}]$ and is evaluated at the location of $\{\mathbf{m}\}_i$. Derivatives are needed to calculate rotations $[\mathbf{N}']_i\{\mathbf{d}\}$, through which moments $\{\mathbf{m}\}_i$ act in doing work equal to $\{\mathbf{d}\}^T\{\mathbf{r}_e\}$. An example of loads that result from Eq. 4.3-5 appears in Fig. 4.3-6*c*.

Example. Bar Element. The bar element in Fig. 4.3-2 carries a distributed axial load q, having dimensions of force per unit length, which varies linearly from intensity q_1 at $x = 0$ to intensity q_2 at $x = L$. In addition, a concentrated axial force F acts at $x = 2L/3$. Consistent nodal loads $\{\mathbf{r}_e\}$ are required.

To account for force F we use Eq. 4.3-4. To account for load q we can use the last integral in Eq. 4.1-6, writing the force increment as $q\,dx$ instead of $\{\mathbf{\Phi}\}\,dS$. Thus

$$\underset{2\times 1}{\{\mathbf{r}_e\}} = \int_0^L \lfloor \mathbf{N} \rfloor^T q\,dx + \lfloor \mathbf{N} \rfloor_{2L/3}^T F \tag{4.3-6}$$

from which, with $\lfloor \mathbf{N} \rfloor = \left\lfloor \frac{L-x}{L} \quad \frac{x}{L} \right\rfloor$ and q as given in Fig. 4.3-2*a*,

$$\{\mathbf{r}_e\} = \frac{L}{6}\begin{Bmatrix} 2q_1 + q_2 \\ q_1 + 2q_2 \end{Bmatrix} + \begin{Bmatrix} F/3 \\ 2F/3 \end{Bmatrix} \tag{4.3-7}$$

We see that if $q_1 = q_2 = q$, a constant, then the total load qL is equally distributed to element nodes. The fraction of force F allocated to each node is dictated by Eq. 4.3-4 and is shown graphically as an ordinate of the appropriate shape function in Fig. 4.3-2. (In civil engineering parlance, N_1 and N_2 are *influence lines* for reactions at the nodes.)

The concept of work-equivalent loads is discussed in connection with Eq. 4.3-3. To show that $\{\mathbf{r}_e\}$ of eq. 4.3-7 is indeed a set of work-equivalent loads, we must show that

$$\lfloor u_1 \quad u_2 \rfloor \{\mathbf{r}_e\} = \int_0^L uq\,dx + Fu_{2L/3} \tag{4.3-8}$$

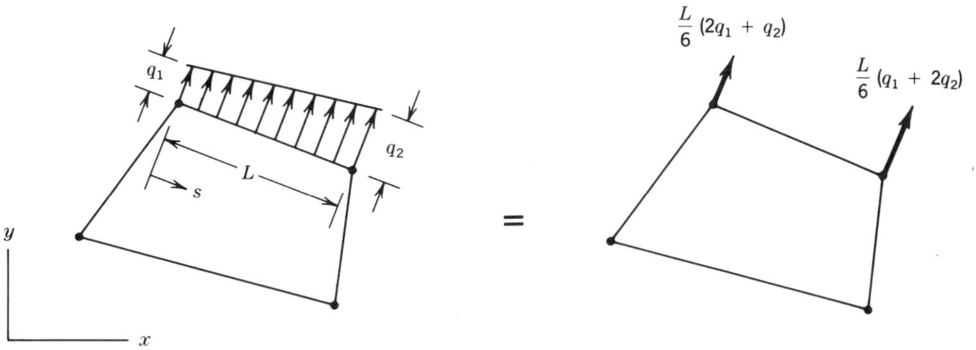

Figure 4.3-3. Linearly varying load on a linear edge, and the consistent nodal loads produced.

where

$$u = \frac{L - x}{L} u_1 + \frac{x}{L} u_2 \qquad \text{and} \qquad q = \frac{L - x}{L} q_1 + \frac{x}{L} q_2 \tag{4.3-9}$$

Substitution of Eqs. 4.3-7 and 4.3-9 into Eq. 4.3-8 shows that Eq. 4.3-8 is indeed satisfied.

Plane Elements. Figure 4.3-3 shows a distributed load of intensity q that acts normal to a linear edge of a plane element. The inclination of the edge with respect to global coordinates xy does not matter. So long as edge-normal displacement varies linearly with edge-tangent coordinate s, loads are allocated to nodes as shown. This is of course the same allocation as seen in Eq. 4.3-7. An edge-*parallel* traction that varies linearly with s would be allocated to nodes in the same proportions.

Figure 4.3-4 shows a uniformly distributed load of intensity q that acts normal to a *quadratic* edge. That is, the edge-normal displacement v is $v = \lfloor \mathbf{N} \rfloor \lfloor v_4 \quad v_7 \quad v_3 \rfloor^T$, where

$$\lfloor \mathbf{N} \rfloor = \left\lfloor \frac{2x^2}{L^2} - \frac{x}{L} \qquad 1 - \frac{4x^2}{L^2} \qquad \frac{2x^2}{L^2} + \frac{x}{L} \right\rfloor \tag{4.3-10}$$

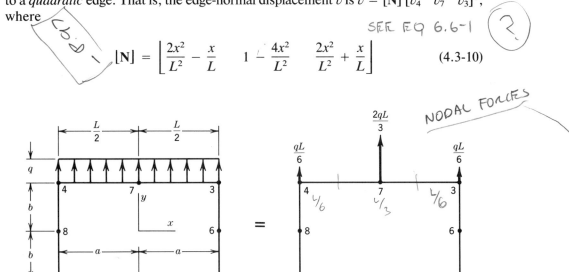

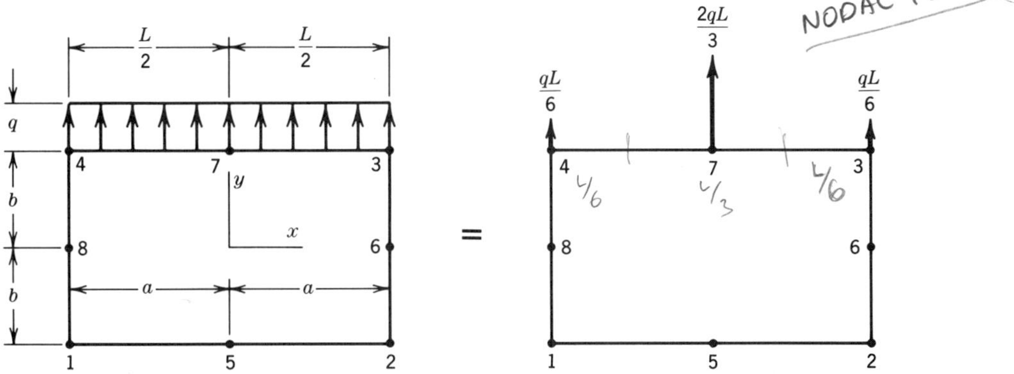

Figure 4.3-4. Uniform traction of intensity q on a quadratic edge, and the consistent nodal loads produced.

The edge is straight and node 7 is at the midpoint. Work-equivalent loads at nodes 4, 7, and 3 are

$$\{\mathbf{r}_e\} = \int_{-L/2}^{L/2} \lfloor \mathbf{N} \rfloor^T q \, dx = \begin{Bmatrix} qL/6 \\ 2qL/3 \\ qL/6 \end{Bmatrix} \tag{4.3-11}$$

We see that a uniform traction does *not* produce the same force at each node on a quadratic edge.

Nodal loads produced by a concentrated force *within* a plane element are evaluated by means of Eq. 4.3-4. Shape functions for the rectangular element of Fig. 4.3-4 are given by Eqs. 6.6-1 if one substitutes $\xi = x/a$ and $\eta = y/b$. If a concentrated force F acts toward (say) the right at the center of this element (where $\xi = \eta = 0$ in Eqs. 6.6-1), Eq. 4.3-4 dictates that rightward forces $F/2$ appear at nodes 5, 6, 7, and 8 and that *leftward* forces $F/4$ appear at nodes 1, 2, 3, and 4. The nodal forces sum to F, as they must, but the appearance of nodal forces whose sense is opposite to that of F is unexpected from the standpoint of "common sense." Note that in a nonstructural problem, the analogue of a concentrated force F is a point source or a sink.

If a uniform body force acts on a plane four-node rectangular element, one-quarter of the total force appears at each node. If the body force is again uniform and the element again plane and rectangular, but now with eight nodes as in Fig. 4.3-4, the total force is allocated to nodes in the proportions shown on the upper face of the element in Fig. 4.3-5b.

Solid Elements. Again we use Eq. 4.1-6. If there is a body force $\{\mathbf{F}\}$, integration spans the entire element volume and all shape functions of the element are involved. If there is a traction $\{\mathbf{\Phi}\}$ on one face, integration spans only the loaded face, and only the shape functions associated with nodes on that face are involved.

Imagine, for example, that a uniform normal stress σ_c acts on a rectangular face that has corner and midside nodes (Fig. 4.3-5). Shape functions for this face may be taken from Eqs. 6.6-1 with $\xi = x/a$ and $\eta = y/b$. The last integral in Eq. 4.1-6 yields the nodal loads shown in Fig. 4.3-5. The total force is $4P + 4Q = \sigma_c A$, as required, but loads at corner nodes have a direction *opposite* to that of σ_c.

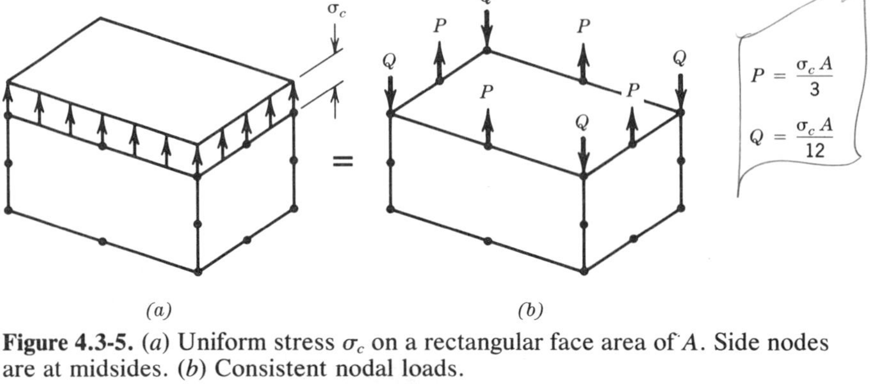

Figure 4.3-5. (*a*) Uniform stress σ_c on a rectangular face area of A. Side nodes are at midsides. (*b*) Consistent nodal loads.

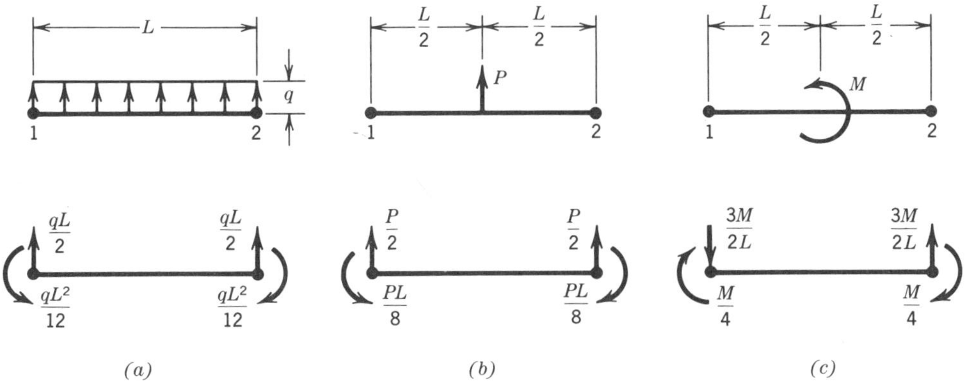

Figure 4.3-6. Consistent nodal loads associated with loads q (constant), P, and M on a standard four-d.o.f. beam element (Eq. 4.2-3).

Beam Elements. Figure 4.3-6 shows nodal loads produced by typical loading patterns on a beam element. Nodal loads are calculated by use of Eqs. 4.1-6, 4.3-4, and 4.3-5. Specifically,

$$\underset{4\times 1}{\{\mathbf{r}_e\}} = \int_0^L \lfloor \mathbf{N} \rfloor^T q\, dx + \lfloor \mathbf{N} \rfloor_{L/2}^T P + \left\lfloor \frac{d\mathbf{N}}{dx} \right\rfloor_{L/2}^T M \tag{4.3-12}$$

Loads $\{\mathbf{r}_e\}$ are of course work-equivalent to the original loads q, P, or M, in the sense defined in connection with Eq. 4.3-3. They are also *statically* equivalent; that is, loads $\{\mathbf{r}_e\}$ produce the same resultant force and the same resultant moment about an arbitrary point as do the original loads, as the reader can easily show.

Error Produced by Lumping. The following example shows the merit of using consistent nodal loads rather than a lumping. Consider the uniformly loaded cantilever beam of Fig. 4.3-7*a*. Consistent nodal loads for a one-element model are shown in Fig. 4.3-7*b*. Taking $[\mathbf{k}]$ from Eq. 4.2-5 and fixing the left end of the beam, we arrive at the following set of equations to be solved for w_2 and θ_2,

$$\frac{EI}{L_T^3}\begin{bmatrix} 12 & -6L_T \\ -6L_T & 4L_T^2 \end{bmatrix}\begin{Bmatrix} w_2 \\ \theta_2 \end{Bmatrix} = \begin{Bmatrix} qL_T/2 \\ -qL_T^2/12 \end{Bmatrix} \tag{4.3-13}$$

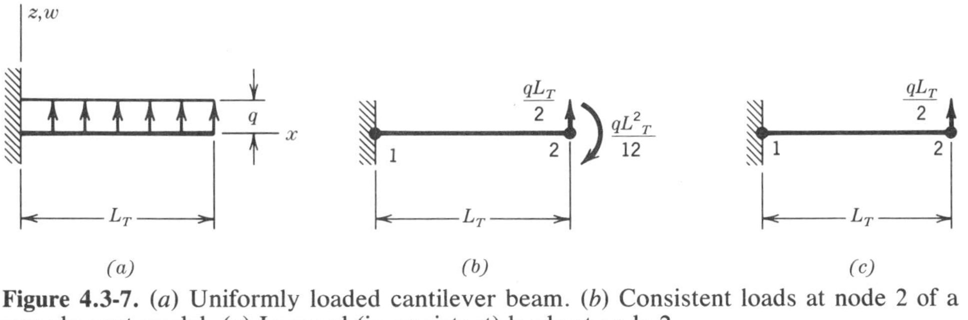

Figure 4.3-7. (*a*) Uniformly loaded cantilever beam. (*b*) Consistent loads at node 2 of a one-element model. (*c*) Lumped (inconsistent) loads at node 2.

from which

$$w_2 = \frac{qL_T^4}{8EI} \quad \text{and} \quad \theta_2 = \frac{qL_T^3}{6EI} \tag{4.3-14}$$

These are the exact values of w_2 and θ_2. (Values of w for $0 < x < L_T$ are not exact. The approximating field $w = \Sigma\, N_i d_i$ is cubic in x, but the exact field for a uniformly distributed load is quartic in x.)

If the beam is divided into two or more elements of equal length L, moment loads cancel at all interior nodes. Accordingly, lumped loading differs from consistent loading only in that lumped loading omits the clockwise moment $qL^2/12$ at the beam tip, thus causing tip deflection and tip rotation to be overestimated. With n the number of equal-length elements, lumped loading yields the following percentage errors in deflection and rotation at the right end.

	$n = 1$	$n = 2$	$n = 3$	$n = 4$
Deflection error	33.3%	8.3%	3.7%	2.1%
Rotation error	50.0%	12.5%	5.6%	3.1%

Consistent loading produces exact values of end deflection and end rotation for all values of n.

4.4 EQUILIBRIUM AND COMPATIBILITY IN THE SOLUTION

In an exact solution, according to the theory of elasticity, every differential element of a continuum is in static equilibrium, and compatibility prevails everywhere. An approximate finite element solution does not fulfill these requirements in every sense. In the present section we note the extent to which equilibrium and compatibility conditions may be satisfied at nodes, across interelement boundaries, and within individual elements.

1. *Equilibrium of nodal forces and moments is satisfied.* The structural equations $\{\mathbf{R}\} - [\mathbf{K}]\{\mathbf{D}\} = \{\mathbf{0}\}$ are nodal equilibrium equations. Therefore, the solution vector $\{\mathbf{D}\}$ is such that nodal forces and moments have a zero resultant at every node.
2. *Compatibility prevails at nodes.* Loosely speaking, elements connected to one another have the same displacements at the connection point. More precisely, elements are compatible at nodes to the extent of nodal d.o.f. they share. The latter statement allows the modeling of a physical hinge or roller between adjacent nodes that would otherwise be fully connected; one then connects some but not all nodal d.o.f. For example, if adjacent beam elements meet at a node where they share only translational d.o.f., a hinge connection is created.

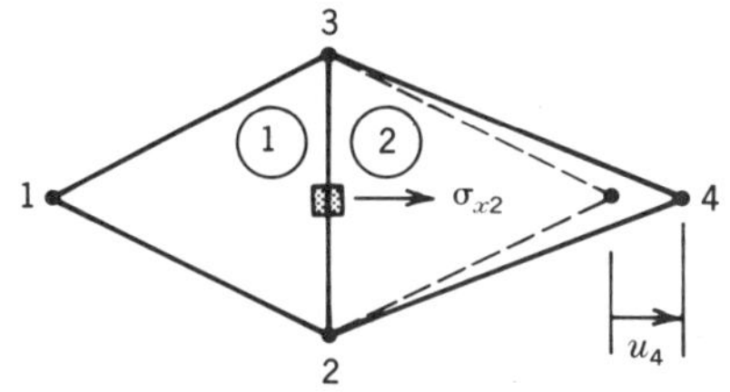

Figure 4.4-1. Differential element (shaded) that spans an interelement boundary.

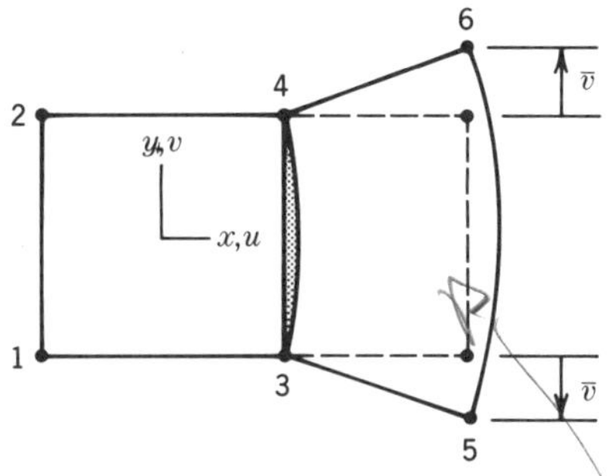

Figure 4.4-2. Adjacent incompatible plane elements. All nodes but 5 and 6 have zero displacement.

3. *Equilibrium is usually not satisfied across interelement boundaries.* Figure 4.4-1 provides a simple example. Imagine that the elements are constant-strain triangles (Eq. 4.2-9) and that node 4 is the only node displaced, as shown. Then σ_{x2} is the only nonzero stress, and the shaded differential element is not in equilibrium. Other types of finite elements behave similarly. (Interelement continuity of stresses may be displayed by elements that include strains among their nodal d.o.f., but such elements are not commonly used.)

   When examining finite element stress output, one should not expect that stresses in adjacent elements will be the same along the common edge, that stresses at nodes will be the same in all elements that share the node, or that stress boundary conditions will be exactly satisfied (e.g., in Fig. 1.1-2*b* we will not compute precisely $\sigma_x = \tau_{xy} = 0$ along the right edge). For a properly constructed mesh these discrepancies will be small and will become smaller with mesh refinement.

4. *Compatibility may or may not be satisfied across interlement boundaries.* Interelement compatibility is satisfied by all elements we have discussed thus far. For example, with both the constant-strain triangle and the plane bilinear element (both discussed in Section 4.2), compatibility is guaranteed because element sides remain straight even after the element is deformed. (That sides remain straight can be shown by noting that edges are initially straight and displacements along a side are linear functions of the coordinates.)

   Other elements, not yet discussed, may be incompatible. Several successful plate elements are incompatible in rotation about an interlement boundary. Figure 4.4-2 is another case in point. Stretching of the right edge causes vertical edges of the right element to bend, and the interelement gap (shaded) appears. (This element, called either *incompatible* or *nonconforming,* is discussed in Section 8.3.)

   Incompatibilities between elements should tend toward zero as more and more elements are used to model a structure. Indeed, this *must* happen if an incompatible element is to be considered reliable enough for general use.

5. *Equilibrium is usually not satisfied within elements.* In the absence of body forces, the differential equations of equilibrium (Eqs. 1.6-2) are exactly satisfied by the constant-strain triangle (Eq. 4.2-9) but not by the bilinear rec-

tangle (Eq. 4.2-12) unless $a_4 = a_8 = 0$. In general, satisfaction of the differential equations of equilibrium at every point in an element demands a relation among element d.o.f. that usually does not result from solution of the global finite element equations $[\mathbf{K}]\{\mathbf{D}\} = \{\mathbf{R}\}$.

If exact results are to be approached as a mesh is refined, then satisfaction of the equilibrium equations must be approached throughout each element. For elements having few d.o.f., this means that a constant-strain condition must be approached within each element. Higher-order elements, such as the quadratic element of Fig. 4.3-4, can satisfy the differential equations of equilibrium while displaying either constant or linear-strain fields. However, an important property of *any* reliable element is that its displacement field be capable of representing all possible states of constant strain.[1]

6. *Compatibility is satisfied within elements.* We require only that the element displacement field be continuous and single-valued. These properties are automatically provided by polynomial fields.

4.5 CONVERGENCE REQUIREMENTS

If a particular problem is repeatedly analyzed, each time using a finer mesh of elements, we generate a sequence of approximate solutions. How can we be assured that the sequence converges to the theoretically exact result? In the following, requirements for convergence are stated in general terms, then interpreted in terms appropriate to structural mechanics and elements based on displacement fields.

General. Let the field variable be $\phi = \phi(x,y,z)$, and let there be a functional $\Pi = \Pi(\phi)$ that yields the governing differential equation of the physical problem from the stationary condition $d\Pi = 0$. Assume that Π contains derivatives of ϕ through order m. If the exact ϕ is to be approached as the mesh is refined, then:

1. Within each element, the assumed field for ϕ must contain a complete polynomial of degree m. (Completeness is discussed in Section 3.6.)
2. Across boundaries between elements, there must be continuity of ϕ and its derivatives through order $m - 1$.
3. Let elements be used in a mesh (rather than tested individually), and let boundary conditions on the mesh be appropriate to a constant value of any of the mth derivatives of ϕ. Then, as the mesh is refined, each element must come to display that constant value.

For example, if $\phi = \phi(x,y)$ and Π contains first derivatives of ϕ, then the lowest-order acceptable field has the form $\phi = a_1 + a_2x + a_3y$ in each element, only ϕ itself need be continuous across interelement boundaries, and each element of an appropriately loaded mesh must display a constant value of $\phi_{,x}$ (or of $\phi_{,y}$ for other appropriate loading), at least as the mesh is refined.

[1]This suggests the following criterion for modeling. In a mesh of low-order elements (such as the bilinear element), the ratio of stress variation across the element to mean stress within the element should be small.

Requirement 1 ensures that ϕ will be continuous within elements and is necessary (but not always sufficient) in order for Requirement 3 to be satisfied.

Requirement 2 is met at all stages of mesh refinement by compatible elements. Incompatible elements must *become* compatible as the mesh is refined ad infinitum.

Requirement 3, which must be met in the limit of mesh refinement, is also met even in a coarse mesh by most commonly used elements.

The order of differentiation m can be determined by examination of either the governing differential equation or its associated functional Π. The correspondence between differential equation and Π is as follows: if derivatives of ϕ of order $2m$ appear in the differential equation, then derivatives of ϕ of order m appear in Π. If $2m$ is odd, so that no variational principle exists, one can yet obtain a finite element formulation by weighted residual methods such as the Galerkin method. Then one regards m as the highest-order derivative of ϕ to be found in integral expressions used to generate the finite element matrices. Usually these expressions result from integrations by parts that reduce m as much as possible.

Satisfaction of Requirements 1, 2, and 3 guarantees convergence to correct results, but says nothing about accuracy in a coarse mesh or the rate of convergence with mesh refinement. However, if the requirements are met at all stages of mesh refinement, and if each refinement is achieved by dividing the elements of the previous mesh into two or more elements, then convergence is monotonic [4.3]. This manner of subdivision means that each new mesh contains the old mesh, especially in the mathematical sense of having the old trial space embedded in the new.

Structural Mechanics. When elements are based on displacement fields, there is often more than one field required (e.g., fields for both u and v are needed in a plane problem). The order of differentiation, m, is determined from the strain energy term in Π_p. Sometimes m has two values for one element. A case in point is a thin flat element that must both stretch and bend. Here $m = 1$ for displacements u and v tangent to the element midsurface and $m = 2$ for displacement w normal to the element midsurface. Interelement compatibility of u, v, w, $w_{,x}$ and $w_{,y}$ is required, at least as the mesh is refined ad infinitum.

Together, Requirements 1 and 3 say that a mesh of elements must, when given appropriate boundary conditions, display rigid-body motion or a state of constant strain. For the flat element mentioned in the preceding paragraph, we must find that strains $\{\boldsymbol{\epsilon}\} = [\mathbf{B}]\{\mathbf{d}\}$ are zero when $\{\mathbf{d}\}$ represents translation along (or small rotation about) any coordinate axis, and that the mesh gives constant values of strains $\epsilon_x = u_{,x}$, $\epsilon_y = v_{,y}$, and $\gamma_{xy} = u_{,y} + v_{,x}$ when appropriate stretching loads are applied to boundaries of the mesh. When bending or twisting loads are applied, a constant-strain state must be observed in a layer parallel to the element midsurface, which means that the mesh must be able to display constant curvatures $w_{,xx}$ and $w_{,yy}$ and the constant twist $w_{,xy}$.

As an example of rigid-body motion, consider the constant-strain triangle, Eq. 4.2-9. The motion $u = a_1$ is a translation in the x direction, in which all points have displacement a_1. Similarly, $v = a_4$ is a translation in the y direction. With $a_6 = -a_3$, the motion $u = a_3 y$ and $v = -a_3 x$ is a small clockwise rigid-body rotation through an angle a_3 about the point $x = y = 0$ (Fig. 4.5-1). One easily checks that $\epsilon_x = \epsilon_y = \gamma_{xy} = 0$ for this rotation.

All elements discussed in Section 4.2 can display rigid-body translation and

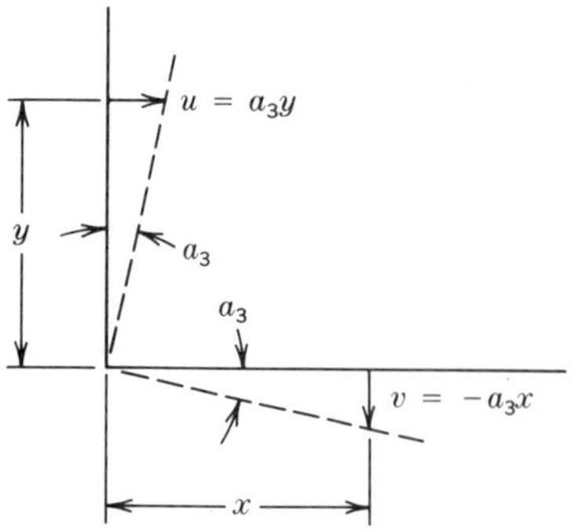

Figure 4.5-1. Rigid-body rotation through a small angle a_3.

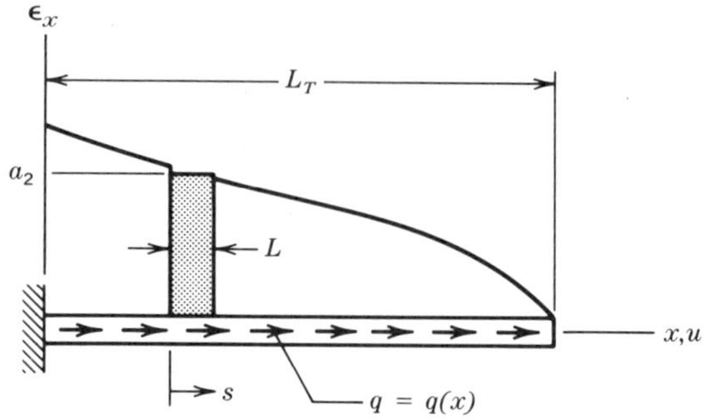

Figure 4.5-2. Axial strain ϵ_x in a bar under axial load q. A typical element has length L.

small rigid-body rotation without strain. An example of an element that *cannot,* at least in a coarse mesh, is an element used to model a shell of revolution—for example, a spherical shell [4.4]. The element displacement field is written in curvilinear coordinates rather than in Cartesian coordinates. When element nodal d.o.f. {**d**} represent a rigid-body translation along the axis of revolution, this shell element yields zero strains only if the arc subtended by the element meridian approaches zero. As the mesh is refined, the arc subtended by each element decreases, and convergence to correct results is obtained.

A physical explanation of Requirement 3 can be given with reference to Fig. 4.5-2. The structure is modeled by standard bar elements whose displacement field is of the form $u = a_1 + a_2 s$ and whose strain is therefore $\epsilon_x = a_2$. If $L_T >> L$, the actual strain distribution over length L departs very little from the constant value $\epsilon_x = a_2$. Clearly, as L shrinks, the actual curve can be matched arbitrarily closely in stairstep fashion. One could *not* converge to an arbitrarily close match by using an element based on the field $u = a_1 + a_2 s^2$. Here ϵ_x is not constant; it is $\epsilon_x = 2a_2 s$, for which $\epsilon_x = 0$ at the left end of every element. Such a defect is not correctible by mesh refinement.

Requirement 3 is stated in terms of a mesh of elements rather than in terms of a single element for the following reason. It is possible for an element to be based on a polynomial field that contains constant-strain terms, yet fail to display constant strain when a mesh of arbitrarily shaped elements is appropriately supported and loaded. None of the elements discussed in Section 4.2 exhibits this difficulty.

Most elements meet Requirement 3 when used as a coarse mesh, but some elements require mesh refinement. This matter is discussed in Section 4.6, in connection with the "weak" patch test.

Geometric Isotropy. Computed results for a given structure should not depend on how the mesh is oriented in global coordinates. For example, the computed displacement of load P in Fig. 4.5-3 should be the same whether the element has orientation (*a*) or orientation (*b*). Elements that are well behaved in this regard are called *geometrically isotropic* (also called *geometrically invariant* and *spatially isotropic*). Although geometric invariance is not required for convergence with mesh refinement, it is very desirable that elements have no "preferred directions" so that the user will not encounter results that seem puzzling or even alarming.

If a plane element is to be geometrically isotropic, it is necessary that its displacement expansions for u and v have the same form and include terms sym-

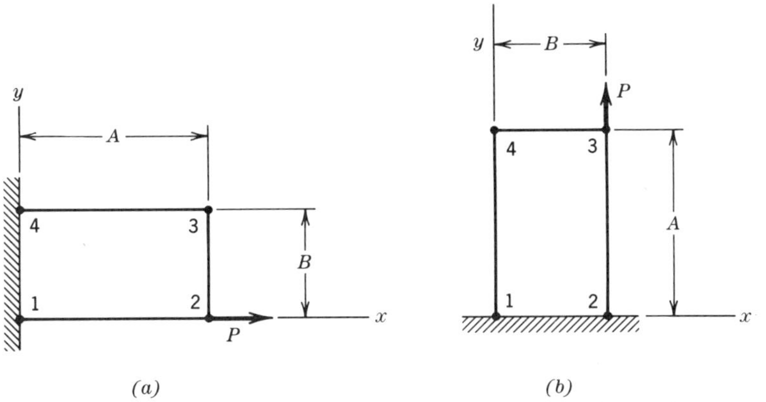

Figure 4.5-3. A rectangular plane element, loaded by force P at one corner.

metric about the axis of a Pascal triangle (Fig. 3.6-1). The constant-strain triangle (Eq. 4.2-9) and the bilinear element (Eq. 4.2-12) both satisfy these requirements. In particular, to produce a bilinear element one should not supplement the complete linear fields of Eq. 4.2-9 with (say) $u = a_4x^2$ and $v = a_8x^2$, as this would destroy geometric isotropy. One selects instead the modes $u = a_4xy$ and $v = a_8xy$, which favor neither x nor y. If used in the problem of Fig. 4.5-3, the bilinear element yields the same displacement of load P for orientation (*a*) and for orientation (*b*). (We have not yet explained how to deal with *arbitrary* orientations of this element; see Chapter 6.)

Similar remarks apply to solid elements. The trilinear element, Eq. 4.2-16, uses all linear terms, a balanced selection of quadratic terms, and a single cubic term that favors none of the coordinate directions over another.

4.6 THE PATCH TEST

The patch test was originated by Irons [4.5,4.6]. It is a simple test that can be performed numerically, so as to check the validity of an element formulation and its programmed implementation. We assume that the element is stable in the sense described below. Then, if the element passes the patch test, we have assurance that all convergence criteria noted in Section 4.5 are met. Therefore, when this type of element is used to model any other structure, mesh refinement will produce a sequence of approximate solutions that converges to the exact solution. In other words, the patch test serves as a necessary and sufficient condition for correct convergence of a finite element formulation. The test can be described in a general way, but we will describe it in terms appropriate to structural mechanics.

Procedure. One assembles a small number of elements into a "patch," taking care to place at least one node within the patch, so that the node is shared by two or more elements, and so that one or more interelement boundaries exist. Figure 4.6-1 shows an acceptable two-dimensional patch, built of four-node elements of a type we have not yet discussed. Boundary nodes of the patch are loaded by consistently derived nodal loads appropriate to a state of constant stress. Internal nodes are neither loaded nor restrained. The patch is provided with just

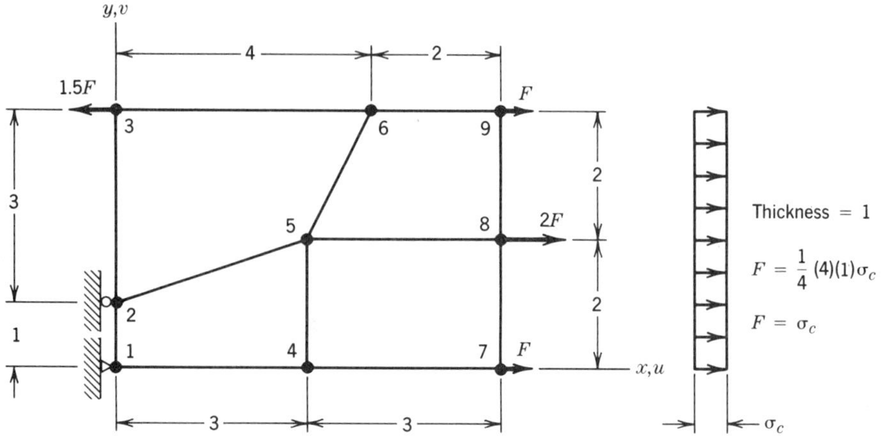

Figure 4.6-1. Patch of four-node elements, loaded by forces F consistent with the uniform stress state $\sigma_x = \sigma_c$, $\sigma_y = \tau_{xy} = 0$.

enough supports to prevent rigid-body motion. One next executes a standard solution and examines the computed stresses. If, *throughout the element,* computed stresses agree with exact stresses for the physical problem modeled, then the patch test is passed. By "agree" we mean *exact* agreement, allowing only for numerical noise associated with the finite length of computer words.

The patch test must be repeated for all other constant-stress states demanded of the element being tested. In Fig. 4.6-1, a test for constant σ_x is depicted; we must test the patch again for constant σ_y, and again for constant τ_{xy} (each time using appropriate nodal loads). Solid elements must be patch-tested for constant states of σ_x, σ_y, σ_z, τ_{xy}, τ_{yz}, and τ_{zx}. A patch of plate-bending elements must display constant bending moments M_x and M_y and constant twisting moment M_{xy}.

It is only for convenience that we examine stress states rather than strain states: typically, a computer program outputs stresses rather than strains. Computed nodal d.o.f. can also be examined; if they are incorrect, strains and stresses will also be incorrect. If displacements are correct but stresses are incorrect, one suspects that the stress calculation subroutine may be in error.

Support conditions must not prevent the constant state from occurring. In Fig. 4.6-1, complete fixity of nodes 1, 2, and 3 would be fatal to the patch test, as Poisson contraction in the y direction would be prevented along the left edge. It would be acceptable to impose the boundary condition $u_3 = 0$ rather than a load at node 3—that is, to replace the force $1.5F$ by a roller support.

Stability. At the outset we assumed that the element to be patch-tested is stable. A stable element is one that admits no zero-energy deformation states when adequately supported against rigid-body motion. This matter is discussed in detail in Section 6.12. Unstable elements should be used with caution. They can produce an unstable mesh, whose displacements are excessive and quite unrepresentative of the actual structure.

Instabilities can be detected by an eigenvalue test (see Section 18.8). They can also be detected by a "perturbed" patch test, as follows [4.6]. Let the patch be just adequately supported, and add a small amount to one of the consistently derived nodal loads (e.g., apply an additional load equal to an existing load times the square root of the round-off limit, 10^{-6} for round-offs of order 10^{-12}). If computed displacements change a large amount as a result, the patch is unstable. If this test is applied to a single element, one can detect if the element is unstable. (An unstable element does not necessarily produce an unstable mesh.)

"Weak" Patch Test. An element that fails to display constant stress in a patch of large elements has not necessarily failed the patch test. If, as the mesh is repeatedly subdivided, elements come to display the expected state of constant stress, then the element is said to have passed a "weak" patch test, and convergence to correct results is assured.

For example, a certain incompatible plane element fails the patch test if of arbitrary quadrilateral shape, but passes if each element of the patch is a parallelogram. If a mesh of arbitrary quadrilaterals is subdivided again and again, the newly created elements approach parallelograms in shape. Thus the element is found to pass a weak patch test. This result might be anticipated by applying the standard patch test to a mesh of parallelograms, Fig. 4.6-2, using nodal loads computed by means of Eqs. 1.6-3 and 4.1-6.

As another example, consider an axially symmetric geometry. One uses elements whose strain fields include the circumferential strain $\epsilon_\theta = u/r$ (Fig. 4.6-3). As a mesh is refined, h/a and θ_e become small for each element. Accordingly, one applies the weak patch test to see whether ϵ_θ becomes constant over each element as h/a and θ_e approach zero.

Remarks. It is possible for an element that fails the patch test to give better answers in a coarse mesh than an equal number of elements that pass. Nevertheless, an element that fails cannot be trusted. An element that fails may provide convergence, but may converge to an incorrect result. This behavior is observed in certain plate-bending problems when the elements used do not permit a state of constant twist.

"Higher-order" patch tests are possible. For example, a plane element whose displacement field includes a complete quadratic expansion for u and v (Fig. 3.6-1) should be able to represent exactly a field of pure bending. "Robustness" can also be checked [4.6]. For example, Poisson's ratio ν should have no effect on pure bending of a plane mesh; by computation, one discovers to what extent an element is insensitive to ν.

Apart from their use in testing elements, patch tests provide good example problems for learning to use an unfamiliar computer program.

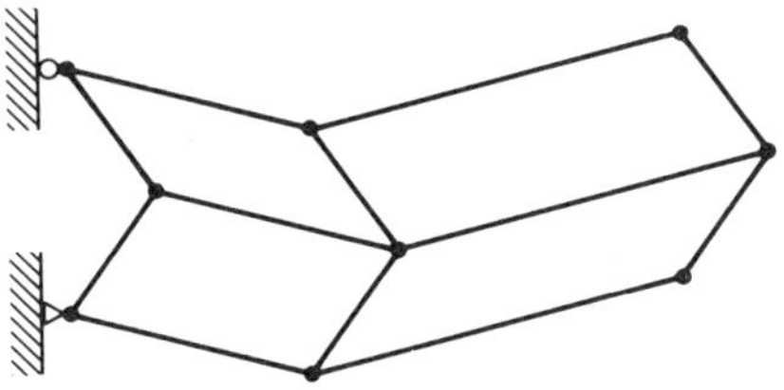

Figure 4.6-2. Patch of parallelogram elements, ready for application of patch-test loadings.

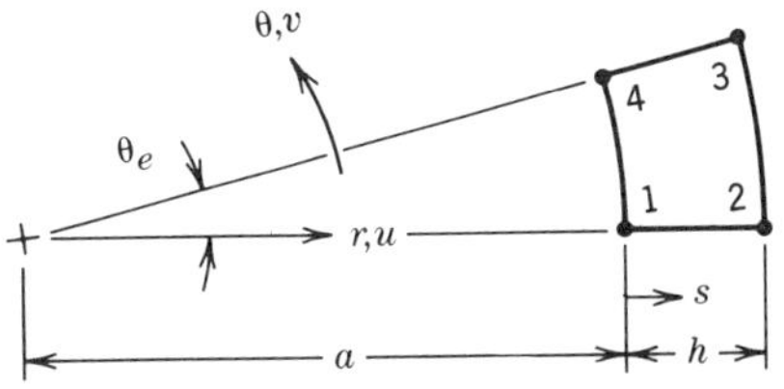

Figure 4.6-3. A plane element in polar coordinates. Displacements u and v are respectively radial and circumferential.

4.7 STRESS CALCULATION

Stress $\{\boldsymbol{\sigma}\}$ in an element can be calculated when its nodal d.o.f. $\{\mathbf{d}\}$ are known. These d.o.f. are available after the structural equations $[\mathbf{K}]\{\mathbf{D}\} = \{\mathbf{R}\}$ have been solved. Equation 1.7-8, repeated here, is

$$\{\boldsymbol{\sigma}\} = [\mathbf{E}](\{\boldsymbol{\epsilon}\} - \{\boldsymbol{\epsilon}_0\}) + \{\boldsymbol{\sigma}_0\} \tag{4.7-1}$$

in which mechanical strains $\{\boldsymbol{\epsilon}\} = [\mathbf{B}]\{\mathbf{d}\}$ are produced by displacements of the nodes. Typically, $\{\boldsymbol{\sigma}_0\}$ is omitted and $\{\boldsymbol{\epsilon}_0\}$ is used to account for thermal strains. Thus, in a plane problem with isotropic material, with T the temperature relative to a stress-free temperature and α the coefficient of thermal expansion,

$$\begin{Bmatrix} \sigma_x \\ \sigma_y \\ \tau_{xy} \end{Bmatrix} = \frac{E}{1 - \nu^2} \begin{bmatrix} 1 & \nu & 0 \\ \nu & 1 & 0 \\ 0 & 0 & \dfrac{1 - \nu}{2} \end{bmatrix} \left([\mathbf{B}]\{\mathbf{d}\} - \begin{Bmatrix} \alpha T \\ \alpha T \\ 0 \end{Bmatrix} \right) \tag{4.7-2}$$

Matrix $[\mathbf{B}]$ is a function of the coordinates and must be evaluated at the location in the element where stresses are desired.

The calculation $\{\boldsymbol{\epsilon}\} = [\mathbf{B}]\{\mathbf{d}\}$ involves differentiation of the displacement field. Accordingly, one expects that stresses will be less accurate than displacements. In low-order elements stresses are often most accurate at the element centroid, less accurate at midsides, and least accurate at corners. Elements of higher order usually display multiple points of optimal accuracy for stresses. The locations of these points depend on the element geometry and the displacement field, and can often be predicted before doing numerical calculations. Stresses at other locations are usually best found by extrapolation from the optimal points (Section 6.13).

The most commonly used elements display only the minimum acceptable degree of interelement continuity. C^0 elements are popular for determining displacements in plane and solid elasticity, and C^1 elements are popular for determining displacements and slopes in plate bending. Stresses depend on displacement derivatives in C^0 elements (or on curvatures in commonly used C^1 elements). Therefore, unless a state of constant stress prevails, stresses are discontinuous across boundaries between C^0 elements, and C^0 elements that share a node do not all display the same state of stress at that node. A substantial difference in stress between adjacent elements suggests a need for mesh refinement.

The average stress at a node is more to be trusted than the nodal stress in any one element attached to the node. A weighted nodal average—for example, with each element contribution weighted in proportion to its interior corner angle at the shared node—may be more reliable than a simple nodal average. However, averaging of stresses on either side of a physical discontinuity such as a step change in thickness should be avoided. Nodal average stresses may be used by a postprocessor in graphic display of results. A caution: a smooth stress field produced by a postprocessor may hide large stress differences in adjacent elements that indicate a need for mesh refinement.

Imagine that the average strain along a line joining any two nodes is desired. The following trick is useful. Connect a straight bar element to the two nodes, but make its cross-sectional area very small, so that its stiffness is negligible in comparison with the stiffness of the existing structure. The computer program will calculate the bar stress, which the user can divide by E to obtain the strain desired. Effectively, we have attached a strain gage to the structure.

Body Forces and Surface Tractions. In the absence of $\{\boldsymbol{\epsilon}_0\}$ and $\{\boldsymbol{\sigma}_0\}$, stress is due entirely to mechanical loads. Imagine that the mechanical load is entirely body force, as in the axially loaded bar of Fig. 4.7-1. If a single bar element is used, both of its nodes are fixed, and we compute $\{\mathbf{d}\} = \{\mathbf{0}\}$ and $\{\boldsymbol{\sigma}\} = \{\mathbf{0}\}$. Two-element and four-element models yield increasingly better results, as expected. The *exact* stress variation can be obtained from any of these models if the effect of body force within each separate element is taken into account. For example, with a one-element model of weight W, we add the stress

$$\sigma_y = \frac{W}{A}\left(\frac{y}{L_T} - \frac{1}{2}\right) \tag{4.7-3}$$

to the stresses $\{\boldsymbol{\sigma}\} = [\mathbf{E}][\mathbf{B}]\{\mathbf{d}\}$, which are zero for a one-element model. A similar formula can be written for σ_y in an arbitrary element of a multi-element model.

Analogous refinement can be made when calculating bending moments in a beam element. The bending moment $M = EI\lfloor\mathbf{B}\rfloor\{\mathbf{d}\}$ can be augmented by the bending moment produced by lateral loads on an element whose ends are completely fixed. This refinement may be worthwhile because beams are frequently analyzed and their loadings are comparatively easy to categorize and incorporate in a computer program.

For other elements, such adjustments are ignored in practice. The adjustments are less important for plane and solid elements than for beam elements. Moreover, they are not easily formulated for an element of general shape.

Thermal Stress. As a usual rule, for analysis of thermal stress one should construct a finite element model that permits the strain field to have about the same level of complexity as seen in the temperature field. Thus, one avoids a mismatch between $\{\boldsymbol{\epsilon}\}$ and $\{\boldsymbol{\epsilon}_0\}$ in Eq. 4.7-1. A mismatch can produce an unreliable stress prediction, especially among simpler elements, as the following example shows.

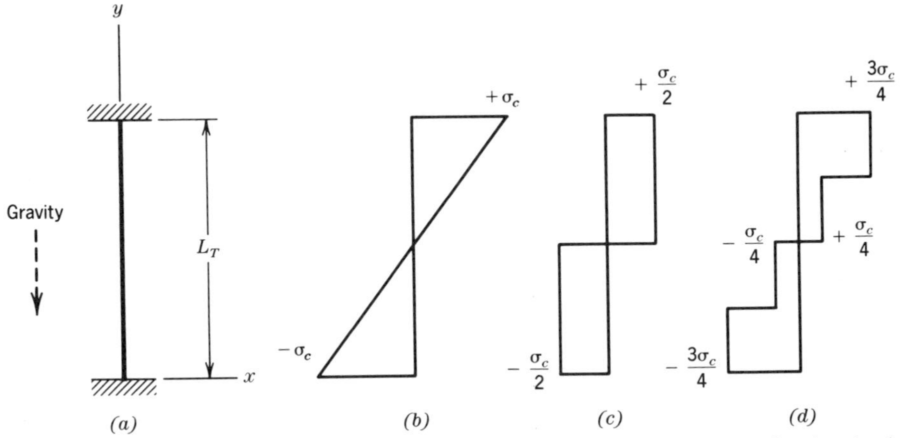

Figure 4.7-1. (*a*) Uniform bar of weight W and cross-sectional area A, fixed at both ends. (*b*) Axial stress variation, where $\sigma_c = W/2A$. (*c,d*) Stresses predicted by two-element and four-element models.

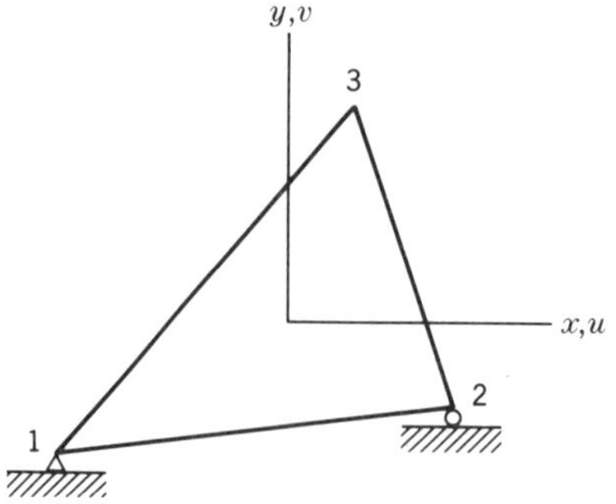

Figure 4.7-2. A triangular element.

Let the triangular element of Fig. 4.7-2 have the linear temperature field

$$T = c_1 x + c_2 y \tag{4.7-4}$$

where c_1 and c_2 are constants. If the centroid of the triangle is at $x = y = 0$, and if the element is an isotropic constant-strain triangle, then Eq. 4.1-6 shows that nodal loads $\{\mathbf{r}_e\}$ produced by the temperature field are all zero. Therefore, $\{\mathbf{d}\} = \{\mathbf{0}\}$, and, by Eq. 4.7-2,

$$\begin{Bmatrix} \sigma_x \\ \sigma_y \\ \tau_{xy} \end{Bmatrix} = -\frac{E\alpha}{1-\nu}(c_1 x + c_2 y)\begin{Bmatrix} 1 \\ 1 \\ 0 \end{Bmatrix} \tag{4.7-5}$$

These stresses can be large. However, they are spurious: theory of elasticity shows that a linear temperature field produces deformation of the triangle but *zero stresses*. Equation 4.7-5 would indeed yield zero stresses if evaluated at the centroid of the triangle, $x = y = 0$. This amounts to using a constant temperature field in the constant-strain triangle when calculating stresses. Generalizing, we infer that, for the purpose of stress analysis, one should reduce (if necessary) the order of the element temperature field to the same order as the element strain field. Thus, if Fig. 4.7-2 is regarded as a *quadratic* triangle (three corner nodes and three midside nodes), the element has a linear strain field. The temperature field of Eq. 4.7-4 is also linear, so the element correctly yields $\sigma_x = \sigma_y = \tau_{xy} = 0$.

As a counterexample [4.7, 4.8], imagine that a uniform beam is fixed between rigid walls (e.g., for a one-element model, set $u_1 = u_2 = u_3 = u_4 = 0$ in Fig. 4.2-4). Let $\nu = 0$ and let the temperature vary linearly with distance y from the midsurface of the beam, $T = T_0 y$. Use of the actual temperature field $T = T_0 y$ yields $\sigma_x = \sigma_y = -E\alpha T$, which is correct for σ_x but wrong for σ_y. Use of the reduced temperature field ($T = 0$ in this case) yields $\sigma_x = \sigma_y = 0$, which is wrong for σ_x but correct for σ_y.

Accordingly, we cannot say whether it is always better to accept a complicated temperature field or to smooth it in element by element fashion, as neither strategy will be best in all cases [4.7, 4.8].

Iterative Improvement. Computed stresses can be iteratively improved as follows [4.9, 4.10]. First, solve the problem in standard fashion, and apply Eq. 4.7-1 to obtain stresses in each element. Next, at each node, use stresses from the surrounding elements to produce nodal average stresses. In a typical element, interpolate to find stresses $\{\boldsymbol{\sigma}\}$ within the element,

$$\{\boldsymbol{\sigma}\} = [\mathbf{N}]\{\overline{\boldsymbol{\sigma}}\} \tag{4.7-6}$$

where $[\mathbf{N}]$ is a shape function matrix and $\{\overline{\boldsymbol{\sigma}}\}$ contains nodal average stresses for the element at hand. Terms may be added to the right-hand side of Eq. 4.7-6 in order to involve stresses within the element as well [4.10]. Known stress boundary conditions can be accounted for in Eq. 4.7-6. We recall from Eq. 4.1-6 that the integral of $[\mathbf{B}]^T\{\boldsymbol{\sigma}_0\}$ is a vector of nodal loads $\{\mathbf{r}_e\}$ that is statically equivalent to the stress distribution $\{\boldsymbol{\sigma}_0\}$. If, instead of $\{\boldsymbol{\sigma}_0\}$, we use the *total* stress $\{\boldsymbol{\sigma}\}$ from Eq. 4.7-6, the sum of nodal loads $\{\mathbf{r}_e\}$ over all *numel* elements of the structure should be statically equivalent to the entire structure load vector $\{\mathbf{R}\}$. So we write

$$[\mathbf{K}]\{\Delta\mathbf{D}\} = \{\mathbf{R}\} - \sum_{n=1}^{\text{numel}} \left(\int_{V_e} [\mathbf{B}]^T\{\boldsymbol{\sigma}\}\, dV \right)_n \tag{4.7-7}$$

where $\{\boldsymbol{\sigma}\}$ comes from Eq. 4.7-6. If $\{\boldsymbol{\sigma}\}$ is exact, the right-hand side of Eq. 4.7-7 is zero. Otherwise, it is a load imbalance that drives the solution toward a configuration that reduces the imbalance. We compute increments $\{\Delta\mathbf{D}\}$ from Eq. 4.7-7. The new configuration is $\{\mathbf{D}\}_{\text{new}} = \{\mathbf{D}\}_{\text{old}} + \{\Delta\mathbf{D}\}$. New stresses are computed based on $\{\mathbf{D}\}_{\text{new}}$, Eqs. 4.7-6 and 4.7-7 are applied to the new stresses, and another $\{\Delta\mathbf{D}\}$ is determined. The process repeats until convergence. Note that Eq. 4.7-7 does not require repeated construction and reduction of $[\mathbf{K}]$.

Example applications appear in Fig. 4.7-3 and Table 4.7-1. The four-node elements used are described in Section 6.3. Symmetry is exploited by analyzing only one quadrant of the object. Stress at point *A* is obtained by extrapolation of stresses at four integration stations in the adjacent element. The "standard method" in Table 4.7-1 is the usual single-pass analysis.

Stress Concentrations. A stress raiser, such as a hole or a notch, might be analyzed by a brute-force approach—that is, by using a profusion of elements to surround

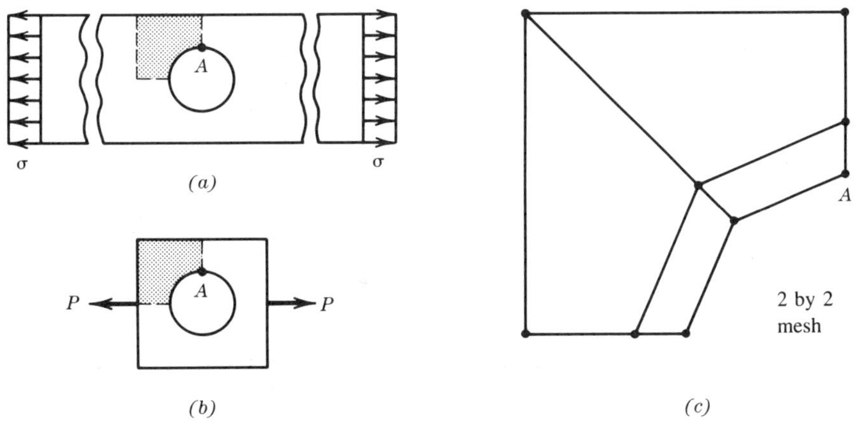

Figure 4.7-3. Flat plate with a central hole of diameter equal to half the plate width. (*a*) Uniform tensile load on portions distant from the hole. (*b*) Point forces on a square plate. (*c*) Typical mesh used on quadrant shaded in parts *a* and *b*.

TABLE 4.7-1. HORIZONTAL STRESS AT POINT *A* IN FIG. 4.7-3 USING THE STANDARD METHOD AND THE METHOD OF EQ. 4.7-7 (SIX ITERATIONS) [4.12].

	2 by 2 Mesh in Quadrant			4 by 4 Mesh in Quadrant		
Loading Case	Standard	Six Iterations	Exact	Standard	Six Iterations	Exact
Fig. 4.7-3*a*	30.9	38.1	43.2	36.6	39.5	43.2
Fig. 4.7-3*b*	73.6	85.4	112.3[a]	95.3	96.9	112.3[a]

[a]From an 8 by 8 mesh of nine-node elements.

the discontinuity. This is both tedious and expensive. An alternative that uses a coarse mesh may be available, as follows [4.13].

Analyze the problem of interest using the coarsest mesh that will model the geometry. Let the computed stress at the point of interest be called σ_a. Next, use a locally identical mesh to analyze the most closely related condition for which results are already available—for example, in a table of stress concentration factors. For this "secondary" case, call the computed stress σ_r and the tabulated stress σ_t. The ratio σ_t/σ_r is regarded as a correction factor that can be applied to closely related geometries. Thus the stress in the case of interest is now estimated to be $(\sigma_t/\sigma_r)\sigma_a$ rather than σ_a.

The success of this method depends on the availability of tabulated cases and on the skill of the analyst in selecting an *appropriate* secondary case.

As an example, consider the determination of the peak stress at *A* in the point-loaded plate of Fig. 4.7-3*b*. As the secondary case we take the tensile strip of Fig. 4.7-3*a*. The tabulated stress concentration factor for the tensile strip is 2.16, which yields the stress at point *A* as 43.2 for the load applied. Thus, from the 2 by 2 and 4 by 4 mesh results in Table 4.7-1, respectively, we estimate that the stress at point *A* in the point-loaded plate is

$$\frac{43.2}{30.9}73.6 = 103 \qquad \text{and} \qquad \frac{43.2}{36.6}95.3 = 112 \tag{4.7-8}$$

Both of these results are better than the results given by six iterations in Table 4.7-1.

If the case of interest and the secondary case are identical, one of course obtains the tabulated stress. This means only that a finite element analysis is unnecessary, except perhaps as used for the entire structure in order to obtain the load applied to the portion of interest or the average stress field in a certain region (e.g., to obtain stress σ in Fig. 4.7-3*a*).

4.8 OTHER FORMULATION METHODS

The present chapter is devoted to elements whose properties are based on assumed displacement fields. This type of element is the most popular. The reader should be aware that several other element types are possible and are in use. If an alternatively derived element has displacement d.o.f., it can be used in combination with displacement-based elements. Indeed, the user of an analysis program may be unaware that some elements in the program are not displacement-based.

Displacement fields and strain energy expressions provide only one of many approaches to formulation of finite elements.

A type of "hybrid" element makes use of an assumed stress distribution within the element and assumed displacements along its edges. The formulation procedure yields an element with displacement d.o.f. A "mixed" element has both displacement d.o.f. and force d.o.f. Its characteristic matrix is not a stiffness matrix; rather, it contains a flexibility submatrix and a submatrix that couples the displacement and force d.o.f. Both hybrid elements and mixed elements can be formulated from appropriate functionals.

For some physical problems, such as certain problems in fluid mechanics, no variational principle exits. Finite element formulations can yet be developed by *weighted residual methods,* chief among which is the Galerkin method.

PROBLEMS

Section 4.1

4.1 If element d.o.f. are given virtual (i.e., small imaginary) displacements $\{\delta \mathbf{d}\}$, strains are changed in the amount $\{\delta \epsilon\} = [\mathbf{B}]\{\delta \mathbf{d}\}$. Loads acting on the structure do work that is stored as the strain energy $\delta U = \int \{\delta \epsilon\}^T\{\sigma\}\, dV$. Complete this virtual work argument to obtain the structure equations $[\mathbf{K}]\{\mathbf{D}\} = \{\mathbf{R}\}$. In the process, identify formulas for $[\mathbf{k}]$ and $\{\mathbf{r}_e\}$.

4.2 Derive the stiffness matrix of a bar, Fig. 4.2-1, by use of the a-basis method, Eqs. 4.1-17 and 4.1-18.

Section 4.2

4.3 The three-node bar element shown is uniform and is allowed only axial displacement u. Its nodal d.o.f. are u_1, u_2, and u_3. Its shape functions are given by Eq. 4.3-10. Determine the element stiffness matrix.

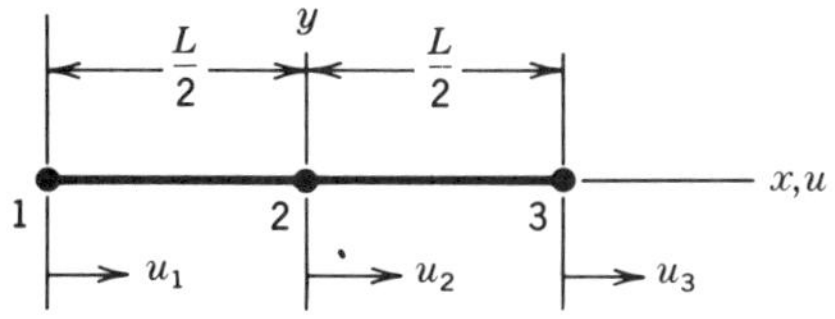

Problem 4.3

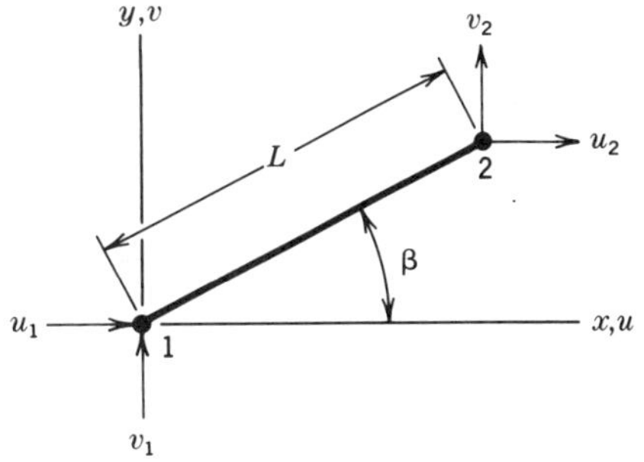

Problem 4.4

4.4 For the plane truss element shown, write an expression for axial displacement as a function of the four nodal d.o.f., angle β, and an axial coordinate along the bar. Use this expression to determine the 4 by 4 element stiffness matrix. Check your result against Eq. 2.4-3.

4.5 Imagine that a beam element has positive directions for nodal d.o.f. and nodal load as shown in the sketch, rather than the directions shown in Fig. 4.2-2. How does this sign convention change $[\mathbf{k}]$ of Eq. 4.2-5? What is awkward about this sign convention?

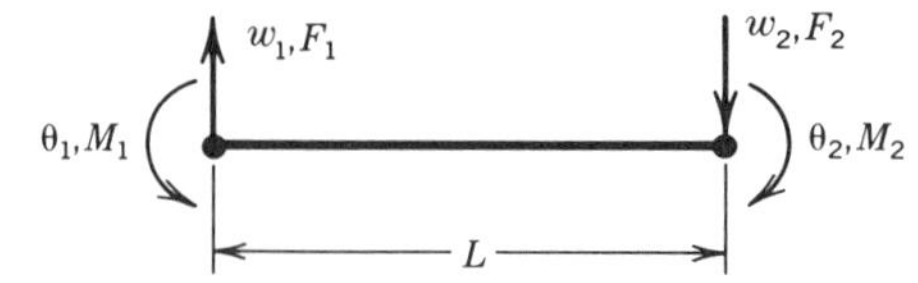

Problem 4.5

4.6 (a) Verify the correctness of the beam element stiffness matrix, Eq. 4.2-5. Use either Eq. 4.2-5 or Eq. 4.2-7.

(b) Verify that [**B**]{**d**} and [**k**]{**d**} are both zero for a small rigid-body rotation of the beam element about its left end.

4.7 Imagine that a pin-jointed plane truss is modeled by plane frame elements. If rotational d.o.f. θ are suppressed at all nodes, is the truss correctly modeled? Explain.

4.8 (a) Imagine that the uniform bar element of Fig. 4.2-1 is to have *two* d.o.f. at each node—namely, axial displacement u and axial strain ϵ_x. Thus {**d**} $= \lfloor u_1 \quad \epsilon_{x1} \quad u_2 \quad \epsilon_{x2} \rfloor^T$. Derive the 4 by 4 element stiffness matrix.

(b) Should this element be used to model a bar that has abrupt changes in cross section, as in Fig. 1.1-1*b*? Explain.

4.9 Consider a plane grillage, which consists of bars that all occupy the xy plane. Let each bar be uniform and parallel to either the x axis or the y axis. Each bar can resist bending, in either the yz or the zx plane, and torsion about its axis. Write the 6 by 6 stiffness matrix of an element that lies parallel to the y axis. Express your answer in terms of E, I_x, G, J, and element length L. Indicate the d.o.f. on which [**k**] operates, arrayed in order consistent with [**k**].

4.10 The cantilever beam shown is modeled by an extremely coarse mesh of constant-strain triangles. The beam is loaded by an end moment. Qualitatively plot the variation of σ_x versus x along the x axis.

4.11 (a) For the particular triangular element shown, evaluate the strain–displacement matrix [**B**] in terms of dimensions a and b. Let d.o.f. in {**d**} have the ordering $\lfloor u_1 \quad v_1 \quad u_2 \quad v_2 \quad u_3 \quad v_3 \rfloor^T$.

(b) Let {**d**} contain displacements consistent with Eq. 4.2-9—that is, $u_1 = a_1$, $u_2 = a_1 + a_2 a$, and so on. Show that [**B**]{**d**} yields the strains of Eq. 4.2-10.

(c) Determine [**k**] in terms of a, b, E, and t if thickness t is constant and Poisson's ratio is zero.

(d) Fix nodes 1 and 2, apply a y-direction force P to node 3, and solve for u_3 and v_3. Also compute stresses in the element.

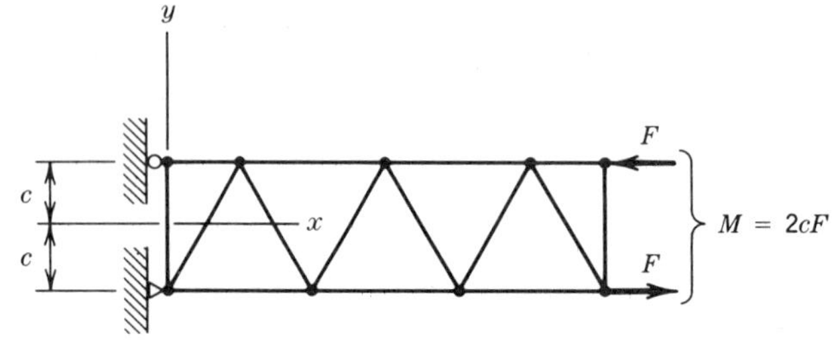

Problem 4.10

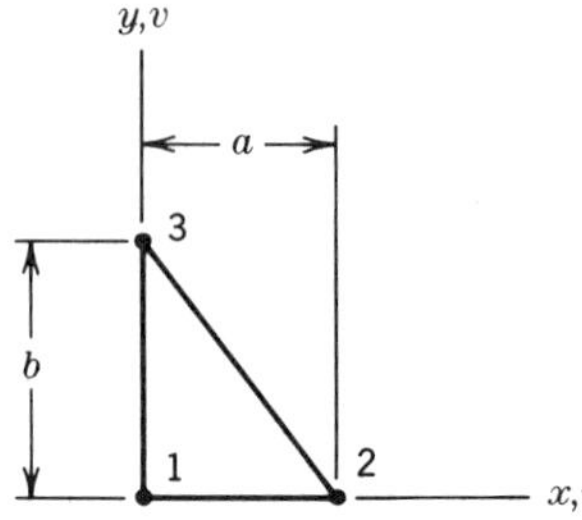

Problem 4.11

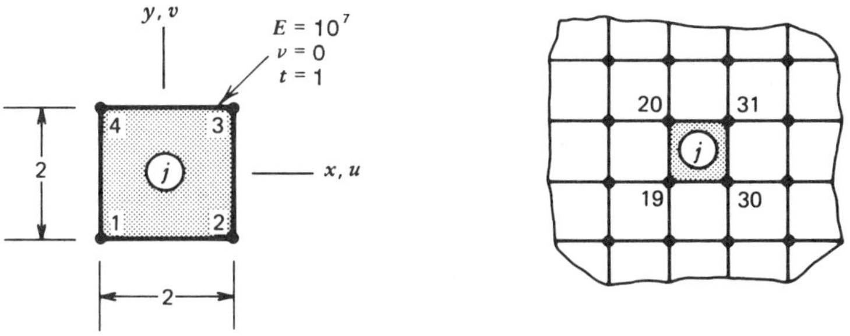

Problem 4.12

4.12 Element j is of the type shown in Fig. 4.2-4. It has stiffness matrix [**k**] and nodal displacement vector $\{\mathbf{d}\} = \lfloor u_1 \quad v_1 \quad u_2 \quad v_2 \; . \; . \; . \; v_4 \rfloor^T$. Element j is to be attached to nodes 19, 20, 30, and 31 of the structure, a fragment of which is shown. The structure stiffness matrix [**K**] is to be stored in full (not banded) format, and no d.o.f. are as yet removed by imposing support conditions. What is the numerical contribution of element j to the single coefficient in [**K**] at the intersection of
(a) row 48 and column 39?
(b) row 37 and column 37?
(c) row 59 and column 61?

4.13 Consider the plane bilinear element of Fig. 4.2-4 and the linear displacement field $u = a_1 + a_2x + a_3y$, $v = a_4 + a_5x + a_6y$. Let nodal d.o.f. {**d**} be consistent with this field; that is, let $u_1 = a_1 - a_2a - a_3b$, and so on.
(a) Show that Eq. 4.2-13 yields the given u and v fields.
(b) Show that Eq. 4.2-14 yields the constant-strain state $\epsilon_x = a_2$, $\epsilon_y = a_6$, and $\gamma_{xy} = a_3 + a_5$.

4.14 The best approximation of a state of pure bending that a plane bilinear element can display is $u = \bar{u}xy/ab$ and $v = 0$, where $\bar{u}$ is the magnitude of a corner displacement.
(a) Sketch the deformed element. Show nodal forces F that produce the deformation state.
(b) If Poisson's ratio is taken as zero, strain energy per unit volume is $(E\epsilon_x^2 + E\epsilon_y^2 + G\gamma_{xy}^2)/2$. Use this information to determine F as a function of a, b, E, G, $\bar{u}$, and element thickness t.
(c) What is the correct value of F, according to elementary beam theory?
(d) In parts (b) and (c), one can define a stiffness measure S as $S = F/\bar{u}$. Give a physical explanation as to why $S_{(b)} > S_{(c)}$.
(e) Show that the ratio $S_{(b)}/S_{(c)}$ approaches unity only as a/b approaches zero. What happens as a/b becomes large?

4.15 For the rectangular solid element of Fig. 4.2-5, write out the first three columns of [**N**] and the first three columns of [**B**].

4.16 The block of material shown is loaded by axial force $P = \sigma_c bt$, which produces axial deflection D. Axial stiffness is $k = P/D$.
(a) Show that k is inversely proportional to L if cross-sectional area $A = bt$ remains constant.
(b) Show that k is independent of b and L if t remains constant and the aspect ratio b/L is not changed.

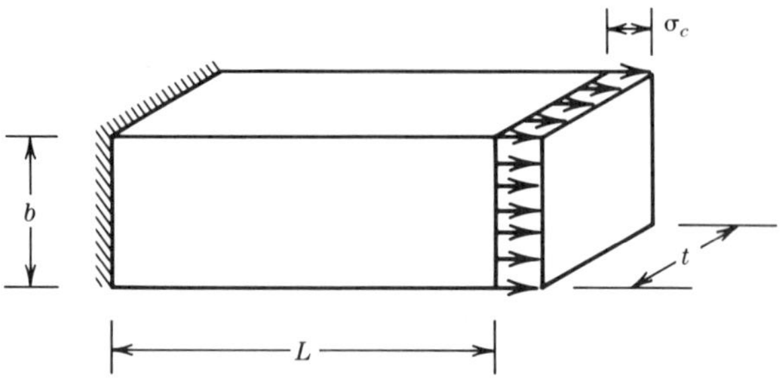

Problem 4.16

(c) Show that k is directly proportional to a linear dimension if the shape of the element is not changed.

(These behaviors are in fact observed in axial, plane, and solid elements, respectively.)

Section 4.3

4.17 Show that integrals containing $\{\epsilon_0\}$ and $\{\sigma_0\}$ in Eq. 4.1-6 yield nodal loads $\{\mathbf{r}_e\}$ that are self-equilibrating—that is, they sum to zero net force on an element. For simplicity, restrict your argument to a C^0 element, such as a plane bilinear element.

4.18 Write a work equation analogous to Eq. 4.3-3, but include $\{\epsilon_0\}$ and $\{\sigma_0\}$ rather than $\{\Phi\}$. Interpret the result: that is, can loads $\{\mathbf{r}_e\}$ still be called work-equivalent?

4.19 (a) A two-node bar element is uniformly heated an amount T. The cross-sectional area of the bar varies linearly from A_1 at $x = 0$ to A_2 at $x = L$. What loads $\{\mathbf{r}_e\}$ are predicted by Eq. 4.1-6? Continue to use the linear shape functions given in Fig. 4.3-2.

(b) Use the methods of elementary mechanics of materials to compute the forces this bar would apply to rigid walls if the bar were placed between the walls and uniformly heated an amount T. Why do these forces differ from loads $\{\mathbf{r}_e\}$ of part (a)?

4.20 A three-node bar element is subjected to a temperature change that varies linearly with x, as shown. What axially directed nodal loads appear? Shape functions are given by Eq. 4.3-10.

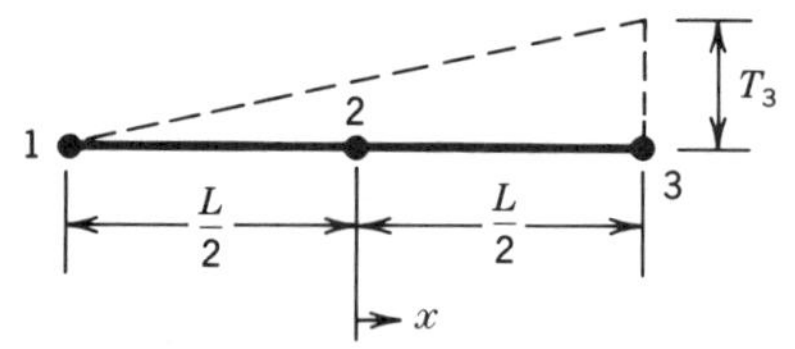

Problem 4.20

4.21 Let temperature T in the plane bilinear element of Fig. 4.2-4 vary as $T = T_0 x$, where T_0 is a constant. Material properties E, ν, and α are constant over the element, as is thickness t. What nodal loads $\{\mathbf{r}_e\}$ are produced by T?

4.22 Imagine that a uniform prestress $\sigma_x = \bar{\sigma}$ exists throughout the trilinear element of Fig. 4.2-5. All other stresses are zero. What nodal loads $\{\mathbf{r}_e\}$ are produced?

4.23 Consider the uniform, three-node bar element of Problem 4.3. Let the element be fixed at the left end and loaded by a uniformly distributed axial load of intensity q. The respective rows of $[\mathbf{k}]$ are $\lfloor 7, -8, 1 \rfloor$, $\lfloor -8, 16, -8 \rfloor$, and $\lfloor 1, -8, 7 \rfloor$, each times $AE/3L$. Calculate u_2 and u_3 if nodal loads are
(a) calculated in consistent fashion.
(b) of magnitude $qL/3$ at each of the three nodes.

4.24 Verify the correctness of the three nodal loads shown in Fig. 4.3-4.

4.25 (a) Compute loads allocated to nodes 4, 7, and 3 in Fig. 4.3-4 by an edge-normal traction that varies linearly with x, from intensity $-\bar{q}$ (compression) at node 4 to intensity $+\bar{q}$ (tension) at node 3.
(b) Combine the result of part (a) with the loads of Eq. 4.3-11 to determine $\{\mathbf{r}_e\}$ for a traction that varies linearly from intensity q_4 at node 4 to intensity q_3 at node 3.

4.26 (a) Force F acts on one edge of the plane bilinear element at $y = b/2$, as shown. What 8 by 1 load vector $\{\mathbf{r}_e\}$ results?
(b) What would be the three nonzero nodal loads if the left edge had three uniformly spaced nodes? (See Eq. 4.3-10.)

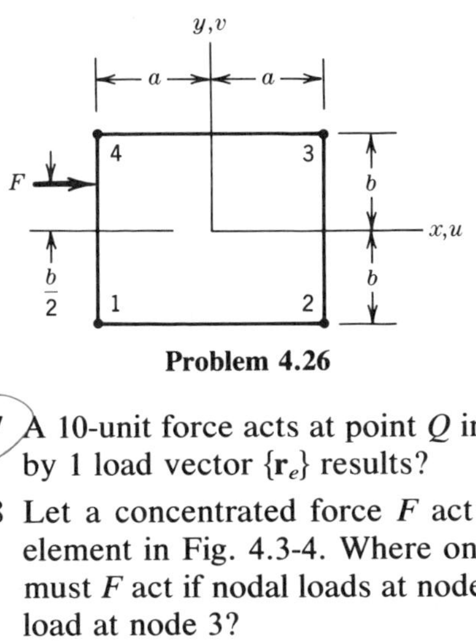

Problem 4.26

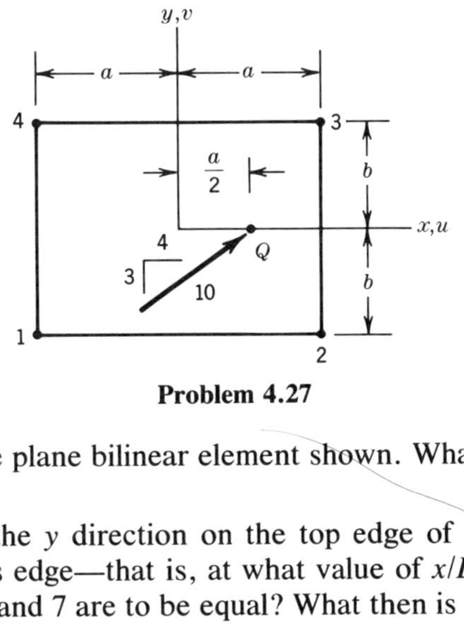

Problem 4.27

4.27 A 10-unit force acts at point Q in the plane bilinear element shown. What 8 by 1 load vector $\{\mathbf{r}_e\}$ results?

4.28 Let a concentrated force F act in the y direction on the top edge of the element in Fig. 4.3-4. Where on this edge—that is, at what value of x/L—must F act if nodal loads at nodes 4 and 7 are to be equal? What then is the load at node 3?

4.29 A uniform body force F_x acts in the positive x direction on a plane rectangular bilinear element (Fig. 4.2-4). Use Eq. 4.1-6 to compute values of the nodal loads in terms of F_x, a, b, and uniform element thickness t.

4.30 Verify the nodal loads shown in Fig. 4.3-5b by integration of $[\mathbf{N}]^T\{\mathbf{\Phi}\}$ over a rectangular face. Shape functions are given in Eqs. 6.6-1. For simplicity you may use $\xi = x$ and $\eta = y$ as coordinates whose origin is at the center of a square face two units on a side.

4.31 Imagine that the temperature of a uniform beam element varies linearly through its depth, from $-T$ along the lower surface to $+T$ along the upper surface. Nodal moments, but not nodal forces, appear in $\{\mathbf{r}_e\}$. Compute these moments by use of mechanics of materials arguments rather than by use of Eq. 4.1-6.

4.32 Use the virtual work concept to compute the moment $M_1 = qL^2/12$ at node 1 in Fig. 4.3-6*a*. That is, compute work done by load q as it moves through displacements created by virtual rotation $\delta\theta_1$ (with other nodal d.o.f. kept at zero), and equate it to work done by M_1 in moving through displacement $\delta\theta_1$.

4.33 A concentrated lateral force F is applied to a standard beam element at a distance x from the left end. For what value of x between 0 and L will $\{\mathbf{r}_e\}$ contain the largest magnitude of moment at node 1? *Suggestion:* Note the interpretation made in Fig. 4.3-2*b*.

4.34 Let a uniformly distributed load of intensity q act over only the left half of a beam element. Compute the 4 by 1 load vector $\{\mathbf{r}_e\}$. Check that loads in $\{\mathbf{r}_e\}$ are statically equivalent to the original load q.

4.35 The uniform cantilever beam shown carries a concentrated lateral force P. Model the beam by a single beam element.

(a) Calculate w_2 (the lateral deflection at the right end). Use the consistent load vector at node 2. Express your answer in terms of P, L, E, I, and x.

(b) Again calculate w_2, but use load lumping: let the lateral force at node 2 be Px/L and ignore the moment load at node 2.

(c) Compute the exact w_2 according to elementary beam theory.

(d) Compute the ratio of the finite element w_2 to the exact w_2. Do this for part (a) and for part (b). Plot these ratios versus x/L.

(e) Compute the bending moment at the left end, first as given by the consistent loads, then as given by lumped loads. Compute the ratio of each moment to the exact value, and plot the two ratios versus x for $0 < x < L$.

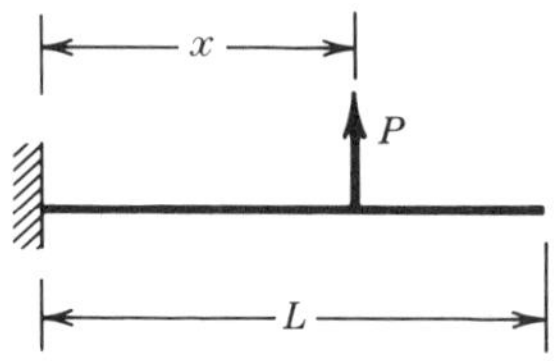

Problem 4.35

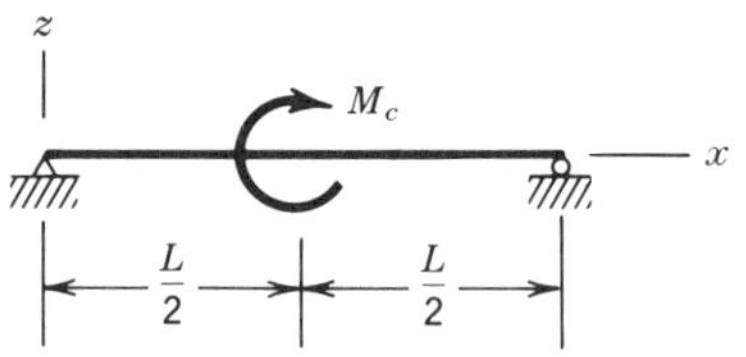

Problem 4.36

4.36 A uniform simply supported beam is loaded by moment M_c at midspan, as shown. If the entire beam is modeled by one element, what rotation at midspan is computed? (The exact answer is $\theta = -ML/12EI$.)

4.37 Verify the percentage errors listed at the end of Section 4.3. *Suggestion:* Use beam theory rather than equations $[\mathbf{K}]\{\mathbf{D}\} = \{\mathbf{R}\}$.

4.38 Consider the bending moment M at the left end of the uniformly loaded cantilever beam element in Fig. 4.3-7*a*. Compute M from the equation $M = EI\lfloor\mathbf{B}\rfloor\{\mathbf{d}\}$, where $\lfloor\mathbf{B}\rfloor$ is given by Eq. 4.2-4 and nonzero d.o.f. are those used in Eq. 4.3-13. What percentage errors in M are given by the consistent loading and by the lumped loading in Fig. 4.3-7?

Section 4.4

4.39 Imagine that the right-hand element in Fig. 4.4-2 is not an incompatible element, but rather the bilinear element of Fig. 4.2-4. Let vertical displacements $\bar{v}$ be imposed as shown at nodes 5 and 6. The remaining d.o.f. are zero.

(a) Show that edges 3–5 and 4–6 remain straight.

(b) Compute the stresses on both sides of interelement boundary 3–4. Express your answers in terms of E, ν, $\bar{v}$, and the coordinates.

4.40 A uniform beam is modeled by bilinear plane elements, as shown. For *each* of the loadings (1), (2), and (3), answer the following questions about stresses displayed by the finite element solution. In parts (c) and (d), plot stresses qualitatively (without numerical calculation).

(a) Is σ_x continuous across the interelement boundary at A? Explain.

(b) Is σ_y zero at B? Explain.

(c) Plot τ_{xy} along the line $y = 0$.

(d) Plot σ_x along the line $y = 0$.

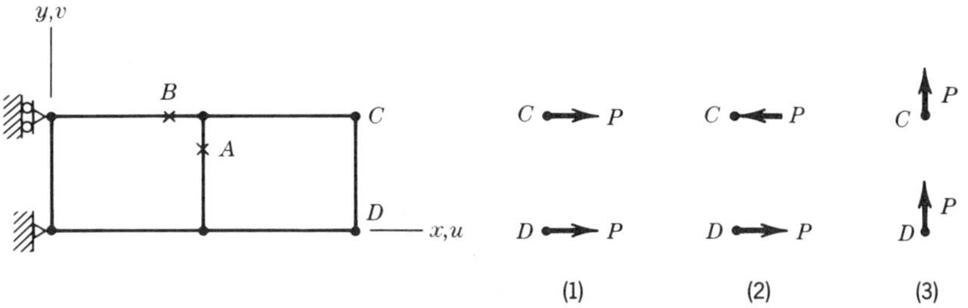

Problem 4.40

4.41 Apply the displacement field of the bilinear element (Eq. 4.2-12) to the trapezoidal element shown. Replace the a_i by nodal d.o.f. u_i, then evaluate u along the line $x = y$. Hence, demonstrate that this element is incompatible. (The *isoparametric formulation,* Chapter 6, can produce a compatible element of this shape.)

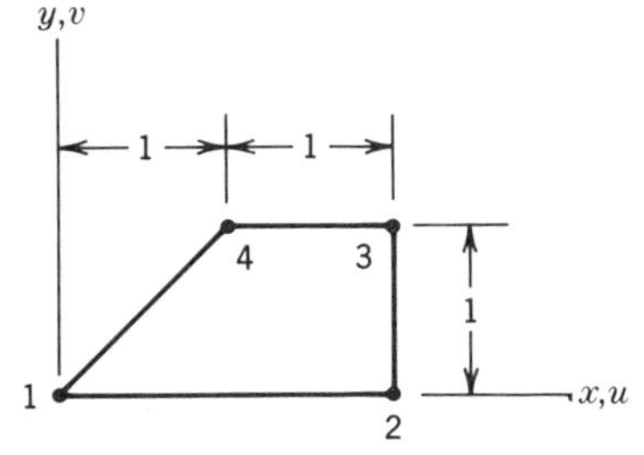

Problem 4.41

4.42 Assume that a plane body is isotropic and that body forces F_x and F_y are constant. In terms of F_x, F_y, E, and ν, what must be the values of a_4 and a_8 in Eq. 4.2-12 if the differential equations of equilibrium are to be satisfied?

Section 4.5

4.43 The bar element shown is uniform and has nodes 1 and 2. Let the assumed axial displacement field have the form $u = \lfloor 1 \quad x^2 \rfloor \lfloor a_1 \quad a_2 \rfloor^T$. First, replace

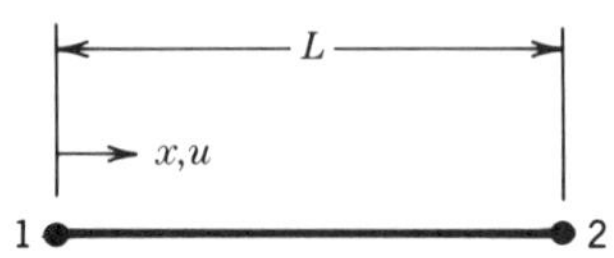

Problem 4.43

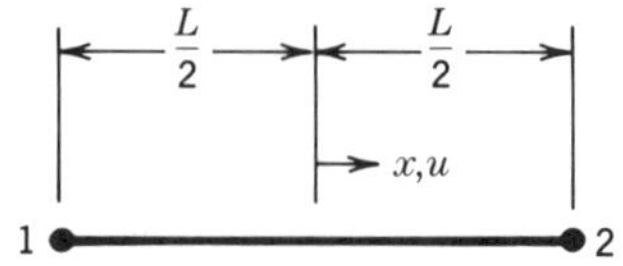

Problem 4.44

d.o.f. a_1 and a_2 by nodal d.o.f. u_1 and u_2. Then determine the strain–displacement matrix $\lfloor \mathbf{B} \rfloor$ and the element stiffness matrix $[\mathbf{k}]$, in terms of A, E, and L. What defects do you see in these results, and what is their source?

4.44 Repeat Problem 4.43 with reference to the element shown (nodes at $x = \pm L/2$) and the assumed axial displacement field $u = \lfloor x \quad x^2 \rfloor \lfloor a_1 \quad a_2 \rfloor^T$.

4.45 (a) Compute the axial deflection at $x = 2L$ and the axial stress at $x = 0$ in the uniform two-element model shown. Use the element stiffness matrix derived in Problem 4.43 and evaluate stress by the calculation $\sigma_x = E \lfloor \mathbf{B} \rfloor \{\mathbf{d}\}$. Are the results correct?

(b) Repeat part (a), but use the element stiffness matrix derived in Problem 4.44.

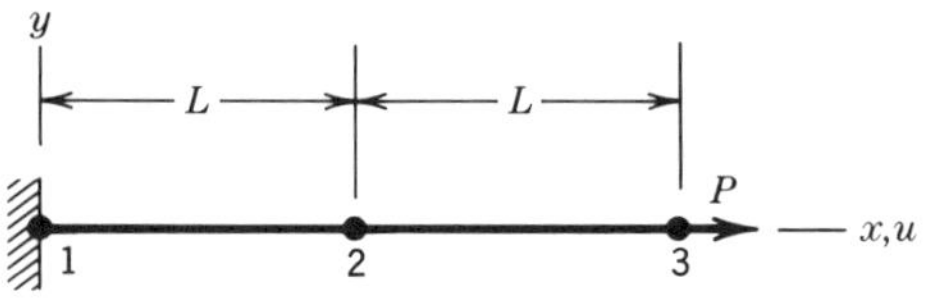

Problem 4.45

4.46 If each of the two coefficients xy in Eq. 4.2-12 is replaced by $x^2 + y^2$, the element becomes incompatible. Why? *Suggestion:* Let two adjacent elements have two corner nodes in common. Along the boundary between elements, d.o.f. of these corner nodes must produce the same boundary displacement in each element if the elements are to be compatible. But how many d.o.f. are needed to define a quadratic curve? And what does this imply?

4.47 It is proposed that a beam element be based on a cubic polynomial but that d.o.f. are to be only lateral displacements w_i, where $i = 1, 2, 3, 4$. Nodes are to be at either end and at the third points. What convergence criterion is violated by this element?

4.48 What quadratic terms are omitted from the displacement field of the solid trilinear element (Eq. 4.2-16)? What cubic terms are omitted?

4.49 In three dimensions, which of the cubic terms would you add to a complete quadratic field (10 terms), if there are to be no preferred directions and the total number of terms in the expansion is to be (a) 11, (b) 13, (c) 14, (d) 16, (e) 17, (f) 19?

Section 4.6

4.50 Imagine that the mesh of Fig. 4.6-1 passes the patch test for constant values of σ_x, σ_y, and τ_{xy}. If now all nine nodes are assigned displacements consistent with a field of constant strain (e.g., Eq. 4.2-9), what loads at node 5 should result from the calculation $[\mathbf{K}]\{\mathbf{D}\}$, and why?

4.51 For the four-element patch shown in Fig. 4.6-1, determine the nodal loads appropriate to the following patch tests. In each case show the loads on a sketch.
(a) Uniaxial stress $\sigma_y = \sigma_c$.
(b) Shear stress $\tau_{xy} = \tau_c$.

4.52 Sketch an assembly of hexahedral solid elements, suitable for use as a patch test mesh. Let the elements have corner nodes only and some corner angles other than 90°. Show supports and nodal loads appropriate to a test for uniform tensile stress σ_z.

4.53 For the beam element of Fig. 4.2-2, consider the lateral displacement field

$$w = \frac{L - x}{L} w_1 + \frac{x}{2}\left(\frac{L - x}{L}\right) \theta_1 + \frac{x}{L} w_2 - \frac{x}{2}\left(\frac{L - x}{L}\right) \theta_2$$

(a) Show that this field includes the required rigid-body motion capability.
(b) If nodal d.o.f. consistent with a state of constant curvature are prescribed, is the correct $w_{,xx}$ obtained?
(c) Determine the element stiffness matrix based on the given field. What defects does it have?

4.54 Divide the element shown into four elements by connecting midpoints of opposite sides. Subdivide the new elements again in the same way. In the limit, what shape does each element approach?

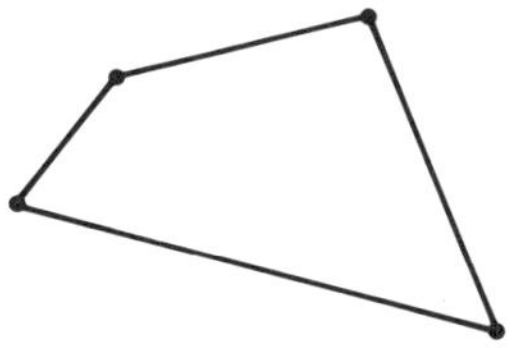

Problem 4.54

4.55 Imagine that the four-element patch of Fig. 4.6-1 is to be tested to see how well it models the pure bending states given in parts (a) and (b). For each of these states, what should be the nodal loads? Rearrange support conditions as may be convenient and appropriate.
(a) σ_x varies linearly from -6 at $y = 0$ to $+6$ at $y = 4$.
(b) σ_y varies linearly from -4 at $x = 0$ to $+4$ at $x = 6$.

4.56 In Fig. 4.6-3, let u vary linearly with s along edge 1–2. Interpolate u from nodal d.o.f. u_1 and u_2. Assume that all material points in the element move only radially, so that $v = 0$ and $u = u(r)$. Show that circumferential strain ϵ_θ becomes independent of s as a becomes much larger than h.

Section 4.7

4.57 Nodal d.o.f. of (say) plane elements need not be restricted to displacements u and v. One might also use all four first derivatives of u and v, for a total of six d.o.f. per node. Such an approach has both advantages and disadvantages. What do you think they are?

4.58 Model the bar of Fig. 4.7-1 by two elements, each of length $L_T/2$. Solve for the nodal d.o.f. and the resulting stress in the upper element. Augment this stress by the appropriate form of Eq. 4.7-3. Is the exact stress field obtained?

4.59 Consider the bending moment at the center of the uniformly loaded cantilever beam of Fig. 4.3-7.
(a) What is the exact value?
(b) What value is given by $M = EI\lfloor \mathbf{B} \rfloor \{\mathbf{d}\}$ and the exact values of d.o.f. w_2 and θ_2? Use a one-element model.
(c) What value is given by adding the result of part (b) to the bending moment at the center of a uniformly loaded clamped–clamped beam of length L?

4.60 In Problem 4.35*e*, nodal d.o.f. calculated by use of consistent nodal loads produce a bending moment $M = (2Lx^2 - x^3)P/L^2$ at the fixed end of the cantilever beam. If this M is augmented by the bending moment at the left end of a clamped–clamped beam under load P at position x, is the exact moment produced?

4.61 Calculate the bending moment at $x = L/4$ in the one-element simply supported beam of Problem 4.36. Compute this moment by adding (a) M caused by nodal d.o.f. θ_1 and θ_2, and (b) M in a fixed–fixed beam of length L with M_c applied at midspan.

4.62 The uniform bar element shown is given the temperature variation $T = cx/L$, where c is a constant.
(a) Solve for u_2, then for axial stress $\sigma_x = E(\epsilon_x - \epsilon_{x0})$.
(b) Repeat part (a), but for stress calculation use an appropriate constant temperature T_0 rather than $T = cx/L$.
(c) Why do the results of parts (a) and (b) differ? Is either correct?

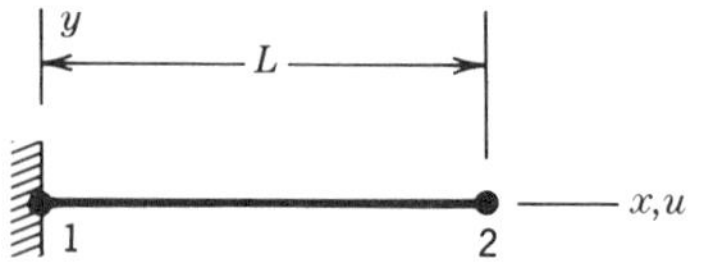

Problem 4.62

4.63 The sketch shows a uniform bar of length $2L$, loaded by a uniformly distributed axial load of intensity q. Also shown is a two-element model and the consistently derived loads at nodes 2 and 3. Carry out two cycles of iterative improvement (see Eq. 4.7-7). To what result does the solution appear to be converging, and what then are the nodal average stresses?

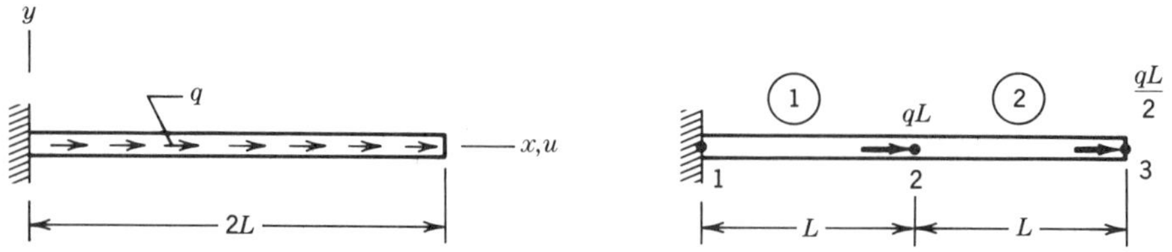

Problem 4.63

CHAPTER 5

STRAIGHT-SIDED TRIANGLES AND TETRAHEDRA

Natural coordinates are introduced. Triangles and tetrahedra are discussed, with the emphasis on triangles. Shape functions are general, but integration formulas and element matrices presented are restricted to elements having straight sides and evenly spaced side nodes. These restrictions are removed in Chapter 6.

5.1 NATURAL COORDINATES (LINEAR)

Natural coordinates are dimensionless. They are defined with reference to the element rather than with reference to the global coordinate system in which the element resides. They are used in preference to Cartesian coordinates because they simplify the process of formulating element matrices. The simplest instance, that of natural coordinates for a straight line, is considered in the present section. Similarly defined natural coordinates for triangles and tetrahedra are considered in Section 5.2. Differently defined natural coordinates for quadrilaterals and hexahedra are used in Chapter 6.

Natural coordinates ξ_1 and ξ_2 in Fig. 5.1-1 are defined as ratios of lengths:

$$\xi_1 = \frac{L_1}{L} \qquad \text{and} \qquad \xi_2 = \frac{L_2}{L} \tag{5.1-1}$$

Since $L_1 + L_2 = L$, coordinates ξ_1 and ξ_2 are not independent. They satisfy the constraint relation

$$\xi_1 + \xi_2 = 1 \tag{5.1-2}$$

Note that ξ_1 and ξ_2 are each either zero or unity at end points 1 and 2. Definitions

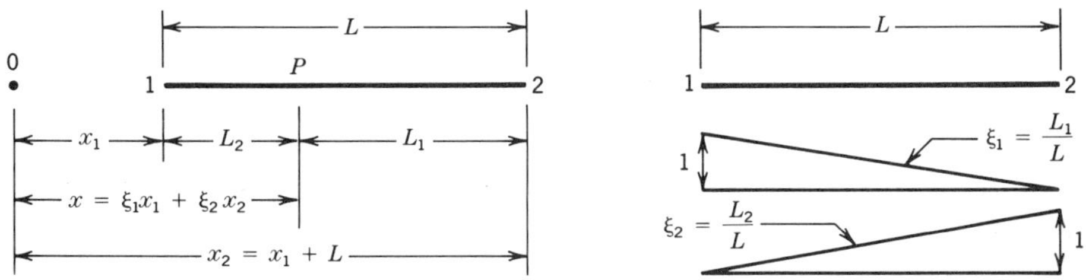

Figure 5.1-1. Natural coordinates ξ_1 and ξ_2 along a straight line. Point *P* is arbitrarily located; that is, *x* can have any value.

5.1-1 are independent of global coordinate x. Even so, ξ_1 and ξ_2 can be used to state the position of the arbitrary point P in terms of x_1 and x_2:

$$x = \xi_1 x_1 + \xi_2 x_2 \tag{5.1-3}$$

For example, the centroid of line 1–2 is at $\xi_1 = \xi_2 = \frac{1}{2}$, where $x = (x_1 + x_2)/2$.

Equations 5.1-2 and 5.1-3, and the inverse relations that state ξ_1 and ξ_2 in terms of x, are

$$\begin{Bmatrix} 1 \\ x \end{Bmatrix} = \begin{bmatrix} 1 & 1 \\ x_1 & x_2 \end{bmatrix} \begin{Bmatrix} \xi_1 \\ \xi_2 \end{Bmatrix} \quad \text{and} \quad \begin{Bmatrix} \xi_1 \\ \xi_2 \end{Bmatrix} = \frac{1}{L} \begin{bmatrix} x_2 & -1 \\ -x_1 & 1 \end{bmatrix} \begin{Bmatrix} 1 \\ x \end{Bmatrix} \tag{5.1-4}$$

Equations 5.1-4 provide a linear mapping between the x and ξ coordinate systems.

Interpolation along line 1–2 can be done in natural coordinates. *Linear* interpolation of a function ϕ, from nodal values ϕ_1 and ϕ_2, is

$$\phi = \lfloor \mathbf{N} \rfloor \begin{Bmatrix} \phi_1 \\ \phi_2 \end{Bmatrix}, \qquad \text{where} \quad \lfloor \mathbf{N} \rfloor = \lfloor \xi_1 \quad \xi_2 \rfloor \tag{5.1-5}$$

In this instance the individual shape functions are $N_1 = \xi_1$ and $N_2 = \xi_2$. *Quadratic* interpolation from nodal values ϕ_1 at $\xi_1 = 1$, ϕ_2 at $\xi_2 = 1$, and ϕ_3 at $\xi_1 = \xi_2 = \frac{1}{2}$, is

$$\phi = \xi_1(2\xi_1 - 1)\phi_1 + \xi_2(2\xi_2 - 1)\phi_2 + 4\xi_1\xi_2\phi_3 \tag{5.1-6}$$

The three quadratic shape functions are shown in Fig. 5.1-2. Equation 5.1-6 can be derived by starting with $\phi = a_1\xi_1^2 + a_2\xi_2^2 + a_3\xi_1\xi_2$ (which is quadratic in x), evaluating the a_i by substitutions such as $\phi = \phi_1$ at $\xi_1 = 1$, and using Eq. 5.1-2. Alternatively, one can use insight: by comparing curves in Figs. 5.1-1 and 5.1-2, we see that $N_1 = \xi_1 - N_3/2$ and $N_2 = \xi_2 - N_3/2$.

Stiffness matrices can be formulated by use of natural coordinates. For example, imagine that line 1–2 in Fig. 5.1-1 is a two-node bar element of cross-sectional area A and elastic modulus E. From Eq. 5.1-5 with $\phi = u$, axial displacement is $u = \xi_1 u_1 + \xi_2 u_2$. Therefore, the axial strain is

$$\epsilon_x = \frac{du}{dx} = \frac{\partial u}{\partial \xi_1}\frac{\partial \xi_1}{\partial x} + \frac{\partial u}{\partial \xi_2}\frac{\partial \xi_2}{\partial x} = u_1\left(-\frac{1}{L}\right) + u_2\left(\frac{1}{L}\right) \tag{5.1-7}$$

in which the factors $\pm 1/L$ are obtained from the second of Eqs. 5.1-4; that is,

$$\frac{\partial \xi_1}{\partial x} = \frac{\partial}{\partial x}\left(\frac{x_2 - x}{L}\right) = -\frac{1}{L} \quad \text{and} \quad \frac{\partial \xi_2}{\partial x} = \frac{\partial}{\partial x}\left(\frac{x - x_1}{L}\right) = \frac{1}{L} \tag{5.1-8}$$

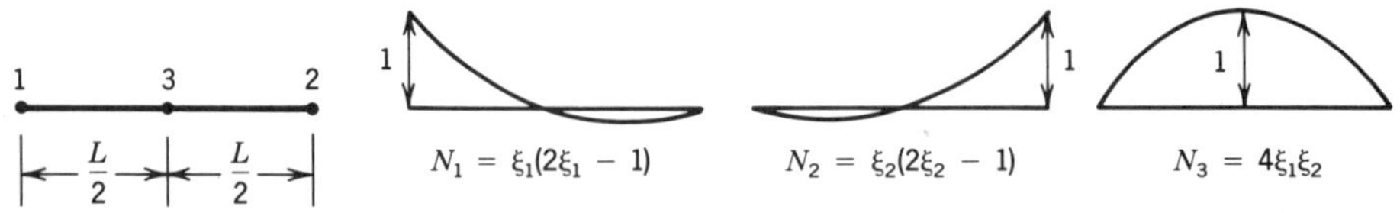

Figure 5.1-2. Quadratic shape functions over a span L in natural coordinates ξ_1 and ξ_2, with 3 a central node.

The element stiffness matrix that operates on u_1 and u_2 is

$$[\mathbf{k}] = \int_L AE\, \lfloor \mathbf{B} \rfloor^T \lfloor \mathbf{B} \rfloor \, dL, \qquad \text{where} \quad \lfloor \mathbf{B} \rfloor = \left\lfloor -\frac{1}{L} \quad \frac{1}{L} \right\rfloor \qquad (5.1\text{-}9)$$

The familiar result, Eq. 2.4-5, follows immediately.

For elements of quadratic or higher order, the integrand of [k] contains functions of ξ_1 and ξ_2. Integration of polynomials in ξ_1 and ξ_2 can be done by the formula

$$\int_L \xi_1^k \xi_2^\ell \, dL = L\, \frac{k!\ell!}{(1 + k + \ell)!} \qquad (5.1\text{-}10)$$

where k and ℓ are nonnegative integers and L is the distance between the end points $\xi_1 = 1$ and $\xi_2 = 1$. When it appears, the factorial 0! is defined as unity. Integration in Eq. 5.1-9 involves integration of only dL. Therefore, $k = \ell = 0$ in Eq. 5.1-10, and integration yields L. As additional examples of the application of Eq. 5.1-10,

$$\int_L x \, dL = \int_L (\xi_1 x_1 + \xi_2 x_2) \, dL = \frac{L}{2} (x_1 + x_2) \qquad (5.1\text{-}11)$$

$$\int_L \xi_1 \xi_2^2 \, dL = L\, \frac{2}{4!} = \frac{L}{12} \qquad (5.1\text{-}12)$$

$$\int (\xi_1 A_1 + \xi_2 A_2)\, dL = \frac{L}{2}(A_1 + A_2)$$

5.2 NATURAL COORDINATES (AREA AND VOLUME)

As defined in Section 5.1, natural coordinates for a line are ratios of lengths. Analogous natural coordinates for triangles and tetrahedra are respectively defined as ratios of areas and ratios of volumes. In the present section we assume that sides of triangles and edges of tetrahedra are straight.

Area Coordinates. In Fig. 5.2-1, an arbitrarily located point P divides a triangle 1–2–3 into three subareas A_1, A_2, and A_3. Area coordinates are defined as ratios of areas[1]:

$$\xi_1 = \frac{A_1}{A} \qquad \xi_2 = \frac{A_2}{A} \qquad \xi_3 = \frac{A_3}{A} \qquad (5.2\text{-}1)$$

where A is the area of triangle 1–2–3. Since $A = A_1 + A_2 + A_3$, the ξ_i are not independent. They satisfy the constraint equation

$$\xi_1 + \xi_2 + \xi_3 = 1 \qquad (5.2\text{-}2)$$

[1]Area coordinates can also be regarded as ratios of lengths. For example, if L_1 is the length of side 1 and a distance s is measured normal to side 1, then $\xi_1 = A_1/A = (L_1 s/2)/(L_1 h/2) = s/h$, where h is the height of the triangle measured normal to side 1. Similarly, volume coordinates can be regarded as ratios of lengths.

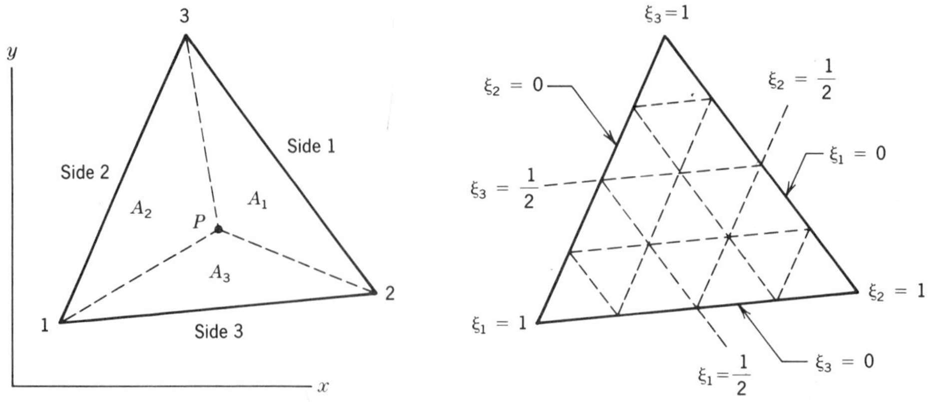

Figure 5.2-1. Natural (area) coordinates for a triangle.

The centroid of a straight-sided triangle is at $\xi_1 = \xi_2 = \xi_3 = \frac{1}{3}$.

The constraint equation and the linear relation between Cartesian and area coordinates are expressed by equations analogous to Eqs. 5.1-4,

$$\begin{Bmatrix} 1 \\ x \\ y \end{Bmatrix} = [\mathbf{A}] \begin{Bmatrix} \xi_1 \\ \xi_2 \\ \xi_3 \end{Bmatrix} \quad \text{and} \quad \begin{Bmatrix} \xi_1 \\ \xi_2 \\ \xi_3 \end{Bmatrix} = [\mathbf{A}]^{-1} \begin{Bmatrix} 1 \\ x \\ y \end{Bmatrix} \tag{5.2-3}$$

where, with $x_{ij} = x_i - x_j$ and $y_{ij} = y_i - y_j$,

$$[\mathbf{A}] = \begin{bmatrix} 1 & 1 & 1 \\ x_1 & x_2 & x_3 \\ y_1 & y_2 & y_3 \end{bmatrix} \quad \text{and} \quad [\mathbf{A}]^{-1} = \frac{1}{2A} \begin{bmatrix} (x_2 y_3 - x_3 y_2) & y_{23} & x_{32} \\ (x_3 y_1 - x_1 y_3) & y_{31} & x_{13} \\ (x_1 y_2 - x_2 y_1) & y_{12} & x_{21} \end{bmatrix} \tag{5.2-4}$$

That the first of Eqs. 5.2-3 is correct may be checked by noting that it yields the correct x and y values at the vertices (and at other points, such as midsides), and that x and y vary linearly elsewhere.

Twice the area of triangle 1–2–3 is

$$2A = \det [\mathbf{A}] = x_{21} y_{31} - x_{31} y_{21} \tag{5.2-5}$$

where $x_{21} = x_2 - x_1$, and so on. The latter expression for $2A$ is neither unique nor obvious but can be obtained by manipulation of the determinant. If labels 1–2–3 are reversed, so that the order 1–2–3 is clockwise around the triangle, Eq. 5.2-5 gives a negative number for $2A$.

The first of Eqs. 5.2-3 allows one to find the Cartesian coordinates of a point when its area coordinates are given. A point in the triangle is uniquely located by specifying any two of its area coordinates. If all three are specified they must satisfy the constraint equation, Eq. 5.2-2. Further discussion of interpolation in area coordinates appears in subsequent sections, where properties of specific elements are formulated.

Formulation of element matrices requires that a function ϕ, expressed in terms of area coordinates, be differentiated with respect to Cartesian coordinates. By the chain rule, with $\phi = \phi(\xi_1, \xi_2, \xi_3)$,

$$\frac{\partial \phi}{\partial x} = \frac{\partial \phi}{\partial \xi_1}\frac{\partial \xi_1}{\partial x} + \frac{\partial \phi}{\partial \xi_2}\frac{\partial \xi_2}{\partial x} + \frac{\partial \phi}{\partial \xi_3}\frac{\partial \xi_3}{\partial x} \qquad (5.2\text{-}6a)$$

$$\frac{\partial \phi}{\partial y} = \frac{\partial \phi}{\partial \xi_1}\frac{\partial \xi_1}{\partial y} + \frac{\partial \phi}{\partial \xi_2}\frac{\partial \xi_2}{\partial y} + \frac{\partial \phi}{\partial \xi_3}\frac{\partial \xi_3}{\partial y} \qquad (5.2\text{-}6b)$$

From the second of Eqs. 5.2-3,

$$A_1 = \frac{y_{23}\,x_{32}}{2}$$

$$\frac{\partial \xi_1}{\partial x} = \frac{y_{23}}{2A} \qquad \frac{\partial \xi_2}{\partial x} = \frac{y_{31}}{2A} \qquad \frac{\partial \xi_3}{\partial x} = \frac{y_{12}}{2A} \qquad (5.2\text{-}7a)$$

$$\frac{\partial \xi_1}{\partial y} = \frac{x_{32}}{2A} \qquad \frac{\partial \xi_2}{\partial y} = \frac{x_{13}}{2A} \qquad \frac{\partial \xi_3}{\partial y} = \frac{x_{21}}{2A} \qquad (5.2\text{-}7b)$$

where $y_{23} = y_2 - y_3$, and so on. Equations 5.2-7 apply to straight-sided triangles whose side nodes (if any) are evenly spaced.

Integration of a polynomial in area coordinates over the triangle area is accomplished by a formula analogous to Eq. 5.1-10. If k, ℓ, and m are nonnegative integers, then

$$\int_A \xi_1^k \xi_2^\ell \xi_3^m \, dA = 2A \frac{k!\ell!m!}{(2 + k + \ell + m)!} \qquad (5.2\text{-}8)$$

where A is the entire area of the triangle in Fig. 5.2-1. Integration along a side, such as side 3 of the triangle, where $\xi_3 = 0$, is accomplished by Eq. 5.1-10.

The following form of the area integration formula is often useful [5.1]. It allows the integrand to be expressed in terms of Cartesian coordinates rather than area coordinates. Let x and y be *centroidal axes,* so that vertex coordinates satisfy the equations $x_1 + x_2 + x_3 = 0$ and $y_1 + y_2 + y_3 = 0$. Then, if r and s are nonnegative integers,

$$\int_A x^r y^s \, dA = C_{r+s} A(x_1^r y_1^s + x_2^r y_2^s + x_3^r y_3^s) \qquad (5.2\text{-}9)$$

where

$r + s$	1	2	3	4	5
C_{r+s}	0	1/12	1/30	1/30	2/105

For example,

$$\int_A x^2 \, dA = \frac{A}{12}(x_1^2 + x_2^2 + x_3^2) \qquad (5.2\text{-}10)$$

where, from Eq. 5.2-5, $A = x_{21}y_{31} - x_{31}y_{21}$. Equation 5.2-9 can be derived from Eqs. 5.2-3, 5.2-4, and 5.2-8, but the manipulations are tedious.

Numerical integration formulas for triangles appear in Section 6.8.

Volume Coordinates. Volume coordinates for a tetrahedron are a direct extension of area coordinates for a triangle, so we will be brief.

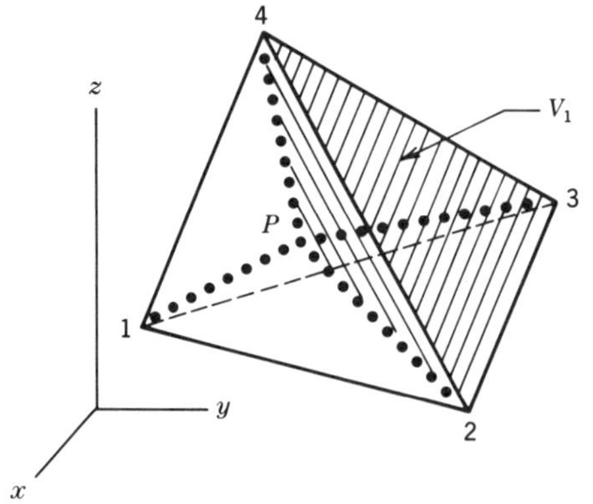

Figure 5.2-2. Tetrahedron 1–2–3–4. Volume V_1 refers to subtetrahedron P–2–3–4, which is identified by hatching.

Point P is the common vertex of four subtetrahedra (Fig. 5.2-2). Volume coordinates are $\xi_i = V_i/V$, where the volume of tetrahedron 1–2–3–4 is $V = \Sigma V_i$ and i runs from 1 to 4. Cartesian and volume coordinates have the relation

$$\begin{Bmatrix} 1 \\ x \\ y \\ z \end{Bmatrix} = \begin{bmatrix} 1 & 1 & 1 & 1 \\ x_1 & x_2 & x_3 & x_4 \\ y_1 & y_2 & y_3 & y_4 \\ z_1 & z_2 & z_3 & z_4 \end{bmatrix} \begin{Bmatrix} \xi_1 \\ \xi_2 \\ \xi_3 \\ \xi_4 \end{Bmatrix} \tag{5.2-11}$$

The determinant of the square matrix is $6V$, where V is positive if nodes are numbered so that the sequence 1–2–3 runs counterclockwise when viewed from node 4. Relations analogous to Eqs. 5.2-7 contain terms from the inverse of the square matrix in Eq. 5.2-11.

The integration formula in volume coordinates is

$$\int_V \xi_1^k \xi_2^\ell \xi_3^m \xi_4^n \, dV = 6V \frac{k!\ell!m!n!}{(3 + k + \ell + m + n)!} \tag{5.2-12}$$

A formula analgous to Eq. 5.2-9, but for a special case, is

$$\int_V x^r y^s z^t \, dV = \frac{V}{20} \sum_{i=1}^{4} x_i^r y_i^s z_i^t \qquad (\text{if } r + s + t = 2) \tag{5.2-13}$$

provided that the centroid of the tetrahedron is at $x = y = z = 0$.

Remarks. Triangles and tetrahedra are instances of a *simplex,* which is a figure in n-dimensional space that has $n + 1$ vertices and is bounded by $n + 1$ surfaces of dimensionality $n - 1$. Other names for area coordinates are *areal, triangular,* and *trilinear* coordinates. Volume coordinates may also be called *tetrahedronal* coordinates. Area and volume coordinates are known to mathematicians as *simplex* or *barycentric* coordinates. They are not new [5.2,5.3] but seem to have been independently devised when finite element theory found a need for them.

5.3 INTERPOLATION FIELDS FOR PLANE TRIANGLES

Triangular elements allow a complete polynomial in Cartesian coordinates to be used for the field quantity (e.g., for temperature or for displacement). In other words, in Fig. 5.3-1, with internal nodes present for cubic and higher-order elements, *all* terms of a truncated Pascal triangle are used in the shape functions: through the second row for a linear element, through the third row for a quadratic element, and so on.

In order to generate finite elements, we seek shape functions $N_i = N_i(\xi_1,\xi_2,\xi_3)$ in the relation $\phi = \Sigma N_i\phi_i$, where the ϕ_i are nodal d.o.f. One way to generate shape functions N_i is the usual way of starting with a polynomial that contains constants a_i that must be determined. Consider a function $\phi = \phi(\xi_1,\xi_2,\xi_3) = \phi(x,y)$, where ϕ is given by the expansion

$$\phi = \sum_{i=1}^{n} a_i \xi_1^q \xi_2^r \xi_3^s \tag{5.3-1}$$

in which q, r, and s are nonnegative integers that range over the n possible combinations for which $q + r + s = p$. Thus ϕ is a complete polynomial of degree p in Cartesian coordinates [5.4]. For example, for the quadratic triangle in Fig. 5.3-2*b*, $n = 6$, $p = 2$, and

$$\phi = a_1\xi_1^2 + a_2\xi_2^2 + a_3\xi_3^2 + a_4\xi_1\xi_2 + a_5\xi_2\xi_3 + a_6\xi_3\xi_1 \tag{5.3-2}$$

which, for a straight-sided triangle, is equivalent to

$$\phi = b_1 + b_2x + b_3y + b_4x^2 + b_5xy + b_6y^2 \tag{5.3-3}$$

where the b_i are constants related to constants a_i of Eq. 5.3-2.

To obtain shape functions N_i from Eq. 5.3-1, we express the a_i in terms of nodal d.o.f. ϕ_i. Consider again the quadratic triangle. Side node 4 is at $\xi_1 = \xi_2 = \frac{1}{2}$ and $\xi_3 = 0$, side node 5 is at $\xi_2 = \xi_3 = \frac{1}{2}$ and $\xi_1 = 0$, and side node 6 is at $\xi_3 = \xi_1 = \frac{1}{2}$ and $\xi_2 = 0$. In Eq. 5.3-2 we set $\phi = \phi_1$ for $\xi_1 = 1$ and $\xi_2 = \xi_3 = 0$, $\phi = \phi_2$ for $\xi_2 = 1$ and $\xi_3 = \xi_1 = 0$, and so on through $\phi = \phi_6$ for $\xi_3 = \xi_1 = \frac{1}{2}$ and $\xi_2 = 0$. After we have solved for the a_i, the coefficients of the ϕ_i are identified as shape functions.

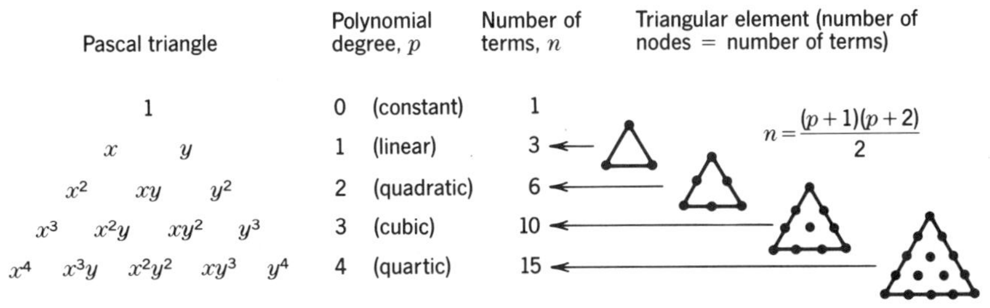

Figure 5.3-1. Relation between type of plane triangular element and number of polynomial coefficients used for interpolation.

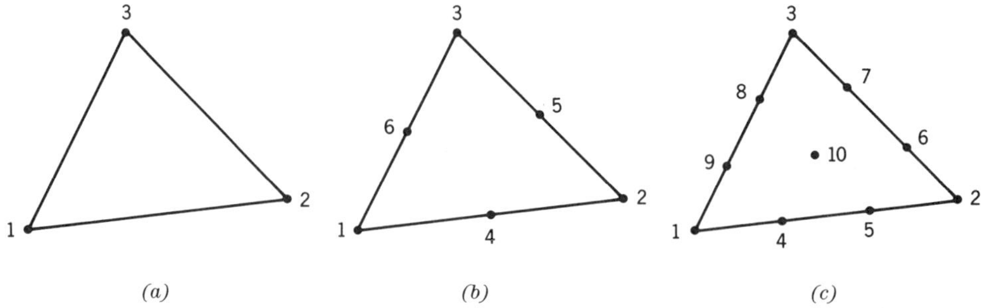

Figure 5.3-2. Triangular elements. (*a*) Linear. (*b*) Quadratic. (*c*) Cubic. Node 10 is at the centroid, $\xi_1 = \xi_2 = \xi_3 = \frac{1}{3}$.

An alternative way of determining the shape functions is available. With nodes permitted in the element interior, but no derivative d.o.f. permitted at any node, the elements can be identified as Lagrangian. Hence, the Lagrange interpolation formula produces shape functions N_i directly. The procedure is not difficult but cannot be explained briefly. Details appear in [5.4].

Shape functions for the elements of Fig. 5.3-2 are as follows. Note that (as expected) each N_i is unity at node i but vanishes at all other nodes. For the *linear* triangle, Fig. 5.3-2*a*,

$$N_1 = \xi_1 \qquad N_2 = \xi_2 \qquad N_3 = \xi_3 \tag{5.3-4}$$

Thus, individual shape functions of a linear triangle are the area coordinates themselves. For the *quadratic* triangle, Fig. 5.3-2*b*,

$$\begin{array}{lll} N_1 = \xi_1(2\xi_1 - 1) & N_2 = \xi_2(2\xi_2 - 1) & N_3 = \xi_3(2\xi_3 - 1) \\ N_4 = 4\xi_1\xi_2 & N_5 = 4\xi_2\xi_3 & N_6 = 4\xi_3\xi_1 \end{array} \tag{5.3-5}$$

For the *cubic* triangle, Fig. 5.3-2*c*,

$$\begin{array}{ll} N_i = \frac{1}{2}\,\xi_i(3\xi_i - 1)(3\xi_i - 2) & \text{for } i = 1, 2, 3 \\ N_4 = \frac{9}{2}\,\xi_2\xi_1(3\xi_1 - 1) & N_6 = \frac{9}{2}\,\xi_3\xi_2(3\xi_2 - 1) \\ N_8 = \frac{9}{2}\,\xi_1\xi_3(3\xi_3 - 1) & N_{10} = 27\,\xi_1\xi_2\xi_3 \end{array} \tag{5.3-6}$$

and N_5, N_7, and N_9 are obtained from N_4, N_6, and N_8, respectively, by interchanging subscripts; for example, $N_5 = 9\xi_1\xi_2(3\xi_2 - 1)/2$.

Equations 5.3-5 and 5.3-6 can also be used for triangular elements having curved sides. The procedure is described in Section 6.8.

5.4 THE LINEAR TRIANGLE

Scalar Field Element. For simplicity, we begin with a scalar field problem. Each node has a single d.o.f. An example application is heat conduction, for which ϕ represents temperature (see Section 3.10 or Chapter 16). For the three-node triangle, from Eq. 5.3-4,

$$\phi = \lfloor \mathbf{N} \rfloor \begin{Bmatrix} \phi_1 \\ \phi_2 \\ \phi_3 \end{Bmatrix}, \qquad \text{where} \quad \lfloor \mathbf{N} \rfloor = \lfloor \xi_1 \quad \xi_2 \quad \xi_3 \rfloor \tag{5.4-1}$$

In order to generate the characteristic matrix [**k**], derivatives of ϕ with respect to x and y are needed. Using Eqs. 5.2-6 and 5.4-1, we have

$$\frac{\partial \phi}{\partial x} = \sum_{i=1}^{3} \frac{\partial N_i}{\partial x} \phi_i = \sum_{i=1}^{3} \frac{\partial \xi_i}{\partial x} \phi_i \tag{5.4-2}$$

and similarly for $\partial\phi/\partial y$, where derivatives $\partial\xi_i/\partial x$ and $\partial\xi_i/\partial y$ are given by Eqs. 5.2-7. Hence,

$$\begin{Bmatrix} \phi_{,x} \\ \phi_{,y} \end{Bmatrix} = \begin{bmatrix} \mathbf{N}_{,x} \\ \mathbf{N}_{,y} \end{bmatrix} \begin{Bmatrix} \phi_1 \\ \phi_2 \\ \phi_3 \end{Bmatrix} = [\mathbf{B}] \begin{Bmatrix} \phi_1 \\ \phi_2 \\ \phi_3 \end{Bmatrix}, \qquad \text{where} \quad [\mathbf{B}] = \frac{1}{2A} \begin{bmatrix} y_{23} & y_{31} & y_{12} \\ x_{32} & x_{13} & x_{21} \end{bmatrix} \tag{5.4-3}$$

and A is the area of the triangle. The element characteristic matrix is

$$[\mathbf{k}] = \int_{V_e} [\mathbf{B}]^T k \, [\mathbf{B}] \, dV \tag{5.4-4}$$

where k is a material property. If k and element thickness t are constant over the element, then $[\mathbf{k}] = ktA[\mathbf{B}]^T[\mathbf{B}]$.

Constant-Strain Triangle. In plane stress analysis the manipulations are quite similar to those presented in Eqs. 5.4-1 to 5.4-4. Displacement fields $u = u(x,y)$ and $v = v(x,y)$ are each interpolated from nodal d.o.f. u_i and v_i. The linear element has six displacements in the vector of nodal d.o.f. {**d**}. With $\{\mathbf{d}\} = \lfloor u_1 \quad v_1 \quad u_2 \quad v_2 \quad u_3 \quad v_3 \rfloor^T$,

$$\begin{Bmatrix} u \\ v \end{Bmatrix} = [\mathbf{N}]\{\mathbf{d}\}, \qquad \text{where} \quad [\mathbf{N}] = \begin{bmatrix} \xi_1 & 0 & \xi_2 & 0 & \xi_3 & 0 \\ 0 & \xi_1 & 0 & \xi_2 & 0 & \xi_3 \end{bmatrix} \tag{5.4-5}$$

Strains $\{\boldsymbol{\epsilon}\} = [\mathbf{B}]\{\mathbf{d}\}$ are given by Eqs. 1.5-6 and 1.5-8. Using these equations, and also Eqs. 5.2-6 and 5.2-7, we obtain

$$[\mathbf{B}] = [\partial][\mathbf{N}] = \frac{1}{2A} \begin{bmatrix} y_{23} & 0 & y_{31} & 0 & y_{12} & 0 \\ 0 & x_{32} & 0 & x_{13} & 0 & x_{21} \\ x_{32} & y_{23} & x_{13} & y_{31} & x_{21} & y_{12} \end{bmatrix} \tag{5.4-6}$$

With [**E**] given by Eq. 1.7-4 or Eq. 1.7-5, the stiffness matrix of the linear triangle is

$$[\mathbf{k}] = \int_{V_e} [\mathbf{B}]^T[\mathbf{E}][\mathbf{B}] \, dV \tag{5.4-7}$$

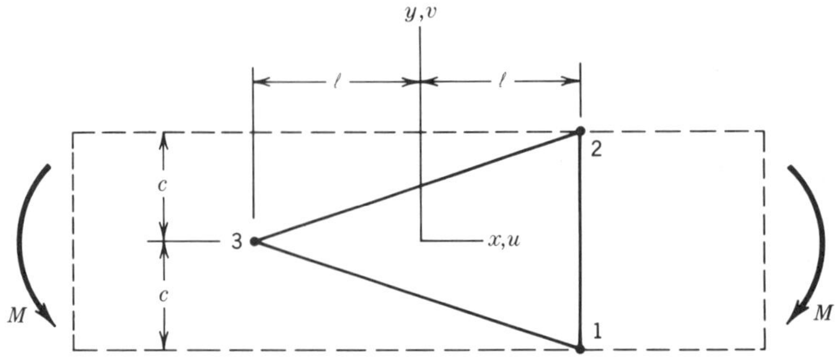

Figure 5.4-1. A linear triangle in a beam subjected to pure bending.

If [E] and element thickness t are constant, then $[\mathbf{k}] = tA[\mathbf{B}]^T[\mathbf{E}][\mathbf{B}]$.

The element associated with Eq. 5.4-5 is called the *constant-strain triangle* (CST) because its strain field contains only constants. Because x and y terms are absent, the CST element behaves poorly in bending. Consider, for example, Fig. 5.4-1. Let the origin $x = y = 0$ be motionless, and let bending moment M be of such magnitude that $u = \bar{u}$ at $x = \ell$ on top of the beam. Hence, the correct displacement field associated with pure bending, and the resulting strain field, are

$$\begin{aligned} u &= \frac{\bar{u}}{c\ell}xy \qquad v = \frac{\bar{u}}{2c\ell}(-x^2 - \nu y^2) \\ \epsilon_x &= \frac{\bar{u}}{c\ell}y \qquad \epsilon_y = -\nu\frac{\bar{u}}{c\ell}y \qquad \gamma_{xy} = 0 \end{aligned} \tag{5.4-8}$$

To see how the CST represents these bending strains, we impose on the CST nodal displacements {d} consistent with the bending field of Eq. 5.4-8 ($u_1 = -\bar{u}$, $u_2 = \bar{u}$, $u_3 = 0$, and similar prescriptions of v_1, v_2, and v_3 from Eqs. 5.4-8). Hence,

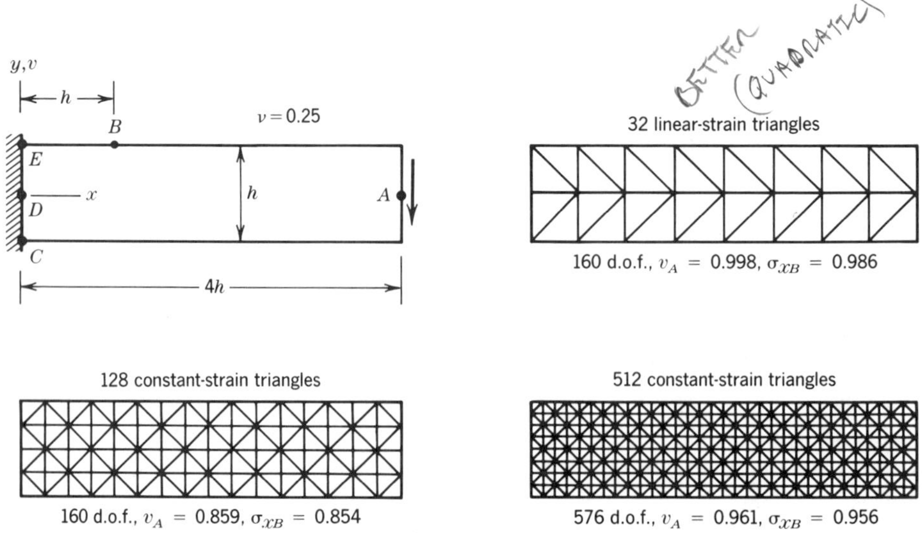

Figure 5.4-2. Tip deflection v_A and flexural stress σ_{xB} in an isotropic cantilever beam of constant thickness [5.5]. The transverse load is parabolically distributed over the right end. Values reported are *ratios* of computed values to values predicted by theory of elasticity.

from $\{\boldsymbol{\epsilon}\} = [\mathbf{B}]\{\mathbf{d}\}$, with $[\mathbf{B}]$ given by Eq. 5.4-6, we obtain strains in the CST of Fig. 5.4-1. They are

$$\epsilon_x = 0 \qquad \epsilon_y = 0 \qquad \gamma_{xy} = \left(\frac{1}{c} - \nu \frac{c}{4\ell^2}\right) \bar{u} \tag{5.4-9}$$

Clearly these strains are quite wrong. Despite their poor performance in bending, CST elements can adequately model a beam in bending if a great many elements are used through the depth of the beam.

Figure 5.4-2 gives numerical evidence of how the CST behaves. Theoretical values allow for transverse shear deformation and assume that the end at $x = 0$ is free to warp while points C, D, and E remain on a vertical line.

Elements better than the CST are available. Routine use of the CST in stress analysis is not recommended.

5.5 THE QUADRATIC TRIANGLE

The quadratic triangle with straight sides and midside nodes, Fig. 5.5-1, dates from 1964. It is an excellent element for stress analysis. Its strain field contains a complete linear polynomial for ϵ_x, ϵ_y, and γ_{xy}. Accordingly, it is also known as the *linear-strain triangle*. Its sides can deform into quadratic curves, which means that a uniform edge traction is allocated to nodal loads $\{\mathbf{r}_e\}$ in the 1–4–1 proportion seen in Fig. 4.3-4.

In the present section we formulate the element stiffness matrix under the restrictions that sides are straight and side nodes are at midsides. In Section 6.8 these restrictions are removed.

For convenience of notation in the following development, we arrange nodal d.o.f. in the order

$$\{\mathbf{d}\} = \lfloor u_1 \quad u_2 \quad u_3 \quad u_4 \quad u_5 \quad u_6 \quad v_1 \quad v_2 \quad v_3 \quad v_4 \quad v_5 \quad v_6 \rfloor^T \tag{5.5-1}$$

The x- and y-direction displacement fields are

$$u = \sum_{i=1}^{6} N_i u_i \qquad \text{and} \qquad v = \sum_{i=1}^{6} N_i v_i \tag{5.5-2}$$

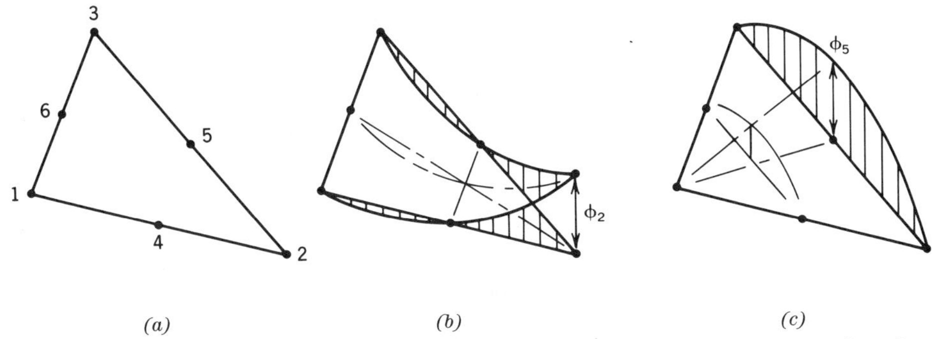

Figure 5.5-1. Quadratic triangle. (*a*) Node numbering. (*b*) A vertex-node shape function. (*c*) A side-node shape function.

where the N_i are given by Eqs. 5.3-5. Next

$$\epsilon_x = \frac{\partial u}{\partial x} = \sum_{i=1}^{6} \frac{\partial N_i}{\partial x} u_i \quad \text{and} \quad \frac{\partial N_i}{\partial x} = \sum_{j=1}^{3} \frac{\partial N_i}{\partial \xi_j} \frac{\partial \xi_j}{\partial x} \tag{5.5-3}$$

where the latter equation is a restatement of Eq. 5.2-6a. Thus, from Eqs. 5.2-7a, 5.3-5, and 5.5-3,

$$\epsilon_x = \frac{1}{2A} \lfloor (4\xi_1 - 1)y_{23} \quad (4\xi_2 - 1)y_{31} \quad (4\xi_3 - 1)y_{12} \quad 4(\xi_2 y_{23} + \xi_1 y_{31}) \quad 4(\xi_3 y_{31} + \xi_2 y_{12}) \quad 4(\xi_1 y_{12} + \xi_3 y_{23}) \rfloor \begin{Bmatrix} u_1 \\ u_2 \\ u_3 \\ u_4 \\ u_5 \\ u_6 \end{Bmatrix} \tag{5.5-4}$$

Expressions for strains ϵ_y and γ_{xy} are developed similarly. The complete result is

$$\{\boldsymbol{\epsilon}\} = \begin{Bmatrix} \epsilon_x \\ \epsilon_y \\ \gamma_{xy} \end{Bmatrix} = [\mathbf{B}]\{\mathbf{d}\}, \quad \text{where} \quad \underset{3\times 12}{[\mathbf{B}]} = \begin{bmatrix} \mathbf{B}_x & \mathbf{0} \\ \mathbf{0} & \mathbf{B}_y \\ \mathbf{B}_y & \mathbf{B}_x \end{bmatrix} \tag{5.5-5}$$

$\mathbf{B}_x$ is the row matrix in Eq. 5.5-4 and $\mathbf{B}_y$ is a similar row matrix.

Because each strain in $\{\boldsymbol{\epsilon}\}$ is a complete linear field, the following convenient trick can be used [5.5]. We interpolate strains $\{\boldsymbol{\epsilon}\}$ from strains $\{\boldsymbol{\epsilon}_c\}$ at corner nodes 1, 2, and 3,

$$\{\boldsymbol{\epsilon}\} = [\mathbf{Q}]\{\boldsymbol{\epsilon}_c\} = [\mathbf{Q}]\lfloor \epsilon_{x1} \quad \epsilon_{x2} \quad \epsilon_{x3} \quad \epsilon_{y1} \quad \epsilon_{y2} \quad \epsilon_{y3} \quad \gamma_{xy1} \quad \gamma_{xy2} \quad \gamma_{xy3} \rfloor^T \tag{5.5-6}$$

where ϵ_{x1} is ϵ_x at node 1, and so on, and, making use of Eq. 5.3-4,

$$[\mathbf{Q}] = \begin{bmatrix} \xi_1 & \xi_2 & \xi_3 & 0 & 0 & 0 & 0 & 0 & 0 \\ 0 & 0 & 0 & \xi_1 & \xi_2 & \xi_3 & 0 & 0 & 0 \\ 0 & 0 & 0 & 0 & 0 & 0 & \xi_1 & \xi_2 & \xi_3 \end{bmatrix} \tag{5.5-7}$$

Corner strains $\{\boldsymbol{\epsilon}_c\}$ are written in terms of nodal d.o.f. $\{\mathbf{d}\}$ by evaluating Eq. 5.5-5 at nodes 1, 2, and 3. Thus

$$\{\boldsymbol{\epsilon}_c\} = [\mathbf{H}]\{\mathbf{d}\} \quad \text{and} \quad \{\boldsymbol{\epsilon}\} = [\mathbf{Q}][\mathbf{H}]\{\mathbf{d}\} \tag{5.5-8}$$

where $[\mathbf{H}]$ is a 9 by 12 matrix of element dimensions: $H_{11} = 3y_{23}/2A$, $H_{12} = -y_{31}/2A$, and so on. Combining Eqs. 5.5-6 and 5.5-8, we obtain

$$\{\boldsymbol{\epsilon}\} = [\mathbf{B}]\{\mathbf{d}\}, \quad \text{where} \quad \underset{3\times 12}{[\mathbf{B}]} = \underset{3\times 9}{[\mathbf{Q}]}\ \underset{9\times 12}{[\mathbf{H}]} \tag{5.5-9}$$

The element stiffness matrix is

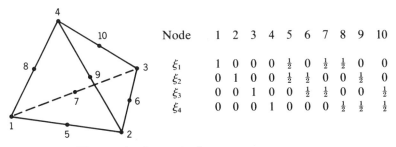

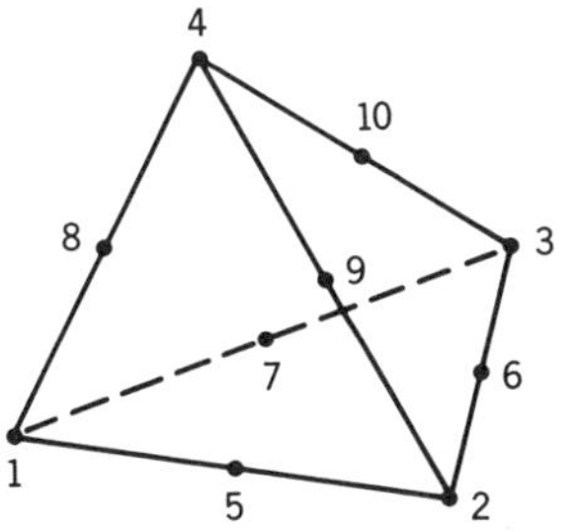

Node	1	2	3	4	5	6	7	8	9	10
ξ_1	1	0	0	0	$\frac{1}{2}$	0	$\frac{1}{2}$	$\frac{1}{2}$	0	0
ξ_2	0	1	0	0	$\frac{1}{2}$	$\frac{1}{2}$	0	0	$\frac{1}{2}$	0
ξ_3	0	0	1	0	0	$\frac{1}{2}$	$\frac{1}{2}$	0	0	$\frac{1}{2}$
ξ_4	0	0	0	1	0	0	0	$\frac{1}{2}$	$\frac{1}{2}$	$\frac{1}{2}$

Figure 5.6-1. The quadratic tetrahedron.

$$[\mathbf{k}] = \int_{V_e} [\mathbf{B}]^T[\mathbf{E}][\mathbf{B}]\, dV = [\mathbf{H}]^T \int_{V_e} [\mathbf{Q}]^T[\mathbf{E}][\mathbf{Q}]\, dV\, [\mathbf{H}] \tag{5.5-10}$$

where $dV = t\, dA$ and t is the element thickness. Again, note the ordering of nodal d.o.f. assumed in Eq. 5.5-1. The latter form of Eq. 5.5-10 can be used to advantage in programming. If t is constant or a polynomial in area coordinates, integrations are easily done by means of Eq. 5.2-8.

Behaviors of the constant- and linear-strain triangles are compared in Fig. 5.4-2. We see that the linear-strain triangle is much better in this problem despite using fewer d.o.f.

Fortran coding for straight-sided quadratic triangles of constant thickness may be found in [5.5]. Alternative compact coding for linear, quadratic, and cubic triangles is discussed in [5.6].

5.6 THE QUADRATIC TETRAHEDRON

Tetrahedral elements are a straightforward extension of triangular elements. Accordingly, we cite only the quadratic element, Fig. 5.6-1, whose shape functions are

$$\begin{aligned} &N_i = \xi_i(2\xi_i - 1) \quad \text{for } i = 1, 2, 3, 4 \\ &N_5 = 4\xi_1\xi_2 \qquad N_6 = 4\xi_2\xi_3 \qquad N_7 = 4\xi_3\xi_1 \\ &N_8 = 4\xi_1\xi_4 \qquad N_9 = 4\xi_2\xi_4 \qquad N_{10} = 4\xi_3\xi_4 \end{aligned} \tag{5.6-1}$$

Equation 5.5-10 again describes the element stiffness matrix, where $[\mathbf{k}]$ is now 30 by 30, $[\mathbf{E}]$ is 6 by 6, and $[\mathbf{B}]$ is 6 by 30.

PROBLEMS

Section 5.1

5.1 (a) Derive Eq. 5.1-6 by using the substitution procedure suggested below that equation.

(b) Similarly, derive shape functions for cubic interpolation along a line. Nodes are at $\xi_1 = 1$, $\xi_1 = \frac{2}{3}$, $\xi_1 = \frac{1}{3}$, and $\xi_1 = 0$. Start with $\phi = a_1\xi_1^3 + a_2\xi_1^2\xi_2 + a_3\xi_1\xi_2^2 + a_4\xi_2^3$.

5.2 Use Eq. 5.1-10 to integrate the following functions over span L in Fig. 5.1-1.
(a) x^2
(b) $\xi_1^3 \xi_2^2$
(c) $\xi_1 x$

5.3 In Eq. 5.1-9, let A vary linearly from A_1 at node 1 to A_2 at node 2. Determine the resulting stiffness matrix [**k**].

5.4 In Eq. 5.1-6, let ϕ represent axial displacement u of a uniform bar element. Determine the 3 by 3 element stiffness matrix. Let node 3 lie at the center of the bar (hence, ξ_1 and ξ_2 are linear functions of x).

Section 5.2

5.5 Let x_1 and y_1 in [**A**] of Eq. 5.2-4 be replaced by x and y. Then, Eq. 5.2-5 yields $2A_1 = \det[\mathbf{A}]$. Similar expressions for $2A_2$ and $2A_3$ can be written. Hence, $\lfloor \xi_1 \quad \xi_2 \quad \xi_3 \rfloor^T = \lfloor 2A_1 \quad 2A_2 \quad 2A_3 \rfloor^T / 2A = [\mathbf{A}]^{-1} \lfloor 1 \quad x \quad y \rfloor^T$. In this way verify the expression for $[\mathbf{A}]^{-1}$ in Eq. 5.2-4.

5.6 (a) Verify that $[\mathbf{A}]^{-1}$ is correctly stated in Eq. 5.2-4 by forming the product $[\mathbf{A}][\mathbf{A}]^{-1}$.
(b) Show that the equation of side 1 of the triangle is $x_2 y_3 - x_3 y_2 + y_{23} x + x_{32} y = 0$.
(c) Derive the latter expression for $2A$ in Eq. 5.2-5 from det[**A**].
(d) In Fig. 5.2-1, drop lines from the triangle vertices to the x axis. One can now identify three trapezoids, whose upper sides are 1–2, 2–3, and 3–1. Two trapezoidal areas minus a third equals the triangle area. Hence, derive the expression $2A = x_{21} y_{31} - x_{31} y_{21}$ in Eq. 5.2-5.
(e) Is it also true that $2A = x_{32} y_{12} - x_{12} y_{32}$? Explain.

5.7 Let the function $\phi = 27 \xi_1 \xi_2 \xi_3$ be defined over the triangle in Fig. 5.2-1.
(a) Sketch this function in isometric view (analogous to Fig. 1.1-3).
(b) Integrate ϕ over the triangle area A.

5.8 Let a function ϕ vary linearly over face 1–2–3 of a tetrahedron. Values of ϕ at corner nodes on this face are ϕ_1, ϕ_2, and ϕ_3. The integral of ϕ over face 1–2–3 is $A_{123}(\phi_1 + \phi_2 + \phi_3)/3$. Derive this result by formal integration.

5.9 Show that Eq. 5.2-10 is produced by Eq. 5.2-8.

5.10 For a triangle having the dimensions shown, use formulas for area moments and products of inertia to numerically verify Eq. 5.2-9 for the case $r + s = 2$.

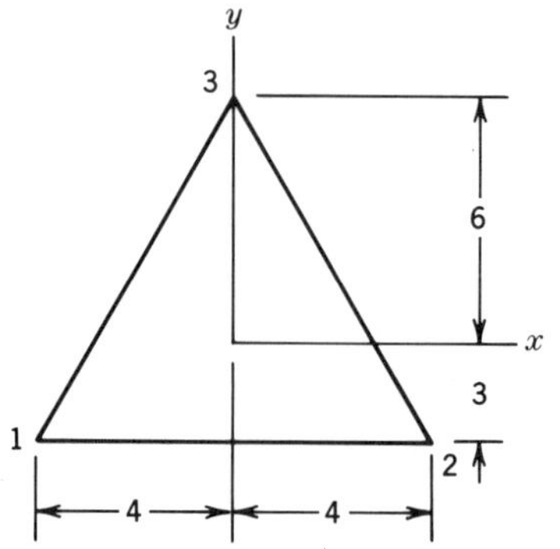

Problem 5.10

5.11 For $r + s = 1$ and again for $r + s = 2$, write integration formulas analogous to Eq. 5.2-9 if the origin of coordinates need not be at the centroid of the triangle. (Let the centroid have coordinates $x = x_c$ and $y = y_c$.)

Section 5.3

5.12 For a cubic triangle (10 nodes), write an equation analogous to Eq. 5.3-2.

5.13 From Eq. 5.3-2, derive the shape functions of Eqs. 5.3-5.

Section 5.4

5.14 (a) For a linear triangle, the equation that corresponds to Eq. 5.3-3 is $\phi = b_1 + b_2x + b_3y$. From this, and without using area coordinates, obtain shape functions N_1, N_2, and N_3 in terms of x, y, and Cartesian coordinates x_i and y_i of the vertex nodes.
(b) Show that the N_i of part (a) yield the N_i of Eq. 5.3-4.
(c) Obtain $[\mathbf{B}]$ of Eq. 5.4-6 from the N_i of part (a).

5.15 Verify the results given in Eqs. 5.4-9. Start with Eqs. 5.4-8.

Section 5.5

5.16 Write out the first three rows of matrix $[\mathbf{H}]$ in Eq. 5.5-8. Then show that the product $[\mathbf{Q}][\mathbf{H}]$ in Eq. 5.5-9 yields the row matrix $\mathbf{B}_x$ defined by Eq. 5.5-4 (check at least the first row of the product).

5.17 Consider use of the quadratic triangle for a scalar field problem (similar to Eqs. 5.4-1 through 5.4-3 for the linear triangle). Redefine matrices $[\mathbf{Q}]$ and $[\mathbf{H}]$ of the quadratic triangle as may be necessary in order to obtain a result for a scalar field problem analogous to the latter form of Eq. 5.5-10. If material property k and thickness t are constant, determine c and $[\mathbf{M}]$ in the form $[\mathbf{k}] = c[\mathbf{H}]^T[\mathbf{M}][\mathbf{H}]$, where integrations have been completed to yield $[\mathbf{M}]$, and c represents constants.

5.18 Impose nodal displacements associated with u and v of Eqs. 5.4-8 on the element of Fig. 5.4-1, but consider that the element is a quadratic triangle (six nodes; Eqs. 5.5-2). For simplicity, consider the special case $\nu = 0$. Does the quadratic element display the correct strains?

5.19 Let an edge of a quadratic triangle lie parallel to the x axis. Determine the consistent nodal loads that result from the following distributed loads on this edge.
(a) Uniform traction in the y direction.
(b) Shear stress parallel to the edge, varying parabolically from a maximum at midedge to zero at the adjacent vertices.
(c) A traction in the y direction that varies linearly from $+\overline{\sigma}$ at one vertex to $-\overline{\sigma}$ at the adjacent vertex.

5.20 (a) Evaluate nodal loads $\{\mathbf{r}_e\}$ that result from heating of an isotropic quadratic triangle. Let the temperature vary linearly over the element and be defined by corner temperatures T_1, T_2, and T_3. Express $\{\mathbf{r}_e\}$ in terms of E, α, ν, and corner coordinates and temperatures.
(b) Show that the nodal forces of part (a) are self-equilibrating—that is, show $\Sigma\ (r_e)_i = 0$.

Section 5.6

5.21 The strain–displacement relation for a tetrahedron can be written in the form $\{\epsilon\} = [\mathbf{Q}][\mathbf{H}]\{\mathbf{d}\}$ (analogous to Eqs. 5.5-9). Write out matrices $[\mathbf{Q}]$ and $[\mathbf{H}]$ for a quadratic tetrahedron. Use the symbol a_{ij} to denote coefficients of the inverse of the matrix in Eq. 5.2-11, but do not bother to compute the inverse.

5.22 Let a uniform pressure p act normal to one face of a quadratic tetrahedron. What are the consistent nodal loads?

CHAPTER 6

THE ISOPARAMETRIC FORMULATION

The "isoparametric" formulation can be used to produce many types of useful elements. Plane isoparametric elements are emphasized in the present chapter. Numerical integration, used in element formulation, is described and its possible pitfalls are discussed.

6.1 INTRODUCTION

The isoparametric formulation makes it possible to generate elements that are nonrectangular and have curved sides. These shapes have obvious uses in grading a mesh from coarse to fine, in modeling arbitrary shapes, and in modeling curved boundaries (Fig. 6.1-1). The isoparametric family includes elements for plane, solid, plate, and shell problems. There are also special elements for fracture mechanics and elements for nonstructural problems.

In formulating isoparametric elements, natural coordinate systems must be used (systems $\xi\eta$ and $\xi\eta\zeta$ in Fig. 6.1-2). Displacements are expressed in terms of natural

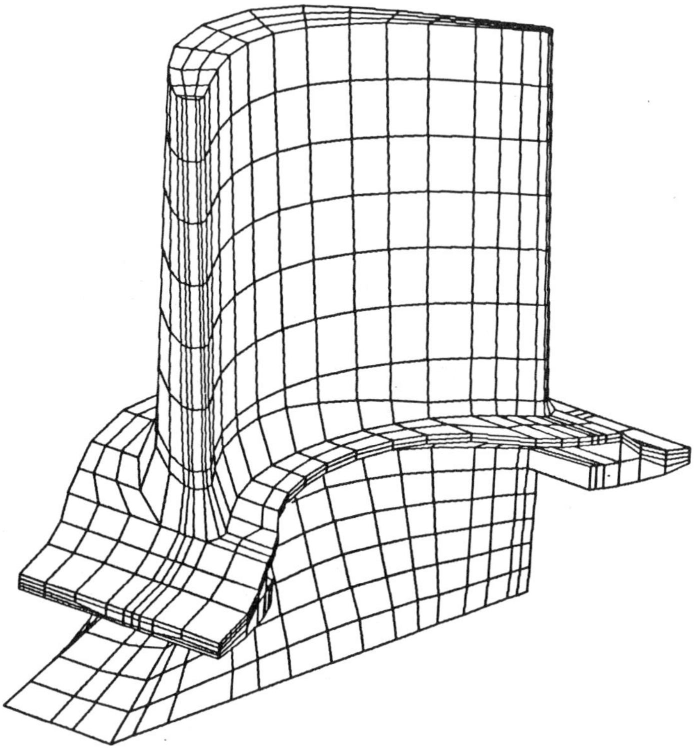

Figure 6.1-1. Turbine blade, modeled by solid elements. (*Courtesy of NASA Lewis Research Center, Cleveland, Ohio.*)

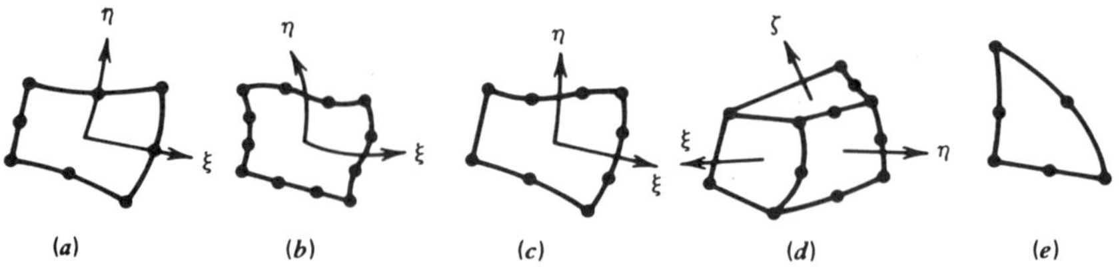

Figure 6.1-2. Example isoparametric elements. (*a*) Quadratic plane element. (*b*) Cubic plane element. (*c*) A "degraded" cubic element. The left and lower sides can be joined to linear and quadratic elements. (*d*) Quadratic solid element with some linear edges. (*e*) A quadratic plane triangle.

coordinates, but must be differentiated with respect to global coordinates x, y, and z. Accordingly, a transformation matrix, called **[J]**, must be invoked. In addition, integrations must be done numerically rather than analytically if elements are nonrectangular. Closed-form integrations are possible in some special cases, but expressions tend to be lengthy, tedious to work out, and therefore more subject to errors of algebra and coding than numerical integration.

The term "isoparametric" means "same parameters" and is explained as follows. Because either displacements or coordinates can be interpolated from nodal values,

1. Nodal d.o.f. $\{\mathbf{d}\}$ define displacements $\lfloor u \quad v \quad w \rfloor$ of a point in the element; that is, $\lfloor u \quad v \quad w \rfloor^T = [\mathbf{N}]\{\mathbf{d}\}$.
2. Nodal coordinates $\{\mathbf{c}\}$ define coordinates $\lfloor x \quad y \quad z \rfloor$ of a point in the element; that is, $\lfloor x \quad y \quad z \rfloor^T = [\tilde{\mathbf{N}}]\{\mathbf{c}\}$.

Shape function matrices $[\mathbf{N}]$ and $[\tilde{\mathbf{N}}]$ are functions of ξ, η, and ζ. An element is *isoparametric* if $[\mathbf{N}]$ and $[\tilde{\mathbf{N}}]$ are identical. If $[\mathbf{N}]$ is of higher degree than $[\tilde{\mathbf{N}}]$, the element is called *subparametric*, but if $[\mathbf{N}]$ is of lower degree than $[\tilde{\mathbf{N}}]$, the element is called *superparametric*.

Isoparametric elements were developed by Taig in 1958 [1.10], but no work was published until 1966 [6.1].

6.2 AN ISOPARAMETRIC BAR ELEMENT

As a simple introduction to isoparametric elements, consider a straight, three-node element, Fig. 6.2-1*a*. Coordinate ξ is a *natural* or *intrinsic* coordinate: ends of the bar lie at $\xi = \pm 1$, regardless of the physical length L of the bar. Moreover, ξ is attached to the bar and remains an axial coordinate regardless of how the bar

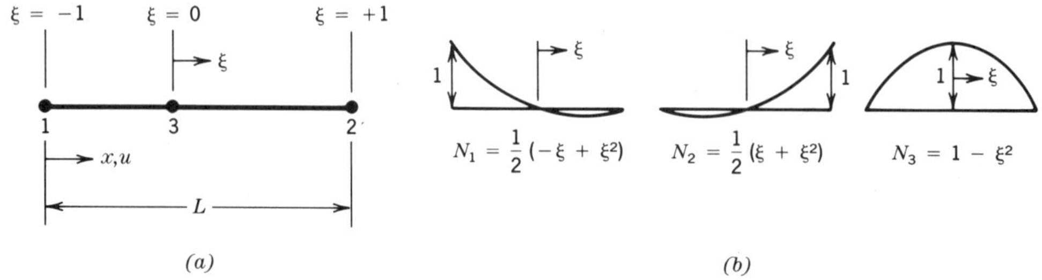

Figure 6.2-1. (*a*) Three-node (quadratic) bar element with natural coordinate ξ. (*b*) The three shape functions.

is oriented in global coordinates xyz. For convenience, not necessity, ξ and x are collinear in the present example. Node 3 is at $\xi = 0$, but need not be at the physical center of the bar.

Coordinate ξ in Fig. 6.2-1 differs from the natural coordinates ξ_1 and ξ_2 used in Section 5.1; ξ is chosen in preference to ξ_1 and ξ_2 because it is more closely related to coordinates $\xi\eta$ used for plane quadrilateral elements in subsequent discussions.

We begin with an assumed field, written in terms of the natural coordinate ξ, for both coordinate x and axial displacement u:

$$x = \lfloor 1 \quad \xi \quad \xi^2 \rfloor \begin{Bmatrix} a_1 \\ a_2 \\ a_3 \end{Bmatrix} \qquad \text{and} \qquad u = \lfloor 1 \quad \xi \quad \xi^2 \rfloor \begin{Bmatrix} a_4 \\ a_5 \\ a_6 \end{Bmatrix} \tag{6.2-1}$$

where the a_i are generalized coordinates. One can now determine shape functions by following the formal substitution procedure detailed in Eqs. 3.8-5 through 3.8-8. Another way to determine shape functions is to use Lagrange's interpolation formula, Eq. 3.12-3, replacing x's by ξ's. Or, finally, one can proceed largely by inspection. For example, note that the linear ramps $\frac{1}{2}(1 - \xi)$ and $\frac{1}{2}(1 + \xi)$ have magnitude $\frac{1}{2}$ at $\xi = 0$ (Fig. 3.12-1*a*). We require that N_1 and N_2 both vanish at $\xi = 0$. Accordingly, if $N_3 = 1 - \xi^2$ is known, we obtain $N_1 = \frac{1}{2}(1 - \xi) - \frac{1}{2}N_3$ and $N_2 = \frac{1}{2}(1 + \xi) - \frac{1}{2}N_3$. By any of these procedures, we arrive at

$$x = \lfloor \mathbf{N} \rfloor \lfloor x_1 \quad x_2 \quad x_3 \rfloor^T \qquad \text{and} \qquad u = \lfloor \mathbf{N} \rfloor \lfloor u_1 \quad u_2 \quad u_3 \rfloor^T \tag{6.2-2}$$

where

$$\lfloor \mathbf{N} \rfloor = \lfloor \tfrac{1}{2}(-\xi + \xi^2) \qquad \tfrac{1}{2}(\xi + \xi^2) \qquad 1 - \xi^2 \rfloor \tag{6.2-3}$$

The element is isoparametric because the same $\lfloor \mathbf{N} \rfloor$ is used for interpolation of both x and u. To evaluate x or u at any point on the bar, we substitute the ξ coordinate of that point into Eq. 6.2-2.

Construction of the element stiffness matrix requires that the strain–displacement relation be known. Axial strain ϵ_x is

$$\epsilon_x = \frac{du}{dx} = \left(\frac{d}{dx} \lfloor \mathbf{N} \rfloor \right) \begin{Bmatrix} u_1 \\ u_2 \\ u_3 \end{Bmatrix}, \qquad \text{where} \qquad \frac{d}{dx} = \frac{d\xi}{dx}\frac{d}{d\xi} \tag{6.2-4}$$

The chain rule for d/dx must be invoked because $\lfloor \mathbf{N} \rfloor$ is expressed in terms of ξ rather than in terms of x. Unfortunately, $d\xi/dx$ is not immediately available. We must first calculate its *inverse*, $dx/d\xi$, from the first of Eqs. 6.2-2. Let $J = dx/d\xi$. Then

$$J = \frac{d}{d\xi} \lfloor \mathbf{N} \rfloor \begin{Bmatrix} x_1 \\ x_2 \\ x_3 \end{Bmatrix} = \lfloor \tfrac{1}{2}(-1 + 2\xi) \qquad \tfrac{1}{2}(1 + 2\xi) \qquad -2\xi \rfloor \begin{Bmatrix} x_1 \\ x_2 \\ x_3 \end{Bmatrix} \tag{6.2-5}$$

J is called a *Jacobian*. It can be regarded as a scale factor that describes the physical length dx associated with a reference length $d\xi$; that is, $dx = J\, d\xi$. The element stiffness matrix is

$$[\mathbf{k}] = \int_0^L \lfloor \mathbf{B} \rfloor^T AE \lfloor \mathbf{B} \rfloor \, dx = \int_{-1}^{1} \lfloor \mathbf{B} \rfloor^T AE \lfloor \mathbf{B} \rfloor \, J \, d\xi \tag{6.2-6}$$

where, from Eqs. 6.2-4 and 6.2-5,

$$\lfloor \mathbf{B} \rfloor = \frac{1}{J} \frac{d}{d\xi} \lfloor \mathbf{N} \rfloor = \frac{1}{J} \lfloor \tfrac{1}{2}(-1 + 2\xi) \quad \tfrac{1}{2}(1 + 2\xi) \quad -2\xi \rfloor \tag{6.2-7}$$

Only if node 3 is at the middle of the bar does J reduce to the constant value $J = L/2$. The specific form of J depends on the numerical values assigned to x_1, x_2, and x_3 in Eq. 6.2-5. In general, J is a function of ξ. Accordingly, $\lfloor \mathbf{B} \rfloor$ contains ξ in both numerator and denominator of every term. Therefore, Eq. 6.2-6 cannot be conveniently integrated in closed form. In practice, numerical integration is used instead.

The preceding example illustrates some of the concepts and manipulations associated with isoparametric elements, but shows none of their versatility. For this we must consider plane and solid elements.

6.3 PLANE BILINEAR ISOPARAMETRIC ELEMENT

The following development generalizes the four-node element of Fig. 4.2-4 from a rectangle to arbitrary quadrilateral shape. For a rectangular element of side lengths $2a$ and $2b$, with $x = 0$ and $y = 0$ at the element center, isoparametric coordinates ξ and η can be regarded as dimensionless Cartesian coordinates $\xi = x/a$ and $\eta = y/b$. This special case can be used as a study aid in the following discussion.

Isoparametric coordinates in a plane are shown in Fig. 6.3-1*a*. For a four-node element, axes ξ and η pass through midpoints of opposite sides. Axes ξ and η need not be orthogonal, and neither need be parallel to the x axis or the y axis. Sides of the element are at $\xi = \pm 1$ and at $\eta = \pm 1$. Coordinates x and y within the element are defined by

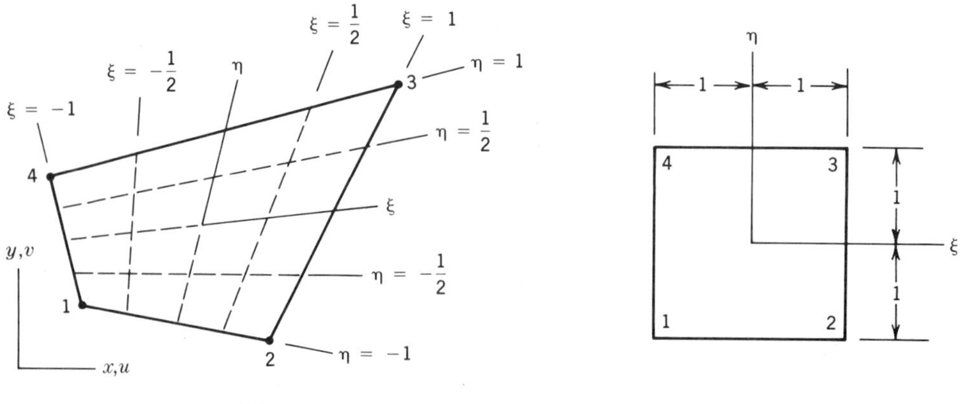

(*a*) (*b*)

Figure 6.3-1. (*a*) Four-node plane isoparametric element in xy space. (*b*) Plane isoparametric element in $\xi\eta$ space.

$$x = \sum N_i x_i \qquad \text{and} \qquad y = \sum N_i y_i \tag{6.3-1}$$

Summations run from 1 to 4. Individual shape functions are

$$\begin{aligned} N_1 &= \tfrac{1}{4}(1 - \xi)(1 - \eta) \qquad & N_2 &= \tfrac{1}{4}(1 + \xi)(1 - \eta) \\ N_3 &= \tfrac{1}{4}(1 + \xi)(1 + \eta) \qquad & N_4 &= \tfrac{1}{4}(1 - \xi)(1 + \eta) \end{aligned} \tag{6.3-2}$$

These N_i are clearly similar to those in Eq. 3.12-10; indeed they are identical for the special case $\xi = x/a$ and $\eta = y/b$. Equations 6.3-2 can be established by any of the usual methods: formal substitution, Lagrange's interpolation formula, or inspection and trial.

For a given element geometry, the orientation of $\xi\eta$ axes with respect to xy axes is dictated by Eqs. 6.3-2 and the node numbers assigned to the element at hand. For example, in Fig. 6.3-1a, a cyclic change in node numbers (2 changed to 1, 3 changed to 2, etc.) would place the ξ axis where the η axis is now shown and would place the η axis in the present $-\xi$ direction. Figure 6.3-1b would not be changed: because of Eqs. 6.3-2, node 1 is always at $\xi = \eta = -1$, node 2 always at $\xi = 1$ and $\eta = -1$, and so on.

The point $\xi = \eta = 0$ can be regarded as the center of the element, but it is not in general the centroid of the element area.

Scalar Field Element. Let the field quantity be ϕ, where $\phi = \phi(x,y)$ or $\phi = \phi(\xi,\eta)$. For example, in heat conduction analysis, ϕ represents temperature. The simplest element has one d.o.f. per node. Within a four-node element, ϕ is interpolated from nodal values $\{\boldsymbol{\phi}_e\} = \lfloor \phi_1 \quad \phi_2 \quad \phi_3 \quad \phi_4 \rfloor^T$,

$$\phi = \lfloor \mathbf{N} \rfloor \{\boldsymbol{\phi}_e\} \qquad \text{or} \qquad \phi = \sum N_i \phi_i \tag{6.3-3}$$

where, to make the element isoparametric, the N_i are taken from Eq. 6.3-2. The following derivatives of ϕ are needed in element formulation:

$$\begin{Bmatrix} \phi_{,x} \\ \phi_{,y} \end{Bmatrix} = [\mathbf{B}]\{\boldsymbol{\phi}_e\}, \qquad \text{where} \quad [\mathbf{B}] = \begin{bmatrix} N_{1,x} & N_{2,x} & N_{3,x} & N_{4,x} \\ N_{1,y} & N_{2,y} & N_{3,y} & N_{4,y} \end{bmatrix} \tag{6.3-4}$$

For the element characteristic matrix we refer to Eq. 3.10-9. With k a material property and t the element thickness,

$$[\mathbf{k}] = \int\!\!\int [\mathbf{B}]^T k [\mathbf{B}]\, t\, dx\, dy = \int_{-1}^{1}\int_{-1}^{1} [\mathbf{B}]^T k [\mathbf{B}]\, tJ\, d\xi\, d\eta \tag{6.3-5}$$

in which J arises because of the change of coordinates.[1] In the latter form of Eq. 6.3-5, $[\mathbf{B}]$ is a function of ξ and η, and is determined as follows.

Because ϕ is expressed in terms of ξ and η, not x and y, derivatives needed in Eq. 6.3-4 are not immediately available. We therefore begin by taking derivatives with respect to ξ and η instead:

[1]The Jacobian of the transformation, J, is familiar in elementary calculus: for example, when an integral is changed from Cartesian coordinates to polar coordinates, $dx\, dy$ is replaced by $r\, dr\, d\theta$, in which $J = r$.

$$\begin{Bmatrix} \phi_{,\xi} \\ \phi_{,\eta} \end{Bmatrix} = [\mathbf{D}_N]\{\boldsymbol{\phi}_e\}, \qquad \text{where} \quad [\mathbf{D}_N] = \begin{bmatrix} N_{1,\xi} & N_{2,\xi} & N_{3,\xi} & N_{4,\xi} \\ N_{1,\eta} & N_{2,\eta} & N_{3,\eta} & N_{4,\eta} \end{bmatrix} \tag{6.3-6}$$

From Eqs. 6.3-2, $N_{1,\xi} = -(1 - \eta)/4$, $N_{1,\eta} = -(1 - \xi)/4$, and so on. Next we must relate $\phi_{,\xi}$ and $\phi_{,\eta}$ to $\phi_{,x}$ and $\phi_{,y}$. The necessary relation is derived subsequently and has the form

$$\begin{Bmatrix} \phi_{,x} \\ \phi_{,y} \end{Bmatrix} = [\boldsymbol{\Gamma}] \begin{Bmatrix} \phi_{,\xi} \\ \phi_{,\eta} \end{Bmatrix} \tag{6.3-7}$$

which provides the **[B]** matrix as

$$[\mathbf{B}] = [\boldsymbol{\Gamma}][\mathbf{D}_N] \tag{6.3-8}$$

Jacobian Matrix [J]. We now seek an expression for $[\boldsymbol{\Gamma}]$ in Eq. 6.3-7. By the chain rule,

$$\frac{\partial \phi}{\partial x} = \frac{\partial \phi}{\partial \xi}\frac{\partial \xi}{\partial x} + \frac{\partial \phi}{\partial \eta}\frac{\partial \eta}{\partial x} \qquad \text{and} \qquad \frac{\partial \phi}{\partial y} = \frac{\partial \phi}{\partial \xi}\frac{\partial \xi}{\partial y} + \frac{\partial \phi}{\partial \eta}\frac{\partial \eta}{\partial y} \tag{6.3-9}$$

Thus $\Gamma_{11} = \xi_{,x}$, $\Gamma_{12} = \eta_{,x}$, $\Gamma_{21} = \xi_{,y}$ and $\Gamma_{22} = \eta_{,y}$. Unfortunately, the partial derivatives of ξ and η with respect to x and y are not directly available from our equations. Therefore, we must write the *inverse* of Eq. 6.3-7 first, which is easily done. We write

$$\begin{aligned} \frac{\partial \phi}{\partial \xi} &= \frac{\partial \phi}{\partial x}\frac{\partial x}{\partial \xi} + \frac{\partial \phi}{\partial y}\frac{\partial y}{\partial \xi} \\ \frac{\partial \phi}{\partial \eta} &= \frac{\partial \phi}{\partial x}\frac{\partial x}{\partial \eta} + \frac{\partial \phi}{\partial y}\frac{\partial y}{\partial \eta} \end{aligned} \qquad \text{or} \qquad \begin{Bmatrix} \phi_{,\xi} \\ \phi_{,\eta} \end{Bmatrix} = [\mathbf{J}] \begin{Bmatrix} \phi_{,x} \\ \phi_{,y} \end{Bmatrix} \tag{6.3-10}$$

where **[J]** is called the *Jacobian matrix:*

$$[\mathbf{J}] = \begin{bmatrix} x_{,\xi} & y_{,\xi} \\ x_{,\eta} & y_{,\eta} \end{bmatrix} = \begin{bmatrix} \sum N_{i,\xi} x_i & \sum N_{i,\xi} y_i \\ \sum N_{i,\eta} x_i & \sum N_{i,\eta} y_i \end{bmatrix} \tag{6.3-11}$$

Equation 6.3-11 is valid for *all* plane isoparametric elements, where i ranges over the number of nodes (and shape functions) used to define element geometry. For the four-node element at hand, i runs from 1 to 4, so, from Eqs. 6.3-6 and 6.3-11,

$$[\mathbf{J}] = [\mathbf{D}_N] \begin{bmatrix} x_1 & y_1 \\ x_2 & y_2 \\ x_3 & y_3 \\ x_4 & y_4 \end{bmatrix} \tag{6.3-12}$$

in which, for the bilinear element,

$$[\mathbf{D}_N] = \frac{1}{4} \begin{bmatrix} -(1 - \eta) & (1 - \eta) & (1 + \eta) & -(1 + \eta) \\ -(1 - \xi) & -(1 + \xi) & (1 + \xi) & (1 - \xi) \end{bmatrix} \tag{6.3-13}$$

Matrix $[\mathbf{\Gamma}]$ is the inverse of $[\mathbf{J}]$,

$$[\mathbf{\Gamma}] = [\mathbf{J}]^{-1} = \frac{1}{J}\begin{bmatrix} J_{22} & -J_{12} \\ -J_{21} & J_{11} \end{bmatrix} \tag{6.3-14}$$

where J is the determinant of the Jacobian matrix

$$J = \det[\mathbf{J}] = J_{11}J_{22} - J_{21}J_{12} \tag{6.3-15}$$

Jacobian J can be regarded as a scale factor that yields area $dx\,dy$ from $d\xi\,d\eta$. In general, J is a function of ξ and η, but for rectangles and parallelograms it is constant.

All ingredients are now at hand for evaluation of $[\mathbf{k}]$ according to the second form of Eqs. 6.3-5.

Plane Stress Element. There are now two fields—namely, the displacements

$$u = \sum N_i u_i \qquad \text{and} \qquad v = \sum N_i v_i \tag{6.3-16}$$

where the N_i are again the same as the functions used to define shape, Eqs. 6.3-1 and 6.3-2. Displacements u and v are x-parallel and y-parallel; they are *not* ξ-parallel and η-parallel.

The strain–displacement relation is $\{\boldsymbol{\epsilon}\} = [\mathbf{B}]\{\mathbf{d}\}$, where $\{\mathbf{d}\} = \lfloor u_1 \quad v_1 \quad u_2 \quad \cdots \quad v_4 \rfloor^T$ and $[\mathbf{B}]$ is the product of the rectangular matrices in the following three equations, which respectively state the strain–displacement relation (Eq. 1.5-6), an expanded form of Eq. 6.3-7, and an expanded form of Eq. 6.3-6:

$$\{\boldsymbol{\epsilon}\} = \begin{Bmatrix} \epsilon_x \\ \epsilon_y \\ \gamma_{xy} \end{Bmatrix} = \begin{bmatrix} 1 & 0 & 0 & 0 \\ 0 & 0 & 0 & 1 \\ 0 & 1 & 1 & 0 \end{bmatrix} \begin{Bmatrix} u_{,x} \\ u_{,y} \\ v_{,x} \\ v_{,y} \end{Bmatrix} \tag{6.3-17}$$

$$\begin{Bmatrix} u_{,x} \\ u_{,y} \\ v_{,x} \\ v_{,y} \end{Bmatrix} = \begin{bmatrix} \Gamma_{11} & \Gamma_{12} & 0 & 0 \\ \Gamma_{21} & \Gamma_{22} & 0 & 0 \\ 0 & 0 & \Gamma_{11} & \Gamma_{12} \\ 0 & 0 & \Gamma_{21} & \Gamma_{22} \end{bmatrix} \begin{Bmatrix} u_{,\xi} \\ u_{,\eta} \\ v_{,\xi} \\ v_{,\eta} \end{Bmatrix} \tag{6.3-18}$$

$$\begin{Bmatrix} u_{,\xi} \\ u_{,\eta} \\ v_{,\xi} \\ v_{,\eta} \end{Bmatrix} = \begin{bmatrix} N_{1,\xi} & 0 & N_{2,\xi} & 0 & N_{3,\xi} & 0 & N_{4,\xi} & 0 \\ N_{1,\eta} & 0 & N_{2,\eta} & 0 & N_{3,\eta} & 0 & N_{4,\eta} & 0 \\ 0 & N_{1,\xi} & 0 & N_{2,\xi} & 0 & N_{3,\xi} & 0 & N_{4,\xi} \\ 0 & N_{1,\eta} & 0 & N_{2,\eta} & 0 & N_{3,\eta} & 0 & N_{4,\eta} \end{bmatrix} \underset{8\times 1}{\{\mathbf{d}\}} \tag{6.3-19}$$

Coefficients Γ_{ij} are given by Eqs. 6.3-11 and 6.3-14.

The element stiffness matrix, Eq. 4.1-5, is

$$\underset{8\times 8}{[\mathbf{k}]} = \int\!\!\int \underset{8\times 3}{[\mathbf{B}]^T}\,\underset{3\times 3}{[\mathbf{E}]}\,\underset{3\times 8}{[\mathbf{B}]}\,t\,dx\,dy = \int_{-1}^{1}\int_{-1}^{1} [\mathbf{B}]^T\,[\mathbf{E}]\,[\mathbf{B}]\,tJ\,d\xi\,d\eta \tag{6.3-20}$$

where t is the element thickness. Contributions to the element load vector $\{\mathbf{r}_e\}$, Eq. 4.1-6, include the terms

$$\int_{-1}^{1}\int_{-1}^{1} ([\mathbf{B}]^T[\mathbf{E}]\{\boldsymbol{\epsilon}_0\} - [\mathbf{B}]^T\{\boldsymbol{\sigma}_0\} + [\mathbf{N}]^T\{\mathbf{F}\})\, tJ\, d\xi\, d\eta \tag{6.3-21}$$

For convenience in computer programming, contributions to $\{\mathbf{r}_e\}$ from surface tractions $\{\boldsymbol{\Phi}\}$ are evaluated separately. For this, the manipulations associated with isoparametric coordinates may be unnecessary. For example, linearly varying traction on a straight edge is allocated to nodes as shown in Fig. 4.3-3.

Remarks. Isoparametric elements are geometrically isotropic. Thus, for the element of Fig. 6.3-1, the numerical values of coefficients in $[\mathbf{k}]$ do not depend on whether element nodes are labeled 1–2–3–4, 2–3–4–1, 3–4–1–2, or 4–1–2–3. However, the cyclic order must be maintained and must run counterclockwise if J is not to become negative over part or all of the element.

From Eqs. 6.3-8 and 6.3-14 we see that J appears in the denominator of each coefficient B_{ij}. Again, J is in general a function of ξ and η. Therefore, the denominator of each term to be integrated in Eqs. 6.3-5 and 6.3-20 will in general contain a polynomial of the form $a_1 + a_2\xi + a_3\eta + a_4\xi\eta$. (A polynomial of higher degree appears for elements having more than four nodes.) Closed-form expressions for integrals of terms in Eq. 6.3-20 would be lengthy. Accordingly, integration is done numerically instead, usually by formulas known as Gauss quadrature.

6.4 SUMMARY OF GAUSS QUADRATURE

"Quadrature" is the name applied to evaluating an integral numerically, rather than analytically as is done in tables of integrals. There are many quadrature rules. The reader may have encountered the Newton–Cotes rules such as Simpson's rule. Here we discuss only the Gauss rules, as they are most appropriate for elements discussed in this chapter.

One Dimension. An integral having arbitrary limits can be transformed so that its limits are from -1 to $+1$. With $f = f(x)$, and with the substitution $x = \frac{1}{2}(1 - \xi)x_1 + \frac{1}{2}(1 + \xi)x_2$,

$$I = \int_{x_1}^{x_2} f\, dx \qquad \text{becomes} \qquad I = \int_{-1}^{1} \phi\, d\xi \tag{6.4-1}$$

Thus the integrand is changed from $f = f(x)$ to $\phi = \phi(\xi)$, where ϕ incorporates the Jacobian of the transformation, $J = dx/d\xi = \frac{1}{2}(x_2 - x_1)$. The latter form of Eq. 6.4-1 makes it possible to write convenient quadrature formulas.

The foregoing linear transformation suffices to make arbitrary limit changes. We can always consider a convenient reference interval such as -1 to $+1$. In practice the limit change is done automatically by the isoparametric transformation and J is usually more complicated than $\frac{1}{2}(x_2 - x_1)$.

To approximate the integral in the simplest way, one can sample (evaluate) ϕ at the midpoint $\xi = 0$ and multiply by the length of the interval (Fig. 6.4-1*a*). Thus we approximate the shaded area by a rectangular area of height ϕ_1 and length

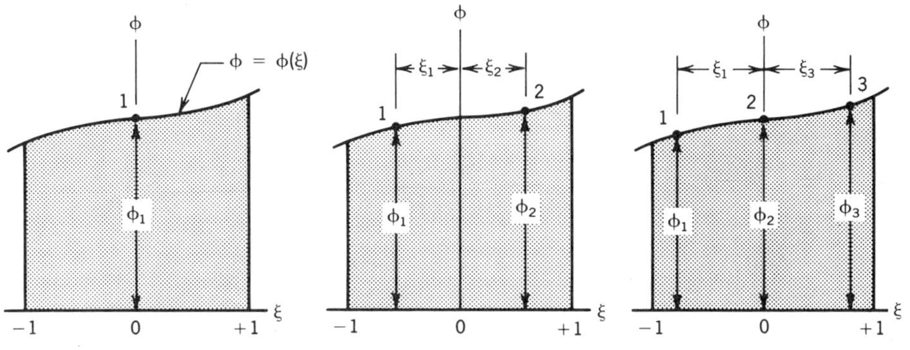

(*a*) (*b*) (*c*)

Figure 6.4-1. Gauss quadrature to compute the shaded area under the curve $\phi = \phi(\xi)$, using (*a*) one, (*b*) two, and (*c*) three sampling points (also called Gauss points).

2, so that $I \approx 2\phi_1$. This result is exact if $\phi = \phi(\xi)$ happens to describe a straight line of any finite slope.

Generalization of the foregoing procedure leads to the quadrature formula

$$I = \int_{-1}^{1} \phi \, d\xi \approx W_1\phi_1 + W_2\phi_2 + \cdots + W_n\phi_n \tag{6.4-2}$$

Thus, to approximate I, we evaluate $\phi = \phi(\xi)$ at each of several locations ξ_i to obtain ordinates ϕ_i, multiply each ϕ_i by an appropriate weight W_i, and add. In the preceding one-point example, where $I \approx 2\phi_1$, we have $n = 1$ and $W_1 = 2$. Gauss was able to prescribe the locations ξ_i and weights W_i such that greatest accuracy is achieved for a given n.

Sampling points are located symmetrically with respect to the center of the integration interval. Symmetrically paired points have the same weight W_i. Data appear in Table 6.4-1. These data are sometimes called Gauss–Legendre coefficients because sampling point locations happen to be roots of Legendre polynomials. Much more extensive tabulations are available [6.2]. In programming,

TABLE 6.4-1. Sampling points and weights for Gauss quadrature over the interval $\xi = -1$ to $\xi = +1$.

Order n	Location ξ_i of Sampling Point	Weight Factor W_i
1	0.	2.
2	$\pm 0.57735\ 02691\ 89626 = \pm 1/\sqrt{3}$	1.
3	$\pm 0.77459\ 66692\ 41483 = \pm\sqrt{0.6}$	$0.55555\ 55555\ 55555 = \dfrac{5}{9}$
	0.	$0.88888\ 88888\ 88888 = \dfrac{8}{9}$
4	$\pm 0.86113\ 63115\ 94053 = \pm\left[\dfrac{3 + 2r}{7}\right]^{1/2}$	$0.34785\ 48451\ 37454 = \dfrac{1}{2} - \dfrac{1}{6r}$
	$\pm 0.33998\ 10435\ 84856 = \pm\left[\dfrac{3 - 2r}{7}\right]^{1/2}$ where $r = \sqrt{1.2}$	$0.65214\ 51548\ 62546 = \dfrac{1}{2} + \dfrac{1}{6r}$

we first check the correctness of tabulated data, then code the ξ_i and W_i with as many digits as the computer allows, in order to avoid unnecessary rounding error.

Example. Consider the polynomial $\phi = a_1 + a_2\xi + a_3\xi^2 + a_4\xi^3$, where the a_i are constants. The exact integral is

$$I = \int_{-1}^{1} \phi \, d\xi = 2a_1 + \tfrac{2}{3}a_3 \tag{6.4-3}$$

The approximate integral given by a one-point rule is

$$I_1 \approx 2a_1 \tag{6.4-4}$$

The integral given by a two-point rule is, with $\xi_1 = -p$, $\xi_2 = p$, and $p = 1/\sqrt{3}$,

$$\begin{aligned} I_2 &= 1.0\,(a_1 - a_2p + a_3p^2 - a_4p^3) + 1.0\,(a_1 + a_2p + a_3p^2 + a_4p^3) \\ I_2 &= 2a_1 + \tfrac{2}{3}a_3 \end{aligned} \tag{6.4-5}$$

In the foregoing example we see an instance of a general rule: *a polynomial of degree 2n − 1 is integrated exactly by n-point Gauss quadrature*. Use of more than n points will still produce the exact result. The *degree of precision* of a quadrature rule is the degree of the highest-order polynomial that is exactly integrated. Thus a second-order Gauss rule has degree of precision 3.

If the function $\phi = \phi(\xi)$ is not a polynomial, Gauss quadrature is inexact, but becomes more accurate as more points are used. Here we refer to the accuracy of integration, not to the accuracy of the results of finite element analysis. We will see, for example, that some elements used for stress analysis are improved by using a Gauss rule of lower order than would be chosen if the goal were accurate integration. The question of what quadrature order is best is addressed in Sections 6.11 and 6.12.

It is important to realize that the *ratio* of two polynomials is in general not a polynomial, and therefore will not be integrated exactly by Gauss quadrature.

Two and Three Dimensions. Multidimensional Gauss rules, called Gaussian product rules, are formed by successive application of one-dimensional Gauss rules. In two dimensions, consider the function $\phi = \phi(\xi,\eta)$. We elect to integrate first with respect to ξ and then with respect to η:

$$\begin{aligned} I &= \int_{-1}^{1}\int_{-1}^{1} \phi(\xi,\eta)\,d\xi\,d\eta \approx \int_{-1}^{1}\left[\sum_i W_i\phi(\xi_i,\eta)\right]d\eta \\ &\approx \sum_j W_j\left[\sum_i W_i\phi(\xi_i,\eta_j)\right] = \sum_i\sum_j W_iW_j\phi(\xi_i,\eta_j) \end{aligned} \tag{6.4-6}$$

For the four-point rule depicted in Fig. 6.4-2a, $W_iW_j = 1$, and Eq. 6.4-6 becomes

$$I \approx \phi_1 + \phi_2 + \phi_3 + \phi_4 \tag{6.4-7}$$

where ϕ_i is the numerical value of ϕ at the ith Gauss point. For the nine-point rule depicted in Fig. 6.4-2b, Eq. 6.4-6 yields

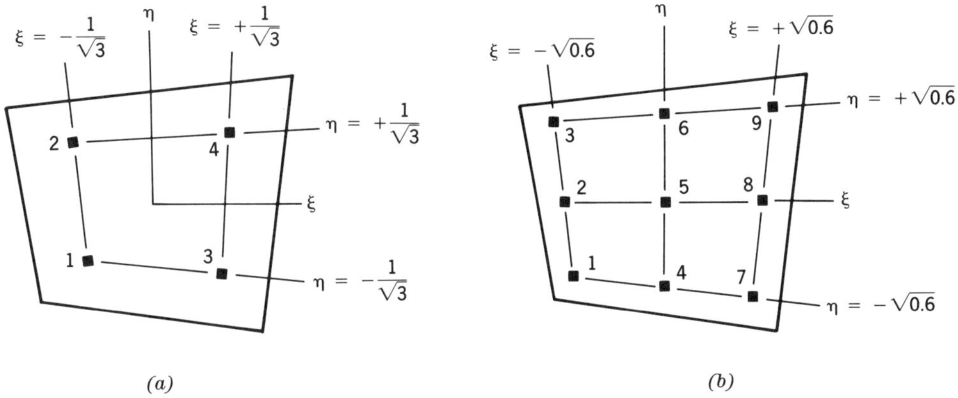

Figure 6.4-2. Gauss point locations in a quadrilateral element using (*a*) four points (order 2 rule), and (*b*) nine points (order 3 rule).

$$I \approx \frac{25}{81}(\phi_1 + \phi_3 + \phi_7 + \phi_9) + \frac{40}{81}(\phi_2 + \phi_4 + \phi_6 + \phi_8) + \frac{64}{81}\phi_5 \quad (6.4\text{-}8)$$

In three dimensions, the Gauss quadrature rule has the form

$$I = \int_{-1}^{1}\int_{-1}^{1}\int_{-1}^{1} \phi(\xi,\eta,\zeta)\, d\xi\, d\eta\, d\zeta \approx \sum_i \sum_j \sum_k W_i W_j W_k \phi(\xi_i,\eta_j,\zeta_k) \quad (6.4\text{-}9)$$

In an equation such as Eq. 6.3-20, each coefficient of the integrand $[\mathbf{B}]^T[\mathbf{E}][\mathbf{B}]tJ$ is in general a function of ξ and η. There are 64 coefficients (or, 36 *different* coefficients, owing to symmetry), and each must be integrated as ϕ is integrated in Eq. 6.4-6: by evaluation at specific points, multiplication by weight factors, and addition. It is not necessary to use the same Gauss rule in both directions, but doing so is most common.

6.5 COMPUTER SUBROUTINES FOR THE BILINEAR ISOPARAMETRIC ELEMENT

The stiffness matrix of Eq. 6.3-20 and the load vector of Eq. 6.3-21 are evaluated by the subroutines presented here. Gauss quadrature is used. Of course, the coding is not unique: other procedures may be more compact, more general, or more efficient [6.3,6.4]. Nevertheless, the subroutines show precisely what must be done. A thorough understanding of these subroutines makes it easier to understand the isoparametric formulation in general.

The notation, assumptions, and procedures are as follows. Shape functions of Eqs. 6.3-2 and their derivatives can be written in the form

$$N_i = (1 + \xi\xi_i)(1 + \eta\eta_i)/4 \quad (6.5\text{-}1a)$$

$$N_{i,\xi} = \xi_i(1 + \eta\eta_i)/4 \quad (6.5\text{-}1b)$$

$$N_{i,\eta} = \eta_i(1 + \xi\xi_i)/4 \quad (6.5\text{-}1c)$$

```
      SUBROUTINE SHAPE (PXI,PET,XL,YL,EXI,EYI,ESI,THC,THK,DETJAC)
      IMPLICIT DOUBLE PRECISION (A-H,O-Z)
      DOUBLE PRECISION NXI(4),NET(4),JAC
      DIMENSION XII(4),ETI(5)
      DIMENSION DUP(1),XL(4),YL(4),EXI(4),EYI(4),ESI(4),THC(4)
      COMMON /Q4/ EN(4),JAC(2,2),B(3,9),E(3,3),SE(8,8),RE(8)
      EQUIVALENCE (XII(1),ETI(2)),(DUP(1),JAC(1))
      DATA ETI /-1.D0,-1.D0,+1.D0,+1.D0,-1.D0/
C --- Find shape functions (EN) and their derivatives (NXI,NET).
      DO 10 L=1,4
      DUM1 = (1. + XII(L)*PXI)/4.
      DUM2 = (1. + ETI(L)*PET)/4.
      EN(L) = 4.*DUM1*DUM2
      NXI(L) = XII(L)*DUM2
   10 NET(L) = ETI(L)*DUM1
C --- Clear arrays JAC and B (words in common are stored in sequence).
      DO 20 L=1,31
   20 DUP(L) = 0.
C --- Find Jacobian JAC and its determinant.  Replace JAC by its inverse.
      DO 30 L=1,4
      JAC(1,1) = JAC(1,1) + NXI(L)*XL(L)
      JAC(1,2) = JAC(1,2) + NXI(L)*YL(L)
      JAC(2,1) = JAC(2,1) + NET(L)*XL(L)
   30 JAC(2,2) = JAC(2,2) + NET(L)*YL(L)
      DETJAC = JAC(1,1)*JAC(2,2) - JAC(2,1)*JAC(1,2)
      DUM1   = JAC(1,1)/DETJAC
      JAC(1,1) =  JAC(2,2)/DETJAC
      JAC(1,2) = -JAC(1,2)/DETJAC
      JAC(2,1) = -JAC(2,1)/DETJAC
      JAC(2,2) =  DUM1
C --- Form strain-displacement matrix [B] (zero entries are already set).
      DO 40 J=1,4
      L = 2*J
      K = L-1
      B(1,K) = JAC(1,1)*NXI(J) + JAC(1,2)*NET(J)
      B(2,L) = JAC(2,1)*NXI(J) + JAC(2,2)*NET(J)
      B(3,K) = B(2,L)
   40 B(3,L) = B(1,K)
C --- Interpolate initial strains from corner values, and store as column
C --- 9 of [B] (already cleared to zero).  Similarly, get thickness THK.
      THK = 0.
      DO 50 L=1,4
      B(1,9) = B(1,9) + EN(L)*EXI(L)
      B(2,9) = B(2,9) + EN(L)*EYI(L)
      B(3,9) = B(3,9) + EN(L)*ESI(L)
   50 THK    = THK    + EN(L)*THC(L)
      RETURN
      END
```

Figure 6.5-1. Fortran subroutine SHAPE. For the plane stress element of Section 6.3, it calculates the shape functions and their derivatives, the Jacobian matrix, its inverse and determinant, matrix [**B**], and initial strains and element thickness, all at the point whose ξ and η coordinates are PXI and PET.

where i is the number of the shape function, and

$$\begin{aligned} \xi_i &= -1., 1., 1., -1. \quad &\text{for} \quad i = 1, 2, 3, 4 \\ \eta_i &= -1., -1., 1., 1. \quad &\text{for} \quad i = 1, 2, 3, 4 \end{aligned} \tag{6.5-2}$$

In Fig. 6.5-1, ξ_i and η_i are placed in arrays XII and ETI by DATA and EQUIVALENCE statements. N_i, $N_{i,\xi}$ and $N_{i,\eta}$ are computed and stored in arrays EN, NXI, and NET. Because of the EQUIVALENCE statement containing DUP(1), the loop on statement 20 neatly initializes arrays JAC and B to zero. Coordinates ξ and η in Eqs. 6.5-1 are called PXI and PET in Fig. 6.5-1 and are transmitted as formal parameters. PXI and PET are Gauss point coordinates if SHAPE is called by QUAD4 (Fig. 6.5-2), but other coordinates could be prescribed by a subsequent calling routine (as, for example, when evaluating [**B**] for use in stress calculation by means of Eq. 4.7-1).

```
      SUBROUTINE QUAD4 (NGAUSS,XL,YL,EXI,EYI,ESI,THC,PLACE,WGT,
     1                 BODYFX,BODYFY)
      IMPLICIT DOUBLE PRECISION (A-H,O-Z)
      DOUBLE PRECISION JAC
C  E = material property matrix. Data at element nodes is as follows:
C  XY,YL       = Cartesian coordinates.         (Used in subroutine SHAPE)
C  EXI,EYI,ESI = initial strains (x,y,shear).  (Used in subroutine SHAPE)
C  THC         = thicknesses in z direction.   (Used in subroutine SHAPE)
C  The calling program must supply the following data in 3 by 3 arrays:
C              | 0.  -.57735---  -.77459--- |          | 2. 1. .555--- |
C  [PLACE] =   | 0.  +.57735---  0.         | , [WGT] = | 0. 1. .888--- |
C              | 0.  0.          +.77459--- |          | 0. 0. .555--- |
      COMMON /Q4/ EN(4),JAC(2,2),B(3,9),E(3,3),SE(8,8),RE(8)
      DIMENSION XL(4),YL(4),EXI(4),EYI(4),ESI(4),THC(4),BTE(8,3),
     1          PLACE(3,3),WGT(3,3)
C --- Clear load vector {r} and upper triangle of stiffness matrix [k].
      DO 10 K=1,8
      RE(K) = 0.
      DO 10 L=K,8
   10 SE(K,L) = 0.
C --- Start Gauss quadrature loop. Use NGAUSS by NGAUSS rule.
      DO 90 NA = 1,NGAUSS
      PXI = PLACE(NA,NGAUSS)
        DO 80 NB = 1,NGAUSS
        PET = PLACE(NB,NGAUSS)
        CALL SHAPE (PXI,PET,XL,YL,EXI,EYI,ESI,THC,THK,DETJAC)
        DV = WGT(NA,NGAUSS)*WGT(NB,NGAUSS)*THK*DETJAC
C ------- Store [B]-transpose times [E] in 8 by 3 work array [BTE].
          DO 30 J=1,4
          L = 2*J
          K = L-1
C --------- Do only multiplications that give a nonzero product.
            DO 20 N=1,3
            BTE(K,N) = B(1,K)*E(1,N) + B(3,K)*E(3,N)
   20       BTE(L,N) = B(2,L)*E(2,N) + B(3,L)*E(3,N)
C ------- Add contribution of body forces to nodal load array {r}.
          RE(K) = RE(K) + EN(J)*BODYFX*DV
   30     RE(L) = RE(L) + EN(J)*BODYFY*DV
C --------- Loop on rows of [k] (array SE) and {r} (array RE).
            DO 70 NROW=1,8
C ----------- Add contribution of initial strains to load array {r}.
              DO 40 J=1,3
   40         RE(NROW) = RE(NROW) + BTE(NROW,J)*B(J,9)*DV
C --------- Loop to add contribution to element stiffness matrix [k].
            DO 60 NCOL=NROW,8
            DUM = 0.
C ----------- Loop for product [B]T*[E]*[B]. Zeros in [B] not skipped.
              DO 50 J=1,3
   50         DUM = DUM + BTE(NROW,J)*B(J,NCOL)
   60       SE(NROW,NCOL) = SE(NROW,NCOL) + DUM*DV
   70     CONTINUE
   80   CONTINUE
   90 CONTINUE
C --- Fill in lower triangle of element stiffness matrix by symmetry.
      DO 100 K=1,7
      DO 100 L=K,8
  100 SE(L,K) = SE(K,L)
      RETURN
      END
```

Figure 6.5-2. Fortran subroutine QUAD4. It generates [**k**] and $\{\mathbf{r}_e\}$ for the plane stress element of Section 6.3 by Gauss quadrature. We store [**k**] in array SE and $\{\mathbf{r}_e\}$ in array RE.

Through statement 40, subroutine SHAPE follows exactly the development in Section 6.3. In the DO 50 loop, initial strains (ϵ_{x0} = EXI, ϵ_{y0} = EYI, γ_{xy0} = ESI) and element thickness (t = THC) are prescribed at the four nodes. Values of these quantities at coordinates PXI and PET are found by interpolation (write $t = \Sigma N_i t_i$, analogous to Eqs. 6.3-1). Initial stresses $\{\boldsymbol{\sigma}_0\}$ are not included but can be added as an exercise.

The foregoing calculations in SHAPE must be carried out at every Gauss point used by QUAD4 (Fig. 6.5-2).

QUAD4 requires as input data the global nodal coordinates XL and YL, nodal thicknesses and initial strains, the material property matrix [**E**] (presumed full, as for a general material, and constant over the element), body forces F_x = BODYFX and F_y = BODYFY, and the quadrature order NGAUSS. As indicated by comments in the listing, Gauss quadrature data must be coded before the subroutine is used. Although perhaps less obvious than subroutine SHAPE, subroutine QUAD4 is a straightforward application of Gauss quadrature to Eqs. 6.3-20 and 6.3-21. DV represents the product $W_i W_j tJ$, which is a common multiplier of each coefficient to be integrated—that is, of each coefficient in $[\mathbf{B}]^T[\mathbf{E}][\mathbf{B}]$, in $[\mathbf{B}]^T[\mathbf{E}]\{\boldsymbol{\epsilon}_0\}$, and in $[\mathbf{N}]^T\{\mathbf{F}\}$.

Cost is reduced by using quadrature to generate only the upper triangle of [**k**], leaving the lower triangle to be completed by symmetry as the last step. Coding of the DO 20 loop is an attempt to exploit the sparsity of [**B**]. Additional economies have been proposed [6.3,6.4].

6.6 QUADRATIC PLANE ELEMENTS

One, two, or more nodes can be placed on each side of a four-node quadrilateral. The resulting elements are called quadratic, cubic, and so on. Here we discuss the quadratic quadrilateral element (Fig. 6.6-1). Quadratic triangular elements are discussed in Section 6.8.

As with the bilinear element, sides of the quadratic element are at $\xi = \pm 1$ and at $\eta = \pm 1$. Two of the side nodes are at $\xi = 0$ and two are at $\eta = 0$. Axes ξ and η may be curved in a quadratic element. As shown in Fig. 6.6-1, sides of an undeformed element may be straight lines or quadratic curves. Similarly, displacements may be linear or quadratic. (A side capable of deforming quadratically need not actually do so; it would deform only linearly in passing a constant-strain patch test.)

Shape functions of a quadratic element can be generated systematically [3.4], or by inspection and trial, as follows. In Fig. 6.6-2*a*, one obtains N_5 by interpolating quadratically in ξ and linearly in η, taking care that $N_5 = 1$ at node 5 and $N_5 = 0$ at all other nodes. Similarly, N_8 is obtained in Fig. 6.6-2*b*. Next, one observes that $N_{(c)}$ (which is N_1 of a bilinear element) has ordinate 0.5 at nodes 5 and 8. The function $N_{(c)} - \frac{1}{2}N_5 - \frac{1}{2}N_8$ is therefore zero at all nodes but node 1, where

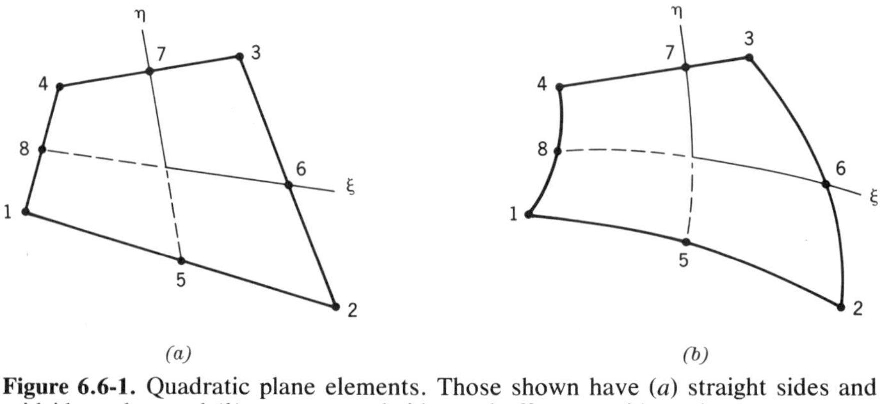

Figure 6.6-1. Quadratic plane elements. Those shown have (*a*) straight sides and midside nodes, and (*b*) some curved sides and off-center side nodes.

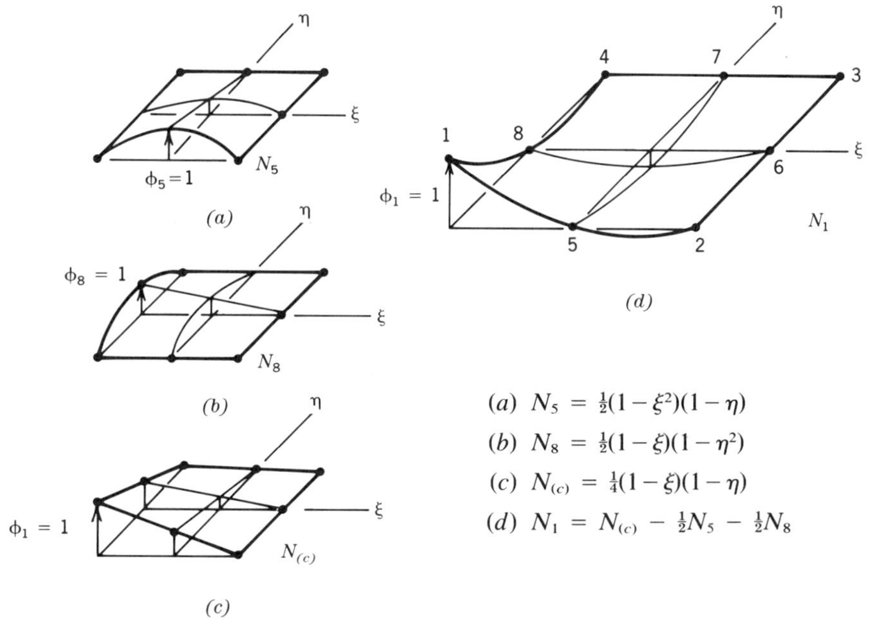

Figure 6.6-2. Selected shape functions for the quadratic element in Fig. 6.6-1, shown normal to square elements in $\xi\eta$ coordinates.

it is unity; therefore, it is shape function N_1. The complete set of shape functions is

$$
\begin{aligned}
N_1 &= \tfrac{1}{4}(1-\xi)(1-\eta) - \tfrac{1}{2}(N_8+N_5) & N_5 &= \tfrac{1}{2}(1-\xi^2)(1-\eta) \\
N_2 &= \tfrac{1}{4}(1+\xi)(1-\eta) - \tfrac{1}{2}(N_5+N_6) & N_6 &= \tfrac{1}{2}(1+\xi)(1-\eta^2) \\
N_3 &= \tfrac{1}{4}(1+\xi)(1+\eta) - \tfrac{1}{2}(N_6+N_7) & N_7 &= \tfrac{1}{2}(1-\xi^2)(1+\eta) \\
N_4 &= \tfrac{1}{4}(1-\xi)(1+\eta) - \tfrac{1}{2}(N_7+N_8) & N_8 &= \tfrac{1}{2}(1-\xi)(1-\eta^2)
\end{aligned}
\qquad (6.6\text{-}1)
$$

The foregoing element, and other isoparametric elements having boundary nodes only, are sometimes called "serendipity" elements. Addition of an internal node (node 9) at $\xi = \eta = 0$ in Fig. 6.6-1 makes the element a "Lagrange" quadratic element, Fig. 6.6-3*a*. Element sides may be straight or curved. The name "Lagrange" is used because the element shape functions can be obtained by taking

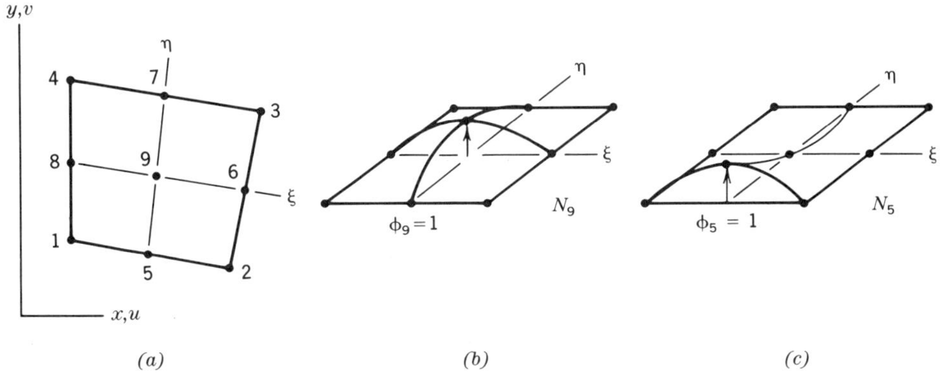

Figure 6.6-3. (*a*) Nine-node Lagrange element in Cartesian coordinates. (*b,c*) Shape functions N_9 and N_5, shown normal to square elements in $\xi\eta$ coordinates.

products of one-dimensional Lagrange interpolants (Eqs. 3.12-6). Computed results are usually most accurate if node 9 is assigned the location given by $\xi = \eta = 0$ in Eqs. 6.6-1—that is,

$$x_9 = \sum N_i x_i = -\tfrac{1}{4}(x_1 + x_2 + x_3 + x_4) + \tfrac{1}{2}(x_5 + x_6 + x_7 + x_8) \qquad (6.6\text{-}2)$$

and similarly for y_9. Thus the geometry of the Lagrange element is completely defined by coordinates of the eight boundary nodes.

The shape function associated with node 9 of the quadratic Lagrange element is

$$N_9 = (1 - \xi^2)(1 - \eta^2) \qquad (6.6\text{-}3)$$

which may be called a "bubble function" because it resembles a bubble blown over a quadrilateral opening in a plate, as shown in Fig. 6.6-3*b*. The first eight shape functions of the Lagrange quadratic element can be obtained by modifying the N_i of Eqs. 6.6-1 so that each is zero at $\xi = \eta = 0$ (compare N_5 in Fig. 6.6-2*a* with N_5 in Fig. 6.6-3*c*). The results are shown in Table 6.6-1, which is explained as follows.

With all nine nodes included, all nine N_i of Table 6.6-1 are used. If only node 9 is omitted, the N_i reduce to those of Eqs. 6.6-1. If nodes 5 through 9 are omitted, the element becomes bilinear and the N_i reduce to those of Eqs. 6.3-2. If nodes 6 through 9 are omitted, the element has three linear sides and one quadratic side, and N_3 and N_4 become bilinear shape functions. (A five-node element is a "transition" element that can be connected to both bilinear and biquadratic elements without incompatibility.) Alternative shape functions for the nine-node element are explained in connection with Eqs. 8.1-4 and 8.1-5.

Let it be required to determine the characteristic matrix [**k**] of a scalar field element (see Eq. 6.3-5). To do so, we express J and [**B**] in terms of ξ and η, then perform Gauss quadrature as described in Section 6.4. [**J**] is given by Eq. 6.3-11, in which index i runs from 1 to n, where n is the number of nodes in the element; that is,

TABLE 6.6-1. SHAPE FUNCTIONS OF A PLANE QUADRILATERAL THAT HAS FROM FOUR TO NINE NODES. EXAMPLE (NOTE THAT N_5 THROUGH N_8 CONTAIN N_9): $N_1 = \frac{1}{4}(1 - \xi)(1 - \eta) - \frac{1}{4}(1 - \xi^2)(1 - \eta) - \frac{1}{4}(1 - \xi)(1 - \eta^2) + \frac{1}{4}N_9$. NODE 9 IS AT $\xi = \eta = 0$.

	Include Only If Node i Is Present in the Element				
	$i = 5$	$i = 6$	$i = 7$	$i = 8$	$i = 9$
$N_1 = \frac{1}{4}(1 - \xi)(1 - \eta)$	$-\frac{1}{2}N_5$			$-\frac{1}{2}N_8$	$-\frac{1}{4}N_9$
$N_2 = \frac{1}{4}(1 + \xi)(1 - \eta)$	$-\frac{1}{2}N_5$	$-\frac{1}{2}N_6$			$-\frac{1}{4}N_9$
$N_3 = \frac{1}{4}(1 + \xi)(1 + \eta)$		$-\frac{1}{2}N_6$	$-\frac{1}{2}N_7$		$-\frac{1}{4}N_9$
$N_4 = \frac{1}{4}(1 - \xi)(1 + \eta)$			$-\frac{1}{2}N_7$	$-\frac{1}{2}N_8$	$-\frac{1}{4}N_9$
$N_5 = \frac{1}{2}(1 - \xi^2)(1 - \eta)$					$-\frac{1}{2}N_9$
$N_6 = \frac{1}{2}(1 + \xi)(1 - \eta^2)$					$-\frac{1}{2}N_9$
$N_7 = \frac{1}{2}(1 - \xi^2)(1 + \eta)$					$-\frac{1}{2}N_9$
$N_8 = \frac{1}{2}(1 - \xi)(1 - \eta^2)$					$-\frac{1}{2}N_9$
$N_9 = (1 - \xi^2)(1 - \eta^2)$					

$$[\mathbf{J}] = [\mathbf{D}_N]\begin{bmatrix} x_1 & y_1 \\ x_2 & y_2 \\ \cdot & \cdot \\ \cdot & \cdot \\ \cdot & \cdot \\ x_n & y_n \end{bmatrix} \tag{6.6-4}$$

where $$[\mathbf{D}_N] = \begin{bmatrix} N_{1,\xi} & N_{2,\xi} & N_{3,\xi} & \cdots & N_{n,\xi} \\ N_{1,\eta} & N_{2,\eta} & N_{3,\eta} & \cdots & N_{n,\eta} \end{bmatrix} \tag{6.6-5}$$

Hence, Eq. 6.3-14 gives $[\boldsymbol{\Gamma}] = [\mathbf{J}]^{-1}$ and Eq. 6.3-15 gives $J = \det[\mathbf{J}]$. Derivatives of ϕ are

$$\begin{Bmatrix} \phi_{,x} \\ \phi_{,y} \end{Bmatrix} = [\boldsymbol{\Gamma}]\begin{Bmatrix} \phi_{,\xi} \\ \phi_{,\eta} \end{Bmatrix} \quad \text{and} \quad \begin{Bmatrix} \phi_{,\xi} \\ \phi_{,\eta} \end{Bmatrix} = [\mathbf{D}_N]\{\boldsymbol{\phi}_e\} \tag{6.6-6}$$

where $\{\boldsymbol{\phi}_e\}$ are nodal values of ϕ. From Eqs. 6.6-6, we obtain

$$\begin{Bmatrix} \phi_{,x} \\ \phi_{,y} \end{Bmatrix} = [\mathbf{B}]\{\boldsymbol{\phi}_e\}, \quad \text{where} \quad [\mathbf{B}] = [\boldsymbol{\Gamma}][\mathbf{D}_N] \tag{6.6-7}$$

The characteristic matrix, to be integrated numerically, is

$$\underset{n\times n}{[\mathbf{k}]} = \int_{-1}^{1}\int_{-1}^{1} \underset{n\times 2}{[\mathbf{B}]^T} k \underset{2\times n}{[\mathbf{B}]}\, tJ\, d\xi\, d\eta \tag{6.6-8}$$

Except that n may now be greater than 4, the foregoing argument is identical to that in Section 6.3.

Remarks. By adding yet more nodes, quadrilateral elements become successively cubic, quartic, and so on. Examination of the polynomial expansions for serendipity elements (which have no internal nodes) shows that serendipity elements "leave out the middle" of a Pascal triangle, Fig. 6.6-4. In contrast, Lagrange elements (which have one or more internal nodes) use a square block of terms, which allows Lagrange elements to have better accuracy.

All isoparametric elements lose accuracy when distorted from a rectangular

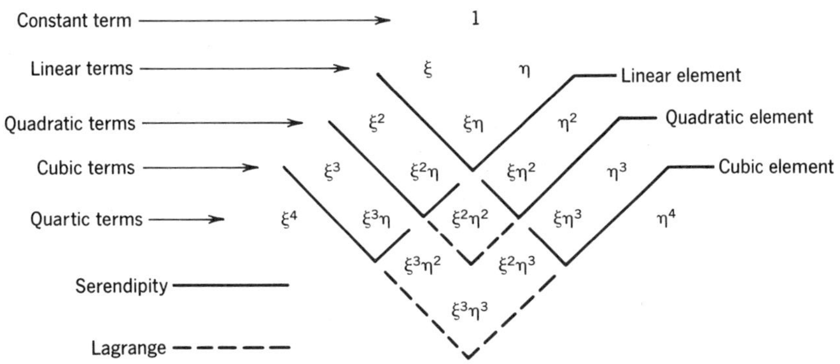

Figure 6.6-4. Polynomial coefficients in plane serendipity elements (boundary nodes only) and plane Lagrange elements (boundary and internal nodes).

shape. However, the nine-node element is much less sensitive than the eight-node element to nonrectangularity, to curvature of sides, and to placing side nodes away from midsides. Indeed, it has been found that a nonrectangular nine-node element can still represent a state of pure bending, provided that sides are straight and side nodes are at midsides as in Fig. 6.6-3*a* [4.6]. The eight-node element does not have this capability.

In dynamic problems, it is often desirable to use lumped (diagonal) mass matrices. All popular lumping schemes result in positive nodal masses for the nine-node element. Such is not the case for the eight-node element, which may have some negative nodal masses (see Chapter 13).

It is easy to alter the subroutines in Section 6.5 to accommodate a quadratic element. The major change is revision of the shape functions in subroutine SHAPE. Otherwise, one increases the size of some arrays and increases the range of most DO loops. The Jacobian matrix remains 2 by 2. In forming **[B]** from Eqs. 6.3-17 through 6.3-19, the rectangular matrix in Eq. 6.3-19 contains 16 or 18 columns, depending on whether node 9 is omitted or included.

6.7 HEXAHEDRAL (SOLID) ISOPARAMETRIC ELEMENTS

In three dimensions, the isoparametric procedure closely resembles the two-dimensional development in Section 6.3. The general procedures used for isoparametric elements have little to do with the specific shape functions of a particular element.

For element geometry and the field quantity ϕ of a solid isoparametric element, we write

$$x = \sum N_i x_i \qquad y = \sum N_i y_i \qquad z = \sum N_i z_i \qquad \phi = \sum N_i \phi_i \tag{6.7-1}$$

where i ranges over the number of nodes in the element. Shape functions N_i are functions of isoparametric coordinates ξ, η, and ζ. Faces of the element lie at $\xi = \pm 1$, $\eta = \pm 1$, and $\zeta = \pm 1$. The Jacobian matrix is defined analogously to Eq. 6.3-11, but is now 3 by 3:

$$[\mathbf{J}] = \begin{bmatrix} x_{,\xi} & y_{,\xi} & z_{,\xi} \\ x_{,\eta} & y_{,\eta} & z_{,\eta} \\ x_{,\zeta} & y_{,\zeta} & z_{,\zeta} \end{bmatrix} = \sum \begin{bmatrix} N_{i,\xi}x_i & N_{i,\xi}y_i & N_{i,\xi}z_i \\ N_{i,\eta}x_i & N_{i,\eta}y_i & N_{i,\eta}z_i \\ N_{i,\zeta}x_i & N_{i,\zeta}y_i & N_{i,\zeta}z_i \end{bmatrix} \tag{6.7-2}$$

With $[\boldsymbol{\Gamma}] = [\mathbf{J}]^{-1}$,

$$\lfloor \phi_{,x} \quad \phi_{,y} \quad \phi_{,z} \rfloor^T = [\boldsymbol{\Gamma}] \lfloor \phi_{,\xi} \quad \phi_{,\eta} \quad \phi_{,\zeta} \rfloor^T \tag{6.7-3}$$

For a scalar field problem in three dimensions, with n the number of nodes per element, Eq. 6.3-5 becomes

$$\underset{n\times n}{[\mathbf{k}]} = \int_{-1}^{1}\int_{-1}^{1}\int_{-1}^{1} \underset{n\times 3}{[\mathbf{B}]^T} k \underset{3\times n}{[\mathbf{B}]} \; J \, d\xi \, d\eta \, d\zeta \tag{6.7-4}$$

The Jacobian determinant $J = \det[\mathbf{J}]$ expresses the ratio of volume $dx\,dy\,dz$ to $d\xi\,d\eta\,d\zeta$. In general, **[J]** and **[B]** are functions of ξ, η, and ζ.

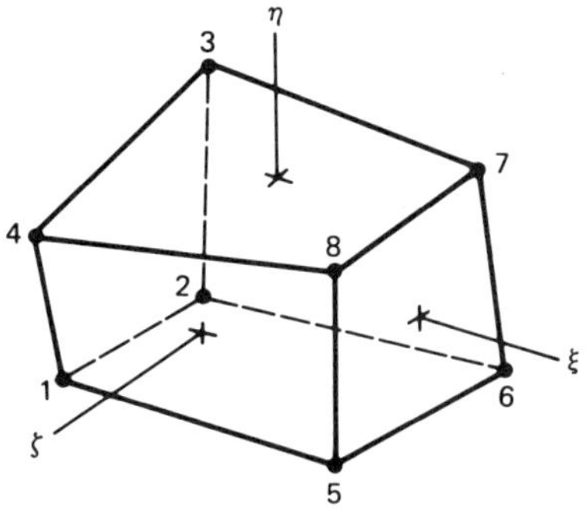

Figure 6.7-1. Linear solid (eight-node brick) element. Also called a "trilinear" element.

In structural mechanics we deal with displacements u, v, and w, which are parallel to x, y, and z directions, respectively. Matrix [**B**] is evaluated by expanded forms of Eqs. 6.3-17 through 6.3-19. As expanded, Eq. 6.3-19 has nine rows, Eq. 6.3-18 contains a 9 by 9 square matrix, and Eq. 6.3-17 states the strain–displacement relations (Eqs. 1.5-6) as

$$\begin{Bmatrix} \epsilon_x \\ \epsilon_y \\ \epsilon_z \\ \gamma_{xy} \\ \gamma_{yz} \\ \gamma_{zx} \end{Bmatrix} = [\mathbf{H}] \begin{Bmatrix} u_{,x} \\ u_{,y} \\ u_{,z} \\ v_{,x} \\ \vdots \\ w_{,z} \end{Bmatrix}, \qquad \text{where} \quad \underset{6\times 9}{[\mathbf{H}]} = \begin{bmatrix} 1 & 0 & 0 & 0 & 0 & 0 & 0 & 0 & 0 \\ 0 & 0 & 0 & 0 & 1 & 0 & 0 & 0 & 0 \\ 0 & 0 & 0 & 0 & 0 & 0 & 0 & 0 & 1 \\ 0 & 1 & 0 & 1 & 0 & 0 & 0 & 0 & 0 \\ 0 & 0 & 0 & 0 & 0 & 1 & 0 & 1 & 0 \\ 0 & 0 & 1 & 0 & 0 & 0 & 1 & 0 & 0 \end{bmatrix} \tag{6.7-5}$$

Shape functions of a linear solid element (also called an eight-node brick element), Fig. 6.7-1, are

$$N_i = \tfrac{1}{8}(1 \pm \xi)(1 \pm \eta)(1 \pm \zeta) \tag{6.7-6}$$

in which $i = 1, 2, \ldots, 8$ and the choice of algebraic signs should be obvious to an adequately prepared reader.

A quadratic "serendipity" solid element has corner nodes and a node on each edge, for a total of 20 nodes. A quadratic "Lagrange" solid element also includes midface nodes and a node at $\xi = \eta = \zeta = 0$, for a total of 27 nodes. Shape functions of these quadratic elements are analogous to those in Eqs. 6.6-1 and Table 6.6-1 and may be found in [2.1]. Stiffness matrices of these elements are, respectively, 60 by 60 and 81 by 81. Quadratic elements may be good enough to permit the structure of Fig. 6.7-2 to be modeled by a single layer of elements, except near the fillet where stress gradients may be large.

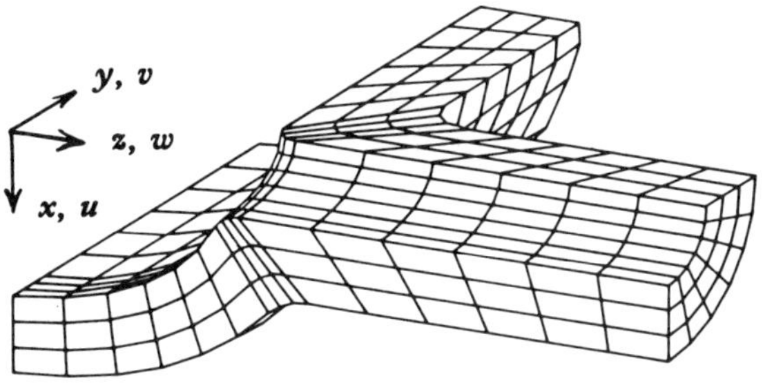

Figure 6.7-2. Model of one octant of a cylinder-to-cylinder intersection.

6.8 TRIANGULAR ISOPARAMETRIC ELEMENTS

In Chapter 5, we discussed triangular elements that are restricted to have straight sides and evenly spaced side nodes (Fig. 6.8-1*a*). In the present section these restrictions are removed, to allow triangles such as that in Fig. 6.8-1*b*, where sides need not be straight and side nodes need not be at midsides. We will see that the formulation procedure is essentially the same as that used for quadrilateral isoparametric elements.

Let ϕ be a scalar field, interpolated from nodal values ϕ_i. Similarly, x and y are coordinates interpolated from nodal values of x_i and y_i:

$$\phi = \sum N_i\phi_i \qquad x = \sum N_i x_i \qquad y = \sum N_i y_i \tag{6.8-1}$$

The element is isoparametric if the same shape functions N_i are used in all three summations. For a triangle, the N_i are expressed in terms of area coordinates ξ_1, ξ_2, and ξ_3, as explained in Section 5.3.

Imagine that the element characteristic matrix $[\mathbf{k}]$ for a scalar field problem is required. Equations 6.6-4 through 6.6-7 may be used exactly as written, without even a change in symbols. We argue as follows.

Area coordinates are not independent. They satisfy the constraint relation $\xi_1 + \xi_2 + \xi_3 = 1$. Accordingly, only two need be given to uniquely locate a point. For example, if ξ_1 and ξ_2 are given and the constraint relation is invoked, Eqs. 6.8-1 can be evaluated. We therefore define

$$\begin{aligned} \xi_1 &= \xi \\ \xi_2 &= \eta \\ \xi_3 &= 1 - \xi - \eta \end{aligned} \tag{6.8-2}$$

To evaluate the shape function derivatives seen in Eq. 6.6-5, we invoke the chain rule:

$$\frac{\partial N_i}{\partial \xi} = \frac{\partial N_i}{\partial \xi_1}\frac{\partial \xi_1}{\partial \xi} + \frac{\partial N_i}{\partial \xi_2}\frac{\partial \xi_2}{\partial \xi} + \frac{\partial N_i}{\partial \xi_3}\frac{\partial \xi_3}{\partial \xi} = \frac{\partial N_i}{\partial \xi_1} - \frac{\partial N_i}{\partial \xi_3} \tag{6.8-3}$$

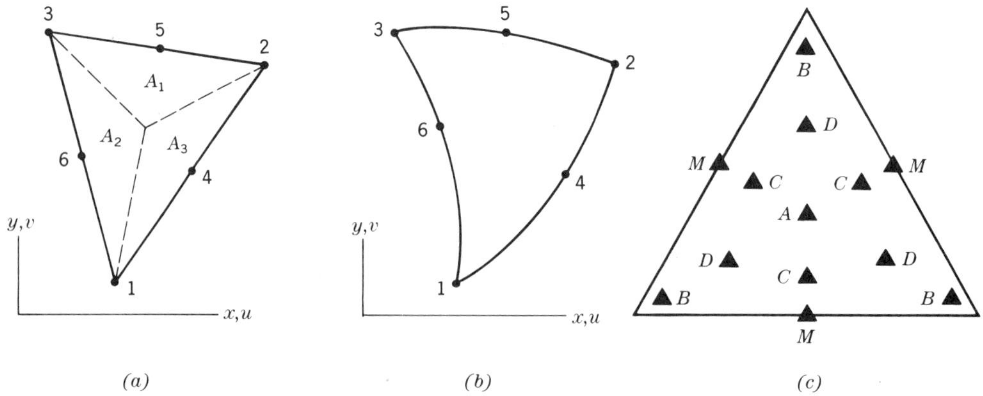

Figure 6.8-1. (*a*) Quadratic triangle with straight sides and midside nodes, showing subareas A_i that define area coordinates ξ_i. (*b*) Quadratic triangle of arbitrary shape. (*c*) Sampling points used in quadrature formulas of Table 6.8-1.

Similarly

$$\frac{\partial N_i}{\partial \eta} = \frac{\partial N_i}{\partial \xi_2} - \frac{\partial N_i}{\partial \xi_3} \tag{6.8-4}$$

For example, using the quadratic shape functions of Eqs. 5.3-5, we obtain for Eq. 6.6-5 the expression

$$\underset{2\times 6}{[\mathbf{D}_N]} = \begin{bmatrix} 4\xi_1 - 1 & 0 & -4\xi_3 + 1 & 4\xi_2 & -4\xi_2 & 4(\xi_3 - \xi_1) \\ 0 & 4\xi_2 - 1 & -4\xi_3 + 1 & 4\xi_1 & 4(\xi_3 - \xi_2) & -4\xi_1 \end{bmatrix} \tag{6.8-5}$$

$[\mathbf{J}]$ is produced by letting $[\mathbf{D}_N]$ premultiply a 6 by 2 array of nodal coordinates (let $n = 6$ in Eq. 6.6-4). Hence, $[\mathbf{B}] = [\mathbf{J}]^{-1}[\mathbf{D}_N]$, as in Eq. 6.6-7.

Equation 6.6-8 is a valid expression for $[\mathbf{k}]$ of an n-node isoparametric triangle if the limits are 0 to 1 on the first integral and 0 to $1 - \eta$ on the second. Alternatively, we can write the following form, which is more appropriate to subsequent numerical integration:

$$\underset{n\times n}{[\mathbf{k}]} = \int_A \underset{n\times 2}{[\mathbf{B}]^T} k \underset{2\times n}{[\mathbf{B}]}\ t\, dA \tag{6.8-6}$$

Here t is element thickness, A is the area of the triangle, and $[\mathbf{B}]$ is a function of area coordinates ξ_1, ξ_2, and ξ_3.

Numerical Integration for Triangles. Let ϕ be a function of area coordinates ξ_1, ξ_2, and ξ_3. The quadrature rule is

$$\int_A \phi\, dA = \frac{1}{2} \sum_{i=1}^{n} W_i J_i \phi_i \tag{6.8-7}$$

in which ϕ_i is the value of ϕ at a specific point in the triangle, W_i is the weight appropriate to this point, and n is the number of sampling points used. In going from Eq. 6.8-6 to Eq. 6.8-7 we set $dA = J\, d\xi\, d\eta$. The factor of $\frac{1}{2}$ appears in Eq. 6.8-7 because the area of a reference triangle in area coordinates is $\frac{1}{2}$. For an undistorted triangle of unit area in Cartesian coordinates, $J = 2$ throughout. Hence, since $\Sigma W_i = 1$ (see data in Table 6.8-1), we obtain $\int dA = 1$ when $\phi = 1$ in Eq. 6.8-7.

Equation 6.8-7 must be applied to each term in the integrand of Eq. 6.8-6.

Equation 6.8-7 is similar in form to Eq. 6.4-6, but in Eq. 6.8-7 weights are expressed directly rather than as the product of one-dimensional weights. Data appear in Table 6.8-1 [6.5]. For points of multiplicity 3 (points B, C, D, and M) there are three points that have the same weight W_i. Area coordinates are given for one of the three; the other two are given by cyclic permutation. For example, using the first three-point formula in Table 6.8-1, we evaluate function ϕ at the three locations

$$\begin{array}{ccc} (1) & (2) & (3) \\ \xi_1 = \frac{2}{3} & \xi_2 = \frac{2}{3} & \xi_3 = \frac{2}{3} \\ \xi_2 = \xi_3 = \frac{1}{6} & \xi_3 = \xi_1 = \frac{1}{6} & \xi_1 = \xi_2 = \frac{1}{6} \end{array} \tag{6.8-8}$$

TABLE 6.8-1. Gauss quadrature formulas for integration over a triangle according to Eq. 6.8-7. Approximate locations of entries in the "points" column are shown in Fig. 6.8-1c. *ns* = points of multiplicity 6, not shown in Fig. 6.8-1c.

Points	Multiplicity	Area Coordinates ξ_1, ξ_2, ξ_3			Weights W_i
		1-point formula	degree of precision 1		
A	1	0.33333 33333 33333	0.33333 33333 33333	0.33333 33333 33333	1.00000 00000 00000
		3-point formula	degree of precision 2		
B	3	0.66666 66666 66667	0.16666 66666 66667	0.16666 66666 66667	0.33333 33333 33333
		3-point formula	degree of precision 2		
M	3	0.50000 00000 00000	0.50000 00000 00000	0.00000 00000 00000	0.33333 33333 33333
		4-point formula	degree of precision 3		
A	1	0.33333 33333 33333	0.33333 33333 33333	0.33333 33333 33333	−0.56250 00000 00000
B	3	0.60000 00000 00000	0.20000 00000 00000	0.20000 00000 00000	0.52083 33333 33333
		6-point formula	degree of precision 4		
B	3	0.81684 75729 80459	0.09157 62135 09771	0.09157 62135 09771	0.10995 17436 55322
C	3	0.10810 30181 68070	0.44594 84909 15965	0.44594 84909 15965	0.22338 15896 78011
		7-point formula	degree of precision 5		
A	1	0.33333 33333 33333	0.33333 33333 33333	0.33333 33333 33333	0.22500 00000 00000
B	3	0.79742 69853 53087	0.10128 65073 23456	0.10128 65073 23456	0.12593 91805 44827
C	3	0.47014 20641 05115	0.47014 20641 05115	0.05971 58717 89770	0.13239 41527 88506
		12-point formula	degree of precision 6		
B	3	0.87382 19710 16996	0.06308 90144 91502	0.06308 90144 91502	0.05084 49063 70207
D	3	0.50142 65096 58179	0.24928 67451 70910	0.24928 67451 70910	0.11678 62757 26379
ns	6	0.63650 24991 21399	0.31035 24510 33784	0.05314 50498 44817	0.08285 10756 18374
		13-point formula	degree of precision 7		
A	1	0.33333 33333 33333	0.33333 33333 33333	0.33333 33333 33333	−0.14957 00444 67682
D	3	0.47930 80678 41920	0.26034 59660 79040	0.26034 59660 79040	0.17561 52574 33208
B	3	0.86973 97941 95568	0.06513 01029 02216	0.06513 01029 02216	0.05334 72356 08838
ns	6	0.63844 41885 69810	0.31286 54960 04874	0.04869 03154 25316	0.07711 37608 90257

and apply weight $W = \frac{1}{3}$ to each. Center point A has multiplicity 1. Points ns (not shown) have multiplicity 6. The formulas in Table 6.8-1 are symmetric in the area coordinates. These formulas were not derived by Gauss but are called Gaussian because sampling points are optimally placed rather than simply being located in a uniform pattern.

The "degree of precision" in Table 6.8-1 refers to the degree of the highest-order complete polynomial in Cartesian coordinates that is integrated exactly by a formula. If ϕ in Eq. 6.8-7 is not a polynomial, numerical integration is not exact but becomes more accurate as more sampling points are used. Inexact integration is expected for elements such as that in Fig. 6.8-1*b*, because geometric distortion produces a $[\mathbf{B}]$ matrix whose terms are not polynomials but rather the *ratio* of two polynomials.

Remarks. Like triangles, tetrahedra can be distorted from regular shapes such as that in Fig. 5.6-1. Triangles and tetrahedra decline in accuracy as sides become curved or side nodes become unevenly spaced. There is evidence that the ten-node tetrahedron is more sensitive to distortion than the six-node triangle.

If elements have straight sides and evenly spaced side nodes, one has the option of writing explicit expressions for stiffness coefficients rather than using numerical integration. The amount of algebra required to explicitly formulate a stiffness matrix in area or volume coordinates can be large. Fortunately, symbol-processing programs are available, of which MACSYMA may be the best known. Operations on polynomials, such as multiplication, factoring, differentiation, and integration can be performed *symbolically,* with output in the form of Fortran statements if so desired. The potential for saving time and reducing errors is obvious [6.6].

6.9 CONSISTENT ELEMENT NODAL LOADS $\{\mathbf{r}_e\}$

Consistent element nodal loads caused by initial strains $\{\boldsymbol{\epsilon}_0\}$, initial stresses $\{\boldsymbol{\sigma}_0\}$, body forces $\{\mathbf{F}\}$, and surface tractions $\{\boldsymbol{\Phi}\}$ are computed according to Eq. 4.1-6. This matter is thoroughly discussed in Section 4.3. Results presented there remain valid if sides are straight and elements are rectangular; it does not matter that the elements may now be called isoparametric. For curved sides and general element shapes, results will differ and will be affected by the amount of geometric distortion.

Loads $\{\mathbf{r}_e\}$ caused by $\{\boldsymbol{\epsilon}_0\}$, $\{\boldsymbol{\sigma}_0\}$, and $\{\mathbf{F}\}$ are given by Eq. 6.3-21 for plane quadrilateral elements having any number of nodes. For solid hexahedra a triple integral is used, in which $tJ\,d\xi\,d\eta$ is replaced by $J\,d\xi\,d\eta\,d\zeta$. For triangular elements integration spans the triangle area A, $tJ\,d\xi\,d\eta$ is replaced by $t\,dA$, and formulas discussed in Section 6.8 are used.

A result of interest for the quadratic triangle is easy to show. Let the triangle be of constant thickness and have straight sides and midside nodes, as in Fig. 6.8-1*a*, and let there be a *uniform* body force in (say) the $+x$ direction. Taking shape functions from Eq. 5.3-5, and using Eq. 5.2-8 for integrations, we obtain the following x-direction nodal loads:

$$\int_A \underset{6\times 1}{\lfloor \mathbf{N} \rfloor^T} F_x t\,dA = \frac{F_x A t}{3} \lfloor 0 \quad 0 \quad 0 \quad 1 \quad 1 \quad 1 \rfloor^T \tag{6.9-1}$$

Thus the entire force F_xAt is apportioned equally to only the midside nodes, regardless of the shape of the triangle (so long as sides remain straight and side nodes remain at midsides). This division also applies to the load produced by a uniform pressure against area A of the triangle. (Such a pressure loading may act on one six-node face of a ten-node tetrahedron, or on a six-node triangle formulated as a plate bending element.) If element sides are curved, the functions to be integrated are not polynomials: then numerical integration is required in Eq. 6.9-1, and the resulting nodal loads are not the same as those in Eq. 6.9-1.

Procedures for evaluating consistent loads $\{\mathbf{r}_e\}$ caused by tractions on curved edges and warped surfaces are discussed in Section 5.8 of the second edition of this book.

6.10 THE VALIDITY OF ISOPARAMETRIC ELEMENTS

The critical test of element validity is the patch test, discussed in Section 4.6. In the present section we will stop short of the patch test, but will argue that the isoparametric formulation endows elements with all characteristics needed for convergence as discussed in Section 4.5. Elements discussed in the present chapter do in fact pass the patch test.

Isoparametric elements are well suited to problems whose functional Π contains first spatial derivatives (or whose governing differential equations contain second spatial derivatives). Examples include heat conduction, plane stress analysis, and stress analysis of solids. We wish to show that isoparametric elements (a) provide C^0 continuity (interelement continuity of the primary field variable), and (b) contain a complete linear polynomial in Cartesian coordinates [6.7].

Continuity. Interelement continuity of a field variable ϕ can be demonstrated with the aid of Fig. 6.10-1. Along the common edge ABC, adjacent elements display the same edge-tangent coordinate: $\eta_1 = \eta_2$. In addition, along edge ABC, shape functions of the two elements are identical functions of η and operate on d.o.f. of nodes A, B, and C only. Thus, whether viewed from element 1 or from element 2, the field variable along ABC is the same function. This argument also shows that elements match geometrically, because coordinates of A, B, and C define a unique quadratic curve.

Completeness. Imagine that $\phi = \phi(x,y,z)$ is a polynomial field. If element nodes are attached to this field, so that nodal d.o.f. are $\phi_i = \phi(x_i,y_i,z_i)$, does shape function interpolation from nodal ϕ_i yield the original field ϕ throughout the element? The answer is yes if ϕ is linear and the element is isoparametric. Specifically, we propose to show that ϕ within an isoparametric element is a complete linear polynomial—that is, that

$$\phi = a_1 + a_2x + a_3y + a_4z \tag{6.10-1}$$

when nodal d.o.f. ϕ_i are consistent with this field and ϕ within the element is evaluated by the interpolation

$$\phi = \sum N_i\phi_i \tag{6.10-2}$$

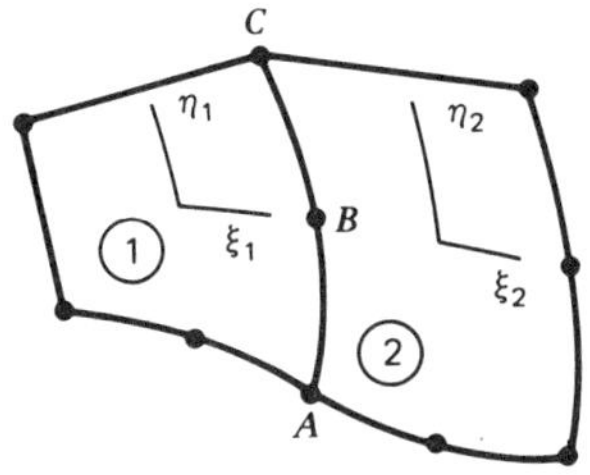

Figure 6.10-1. Adjacent elements cited in compatibility arguments.

To do so we first evaluate Eq. 6.10-1 at each node:

$$\phi_i = a_1 + a_2 x_i + a_3 y_i + a_4 z_i \tag{6.10-3}$$

Here x_i, y_i, and z_i are Cartesian coordinates of node i, and i ranges over all nodes of the element, however many nodes there may be. Equations 6.10-2 and 6.10-3 yield

$$\phi = a_1 \sum N_i + a_2 \sum N_i x_i + a_3 \sum N_i y_i + a_4 \sum N_i z_i \tag{6.10-4}$$

But, if the element is isoparametric, coordinates are interpolated in the same way as ϕ—that is,

$$x = \sum N_i x_i \qquad y = \sum N_i y_i \qquad z = \sum N_i z_i \tag{6.10-5}$$

Accordingly, if $\Sigma N_i = 1$, Eq. 6.10-4 reduces to Eq. 6.10-1, and the proposition is proved.

To show that $\Sigma N_i = 1$, we note that Eqs. 6.10-5 can be used regardless of the location of the origin of Cartesian coordinates with respect to the element. Imagine that there is a second system XYZ, translated a distance h along the x axis so that $x = X + h$. Hence

$$x = \sum N_i x_i = \sum N_i X_i + h \sum N_i = X + h \sum N_i = (x - h) + h \cdot \sum N_i \tag{6.10-6}$$

from which we obtain $h = h \, \Sigma N_i$ and therefore conclude that $\Sigma N_i = 1$.

Furthermore, from $\Sigma N_i = 1$ we obtain

$$\sum N_{i,\xi} = \sum N_{i,\eta} = \sum N_{i,\zeta} = 0 \tag{6.10-7}$$

Equations 6.10-7 can be helpful in checking derivations and coding. Area coordinates for a triangle contain a redundant coordinate, which must be eliminated by means of Eqs. 6.8-2 in order for Eqs. 6.10-7 to be valid.

The foregoing arguments show that isoparametric elements have properties necessary to the passing of a patch test. They say nothing about accuracy in a coarse mesh, convergence rate with mesh refinement, or how accuracy declines as element geometry is distorted from a compact and regular shape (as in going from Fig. 6.6-1a to Fig. 6.6-1b). As sides become curved and as side nodes become unevenly spaced, an element tends to lose its ability to represent quadratic and higher polynomials in xyz coordinates. For example, an element that contains a

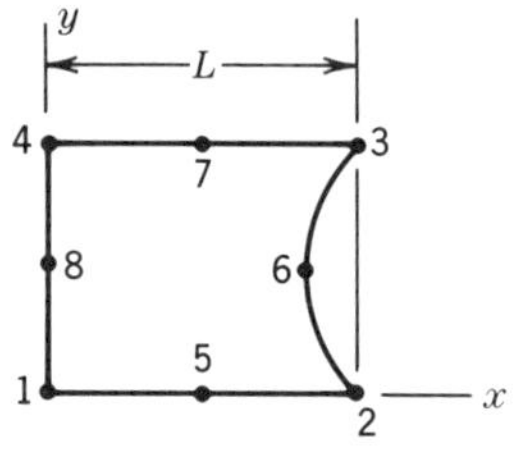

Figure 6.10.2. Superparametric element, in which $\phi = \phi(\phi_1, \phi_2, \phi_3, \phi_4)$.

complete quadratic polynomial in $\xi\eta\zeta$ coordinates may not contain a complete quadratic polynomial in xyz coordinates, depending on the type of distortion from a regular shape. The argument of Eqs. 6.10-1 to 6.10-5 shows only that in spite of geometric distortion, a complete linear polynomial in xyz coordinates will remain.

Subparametric and Superparametric Elements. Such elements are defined in Section 6.1. For example, with straight sides and midside nodes, the element of Fig. 6.6-1*a* is subparametric, as its shape is then defined by the coordinates of nodes 1 through 4. The field quantity ϕ is still defined by the ϕ_i of all eight nodes. For such an element, the foregoing completeness argument remains valid. Specifically, for the element of Fig. 6.6-1*a*, summations in Eqs. 6.10-4 and 6.10-5 run from 1 to 8, but in Eqs. 6.10-5 the interpolations reduce to those of the bilinear element, Eqs. 6.3-1 and 6.3-2, provided that $x_5 = (x_1 + x_2)/2$, $y_5 = (y_1 + y_2)/2$, and so on. Thus the linear field of Eq. 6.10-1 is present when geometry is defined by only the four corner nodes.

Superparametric elements are usually not valid (however, with certain restrictions, valid superparametric elements for beams, plates, and shells are possible). Consider Fig. 6.10-2. Let the shape be defined by all eight nodes, that is, in a more general way than is the field quantity ϕ, which is defined by ϕ_i at nodes 1, 2, 3, and 4 only. Imagine that $\phi_1 = \phi_4 = 0$ while $\phi_2 = \phi_3 = c$, a constant. Thus we expect the element to display the constant gradient $\phi_{,x} = c/L$. However, gradient calculation according to Eq. 6.6-7 makes $\phi_{,x}$ a function of ξ rather than constant, owing to [**J**] in Eq. 6.6-4, in which $x_6 < L$. The superparametric element of Fig. 6.10-2 would fail a patch test.

6.11 APPROPRIATE ORDER OF QUADRATURE

For numerically integrated elements, we define "full integration" as a quadrature rule sufficient to provide the exact integrals of all terms k_{ij} in the element stiffness matrix if the element is undistorted (e.g., if a quadratic element has straight sides and midside nodes). The same "full integration" rule will not exactly integrate all k_{ij} if sides are curved or if side nodes are offset from the midpoints, for then J is not constant throughout the element.

For example, in Eq. 6.2-6, $\lfloor \mathbf{B} \rfloor$ is linear in ξ and J is constant if node 3 is centered. Therefore, the integrand contains terms up to ξ^2, which are integrated exactly by two Gauss points. Accordingly, for this element, even if node 3 is not centered, two-point Gauss quadrature is considered "full integration."

Use of full integration is the only sure way to avoid pitfalls such as mesh instabilities, which are discussed in Section 6.12.

However, a lower-order quadrature rule, called "reduced integration," may be desirable for two reasons. First, since the expense of generating a matrix [**k**] by numerical integration is proportional to the number of sampling points, using fewer sampling points means lower cost. Second, a low-order rule tends to soften an element, thus countering the overly stiff behavior associated with an assumed displacement field. Softening comes about because certain higher-order polynomial terms happen to vanish at Gauss points of a low-order rule, so that these terms make no contribution to strain energy. In other words, with fewer sampling points, some of the more complicated displacement modes offer less resistance to deformation. In sum, our argument is that reduced integration may be able to simultaneously reduce cost, reduce accuracy in the evaluation of integral expressions, and *increase* the accuracy of a finite element analysis. Reduced integration should not be used if cost reduction is the only motivation.

The number of Gauss points has a lower limit because in the limit of mesh refinement, element volume must be integrated exactly. We argue as follows. As a mesh is indefinitely refined, a constant-strain condition is approached in each element, provided that the element is valid in the patch test sense. Thus strain energy density U_0 becomes constant throughout each element. Strain energy in an element, for plane and solid problems, respectively, is

$$U_e = \int\int U_0 tJ \, d\xi \, d\eta \qquad \text{or} \qquad U_e = \int\int\int U_0 J \, d\xi \, d\eta \, d\zeta \qquad (6.11\text{-}1)$$

If U_0 is constant, then U_e will be correct if volume $dV = tJ \, d\xi \, d\eta$ (or $dV = J \, d\xi \, d\eta \, d\zeta$) is correctly integrated. In practice, we prefer to use exact volume integration for any shape and size of element.

From Eq. 4.1-10 we see that integrals in Eq. 6.11-1 produce terms in the element stiffness matrix [**k**]. Accordingly, if [**k**] is produced by an integration rule adequate to compute element volume exactly, the element will be able to provide the correct strain energy in a constant-strain deformation mode.

Thus, for an element of arbitrary geometry, the minimum quadrature requirement is a rule that exactly integrates tJ (plane case) or J (solid case). In a plane bilinear element of constant thickness, tJ is linear in ξ and in η, so one Gauss point is required. In a plane quadratic element of constant thickness, tJ contains ξ^3 and η^3, so a 2 by 2 Gauss rule is required. The eight-node solid also requires an order 2 Gauss rule (8 points).

However, with rare practical exceptions, indefinitely repeated subdivision of a mesh yields elements that become straight-sided parallelograms of constant thickness. Thus t and J cease to be functions of the coordinates and, *in the limit,* a single Gauss point yields the correct element volume.

For an isoparametric element based on an assumed displacement field, the best quadrature rule is usually the lowest-order rule that computes volume correctly and does not produce instability. Numerical testing of any proposed rule is mandatory. Solution accuracy may be mesh-dependent and problem-dependent, but usually one quadrature rule will be clearly superior to others. For bilinear and eight-node plane elements, and for the eight-node linear solid element, an order 2 Gauss rule is favored (four and eight points for plane and solid elements, respectively). The quadratic serendipity solid, having eight corner nodes and twelve edge nodes, can be integrated with an order 3 rule (27 points), but a special 14-point rule may be preferred, especially if the element is made very thin in one direction [6.8-6.10].

6.12 ELEMENT AND MESH INSTABILITIES

An *instability* may also be called a *spurious singular mode*. In structural mechanics, an instability may be known as a *mechanism*, a *kinematic mode*, an *hourglass mode*, or a *zero-energy mode*. The term "zero-energy mode" refers to a nodal displacement vector $\{\mathbf{D}\}$ that is not a rigid-body motion but nevertheless produces zero strain energy $\{\mathbf{D}\}^T[\mathbf{K}]\{\mathbf{D}\}/2$. Instabilities arise because of shortcomings in the element formulation process, such as use of a low-order Gauss quadrature rule. In the present context, an instability has nothing to do with buckling problems of structures.

A structure that appears adequately constrained may yet have an instability that makes $[\mathbf{K}]$ singular. Or, unstable elements may combine to form a structure that is stable but unduly susceptible to certain load patterns, so that computed displacements are excessive.

To explain the term "zero-energy mode" further and show how such a mode may arise, we substitute the relation $\{\epsilon\} = [\mathbf{B}]\{\mathbf{d}\}$ into the expression for strain energy in an element, U_e. From Eq. 4.1-10 and the standard expression for $[\mathbf{k}]$, Eq. 4.1-5, we obtain

$$U_e = \tfrac{1}{2}\{\mathbf{d}\}^T[\mathbf{k}]\{\mathbf{d}\} = \tfrac{1}{2}\{\mathbf{d}\}^T \int_{V_e} [\mathbf{B}]^T[\mathbf{E}][\mathbf{B}]\, dV\{\mathbf{d}\} = \tfrac{1}{2}\int_{V_e} \{\epsilon\}^T[\mathbf{E}]\{\epsilon\}\, dV \qquad (6.12\text{-}1)$$

When $[\mathbf{k}]$ is formed by numerical integration, it contains only the information that can be sensed at the sampling points of the quadrature rule. If it happens that strains $\{\epsilon\} = [\mathbf{B}]\{\mathbf{d}\}$ are zero at all sampling points for a certain mode $\{\mathbf{d}\}$, then U_e will vanish for that $\{\mathbf{d}\}$, and, according to Eq. 6.12-1, $[\mathbf{k}]$ will be a zero-stiffness matrix in the sense that strain energy $U_e = \{\mathbf{d}\}^T[\mathbf{k}]\{\mathbf{d}\}/2$ is zero for this particular $\{\mathbf{d}\}$. We expect that $U_e = 0$ if $\{\mathbf{d}\}$ is a rigid-body motion. If $U_e = 0$ when $\{\mathbf{d}\}$ is *not* a rigid-body motion, then an instability is present.

An element that displays a mechanism is said to be *rank deficient*. That is, the rank of $[\mathbf{k}]$ is less than the number of element d.o.f. minus the number of rigid-body modes.

An instability in an existing $[\mathbf{k}]$ can be detected by means of an eigenvalue test (Section 18.8). In the present section we give examples of instabilities and briefly discuss their prevention.

Examples. Consider the four-node plane (bilinear) element, whose stiffness matrix is 8 by 8. Eight independent displacement modes $\{\mathbf{d}\}$ can be identified (Fig. 6.12-1). The first three are rigid-body modes, for which $U_e = 0$, as is correct, regardless of the quadrature rule used. The next three modes, numbers 4, 5, and 6, are constant-strain modes, for which $U_e > 0$, regardless of the quadrature rule used. Modes 7 and 8 are bending modes. An order 1 rule, whose single Gauss point is at the element center, does not sense these modes, as $\epsilon_x = \epsilon_y = \gamma_{xy} = 0$ at the center. Accordingly, $U_e = 0$ for modes 7 and 8, and the element displays two mechanisms. These two spurious modes disappear if the Gauss rule is order 2 or greater.

The foregoing mechanisms can appear in a mesh of elements as well as in a single element (Fig. 6.12-2). In Fig. 6.12-2*d*, modes 7 and 8 of Fig. 6.12-1 are combined with a rigid-body rotation of each element. The mechanisms of Fig.

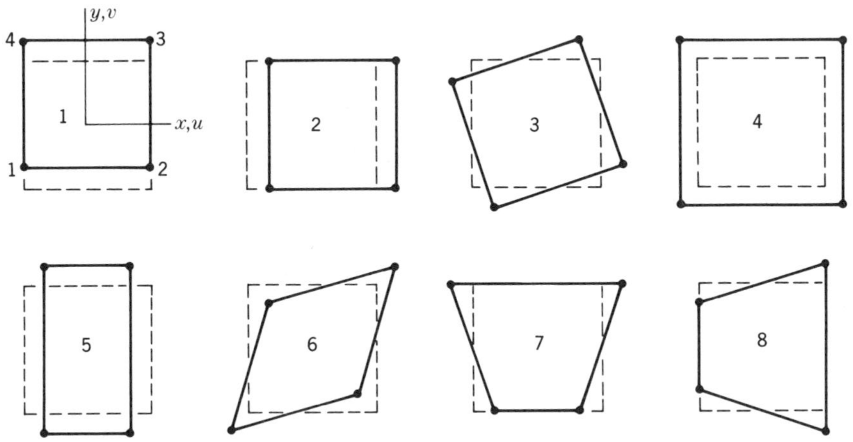

Figure 6.12-1. Independent displacement modes of a bilinear element.

6.12-2 are called *hourglass modes* because of their physical shape. Each of these distortions, as well as each straining mode in Fig. 6.12-1, would be considered the same mode if nodal d.o.f. were reversed—that is, if {**d**} of that mode were replaced by $-\{\mathbf{d}\}$.

Elements need not be rectangular in order to display a mechanism. Imagine, for example, that displacements are $u = a_1\xi\eta$ and $v = a_2\xi\eta$, where a_1 and a_2 are constants. Then, at $\xi = \eta = 0$, we have $u_{,\xi} = u_{,\eta} = v_{,\xi} = v_{,\eta} = 0$; hence, according to Eqs. 6.3-17 and 6.3-18, $\epsilon_x = \epsilon_y = \gamma_{xy} = 0$ at the Gauss point of an order 1 rule, regardless of the shape of the element.

Consider next the quadratic plane element, having either eight or nine nodes, and integrated with a 2 by 2 Gauss rule (Fig. 6.12-3). Displacements in the nine-node element of Fig. 6.12-3*b* are [6.11]

$$u = 3\xi^2\eta^2 - \xi^2 - \eta^2 \qquad v = 0 \tag{6.12-2}$$

At the Gauss points of a 2 by 2 rule—that is, where ξ and η are $\pm 1/\sqrt{3}$—one finds $u_{,\xi} = u_{,\eta} = v_{,\xi} = v_{,\eta} = 0$. Therefore, according to Eqs. 6.3-17 and 6.3-18, strains are zero at these points, for any geometric shape of the undeformed ele-

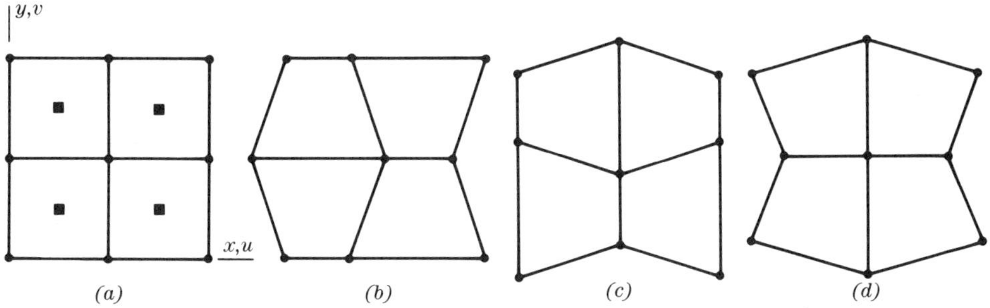

Figure 6.12-2. (*a*) Mesh of four bilinear elements, showing Gauss points of an order 1 rule in each element (squares). (*b,c,d*) Possible mechanisms ("hourglass" modes).

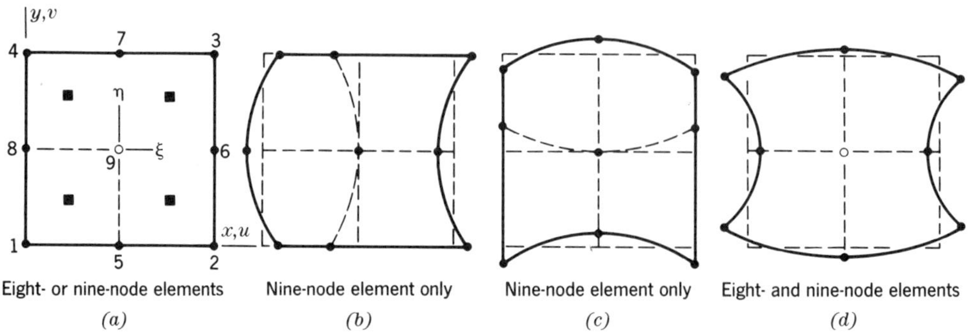

Figure 6.12-3. Possible mechanisms ("hourglass" modes) in quadratic elements integrated by an order 2 rule. Gauss points are shown by squares.

ment. A similar v field is possible (Fig. 6.12-3*c*). Thus we have identified two mechanisms.

The foregoing two mechanisms are not possible in the eight-node element because the $\xi^2\eta^2$ term is not present (see Eqs. 6.6-1). However, yet another mechanism is possible in both eight-node and nine-node elements (Fig. 6.12-3*d*). Its displacement field is simple to state for a square element; it is

$$u = \xi(3\eta^2 - 1) \qquad \text{and} \qquad v = \eta(1 - 3\xi^2) \tag{6.12-3}$$

Again, strains are zero at the Gauss points of a 2 by 2 rule. This mechanism is usually not of great concern because two adjacent elements cannot both have such a mode, as may be seen by trying to connect two deformed elements. Thus an instability present in individual elements is not present in the mesh.

Summing up, Fig. 6.12-3 identifies three element instabilities in quadratic elements arising from a 2 by 2 Gauss quadrature rule. The element stiffness matrix has rank 12 for both eight-node and nine-node elements (rank equals order less the number of rigid-body and instability modes). None of these instabilities exists if the Gauss rule is 3 by 3 or greater.

A mesh that has no mechanisms may yet behave badly because restraints on the mechanisms are weak. Consider Fig. 6.12-4*a*. Elements may be the four-node elements of Fig. 6.12-2 or the nine-node elements of Fig. 6.12-3, respectively integrated by one-point and four-point rules. Load P is concentrated and applied centrally rather than being distributed across the right end. Mechanisms are not possible because all nodes at the left support are fixed. However, this restraint

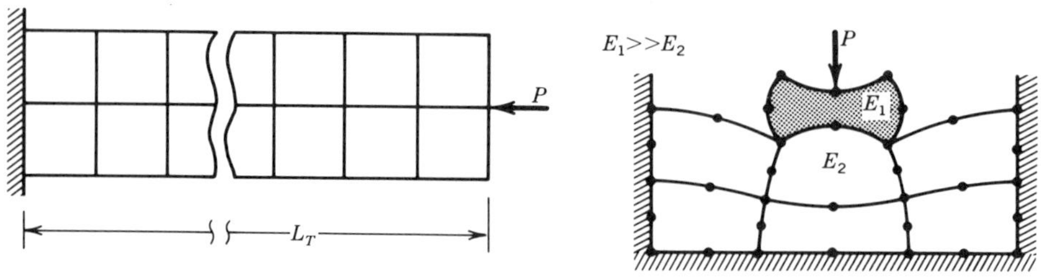

Figure 6.12-4. Problems that involve "near mechanisms," if reduced integration is used. In (*b*), elements that were initially rectangular are here shown deformed.

becomes weaker with increasing distance from the support. Near the load, distortions of the type shown in Figs. 6.12-2*b* and 6.12-3*b* become pronounced. Indeed, for a 2 by 24 mesh, the computed displacement of load P may be over 500 times the displacement predicted by the elementary formula PL/AE [6.12].

A similar situation is depicted in Fig. 6.12-4*b* [4.6]. A 2 by 2 Gauss rule is used to integrate [**k**] of each element. The stiff element, shown shaded, is weakly restrained by soft elements connected to the fixed boundary, allowing the mode of Fig. 6.12-3*d* (with signs of {**d**} reversed) to become pronounced, although not unbounded.

Elements for solids, for plate bending, and for nonstructural problems can also suffer from instabilities. Methods for detecting and controlling these modes are similar to methods used for plane elements.

A conservative analyst will avoid using any element that contains a possible instability because its dangers may not be foreseen.

Control of Instabilities. Various control methods have been proposed. Their goal is to eliminate instability by providing restraint, but without simultaneously stiffening the element's response to "legitimate" modes that are already working well. In what follows we summarize an effective method, with particular reference to a rectangular bilinear element. The method adds "hourglass stiffness" to an element integrated by one-point quadrature. The resulting element is inexpensive to formulate and works very well.

For simplicity, consider only the x-direction nodal displacements $\{\mathbf{d}_x\}$ of the eight modes shown in Fig. 6.12-1. For modes 1 and 8, $\{\mathbf{d}_x\} = \{\mathbf{0}\}$. An arbitrary combination of modes 2 through 6 is

$$\{\mathbf{d}_x\} = a_2\begin{Bmatrix}1\\1\\1\\1\end{Bmatrix} + a_3\begin{Bmatrix}1\\1\\-1\\-1\end{Bmatrix} + a_4\begin{Bmatrix}-1\\1\\1\\-1\end{Bmatrix} + a_5\begin{Bmatrix}1\\-1\\-1\\1\end{Bmatrix} + a_6\begin{Bmatrix}-1\\-1\\1\\1\end{Bmatrix} \tag{6.12-4}$$

where the a_i are constants. Mode 7 is

$$\{\mathbf{d}_x\}_7 = a_7 \lfloor 1 \quad -1 \quad 1 \quad -1 \rfloor^T \tag{6.12-5}$$

To provide mode 7 with the stiffness it lacks under one-point quadrature, we form the "stabilization matrix"

$$[\mathbf{k}]_7 = \{\mathbf{d}_x\}_7\{\mathbf{d}_x\}_7^T \tag{6.12-6}$$

A similar matrix $[\mathbf{k}]_8$, containing a constant a_8, serves to restrain mode 8. To the stiffness matrix computed by one-point quadrature, we now add $[\mathbf{k}]_7$ and $[\mathbf{k}]_8$. It is possible to choose values of a_7 and a_8 such that a rectangular element displays the exact strain energy in states of pure bending.

Note that mode 7 is orthogonal to all other modes—that is,

$$\{\mathbf{d}_x\}_7^T\{\mathbf{d}_x\}_i = 0 \qquad \text{for } i = 1, 2, 3, 4, 5, 6, 8 \tag{6.12-7}$$

Orthogonality prevents $[\mathbf{k}]_7$ from stiffening modes other than mode 7. That this is so may be seen by computing nodal forces $\{\bar{\mathbf{r}}\}_i$ associated with matrix $[\mathbf{k}]_7$,

$$\{\bar{\mathbf{r}}\}_i = [\mathbf{k}]_7\{\mathbf{d}_x\}_i = \{\mathbf{d}_x\}_7\{\mathbf{d}_x\}_7^T\{\mathbf{d}_x\}_i = \{\mathbf{d}_x\}_7(0) = \{\mathbf{0}\} \tag{6.12-8}$$

for $i = 1$ through 6 and for $i = 8$.

The foregoing control method can be generalized to elements having more than four nodes and to elements of arbitrary shape [6.11, 6.13, 13.49, 13.52-13.54].

6.13 REMARKS ON STRESS COMPUTATION

Element stresses follow from Eq. 4.7-1, with the substitution $\{\boldsymbol{\epsilon}\} = [\mathbf{B}]\{\mathbf{d}\}$:

$$\{\boldsymbol{\sigma}\} = [\mathbf{E}]([\mathbf{B}]\{\mathbf{d}\} - \{\boldsymbol{\epsilon}_0\}) + \{\boldsymbol{\sigma}_0\} \tag{6.13-1}$$

Here, in isoparametric elements, $[\mathbf{B}]$ is a function of the natural coordinates and $\{\boldsymbol{\sigma}\}$ contains stresses referred to the global coordinate system xyz. Where in the element should stresses be calculated? For isoparametric elements, it often happens that stresses (especially *shear* stresses) are most accurate at Gauss points of a quadrature rule one order less than that required for full integration of the element stiffness matrix.

Consider Fig. 6.13-1. Sides of a bilinear element remain straight during deformation. A typical element, deformed by bending moment but with rigid-body motion removed, is shown in Fig. 6.13-1*b*. Displacements in the element are $u = -a_1\xi\eta$ and $v = 0$, where a_1 is a positive constant. Thus shear strain γ_{xy} is proportional to ξ. On the neutral surface of bending, γ_{xy} displays the sawtooth pattern seen in Fig. 6.13-1*c*. Only at $\xi = 0$ in each element is γ_{xy} correctly computed (as zero) under pure bending deformation. In a general problem of plane stress analysis, where bending can occur in both directions (modes 7 and 8 of Fig. 6.12-1 simultaneously), the best computation point for γ_{xy} in a bilinear element is at $\xi = \eta = 0$. This is the Gauss point location of an order 1 rule, which is one order less than the order 2 rule of full integration.

A similar circumstance occurs with the eight-node and nine-node quadratic elements. In the beam of Fig. 6.13-2, the exact γ_{xy} is constant along the x axis. In the quadratic element, γ_{xy} along the x axis displays the parabolic distributions shown. However, one finds that the quadratic element displays the correct γ_{xy} at the Gauss points of a 2 by 2 quadrature rule. In other problems of stress analysis, normal strains can also display parabolic variations, and again the most accurate strains are to be found at the Gauss points of an order 2 rule.

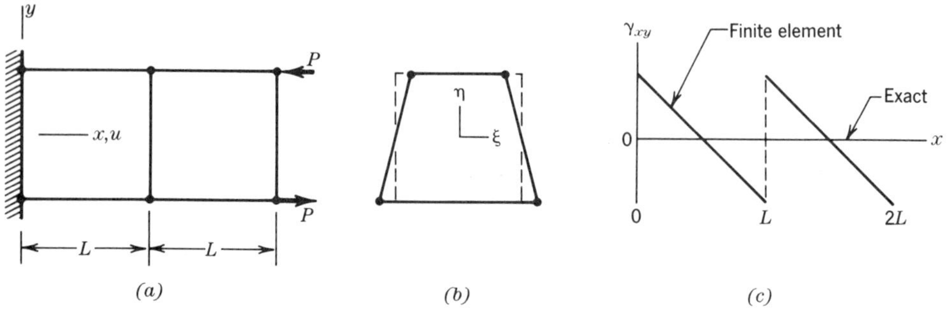

Figure 6.13-1. (*a*) Beam loaded in bending. (*b*) Bending distortion of a typical bilinear element. (*c*) Shear strain along the x axis.

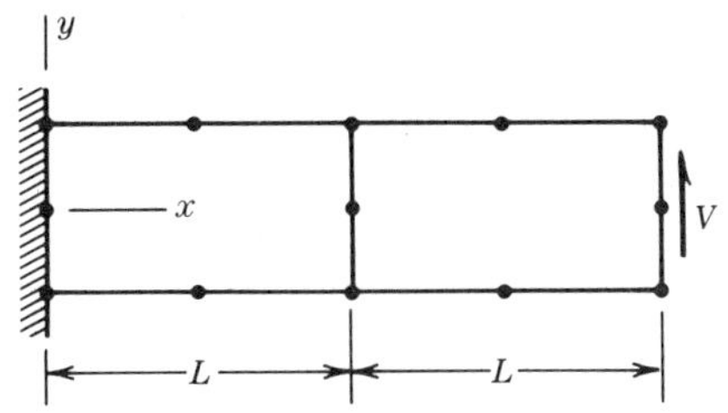

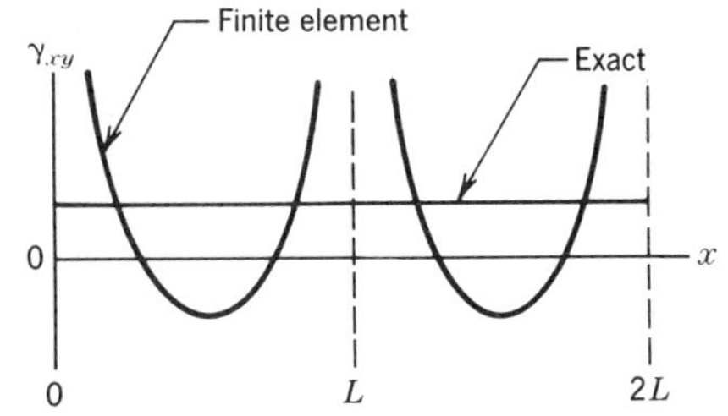

Figure 6.13-2. (*a*) Beam loaded by transverse tip force V. (*b*) Shear strain along the x axis.

In elements based on displacement fields, one expects stresses to be less accurate than displacements, as explained in Section 3.5. However, in the foregoing examples, stresses are "superaccurate" or "superconvergent" at the Gauss points because there they have the same degree of accuracy as displacements. Indeed, in unusual situations it may happen that stresses are *more* accurate than displacements. For example, in Fig. 6.12-4*a*, stresses may be substantially correct at Gauss points (of an order 1 or order 2 rule, for four- and nine-node elements, respectively), although displacements are grossly in error. This is possible because the modes that permit excessive displacements produce zero strain at the Gauss points.

The theory of locating error-minimal points for stress computation is explained elsewhere [6.14,6.15]. One discovers that these points are Gauss points: at $\xi = \eta = 0$ in bilinear (plane) and trilinear (solid) elements, and where ξ, η, and ζ are $\pm 1/\sqrt{3}$ in eight- or nine-node quadratic (plane) and 20- or 21-node quadratic (solid) elements. These conclusions are rigorously true for rectangular elements. For distorted elements, Gauss points may not be optimal locations but they remain very good choices.

Stresses at Gauss points can be interpolated or extrapolated to other points in the element. The result obtained is usually more accurate than the result of evaluating Eq. 6.13-1 directly at the point of interest. The interpolation–extrapolation process is explained as follows.

Imagine that stresses have been computed at the four Gauss points of a plane element (points 1, 2, 3, and 4 in Fig. 6.13-3). We now wish to interpolate or extrapolate these stress values to other points in the element. In Fig. 6.13-3, coordinate r is proportional to ξ and s is proportional to η. At (say) point 3, $r = s = 1$ and $\xi = \eta = 1/\sqrt{3}$. Therefore, the factor of proportionality is $\sqrt{3}$; that is,

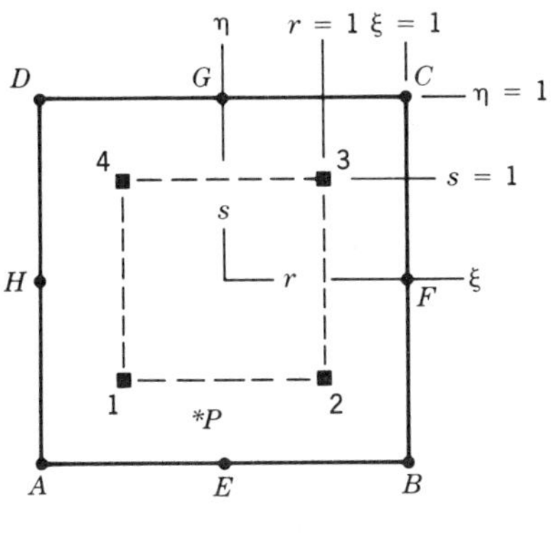

Figure 6.13-3. Natural coordinate systems used in extrapolation of stresses from Gauss points.

$$r = \xi\sqrt{3} \qquad \text{and} \qquad s = \eta\sqrt{3} \tag{6.13-2}$$

Stresses at any point P in the element are found by the usual shape functions,

$$\sigma_P = \sum N_i \sigma_i \qquad \text{for} \quad i = 1, 2, 3, 4 \tag{6.13-3}$$

where σ is σ_x, σ_y, or τ_{xy}. The N_i are the bilinear shape functions given by Eq. 6.3-2, but now written in terms of r and s rather than ξ and η; that is,

$$N_i = \tfrac{1}{4}(1 \pm r)(1 \pm s) \tag{6.13-4}$$

In Eq. 6.13-3, the N_i are evaluated at the r and s coordinates of point P. For example, let point P coincide with corner A. To calculate stress σ_{xA} at corner A from σ_x values at the four Gauss points, we substitute $r = s = -\sqrt{3}$ into the shape functions, and obtain

$$\sigma_{xA} = 1.866\sigma_{x1} - 0.500\sigma_{x2} + 0.134\sigma_{x3} - 0.500\sigma_{x4} \tag{6.13-5}$$

For solids, an interpolation–extrapolation formula similar to Eq. 6.13-3 is based on stresses at eight Gauss points and the trilinear N_i of Eq. 6.7-6.

In Section 4.7 we advised that usually the temperature field used for thermal stress analysis should have the same competence as the element strain field. Accordingly, if element stresses are based on Gauss point values, thermal strains $\{\boldsymbol{\epsilon}_0\}$ in Eq. 6.13-1 should also be based on Gauss point values.

6.14 EXAMPLES. EFFECT OF ELEMENT GEOMETRY

Simple test problems show how accuracy is affected by element distortion, changes in Gauss quadrature rule, and changes in element aspect ratio. Our examples are two-dimensional, but the trends displayed pertain to three-dimensional elements as well.

Example Problems. Table 6.14-1 illustrates the behavior of the bilinear element when its [**k**] is formed by four-point Gauss quadrature. Results are expressed as

TABLE 6.14-1. STRESSES AND DEFLECTIONS IN CANTILEVER BEAMS OF CONSTANT THICKNESS UNDER TRANSVERSE TIP LOAD P. LENGTH = 10, DEPTH = 2, ν = 0.25. VALUES BY BEAM THEORY = 1.000, OF WHICH 3% OF v_A IS DUE TO TRANSVERSE SHEAR DEFORMATION.

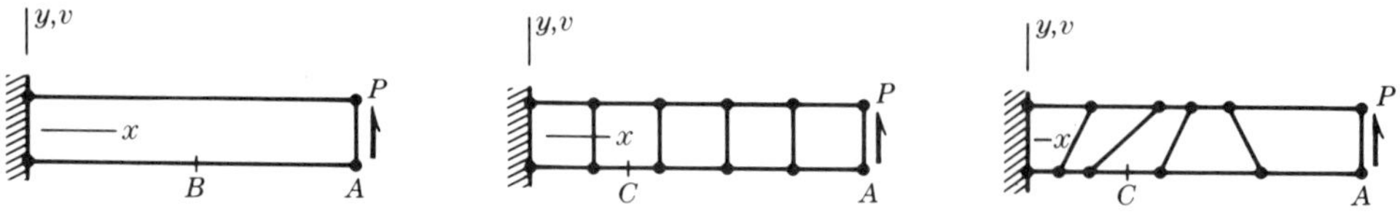

σ_{xB}	v_A	σ_{xC}	v_A	σ_{xC}	v_A
0.096	0.091	0.727	0.682	0.301	0.494

the ratio of computed value to the value given by beam theory. We see that square elements are better than elongated elements, and that geometric distortion stiffens the element and makes answers less accurate.

The principal failing of the bilinear element is that under pure bending loads, for which γ_{xy} should be zero, the element displays substantial values of γ_{xy} except at its center, as noted in connection with Fig. 6.13-1. This defect, known as *parasitic shear,* makes the element too stiff in bending. An improved form of the element discussed in Section 8.3.

Table 6.14-2 illustrates the behavior of eight-node and nine-node versions of the quadratic element [6.12]. All nodes at the left end are fixed. Load P on the right end is allotted to nodes in the proportion 1–8–1, which is consistent with a parabolic distribution of shear stress. Side nodes are midway along the sides. Point B is a Gauss point of a 2 by 2 quadrature rule. Results are expressed as the ratio of computed value to the value given by beam theory.

When elements are rectangular, we see that eight-node and nine-node elements have comparable accuracy. Both become stiffer if the quadrature rule used to generate [**k**] is changed from 2 by 2 to 3 by 3.

Next in Table 6.14-2, elements are made trapezoidal by moving nodes C and D horizontally to positions $L/4$ and $3L/4$, where L is the length of the beam. The final mesh in Table 6.14-2 introduces one curved interelement boundary by moving node E left of center an amount $L/20$. We see that 2 by 2 is the preferred integration rule, and that the eight-node element is much more sensitive to geometric distortion than the nine-node element. In one case, stress σ_{xB} in the eight-node element is not even of the correct sign.

The obvious lesson is that an ideal element is compact, straight-sided, and has equal corner angles. Of course, elements must be distorted to some extent in modeling an actual structure, but gratuitous distortion is to be avoided. In particular, if an element side is curved to model the curved boundary of a structure, other element sides that form interelement boundaries should be straight.

Quadratic triangles (Sections 5.5 and 6.8) have approximately the same accuracy as the nine-node element. For example, in the second case in Table 6.14-2, let

TABLE 6.14-2. STRESSES AND DEFLECTIONS IN TWO-ELEMENT CANTILEVER BEAMS OF CONSTANT THICKNESS UNDER TRANSVERSE TIP LOAD P. LENGTH = 100, DEPTH = 10, ν = 0.30. VALUES BY BEAM THEORY = 1.000 (IN WHICH THE TRANSVERSE-SHEAR CONTRIBUTION TO v_A IS NEGLECTED). SKETCHES ARE NOT TO SCALE.

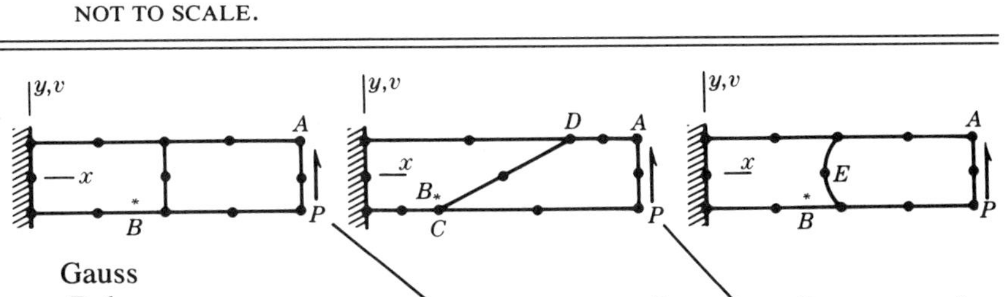

Element Type	Gauss Rule	σ_{xB}	v_A	σ_{xB}	v_A	σ_{xB}	v_A
8 node	2 × 2	1.000	0.968	0.051	0.362	−0.048	0.430
8 node	3 × 3	1.129	0.930	0.048	0.161	0.050	0.221
9 node	2 × 2	1.000	1.006	1.125	1.109	0.958	0.955
9 node	3 × 3	1.141	0.954	0.687	0.791	0.705	0.737

each quadrilateral be divided along its shorter diagonal, to which a midside node is added. Thus we produce four straight-sided quadratic triangles, which yield $v_A = 0.796$.

Geometric Distortion: Examples and Tests. Elements in Fig. 6.14-1 have very poor geometry. Such elements should not be used. But if used, and if surrounded by elements of acceptable geometry, stresses will be poor in and very near the distorted element, but reasonable at some distance away because of Saint-Venant's principle.

In Fig. 6.14-1, Gauss points of a 2 by 2 rule lie at centers of the small black quadrilaterals. Dashed lines are lines of constant ξ and constant η. Dashed lines would be parallel if the element were rectangular (or if the elements were sketched in $\xi\eta$ space, where each element is square). The first element in Fig. 6.14-1 could be a bilinear element or a quadratic element with midside nodes. The remaining three elements are distortions of quadratic rectangles with midside nodes, respectively created by moving one corner node, one side node, and two side nodes.

In Fig. 6.14-1a there is a singularity at node 3, where the Jacobian determinant J is zero. Elsewhere in the element, $J > 0$. If the $\xi\eta$ system were made left-handed, by numbering nodes clockwise around the element but leaving shape

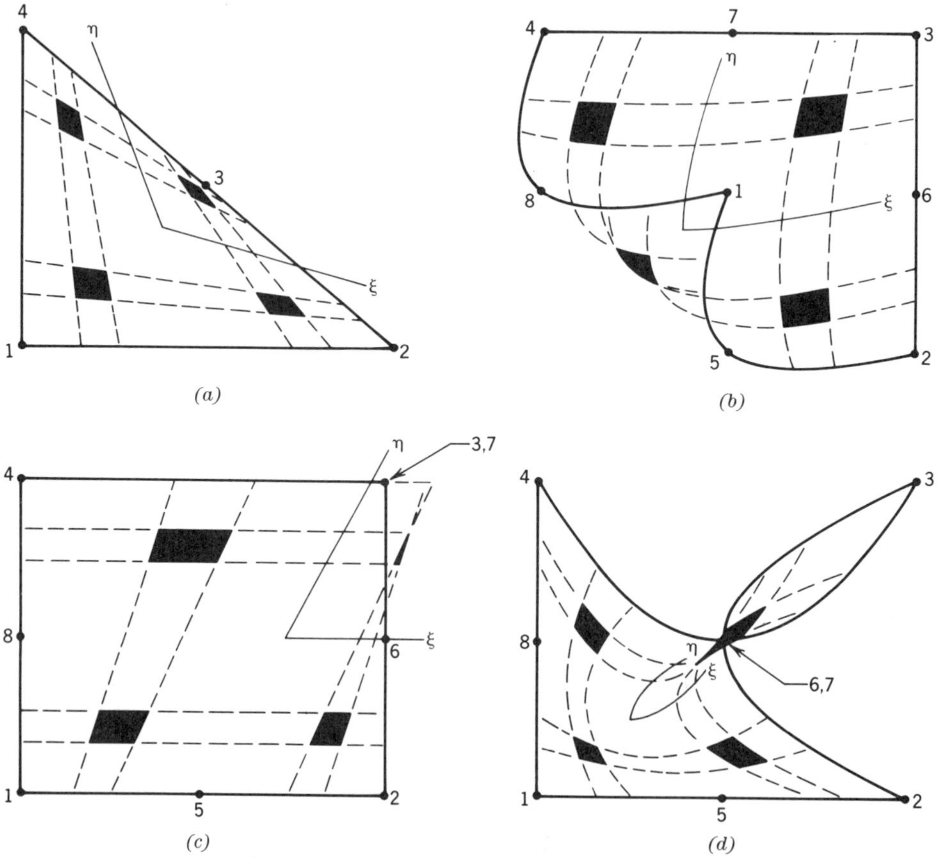

Figure 6.14-1. Badly shaped elements, showing Gauss points bounded by lines of constant ξ and constant η. (a) Interior angle at node 3 is 180°. (b) Node 1 moved to the center of the original rectangle. (c) Node 7 moved from top midside to corner 3. (d) Two side nodes moved to center of the original rectangle.

functions and the xy system unchanged, we would find $J < 0$ within the element, and all diagonal coefficients of the element [**k**] would be negative.

In the latter three elements of Fig. 6.14-1, part of each element falls outside the intended element boundaries and $J < 0$ at one of the Gauss points. These distortions do not prevent the elements from passing constant-strain patch tests, but drastically reduce the ability of the elements to represent more complicated states of deformation [6.16].

A user-oriented finite element program performs tests on the geometry of elements. Clearly it is easy to check that interior corner angles of quadrilaterals are not far from 90°, that side nodes are not far from the midpoint of a straight line between adjacent corner nodes, and that J is positive at each Gauss point and not greatly different from the value of J at other Gauss points. It is extra trouble, but perhaps advisable, to check that J is also positive at vertex nodes [6.16].

PROBLEMS

Section 6.2

6.1 (a) Determine $\lfloor \mathbf{N} \rfloor$ of Eq. 6.2-3 by following the formal procedure suggested below Eq. 6.2-1.

(b) Determine $\lfloor \mathbf{N} \rfloor$ of Eq. 6.2-3 by use of Lagrange's formula, as suggested below Eq. 6.2-1.

6.2 (a) Show that if x_3 is at the middle of a bar of length L (Fig. 6.2-1*a*), then $J = L/2$ in Eq. 6.2-5. Let node 1 have the arbitrary value x_1.

(b) How far from the center of the bar can node 3 be placed if, according to Eq. 6.2-4, strain ϵ_x is to remain finite at the ends of the bar for arbitrary values of u_1, u_2, and u_3?

6.3 Determine the element stiffness matrix [**k**] if $x_1 = 0$, $x_2 = L$, and $x_3 = L/2$ in Fig. 6.2-1*a*. Let A and E be constant and do integrations explicitly.

6.4 Omit node 3 in Fig. 6.2-1*a*, so that the bar becomes a linear element with end nodes only. Derive the 2 by 2 stiffness matrix [**k**] by using the natural coordinate ξ.

6.5 The bar shown is fixed at both ends. It is modeled by one three-node element, whose shape functions are given by Eq. 6.2-3. Show that if the bar is uniform and loaded axially by its own weight, the exact stress distribution is obtained.

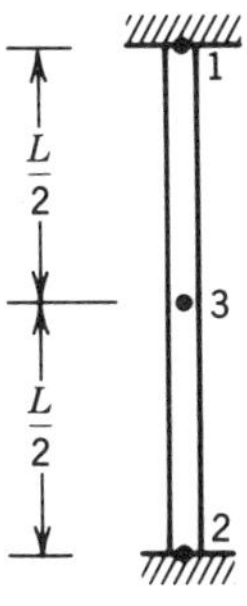

Problem 6.5

Section 6.3

6.6 With reference to Fig. 6.3-1*a*, let $x = \lfloor 1 \quad \xi \quad \eta \quad \xi\eta \rfloor \lfloor a_1 \quad a_2 \quad a_3 \quad a_4 \rfloor^T$.

(a) Hence, write [**A**] in the relation $\lfloor x_1 \quad x_2 \quad x_3 \quad x_4 \rfloor^T = [\mathbf{A}] \lfloor a_1 \quad a_2 \quad a_3 \quad a_4 \rfloor^T$.

(b) By inspection of Eqs. 6.3-2, write $[\mathbf{A}]^{-1}$ in the relation $x = \lfloor 1 \quad \xi \quad \eta \quad \xi\eta \rfloor [\mathbf{A}]^{-1} \lfloor x_1 \quad x_2 \quad x_3 \quad x_4 \rfloor^T$.

(c) Check your answers by seeing if $[\mathbf{A}][\mathbf{A}]^{-1} = [\mathbf{I}]$.

6.7 Sketch a quadrilateral, with corners properly lettered and $\xi\eta$ axes properly oriented, if shape functions are written as

$$N_A = \tfrac{1}{4}(1 - \xi)(1 + \eta) \qquad N_C = \tfrac{1}{4}(1 - \xi)(1 - \eta)$$
$$N_B = \tfrac{1}{4}(1 + \xi)(1 + \eta) \qquad N_D = \tfrac{1}{4}(1 + \xi)(1 - \eta)$$

6.8 The choice of natural coordinates made in Fig. 6.3-1 is not unique. As an alternative one could adopt natural coordinates r and s, as shown in the sketch for this problem. Write shape functions of the bilinear element in terms of r and s.

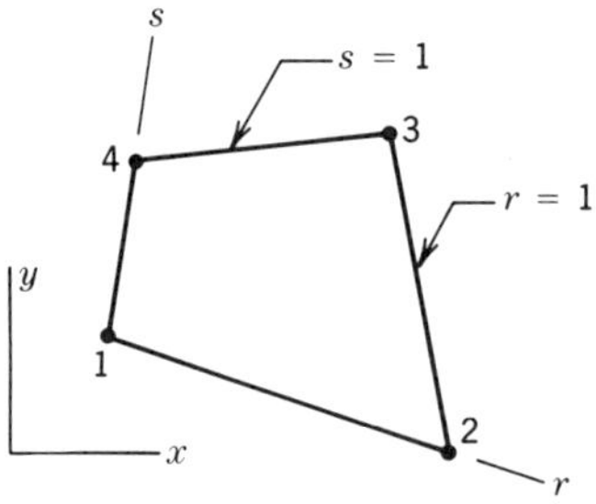

Problem 6.8

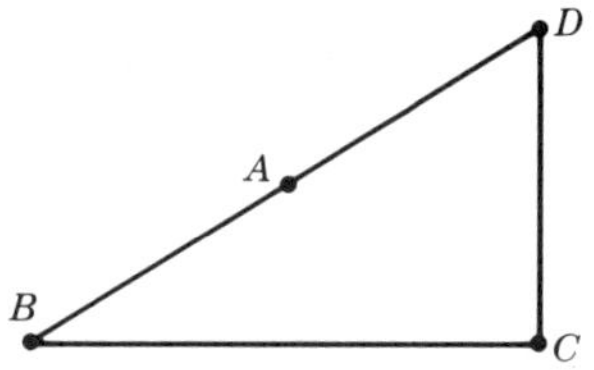

Problem 6.9

6.9 For the element shown, sketch the lines $\xi = -0.5$, 0.0, and 0.5 and the lines $\eta = -0.5$, 0.0, and 0.5. The N_i are given by Eq. 6.3-2. Let $\xi = \eta = -1$ at (a) point A, (b) point B, (c) point C, and (d) point D.

6.10 Show that $y_{,\eta} = J\xi_{,x}$. Also write the remaining three similar relationships among the ξ and η derivatives of x and y and the x and y derivatives of ξ and η.

6.11 Sketch a bilinear element for which J is a function of ξ but not of η.

6.12 Both elements shown are square and two units on a side. Both are improperly numbered. For each, determine $[\mathbf{J}]$ and J, using the N_i of Eq. 6.3-2. What do the given numberings imply about the $\xi\eta$ axes of the first element and the actual shape of the second element?

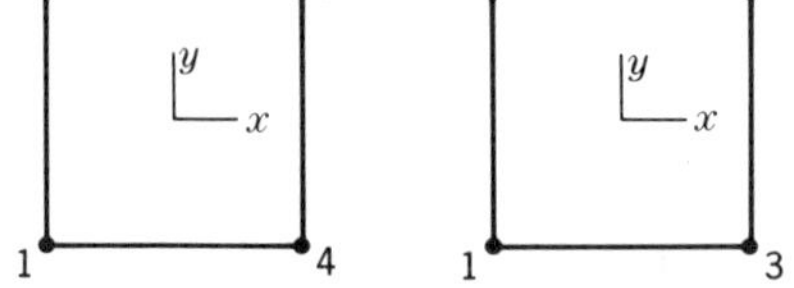

(a) (b) **Problem 6.12**

6.13 Evaluate $[\mathbf{J}]$ and J for each of the four elements shown. Also compute the ratio of element area to the area of a square two units on a side. How is this ratio related to J, and why?

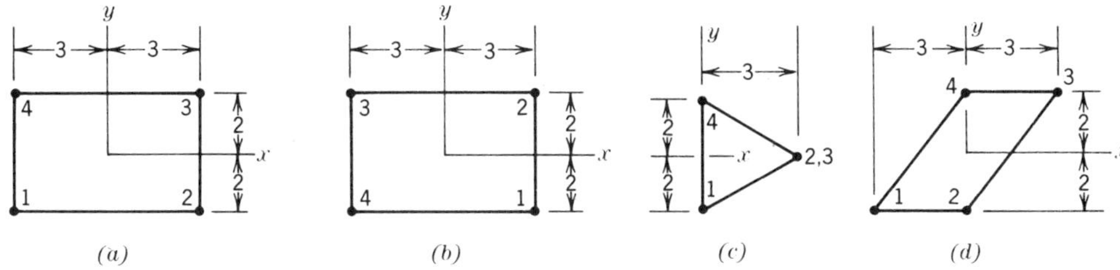

Problem 6.13

Section 6.4

6.14 Derive the locations and weights of an order 2 Gauss rule by requiring that it integrate exactly the polynomial $\phi = a_1 + a_2\xi + a_3\xi^2 + a_4\xi^3$ in the range $-1 \leq \xi \leq 1$. Assume that weights and points are symmetric with respect to the axis $\xi = 0$.

6.15 Use one-, two-, and three-point Gauss quadrature to integrate each of the following functions. Compare these answers with the exact answers.
(a) $\phi = \cos \pi x/2$ between $x = -1$ and $x = 1$.
(b) $\phi = (2 - x)/(2 + x)$ between $x = -1$ and $x = 1$.
(c) $\phi = 1/(x^2 - 3x + 4)$ between $x = -1$ and $x = 1$.
(d) $\phi = 1/x$ between $x = 1$ and $x = 7$.

6.16 Write an expression for I, analogous to Eq. 6.4-8, for (a) a 2 by 3 quadrature rule, and (b) a 3 by 4 quadrature rule.

6.17 For any of the quadrature rules in Table 6.4-1, weights W_i sum to 2 in one dimension, weight products W_iW_j sum to 4 in two dimensions, and weight products $W_iW_jW_k$ sum to 8 in three dimensions. Why? Check this behavior in Eq. 6.4-8.

6.18 Use a 2 by 2 Gauss rule to approximate I over the rectangular region shown.

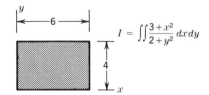

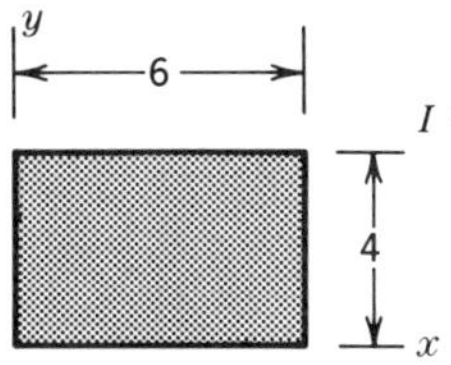

$$I = \iint \frac{3 + x^2}{2 + y^2}\, dx\, dy$$

Problem 6.18

6.19 In Problem 6.8, what are the r and s coordinates of the Gauss points of an order 2 rule? And what are the corresponding weights W_i? (Integration is from 0 to 1 for both r and s.)

6.20 (a) Determine the element stiffness matrix $[\mathbf{k}]$ of a two-node uniform bar element of length L by use of an order 2 Gauss quadrature rule. Check your result against Eq. 2.4-5.
(b) Repeat part (a), but let the cross-sectional area vary linearly from A_1 at node 1 to A_2 at node 2.
(c) Repeat part (b), but use one-point Gauss quadrature.

6.21 Determine the 3 by 3 element stiffness matrix $[\mathbf{k}]$ if node 3 in Fig. 6.2-1a is at the middle of the bar and AE is constant. Use an order 2 Gauss quadrature rule.

Section 6.5

6.22 Subroutine QUAD4 (Fig. 6.5-2) would be more efficient if DV were removed from statement 60 and placed elsewhere. Where? And why?

6.23 Using Figs. 6.5-1 and 6.5-2 as a guide, write Fortran statements that will generate the stiffness matrix of the bar element in Fig. 6.2-1 by three-point Gauss quadrature. Let cross-sectional area A be linearly interpolated from known values at node 1 and node 2. Node 3 is not necessarily at the midpoint.

6.24 Using Figs. 6.5-1 and 6.5-2 as a guide, write Fortran statements that will generate the stiffness matrix of a uniform beam element (Fig. 4.2-2 and Eq. 4.2-5) by two-point Gauss quadrature.

Section 6.6

6.25 For the element described by Eqs. 6.6-1, write the eight-term displacement function $\phi = a_1 + a_2\xi + \cdots + a_8\xi\eta^2$. Then write the 8 by 8 matrix $[\mathbf{A}]$ that arises in exchanging the a_i for nodal d.o.f. (in the manner of Eq. 3.8-6).

6.26 The element of Fig. 6.6-1*a* can be called subparametric. Why?

6.27 (a) Sketch a *rectangular* eight-node element for which J is a function of η but not of ξ.
(b) Sketch a nonrectangular eight-node element with midside nodes for which J is a function of η but not of ξ.

6.28 What changes would be needed in Figs. 6.5-1 and 6.5-2 to convert these subroutines so that they apply to the eight-node element described by Eqs. 6.6-1? Do not code the shape functions and their derivatives, but otherwise describe the changes precisely.

6.29 Identify the defects associated with connecting four-node and eight-node elements in the pattern shown.

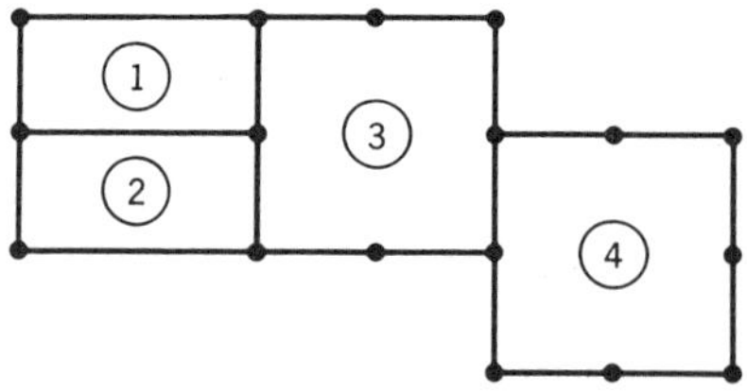

Problem 6.29

6.30 For the quadratic Lagrange element (nine nodes), sketch shape function N_1 in the manner of Fig. 6.6-2*d*. Decide whether N_1 is positive or negative in each quadrant by evaluating N_1 at $\xi = \pm\frac{1}{2}$ and $\eta = \pm\frac{1}{2}$.

6.31 Consider the nine-node element whose shape functions are given by Table 6.6-1. Any of nodes 5 through 9 can be omitted. In similar fashion, could node 1 be omitted, as shown, so as to produce a valid element with two straight edges, whose displacements are governed by nodes 2, 4, 5, and 8? *Suggestion:* Consider the horizontal (or the vertical) displacement at the lower left corner.

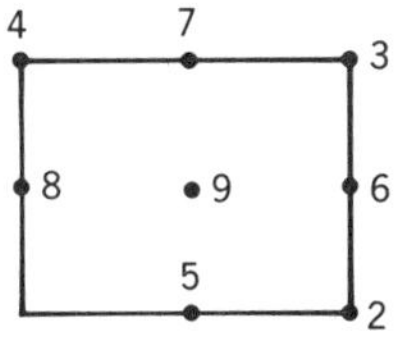

Problem 6.31

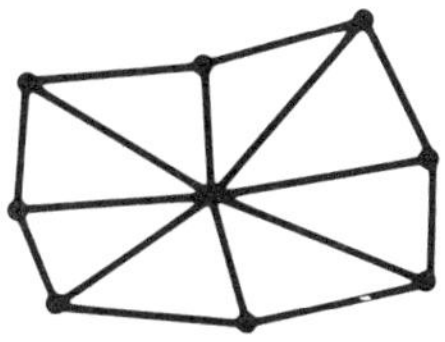

Problem 6.32

6.32 As an alternative to the nine-node element whose shape functions appear in Table 6.6-1, a nine-node element can be formed by combining eight linear triangles, as shown. What are comparative advantages and disadvantages of these two alternatives?

Section 6.7

6.33 What changes would be needed in Figs. 6.5-1 and 6.5-2 to convert these subroutines so that they apply to the eight-node solid of Fig. 6.7-1? Describe the changes precisely, including the coding of new shape functions and their derivatives. Assume that a subroutine can be called to invert **[J]** and compute J.

Section 6.8

6.34 Show that the bilinear element (Fig. 6.3-1a) becomes a constant-strain triangle if nodes 1 and 4 coalesce. For simplicity, use the particular geometry shown in the sketch.

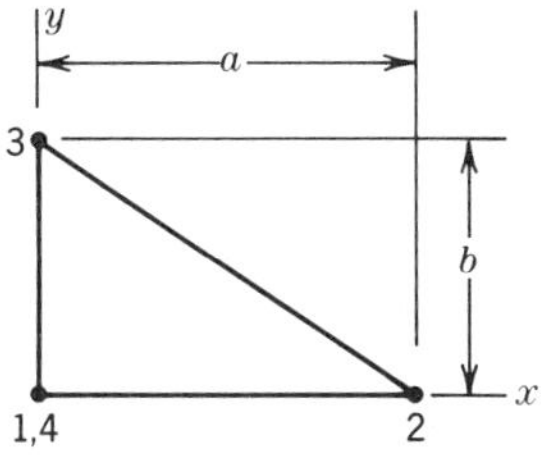

Problem 6.34

6.35 Use the arguments of Eqs. 6.8-1 to 6.8-5 to evaluate **[B]** for a linear (three-node) triangle. Compare your result with Eqs. 5.2-5 and 5.4-3.

6.36 If ϕ is constant, the product $[\mathbf{B}]\{\boldsymbol{\phi}_e\}$ must be zero. Use Eq. 6.8-5 to show that this is so.

6.37 As an alternative to Eqs. 6.8-2 through 6.8-4, one can eliminate (say) ξ_3 from shape functions N_i by use of the constraint relation $\xi_1 + \xi_2 + \xi_3 = 1$, then take the derivatives $\partial N_i/\partial\xi_1$ and $\partial N_i/\partial\xi_2$. Verify that this procedure also yields Eq. 6.8-5.

6.38 Consider a six-noded triangle. Imagine that we wish to move the side nodes to the positions shown, where a and b are dimensionless fractions of edge length (so that $a + b = 1$). We can accomplish this positioning in isoparametric fashion, using Eqs. 5.3-5, but the element displacement field is then not a complete quadratic in Cartesian coordinates. Show that we can accomplish the positioning in subparametric fashion, thus retaining the quad-

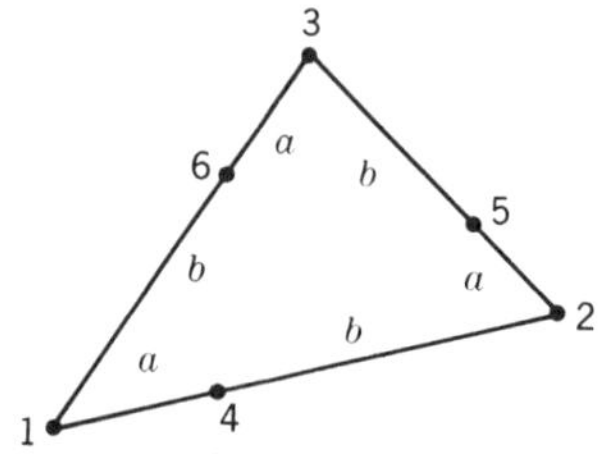

Problem 6.38

ratic field. *Suggestion:* Abandon uniform side-node spacing on the reference element, and apply Eq. 5.3-2 to element geometry. Obtain shape functions N_i, and show that they reduce to Eqs. 5.3-5 for $a = b = \frac{1}{2}$.

6.39 (a) If $\phi = 1$ in Eq. 6.8-7, one concludes that $\Sigma W_i = 1$. Verify this property in Table 6.8-1.

(b) All points listed in Table 6.8-1 should satisfy the constraint relation $\xi_1 + \xi_2 + \xi_3 = 1$. Verify that this is so for the 13-point formula.

6.40 (a) Integrate the function $\phi = (1 + \xi_1\xi_2)^{-1}$, using each of the first four integration formulas in Table 6.8-1. Let $A = 1$ and $J = 2$.

(b) Integrate the function $\phi = \xi_1\xi_2\xi_3$ by use of the appropriate rule in Table 6.8-1. Verify your result by use of Eq. 5.2-8. Let $A = 1$ and $J = 2$.

6.41 Evaluate Eq. 5.2-10 for the triangle shown. Evaluate this same integral by use of the first three-point formula in Table 6.8-1, and compare results.

6.42 In the wedge-shaped elements shown, let ζ be a coordinate that has values $+1$ and -1, respectively, on the top and bottom triangular faces. Write shape functions N_i for (a) the 6-node element, and (b) the 15-node element.

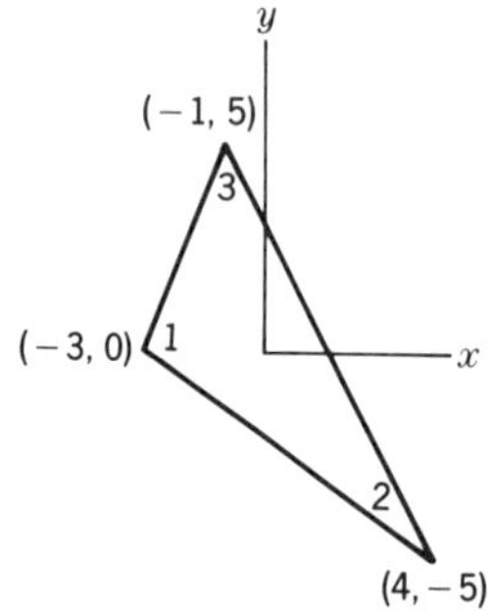

Problem 6.41

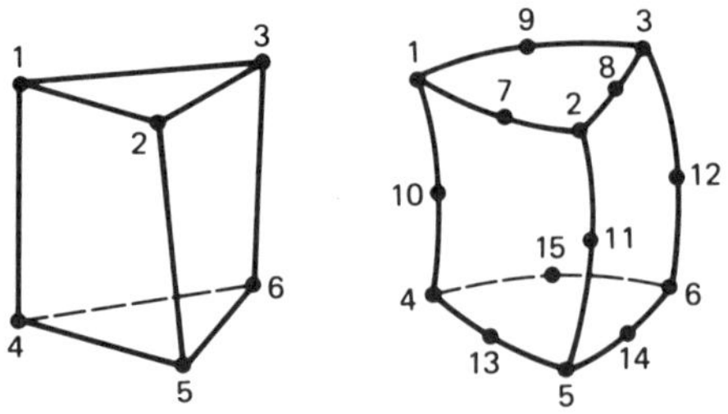

Problem 6.42

Section 6.9

6.43 Verify the results given in Eq. 6.9-1.

6.44 Let the centroid of the triangle of Fig. 6.8-1a lie at $x = y = 0$. Imagine that pressure $p = cx$, where c is a constant, acts normal to area A of the triangle. Evaluate the consistent nodal loads produced by p, in terms of c, A, x_1, x_2, and x_3.

6.45 Let a uniform traction act on the surface of a nine-node quadratic element of rectangular shape. In the consistent load vector $\{\mathbf{r}_e\}$, what fraction of the total force appears at each node? (See Fig. 4.3-5 for the corresponding eight-node case.)

Section 6.10

6.46 Use shape functions of Table 6.6-1 to demonstrate the interelement compatibility argument made in Section 6.10 with reference to Fig. 6.10-1.

6.47 Verify that $\Sigma N_i = 1$ for the N_i of (a) Eqs. 5.1-5, (b) Eqs. 5.1-6, (c) Eqs. 5.3-5, (d) Eqs. 6.3-2, and (e) Eqs. 6.6-1.

6.48 Verify that Eqs. 6.10-7 are satisfied for the N_i of (a) Eqs. 5.1-5, (b) Eqs. 5.1-6, (c) Eqs. 5.4-1, and (d) Eqs. 6.3-2.

6.49 Let $u_1 = 0$ and $u_2 > 0$ in the three-node bar element shown. For $0 < x < L$, interpolate axial displacement u linearly, so that the element is superparametric. Calculate axial strain $\epsilon_x = \lfloor \mathbf{B} \rfloor \{\mathbf{u}\}$, where $\{\mathbf{u}\} = \lfloor u_1 \quad u_2 \rfloor^T$ and J (the denominator of $\lfloor \mathbf{B} \rfloor$) is given by Eq. 6.2-5. Hence, show that the element fails unless $x_3 = L/2$.

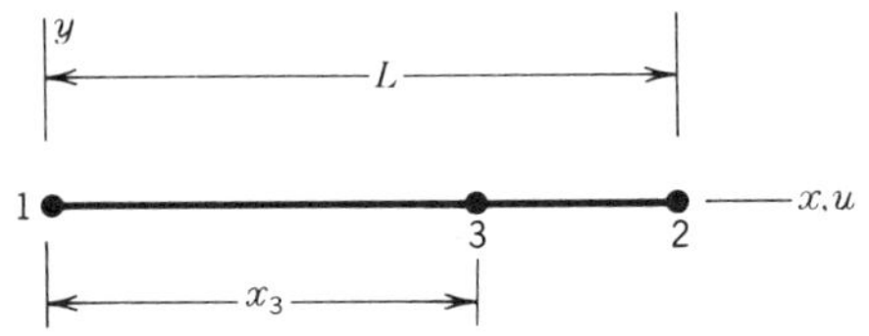

Problem 6.49

6.50 Write out the formula $x = \Sigma N_i x_i$ for the element of Fig. 6.6-1*a*. Hence, show the correctness of the completeness argument in Section 6.10 for this subparametric element.

Section 6.11

6.51 Let u_1, u_2, and u_3 be prescribed in the three-node bar of Fig. 6.2-1. Let AE be constant. What order of Gauss rule is needed to calculate strain energy in the element if (a) $x_3 = L/2$, and (b) $x_3 \neq L/2$?

6.52 If element thickness t can vary and is computed as $t = \Sigma N_i t_i$ from nodal values t_i, what order of Gauss quadrature is needed to compute the exact volume of (a) a bilinear element (four nodes), and (b) a quadratic element (eight nodes)?

6.53 Show that the volume of a trilinear solid element is correctly computed by an order 2 Gauss rule.

6.54 Let the following elements be rectangular in geometry, with side nodes evenly spaced and thicknesses constant. What order of Gauss quadrature is needed to obtain the exact *stiffness matrix*—that is, to integrate each k_{ij} exactly?

(a) Plane bilinear element (four nodes).
(b) Plane quadratic element (eight nodes).
(c) Solid trilinear element (eight nodes).
(d) Plane quadratic triangle (six nodes).

6.55 Repeat Problem 6.54 if element thickness t is variable and interpolated from nodal values, $t = \Sigma N_i t_i$.

6.56 A 2 by 2 Gauss rule is used to form $[\mathbf{k}]$ for each element in Table 6.14-1. What will the qualitative change in deflection v_A in each of the three cases be if the Gauss rule is changed to 3 by 3? Why?

Section 6.12

6.57 The plane structure shown is built of four bilinear elements, each integrated by one-point Gauss quadrature.

(a) Sketch the possible mechanisms of the structure.

(b) If you ask a computer program to solve for the displacement of load P, what do you think will happen?

(c) Add one roller support that will prevent an instability.

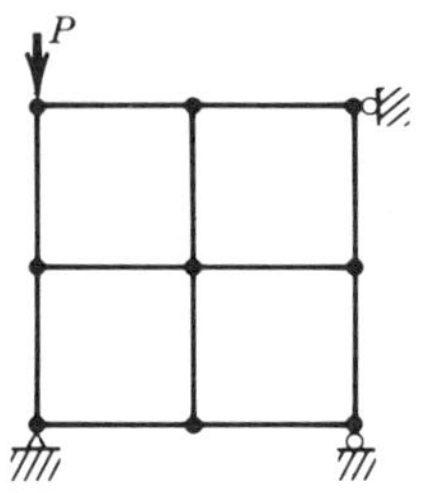

Problem 6.57

6.58 Verify that Eqs. 6.12-3 yield zero strains at Gauss points of an order 2 rule.

6.59 For each of the following elements, write (if possible) a vector {**d**} of nodal d.o.f. that represents an instability mode under one-point Gauss quadrature.

(a) The three-node bar element of Fig. 6.2-1.

(b) The standard four-d.o.f. beam element, Fig. 4.2-2.

6.60 Imagine that the bilinear element of Fig. 6.12-1 is integrated with a 2 by 1 Gauss rule. What is the rank of the element stiffness matrix?

6.61 (a) Consider the quadratic serendipity solid element (a hexahedron having eight corner nodes and twelve side nodes). If integrated by an order 2 Gauss rule, what do you anticipate will be the rank of its stiffness matrix? State your reason.

(b) Can these elements be put together so that the mesh has a mechanism?

6.62 There exists a six-point quadrature rule for hexahedra that uses a sampling point at the middle of each face [6.9]. What mechanisms are possible for a rectangular eight-node element whose stiffness matrix is formed by this rule? Can a mesh of elements also display these mechanisms?

6.63 Consider the upper right-hand element in Fig. 6.12-2*d*. Let the undeformed element be square, two units on a side.

(a) Show that strains are zero at the center of the element. Let all nonzero d_i in {**d**} have magnitude c.

(b) Show that {**d**} of the deformed element can be obtained by combining modes 7 and 8 of Fig. 6.12-1 with a rigid-body rotation.

6.64 (a) Let a vector {**d**} contain nodal d.o.f. of an arbitrary plane element, with all u_i in the upper half and all v_i in the lower half. Consider the following similarly-partitioned vectors of 0's, 1's, and nodal coordinates x_i and y_i:

$$\begin{Bmatrix} \mathbf{1} \\ \mathbf{0} \end{Bmatrix}, \begin{Bmatrix} \mathbf{x} \\ \mathbf{0} \end{Bmatrix}, \begin{Bmatrix} \mathbf{y} \\ \mathbf{0} \end{Bmatrix}, \begin{Bmatrix} \mathbf{0} \\ \mathbf{1} \end{Bmatrix}, \begin{Bmatrix} \mathbf{0} \\ \mathbf{x} \end{Bmatrix}, \begin{Bmatrix} \mathbf{0} \\ \mathbf{y} \end{Bmatrix}$$

Identify these vectors, singly or in combination, with three rigid-body modes and three constant-strain modes.

(b) Let the element be square, two units on a side and have nine nodes. Write {**d**} for the mode of Eq. 6.12-3b, and show that it is orthogonal to the modes of part (a).

6.65 Let the trilinear solid element of Fig. 6.7-1 be rectangular and two units on a side, so that $\xi = x$, $\eta = y$ and $\zeta = z$.

(a) What is the rank of the element stiffness matrix if it is integrated by use of a single Gauss point?

(b) Consider only x-direction displacements u. Let $\{\mathbf{d}_x\}$ represent the u_i of the eight nodes. Write a $\{\mathbf{d}_x\}$ of arbitrary magnitude for each zero-energy mode that involves only the u_i.

(c) Similarly, write a $\{\mathbf{d}_x\}$ for a total of four independent rigid-body and constant-strain modes that involve only the u_i.

(d) Show that the modes of part (b) are orthogonal to those of part (c).

6.66 Determine a_7 in Eq. 6.12-5 so that a rectangular element of uniform thickness has the exact strain energy in mode 7 (a pure bending mode). Express your answer in terms of the elastic modulus and element dimensions.

Section 6.13

6.67 Use the bending-deformation mode of Fig. 6.13-1*b* and the N_i of Eq. 6.3-2 to show that $\gamma_{xy} = -c\xi$ on $\eta = 0$, where c is a positive constant. For simplicity, assume that elements are square.

6.68 (a) Verify the numerical factors in Eq. 6.13-5.

(b) Apply Eq. 6.13-3 to nodes B, C, and D in Fig. 6.13-3. (Obtain numerical factors, as in Eq. 6.13-5.)

(c) Apply Eq. 6.13-3 to nodes E, F, G, and H in Fig. 6.13-3. (Obtain numerical factors, as in Eq. 6.13-5.)

6.69 Write a formula analogous to Eq. 6.13-3 that uses σ_i at the eight Gauss points of an order 2 rule in a solid element. Use it to write expressions for σ_x, analogous to Eq. 6.13-5, at (a) node 8 in Fig. 6.7-1, and (b) the point where axis ξ pierces the right-hand face in Fig. 6.7-1.

6.70 In the bilinear element (four nodes), stresses calculated directly at nodes agree exactly with stress extrapolated to nodes from four Gauss points, if the element is a parallelogram. Results disagree if the element is an arbitrary quadrilateral. Why?

6.71 Imagine that the bilinear element is not of constant thickness. What role does the thickness variation play in stress calculation according to Eq. 6.13-1? Suggest an ad hoc adjustment for thickness variation that might improve the accuracy of computed stresses.

Section 6.14

6.72 In Fig. 6.14-1*b*, locate the points described by the following coordinates.

(a) $\xi = -1$ and $\xi = -1/\sqrt{3}$.

(b) $\eta = 0$ and $\eta = -1$.

6.73 Consider an isosceles triangle, created by moving nodes 3 and 4 of a rectangular bilinear element (Fig. 4.2-4) so that they coincide on the η axis.

(a) Sketch the element and the Gauss points of a 2 by 2 rule, in the manner of Fig. 6.14-1.

(b) If J is computed at each Gauss point, what is the ratio J_{max}/J_{min}?

Summary Questions

6.74 Each of the structures shown may be analyzed as two-dimensional. Greatest stresses and greatest deflections are desired. Imagine that an initial (coarse mesh) analysis is to be undertaken, so that errors will be roughly 10% or less. For each structure, sketch a suitable mesh, first using linear elements and then using quadratic elements. State your assumptions and approximations regarding how loads and supports are specified, the use of symmetry, the quadrature rule needed, treatment of stress concentrations, and so on.

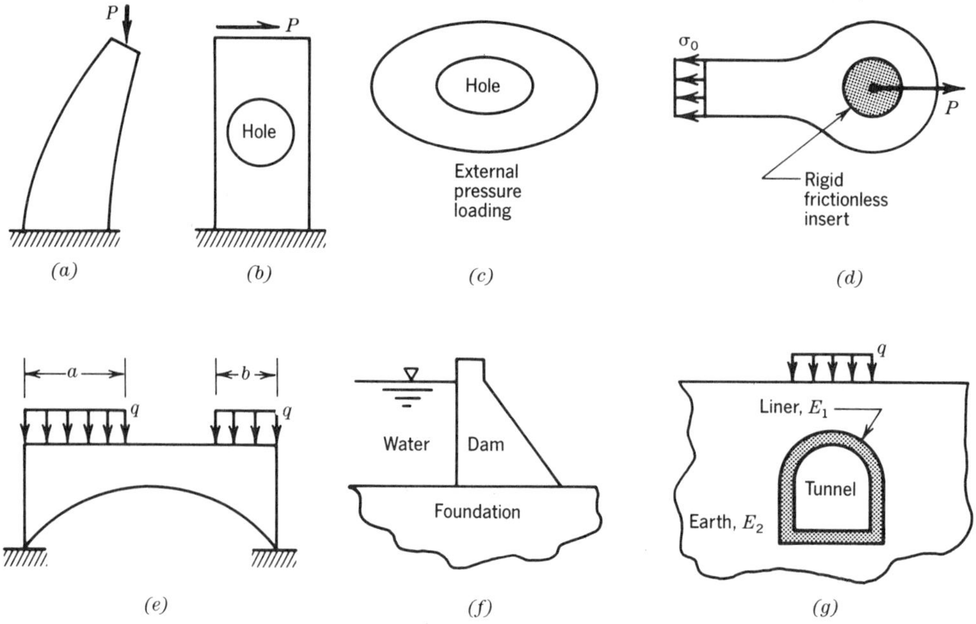

Problem 6.74

COORDINATE TRANSFORMATION

Uses of coordinate transformations in structural mechanics are described, with emphasis on transformation of stiffness properties.

7.1 INTRODUCTION

SEE P. 210 FIG 7.2-1

Coordinate transformation permits the calculation of elastic property matrix $[E']$ and stiffness matrix $[k']$ in one coordinate system with subsequent transformation to matrices $[E]$ and $[k]$ in another coordinate system. Other uses of coordinate transformation include condensation techniques in structural dynamics and imposition of constraints. Constraints are discussed in detail in Chapter 9.

The form $[Q] = [T]^T[Q'][T]$ appears repeatedly. Here $[Q']$ is the matrix to be transformed and $[T]$ is the transformation matrix. The transformed matrix $[Q]$ is symmetric if $[Q']$ is symmetric. Matrix $[T]$ may be rectangular or square. If square it may not be orthogonal. *IT MAY BE OR MAY NOT BE* The specific form of $[T]$ depends on the problem at hand.

One often has the option of taking $[Q']$ as either an element matrix or the corresponding structure matrix. Computer programming is usually easiest when transformations are done *before* elements are assembled, even though we must then transform several small matrices instead of one large one.

Formal matrix multiplication to produce $[T]^T[Q'][T]$ is often wasteful because $[T]$ is often sparse. Sparsity should be exploited, or terms in the product should be hand-calculated and then coded.

Caution. Transformations modify stiffness matrices. Errors and inconsistencies in stiffness matrices can lead to numerical difficulties and seriously degrade accuracy. It matters little if errors in $[T]$ produce only a slightly different geometry than intended. But damage is done if errors in $[T]$ act to falsify equilibrium equations. To avoid damage, we should state and manipulate transformation matrices and constraint equations with as much precision as is granted to stiffness coefficients K_{ij}.

7.2 TRANSFORMATION OF VECTORS

Consider a vector **V** whose scalar components in the x, y, and z directions are u, v, and w (Fig. 7.2-1). Components of **V** in the x', y' and z' directions are u', v', and w'. We wish to express u', v', and w' in terms of u, v, w, and the cosines of

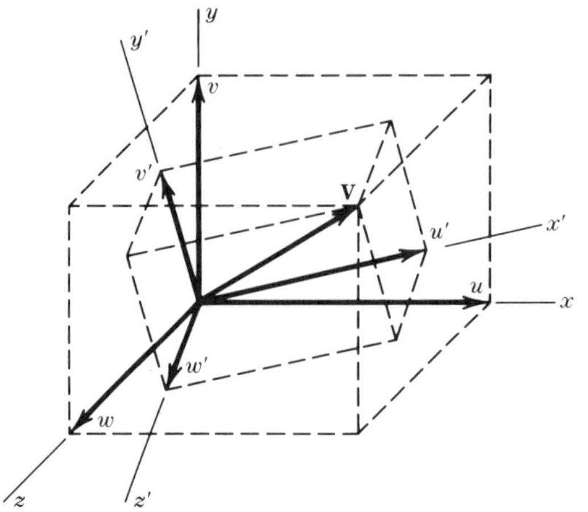

Direction cosines of axes:

	x	y	z
x'	ℓ_1	m_1	n_1
y'	ℓ_2	m_2	n_2
z'	ℓ_3	m_3	n_3

Figure 7.2-1. Coordinate systems xyz and $x'y'z'$, with table of direction cosines of angles between axes; for example, ℓ_1 is the cosine of the angle between axes x and x'. Components of a vector **V** can be expressed in either coordinate system.

angles between axes $x'y'z'$ and xyz. Vector **V** can be regarded as a position vector, a nodal force or moment vector, or a nodal displacement or rotation vector.[1]

Component u' can be regarded as the sum of components of displacements u, v, and w parallel to the x' axis. That is, $u' = \ell_1 u + m_1 v + n_1 w$. Components v' and w' can be written similarly. In matrix format, these relations are

$$\begin{Bmatrix} u' \\ v' \\ w' \end{Bmatrix} = [\mathbf{\Lambda}] \begin{Bmatrix} u \\ v \\ w \end{Bmatrix}, \qquad \text{where} \quad [\mathbf{\Lambda}] = \begin{bmatrix} \ell_1 & m_1 & n_1 \\ \ell_2 & m_2 & n_2 \\ \ell_3 & m_3 & n_3 \end{bmatrix} \tag{7.2-1}$$

Matrix $[\mathbf{\Lambda}]$ is called the *rotation matrix*. Equation 7.2-1 relates vectorial components in two systems. Matrix $[\mathbf{\Lambda}]$ is *orthogonal* (i.e., its inverse is equal to its transpose). Therefore, the inverse of the transformation in Eq. 7.2-1 is

$$\begin{Bmatrix} u \\ v \\ w \end{Bmatrix} = [\mathbf{\Lambda}]^T \begin{Bmatrix} u' \\ v' \\ w' \end{Bmatrix} \tag{7.2-2}$$

Vector **V** in Fig. 7.2-1 may represent a displacement whose components can be expressed as $\{\mathbf{d}\} = \lfloor u \quad v \quad w \rfloor^T$ or as $\{\mathbf{d}'\} = \lfloor u' \quad v' \quad w' \rfloor^T$. Or, **V** may represent a force whose components can be expressed as $\{\mathbf{r}\} = \lfloor f_x \quad f_y \quad f_z \rfloor^T$ or as $\{\mathbf{r}'\} = \lfloor f_x' \quad f_y' \quad f_z' \rfloor^T$. Accordingly, components of displacements and forces obey the transformation rules

$$\{\mathbf{d}'\} = [\mathbf{\Lambda}]\{\mathbf{d}\} \quad \text{and} \quad \{\mathbf{d}\} = [\mathbf{\Lambda}]^T\{\mathbf{d}'\} \tag{7.2-3}$$

$$\{\mathbf{r}'\} = [\mathbf{\Lambda}]\{\mathbf{r}\} \quad \text{and} \quad \{\mathbf{r}\} = [\mathbf{\Lambda}]^T\{\mathbf{r}'\} \tag{7.2-4}$$

[1]Provided that rotation is small, as is usual. Finite (large) rotations do not combine vectorially.

Similarly, $[\mathbf{\Lambda}]$ and $[\mathbf{\Lambda}]^T$ can be used to transform (small) rotations $\{\mathbf{d}\} = \lfloor \theta_x \quad \theta_y \quad \theta_z \rfloor^T$ and moments $\{\mathbf{r}\} = \lfloor M_x \quad M_y \quad M_z \rfloor^T$.

In Eqs. 7.2-3 and 7.2-4 we require only that xyz and $x'y'z'$ each be a set of three mutually perpendicular directions. Neither system need be a Cartesian system. For example, $x'y'z'$ might be a cylindrical system (in which $x = r$, $y = \theta$, and $z = z$).

General Transformations. If $\{\mathbf{d}'\} = [\mathbf{T}]\{\mathbf{d}\}$, where $[\mathbf{T}]$ is not necessarily orthogonal and may not even be square, it remains true that $\{\mathbf{r}\} = [\mathbf{T}]^T \{\mathbf{r}'\}$. The proof is as follows. We argue that since $\{\mathbf{r}\}$ and $\{\mathbf{r}'\}$ describe the same resultant force, work done by the force during a prescribed virtual displacement must be independent of the coordinate system in which the work is computed. Let $\{\delta\mathbf{d}\}$ and $\{\delta\mathbf{d}'\}$ be two ways to describe the same virtual displacement (i.e., $\{\delta\mathbf{d}'\} = [\mathbf{T}]\{\delta\mathbf{d}\}$). Writing the work equality and using the relation $\{\delta\mathbf{d}'\}^T = \{\delta\mathbf{d}\}^T[\mathbf{T}]^T$, we obtain

$$\{\delta\mathbf{d}\}^T\{\mathbf{r}\} = \{\delta\mathbf{d}'\}^T\{\mathbf{r}'\} \qquad \text{or} \qquad \{\delta\mathbf{d}\}^T\{\mathbf{r}\} = \{\delta\mathbf{d}\}^T[\mathbf{T}]^T\{\mathbf{r}'\} \tag{7.2-5}$$

from which

$$\{\delta\mathbf{d}\}^T(\{\mathbf{r}\} - [\mathbf{T}]^T\{\mathbf{r}'\}) = 0, \qquad \text{and therefore} \quad \{\mathbf{r}\} = [\mathbf{T}]^T\{\mathbf{r}'\} \tag{7.2-6}$$

The latter equation may be written because the equation before it must be true for *any* virtual displacement $\{\delta\mathbf{d}\}$. Only when $[\mathbf{T}]$ is orthogonal does the second of Eqs. 7.2-3 result from $\{\mathbf{d}'\} = [\mathbf{T}]\{\mathbf{d}\}$ and the first of Eqs. 7.2-4 result from Eq. 7.2-6.

7.3 TRANSFORMATION OF STRESS, STRAIN, AND MATERIAL PROPERTIES

Transformation of stresses $\{\boldsymbol{\sigma}\}$ and strains $\{\boldsymbol{\epsilon}\}$ in two dimensions leads to the familiar Mohr's circle calculations. In this section we consider the problem in three dimensions. We also consider the transformation of material properties $[\mathbf{E}]$. Analogous transformations related to plate bending appear in Section 11.1.

Strains. Strain transformations are essentially transformations of displacement derivatives. That is, to relate ϵ_x' in coordinates $x'y'z'$ to ϵ_x in coordinates xyz, we must relate $\partial u'/\partial x'$ to $\partial u/\partial x$ and to other derivatives of u, v, and w. From Eq. 7.2-1,

$$\frac{\partial u'}{\partial x'} = \ell_1 \frac{\partial u}{\partial x'} + m_1 \frac{\partial v}{\partial x'} + n_1 \frac{\partial w}{\partial x'}, \text{ and so on} \tag{7.3-1}$$

By chain rule differentiation, with $\partial x/\partial x' = \ell_1$, $\partial y/\partial x' = m_1$, and $\partial z/\partial x' = n_1$,

$$\frac{\partial u}{\partial x'} = \ell_1 \frac{\partial u}{\partial x} + m_1 \frac{\partial u}{\partial y} + n_1 \frac{\partial u}{\partial z} \tag{7.3-2}$$

By this process we obtain

$$\underset{9\times 1}{\left\lfloor \frac{\partial u'}{\partial x'} \quad \frac{\partial u'}{\partial y'} \quad \frac{\partial u'}{\partial z'} \cdots \frac{\partial w'}{\partial z'} \right\rfloor^T} = \begin{bmatrix} \ell_1\Lambda & m_1\Lambda & n_1\Lambda \\ \ell_2\Lambda & m_2\Lambda & n_2\Lambda \\ \ell_3\Lambda & m_3\Lambda & n_3\Lambda \end{bmatrix} \underset{9\times 1}{\lfloor u_{,x} \quad u_{,y} \quad u_{,z} \quad \cdots \quad w_{,z} \rfloor^T} \tag{7.3-3}$$

where $[\Lambda]$ is given by Eq. 7.2-1. The 9 by 9 square matrix in Eq. 7.3-3 is orthogonal.

A state of strain can be expressed as $\{\epsilon'\}$ in $x'y'z'$ coordinates or as $\{\epsilon\}$ in xyz coordinates. One now introduces the strain–displacement relations (Eqs. 1.5-6) into Eq. 7.3-3. After straightforward but tedious expansion and gathering of terms, one obtains the relation between $\{\epsilon'\}$ and $\{\epsilon\}$ as

$$\{\epsilon'\} = [\mathbf{T}_\epsilon]\{\epsilon\} \tag{7.3-4}$$

where

$$[\mathbf{T}_\epsilon] = \left[\begin{array}{ccc|ccc} \ell_1^2 & m_1^2 & n_1^2 & \ell_1 m_1 & m_1 n_1 & n_1 \ell_1 \\ \ell_2^2 & m_2^2 & n_2^2 & \ell_2 m_2 & m_2 n_2 & n_2 \ell_2 \\ \ell_3^2 & m_3^2 & n_3^2 & \ell_3 m_3 & m_3 n_3 & n_3 \ell_3 \\ \hline 2\ell_1\ell_2 & 2m_1 m_2 & 2n_1 n_2 & \ell_1 m_2 + \ell_2 m_1 & m_1 n_2 + m_2 n_1 & n_1 \ell_2 + n_2 \ell_1 \\ 2\ell_2\ell_3 & 2m_2 m_3 & 2n_2 n_3 & \ell_2 m_3 + \ell_3 m_2 & m_2 n_3 + m_3 n_2 & n_2 \ell_3 + n_3 \ell_2 \\ 2\ell_3\ell_1 & 2m_3 m_1 & 2n_3 n_1 & \ell_3 m_1 + \ell_1 m_3 & m_3 n_1 + m_1 n_3 & n_3 \ell_1 + n_1 \ell_3 \end{array}\right] \tag{7.3-5}$$

Strains in $\{\epsilon'\}$ and $\{\epsilon\}$ are ordered as in Eqs. 1.5-6, and the engineering definition of shear strain is used (e.g., $\gamma_{xy} = u_{,y} + v_{,x}$). Partitioning seen in Eq. 7.3-5 is used in what follows.

Stresses. A stress transformation relates stresses $\{\sigma\}$ in xyz coordinates to stresses $\{\sigma'\}$ in $x'y'z'$ coordinates. To determine the form of this transformation, we consider internal virtual work per unit volume, done by stresses during a prescribed virtual displacement. This work must be the same whether it is computed in the xyz system or in the $x'y'z'$ system. Therefore, writing the work equality and using Eq. 7.3-4, we obtain

$$\{\delta\epsilon\}^T\{\sigma\} = \{\delta\epsilon'\}^T\{\sigma'\} \quad \text{or} \quad \{\delta\epsilon\}^T\{\sigma\} = \{\delta\epsilon\}^T[\mathbf{T}_\epsilon]^T\{\sigma'\} \tag{7.3-6}$$

Equation 7.3-6 must be true for any virtual strain state $\{\delta\epsilon\}$. Hence

$$\{\sigma\} = [\mathbf{T}_\epsilon]^T\{\sigma'\} \quad \text{or} \quad \{\sigma'\} = [\mathbf{T}_\epsilon]^{-T}\{\sigma\} \tag{7.3-7}$$

Coefficients in $\{\sigma\}$ and $\{\sigma'\}$ are ordered as in Eq. 1.7-1.

The inverse-transpose matrix in Eq. 7.3-7 is easy to compute. After assigning labels $\mathbf{T}_{11}$, $\mathbf{T}_{12}$, $\mathbf{T}_{21}$, and $\mathbf{T}_{22}$ to the partitions in Eq. 7.3-5, one discovers that

$$\text{if} \quad [\mathbf{T}_\epsilon] = \begin{bmatrix} \mathbf{T}_{11} & \mathbf{T}_{12} \\ \mathbf{T}_{21} & \mathbf{T}_{22} \end{bmatrix} \quad \text{then} \quad [\mathbf{T}_\epsilon]^{-T} = \begin{bmatrix} \mathbf{T}_{11} & 2\mathbf{T}_{12} \\ \frac{1}{2}\mathbf{T}_{21} & \mathbf{T}_{22} \end{bmatrix} \tag{7.3-8}$$

For 2-D

$$\Lambda = \begin{bmatrix} \cos \beta & \sin \beta \\ -\sin \beta & \cos \beta \end{bmatrix}$$

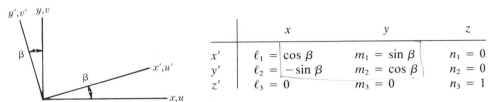

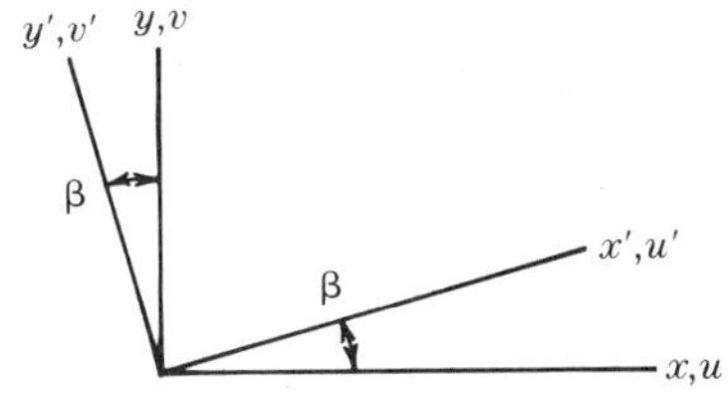

	x	y	z
x'	$\ell_1 = \cos\beta$	$m_1 = \sin\beta$	$n_1 = 0$
y'	$\ell_2 = -\sin\beta$	$m_2 = \cos\beta$	$n_2 = 0$
z'	$\ell_3 = 0$	$m_3 = 0$	$n_3 = 1$

Figure 7.3-1. The two-dimensional case. Coordinate systems xy and $x'y'$, with table of direction cosines between axes.

Thus $[\mathbf{T}_\epsilon]^{-T}$ is obtained from $[\mathbf{T}_\epsilon]$ by shifting factors of 2 in $[\mathbf{T}_\epsilon]$ symmetrically about the diagonal.

Material Properties. A single stress–strain relation can be written as $\{\boldsymbol{\sigma}\} = [\mathbf{E}]\{\boldsymbol{\epsilon}\}$ in the xyz coordinate system or as $\{\boldsymbol{\sigma'}\} = [\mathbf{E'}]\{\boldsymbol{\epsilon'}\}$ in the $x'y'z'$ coordinate system. Imagine that $[\mathbf{E'}]$ is known and $[\mathbf{E}]$ is desired. By substitution from Eqs. 7.3-4, 7.3-7, and the relation $\{\boldsymbol{\sigma'}\} = [\mathbf{E'}]\{\boldsymbol{\epsilon'}\}$,

$$\{\boldsymbol{\sigma}\} = [\mathbf{T}_\epsilon]^T\{\boldsymbol{\sigma'}\} = [\mathbf{T}_\epsilon]^T[\mathbf{E'}]\{\boldsymbol{\epsilon'}\} = [\mathbf{T}_\epsilon]^T[\mathbf{E'}][\mathbf{T}_\epsilon]\{\boldsymbol{\epsilon}\} \qquad (7.3\text{-}9)$$

from which

$$[\mathbf{E}] = [\mathbf{T}_\epsilon]^T[\mathbf{E'}][\mathbf{T}_\epsilon] \qquad (7.3\text{-}10)$$

This transformation concerns conditions at a point. Therefore, it is not necessary that xyz and $x'y'z'$ be Cartesian systems. For example, one coordinate system might be Cartesian and the other cylindrical.

Plane Problems. A two-dimensional problem is a special case in which $n_3 = 1$ and $\ell_3 = m_3 = n_1 = n_2 = 0$ (see Fig. 7.3-1). In the xy plane, $\{\boldsymbol{\epsilon}\} = \lfloor \epsilon_x \quad \epsilon_y \quad \gamma_{xy} \rfloor^T$, $\{\boldsymbol{\sigma}\} = \lfloor \sigma_x \quad \sigma_y \quad \tau_{xy} \rfloor^T$, $[\mathbf{E}]$ is 3 by 3, and

$$[\mathbf{T}_\epsilon] = \begin{bmatrix} c^2 & s^2 & cs \\ s^2 & c^2 & -cs \\ -2cs & 2cs & c^2 - s^2 \end{bmatrix} \quad \text{and} \quad [\mathbf{T}_\epsilon]^{-T} = \begin{bmatrix} c^2 & s^2 & 2cs \\ s^2 & c^2 & -2cs \\ -cs & cs & c^2 - s^2 \end{bmatrix}$$

$$(7.3\text{-}11)$$

where $c = \cos \beta$ and $s = \sin \beta$. Hence, one can recognize Eqs. 7.3-7 as the familiar Mohr's circle relations used in elementary mechanics of materials.

7.4 TRANSFORMATION OF STIFFNESS MATRICES

In two coordinate systems such as xyz and $x'y'z'$, the element stiffness relation can be written as

$$[\mathbf{k}]\{\mathbf{d}\} = \{\mathbf{r}\} \qquad \text{or as} \qquad [\mathbf{k'}]\{\mathbf{d'}\} = \{\mathbf{r'}\} \qquad (7.4\text{-}1)$$

The stiffness matrix of a given element can be expressed as either **[k]** or **[k′]**. The two matrices differ because they operate on different nodal d.o.f.—namely, {**d**} and {**d′**}. We imagine here that **[k′]** is known and **[k]** is desired. The necessary transformation is now derived.

A review of the argument associated with Eqs. 7.2-5 and 7.2-6 shows that no special form need be assumed for the matrix that relates {**d′**} and {**d**}. It is required only that the relation be known. We will call the relational matrix **[T]**. The argument of Eqs. 7.2-5 and 7.2-6 is that

$$\text{if} \qquad \{\mathbf{d}'\} = [\mathbf{T}]\{\mathbf{d}\} \qquad \text{then} \qquad \{\mathbf{r}\} = [\mathbf{T}]^T\{\mathbf{r}'\} \tag{7.4-2}$$

Examples will follow. For now we remark only that {**d**} and {**d′**} need not be the same size and need not even contain the same kind of d.o.f.

Hence, the stiffness transformation is easy to derive. By substitution of Eqs. 7.4-2 into Eq. 7.4-1,

$$[\mathbf{k}]\{\mathbf{d}\} = \{\mathbf{r}\} = [\mathbf{T}]^T\{\mathbf{r}'\} = [\mathbf{T}]^T[\mathbf{k}']\{\mathbf{d}'\} = [\mathbf{T}]^T[\mathbf{k}'][\mathbf{T}]\{\mathbf{d}\} \tag{7.4-3}$$

from which

$$[\mathbf{k}] = [\mathbf{T}]^T[\mathbf{k}'][\mathbf{T}] \tag{7.4-4}$$

Equation 7.4-4 does not change the orientation of the element in fixed global coordinates or alter element properties; rather, this transformation alters the *formal expression* of element properties to agree with a change of d.o.f. from {**d′**} to {**d**}.

For future reference, we note that mass and damping matrices used in dynamics transform in the same way. That is, $[\mathbf{m}] = [\mathbf{T}]^T[\mathbf{m}'][\mathbf{T}]$ and $[\mathbf{c}] = [\mathbf{T}]^T[\mathbf{c}'][\mathbf{T}]$.

7.5 EXAMPLES: TRANSFORMATION OF STIFFNESS MATRICES

Plane Truss Element. Imagine that the stiffness matrix of the bar in Fig. 7.5-1*a* in local coordinates $x'y'$ is called **[k′]** and is known. From it, **[k]** is to be determined, where **[k]** is the stiffness matrix of the bar referred to global coordinates xy. Thus **[k′]** and **[k]** describe the same bar but use different d.o.f. to do so. We have

$$[\mathbf{k}'] = \frac{AE}{L}\begin{bmatrix} 1 & 0 & -1 & 0 \\ 0 & 0 & 0 & 0 \\ -1 & 0 & 1 & 0 \\ 0 & 0 & 0 & 0 \end{bmatrix} \qquad \text{and} \qquad \begin{Bmatrix} u_1' \\ v_1' \\ u_2' \\ v_2' \end{Bmatrix} = \begin{bmatrix} c & s & 0 & 0 \\ -s & c & 0 & 0 \\ 0 & 0 & c & s \\ 0 & 0 & -s & c \end{bmatrix} \begin{Bmatrix} u_1 \\ v_1 \\ u_2 \\ v_2 \end{Bmatrix} \tag{7.5-1}$$

where $c = \cos\beta$ and $s = \sin\beta$. The square matrix of sines and cosines is **[T]**. It is built from matrices **[Λ]** of Eq. 7.2-1, specialized to two dimensions (Fig. 7.3-1). We find that $[\mathbf{k}] = [\mathbf{T}]^T[\mathbf{k}'][\mathbf{T}]$ is the stiffness matrix given by Eq. 2.4-3, as expected.

However, the foregoing procedure involves unnecessary effort. Terms in rows

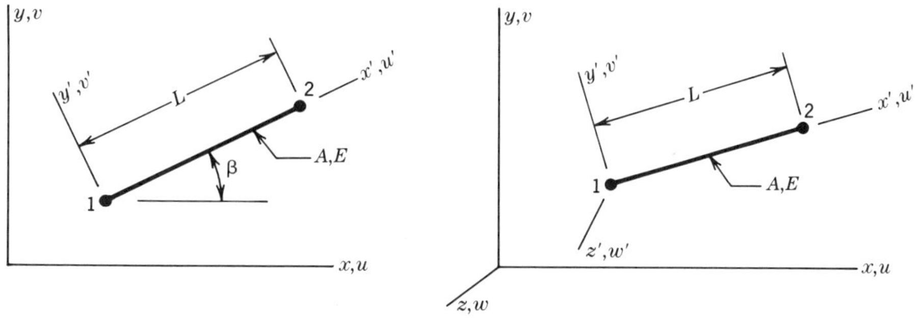

Figure 7.5-1. A uniform two-force (bar or truss) element in local and global reference frames. (*a*) Two-dimensional case. (*b*) Three-dimensional case.

2 and 4 of [**T**], which pertain to v_1' and v_2', are always multiplied by zero. This is physically reasonable, as axis x' completely defines the orientation of the element. Rather than use Eqs. 7.5-1, it is more efficient to use for [**k**′] the 2 by 2 matrix in Eq. 2.4-5. Thus

$$[\mathbf{k}'] = \frac{AE}{L}\begin{bmatrix} 1 & -1 \\ -1 & 1 \end{bmatrix} \quad \text{and} \quad \begin{Bmatrix} u_1' \\ u_2' \end{Bmatrix} = [\mathbf{T}]\begin{Bmatrix} u_1 \\ v_1 \\ u_2 \\ v_2 \end{Bmatrix} \tag{7.5-2}$$

where, with $c = \cos\beta$ and $s = \sin\beta$,

$$\underset{2\times 4}{[\mathbf{T}]} = \begin{bmatrix} c & s & 0 & 0 \\ 0 & 0 & c & s \end{bmatrix} \tag{7.5-3}$$

With [**k**′] and [**T**] thus defined, the operation $[\mathbf{k}] = [\mathbf{T}]^T[\mathbf{k}'][\mathbf{T}]$ again produces the expected 4 by 4 matrix of Eq. 2.4-3.

Space Truss Element. With [**k**′] again defined as in Eq. 7.5-2, we wish to obtain from it the 6 by 6 matrix [**k**] for the element in Fig. 7.5-1*b*, which operates on nodal displacements parallel to x, y, and z axes. Vectors of local and global d.o.f. for this element are

$$\{\mathbf{d}'\} = \lfloor u_1' \quad u_2' \rfloor^T \quad \text{and} \quad \{\mathbf{d}\} = \lfloor u_1 \quad v_1 \quad w_1 \quad u_2 \quad v_2 \quad w_2 \rfloor^T \tag{7.5-4}$$

The transformation is $\{\mathbf{d}'\} = [\mathbf{T}]\{\mathbf{d}\}$, where

$$\underset{2\times 6}{[\mathbf{T}]} = \begin{bmatrix} \ell_1 & m_1 & n_1 & 0 & 0 & 0 \\ 0 & 0 & 0 & \ell_1 & m_1 & n_1 \end{bmatrix} \tag{7.5-5}$$

and ℓ_1, m_1, and n_1 are direction cosines of axis x'. The desired result is $[\mathbf{k}] = [\mathbf{T}]^T[\mathbf{k}'][\mathbf{T}]$.

Plane Frame Element. This element is a plane beam but with axial deformation permitted. We first write the stiffness matrix [**k**′] in local coordinates $x'y'$, Fig.

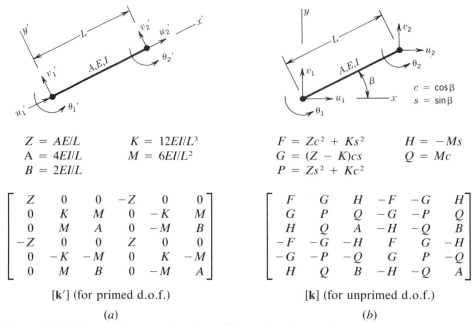

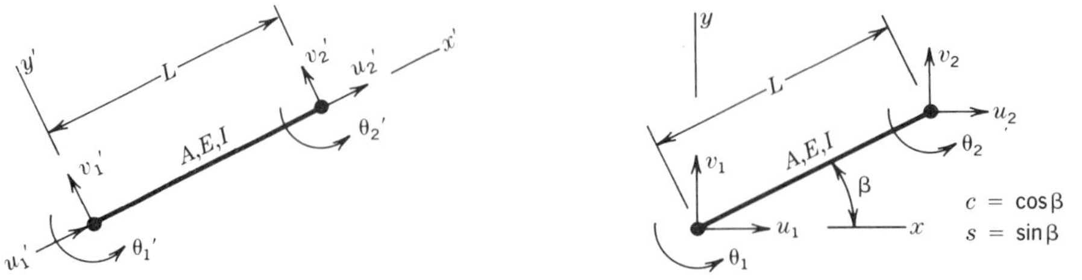

$$Z = AE/L \qquad K = 12EI/L^3 \qquad F = Zc^2 + Ks^2 \qquad H = -Ms$$
$$A = 4EI/L \qquad M = 6EI/L^2 \qquad G = (Z - K)cs \qquad Q = Mc$$
$$B = 2EI/L \qquad P = Zs^2 + Kc^2$$

$$\begin{bmatrix} Z & 0 & 0 & -Z & 0 & 0 \\ 0 & K & M & 0 & -K & M \\ 0 & M & A & 0 & -M & B \\ -Z & 0 & 0 & Z & 0 & 0 \\ 0 & -K & -M & 0 & K & -M \\ 0 & M & B & 0 & -M & A \end{bmatrix} \qquad \begin{bmatrix} F & G & H & -F & -G & H \\ G & P & Q & -G & -P & Q \\ H & Q & A & -H & -Q & B \\ -F & -G & -H & F & G & -H \\ -G & -P & -Q & G & P & -Q \\ H & Q & B & -H & -Q & A \end{bmatrix}$$

[**k**'] (for primed d.o.f.) [**k**] (for unprimed d.o.f.)

(*a*) (*b*)

Figure 7.5-2. The stiffness matrix of a uniform plane frame element.

7.5-2*a*. The element can both stretch and bend in the xy (or the $x'y'$) plane. Element stiffness matrix [**k**'] operates on the d.o.f.

$$\{\mathbf{d}'\} = \lfloor u_1' \quad v_1' \quad \theta_1' \quad u_2' \quad v_2' \quad \theta_2' \rfloor^T \tag{7.5-6}$$

D.o.f. u_1' and u_2' are associated with axial stiffness AE/L. The remaining d.o.f. are associated with bending. Axial and bending effects do not interact (unless a large axial load produces "beam–column" action). We therefore create the 6 by 6 matrix [**k**'] of the frame element by taking terms from the beam element matrix (Eq. 4.2-5) and the truss element matrix (Eq. 7.5-1). The resulting [**k**'] appears in Fig. 7.5-2*a*.

To generate [**k**] in global coordinates xy we apply Eq. 7.4-4. The transformation matrix is

$$\underset{6\times 6}{[\mathbf{T}]} = \begin{bmatrix} \mathbf{T}_n & \mathbf{0} \\ \mathbf{0} & \mathbf{T}_n \end{bmatrix}, \qquad \text{where} \quad [\mathbf{T}_n] = \begin{bmatrix} \cos\beta & \sin\beta & 0 \\ -\sin\beta & \cos\beta & 0 \\ 0 & 0 & 1 \end{bmatrix} \tag{7.5-7}$$

The "1" appears in [$\mathbf{T}_n$] because the rotation vectors do not change in direction: $\theta_1' = \theta_1$ and $\theta_2' = \theta_2$. The [**k**] that results from transformation, and the d.o.f. on which it operates, are shown in Fig. 7.5-2*b*.

7.6 INCLINED SUPPORT

Consider a structure that has translational nodal d.o.f. directed along the coordinate axes xyz. It may happen that a certain node is allowed to move only in a

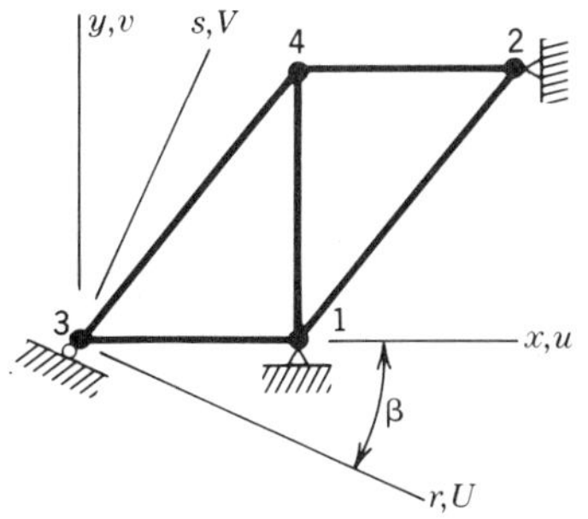

Figure 7.6-1. A plane truss or plane frame in which node 3 is allowed to move in only the r direction.

plane that is not parallel to a coordinate plane. In other words, displacement is *prohibited* in a direction that is not parallel to x, y, or z axes. A way to treat this boundary condition is now illustrated by using a plane structure.

In Fig. 7.6-1, the inclined support requires that $v_3 = -u_3 \tan\beta$ (or, in terms of other d.o.f., it requires that $V_3 = 0$ while U_3 is unrestrained). It is easier to deal with the constraint $V_3 = 0$ than with the constraint $v_3 = -u_3 \tan\beta$. The procedure described in the following replaces u_3 and v_3 by U_3 and V_3 without changing other d.o.f. of the structure. One then uses a standard method to set $V_3 = 0$ (see Section 2.10). U_3 remains active and is computed as part of the solution vector $\{\mathbf{D}\}$ in the usual way.

Before any support conditions are imposed in Fig. 7.6-1, the structure stiffness equations $[\mathbf{K}]\{\mathbf{D}\} = \{\mathbf{R}\}$, partitioned by node, are

$$\begin{bmatrix} \mathbf{K}_{11} & \mathbf{K}_{12} & \mathbf{K}_{13} & \mathbf{K}_{14} \\ \mathbf{K}_{21} & \mathbf{K}_{22} & \mathbf{0} & \mathbf{K}_{24} \\ \mathbf{K}_{31} & \mathbf{0} & \mathbf{K}_{33} & \mathbf{K}_{34} \\ \mathbf{K}_{41} & \mathbf{K}_{42} & \mathbf{K}_{43} & \mathbf{K}_{44} \end{bmatrix} \begin{Bmatrix} \mathbf{D}_1 \\ \mathbf{D}_2 \\ \mathbf{D}_3 \\ \mathbf{D}_4 \end{Bmatrix} = \begin{Bmatrix} \mathbf{R}_1 \\ \mathbf{R}_2 \\ \mathbf{R}_3 \\ \mathbf{R}_4 \end{Bmatrix} \tag{7.6-1}$$

where, depending on whether the structure is a plane truss or a plane frame,

$$\{\mathbf{D}_i\} = \lfloor u_i \quad v_i \rfloor^T \qquad \text{or} \qquad \{\mathbf{D}_i\} = \lfloor u_i \quad v_i \quad \theta_i \rfloor^T \tag{7.6-2}$$

To replace u_3 and v_3 by U_3 and V_3, we write the transformation relation

$$\begin{Bmatrix} u_3 \\ v_3 \end{Bmatrix} = [\mathbf{T}_3] \begin{Bmatrix} U_3 \\ V_3 \end{Bmatrix} \qquad \text{or} \qquad \begin{Bmatrix} u_3 \\ v_3 \\ \theta_3 \end{Bmatrix} = [\mathbf{T}_3] \begin{Bmatrix} U_3 \\ V_3 \\ \theta_3 \end{Bmatrix} \tag{7.6-3}$$

where, with $c = \cos\beta$ and $s = \sin\beta$,

$$[\mathbf{T}_3] = \begin{bmatrix} c & s \\ -s & c \end{bmatrix} \qquad \text{or} \qquad [\mathbf{T}_3] = \begin{bmatrix} c & s & 0 \\ -s & c & 0 \\ 0 & 0 & 1 \end{bmatrix} \tag{7.6-4}$$

for plane truss and plane frame, respectively. The transformation matrix $[\mathbf{T}]$ for the entire structure is a unit matrix except for $[\mathbf{T}_3]$ on the diagonal. With $\lceil \mathbf{I} \rfloor$ a 2 by 2 or a 3 by 3 unit matrix, $[\mathbf{T}]$ is

$$[\mathbf{T}] = \lceil \mathbf{I} \quad \mathbf{I} \quad \mathbf{T}_3 \quad \mathbf{I} \rfloor \tag{7.6-5}$$

After Eq. 7.6-1 is transformed, $\lfloor U_3 \quad V_3 \rfloor^T$ or $\lfloor U_3 \quad V_3 \quad \theta_3 \rfloor^T$ replaces $\{\mathbf{D}_3\}$, $[\mathbf{T}_3]^T\{\mathbf{R}_3\}$ replaces $\{\mathbf{R}_3\}$, and the structure stiffness matrix becomes

$$[\mathbf{T}]^T[\mathbf{K}][\mathbf{T}] = \begin{bmatrix} \mathbf{K}_{11} & \mathbf{K}_{12} & \mathbf{K}_{13}\mathbf{T}_3 & \mathbf{K}_{14} \\ \mathbf{K}_{21} & \mathbf{K}_{22} & \mathbf{0} & \mathbf{K}_{24} \\ \mathbf{T}_3^T\mathbf{K}_{31} & \mathbf{0} & \mathbf{T}_3^T\mathbf{K}_{33}\mathbf{T}_3 & \mathbf{T}_3^T\mathbf{K}_{34} \\ \mathbf{K}_{41} & \mathbf{K}_{42} & \mathbf{K}_{43}\mathbf{T}_3 & \mathbf{K}_{44} \end{bmatrix} \tag{7.6-6}$$

Transformed arrays $[\mathbf{K}]$ and $\{\mathbf{R}\}$ can be transformed again if there is another skew support. Conceivably, all nodes of the truss or frame could be skew and all translational d.o.f. in $\{\mathbf{D}\}$ could have different directions. If n successive transformations are used so that $\{\mathbf{D}'\} = [\mathbf{T}_1]\{\mathbf{D}\}$, $\{\mathbf{D}''\} = [\mathbf{T}_2]\{\mathbf{D}'\}$, and so on, original d.o.f. $\{\mathbf{D}^n\}$ are related to final d.o.f. $\{\mathbf{D}\}$ by the equation

$$\{\mathbf{D}^n\} = [\mathbf{T}_n][\mathbf{T}_{n-1}] \cdots [\mathbf{T}_1]\{\mathbf{D}\} \tag{7.6-7}$$

In the preceding explanation, transformation is done at the structure level. This approach requires that we construct and use $[\mathbf{T}]$ in a manner consistent with whatever compact storage format has been adopted for the structure stiffness matrix. It also requires that a d.o.f. to be suppressed (e.g., V_3 in Fig. 7.6-1) remain present until transformation is complete. If, instead, the separate element matrices are transformed *before* assembly, the scheme of Figs. 2.10-4 and 2.10-5 can be used to exclude from $\{\mathbf{D}\}$ the d.o.f. to be suppressed. The required transformation matrix for a plane frame element, with all six of its d.o.f. included, is

$$\underset{6\times 6}{[\mathbf{T}]} = \begin{bmatrix} \mathbf{T}_3 & \mathbf{0} \\ \mathbf{0} & \mathbf{I} \end{bmatrix} \quad \text{or} \quad \underset{6\times 6}{[\mathbf{T}]} = \begin{bmatrix} \mathbf{I} & \mathbf{0} \\ \mathbf{0} & \mathbf{T}_3 \end{bmatrix} \tag{7.6-8}$$

depending on which node of the element coincides with the affected node of the frame. This transformation must be applied to every element that is attached to the affected node (node 3 in Fig. 7.6-1).

7.7 JOINING DISSIMILAR ELEMENTS TO ONE ANOTHER

An element match termed "dissimilar" is depicted in Fig. 7.7-1*a*. The left end of a plane frame element is to be attached at an arbitrary location along an edge of a plane four-node quadrilateral element. Node 5 of the frame element does not coincide with a node of the quadrilateral. Moreover, rotational d.o.f. appear at nodes 5 and 6, but nodes 1 through 4 have only translational d.o.f. A method of connecting these two elements is now described.

The frame element stiffness relation is $[\mathbf{k}']\{\mathbf{d}'\} = \{\mathbf{r}'\}$, where

$$\{\mathbf{d}'\} = \lfloor u_5 \quad v_5 \quad \theta_5 \quad u_6 \quad v_6 \quad \theta_6 \rfloor^T \tag{7.7-1}$$

We seek modified matrices $[\mathbf{k}]$ and $\{\mathbf{r}\}$ for the frame element, where

$$[\mathbf{k}] = [\mathbf{T}]^T[\mathbf{k}'][\mathbf{T}] \quad \text{and} \quad \{\mathbf{r}\} = [\mathbf{T}]^T\{\mathbf{r}'\} \tag{7.7-2}$$

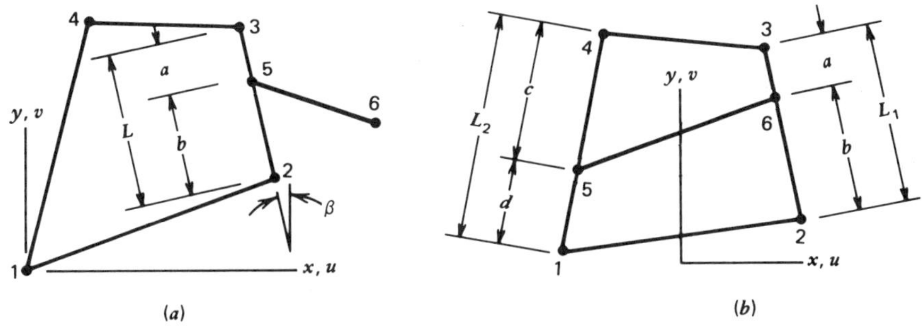

Figure 7.7-1. (*a*) A standard plane frame element connected to a four-node plane element. (*b*) A two-force (bar) element connected to a four-node plane element.

New d.o.f. of the frame element are to be

$$\{\mathbf{d}\} = \lfloor u_2 \quad v_2 \quad u_3 \quad v_3 \quad u_6 \quad v_6 \quad \theta_6 \rfloor^T \tag{7.7-3}$$

In the expression $\{\mathbf{d}'\} = [\mathbf{T}]\{\mathbf{d}\}$, transformation matrix $[\mathbf{T}]$ is written by saying that translational motion of node 5 is linearly interpolated along edge 2–3 from translational d.o.f. at nodes 2 and 3, and that rotation θ_5 is the same as the rotation of edge 2–3. Thus, with $c = \cos\beta$ and $s = \sin\beta$,

$$\underset{6\times 7}{[\mathbf{T}]} = \begin{bmatrix} \mathbf{T}_3 & \mathbf{0} \\ \mathbf{0} & \mathbf{I} \end{bmatrix}, \qquad \text{where} \quad [\mathbf{T}_3] = \frac{1}{L}\begin{bmatrix} a & 0 & b & 0 \\ 0 & a & 0 & b \\ c & s & -c & -s \end{bmatrix} \tag{7.7-4}$$

and $\lceil \mathbf{I} \rfloor = \lceil 1 \quad 1 \quad 1 \rfloor$. We see that $[\mathbf{k}]$ is a 7 by 7 matrix. Quadrilateral and frame elements can now be assembled to one another or assembled into the rest of the structure. Node 5 and its d.o.f. do not appear in the assembled structure. Node 5 may be called a "slave" node because its d.o.f. are completely determined by d.o.f. of "master" nodes 2 and 3.

As a second example, consider the problem of Fig. 7.7-1*b*. A two-force member, perhaps a reinforcing bar in concrete, is to be connected to points arbitrarily located along edges of a plane four-node element. Matrices $\{\mathbf{r}'\}$ and $[\mathbf{k}']$ of the bar are associated with d.o.f. u_5, v_5, u_6, and v_6. By transformation, $\{\mathbf{r}'\}$ and $[\mathbf{k}']$ are to be converted to $\{\mathbf{r}\}$ and $[\mathbf{k}]$, which are associated with d.o.f. u_i and v_i of the four corner nodes of the quadrilateral, $i = 1, 2, 3, 4$. Thus

$$\underset{4\times 1}{\{\mathbf{r}'\}} \text{ becomes } \underset{8\times 1}{\{\mathbf{r}\}} \text{ and } \underset{4\times 4}{[\mathbf{k}']} \text{ becomes } \underset{8\times 8}{[\mathbf{k}]} \tag{7.7-5}$$

The displacement transformation for d.o.f. $\{\mathbf{d}'\}$ of the bar is $\{\mathbf{d}'\} = [\mathbf{T}]\{\mathbf{d}\}$, and the required transformation matrix $[\mathbf{T}]$ is 4 by 8. With displacements linearly interpolated along edges of the quadrilateral, $[\mathbf{T}]$ contains terms like those in the first two rows of $[\mathbf{T}_3]$ in Eq. 7.7-4. After transformation, $\{\mathbf{r}\}$ and $[\mathbf{k}]$ can be directly added to the corresponding arrays of the quadrilateral or assembled into the structure. Nodes 5 and 6 and their d.o.f. are then not explicitly present.

One can say that d.o.f. of the frame and bar elements in Fig. 7.7-1 are constrained to follow d.o.f. of the quadrilateral.[2]

[2]Constraints are discussed in detail in Chapter 9.

7.8 RIGID LINKS. RIGID ELEMENTS

Rigid members impose relationships among d.o.f. This circumstance is sometimes called a multipoint constraint.[3] In the present section we consider rigid members as an application of coordinate transformation.

Rigid Links. Imagine that a plate is to be reinforced by a beam (Fig. 7.8-1). Nodes of the beam do not coincide with nodes of the plate. (If nodes were coincident, the beam–plate connection would be easy; one would simply assemble elements in the usual way.) Even with an offset beam, it is still possible to connect beam and plate in such a way that d.o.f. of only the plate appear in the assembled structure. The procedure for doing so is now described.

The procedure invokes a transformation that makes beam d.o.f. at nodes 3 and 4 "slave" to "master" d.o.f. at nodes 1 and 2 in the plate. This is accomplished by adding imaginary, weightless, rigid links—one between nodes 1 and 3 and another between nodes 2 and 4. We assume that the beam has bending stiffness (associated with d.o.f. w_3, θ_3, w_4, and θ_4) and axial stiffness (associated with d.o.f. u_3 and u_4). These six d.o.f. must be incorporated in the transformation relation. At the left end, the transformation is

$$\begin{Bmatrix} u_3 \\ w_3 \\ \theta_3 \end{Bmatrix} = [\mathbf{T}_\ell] \begin{Bmatrix} u_1 \\ w_1 \\ \theta_1 \end{Bmatrix}, \qquad \text{where} \quad [\mathbf{T}_\ell] = \begin{bmatrix} 1 & 0 & b \\ 0 & 1 & 0 \\ 0 & 0 & 1 \end{bmatrix} \tag{7.8-1}$$

A similar transformation is written at the right end by replacing subscripts 1 by 2 and 3 by 4. We see that d.o.f. u_3 and u_4 are activated by θ_1 and θ_2. Thus, because of the rigid links, axial stiffness of the beam is seen as bending stiffness by the plate d.o.f.

Let $\{\mathbf{r}'\}$ and $[\mathbf{k}']$ be beam element matrices associated with d.o.f. at nodes 3 and 4 (see Fig. 7.5-2*a* for $[\mathbf{k}']$). Transformed arrays $\{\mathbf{r}\}$ and $[\mathbf{k}]$, associated with d.o.f. at plate nodes 1 and 2, are

$$\begin{aligned} \{\mathbf{r}\} &= [\mathbf{T}]^T\{\mathbf{r}'\} \\ [\mathbf{k}] &= [\mathbf{T}]^T[\mathbf{k}'][\mathbf{T}] \end{aligned} \qquad \text{where} \quad \underset{6\times 6}{[\mathbf{T}]} = \begin{bmatrix} \mathbf{T}_\ell & \mathbf{0} \\ \mathbf{0} & \mathbf{T}_\ell \end{bmatrix} \tag{7.8-2}$$

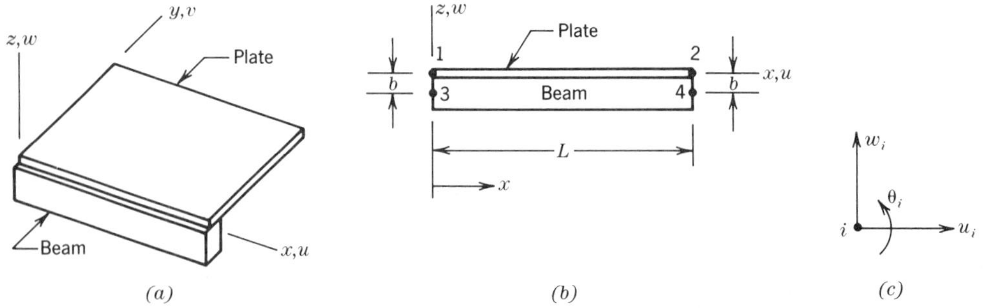

Figure 7.8-1. (*a*) A reinforcing beam joined to one edge of a plate element. (*b*) Side view. (*c*) Typical node i (i = 1, 2, 3, 4), showing d.o.f. considered in the coordinate transformation.

[3] Constraints are discussed in detail in Chapter 9.

Clearly this procedure can be extended to deal with a stiffener that is arbitrarily oriented in space, and with rigid links that are not perpendicular to the stiffener.

The foregoing transformation introduces an error that can cause displacements to be significantly overestimated [7.1]. The error can be attributed to incomplete coupling between beam and plate. Axial displacement in the beam should be

$$u_{\text{beam}} = u_{\text{plate}} + b\theta_{\text{plate}} \tag{7.8-3}$$

Imagine, for example, that all d.o.f. of the plate in Fig. 7.8-1*b* are zero but w_2. Then w_{plate} is cubic in x and θ_{plate} is quadratic in x. Hence, according to Eq. 7.8-3, u_{beam} should be quadratic in x. However, Eq. 7.8-1 yields $u_3 = u_4 = 0$; hence, $u_{\text{beam}} = 0$. Thus, for this deformation mode, beam and plate bending stiffnesses are simply added rather than being combined in a way that recognizes a common neutral axis. For a test case in which a uniform cantilever was loaded by a transverse tip force, with n plate elements along the length, tip displacement was overestimated by 69% for $n = 1$, 17% for $n = 2$, and 4.3% for $n = 4$ [7.1]. The error tends toward zero as each element approaches a state of constant curvature.

A method that eliminates the error was suggested by Miller [7.2]. He introduces axial displacement d.o.f. at $x = L/2$, say u_5 in the plate and u_6 in the beam. Axial displacement in the beam is now quadratic in x, as is desired. The axial stiffness portion of $[\mathbf{k}']$ is 3 by 3 and is associated with u_3, u_4, and u_6 (see Section 6.2). The transformation is essentially that of Eq. 7.8-2, augmented by

$$u_6 = u_5 + b\left(\frac{dw}{dx}\right)_{x=L/2} \tag{7.8-4}$$

where plate rotation dw/dx depends on w_1, θ_1, w_2, and θ_2. Transformation causes the 7 by 7 beam element stiffness matrix to operate on d.o.f. u_1, w_1, θ_1, u_2, w_2, θ_2, and u_5. This matrix is then combined with the plate element stiffness matrix (whose row and column corresponding to u_5 are null). Finally, condensation removes u_5, thus producing a combined $[\mathbf{k}]$ that operates on the usual plate element d.o.f.

Another difficulty, encountered in dynamic problems, is that the transformation converts a diagonal beam mass matrix $[\mathbf{m}']$ to a nondiagonal mass matrix $[\mathbf{m}]$. Ad hoc adjustments of $[\mathbf{m}]$ can make it diagonal again.

Rigid Elements. A rigid element might be used to model part of a linkage mechanism that couples elastic bodies. Or a particular element might be of much higher modulus than surrounding elements. In the latter case, errors of the type discussed in Section 18.2 are likely, and it is better to make the element perfectly rigid rather than very stiff.

Imagine that the triangle of Fig. 7.8-2 is to be idealized as perfectly rigid.

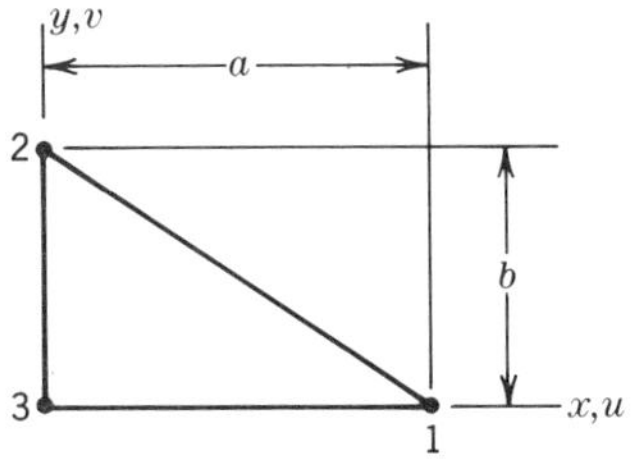

Figure 7.8-2. A plane triangle. Other elements of the structure are connected to it but are not shown.

Therefore, its motion is completely described by three d.o.f., say u_1, v_1, and u_2. These d.o.f. are related to the original six d.o.f. by the transformation

$$\{\mathbf{d}'\} = [\mathbf{T}]\{\mathbf{d}\} \quad \text{or} \quad \begin{Bmatrix} u_1 \\ v_1 \\ u_2 \\ v_2 \\ u_3 \\ v_3 \end{Bmatrix} = \begin{bmatrix} 1 & 0 & 0 \\ 0 & 1 & 0 \\ 0 & 0 & 1 \\ -a/b & 1 & a/b \\ 1 & 0 & 0 \\ -a/b & 1 & a/b \end{bmatrix} \begin{Bmatrix} u_1 \\ v_1 \\ u_2 \end{Bmatrix} \tag{7.8-5}$$

in which $u_3 = u_1$ and $v_2 = v_3 = v_1 - \theta a$, where $\theta = (u_1 - u_2)/b$ is a small rigid-body rotation. Transformation according to Eq. 7.8-5 is applied to all elements of the structure that contain any of the d.o.f. v_2, u_3, and v_3. Thus, v_2, u_3, and v_3 no longer appear as d.o.f. in $\{\mathbf{D}\}$. The particular stiffness coefficients of triangle 1–2–3 do not matter; they are overridden by the rigid-body constraint.

The choice $\{\mathbf{d}\} = \lfloor u_1 \quad v_1 \quad u_2 \rfloor^T$ is not unique, and would be unacceptable if node numbers were rearranged so that $y_2 - y_1 = b = 0$. Not only would there be a division by zero in Eq. 7.8-5, but the use of u_1 and u_2 as independent d.o.f. would contradict the assumption that the triangle is rigid.

PROBLEMS

Section 7.2

7.1 If a vector $\mathbf{V}$ has length L, then $\mathbf{V}\cdot\mathbf{V} = L^2$ regardless of the coordinate system in which $\mathbf{V}$ resides. Hence, using Eq. 7.2-1, show that $\Sigma\, \ell_i = 1$, $\Sigma\, \ell_i m_i = 0$, and so on (six such relations altogether).

7.2 (a) Let $x' = -x$ and $y' = -y$. What is $[\boldsymbol{\Lambda}]$ in Eq. 7.2-1 if both coordinate systems are right-handed?
(b) Similarly, what is $[\boldsymbol{\Lambda}]$ if $x' = y$ and $z' = z$?

7.3 (a) If $z = z'$ and x' is located at a counterclockwise angle θ from x, what is $[\boldsymbol{\Lambda}]$ in Eq. 7.2-1?
(b) For this $[\boldsymbol{\Lambda}]$, show that $[\boldsymbol{\Lambda}]^{-1} = [\boldsymbol{\Lambda}]^T$.

Section 7.3

7.4 Let $[\mathbf{E}']$ be 3 by 3, as for a plane stress problem. Show that Eq. 7.3-10 yields $[\mathbf{E}'] = [\mathbf{E}]$ if the material is isotropic.

7.5 Let an orthotropic material have principal directions x', y', and z (i.e., axes z' and z coincide). Write the 6 by 6 matrix $[\mathbf{T}_\epsilon]$ for this situation. Express your answer in terms of $\sin\beta$ and $\cos\beta$.

7.6 Consider a plane problem for which the 3 by 3 matrix $[\mathbf{E}']$ is diagonal, with $E'_{11} = E_a$, $E'_{22} = E_b$, and $E'_{33} = G$. What is $[\mathbf{E}]$ for an arbitrary angle β in Fig. 7.3-1? As a partial check on your answer, try the case $\beta = \pi/2$.

7.7 Is $[\mathbf{T}_\epsilon]$ of Eq. 7.3-11 an orthogonal matrix?

Section 7.4

7.8 For a given distortion, strain energy in an element (Eq. 4.1-10) must be independent of the coordinate system in which it is computed. Use this argument to derive Eq. 7.4-4.

Section 7.5

7.9 Two forms of [**k**′] for a truss element are given in Section 7.5, one by Eq. 7.5-1 and the other by Eq. 7.5-2. Verify that appropriate transformation of each form produces the stiffness matrix of Eq. 2.4-3.

7.10 (a) Verify the [**k**] determined in Problem 2.16 by coordinate transformation of Eq. 2.4-3.

(b) Obtain the same result by coordinate transformation of [**k**′] in Eq. 7.5-1.

(c) Obtain the same result by coordinate transformation of [**k**′] in Eq. 7.5-2.

7.11 Write a compact set of Fortran statements that will generate [**k**] of a space truss element (see Eq. 7.5-5).

7.12 Verify that [**k**] in Fig. 7.5-2 follows from [**k**′] by application of Eqs. 7.4-4 and 7.5-7.

7.13 A *plane grillage* is a plane network of straight members that carries loads normal to its plane. Thus the grillage resembles a plane frame, but carries lateral loads. A typical member resists bending and torsional deformation and has six d.o.f., as shown. Write, in terms of angle α in the xy plane, the transformation matrix that would be used to convert [**k**′] to [**k**], where [**k**] operates on d.o.f. w (lateral deflection), θ_x (rotation about the x axis), and θ_y (rotation about the y axis) at each node.

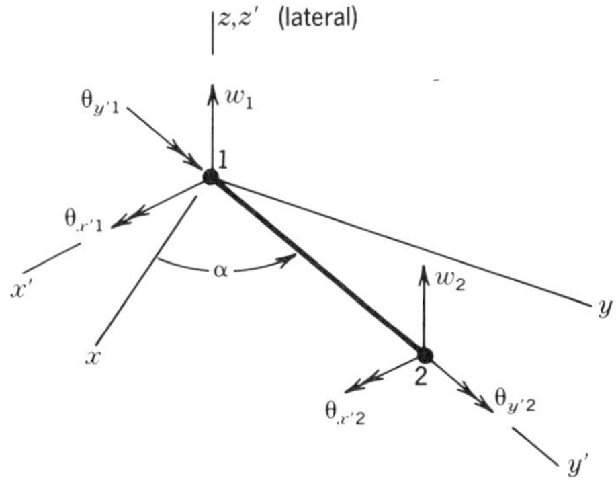

Problem 7.13

7.14 A *space beam* can resist axial load, twisting about its axis, bending about either principal axis of its cross section, and transverse loads. Assume that the beam is straight, uniform, and has six d.o.f. at each end.

(a) Let the beam lie on the x' axis and let y' and z' be parallel to principal axes of the cross section. Write the 12 by 12 stiffness matrix [**k**′] that operates on d.o.f. $\{\mathbf{d}'\} = \lfloor u_1' \quad v_1' \quad w_1' \quad \theta_{x1}' \cdots \theta_{y2}' \quad \theta_{z2}' \rfloor^T$. Let θ vectors point in the positive coordinate directions.

(b) Now consider that the beam is arbitrarily oriented in xyz coordinates.

Stiffness matrix **[k]** is desired, where **[k]** operates on d.o.f. {**d**} that are parallel to x, y, and z axes. Write the transformation matrix **[T]**.

7.15 The size, shape, and orientation in space of a plane triangular element are defined by the known global coordinates of its three corner nodes.

(a) Let node 3 be at the origin of local coordinates $x'y'z'$. In addition, let nodes 3 and 1 define the x' axis and let the plane of the element define the $x'y'$ plane. Describe how to compute the direction cosines of Fig. 7.2-1 from the given information.

(b) If the element is a constant-strain triangle, write the matrix **[T]** that will produce the 9 by 9 global matrix **[k]** from the 6 by 6 local matrix **[k′]**.

7.16 The bar shown is *rigid* and is supported by two linear springs of stiffness k_1 and k_2. Only vertical displacement is permitted.

(a) Write the stiffness matrix that operates on d.o.f. v_1 and v_2. Then transform this matrix so that you obtain a stiffness matrix that operates on v_1 and θ_1, where θ_1 is a small rotation of the bar with respect to the horizontal and about the left end.

(b) Redefine v_2 so that it is the vertical displacement at the midpoint of the bar. Do not change v_1. Then repeat part (a).

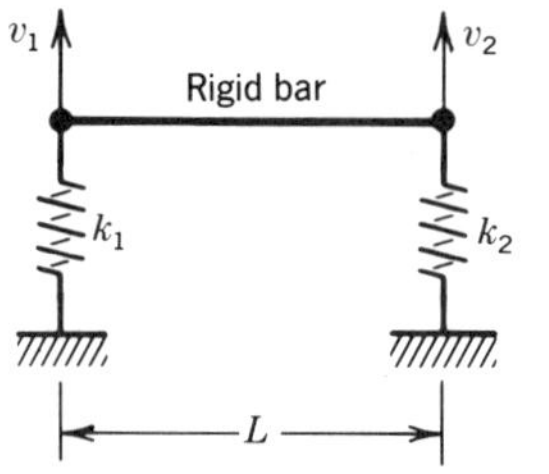

Problem 7.16a

7.17 Work Problem 7.16 in reverse. That is, start with the final stiffness matrix that operates on v_1 and θ_1. By transformation, obtain from it the matrix in Problem 7.16a that operates on v_1 and v_2. Similarly, obtain the matrix in Problem 7.16b that operates on v_1 and the midpoint v_2.

7.18 Let a standard bar element of axial stiffness $k = AE/L$ be restricted to motion along its axis. Its **[k]** is 2 by 2 and operates on nodal d.o.f. u_1 and u_2. Transform **[k]** so that it operates on nodal d.o.f. u_1 and u_r, where u_r is the displacement of node 2 relative to node 1.

7.19 A three-node bar element and its shape functions are shown in Fig. 6.2-1. Imagine that d.o.f. u_3 is to be replaced by u_r, where u_r is the displacement at $\xi = 0$ *relative* to the displacement at $\xi = 0$ dictated by u_1 and u_2. Thus, $u_3 = u_r + \frac{1}{2}(u_1 + u_2)$. Write the transformation matrix and use it to determine the new shape functions.

Section 7.6

7.20 Let loads F_x and F_y act at node 3 in Fig. 7.6-1. Verify that the operation $[\mathbf{T}]^T\{\mathbf{r}'\}$ transforms F_x and F_y to the correct r and s components.

7.21 Let Fig. 7.6-1 represent a plane truss for which axial stiffness $k = AE/L$ is the same for each bar. Also let the three interior angles in each panel be 45°, 45°, and 90°. Apply a downward load P at node 4 and set $u_4 = 0$. If $\beta = \arctan 0.75$, what is the force in bar 3–1 in terms of P?

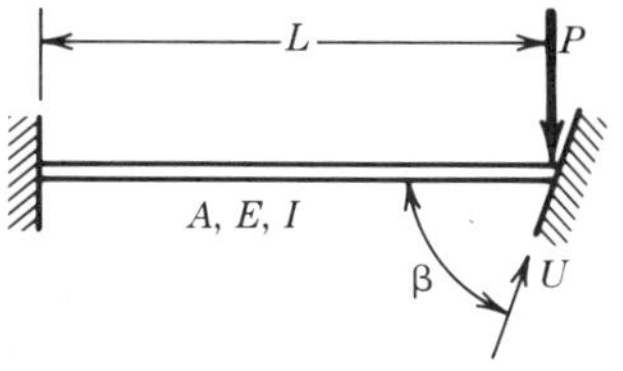

Problem 7.22

7.22 The right end of the cantilever beam slides without friction on a rigid wall, as shown. Represent the cantilever as a single element with axial, transverse, and rotational d.o.f. at the right end.

(a) Transform and impose the boundary conditions. Thus, obtain a 2 by 2 matrix **[K]** that operates on tangential displacement U and rotation θ at the right end.

(b) In addition, let the condition $\theta = 0$ be imposed. Solve for U.

7.23 Imagine that, at a certain node of a space truss, motion is to be prohibited along a line whose direction cosines are ℓ_1, ℓ_2, and ℓ_3. Motion is permitted in all directions normal to the line. Original nodal d.o.f. are displacements in coordinate directions x, y, and z.

(a) Explain precisely how to define suitable new directions for d.o.f. at the node, and write the transformation matrix at the node (analogous to $[\mathbf{T}_3]$ in Eq. 7.6-4).

(b) Check your result for the special case $\ell_2 = 1$.

Section 7.7

7.24 The element shown is of arbitrary quadrilateral shape and is formulated as a bilinear element (Section 6.3). A constant-strain triangle (six d.o.f.; lettered nodes) is to be attached, so that lettered nodes lie at $\xi = \pm 0.5$ and $\eta = \pm 0.5$. Write **[T]** in the relation $\{\mathbf{d}'\} = [\mathbf{T}]\{\mathbf{d}\}$, where $\{\mathbf{d}'\}$ and $\{\mathbf{d}\}$ contain d.o.f. of lettered nodes ("slaves") and numbered nodes ("masters") respectively.

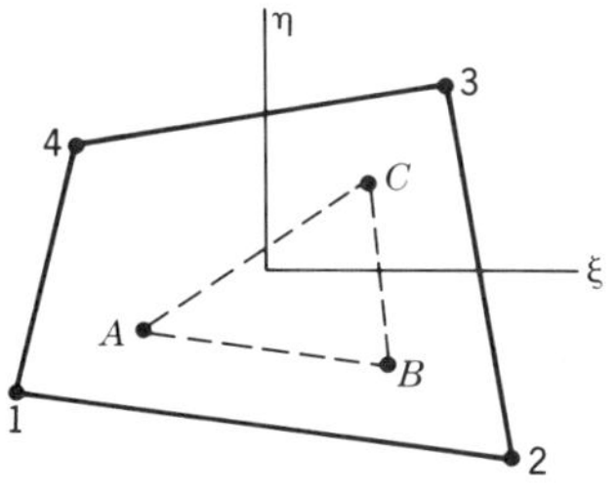

Problem 7.24

7.25 At node 5 of the frame element in Fig. 7.7-1*a*, let forces F_x and F_y and moment M_5 (counterclockwise) be applied. How are these loads distributed to nodes 2 and 3 by the transformation of Eq. 7.7-4? Do the new loads exert the same force and moment resultants as the original loads?

7.26 Write the transformation matrix for the problem described in connection with Eq. 7.7-5. What is $[\mathbf{k}']$ for this problem?

7.27 Plane element 1 in the sketch is bilinear (Section 6.3). It has the usual two d.o.f. per node. Plane *frame* element 2 has the usual three d.o.f. per node (u_i, w_i, θ_i). Consider the element stiffness matrices $[\mathbf{k}_1]$ and $[\mathbf{k}_2]$.

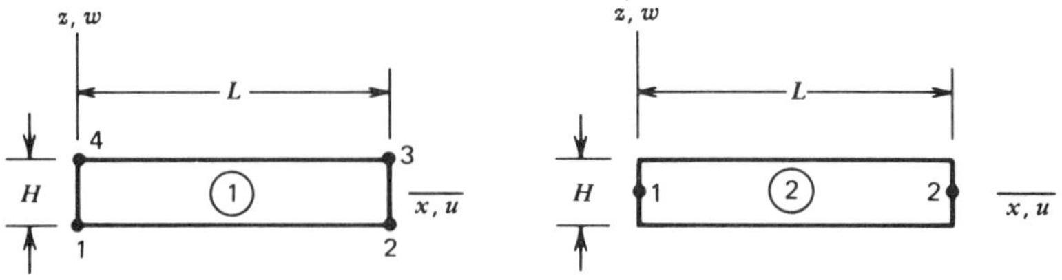

Problem 7.27

(a) Write a transformation matrix $[\mathbf{T}_1]$ that could be used to convert $[\mathbf{k}_1]$ so that it operates on the d.o.f. of element 2.
(b) Write a transformation matrix $[\mathbf{T}_2]$ that could be used to convert $[\mathbf{k}_2]$ so that it operates on the d.o.f. of element 1.
(c) Should $[\mathbf{T}_1][\mathbf{T}_2]$ and $[\mathbf{T}_2][\mathbf{T}_1]$ be unit matrices? Find an argument that says so.
(d) Evaluate the products $[\mathbf{T}_1][\mathbf{T}_2]$ and $[\mathbf{T}_2][\mathbf{T}_1]$. How can the results be explained?

Section 7.8

7.28 Element *ij* is a plane frame element (see sketch). Imagine that d.o.f. at nodes *i* and *j* are to be made slave to d.o.f. at nodes 1 and 2 via rigid links *i*1 and *j*2. Write the 6 by 6 transformation matrix [**T**].

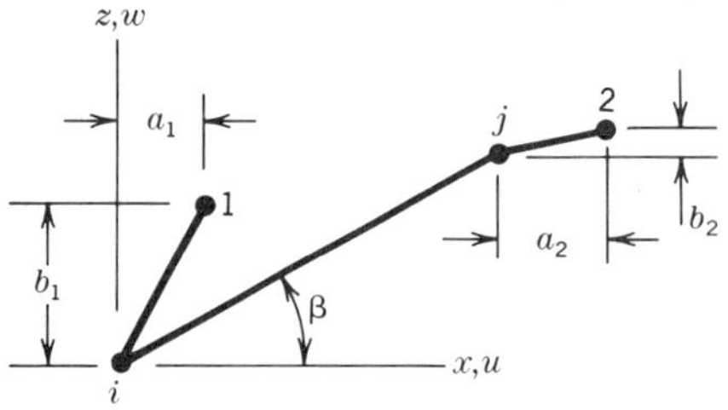

Problem 7.28

7.29 Consider a frame element *ij*, arbitrarily oriented in space. D.o.f. at node *i* are $\lfloor u_i \quad v_i \quad w_i \quad \theta_{xi} \quad \theta_{yi} \quad \theta_{zi} \rfloor$, where rotational d.o.f. vectors point in positive coordinate directions. D.o.f. at node *j* are similar. The element is to be made slave to d.o.f. at some other nodes (nodes 1 and 2, say) via rigid links *i*1 and *j*2, which are arbitrarily oriented. Write the 12 by 12 transformation matrix [**T**].

7.30 (a) Evaluate Eq. 7.8-4. That is, express u_6 in terms of b, u_5, w_1, θ_1, w_2, and θ_2 (see Eq. 4.2-3 and Fig. 3.13-2).
(b) Write the 7 by 7 transformation matrix [**T**] in $\{\mathbf{d}'\} = [\mathbf{T}]\{\mathbf{d}\}$, where $\{\mathbf{d}'\} = \lfloor u_3 \quad w_3 \quad \theta_3 \quad u_4 \quad w_4 \quad \theta_4 \quad u_6 \rfloor^T$.

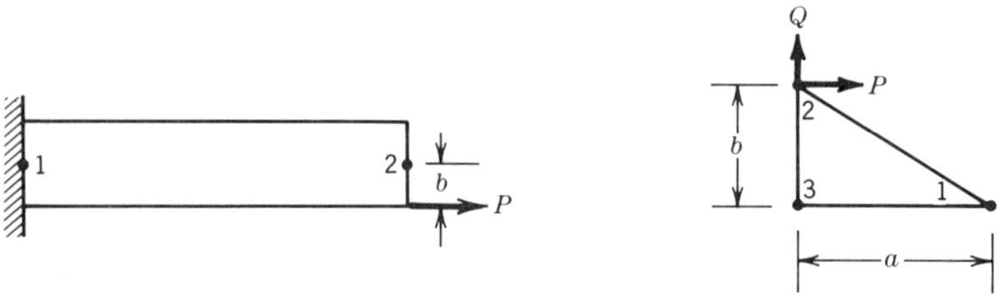

Problem 7.31

7.31 Obtain nodal loads $\{\mathbf{r}\} = [\mathbf{T}]^T\{\mathbf{r}'\}$ for the elements and loads shown. Use $[\mathbf{T}]$ from Eq. 7.8-1 and Eq. 7.8-5 for the respective elements. Sketch $\{\mathbf{r}\}$, and argue why $\{\mathbf{r}\}$ is reasonable or unreasonable.

7.32 Rewrite Eq. 7.8-5 if the triangle is of arbitrary shape, with nodal coordinates x_i and y_i $(i = 1, 2, 3)$.

7.33 Rewrite Eq. 7.8-5 if the "master" d.o.f. are changed from u_1, v_1, and u_2 to u_2, u_3, and v_3.

8 CHAPTER

TOPICS IN STRUCTURAL MECHANICS

Miscellaneous elements, procedures, and remarks are presented. Some topics are of general interest while others pertain to structural mechanics.

8.1 D.O.F. WITHIN ELEMENTS. CONDENSATION

Occasionally, the basic building block of a finite element mesh is a *macroelement*—that is, a "patch" that consists of two or more elements coupled together. A macroelement can be regarded as a small structure. Its component elements are called subelements. Two examples appear in Fig. 8.1-1. Both macroelements are built of triangular subelements. In both cases the user of a computer program need define only the boundary nodes (which are numbered in the sketch). The program itself can automatically locate the internal nodes, generate and combine matrices of the subelements, and produce a stiffness matrix and load vector associated with only the boundary nodes. There is no limit to the number of subelements or the number of internal d.o.f. This observation leads to *substructuring,* discussed in Section 8.14.

D.o.f. of internal nodes are coupled only to d.o.f. of other internal nodes and to d.o.f. of nodes on the macroelement boundary. There is no coupling of internal d.o.f. to d.o.f. of nodes outside the macroelement. Accordingly, equations associated with internal d.o.f. can be processed separately from other equations of the structure. Separate processing can be both efficient and convenient for users, as will be seen subsequently. In the present section we emphasize the processing procedures.

Condensation. *Condensation* is the process of reducing the number of d.o.f. by substitution, for example, by starting a Gauss elimination solution of equations for unknowns but stopping before the stiffness matrix has been fully reduced. Condensation by elimination is also called *static condensation*. Condensation in dynamics is usually called *reduction* and introduces an approximation. Static condensation, described as follows, is strictly a manipulation and introduces no approximation.

Let the equations $[\mathbf{k}]\{\mathbf{d}\} = \{\mathbf{r}\}$ represent a portion of the entire structure. This portion might be a macroelement built of subelements or a single element that has "nodeless" d.o.f. (such an element will be described in the following). Let d.o.f. $\{\mathbf{d}\}$ be partitioned so that $\{\mathbf{d}\} = \lfloor \mathbf{d}_r \quad \mathbf{d}_c \rfloor^T$, where $\{\mathbf{d}_r\}$ are boundary d.o.f. to be retained and $\{\mathbf{d}_c\}$ are internal d.o.f. to be eliminated by condensation. Thus $[\mathbf{k}]\{\mathbf{d}\} = \{\mathbf{r}\}$ becomes

$$\begin{bmatrix} \mathbf{k}_{rr} & \mathbf{k}_{rc} \\ \mathbf{k}_{cr} & \mathbf{k}_{cc} \end{bmatrix} \begin{Bmatrix} \mathbf{d}_r \\ \mathbf{d}_c \end{Bmatrix} = \begin{Bmatrix} \mathbf{r}_r \\ \mathbf{r}_c \end{Bmatrix} \tag{8.1-1}$$

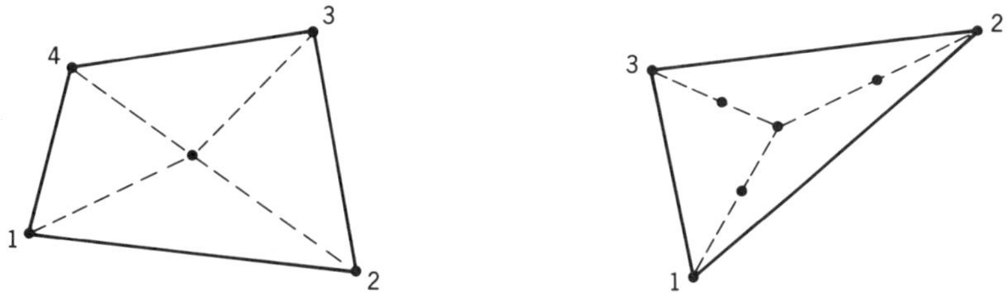

(*a*) (*b*)

Figure 8.1-1. Elements having internal nodes. Boundary nodes are numbered; internal nodes are not. (*a*) A quadrilateral built of four triangles. (*b*) A triangle built of three triangles.

The lower partition is solved for $\{\mathbf{d}_c\}$:

$$\{\mathbf{d}_c\} = -[\mathbf{k}_{cc}]^{-1}([\mathbf{k}_{cr}]\{\mathbf{d}_r\} - \{\mathbf{r}_c\}) \tag{8.1-2}$$

Next, $\{\mathbf{d}_c\}$ is substituted into the upper partition of Eqs. 8.1-1. Thus

$$\underbrace{([\mathbf{k}_{rr}] - [\mathbf{k}_{rc}][\mathbf{k}_{cc}]^{-1}[\mathbf{k}_{cr}])}_{\text{condensed } [\mathbf{k}]}\{\mathbf{d}_r\} = \underbrace{\{\mathbf{r}_r\} - [\mathbf{k}_{rc}][\mathbf{k}_{cc}]^{-1}\{\mathbf{r}_c\}}_{\text{condensed } \{\mathbf{r}\}} \tag{8.1-3}$$

The element is now treated in standard fashion; that is, the condensed $[\mathbf{k}]$ and the condensed $\{\mathbf{r}\}$ are assembled into the structure, boundary conditions are imposed, and structural d.o.f. $\{\mathbf{D}\}$ are computed. Thus $\{\mathbf{d}_r\}$ becomes known, and $\{\mathbf{d}_c\}$ (which may be needed in stress calculation) follows from Eq. 8.1-2. Computation of $\{\mathbf{d}_c\}$ is called *recovery* of internal d.o.f. Computer algorithms for condensation and recovery are discussed in Section 8.2.

Equation 8.1-3 is Gauss elimination, carried out on d.o.f. $\{\mathbf{d}_c\}$ only (compare with Eq. B.2-2, Appendix B). Completion of the elimination process, and solution for $\{\mathbf{d}_r\}$, awaits assembly of all remaining elements of the structure. Thus condensation is simply the first set of eliminations in a solution of the structure equations $[\mathbf{K}]\{\mathbf{D}\} = \{\mathbf{R}\}$. The same solution vector $\{\mathbf{D}\}$ would result if internal d.o.f. were eliminated later. The advantage of eliminating them first, at the element level and before assembly, is that the order of the structure stiffness matrix is reduced because d.o.f. $\{\mathbf{d}_c\}$ are not carried into the global set of equations.

The partitioning used in Eq. 8.1-1 is a conceptual convenience rather than a computational necessity. D.o.f. to be condensed can appear anywhere in $\{\mathbf{d}\}$, and can be processed serially rather than simultaneously. After a d.o.f. d_k is condensed, rows $i \neq k$ and columns $j \neq k$ comprise the condensed $[\mathbf{k}]$.

Nodeless D.o.f. Internal d.o.f. need not be associated with a node. The 18 d.o.f. plane element of Table 6.6-1 can be restated in terms of nodeless internal d.o.f., as we now describe. For $i = 1$ to 8, let shape functions N_i be those of the 16 d.o.f. element, Eq. 6.6-1. Then displacements in the 18 d.o.f. element are

$$u = \sum_{i=1}^{8} N_i u_i + N_9 a_1 \quad \text{and} \quad v = \sum_{i=1}^{8} N_i v_i + N_9 a_2 \tag{8.1-4}$$

where, as in Eq. 6.6-3,

$$N_9 = (1 - \xi^2)(1 - \eta^2) \tag{8.1-5}$$

Mode N_9 is called a "bubble function" mode, and a_1 and a_2 are nodeless d.o.f. to be condensed, $\{\mathbf{d}_c\} = \lfloor a_1 \quad a_2 \rfloor^T$. Physically, a_1 and a_2 represent the displacement components at $\xi = \eta = 0$ *relative* to the displacement components $\Sigma\, N_i u_i$ and $\Sigma\, N_i v_i$ at $\xi = \eta = 0$ dictated by d.o.f. at the eight boundary nodes. It is not necessary to assign such a physical meaning, or to calculate actual displacements at $\xi = \eta = 0$, because d.o.f. a_1 and a_2 are not connected to other elements—that is, a_1 and a_2 in an element are not d.o.f. of another element as well.

In processing, nodeless d.o.f. are treated no differently than any other d.o.f. Thus, for the 18 d.o.f. element, it does not matter whether the N_i are given by Table 6.6-1 or by Eqs. 8.1-4: if formulation procedures of preceding chapters are used consistently, then, from either starting point, identical 16 by 16 condensed matrices $[\mathbf{k}]$ and 16 by 1 consistent load vectors $\{\mathbf{r}_e\}$ appear after condensation of the two internal d.o.f. (u_9 and v_9 or a_1 and a_2). Before condensation, loads in $\{\mathbf{r}_e\}$ associated with a_1 and a_2 may appear to be too large, even though correct, because a_1 and a_2 are not actual displacements.

When a_1 and a_2 are used as internal d.o.f., element geometry (e.g., the Jacobian matrix $[\mathbf{J}]$ of Eq. 6.6-4) is defined by the eight N_i of Eqs. 6.6-1 and the coordinates of the eight boundary nodes. Use of Eq. 6.6-2 and the N_i of Table 6.6-1 would yield the same geometry but with slightly more computational effort.

Releases. A "release" is a lack of complete connection between nodes that would usually be fully connected. The plane frame of Fig. 8.1-2 is a case in point. The structure displacement vector $\{\mathbf{D}\}$ contains three d.o.f. per node. At node A, where there is a hinge, the two frames are not to share the same nodal rotation θ_A, as this would imply a rigid connection rather than a hinge. One way to model the hinge is to condense θ at A in (say) the left frame, then fill the row and column just condensed with zeros so that no rotational stiffness at A will be contributed to the structure by the left frame. Thus assembly makes the left and right frames share only u_A and v_A, and θ_A in $\{\mathbf{D}\}$ now represents the rotation at A in the right frame. Rotation at A in the left frame is treated as an internal d.o.f. to be recovered after $\{\mathbf{D}\}$ is known.

Another way to treat the hinge at A in Fig. 8.1-2 is to define two separate nodes at A, one in the left frame and one in the right but having the same location. Thus there are a total of six d.o.f. at A. Next, one joins only translational d.o.f. of the two nodes by means of a constraint technique (see Chapter 9). A similar treatment could be used at an interface between elastic bodies that may slide on one another.

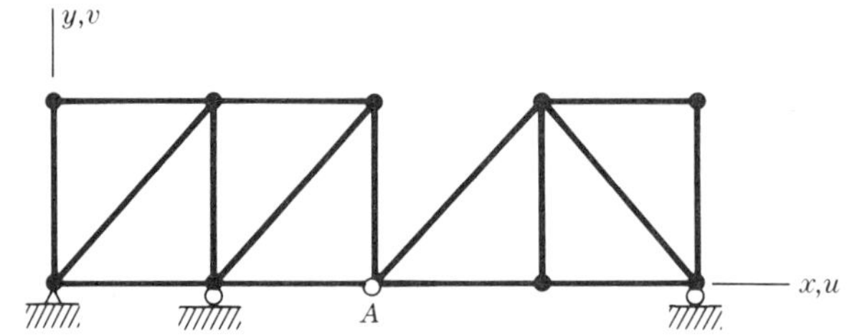

Figure 8.1-2. Two frames with a hinge connection at A.

```
C---    Do condensation operations on lower triangle of SE.
        DO 30 K=1,NUM
        LL = NSIZE - K
        KK = LL + 1
        DO 20 L=1,LL
        DUM = SE(KK,L)/SE(KK,KK)
        DO 10 M=1,L
     10 SE(L,M) = SE(L,M)  - SE(KK,M)*DUM
        RE(L)   = RE(L)    - RE(KK)  *DUM
     20 CONTINUE
     30 CONTINUE
C---    Fill in the upper triangle of SE by symmetry.
        DO 40 K=1,LL
        DO 40 L=1,K
     40 SE(L,K) = SE(K,L)
```

Figure 8.2-1. A Fortran condensation algorithm that accepts a *symmetric* matrix, stored in full rather than banded format, does Gauss elimination from the bottom with diagonal pivots and yields the condensed [**k**] and {**r**} of Eq. 8.1-3.

8.2 CONDENSATION AND RECOVERY ALGORITHMS

Our algorithm is a straightforward coding of Gauss elimination to produce the condensed arrays [**k**] and {**r**} of Eqs. 8.1-3 from the uncondensed arrays of Eq. 8.1-1. Let there be NSIZE d.o.f. in the uncondensed system, and let it be required to condense the last NUM of these d.o.f. In Fig. 8.2-1 condensation is done in place, so that after completion of the condensation algorithm, the condensed [**k**] resides in the upper left NSIZE-NUM rows and columns of array SE, and the condensed {**r**} resides in the upper NSIZE-NUM rows of array RE.

After structural equations have been solved, d.o.f. $\{\mathbf{d}_r\}$ reside in the structure solution vector {**D**}. It remains to recover $\{\mathbf{d}_c\}$ in Eq. 8.1-2. This can be done by back-substitution, thus completing the Gauss elimination solution that was begun by condensation. Figure 8.2-2 gives the Fortran coding. In Fig. 8.2-2, arrays SM and RM contain the last NUM rows of arrays SE and RE exactly as they stand after completion of the condensation routine. In Fig. 8.2-1 these rows *follow* the condensed matrices, and are numbered NUM+1 through NSIZE. In Fig. 8.2-2 these rows are duplicated in arrays SM and RM, where they occupy rows 1 through NUM. In Fig. 8.2-2 the first NSIZE-NUM rows of array DE contain the known d.o.f. $\{\mathbf{d}_r\}$. Internal d.o.f. $\{\mathbf{d}_c\}$ are computed and stored in the last NUM rows of array DE.

In the foregoing algorithms, mass storage would probably be used to store the last NUM rows of arrays SE and RE from Fig. 8.2-1. Node point coordinates, and $\{\boldsymbol{\epsilon}_0\}$ and $\{\boldsymbol{\sigma}_0\}$, would also be stored. After {**D**} is known, these data would be recalled and Fig. 8.2-2 used to compute $\{\mathbf{d}_c\}$. At each point where stresses are needed, [**B**] can be reconstructed from the node point coordinates. Finally, strains are $\{\boldsymbol{\epsilon}\} = [\mathbf{B}]\lfloor \mathbf{d}_r \quad \mathbf{d}_c \rfloor^T - \{\boldsymbol{\epsilon}_0\}$ and stresses are $\{\boldsymbol{\sigma}\} = [\mathbf{E}]\{\boldsymbol{\epsilon}\} + \{\boldsymbol{\sigma}_0\}$.

```
        DO 60 J=1,NUM
        JJ = NSIZE - NUM + J
        DUM = 0.
        K = JJ - 1
        DO 50 L=1,K
     50 DUM = DUM + SM(J,L)*DE(L)
     60 DE(JJ) = (RM(J) - DUM)/SM(J,JJ)
```

Figure 8.2-2. Recovery of previously condensed d.o.f. $\{\mathbf{d}_c\}$ when $\{\mathbf{d}_r\}$ and $\{\mathbf{r}_c\}$ are known.

Alternative Method. In an alternative method [8.1], explicit recovery of $\{\mathbf{d}_c\}$ is avoided. Instead, as part of the process of generating the element stiffness matrix, an element stress matrix [**S**] and stress vector $\{\boldsymbol{\rho}\}$ are also generated. After element d.o.f. $\{\mathbf{d}_r\}$ are known, element stresses $\{\boldsymbol{\sigma}\}$ are computed by the equation

$$\{\boldsymbol{\sigma}\} = [\mathbf{S}]\{\mathbf{d}_r\} + \{\boldsymbol{\rho}\} \tag{8.2-1}$$

in which previously condensed d.o.f. $\{\mathbf{d}_c\}$ do not appear.

The most significant differences between the two methods are as follows. The first method, Figs. 8.2-1 and 8.2-2, explicitly recovers $\{\mathbf{d}_c\}$ and reconstructs a (somewhat sparse) matrix [**B**] at each stress point. The alternative method generates, condenses, and stores an array [**S**] that is smaller than [**B**] but not sparse. Both methods yield the same stresses. The relative cost of the two methods depends on billing charges for computing and for mass storage, the size of $\{\mathbf{d}_r\}$ in relation to $\{\mathbf{d}_c\}$, the number of load conditions, the number of stress points, and other less important factors. The method of Figs. 8.2-1 and 8.2-2 tends to be cheaper if $\{\mathbf{d}_c\}$ is small in relation to $\{\mathbf{d}_r\}$, if the number of load cases is small, or if the number of stress points is large. Further comparison appears in [8.2].

8.3 PARASITIC SHEAR. INCOMPATIBLE ELEMENTS

Parasitic Shear. Bilinear elements, discussed in Section 6.3, are attractive because they are simple and have only corner nodes. Unfortunately, they are too stiff in bending, whether the element is a rectangle or an arbitrary quadrilateral. We illustrate the point with reference to the rectangular element in Fig. 8.3-1. Here ξ and η are dimensionless Cartesian coordinates, $\xi = x/a$ and $\eta = y/b$. Let bending moment M_1 be applied, so that nodal displacements $\bar{u}$ arise in response, as shown in Fig. 8.3-1*b*. According to Eqs. 6.3-2 and 6.3-16, the element deformation field is

$$u = \xi\eta\bar{u} \qquad \text{and} \qquad v = 0 \tag{8.3-1}$$

Thus top and bottom edges $\eta = \pm 1$ remain straight, and strains in the element are

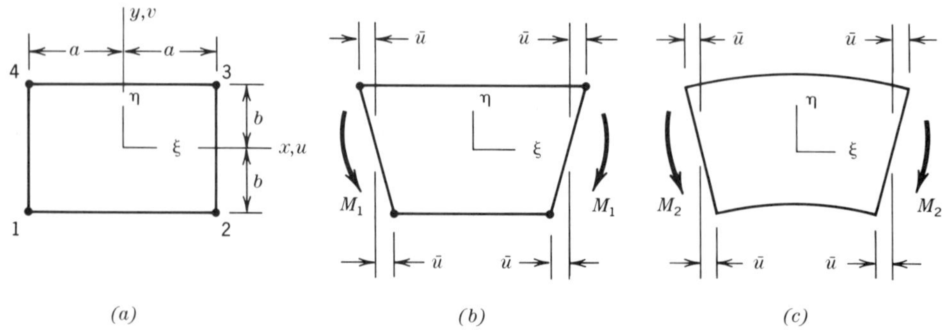

Figure 8.3-1. (*a*) A rectangular bilinear element. (*b*) The bilinear element deformed by bending moment M_1. (*c*) Correct deformed geometry for pure bending under bending moment M_2.

$$\epsilon_x = \eta \frac{\overline{u}}{a} \qquad \epsilon_y = 0 \qquad \gamma_{xy} = \xi \frac{\overline{u}}{b} \tag{8.3-2}$$

The *correct* shape under pure bending, Fig. 8.3-1*c*, is

$$u = \xi\eta\overline{u} \qquad \text{and} \qquad v = (1 - \xi^2)\frac{a\overline{u}}{2b} + (1 - \eta^2)\,\nu\frac{b\overline{u}}{2a} \tag{8.3-3}$$

where ν is Poisson's ratio. From Eqs. 8.3-3, the correct strains under pure bending are

$$\epsilon_x = \eta \frac{\overline{u}}{a} \qquad \epsilon_y = -\nu\eta \frac{\overline{u}}{a} \qquad \gamma_{xy} = 0 \tag{8.3-4}$$

Upon comparing Eqs. 8.3-2 and 8.3-4, we see that if bending displacements $\overline{u}$ are imposed, the correct behavior gives rise to storage of strain energy caused by normal strains alone, but the bilinear element stores strain energy caused by normal strain ϵ_x *and* a spurious shear strain γ_{xy}. Thus, for the same deformation, $M_1 > M_2$ in Fig. 8.3-1. Specifically, by computing the ratio of strain energies in the two cases, we obtain

$$\frac{M_1}{M_2} = \frac{1}{1 + \nu}\left[\frac{1}{1 - \nu} + \frac{1}{2}\left(\frac{a}{b}\right)^2\right] \tag{8.3-5}$$

The unwanted shear strain that produces $M_1 > M_2$ is called *parasitic shear.* Equation 8.3-5 shows that its effect is disastrous if a/b is large; that is, for large a/b the mesh "locks." Locking is discussed in more detail in Section 9.4.

Incompatible Elements. Upon comparing Eqs. 8.3-1 and 8.3-3, we see that the bilinear element errs by omitting from v the displacement modes associated with $(1 - \xi^2)$ and $(1 - \eta^2)$. In the bending mode where $v = \overline{v}\xi\eta$, similar modes are omitted from u. The eight-node trilinear solid element (Section 6.7) suffers from the same defects. These elements, whether rectangular or not, can be improved by adding the missing modes as internal freedoms. We write [8.3]

Eight-node solid element →
Four-node plane element →

$$\begin{aligned} u &= \Sigma N_i u_i + (1 - \xi^2)a_1 + (1 - \eta^2)a_2 \quad + (1 - \zeta^2)a_7 \\ v &= \Sigma N_i v_i + (1 - \xi^2)a_3 + (1 - \eta^2)a_4 \quad + (1 - \zeta^2)a_8 \\ w &= \Sigma N_i w_i + (1 - \xi^2)a_5 + (1 - \eta^2)a_6 \quad + (1 - \zeta^2)a_9 \end{aligned} \tag{8.3-6}$$

where the a_i are nodeless d.o.f. For the plane element, $i = 1, 2, 3, 4$ and the N_i are given by Eq. 6.3-2. For the solid element, $i = 1, 2, \ldots, 8$ and the N_i are given by Eq. 6.7-6. The plane element associated with Eqs. 8.3-6 is usually called the Q6 element. If rectangular, it models pure bending exactly regardless of element aspect ratio. In programming this element—for example, by modifying Figs. 6.5-1 and 6.5-2—one adds four columns to array B in order to accommodate the four additional d.o.f., expands other arrays and loop indexes as required, but computes the Jacobian matrix as before (as though the element had only the basic nodal d.o.f.).

The Q6 element is *incompatible* or *nonconforming*. For example, as suggested by Fig. 8.3-2, the mode $u = (1 - \eta^2)a_2$ might be activated in one element but not in its neighbors to the left and right, thus producing a gap on one side and an overlap on the other. But incompatible elements are still valid if incompatibilities disappear and a constant-strain state is approached as the mesh is refined. That is, the element is valid if it passes the patch test. If an element of general shape is to pass the patch test, a modified integration scheme is needed; it will be described subsequently.

Incompatible elements often yield results of high quality. Interelement gaps and overlaps tend to soften a structure. Softening counters the inherent overstiffness of an assumed-displacement approximation. A good balance of the two effects leads to good results with a coarse mesh. However, the upper-bound nature of the approximation is lost: there is no guarantee that a mesh of incompatible elements will be stiffer than the actual structure. Moreover, in problems that should be independent of Poisson's ratio ν, a coarse mesh of incompatible elements may display a dependence on ν.

After formulation of element matrices, condensation removes the a_i of Eq. 8.3-6. Thus, for the four-node plane element, a_1 through a_4 are eliminated, leaving an 8 by 8 condensed matrix $[\mathbf{k}]$. *For the special case of a rectangular element,* as in Fig. 8.3-1, this condensed $[\mathbf{k}]$ is the same as the 8 by 8 $[\mathbf{k}]$ produced *directly* by the displacement field [8.4],

$$\begin{Bmatrix} u \\ v \end{Bmatrix} = \begin{bmatrix} N_1 & N_x & N_2 & -N_x & N_3 & N_x & N_4 & -N_x \\ N_y & N_1 & -N_y & N_2 & N_y & N_3 & -N_y & N_4 \end{bmatrix} \begin{Bmatrix} u_1 \\ v_1 \\ \vdots \\ u_4 \\ v_4 \end{Bmatrix} \tag{8.3-7}$$

where N_1 through N_4 are given by Eqs. 6.3-2, and

$$\begin{aligned} N_x &= (1 - \xi^2)\, \nu \frac{a}{8b} + (1 - \eta^2) \frac{b}{8a} \\ N_y &= (1 - \xi^2) \frac{a}{8b} + (1 - \eta^2)\, \nu \frac{b}{8a} \end{aligned} \tag{8.3-8}$$

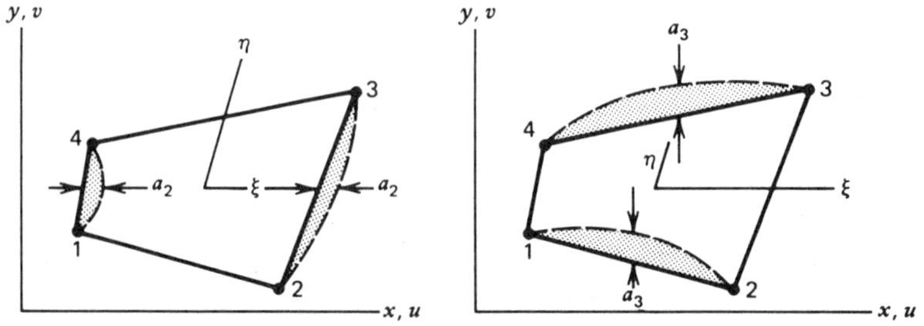

Figure 8.3-2. Dashed lines show edge displacements associated with the incompatible modes $u = (1 - \eta^2)a_2$ and $v = (1 - \xi^2)a_3$ in the plane element described by Eqs. 8.3-6.

Yet a third way to obtain the same [**k**] (for a rectangular element) is to begin with the stress field

$$\sigma_x = \beta_1 + \beta_4 y \qquad \sigma_y = \beta_2 + \beta_5 x \qquad \tau_{xy} = \beta_3 \tag{8.3-9}$$

where the β_i are constants. Using these stress modes, one can invoke the "hybrid" method (Section 8.5), or one can compute the associated displacement field by integration and then proceed in the usual way. Indeed, the latter approach was used very early in finite element history [1.8; see also the discussion cited in Ref. 8.8].

Modified Integration Scheme. The plane and solid elements described by Eq. 8.3-6 pass patch tests only if they are either rectangular or parallelograms and parallelepipeds. A modified integration scheme, here described, corrects this failing [8.3]. Thus modified, the plane element is known as the QM6 element.

In the augmented strain–displacement relation $\{\boldsymbol{\epsilon}\} = [\mathbf{B}]\{\mathbf{d}\}$, let $[\mathbf{B}_a]$ represent the latter columns of [**B**], that is, the portion of [**B**] associated with nodeless d.o.f. a_i. Hence, from Eq. 4.1-6, the contribution of the a_i to the consistent element nodal load vector $\{\mathbf{r}_e\}$ is

$$\{\mathbf{r}_{ea}\} = -\int_{V_e} [\mathbf{B}_a]^T\{\boldsymbol{\sigma}_0\}\, dV \tag{8.3-10}$$

Imagine that instead of representing initial stresses, $\{\boldsymbol{\sigma}_0\}$ represents element stresses produced by nodal displacements $\{\mathbf{d}_r\}$ on the element boundary. The basic isoparametric element, with neither the a_i nor $\{\mathbf{r}_{ea}\}$ present, is able to pass a patch test. In other words, when $\{\boldsymbol{\sigma}_0\}$ is constant and produced by the "essential" d.o.f. $\{\mathbf{d}_r\}$, certain "correct" nodal loads associated with $\{\boldsymbol{\sigma}_0\}$ are applied by an element to its nodes. These loads should not be disturbed if incompatible modes are added. Accordingly, in a patch test no *additional* nodal loads should be associated with the a_i. This means that $\{\mathbf{r}_{ea}\}$ of Eq. 8.3-10 must vanish when $\{\boldsymbol{\sigma}_0\}$ is constant. When $\{\boldsymbol{\sigma}_0\}$ is constant, we see that $\{\mathbf{r}_{ea}\}$ will be zero if, for plane and solid elements, respectively,

$$\int_{-1}^{1}\int_{-1}^{1} [\mathbf{B}_a]^T\, tJ\, d\xi\, d\eta = 0 \qquad \int_{-1}^{1}\int_{-1}^{1}\int_{-1}^{1} [\mathbf{B}_a]^T\, J\, d\xi\, d\eta\, d\zeta = 0 \tag{8.3-11}$$

where t is element thickness and J is the Jacobian determinant. For parallelograms of constant thickness and parallelepipeds, t and J are constant and $[\mathbf{B}_a]$ contains first powers of ξ and η (and ζ for solids), so that Eq. 8.3-11 is satisfied automatically. For elements of general shape, t, J, and $[\mathbf{B}_a]$ are more complicated functions; Eqs. 8.3-11 are not satisfied and the patch test is failed. But we can "artificially" satisfy Eqs. 8.3-11 as follows. In forming $[\mathbf{B}_a]$ and integrating, instead of using the correct $[\mathbf{J}]^{-1}$ and J at the Gauss quadrature points, use the constant values $[\mathbf{J}_0]^{-1}$ and J_0, where $[\mathbf{J}_0]$ and J_0 are the Jacobian matrix and its determinant at $\xi = \eta = \zeta = 0$. In addition, for plane elements, use t_0 rather than t if t varies. (If Figs. 6.5-1 and 6.5-2 are adapted to the QM6 element, these adjustments can all be confined to Fig. 6.5-1, in which only columns 9 through 12 of the augmented [**B**] are affected.)

Elements Q6 and QM6 work almost as well as the quadratic element of Eqs.

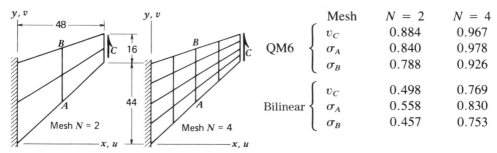

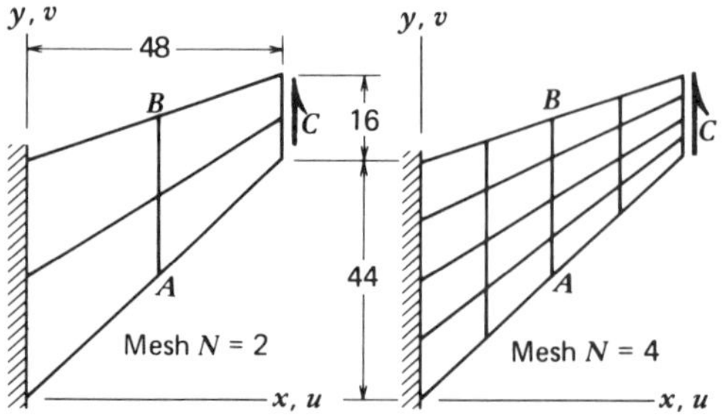

	Mesh	$N = 2$	$N = 4$
QM6	v_C	0.884	0.967
	σ_A	0.840	0.978
	σ_B	0.788	0.926
Bilinear	v_C	0.498	0.769
	σ_A	0.558	0.830
	σ_B	0.457	0.753

Figure 8.3-3. A plane structure with a uniformly distributed load along the right edge and with $E = 1.0$, $\nu = 1/3$. "Bilinear" refers to the element of Eqs. 6.3-2, and v_C = deflection at C, σ_A = maximum stress at A, σ_B = minimum stress at B, all reported as the ratio of computed value to best-known answer [8.3].

6.6-1, but only if rectangular. Nevertheless, the QM6 element is considerably more accurate than the bilinear element, as seen in Fig. 8.3-3. (The same problem is solved, using different elements, in Fig. 8.6-2.)

Stresses are calculated by using all element d.o.f., including the a_i, to evaluate element strains. Calculated stresses are usually more accurate if element nodal loads in $\{\mathbf{r}_e\}$ that are associated with incompatible modes are set to zero during recovery of the associated d.o.f. a_i.

8.4 ROTATIONAL D.O.F. IN PLANE ELEMENTS

The d.o.f. considered are rotations at corner nodes. Such a d.o.f. may also be called a *drilling freedom*. Its vector representation is normal to the plane of the element. The usual translational d.o.f. at nodes are retained. Thus, a plane triangle with only corner nodes has nine d.o.f. [8.5].

A good reason for use of drilling d.o.f. is found in the modeling of shells as an assembly of flat elements (discussed in detail in Section 12.3). Each element can model bending and stretching actions. Typically, in a general-purpose computer program, d.o.f. allowed at each structure node consist of three displacements and three rotations. Thus drilling d.o.f. are present among structural d.o.f. $\{\mathbf{D}\}$ whether or not they are present among element d.o.f. $\{\mathbf{d}\}$. If flat shell elements connected to a certain node all happen to be coplanar, but elements do not include drilling d.o.f., then the drilling d.o.f. in $\{\mathbf{D}\}$ at that node is not resisted, and $[\mathbf{K}]$ is singular. This difficulty is neatly avoided by including drilling d.o.f. in $\{\mathbf{d}\}$. Simultaneously, all of the six d.o.f. available at a node are exploited.

In what follows we presume that drilling d.o.f. are associated with parabolic displaced shapes of element sides. In Fig. 8.4-1a, drilling d.o.f. ω_i and ω_j appear at nodes i and j of a typical element side of length L. At midside, ω_i and ω_j produce the edge-normal displacement δ:

$$\delta = \frac{L}{8}(\omega_j - \omega_i) \tag{8.4-1}$$

Thus, if $\omega_i = \omega_j$, the edge remains straight. If $\omega_i \neq \omega_j$, the side assumes a parabolic shape. If $\omega_i = -\omega_j$, one can regard δ as the midspan deflection of a simply

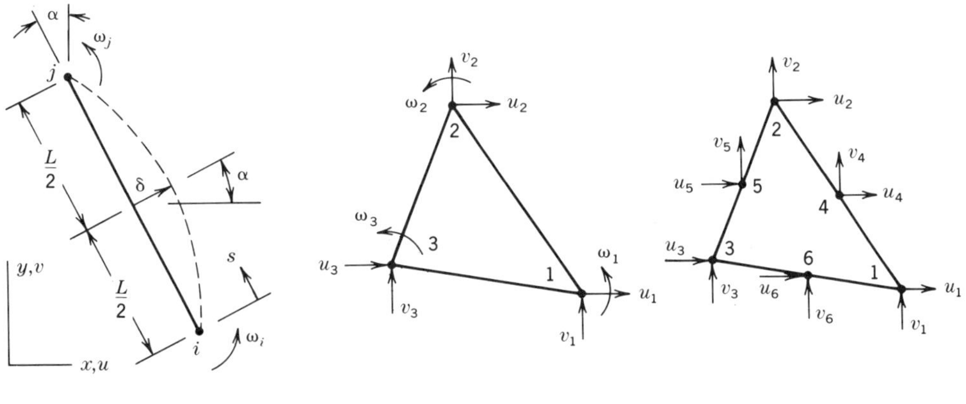

Figure 8.4-1. (*a*) Side displacement produced by drilling freedoms ω_i and ω_j. (*b*) A nine-d.o.f. plane triangle. (*c*) The linear-strain triangle.

supported beam of length L under a pure bending moment that produces end rotations $|\omega_i| = |\omega_j|$. Extending this beam analogy, imagine that supports of the beam have translational displacement components u_i, v_i, u_j, and v_j. Thus edge displacement is the sum of two parts: (a) a straight shape associated with nodal translational d.o.f., and (b) the parabolic shape associated with δ. The x and y components of δ are $\delta \cos \alpha$ and $\delta \sin \alpha$. Therefore, the total displacements u and v of a typical point on the edge are

$$\begin{Bmatrix} u \\ v \end{Bmatrix} = \frac{L - s}{L} \begin{Bmatrix} u_i \\ v_i \end{Bmatrix} + \frac{s}{L} \begin{Bmatrix} u_j \\ v_j \end{Bmatrix} + \frac{(L - s)s}{2L} (\omega_j - \omega_i) \begin{Bmatrix} \cos \alpha \\ \sin \alpha \end{Bmatrix} \tag{8.4-2}$$

Similar expressions may be written for all other sides of the element. Hence, one can devise shape functions N_i for a triangle, a quadrilateral, and so on. In the triangle of Fig. 8.4-1*b*, contributions to u and v from nodal translations are interpolated linearly over the element, and contributions associated with drilling d.o.f. are interpolated using products of area coordinates, such as $\xi_1 \xi_2$ for δ on side 1–2. Finally the usual process of element formulation is pursued (Eqs. 4.1-5 and 4.1-6).

One can also begin with an element that has straight sides and midside nodes, and convert it to an element that has corner nodes only, each with two translational d.o.f. and one drilling freedom. The conversion process relates midside δ's to nodal ω's and constrains each edge-tangent displacement component to vary linearly with edge-tangent coordinate s. The element may have any number of sides. As applied to a linear-strain triangle, Fig. 8.4-1*c*, the transformation procedure is outlined as follows.

Consider, for example, side 1–4–2 of the linear-strain triangle, Fig. 8.4-1*c*. D.o.f. at node 4 are related to d.o.f. at nodes 1 and 2 of the new element, Fig. 8.4-1*b*, by evaluating Eq. 8.4-2 with $s = L/2$. With $i = 1, j = 2$, $L \cos \alpha = y_2 - y_1$, and $L \sin \alpha = x_1 - x_2$, Eq. 8.4-2 yields

$$\begin{Bmatrix} u_4 \\ v_4 \end{Bmatrix} = \frac{1}{2} \begin{Bmatrix} u_1 \\ v_1 \end{Bmatrix} + \frac{1}{2} \begin{Bmatrix} u_2 \\ v_2 \end{Bmatrix} + \frac{\omega_2 - \omega_1}{8} \begin{Bmatrix} y_2 - y_1 \\ x_1 - x_2 \end{Bmatrix} \tag{8.4-3}$$

After doing the same for d.o.f. at nodes 5 and 6, we can relate d.o.f. in Figs. 8.4-1*b* and 8.4-1*c* by the transformation

$$\lfloor u_1 \quad v_1 \quad u_2 \cdots u_6 \quad v_6 \rfloor^T = \underset{12\times 9}{[\mathbf{T}]} \lfloor u_1 \quad v_1 \quad \omega_1 \quad u_2 \quad v_2 \quad \omega_2 \quad u_3 \quad v_3 \quad \omega_3 \rfloor^T \tag{8.4-4}$$

If desired, we can obtain the new element by transforming the stiffness matrix $[\mathbf{k}']$ of the linear-strain triangle:

$$\underset{9\times 9}{[\mathbf{k}]} = [\mathbf{T}]^T \underset{12\times 12}{[\mathbf{k}']} [\mathbf{T}] \tag{8.4-5}$$

More efficient coding will result if $[\mathbf{k}]$ is formulated directly from shape functions $[\mathbf{N}]$ appropriate to the nine-d.o.f. element. These may be obtained by transforming shape functions $[\mathbf{N}']$ of the linear-strain triangle:

$$\begin{Bmatrix} u \\ v \end{Bmatrix} = \underset{2\times 12}{[\mathbf{N}']} \underset{12\times 1}{\{\mathbf{d}'\}} = [\mathbf{N}'][\mathbf{T}] \underset{9\times 1}{\{\mathbf{d}\}} = \underset{2\times 9}{[\mathbf{N}]} \{\mathbf{d}\} \tag{8.4-6}$$

Shape functions N_i in Eq. 8.4-6 agree with those obtained by the procedure described below Eq. 8.4-2.

Deformations are everywhere zero if all nodal rotations in the mesh have the same value. This mechanism can be suppressed by prescribing the value of one nodal rotation in the mesh, for example, by setting $\omega_1 = 0$.

The true rotation at a node i is defined as

$$\theta_i = \tfrac{1}{2}(v_{,x} - u_{,y})_i \tag{8.4-7}$$

For rigid-body rotation in the plane of the element, $\theta_i = \omega_i$. Otherwise, equality may not prevail [8.5]. Interelement continuity of true rotations is not in general provided, nor is it necessary for proper convergence of the finite element solution.

A plane quadrilateral with drilling d.o.f. is described in [8.37]. It incorporates various improvements, including a device to control the aforementioned mechanism. This device is summarized as follows. Let the strain energy in each element be augmented by U_ω,

$$U_\omega = \tfrac{1}{2} V_e G(2\gamma) \left(\theta_0 - \frac{1}{n} \sum_{i=1}^{n} \omega_i \right)^2 \tag{8.4-8}$$

where V_e = element volume, G = shear modulus, γ = dimensionless constant ($\gamma = 10^{-6}$ is recommended), $\theta_0 = (v_{,x} - u_{,y})_0/2$ is the rotation at the element center, and n = number of nodes where rotational d.o.f. ω_i are used. By inserting shape functions, we can write the parenthetic expression in Eq. 8.4-8 in matrix format; that is, $(\cdots) = \lfloor \mathbf{Q} \rfloor \{\mathbf{d}\}$, where $\lfloor \mathbf{Q} \rfloor$ is a row matrix. Hence

$$U_\omega = \tfrac{1}{2} \{\mathbf{d}\}^T [\mathbf{k}_\omega]\{\mathbf{d}\} \qquad \text{in which} \qquad [\mathbf{k}_\omega] = 2\gamma V_e G \lfloor \mathbf{Q} \rfloor^T \lfloor \mathbf{Q} \rfloor \tag{8.4-9}$$

where $[\mathbf{k}_\omega]$ is a rank 1 "stabilization matrix" that is added to the existing element stiffness matrix. Matrix $[\mathbf{k}_\omega]$ has no effect on the ability of the element to represent constant-strain states and rigid-body modes.

Numerical examples are reported in Fig. 8.6-2 and in Refs. 8.5, 8.6, and 8.37.

8.5 ASSUMED-STRESS HYBRID FORMULATION

The assumed-stress hybrid method is a way of formulating a stiffness matrix by use of *independent* assumptions of (a) an equilibrium stress field within the element, and (b) interelement-compatible displacement modes on the element boundary. Mathematically, the method can be stated as a modified complementary energy principle. The principle of stationary complementary energy states that: among all stress fields that satisfy the differential equations of equilibrium, the stress field that also satisfies compatibility conditions makes the complementary energy stationary with respect to small variations of stress. For a linearly elastic material, strain energy per unit volume can be written as

$$U_0 = \tfrac{1}{2}\{\boldsymbol{\epsilon}\}^T[\mathbf{E}]\{\boldsymbol{\epsilon}\} \quad \text{or as} \quad U_0 = \tfrac{1}{2}\{\boldsymbol{\sigma}\}^T[\mathbf{E}]^{-1}\{\boldsymbol{\sigma}\} \tag{8.5-1}$$

(potential energy) (complementary energy)

Starting with an expression for U_0, one can write various functionals. The functional for potential energy is Π_p, Eq. 4.1-1, which yields the stiffness matrix of an element based on an assumed displacement field. Analogously, one can write a complementary energy functional that yields the stiffness matrix of an assumed-stress hybrid element [8.7]. Rather than discuss the functional, we consider the following more direct method, which is the method by which assumed-stress hybrid elements were first derived [8.8]. Although the hybrid method is general, the discussion that follows is oriented toward plane problems without body forces.

One begins with a stress field that satisfies the differential equations of equilibrium, Eqs. 1.6-2 and 1.6-4. Symbolically,

$$\{\boldsymbol{\sigma}\} = [\mathbf{P}]\{\boldsymbol{\beta}\} \tag{8.5-2}$$

where, for plane problems, $\{\boldsymbol{\sigma}\} = \lfloor \sigma_x \quad \sigma_y \quad \tau_{xy} \rfloor^T$, and $\{\boldsymbol{\beta}\}$ contains constants β_i that are yet to be determined. Equations 8.3-9 are a 5-β example of such an equilibrium stress field. From Eqs. 8.5-1 and 8.5-2, the complementary strain energy in an element of volume V_e is

$$U = \int_{V_e} U_0 \, dV = \tfrac{1}{2}\{\boldsymbol{\beta}\}^T[\mathbf{H}]\{\boldsymbol{\beta}\} \tag{8.5-3}$$

where

$$[\mathbf{H}] = \int_{V_e} [\mathbf{P}]^T[\mathbf{E}]^{-1}[\mathbf{P}] \, dV \tag{8.5-4}$$

Let $\{\boldsymbol{\Phi}\}$ represent tractions at the element boundary S_e, obtained by evaluating Eq. 8.5-2 on the boundary. Also let boundary displacements $\{\mathbf{u}_b\}$ be interpolated from nodal d.o.f. $\{\mathbf{d}\}$. (An example will follow.) Thus

$$\{\boldsymbol{\Phi}\} = [\mathbf{R}]\{\boldsymbol{\beta}\} \quad \text{and} \quad \{\mathbf{u}_b\} = [\mathbf{L}]\{\mathbf{d}\} \tag{8.5-5}$$

The total complementary energy in the element is U minus work done by tractions $\{\boldsymbol{\Phi}\}$ in moving through displacements $\{\mathbf{u}_b\}$; that is,

$$\Pi_c = U - \int_{S_e} \{\Phi\}^T\{\mathbf{u}_b\}\, dS = \tfrac{1}{2}\{\boldsymbol{\beta}\}^T[\mathbf{H}]\{\boldsymbol{\beta}\} - \{\boldsymbol{\beta}\}^T[\mathbf{G}]\{\mathbf{d}\} \tag{8.5-6}$$

where

$$[\mathbf{G}] = \int_{S_e} [\mathbf{R}]^T[\mathbf{L}]\, dS \tag{8.5-7}$$

Making Π_c stationary with respect to variations of stress, we write

$$\frac{\partial \Pi_c}{\partial \beta_i} = 0 \quad \text{for } i = 1, 2, \ldots, n \quad \text{or} \quad \left\{\frac{\partial \Pi_c}{\partial \boldsymbol{\beta}}\right\} = \{\mathbf{0}\} \tag{8.5-8}$$

from which

$$[\mathbf{H}]\{\boldsymbol{\beta}\} = [\mathbf{G}]\{\mathbf{d}\} \quad \text{or} \quad \{\boldsymbol{\beta}\} = [\mathbf{H}]^{-1}[\mathbf{G}]\{\mathbf{d}\} \tag{8.5-9}$$

At this point one can say that we have asked for the stress field within an element when displacements on its boundary are prescribed, and answered by finding values of $\{\boldsymbol{\beta}\}$ that define the best stress field $\{\boldsymbol{\sigma}\}$ that is contained in the approximation $\{\boldsymbol{\sigma}\} = [\mathbf{P}]\{\boldsymbol{\beta}\}$. Substitution of $\{\boldsymbol{\beta}\}$ from Eq. 8.5-9 into Eq. 8.5-3 yields

$$U = \tfrac{1}{2}\{\mathbf{d}\}^T[\mathbf{k}]\{\mathbf{d}\}, \quad \text{where} \quad [\mathbf{k}] = [\mathbf{G}]^T[\mathbf{H}]^{-1}[\mathbf{G}] \tag{8.5-10}$$

in which $[\mathbf{k}]$ is recognized as a stiffness matrix because its form matches that of Eq. 4.1-10.

Example. Consider a typical straight edge ij of a plane element, Fig. 8.5-1a. Tractions Φ_x and Φ_y are related to stresses σ_x, σ_y, and τ_{xy} by direction cosines ℓ and m of the outward normal to the edge. From Eq. 1.6-3,

$$\begin{aligned} \Phi_x &= \ell\sigma_x + m\tau_{xy} \\ \Phi_y &= \ell\tau_{xy} + m\sigma_y \end{aligned} \quad \text{where} \quad \begin{aligned} \ell &= \cos\alpha = (y_j - y_i)/L_{ij} \\ m &= \sin\alpha = (x_i - x_j)/L_{ij} \end{aligned} \tag{8.5-11}$$

For the particular case of a 5–β rectangular element, Eqs. 8.3-9 and Fig. 8.5-1b, arrays $[\mathbf{P}]$ and $[\mathbf{R}]$ are

$$[\mathbf{P}] = \begin{bmatrix} 1 & 0 & 0 & y & 0 \\ 0 & 1 & 0 & 0 & x \\ 0 & 0 & 1 & 0 & 0 \end{bmatrix} \quad \text{and} \quad [\mathbf{R}] = \begin{bmatrix} 0 & 0 & -1 & 0 & 0 \\ 0 & -1 & 0 & 0 & -x \\ 1 & 0 & 0 & y & 0 \\ 0 & 0 & 1 & 0 & 0 \\ 0 & 0 & 1 & 0 & 0 \\ 0 & 1 & 0 & 0 & x \\ -1 & 0 & 0 & -y & 0 \\ 0 & 0 & -1 & 0 & 0 \end{bmatrix} \tag{8.5-12}$$

where $[\mathbf{R}]$ is obtained from $[\mathbf{P}]$ and Eq. 8.5-11. The first two rows of $[\mathbf{R}]$ pertain to Φ_x and Φ_y along side 1–2 (where $\ell = 0$ and $m = -1$), the third and fourth rows pertain to Φ_x and Φ_y along side 2–3 (where $\ell = 1$ and $m = 0$), and so on. Matrix $[\mathbf{L}]$ of Eq. 8.5-5 is 8 by 8 for the element of Fig. 8.5-1b, and relates u and v displacement components along all four edges to nodal d.o.f. $\{\mathbf{d}\}$. That is,

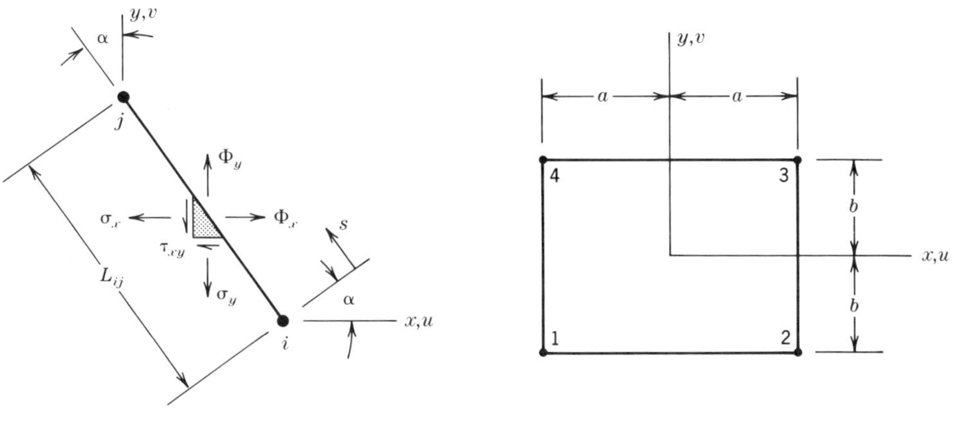

Figure 8.5-1. (*a*) Stresses and surface tractions at a typical edge *ij*. (*b*) A rectangular element.

$$\lfloor u_{12} \quad v_{12} \quad u_{23} \cdots u_{41} \quad v_{41} \rfloor^T = [\mathbf{L}] \lfloor u_1 \quad v_1 \cdots u_4 \quad v_4 \rfloor^T \tag{8.5-13}$$

If u and v along an edge are linearly interpolated from nodal d.o.f. on that edge, the rows of [**L**] are

$$\begin{aligned} &\text{row 1:} \quad \frac{a-x}{2a},\ 0,\ \frac{a+x}{2a},\ 0,\ 0,\ 0,\ 0,\ 0 \\ &\text{row 2:} \quad 0,\ \frac{a-x}{2a},\ 0,\ \frac{a+x}{2a},\ 0,\ 0,\ 0,\ 0 \end{aligned} \tag{8.5-14}$$

and so on. All ingredients are now at hand: the element stiffness matrix is obtained by application of Eqs. 8.5-4, 8.5-7, and 8.5-10. In integration of Eq. 8.5-4, dV becomes $t\,dA = t\,dx\,dy$, where t is the element thickness. In integration of Eq. 8.5-7, $dS = t\,ds$, where $ds = dx$ or $ds = dy$ for sides parallel to x and y axes, respectively. Terms that contain x are integrated from $-a$ to $+a$ and terms that contain y are integrated from $-b$ to $+b$.

Remarks. After {**d**} is known, element stresses are recovered by use of Eqs. 8.5-2 and 8.5-9. Thus

$$\{\boldsymbol{\sigma}\} = [\mathbf{P}][\mathbf{H}]^{-1}[\mathbf{G}]\{\mathbf{d}\} \tag{8.5-15}$$

Assumed-stress hybrid elements can be joined to displacement-based elements because both use displacement quantities as nodal d.o.f. The user of a computer program may be unaware that some of its elements are hybrid elements.

Assumed-stress hybrid elements become stiffer as $\{\boldsymbol{\beta}\}$ grows—that is, as more terms are added to the stress expansion. They usually become more flexible as element edges are permitted more complicated displacement patterns. No bound can be set: we cannot say in general that a mesh of hybrid elements will be too stiff or too flexible. If $\{\boldsymbol{\sigma}\} = [\mathbf{P}]\{\boldsymbol{\beta}\}$ is not a complete polynomial, the element will not be geometrically isotropic. For example, the stress field described by [**P**] of Eq. 8.5-12 is not complete. A complete linear stress field in the plane is

$$\begin{Bmatrix} \sigma_x \\ \sigma_y \\ \tau_{xy} \end{Bmatrix} = \begin{bmatrix} 1 & 0 & 0 & y & 0 & x & 0 \\ 0 & 1 & 0 & 0 & x & 0 & y \\ 0 & 0 & 1 & 0 & 0 & -y & -x \end{bmatrix} \begin{Bmatrix} \beta_1 \\ \beta_2 \\ \vdots \\ \beta_7 \end{Bmatrix} \tag{8.5-16}$$

This field contains seven β_i rather than nine in order to satisfy the differential equations of equilibrium. If used with linear edge displacements (e.g., Eq. 8.5-14), the 7–β element has eight displacement d.o.f. but is stiffer than a 5–β element having the same eight displacement d.o.f.

It is possible to relax the requirement that stresses satisfy the differential equations of equilibrium a priori. Procedures suggested by Wolf [8.9] and Pian [8.10] use a functional in which displacements within the element act as Lagrange multipliers of the differential equations of equilibrium. Equilibrium is then satisfied in an average sense rather than at every point.

8.6 A PLANE HYBRID TRIANGLE WITH ROTATIONAL D.O.F.

Hybrid element theory (Section 8.5) and "drilling" d.o.f. (Section 8.4) are combined in the element here described [8.11]. It is a triangle built of three subtriangles. The final macroelement is a nine-d.o.f. triangle (Fig. 8.6-1). The same element can be obtained by standard displacement theory, as explained in the following.

Basic Triangular Subelement. Stresses are assumed constant: $\sigma_x = \beta_1$, $\sigma_y = \beta_2$, and $\tau_{xy} = \beta_3$. Thus **[P]** is a unit matrix, and Eq. 8.5-4 yields

$$\underset{3\times 3}{[\mathbf{H}]} = At[\mathbf{E}]^{-1} \qquad \text{and} \qquad [\mathbf{H}]^{-1} = \frac{1}{At}[\mathbf{E}] \tag{8.6-1}$$

where A = subelement area and t = subelement thickness.

Matrix **[R]** is 6 by 3 and is formed from boundary tractions $\Phi_x = \ell\beta_1 + m\beta_3$ and $\Phi_y = \ell\beta_3 + m\beta_2$. Direction cosines ℓ and m are given by Eq. 8.5-11. For

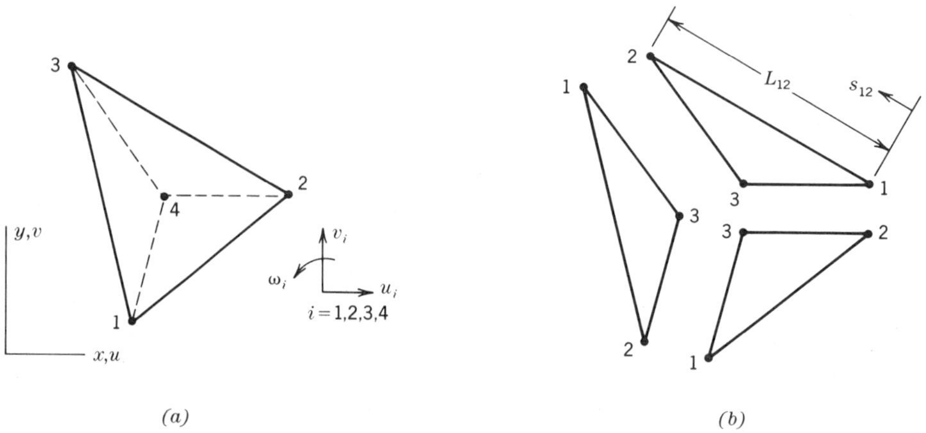

Figure 8.6-1. (*a*) Triangular macroelement built of three subtriangles. D.o.f. of a typical node *i* are shown. (*b*) Node labels for the subtriangles.

example, terms in the first two rows of [**R**] pertain to side 1–2 of the subelement and are

$$\begin{aligned} &\text{row 1:} \quad (y_2 - y_1)/L_{12},\ 0,\ (x_1 - x_2)/L_{12} \\ &\text{row 2:} \quad 0,\ (x_1 - x_2)/L_{12},\ (y_2 - y_1)/L_{12} \end{aligned} \tag{8.6-2}$$

Boundary displacements and nodal d.o.f. are related via matrix [**L**]. The relationship resembles Eq. 8.5-13, but involves one less side, one less node, and three d.o.f. per node. Rows of [**L**] are written by application of Eq. 8.4-2. For example, the first row of [**L**] expresses the relation

$$u_{12} = \frac{L_{12} - s_{12}}{L_{12}} u_1 + \frac{s_{12}}{L_{12}} u_2 + \frac{y_2 - y_1}{L_{12}} \frac{(L_{12} - s_{12})s_{12}}{2L_{12}} (\omega_2 - \omega_1) \tag{8.6-3}$$

where s_{12} is an edge-tangent coordinate, Fig. 8.6-1*b*.

When matrix [**G**] of Eq. 8.5-7 is computed, $dS = t\, ds_{ij}$ and limits of integration are from 0 to L_{ij}, where L_{ij} is L_{12}, L_{23}, or L_{31} for the respective element sides. Let nodal d.o.f. have the ordering $\{\mathbf{d}\} = \lfloor u_1 \quad v_1 \quad \omega_1 \quad u_2 \quad v_2 \quad \omega_2 \quad u_3 \quad v_3 \quad \omega_3 \rfloor^T$. Then, in partitioned form, [**G**] is

$$\underset{3\times 9}{[\mathbf{G}]} = [\mathbf{G}_A \quad \mathbf{G}_B \quad \mathbf{G}_C] \tag{8.6-4}$$

where, with $y_{ij} = y_i - y_j$ and $x_{ij} = x_i - x_j$,

$$[\mathbf{G}_A] = \frac{t}{12} \begin{bmatrix} 6y_{23} & 0 & y_{31}^2 - y_{12}^2 \\ 0 & 6x_{32} & x_{13}^2 - x_{21}^2 \\ 6x_{32} & 6y_{23} & 2(x_{13}y_{31} - x_{21}y_{12}) \end{bmatrix} \tag{8.6-5}$$

Submatrices [$\mathbf{G}_B$] and [$\mathbf{G}_C$] are obtained from [$\mathbf{G}_A$] by advancing the subscripts by 1 and 2 respectively around the loop 1–2–3; for example, x_{32} in [$\mathbf{G}_A$] becomes x_{13} in [$\mathbf{G}_B$] and x_{21} in [$\mathbf{G}_C$].

The foregoing triangular subelement can also be generated from an assumed displacement field: one uses the 2 by 9 matrix [**N**] of Eq. 8.4-6, and integrates $[\mathbf{B}]^T[\mathbf{E}][\mathbf{B}]$ by using a single Gauss point at the centroid of the element (or subelement, in the present context).

Macroelement. The composite element—that is, the macroelement, Fig. 8.6-1*a*—is formed by combining three of the foregoing subelements. Node 4 is at the centroid of the macroelement, that is, at

$$x_4 = \tfrac{1}{3}(x_1 + x_2 + x_3) \qquad \text{and} \qquad y_4 = \tfrac{1}{3}(y_1 + y_2 + y_3) \tag{8.6-6}$$

D.o.f. at node 4 can be eliminated by static condensation, leaving a three-node element having nine d.o.f.

However, such an element is a bit too flexible in many problems. Accordingly, prior to condensation, the rotational d.o.f. at node 4 is constrained to be the average of rotational d.o.f. at nodes 1, 2, and 3. This is accomplished by first transforming the 12 by 12 matrix [**k**] so that it operates on nodal d.o.f. {**d**} in which ω_{4r} replaces ω_4, where

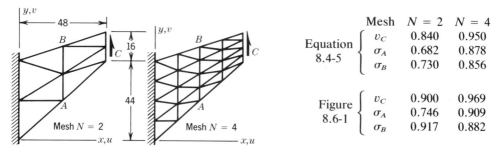

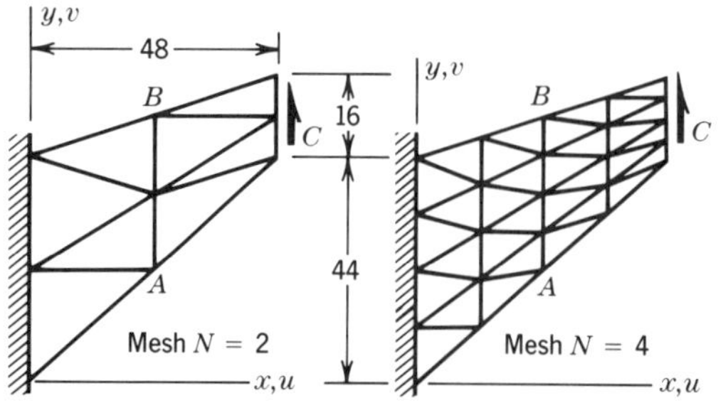

	Mesh	$N = 2$	$N = 4$
Equation 8.4-5	v_C	0.840	0.950
	σ_A	0.682	0.878
	σ_B	0.730	0.856
Figure 8.6-1	v_C	0.900	0.969
	σ_A	0.746	0.909
	σ_B	0.917	0.882

Figure 8.6-2. A plane structure with a uniformly distributed load along the right edge and with $E = 1.0$, $\nu = 1/3$. Results displayed are obtained from elements that have drilling d.o.f. "Equation 8.4-5" refers to the nine-d.o.f. displacement-based element obtainable from the linear-strain triangle [8.5]. "Figure 8.6-1" refers to the composite element with constant-stress subtriangles and $\omega_4 = (\omega_1 + \omega_2 + \omega_3)/3$. Here v_C = deflection at C, σ_A = maximum stress at A, σ_B = minimum stress at B, all reported as the ratio of computed value to best-known answer.

$$\omega_{4r} = \omega_4 - \tfrac{1}{3}(\omega_1 + \omega_2 + \omega_3) \tag{8.6-7}$$

Here ω_{4r} is the rotational d.o.f. at node 4 *relative* to the average of the three vertex rotational d.o.f. The constraint is $\omega_{4r} = 0$, which is enforced by striking out the row and column of the transformed **[k]** associated with ω_{4r}. This leaves an 11 by 11 stiffness matrix, from which u_4 and v_4 are eliminated by static condensation.

The resulting nine-d.o.f. element is geometrically isotropic and has rank 5, corresponding to three rigid-body motions and the mechanism in which all nodal rotations are the same. The patch test is passed. Numerical results are reported as the "Fig. 8.6-1" entries in Fig. 8.6-2. The same problem is solved, using different elements, in Fig. 8.3-3.

8.7 USER-DEFINED ELEMENTS. ELASTIC KERNEL

The user of a computer program may wish to supply an element of a type not provided in the program. If all stiffness coefficients k_{ij} of the new element are supplied directly, rather than as the output of a tested algorithm, there is substantial risk of introducing numerical error. In the following we explain a procedure whereby only *some* of the k_{ij} need be supplied, and we comment on why this procedure is effective in avoiding the possible error.

First, the element stiffness equation $[\mathbf{k}]\{\mathbf{d}\} = \{\mathbf{r}\}$ is partitioned,

$$\begin{bmatrix} \mathbf{k}_{RR} & \mathbf{k}_{RE} \\ \mathbf{k}_{RE}^T & \mathbf{k}_{EE} \end{bmatrix} \begin{Bmatrix} \mathbf{d}_R \\ \mathbf{d}_E \end{Bmatrix} = \begin{Bmatrix} \mathbf{r}_R \\ \mathbf{r}_E \end{Bmatrix} \tag{8.7-1}$$

in which d.o.f. $\{\mathbf{d}_R\}$ are used to define rigid-body motion of the element and d.o.f. $\{\mathbf{d}_E\}$ are used to define straining modes (an example follows). For any element, there is more than one way to partition $\{\mathbf{d}\}$ into $\{\mathbf{d}_R\}$ and $\{\mathbf{d}_E\}$. We wish to describe how the stiffness matrix $[\mathbf{k}_{EE}]$ of an element that is fully (but not redundantly) restrained from rigid-body motion ($\{\mathbf{d}_R\} = \{\mathbf{0}\}$) can be converted to a complete stiffness matrix **[k]**, ready for assembly into the structure.

The lower partition of Eq. 8.7-1 is solved for $\{\mathbf{d}_E\}$. Also, this expression for $\{\mathbf{d}_E\}$ is substituted into the upper partition. Thus

$$\begin{bmatrix} (\mathbf{k}_{RR} - \mathbf{k}_{RE}\mathbf{k}_{EE}^{-1}\mathbf{k}_{RE}^T) & \mathbf{k}_{RE}\mathbf{k}_{EE}^{-1} \\ -\mathbf{k}_{EE}^{-1}\mathbf{k}_{RE}^T & \mathbf{k}_{EE}^{-1} \end{bmatrix} \begin{Bmatrix} \mathbf{d}_R \\ \mathbf{r}_E \end{Bmatrix} = \begin{Bmatrix} \mathbf{r}_R \\ \mathbf{d}_E \end{Bmatrix} \tag{8.7-2}$$

Matrix $[\mathbf{k}_{EE}]$ is symmetric and is invertible because rigid-body motion is prevented. If there is no elastic distortion, then $\{\mathbf{r}_E\} = \{\mathbf{0}\}$ and, therefore,

$$-[\mathbf{k}_{EE}]^{-1}[\mathbf{k}_{RE}]^T\{\mathbf{d}_R\} = \{\mathbf{d}_E\} \tag{8.7-3}$$

With $\{\mathbf{r}_E\} = \{\mathbf{0}\}$, only rigid-body motion is possible. We also know that in rigid-body motion nodal d.o.f. are related strictly by kinematics, expressed by a matrix $[\mathbf{T}]$ of element dimensions:

$$\{\mathbf{d}_E\} = [\mathbf{T}]\{\mathbf{d}_R\} \tag{8.7-4}$$

Typically $\{\mathbf{d}_E\}$ contains more d.o.f. than $\{\mathbf{d}_R\}$, in which case $[\mathbf{T}]$ contains more rows than columns. Because Eqs. 8.7-3 and 8.7-4 must be true for *any* $\{\mathbf{d}_R\}$, we thus conclude, by comparison, that

$$[\mathbf{k}_{RE}]^T = -[\mathbf{k}_{EE}][\mathbf{T}] \tag{8.7-5}$$

Next imagine that $\{\mathbf{r}_E\} \neq \{\mathbf{0}\}$. Then, because $\{\mathbf{d}_R\}$ contains only enough d.o.f. to prevent rigid-body motion, $\{\mathbf{r}_R\}$ can be computed from $\{\mathbf{r}_E\}$ entirely by equations of statics, independently of $\{\mathbf{d}_R\}$. Accordingly, the coefficient of $\{\mathbf{d}_R\}$ in the upper partition of Eq. 8.7-2 must vanish. From this and Eq. 8.7-5 we obtain

$$[\mathbf{k}_{RR}] = [\mathbf{k}_{RE}][\mathbf{k}_{EE}]^{-1}[\mathbf{k}_{RE}]^T = [\mathbf{T}]^T[\mathbf{k}_{EE}][\mathbf{T}] \tag{8.7-6}$$

The stiffness matrix that operates on *all* nodal d.o.f., $\{\mathbf{d}\} = \lfloor \mathbf{d}_R \quad \mathbf{d}_E \rfloor^T$, is therefore

$$[\mathbf{k}] = \begin{bmatrix} \mathbf{T}^T\mathbf{k}_{EE}\mathbf{T} & -\mathbf{T}^T\mathbf{k}_{EE} \\ -\mathbf{k}_{EE}\mathbf{T} & \mathbf{k}_{EE} \end{bmatrix} \tag{8.7-7}$$

Matrix $[\mathbf{k}_{EE}]$ is called the *elastic kernel.*

The user must supply $[\mathbf{k}_{EE}]$ (or supply and then invert the flexibility matrix $[\mathbf{k}_{EE}]^{-1}$) and $[\mathbf{T}]$ to the computer program. Equation 8.7-7 then produces a $[\mathbf{k}]$ that *requires no force to produce rigid-body motion.* If the user were required to prescribe the *entire* $[\mathbf{k}]$, the individual k_{ij} would have to be of full computer-word accuracy to avoid the possibility of introducing serious errors (Section 18.2). By use of Eq. 8.7-7, slight errors in the $(\mathrm{k}_{EE})_{ij}$ produce only slight defects in elastic response; they do not cause rigid-body motion to be misrepresented.

Example. Consider the standard beam element of Fig. 4.2-2. Let the element be fixed at its left end, so that $\{\mathbf{d}_R\} = \lfloor w_1 \quad \theta_1 \rfloor^T$. Then $\{\mathbf{d}_E\} = \lfloor w_2 \quad \theta_2 \rfloor^T$. From Eq. 4.2-4, the curvature

$$w_{,xx} = \left\lfloor \frac{6}{L^2} - \frac{12x}{L^3} \quad -\frac{2}{L} + \frac{6x}{L^2} \right\rfloor \{\mathbf{d}_E\} \tag{8.7-8}$$

is used to construct $[\mathbf{k}_{EE}]$, which is found to be the lower right 2 by 2 submatrix in Eq. 4.2-5. Equation 8.7-4 becomes

$$\begin{Bmatrix} w_2 \\ \theta_2 \end{Bmatrix} = [\mathbf{T}] \begin{Bmatrix} w_1 \\ \theta_1 \end{Bmatrix}, \qquad \text{where} \quad [\mathbf{T}] = \begin{bmatrix} 1 & L \\ 0 & 1 \end{bmatrix} \tag{8.7-9}$$

Equation 8.7-7 then yields $[\mathbf{k}]$ of Eq. 4.2-5.

8.8 HIGHER DERIVATIVES AS NODAL D.O.F.

For the following discussion we define "essential" d.o.f. as the particular nodal d.o.f. needed to achieve the minimally-acceptable degree of interelement compatibility. These are the familiar nodal d.o.f. used in Chapters 1 through 7: for example, u_i and v_i for bars and plane elements, w_i and θ_i for beam elements. We define a "higher derivative" as one that is not needed to define interelement compatibility. Thus, in the stretching of a bar or in plane stress, *all* derivatives of u and v would be considered "higher." In the bending of a beam or a thin plate, higher derivatives are second and greater derivatives of lateral displacement. When used as nodal d.o.f., higher derivatives are also called "extra" or "excessive."

Elements with higher-derivative d.o.f. have certain advantages. They are based on fields having many generalized coordinates, so they provide good accuracy in a coarse mesh. Strains (or curvatures) needed in the calculation of stresses (or bending moments) appear in $\{\mathbf{D}\}$. Thus, being primary unknowns, strains may be computed more accurately than the conventionally computed strains $\{\boldsymbol{\epsilon}\} = [\mathbf{B}]\{\mathbf{d}\}$, which invoke difference operations on essential d.o.f. $\{\mathbf{d}\}$. Moreover, the extra d.o.f. are available at nodes, the very place where conventional strains $\{\boldsymbol{\epsilon}\} = [\mathbf{B}]\{\mathbf{d}\}$ are likely to be least accurate.

However, elements having higher-derivative d.o.f. are sometimes awkward to use. At an elastic–plastic boundary, or where there is an abrupt change in stiffness or material properties, continuity of higher derivative d.o.f. must *not* be enforced. For example, if two beam elements of different stiffness are joined, they have the same moment but different curvature at the node they share. A maneuver appropriate to such a circumstance is to release the curvature d.o.f. in one of the elements before assembly (Section 8.1). But, by doing so, we reduce the benefit of these d.o.f. where it is most needed—near a high-stress gradient.

Release of higher-derivative d.o.f. is again required and the benefit of these d.o.f. is again reduced if elements with derivative d.o.f. must be used in combination with elements that have only essential nodal d.o.f. Indeed, many computer programs allow up to six d.o.f. per node (three translations and three rotations) and so may be unable to accommodate a higher-order element without basic changes.

Boundary conditions may become awkward because the physical meaning of higher-derivative d.o.f. and their associated nodal loads is obscure. For example, if a plane element includes derivatives $u_{,x}$, $u_{,y}$, $v_{,x}$, and $v_{,y}$ as nodal d.o.f., a stress-free boundary dictates a constraint relation among these d.o.f. but does not dictate the numerical value of any of them.

In summary, higher-derivative d.o.f. tend to make the finite element method awkward in application to problems for which it is most powerful—to structures built of different element types and involving thickness changes, stiffeners, and parts that join with sharp angles instead of smooth curves.

8.9 FRACTURE MECHANICS. SINGULARITY ELEMENTS

Fracture mechanics deals with the conditions under which a body can fail owing to the propagation of an existing crack of macroscopic size [8.12]. In analysis, one might ask for the load that will produce failure when a crack of known size is present, or for the allowable size of a crack when a known load must be sustained.

Consider an arbitrarily loaded body that contains a crack. By isolating material in the immediate neighborhood of a crack tip, one can identify the three possible deformation modes shown in Fig. 8.9-1. These modes may appear singly or in arbitrary combination. Formulas exist for stresses and displacements in the immediate neighborhood of a crack tip. For example, if the crack is Mode I and the material is linearly elastic and isotropic, the y-direction stress and displacement are

$$\sigma_y = \frac{K_{\mathrm{I}}}{(2\pi r)^{1/2}}\left(\cos\frac{\theta}{2}\right)\left[1 + \sin\frac{\theta}{2}\sin\frac{3\theta}{2}\right] \tag{8.9-1}$$

$$v = \frac{(2\pi r)^{1/2}}{8G\pi} K_{\mathrm{I}}\left[(2\kappa + 1)\sin\frac{\theta}{2} - \sin\frac{3\theta}{2}\right] \tag{8.9-2}$$

where G = shear modulus, and with ν = Poisson's ratio,

$$\kappa = 3 - 4\nu \quad \text{(plane strain)} \qquad \text{or} \qquad \kappa = \frac{3 - \nu}{1 + \nu} \quad \text{(plane stress)} \tag{8.9-3}$$

Here K_{I} is called the *stress intensity factor* for Mode I. It can be defined as

$$K_{\mathrm{I}} = \lim_{r \to 0}\left[\sigma_y (2\pi r)^{1/2}\right] \qquad \text{for} \quad \theta = 0 \tag{8.9-4}$$

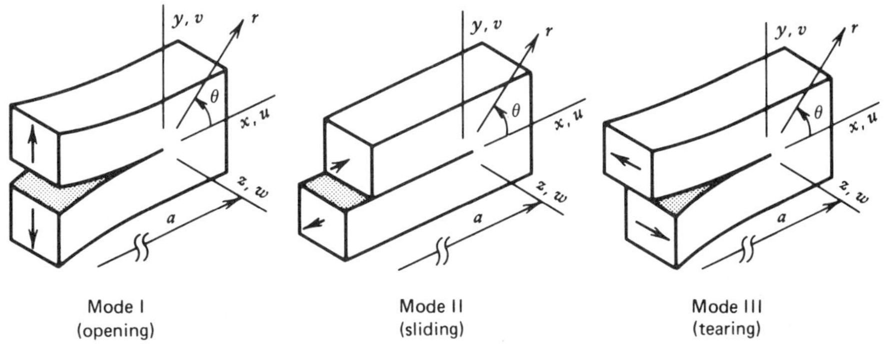

Figure 8.9-1. Deformation modes in the immediate neighborhood of a crack tip.

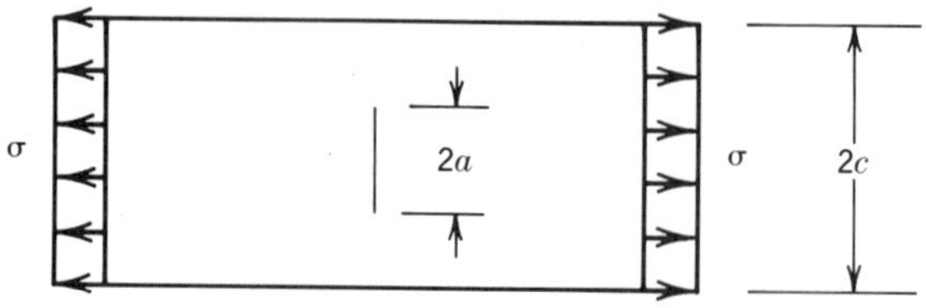

Figure 8.9-2. Flat plate with a central crack of width $2a$. Tensile stress σ is uniform well away from the crack.

A stress intensity factor is *not* a stress concentration factor: a stress intensity factor pertains to a singularity in the stress field, whereas a stress concentration factor pertains to geometries that do not produce infinite stresses. Nevertheless, two factors are analogous in that results are known and tabulated for several geometries and loadings [8.13]. For example, in Fig. 8.9-2,

$$K_I = \sigma(\pi a)^{1/2} \frac{1 - 0.5(a/c) + 0.326(a/c)^2}{[1 - (a/c)]^{1/2}} \tag{8.9-5}$$

There is a value of K_I denoted by K_{IC} and called *fracture toughness*. K_{IC} can be regarded as a material constant[1] for which data are known. If $K_I = K_{IC}$ in Eq. 8.9-5, a "critical" condition exists—that is, fracture impends. Thus, given K_{IC}, a, and c, one can solve for the critical value of σ. Or, given K_{IC}, σ, and c, one can solve for the critical value of crack length $2a$. In Eq. 8.9-5, note that σ is stress on the gross cross-sectional area $2ct$, not stress on the net area $2(c - a)t$.

For complicated geometries and loadings, formulas such as Eq. 8.9-5 are not tabulated. A substitute relation can be determined numerically. Specifically, one can apply an arbitrarily chosen reference load to a finite element model and solve for K_I by methods described in connection with Eqs. 8.9-8. The critical load is then equal to K_{IC}/K_I times the reference load. (The same loads may also produce nonzero values of K_{II} and K_{III}. Unfortunately, for such mixed-mode conditions, the failure load cannot be accurately predicted by existing methods.)

Quarter-Point Elements (QPE). The stress field of Eq. 8.9-1 displays a stress singularity of order $r^{-1/2}$. An element having side nodes can be made to display a $r^{-1/2}$ stress (or strain) singularity by appropriate definition of its geometry. Consider, for example, the three-node bar element discussed in Section 6.2. This element is shown again in Fig. 8.9-3, now with node 3 moved to the quarter point. With $x_1 = 0$, $x_2 = L$, and $x_3 = L/4$, Eq. 6.2-2 yields

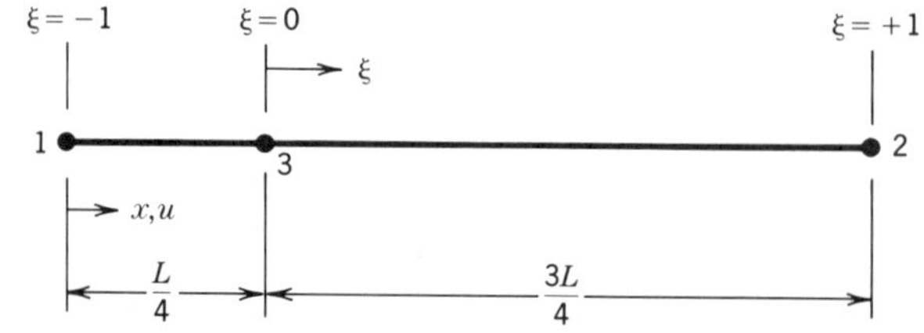

Figure 8.9-3. Three-node quarter-point bar element.

[1]Provided that t and a are both at least $2.5(K_{IC}/Y)^2$, where t = specimen thickness and Y = yield strength in a tension test. For smaller values of t and a, fracture toughness is a function of t and a.

$$x = \frac{L}{4}(1 + \xi)^2 \qquad \text{or} \qquad \xi = 2\left(\frac{x}{L}\right)^{1/2} - 1 \tag{8.9-6}$$

Equations 6.2-5 and 6.2-7 yield J and $\lfloor \mathbf{B} \rfloor$, from which ξ may be eliminated by means of Eq. 8.9-6. One obtains

$$\lfloor \mathbf{B} \rfloor = \left\lfloor \left(\frac{2}{L} - \frac{3}{2(Lx)^{1/2}}\right), \left(\frac{2}{L} - \frac{1}{2(Lx)^{1/2}}\right), \left(-\frac{4}{L} + \frac{2}{(Lx)^{1/2}}\right) \right\rfloor \tag{8.9-7}$$

Accordingly, stress $\sigma_x = E\lfloor \mathbf{B} \rfloor\{\mathbf{d}\}$ varies as $x^{-1/2}$—that is, as $r^{-1/2}$ along the line $\theta = 0$. Stress becomes infinite at $x = 0$ for displacements other than rigid-body motion.

The six-node plane triangle discussed in Section 5.5 can display the $r^{-1/2}$ singularity in its strain field if its side nodes are moved to quarter points near the crack tip, a shown in Fig. 8.9-4 [8.14]. The quarter-point sides should be straight and the side node opposite the crack tip should be at midside.

Another effective singularity element can be formed from a four-sided quadratic element (Fig. 6.6-1*a*) by collapsing one side to produce a triangle: for example, in Fig. 6.6-1*a*, nodes 1, 4, and 8 can be assigned the same coordinates *and* the same displacements. Nodes 5 and 7 are moved to the quarter points near the collapsed side. The side opposite (side 2–6–3 in this example) must be kept straight to avoid significant errors.

Rectangular QPE's—for example, the element of Fig. 6.6-1*a* with node 1 at the crack tip and nodes 5 and 8 at the quarter points—display the $r^{-1/2}$ singularity only along two sides and the diagonal [8.15]. They are less accurate than triangular QPE's, which display the $r^{-1/2}$ singularity along all rays emanating from the crack tip.

In a QPE, the singularity is precisely at the vertex—that is, at $r = 0$ in Fig. 8.9-4*b*. If side nodes are closer to midsides, the singularity moves away from the element (and becomes infinitely distant if side nodes are at midsides). Side nodes need not be precisely at quarter points, as a small error of order e in position produces an error of order e^2 in the stress intensity factor [8.15]. Indeed, one may

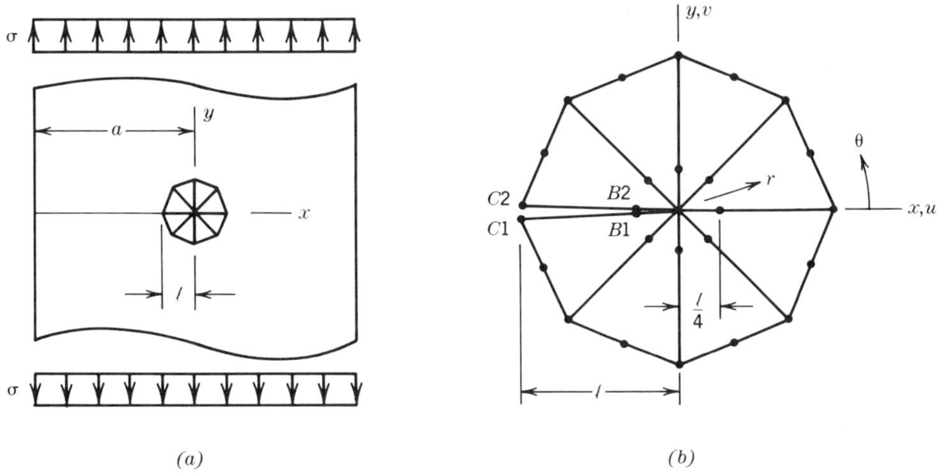

Figure 8.9-4. (*a*) Plate with edge crack of length *a*. Only those elements around the crack tip are shown. (*b*) Mesh of QPE's around the crack tip.

deliberately use an "almost" QPE. In three-dimensional analysis, curvature of the crack front may place the singularity *inside* a QPE. This trouble may be avoided by placing side nodes a bit closer to midside rather than at quarter points.

In a QPE, strains are represented as a constant plus a term proportional to $r^{-1/2}$, as may be seen in Eq. 8.9-7. Accordingly, if ℓ/a in Fig. 8.9-4 is small, the region of the structure in which the singular stress field is represented decreases. But if ℓ/a is large, the nonsingular variations of stress are represented by only the constant term over a larger region of the structure. The best value of ℓ/a in a mesh with a fixed number of elements is problem-dependent. Many analyses have used $\ell/a \approx 0.1$, but the value of ℓ/a is not critical in a body of arbitrary geometry if the mesh is adequate to represent stresses in the body were the crack not present [8.16,8.17]. An additional recommendation is that at least four (in a Mode I problem) or eight (in a mixed-mode problem) elements surround the crack tip [8.18]. There is disagreement as to what quadrature rule is best in generating [**k**] of a QPE.

A common way to calculate stress intensity factors from a finite element analysis is the crack-opening displacement method. From displacements of nodes on $\theta = \pm\pi$ in Fig. 8.9-4 [8.17], it can be shown that QPE's yield the Mode I and Mode II stress intensity factors

$$K_{\mathrm{I}} = \frac{2G}{\kappa + 1}\left(\frac{\pi}{2\ell}\right)^{1/2}[(4v_{B2} - v_{C2}) - (4v_{B1} - v_{C1})] \tag{8.9-8a}$$

$$K_{\mathrm{II}} = \frac{2G}{\kappa + 1}\left(\frac{\pi}{2\ell}\right)^{1/2}[(4u_{B2} - u_{C2}) - (4u_{B1} - u_{C1})] \tag{8.9-8b}$$

where κ is given by Eq. 8.9-3.

QPE's have the appeal of simplicity. Elements having side nodes are available in most programs, and they become QPE's when input data locates their side nodes at quarter points. Stress intensity factors are easily computed from Eqs. 8.9-8. However, other methods are available, as follows.

Other Methods. By various methods, including hybrid methods, the strain or stress field used to formulate an element can be made to contain a singularity without invoking special placement of nodes. Indeed, a stress intensity factor can become a d.o.f. in {**D**}. As alternatives to Eqs. 8.9-8, one can use the virtual crack extension method [8.19] or the J integral.

The simplest alternative is not to use singularity elements at all. Stress intensity factors can be obtained by using ordinary elements to surround the crack tip. Singularity elements merely make possible greater accuracy for a given computational effort.

8.10 ELASTIC FOUNDATIONS

Sometimes one elastic structure is supported by another, but stress analysis is required for only the first of the two. Then it suffices to model the *effect* of the second structure on the first. We need not model the second structure in such detail that stresses within it can be determined. Examples include a rail on a roadbed or a pavement slab on soil. The rail or the slab must be analyzed; the supporting effect of the roadbed or soil must be modeled. Elastic support can be represented by a *foundation stiffness matrix,* $[\mathbf{k}_f]$ for a foundation element and $[\mathbf{K}_f] = \Sigma\, [\mathbf{k}_f]$ for the entire foundation structure. If $[\mathbf{K}_s]$ is the stiffness matrix of the supported structure, then $[\mathbf{K}_s] + [\mathbf{K}_f]$ is the net stiffness matrix of the supported structure on its elastic foundation.

In what follows we presume that $[\mathbf{K}_f]$ is to operate on interface d.o.f. only—that is, on d.o.f. of only the nodes shared by the supported structure and its foundation [8.20,8.21]. Physically, the *j*th column of $[\mathbf{K}_f]$ represents forces (and perhaps moments as well) that must be applied to nodes on the surface of the foundation to cause the *j*th foundation d.o.f. to have unit value while all other d.o.f. on the interface are zero.

For an elastic solid foundation model, $[\mathbf{K}_f]$ is a full matrix, as loads must be applied to *all* interface d.o.f. when only one of these d.o.f. is activated. An approximate foundation model, simple and inexpensive yet often adequate, is the Winkler foundation model.

A Winkler foundation, Fig. 8.10-1*a*, deflects only where load is applied. Adjacent foundation material is utterly unaffected. A Winkler foundation of modulus β applies a vertical pressure βw when deflected vertically an amount w. In this regard the foundation acts exactly like a liquid of density β. However, we assume that, unlike the pressure of bouyancy, foundation pressure can act either upward or downward. If instead part of a structure lifts off the foundation, the problem is nonlinear, as then contact forces and contact geometry are both unknown at the outset.

We define a foundation element as the area on the foundation surface that makes contact with an element (or element face) of the supported structure. Thus a rectangular plate element in a supported paving slab would define a rectangular foundation element of identical shape and size. To determine $[\mathbf{k}_f]$ for a Winkler foundation element, we can use the following strain energy argument. Let dA be an increment of the area A of the foundation element. Then deflection w normal to A produces a force increment $dF = \beta w\, dA$. By analogy with a linear spring, whose strain energy is $F\Delta/2$ when deflected an amount Δ, the strain energy increment in the foundation is $dU = dF(w/2) = \beta w^2\, dA/2$. If w is governed by d.o.f. $\{\mathbf{d}\}$ of the aforementioned plate element, then $w = \lfloor\mathbf{N}\rfloor\{\mathbf{d}\}$, where $\lfloor\mathbf{N}\rfloor$ is the lateral-displacement shape function matrix of the plate element. Hence

$$U = \tfrac{1}{2}\int \beta w^2\, dA = \tfrac{1}{2}\int w^T \beta w\, dA = \tfrac{1}{2}\,\{\mathbf{d}\}^T[\mathbf{k}_f]\{\mathbf{d}\} \qquad (8.10\text{-}1)$$

in which

$$[\mathbf{k}_f] = \int \beta\lfloor\mathbf{N}\rfloor^T\lfloor\mathbf{N}\rfloor\, dA \qquad (8.10\text{-}2)$$

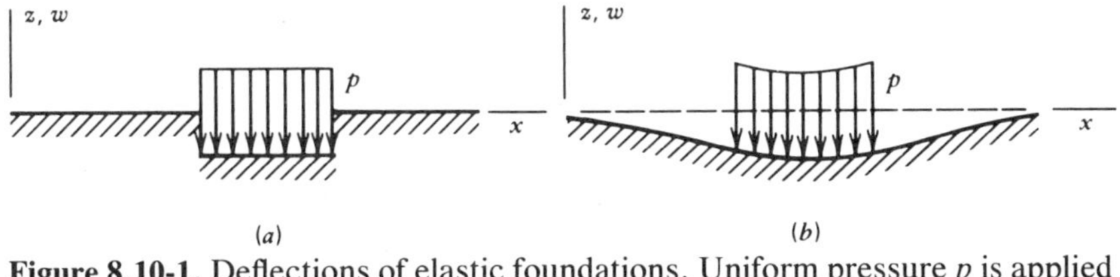

Figure 8.10-1. Deflections of elastic foundations. Uniform pressure p is applied directly to the foundation; no structure is interposed. (*a*) Winkler foundation model. (*b*) Elastic solid foundation model.

is a foundation stiffness matrix that operates on the same d.o.f. as the plate element in contact with the foundation.

If the supported element were a beam rather than a plate, then $\lfloor \mathbf{N} \rfloor$ in Eq. 8.10-1 would contain the cubic shape functions of a beam and $dA = b\,dx$, where b is the width of the beam. However, cubic functions are not exact because the beam element is not loaded only by forces and moments at its end nodes; it is also loaded by distributed foundation pressure. Use of the exact deflected shape [8.22] leads to stiffness coefficients in the combined matrix $[\mathbf{k}_{\text{beam}}] + [\mathbf{k}_f]$. These coefficients involve lengthy expressions, but they are much more accurate in a coarse mesh than coefficients based on a cubic polynomial.

If the supported element is neither a plate nor a beam, but (say) an eight-node hexahedron, then $\{\mathbf{d}\}$ in Eq. 8.10-1 would not contain nodal rotation d.o.f. It would contain only the w d.o.f. of the four corner nodes of the quadrilateral contact area, and the N_i would be those of Eq. 6.3-2. Indeed, one could ignore nodal rotation d.o.f. even if the supported element *is* a plate. Then $[\mathbf{k}_f]$ becomes more sparse and does not resist nodal rotations. Ultimately one can imagine for the supported structure element only a rigid-body lateral translation w, and divide the foundation resisting force $\beta A w$ equally among element nodes in contact with the foundation. Thus, for a supported element that has n contacting nodes, $[\mathbf{k}_f]$ becomes a diagonal matrix whose n nonzero coefficients are $k_{fii} = \beta A/n$. By this "lumping" procedure the foundation is reduced to a set of linear springs at the contacting nodes.

The name *scalar element* is given to a linear or torsional spring that connects a node to a support. A scalar element can resist only a deformation along its axis or a twist about its axis.

8.11 MEDIA OF INFINITE EXTENT

Many physical problems deal with an unbounded medium. Examples include a wing moving through air, diffraction of water around an island, and a load supported by the ground (Fig. 8.11-1*a*). In all these problems a finite element model must be terminated somewhere short of infinity. Simple truncation at a rigid boundary, Fig. 8.11-1*b*, is usually adequate in static problems. However, it is

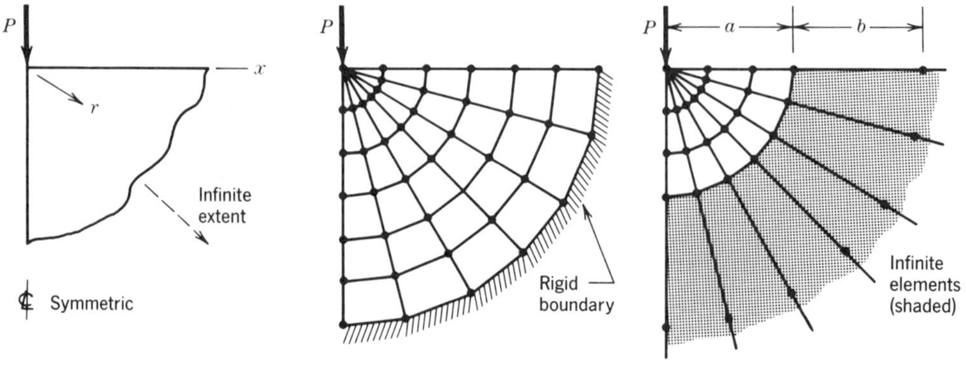

(*a*) (*b*) (*c*)

Figure 8.11-1. (*a*) Load P on plane or axially symmetric body of infinite extent below the x axis. (*b*) Large mesh of conventional elements. (*c*) Smaller mesh, bounded by infinite elements.

unclear where the rigid boundary should be placed, and the analysis may be expensive because many elements are used. In dynamic problems a rigid boundary reflects a wave, regardless of the size of the mesh; therefore, the model misrepresents reality.

Various methods for numerical analysis of unbounded field problems, both static and dynamic, have been devised [8.23]. In what follows we summarize a particular kind of "infinite element" for static analysis that is simple and effective [8.23-8.25]. In Fig. 8.11-1*c*, infinite elements permit satisfactory results to be obtained from fewer elements than would otherwise be required.

Infinite Elements. In stress analysis, infinite elements are analogous to an elastic foundation in that they provide correct or approximately correct support conditions for a region of interest that is modeled by a mesh of standard elements. Stresses in the infinite elements are usually not of interest and may not be accurate.

In formulating an infinite element, one makes use of two sets of shape functions. These are the standard shape functions $[\mathbf{N}]$ and either one of the following: (1) "decay" shape functions $[\mathbf{N}_d]$, which approach zero as a coordinate approaches infinity, or (2) "growth" shape functions $[\mathbf{M}]$, which grow without limit as a coordinate approaches infinity. In the first method $[\mathbf{N}]$ is applied to geometry and $[\mathbf{N}_d]$ to the field variable, so that the element remains of finite size while the field variable decays. In the second method $[\mathbf{N}]$ is applied to the field variable and $[\mathbf{M}]$ to geometry, so that the element grows to infinite size. The second method yields what are called "mapped" infinite elements. They are easy to implement and are described as follows.

In order to illustrate concepts and introduce procedures, we consider a one-dimensional element—namely, element 1–2–3 in Fig. 8.11-2 [8.25]. Distance a between nodes 1 and 2 may be considered a characteristic length of the element. Point 0, a distance a to the left of node 1, is not a node; it is a "pole" whose significance is discussed subsequently. Geometry of the element is interpolated according to

$$x = M_1 x_1 + M_2 x_2, \quad \text{where} \quad \begin{aligned} M_1 &= -\frac{2\xi}{1-\xi} \\ M_2 &= \frac{1+\xi}{1-\xi} \end{aligned} \tag{8.11-1}$$

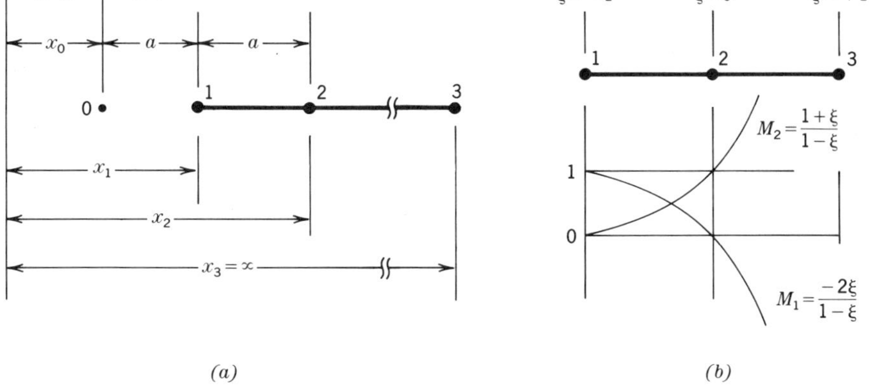

Figure 8.11-2. (*a*) One-dimensional infinite element in physical space. (*b*) The same element in natural-coordinate space.

which yields $x = x_1$ at $\xi = -1$ and $x = x_2$ at $\xi = 0$. As for x_3, from Eq. 8.11-1,

$$x_3 = \lim_{\xi \to 1} \frac{-2\xi x_1 + (1 + \xi)x_2}{1 - \xi} = \infty \tag{8.11-2}$$

Accordingly, the mapping of Eq. 8.11-1 automatically places node 3 at infinity, and node 3 need not be explicitly present in Eq. 8.11-1. A field variable ϕ can be interpolated by standard shape functions. For the present three-node line element, from Eq. 6.2-2, the field interpolation $\phi = \lfloor \mathbf{N} \rfloor \{\boldsymbol{\phi}_e\}$ is the usual quadratic

$$\phi = \left\lfloor -\frac{\xi + \xi^2}{2} \quad (1 - \xi^2) \quad \frac{\xi + \xi^2}{2} \right\rfloor \begin{Bmatrix} \phi_1 \\ \phi_2 \\ \phi_3 \end{Bmatrix} \tag{8.11-3}$$

Typically, ϕ_3 is set to a constant value (usually zero) as a boundary condition. Formulation of the element stiffness matrix, Eq. 6.2-6, proceeds in standard fashion except that mapping functions M_1 and M_2 of Eq. 8.11-1 are used to form the Jacobian J. Specifically, in Eq. 6.2-6 we require the strain–displacement matrix $[\mathbf{B}]$ and the Jacobian J, which for the infinite line element are

$$\lfloor \mathbf{B} \rfloor = \frac{1}{J} \left\lfloor \frac{d}{d\xi} \mathbf{N} \right\rfloor \qquad \text{and} \qquad J = M_{1,\xi} x_1 + M_{2,\xi} x_2 \tag{8.11-4}$$

where $J = dx/d\xi$ is obtained from Eq. 8.11-1 and $\lfloor \mathbf{N} \rfloor$ is given by Eq. 8.11-3.

To show how the foregoing infinite element represents field quantity ϕ, we first solve Eq. 8.11-1 for ξ. With $x = x_0 + r$ and other dimensions shown in Fig. 8.11-2*a*,

$$\xi = \frac{x - x_2}{x - 2x_1 + x_2} = 1 - \frac{2a}{r} \tag{8.11-5}$$

Substitution of Eq. 8.11-5 into Eq. 8.11-3 yields

$$\phi = \phi_3 + (-\phi_1 + 4\phi_2 - 3\phi_3)\frac{a}{r} + (2\phi_1 - 4\phi_2 + 2\phi_3)\frac{a^2}{r^2} \tag{8.11-6}$$

We see that as r approaches infinity, ϕ approaches ϕ_3 (which may be set to zero as a boundary condition). The constant value $\phi = c$ prevails if $\phi_1 = \phi_2 = \phi_3 = c$, but linear variations of ϕ with r are not represented. In general, the two parenthetic expressions in Eq. 8.11-6 do not vanish, so ϕ becomes infinite at point 0 because $r = 0$ at point 0. Point 0 is therefore a pole or singular point about which field quantity ϕ decays. This suggests that in a problem such as that of Fig. 8.11-1*c*, in which there is indeed a singularity at $r = 0$, one should use $a = b$.

It is not necessary that the mapping and the field interpolation rely on identical sets of nodes. For example, we can use the three-node mapping of Fig. 8.11-2 and Eq. 8.11-1, but replace Eq. 8.11-3 by a linear field interpolation between nodes 1 and 3,

$$\phi = \left\lfloor \frac{1 - \xi}{2} \quad \frac{1 + \xi}{2} \right\rfloor \begin{Bmatrix} \phi_1 \\ \phi_3 \end{Bmatrix} \tag{8.11-7}$$

This is perhaps the simplest possible infinite element.

Equation 8.11-7 offers the following physical interpretation. Let ϕ be axial displacement u and let node 3 be fixed. Then, from Eqs. 8.11-1 and 8.11-7, axial strain is

$$\epsilon_x = \frac{1}{J}\left\lfloor \frac{d}{d\xi}\mathbf{N} \right\rfloor \begin{Bmatrix} u_1 \\ 0 \end{Bmatrix} = \frac{(1-\xi)^2}{2a}\left(-\frac{1}{2}\right) u_1 = -\frac{u_1}{2a}\frac{(1-\xi)^2}{2} \tag{8.11-8}$$

We see that for an imagined element of physical length $2a$ between nodes 1 and 3, axial strain decays parabolically from $\epsilon_x = -u_1/a$ at end $\xi = -1$ to $\epsilon_x = 0$ at end $\xi = +1$, rather than being the constant value $\epsilon_x = -u_1/2a$ throughout as would be the case for a standard two-node element of length $2a$.

For analysis of plane and axially symmetric bodies, one needs infinite elements that are mathematically two-dimensional. Such an element is shown in Fig. 8.11-3. It extends to infinity in the ξ direction and is directly analogous to the element of Fig. 8.11-2. If the field variable ϕ is set to zero at element nodes 5 and 6, one need not use N_5 and N_6 in element formulation, and d.o.f. ϕ_5 and ϕ_6 need not appear in {**D**}. However, nodal d.o.f. ϕ_i on outer edges of infinite elements may be left unspecified, as d.o.f. to be determined, if unrestrained outer boundaries do not imply the possibility of rigid-body motion. An axially symmetric plane problem, in which only axially symmetric deformations are allowed, is a case in point.

Computer programming of mapped infinite elements is straightforward. In terms of Figs. 6.5-1 and 6.5-2, the essential change is alteration of the loop on statement 30 in Fig. 6.5-1: mapping functions [**M**] must be used to generate the Jacobian matrix, its inverse, and its determinant. Throughout the remainder of the subroutine one uses shape functions [**N**] and shape function derivatives (appropriate to the number of element nodes used for the field variable) in the manner already programmed.

Boundary Element Method (BEM). The BEM is an alternative to the finite element method (FEM). BEM can be applied to bounded or unbounded domains, but seems best suited to the latter. Like FEM, BEM uses nodes and elements, but only on the boundary. Thus, as compared with FEM, dimensionality is reduced by one; for example, a solid analyzed by BEM uses a two-dimensional mesh that covers only its surface. BEM and FEM can be coupled, so that BEM might replace infinite elements as the supporting medium for a structure modeled by FEM. BEM accurately models response in the domain bounded by its mesh (unlike infinite

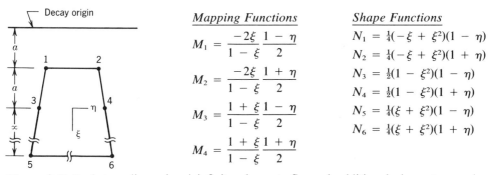

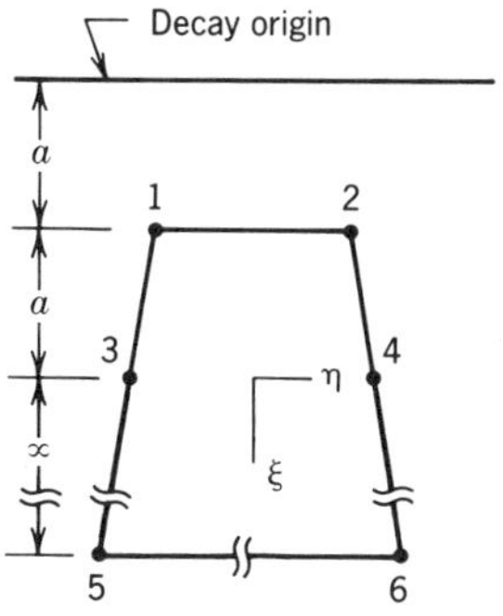

Mapping Functions

$$M_1 = \frac{-2\xi}{1-\xi}\,\frac{1-\eta}{2}$$

$$M_2 = \frac{-2\xi}{1-\xi}\,\frac{1+\eta}{2}$$

$$M_3 = \frac{1+\xi}{1-\xi}\,\frac{1-\eta}{2}$$

$$M_4 = \frac{1+\xi}{1-\xi}\,\frac{1+\eta}{2}$$

Shape Functions

$$N_1 = \tfrac{1}{4}(-\xi + \xi^2)(1 - \eta)$$
$$N_2 = \tfrac{1}{4}(-\xi + \xi^2)(1 + \eta)$$
$$N_3 = \tfrac{1}{2}(1 - \xi^2)(1 - \eta)$$
$$N_4 = \tfrac{1}{2}(1 - \xi^2)(1 + \eta)$$
$$N_5 = \tfrac{1}{4}(\xi + \xi^2)(1 - \eta)$$
$$N_6 = \tfrac{1}{4}(\xi + \xi^2)(1 + \eta)$$

Figure 8.11-3. A two-dimensional infinite element. Several additional elements are described in [8.25].

elements, which provide support but do not offer internal accuracy). However, the computational expense of BEM increases quickly if the response at several interior locations is needed. Although [**K**] of FEM is usually large, sparse, and symmetric, the analogous matrix of BEM is small, full, and unsymmetric. With an increase in the ratio of surface to volume, BEM becomes a less attractive alternative to FEM, because a mesh must be supplied for each boundary (each surface, hole, joint plane, or other discontinuity).

The theory of BEM is not easily explained. The mathematics required is more advanced than that needed for FEM. The interested reader will find several texts, conference proceedings, journal articles, and surveys [8.26,8.27].

8.12 FINITE ELEMENTS AND FINITE DIFFERENCES

Both the finite element method and the older finite difference method discretize a continuum, and both generate simultaneous algebraic equations to be solved for nodal d.o.f. Otherwise, the methods are superficially different. Finite difference stencils overlap one another and sometimes have nodes outside the structure boundary. Finite elements do not overlap and have no nodes outside the structure boundary. Finite differences are usually explained as a way to solve differential equations; finite elements are usually explained as a way to minimize a functional.

But a finite difference model *can* be derived from a functional [8.28]. For example, if Π_p is the functional and $\{\mathbf{D}\}$ are nodal d.o.f., we can write finite difference expressions for the derivatives in Π_p and generate algebraic equations from the stationary condition $\{\partial\Pi_p/\partial\mathbf{D}\} = \{\mathbf{0}\}$. This procedure is called the *finite difference energy method*. It produces a symmetric coefficient matrix if the finite element method produces a symmetric coefficient matrix for the same physical problem.

Thus the finite difference and finite element methods differ only in the choice of d.o.f. and in the location of nodes. Indeed, we can say that finite elements are a device for generating finite difference equations. Sometimes the two methods produce identical equations.

Both methods have about the same accuracy. Computer cost is often less when finite differences are used. Inevitably, cost comparisons depend on the type of problem and program organization as well as on the analysis method.

The finite difference energy method is well suited to shells of revolution [8.28]. It is also suited to "pure" continua, where there is only one medium, such as a homogeneous solid or fluid. It is not well suited to a structure with a complicated boundary shape or to a structure that must be modeled by a mixture of materials or a mixture of forms, such as a vehicle that combines bar, beam, plate, and shell components. For such a problem the finite element method has no rival.

8.13 REANALYSIS METHODS

Imagine that an initial solution has been obtained. Then the structure is altered: by changing member sizes, changing materials, or otherwise altering the finite element mesh. Loads on the structure are not changed.[2] Symbolically, we have

$$\text{Initial system:} \quad [\mathbf{K}]\{\mathbf{D}\} = \{\mathbf{R}\} \tag{8.13-1}$$

$$\text{Altered system:} \quad [\mathbf{K}^*]\{\mathbf{D}^*\} = \{\mathbf{R}\} \tag{8.13-2}$$

where

$$[\mathbf{K}^*] = [\mathbf{K}] + [\Delta\mathbf{K}] \qquad \text{and} \qquad \{\mathbf{D}^*\} = \{\mathbf{D}\} + \{\Delta\mathbf{D}\} \tag{8.13-3}$$

D.o.f. $\{\mathbf{D}^*\}$ are desired. The obvious approach is complete re-solution: that is, solve Eq. 8.13-2. Alternatives, called *reanalysis methods,* intend to obtain $\{\mathbf{D}^*\}$ with less computational effort than complete re-solution, by using information available from the previously obtained solution of Eq. 8.13-1. In vibration analysis, the analogous problem is to obtain modified frequencies without redoing the eigenvalue extraction.

Many methods of reanalysis have been proposed [8.29]. They have been categorized as follows [8.30].

1. *Direct methods* require a finite and predictable number of steps. They produce the exact $\{\mathbf{D}^*\}$ and work best when only a small portion of the structure is altered. If $[\mathbf{K}]$ has semibandwidth b, and less than roughly b rows of $[\mathbf{K}]$ are altered, then a direct method of reanalysis may be more efficient than complete re-solution.
2. *Iterative methods* converge from $\{\mathbf{D}\}$ toward $\{\mathbf{D}^*\}$ at a rate that is case-dependent. Iterative methods work best when alterations are small. Large differences between $[\mathbf{K}]$ and $[\mathbf{K}^*]$ make the iterations converge slowly or even diverge.
3. *Approximate methods* are usually based on a truncated series expansion or on a reduced set of structural equations. They are best suited to problems where exact results are not needed, for example, in intermediate stages of design or optimization.

To do reanalysis, one must choose among the many methods, and revise and enlarge the computer program. The option of complete re-solution is easier and may also be more efficient in many problems.

However, the option of substructuring should be noted. Substructuring (Section 8.14) is done for various reasons. One of its benefits is that the effect of alterations in a single substructure is efficiently computed. In this regard, substructuring is a direct method of reanalysis.

8.14 SUBSTRUCTURING

Mathematically, a substructure is a partially solved portion of the complete set of structural equations. Physically, a substructure is one of two or more parts

[2]Hence, the same $\{\mathbf{R}\}$ appears in Eqs. 8.13-1 and 8.13-2. Strictly, this requires that structural alterations have no effect on element loads $\{\mathbf{r}_e\}$, Eq. 4.1-6. If $\{\mathbf{r}_e\}$ is changed—for example, by resizing elements—$\{\mathbf{R}\}$ will be slightly changed but will be statically equivalent to its previous value.

into which a structure or a finite element mesh is divided. Multilevel substructuring is possible (Fig. 8.14-1). Substructuring has other names in other contexts: *blocking* or *dissection* when used by numerical analysts, and *diakoptics* or *tearing* when used by electrical engineers. We will describe the procedure in structural terms [8.31,8.32].

Procedure. In brief, a substructure is a "superelement," that is, a single element with many nodes on its boundary and many interior d.o.f. The name "macroelement" is also appropriate. The process is that of condensation and recovery, as described in Sections 8.1 and 8.2. Indeed, elements in Fig. 8.1-1 are substructures having few d.o.f. After the division of a structure into substructures has been selected, static analysis proceeds as follows.

1. Evaluate [**k**] and {**r**} for each substructure, where [**k**] and {**r**} pertain to all d.o.f. of the substructure. Eliminate internal d.o.f. by condensation; that is, apply Eq. 8.1-3. The condensed [**k**] and {**r**} pertain to only the boundary d.o.f. $\{\mathbf{d}_r\}$ of the substructure, which may be called "attachment" d.o.f.
2. Assemble substructures by connecting attachment nodes (i.e., nodes shared by substructures). Thus generate structural equations $[\mathbf{K}_m]\{\mathbf{D}_m\} = \{\mathbf{R}_m\}$, in which $\{\mathbf{D}_m\}$ contains the attachment d.o.f. of all substructures. (Attachment nodes on mating boundaries of adjacent substructures must match in physical placement and in orientation of their d.o.f.) Solve for $\{\mathbf{D}_m\}$.
3. For each substructure, extract from $\{\mathbf{D}_m\}$ the attachment d.o.f. $\{\mathbf{d}_r\}$ of that substructure. Use Eq. 8.1-2 to compute interior d.o.f. $\{\mathbf{d}_c\}$. Now all d.o.f. of the substructure are known. Hence, stress calculation proceeds in the usual way.

Clearly this is a finite element process in which elements have many internal d.o.f. and are given the name "substructures." It differs from a standard finite element process in that one does not form a single stiffness matrix that operates on *all*

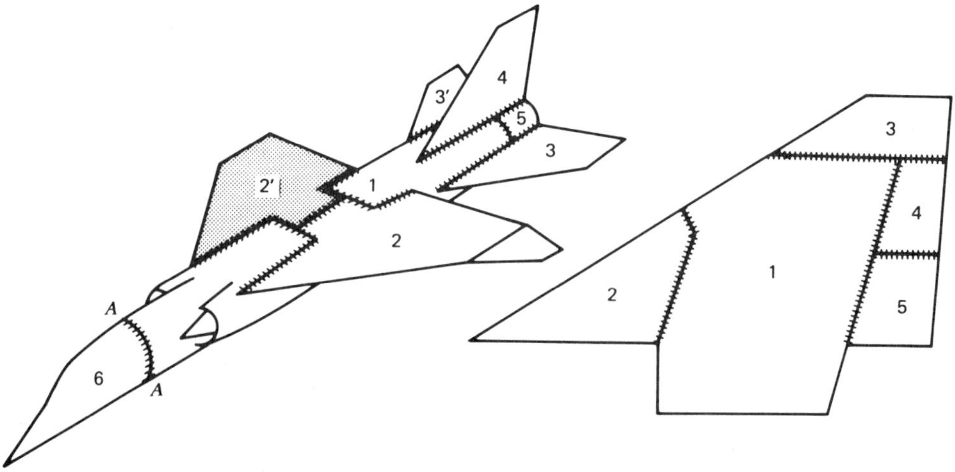

Figure 8.14-1. (*a*) An aircraft divided into substructures 1, 2, 2′, and so on. (*b*) Division of substructure 2′ into further substructures.

d.o.f. of the structure. Instead, there are several stiffness matrices, one for each substructure, and there is information about how to connect them so as to form $[\mathbf{K}_m]$.

Remarks. The names "masters" and "slaves" are sometimes used for retained and condensed d.o.f., respectively. In static analysis, all attachment d.o.f. $\{\mathbf{D}_m\}$ are masters and all d.o.f. interior to the substructures are slaves. In dynamic analysis, most interior d.o.f. are slaves but some may be masters. (Condensation in dynamics is discussed in Section 13.7.) Thus, in dynamic analysis, master d.o.f. $\{\mathbf{D}_m\}$ may not consist entirely of attachment d.o.f. shared by substructures. In static analysis *no approximation is introduced by substructuring.* In dynamic analysis some loss of accuracy is produced by substructuring.

In both static and dynamic analysis one desires that boundaries between substructures be small, so that the ratio of masters to slaves is small and $[\mathbf{K}_m]$ is kept to more manageable size. Accordingly, some structures are more amenable to substructuring than others; for example, a long cylinder can be more effectively substructured than a sphere.

No two substructures need be alike. However, there is special advantage if a structure contains many repetitions of the same form, particularly if the ratio of masters to slaves is small. Consider Fig. 8.14-2*a*. After the first analysis step has produced the condensed substructure $[\mathbf{k}]$ that operates on d.o.f. along boundaries *AB* and *CD*, this $[\mathbf{k}]$ need only be replicated with different node numberings to form the structure matrix $[\mathbf{K}_m] = \Sigma\,[\mathbf{k}]$. (Indeed, substructures within *ABCD* can be identified: the condensed $[\mathbf{k}]$ of the shaded substructure in Fig. 8.14-2*b* can be reflected about vertical and horizontal centerlines of *ABCD*, after which condensation of d.o.f. along these centerlines produces the condensed $[\mathbf{k}]$ of *ABCD*.) Thus substructuring is computationally efficient, as internal d.o.f. are processed only once, in forming the condensed $[\mathbf{k}]$ of the typical substructure. Without substructuring, the analysis would take longer, even though the total number of d.o.f. is unchanged.

Other advantages of substructuring include the following. There is a managerial advantage in breaking a large problem into smaller and more tractable parts. Different substructures can be studied simultaneously by different design groups. The work of one group can be almost independent of the others if the interaction between substructures is small. Design changes or analysis of nonlinearities, if

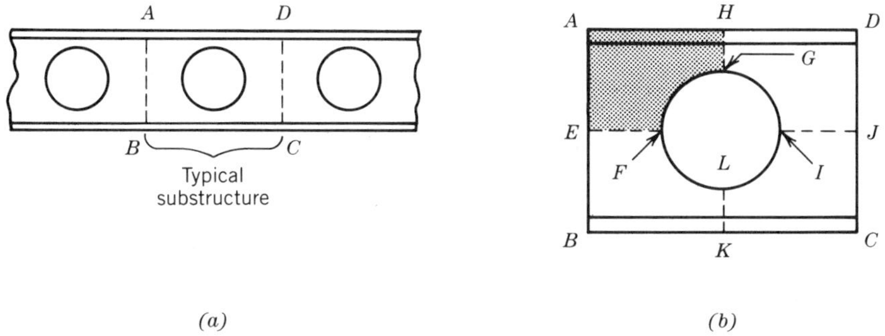

Figure 8.14-2. (*a*) Typical repeating substructure *ABCD* in an I beam with holes in its web. (*b*) A possible substructure of *ABCD* is shown shaded.

confined to a single substructure, leave matrices of all other substructures unchanged. The results of substructure analyses can be checked separately and revised if necessary before substructures are combined to form the complete structure.

Disadvantages of substructuring include the following. Substructuring replaces one long computer run by several shorter runs. Although this can be an advantage, it is a disadvantage if turnaround is slow. The computer program is more complicated because of increased file handling and data transfers, the need for efficient data structures, and user conveniences. Thus bookkeeping and overhead expense increase. If only one analysis is to be performed and there are no repeating substructures, it is cheaper to analyze the structure entire than to use substructuring. (In practice, design changes are expected, so it is unlikely that there will be but a single analysis.)

In vibration analysis it is possible to compute modes and frequencies of a structure from modes of its component substructures (see Section 13.8).

8.15 STRUCTURAL SYMMETRY

Figure 8.15-1*a* represents a thin square plate under lateral load. Imagine that the plate is homogeneous, isotropic, uniformly loaded, and has all four edges simply supported. Axes x, y, s, and t are all axes of symmetry. Accordingly, in static analysis, one need not analyze the entire plate: analysis of a single quadrant or a single octant—for example, one of those shown shaded in Fig. 8.15-1*a*—tells all that there is to know. As compared with analysis of the entire plate, data preparation time and computational expense are reduced.

How can symmetry be recognized? To be symmetric, a structure must have symmetry of shape, material properties, and support conditions. Symmetry can be classed as reflective with respect to an axis or to a plane, or rotational with respect to an axis. A symmetric structure is one for which one or more reflections and/or rotations brings the structure to a configuration indistinguishable from the

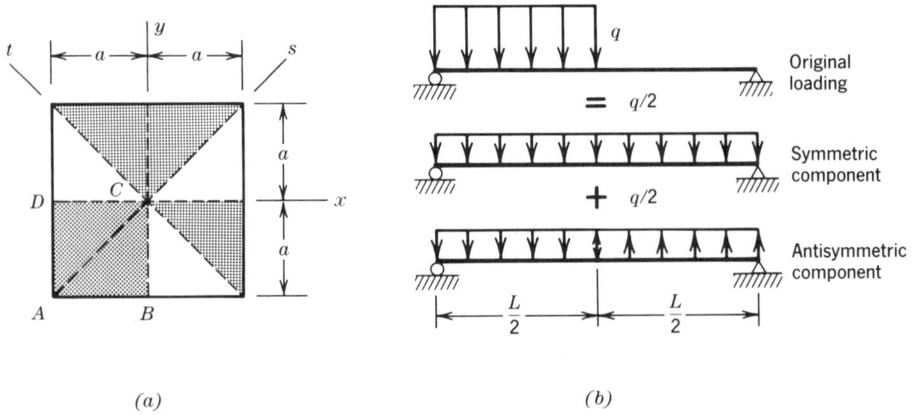

Figure 8.15-1. (*a*) A laterally loaded square plate with simply supported edges. (*b*) A uniform beam whose loading is broken into symmetric and antisymmetric components.

original configuration with respect to shape, material properties, and support conditions. In the plate of Fig. 8.15-1*a*, each dashed line is an axis of reflective symmetry. An axis normal to the plate through its center is an axis of rotational symmetry, because successive 90° rotations bring the structure into coincidence with itself. Other examples of rotational symmetry include solids of revolution (Chapter 10) and cyclic symmetry (Section 8.16).

A symmetric structure may carry symmetric or antisymmetric loads. Antisymmetry of loads exists if a single reflection of the structure, *with* its loads, followed by reversal of all loads, results in self-coincidence. An example appears in Fig. 8.15-1*b*, in which the plane of reflection is normal to the beam and intersects its center.

Symmetric loads, when acting on symmetric structures, produce symmetric effects in linear static analysis [8.33]. To exploit this rule, we analyze a part of the structure with appropriate boundary conditions. At a plane of geometric symmetry, displacement boundary conditions for *symmetric* loading are

1. No translational motion is perpendicular to a plane of geometric symmetry.
2. Rotation vectors have no component in a plane of geometric symmetry.

At a plane of geometric symmetry, displacement boundary conditions for *antisymmetric* loading are

1. Translational motion has no component in a plane of geometric symmetry.
2. Rotation vectors have no component perpendicular to a plane of geometric symmetry.

For symmetric and antisymmetric loads acting on symmetric structures, the resulting displacements and stresses are respectively symmetric and antisymmetric.

Consider, for example, the plate of Fig. 8.15-1*a*. Let finite elements (not shown) have as d.o.f. lateral translation w and rotations θ_y and θ_x about x and y axes, respectively (e.g., $\theta_y = w_{,y}$). Under uniform load, the loading has symmetry with respect to x and y axes. Quadrant *ABCD* can be analyzed with boundary conditions $w = 0$ along *AB* and *AD*, $\theta_x = 0$ along *AB* and *BC*, and $\theta_y = 0$ along *AD* and *CD*. If the load is uniform but alternately up and down over the four quadrants in checkerboard fashion, the loading has antisymmetry with respect to x and y axes. Quadrant *ABCD* can be analyzed with boundary conditions $w = 0$ along *AB*, *BC*, *CD* and *DA*, $\theta_y = 0$ along *AD* and *BC*, and $\theta_x = 0$ along *AB* and *CD*.

Remarks. In vibration analysis, symmetry must be exploited with caution, as symmetry of geometry does not imply symmetry of all vibration modes. Similarly, caution is needed if buckling or other nonlinear behavior arises, as symmetries present in an initial linear analysis may subsequently disappear.

It may be expedient to express a load as the sum of symmetric and antisymmetric components (Fig. 8.15-1*b*). Thus, rather than analyzing the entire structure once, one analyses half the structure twice [8.34].

Additional types of symmetries, known as skew-symmetric and skew-antisymmetric, can be identified and exploited [8.35].

Before a large model is analyzed, a model with few d.o.f. can be studied to check that anticipated symmetries indeed exist, and perhaps discover unanticipated symmetries.

8.16 CYCLIC SYMMETRY

A structure such as an impeller in a centrifugal pump is not a solid of revolution and cannot be analyzed as such. Nor is there a plane of reflective symmetry. Yet one can recognize a repetition of geometry and loading (Fig. 8.16-1). This circumstance is called *cyclic symmetry, sectorial symmetry,* or *rotational periodicity.* It is possible to analyze a representative substructure rather than the entire structure. The procedure is as follows [8.36].

For a typical substructure *AABB*, Fig. 8.16-1*b*, let $[\mathbf{K}]\{\mathbf{D}\} = \{\mathbf{R}\}$ represent the substructure equations, where $\{\mathbf{D}\}$ includes all substructure d.o.f. In partitioned form, these equations are

$$\begin{bmatrix} \mathbf{K}_{II} & \mathbf{K}_{IA} & \mathbf{K}_{IB} \\ \mathbf{K}_{IA}^T & \mathbf{K}_{AA} & \mathbf{K}_{AB} \\ \mathbf{K}_{IB}^T & \mathbf{K}_{AB}^T & \mathbf{K}_{BB} \end{bmatrix} \begin{Bmatrix} \mathbf{D}_I \\ \mathbf{D}_A \\ \mathbf{D}_B \end{Bmatrix} = \begin{Bmatrix} \mathbf{R}_I \\ \mathbf{R}_A \\ \mathbf{0} \end{Bmatrix} + \begin{Bmatrix} \mathbf{0} \\ \mathbf{F}_A \\ \mathbf{F}_B \end{Bmatrix} \tag{8.16-1}$$

where $\{\mathbf{D}_A\}$ and $\{\mathbf{D}_B\}$ contain d.o.f. on interface boundaries *AA* and *BB*, respectively, and $\{\mathbf{D}_I\}$ contains all remaining d.o.f. (from nodes within the substructure and noninterface nodes on boundaries $r = a$ and $r = b$). Loads $\{\mathbf{F}_A\}$ and $\{\mathbf{F}_B\}$ result from elastic deformations and are applied along *AA* and *BB* by neighboring substructures. Loads $\{\mathbf{R}_I\}$ and $\{\mathbf{R}_A\}$ represent imposed loads, typically caused by rotation and uneven (but cyclically symmetric) heating. Loads $\{\mathbf{R}_B\}$ are absent because imposed loads on interface boundaries must appear on only one interface boundary; if mistakenly placed on both, the substructure receives twice the load intended.

All repeating substructures are identical. Therefore, subject to a subsequent caution,

$$\{\mathbf{D}_B\} = \{\mathbf{D}_A\} \qquad \text{and} \qquad \{\mathbf{F}_B\} = -\{\mathbf{F}_A\} \tag{8.16-2}$$

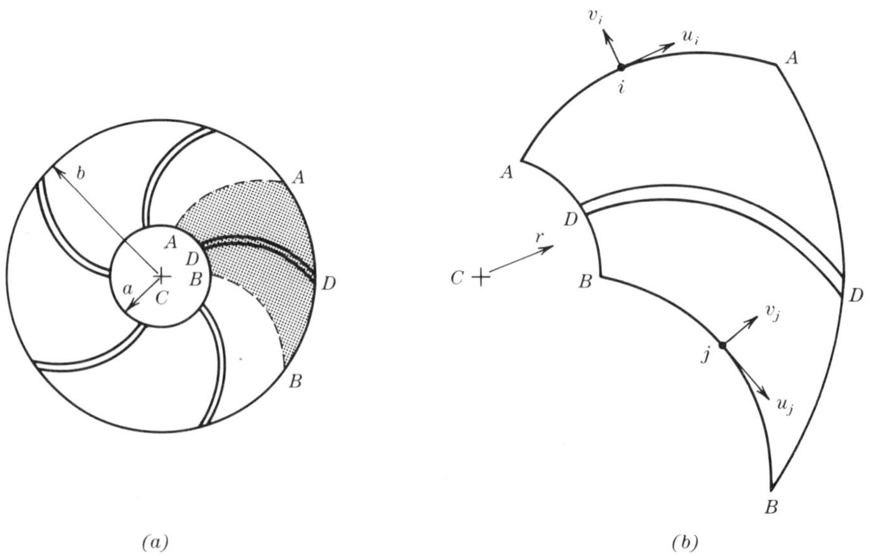

Figure 8.16-1. (*a*) A hypothetical pump impeller, viewed along its axis of rotation. Vanes such as *DD*, seen here in edge view, are mounted on a circular disk. (*b*) A typical repeating substructure.

Using transformation procedures explained in Chapter 7, we write

$$\begin{Bmatrix} \mathbf{D}_I \\ \mathbf{D}_A \\ \mathbf{D}_B \end{Bmatrix} = [\mathbf{T}] \begin{Bmatrix} \mathbf{D}_I \\ \mathbf{D}_A \end{Bmatrix}, \qquad \text{where} \quad [\mathbf{T}] = \begin{bmatrix} \mathbf{I} & \mathbf{0} \\ \mathbf{0} & \mathbf{I} \\ \mathbf{0} & \mathbf{I} \end{bmatrix} \tag{8.16-3}$$

and $[\mathbf{I}]$ is a unit matrix. The transformations $[\mathbf{T}]^T[\mathbf{K}][\mathbf{T}]$ and $[\mathbf{T}]^T\{\mathbf{R}\}$, applied to $[\mathbf{K}]$ and $\{\mathbf{R}\}$ of Eq. 8.16-1, yield

$$\begin{bmatrix} \mathbf{K}_{II} & \mathbf{K}_{IA} + \mathbf{K}_{IB} \\ \mathbf{K}_{IA}^T + \mathbf{K}_{IB}^T & \mathbf{K}_{AA} + \mathbf{K}_{AB} + \mathbf{K}_{AB}^T + \mathbf{K}_{BB} \end{bmatrix} \begin{Bmatrix} \mathbf{D}_I \\ \mathbf{D}_A \end{Bmatrix} = \begin{Bmatrix} \mathbf{R}_I \\ \mathbf{R}_A \end{Bmatrix} \tag{8.16-4}$$

in which $\{\mathbf{F}_A\}$ and $\{\mathbf{F}_B\}$ do not appear because of Eq. 8.16-2. Solution for nodal d.o.f. and stresses now proceeds in the usual way.

Caution. The number and location of nodes along *AA* and *BB* must correspond exactly, and d.o.f. at corresponding nodes must have the same orientation with respect to the interface boundary. For example, in Fig. 8.16-1*b*, if *i* and *j* are corresponding nodes (e.g., both the *k*th node on their respective boundaries), then we must have $r_i = r_j$, and d.o.f. at *i* and *j* must have directions such as those shown, where u and v are respectively tangent and normal to each interface boundary. Directions u (radial) and v (tangential) in polar coordinates with point *C* as pole are also acceptable, but directions u and v in Cartesian coordinates are not acceptable.

Equation 8.16-4 can be produced *automatically* by the assembly process, thus avoiding the transformation defined by Eq. 8.16-3. The trick is to assign the same node number to corresponding nodes along *AA* and *BB*; for example, nodes *i* and *j* cited in the preceding paragraph would *both* be given the number (say) 125. Thus, the additions seen in Eq. 8.16-4 are produced automatically when elements are assembled. One must of course use actual node point coordinates in the formulation of element matrices.

PROBLEMS

Section 8.1

8.1 Apply Eq. 8.1-3 to the problem in Fig. 2.11-1. Specifically, eliminate u_2 and u_3 in Fig. 2.11-1*b*, and obtain the condensed equation $2u_4 = 8$ seen in Fig. 2.11-1*d*.

8.2 What values of the 18 d.o.f. in Eq. 8.1-4 are associated with rigid-body translation of magnitude $\bar{u}$ in the $+x$ direction?

8.3 A 9 by 9 transformation matrix $[\mathbf{T}]$ can be applied to the shape function matrix $\lfloor\mathbf{N}\rfloor$ associated with Table 6.6-1 to yield the shape function matrix associated with Eq. 8.1-4. Write this matrix $[\mathbf{T}]$ and the equation that relates the two shape function matrices.

8.4 A three-node bar element with a central node 3 is shown, along with its stiffness matrix, which operates on d.o.f. $\lfloor u_1 \quad u_2 \quad u_3 \rfloor$.

(a) Determine the 2 by 2 matrix $[\mathbf{k}]$ produced by condensation of u_3.

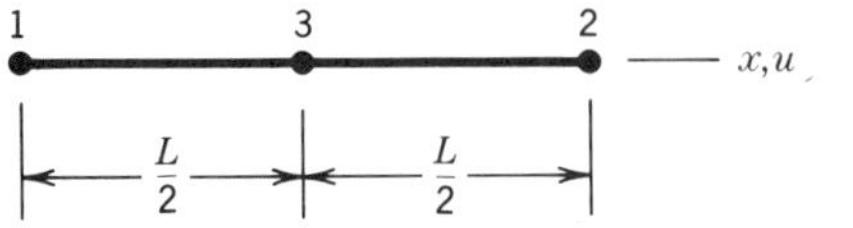

$$[\mathbf{k}] = \frac{AE}{3L}\begin{bmatrix} 7 & 1 & -8 \\ 1 & 7 & -8 \\ -8 & -8 & 16 \end{bmatrix}$$

Problem 8.4

(b) Apply a uniformly distributed axial load of intensity q. From the consistent load vector $\{\mathbf{r}_e\}$, obtain a condensed load vector associated with u_1 and u_2.

8.5 A stiffness matrix and a consistent load vector can be formulated for the three-node bar element of Problem 8.4 by use of the displacement field

$$u = \frac{L-x}{L}u_1 + \frac{x}{L}u_2 + x(L-x)a_1$$

where a_1 is a nodeless d.o.f.

(a) Determine the 3 by 3 stiffness matrix $[\mathbf{k}]$ dictated by the given u field. Let the element be uniform.

(b) Under what circumstances do you think the added mode $x(L-x)a_1$ will improve the results given by the basic linear element? In what stage of a finite element stress analysis does a_1 have an effect?

8.6 Imagine that d.o.f. θ_1 and θ_2 of the standard four-d.o.f. beam element (Fig. 4.2-2 and Eq. 4.2-5) are to be eliminated. What do you think the 2 by 2 condensed matrix $[\mathbf{k}]$ will be? Verify your prediction.

8.7 Let a plane frame element be joined to a rotational spring at each end, with respective spring stiffnesses k_1 and k_2 (moment per radian). Let β_1 and β_2 be *structure* node rotations. Rotational d.o.f. θ_1 and θ_2 of the frame element are to be connected to structure nodes through the rotational springs, so that in Fig. 4.2-2 $\theta_1 \neq \beta_1$ and $\theta_2 \neq \beta_2$ unless k_1 and k_2 approach infinity. Translational d.o.f. are to be connected directly, as usual. Beginning with an 8 by 8 stiffness matrix that operates on d.o.f. $\lfloor u_1 \quad w_1 \quad \theta_1 \quad u_2 \quad w_2 \quad \theta_2 \quad \beta_1 \quad \beta_2 \rfloor^T$, describe how to determine a 6 by 6 matrix $[\mathbf{k}]$ that operates on d.o.f. $\lfloor u_1 \quad w_1 \quad \beta_1 \quad u_2 \quad w_2 \quad \beta_2 \rfloor^T$ and is a function of A, E, I, L, k_1, and k_2.

8.8 Addition to an element of internal d.o.f., such as a_1 and a_2 in Eq. 8.1-4, can be regarded as a device that permits better approximation of equilibrium equations within the element, without affecting interelement compatibility. Accordingly, do you think the constant-strain triangle (Section 5.4) would be improved by addition of the bubble function modes $u = \xi_1\xi_2\xi_3 a_1$ and $v = \xi_1\xi_2\xi_3 a_2$? Why or why not?

8.9 Consider the frame of Fig. 8.1-2. Imagine that, before assembly, rotation θ_A is condensed in all four elements that meet at node A. What do you think will be the effect of these condensations, both physically and in the numerical process?

8.10 Cantilever beams AB and BC are identical and are connected by a hinge at B, as shown. Use condensation, as described in connection with Fig. 8.1-2, to evaluate the rotation in both beams at B. Verify your result by elementary beam theory.

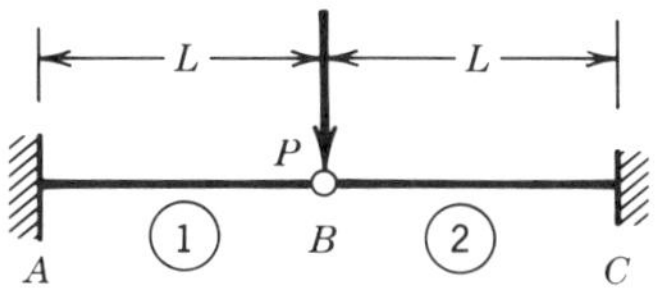

Problem 8.10

8.11 In Problem 8.10, how would you determine the value of P needed to produce a prescribed amount of relative rotation between the beams at B?

Section 8.2

8.12 Modify Figs. 8.2-1 and 8.2-2 to allow for NL load cases rather than only one.

Section 8.3

8.13 (a) For the elements shown in Figs. 8.3-1*b* and 8.3-1*c*, compute the ratio of element strain energies, U_1/U_2.
(b) Use this result to verify the correctness of Eq. 8.3-5.

8.14 Consider the two beams built of rectangular elements in Table 6.14-1 (one-element case and the first five-element case). If one assumes that Eq. 8.3-5 is approximately true for these beams, what end deflections would be expected? Compare these results with those in Table 6.14-1.

8.15 (a) Do the a_i of Eqs. 8.3-6 represent relative or absolute motions?
(b) If, after computation of nodal d.o.f. in a mesh of QM6 elements, the nodeless d.o.f. a_i are omitted from stress computation, what consequences do you expect? Consider, for example, the rectangular-element test cases in Table 6.14-1.

8.16 The sketch shows a cantilever beam modeled by QM6 elements. For the loading shown, will exact values of stresses σ_x be computed? Why or why not?

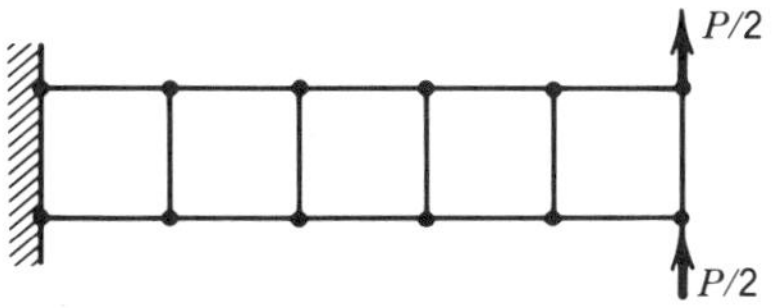

Problem 8.16

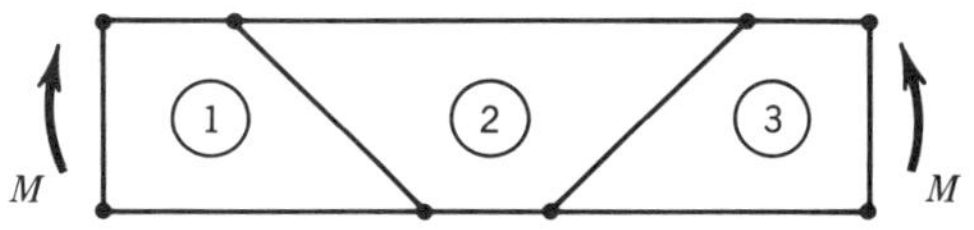

Problem 8.17

8.17 All three elements in the beam shown are plane QM6 elements. Examine displacements along sides of element 2 under the moment loading shown. Hence, show that pure bending is not modeled exactly by nonrectangular QM6 elements.

8.18 For the nodal d.o.f. $\bar{u}$ applied in Fig. 8.3-1, show that Eq. 8.3-7 yields Eqs. 8.3-3.

8.19 Imagine that Figs. 6.5-1 and 6.5-2 are to be modified so that they will apply to the QM6 element. Clearly state what changes and additions are required, and supply new coding where needed in Fig. 6.5-1.

Section 8.4

8.20 Show that δ in Eq. 8.4-1 can be regarded as a beam midspan deflection, as noted below Eq. 8.4-1.

8.21 Following the procedure suggested below Eq. 8.4-2, write shape functions for the nine-d.o.f. triangle of Fig. 8.4-1*b*.

8.22 (a) Establish the contents of matrix **[T]** in Eq. 8.4-4.
(b) Use this **[T]** to determine shape functions N_i of a nine-d.o.f. triangle from shape functions N_i' of a linear-strain triangle.

8.23 Imagine that lateral deflection w of a uniform beam element is defined by three nodal values, as shown.
(a) Establish the 3 by 4 transformation matrix **[T]** that will convert this element to one that operates on the standard d.o.f. w_1, θ_1, w_2, and θ_2.
(b) Hence, establish the new shape functions N_1, N_2, N_3, and N_4.
(c) What property does the element have that may pose a difficulty?

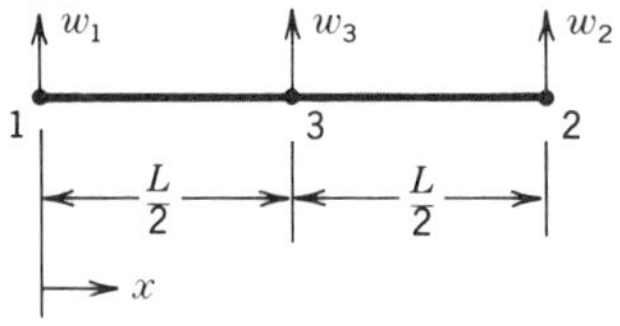

$$w = \frac{2x^2 - 3Lx + L^2}{L^2} w_1 + \frac{2x^2 - Lx}{L^2} w_2 + \frac{4x(L-x)}{L^2} w_3$$

Problem 8.23

Section 8.5

8.24 Complete the steps of generating **[k]** for the element of Fig. 8.5-1*b*, as follows. Use **[P]** and **[R]** from Eq. 8.5-12.
(a) Complete matrix **[L]**, begun in Eqs. 8.5-14.
(b) Generate matrix **[G]**, Eq. 8.5-7.
(c) Generate matrix **[H]**, Eq. 8.5-4. For simplicity, let $\nu = 0$, so that $[\mathbf{E}] = E\lceil 1 \quad 1 \quad \tfrac{1}{2} \rfloor$.
(d) Generate **[k]**, Eq. 8.5-10, again for $\nu = 0$.

8.25 Use the assumed-stress hybrid method to evaluate **[k]** for the six-d.o.f. plane triangle shown. Use $\{\boldsymbol{\beta}\} = \lfloor \beta_1 \quad \beta_2 \quad \beta_3 \rfloor^T$. For simplicity, take Poisson's ratio as zero. (This **[k]** should agree with the **[k]** obtained in Problem 4.11c.)

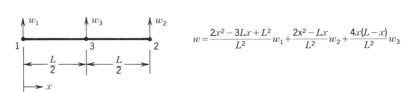

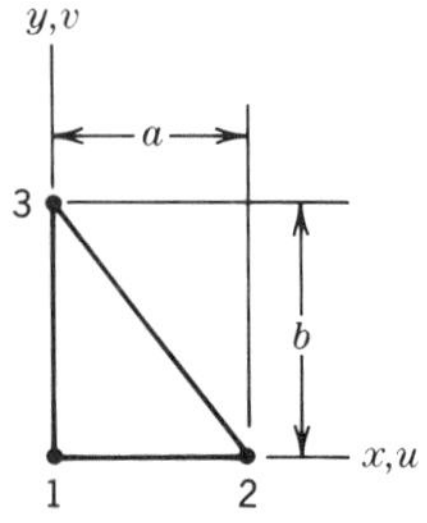

Problem 8.25

8.26 (a) Write the equation $\{\boldsymbol{\sigma}\} = [\mathbf{P}]\{\boldsymbol{\beta}\}$ for a plane element if $\sigma_x = \beta_1 + \beta_4 x$, $\sigma_y = \beta_2 + \beta_5 y$, and $\tau_{xy} = \beta_3$. Do you think such an element would be a good one?

(b) If $\beta_4 = \beta_5 = 0$ in part (a), so that $\{\boldsymbol{\beta}\} = \lfloor \beta_1 \quad \beta_2 \quad \beta_3 \rfloor^T$, what defect would you expect to see in the stiffness matrix of a plane eight-d.o.f. rectangular element?

8.27 The beam element shown is to include the effects of transverse shear deformation. If bending moment M is taken as $M = \beta_1 + \beta_2 x$, then the shear force $V = \beta_2$ satisfies the equilibrium equation $dM/dx = V$. With $\{\boldsymbol{\sigma}\} = \lfloor M \quad V \rfloor^T$, strain energy in the element is $U = \frac{1}{2} \int \{\boldsymbol{\sigma}\}^T \begin{bmatrix} \frac{1}{EI} & \frac{f}{AG} \end{bmatrix} \{\boldsymbol{\sigma}\}\, dx$, where f is a "form factor" ($f = 1.2$ for a rectangular cross section). [**R**] relates nodal moments and shear forces to $\{\boldsymbol{\beta}\}$, [**L**] is a unit matrix, and $[\mathbf{R}]^T[\mathbf{L}]$ requires no integration. Derive [**k**] and show that it reduces to Eq. 4.2-5 as shear modulus G becomes large [4.11].

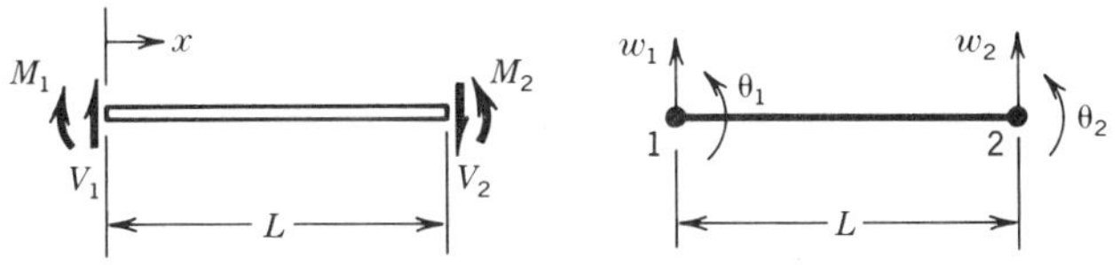

Problem 8.27

Section 8.7

8.28 (a) Following the example of Eqs. 8.7-8 and 8.7-9, determine [**k**] for a standard four-d.o.f. beam element. However, use $\{\mathbf{d}_R\} = \lfloor w_1 \quad w_2 \rfloor^T$.

(b) Imagine that the leading diagonal coefficient of $[\mathbf{k}_{EE}]$ in part (a) is in error by an amount e. Show that [**k**] still represents rigid-body motion correctly.

8.29 The two-spring structure shown is allowed axial nodal displacements u_1, u_2, and u_3. If $\{\mathbf{d}_R\} = u_1$, write $[\mathbf{k}_{EE}]$, and from it determine [**k**].

8.30 (a) A bar element of axial stiffness $k = AE/L$ is permitted only axial nodal displacements u_1 and u_2. Write $[\mathbf{k}_{EE}]$, and from it determine [**k**].

(b) Repeat part (a), but let there be four d.o.f. $\{\mathbf{d}\} = \lfloor u_1 \quad v_1 \quad u_2 \quad v_2 \rfloor^T$, as in Eq. 2.4-3, so that plane motion is possible.

8.31 A flat elastic disk has inside and outside radii r_1 and r_2, as shown. Nodal d.o.f. are circumferential displacements v_1 and v_2. When the disk is fixed at $r = r_1$, the ratio of torque T_2 on edge $r = r_2$ to the resulting angle of twist θ_2 is a number C. Determine the stiffness matrix [**k**] that operates on d.o.f. v_1 and v_2, in terms of C, r_1, and r_2. Verify that $[\mathbf{k}]\{\mathbf{d}\} = \{\mathbf{0}\}$ if $\{\mathbf{d}\}$ represents rigid-body motion.

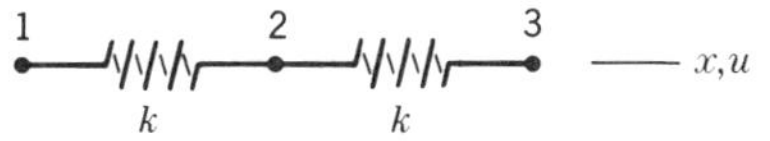

Problem 8.29

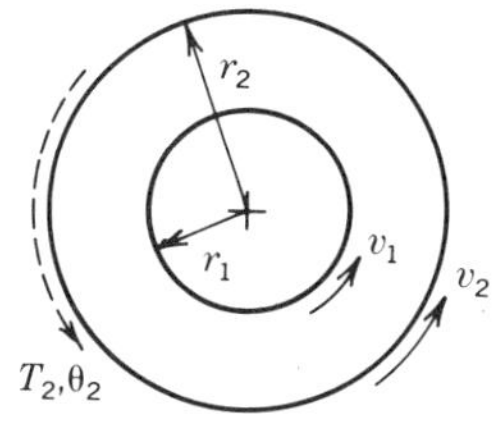

Problem 8.31

Section 8.8

8.32 The structure shown is built of plane elements. D.o.f. (at each corner node) consist of u, v, $u_{,x}$, $v_{,x}$, $u_{,y}$, and $v_{,y}$. Pressure p acts along edge AB. Edge BC is fixed. What boundary conditions should be imposed on nodal d.o.f. along edges AB, BC, CD, and DA? Assume that the material is isotropic. What is different if the material is anisotropic?

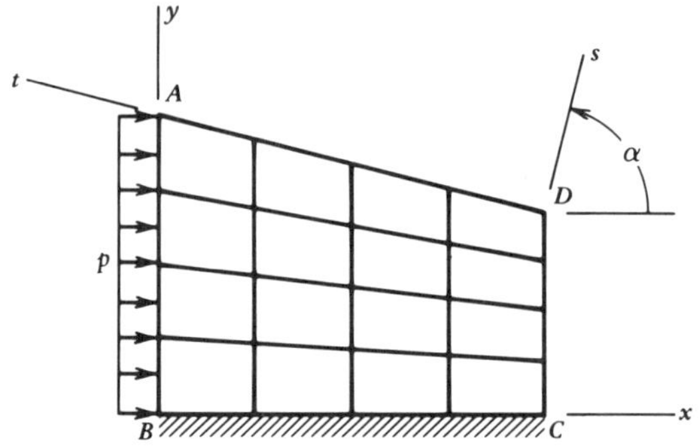

Problem 8.32

Section 8.9

8.33 A long bar, 100 mm wide and 20 mm thick, is loaded in tension by an axial force P.

(a) If the yield strength is $Y = 1150$ MPa and $K_{IC} = 77$ MPa$\sqrt{\text{m}}$, and a central crack 15 mm long is present, what force P will fracture the bar?

(b) If the yield strength is $Y = 1410$ MPa and $K_{IC} = 50$ MPa$\sqrt{\text{m}}$, what is the critical crack length if the force P determined in part (a) is applied?

8.34 Rather than use Eq. 8.9-8a to determine K_I, one can determine K_I from displacements of points $B1$ and $B2$ alone in Fig. 8.9-4. Derive the appropriate formula.

8.35 Consider the bar element of Fig. 8.9-3, but place node 3 at the third point rather than at the quarter point. At what value of x/L is a stress singularity indicated?

8.36 Let quarter-point elements be used to solve a certain crack problem (e.g., Fig. 8.9-4). Imagine that the problem is solved again, this time using quarter-point elements of smaller size. Now the computed results are found to be less accurate than before. Explain how this is possible.

Section 8.10

8.37 The beam element shown has the usual d.o.f. $\{\mathbf{d}\} = \lfloor w_1 \quad \theta_1 \quad w_2 \quad \theta_2 \rfloor^T$. The element has width b and rests on a Winkler foundation of modulus β. Determine the foundation matrix $[\mathbf{k}_f]$ defined by each of the following approximations.

(a) Deflection w is cubic in x, as in the standard beam element.

(b) Deflection w is quadratic in x (see Eq. 8.4-2).

(c) Deflection w is linear in x (and independent of θ_1 and θ_2).

(d) Deflection w is constant.

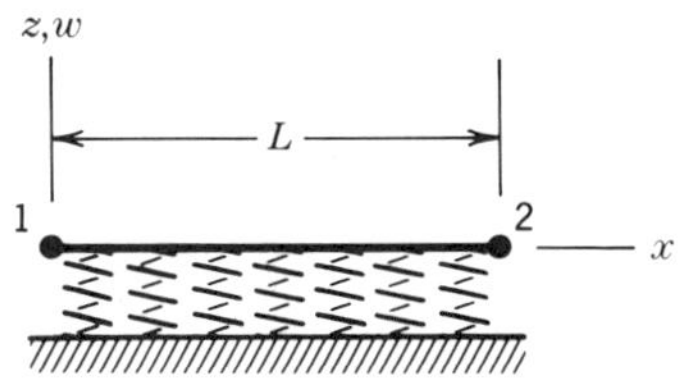

Problem 8.37

8.38 The sketches represent top views of triangular elements that rest on a Winkler foundation of modulus β. Assume that vertical deflection w depends only on nodal values of w. Determine an expression for $[\mathbf{k}_f]$ of
(a) the three-node element (Eqs. 5.3-4).
(b) the six-node element (Eqs. 5.3-5), if sides are straight and side nodes are at midsides.

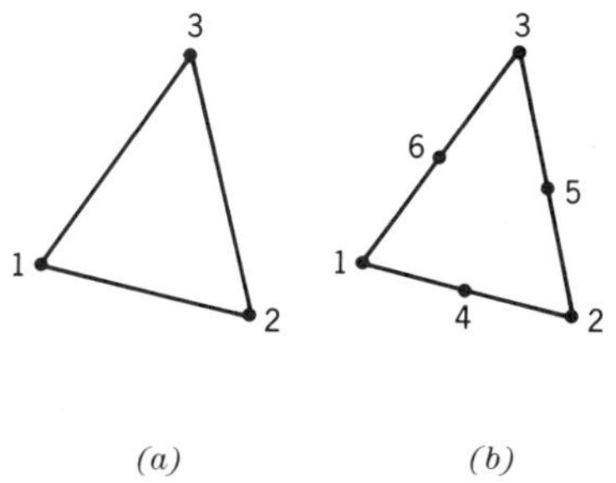

Problem 8.38

8.39 Imagine that separation is possible between a beam and its Winkler elastic foundation. Outline a solution algorithm for such a problem. In this exercise, do not be concerned with computational efficiency.

8.40 Imagine that a Winkler elastic foundation, which has translational modulus β, is augmented by a rotational modulus α (whose units are force divided by length). For an element on such a composite foundation, what formula for $[\mathbf{k}_f]$ replaces Eq. 8.10-2? (A symbolic result is desired, with terms defined, rather than specifics of a particular element.)

Section 8.11

8.41 Show that Eqs. 8.11-5 and 8.11-6 indeed result from the manipulations described.

8.42 Use ϕ from Eq. 8.11-7 and the mapping of Eq. 8.11-1 to determine the following:
(a) ϕ as a function of r (analogous to Eq. 8.11-6).
(b) J as a function of ξ and a.
(c) Element matrix $[\mathbf{k}]$. Use exact integration.
(d) Element matrix $[\mathbf{k}]$. Use a two-point Gauss rule.

8.43 (a) For the infinite element shown, let field variable ϕ depend on nodal values ϕ_1, ϕ_2, ϕ_5, and ϕ_6 only (not on ϕ_3 and ϕ_4). Write mapping functions and shape functions, in the manner of Fig. 8.11-3.

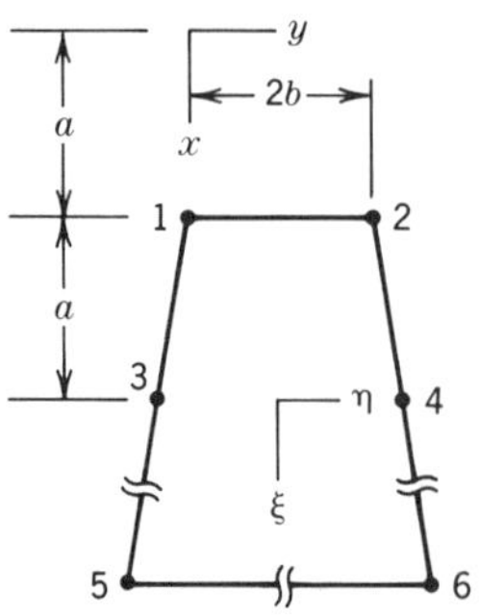

Problem 8.43

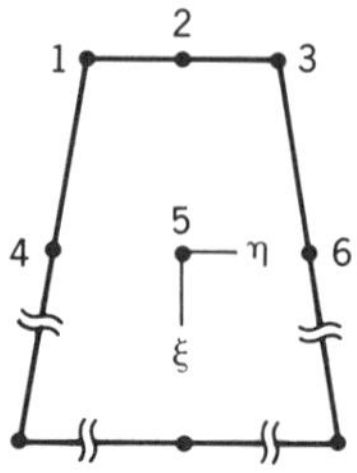

Problem 8.44

(b) Let sides 1–3–5 and 2–4–6 be parallel. Evaluate **[J]** and J.

(c) If $\phi_5 = \phi_6 = 0$, what 2 by 2 element characteristic matrix **[k]** operates on ϕ_1 and ϕ_2? Again let sides 1–3–5 and 2–4–6 be parallel. (See Eq. 6.3-5, and let t = element thickness and k = material characteristic, both uniform over the element.)

8.44 Write mapping functions for the infinite plane element shown.

Section 8.13

8.45 If $\{\mathbf{R}\}$ is unchanged and structural alterations are minor, then $\{\Delta\mathbf{D}\} \approx -[\mathbf{K}]^{-1}([\Delta\mathbf{K}]\{\mathbf{D}\})$. Derive this expression for $\{\Delta\mathbf{D}\}$. What are advantages and disadvantages of this method?

8.46 Equation 8.13-2 can be cast in the iterative form $[\mathbf{K}]\{\mathbf{D}^*\}_{i+1} = \{\mathbf{R}\} - [\Delta\mathbf{K}]\{\mathbf{D}^*\}_i$. Consider the application of this equation to single-d.o.f. problems as follows.

(a) Let $K = 0.5$, $K^* = 0.8$, and $R = 2$. Starting with $D_0^* = D = 4.0$, compute D_5^* (i.e., apply five iterative cycles).

(b) For what range of values of $\Delta K/K$ does this iterative method converge?

Section 8.14

8.47 In Fig. 8.14-2*b*, imagine that the reduced **[k]** for substructure *AEFGH* is known. How can one transform this **[k]** so that it pertains to substructure *HGIJD*, ready for assembly with substructure *AEFGH*? For brevity, consider only translational d.o.f. u_i and v_i at the lettered corners.

Section 8.15

8.48 Let the plate of Fig. 8.15-1*a* be uniformly loaded. Imagine that octant *ABC* is modeled by square elements, as shown in the sketch for this problem, so that some elements straddle the symmetry axis *AC*. What boundary conditions should be applied to these elements, for example, to typical element 1–2–3–4? State these conditions with reference to (a) *st* axes, and (b) *xy* axes.

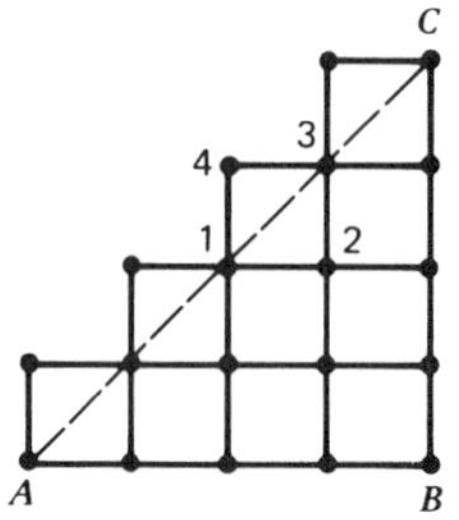

Problem 8.48

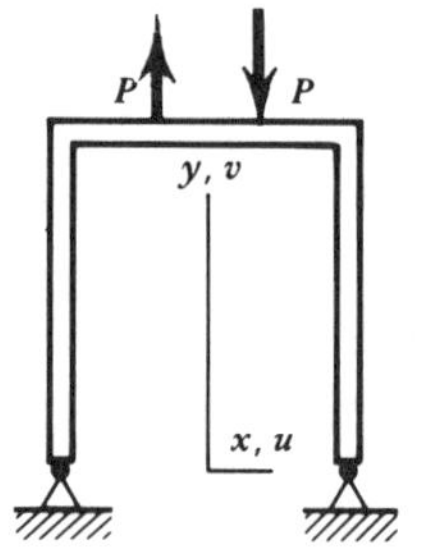

Problem 8.49

8.49 For the frame shown, geometry is symmetric but loads P are antisymmetric with respect to the y axis.

(a) By considering a reflection and a load reversal, show that vertical support reactions are equal in magnitude and horizontal support reactions are zero.

(b) If half the frame is analyzed, what support condition should be used at the point where the y axis crosses the frame?

8.50 By superposing results from symmetric and antisymmetric loadings in Fig. 8.15-1b, and using formulas from beam theory, determine (a) the deflection at midspan, and (b) the rotation at the left end.

Section 8.16

8.51 The three-node truss shown carries radial loads P and contains three identical bars, each of axial stiffness $k = AE/L$. Use cyclic symmetry methods to determine the radial displacement of a typical node.

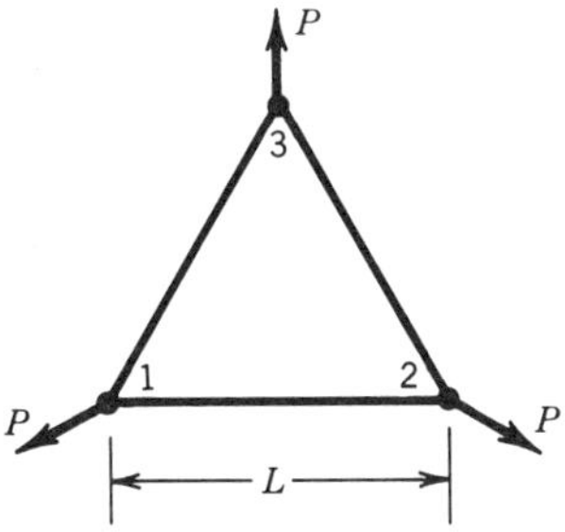

Problem 8.51

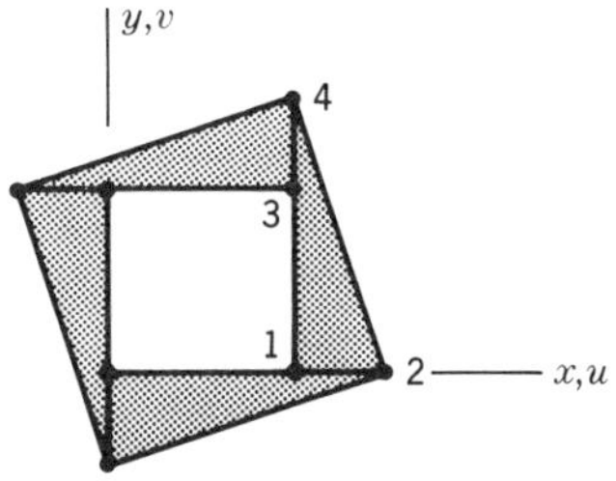

Problem 8.52

8.52 The plane structure shown consists of four identical triangular areas, which form a square outer boundary and a square opening. Imagine that there exists an 8 by 8 stiffness matrix $[\mathbf{k}']$ for one triangle, which operates on d.o.f. u_i and v_i at nodes $i = 1, 2, 3, 4$.

(a) Construct a transformation matrix $[\mathbf{T}_a]$ for the operation $[\mathbf{k}] = [\mathbf{T}_a]^T[\mathbf{k}'][\mathbf{T}_a]$, in which $[\mathbf{k}]$ operates on d.o.f. suitable for exploitation of cyclic symmetry.

(b) Construct an 8 by 4 constraint transformation matrix $[\mathbf{T}]$ appropriate to this exercise (see Eq. 8.16-3).

(c) What is the appropriate transformation matrix if parts (a) and (b) are to be done as a single transformation?

8.53 A long, uniform beam is supported and loaded in a repetitive pattern, as shown. Use cyclic symmetry methods to determine nodal d.o.f. w_1, θ_1, and θ_2 in terms of P, a, E, and I.

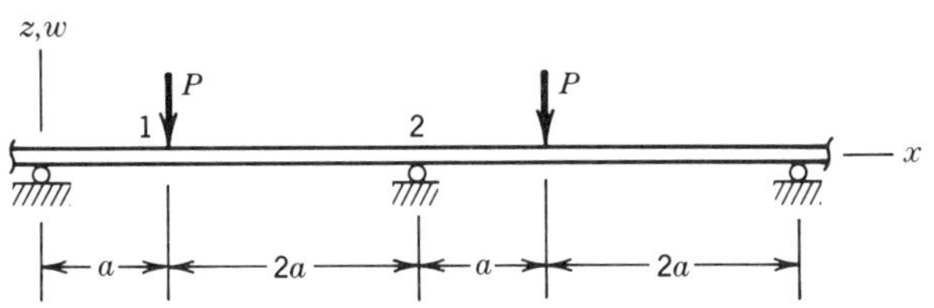

Problem 8.53

9 CHAPTER

CONSTRAINTS

Constraints enforce a relationship among d.o.f. Procedures for imposing a constraint include transformation, Lagrange multipliers, and penalty functions. Naturally arising constraints, constraint counting, and integration rules for incompressible materials are also discussed.

9.1 CONSTRAINTS. TRANSFORMATIONS

Constraints. A *constraint* either prescribes the value of a d.o.f. (as in imposing a support condition) or prescribes a relationship among d.o.f. In common terminology, a *single-point constraint* sets a single d.o.f. to a known value (often zero), and a *multipoint constraint* imposes a relationship between two or more d.o.f. Thus support conditions in the three-bar truss of Fig. 2.2-1 invoke three single-point constraints. Rigid links and rigid elements, discussed in Section 7.8, each invoke a multipoint constraint.

Figure 9.1-1 shows an example in which constraints could be imposed. In a typical frame, axial deformation of a member can usually be ignored; only bending deformation is significant. Accordingly, in Fig. 9.1-1, one could impose the single-point constraints $v_A = 0$ and $v_B = 0$, and the multipoint constraint $u_A = u_B$, after which the active d.o.f. consist of only θ_A, θ_B, and either u_A or u_B. (Failure to impose the constraint $u_A = u_B$ invites numerical difficulty; see Section 18.2.) Special-purpose computer programs for tall buildings may incorporate constraints of this type by allowing only three d.o.f. per floor, these being the rotation θ_z of a floor about a vertical z axis and the horizontal displacement components u and v.

For each equation of constraint, one d.o.f. can be eliminated from the vector of structural d.o.f. $\{\mathbf{D}\}$. However, doing so may involve appreciable manipulation and typically increases the bandwidth (or the frontwidth) of the structural equations. The Lagrange multiplier method of treating constraints, discussed subsequently, *adds* to the number of equations but requires less manipulation.

Transformation Equations. Constraint equations that couple d.o.f. in $\{\mathbf{D}\}$ can be written in the form

$$[\mathbf{C}]\{\mathbf{D}\} = \{\mathbf{Q}\} \tag{9.1-1}$$

where $[\mathbf{C}]$ and $\{\mathbf{Q}\}$ contain constants. There are more d.o.f. in $\{\mathbf{D}\}$ than constraint equations, so $[\mathbf{C}]$ has more columns than rows. We now consider the common case $\{\mathbf{Q}\} = \{\mathbf{0}\}$. Let Eq. 9.1-1 be partitioned so that

$$[\mathbf{C}_r \quad \mathbf{C}_c] \begin{Bmatrix} \mathbf{D}_r \\ \mathbf{D}_c \end{Bmatrix} = \{\mathbf{0}\} \tag{9.1-2}$$

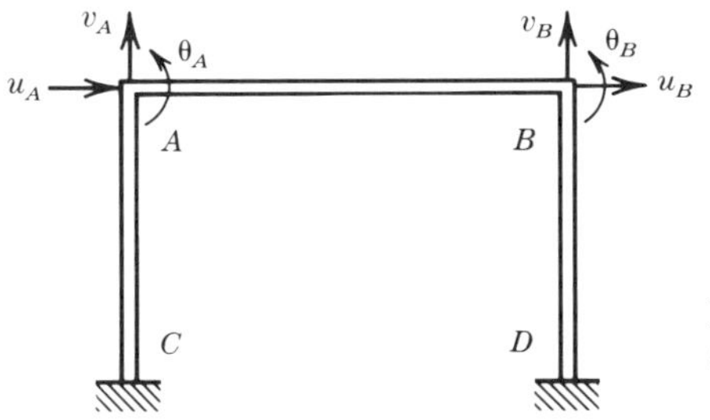

Figure 9.1-1. A three-element plane frame, fixed at nodes C and D. D.o.f. at nodes A and B are shown.

where $\{\mathbf{D}_r\}$ and $\{\mathbf{D}_c\}$ are, respectively, d.o.f. to be retained and d.o.f. to be eliminated or "condensed out." Because there are as many d.o.f. $\{\mathbf{D}_c\}$ as there are independent equations of constraint in Eq. 9.1-2, matrix $[\mathbf{C}_c]$ is square and nonsingular. Solution for $\{\mathbf{D}_c\}$ yields

$$\{\mathbf{D}_c\} = [\mathbf{C}_{rc}]\{\mathbf{D}_r\}, \qquad \text{where} \quad [\mathbf{C}_{rc}] = -[\mathbf{C}_c]^{-1}[\mathbf{C}_r] \tag{9.1-3}$$

We now write as one relation the identity $\{\mathbf{D}_r\} = \{\mathbf{D}_r\}$ and Eq. 9.1-3:

$$\begin{Bmatrix} \mathbf{D}_r \\ \mathbf{D}_c \end{Bmatrix} = [\mathbf{T}]\{\mathbf{D}_r\}, \qquad \text{where} \quad [\mathbf{T}] = \begin{bmatrix} \mathbf{I} \\ \mathbf{C}_{rc} \end{bmatrix} \tag{9.1-4}$$

With the transformation matrix $[\mathbf{T}]$ now defined, the familiar transformations $\{\mathbf{R}\} = [\mathbf{T}]^T\{\mathbf{R}'\}$ and $[\mathbf{K}] = [\mathbf{T}]^T[\mathbf{K}'][\mathbf{T}]$ of Eqs. 7.4-2 and 7.4-4 can be applied to the structural equations $[\mathbf{K}']\{\mathbf{D}'\} = \{\mathbf{R}'\}$, which are partitioned as

$$\begin{bmatrix} \mathbf{K}_{rr} & \mathbf{K}_{rc} \\ \mathbf{K}_{cr} & \mathbf{K}_{cc} \end{bmatrix} \begin{Bmatrix} \mathbf{D}_r \\ \mathbf{D}_c \end{Bmatrix} = \begin{Bmatrix} \mathbf{R}_r \\ \mathbf{R}_c \end{Bmatrix} \tag{9.1-5}$$

The condensed system is

$$[\mathbf{K}_{rr} + \mathbf{K}_{rc}\mathbf{C}_{rc} + \mathbf{C}_{rc}^T\mathbf{K}_{cr} + \mathbf{C}_{rc}^T\mathbf{K}_{cc}\mathbf{C}_{rc}]\{\mathbf{D}_r\} = \{\mathbf{R}_r + \mathbf{C}_{rc}^T\mathbf{R}_c\} \tag{9.1-6}$$

After Eq. 9.1-6 is solved for $\{\mathbf{D}_r\}$, Eq. 9.1-3 yields $\{\mathbf{D}_c\}$. If $\{\mathbf{Q}\} \neq \{\mathbf{0}\}$ in Eq. 9.1-1, additional terms appear on the right-hand side of Eq. 9.1-6.

If Eq. 9.1-2 simply sets certain d.o.f. $\{\mathbf{D}_c\}$ to zero, then $[\mathbf{C}_r] = [\mathbf{0}]$ and $[\mathbf{C}_c] = [\mathbf{I}]$, hence $[\mathbf{C}_{rc}] = [\mathbf{0}]$, and Eq. 9.1-6 is equivalent to discarding rows and columns associated with $\{\mathbf{D}_c\}$. Otherwise, the choice of which d.o.f. to place in $\{\mathbf{D}_c\}$ is not unique, so the choice of $[\mathbf{C}_c]$ is not unique. One might then define $[\mathbf{C}_c]$ to be the last c linearly independent columns of $[\mathbf{C}]$.

It is possible to avoid the reordering, partitioning, and matrix multiplications implied by Eq. 9.i-6 by applying individual constraint equations serially and retaining all d.o.f. of $\{\mathbf{D}_r\}$ and $\{\mathbf{D}_c\}$ in the transformed equations [6.1]. The transformed coefficient matrix may not be positive definite.

Example. Consider the three-element structure of Fig. 9.1-2. With only axial deformation allowed, and after the support condition $u = 0$ is imposed at $x = 0$, the structural equations are

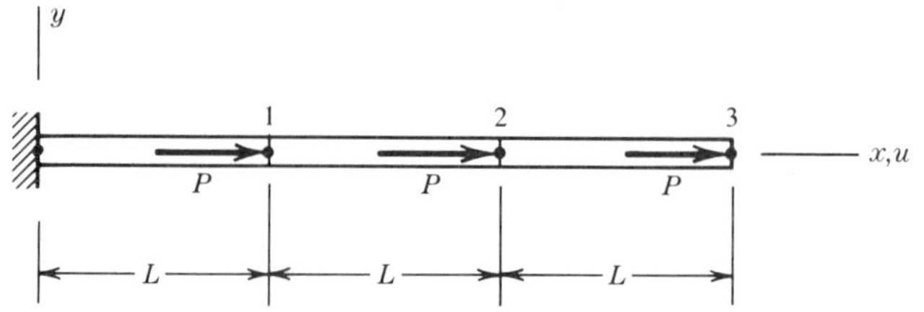

Figure 9.1-2. Three identical bar elements, each of axial stiffness $k = AE/L$.

$$\begin{bmatrix} 2k & -k & 0 \\ -k & 2k & -k \\ 0 & -k & k \end{bmatrix} \begin{Bmatrix} u_1 \\ u_2 \\ u_3 \end{Bmatrix} = \begin{Bmatrix} P \\ P \\ P \end{Bmatrix} \tag{9.1-7}$$

Imagine that the constraint $u_2 = u_3$ is to be imposed. With the choice $D_c = u_3$, Eqs. 9.1-2 and 9.1-3 become

$$[0 \quad 1 \mid -1] \begin{Bmatrix} u_1 \\ u_2 \\ u_3 \end{Bmatrix} = 0 \quad \text{and} \quad [\mathbf{C}_{rc}] = [0 \quad 1] \tag{9.1-8}$$

The transformation matrix of Eq. 9.1-4 and the reduced system of Eq. 9.1-6 are

$$[\mathbf{T}] = \begin{bmatrix} 1 & 0 \\ 0 & 1 \\ 0 & 1 \end{bmatrix} \quad \text{and} \quad \begin{bmatrix} 2k & -k \\ -k & k \end{bmatrix} \begin{Bmatrix} u_1 \\ u_2 \end{Bmatrix} = \begin{Bmatrix} P \\ 2P \end{Bmatrix} \tag{9.1-9}$$

Equation 9.1-9 yields $u_1 = 3P/k$ and $u_2 = 5P/k$. Hence, Eq. 9.1-8 yields $u_3 = 5P/k$.

In Section 8.16, we note that two different nodes can be forced to have the same d.o.f. in $\{\mathbf{D}\}$ by giving them the same node number. (Actual nodal coordinates are still used in the generation of element matrices.) Thus a node whose d.o.f. would all appear in $\{\mathbf{D}_c\}$ can be assigned a node number associated with $\{\mathbf{D}_r\}$ instead of using the transformation, Eq. 9.1-6. Any externally applied loads on d.o.f. $\{\mathbf{D}_c\}$ must be transferred to d.o.f. in $\{\mathbf{D}_r\}$. In applying this method to the foregoing example problem, one assigns the number 2 to the rightmost two nodes. This causes addition of the four coefficients in $[\mathbf{k}]$ of the right element, for a sum of zero at node 3, effectively removing the right element (but not its load) from the structure, and producing Eq. 9.1-9 upon assembly of the remaining two elements.

The condensed system in Eq. 9.1-6 is different from the system obtained by static condensation, Eq. 8.1-3. In Eq. 8.1-3, condensed d.o.f. are related to retained d.o.f. by equilibrium equations already present in the system $[\mathbf{K}]\{\mathbf{D}\} = \{\mathbf{R}\}$. In Eq. 9.1-6, condensed d.o.f. $\{\mathbf{D}_c\}$ are related to retained d.o.f. $\{\mathbf{D}_r\}$ by *supplementary* equations of constraint that replace certain equilibrium equations. Accordingly, constraints may appear to falsify certain equilibrium equations. Figure 9.1-3 is a case in point. The original system, and the system that results from the constraint $v_1 = v_2$, are respectively

$$\begin{bmatrix} k & 0 \\ 0 & k \end{bmatrix} \begin{Bmatrix} v_1 \\ v_2 \end{Bmatrix} = \begin{Bmatrix} P \\ 0 \end{Bmatrix} \quad \text{and} \quad (2k)v_1 = P \tag{9.1-10}$$

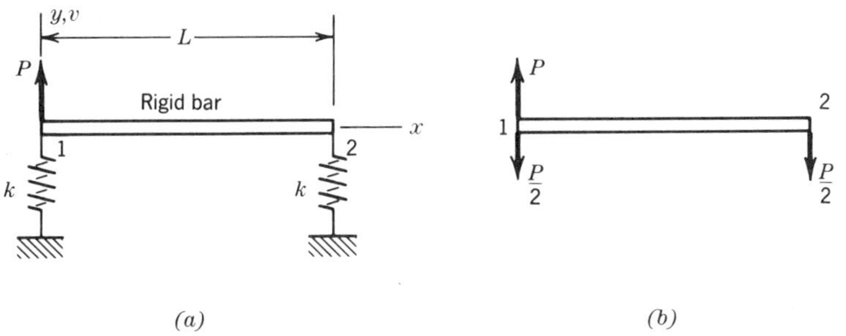

Figure 9.1-3. (*a*) A rigid bar supported by two springs. (*b*) External and elastic forces applied to the bar if the constraint $v_1 = v_2$ is imposed. Forces of constraint are not shown.

Hence, $v_1 = v_2 = P/2k$, and forces carried by the springs are $kv_1 = kv_2 = P/2$. Net forces applied to the bar, Fig. 9.1-3*b*, satisfy equilibrium of *y*-direction forces but not moment equilibrium. Of course, the condensed structure is not that of Fig. 9.1-3*b*; it is a single spring of stiffness $2k$, loaded by force P.

9.2 LAGRANGE MULTIPLIERS

Lagrange's method of undetermined multipliers is used to find the maximum or minimum of a function whose variables are not independent but have some prescribed relation. In structural mechanics the function is potential energy Π_p and the variables are d.o.f. in $\{\mathbf{D}\}$. System unknowns become $\{\mathbf{D}\}$ *and* the Lagrange multipliers.

The theory is easy to describe. We write the constraint equation (Eq. 9.1-1) as the homogeneous equation $[\mathbf{C}]\{\mathbf{D}\} - \{\mathbf{Q}\} = \{\mathbf{0}\}$ and multiply its left-hand side by a row vector $\{\boldsymbol{\lambda}\}^T$ that contains as many Lagrange multipliers λ_i as there are constraint equations. Next we add the result to the potential expression, Eq. 4.1-7:

$$\Pi_p = \frac{1}{2}\{\mathbf{D}\}^T[\mathbf{K}]\{\mathbf{D}\} - \{\mathbf{D}\}^T\{\mathbf{R}\} + \{\boldsymbol{\lambda}\}^T([\mathbf{C}]\{\mathbf{D}\} - \{\mathbf{Q}\}) \tag{9.2-1}$$

The expression in parentheses is zero, so we have added nothing to Π_p. Next we make Π_p stationary by writing the equations $\{\partial\Pi_p/\partial\mathbf{D}\} = \{\mathbf{0}\}$ and $\{\partial\Pi_p/\partial\boldsymbol{\lambda}\} = \{\mathbf{0}\}$, following differentiation rules stated in Appendix A. The result is

$$\begin{bmatrix} \mathbf{K} & \mathbf{C}^T \\ \mathbf{C} & \mathbf{0} \end{bmatrix} \begin{Bmatrix} \mathbf{D} \\ \boldsymbol{\lambda} \end{Bmatrix} = \begin{Bmatrix} \mathbf{R} \\ \mathbf{Q} \end{Bmatrix} \tag{9.2-2}$$

The lower partition of Eqs. 9.2-2 is Eq. 9.1-1, the equation of constraint. Equations 9.2-2 are solved for both $\{\mathbf{D}\}$ and $\{\boldsymbol{\lambda}\}$. The λ_i may be interpreted as forces of constraint (see the following example problem).

Strict partitioning—that is, $\{\mathbf{D}\}$ followed by $\{\boldsymbol{\lambda}\}$ in Eq. 9.2-2—increases bandwidth to the maximum. If instead the D_i and λ_i are interlaced, bandwidth can be much less, although not as small as when the λ_i are absent. However, in a Gauss elimination solution with pivoting on the diagonal, a zero pivot appears if a constraint equation is processed before any of the d.o.f. to which it is coupled. Otherwise, the null submatrix fills in and the solution proceeds normally if the stiffness matrix $[\mathbf{K}]$ is by itself positive definite.

The Lagrange multiplier method is more attractive than the transformation method of Section 9.1 if there are few constraint equations that couple many d.o.f. However, Lagrange multipliers are active at the structure level, but transformation equations can be applied at either the structure level or element by element. The latter has the appeal of disposing of constraints at an early stage, when the matrices are small and more manageable.

Example. Again we solve the example problem of Fig. 9.1-2. The constraint equation is Eq. 9.1-8. Equation 9.2-2 assumes the form

$$\begin{bmatrix} 2k & -k & 0 & 0 \\ -k & 2k & -k & 1 \\ 0 & -k & k & -1 \\ 0 & 1 & -1 & 0 \end{bmatrix} \begin{Bmatrix} u_1 \\ u_2 \\ u_3 \\ \lambda \end{Bmatrix} = \begin{Bmatrix} P \\ P \\ P \\ 0 \end{Bmatrix} \tag{9.2-3}$$

The solution of Eq. 9.2-3 is

$$\lfloor u_1 \quad u_2 \quad u_3 \quad \lambda \rfloor = \left\lfloor \frac{3P}{k} \quad \frac{5P}{k} \quad \frac{5P}{k} \quad -P \right\rfloor \tag{9.2-4}$$

The result $\lambda = -P$ can be regarded as the force of constraint applied through the now rigid link 2–3. The algebraic sign of λ is not significant: had we written $[\mathbf{C}] = [0 \quad -1 \quad 1]$ in Eq. 9.1-8, we would obtain $\lambda = +P$ but the same values of u_1, u_2, and u_3.

9.3 PENALTY FUNCTIONS

If the constraint equation $[\mathbf{C}]\{\mathbf{D}\} = \{\mathbf{Q}\}$, Eq. 9.1-1, is written in the form

$$\{\mathbf{t}\} = [\mathbf{C}]\{\mathbf{D}\} - \{\mathbf{Q}\} \tag{9.3-1}$$

then $\{\mathbf{t}\} = \{\mathbf{0}\}$ implies satisfaction of the constraints. The usual potential Π_p of a structural system can be augmented by a *penalty function* $\{\mathbf{t}\}^T[\boldsymbol{\alpha}]\{\mathbf{t}\}/2$, where $[\boldsymbol{\alpha}]$ is a diagonal matrix of "penalty numbers" α_i. Thus

$$\Pi_p = \tfrac{1}{2}\{\mathbf{D}\}^T[\mathbf{K}]\{\mathbf{D}\} - \{\mathbf{D}\}^T\{\mathbf{R}\} + \tfrac{1}{2}\{\mathbf{t}\}^T[\boldsymbol{\alpha}]\{\mathbf{t}\} \tag{9.3-2}$$

If $\{\mathbf{t}\} = \{\mathbf{0}\}$ the constraints are satisfied and we have added nothing to Π_p. If $\{\mathbf{t}\} \neq \{\mathbf{0}\}$ the penalty of constraint violation becomes more prominent as $[\boldsymbol{\alpha}]$ increases.

Next we substitute Eq. 9.3-1 into Eq. 9.3-2 and write the minimum condition $\{\partial\Pi_p/\partial\mathbf{D}\} = \{\mathbf{0}\}$. Thus, from Eqs. 9.3-1 and 9.3-2,

$$([\mathbf{K}] + [\mathbf{C}]^T[\boldsymbol{\alpha}][\mathbf{C}])\{\mathbf{D}\} = \{\mathbf{R}\} + [\mathbf{C}]^T[\boldsymbol{\alpha}]\{\mathbf{Q}\} \tag{9.3-3}$$

in which $[\mathbf{C}]^T[\boldsymbol{\alpha}][\mathbf{C}]$ can be called the *penalty matrix*. If $[\boldsymbol{\alpha}] = [\mathbf{0}]$, the constraints are ignored. As $[\boldsymbol{\alpha}]$ grows, $\{\mathbf{D}\}$ changes in such a way that the constraint equations are more nearly satisfied. The analyst is responsible for selecting appropriate numerical values of the α_i.

Preferably, for a reason that will subsequently be explained, penalty numbers

α_i are dimensionless. Equations 9.3-1 can easily be written in such a way that the α_i are dimensionless if d.o.f. coupled by the constraint equation are all of the same type, for example, all translations or all rotations. If d.o.f. are different types, some types can be redefined to agree with the others (e.g., $L\theta_i$ can replace θ_i); however, the labor of making such a change may outweigh its benefits.

The method of imposing a prescribed zero or nonzero d.o.f. D_i by adding large numbers to K_{ii} and R_i is a penalty method. For example, in Fig. 2.10-6 the penalty matrix added to [**K**] contains a single coefficient—namely, the large spring stiffness k_s. In this example, α is dimensionless if we define $k_s = \alpha k$ and the constraint as $\sqrt{k}\, v_1 = 0$, where k is a spring stiffness whose magnitude is approximately the same as a typical K_{ij} already present in [**K**].

Example. Imagine that the constraint $u_1 = u_2$ is to be imposed on the structure of Fig. 9.3-1.

There is no unique way to write the constraint relation. We will write [**C**] in such a way that the penalty numbers are dimensionless. Thus for Eq. 9.3-1 we elect to write

$$[\mathbf{C}] = \lfloor \sqrt{k} \quad -\sqrt{k} \rfloor \qquad \{\mathbf{D}\} = \begin{Bmatrix} u_1 \\ u_2 \end{Bmatrix} \qquad \{\mathbf{Q}\} = \{\mathbf{0}\} \tag{9.3-4}$$

where $k = AE/L$. With but one constraint, $\lfloor \boldsymbol{\alpha} \rfloor = \alpha$, a scalar. Equation 9.3-3 becomes

$$\left(\begin{bmatrix} 2k & -k \\ -k & k \end{bmatrix} + \alpha \begin{bmatrix} k & -k \\ -k & k \end{bmatrix} \right) \begin{Bmatrix} u_1 \\ u_2 \end{Bmatrix} = \begin{Bmatrix} P \\ P \end{Bmatrix} \tag{9.3-5}$$

which has the solution

$$u_1 = \frac{2P}{k} \qquad \text{and} \qquad u_2 = \frac{3 + 2\alpha}{1 + \alpha} \frac{P}{k} \tag{9.3-6}$$

If $\alpha = 0$, then $u_2 = 3P/k$, as expected. As α becomes large, u_2 approaches the value $2P/k$, which is correct for the constrained system. Note that $u_1 - u_2 = -P/k(1 + \alpha)$ and $\{\mathbf{t}\} = t = \sqrt{k}\,(u_1 - u_2)$, so that the coefficient of α in the penalty function approaches zero as α approaches infinity.

The symbolic manipulations that produce Eq. 9.3-6 from Eq. 9.3-5 obscure a difficulty that may arise if the manipulations are done numerically. The second square matrix in Eq. 9.3-5 is recognized as the stiffness of a bar element that spans nodes 1 and 2 (i.e., a "constraint" bar in parallel with the bar already there).

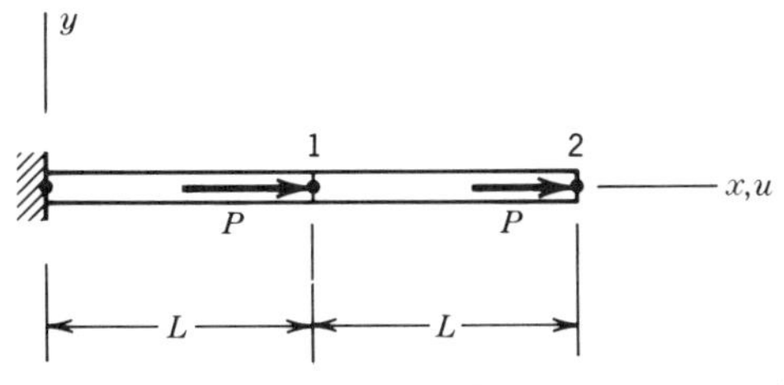

Figure 9.3-1. Two identical bar elements, each of axial stiffness $k = AE/L$.

As α grows, the structure becomes the error-prone case of a stiff region supported by a flexible region (see Section 18.2).

If constraints do not couple all d.o.f. in $\{\mathbf{D}\}$, then $[\mathbf{C}]$ has more columns than rows, and $[\mathbf{C}]^T[\boldsymbol{\alpha}][\mathbf{C}]$ is certain to be a singular matrix. In some important problems, discussed in Section 9.4, constraints *do* couple all d.o.f. in $\{\mathbf{D}\}$, and singularity of $[\mathbf{C}]^T[\boldsymbol{\alpha}][\mathbf{C}]$ is not guaranteed. However, we *want* this matrix to be singular, as the following argument illustrates.

For simplicity let all α_i in $[\boldsymbol{\alpha}]$ be the same number, say α. In addition, let $\{\mathbf{Q}\} = \{\mathbf{0}\}$ in Eq. 9.3-3. Then, as α becomes large, Eq. 9.3-3 becomes

$$[\mathbf{C}]^T[\mathbf{C}]\{\mathbf{D}\} \approx \frac{1}{\alpha}\{\mathbf{R}\} \tag{9.3-7}$$

Equation 9.3-7 shows that if $[\mathbf{C}]^T[\mathbf{C}]$ is nonsingular, then as α grows the solution vector $\{\mathbf{D}\}$ approaches zero. In other words, the mesh "locks." Only if $[\mathbf{C}]^T[\mathbf{C}]$ is singular can $\{\mathbf{D}\}$ be nonzero. Then the number of independent nonzero $\{\mathbf{D}\}$'s that satisfy Eq. 9.3-7 is equal to the difference between the order of $[\mathbf{C}]^T[\mathbf{C}]$ and its rank. The practical significance of this argument is discussed in subsequent sections.

In comparison with Lagrange multipliers, penalty functions have the advantage of introducing no new variables. However, the penalty matrix may significantly increase the bandwidth (or wave front) of the structural equations, depending on how d.o.f. are numbered and what d.o.f. are coupled by the constraint equation. Implementation of a penalty function can be as easy as assigning a high modulus to an element already in the structure. Penalty functions have the disadvantage that penalty numbers must be chosen in an allowable range: large enough to be effective but not so large as to provoke numerical difficulties [9.2,9.3].

9.4 NATURALLY ARISING PENALTY FORMULATIONS. NUMERICAL INTEGRATION AND CONSTRAINTS

In Section 9.3, constraints are imposed by explicitly adding a penalty matrix to an existing stiffness matrix $[\mathbf{K}]$. In some situations $[\mathbf{K}]$ *already contains* a contribution that can be identified as a penalty matrix. Thus some applications of penalty methods arise *naturally* in the sense that large stiffnesses with respect to particular deformations can be interpreted as penalty numbers. Here we discuss two of these applications—transverse shear in beams and material incompressibility—which can easily lead to "locking" difficulties unless care is taken in the choice of element type and integration rule. Physical interpretation of the constraints makes it possible to understand the reasons for locking and leads to guidelines that can be used to avoid the difficulty. The guideline presented subsequently is called "constraint counting." It does not provide a rigorous guarantee of success but can be quite effective in practice. In the present section we identify the constraints precisely, and in the next section we show how to count them.

Mindlin Beam Element. The beam element shown in Fig. 9.4-1*a* allows transverse shear deformation. The element has four d.o.f., as is usual. However, rotational d.o.f. θ_1 and θ_2 are *not* values of dw/dx at the nodes, as is the case with the standard beam element of Eq. 4.2-5. A plane initially normal to the midsurface

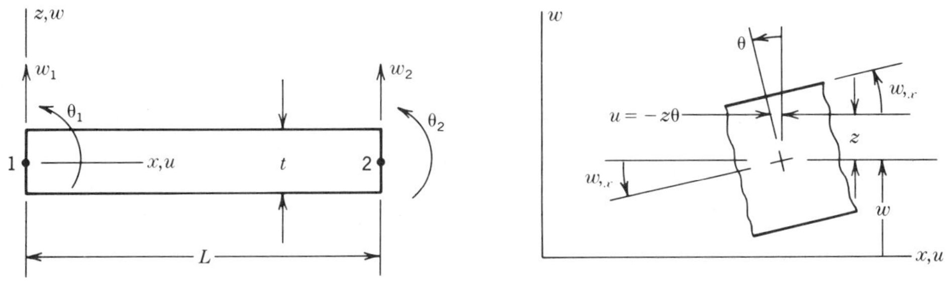

(*a*) (*b*)

Figure 9.4-1. (*a*) A "Mindlin" beam element. (*b*) Displacements and rotations.

remains plane but not necessarily normal. In Fig. 9.4-1*a*, θ represents the rotation of a line that was initially normal to the undeformed longitudinal axis of the beam. We will interpolate θ and w *independently,* so that the beam can represent transverse shear deformation. This element is called a Mindlin beam element [9.4]. It can model constant bending moment but not linearly varying bending moment. The Mindlin beam element generalizes to *Mindlin plate elements,* discussed in Section 11.3. The standard beam element, also known as an *Euler beam element,* can model linearly varying bending moment but does not account for transverse shear deformation.[1]

Axial normal strain ϵ_x and transverse shear strain γ_{zx} in a Mindlin beam element are, from Fig. 9.4-1*b*,

$$\epsilon_x = u_{,x} = -z\theta_{,x} \qquad \text{and} \qquad \gamma_{zx} = w_{,x} - \theta \tag{9.4-1}$$

where θ is a small angle of rotation. Thus γ_{zx} is taken as constant over the depth t. Nonzero stresses are assumed to consist only of axial normal stress σ_x and transverse shear stress τ_{zx}. Accordingly, strain energy in the Mindlin beam element is $U = U_b + U_s$, where U_b and U_s are strain energies of bending and shear, respectively:

$$U_b = \frac{1}{2}\int_{V_e} \frac{\sigma_x^2}{E}\, dV = \frac{1}{2}\int_{V_e} E\epsilon_x^2\, dV = \frac{1}{2}\frac{Ebt^3}{12}\int_0^L \theta_{,x}^2\, dx \tag{9.4-2a}$$

$$U_s = \frac{1}{2}\int_{V_e} \frac{\tau_{zx}^2}{G}\, dV = \frac{1}{2}\int_{V_e} G\gamma_{zx}^2\, dV = \frac{1}{2}\frac{Gbt}{1.2}\int_0^L (w_{,x} - \theta)^2\, dx \tag{9.4-2b}$$

Here E = elastic modulus, G = shear modulus, b is the width of the beam, and 1.2 is the "form factor" that accounts for a parabolic distribution of τ_{zx} over a rectangular cross section.[2]

[1]The standard beam element, Eq. 4.2-5, can be modified to account for transverse shear deformation, while retaining the ability to model both constant and linearly varying bending moment [4.2,4.11].

[2]Let a beam have a rectangular cross section of dimensions b by t. If P is the transverse shear force, then $\tau_{zx} = (3P/2bt^3)(t^2 - 4z^2)$, where $z = 0$ at the neutral axis. Equation 9.4-2b then yields $U_s = 1.2(P^2L/btG)/2$. This result suggests the view that a uniform stress $\tau_{zx} = P/bt$ acts over a modified area $A = bt/1.2$, so that the same U_s results. A uniform stress $\tau_{zx} = G\gamma_{zx}$ is provided by Eq. 9.4-1.

Displacement w and rotation θ are interpolated linearly between nodes:

$$w = \frac{L - x}{L} w_1 + \frac{x}{L} w_2 \qquad \text{and} \qquad \theta = \frac{L - x}{L} \theta_1 + \frac{x}{L} \theta_2 \tag{9.4-3}$$

In the usual way, from Eqs. 9.4-2 and 9.4-3 we obtain a 4 by 4 element stiffness matrix $[\mathbf{k}] = [\mathbf{k}_b] + [\mathbf{k}_s]$, where $[\mathbf{k}_b]$ resists bending strain ϵ_x and $[\mathbf{k}_s]$ resists shear strain γ_{zx}. With this notation, for a structure,

$$([\mathbf{K}_b] + [\mathbf{K}_s])\{\mathbf{D}\} = \{\mathbf{R}\} \tag{9.4-4}$$

D.o.f. w_i and θ_i in $\{\mathbf{D}\}$ are coupled by terms in $[\mathbf{K}_s]$ but not by terms in $[\mathbf{K}_b]$.

Deflections $\{\mathbf{D}\}$ of a *thin* beam should be governed by only $[\mathbf{K}_b]$ because transverse shear deformation is negligible. In other words, if $Gbt/1.2$ becomes much larger than $Ebt^3/12$ in Eq. 9.4-2, then $[\mathbf{K}_s]$, which arises from U_s, should enforce the constraint $\gamma_{zx} = 0$. But, as a beam becomes slender, $[\mathbf{K}_s]$ grows in relation to $[\mathbf{K}_b]$. So $[\mathbf{K}_s]$ acts as a penalty matrix that causes Eq. 9.4-4 to yield $\{\mathbf{D}\} = \{\mathbf{0}\}$—*unless* $[\mathbf{K}_s]$ *is singular*. In other words, unless $[\mathbf{K}_s]$ is singular, the computed deflection of a very slender beam is almost zero. A singular $[\mathbf{K}_s]$ can enforce the $\gamma_{zx} = 0$ constraint without locking. As discussed subsequently, a singular $[\mathbf{K}_s]$ can be achieved by reduced integration.

As a simple example of locking, let the beam element of Fig. 9.4-1*a* be fixed at the left end and loaded by a transverse force P at the right end. Then $w = w_2 x/L$, $\theta = \theta_2 x/L$, and $\Pi_p = U - Pw_2$. For simplicity let $\nu = 0$, so that $E = 2G$. Then the equilibrium equations $\partial\Pi_p/\partial w_2 = 0$ and $\partial\Pi_p/\partial\theta_2 = 0$ yield

$$w_2 = \frac{12(t/L)^2 + 20}{12(t/L)^2 + 5}\left(1.2\frac{PL}{GA}\right) \tag{9.4-5}$$

where $A = bt$. For a very short beam, we obtain $w_2 \approx 1.2PL/GA$, which is the correct expression for the portion of end deflection that is due to transverse shear deformation. For a slender beam, $L >> t$, we obtain $w_2 \approx 4.8PL/GA$. This value includes no bending deformation and is far too small; that is, the beam locks as L/t increases.

Incompressible Materials. As Poisson's ratio ν approaches 0.5, a material becomes incompressible. Values of ν near 0.5 occur in rubberlike materials and in materials that flow, such as fluids and plastic solids. Unless the problem is one of plane stress, the value $\nu = 0.5$ is forbidden because denominators become zero in material property matrices $[\mathbf{E}]$ (Eqs. 1.7-3 and 1.7-4, for example). It is tempting to approximate incompressibility by using (say) $\nu = 0.49$. But, near $\nu = 0.5$, stresses are strongly dependent on ν. Also, structural equations become ill conditioned as ν approaches 0.5, for reasons explained next.

The shear modulus G and bulk modulus B of an isotropic material are

$$G = \frac{E}{2(1 + \nu)} \qquad B = \frac{E}{3(1 - 2\nu)} \tag{9.4-6}$$

In terms of G and B, the material property matrix $[\mathbf{E}]$ of Eq. 1.7-3 is

$$[\mathbf{E}] = G\begin{bmatrix} 4/3 & -2/3 & -2/3 & 0 & 0 & 0 \\ -2/3 & 4/3 & -2/3 & 0 & 0 & 0 \\ -2/3 & -2/3 & 4/3 & 0 & 0 & 0 \\ 0 & 0 & 0 & 1 & 0 & 0 \\ 0 & 0 & 0 & 0 & 1 & 0 \\ 0 & 0 & 0 & 0 & 0 & 1 \end{bmatrix} + B\begin{bmatrix} 1 & 1 & 1 & 0 & 0 & 0 \\ 1 & 1 & 1 & 0 & 0 & 0 \\ 1 & 1 & 1 & 0 & 0 & 0 \\ 0 & 0 & 0 & 0 & 0 & 0 \\ 0 & 0 & 0 & 0 & 0 & 0 \\ 0 & 0 & 0 & 0 & 0 & 0 \end{bmatrix} \tag{9.4-7}$$

or, abbreviated, $[\mathbf{E}] = G[\mathbf{E}_G] + B[\mathbf{E}_B]$. The element stiffness matrix (Eq. 4.1-5) becomes

$$[\mathbf{k}] = G\int_{V_e}[\mathbf{B}]^T[\mathbf{E}_G][\mathbf{B}]\,dV + B\int_{V_e}[\mathbf{B}]^T[\mathbf{E}_B][\mathbf{B}]\,dV \tag{9.4-8}$$

Therefore, structural equations have the form

$$(G[\mathbf{K}_G] + B[\mathbf{K}_B])\{\mathbf{D}\} = \{\mathbf{R}\} \tag{9.4-9}$$

As ν approaches 0.5, bulk modulus B approaches infinity. Therefore, $B[\mathbf{K}_B]$ acts as a penalty matrix that enforces the constraint of incompressibility. As ν approaches 0.5, numerical trouble becomes more likely, and finally the mesh "locks"—*unless* $[\mathbf{K}_B]$ *is singular.*

Remarks. In the notation of Eq. 9.3-3, Eqs. 9.4-4 and 9.4-9 correspond to *homogeneous* constraints, $\{\mathbf{Q}\} = \{\mathbf{0}\}$. Therefore, Eqs. 9.4-4 and 9.4-9 arise naturally from minimization of a potential Π_p that can be stated in the form

$$\Pi_p = \beta\int_V (F + \alpha H)\,dV + P \tag{9.4-10}$$

Here F and H are proportional to strain energy densities, β is a common factor of F and H, α is the penalty number, and P represents work done by loads of all types (P need not be detailed here). There is no unique way to choose β, but if possible it should be chosen in such a way that α is *dimensionless*. A dimensionless α makes it easier to select a numerical value of α such that ill-conditioning is avoided (see guidelines at the end of this section).

For the Mindlin beam, the correspondence between Eqs. 9.4-2 and 9.4-10 can be as follows. Replace dV by dx in Eq. 9.4-10, and let

$$\begin{aligned} F &= \frac{E}{2}\,\theta_{,x}^2 \qquad & H &= \frac{E}{2L_T^2}(w_{,x} - \theta)^2 \\ \beta &= \frac{bt^3}{12} \qquad & \alpha &= \frac{10\,L_T^2 G}{t^2 E} = \frac{5}{1+\nu}\left(\frac{L_T}{t}\right)^2 \end{aligned} \tag{9.4-11}$$

where t is the beam depth (thickness) and $L_T = \Sigma\, L_i$ is the total length of the beam after elements of length L_i have been assembled.[3] We know that beam deflections depend more strongly on L_T and t than on the beam width b. Accordingly, b is excluded from F, H, and α, so that the effect of the L_T/t ratio is apparent.

[3]If t is not constant throughout the beam, one can use a typical value of t, because a practical beam will not display wild variations of t.

Since the factor $5/(1 + \nu)$ varies little for valid choices of ν, we see that α depends strongly on $(L_T/t)^2$. When $(L_T/t)^2$ is large, the penalty number α is large. (Note that when L_T/t approaches infinity so does L_i/t, so the foregoing arguments could be restated using element length rather than L_T.)

The incompressible case, Eq. 9.4-9, can be obtained from Eq. 9.4-10 if

$$F = \frac{E}{4(1 + \nu)} \{\epsilon\}^T[\mathbf{E}_G]\{\epsilon\} \qquad H = \frac{E}{2} \{\epsilon\}^T[\mathbf{E}_B]\{\epsilon\}$$
$$\beta = 1 \qquad \alpha = \frac{1}{3(1 - 2\nu)} \tag{9.4-12}$$

The penalty number α becomes large as ν approaches 0.5.

Constraints and Quadrature Points. We wish to show that the number of penalty function constraints is proportional to the number of sampling points used to integrate element matrices.

Numerical integration of stiffness matrices (such as $[\mathbf{K}_b]$, $[\mathbf{K}_s]$, $[\mathbf{K}_G]$, and $[\mathbf{K}_B]$ in Eqs. 9.4-4 and 9.4-9) corresponds to evaluation of energy Π_p by numerical integration. Thus Eq. 9.4-10 can be written as

$$\Pi_p \approx \beta \sum_{I=1}^{\text{numel}} \left[\sum_{i=1}^{n} (F_I J_I)_i W_i + \alpha \sum_{j=1}^{m} (H_I J_I)_j T_j \right] + P \tag{9.4-13}$$

where $(F_I J_I)_i$ and $(H_I J_I)_j$ are values of functions F and H times the Jacobian J, evaluated at the ith or jth sampling point, and W_i and T_j are positive weights (or weight products) appropriate to the integration rule. If $n = m$ and sampling points i and j are the same, the integration scheme is *uniform;* otherwise it is *selective*. The integration scheme is called *full* if enough sampling points are used to provide exact integration of all stiffness coefficients of an undistorted element (e.g., a rectangular element). If fewer sampling points are used—that is, if either n or m is reduced—the integration scheme is called *reduced* and Π_p is said to be *under-integrated.* In what follows we assume that kinematic modes of deformation are either impossible or are suppressed by boundary conditions (kinematic modes are discussed in Section 6.12).

Now consider the effect of letting α become very large in Eq. 9.4-10. If the correct Π_p is to be closely approximated, a large α must be associated with a zero or near-zero value of the second summation in Eq. 9.4-13. Since $(J_I)_j > 0$ and $T_j > 0$, the desired condition is that $(H_I)_j = 0$. From equations such as Eqs. 9.4-11 and 9.4-12, we see that $H = 0$ implies satisfaction of the constraint (zero shear strain and zero volume change in the respective cases). Thus, in these examples, *each integration point used to evaluate the penalty matrix imposes a constraint,* and the total number of constraints in the structure is the number of elements times the number of penalty integration points per element.

The preceding discussion suggests that locking difficulties may be avoided by applying reduced integration to terms that yield a penalty matrix. Thus, for example, we would use a one-point rule to evaluate U_s of Eq. 9.4-2b when the shape functions of Eq. 9.4-3 are used (for which two-point integration would be exact). Further discussion and other examples appear in Section 9.5.

In the foregoing examples, material property matrices used in generating the penalty matrix have rank one (E in Eq. 9.4-11, $[\mathbf{E}_B]$ in Eq. 9.4-12). In other problems, the analogous matrices may have rank greater than one. Then the number of penalty constraints per element may be greater than the number of integration points [9.8]. This happens in certain plate-bending formulations, where an integration point constrains *two* transverse shear strains, namely γ_{yz} and γ_{zx} (see Section 11.3). In some problems it is conceivable that the number of constraints will exceed the number of d.o.f. This circumstance implies that some of the constraints are redundant.

Guideline for Choice of α. If computer words carry approximately p decimal digits, experience has shown that α should not exceed $10^{p/2}$ if ill-conditioning and numerical difficulty are to be avoided. If this guideline is followed, coefficients of $[\mathbf{K}]$ in Eq. 9.3-3 influence the latter $p/2$ digits in computer words used to store the complete matrix $[\mathbf{K} + \mathbf{C}^T\boldsymbol{\alpha}\mathbf{C}]$. Typically $10^{p/2}$ is 10^3 to 10^4 in single precision and 10^6 to 10^7 in double precision. If material properties yield a larger value of α, it is best to lower L_T/t (Eq. 9.4-11) or ν (Eq. 9.4-12) artificially so that the guideline is satisfied.

The foregoing choice of α is made *after* one has chosen an integration rule that avoids locking of the mesh. Even without locking, the penalty matrix enforces constraints, and it is to avoid numerical difficulty associated with these remaining constraints that one takes care in the choice of α.

9.5 CONSTRAINT COUNTING

We seek a guideline for choosing a suitable numerical integration formula in problems where penalty constraints arise naturally. Specifically, if the number of constraints is proportional to the number of sampling points used to integrate the penalty matrix, how many points per element should be used?

In what follows we continue to assume that all weight factors in quadrature rules are positive. Otherwise, it is conceivable that terms of a summation will cancel one another. This would confuse the counting rule.

Mesh Locking. Let the beam in Fig. 9.5-1a be built of the shear-flexible beam elements of Fig. 9.4-1. Support conditions suppress w and θ d.o.f. at the fixed

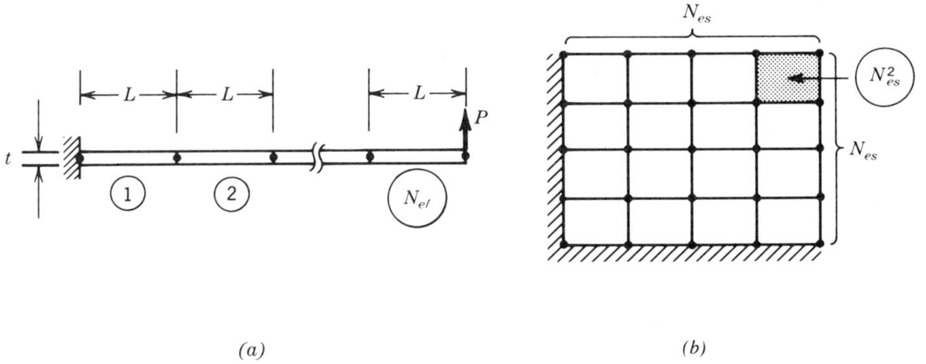

Figure 9.5-1. (a) Cantilever beam built of $N_{e\ell}$ elements. (b) An N_{es} by N_{es} mesh of bilinear elements. The case $N_{es} = 4$ is shown.

end, leaving $2N_{e\ell}$ active d.o.f. in $\{\mathbf{D}\}$. Now imagine that $[\mathbf{K}_s]$ of Eq. 9.4-4 is generated by use of two sampling points per element (thus, $[\mathbf{K}_s]$ is integrated exactly). Then, if the beam is slender, there are $2N_{e\ell}$ penalty constraints. All d.o.f. $\{\mathbf{D}\}$ are now used to satisfy the constraint $w_{,x} - \theta \approx 0$, and the computed deflection of load P is nearly zero. This conclusion is unchanged by changing the supports. With *no* supports, rigid-body motion of the "locked" beam becomes possible.

The same situation prevails if the beam is built of bilinear plane elements (four nodes, eight d.o.f., as in Table 6.14-1) integrated with a 2 by 2 Gauss rule. Now the beam contains $4N_{e\ell}$ active d.o.f. However, owing to the "parasitic shear" discussed in Section 8.3, there are $4N_{e\ell}$ constraints if the beam is thin. Again the mesh locks as the length-to-thickness ratio becomes large.

In Fig. 9.5-1*b* there are two d.o.f. per node and therefore $2N_{es}^2$ active d.o.f. Let the material be nearly incompressible; that is,

$$\epsilon_V = u_{,x} + v_{,y} + w_{,z} \approx 0 \tag{9.5-1}$$

where ϵ_V is the volumetric strain. If the plane strain condition $w_{,z} = 0$ prevails, then the penalty constraint $u_{,x} + v_{,y} \approx 0$ is enforced at each integration point. With a 2 by 2 integration rule there are $4N_{es}^2$ constraints, and the mesh is locked.

In the foregoing examples, the ratio of number of d.o.f. to number of penalty constraints is 1/1 in Fig. 9.5-1*a* and 1/2 in Fig. 9.5-1*b*. Note that these same ratios can be determined from a single element. Consider the addition to the mesh of a single element, such as the element shaded in Fig. 9.5-1*b*. It brings two additional d.o.f. to the mesh, and under a 2 by 2 Gauss rule it also brings four additional constraints, for a d.o.f.-to-constraint ratio of 1/2, as previously determined. For the mesh as a whole, the 1/2 ratio would still be approximately correct if support conditions are changed, provided that N_{es} is large. Accordingly, we will henceforth presume that meshes contain many elements, and do our "constraint counting" by examining the *additional* d.o.f. and constraints that are brought to a mesh by adding a single element (or perhaps a single "macroelement" built of subelements).

Desirable Constraint Ratios. We define the constraint ratio r as the ratio of the number of active d.o.f. in $\{\mathbf{D}\}$ to the number of penalty constraints. Locking occurs if $r \leq 1$. If r is *slightly* greater than unity, the mesh does not lock, but poor results are likely because most d.o.f. are occupied in satisfying penalty constraints; few d.o.f. are left to model the elastic behavior of the system. Extensive numerical testing has shown that near-optimal constraint ratios are $r = 2/1$ for two-dimensional problems and $r = 3/1$ for three-dimensional problems. In each case these ratios correspond to the number of differential equations of equilibrium (two and three for plane and solid problems, respectively) divided by the number of constraint conditions on the system of governing differential equations (one constraint; Eq. 9.5-1 for incompressibility or $\gamma_{zx} = 0$ for Mindlin beams).

A favorable constraint ratio is usually achieved by underintegration or by *selective reduced integration*—that is, by use of $m < n$ in Eq. 9.4-13. In the context of Eq. 9.4-9 this means using a lower-order Gauss quadrature rule for $[\mathbf{K}_B]$ than for $[\mathbf{K}_G]$.[4]

[4]If quadrature points are unsymmetrically distributed, such an element may not be geometrically isotropic. This should be of no consequence, because lack of geometric isotropy is annoying only in a coarse mesh, and a rather fine mesh is needed to produce accurate results if the material is nearly incompressible.

Further Examples. With the Mindlin beam element of Fig. 9.4-1, it suffices to use uniform integration with $m = n = 1$—that is, a single Gauss point for both $[\mathbf{k}_b]$ and $[\mathbf{k}_s]$. This integration is exact for $[\mathbf{k}_b]$ because $\theta_{,x}$ is not a function of x. However, $w_{,x} - \theta$ is linear in x, so $[\mathbf{k}_s]$ is underintegrated by a single Gauss point. Thus two d.o.f. and one constraint are added to the mesh by each element, for an ideal constraint ratio of $r = 2/1$. As an exercise, one can show that in place of Eq. 9.4-5, one now obtains a far more accurate result for a tip-loaded cantilever element. Additional discussion of transverse shear constraints can be found in [9.7,9.8] and in Section 11.3.

Figure 9.5-2 shows elements that might be used for nearly incompressible media (Eq. 9.4-9). Plane and solid elements have respectively two and three d.o.f. per node. For the bilinear and trilinear elements, it is appropriate to use one-point integration to obtain $[\mathbf{K}_B]$: thus, for Figs. 9.5-2*a* and 9.5-2*b*, we obtain $r = 2/1$ and $r = 3/1$, respectively, which are the optimal ratios. In Fig. 9.5-2*c*, depending on whether the element has eight or nine nodes, uniform 3 by 3 integration gives $r = 6/9$ or $r = 8/9$, respectively, and selective reduced integration with a 2 by 2 rule for the penalty matrix gives $r = 6/4$ or $r = 8/4$. The latter is optimal.

For convenience, triangles in Fig. 9.5-2*d* are considered as a two-element patch that brings eight d.o.f. to a mesh. No integration rule for the penalty matrix is entirely satisfactory: three points per triangle gives $r = 8/6$ (too low), and one point per triangle gives $r = 8/2$ (too high). Indeed, were the triangles to have vertex nodes only, even one point per triangle would be too high: thus $r = 2/2$, which means that a mesh would lock. These difficulties with triangles for plane problems suggest that tetrahedra for solid problems would not work well. However, the question of what approach is best for solid problems is not yet settled.

In distorted meshes, one may find that reduced integration is not adequate to represent element volume exactly. This difficulty disappears with mesh refinement if subdivision causes elements to become parallelograms or parallelepipeds. The difficulty can be avoided by use of the *consistent penalty method,* which has additional attributes to recommend it (see Section 9.6).

9.6 ADDITIONAL TECHNIQUES FOR INCOMPRESSIBLE MEDIA

Deviatoric–Dilatational Splitting. It is convenient to regard stresses $\{\boldsymbol{\sigma}\}$ as being composed of a *deviatoric* state $\{\boldsymbol{\sigma}_D\}$ (which produces no change of volume) and a *dilatational* state $\{\boldsymbol{\sigma}_V\}$ (which produces no change of shape). Thus, from Eq. 9.4-7,

$$\{\boldsymbol{\sigma}\} = \{\boldsymbol{\sigma}_D\} + \{\boldsymbol{\sigma}_V\} = G[\mathbf{E}_G]\{\boldsymbol{\epsilon}\} + B[\mathbf{E}_B]\{\boldsymbol{\epsilon}\} \tag{9.6-1}$$

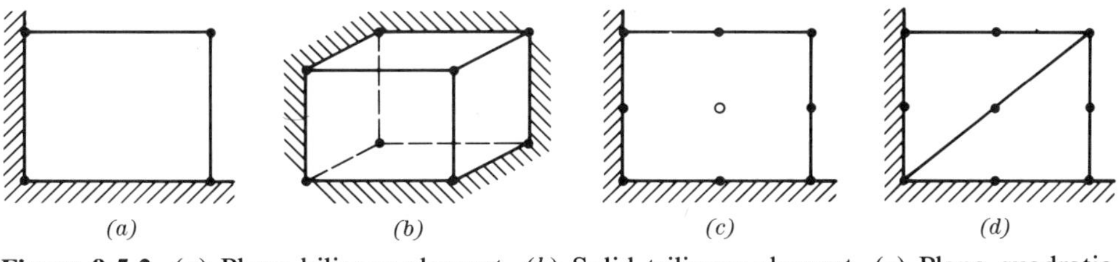

Figure 9.5-2. (*a*) Plane bilinear element. (*b*) Solid trilinear element. (*c*) Plane quadratic element. (*d*) Two plane triangular elements.

We discover that $\{\boldsymbol{\sigma}_V\}$ contains three equal normal stresses and no shear stresses. Specifically, in terms of the volumetric strain ϵ_V stated in Eq. 9.5-1, nonzero stresses in $\{\boldsymbol{\sigma}_V\}$ are

$$\sigma_{xV} = \sigma_{yV} = \sigma_{zV} = \lambda, \qquad \text{where} \quad \lambda = B\epsilon_V \tag{9.6-2}$$

in which λ is called the *hydrostatic pressure function* and B is the bulk modulus, defined in Eq. 9.4-6. Equations 9.6-1 and 9.6-2 yield $\lambda = (\sigma_x + \sigma_y + \sigma_z)/3$. (In the theory of plasticity, where $\{\boldsymbol{\sigma}_D\}$ and $\{\boldsymbol{\sigma}_V\}$ are also used, λ is called the mean stress.) With λ from Eq. 9.6-2, Eq. 9.6-1 assumes the form

$$\{\boldsymbol{\sigma}\} = \{\boldsymbol{\sigma}_D\} + \lambda\lfloor 1 \quad 1 \quad 1 \quad 0 \quad 0 \quad 0\rfloor^T \tag{9.6-3}$$

where $\{\boldsymbol{\sigma}_D\} = G[\mathbf{E}_G]\{\boldsymbol{\epsilon}\}$ and $G[\mathbf{E}_G]$ is stated in Eq. 9.4-7. For a completely incompressible material λ is a "volumetric stress," which can be regarded as a system of pressures that keeps the body from dilatating.

One may argue that if Poisson's ratio approaches 0.5 while all other parameters of the problem are held fixed, the incompressible pressure field is, at each material point, the limit of the slightly compressible pressure λ computed from Eq. 9.6-2 [9.9]. The penalty method provides a way to "perturb" the exactly incompressible solution slightly and thus obtain λ as a good approximation of the exactly incompressible pressure.

Pressure Calculation in the Penalty Method. In the penalty method, one can calculate pressure λ by evaluating ϵ_V from the displacement field and then using Eq. 9.6-2. The error in λ associated with use of a penalty constraint rather than an exact constraint is of order $10^{-p/2}$ if α is chosen as $10^{p/2}$, as suggested at the end of Section 9.4.[5] However, this error estimate is valid only if λ is calculated at locations where the constraint is enforced—that is, at the Gauss points used to evaluate element matrices $[\mathbf{k}_B]$. For the penalty method, then, the (reduced) volumetric integration points of Eq. 9.4-13 play a three-part role: that of "pressure points," that of constraint points, and that of volumetric integration points.

If pressures are desired at other points, techniques of Section 6.13 can be used to extrapolate pressures to displacement nodes or to any other points in an element. Expressions analogous to Eq. 6.13-5 can be developed for any of the elements discussed in Section 9.5. However, for many of those elements, pressures are susceptible to an instability akin to the hourglassing discussed in Section 6.12. Unfortunately, this is particularly true of elements such as the bilinear and quadratic elements of Fig. 9.5-2 when the standard one-point or four-point reduced formulas are selectively applied to the volumetric terms of Eq. 9.4-13. One way to avoid this is by postprocessing the computed pressures using an "averaging" or "smoothing" scheme to smooth out the penalty pressures. This can be rigorously justified in terms of error analysis [9.11-9.13], and is not difficult to implement [9.7,9.14].

[5]The value $\alpha \approx 10^{p/3}$ can be rigorously justified, but in practice $10^{p/3}$ appears to be sufficiently pessimistic that a value of α approaching $10^{p/2}$ can usually be used.

Consistent Penalty Method. The consistent penalty method provides an alternative way to calculate $[\mathbf{K}_B]$. Reduced integration is not required, and incompressibility constraints are imposed at certain "pressure points" rather than at integration points. Thus, as compared with the preceding penalty method, constraint points are divorced from integration points, and the choice of integration rule for $[\mathbf{K}_B]$ is not dictated by constraint counting. Constraint counting is still used, but now to achieve a balance between the number of pressure points and the number of displacement d.o.f. If pressure points are well chosen, the aforementioned pressure instabilities will not arise. A full description of the method is beyond the scope of this book: it is a special case of the "$[\overline{\mathbf{B}}]$ method" of [9.7], and details appear in [9.10]. The following summary is offered.

Let λ_i represent hydrostatic pressure at the pressure points, such as the three shown in Fig. 9.6-1. Pressure λ over the element is interpolated from these λ_i. Thus λ is interpolated independently of d.o.f. at element nodes. The formulation procedure leads to equations like Eq. 9.2-2, except that (a) $\{\mathbf{Q}\} = \{\mathbf{0}\}$, and (b) the lower-right submatrix $[\mathbf{0}]$ is replaced by a square nonsingular matrix that represents slight compressibility.

An important feature of pressure points is that they are not shared by adjacent elements, even if pressure points are placed on element boundaries. Thus the pressure points λ_i are internal d.o.f. that can be eliminated by condensation before elements are assembled (see Eq. 8.1-1, and let $\{\mathbf{d}_c\} = \{\boldsymbol{\lambda}\}$). The result of condensation and assembly is a set of equations like Eq. 9.4-9, obtained with computational efficiency comparable to that of the reduced-integration penalty method, and with an improved $[\mathbf{K}_B]$.

The nine-node, three-pressure-point element of Fig. 9.6-1, in consistent penalty form and with $\alpha = 10^{p/2}$, is the best element known for two-dimensional incompressible elasticity and fluid flow. That the constraint ratio is $r = 8/3$, rather than the optimal $r = 2/1$, evidently causes no ill effects. There are rigorous error bounds for the element, showing that the pressure is as accurate as strains in compressible elasticity [9.10]. There are no difficulties with spurious pressure modes.

In three-dimensional problems, the situation is not nearly as clear. There appears to be no three-dimensional analogue of the two-dimensional quadratic element with three pressure points. At present, we recommend the eight-node brick of Fig. 9.5-2*b* with one pressure point, using the consistent formulation or the selective/reduced formulation (if the elements are not severely distorted by isoparametric transformations). For undistorted elements these two formulations are identical [9.7,9.10]. Unfortunately, this element has spurious modes, and pressure smoothing is required. We recommend the pressure-smoothing scheme described in [9.14]. Development of good elements for incompressible media in three dimensions is an active area of current finite element research.

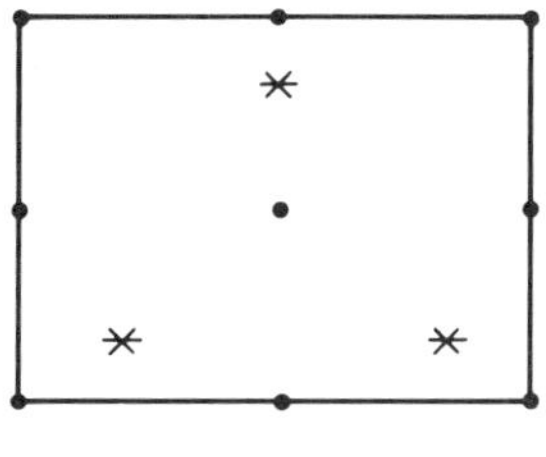

Figure 9.6-1. A plane element with nine nodes and three pressure points, used in a consistent penalty method [9.10].

PROBLEMS

Section 9.1

9.1 In Fig. 9.1-1, let members CA, AB, and BD be identical. Also assume that conditions $v_A = v_B = 0$ have already been imposed, so that [**K**] operates on d.o.f. u_A, θ_A, u_B, and θ_B. Write [**K**], then condense it to a 3 by 3 matrix by imposing the constraint $u_A = u_B$, so that {**D**} becomes $\{\mathbf{D}\} = \lfloor u_A \quad \theta_A \quad \theta_B \rfloor^T$.

9.2 Let the quadratic element of Eqs. 6.6-1 have straight sides and midside nodes. Imagine that the displacement of each side node is to be the average of displacements of the two adjacent corner nodes. Write the appropriate form of Eq. 9.1-4. (You may wish to consider the u_i separately from the v_i.) What kind of element will be produced by applying these constraints? Verify your prediction.

9.3 Write Eq. 7.8-5 in the form of Eq. 9.1-2. How many equations of constraint are there?

9.4 Write the specific form of Eq. 9.1-2 appropriate to Fig. 8.1-2; that is, write the equation that joins the two frames at A with a hinge connection. For simplicity, include in your equations only the three d.o.f. of each frame at A (six d.o.f. altogether). Identify matrices $[\mathbf{C}_r]$ and $[\mathbf{C}_c]$ in your formulation.

9.5 The bar element shown has axial stiffness $k = AE/L$ and axially directed d.o.f. u_1 and u_2. Using the procedure of Section 9.1, solve for u_1 if the constraint $u_2 = 0$ is imposed.

9.6 (a) Let {**Q**} be nonzero in Eq. 9.1-1. Hence, derive the equation analogous to Eq. 9.1-6.
(b) Let the constraint $u_2 = \bar{u}$ be applied in Problem 9.5. Use the method of part (a) to determine u_1.

9.7 Element 2 is to be connected to element 1, as shown. Explain in detail how to treat the stiffness matrix of element 2 before assembly so that node 3 is constrained to lie on *linear* edge 1–2. (D.o.f. of nodes 1 and 2, but not of node 3, are to appear in the structural equations.)

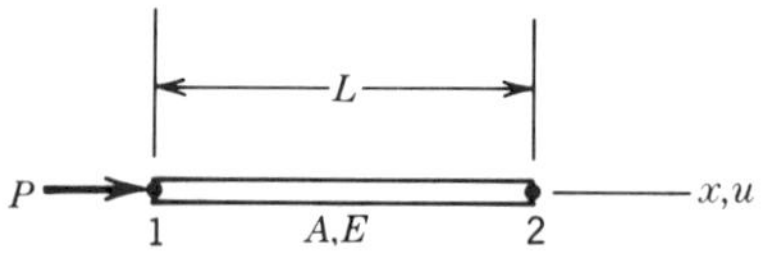

Problem 9.5

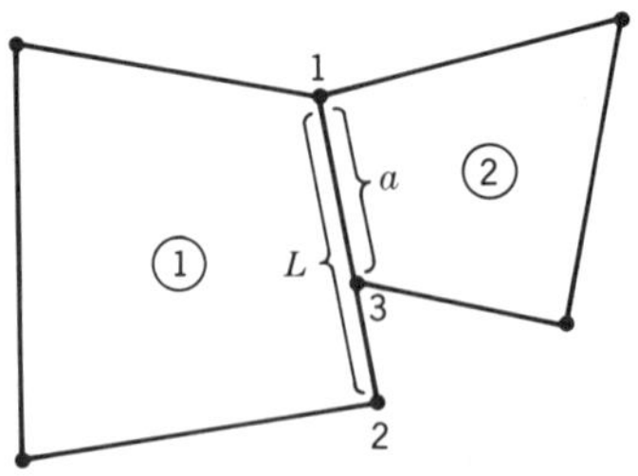

Problem 9.7

9.8 Three nodes lie on an x axis at coordinates x_1, x_2, and x_3. Write a relation in the form of Eq. 9.1-1 that constrains their x-direction displacements to be directly proportional to x.

9.9 If Fig. 9.1-2 were to represent three identical *beam* elements under *lateral* load, what constraint among d.o.f. would be enforced by the "same node number" device (described below Eq. 9.1-9) applied to nodes 2 and 3? Would [**k**] of the right element contribute to stiffness of the structure? If so, in what way?

9.10 The two rigid links AB and BC are connected by a hinge at B and are supported by identical springs at A, B, and C, as shown. Write the structural equations that use v_A, v_B, and v_C as d.o.f. Then impose the constraint that the hinge does not allow relative rotation between the two links, and solve for v_A, v_B, and v_C.

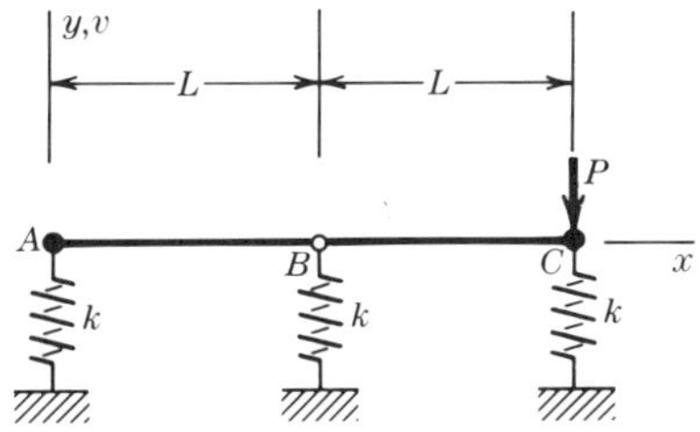

Problem 9.10

9.11 The two-element uniform cantilever beam shown is built of standard beam elements (Eq. 4.2-5). Imagine now that element 2–3 is to be made rigid. Impose this constraint, eliminate w_3 and θ_3, solve for w_2 and θ_2, and compare these results with the prediction of elementary beam theory.

9.12 Two identical standard beam elements (Eq. 4.2-5) are joined at node 2 and are simply supported at nodes 1, 2, and 3, as shown. Node 3 is loaded by moment M_0.

(a) Impose the constraint $\theta_2 = \theta_3$ and solve for all three rotational d.o.f. in terms of M_0, L, E, and I.

(b) Sketch a free-body diagram of the constrained two-element structure.

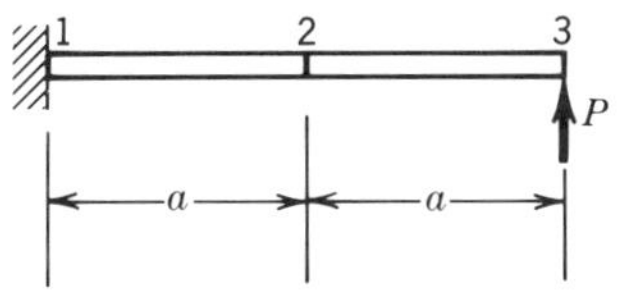

Problem 9.11

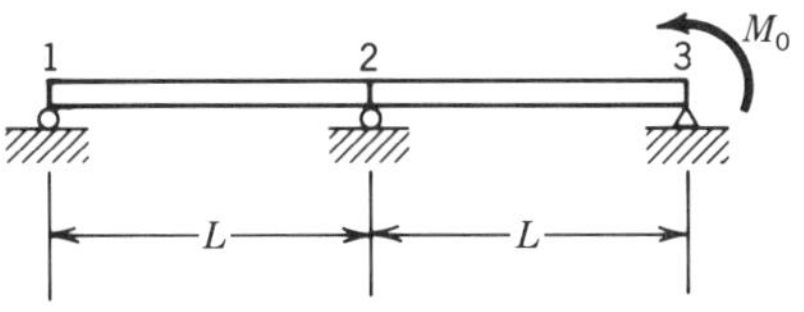

Problem 9.12

Section 9.2

9.13 The following question is strictly mathematical, and serves as a review of the Lagrange multiplier method. What is the area of the largest rectangle that can be inscribed in the ellipse $(x/a)^2 + (y/b)^2 - 1 = 0$?

9.14 Solve the problem of Fig. 9.1-3 (i.e., impose $v_1 = v_2$) by use of a Lagrange multiplier.

9.15 A bar of axial stiffness $k = AE/L$ lies along the x axis and is allowed only axial displacements u. Its right end carries a force $P = 3$ in the $+x$ direction. Its left end (node 1) is to be displaced two units to the right. Impose displacement $u_1 = 2$ and solve for u_2 by use of a Lagrange multiplier.

9.16 To model the uniform cantilever beam shown, use a single standard beam element (Eq. 4.2-5). Use the Lagrange multiplier method to determine the deflection of load P under the following constraint conditions.

(a) The right end is to remain tangent to a straight line between nodes 1 and 2.

(b) The right end is to rotate half as much as the midpoint of the beam, but in the opposite direction.

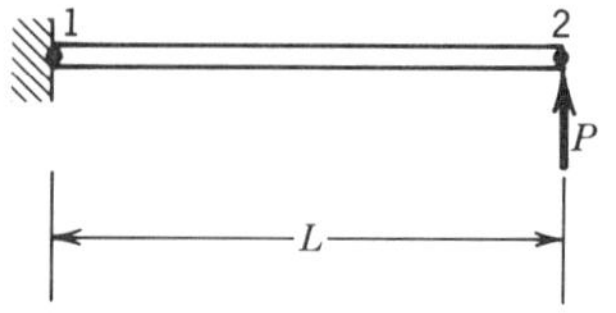

Problem 9.16

Section 9.3

9.17 (a) Derive Eq. 9.3-3 from Eqs. 9.3-1 and 9.3-2.
(b) Revise the argument associated with Eq. 9.3-7: do not make the simplifying assumptions $\lceil \boldsymbol{\alpha} \rfloor = \lceil \mathbf{I} \rfloor \alpha$ and $\{\mathbf{Q}\} = \{\mathbf{0}\}$.

9.18 Solve the problem of Fig. 9.1-3 (i.e., impose $v_1 = v_2$) by use of a penalty number.

9.19 Repeat Problem 9.15, but use a penalty number α instead of a Lagrange multiplier. For $k = 1$, tabulate u_1 and u_2 for the values $\alpha = 1, 4, 10$, and 100.

9.20 Use the penalty method to solve
(a) Problem 9.16a.
(b) Problem 9.16b.

Section 9.4

9.21 Derive Eq. 9.4-5.

9.22 Use one-point quadrature to evaluate U_s in Eq. 9.4-2. Hence, use the equations $\partial \Pi_p / \partial w_2 = 0$ and $\partial \Pi_p / \partial \theta_2 = 0$ to derive expressions for w_2 and θ_2 for the problem posed in connection with Eq. 9.4-5. Compare these results with exact values.

9.23 (a) Let a Mindlin beam element (Fig. 9.4-1*a*) be simply supported at nodes 1 and 2 and loaded in pure bending. Use Eqs. 9.4-2 to show that exact integration gives an element strain energy U consistent with an effective moment of inertia $I_e = I(1 + GL^2/1.2Et^2)$, where $I = bt^3/12$ for the rectangular cross section.
(b) Show that if U_s is integrated by one-point quadrature, the resulting U is consistent with the exact moment of inertia.

9.24 Use Eqs. 9.4-2 and 9.4-3 to evaluate the stiffnesses $[\mathbf{k}_b]$ and $[\mathbf{k}_s]$ of the Mindlin beam element in the following ways, and determine the rank of each matrix.
(a) Evaluate $[\mathbf{k}_b]$ by one-point Gauss quadrature.
(b) Evaluate $[\mathbf{k}_s]$ by one-point Gauss quadrature. Call the result $[\mathbf{k}_s]_1$.
(c) Evaluate $[\mathbf{k}_s]$ by two-point Gauss quadrature. Call the result $[\mathbf{k}_s]_2$.

9.25 (a) Use $[\mathbf{k}_b]$ and $[\mathbf{k}_s]_1$ from Problem 9.24 to model a one-element cantilever beam fixed at node 1 (the left end). Load node 2 by moment M only. Solve for w_2 and θ_2. Investigate what happens as L becomes much larger than t.
(b) Repeat part (a) but use $[\mathbf{k}_s]_2$ rather than $[\mathbf{k}_s]_1$.

9.26 In Problem 9.25, compute the ratio of w_2 obtained from use of $[\mathbf{k}_s]_1$ to w_2 obtained from use of $[\mathbf{k}_s]_2$. If $\nu = 0$, for what value of L/t is w_2 only 10% in error?

9.27 A rectangular bilinear element (four nodes) is subjected to the constraint $\iint (\epsilon_x + \epsilon_y)\, dx\, dy = 0$. Show that the same constraint is produced by setting $\epsilon_x + \epsilon_y = 0$ in a one-point Gauss quadrature rule.

Section 9.5

9.28 Give a detailed explanation of why the constraint ratio r is the same for a one-element test mesh as it is for a square test mesh with N_{es} elements per side.

9.29 The sketch shows a mesh of plane constant-strain triangular elements. The material is incompressible, and plane strain conditions are enforced. All nodes i in contact with the supports are fixed ($u_i = v_i = 0$).
(a) What is the constraint ratio? Will the mesh lock?
(b) Can the constraint ratio be increased by use of reduced integration?
(c) What is the constraint ratio if the number of elements is greatly increased?
(d) Consider elements 1 and 2, and show that the incompressibility constraint implies $u_j = v_j = 0$ at node j. Extend this argument to show that the mesh is completely locked.
(e) Try some other arrangements of constant-strain triangles to see whether the same conclusion holds.
(f) Revise the support conditions so that part (a) of this problem will yield the constraint ratio 1/1.

9.30 What are the constraint ratios r (in a mesh of many elements) for the plane and solid Q6 elements of Section 8.3 if integration is done using an order 2 Gauss rule and the material is nearly incompressible?

9.31 The elements shown are (a) a cubic triangle, and (b) a quartic triangle. The stiffness matrix of the cubic triangle can be *exactly* integrated by the six-point formula of Table 6.8-1 (when material properties and element thickness are constant). Similarly, there is a ten-point formula that integrates the quartic element stiffness matrix exactly. In a refined mesh, for an incompressible material, what is the constraint ratio for each of these exactly integrated elements?

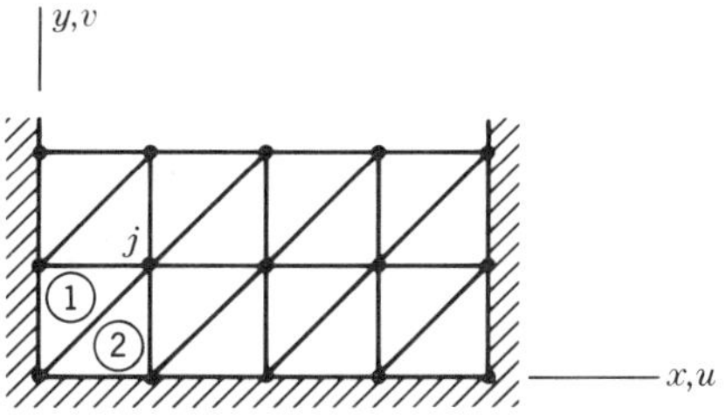

Problem 9.29

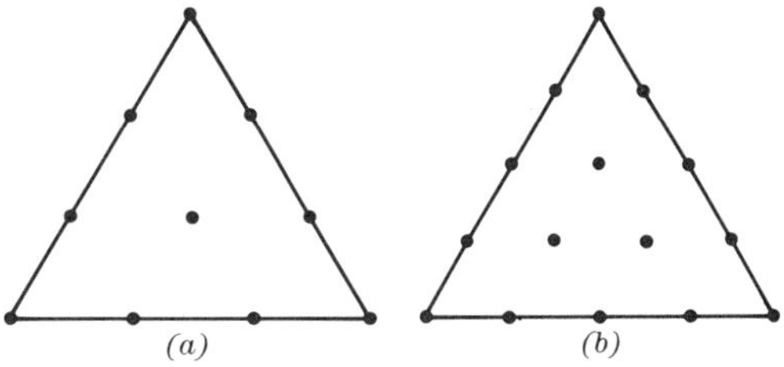

Problem 9.31

Section 9.6

9.32 (a) Show that $\lambda = (\sigma_x + \sigma_y + \sigma_z)/3$, as stated below Eq. 9.6-2.
(b) Show that $\beta\alpha H$ in Eq. 9.4-10 can be expressed as $B\epsilon_V^2/2$.

9.33 Consider the element shown, which is proposed as a higher-order solid element for incompressible media. It has 20 nodes on the edges and a "bubble" shape function added at $\xi = \eta = \zeta = 0$. (Shape functions N_1 through N_{20} are not needed in this problem but may be found in Ref. 2.1, p. 201.)
(a) Give an expression for the shape function $N_{21}(\xi,\eta,\zeta)$ that assures interelement compatibility.
(b) What is the constraint ratio when four internal pressure points (at the nodes of a tetrahedron) are used? (*Note:* This is an attempt to generalize

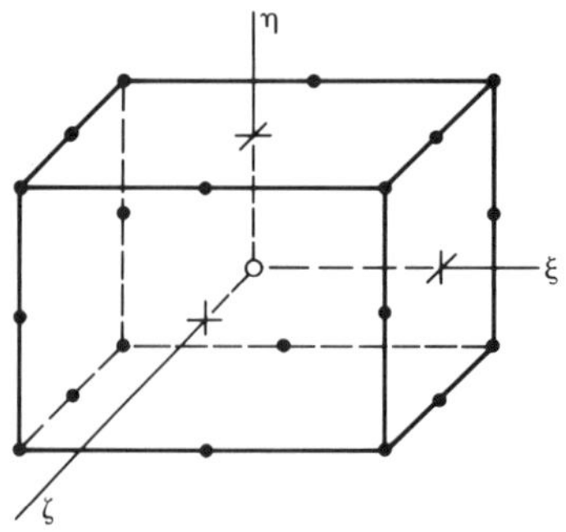

Problem 9.33

Fig. 9.6-1 to a solid element. Little is currently known of the behavior of this element.)

CHAPTER 10

SOLIDS OF REVOLUTION

Analysis methods for axially symmetric bodies are described. Loads may be with or without axial symmetry. Loads without axial symmetry are treated by superposition, using Fourier series.

10.1 INTRODUCTION

A solid of revolution is generated by revolving a plane figure about an axis, and is most easily described in cylindrical coordinates r, θ, and z (Fig. 10.1-1). The geometry is axially symmetric, and if material properties and loads are also axially symmetric, the problem is mathematically two-dimensional. That is, if geometry, support conditions, loads $\{\mathbf{R}\}$, and material property matrix $[\mathbf{E}]$ are all independent of θ, and if the material either is isotropic or has θ as a principal material direction, then static displacements and stresses are independent of θ: circumferential displacement v is zero, material points have only u (radial) and w (axial) displacement components, and the nonzero stresses are those shown in Fig. 10.1-1*a*. The analysis procedure for static problems having axial symmetry is very similar to the procedure used for static problems of plane stress or plane strain. (In a vibration or buckling problem, symmetric *and unsymmetric* modes should be expected, even if geometry, support conditions, and elastic properties are all axially symmetric. A vibration or buckling analysis that assumes θ independence would miss all modes that are not axially symmetric.)

If the solid is axially symmetric but the loading is not, displacements and stresses are three-dimensional rather than axially symmetric. A Fourier series method can then be used. The given loading is expressed as the sum of several component loadings, and an analysis is done for each load component. According to the principle of superposition, the original problem is solved by adding the solutions of the component problems. Each component analysis remains mathematically two-dimensional. Thus the original three-dimensional problem is exchanged for a series of two-dimensional problems. The exchange is usually worthwhile because three-dimensional problems are expensive to set up and run.

A finite element model of a solid of revolution has nodal circles rather than nodal points (Fig. 10.1-1). So does a *shell* of revolution, which is an effective model if the body is thin-walled (Section 12.4). If a body of revolution (having nodal circles) must be attached to a solid body (having nodal points), there is some difficulty in making the connection.

Finite element analysis for axially symmetric solids was first published in 1965 [10.1]. Computer programs are readily available [10.2]. Indeed, minor additions

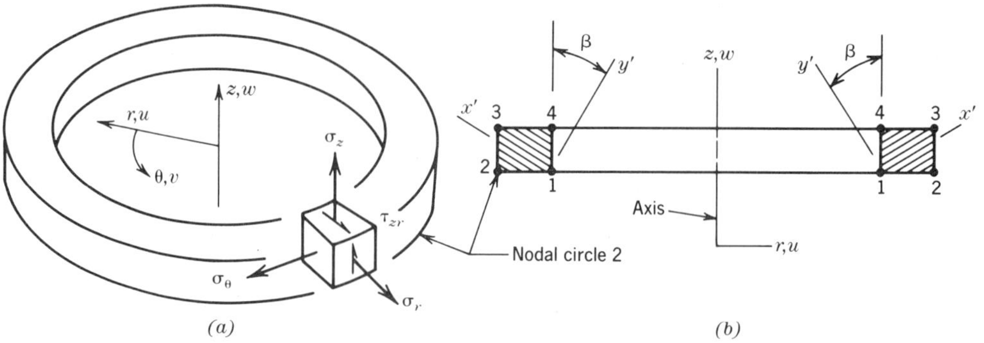

Figure 10.1-1. An axially symmetric finite element of rectangular cross section. (*a*) Isometric view, showing stresses produced by axially symmetric loading. (*b*) Cross section containing the z axis. Hatching suggests an orthotropic material whose principal directions are x', y', and θ.

to a program for plane problems makes the program capable of analyzing solids of revolution as well.

10.2 ELASTICITY RELATIONS FOR AXIAL SYMMETRY

If the analysis problem is axially symmetric, then (see Fig. 10.1-1)

$$v = 0 \text{ and } \tau_{r\theta} = \tau_{\theta z} = \gamma_{r\theta} = \gamma_{\theta z} = 0 \qquad (10.2\text{-}1)$$

Equations 10.2-1 prevail if geometry, support conditions, and loading are all axially symmetric, θ is a principal material direction, and β in Fig. 10.1-1*b* is independent of θ. Thus the material may be isotropic. Or, if orthotropic, principal material axes x' and y' must not change direction with θ, and the third principal material axis must not form a helix. Accordingly, the most general stress–strain relation allowed has the form

$$\begin{Bmatrix} \sigma_r \\ \sigma_\theta \\ \sigma_z \\ \tau_{zr} \end{Bmatrix} = \begin{bmatrix} E_{11} & E_{12} & E_{13} & E_{14} \\ & E_{22} & E_{23} & E_{24} \\ & & E_{33} & E_{34} \\ \text{symm.} & & & E_{44} \end{bmatrix} \left(\begin{Bmatrix} \epsilon_r \\ \epsilon_\theta \\ \epsilon_z \\ \gamma_{zr} \end{Bmatrix} - \{\epsilon_0\} \right) \qquad (10.2\text{-}2)$$

in which $\{\epsilon_0\}$ represents initial strains, and the trivial relations $\tau_{r\theta} = 0$ and $\tau_{\theta z} = 0$ are simply not written. If, in addition to θ, r and z are also principal material directions ($\beta = 0$ in Fig. 10.1-1), then $E_{14} = E_{24} = E_{34} = 0$. For the special case of isotropy and thermal loading, Eq. 10.2-2 becomes

$$\begin{Bmatrix} \sigma_r \\ \sigma_\theta \\ \sigma_z \\ \tau_{zr} \end{Bmatrix} = \frac{(1-\nu)E}{(1+\nu)(1-2\nu)} \begin{bmatrix} 1 & f & f & 0 \\ & 1 & f & 0 \\ & & 1 & 0 \\ \text{symm.} & & & g \end{bmatrix} \left(\begin{Bmatrix} \epsilon_r \\ \epsilon_\theta \\ \epsilon_z \\ \gamma_{zr} \end{Bmatrix} - \begin{Bmatrix} \alpha T \\ \alpha T \\ \alpha T \\ 0 \end{Bmatrix} \right) \qquad (10.2\text{-}3a)$$

in which

$$f = \frac{\nu}{1 - \nu} \qquad \text{and} \qquad g = \frac{1 - 2\nu}{2(1 - \nu)} \tag{10.2-3b}$$

and α = coefficient of thermal expansion and T = temperature relative to a reference temperature at which the body is free of stress. Thus $E_{44} = G$, the shear modulus.

The strain–displacement relations are

$$\begin{aligned} \epsilon_r &= u_{,r} \qquad & \epsilon_\theta &= \frac{2\pi(r + u) - 2\pi r}{2\pi r} = \frac{u}{r} \\ \epsilon_z &= w_{,z} \qquad & \gamma_{zr} &= u_{,z} + w_{,r} \end{aligned} \tag{10.2-4}$$

In matrix format, Eqs. 10.2-4 are

$$\begin{Bmatrix} \epsilon_r \\ \epsilon_\theta \\ \epsilon_z \\ \gamma_{zr} \end{Bmatrix} = [\partial] \begin{Bmatrix} u \\ w \end{Bmatrix}, \qquad \text{where} \quad [\partial] = \begin{bmatrix} \partial/\partial r & 0 \\ 1/r & 0 \\ 0 & \partial/\partial z \\ \partial/\partial z & \partial/\partial r \end{bmatrix} \tag{10.2-5}$$

Or, in alternative format, the same relations are

$$\begin{Bmatrix} \epsilon_r \\ \epsilon_\theta \\ \epsilon_z \\ \gamma_{zr} \end{Bmatrix} = [\mathbf{H}] \begin{Bmatrix} u_{,r} \\ u_{,z} \\ w_{,r} \\ w_{,z} \\ u \end{Bmatrix}, \qquad \text{where} \quad [\mathbf{H}] = \begin{bmatrix} 1 & 0 & 0 & 0 & 0 \\ 0 & 0 & 0 & 0 & 1/r \\ 0 & 0 & 0 & 1 & 0 \\ 0 & 1 & 1 & 0 & 0 \end{bmatrix} \tag{10.2-6}$$

10.3 FINITE ELEMENTS FOR AXIAL SYMMETRY

One may follow the standard formulation procedure, which is contained in Eqs. 4.1-5 and 4.1-6. Consider, for example, an eight-d.o.f. element of rectangular cross section, shown in Fig. 10.3-1. Its displacement field $\{\mathbf{u}\} = [\mathbf{N}]\{\mathbf{d}\}$ is

$$\begin{Bmatrix} u \\ w \end{Bmatrix} = \begin{bmatrix} N_1 & 0 & N_2 & 0 & N_3 & 0 & N_4 & 0 \\ 0 & N_1 & 0 & N_2 & 0 & N_3 & 0 & N_4 \end{bmatrix} \begin{Bmatrix} u_1 \\ w_1 \\ u_2 \\ \vdots \\ w_4 \end{Bmatrix} \tag{10.3-1}$$

where shape functions N_1 through N_4 are stated in Eq. 3.12-10, except that z replaces y and $r - r_m$ replaces x, where r_m is the mean radius $(r_1 + r_2 + r_3 + r_4)/4$. Thus

$$N_i = \frac{[a \pm (r - r_m)](b \pm z)}{4ab} \tag{10.3-2}$$

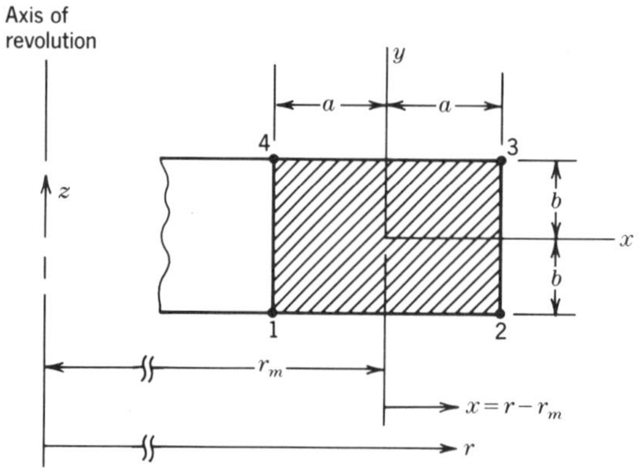

Figure 10.3-1. Geometry of an element of rectangular cross section.

in which $z = 0$ at the element center. The element stiffness matrix is

$$\underset{8\times 8}{[\mathbf{k}]} = \int_A \int_{-\pi}^{\pi} [\mathbf{B}]^T[\mathbf{E}][\mathbf{B}]\ r\ d\theta\ dA \tag{10.3-3}$$

where A = cross-sectional area of the element and $dA = drdz$. From Eqs. 10.2-5 and 10.3-1, $[\mathbf{B}] = [\partial][\mathbf{N}]$. However, since $\partial/\partial r = \partial/\partial x$ and $r = r_m + x$, we can also write

$$[\mathbf{B}] = \begin{bmatrix} \partial/\partial x & 0 \\ 1/(r_m + x) & 0 \\ 0 & \partial/\partial z \\ \partial/\partial z & \partial/\partial x \end{bmatrix} [\mathbf{N}], \qquad \text{where} \quad N_i = \frac{(a \pm x)(b \pm z)}{4ab} \tag{10.3-4}$$

and where $x = 0$ at $r = r_m$, the element center.

Equation 10.3-4 yields the same $[\mathbf{B}]$ as does $[\mathbf{B}] = [\partial][\mathbf{N}]$ with the N_i given by Eq. 10.3-2. Equation 10.3-4 shows that, as compared with a plane problem, the *only change* in $[\mathbf{B}]$ is the added row that computes $\epsilon_\theta = u/r$ (compare Eqs. 10.3-4 and 4.2-14). Moreover, $[\mathbf{k}]$ is the same size; for example, it is 8 by 8 for the foregoing four-node bilinear element, whether the problem is plane or axially symmetric.

If the element is isoparametric, shape functions N_i are functions of ξ and η, and we must use the usual coordinate transformation to relate derivatives,

$$\begin{Bmatrix} u_{,r} \\ u_{,z} \\ w_{,r} \\ w_{,z} \\ u \end{Bmatrix} = \underset{5\times 5}{[\mathbf{\Gamma}]} \begin{Bmatrix} u_{,\xi} \\ u_{,\eta} \\ w_{,\xi} \\ w_{,\eta} \\ u \end{Bmatrix}, \qquad \text{where} \quad [\mathbf{\Gamma}] = \begin{bmatrix} \mathbf{J}^{-1} & \mathbf{0} & \mathbf{0} \\ \mathbf{0} & \mathbf{J}^{-1} & \mathbf{0} \\ \mathbf{0} & \mathbf{0} & 1 \end{bmatrix} \tag{10.3-5}$$

Jacobian matrix $[\mathbf{J}]$ is unchanged: it is still 2 by 2 and is as described in Section 6.3. The ξ and η derivatives in Eq. 10.3-5 are related to nodal d.o.f. by the equation

$$\lfloor u_{,\xi} \quad u_{,\eta} \quad w_{,\xi} \quad w_{,\eta} \quad u \rfloor^T = \underset{5\times 8}{[\mathbf{Q}]} \lfloor u_1 \quad w_1 \quad u_2 \cdots w_4 \rfloor^T \tag{10.3-6}$$

The first four rows of [**Q**] appear in Eq. 6.3-19. The fifth row is $\lfloor N_1 \quad 0 \quad N_2 \quad 0 \quad N_3 \quad 0 \quad N_4 \quad 0 \rfloor$. Shape functions N_i of a four-node isoparametric element are given by Eqs. 6.3-2. From Eqs. 10.2-6, 10.3-5, and 10.3-6, [**B**] = [**H**][**Γ**][**Q**]. The element stiffness matrix of an isoparametric element is

$$[\mathbf{k}] = \int_{-1}^{1} \int_{-1}^{1} \int_{-\pi}^{\pi} [\mathbf{B}]^T[\mathbf{E}][\mathbf{B}] r \, d\theta \, J \, d\xi \, d\eta \tag{10.3-7}$$

where

$$r = \sum N_i r_i \qquad \text{or} \qquad r = r_m + \sum N_i x_i \tag{10.3-8}$$

In numerical integration, either of Eqs. 10.3-8 may be used to determine r at quadrature points.

Element nodal loads $\{\mathbf{r}_e\}$—for example, from heating or from centrifugal force—are calculated in straightforward fashion from Eq. 4.1-6. Here $dV = r \, d\theta \, dA$ or $dV = r \, d\theta \, J \, d\xi \, d\eta$, and $dS = r \, d\theta \, d\ell$, where $d\ell$ is an increment of meridional length.

A uniform line load q (units N/m) on a nodal circle of radius r_i produces the nodal load $2\pi r_i q$. The net static force is also $2\pi r_i q$ if q acts axially, but is zero if q acts radially. Nevertheless, radial load q produces deformations and stresses.

Remarks. The preceding formulation produces 2π as a multiplier of every K_{ij} and every R_i in the structural equations [**K**]{**D**} = {**R**}. This superfluous multiplier can be avoided by letting integrals for [**k**] and $\{\mathbf{r}_e\}$ have theta limits of from zero to one radian. With this approach, externally applied loads must also pertain to a one-radian segment.

Some coefficients in the integrands of Eqs. 10.3-3 and 10.3-7 have $1/r$ as a multiplier. With Gauss quadrature these terms remain finite because there are no Gauss points at $r = 0$. If [**k**] is formed explicitly, we can produce "core" elements by using a displacement field for which $u = 0$ for $r = 0$ and evaluating indeterminate forms 0/0 that appear in the formulation by the L'Hôpital rule. If numerical integration is used instead, for acceptable accuracy core elements may require more integration points in the radial direction than are used for elements distant from the axis of revolution.

During stress computation, the indeterminate form $\epsilon_\theta = u/r = 0/0$ arises for points on the z axis. We can avoid this trouble by calculating ϵ_θ slightly alway from the axis or by extrapolating strains at Gauss points to the axis. Another option is to exploit the theoretical requirement that $\epsilon_r = \epsilon_\theta$ at $r = 0$. Thus, for stress computation at $r = 0$, we merely replace the ϵ_θ row of [**B**] by the ϵ_r row.

Because of axial symmetry, z-direction translation is the only possible rigid-body motion. It can be restrained by prescribing w on a single nodal circle. The radial displacement $u = 0$ should be prescribed at all nodes that lie on the z axis.

Valid elements for solids of revolution must pass a weak patch test (see Section 4.6). Consider, for example, elements Q6 and QM6 (Section 8.3). These elements develop a spurious radial bulge because internal d.o.f. are activated. The bulge creates spurious shear strain γ_{zr} except at $\xi = \eta = 0$. Nevertheless, the element is valid because the bulge tends to vanish as element cross-sectional dimensions become small in comparison with the mean radius of the element. In general use, stresses predicted by the QM6 element may be more reliable if γ_{zr} is evaluated only at $\xi = \eta = 0$.

10.4 FOURIER SERIES

The response of an axially symmetric body to asymmetric loads can be analyzed by superposing component analyses, each of which represents the response attributable to one component of the total load. The method relies on Fourier series, which is summarized as follows without reference to bodies of revolution.

Fourier series represent functions that are periodic. A Fourier series for a dependent variable $\phi = \phi(\theta)$ can be written

$$\phi = \sum_{n=0}^{\infty} p_n \cos n\theta + \sum_{n=1}^{\infty} q_n \sin n\theta \tag{10.4-1}$$

where n is an integer. The period of ϕ is 2π, for example, from $\theta = -\pi$ to $\theta = \pi$.

Sine terms are called *odd* or *antisymmetric,* as $\phi(\theta) = -\phi(-\theta)$. Cosine terms are called *even* or *symmetric,* as $\phi(\theta) = \phi(-\theta)$. Coefficients p_n and q_n are functions of n but not of θ. The following integrals, where m and n are integers, are useful:

$$\int_{-\pi}^{\pi} \sin m\theta \sin n\theta \, d\theta = \begin{cases} \pi & \text{for } m = n \neq 0 \\ 0 & \text{for } m \neq n \text{ and for } m = n = 0 \end{cases} \tag{10.4-2a}$$

$$\int_{-\pi}^{\pi} \cos m\theta \cos n\theta \, d\theta = \begin{cases} 2\pi & \text{for } m = n = 0 \\ \pi & \text{for } m = n \neq 0 \\ 0 & \text{for } m \neq n \end{cases} \tag{10.4-2b}$$

$$\int_{-\pi}^{\pi} \sin m\theta \cos n\theta \, d\theta = 0 \qquad \text{for all } m \text{ and } n \tag{10.4-2c}$$

Imagine that a certain periodic function $\phi = \phi(\theta)$ is known, but not expressed as a Fourier series. An equivalent Fourier series representation of ϕ requires that p_n and q_n in Eq. 10.4-1 be determined. To do so, we integrate the function, then multiply it by the single term cos $n\theta$ and integrate, then multiply it by the single term sin $n\theta$ and integrate. Equation 10.4-1 is similarly integrated, making use of Eqs. 10.4-2. Thus equations that determine p_0, p_n, and q_n are

$$\int_{-\pi}^{\pi} \phi \, d\theta = 2\pi p_0 \qquad \int_{-\pi}^{\pi} \phi \cos n\theta \, d\theta = \pi p_n \qquad \int_{-\pi}^{\pi} \phi \sin n\theta \, d\theta = \pi q_n \tag{10.4-3}$$

Integrals in Eq. 10.4-3 may be evaluated analytically, numerically, or even graphically.

Example. The square wave in Fig. 10.4-1 is to be represented by a Fourier series. Here $\theta = \pi x/L$, and $\phi = \phi_0$ can be regarded as a uniformly distributed load of intensity ϕ_0 on a span of length L. In Eqs. 10.4-3 we use $\phi = -\phi_0$ for $-L < x < 0$ and $\phi = +\phi_0$

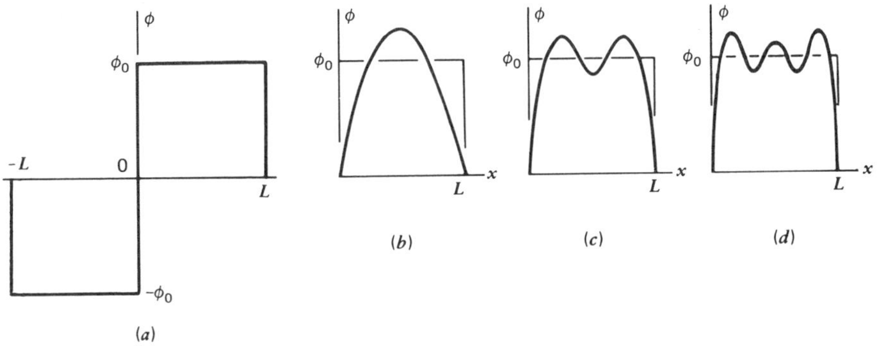

Figure 10.4-1. (*a*) Square wave and its representation by truncated Fourier series, using terms (*b*) $n = 1$, (*c*) $n = 1$ and 3, and (*d*) $n = 1$, 3, and 5.

for $0 < x < L$. Results of the respective integrations in Eq. 10.4-3 are 0, 0, and 0 (n even) or $4\phi_0/n$ (n odd). Hence, $p_0 = 0$, $p_n = 0$, and $q_n = 0$ (n even). For n odd,

$$q_n = \frac{4\phi_0}{n\pi} \quad \text{and} \quad \phi = \sum_{n=1,3,\ldots} \frac{4\phi_0}{n\pi} \sin \frac{n\pi x}{L} \tag{10.4-4}$$

As suggested by Fig. 10.4-1, the square wave can be modeled arbitrarily closely by using enough series terms.

Example. Concentrated loads P produce an interesting result. In the coordinates of Fig. 10.4-1*a*, consider a load P downward at $x = -L/2$ and a load P upward at $x = +L/2$. Here $\theta = \pi x/L$, and P can be regarded as a concentrated center load on a beam that extends from $x = 0$ to $x = L$. Equations 10.4-3 yield $p_0 = 0$, $p_n = 0$, and

$$q_n = \frac{2P}{L} \sin \frac{n\pi}{2} \qquad \phi = \sum_{n=1,2,3,\ldots} \left(\frac{2P}{L} \sin \frac{n\pi}{2} \right) \sin \frac{n\pi x}{L} \tag{10.4-5}$$

This series for ϕ does not converge. However, when used to load the aforementioned beam, convergent results are obtained for displacement and stress.

A Simple Application in Stress Analysis. The beam problem depicted in Fig. 10.4-2 illustrates features of series analysis that also appear in series analysis of solids (and shells) of revolution. For a beam, the equilibrium and moment–curvature relations are

$$V_{,x} = \phi \qquad M_{,x} = V \qquad EIw_{,xx} = M \tag{10.4-6}$$

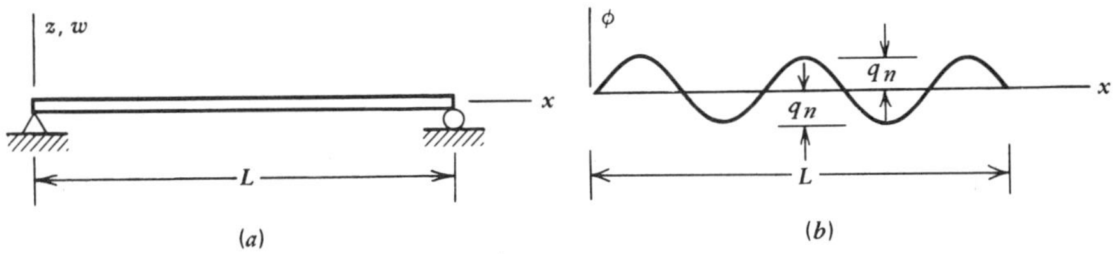

Figure 10.4-2. (*a*) Simply supported beam. (*b*) The sine wave loading $\phi = q_n \sin(n\pi x/L)$, where q_n is the amplitude of ϕ. The case $n = 5$ is depicted.

where V is transverse shear force, M is bending moment, and ϕ is distributed load. When combined, Eqs. 10.4-6 yield, for constant bending stiffness EI,

$$EIw_{,xxxx} = \phi \tag{10.4-7}$$

Consider the loading of Fig. 10.4-2*b*, which corresponds to one term of the second summation in the Fourier series of Eq. 10.4-1:

$$\phi = q_n \sin \frac{n\pi x}{L} \tag{10.4-8}$$

Here q_n does not depend on x. Assume that the displacement is the admissible function

$$w = w_n \sin \frac{n\pi x}{L} \tag{10.4-9}$$

where w_n does not depend on x. We substitute w and ϕ into Eq. 10.4-7 and obtain

$$\left[EI \left(\frac{n\pi}{L} \right)^4 w_n - q_n \right] \sin \frac{n\pi x}{L} = 0 \tag{10.4-10}$$

This can be true for all x only if the bracketed expression vanishes. Hence

$$w_n = \frac{q_n}{EI} \left(\frac{L}{n\pi} \right)^4 \tag{10.4-11}$$

Substitution of Eq. 10.4-11 into Eq. 10.4-9 defines the correct and unique solution for w, since it satisfies all requirements of equilibrium, compatibility, and boundary conditions (Section 1.6). Note that a sine wave of loading produces corresponding sine waves of deflection and bending moment, regardless of n. That is, the various harmonics are uncoupled—the nth wave does not interact with the mth wave.

Now that w_n is known, Eq. 10.4-9 defines w for any x and for any n. If two or more sine wave loadings act simultaneously, each associated with a different n, the net deflection and the net bending moment are determined by superposition; that is,

$$w = \sum_n q_n \frac{L^4}{EIn^4\pi^4} \sin \frac{n\pi x}{L} \quad \text{and} \quad M = -\sum_n q_n \frac{L^2}{n^2\pi^2} \sin \frac{n\pi x}{L} \tag{10.4-12}$$

where $M = EIw_{,xx}$. A particular loading requires a particular q_n. For example, to analyze the effect of a uniformly distributed loading of intensity ϕ_0, we substitute $q_n = 4\phi_0/n\pi$ from Eq. 10.4-4 into Eqs. 10.4-12. For a concentrated load at mid-span, we substitute q_n from Eq. 10.4-5.

What we have done is find the response of the beam to a single load component from a single equation (Eq. 10.4-11) that says nothing about how w varies with x. We pay for this simplicity by having to solve the equation several times, once for each Fourier component of loading.

In dealing with bodies of revolution, we replace Eq. 10.4-11 by a set of simultaneous algebraic equations, which must be solved for each Fourier component of loading.

10.5 LOADS WITHOUT AXIAL SYMMETRY: INTRODUCTION

In solids (and shells) of revolution under loads without axial symmetry, we can arrange to compute structural response to each Fourier harmonic of loading from a set of comparatively simple equations that makes no reference to how deformations vary with circumferential coordinate θ. This set of equations must be solved several times, once for each harmonic of loading. Response attributable to the given loading is found by superposing the separate analyses and, in general, displays deformations that vary with θ as well as with r and z. The superposition method is much less expensive than a single three-dimensional analysis if only a few Fourier harmonics are needed to represent the load. As examples, wind loading requires few harmonics but a concentrated force requires many.

With θ a principal material direction, the most general stress–strain relation $\{\boldsymbol{\sigma}\} = [\mathbf{E}]\{\boldsymbol{\epsilon}\}$ has the form

$$\begin{Bmatrix} \sigma_r \\ \sigma_\theta \\ \sigma_z \\ \tau_{zr} \\ \tau_{r\theta} \\ \tau_{\theta z} \end{Bmatrix} = \begin{bmatrix} E_{11} & E_{12} & E_{13} & E_{14} & 0 & 0 \\ & E_{22} & E_{23} & E_{24} & 0 & 0 \\ & & E_{33} & E_{34} & 0 & 0 \\ & & & E_{44} & 0 & 0 \\ \text{symmetric} & & & & E_{55} & E_{56} \\ & & & & & E_{66} \end{bmatrix} \begin{Bmatrix} \epsilon_r \\ \epsilon_\theta \\ \epsilon_z \\ \gamma_{zr} \\ \gamma_{r\theta} \\ \gamma_{\theta z} \end{Bmatrix} \tag{10.5-1}$$

If r and z are also principal material directions, or if the material is isotropic, then $E_{14} = E_{24} = E_{34} = E_{56} = 0$. If the material is isotropic, one uses Eq. 10.2-3 and the values $E_{44} = E_{55} = E_{66} = G$, where G is the shear modulus, $G = 0.5\,E/(1 + \nu)$.

Let loading of the body be expressed as Fourier series: for example, the radially directed body force is $F_r = \Sigma\,\overline{F}_{rn} \cos n\theta$, where $\overline{F}_{rn}$ is an amplitude, like p_n in Eq. 10.4-1. Thus

$$\lfloor F_r \quad F_z \quad \Phi_r \quad \Phi_z \quad T \rfloor = \sum_n \lfloor \overline{F}_{rn} \quad \overline{F}_{zn} \quad \overline{\Phi}_{rn} \quad \overline{\Phi}_{zn} \quad \overline{T}_n \rfloor \cos n\theta \tag{10.5-2a}$$

$$\lfloor F_\theta \quad \Phi_\theta \rfloor = \sum_n \lfloor \overline{F}_{\theta n} \quad \overline{\Phi}_{\theta n} \rfloor \sin n\theta \tag{10.5-2b}$$

where T is temperature, and the F's and Φ's are, respectively, body forces per unit volume and surface tractions in the r, θ, and z directions. Equations 10.5-2 represent a state of symmetry with respect to the plane $\theta = 0$ (antisymmetric Fourier terms are considered subsequently).

We will show that when loads are described by Eqs. 10.5-2, displacements are described by

$$\text{Radial displacement} = u = \sum_n \overline{u}_n \cos n\theta \tag{10.5-3a}$$

$$\text{Circumferential displacement} = v = \sum_n \overline{v}_n \sin n\theta \tag{10.5-3b}$$

$$\text{Axial displacement} = w = \sum_n \overline{w}_n \cos n\theta \tag{10.5-3c}$$

All three displacements are needed because the physical problem is three-dimensional. In Eqs. 10.5-2 and 10.5-3, n is an integer, and all barred quantities are functions of r, z, and n but not of θ. Thus the barred terms are amplitudes.

The strain–displacement relations in cylindrical coordinates are $\{\boldsymbol{\epsilon}\} = [\partial]\{\mathbf{u}\}$; that is,

$$\begin{Bmatrix} \epsilon_r \\ \epsilon_\theta \\ \epsilon_z \\ \gamma_{zr} \\ \gamma_{r\theta} \\ \gamma_{\theta z} \end{Bmatrix} = \begin{bmatrix} \partial/\partial r & 0 & 0 \\ 1/r & \partial/(r\,\partial\theta) & 0 \\ 0 & 0 & \partial/\partial z \\ \partial/\partial z & 0 & \partial/\partial r \\ \partial/(r\,\partial\theta) & (\partial/\partial r - 1/r) & 0 \\ 0 & \partial/\partial z & \partial/(r\,\partial\theta) \end{bmatrix} \begin{Bmatrix} u \\ v \\ w \end{Bmatrix} \tag{10.5-4}$$

These relations are independent of material properties and of whether or not u, v, and w are described by series.

Consider now a typical single harmonic of displacement, say the nth. If we substitute Eqs. 10.5-3 into Eq. 10.5-4 and the resulting strains into Eq. 10.5-1, we find that stresses of the nth harmonic have the form

$$\lfloor \sigma_{rn} \quad \sigma_{\theta n} \quad \sigma_{zn} \quad \tau_{zrn} \rfloor = \lfloor \bar{\sigma}_{rn} \quad \bar{\sigma}_{\theta n} \quad \bar{\sigma}_{zn} \quad \bar{\tau}_{zrn} \rfloor \cos n\theta \tag{10.5-5a}$$

$$\lfloor \tau_{r\theta n} \quad \tau_{\theta zn} \rfloor = \lfloor \bar{\tau}_{r\theta n} \quad \bar{\tau}_{\theta zn} \rfloor \sin n\theta \tag{10.5-5b}$$

where the barred quantities are functions of r, z, and n but not of θ. If Eqs. 10.5-2 and 10.5-5 are substituted into the three differential equations of equilibrium, Eqs. 1.6-4, we find that these three equations assume the forms

$$Q_1 \cos n\theta = 0 \qquad Q_2 \cos n\theta = 0 \qquad Q_3 \sin n\theta = 0 \tag{10.5-6}$$

where the Q_i are functions of r, z, and n but not of θ. Equations 10.5-6 are analogous to Eq. 10.4-10. Equations 10.5-6 must prevail for all θ, so $Q_1 = Q_2 = Q_3 = 0$. As will be seen in Section 10.6, in a finite element context the equations $Q_1 = Q_2 = Q_3 = 0$ produce the equilibrium equations of the nth harmonic

$$[\mathbf{K}]_n\{\mathbf{D}\}_n - \{\mathbf{R}\}_n = \{\mathbf{0}\} \tag{10.5-7}$$

Equations 10.5-7 are analogous to Eq. 10.4-11. Their solution yields the nodal d.o.f. $\{\mathbf{D}\}_n = \lfloor \bar{u}_{1n} \quad \bar{v}_{1n} \quad \bar{w}_{1n} \quad \bar{u}_{2n} \ \ldots \rfloor^T$, which are displacement amplitudes of the nodal circles in the nth harmonic. Matrix $[\mathbf{K}]_n$ depends on n. The load coefficients in $\{\mathbf{R}\}_n$ correspond to the q_n of Eq. 10.4-1.

We see that n circumferential waves of loading are associated with n circumferential waves of stress and of displacement. *The Fourier harmonics are not coupled.* Different numerical values of n present different problems that do not interact. Thus the need for a division into finite elements in the circumferential direction is replaced by the need to superpose separate solutions for a structure divided into finite elements in only its cross section. A single mesh is used for all the separate solutions. In most practical problems only a few load harmonics need be analyzed. A computer program can automatically cycle through a user-specified number of harmonics and superpose the separate solutions.

Remarks. The preceding discussion invokes only loads and displacements that have $\theta = 0$ as a plane of symmetry. In general, antisymmetric terms are also present. Thus Eqs. 10.5-2 are augmented to read

$$\lfloor F_r \quad F_z \quad \Phi_r \quad \Phi_z \quad T \rfloor = \sum_n \lfloor \bar{\mathbf{L}}_{cn} \rfloor \cos n\theta + \sum_n \lfloor \bar{\bar{\mathbf{L}}}_{sn} \rfloor \sin n\theta \qquad (10.5\text{-}8a)$$

$$\lfloor F_\theta \quad \Phi_\theta \rfloor = \sum_n \lfloor \bar{\mathbf{L}}_{sn} \rfloor \sin n\theta + \sum_n \lfloor \bar{\bar{\mathbf{L}}}_{cn} \rfloor \cos n\theta \qquad (10.5\text{-}8b)$$

where $\lfloor \bar{\mathbf{L}}_{cn} \rfloor$ and $\lfloor \bar{\mathbf{L}}_{sn} \rfloor$ represent the symmetric load amplitudes, already present in Eqs. 10.5-2, and $\lfloor \bar{\bar{\mathbf{L}}}_{sn} \rfloor$ and $\lfloor \bar{\bar{\mathbf{L}}}_{cn} \rfloor$ represent additional antisymmetric load amplitudes. Similarly, the symmetric displacement field, Eqs. 10.5-3, is augmented by antisymmetric terms and becomes

$$u = \sum_n \bar{u}_n \cos n\theta + \sum_n \bar{\bar{u}}_n \sin n\theta \qquad (10.5\text{-}9a)$$

$$v = \sum_n \bar{v}_n \sin n\theta - \sum_n \bar{\bar{v}}_n \cos n\theta \qquad (10.5\text{-}9b)$$

$$w = \sum_n \bar{w}_n \cos n\theta + \sum_n \bar{\bar{w}}_n \sin n\theta \qquad (10.5\text{-}9c)$$

The motivation for the arbitrarily chosen negative sign in the v series is explained in the subsection that follows Eq. 10.6-9.

Axially symmetric problems are represented by the $n = 0$ terms of the single-barred series. For $n = 1, 2, 3, \ldots$, loads and displacements of the single-barred series represent symmetry about the plane $\theta = 0$. For $n = 0, 2, 4, 6, \ldots$, loads and deformations have both $\theta = 0$ and $\theta = \pi/2$ as planes of symmetry. Example symmetric loads appear in Fig. 10.5-1*c*.

Antisymmetric problems (e.g., Fig. 10.5-1*f*) are represented by the double-barred series. Pure torque is represented by the $n = 0$ terms of the double-barred series. Thus, for example, we can study the twist of shafts of variable diameter. In the torsion problem u and w are everywhere zero, so a finite element solution based on a stress function is also possible [10.3].

When $n = 0$, for any node i, nodal d.o.f. $\bar{v}_{ni}$, $\bar{\bar{u}}_{ni}$, and $\bar{\bar{w}}_{ni}$ have no stiffness associated with them, so these d.o.f. must be suppressed to avoid a singular stiffness matrix.

The simplest displacement boundary condition is zero displacement on a nodal circle. This requires that displacement amplitudes on the circle be zero in every harmonic. If loads are symmetric about both the $\theta = 0$ and $\theta = \pi/2$ planes, then u and v are zero at $r = 0$ in all harmonics. Nonzero and asymmetric displacement conditions can be represented as Fourier series and the separate amplitude coefficients used as prescribed displacements in the separate analyses.

Additional constraints on the Fourier displacement amplitudes can be deduced from the condition that strains remain finite at $r = 0$ [10.4]. If these constraints are not imposed, numerical integration makes some stiffness coefficients significantly larger than others. This circumstance is not likely to be troublesome in static analysis, but it may require a very small time step if explicit integration is applied to transient problems.

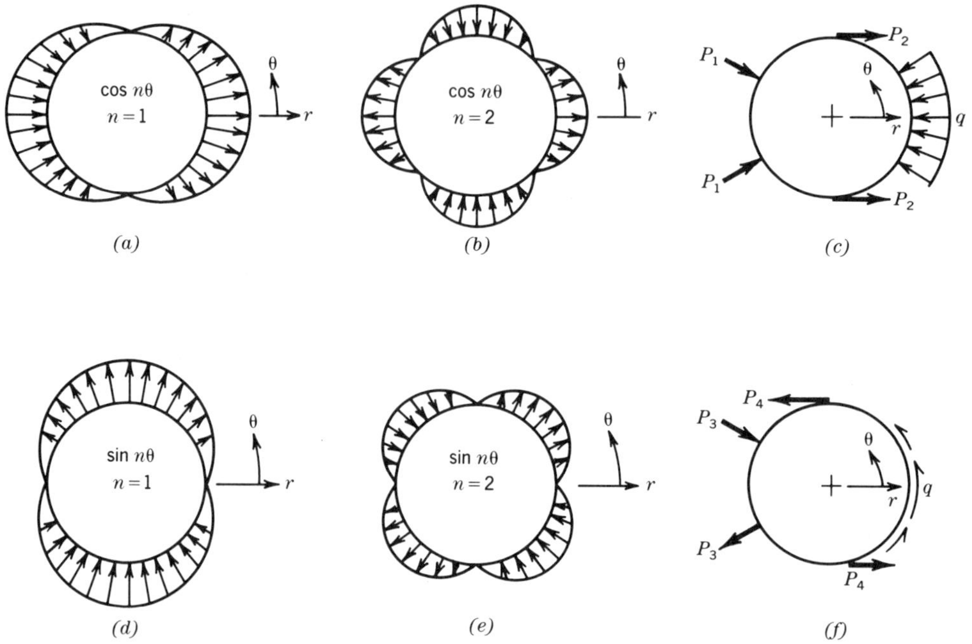

Figure 10.5-1. (*a,b*) Example cosine terms. (*c*) Possible symmetric loads in an $r\theta$ plane. (*d,e*) Example sine terms. (*f*) Possible antisymmetric loads in an $r\theta$ plane.

10.6 LOADS WITHOUT AXIAL SYMMETRY: ELEMENT MATRICES

Within an element, one can interpolate amplitudes $\bar{u}_n$, $\bar{v}_n$, and $\bar{w}_n$ of Eqs. 10.5-3 from *nodal* amplitudes $\bar{u}_{in}$, $\bar{v}_{in}$, and $\bar{w}_{in}$. Consider, for example, the four-node element in Fig. 10.1-1*b*. Displacement amplitudes $\{\bar{\mathbf{u}}\}_n$ in harmonic n are $\{\bar{\mathbf{u}}\}_n = [\bar{\mathbf{N}}]\{\bar{\mathbf{d}}\}_n$, in which nodal displacement amplitudes $\{\bar{\mathbf{d}}\}_n$ pertain to the nth harmonic and are independent of θ. Written out, this relation is

$$\begin{Bmatrix} \bar{u}_n \\ \bar{v}_n \\ \bar{w}_n \end{Bmatrix} = \Big[\underbrace{\begin{matrix} N_1 & 0 & 0 \\ 0 & N_1 & 0 \\ 0 & 0 & N_1 \end{matrix}}_{1} \Big| \underbrace{\begin{matrix} N_2 & 0 & 0 \\ 0 & N_2 & 0 \\ 0 & 0 & N_2 \end{matrix}}_{2} \Big| \underbrace{\begin{matrix} \cdots \\ \cdots \\ \cdots \end{matrix}}_{3} \Big| \underbrace{\begin{matrix} \cdots \\ \cdots \\ \cdots \end{matrix}}_{4} \Big] \{\bar{\mathbf{d}}\}_n \tag{10.6-1a}$$

in which element nodal displacement amplitudes are

$$\{\bar{\mathbf{d}}\}_n = \lfloor \bar{u}_{1n} \quad \bar{v}_{1n} \quad \bar{w}_{1n} \quad \bar{u}_{2n} \quad \bar{v}_{2n} \quad \bar{w}_{2n} \quad \ldots \rfloor^T \tag{10.6-1b}$$

For a rectangular four-node element, shape functions N_i are as stated in Eq. 10.3-2 (or in Eq. 10.3-4). For an isoparametric four-node element, shape functions N_i are as stated in Eqs. 6.3-2. An element with more nodes will display more partitions in Eq. 10.6-1a and more nodal amplitudes in Eq. 10.6-1b. The same shape functions N_i can be used for all harmonics.

The same interpolation is used for the double-barred series in Eqs. 10.5-9: thus, in Eqs. 10.6-1, single-barred quantities become double-barred quantities. In what follows we discuss the single-barred series. The double-barred series is treated similarly.

Summation of the various Fourier harmonics yields displacements in an element is stated by Eqs. 10.5-3. This same displacement field can be written in matrix format by attaching cos $n\theta$ to rows 1 and 3 in Eq. 10.6-1a and sin $n\theta$ to row 2, then summing the various harmonics. Thus

$$\begin{Bmatrix} u \\ v \\ w \end{Bmatrix} = \sum_n \begin{Bmatrix} \bar{u}_n \cos n\theta \\ \bar{v}_n \sin n\theta \\ \bar{w}_n \cos n\theta \end{Bmatrix} = \sum_n \underbrace{\begin{bmatrix} N_1 \cos n\theta & 0 & 0 & \cdots \\ 0 & N_1 \sin n\theta & 0 & \cdots \\ 0 & 0 & N_1 \cos n\theta & \cdots \end{bmatrix}}_{[\mathbf{N}]_n} \{\bar{\mathbf{d}}\}_n \tag{10.6-2}$$

or, with the summation written out,

$$\begin{Bmatrix} u \\ v \\ w \end{Bmatrix} = \underbrace{[\mathbf{N}_{n=0} \quad \mathbf{N}_{n=1} \quad \mathbf{N}_{n=2} \quad \ldots]}_{[\mathbf{N}]}\{\bar{\mathbf{d}}\} \tag{10.6-3a}$$

in which $\{\bar{\mathbf{d}}\}$ lists amplitudes from *all* harmonics, that is,

$$\{\bar{\mathbf{d}}\} = \lfloor \{\bar{\mathbf{d}}\}_{n=0} \quad \{\bar{\mathbf{d}}\}_{n=1} \quad \{\bar{\mathbf{d}}\}_{n=2} \quad \ldots \rfloor^T \tag{10.6-3b}$$

Note that $[\mathbf{N}]_n$ depends on n only because of the cos $n\theta$ and sin $n\theta$ terms. By applying the operator matrix $[\partial]$ in Eq. 10.5-4, one obtains the strain–displacement relation:

$$\{\boldsymbol{\epsilon}\} = [\partial] \begin{Bmatrix} u \\ v \\ w \end{Bmatrix} = \underbrace{[\partial]\lfloor \mathbf{N}_{n=0} \quad \mathbf{N}_{n=1} \quad \mathbf{N}_{n=2} \quad \ldots \rfloor}_{[\mathbf{B}] = [\mathbf{B}_{n=0} \quad \mathbf{B}_{n=1} \quad \ldots]} \{\bar{\mathbf{d}}\} \tag{10.6-4}$$

For example, one determines that the contribution of the nth harmonic to strains, $\{\boldsymbol{\epsilon}\}_n = [\mathbf{B}]_n\{\mathbf{d}\}_n$, is

$$\begin{Bmatrix} \epsilon_{rn} \\ \epsilon_{\theta n} \\ \epsilon_{zn} \\ \gamma_{zrn} \\ \gamma_{r\theta n} \\ \gamma_{\theta zn} \end{Bmatrix} = \Bigg[\underbrace{\begin{matrix} N_{1,r} \cos n\theta & 0 & 0 \\ \dfrac{N_1}{r} \cos n\theta & \dfrac{nN_1}{r} \cos n\theta & 0 \\ 0 & 0 & N_{1,z} \cos n\theta \\ N_{1,z} \cos n\theta & 0 & N_{1,r} \cos n\theta \\ -\dfrac{nN_1}{r} \sin n\theta & \left(N_{1,r} - \dfrac{N_1}{r}\right) \sin n\theta & 0 \\ 0 & N_{1,z} \sin n\theta & -\dfrac{nN_1}{r} \sin n\theta \end{matrix}}_{1} \Bigg| \underbrace{\begin{matrix} \cdots \\ \cdots \\ \cdots \\ \cdots \\ \cdots \\ \cdots \end{matrix}}_{2,3,4,\ldots} \Bigg] \begin{Bmatrix} \bar{u}_{1n} \\ \bar{v}_{1n} \\ \bar{w}_{1n} \\ \hline \bar{u}_{2n} \\ \bar{v}_{2n} \\ \vdots \end{Bmatrix} \tag{10.6-5}$$

where the partitioning corresponds to that used in Eq. 10.6-1a. If the element is isoparametric, the usual transformation of derivatives must be included (see e.g. Eq. 6.3-7):

$$N_{i,r} = \Gamma_{11}N_{i,\xi} + \Gamma_{12}N_{i,\eta} \qquad \text{and} \qquad N_{i,z} = \Gamma_{21}N_{i,\xi} + \Gamma_{22}N_{i,\eta} \tag{10.6-6}$$

Shape functions N_i depend on r and z. Therefore, we see from Eq. 10.6-5 that $[\mathbf{B}]$ is a function of r, z, n, and θ. The element stiffness matrix is given by Eq. 10.3-3 or Eq. 10.3-7. Let there be J nodes per element and M harmonics included in the summation. Then the integrand matrix $[\mathbf{B}]^T[\mathbf{E}][\mathbf{B}]$ is a full matrix of size $3JM$ by $3JM$. It is composed of an M by M array of $3J$ by $3J$ submatrices. The off-diagonal submatrices contain $\sin m\theta \sin n\theta$ or $\cos m\theta \cos n\theta$ in every term, where m and n are *different* integers. According to Eqs. 10.4-2, these terms integrate to zero. We are left with only M on-diagonal submatrices, each $3J$ by $3J$ and containing $\sin^2 n\theta$ or $\cos^2 n\theta$ in every term. After integration according to Eqs. 10.4-2, the common factor π (or 2π for $n = 0$) appears in every term. Integration with respect to r and z (or ξ and η) is done as though the problem were axially symmetric. After integration is complete, terms in each $3J$ by $3J$ submatrix have the form $A + Bn^2$ or the form Cn, where A, B, and C depend on material properties and element geometry but are independent of n and θ. Accordingly, the various expressions here symbolized by A, B, and C need be generated only once, regardless of the number of harmonics used. Element equations, and structural equations after assembly of elements, have the respective forms

$$\begin{bmatrix} \mathbf{k}_0 & & & \\ & \mathbf{k}_1 & & \\ & & \cdot & \\ & & & \cdot \\ & & & & \cdot \end{bmatrix} \begin{Bmatrix} \bar{\mathbf{d}}_0 \\ \bar{\mathbf{d}}_1 \\ \cdot \\ \cdot \\ \cdot \end{Bmatrix} = \begin{Bmatrix} \bar{\mathbf{r}}_0 \\ \bar{\mathbf{r}}_1 \\ \cdot \\ \cdot \\ \cdot \end{Bmatrix} \quad \text{and} \quad \begin{bmatrix} \mathbf{K}_0 & & & \\ & \mathbf{K}_1 & & \\ & & \cdot & \\ & & & \cdot \\ & & & & \cdot \end{bmatrix} \begin{Bmatrix} \bar{\mathbf{D}}_0 \\ \bar{\mathbf{D}}_1 \\ \cdot \\ \cdot \\ \cdot \end{Bmatrix} = \begin{Bmatrix} \bar{\mathbf{R}}_0 \\ \bar{\mathbf{R}}_1 \\ \cdot \\ \cdot \\ \cdot \end{Bmatrix} \tag{10.6-7}$$

where each $[\mathbf{k}]_n$ is of size $3J$ by $3J$, and subscripts 0, 1, and so on indicate the number of the Fourier harmonic. *The M separate harmonics are not coupled.* In practice, matrices are not built for all harmonics at once (as Eqs. 10.6-7 seem to imply). Instead, as suggested by Eq. 10.5-7, separate harmonics of loading are analyzed serially, with results stored for subsequent superposition.

Element loads (Eq. 4.1-6) include contributions such as

$$\{\mathbf{r}\} = \int_{V_e} \underset{3JM\times 3}{[\mathbf{N}]^T} \underset{3\times 1}{\{\mathbf{F}\}}\, dV + \int_{V_e} \underset{3JM\times 6}{[\mathbf{B}]^T} \underset{6\times 6}{[\mathbf{E}]} \underset{6\times 1}{\{\boldsymbol{\epsilon}_0\}}\, dV \tag{10.6-8}$$

where $[\mathbf{N}]$ and $[\mathbf{B}]$ are given by Eqs. 10.6-3a and 10.6-4. Body forces $\{\mathbf{F}\}$ are, from Eq. 10.5-8,

$$\{\mathbf{F}\} = \begin{Bmatrix} \bar{F}_{r0} + \bar{F}_{r1}\cos\theta + \bar{F}_{r2}\cos 2\theta + \cdots \\ 0 \quad + \bar{F}_{\theta 1}\sin\theta + \bar{F}_{\theta 2}\sin 2\theta + \cdots \\ \bar{F}_{z0} + \bar{F}_{z1}\cos\theta + \bar{F}_{z2}\cos 2\theta + \cdots \end{Bmatrix} \tag{10.6-9}$$

Initial strains $\{\boldsymbol{\epsilon}_0\}$ are written similarly. Integration of Eq. 10.6-8 according to Eqs. 10.4-2 shows that $\{\bar{\mathbf{r}}\}_0$ in Eq. 10.6-7 contains only the zero-harmonic (axially symmetric) load terms, $\{\bar{\mathbf{r}}\}_1$ contains only the first-harmonic load terms, and so on. Thus again we see the uncoupling of harmonics.

In the pure torsion harmonic $n = 0$, nodal d.o.f. $\bar{\bar{u}}_{ni}$ and $\bar{\bar{w}}_{ni}$ have no stiffness associated with them. In the axially symmetric harmonic $n = 0$, nodal d.o.f. $\bar{v}_{ni}$ have no stiffness associated with them. These d.o.f. must be suppressed in Eqs. 10.6-7.

Stresses in an element are computed in the usual way, that is, by the equation $\{\boldsymbol{\sigma}\} = [\mathbf{E}]([\mathbf{B}]\{\mathbf{d}\} - \{\boldsymbol{\epsilon}_0\})$, in which $[\mathbf{B}]$ is as stated in Eq. 10.6-4. Thus stresses from the various harmonics are superposed.

Antisymmetric Harmonics. When the double-barred terms in Eqs. 10.5-8 and 10.5-9 are used, the preceding arguments are almost unchanged. One finds that sin $n\theta$ and cos $n\theta$ are interchanged in Eqs. 10.6-2, 10.6-5, and 10.6-9. In addition, algebraic signs change in the last two rows of $[\mathbf{B}]_n$ in Eq. 10.6-5. However, one finds that stiffness matrices $[\mathbf{k}]_0$, $[\mathbf{k}]_1$, and so on, in Eq. 10.6-7 are *identical* to those obtained in the symmetric case. This convenience is the motivation for the arbitrarily chosen negative sign in Eqs. 10.5-9: if the sign were positive instead, the $[\mathbf{k}]_i$ for a given i would differ between symmetric and antisymmetric cases.

More General Elastic Properties. If θ is not a principal material direction, $[\mathbf{E}]$ in Eq. 10.5-1 becomes a full matrix, and each stress in Eq. 10.5-5 depends on both sin $n\theta$ and cos $n\theta$. Symmetric and antisymmetric terms are now coupled in each harmonic, but different harmonics are uncoupled. Thus Eqs. 10.6-7 are still valid, but each $[\mathbf{k}]_n$ is now $6J$ by $6J$ in size. Details of these arguments appear in [10.5].

The problem is more difficult if elastic properties depend on θ. One physical cause of this circumstance is the combination of temperature-dependent moduli and a θ-dependent temperature field. A Fourier series attack can again be used, but all harmonics are coupled [10.6,10.7].

10.7 RELATED PROBLEMS

The Fourier series treatment described in Sections 10.5 and 10.6 is also known as the *semianalytical method* and the *separation of variables method*. When used for plates, it is called the *finite strip method*.

Besides its application to plates and to solids and shells of revolution, the Fourier series method can be applied to prismatic solids [10.8,10.9]. Then the name *finite prism method* may be used. The solid, and its elements, are prismatic (Fig. 10.7-1). The displacement field is again Eq. 10.5-9, except that $\pi y/L$ replaces θ. If only the single-barred series are used, deformation and loading are symmetric about the xz plane, with $v = 0$ at $y = 0$ and at $y = \pm L$. As usual, arbitrary loads and displacements are treated by determining their Fourier coefficients and making a separate analysis for each, then superposing results. Problems such as that of Fig. 10.7-1 may require 9 to 19 Fourier coefficients.

In a physical sense, what has been done in Fig. 10.7-1 is to take a toroidal solid that extends from $-\pi$ to π and straighten it out to form a prismatic solid that extends from $-L$ to L. The straightening can be "faked" by moving the z axis

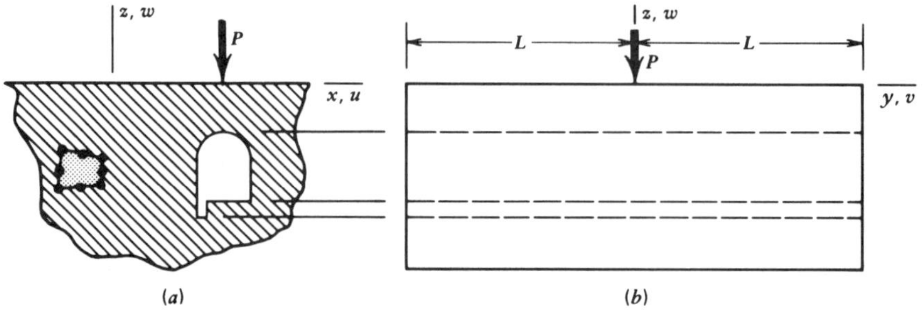

Figure 10.7-1. A point load P over a long tunnel. A suitably large portion of the surrounding earth or rock is modeled by finite elements, one of which is shown and shaded. (*a*) Front view. (*b*) Right-side view.

in Fig. 10.7-1*a* far to the left and making it an axis of revolution. Then the almost-prismatic solid can be analyzed by a computer program for solids of revolution.

The problem of a curved beam bent by a moment M_0 is axially symmetric in geometry and material properties. But it is not obvious that an axisymmetric analysis can deal with a moment loading. Reference 10.10 describes how. The trick is to use a thermal load to simulate the strains produced by M_0. Note that if the curved beam is a thin-walled pipe elbow, its cross section becomes oval in response to M_0 and therefore is more flexible than a pipe whose cross section remains circular. Pipe elbow elements that include this effect have been developed [10.11].

Some geometries are ''almost'' axially symmetric, for example, the geometry or material properties have modest departures from θ independence, or an axially symmetric body is attached to a body that has no symmetry. Aspects of such problems are discussed in [10.7,10.12,10.13].

PROBLEMS

Section 10.2

10.1 If a problem is to be mathematically two-dimensional, θ independence is required of all dependent variables. Explain by example why this requires that θ be a principal direction of an orthotropic material. *Suggestion:* Consider axial load on a cylinder.

Section 10.3

10.2 Imagine that Fig. 6.12-1 depicts displacements and deformations of the square cross section of an axially symmetric four-node element. The axis of revolution is to the left of each cross section. For parts (a) and (b), identify each of the eight modes: that is, is it a rigid-body mode, a zero-energy deformation mode, or a straining mode?

(a) Let [**k**] be generated by one-point Gauss quadrature.

(b) Let [**k**] be generated by four-point Gauss quadrature.

10.3 Arguments are presented in Section 6.11 regarding the number of Gauss points needed for correct volume calculation and correct convergence of computed results. Reference there is to plane elements. How should these arguments be amended if reference is to axially symmetric elements instead?

10.4 Revise Figs. 6.5-1 and 6.5-2 to deal with an axially symmetric problem; that is, describe precisely what changes and additions are necessary.

10.5 Consider an axially symmetric element whose cross section is a three-node triangle. The displacement field has the form seen in Eq. 4.2-9. Determine stiffness matrix $[\mathbf{k}_a]$, which is defined by Eq. 4.1-18, to the extent of writing the integrand in detail and computing the product $[\mathbf{B}_a]^T[\mathbf{E}][\mathbf{B}_a]$. Use $[\mathbf{E}]$ from Eq. 10.2-2.

10.6 The sketch shows the cross section of a flat element shaped like a metal washer. D.o.f. are radial displacements u_1 and u_2 at nodal circles 1 and 2. The material is isotropic.

(a) Formulate matrices $[\mathbf{N}]$ and $[\mathbf{B}]$.

(b) Let $\nu = 0$, and generate $[\mathbf{k}]$ for a one-radian segment by explicit integration.

(c) Let $L = r_2 - r_1$ and $r_m = (r_1 + r_2)/2$. Simplify integration by assuming that $r = r_m$. Hence, determine $[\mathbf{k}]$ (for a nonzero Poisson's ratio). For what geometry is this $[\mathbf{k}]$ a good approximation?

(d) For $\nu = 0$, show that the $[\mathbf{k}]$'s of parts (b) and (c) agree for $r_m >> L$.

(e) From part (b), obtain $[\mathbf{k}]$ for the special case $\nu = r_1 = u_1 = 0$.

(f) From part (c), obtain $[\mathbf{k}]$ for the special case $\nu = r_1 = u_1 = 0$.

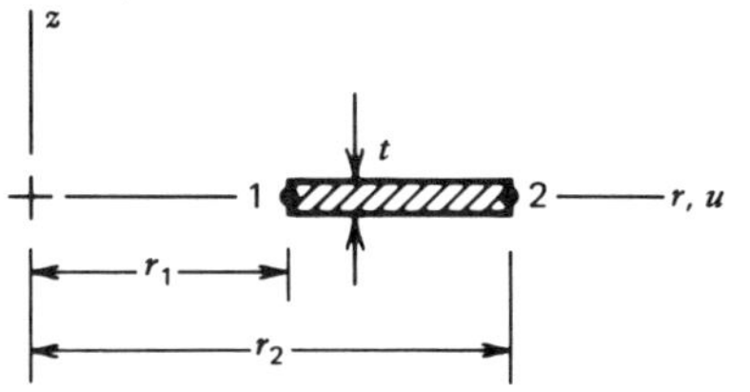

Problem 10.6

10.7 The four-node axisymmetric element shown has a rectangular cross section of dimensions $2a$ and $2b$. The material is homogeneous, isotropic, and has mass density ρ. Evaluate element nodal loads $\{\mathbf{r}_e\}$ produced by the following actions.

(a) Uniform radial pressure p_1 (tensile).

(b) Uniform radial pressure p_2 (tensile).

(c) Line load q_1 (units N/m), which acts at $x = y = 0$.

(d) Line load q_2 (units N/m), which acts at $x = 0$, $y = b$.

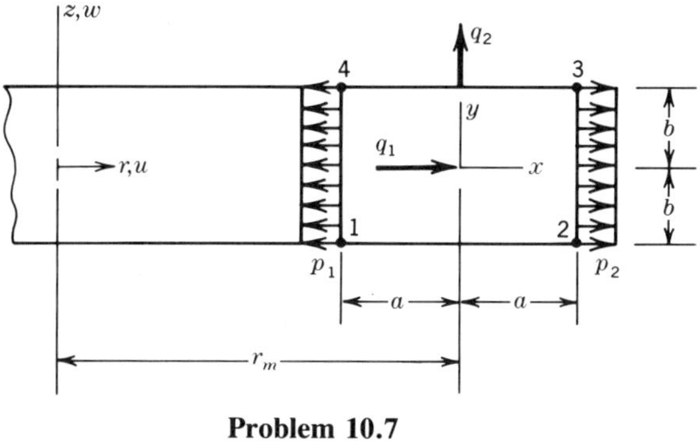

Problem 10.7

(e) Uniform heating an amount T. Stop after setting up a triple integral over $d\theta\, dx\, dy$.

(f) Rotation at constant angular velocity ω. Stop after setting up a triple integral over $d\theta\, dx\, dy$.

10.8 The sketch represents three nodes on a z = constant face of an axisymmetric quadratic element. Node 7 is at midside. Determine the consistent nodal load vector for these three nodes if z-direction surface traction Φ_z is applied as follows.

(a) Φ_z is the constant value p over the face.

(b) $\Phi_z = (\xi^2 - \xi)p_4/2 + (1 - \xi^2)p_7 + (\xi^2 + \xi)p_3/2$, which is a parabolic variation based on nodal values p_4, p_7, and p_3.

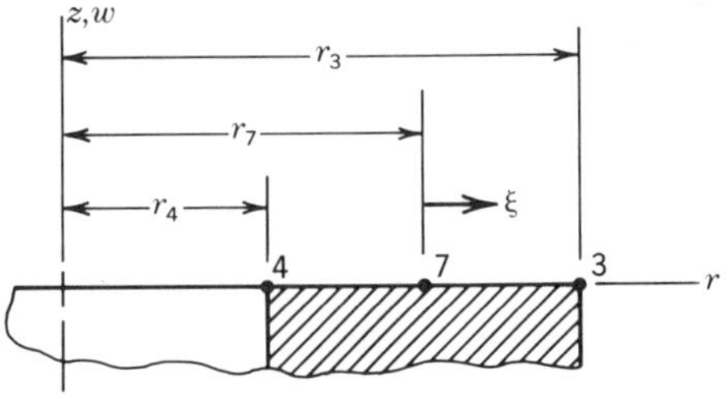

Problem 10.8

10.9 In the sketch for Problem 10.8, imagine that $r_4 = 0$, and that $w_4 > 0$ and $w_7 = w_3 = 0$. Is such a deformation mode reasonable? What do you conclude about γ_{zr} at $r = 0$?

10.10 Show that $\epsilon_r = \epsilon_\theta$ at $r = 0$ in an axially symmetric problem.

Section 10.4

10.11 (a) Use Eqs. 10.4-2 to verify the right-hand sides of Eqs. 10.4-3.

(b) Verify the expression for q_n in Eq. 10.4-4.

(c) Verify the expression for q_n in Eq. 10.4-5. *Suggestion:* Imagine that P is generated by a distributed load of large intensity acting over a small length.

(d) Determine a Fourier series that represents two radial loads P on a disk, one outward at $\theta = -\pi/2$ and another outward at $\theta = +\pi/2$.

10.12 Use Eqs. 10.4-12 to compute the center deflection and center bending moment in a uniform simply supported beam. Use one, then two, then three series terms, each time computing the percentage error of the approximation.

(a) Consider a uniformly distributed upward load.

(b) Consider a concentrated downward load at midspan.

10.13 A uniform, simply supported beam is loaded by a linearly varying distributed load that has intensity q_L at $x = L$, as shown.

(a) Determine a Fourier series representation of the load.

(b) Hence, determine the deflection and the bending moment at $x = L/2$, using 1, 2, and 3 series terms. Compute the percentage error of each of these approximations.

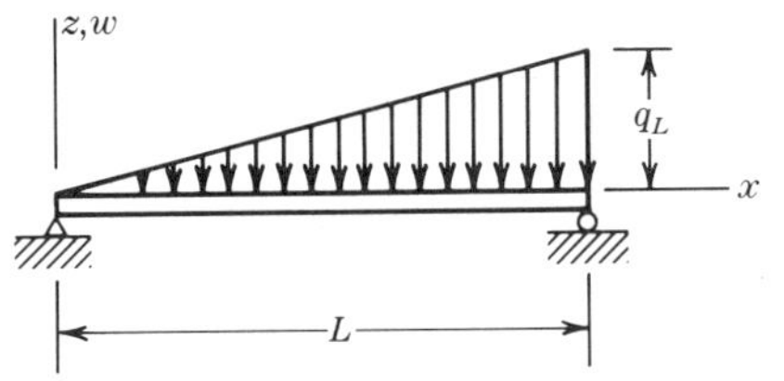

Problem 10.13

Section 10.5

10.14 Assume that a certain axially symmetric pressure vessel can be adequately modeled by four-node elements (as in Fig. 10.1-1*b*), arranged in a 20 by 1 mesh, so that one element spans the wall thickness of the vessel and 20 elements span the axial dimension. Also assume that the loading is described by Eq. 10.5-2, with harmonics $n = 1, 2, 3, 4, 5$, and 6. Alternatively, one could contemplate a fully three-dimensional analysis with a 20 by 1 by m mesh of eight-node solid elements, where m is the number of elements around the circumference.

(a) Estimate m so that the three-dimensional model would be of adequate accuracy. Assume that three elements per half-wave of displacement are acceptable.

(b) Make an estimate of the cost ratio of the three-dimensional solution to the series solution. Base your estimate on the expense of generating stiffness matrices with an order 2 Gauss rule.

(c) Repeat part (b), but base your estimate on the expense of solving equations with a banded equation solver (see Appendix B).

10.15 If $\{\boldsymbol{\epsilon}\} = \{\mathbf{0}\}$, Eqs. 10.5-4 have the solution

$$\begin{Bmatrix} u \\ v \\ w \end{Bmatrix} = \begin{bmatrix} 0 & \cos\theta & z\cos\theta & 0 & \sin\theta & z\sin\theta \\ 0 & -\sin\theta & -z\sin\theta & r & \cos\theta & z\cos\theta \\ 1 & 0 & -r\cos\theta & 0 & 0 & -r\sin\theta \end{bmatrix} \begin{Bmatrix} a_1 \\ a_2 \\ \cdot \\ \cdot \\ \cdot \\ a_6 \end{Bmatrix}$$

where the a_i are constants. A displacement field must contain these terms if there is to be rigid-body motion without strain.

(a) Show that this field does in fact yield $\{\boldsymbol{\epsilon}\} = \{\mathbf{0}\}$.

(b) Compare this field with Eqs. 10.5-9: identify columns of the rectangular matrix as to the value of n and as to belonging to the single-barred or the double-barred series.

(c) For each of the six columns of the rectangular matrix, describe the physical meaning of the displacement mode it represents.

10.16 Specialize Eqs. 10.5-9 to represent the following rigid-body motions.

(a) Axial translation.

(b) Translation perpendicular to the z axis in the plane $\theta = 0$.

(c) Translation perpendicular to the z axis in the plane $\theta = \pi/2$.

(d) Rotation about the z axis.

(e) Rotation about the line $\theta = z = 0$.

(f) Rotation about the line $\theta = \pi/2$, $z = 0$.

10.17 Seven forces are applied to one end of a cylindrical bar of outer radius c, as shown. Forces P_4 form the couple $2P_4c$. Stresses at midheight $z = h/2$ could be calculated by elementary formulas $\sigma = Mc/I$, and so on. However, imagine that, as an exercise, finite elements are to be used instead, so loads must be expressed in the form of Eqs. 10.5-8. For each of the six different loadings, use Eqs. 10.5-8 to write expressions for surface tractions on end $z = h$ that are statically equivalent to the original loading. (This can be done using only terms for which $n = 0$ and/or $n = 1$). Express your answers in terms of c, the P_i, $\sin \theta$, and $\cos \theta$.

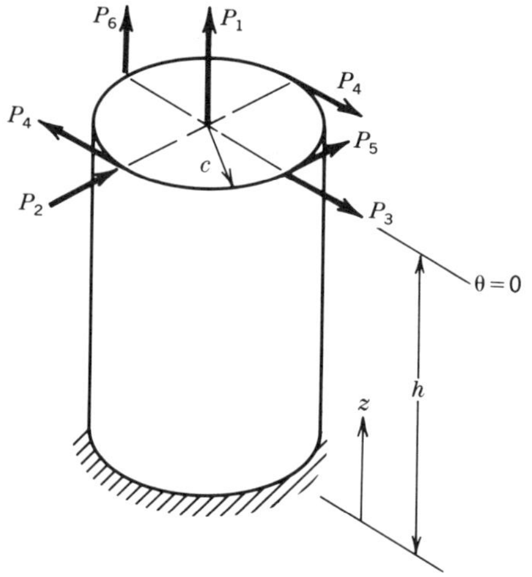

Problem 10.17

10.18 A flat plate containing a circular hole of radius R is loaded by axial force P and bending moment M, as shown. The region enclosed by the dashed line of radius c is to be isolated and analyzed as a solid of revolution. If t = plate thickness and $c >> R$, what load terms from Eqs. 10.5-8 should be used in analysis? Express your answers in terms of P, M, h, t, n, and θ.

(a) Consider P only (let $M = 0$).
(b) Consider M only (let $P = 0$).

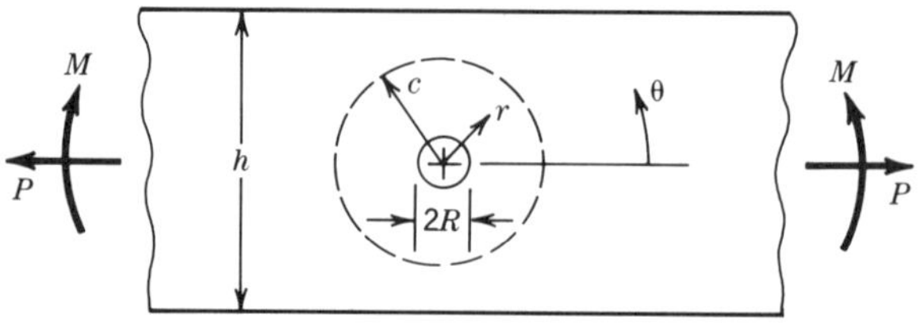

Problem 10.18

Section 10.6

10.19 Assume that [E] has the form shown in Eq. 10.5-1. Use the first partition of $[\mathbf{B}]_n$, as stated in Eq. 10.6-5, to demonstrate the following about the matrix product $[\mathbf{P}] = [\mathbf{B}]^T[\mathbf{E}][\mathbf{B}]$. (*Note:* It is not necessary to write out details of terms extraneous to the question posed.)

(a) Show that [**P**] contains $\sin^2 n\theta$ or $\cos^2 n\theta$ in each term of an on-diagonal submatrix.
(b) Show that [**P**] contains either $\sin m\theta \sin n\theta$ or $\cos m\theta \cos n\theta$, where $m \neq n$, in each term of an off-diagonal submatrix.
(c) Show that each term of an on-diagonal submatrix has the form $A + Bn^2$ or the form Cn.

10.20 (a) Write the form of $[\mathbf{B}]_n$ in Eq. 10.6-5 appropriate to antisymmetric terms (the double-barred series in Eq. 10.5-9).
(b) Show that $[\mathbf{k}]_0$, $[\mathbf{k}]_1$, and so on, are identical to the corresponding matrices obtained for symmetric terms. *Suggestion:* See the note in Problem 10.19.
(c) Show that the conclusion reached in part (b) would not be true if the negative sign in Eq. 10.5-9b were changed to positive.

10.21 Consider the flat element analyzed in Problem 10.6. However, now allow circumferential displacement v as well as radial displacement u, so that loads without axial symmetry can be treated. Element nodal d.o.f. are now u_1, v_1, u_2, and v_2. Let $\theta = 0$ be a plane of symmetry. Formulate $[\mathbf{B}]_n$ for this element (analogous to $[\mathbf{B}]_n$ in Eq. 10.6-5, but including all partitions).

10.22 The element described in Problem 10.21 can be used to solve problems of disks and rings under concentrated loads (see sketch). Why do Fourier harmonics for displacement and stress form convergent series, although the series for concentrated loads are not convergent?

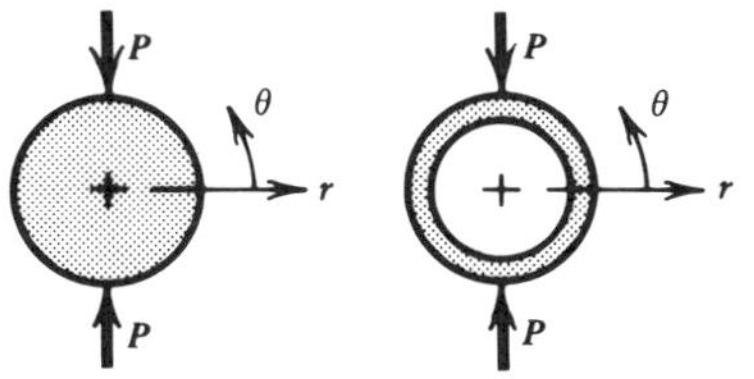

Problem 10.22

11 CHAPTER

BENDING OF FLAT PLATES

Concepts and equations related to the bending of flat plates are reviewed. Elements for thin plates and plates having transverse shear deformation are discussed. Test problems for plate elements are presented.

11.1 PLATE-BENDING THEORY

Loads, Stresses, and Moments. A flat plate, like a straight beam, supports transverse loads by bending action. Figure 11.1-1*a* shows stresses that act on cross sections of a plate whose material is homogeneous and linearly elastic. Normal stresses σ_x and σ_y vary linearly with z and are associated with bending moments M_x and M_y. Shear stress τ_{xy} also varies linearly with z and is associated with twisting moment M_{xy}. Normal stress σ_z is considered negligible in comparison with σ_x, σ_y, and τ_{xy}. Transverse shear stresses τ_{yz} and τ_{zx} vary quadratically with z. Lateral load q includes surface load and body force, both in the z direction. Unless stated otherwise, "plate bending" means that external loads have no components parallel to the xy plane and that $\sigma_x = \sigma_y = \tau_{xy} = 0$ on the midsurface $z = 0$. Excepting stress τ_{xy}, the foregoing stress patterns are a direct extension of beam theory from one dimension to two.

Stresses in Fig. 11.1-1 produce the following bending moments M and transverse shear forces Q:

$$M_x = \int_{-t/2}^{t/2} \sigma_x z\, dz \qquad M_y = \int_{-t/2}^{t/2} \sigma_y z\, dz \qquad M_{xy} = \int_{-t/2}^{t/2} \tau_{xy} z\, dz \tag{11.1-1a}$$

$$Q_x = \int_{-t/2}^{t/2} \tau_{zx}\, dz \qquad Q_y = \int_{-t/2}^{t/2} \tau_{yz}\, dz \tag{11.1-1b}$$

The M's are moments *per unit length* and the Q's are forces *per unit length.* Differential *total* moments and forces are $M_x\, dy$, $Q_x\, dy$, and so on, as shown in Fig. 11.1-1*b*. Stresses σ_x, σ_y, and τ_{xy} are largest at the surfaces $z = \pm t/2$, where they have the respective magnitudes $6M_x/t^2$, $6M_y/t^2$, and $6M_{xy}/t^2$. At arbitrary values of z,

$$\sigma_x = \frac{M_x z}{t^3/12} \qquad \sigma_y = \frac{M_y z}{t^3/12} \qquad \tau_{xy} = \frac{M_{xy} z}{t^3/12} \tag{11.1-2}$$

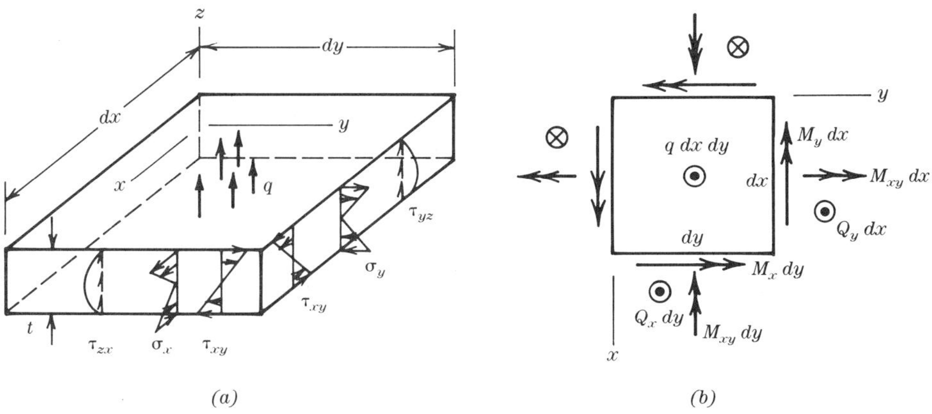

Figure 11.1-1. (*a*) Stresses that act on a differential element of a homogeneous, linearly elastic plate. The distributed lateral load is q (force per unit area). (*b*) The same differential element, viewed normal to the plate. Forces ⊙ and ⊗ act in the positive and negative z directions, respectively.

as may be verified by substituting Eqs. 11.1-2 into Eqs. 11.1-1a. Transverse shear stresses are usually small in comparison with σ_x, σ_y, and τ_{xy}. They have greatest magnitude at $z = 0$, where $\tau_{yz} = 1.5Q_y/t$ and $\tau_{zx} = 1.5Q_x/t$.

Deformations (Kirchhoff Theory). Points on the midsurface $z = 0$ move in only the z direction as the plate deforms in bending. A line that is straight and normal to the midsurface before loading is assumed to remain straight and normal to the midsurface after loading (see line OP in Fig. 11.1-2). Thus transverse shear deformation is assumed to be zero. A point not on the midsurface has displacement components u and v in the x and y directions, respectively. From Fig. 11.1-2, with $w_{,x}$ and $w_{,y}$ small angles of rotation,

$$u = -zw_{,x} \qquad v = -zw_{,y} \qquad \text{hence} \qquad \begin{aligned} \epsilon_x &= u_{,x} &&= -zw_{,xx} \\ \epsilon_y &= v_{,y} &&= -zw_{,yy} \\ \gamma_{xy} &= u_{,y} + v_{,x} &&= -2zw_{,xy} \end{aligned} \qquad (11.1\text{-}3)$$

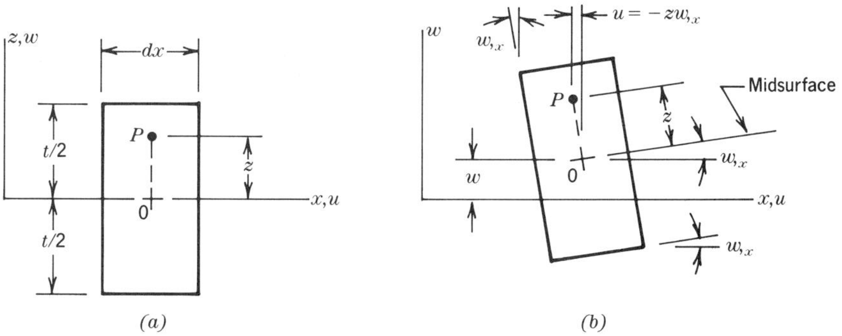

Figure 11.1-2. (*a*) Differential element of a thin plate before loading. (*b*) After loading: deformations associated with Kirchhoff plate theory. Point P displaces w units up and $zw_{,x}$ units leftward because of midsurface displacement w and small rotation $w_{,x}$.

These are the strain–displacement relations of Kirchhoff plate theory, which is applicable to a thin plate.

Deformations (Mindlin Theory). A line that is straight and normal to the midsurface before loading is assumed to remain straight but not necessarily normal to the midsurface after loading. Thus, transverse shear deformation is allowed. The motion of a point not on the midsurface is not governed by slopes $w_{,x}$ and $w_{,y}$ as in Kirchhoff theory. Rather, its motion depends on rotations θ_x and θ_y of lines that were normal to the midsurface of the undeformed plate (Fig. 11.1-3). Thus, with θ_x and θ_y small angles of rotation,

$$\begin{aligned} u &= -z\theta_x \\ v &= -z\theta_y \end{aligned} \qquad \begin{aligned} \epsilon_x &= -z\theta_{x,x} \\ \epsilon_y &= -z\theta_{y,y} \end{aligned} \qquad \begin{aligned} \gamma_{xy} &= -z(\theta_{x,y} + \theta_{y,x}) \\ \gamma_{yz} &= w_{,y} - \theta_y \\ \gamma_{zx} &= w_{,x} - \theta_x \end{aligned} \tag{11.1-4}$$

The foregoing expressions for strain are obtained by straightforward application of Eqs. 1.5-4 and 1.5-5. Equations 11.1-4 are the strain–displacement relations of Mindlin plate theory. This theory accounts for transverse shear deformation and is therefore especially suited to the analysis of thick plates and sandwich plates.

Moment–Curvature Relations (Kirchhoff Theory). We begin with stress–strain relations. Let x and y be principal directions of an orthotropic material. Stress σ_z is considered negligible in comparison with σ_x, σ_y, and τ_{xy}. Transverse shear strains are also considered neligible, so stress–strain relations that involve them need not be written. What remains is the plane stress–strain relation $\{\boldsymbol{\sigma}\} = [\mathbf{E}](\{\boldsymbol{\epsilon}\} - \{\boldsymbol{\epsilon}_0\})$; that is [11.1],

$$\begin{Bmatrix} \sigma_x \\ \sigma_y \\ \tau_{xy} \end{Bmatrix} = \begin{bmatrix} E_x' & E'' & 0 \\ E'' & E_y' & 0 \\ 0 & 0 & G \end{bmatrix} \left(\begin{Bmatrix} \epsilon_x \\ \epsilon_y \\ \gamma_{xy} \end{Bmatrix} - \begin{Bmatrix} \alpha_x T \\ \alpha_y T \\ 0 \end{Bmatrix} \right) \tag{11.1-5}$$

where initial strains $\{\boldsymbol{\epsilon}_0\}$ are presumed caused by thermal expansion with principal expansion coefficients α_x and α_y. For an isotropic material, with E = elastic modulus and ν = Poisson's ratio,

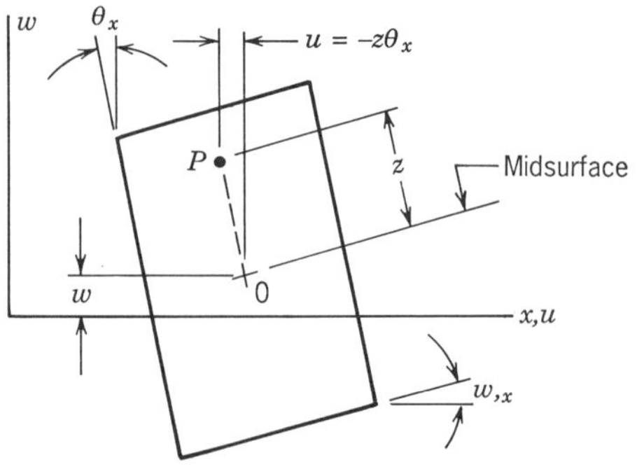

Figure 11.1-3. Differential plate element after loading, analogous to Fig. 11.1-2*b*, but with transverse shear deformation allowed ($w_{,x} \neq \theta_x$, so that $\gamma_{zx} = w_{,x} - \theta_x \neq 0$).

$$E'_x = E'_y = \frac{E''}{\nu} = \frac{E}{1 - \nu^2} \quad \text{and} \quad G = \frac{E}{2(1 + \nu)} \tag{11.1-6}$$

The moment–curvature relation is obtained by substitution of Eqs. 11.1-3 into Eq. 11.1-5 and the result into Eqs. 11.1-1a. This process yields

$$\{\mathbf{M}\} = -[\mathbf{D}_K](\{\boldsymbol{\kappa}\} - \{\boldsymbol{\kappa}_0\}) \tag{11.1-7}$$

where moments and curvatures are

$$\{\mathbf{M}\} = \lfloor M_x \quad M_y \quad M_{xy} \rfloor^T \quad \text{and} \quad \{\boldsymbol{\kappa}\} = \lfloor w_{,xx} \quad w_{,yy} \quad 2w_{,xy} \rfloor^T \tag{11.1-8}$$

In $[\mathbf{D}_K]$ we have $D_{K13} = D_{K31} = D_{K23} = D_{K32} = 0$ and the nonzero terms

$$D_{K11} = \frac{E'_x t^3}{12} \qquad D_{K12} = D_{K21} = \frac{E'' t^3}{12} \qquad D_{K22} = \frac{E'_y t^3}{12} \qquad D_{K33} = \frac{G t^3}{12} \tag{11.1-9}$$

If the material is isotropic, then

$$[\mathbf{D}_K] = \begin{bmatrix} D & \nu D & 0 \\ \nu D & D & 0 \\ 0 & 0 & (1 - \nu)D/2 \end{bmatrix}, \quad \text{where} \quad D = \frac{Et^3}{12(1 - \nu^2)} \tag{11.1-10}$$

D is called "flexural rigidity" and is analogous to bending stiffness EI of a beam. Indeed, if the plate has unit width and $\nu = 0$, then $D = EI = Et^3/12$.

As a particular example of initial curvatures $\{\boldsymbol{\kappa}_0\}$, consider a temperature gradient $T = -2zT_0/t$. This is a linear temperature variation from T_0 at $z = -t/2$ to $-T_0$ at $z = t/2$. Thus the process that yields Eq. 11.1-7 gives the initial curvatures

$$\{\boldsymbol{\kappa}_0\} = \lfloor 2\alpha_x T_0/t \quad 2\alpha_y T_0/t \quad 0 \rfloor^T \tag{11.1-11}$$

Equation 11.1-7 shows that actions in the x and y directions are coupled, even for an isotropic plate. In Fig. 11.1-4*a*, $w_{,yy}$ is constant and $w_{,xx} = w_{,xy} = 0$ in the central portion, but M_x is nonzero because of the Poisson effect. But $M_x = 0$ at the free edges $x = \pm a$, so these edges curl a bit (Fig. 11.1-4*b*). Only if $a \approx t$, so that $M_x \approx 0$ throughout, does the plate act like a beam, displaying the familiar anticlastic surface. The pure twist of Fig. 11.1-4c is associated with moments $-M_{xy}$ alone ($M_x = M_y = 0$) if the plate is isotropic.

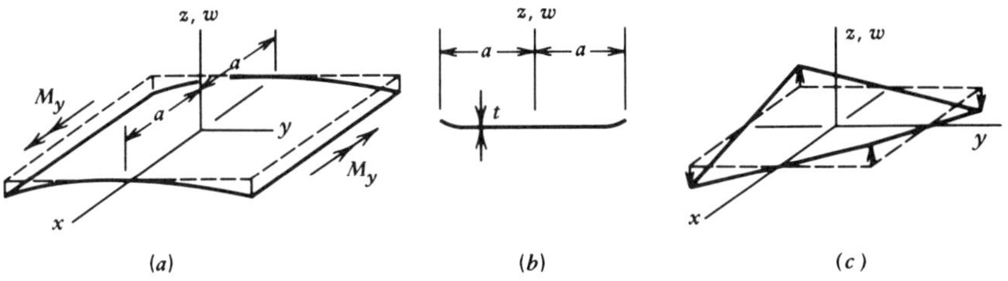

Figure 11.1-4. (*a*) Bending to a cylindrical surface by moments M_y on the edges y = constant. (*b*) Cross section cut by the xz plane. (*c*) The $w = xy$ state of pure twist: $w_{,xx} = w_{,yy} = 0$, $w_{,xy} > 0$.

Moment–Curvature Relations (Mindlin Theory). Again let x and y be principal material directions. The moment–curvature relations of Mindlin plate theory are obtained by essentially the same procedure as used to obtain Eq. 11.1-7. However, we must use Eqs. 11.1-4 instead of Eqs. 11.1-3 and include the shear stress–strain relations $\tau_{yz} = G_{yz}\gamma_{yz}$ and $\tau_{zx} = G_{zx}\gamma_{zx}$. The resulting moment-curvature relation is abbreviated as $\{\mathbf{M}\} = -[\mathbf{D}_M](\{\boldsymbol{\kappa}\} - \{\boldsymbol{\kappa}_0\})$. Written out, this relation is

$$\begin{Bmatrix} M_x \\ M_y \\ M_{xy} \\ Q_y \\ Q_x \end{Bmatrix} = -\underbrace{\begin{bmatrix} & & & 0 & 0 \\ & [\mathbf{D}_K]_{3\times 3} & & 0 & 0 \\ & & & 0 & 0 \\ 0 & 0 & 0 & G_{yz}t & 0 \\ 0 & 0 & 0 & 0 & G_{zx}t \end{bmatrix}}_{[\mathbf{D}_M]} \left(\underbrace{\begin{Bmatrix} \theta_{x,x} \\ \theta_{y,y} \\ \theta_{x,y} + \theta_{y,x} \\ \theta_y - w_{,y} \\ \theta_x - w_{,x} \end{Bmatrix}}_{\{\boldsymbol{\kappa}\}} - \{\boldsymbol{\kappa}_0\} \right) \tag{11.1-12}$$

where $[\mathbf{D}_K]$ is the same as in Eq. 11.1-7. The shear stiffness terms $G_{yz}t$ and $G_{zx}t$ in Eq. 11.1-12 may be replaced by $G_{yz}t/1.2$ and $G_{zx}t/1.2$ to permit the parabolic distributions of τ_{yz} and τ_{zx} (shown in Fig. 11.1-1*a*) to be replaced by uniform distributions, as explained in Section 9.4. If represented as rotation vectors by the right-hand rule, θ_x and θ_y point in the $-y$ and $+x$ directions, respectively. Initial curvatures $\{\boldsymbol{\kappa}_0\}$ are those of Kirchhoff theory, augmented by zeros in positions 4 and 5.

If the plate is isotroptic, then $G_{yz} = G_{zx} = G$ and Eqs. 11.1-10 apply to submatrix $[\mathbf{D}_K]$ in Eq. 11.1-12. For an isotropic sandwich plate, Fig. 11.1-5, with thin facings, G the shear modulus of the core, and E and ν the elastic modulus and Poisson ratio of each facing,

$$\begin{aligned} D_{M11} &= D_{M22} = \frac{D_{M12}}{\nu} = \frac{D_{M21}}{\nu} = \frac{Eh(c+h)^2}{2(1-\nu^2)} \\ D_{M33} &= \frac{Eh(c+h)^2}{4(1+\nu)} \qquad D_{M44} = D_{M55} = \frac{G(c+h)^2}{c} \end{aligned} \tag{11.1-13}$$

and all other entries in the 5 by 5 matrix $[\mathbf{D}_M]$ of Eq. 11.1-12 are zero [11.2].

If principal material directions are x' and y' rather than x and y, as in Fig. 7.3-1, then coordinate transformation is required. Arrays in Eq. 11.1-7 transform as $\{\mathbf{M}\} = [\mathbf{T}_\epsilon]^T\{\mathbf{M}'\}$, $\{\boldsymbol{\kappa}'\} = [\mathbf{T}_\epsilon]\{\boldsymbol{\kappa}\}$, and $[\mathbf{D}_K] = [\mathbf{T}_\epsilon]^T[\mathbf{D}_K'][\mathbf{T}_\epsilon]$, where $[\mathbf{T}_\epsilon]$ is given by Eq. 7.3-11. Coefficients in the southeast corner of $[\mathbf{D}_M]$ in Eq. 11.1-12 are $D_{M44} = m_1^2G_{z'x'}t + m_2^2G_{y'z'}t$, $D_{M55} = \ell_1^2G_{z'x'}t + \ell_2^2G_{y'z'}t$, and $D_{M45} = D_{M54} = \ell_1m_1G_{z'x'}t + \ell_2m_2G_{y'z'}t$, where the ℓ's and m's are given in Fig. 7.3-1.

Remarks. A plate can be loaded by distributed lateral load of intensity q and by initial curvatures $\{\boldsymbol{\kappa}_0\}$, as just discussed. Concentrated forces and line loads may

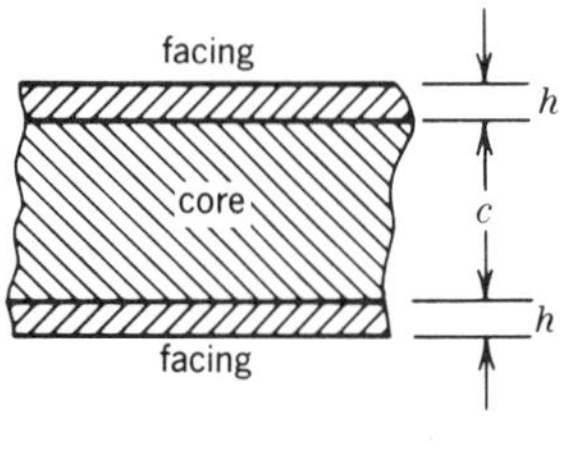

Figure 11.1-5. Cross section of a sandwich plate. Typically the core resists little but transverse shear strains, so almost all bending stiffness is provided by membrane action in thin facings.

also be present. Edge moments M and transverse shears Q may be applied as known loads or as support reactions. Except for line loads, these loads are analogous to loads present in beam theory. Nodal equivalents of loads q and $\{\kappa_0\}$ can be computed by means of Eq. 4.1-6.

In finite element analysis of plates, whether by Kirchhoff or Mindlin theory, d.o.f. at a node i are typically one lateral displacement (w_i) and two rotations ($w_{,xi}$ and $w_{,yi}$ or θ_{xi} and θ_{yi}). At a free edge none of the three d.o.f. is restrained. At a clamped edge all d.o.f. are restrained. Further discussion of boundary conditions appears in Section 11.5.

Full compatibility of interelement displacements requires that, in any z = constant layer, displacements u, v, and w be the same in adjacent elements where the elements meet. Accordingly, from Eqs. 11.1-3, compatible Kirchhoff elements are C^1 elements, as they must display interelement continuity of w, $w_{,x}$, and $w_{,y}$. Note that along (say) a y-parallel interelement boundary, continuity of w ensures continuity of $w_{,y}$ but not continuity of the boundary-normal slope $w_{,x}$. From Eqs. 11.1-4, compatible Mindlin elements are C^0 elements, as the fields, w, θ_x, and θ_y (but not their derivatives) must be interelement-continuous. Note that along (say) a y-parallel interelement boundary, continuity of θ_x does not imply continuity of $w_{,x}$ unless the plate is so thin that $\gamma_{zx} = 0$.

We have tacitly assumed that material properties are either independent of z or symmetric with respect to the midsurface $z = 0$. If not, bending may produce forces in the xy plane so that the midsurface is not a surface where $\sigma_x = \sigma_y = \tau_{xy} = 0$. This effect is pronounced in two-layer laminated plates [11.3].

Appreciable in-plane forces may also arise if deflections w are more than a few tenths of the plate thickness. This happens even when supports apply no in-plane forces, because the deflected shape of the plate requires stretching or shortening in the midsurface (unless deflections are small or the deflected shape is cylindrical or conical). In-plane forces act to support part of the load. Thus the stiffness of a plate effectively increases as deflection increases, which makes the problem nonlinear. In some problems linear theory may overestimate displacements by 50% if deflection w equals thickness t [11.1].

11.2 FINITE ELEMENTS FOR PLATES

A great many finite elements for plates have been proposed: an incomplete survey lists 154 references and 88 different elements [11.4]. In what follows we briefly consider some options in element formulation. Further details may be found in Ref. 11.5 and in papers cited by Ref. 11.4.

Kirchhoff Elements. Kirchhoff theory is applicable to thin plates, in which transverse shear deformation is neglected. Strain energy in the plate is determined entirely by in-plane strains ϵ_x, ϵ_y and γ_{xy}. In turn, strains are determined entirely by the lateral displacement field $w = w(x,y)$, as shown by Eqs. 11.1-3.

The starting point for formulating an element stiffness matrix is the strain energy term of Eq. 4.1-1,

$$U = \int_V \tfrac{1}{2}\{\epsilon\}^T[\mathbf{E}]\{\epsilon\}\, dV, \qquad \text{where} \quad \{\epsilon\}^T = \lfloor -zw_{,xx} \quad -zw_{,yy} \quad -2zw_{,xy} \rfloor \tag{11.2-1}$$

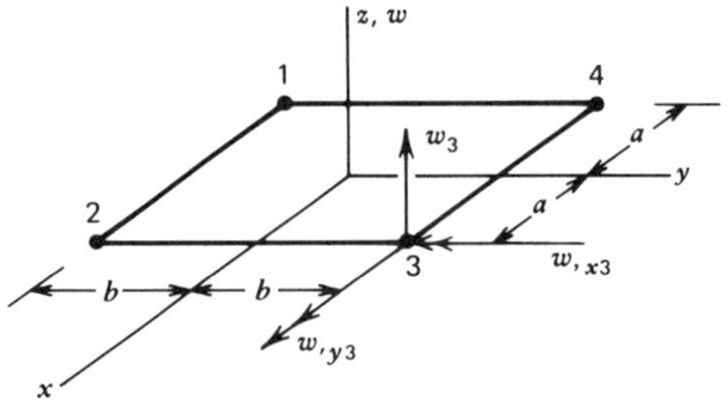

Figure 11.2-1. Twelve-d.o.f. rectangular Kirchhoff plate element, with typical d.o.f. shown at node 3.

and [**E**] is given by Eq. 11.1-5. With $dV = dz\,dA$, where $dA = dx\,dy$ is an increment of midsurface area A, integration through thickness t yields

$$U = \int_A \tfrac{1}{2}\{\boldsymbol{\kappa}\}^T[\mathbf{D}_K]\{\boldsymbol{\kappa}\}\,dA, \qquad \text{where} \quad \{\boldsymbol{\kappa}\}^T = \lfloor w,_{xx} \quad w,_{yy} \quad 2w,_{xy}\rfloor \tag{11.2-2}$$

and [$\mathbf{D}_K$] is the moment–curvature relation of Eq. 11.1-7. An interpolation of w from element nodal d.o.f. {**d**} is devised, then differentiated to yield curvatures {$\boldsymbol{\kappa}$}. For an element having N nodes,

$$w = \underset{1\times 3N}{\lfloor \mathbf{N} \rfloor}\{\mathbf{d}\} \qquad \text{hence} \qquad \{\boldsymbol{\kappa}\} = \underset{3\times 3N}{[\mathbf{B}]}\{\mathbf{d}\} \tag{11.2-3}$$

D.o.f. of a Kirchhoff element are $\{\mathbf{d}\} = \lfloor w_1 \quad w,_{x1} \quad w,_{y1} \ldots w_N \quad w,_{xN} \quad w,_{yN}\rfloor^T$. Finally, by substitution of Eq. 11.2-3 into Eq. 11.2-2, the element stiffness matrix [**k**] appears:

$$U = \tfrac{1}{2}\{\mathbf{d}\}^T[\mathbf{k}]\{\mathbf{d}\}, \qquad \text{where} \quad \underset{3N\times 3N}{[\mathbf{k}]} = \int_A [\mathbf{B}]^T[\mathbf{D}_K][\mathbf{B}]\,dA \tag{11.2-4}$$

For example, consider the twelve-d.o.f. rectangular element of Fig. 11.2-1. Typical d.o.f. $w,_{x3}$ and $w,_{y3}$ are slopes (i.e., rotations) of the plate midsurface at node 3. Their vector representations, shown in Fig. 11.2-1, are determined according to the right-hand rule. Lateral displacement w of this element has the form [11.5]

$$w = \lfloor 1, x, y, x^2, xy, y^2, x^3, x^2y, xy^2, y^3, x^3y, xy^3\rfloor\{\mathbf{a}\} \tag{11.2-5}$$

This element does not preserve interelement continuity of boundary-normal slopes. Vector {**a**} contains twelve generalized coordinates, which must be exchanged for the twelve nodal d.o.f. {**d**} by the usual process (e.g., Eqs. 3.13-3). Thus Eqs. 11.2-3 are established, and [**k**] follows from Eq. 11.2-4.

Early efforts to formulate *triangular* Kirchhoff elements in the same way met with unexpected difficulties. A nine-term field for w is appropriate to the element of Fig. 11.2-2a. Unfortunately, as seen in Eq. 11.2-5, a complete cubic contains 10 terms. Candidate nine-term fields include

$$w = \lfloor 1, x, y, x^2, y^2, x^3, x^2y, xy^2, y^3\rfloor\{\mathbf{a}\} \tag{11.2-6a}$$

$$w = \lfloor 1, x, y, x^2, xy, y^2, x^3, x^2y + xy^2, y^3\rfloor\{\mathbf{a}\} \tag{11.2-6b}$$

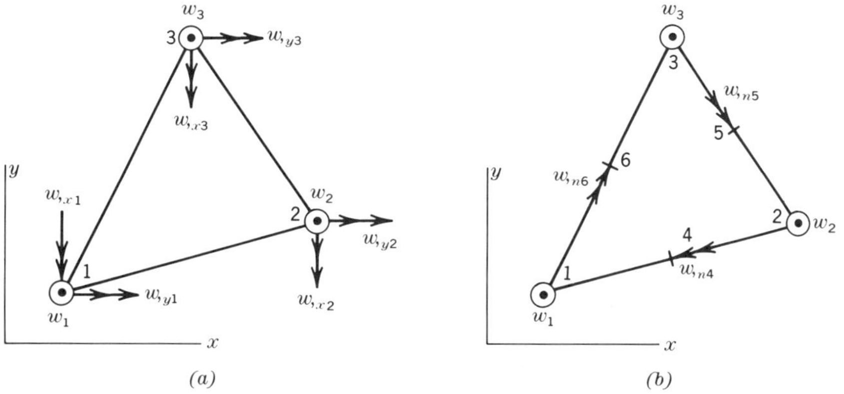

Figure 11.2-2. Triangular Kirchhoff plate elements. (*a*) Nine-d.o.f. element. (*b*) Six-d.o.f., constant-curvature element. The symbol ⊙ indicates an arrow directed out of the paper.

Equation 11.2-6a omits the xy term. The resulting element is therefore incapable of passing a constant-twist patch test, and is considered unacceptable. Equation 11.2-6b leads to an element that lacks geometric isotropy, has poor convergence properties, and for certain element shapes has a singular transformation matrix $[\mathbf{A}]$ in the relation $\{\mathbf{d}\} = [\mathbf{A}]\{\mathbf{a}\}$.

Eventually such difficulties were overcome, often by using subelements, by abandoning strict Kirchhoff plate theory, or by using variational principles other than stationary potential energy. An interesting result of these efforts is the element of Fig. 11.2-2*b*, which appears to be the simplest possible Kirchhoff element. It can represent only rigid-body motion (constant w, $w_{,x}$, or $w_{,y}$) and constant-curvature states (constant $w_{,xx}$, $w_{,yy}$, or $w_{,xy}$). Its six d.o.f. are translations w at corners and normal slopes $w_{,n}$ at midsides, where n is a direction normal to the side.

After nodal d.o.f. have been obtained by solution of the structural equations, element curvatures $\{\boldsymbol{\kappa}\}$ are obtained from Eq. 11.2-3, then bending moments from Eq. 11.1-7, and finally stresses from Eqs. 11.1-2.

Mindlin Elements. Nodal d.o.f. consist of lateral deflections w_i and rotations θ_{xi} and θ_{yi} of midsurface normals. The corresponding deflections and rotations within an element are obtained by independent shape function interpolations:

$$w = \sum N_i w_i \qquad \theta_x = \sum N_i \theta_{xi} \qquad \theta_y = \sum N_i \theta_{yi} \tag{11.2-7}$$

Equations 11.1-4 and 11.2-7 yield strains. We see that the field for w is coupled to the fields for θ_x and θ_y only via the shear strains γ_{yz} and γ_{zx}. Using the strains of Eq. 11.1-4, one can write an expression for strain energy and from it obtain an element stiffness matrix. Mindlin elements are considered in more detail in Section 11.3.

Discrete Kirchhoff Elements. Initially, Eqs. 11.2-7 are again used to express displacements and rotations within an element. However, more d.o.f. are included than are to appear in the final element. Strains of importance in Kirchhoff theory

(ϵ_x, ϵ_y, and γ_{xy} only) are evaluated from Eqs. 11.1-4. Accordingly, element strains and strain energy depend on θ_x and θ_y but are independent of w. Next, using $\gamma_{yz} = \gamma_{zx} = 0$ in Eqs. 11.1-4, we impose the "Kirchhoff constraints" $w_{,y} = \theta_y$ and $w_{,x} = \theta_x$ at certain points. These points are sufficient in number to eliminate the "excess" d.o.f. Thus w becomes coupled to the rotations θ_x and θ_y, and only nodal d.o.f. used for interelement connections remain. Discrete Kirchhoff elements are considered further in Section 11.4.

Finite Strips. The finite strip method [11.7] exploits the "semianalytical" method described in Section 10.5. Imagine that the plate of Fig. 11.2-3 is simply supported along edges $y = 0$ and $y = b$. Lateral displacement w can be taken as

$$w = \sum_n \lfloor \mathbf{N} \rfloor \{\bar{\mathbf{d}}\}_n \sin \frac{n\pi y}{b} \tag{11.2-8}$$

where $\lfloor \mathbf{N} \rfloor$ represents standard cubic shape functions, Fig. 3.13-2, and $\{\bar{\mathbf{d}}\}_n$ contains amplitudes of displacement and rotation along nodal lines 1–1 and 2–2 for mode n. Thus Eq. 11.2-8 expresses the superposition of several solutions, each associated with a single Fourier harmonic of loading. This superposition method replaces division of the plate into elements in the y direction. The finite strip method can also be applied to folded plates and to box beams, either straight or curved. Advantages include modest needs for computer resources and input data. Disadvantages include an inability to cope with arbitrary shapes and arbitrary boundary conditions.

Nodal Loads. Consistent element nodal loads $\{\mathbf{r}_e\}$, Eq. 4.1-6, include terms such as

$$\int_A \lfloor \mathbf{N} \rfloor^T q \, dA \qquad \int \lfloor \mathbf{N}_{,x} \rfloor^T \overline{M}_x \, dy \qquad \text{and} \qquad \int \lfloor \mathbf{N}_{,x} \rfloor^T \overline{M}_{xy} \, dx \tag{11.2-9}$$

where q is the intensity of distributed lateral load, and $\overline{M}_x$ and $\overline{M}_{xy}$ are prescribed boundary values of moment loads. The respective shape function matrices in Eqs. 11.2-9 are associated with lateral displacement, rotation about the y axis along a y-parallel edge that carries $\overline{M}_x$, and rotation about the y axis along an x-parallel edge that carries $\overline{M}_{xy}$.

As a simpler but less accurate alternative to Eq. 4.1-6, a distributed load on a plate element can be "lumped" by assigning equal fractions of the total force on the element to its translational d.o.f. Correct answers are approached with mesh refinement.

For Kirchhoff elements, the first of Eqs. 11.2-9 produces nodal moments as

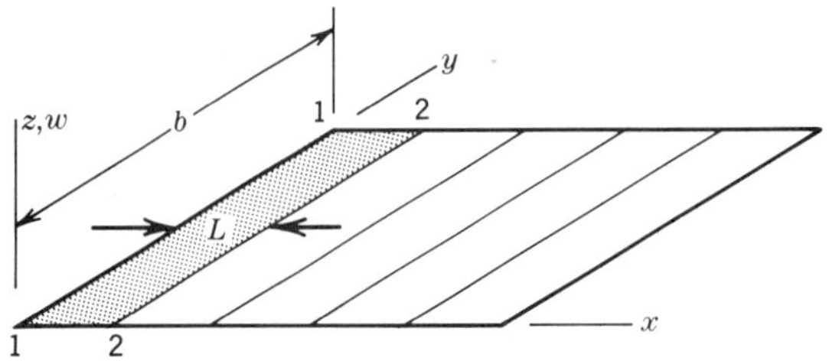

Figure 11.2-3. A thin rectangular plate divided into finite strips. A typical strip is shaded.

well as nodal forces, because $w = \lfloor \mathbf{N} \rfloor \{\mathbf{d}\}$ describes a field dependent on nodal rotations as well as nodal translations. For Mindlin elements, from the first of Eqs. 11.2-7, $w = \lfloor \mathbf{N} \rfloor \{\mathbf{w}\}$ describes a field dependent on only nodal translations, so Eq. 11.2-9 yields only nodal forces. If elements have no internal nodes, a one-element, simply supported, uniformly loaded Mindlin plate would then yield zero deflection, which is unreasonable. Engineers may therefore be willing to forgo mathematical consistency and use in Eq. 11.2-9 a substitute field that provides nodal moments, or they may adopt some other ad hoc strategy in order to improve accuracy in a coarse mesh.

In Kirchhoff plate theory, nodal loads produced by initial curvatures $\{\boldsymbol{\kappa}_0\}$ (e.g., Eq. 11.1-11) are

$$\{\mathbf{r}_e\} = \int_A [\mathbf{B}]^T[\mathbf{D}_K]\{\boldsymbol{\kappa}_0\}\, dA \tag{11.2-10}$$

In Mindlin plate theory, the last two entries in the product $[\mathbf{D}_M]\{\boldsymbol{\kappa}_0\}$ are zero. Accordingly, $[\mathbf{B}]$ can be truncated to three rows, so that $[\mathbf{B}]\{\mathbf{d}\}$ yields only $\theta_{x,x}$, $\theta_{y,y}$, and $\theta_{x,y} + \theta_{y,x}$ (see Eq. 11.3-4). Then Eq. 11.2-10 yields nodal moments produced by $\{\boldsymbol{\kappa}_0\}$, but no z-direction nodal forces.

11.3 MINDLIN PLATE ELEMENTS

Mindlin plate elements account for bending deformation and for transverse shear deformation. Accordingly, they may be used to analyze thick plates as well as thin plates. When used for thin plates, however, they may be less accurate than Kirchhoff elements, which do not allow transverse shear deformation.

Typical Mindlin plate elements are shown in Fig. 11.3-1. For convenience of notation, but not because of any demand of finite element plate theory, we will assume that all three d.o.f. shown in Fig. 11.3-1*c* are present at every node. Rotations θ_x and θ_y are shown by two-headed arrows according to the right-hand rule. These are rotations of a line that was normal to the midsurface of the undeformed plate. Note that $\theta_x \neq w_{,x}$ and $\theta_y \neq w_{,y}$ unless we approach the thin-plate limit, in which case $\gamma_{zx} = \gamma_{yz} = 0$ (see Fig. 11.1-3). A special form of Mindlin plate element is the Mindlin beam element, which the reader may wish to review (see Eqs. 9.4-1 and 9.4-2 and Fig. 9.4-1).

Stiffness Matrix. The starting point for formulating an element stiffness matrix is an expression for strain energy U. With A the area of the plate midsurface,

$$U = \tfrac{1}{2} \int_A \int_{-t/2}^{t/2} \{\boldsymbol{\epsilon}\}^T[\mathbf{E}]\{\boldsymbol{\epsilon}\}\, dz\, dA \tag{11.3-1}$$

where $\{\boldsymbol{\epsilon}\}^T = \lfloor \epsilon_x \quad \epsilon_y \quad \gamma_{xy} \quad \gamma_{yz} \quad \gamma_{zx} \rfloor$, and the individual strains are stated in terms of displacements by Eqs. 11.1-4. Integration through the thickness yields

$$U = \tfrac{1}{2} \int_A \{\boldsymbol{\kappa}\}^T[\mathbf{D}_M]\{\boldsymbol{\kappa}\}\, dA \tag{11.3-2}$$

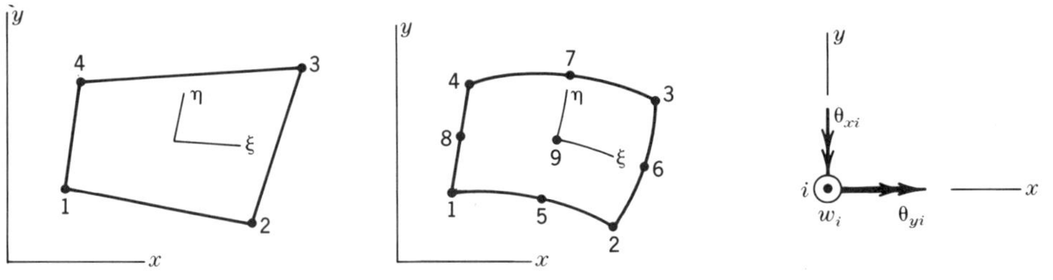

Figure 11.3-1. (*a*) Bilinear element, top view. (*b*) Quadratic element, top view. (*c*) Notation and sign convention for d.o.f. at a typical node *i*. The symbol ⊙ indicates an arrow directed out of the paper.

where $[\mathbf{D}_M]$ and $\{\boldsymbol{\kappa}\}$ are defined in Eq. 11.1-12. If the d.o.f. of Fig. 11.3-1*c* are present at every node, the same shape functions N_i are used to interpolate w, θ_x, and θ_y from nodal values of these quantities, that is,

$$\begin{Bmatrix} w \\ \theta_x \\ \theta_y \end{Bmatrix} = \sum_{i=1}^{N} \begin{bmatrix} N_i & 0 & 0 \\ 0 & N_i & 0 \\ 0 & 0 & N_i \end{bmatrix} \begin{Bmatrix} w_i \\ \theta_{xi} \\ \theta_{yi} \end{Bmatrix} \quad \text{or} \quad \{\mathbf{u}\} = \underset{3\times 3N}{[\mathbf{N}]}\{\mathbf{d}\} \tag{11.3-3}$$

where N is the number of nodes per element, and $\{\mathbf{d}\} = \lfloor w_1 \quad \theta_{x1} \quad \theta_{y1} \; \ldots \; w_N \quad \theta_{xN} \quad \theta_{yN} \rfloor^T$. Curvatures $\{\boldsymbol{\kappa}\}$ stated in Eq. 11.1-12 are

$$\{\boldsymbol{\kappa}\} = \begin{Bmatrix} \theta_{x,x} \\ \theta_{y,y} \\ \theta_{x,y} + \theta_{y,x} \\ \theta_y - w_{,y} \\ \theta_x - w_{,x} \end{Bmatrix} = [\partial]\{\mathbf{u}\}, \quad \text{where} \quad [\partial] = \begin{bmatrix} 0 & \partial/\partial x & 0 \\ 0 & 0 & \partial/\partial y \\ 0 & \partial/\partial y & \partial/\partial x \\ -\partial/\partial y & 0 & 1 \\ -\partial/\partial x & 1 & 0 \end{bmatrix} \tag{11.3-4}$$

Equations 11.3-3 and 11.3-4 yield

$$\{\boldsymbol{\kappa}\} = \underset{5\times 3N}{[\mathbf{B}]}\{\mathbf{d}\}, \quad \text{where} \quad [\mathbf{B}] = [\partial][\mathbf{N}] = \begin{bmatrix} 0 & N_{1,x} & 0 & \ldots \\ 0 & 0 & N_{1,y} & \ldots \\ 0 & N_{1,y} & N_{1,x} & \ldots \\ -N_{1,y} & 0 & N_1 & \ldots \\ -N_{1,x} & N_1 & 0 & \ldots \end{bmatrix} \tag{11.3-5}$$

And finally, from Eqs. 11.3-2 and 11.3-5,

$$U = \tfrac{1}{2}\{\mathbf{d}\}^T[\mathbf{k}]\{\mathbf{d}\}, \quad \text{where} \quad \underset{3N\times 3N}{[\mathbf{k}]} = \int_A [\mathbf{B}]^T[\mathbf{D}_M][\mathbf{B}]\, dA \tag{11.3-6}$$

If the plate is rectangular, shape functions N_i can be expressed in terms of x and y. Then $dA = dx\, dy$. If the plate is of more general shape, as shown in Fig. 11.3-1, the N_i can be expressed in terms of isoparametric coordinates ξ and η. Then $dA = J\, d\xi\, d\eta$, where J is the Jacobian determinant. For bilinear and quadratic elements respectively, shape functions N_i are given by Eqs. 6.3-2 and Table

6.6-1. Shape function derivatives, needed in [**B**], are determined by the usual transformation (e.g, Eq. 6.3-7):

$$N_{i,x} = \Gamma_{11} N_{i,\xi} + \Gamma_{12} N_{i,\eta} \quad \text{and} \quad N_{i,y} = \Gamma_{21} N_{i,\xi} + \Gamma_{22} N_{i,\eta} \tag{11.3-7}$$

Mindlin plate elements can be regarded as special forms of solid elements. For example, the bilinear plate element of Fig. 11.3-1*a* closely resembles the trilinear solid element of Fig. 6.7-1, if the solid element is made thin in one direction. The solid element has twice as many d.o.f. as the plate element. In addition, nodes of the solid that lie on a midsurface-normal line define the thickness-direction strain, ϵ_z, which is ignored in plate-bending theory. If ϵ_z were included in the formulation, nodes that span the thickness would be coupled by stiffness coefficients that become very large in comparison with bending stiffnesses as the plate becomes thin. The discrepancy may lead to numerical difficulty of a type discussed in Section 18.2. In summary, considerations of economy and robustness indicate that solid elements should not be used to model plates.

Quadrature Rule and Locking. One can regard the stiffness matrix of a Mindlin plate element as being composed of a bending stiffness $[\mathbf{k}_b]$ and a transverse shear stiffness $[\mathbf{k}_s]$. From Eq. 11.3-6, with $[\mathbf{B}] = [\mathbf{B}_b] + [\mathbf{B}_s]$,

$$[\mathbf{k}] = \underbrace{\int_A [\mathbf{B}_b]^T[\mathbf{D}_M][\mathbf{B}_b]\, dA}_{[\mathbf{k}_b]} + \underbrace{\int_A [\mathbf{B}_s]^T[\mathbf{D}_M][\mathbf{B}_s]\, dA}_{[\mathbf{k}_s]} \tag{11.3-8}$$

$[\mathbf{B}_b]$ is associated with in-plane strains ϵ_x, ϵ_y, and γ_{xy} and is obtained by setting rows 4 and 5 of [**B**] to zero. $[\mathbf{B}_s]$ is associated with transverse shear strains γ_{yz} and γ_{zx}, and is obtained by setting rows 1, 2, and 3 of [**B**] to zero. The cross product terms $[\mathbf{B}_b]^T[\mathbf{D}_M][\mathbf{B}_s]$ and $[\mathbf{B}_s]^T[\mathbf{D}_M][\mathbf{B}_b]$ are zero because of the distribution of zeros in $[\mathbf{B}_b]$, $[\mathbf{B}_s]$, and $[\mathbf{D}_M]$. Bending stiffness $[\mathbf{k}_b]$ mobilizes only the $[\mathbf{D}_K]$ portion of $[\mathbf{D}_M]$, and transverse shear stiffness $[\mathbf{k}_s]$ mobilizes only the $G_{yz}t$ and $G_{zx}t$ terms in $[\mathbf{D}_M]$.

The splitting of [**k**] into components $[\mathbf{k}_b]$ and $[\mathbf{k}_s]$ is also seen in Eq. 9.4-4 of Section 9.4, where it is argued that each integration point used to evaluate $[\mathbf{k}_s]$ imposes one constraint on transverse shear strain γ_{zx} of the two-node Mindlin beam element, and may produce locking of the mesh if the beam is thin and too many integration points are used to evaluate $[\mathbf{k}_s]$. Similar considerations apply to Mindlin plate elements. However, each integration point used for $[\mathbf{k}_s]$ brings *two* constraints to a Mindlin plate element, one associated with γ_{yz} and the other with γ_{zx}. Locking of Mindlin plate elements caused by too many transverse shear constraints can be avoided by adopting a reduced or selective integration rule to generate [**k**]. Or, one can redefine the transverse shear interpolation; see [11.17].

Various Mindlin plate elements are possible. Some are summarized in Table 11.3-1. Typical behavior is reported in Fig. 11.3-2. "Full integration" is sufficient to avoid element mechanisms. The stiffness matrix of a rectangular element with midside nodes is integrated exactly by full integration.

The bilinear element responds properly to pure bending with either reduced or selective integration. With full (2 by 2) integration and pure bending, parasitic shear strains appear at the Gauss points (Fig. 11.3-3*a*). As the element becomes

TABLE 11.3-1. DATA FOR SELECTED MINDLIN PLATE ELEMENTS.

Element Type		Integration Rule			Shear Constraints	Number of Mechanisms
		Type	$[\mathbf{k}_b]$	$[\mathbf{k}_s]$		
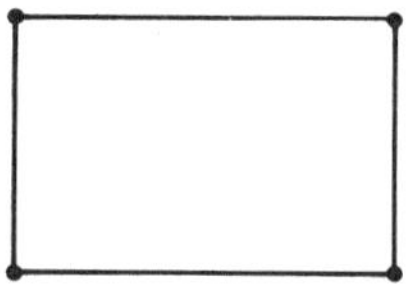	Bilinear: 4 nodes, 12 d.o.f.	Reduced	1 × 1	1 × 1	2	4
		Selective	2 × 2	1 × 1	2	2
		Full	2 × 2	2 × 2	8	0
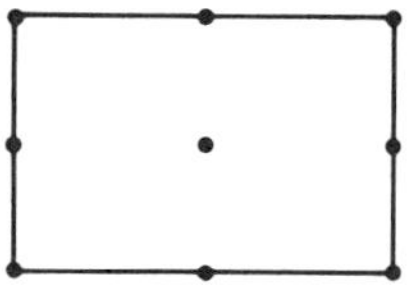	Quadratic: 9 nodes, 27 d.o.f.	Reduced	2 × 2	2 × 2	8	4
		Selective	3 × 3	2 × 2	8	1
		Full	3 × 3	3 × 3	18	0
	Serendipity: 8 nodes, 24 d.o.f.	Reduced	2 × 2	2 × 2	8	1
		Selective	3 × 3	2 × 2	8	0
		Full	3 × 3	3 × 3	18	0
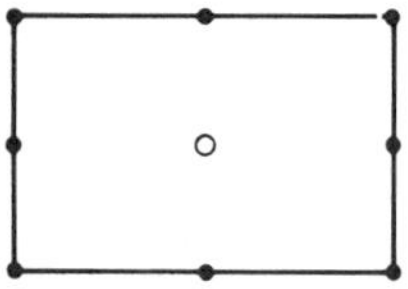	Heterosis: 9 nodes, 26 d.o.f.	Selective	3 × 3	2 × 2	8	0
		(Ref. 11.8. D.o.f. at the center node are θ_x and θ_y only.)				

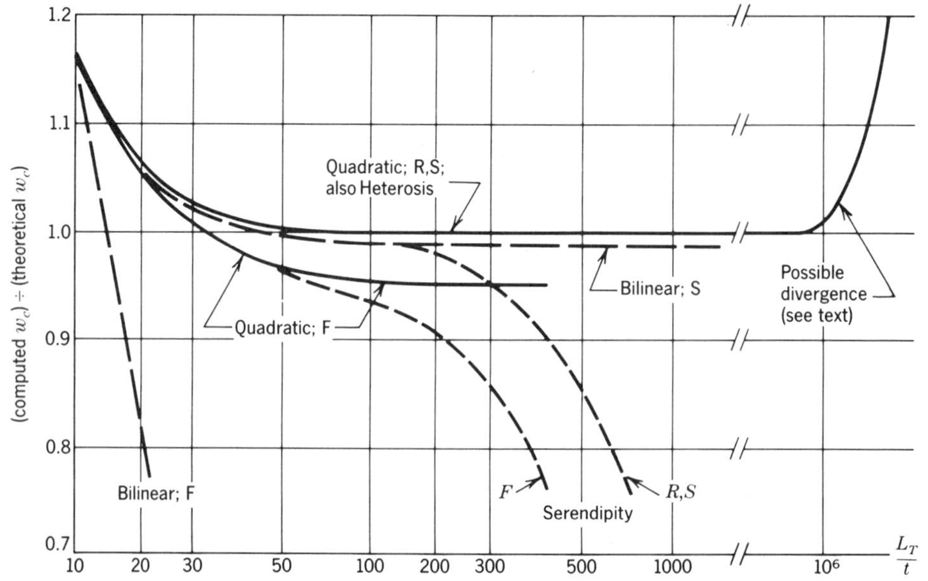

Figure 11.3-2. Center deflection of a uniformly loaded clamped square plate of side length L_T and thickness t. An 8 by 8 mesh is used in all cases. Thin plates correspond to large L_T/t. Transverse shear deformation becomes significant for small L_T/t. Integration rules, from Table 11.3-1, are reduced (R), selective (S), and full (F) [11.9].

thin, its stiffness is due almost entirely to parasitic shear. Thus, if fully integrated, a bilinear Mindlin plate element exhibits almost no bending deformation: that is, the mesh "locks."

The nine-node quadratic element, when integrated with any number of Gauss points, can properly represent pure bending. As seen in Fig. 11.3-3*b*, because lateral deflection w can vary quadratically, zero-shear conditions impose the constraint of *pure* bending rather than *no* bending. The latter element in Fig. 11.3-3*b* displays *linearly varying bending*. Here $w = 0$, not w cubic in x as for a standard beam or a Kirchhoff plate, because the quadratic plate element can display only a quadratic variation of w. Accordingly, linearly varying bending is represented by $w = 0$ and a quadratic variation of θ_x. It can be shown that this state is properly represented only when $[\mathbf{k}_s]$ is integrated with a 2 by 2 rule (Ref. 12.6; see also Problem 11.20).

The "serendipity" element has eight nodes and uses the quadratic shape functions of Eqs. 6.6-1. It is an unreliable element: as shown by Fig. 11.3-2, its accuracy is acceptable only for small values of L_T/t, regardless of quadrature rule.

The upper curves in Fig. 11.3-2 become horizontal, indicating apparent convergence, but may diverge for very large values of L_T/t. This difficulty has nothing to do with locking. Rather, it is caused by the penalty matrix $[\mathbf{k}_s]$ becoming numerically so large that it overwhelms matrix $[\mathbf{k}_b]$, as discussed in Section 9.4. Divergence can be avoided by basing $[\mathbf{k}_s]$ on a value of t that is arbitrarily increased, if necessary, so that

$$\frac{5}{1+\nu}\left(\frac{L_T}{t}\right)^2 < 10^{p/2} \qquad (11.3\text{-}9)$$

where p is the approximate number of decimal digits per computer word (see Eq. 9.4-11). The *actual* value of t is used to compute matrix $[\mathbf{k}_b]$. With this adjustment, transverse shear deformation is misrepresented only when it is so small as not to matter anyway.

The "heterosis" element [11.8] is the best of the elements summarized in Table 11.3-1. It does not exhibit locking, erratic convergence characteristics, or mech-

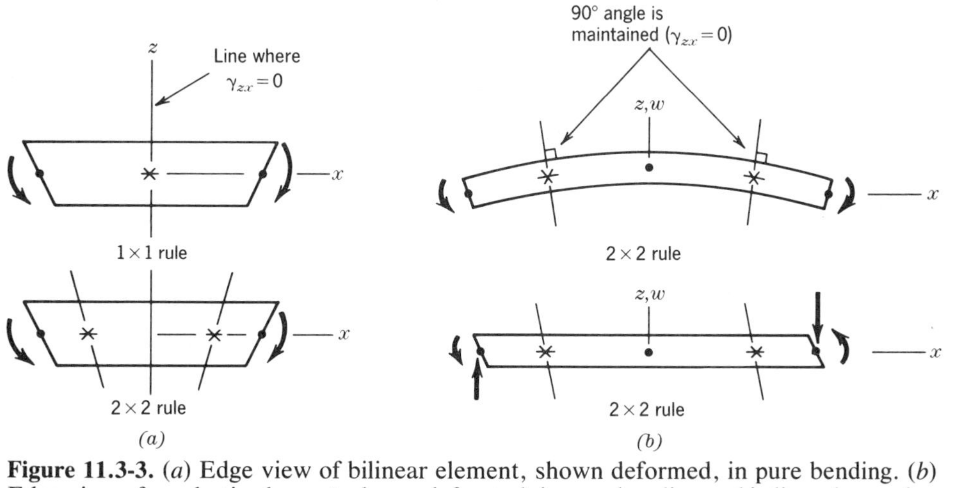

Figure 11.3-3. (*a*) Edge view of bilinear element, shown deformed, in pure bending. (*b*) Edge view of quadratic element, shown deformed, in pure bending and in linearly varying bending.

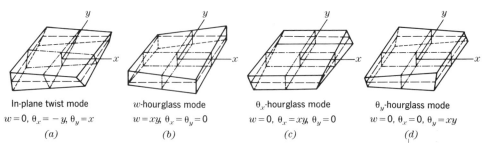

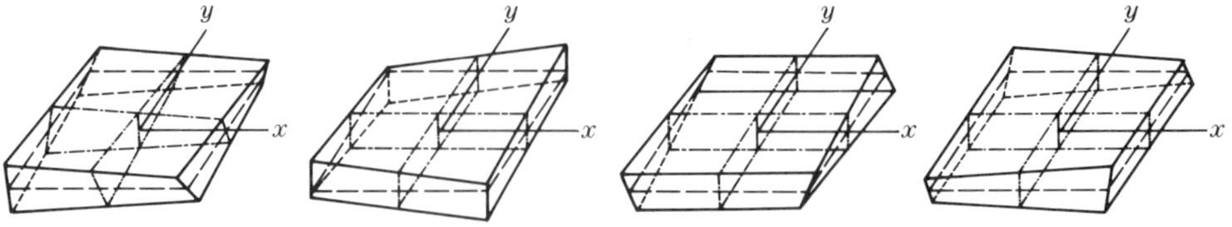

Figure 11.3-4. Mechanisms of the four-node bilinear Mindlin element with reduced integration (one point for all terms) [11.4].

anisms. Other good Mindlin plate elements can be formulated by hybrid methods [9.8].

Mechanisms. An element having one or more mechanisms is not a foolproof element. Occasional numerical disaster is possible in the hands of a user who is unaware or inattentive. Mechanisms of elements in Table 11.3-1 resemble those of the corresponding plane elements, discussed in Section 6.12.

Mechanisms of the bilinear element with reduced integration are shown in Fig. 11.3-4. With selective integration, only two of these mechanisms remain possible: the in-plane twist mode and the w-hourglass mode. The in-plane twist mode, Fig. 11.3-4*a*, is not communicable between adjacent elements, so that a mesh of two or more elements cannot have this mechanism. Control of mechanisms is discussed in [6.11–6.13,13.49,13.52–13.54].

Under both reduced and selective integration, the quadratic and serendipity elements have the mechanism described by Eqs. 6.12-3 (for Mindlin plates, substitute $-z\theta_x$ for u and $-z\theta_y$ for v). This mechanism cannot be communicated between elements. Three additional mechanisms are possible in the nine-node quadratic element under reduced integration. One comes from Eq. 6.12-2, and another from Eq. 6.12-2 with u and v interchanged. The third mechanism is $u = v = 0$, $w = 3\xi^2\eta^2 - \xi^2 - \eta^2$. The latter mechanism is not possible in the heterosis element because the $\xi^2\eta^2$ term is not present in the w field.

Stress Computation. When element nodal d.o.f. $\{\mathbf{d}\}$ are known, Eq. 11.3-5 yields curvatures $\{\boldsymbol{\kappa}\}$, Eq. 11.1-12 yields moments and shears, and Eqs. 11.1-2 yields stresses. Transverse shear stresses may be greatly in error except at Gauss points of the selective integration rule appropriate to $[\mathbf{k}_s]$.(Even at these points, accuracy may be poor unless thickness t used in $[\mathbf{k}_s]$ has been adjusted according to Eq. 11.3-9.) Thus, in the bilinear element, transverse shear stresses τ_{yz} and τ_{zx} should be calculated at the element center, and these values assumed to prevail throughout the element. With the remaining elements in Table 11.3-1, it usually is good strategy to calculate stresses at Gauss points of a 2 by 2 rule, then extrapolate to other locations in the element as required. Extrapolation of stresses is discussed in Section 6.13.

11.4 A TRIANGULAR DISCRETE KIRCHHOFF ELEMENT

The element to be discussed was published in 1969 [11.10]. It was reexamined over ten years later and found to remain among the best elements for analysis of thin plates [11.11]. The element is currently known as DKT, for discrete Kirchhoff triangle. Explicit expressions [11.12] and Fortran coding [11.13] for the element are available. Details of element formulation involve lengthy expressions. The following is a summary. Its essential step, that is, the enforcement of zero transverse shear strain at specific points, is also used in the formulation of other discrete Kirchhoff elements.

The starting point is a straight-sided element with corner and midside nodes (Fig. 11.4-1*a*). Rotations θ_x and θ_y of a midsurface-normal line are each interpolated from nodal rotations θ_{xi} and θ_{yi}, where i runs from 1 to 6, using a complete quadratic polynomial:

$$\theta_x = \sum N_i \theta_{xi} \qquad \text{and} \qquad \theta_y = \sum N_i \theta_{yi} \tag{11.4-1}$$

Here the N_i are given by Eqs. 5.3-5. There are a total of twelve d.o.f. in Eq. 11.4-1. Lateral deflection w along each edge is assumed to be cubic in an edge-tangent coordinate s. Thus, along side 2–3 for example, the rotation $w_{,s}$ at midside node 5 is

$$w_{,s5} = -\frac{3}{2L_{23}} w_2 - \frac{1}{4} w_{,s2} + \frac{3}{2L_{23}} w_3 - \frac{1}{4} w_{,s3} \tag{11.4-2}$$

Two similar equations are written for the remaining two sides. When nodal values of $w_{,s}$ are replaced by nodal values of $w_{,x}$ and $w_{,y}$ by coordinate transformation, there are a total of nine d.o.f. in these three equations for $w_{,s}$ (vertex-node values of w, $w_{,x}$, and $w_{,y}$). Accordingly, in Eq. 11.4-1 and the three rotation equations such as 11.4-2, there are a total of 21 d.o.f.

We seek a nine-d.o.f. element that has the nodal d.o.f. shown in Fig. 11.4-1*b*. Accordingly, the twelve d.o.f. θ_{xi} and θ_{yi} at nodes 1 through 6 must be expressed in terms of w_i, $w_{,xi}$, and $w_{,yi}$ at only the corner nodes. Constraints used for this purpose are as follows.

1. Transverse shear strains γ_{yz} and γ_{zx} vanish at corners 1, 2, and 3. Thus, from Eqs. 11.1-4,

$$\theta_{xi} = w_{,xi} \qquad \text{and} \qquad \theta_{yi} = w_{,yi} \qquad \text{for} \quad i = 1, 2, 3 \tag{11.4-3}$$

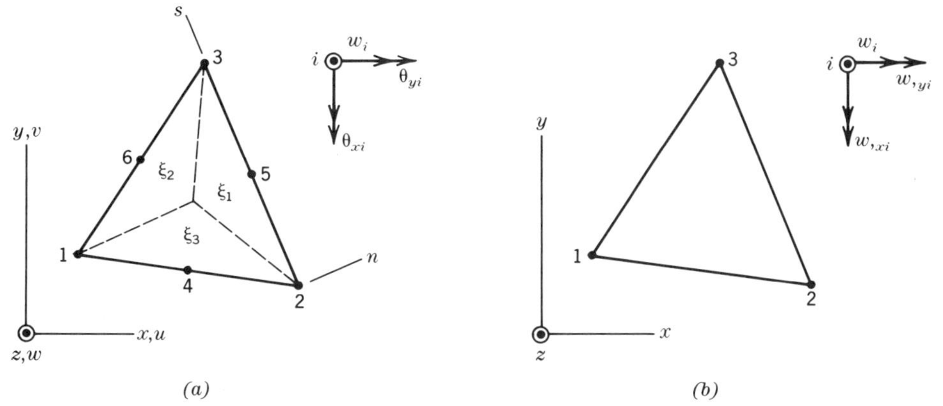

Figure 11.4-1. Development of a discrete Kirchhoff triangle. (*a*) Initial element and its d.o.f. Area coordinates ξ_1, ξ_2, and ξ_3 are shown. (*b*) Final nine-d.o.f. element and its d.o.f.

2. Transverse shear strain γ_{sz} vanishes at nodes 4, 5, and 6, where s is an edge-tangent coordinate. Thus

$$\theta_{si} = w_{,si} \qquad \text{for} \quad i = 4, 5, 6 \tag{11.4-4}$$

Use of Eq. 11.4-4 requires coordinate transformation operations.

3. Normal slopes vary linearly along each edge. Thus

$$\theta_{n4} = \tfrac{1}{2}(w_{,n1} + w_{,n2}) \qquad \theta_{n5} = \tfrac{1}{2}(w_{,n2} + w_{,n3}) \qquad \theta_{n6} = \tfrac{1}{2}(w_{,n3} + w_{,n1}) \tag{11.4-5}$$

After the foregoing constraints have been applied, the twelve nodal d.o.f. θ_{xi} and θ_{yi} in Eqs. 11.4-1 are expressed in terms of the nine nodal d.o.f. w_i, $w_{,xi}$, and $w_{,yi}$ at the corners. Symbolically, these relations are

$$\lfloor \theta_{x1} \quad \theta_{y1} \quad \theta_{x2} \ . \ . \ . \ \theta_{y6} \rfloor^T = \underset{12\times 9}{[\mathbf{T}]} \lfloor w_1 \quad w_{,x1} \quad w_{,y1} \quad w_2 \quad . \ . \ . \quad w_{,y3} \rfloor^T \tag{11.4-6}$$

To generate the element stiffness matrix, we can begin with Eqs. 11.4-1. Strains are stated as in Mindlin plate theory, Eqs. 11.1-4, but now γ_{yz} and γ_{zx} are ignored. Thus one considers only the strains $\epsilon_x = -z\theta_{x,x}$, $\epsilon_y = -z\theta_{y,y}$, and $\gamma_{xy} = -z(\theta_{x,y} + \theta_{y,x})$, that is,

$$\{\boldsymbol{\epsilon}\} = -z[\partial] \begin{Bmatrix} \theta_x \\ \theta_y \end{Bmatrix}, \qquad \text{where} \quad [\partial] = \begin{bmatrix} \partial/\partial x & 0 \\ 0 & \partial/\partial y \\ \partial/\partial y & \partial/\partial x \end{bmatrix} \tag{11.4-7}$$

Equations 11.4-1 and 11.4-7 yield the strains $\{\boldsymbol{\epsilon}\} = \lfloor \epsilon_x \quad \epsilon_y \quad \gamma_{xy} \rfloor^T$ as

$$\{\boldsymbol{\epsilon}\} = \underbrace{-z[\partial] \begin{bmatrix} N_1 & 0 & N_2 & 0 & \cdots & N_6 & 0 \\ 0 & N_1 & 0 & N_2 & \cdots & 0 & N_6 \end{bmatrix}}_{[\mathbf{B}_\theta]} \underbrace{\lfloor \theta_{x1} \quad \theta_{y1} \quad \cdots \quad \theta_{y6} \rfloor^T}_{\{\mathbf{d}_\theta\}} \tag{11.4-8}$$

Equations 5.2-7 must be used in forming $[\mathbf{B}_\theta]$. After integration through the thickness, the strain energy expression $U = \frac{1}{2} \int \{\boldsymbol{\epsilon}\}^T[\mathbf{E}]\{\boldsymbol{\epsilon}\}\, dV$ becomes

$$U = \tfrac{1}{2}\{\mathbf{d}_\theta\}^T[\mathbf{k}_\theta]\{\mathbf{d}_\theta\}, \qquad \text{where} \quad \underset{12\times 12}{[\mathbf{k}_\theta]} = \int_A [\mathbf{B}_\theta]^T[\mathbf{D}_K][\mathbf{B}_\theta]\, dA \tag{11.4-9}$$

Matrix $[\mathbf{D}_K]$ is the rigidity matrix of Kirchhoff plate theory, for example, Eq. 11.1-10. Matrix $[\mathbf{k}_\theta]$ operates on the twelve rotational d.o.f. used in Eq. 11.4-1. It can be converted to a 9 by 9 matrix $[\mathbf{k}]$, which operates on the standard Kirchhoff d.o.f. at the corners, by applying Eq. 11.4-6:

$$[\mathbf{k}] = [\mathbf{T}]^T[\mathbf{k}_\theta][\mathbf{T}] \tag{11.4-10}$$

Transformations of this type are explained in Section 7.4. However, more efficient coding results if Eq. 11.4-6 is substituted into Eq. 11.4-8, so that the strain–displacement relation contains nine columns rather than twelve. Thus $[\mathbf{k}]$ is produced directly.

TABLE 11.4-1. Center deflections of square plates, computed by the DKT element [11.10]. A typical mesh is shown in Fig. 11.4-2. Results are reported as the ratio of computed deflection to exact deflection according to thin-plate theory using $\nu = 0.3$ [11.1]. Simply supported cases use classical boundary conditions (Eqs. 11.5-1).

Mesh Size	Uniformly Loaded		Concentrated Center Load	
	Simply Supported	Clamped	Simply Supported	Clamped
$N_{es} = 1$	1.025	1.500	1.076	1.012
$N_{es} = 2$	0.999	1.228	1.008	1.046
$N_{es} = 4$	1.001	1.069	1.003	1.019
$N_{es} = 8$	1.001	1.021	1.001	1.007

One can visualize the foregoing DKT plate element as a stack of plane linear-strain triangles, pinned together by a rigid thickness-direction rod at each vertex, and with additional constraints that impose Eqs. 11.4-4 and 11.4-5 at midsides.

If the DKT plate element is homogeneous and of constant thickness, $[\mathbf{k}]$ is integrated exactly by a three-point quadrature rule. Explicit formulas for $[\mathbf{k}]$ are also available [11.12,11.13]. After element d.o.f. $\{\mathbf{d}\}$ are known, element strains $\{\boldsymbol{\epsilon}\}$ are computed by successive application of Eqs. 11.4-6 and 11.4-8. Hence, stresses are $\{\boldsymbol{\sigma}\} = [\mathbf{E}](\{\boldsymbol{\epsilon}\} - \{\boldsymbol{\epsilon}_0\})$.

The behavior of the DKT element is reported in Table 11.4-1 [11.10]. Distributed loads were lumped by assigning one-third the total element load to translational d.o.f. at each vetex. Additional results may be found in Refs. 11.11 and 11.12, which show that the element yields accurate bending moments, performs well even at large aspect ratios, and satisfactorily solves the "twisted ribbon" test case.

Fortran coding for the DKT element appears in Fig. 11.4-3. Required input consists of nodal x and y coordinates (X1,X2,X3,Y1,Y2,Y3) and rigidity matrix $[\mathbf{D}_K]$ (the 3 by 3 array D). The stiffness matrix is delivered in array SE. Displacement w is positive in the $+z$ direction. The arrangement of nodal d.o.f. used in this subroutine is

$$\{\mathbf{d}\} = \lfloor w_1 \quad w_{,y1} \quad -w_{,x1} \quad w_2 \quad w_{,y2} \quad -w_{,x2} \quad w_3 \quad w_{,y3} \quad -w_{,x3} \rfloor^T \qquad (11.4\text{-}11)$$

Thus $w_{,y}$ and $-w_{,x}$ are represented by rotation vectors in the $+x$ and $+y$ directions, respectively.

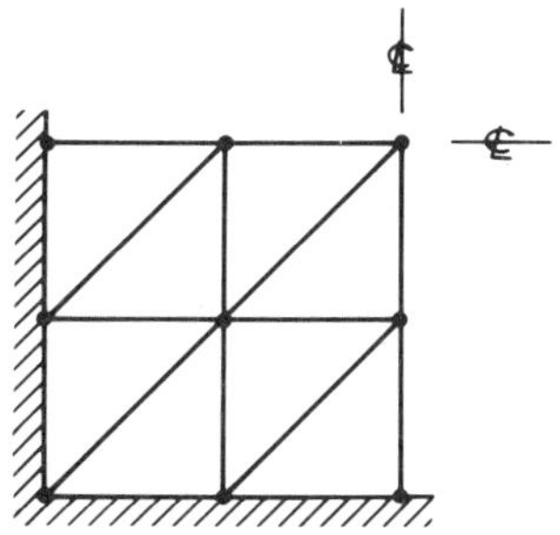

Figure 11.4-2. Typical mesh on one quadrant of a square plate (used for the results reported in Table 11.4-1). There is symmetry about the centerlines.

```
      SUBROUTINE DKT (D,X1,Y1,X2,Y2,X3,Y3,SE)
      IMPLICIT DOUBLE PRECISION (A-H,O-Z)
      DIMENSION D(3,3),DD(9,9),QQ(9,9),PP(3,3),PT(2,3),RS(2,3),Q(3)
      DIMENSION GG(10,9),KOD(2,9),B(3),C(3),ALS(3),PX(3,3),SE(9,9)
      DATA KOD /1,1,2,3,3,2,4,4,5,6,6,5,7,7,8,9,9,8/
      DATA PP /12.D0,4.D0,4.D0,4.D0,2.D0,1.D0,4.D0,1.D0,2.D0/
      B(1)=Y2-Y3
      B(2)=Y3-Y1
      B(3)=Y1-Y2
      C(1)=X3-X2
      C(2)=X1-X3
      C(3)=X2-X1
      DET=(B(1)*C(2)-B(2)*C(1))*24.
      DO 10 I=1,3
      DO 10 J=1,3
   10 PX(I,J)=PP(I,J)/DET
      DO 25 I=1,3
      DO 25 J=1,3
      DO 25 K1=1,3
      II=(I-1)*3+K1
      DO 25 K2=1,3
      JJ=(J-1)*3+K2
   25 DD(II,JJ)=D(I,J)*PX(K1,K2)
      DO 30 I=1,3
      ALS(I)=B(I)*B(I)+C(I)*C(I)
      PT(1,I)=6.*C(I)/ALS(I)
      PT(2,I)=6.*B(I)/ALS(I)
      RS(1,I)=3.*C(I)*C(I)/ALS(I)
      RS(2,I)=3.*B(I)*B(I)/ALS(I)
   30 Q(I)=3.*B(I)*C(I)/ALS(I)
      DO 720 I=1,10
      DO 720 J=1,9
  720 GG(I,J)=0.
      DO 730 I=1,2
      II=(I-1)*5
      P1=PT(I,1)
      P2=PT(I,2)
      P3=PT(I,3)
      R1=RS(I,1)
      R2=RS(I,2)
      R3=RS(I,3)
      GG(II+1,KOD(I,1))=P3
      GG(II+2,KOD(I,1))=-P2
      GG(II+3,KOD(I,1))=-P3
      GG(II+4,KOD(I,1))=P2-P3
      GG(II+5,KOD(I,1))=P2
      GG(II+1,KOD(I,2))=-Q(3)
      GG(II+2,KOD(I,2))=-Q(2)
      GG(II+3,KOD(I,2))=Q(3)
      GG(II+4,KOD(I,2))=Q(2)+Q(3)
      GG(II+5,KOD(I,2))=Q(2)
      GG(II+1,KOD(I,3))=-1.-R3
      GG(II+2,KOD(I,3))=-1.-R2
      GG(II+3,KOD(I,3))=R3
      GG(II+4,KOD(I,3))=R2+R3
      GG(II+5,KOD(I,3))=R2
      GG(II+1,KOD(I,4))=-P3
      GG(II+3,KOD(I,4))=P3
      GG(II+4,KOD(I,4))=P1+P3
      GG(II+1,KOD(I,5))=-Q(3)
      GG(II+3,KOD(I,5))=Q(3)
      GG(II+4,KOD(I,5))=Q(3)-Q(1)
      GG(II+1,KOD(I,6))=1.-R3
      GG(II+3,KOD(I,6))=R3
      GG(II+4,KOD(I,6))=R3-R1
      GG(II+2,KOD(I,7))=P2
      GG(II+4,KOD(I,7))=-P1-P2
      GG(II+5,KOD(I,7))=-P2
      GG(II+2,KOD(I,8))=-Q(2)
      GG(II+4,KOD(I,8))=Q(2)-Q(1)
      GG(II+5,KOD(I,8))=Q(2)
      GG(II+2,KOD(I,9))=1.-R2
      GG(II+4,KOD(I,9))=R2-R1
      GG(II+5,KOD(I,9))=R2
  730 CONTINUE
      DO 850 I=1,9
      QQ(1,I)=B(2)*GG(1,I)+B(3)*GG(2,I)
      QQ(2,I)=2.*B(2)*GG(3,I)+B(3)*GG(4,I)
      QQ(3,I)=B(2)*GG(4,I)+2.*B(3)*GG(5,I)
      QQ(4,I)=-C(2)*GG(6,I)-C(3)*GG(7,I)
      QQ(5,I)=-2.*C(2)*GG(8,I)-C(3)*GG(9,I)
      QQ(6,I)=-C(2)*GG(9,I)-2.*C(3)*GG(10,I)
      QQ(7,I)=C(2)*GG(1,I)+C(3)*GG(2,I)
     1        -B(2)*GG(6,I)-B(3)*GG(7,I)
      QQ(8,I)=2.*C(2)*GG(3,I)+C(3)*GG(4,I)
     1        -2.*B(2)*GG(8,I)-B(3)*GG(9,I)
      QQ(9,I)=C(2)*GG(4,I)+2.*C(3)*GG(5,I)
     1        -B(2)*GG(9,I)-2.*B(3)*GG(10,I)
  850 CONTINUE
      DO 855 I=1,9
      DO 855 J=1,9
      GG(I,J)=0.
      DO 855 K=1,9
  855 GG(I,J)=GG(I,J)+DD(I,K)*QQ(K,J)
      DO 960 L=1,9
      DO 960 J=L,9
      DUM=0.
      DO 900 K=1,9
  900 DUM=DUM+QQ(K,L)*GG(K,J)
      SE(L,J)=DUM
  960 SE(J,L)=DUM
      RETURN
      END
```

Figure 11.4-3. Fortran statements that generate the 9 by 9 stiffness matrix of a DKT plate element [adapted from Ref. 11.13]. See text for notation and order of d.o.f.

11.5 BOUNDARY CONDITIONS AND TEST CASES

Boundary Conditions. Conditions at the edge of a plate are classed as clamped, free, or simply supported. Typically, no single condition prevails along the entire plate boundary. In the notation of Fig. 11.5-1, plate boundary conditions are as follows.

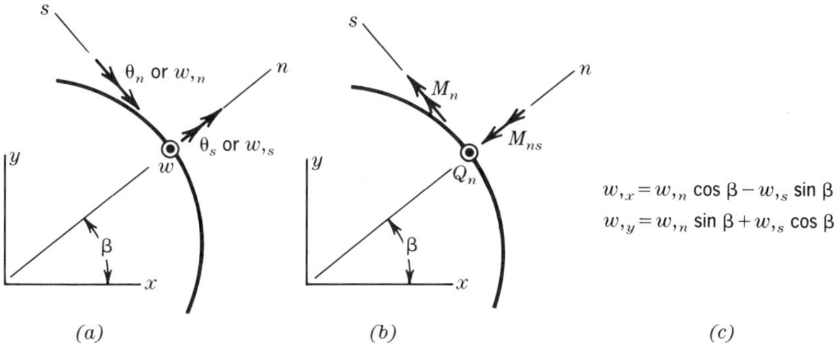

Figure 11.5-1. Coordinates n and s are edge-normal and edge-tangent, respectively. (*a*) Rotations and lateral displacement. (*b*) Moments and transverse shear force. (*c*) Transformation relations for rotations.

clamped	*free*	*simply supported (finite element)*	*simply supported (classical theory)*	
$w = 0$	$Q_n = 0$[1]	$w = 0$	$w = 0$	
$\theta_n = 0$	$M_n = 0$	$M_n = 0$	$M_n = 0$	(11.5-1)
$\theta_s = 0$	$M_{ns} = 0$	$M_{ns} = 0$	$\theta_s = 0$	

Since transverse shear strain is taken as zero in classical thin-plate theory, Eqs. 11.5-1 are modified for Kirchhoff and discrete Kirchhoff elements by replacing θ_n by $w,_n$ and θ_s by $w,_s$.

A clamped edge prevents all motion along the edge. Along a free edge, nodal d.o.f. are unspecified, and remain part of the vector {**D**} of unknown d.o.f. Conditions along a simply supported or "hinged" edge have been found troublesome and require more explanation, as follows.

In classical thin-plate theory, since $\gamma_{zs} = 0$, the boundary condition $w = 0$ necessarily implies the boundary condition $w,_s = 0$ as well. Thus, for a thin simply supported plate, we would expect good numerical results using the "classical theory" conditions in Eqs. 11.5-1, whether the element type is Kirchhoff, discrete Kirchhoff, or Mindlin. Such is indeed the case if boundaries intersect at right angles, as at the four corners of a rectangular plate. However, if the plate is skew, most elements give *poor* results. For example, if the plate of Fig. 11.5-2*a* is modeled by a uniform 14 by 14 mesh, the center displacement may be underestimated by more than 20%. If only the boundary conditions are changed, to the "finite element" simply supported conditions in Eqs. 11.5-1, the error may decline to less than 3%. Apparently, classical simply supported conditions overconstrain the mesh when interior corner angles exceed $\pi/2$. The finite element simply supported conditions produce a plate model that is point-supported at its boundary nodes. Although this may appear to allow too little constraint, especially in a coarse mesh, good results are obtained in practice [11.14].

In Eqs. 11.5-1, the various quantities may have prescribed values other than zero. For example, a line load ($Q_n \neq 0$) could be prescribed along an edge of a

[1]This is the condition that must be used in finite element work. For reasons too lengthy to repeat here, the condition $Q_n + M_{ns,s} = 0$ must be used in classical thin-plate theory [11.1].

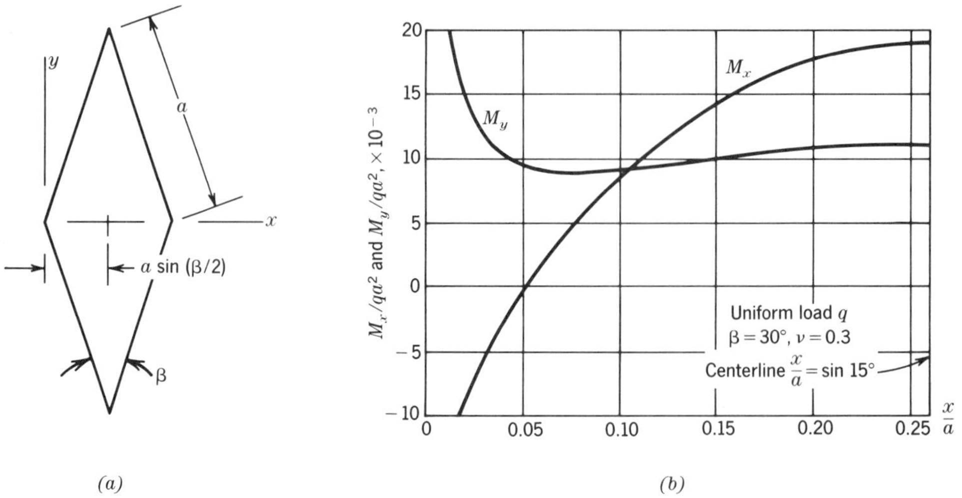

Figure 11.5-2. (*a*) Skew plate with equal side lengths (rhombic plate). (*b*) Bending moments along the *x* axis in a simply supported rhombic plate [11.15].

plate and represented by concentrated lateral forces at nodes along the edge. A nonzero w along the edge may be prescribed instead of Q_n. Thus the edge becomes simply supported or clamped, depending on whether $M_n = 0$ or $\theta_n = 0$.

Test Cases. In Section 4.5 we argue that an element must be able to display a constant-strain state. Plate elements are no exception. Each *layer* $z =$ constant of a plate element must be able to display constant ϵ_x, ϵ_y, and γ_{xy}. Hence, in patch-testing a Kirchhoff or discrete Kirchhoff plate element, we look for constant curvatures $w_{,xx}$ and $w_{,yy}$ and for constant twist $w_{,xy}$. In patch-testing a Mindlin plate element, we look for constant curvatures $\theta_{x,x}$ and $\theta_{y,y}$, constant twist $\theta_{x,y} + \theta_{y,x}$, and constant transverse shear strains $w_{,y} - \theta_y$ and $w_{,x} - \theta_x$. Figure 11.5-3 depicts a patch test for constant $w_{,xx}$ (or for constant $\theta_{x,x}$). One must enforce $w_{,y} = 0$ (or $\theta_y = 0$) at nodes 1 through 4 in order to prevent curling of the edges (unless $\nu = 0$).

Popular test cases include square plates, with various support conditions and loads (see Table 11.4-1). Rectangular plates may be used as well, to test the effect of element aspect ratio. Circular plates can be used to test nonrectangular elements. Exact results are available for many such problems [11.1].

The simply supported rhombic plate, Fig. 11.5-2*a*, is a difficult test case. The obtuse corners are singular points, where moments are theoretically infinite. Some element types fail to show that M_x and M_y are of opposite sign near these corners

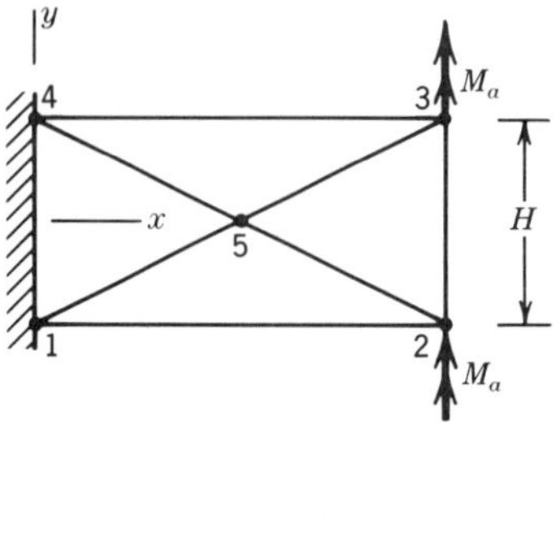

Figure 11.5-3. Patch test for constant curvature, with $M_x = 2M_a/H$, where M_a is a moment load on a node.

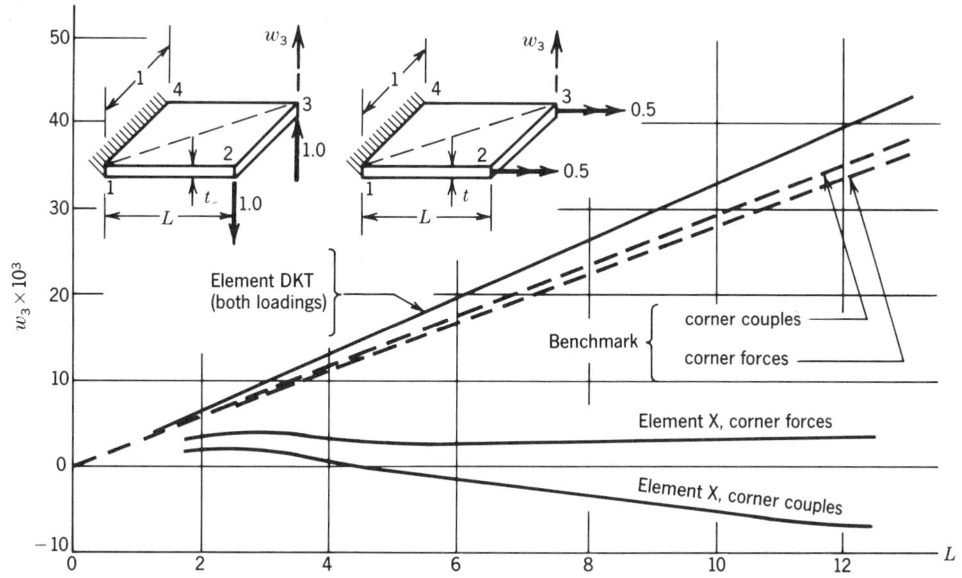

Figure 11.5-4. "Twisted ribbon" test case [11.12,11.16]. Here $E = 10^7$, $\nu = 0.25$, $t = 0.05$, w_3 = deflection of corner indicated.

[11.14–11.16]. Difficulties attendant to use of classical simply supported boundary conditions for this problem have already been noted.

The "twisted ribbon," Fig. 11.5-4, is a test that shows the effect of aspect ratio [11.12,11.16]. The twisting moment may be applied by corner forces or by corner couples, as shown. Usually the entire plate is modeled by one rectangular element or by two triangular elements. "Benchmark" values were obtained from a mesh of 16 rectangular Kirchhoff elements having 16 d.o.f. each. Many types of element fail this test, such as the one identified as "element X," which is too stiff at large aspect ratio, and may fail even to produce a displacement of the correct algebraic sign. For element DKT, the two loadings produce slightly different results, but the difference is scarcely noticeable when plotted. If triangulation is made along diagonal 2–4 rather than diagonal 1–3 as shown, elements DKT and X both produce somewhat different results than shown in Fig. 11.5-4.

PROBLEMS

Section 11.1

11.1 (a) Verify the stress formulas of Eqs. 11.1-2.
(b) Similarly, verify the formulas for τ_{yz} and τ_{zx} given below Eqs. 11.1-2.

11.2 In elementary mechanics of materials, one derives equations for normal and shear stress at an arbitrary angle θ in the xy plane, such as $\sigma_n = \frac{1}{2}(\sigma_x + \sigma_y) + \frac{1}{2}(\sigma_x - \sigma_y)\cos 2\theta + \tau_{xy}\sin 2\theta$. What analogous expressions relate bending and twisting moments M_n and M_{ns} to M_x, M_y, and M_{xy}? *Suggestion:* Use Eqs. 11.1-2.

11.3 (a) In Fig. 11.1-1b presume that the M's and Q's are functions of x and y, so that (for example) $M_x\,dy$ acts along the edge $x = 0$ and $(M_x + M_{x,x}$

$dx)\ dy$ acts along the parallel edge. Show that the equilibrium equations are $Q_{x,x} + Q_{y,y} = -q$, $M_{x,x} + M_{xy,y} = Q_x$, and $M_{xy,x} + M_{y,y} = Q_y$.

(b) Hence, show that $M_{x,xx} + 2M_{xy,xy} + M_{y,yy} + q = 0$.

(c) Use the result of part (b), and Eq. 11.1-7 for isotropic conditions, to show that $\nabla^4 w = q/D$, where ∇^4 is the biharmonic operator.

11.4 A rectangular plate of thickness t has dimensions a and b, as shown. The plate is simply supported along edges AB and CD. Edges BC and DA remain free. If a uniform downward pressure p is applied to the upper surface, what are the principal stresses at the middle of the lower surface, and what is the deflection at the center of the plate?

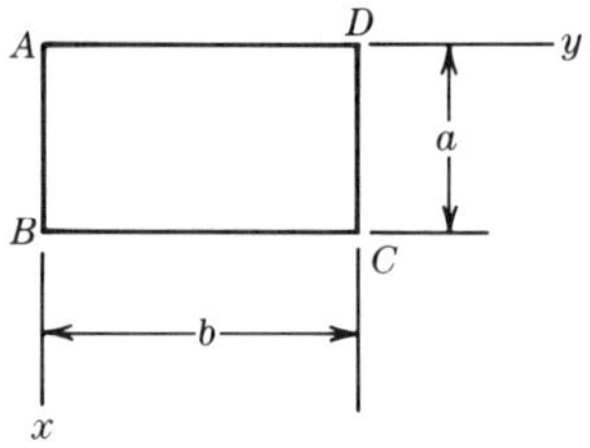

Problem 11.4

11.5 (a) Write Eq. 11.1-5 for an isotropic material. Then derive terms in $[\mathbf{D}_K]$, using the procedure given above Eq. 11.1-7.

(b) Verify the correctness of Eq. 11.1-11.

11.6 Consider an isotropic thin square plate, with edges parallel to x and y axes, loaded only along its edges. Describe the edge loads if the lateral deflection is (a) $w = c_1(x^2 + y^2)$, and (b) $w = c_2(y^2 - x^2)$, where c_1 and c_2 are constants.

11.7 (a) In a sandwich plate, Fig. 11.1-5, the average transverse shear strain γ is related to the core shear strain γ_c by $(c + h)\gamma = c\gamma_c$. Derive this expression. Assume that the facings are much stiffer than the core.

(b) Derive the expression for D_{M44} in Eqs. 11.1-13. *Suggestion:* Consider strain energy.

(c) Work from the bending stiffness EI of a sandwich beam and derive the expression for D_{M11} in Eqs. 11.1-13 (except for the $1 - \nu^2$ factor).

11.8 One sometimes wonders how wide a beam can be before it should be regarded as a plate. How would you decide? Or what would you do if unable to decide?

Section 11.2

11.9 (a) Consider the rectangular plate element of Fig. 11.2-1 and the displacement field of Eq. 11.2-5. Show that interelement compatibility of normal slopes is lacking. For example, show that $w_{,y}$ along $y = b$ does not depend only on $w_{,y}$ at nodes 3 and 4.

(b) Similarly, what can be said about interelement compatibility of w and $w_{,x}$ along edge 3–4?

(c) In Eq. 11.2-5, the terms $a_{11}x^3y$ and $a_{12}xy^3$ might be replaced by $a_{11}x^4$ and $a_{12}y^4$, but it is not wise to do so. Why?

11.10 Imagine that the lateral displacement w of a triangular thin-plate element is taken as a complete quintic (21 terms). For each element shown, and without calculation, allocate d.o.f. to the nodes in a way that seems acceptable. Consider higher-order d.o.f. as needed. Is interelement compatibility achieved? Consider compatibility conditions on the edge $x = 0$.

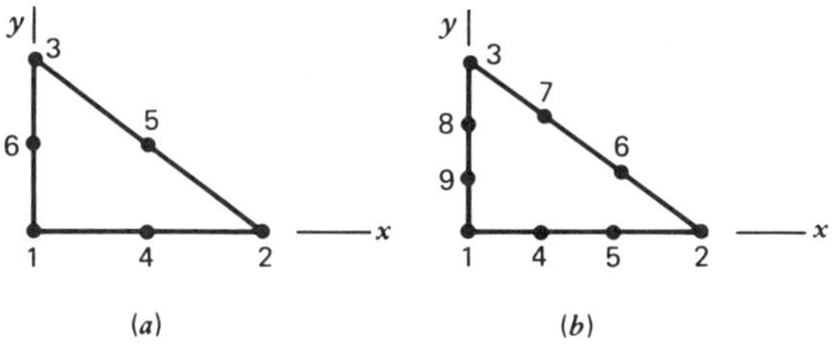

Problem 11.10

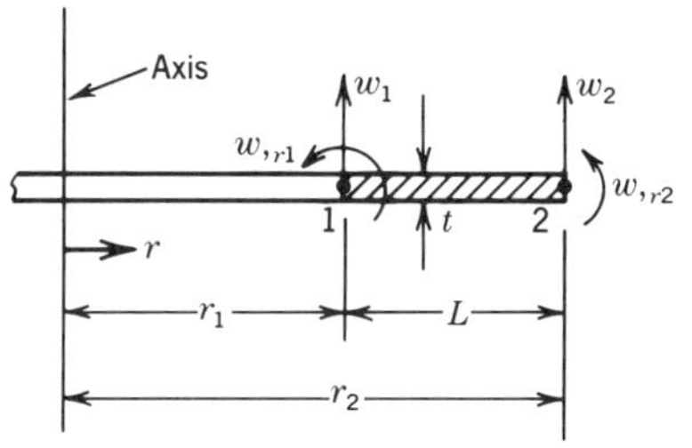

Problem 11.11

11.11 The sketch represents the cross section of an annular (axisymmetric) element for analysis of thin plates ($t << L$). Geometry, loads, and deformations are all axially symmetric. The material is isotropic. Derive the element stiffness matrix [**k**], to the extent of fully defining [**B**], [$\mathbf{D}_K$], and all other terms used in your formula for [**k**], but do not integrate.

11.12 Imagine that the Mindlin beam element of Fig. 9.4-1 is uniform, uniformly loaded, and simply supported at nodes 1 and 2. With one-point quadrature for transverse shear terms, the stiffness matrix that operates of d.o.f. θ_1 and θ_2 is

$$[\mathbf{k}] = \frac{Ebt^3}{12L}\begin{bmatrix} 1 & -1 \\ -1 & 1 \end{bmatrix} + \frac{GbtL}{4.8}\begin{bmatrix} 1 & 1 \\ 1 & 1 \end{bmatrix}$$

where b = element width and t = element depth. Determine the rotation at node 2, and compare it with the exact value, if nodal loads produced by the distributed load are calculated (a) consistently, and (b) from a cubic field for w.

11.13 Imagine that the plate of Fig. 11.2-3 is square, simply supported, and modeled by a single finite strip (i.e., $L = b$). The lateral load is distributed and is described by $q = q_0 \sin(\pi x/b) \sin(\pi y/b)$. Determine the center deflection of the finite strip. *Suggestions:* See Problem 11.3c. Arbitrarily elect to evaluate the equation at the center of the plate. Note, however, that this procedure would not be used in a finite strip computer program.

11.14 Let the rectangular plate element of Fig. 11.2-1 carry a uniformly distributed load q in the $+z$ direction. Determine the resulting nodal moment loads by assuming that the plate element acts like a beam clamped at both ends, spanning first the dimension $2a$ and then the dimension $2b$. Show these loads, properly directed, on a sketch of the plate.

11.15 (a) A constant moment $\overline{M}_{xy}$ is applied along edge 2–3 of the element in Fig. 11.2-1. What nodal loads result? Assume that w along this edge is cubic in y, and governed by w and $w_{,y}$ at nodes 2 and 3.

(b) What would be your answer if the element is instead a Mindlin element, Eqs. 11.2-7, with shape functions N_1 through N_4?

Section 11.3

11.16 Imagine that the four-node plate element of Fig. 11.3-1*a* is to be obtained by specialization of the eight-d.o.f. solid element of Fig. 6.7-1. Describe the steps and substitutions that convert Eqs. 6.7-1 to the equations $w = \Sigma N_i w_i$, $u = -z \Sigma N_i \theta_{xi}$, and $v = -z \Sigma N_i \theta_{yi}$, where $i = 1, 2, 3, 4$.

11.17 Show that Eq. 11.3-8 follows from Eq. 11.3-6 when $[\mathbf{D}_M]$, $[\mathbf{B}_b]$, and $[\mathbf{B}_s]$ are defined as stated in the text.

11.18 A uniform load q acts upward on a rectangular plate element of side lengths $2a$ and $2b$. What are the consistent nodal loads for each of the four elements listed in Table 11.3-1?

11.19 A square plate under concentrated center load is to be analyzed. The boundary is clamped, meaning that all boundary d.o.f. are set to zero. Let the model consist of a single element, which occupies one quadrant. After imposing boundary conditions, how many unknown d.o.f. are left for each of the elements in Table 11.3-1?

11.20 The sketch shows more detail of the latter portion of Fig. 11.3-3*b*. Lateral displacement w is zero at the ends and at the center. Imagine that nothing varies with y; that is, beam action is to be modeled. Under bending moment that varies linearly with x, thin-beam theory shows that end sections rotate an amount θ_b and the middle rotates an amount $\theta_b/2$ in the opposite direction. On this must be superposed rotations θ_s caused by the constant transverse shear force. Thus

$$\gamma_{zx} = w_{,x} - \theta_x = \left(\frac{\theta_b}{2} - \theta_s\right) - \frac{3\theta_b}{2}\left(\frac{x}{a}\right)^2$$

(a) Derive this expression for γ_{zx}, using quadratic shape functions and nodal values of w and θ_x at A, M, and C.

(b) Determine the values of x/a for which γ_{zx} is correctly represented, even when $a >> t$.

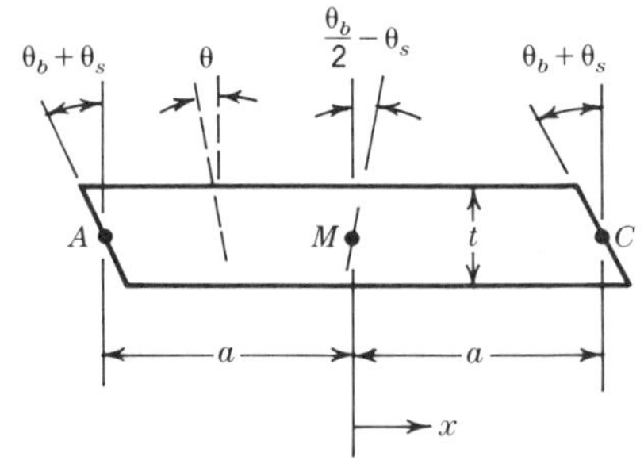

Problem 11.20

11.21 Consider a rectangular quadratic element (Table 11.3-1). Under what circumstances or deformation states will γ_{yz} and γ_{zx} be correctly evaluated along the line $\xi = 0$?

11.22 (a) Sketch a rectangular element in its deformed state if displacements are described by $u = v = 0$, $w = 3\xi^2\eta^2 - \xi^2 - \eta^2$.

(b) Show that this deformation mode yields zero strains at the Gauss points of a 2 by 2 rule.

(c) What kind of loads and support conditions would activate this mode, either for a single element or for a mesh of elements?

11.23 Sketch a 2 by 3 mesh of rectangular bilinear elements. Superposed on this sketch, show the mesh deformed into the w-hourglass mode of Fig. 11.3-4.

11.24 Use Eq. 11.3-4 to evaluate $\{\boldsymbol{\kappa}\}$ for each of the modes in Fig. 11.3-4. Show that only the first two $\{\boldsymbol{\kappa}\}$'s are null for selective integration, and that all four $\{\boldsymbol{\kappa}\}$'s are null for reduced integration.

Section 11.4

11.25 (a) Derive Eq. 11.4-2.

(b) In Eq. 11.4-2, express $w_{,s5}$ in terms of w, $w_{,x}$, and $w_{,y}$ at nodes 2 and 3. Let the n axis in Fig. 11.4-1a make an angle α with the x axis. What is α in terms of the x and y coordinates of nodes 2 and 3?

(c) Similarly, in Eq. 11.4-5, what is θ_{n5} in terms of w, $w_{,x}$, and $w_{,y}$ at nodes 2 and 3 and angle α?

11.26 What can be said about interelement compatibility of the DKT element?

11.27 Why are three Gauss points adequate for exact integration of a DKT element?

11.28 In Table 11.4-1, consider the clamped, uniformly loaded test case with mesh $N = 1$. Do you think the computed result would be more accurate if nodal moments were included in the element load vectors? Why?

11.29 Imagine that the serendipity element of Table 11.3-1 is to be given a "discrete Kirchhoff" treatment by explicitly enforcing zero transverse shear strain at the Gauss points of a 2 by 2 rule. How may d.o.f. do these constraints eliminate? What d.o.f. do you think it appropriate to retain in $\{\mathbf{d}\}$, and why?

Section 11.5

11.30 Demonstrate the free-edge condition $Q_n + M_{ns,s} = 0$, which is stated in the footnote for Eqs. 11.5-1. *Suggestion:* Consider couple forces $M_{ns}\,\Delta s$ and $(M_{ns} + M_{ns,s}\,\Delta s)\,\Delta s$ in adjacent "cells" of length Δs along the edge.

11.31 Consider the mesh shown in Fig. 11.4-2. The mesh models one quadrant of a symmetrically loaded and symmetrically supported square plate. Under each of the following support conditions, how many unknown d.o.f. remain in $\{\mathbf{D}\}$ after boundary conditions have been imposed? Of these, which do you expect will have the same magnitude?

(a) Clamped.

(b) Simply supported (finite element conditions).

(c) Simply supported (classical theory conditions).

CHAPTER 12

SHELLS

The physics of shell behavior is summarized. Advantages and disadvantages of various displacement fields are illustrated by means of singly curved (arch) elements. Element formulations are presented for general shells and for shells of revolution.

12.1 SHELL GEOMETRY AND BEHAVIOR. SHELL ELEMENTS

A shell forms a curved surface in space. Usually a shell is thin in comparison with its span. Geometrically, a shell is described by its thickness t and the shape of the shell midsurface. At every point on the midsurface, one can draw two small arcs that lie in the midsurface, and orient the arcs so as to fit the largest and smallest curvatures of the midsurface at that point. These arcs will be mutually perpendicular. They define the *principal radii of curvature* at that point. In general, principal radii vary from point to point. The centers of curvature lie on a normal to the midsurface.

Examples of shell geometry appear in Fig. 12.1-1. The cylinder and cone are *singly curved*. In addition, they are *developable*, which means that if slit lengthwise they can be unrolled to form flat sheets without having to stretch their midsurfaces. The sphere and hyperboloid are *doubly curved* and are not developable. Radii of curvature are constant in the cylinder and sphere but are variable in the cone and hyperboloid.

Each shell in Fig. 12.1-1 happens to be a *shell of revolution*, meaning that the midsurface is generated by rotating a straight or curved generating line about an axis of revolution in the plane of the generator. A *meridian* is the intersection of the midsurface with a plane that contains the axis of revolution. A *parallel* is the intersection of the midsurface with a plane perpendicular to the axis.

In general, a shell simultaneously displays *bending stresses* and *membrane stresses*. Bending stresses in a shell correspond to bending stresses in a plate (Fig. 11.1-1*a*) and produce bending and twisting moments (Eqs. 11.1-1a). Membrane stresses correspond to stresses in a plane stress problem: they act tangent to the midsurface, and produce midsurface-tangent forces per unit length. These are the membrane forces N_x, N_y, and N_{xy} given by

$$N_x = \int_{-t/2}^{t/2} \sigma_x \, dz \qquad N_y = \int_{-t/2}^{t/2} \sigma_y \, dz \qquad N_{xy} = \int_{-t/2}^{t/2} \tau_{xy} \, dz \qquad (12.1\text{-}1)$$

where x and y are orthogonal coordinates in the midsurface and z is a direction

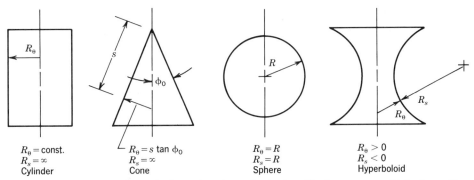

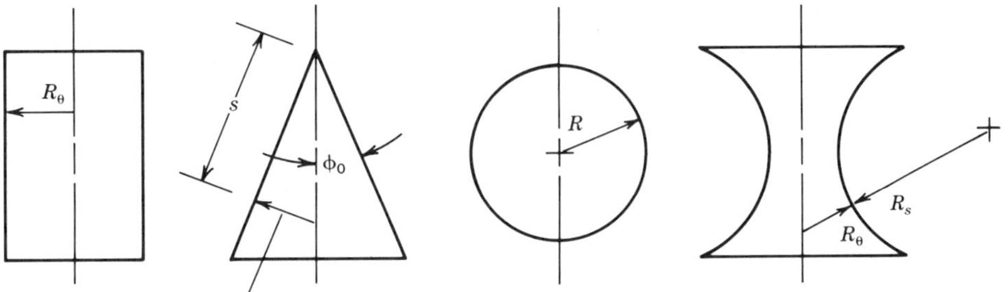

Figure 12.1-1. Shells of revolution, showing principal radii of curvature. R_s is the radius of curvature of the meridian.

normal to the midsurface. Stresses in the shell are composed of a membrane component σ_m and a bending component σ_b; for example, normal stress in the x direction is

$$\sigma_x = \sigma_{mx} + \sigma_{bx} = \frac{N_x}{t} + \frac{M_x z}{t^3/12} \tag{12.1-2}$$

Thus it is assumed that stresses vary linearly through the thickness. Stress on the midsurface $z = 0$ is zero if $N_x = N_y = N_{xy} = 0$.

A shell can carry a large load if membrane action dominates over bending, just as a thin wire can carry a large load in tension but only a small load in bending. Practically, it is not possible to have only membrane action in a shell. Bending action is also present if concentrated loads are applied, if supports apply moments or transverse forces, or if a radius of curvature changes abruptly. As an example of the latter, consider closing a cylindrical pressure vessel with a cap that is hemispherical or ellipsoidal: where the cylinder meets the cap, radius R_s of the meridian changes abruptly from infinite to finite. Typically, bending action is quite localized—that is, bending stresses are large only quite near the load or discontinuity that produces them.

Classical shell theory produces equations that are very difficult to solve. The governing equations in terms of displacements are complicated; they have relatively simple forms only if many approximations are made. Authorities do not agree on what approximations are acceptable, so various shell theories have been proposed (e.g., those of Donnell, Flügge, Sanders, Vlasov, etc.). Like Kirchhoff plate theory, shell theories are limited to small deflections unless higher-order terms are added to account for membrane strains associated with rotation of the shell midsurface.

Classical shell theory is concerned with *thin shells,* in which transverse shear deformation is considered negligible. In practice one may also encounter *thick shells*. Then one must account for transverse shear deformation, and perhaps also for the effects of thickness-direction normal stress.

Shell Elements. Finite elements for shells have been among the most difficult elements to devise. Three approaches to the problem have been pursued:

1. Flat elements, formed by combining a plane membrane element with a plate bending element.
2. Curved elements, formulated by use of a classical shell theory.
3. Mindlin-type elements, similar to Mindlin plate elements described in Section 11.3. Such elements can be regarded as special forms of solid elements, made thin in one direction.

Flat triangular elements model a shell as a faceted surface. Flat elements are easy to formulate. They pass patch tests and do not exhibit strain under rigid-body motion. However, although membrane-bending coupling is present throughout an actual curved shell, it is absent in individual flat elements. In the past, flat elements have not been particularly accurate, but have been useful because of the difficulties of other approaches.

A curved element is necessarily more complicated than a flat element, first because its geometry is more complicated. Then, regardless of what classical shell theory is used, its approximations and complexities are incorporated in the element. Some curved elements cannot display rigid-body motion without strain, either because of defects in the shell theory or because of shortcomings in the element displacement field. Typically, the user of a curved element must supply data in addition to nodal coordinates in order to describe element geometry. Some curved shell elements include derivatives of membrane strain and curvature among their nodal d.o.f.

Mindlin-type or "degenerated solid" elements can be curved and appear to occupy a middle ground between flat elements and curved elements formulated by use of shell theory, both in accuracy and in ease of use. Elements for thin

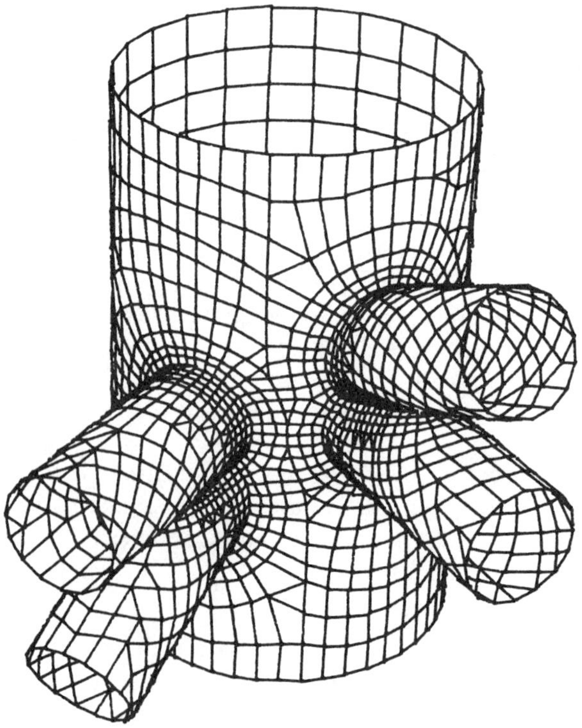

Figure 12.1-2. Intersecting pipes, modeled by quadrilateral and triangular shell elements. (*Courtesy of Algor Interactive Systems, Inc., Pittsburgh, Pennsylvania.*)

shells and for thick shells are available. As with Mindlin plate elements, possible difficulties with locking must be addressed.

Any of the foregoing three approaches might be used to provide elements for a particular shell, such as the shell shown in Fig. 12.1-2.

The membrane stiffness of a thin shell is much larger than its bending stiffness. This is reflected in a large discrepancy between the associated stiffness coefficients in [**K**], regardless of how the shell elements are formulated. Numerical errors of the type discussed in Section 18.2 are possible.

At present it is not clear whether the most cost-effective thin-shell elements will be flat or curved. In what follows, emphasis is placed on flat elements, elements for shells of revolution, and Mindlin elements. Curved elements for thin shells of general shape that are based on a classical shell theory are beyond the scope of this text.

Test Cases for General Shell Elements. If assigned a flat geometry, shell elements can be used to solve problems of plane stress and plate bending. Accordingly, one can begin with patch tests and other commonly used test problems for plane problems and plates (e.g., Fig. 4.6-1, Tables 6.14-1, 6.14-2, and 11.4-1, and proposed test problems in Ref. 12.15). Singly curved shell elements can be tested on arch problems (e.g., Fig. 12.2-1*a*) and on cylindrical shell problems (e.g., Fig. 12.4-4). A good element will have good accuracy on these initial tests and will not have mechanisms. Various additional problems have been used as test cases for general shell elements, including a pinched cylinder, a pinched hemisphere, and a slit cylinder under twisting load. Details may be found in [12.8,12.14–12.16].

12.2 CIRCULAR ARCHES AND ARCH ELEMENTS

The study of arch elements provides insight into various aspects of shell element behavior. In what follows we consider an arch of constant mean radius R, loaded in its own plane.

Equations for Thin Circular Arches. We assume that the arch is thin—that is, that $R \gg t$ in Fig. 12.2-1. A point on the arch midline has s-direction (tangential) displacement u and z-direction (radial) displacement w. Let ϵ_s represent tangential

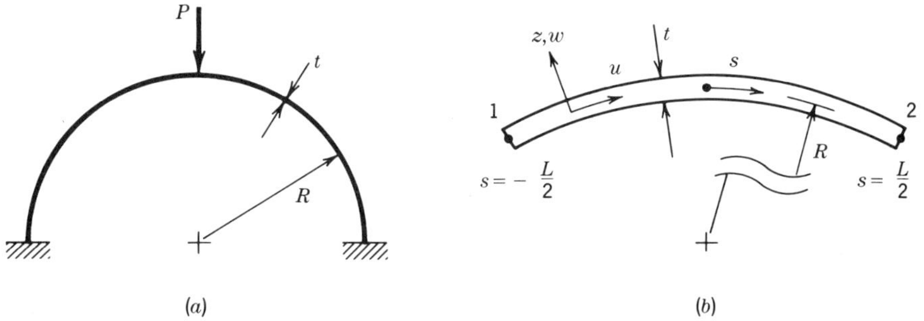

Figure 12.2-1. (*a*) Semicircular arch with clamped ends and concentrated center load. (*b*) Arch element of arc length L. Coordinates s and z are respectively tangent and normal to the arch midline.

strain at an arbitrary point, a distance z from the midline. With the aid of Fig. 12.2-2, and with R constant, we write

$$\epsilon_s = \frac{d}{ds}(\delta_a + \delta_c) + \frac{w}{R} = u_{,s} + \frac{w}{R} + z\left(\frac{u_{,s}}{R} - w_{,ss}\right) \tag{12.2-1}$$

If the approximation $\delta_a \approx u$ is introduced, the term $zu_{,s}/R$ disappears. In alternative notation, Eq. 12.2-1 is

$$\epsilon_s = \epsilon_m + z\kappa, \qquad \text{where} \begin{cases} \epsilon_m = u_{,s} + \dfrac{w}{R} & \text{(12.2-2a)} \\ \kappa = \dfrac{u_{,s}}{R} - w_{,ss} & \text{(12.2-2b)} \end{cases}$$

Membrane strain ϵ_m appears along the arch midline and is associated with s-direction (membrane) force. Curvature change κ is associated with bending moment and is considered positive when the radius of curvature decreases. Transverse shear deformation γ_{zs} is considered zero for a thin arch.

Strain energy U in an arch element is composed of a membrane contribution U_m and a bending contribution U_b. For an element of arc length L,

$$U = U_m + U_b = \int_{-L/2}^{L/2} \frac{EA}{2}\epsilon_m^2\, ds + \int_{-L/2}^{L/2} \frac{EI}{2}\kappa^2\, ds \tag{12.2-3}$$

where E = elastic modulus, A = cross-sectional area, and I = moment of inertia of A about the neutral axis of bending ($I = bt^3/12$ for a rectangular cross section of width b). The expression for U can be derived by integration of strain energy density $E\epsilon_s^2/2$ through the arch thickness t. A term linear in z disappears, and the terms shown in Eq. 12.2-3 remain. Note that membrane stiffness EA becomes much larger than bending stiffness EI as arch thickness t becomes small.

For rigid-body motion, $\epsilon_m = \kappa = 0$. The displacement field for rigid-body motion is therefore

$$u = b_1 \cos\phi + b_2 \sin\phi + b_3 \tag{12.2-4a}$$

$$w = b_1 \sin\phi - b_2 \cos\phi \tag{12.2-4b}$$

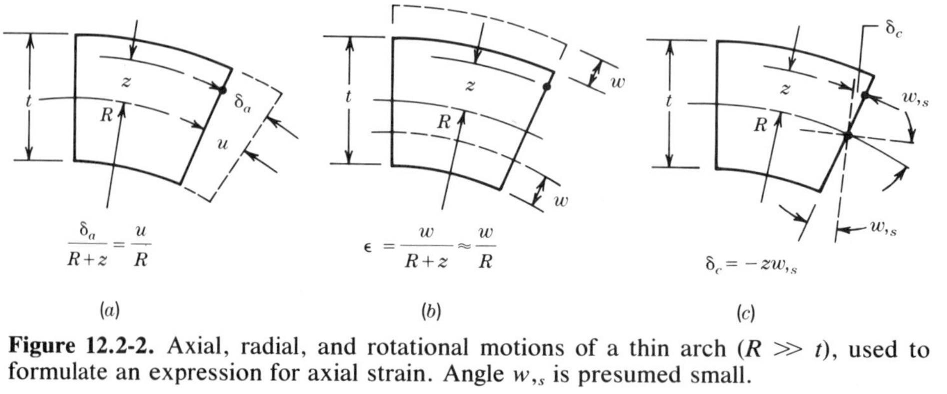

Figure 12.2-2. Axial, radial, and rotational motions of a thin arch ($R \gg t$), used to formulate an expression for axial strain. Angle $w_{,s}$ is presumed small.

where $\phi = s/R$ and the b_i are constants. Here b_1 and b_2 represent translations in mutually perpendicular directions and b_3 represents a rotation about the center of curvature.

A thin arch will bend but will have very little membrane strain. Accordingly, from Eq. 12.2-2, we obtain the *inextensibility condition:*

$$\epsilon_m = 0 \qquad \text{implies} \qquad u_{,s} + \frac{w}{R} = 0 \tag{12.2-5}$$

This condition is satisfied in the limit of thinness as L/t becomes infinite.

Straight Arch Elements. The use of straight elements to model an arch is analogous to the use of flat elements to model a shell. A straight arch element is identical to a plane frame element. To obtain it, one merely combines a standard two-d.o.f. bar element (Eq. 2.4-5) with a standard four-d.o.f. beam element (Eq. 4.2-5). Thus, using d.o.f. in Fig. 12.2-3a, the element stiffness equation $[\mathbf{k}]\{\mathbf{d}\} = \{\mathbf{r}\}$ is

$$\begin{bmatrix} \underset{2\times 2}{[\mathbf{k}_{\text{bar}}]} & \underset{2\times 4}{[\mathbf{0}]} \\ \underset{4\times 2}{[\mathbf{0}]} & \underset{4\times 4}{[\mathbf{k}_{\text{beam}}]} \end{bmatrix} \begin{Bmatrix} u_1 \\ u_2 \\ w_1 \\ \beta_1 \\ w_2 \\ \beta_2 \end{Bmatrix} = \{\mathbf{r}\} \tag{12.2-6}$$

in which $[\mathbf{k}_{\text{bar}}]$ and $[\mathbf{k}_{\text{beam}}]$ come respectively from U_m and U_b in Eq. 12.2-3. Nodal rotations β_1 and β_2 are nodal values of $w_{,s}$.

For assembly with other elements having different orientation, a common set of structural d.o.f. is needed. The choice is not unique. One possibility is to use tangential and radial translational d.o.f. D_s and D_r at each node, as shown in Fig. 12.2-3b. Coordinate transformation (Section 7.4) is used to replace nodal values of u and w by nodal values of D_s and D_r. D.o.f. β_1 and β_2 are unchanged by this transformation. (A similar transformation is described in connection with Eq. 12.4-3.)

The displacement field that produces Eq. 12.2-6 can be written

$$u = a_1 + a_2 s \tag{12.2-7a}$$

$$w = a_3 + a_4 s + a_5 s^2 + a_6 s^3 \tag{12.2-7b}$$

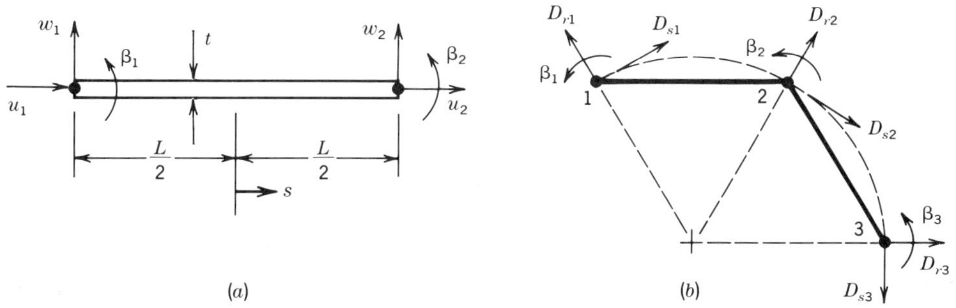

Figure 12.2-3. (a) Straight element, showing d.o.f. in the local coordinate system. (b) Possible choice of global d.o.f, where D_r and D_s are, respectively, radial and tangential.

where u and w are respectively parallel and normal to the straight element, s is a (straight) axial coordinate, and the a_i are constants. Rigid-body motion is accounted for by a_1, a_3, and a_4. Since R is infinite for a straight element, Eqs. 12.2-2 yield $\epsilon_m = a_2$ and $\kappa = -2a_5 - 6a_6 s$. Thus constant-strain states are possible, and these states are not coupled to rigid-body motion. The inextensibility condition, Eq. 12.2-5, is satisfied when $a_2 = 0$. When an additional element is attached to elements already in place, three new d.o.f. are added to the mesh (e.g., the d.o.f. at node 3 in Fig. 12.2-3*b*), but only one inextensibility constraint is added ($a_2 = 0$). Therefore the mesh will not lock, regardless of the values of R, L, and t. (Locking concepts are discussed in Sections 9.4 and 9.5.)

Accordingly, the straight element of Eqs. 12.2-7 is free of major defects. It can be rigorously shown that the straight element is a valid model for a curved arch, and provides convergence to exact answers as the mesh is refined [12.1]. A drawback is that membrane and bending actions are not coupled within a single straight element, as evidenced by the off-diagonal null matrices in Eq. 12.2-6. Another drawback is that a distributed load must be modeled by nodal forces *only*. If nodal moment loads are also applied, as the consistent formulation of Eq. 4.1-6 would dictate, and if elements are of unequal lengths, then spurious bending moments appear. (These moments would be correct if the actual structure were a polygon of straight bars.) And, during stress computation, stresses caused by nodal displacements should *not* be adjusted to account for distributed load on the element (as discussed following Eq. 4.7-3).

In a coarse mesh, accuracy is improved by taking element length L as the arc length between nodes, rather than the chord length as is suggested by Fig. 12.2-3*b*. For example, let $R/t = 40$ for the arch of Fig. 12.2-1*a*, and let the entire arch be modeled by four straight elements, each of length $L = \pi R/4$. The error in the computed displacement of load P is approximately 2%. Use of the chord length $L = 2R \sin 22.5°$ gives an error of approximately 10%.

Curved Arch Elements. A great many types of curved arch element have been proposed, partly because a good curved element proved to be more elusive than first anticipated. Often, curved elements were far less accurate than straight elements: they were much too stiff, especially when applied to an arch that is thin and has a large rise-to-span ratio. Originally the difficulties were blamed on a lack of rigid-body motion capability. Subsequently the difficulties were attributed primarily to *membrane locking* [12.17], which is described as follows.

The simplest curved element has the geometry shown in Fig. 12.2-1*b* and the displacement field of Eqs. 12.2-7, where u and w are now circumferential and radial displacements. Coordinate s follows the arch midline. Element d.o.f. are nodal values of u, w, and $w_{,s}$. (In contrast to straight elements, no coordinate transformation is needed to match d.o.f. prior to assembly of elements.) From Eqs. 12.2-2 and 12.2-7, membrane strain ϵ_m is

$$\epsilon_m = \left(a_2 + \frac{a_3}{R}\right) + \frac{a_4}{R} s + \frac{a_5}{R} s^2 + \frac{a_6}{R} s^3 \qquad (12.2\text{-}8)$$

Now imagine that thickness t of the arch approaches zero. The inextensibility condition, $\epsilon_m = 0$ for all s, demands that

$$a_2 + \frac{a_3}{R} = a_4 = a_5 = a_6 = 0 \qquad (12.2\text{-}9)$$

The condition $a_2 + a_3/R = 0$ implies $\epsilon_m = 0$ at $s = 0$, the element center. This constraint causes no difficulty. The remaining conditions, $a_4 = a_5 = a_6 = 0$, imply that $w_{,s} = w_{,ss} = w_{,sss} = 0$. These are severe and spurious constraints. They produce what is called *membrane locking;* that is, they make the model much too stiff. Numerical evidence supports this conclusion: for example, for the arch of Fig. 12.2-1*a* with $R/t = 40$, four identical elements predict a center displacement that is less than 5% of its correct value [12.2]. Equation 12.2-8 shows that the contribution of w terms to ϵ_m decreases as R becomes large. That is, the tendency to lock decreases as elements become more shallow, and disappears if elements become straight.

One can also conclude that the mesh will lock by doing "constraint counting," which is discussed in Section 9.5. Energy U_m in Eq. 12.2-3 acts as a penalty function that enforces the inextensibility constraint as a curved element becomes thin—that is, as the ratio A/I becomes large. If numerically integrated, U_m (and the membrane stiffness matrix it produces) enforces one constraint for every integration point used to evaluate U_m. Each constraint effectively removes one d.o.f. from its intended role of modeling deformations. Accordingly, if there are three or more integration points per element, *all* d.o.f. of the structure are used to satisfy support conditions and constraints. No d.o.f. are left to model the bending deformations, so a mesh of curved elements tends to lock.

This insight suggests the simple remedy of *selective integration*. One can use a two-point rule to evaluate the stiffness matrix that comes from U_b, but use a one-point rule to evaluate the stiffness matrix that comes from U_m. The single Gauss point is at the element center, $s = 0$. Accordingly, from Eq. 12.2-8, the constraint enforced is

$$a_2 + \frac{a_3}{R} = 0 \qquad (12.2\text{-}10)$$

Now only one d.o.f. per element is used in satisfying the constraint, and the mesh does not lock. Such elements are quite accurate: they have about the same accuracy as the aforementioned straight elements [12.2].

Exactly integrated curved elements can also behave well if element displacement fields are properly designed. For example, consider the fields [12.3]

$$u = a_1 + a_2\phi + a_3\phi^2 + a_4\phi^3 + a_5\phi^4 + a_6\phi^5 \qquad (12.2\text{-}11a)$$

$$w = -a_2 - 2a_3\phi - 3a_4\phi^2 - 4a_5\phi^3 - 5a_6\phi^4 \qquad (12.2\text{-}11b)$$

where $\phi = s/R$. These displacements satisfy the inextensibility condition $\epsilon_m = 0$ for all s. Since $\epsilon_m = 0$, the resulting stiffness matrix comes entirely from the U_b term in Eq. 12.2-3. Computed results are almost exact. We see that if displacement fields are to have enough d.o.f. for a six-d.o.f. element and are also to display $\epsilon_m = 0$, the membrane field must be at least quintic and the lateral displacement field must be one degree lower.

To allow the preceding element to display membrane strain, one can add a constant a_7 to Eq. 12.2-11b. Thus [**k**] becomes 7 by 7 and includes a contribution from U_m in Eq. 12.2-3. D.o.f. a_7 is internal to each element. Again, results are excellent [12.3].

We note that the rigid-body motion field of Eqs. 12.2-4 is not explicitly included in any of the preceding curved elements. One can say that such motion is *implicitly* approximated in Eqs. 12.2-11 because sin ϕ and cos ϕ can be expanded as power series, whose initial terms are contained in Eqs. 12.2-11. Another element that implicitly approximates rigid-body motion is based on a displacement field of the form [12.4]

$$u = a_1 + a_2 s + a_7 \left(1 - \frac{4s^2}{L^2}\right) \tag{12.2-12a}$$

$$w = a_3 + a_4 s + a_5 s^2 + a_6 s^3 \tag{12.2-12b}$$

The mode associated with d.o.f. a_7 is internal or "nodeless"; it vanishes at the element ends $s = \pm L/2$. D.o.f. a_7 can be removed by condensation prior to assembly of elements. As compared with Eqs. 12.2-7, Eqs. 12.2-12 reduce spurious strain energy associated with rigid-body motion by factors of 122,000 and 18,000 for curved elements that subtend arcs of 12° and 20°, respectively [12.4].

A curved element that *explicitly* includes rigid-body motion capability has been suggested. Known as the Cantin–Clough element, it is based on fields of the form [12.2]

$$u = a_1 \cos\phi + a_2 \sin\phi + a_3 + a_4 s \tag{12.2-13a}$$

$$w = a_1 \sin\phi - a_2 \cos\phi + a_5 s^2 + a_6 s^3 \tag{12.2-13b}$$

where $\phi = s/R$. Rigid-body motion, Eqs. 12.2-4, is displayed by the element when $a_4 = a_5 = a_6 = 0$. From Eqs. 12.2-2 and 12.2-13,

$$\epsilon_m = a_4 + \frac{a_5}{R} s^2 + \frac{a_6}{R} s^3 \quad \text{and} \quad \kappa = \frac{a_4}{R} - 2a_5 - 6a_6 s \tag{12.2-14}$$

Inextensibility requires that $a_4 = a_5 = a_6 = 0$, which in turn yields $\kappa = 0$. Accordingly, we expect to encounter locking difficulties. Indeed, for the thin-arch problem of Fig. 12.2-1*a*, the central deflection is almost 50% low when 20 Cantin–Clough elements are used for the entire arch [12.2].

From the foregoing examples we conclude that membrane locking is much more detrimental to thin curved elements than is a lack of explicit rigid-body motion capability.

Mindlin Arch Elements. Displacements of a Mindlin arch element are described by tangential and normal displacements of the midline *and* by rotation of a normal to the midline—that is, by u, w, and β. Thus rotation β may differ from rotation $w_{,s}$. As with Mindlin beam and plate elements discussed in Section 9.4 and Chapter 11, Mindlin arch elements can account for transverse shear deformation, which here is γ_{zs}. Only if $\gamma_{zs} = 0$ does $\beta = w_{,s}$.

Strain energy in a Mindlin arch element is $U = U_m + U_b + U_s$, in which the respective contributions to U are due to membrane strain ϵ_m, curvature change

κ, and transverse shear strain γ_{zs} [12.5]. For an element of length L and constant radius R,

$$U_m = \int_{-L/2}^{L/2} \frac{EA}{2} \epsilon_m^2 \, ds, \qquad \text{where} \quad \epsilon_m = u_{,s} + \frac{w}{R} \tag{12.2-15a}$$

$$U_b = \int_{-L/2}^{L/2} \frac{EI}{2} \kappa^2 \, ds, \qquad \text{where} \quad \kappa = \frac{u_{,s}}{R} - \beta_{,s} \tag{12.2-15b}$$

$$U_s = \int_{-L/2}^{L/2} \frac{GA}{2} \gamma_{zs}^2 \, ds, \qquad \text{where} \quad \gamma_{zs} = w_{,s} - \beta \tag{12.2-15c}$$

To account for a parabolic variation of γ_{zs} through the thickness, we may replace G with $5G/6$. A two-node element can be based on linear interpolations. With a_i = constants and $N_1 = 0.5 - s/L$, $N_2 = 0.5 + s/L$, linear fields are

$$u = a_1 + a_2 s \qquad \text{or} \qquad u = N_1 u_1 + N_2 u_2 \tag{12.2-16a}$$

$$w = a_3 + a_4 s \qquad \text{or} \qquad w = N_1 w_1 + N_2 w_2 \tag{12.2-16b}$$

$$\beta = a_5 + a_6 s \qquad \text{or} \qquad \beta = N_1 \beta_1 + N_2 \beta_2 \tag{12.2-16c}$$

Arguments concerning locking in this element are very similar to arguments made in connection with Eqs. 12.2-8 through 12.2-10 and are summarized as follows. As thickness t approaches zero, all strain energy should be in bending; that is, ϵ_m and γ_{zs} should each vanish for all s, which implies, for a curved element,

$$a_2 + \frac{a_3}{R} = a_4 = a_5 = a_6 = 0 \tag{12.2-17}$$

or which implies, for a straight element ($R = \infty$),

$$a_2 = a_4 - a_5 = a_6 = 0 \tag{12.2-18}$$

In either case, an element added to a thin arch brings three d.o.f. with it, but all d.o.f. are occupied in satisfying constraints, and the mesh locks as t approaches zero. However, reduced integration can produce a workable element. If U_m and U_s are integrated with a one-point rule, only two constraints are imposed per element, one each on ϵ_m and γ_{zs}, and the mesh does not lock.

A curved element based on Eqs. 12.2-16 does not explicitly contain the rigid-body motion field, Eqs. 12.2-4. A straight element ($R = \infty$) can display rigid-body motion; that is, it displays $\epsilon_m = \kappa = \gamma_{zs} = 0$ if $a_2 = a_6 = 0$ and $a_4 = a_5$.

A three-node Mindlin element, having nodes at $s = -L/2$, $s = 0$, and $s = +L/2$, would be called a quadratic element. With $\xi = s/(L/2)$, its displacement field has the form

$$u = a_1 + a_2\xi + a_3\xi^2 \tag{12.2-19a}$$

$$w = a_4 + a_5\xi + a_6\xi^2 \tag{12.2-19b}$$

$$\beta = a_7 + a_8\xi + a_9\xi^2 \tag{12.2-19c}$$

As the arch becomes extremely thin, the inextensibility condition $\epsilon_m = 0$ for all s implies

$$\frac{2a_2}{L} + \frac{a_4}{R} = \frac{4a_3}{L} + \frac{a_5}{R} = a_6 = 0 \qquad (12.2\text{-}20)$$

The constraint $a_6 = 0$ implies membrane locking, as it prevents a thin arch from displaying a constant value of $w,_{ss}$. Reduced integration offers a remedy [12.5,12.6]. Membrane strain can be written in the form

$$\epsilon_m = \left(\frac{2a_2}{L} + \frac{a_4}{R} + \frac{a_6}{3R}\right) + \left(\frac{4a_3}{L} + \frac{a_5}{R}\right)\xi + \frac{a_6}{R}\left[\xi^2 - \tfrac{1}{3}\right] \qquad (12.2\text{-}21)$$

If integration of membrane energy U_m is performed by two-point Gauss quadrature, for which Gauss points are at $\xi = \pm 1/\sqrt{3}$, then the bracketed expression vanishes. Thus no constraint is placed on a_6; rather, $\epsilon_m = 0$ implies only the vanishing of the two parenthetic expressions in Eq. 12.2-21.

A similar argument can be applied to the condition $\gamma_{zs} = 0$, which should prevail in an extremely thin arch. The conclusion is that $\gamma_{zs} = 0$ enforces the constraint $\beta,_{ss} = 0$ in a quadratic element. This is not a locking condition, but it degrades element performance. Again the remedy is to use reduced integration: a two-point Gauss rule should be used to evaluate shear energy U_s. (This advice was previously given in connection with a quadratic plate element, Fig. 11.3-3*b*.)

The quadratic arch element does not explicitly contain the rigid-body motion capability described by Eqs. 12.2-4. However, Eqs. 12.2-4 pertain to a *circular* arch. Imagine now that the element shape is defined by the coordinates of its three nodes. Thus the arch element has parabolic shape. Then, since displacement fields are also second degree, the parabolic element is of the isoparametric family, and arguments given in Section 6.10 demonstrate that the capability for rigid-body motion is present. Similarly, a quadratic *shell* element, adapted from a quadratic isoparametric solid element, is able to display rigid-body motion without strain. (Some cautions about thickness-direction integration should be noted; see Ref. 12.7.)

Remarks. During stress computation in numerically integrated elements, one should evaluate strains at the Gauss points of the reduced quadrature rule appropriate to the element: for example, for Mindlin elements, at the center in linear elements and at $\xi = \pm 1/\sqrt{3}$ in quadratic elements. Large spurious strains may appear at other locations, as shown in Fig. 6.13-2. Figure 6.13-2 depicts transverse shear strain, but membrane strain can display similar behavior. Accurate stress computation in extremely thin elements may require a restriction on thickness t, as noted in the following paragraph.

Use of reduced integration avoids the imposition of spurious constraints, but not all constraints. The constraints that remain may cause numerical difficulty if the element is extremely thin. Difficulty is avoided by simply not allowing thickness t to fall below a certain limit in the computation of stiffness matrix coefficients associated with strain energies U_m and U_s. This matter is discussed at the end of Section 9.4.

12.3 FLAT ELEMENTS FOR SHELLS

A curved shell can be approximated as a faceted surface, formed by connecting flat triangular elements together at vertex nodes. If we elect to use three translational and three rotational d.o.f. per node, each triangular element has 18 d.o.f. Let a typical element lie in the *xy* plane of a local coordinate system *xyz*. Nodal d.o.f. are shown in Fig. 12.3-1. Let these d.o.f. be called $\{\mathbf{d}'\}$ and be arranged in the order[1]

$$\{\mathbf{d}'\} = \lfloor \mathbf{u}_i \quad \mathbf{v}_i \quad \boldsymbol{\theta}_{zi} \quad \mathbf{w}_i \quad \boldsymbol{\theta}_{xi} \quad \boldsymbol{\theta}_{yi} \rfloor^T \tag{12.3-1}$$

where $\lfloor \mathbf{u}_i \rfloor = \lfloor u_1 \quad u_2 \quad u_3 \rfloor$, $\lfloor \mathbf{v}_i \rfloor = \lfloor v_1 \quad v_2 \quad v_3 \rfloor$, and so on. A plane element that includes "drilling freedoms" θ_z among its d.o.f. is described in Section 8.6. Its 9 by 9 stiffness matrix, which we now call $[\mathbf{k}_m]$, models membrane action in the shell. To this element we add a nine-d.o.f. triangular plate element, such as element DKT of Section 11.4. Its 9 by 9 stiffness matrix, which we now call $[\mathbf{k}_b]$, models bending action in the shell. For the composite flat shell element, whose stiffness matrix in local *xyz* coordinates is called $[\mathbf{k}']$, we have

$$[\mathbf{k}']\{\mathbf{d}'\} = \begin{bmatrix} \underset{9\times 9}{[\mathbf{k}_m]} & \underset{9\times 9}{[\mathbf{0}]} \\ \underset{9\times 9}{[\mathbf{0}]} & \underset{9\times 9}{[\mathbf{k}_b]} \end{bmatrix} \begin{Bmatrix} \mathbf{u}_i \\ \mathbf{v}_i \\ \boldsymbol{\theta}_{zi} \\ \mathbf{w}_i \\ \boldsymbol{\theta}_{xi} \\ \boldsymbol{\theta}_{yi} \end{Bmatrix} \tag{12.3-2}$$

This stiffness matrix is clearly analogous to that of a straight arch element, Eq. 12.2-6. Coordinate transformation of $[\mathbf{k}']$ is required prior to assembly of elements so that a common set of d.o.f. is used at each structure node shared by two or

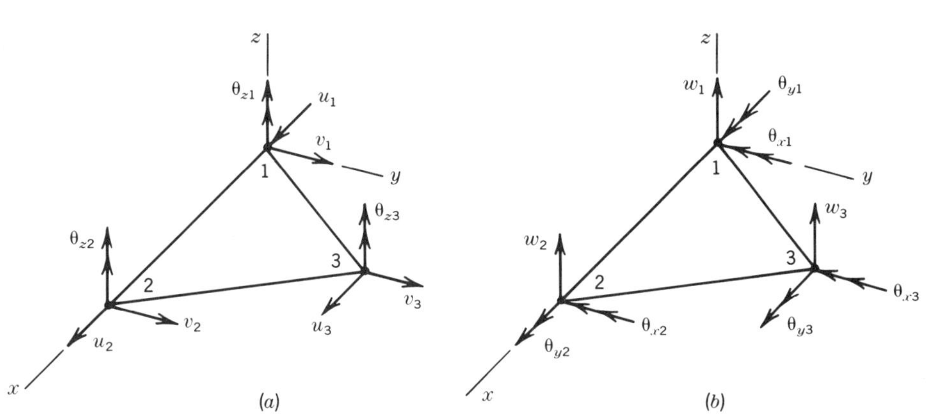

Figure 12.3-1. Triangular element in a local *xy* plane. (*a*) D.o.f. associated with membrane action. (*b*) D.o.f. associated with bending action.

[1]This ordering of d.o.f. is used only for convenience of notation. If this ordering is used, coefficients in $[\mathbf{k}]$ must be appropriately arranged.

more elements. Membrane and bending actions are not coupled within a single element, as evidenced by the 9 by 9 null matrices in Eq. 12.3-2. Nevertheless, the element works well enough to be competitive with curved elements [12.8]. (In Ref. 12.8, $[\mathbf{k}_m]$ is identical to the "basic triangular subelement" discussed in Section 8.6.)

If $[\mathbf{k}_m]$ pertains to the constant-strain triangle (Eq. 5.4-7), which does not use θ_z d.o.f., Eq. 12.3-2 is replaced by

$$[\mathbf{k}']\{\mathbf{d}'\} = \left[\begin{array}{cc|c} \underset{6\times6}{[\mathbf{k}_m]} & \underset{6\times3}{[\mathbf{0}]} & \\ \underset{3\times6}{[\mathbf{0}]} & \underset{3\times3}{[\mathbf{0}]} & \underset{9\times9}{[\mathbf{0}]} \\ \hline \multicolumn{2}{c|}{\underset{9\times9}{[\mathbf{0}]}} & \underset{9\times9}{[\mathbf{k}_b]} \end{array}\right] \begin{Bmatrix} \mathbf{u}_i \\ \mathbf{v}_i \\ \boldsymbol{\theta}_{zi} \\ \mathbf{w}_i \\ \boldsymbol{\theta}_{xi} \\ \boldsymbol{\theta}_{yi} \end{Bmatrix} \tag{12.3-3}$$

Again there is no coupling on the element level between $[\mathbf{k}_m]$ and $[\mathbf{k}_b]$. In addition, no stiffness is associated with θ_z d.o.f. This means that the structure stiffness matrix will be singular if all elements connected to any node happen to be coplanar. This potential difficulty can be avoided by modifying element matrices $[\mathbf{k}']$ as follows. Replace the on-diagonal null matrix in Eq. 12.3-3 by the 3 by 3 matrix in the following equation, which causes element-normal nodal rotations θ_z to produce corresponding moments M_z [12.9],

$$\begin{Bmatrix} M_{z1} \\ M_{z2} \\ M_{z3} \end{Bmatrix} = \alpha EV \begin{bmatrix} 1.0 & -0.5 & -0.5 \\ -0.5 & 1.0 & -0.5 \\ -0.5 & -0.5 & 1.0 \end{bmatrix} \begin{Bmatrix} \theta_{z1} \\ \theta_{z2} \\ \theta_{z3} \end{Bmatrix} \tag{12.3-4}$$

where E is elastic modulus, V is element volume, and α is a number such as 0.3 or less [12.9]. The added matrix provides each θ_z d.o.f. with a fictitious stiffness but offers no resistance to the mode $\theta_{z1} = \theta_{z2} = \theta_{z3}$ or to any other rigid-body motion.

Another way to avoid singularity is to eliminate rotation about a normal to the shell from the list of global d.o.f. at each node. Thus the element has 15 d.o.f. rather than 18.

Numerical tests show that the element of Eq. 12.3-2 is more accurate than the element of Eq. 12.3-3, and that reducing the latter element to 15 d.o.f. further degrades its performance [12.8].

Difficulties with spurious bending moments, noted in connection with straight arch elements, can also occur when flat elements are used to model a shell.

12.4 SHELLS OF REVOLUTION

A shell of revolution resembles a solid of revolution in that elements are symmetric with respect to an axis and node points are cross sections of nodal circles. An element meridional cross section resembles an arch element, whose behavior and pitfalls are discussed in Section 12.2.

Geometry and notation are shown in Fig. 12.4-1. Circumferential displacement v may be nonzero because initially we make no restriction that loads and dis-

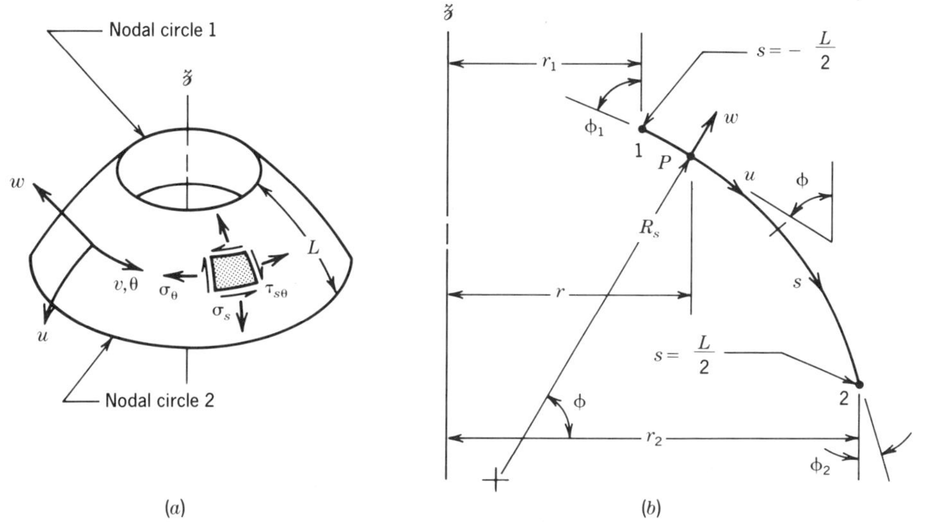

Figure 12.4-1. (*a*) Shell of revolution finite element. (*b*) Meridian of the element. Displacements u, v, and w are mutually orthogonal. R_s is the radius of curvature of the meridian.

placements must be axially symmetric. The following geometric relations can be written:

$$R_\theta = \frac{r}{\cos \phi} \qquad R_s = -\frac{ds}{d\phi} \qquad \sin \phi = \frac{dr}{ds} \qquad \cos \phi = -\frac{dz}{ds} \qquad (12.4\text{-}1)$$

R_θ and R_s are principal radii of curvature. R_s is considered negative if its center lies "outside" the shell (Fig. 12.1-1*d*). The center of an arc $R_\theta \, d\theta$ lies on the axis of revolution, but the center of an arc $R_s \, d\phi$ may not. Both centers lie on the same normal to the shell. Both radii can vary with s. Distance r and angle ϕ can be expressed in terms of s by integrating the second and third of Eqs. 12.4-1. Thus, if R_s is assumed constant over the element [12.10],

$$\phi = \phi_1 - \frac{1}{R_s}\left(\frac{L}{2} + s\right) \qquad \text{and} \qquad r = r_1 + R_s(\cos \phi - \cos \phi_1) \qquad (12.4\text{-}2)$$

The equation for r fails if the meridian is straight, but then r can be linearly interpolated between r_1 and r_2 in terms of s. Stresses shown in Fig. 12.4-1*a* may vary with s and with θ. They may also vary with distance z from the shell mid-surface because of bending action.

Loads without axial symmetry can be treated by superposition; that is, an analysis is performed for each Fourier harmonic of loading and the results of all harmonics are superposed. In essence, the method is that used for solids of revolution, as described in Section 10.5.

Assembly of elements must allow for the possibility that meridians of adjacent elements may meet at a cusp—for example, where a cylindrical shell is joined to a conical cap. Accordingly, it is appropriate to transform from element d.o.f. (Fig. 12.4-2*a*) to a convenient set of global d.o.f. (Fig. 12.4-2*b*). In general, one should infer that all these d.o.f. represent amplitudes of nodal displacements and rota-

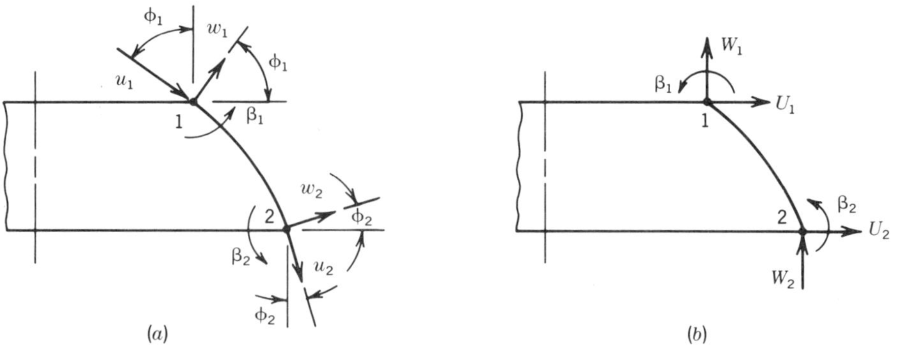

Figure 12.4-2. D.o.f. at nodes, in (*a*) local, and (*b*) global directions. D.o.f. v_1 and v_2, not shown, are perpendicular to the paper.

tions, as in a problem without axial symmetry analyzed by Fourier series. If axial symmetry prevails, circumferential nodal displacements v_i are zero, and the remaining nodal displacements and rotations are θ-independent motions. The transformation of d.o.f. at a typical node i is

$$\begin{Bmatrix} u_i \\ w_i \\ v_i \\ \beta_i \end{Bmatrix} = \begin{bmatrix} \sin\phi_i & -\cos\phi_i & 0 & 0 \\ \cos\phi_i & \sin\phi_i & 0 & 0 \\ 0 & 0 & 1 & 0 \\ 0 & 0 & 0 & 1 \end{bmatrix} \begin{Bmatrix} U_i \\ W_i \\ v_i \\ \beta_i \end{Bmatrix} \tag{12.4-3}$$

D.o.f. v_i and β_i require no transformation. By writing Eq. 12.4-3 for node 1 and then for node 2, one constructs an 8 by 8 transformation matrix $[\mathbf{T}]$, which yields an element matrix $[\mathbf{k}] = [\mathbf{T}]^T[\mathbf{k}'][\mathbf{T}]$, ready for assembly into the structure. If the problem is axially symmetric, d.o.f. v_i do not appear, and $[\mathbf{T}]$ is 6 by 6.

Axial Symmetry. If material properties, support conditions, and loads are all independent of θ, then displacements and stresses are also independent of θ. Thus $v = \tau_{s\theta} = 0$ in Fig. 12.4-1. (Here we exclude pure torsion, for which $v \neq 0$ and $\tau_{s\theta} \neq 0$, and vibration and buckling, whose displacement modes may lack axial symmetry despite axial symmetry of material properties, supports, and loads.)

For axial symmetry, the strain–displacement relations are

$$\begin{aligned} \epsilon_{ms} &= \frac{du}{ds} + \frac{w}{R_s} & \epsilon_{m\theta} &= \frac{u\sin\phi + w\cos\phi}{r} \\ \kappa_s &= \frac{d}{ds}\left(\frac{u}{R_s}\right) - \frac{d^2w}{ds^2} & \kappa_\theta &= \frac{\sin\phi}{r}\left(\frac{u}{R_s} - \frac{dw}{ds}\right) \end{aligned} \tag{12.4-4}$$

where ϵ_{ms} and $\epsilon_{m\theta}$ are membrane strains of the shell midsurface, and κ_s and κ_θ are curvature changes of the midsurface. Subscripts s and θ refer to meridional and circumferential directions, respectively. Transverse shear strain γ_{zs} is assumed to be negligible because the shell is thin. The formulation of a thin shell element proceeds in a way very similar to the formulation of a thin arch element. First, one writes displacement fields for u and w that depend on nodal values of u, w, and rotation $\beta = w_{,s}$. These fields are substituted into Eqs. 12.4-4, and the results into the strain energy expression

$$U = \frac{1}{2}\int_{-L/2}^{L/2} \{\boldsymbol{\epsilon}\}^T \begin{bmatrix} \mathbf{E}_M & \mathbf{0} \\ \mathbf{0} & \mathbf{D}_K \end{bmatrix} \{\boldsymbol{\epsilon}\}\, 2\pi r\, ds \tag{12.4-5}$$

where, with $C = Et/(1 - \nu^2)$ and $D = Et^3/12(1 - \nu^2)$ for an isotropic material,

$$[\mathbf{E}_M] = C\begin{bmatrix} 1 & \nu \\ \nu & 1 \end{bmatrix} \qquad [\mathbf{D}_K] = D\begin{bmatrix} 1 & \nu \\ \nu & 1 \end{bmatrix} \qquad \{\boldsymbol{\epsilon}\} = \begin{Bmatrix} \epsilon_{ms} \\ \epsilon_{m\theta} \\ \kappa_s \\ \kappa_\theta \end{Bmatrix} \tag{12.4-6}$$

Equation 12.4-5 reduces to the form $U = \{\mathbf{d}\}^T[\mathbf{k}]\{\mathbf{d}\}/2$, from which one identifies the element stiffness matrix $[\mathbf{k}]$. After numerical values of nodal d.o.f. have been calculated, strains and curvature changes follow from Eqs. 12.4-4. Membrane forces and bending moments in the shell are

$$\begin{Bmatrix} N_s \\ N_\theta \end{Bmatrix} = \frac{Et}{1 - \nu^2}\begin{bmatrix} 1 & \nu \\ \nu & 1 \end{bmatrix}\begin{Bmatrix} \epsilon_{ms} \\ \epsilon_{m\theta} \end{Bmatrix} \quad \text{and} \quad \begin{Bmatrix} M_s \\ M_\theta \end{Bmatrix} = \frac{Et^3}{12(1 - \nu^2)}\begin{bmatrix} 1 & \nu \\ \nu & 1 \end{bmatrix}\begin{Bmatrix} \kappa_s \\ \kappa_\theta \end{Bmatrix} \tag{12.4-7}$$

Stresses a distance z from the midsurface are obtained by use of Eq. 12.1-2, with $x = s$ for meridional stress and $x = \theta$ for circumferential stress (see Fig. 12.4-3*b*).

Equations 12.4-4 simplify if R_s is infinite; for example, if the element is conical. Use of conical elements to model a doubly curved shell is analogous to use of straight elements to model an arch. If R_s is infinite and $\phi = 0$, the element becomes cylindrical.

A Mindlin Axisymmetric Shell Element. This element is similar to the Mindlin beam element discussed in Section 9.4. We consider a conical element, Fig. 12.4-3. Thus ϕ is independent of s within a single element and R_s is infinite. As before, u and w are midsurface displacements, respectively parallel and normal to a meridian. Let β represent the rotation of a line that was normal to the midsurface of the undeformed shell. Transverse shear strain is $\gamma_{zs} = (dw/ds) -$

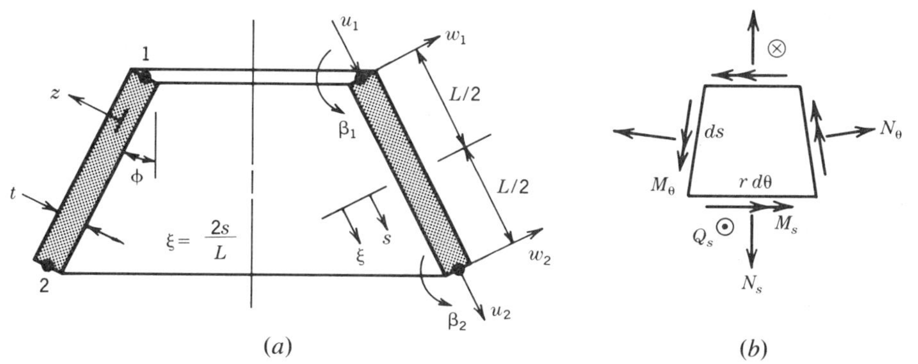

Figure 12.4-3. (*a*) Cross section of a conical shell element. Distance z is measured from the midsurface. (*b*) A differential element, bounded by meridians and parallels, and viewed normal to the shell. Membrane forces N_s and N_θ, transverse shear force Q_s, and bending moments M_s and M_θ are shown.

β. If γ_{zs} is small, then $(dw/ds) \approx \beta$. Thus, with the addition of γ_{zs}, and with R_s infinite, Eqs. 12.4-4 become

$$\{\epsilon\} = \begin{Bmatrix} \epsilon_{ms} \\ \epsilon_{m\theta} \\ \kappa_s \\ \kappa_\theta \\ \gamma_{zs} \end{Bmatrix} = \begin{bmatrix} d/ds & 0 & 0 \\ (\sin\phi)/r & (\cos\phi)/r & 0 \\ 0 & 0 & -d/ds \\ 0 & 0 & -(\sin\phi)/r \\ 0 & d/ds & -1 \end{bmatrix} \begin{Bmatrix} u \\ w \\ \beta \end{Bmatrix} \tag{12.4-8}$$

where, as in Eq. 12.2-15b, β has taken the place of dw/ds.

Equations 12.4-5 and 12.4-6 are again applicable if $\{\epsilon\}$ is taken from Eq. 12.4-8 and the material property matrix in Eq. 12.4-5 is augmented by a shear stiffness term $5Gt/6$—that is, if the material property matrix is

$$\underset{5\times 5}{\begin{bmatrix} \mathbf{E}_M & \mathbf{0} & \mathbf{0} \\ \mathbf{0} & \mathbf{D}_K & \mathbf{0} \\ \mathbf{0} & \mathbf{0} & 5Gt/6 \end{bmatrix}} \tag{12.4-9}$$

in which the factor 5/6 accounts for replacement of the true parabolic variation of γ_{zs} through the thickness by a uniform transverse shear strain.

In what follows we describe a specific element that uses linear interpolations for u and β and a quadratic interpolation for w. Membrane locking is avoided because the meridian is straight. Shear locking is avoided by making γ_{zs} constant over the element, thus invoking only one penalty constraint as the element becomes thin. In the transverse shear constraint, the coefficient $5Gt/6$ in Eq. 12.4-9 plays the role of a penalty number and may require adjustment according to guidelines offered at the end of Section 9.4. The element was suggested by Tessler [12.11].

The displacement field is taken as

$$\begin{Bmatrix} u \\ w \\ \beta \end{Bmatrix} = \begin{bmatrix} N_1 & N_2 & 0 & 0 & 0 & 0 & 0 \\ 0 & 0 & N_1 & N_2 & N_3 & 0 & 0 \\ 0 & 0 & 0 & 0 & 0 & N_1 & N_2 \end{bmatrix} \{\mathbf{d}\} \tag{12.4-10}$$

where $\{\mathbf{d}\} = \lfloor u_1 \quad u_2 \quad w_1 \quad w_2 \quad w_{3r} \quad \beta_1 \quad \beta_2 \rfloor^T$, in which w_{3r} is the transverse displacement at the element center $s = 0$, measured *relative* to the average value $(w_1 + w_2)/2$. With $\xi = 2s/L$, the shape functions are

$$N_1 = \tfrac{1}{2}(1 - \xi) \qquad N_2 = \tfrac{1}{2}(1 + \xi) \qquad N_3 = 1 - \xi^2 \tag{12.4-11}$$

Shear locking could be avoided by use of one Gauss point at $s = 0$ to integrate stiffness terms associated with γ_{zs}. Then internal d.o.f. w_{3r} would be eliminated by static condensation after a 7 by 7 stiffness matrix $[\mathbf{k}]$ is formulated. However, the same result can be produced more efficiently by eliminating w_{3r} at the outset via a requirement that γ_{zs} be constant over the element. It is possible to satisfy this requirement because the w field is one degree higher than the β field. A constant γ_{zs} implies that

$$\frac{d\gamma_{zs}}{ds} = \frac{d^2w}{ds^2} - \frac{d\beta}{ds} = 0 \tag{12.4-12}$$

Equations 12.4-8 through 12.4-12 yield $w_{3r} = (\beta_1 - \beta_2)L/8$. With this substitution, Eq. 12.4-10 becomes

$$\begin{Bmatrix} u \\ w \\ \beta \end{Bmatrix} = \begin{bmatrix} N_1 & N_2 & 0 & 0 & 0 & 0 \\ 0 & 0 & N_1 & N_2 & \overline{N}_3 & -\overline{N}_3 \\ 0 & 0 & 0 & 0 & N_1 & N_2 \end{bmatrix} \{\mathbf{d}\} \tag{12.4-13}$$

where $\{\mathbf{d}\} = \lfloor u_1 \quad u_2 \quad w_1 \quad w_2 \quad \beta_1 \quad \beta_2 \rfloor^T$, and, with $\xi = 2s/L$,

$$N_1 = \tfrac{1}{2}(1 - \xi) \qquad N_2 = \tfrac{1}{2}(1 + \xi) \qquad \overline{N}_3 = \frac{L}{8}(1 - \xi^2) \tag{12.4-14}$$

We see that w is quadratic in s and that the w associated with $\overline{N}_3$ and $\beta_1 = -\beta_2$ is identical to the lateral displacement of a beam whose end rotations are of equal magnitude but opposite sign. One can easily show that Eq. 12.4-13 yields a γ_{zs} that is the same as γ_{zs} at $s = 0$ from Eq. 12.4-10, which is the result desired.

Equations 12.4-8 and 12.4-13 are used to formulate the element stiffness matrix. If the element is cylindrical and its stiffness matrix is to be integrated exactly, membrane contributions require three Gauss points (because ϵ_θ is quadratic in s), curvature contributions require two Gauss points, and the transverse-shear contribution requires one Gauss point. Numerical tests of the element show very good accuracy (e.g., Fig. 12.4-4).

Another Option for Cylindrical Shells. A cylindrical shell that is not circular or not symmetrically loaded can be analyzed by applying Fourier series to a substitute toroidal shell. If the given shell is slightly bent to form a toroidal shell of large radius (like an inner tube for a bicycle tire), the axial coordinate of the cylindrical shell becomes the circumferential coordinate of the toroidal shell. Series analysis of the toroidal shell creates several repetitions of geometry and loading around the circumference. A typical repetition provides a satisfactory model of the original cylindrical shell if enough series terms are used.

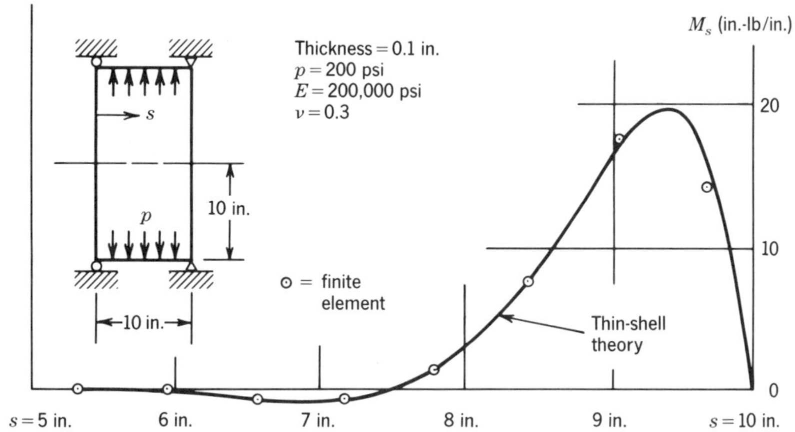

Figure 12.4-4. Meridional bending moment M_s in a cylindrical shell with open ends under uniform internal radial pressure [12.11]. The shell from $s = 5$ in. to $s = 10$ in. is spanned by eight identical elements. Finite element results are shown at element midpoints.

12.5 ISOPARAMETRIC GENERAL SHELL ELEMENTS

Summary. A shell of general shape can be modeled by three-dimensional solid elements that (typically) have a thickness dimension considerably smaller than their other dimensions, as in Fig. 12.5-1*a*. But even for a very thick shell, three nodes along thickness-direction lines supply more d.o.f. than needed. Elimination of the middle nodes yields the element of Fig. 12.5-1*b*, in which thickness-direction strain ϵ_3 is modeled as constant through the thickness. Here the subscript 3 indicates a direction normal to the shell midsurface. As the element becomes even thinner, stiffness coefficients associated with ϵ_3 become far larger than other stiffness coefficients. This circumstance invites numerical difficulties. The difficulty can be avoided by constraining adjacent thickness-direction nodes to have the same thickness-direction displacement. Thus five d.o.f. are associated with each pair of thickness-direction nodes in Fig. 12.5-1*b*. These five d.o.f. can be attached to a single node whose d.o.f. are three translations and two rotations. These five d.o.f. define the motion of a thickness-direction line that remains straight but not necessarily normal to the shell midsurface after deformation. Thus the final element has midsurface nodes only. In formulating the element stiffness matrix, one uses a material property matrix **[E]** that corresponds to the plane stress condition $\sigma_3 = 0$.

The element need not have eight nodes, as in Fig. 12.5-1*c*; other popular forms have four or nine nodes. All are Mindlin-type elements, and are therefore able to account for transverse shear deformation. They can be regarded as more general forms of the plate elements discussed in Section 11.3. As is the case with other plate and shell elements, Mindlin shell elements may encounter problems attributable to mechanisms and locking. These problems may be provoked (or avoided) by the choice of element geometry, type and number of d.o.f., shape functions, and numerical integration scheme. Pertinent references include [12.6,12.7, 12.9,12.12–12.14].

Some details are presented in what follows. No particular element or number of nodes is assumed.

Geometry. At a typical node i, Fig. 12.5-2*a*, one can write a thickness-direction vector $\mathbf{V}_{3i}$,

$$\mathbf{V}_{3i} = t_i \begin{Bmatrix} \ell_{3i} \\ m_{3i} \\ n_{3i} \end{Bmatrix}, \qquad \text{where} \qquad \begin{Bmatrix} \ell_{3i} \\ m_{3i} \\ n_{3i} \end{Bmatrix} = \frac{1}{t_i} \begin{Bmatrix} x_j - x_k \\ y_j - y_k \\ z_j - z_k \end{Bmatrix} \tag{12.5-1}$$

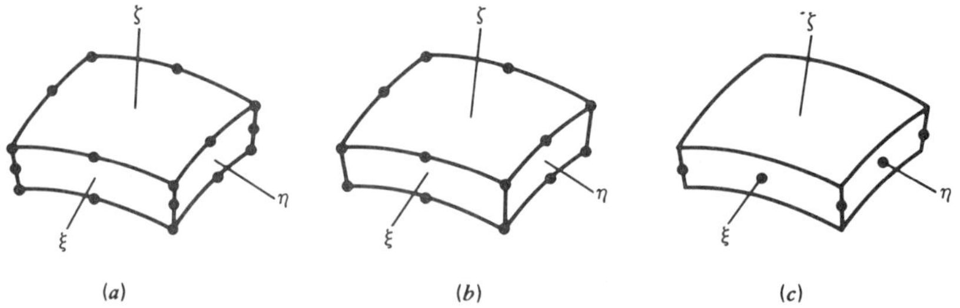

Figure 12.5-1. (*a*) A 20-node, 60-d.o.f. solid element. (*b*) Elimination of four mid-edge nodes yields a 16-node, 48-d.o.f. element. (*c*) Further constraint yields an 8-node, 40-d.o.f. shell element.

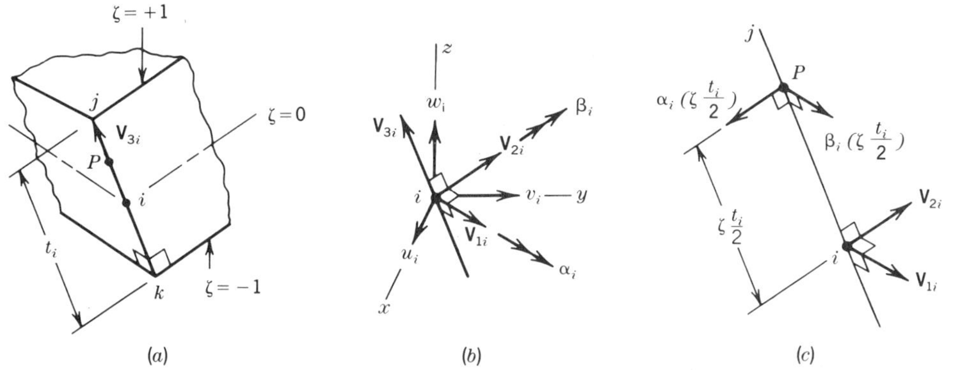

Figure 12.5-2. (*a*) Typical node i, with thickness-direction vector $\mathbf{V}_{3i}$. (*b*) Orthogonal vectors at node i. Directions of rotational d.o.f. α_i and β_i are given by the right-hand rule. Translational d.o.f. u_i, v_i, and w_i are in the Cartesian directions x, y, and z. (*c*) Displacements of an arbitrary point P on $\mathbf{V}_{3i}$ owing to small nodal rotations.

in which ℓ_{3i}, m_{3i}, and n_{3i} are direction cosines of the line kij. The Cartesian coordinates of an arbitrary point in the element are

$$\begin{Bmatrix} x \\ y \\ z \end{Bmatrix} = \sum N_i \begin{Bmatrix} x_i \\ y_i \\ z_i \end{Bmatrix} + \sum N_i \zeta \frac{t_i}{2} \begin{Bmatrix} \ell_{3i} \\ m_{3i} \\ n_{3i} \end{Bmatrix} \tag{12.5-2}$$

where $x_i = (x_j + x_k)/2$, and so on, and shape functions N_i are functions of ξ and η but are independent of ζ. For example, the N_i are given by Eqs. 6.3-2 for a four-node element, by Eqs. 6.6-1 for an eight-node element, and by Table 6.6-1 for a nine-node element. To define element geometry, one can either supply the Cartesian coordinates of all nodes j and k, or supply x_i, y_i, z_i, t_i, and the direction cosines of $\mathbf{V}_{3i}$ for all nodes i.

Vectors $\mathbf{V}_{1i}$ and $\mathbf{V}_{2i}$ in Fig. 12.5-2 are perpendicular to each other and to $\mathbf{V}_{3i}$. Thus $\mathbf{V}_{1i}$ and $\mathbf{V}_{2i}$ are tangent to the midsurface, but they are not required to have a particular relation to Cartesian coordinate directions. $\mathbf{V}_{1i}$ and $\mathbf{V}_{2i}$ are used to define the directions of nodal rotation d.o.f. α_i and β_i, which are shared by all elements that share node i. (It is possible that the set of α_i and β_i directions will differ from node to node.) One can define $\mathbf{V}_{1i}$ as a principal material direction if the material is orthotropic. Or, one can define a midsurface-tangent vector $\mathbf{e}_{1i}$ whose components are $\Delta x = \Sigma N_{i,\xi} x_i \Delta\xi$, $\Delta y = \Sigma N_{i,\xi} y_i \Delta\xi$, and $\Delta z = \Sigma N_{i,\xi} z_i \Delta\xi$, each evaluated at the node i in question, and with $\Delta\xi$ a small number. A similar vector $\mathbf{e}_{2i}$ can be defined using an increment $\Delta\eta$. Then $\mathbf{e}_{3i} = \mathbf{e}_{1i} \times \mathbf{e}_{2i}$ and $\mathbf{V}_{3i} = t_i \mathbf{e}_{3i}/e_{3i}$. Finally, $\mathbf{V}_{1i} = \mathbf{e}_{1i}$ and $\mathbf{V}_{2i} = \mathbf{V}_{3i} \times \mathbf{V}_{1i}$. Direction cosines of $\mathbf{V}_{1i}$ and $\mathbf{V}_{2i}$ are given by dividing each vector by its magnitude. To avoid input data errors, if the three vectors are supplied as data rather than being calculated within the program, one should ensure that $\mathbf{V}_{1i}$, $\mathbf{V}_{2i}$, and $\mathbf{V}_{3i}$ are precisely orthogonal.

For later use, we define the following matrix of direction cosines.

$$[\boldsymbol{\mu}_i] = \begin{bmatrix} -\dfrac{\mathbf{V}_{2i}}{V_{2i}} & \dfrac{\mathbf{V}_{1i}}{V_{1i}} \end{bmatrix} = \begin{bmatrix} -\ell_{2i} & \ell_{1i} \\ -m_{2i} & m_{1i} \\ -n_{2i} & n_{1i} \end{bmatrix} \tag{12.5-3}$$

where V_{1i} and V_{2i} are the magnitudes of $\mathbf{V}_{1i}$ and $\mathbf{V}_{2i}$.

The 3 by 3 Jacobian matrix [**J**], defined by Eq. 6.7-2, contains terms such as

$$
\begin{aligned}
x_{,\xi} &= \sum N_{i,\xi}\,(x_i + \zeta t_i \ell_{3i}/2) \\
x_{,\eta} &= \sum N_{i,\eta}\,(x_i + \zeta t_i \ell_{3i}/2) \\
x_{,\zeta} &= \sum N_i\,(t_i \ell_{3i}/2)
\end{aligned}
\tag{12.5-4}
$$

Displacements and Strains. The displacement of a point P on vector $\mathbf{V}_{3i}$, Fig. 12.5-2, consists of the displacement of node i plus the displacement relative to node i created by rotation of $\mathbf{V}_{3i}$. The relative displacement components, shown in Fig. 12.5-2*c*, must be resolved into x, y, and z components before being added to the displacements of node i. Thus, for example, point P has the x-direction displacement

$$
u_P = u_i - \alpha_i \left(\zeta \frac{t_i}{2}\right) \ell_{2i} + \beta_i \left(\zeta \frac{t_i}{2}\right) \ell_{1i}
\tag{12.5-5}
$$

in which nodal rotations α_i and β_i are presumed small. Displacements of an arbitrary point in the element are

$$
\begin{Bmatrix} u \\ v \\ w \end{Bmatrix} = \sum N_i \begin{Bmatrix} u_i \\ v_i \\ w_i \end{Bmatrix} + \sum N_i \zeta \frac{t_i}{2} [\boldsymbol{\mu}_i] \begin{Bmatrix} \alpha_i \\ \beta_i \end{Bmatrix}
\tag{12.5-6}
$$

Following standard isoparametric procedure, we express strains in terms of displacement derivatives,

$$
\lfloor \epsilon_x \;\; \epsilon_y \;\; \epsilon_z \;\; \gamma_{xy} \;\; \gamma_{yz} \;\; \gamma_{zx} \rfloor^T = [\mathbf{H}] \lfloor u_{,x} \;\; u_{,y} \;\; u_{,z} \;\; v_{,x} \; . \; . \; . \; w_{,z} \rfloor^T
\tag{12.5-7}
$$

$$
\begin{Bmatrix} u_{,x} \\ u_{,y} \\ u_{,z} \\ v_{,x} \\ \vdots \\ w_{,z} \end{Bmatrix} = \begin{bmatrix} \mathbf{J}^{-1} & \mathbf{0} & \mathbf{0} \\ \mathbf{0} & \mathbf{J}^{-1} & \mathbf{0} \\ \mathbf{0} & \mathbf{0} & \mathbf{J}^{-1} \end{bmatrix} \begin{Bmatrix} u_{,\xi} \\ u_{,\eta} \\ u_{,\zeta} \\ v_{,\xi} \\ \vdots \\ w_{,\zeta} \end{Bmatrix}
\tag{12.5-8}
$$

where [**H**] is stated in Eq. 6.7-5 and $[\mathbf{J}]^{-1}$ is the inverse of the 3 by 3 Jacobian matrix [**J**]. All six strains are included in Eq. 12.5-7 because the shell midsurface has no particular orientation with respect to Cartesian coordinates xyz. The condition $\sigma_3 = 0$ will be introduced subsequently via the stress–strain relation. From Eq. 12.5-6 we obtain

$$
\begin{Bmatrix} u_{,\xi} \\ u_{,\eta} \\ u_{,\zeta} \\ v_{,\xi} \\ \vdots \\ w_{,\zeta} \end{Bmatrix} = \sum \begin{bmatrix}
N_{i,\xi} & 0 & 0 & -\zeta t_i N_{i,\xi} \ell_{2i}/2 & \zeta t_i N_{i,\xi} \ell_{1i}/2 \\
N_{i,\eta} & 0 & 0 & -\zeta t_i N_{i,\eta} \ell_{2i}/2 & \zeta t_i N_{i,\eta} \ell_{1i}/2 \\
0 & 0 & 0 & -t_i N_i \ell_{2i}/2 & t_i N_i \ell_{1i}/2 \\
0 & N_{i,\xi} & 0 & -\zeta t_i N_{i,\xi} m_{2i}/2 & \zeta t_i N_{i,\xi} m_{1i}/2 \\
\vdots & \vdots & \vdots & \vdots & \vdots \\
0 & 0 & 0 & -t_i N_i n_{2i}/2 & t_i N_i n_{1i}/2
\end{bmatrix} \begin{Bmatrix} u_i \\ v_i \\ w_i \\ \alpha_i \\ \beta_i \end{Bmatrix}
\tag{12.5-9}
$$

Combination of Eqs. 12.5-7, 12.5-8, and 12.5-9 yields

$$\lfloor \epsilon_x \quad \epsilon_y \quad \epsilon_z \quad \gamma_{xy} \quad \gamma_{yz} \quad \gamma_{zx} \rfloor^T = \sum [\mathbf{B}_i] \lfloor u_i \quad v_i \quad w_i \quad \alpha_i \quad \beta_i \rfloor^T \tag{12.5-10}$$

The complete strain–displacement matrix [**B**] is built of as many 6 by 5 blocks [$\mathbf{B}_i$] as there are nodes in the element.

Stiffness Matrix [k]. The stress–strain relation can be stated as

$$\{\boldsymbol{\sigma}\} = [\mathbf{E}]\{\boldsymbol{\epsilon}\} \qquad \text{or as} \qquad \{\boldsymbol{\sigma}'\} = [\mathbf{E}']\{\boldsymbol{\epsilon}'\} \tag{12.5-11}$$

where $\{\boldsymbol{\sigma}\}$ contains stresses in Cartesian directions xyz and $\{\boldsymbol{\sigma}'\}$ contains stresses in local directions normal and tangent to the shell midsurface. The latter relation is[2]

$$\begin{Bmatrix} \sigma_1 \\ \sigma_2 \\ \sigma_3 \\ \tau_{12} \\ \tau_{23} \\ \tau_{31} \end{Bmatrix} = \underbrace{\begin{bmatrix} E_{11} & E_{12} & 0 & 0 & 0 & 0 \\ E_{12} & E_{22} & 0 & 0 & 0 & 0 \\ 0 & 0 & 0 & 0 & 0 & 0 \\ 0 & 0 & 0 & G_{12} & 0 & 0 \\ 0 & 0 & 0 & 0 & 5G_{23}/6 & 0 \\ 0 & 0 & 0 & 0 & 0 & 5G_{31}/6 \end{bmatrix}}_{[\mathbf{E}']} \begin{Bmatrix} \epsilon_1 \\ \epsilon_2 \\ \epsilon_3 \\ \gamma_{12} \\ \gamma_{23} \\ \gamma_{31} \end{Bmatrix} \tag{12.5-12}$$

where directions 1 and 2 are tangent to the midsurface and direction 3 is normal to it. These directions are presumed to be principal material directions if the material is orthotropic. The factors of 5/6 account for a parabolic variation of transverse shear strain through the thickness. Note that Eq. 12.5-12 is contrived to make the transverse normal stress σ_3 equal to zero. [**E**] is obtained from [**E**′] by the coordinate transformation $[\mathbf{E}] = [\mathbf{T}_\epsilon]^T[\mathbf{E}'][\mathbf{T}_\epsilon]$ (see Eq. 7.3-10). This transformation must be carried out at each Gauss point used in generating [**k**] by numerical integration. Direction cosines needed in [$\mathbf{T}_\epsilon$] are the direction cosines of vectors $\mathbf{V}_1$, $\mathbf{V}_2$, and $\mathbf{V}_3$ at the Gauss point. In turn, these vectors can be found by shape function interpolation from nodal values,

$$\mathbf{V}_1 = \sum N_i \mathbf{V}_{1i} \qquad \mathbf{V}_2 = \sum N_i \mathbf{V}_{2i} \qquad \mathbf{V}_3 = \sum N_i \mathbf{V}_{3i} \tag{12.5-13}$$

in which the N_i are evaluated at the Gauss point in question.

The element stiffness matrix is

$$\underset{5N\times 5N}{[\mathbf{k}]} = \int_{-1}^{1}\int_{-1}^{1}\int_{-1}^{1} \underset{5N\times 6}{[\mathbf{B}]^T} \ \underset{6\times 6}{[\mathbf{E}]} \ \underset{6\times 5N}{[\mathbf{B}]} \ \det[\mathbf{J}] \, d\xi \, d\eta \, d\zeta \tag{12.5-14}$$

where N is the number of nodes per element. If material properties are independent of ζ, and if small errors are acceptable [12.7], then thickness-direction integration can be done explicitly. In doing so one discards terms in [**J**] that depend on ζ, under the assumption that these terms are negligible if the element is not sharply

[2]For isotropy, $E_{11} = E_{22} = E_{12}/\nu = E/(1 - \nu^2)$ and $G_{12} = G_{23} = G_{31} = G = 0.5E/(1 + \nu)$.

curved. Next, [**B**] is split into a part [$\mathbf{B}_0$] that is independent of ζ and a part ζ[$\mathbf{B}_1$] that is linear in ζ, so that [**B**] = [$\mathbf{B}_0$] + ζ[$\mathbf{B}_1$]. Thus terms linear in ζ integrate to zero and Eq. 12.5-14 becomes

$$[\mathbf{k}] = \int_{-1}^{1}\int_{-1}^{1} (2[\mathbf{B}_0]^T[\mathbf{E}][\mathbf{B}_0] + \tfrac{2}{3}[\mathbf{B}_1]^T[\mathbf{E}][\mathbf{B}_1]) \det[\mathbf{J}] \, d\xi \, d\eta \qquad (12.5\text{-}15)$$

in which [**J**] remains 3 by 3 but is evaluated on the midsurface, $\zeta = 0$.

Difficulties arising from shear locking, membrane locking, and mechanisms can be dealt with by selective and reduced integration and other strategies [12.13]. As an element becomes thin, the penalty matrix associated with transverse shear must not be allowed to overwhelm the rest of the stiffness matrix (see the remarks that close Section 9.4).

Element nodal loads (Eq. 4.1-6) come from the usual sources. Those associated with initial strains are, since $\{\boldsymbol{\epsilon}_0'\} = [\mathbf{T}_\epsilon]\{\boldsymbol{\epsilon}_0\}$,

$$\int_{V_e} [\mathbf{B}]^T[\mathbf{E}]\{\boldsymbol{\epsilon}_0\} \, dV = \int_{-1}^{1}\int_{-1}^{1}\int_{-1}^{1} [\mathbf{B}]^T[\mathbf{T}_\epsilon]^T[\mathbf{E}']\{\boldsymbol{\epsilon}_0'\} \det[\mathbf{J}] \, d\xi \, d\eta \, d\zeta \qquad (12.5\text{-}16)$$

Finally, element stresses referred to local directions 1–2–3 are

$$\{\boldsymbol{\sigma}'\} = [\mathbf{E}']([\mathbf{T}_\epsilon][\mathbf{B}]\{\mathbf{d}\} - \{\boldsymbol{\epsilon}_0'\}) \qquad (12.5\text{-}17)$$

Stresses at Gauss points may be more accurate than stresses computed elsewhere in the element, as noted in Section 6.13.

Usually, elements share a common tangent plane at each interelement boundary. Thus d.o.f. α_i and β_i are midsurface-tangent vectors in all elements that share node i. This ideal circumstance would disappear if elements were to form a ridge line where they meet, as in a folded plate. Then $\mathbf{V}_{3i}$ could be defined as an average shell normal vector, with α_i and β_i normal to $\mathbf{V}_{3i}$, but accuracy loss would be expected.

PROBLEMS

Section 12.1

12.1 In terms of R_s and R_θ, how would you qualitatively describe the shape of an American football?

12.2 According to elementary stress formulas and Eqs. 12.1-1, what are N_x, N_y, and N_{xy} for a cylindrical tank under internal pressure?

12.3 The cylindrical tank shown contains a step change in thickness and is capped by a hemispherical shell. Loads consist of axial force Q and internal pressure p.

(a) Explain why load Q cannot be supported by membrane action only.

(b) Free the built-in support condition at CC and divide the vessel into three parts by making circumferential cuts along AA and BB. Then show by a sketch the displaced shape of each part produced by pressure p alone.

(c) Show by a sketch the loads (applied by one part to another) needed to restore continuity of displacements.

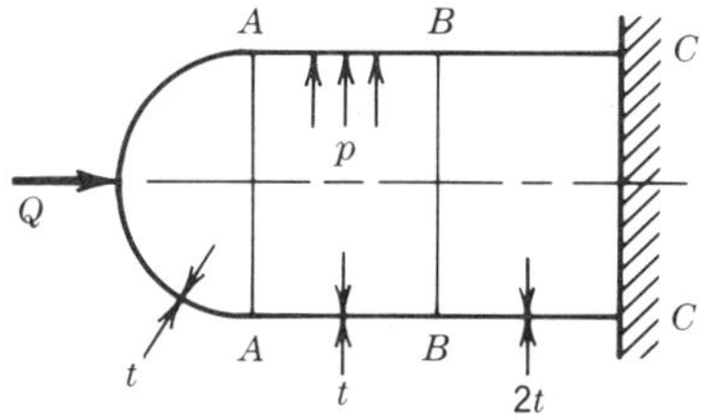

Problem 12.3

12.4 Imagine that a garden hose has an elliptical cross section. Explain why internal pressure causes bending stresses to appear, and sketch their approximate variation over the outside perimeter of the cross section. Also explain why, if internal pressure is to be carried by membrane stresses alone, the cross section must first become circular.

Section 12.2

12.5 Write an expression for strain ϵ_s if the arch radius R is a function of s.

12.6 (a) Derive Eqs. 12.2-4. *Suggestion:* Combine Eqs. 12.2-2 and integrate the resulting differential equation.
(b) Consider a quarter-circle arch that occupies the first quadrant of a Cartesian reference frame. By three separate sketches, show the displacement fields associated with b_1, b_2, and b_3 in Eqs. 12.2-4.

12.7 Write the coordinate transformation of nodal d.o.f. for a straight element; that is, define all terms completely in terms of L and R. Assume that local d.o.f. are ordered as shown in Eq. 12.2-6, and that global d.o.f. in Fig. 12.2-3*b* have the order $\lfloor D_{s1} \quad D_{r1} \quad \beta_1 \quad D_{s2} \quad D_{r2} \quad \beta_2 \rfloor^T$.

12.8 Model a complete circular ring by four straight elements of equal length. The model is therefore a square, as shown. Compute the relative separation of loads P, accounting for bending stiffness only. Take the element length as (a) chord length $L = \sqrt{2}\,R$, and (b) arc length $L = \pi R/2$.

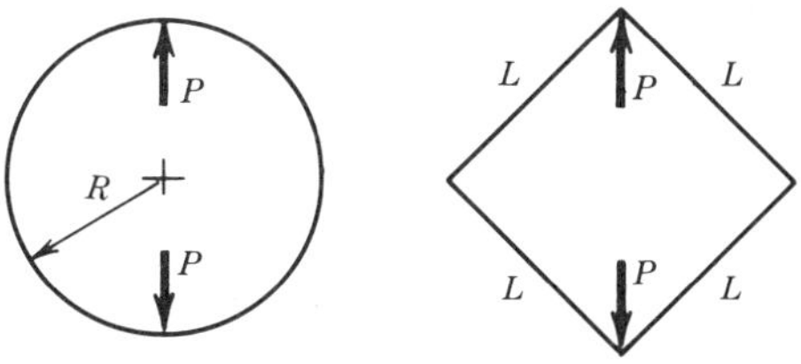

Problem 12.8

12.9 For a curved arch, integrate U_m of Eq. 12.2-3 using the displacement field of Eqs. 12.2-7. Hence, show that the condition $U_m = 0$ implies Eqs. 12.2-9.

12.10 Using the following displacement fields for a curved arch, establish the relation between nodal d.o.f. and the a_i; that is, find [**A**] in the relation

$$\lfloor u_1 \quad w_1 \quad \beta_1 \quad u_2 \quad w_2 \quad \beta_2 \rfloor^T = [\mathbf{A}] \lfloor a_1 \quad a_2 \quad a_3 \quad a_4 \quad a_5 \quad a_6 \rfloor^T$$

where $\beta = w_{,s}$.
(a) Use Eqs. 12.2-7.
(b) Use Eqs. 12.2-11.

12.11 Would the mode associated with d.o.f. a_7 in Eq. 12.2-12a be of any benefit to (a) the straight element of Fig. 12.2-3*a*, or (b) the curved element based on Eqs. 12.2-7? Consider both full and selective integration in your explanation of part (b).

12.12 In the stiffness matrix of the element associated with Eqs. 12.2-11, the diagonal coefficients associated with nodal deflections w_1 and w_2 are each

$$k = \frac{EI}{R^3}\left(\frac{24}{\lambda^3} - \frac{192}{35\lambda} + \frac{16\lambda}{35}\right)$$

where $\lambda = L/2R$ for an element of arc length L. Use this information to solve for the deflection of load P in Fig. 12.2-1*a* from a two-element model.

12.13 In formulating an element stiffness matrix [**k**] from Eqs. 12.2-11, energy U_m makes no contribution to [**k**]. Why cannot U_m simply be discarded in formulating other curved arch elements, for example, those associated with Eqs. 12.2-7 and 12.2-13?

12.14 (a) Verify Eqs. 12.2-17 and 12.2-18.
(b) Determine the analogous equations of constraint that pertain to use of reduced integration for U_m and U_s.

12.15 Investigate the $\gamma_{zs} = 0$ condition and the effect of reduced integration on the three-node Mindlin element (analogous to Eqs. 12.2-20 and 12.2-21).

Section 12.3

12.16 Consider a doubly curved shell, modeled by flat triangular elements, such as those of Eqs. 12.3-2 and 12.3-3. What can you say about interelement compatibility of edge displacements?

12.17 Write an equation analogous to Eq. 12.3-4 but appropriate to a flat element that has four nodes. Do you think this equation should be used if the element is warped rather than flat?

12.18 Imagine that each side of a rectangular box is modeled by a mesh of flat shell elements. Internal pressure is applied. Along the edges where sides intersect, what d.o.f. can probably be set to zero, and why?

Section 12.4

12.19 (a) Derive Eqs. 12.4-2.
(b) Rewrite Eqs. 12.4-2 in a form appropriate to a shell with a straight meridian.

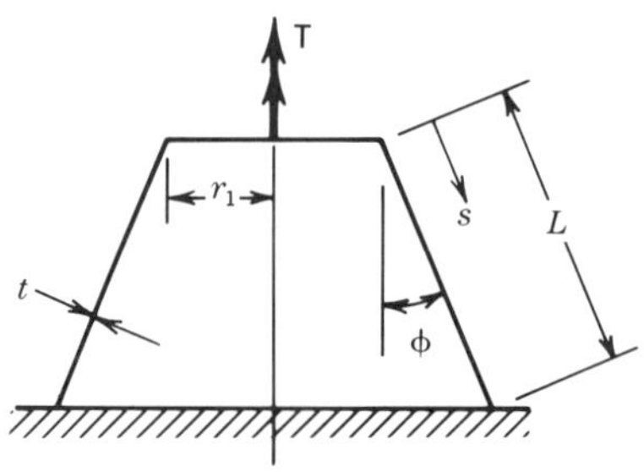

Problem 12.20

12.20 The conical shell shown is fixed at its base and loaded by torque T at the top. Use mechanics of materials concepts, not finite elements, to answer the following questions.
(a) What is shear stress $\tau_{s\theta}$, in terms of T, r_1, t, ϕ, and s?
(b) What is the angle of rotation of the top of the truncated cone relative to the bottom, in terms of T, r_1, t, ϕ, L, and shear modulus G?

12.21 Specialize Eqs. 12.4-4 to the following cases.
(a) A cylindrical shell, with s the only independent variable.
(b) A flat plate, with r the only independent variable.
(c) A sphere, with ϕ the only independent variable.

12.22 Consider a thin cylindrical shell of radius R whose midsurface axial strain ϵ_m is unrestrained. By considering the energy associated with membrane strains ϵ_{ms} and $\epsilon_{m\theta}$, show that displacement w is in effect resisted by an elastic foundation of modulus Et/R^2 (as well as being resisted by bending stiffness).

12.23 Consider a cylindrical shell, thin-walled and symmetrically loaded, but without axial loads. Thus $N_s = 0$, $\epsilon_{ms} = -\nu\epsilon_{m\theta}$, and displacement u need not be considered. Generate the 4 by 4 element stiffness matrix. Use a cubic w field.

12.24 Show that γ_{zs} at $s = 0$ from Eq. 12.4-10 is the same as γ_{zs} for all s from Eq. 12.4-13.

12.25 Show that application of Eq. 12.4-12 converts Eqs. 12.4-10 and 12.4-11 to Eqs. 12.4-13 and 12.4-14.

12.26 Verify the remark made in the sentence following Eq. 12.4-14.

12.27 Let a constant pressure p be applied to the inside surface of the element described by Eq. 12.4-13. Evaluate the consistent element nodal load vector.

12.28 The cylindrical shell shown is modeled by two Mindlin elements, prevented from rotation β at nodal circles 1 and 3, and loaded by a uniform radial pressure on the inside surface. There are no end caps. Consistently computed nodal loads are applied. Will computed results display nonzero meridional curvature κ_s in the following situations? Answer without doing calculations.
(a) $L_1 = L_2$, and w varies linearly between nodes.
(b) $L_1 \neq L_2$, and w varies linearly between nodes.
(c) $L_1 = L_2$, and the element of Eqs. 12.4-13 is used.
(d) $L_1 \neq L_2$, and the element of Eqs. 12.4-13 is used.

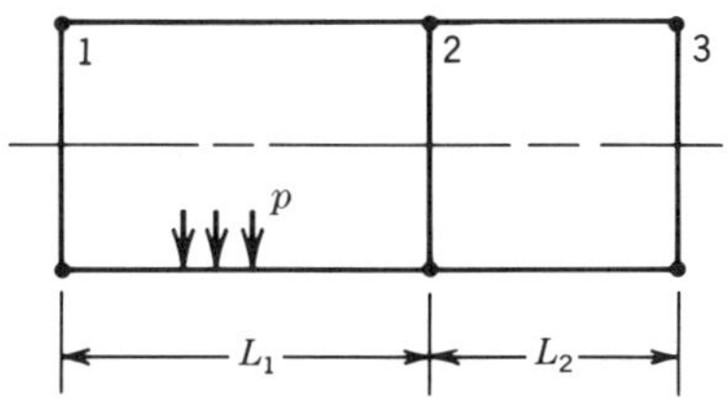

Problem 12.28

Section 12.5

12.29 Write an equation analogous to Eq. 12.5-2 but applicable to the element of Fig. 12.5-1*b*. (Each shape function should depend on ξ and η and should be multiplied by a linear function of ζ.)

12.30 (a) Define terms in the Jacobian matrix **[J]** in terms of ζ, N_i, $N_{i,\xi}$, $N_{i,\eta}$, nodal coordinates, and components of $\mathbf{V}_{3i}$.
(b) Specialize your result for the case of an element that is flat, of constant thickness, and whose midsurface coincides with the *xy* plane.

12.31 (a) Imagine that instead of rotational d.o.f. α_i and β_i in Fig. 12.5-2, we elect to use rotational d.o.f. β_{xi}, β_{yi}, and β_{zi}, which are small rotations about axes x, y, and z. Write the appropriate form of Eq. 12.5-6 and define the terms in the direction cosine matrix $[\boldsymbol{\mu}_i]$ you use.
(b) Check that your formula agrees with Eq. 12.5-6 for the special cases of all vectors $\mathbf{V}_3$ parallel to the x axis, then the y axis, and finally the z axis.
(c) The element now has six d.o.f. per node rather than five. What possible difficulty does this element present?

12.32 (a) How can the transformation of $[\mathbf{E}']$ to $[\mathbf{E}]$ be made more computationally efficient? (Exploit the null row and the null column in $[\mathbf{E}']$.)
(b) Write Eq. 12.5-14 in a form that uses $[\mathbf{E}']$ rather than $[\mathbf{E}]$.

12.33 Let a typical $[\mathbf{B}_i]$ in Eq. 12.5-10 have the form $[\mathbf{B}_i] = [\mathbf{H}][\overline{\mathbf{B}}_i]$, where $[\mathbf{H}]$ is defined by Eq. 6.7-5 and $[\overline{\mathbf{B}}_i]$ is a 9 by 5 matrix. Express $[\overline{\mathbf{B}}_i]$ as a function of ζ, t_i, N_i, $N_{i,\xi}$, $N_{i,\eta}$, direction cosines, and the Γ_{ij} in $[\boldsymbol{\Gamma}] = [\mathbf{J}]^{-1}$.

12.34 Two elements of the type shown by Fig. 12.5-1*c* are to be connected side by side. However, they do not share a common tangent plane; for example, they may be perpendicular, like adjacent sides of a box. How should the connection be accomplished?—that is, how should d.o.f. along the connection line be treated?

12.35 Consider a membrane shell (a shell that has no bending stiffness). Review the formulations presented in Section 12.5, and state how they may be simplified or specialized to deal with a membrane shell.

12.36 The sketch represents an end of an isoparametric *bar* element, whose geometry is defined by the position of nodes along its centerline and vectors $\mathbf{V}_{2i}$ and $\mathbf{V}_{3i}$ that span its rectangular cross section.
(a) Write an equation of geometry analogous to Eq. 12.5-2.
(b) Write an equation of displacement analogous to Eq. 12.5-6.

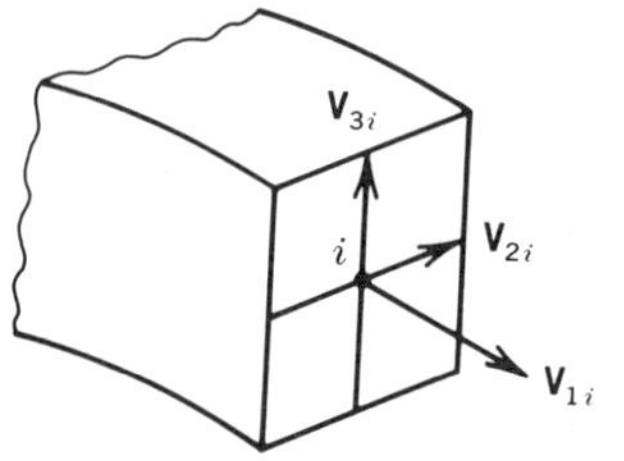

Problem 12.36

CHAPTER 13

FINITE ELEMENTS IN DYNAMICS AND VIBRATIONS

The use of the finite element method for the dynamic analysis of structures is described. Mass and damping matrices are derived. Modal and direct time integration methods of analysis are discussed.

13.1 INTRODUCTION

If the frequency of excitation applied to a structure is less than roughly one-third of the structure's lowest natural frequency of vibration, then the effects of inertia can be neglected and the problem is *quasistatic*. That is, the equations $[\mathbf{K}]\{\mathbf{D}\} = \{\mathbf{R}\}$ are sufficiently accurate even though loads $\{\mathbf{R}\}$, and hence displacements $\{\mathbf{D}\}$, vary (slowly) with time. Loads $\{\mathbf{R}\}$ may result from surface loads and/or body forces. Forces that result from constant or almost constant acceleration are treated in the same manner as gravity forces—that is, by the integral that contains $\{\mathbf{F}\}$ in Eq. 4.1-6.

Inertia becomes important if excitation frequencies are higher than noted above or if the structure vibrates freely. The *mass matrix,* written as $[\mathbf{m}]$ for an element and $[\mathbf{M}]$ for a structure, accounts for inertia and is a discrete representation of the continuous distribution of mass in a structure. The effects of damping, if important, are accounted for by *damping matrices* $[\mathbf{c}]$ and $[\mathbf{C}]$.

Problems of dynamics can be categorized as either *wave propagation problems* or *structural dynamics problems*. In wave propagation problems the loading is often an impact or an explosive blast. The excitation, and hence the structural response, are rich in high frequencies. In such problems we are usually interested in the effects of stress waves. Thus the time duration of analysis is usually short and is typically of the order of a wave traversal time across a structure. A problem that is not a wave propagation problem, but for which inertia is important, is called a structural dynamics problem. In this category, the frequency of excitation is usually of the same order as the structure's lowest natural frequencies of vibration.

Problems of structural dynamics can be subdivided into two broad classifications. In one, we ask for *natural frequencies of vibration* and the corresponding mode shapes. Usually, we wish to compare natural frequencies of the structure with frequencies of excitation. In design, it is usually desirable to assure that these frequencies are well separated. In the other classification, we ask how a structure moves with time under prescribed loads and/or motions of its supports; that is, we ask for a *time-history* analysis. Two popular methods of time-history analysis are *modal methods* and *direct integration methods*. ("Time history" is a commonly used term referring to the record of the variation of a quantity over some interval of time.)

Structural dynamics has an extensive literature and good textbooks [13.1, 13.2,13.45,13.46]. Methods of structural dynamics are largely independent of finite element analysis because these methods presume the availability of stiffness, mass, and damping matrices but do not demand that they arise from a finite element discretization. Indeed, many popular methods were developed before the advent of the finite element method by using matrices resulting from finite difference discretizations. Today, however, matrices are most often obtained from finite element discretizations, and the analysis tools are tailored to fit finite element models.

13.2 DYNAMIC EQUATIONS. MASS AND DAMPING MATRICES

Equations that govern the dynamic response of a structure or medium will be derived by requiring the work of external forces to be absorbed by the work of internal, inertial, and viscous forces for any small kinematically admissible motion (i.e., any small motion that satisfies both compatibility and essential boundary conditions). For a single element, this work balance becomes

$$\int_{V_e} \{\delta\mathbf{u}\}^T\{\mathbf{F}\}\, dV + \int_{S_e} \{\delta\mathbf{u}\}^T\{\mathbf{\Phi}\}\, dS + \sum_{i=1}^{n} \{\delta\mathbf{u}\}_i^T\{\mathbf{p}\}_i$$
$$= \int_{V_e} (\{\delta\boldsymbol{\epsilon}\}^T\{\boldsymbol{\sigma}\} + \{\delta\mathbf{u}\}^T\rho\{\ddot{\mathbf{u}}\} + \{\delta\mathbf{u}\}^T\kappa_d\{\dot{\mathbf{u}}\})\, dV \qquad (13.2\text{-}1)$$

where $\{\delta\mathbf{u}\}$ and $\{\delta\boldsymbol{\epsilon}\}$ are respectively small arbitrary displacements and their corresponding strains, $\{\mathbf{F}\}$ are body forces, $\{\mathbf{\Phi}\}$ are prescribed surface tractions (which typically are nonzero over only a portion of surface S_e), $\{\mathbf{p}\}_i$ are concentrated loads that act at a total of n points on the element, $\{\delta\mathbf{u}\}_i^T$ is the displacement of the point at which load $\{\mathbf{p}\}_i$ is applied, ρ is the mass density of the material, κ_d is a material-damping parameter analogous to viscosity, and volume integration is carried out over the element volume V_e.

Using usual notation, we have for the displacement field $\{\mathbf{u}\}$ (which is a function of both space and time) and its first two time derivatives

$$\{\mathbf{u}\} = [\mathbf{N}]\{\mathbf{d}\} \qquad \{\dot{\mathbf{u}}\} = [\mathbf{N}]\{\dot{\mathbf{d}}\} \qquad \{\ddot{\mathbf{u}}\} = [\mathbf{N}]\{\ddot{\mathbf{d}}\} \qquad (13.2\text{-}2)$$

In Eqs. 13.2-2, shape functions $[\mathbf{N}]$ are functions of space only and nodal d.o.f. $\{\mathbf{d}\}$ are functions of time only. Thus Eqs. 13.2-2 represent a local separation of variables. Combination of Eqs. 13.2-1 and 13.2-2 yields

$$\{\delta\mathbf{d}\}^T \left[\int_{V_e} [\mathbf{B}]^T\{\boldsymbol{\sigma}\}\, dV + \int_{V_e} \rho[\mathbf{N}]^T[\mathbf{N}]\, dV\{\ddot{\mathbf{d}}\} + \int_{V_e} \kappa_d[\mathbf{N}]^T[\mathbf{N}]\, dV\{\dot{\mathbf{d}}\} \right.$$
$$\left. - \int_{V_e} [\mathbf{N}]^T\{\mathbf{F}\}\, dV - \int_{S_e} [\mathbf{N}]^T\{\mathbf{\Phi}\}\, dS - \sum_{i=1}^{n} \{\mathbf{p}\}_i \right] = 0 \qquad (13.2\text{-}3)$$

in which it has been assumed that the locations of concentrated loads $\{\mathbf{p}\}_i$ are coincident with node point locations. Since $\{\delta\mathbf{d}\}$ is arbitrary, Eq. 13.2-3 can be written as

$$[\mathbf{m}]\{\ddot{\mathbf{d}}\} + [\mathbf{c}]\{\dot{\mathbf{d}}\} + \{\mathbf{r}^{\text{int}}\} = \{\mathbf{r}^{\text{ext}}\} \tag{13.2-4}$$

where the element mass and damping matrices are defined as

$$[\mathbf{m}] = \int_{V_e} \rho[\mathbf{N}]^T[\mathbf{N}]\, dV \tag{13.2-5}$$

$$[\mathbf{c}] = \int_{V_e} \kappa_d[\mathbf{N}]^T[\mathbf{N}]\, dV \tag{13.2-6}$$

and the element internal force[1] and external load vectors are defined as

$$\{\mathbf{r}^{\text{int}}\} = \int_{V_e} [\mathbf{B}]^T\{\boldsymbol{\sigma}\}\, dV \tag{13.2-7}$$

$$\{\mathbf{r}^{\text{ext}}\} = \int_{V_e} [\mathbf{N}]^T\{\mathbf{F}\}\, dV + \int_{S_e} [\mathbf{N}]^T\{\boldsymbol{\Phi}\}\, dS + \sum_{i=1}^{n} \{\mathbf{p}\}_i \tag{13.2-8}$$

Equation 13.2-4 is a system of coupled, second-order, ordinary differential equations in time and is called a finite element *semidiscretization* because although displacements $\{\mathbf{d}\}$ are discrete functions of space, they are still continuous functions of time. Methods of dynamic analysis focus on how to solve this equation. Modal methods, discussed in Section 13.6, attempt to uncouple the equations, each of which can then be solved independently of others. Direct integration methods, discussed in Sections 13.9 to 13.13, discretize Eq. 13.2-4 in time to obtain a sequence of simultaneous algebraic equations.

Structure matrices $[\mathbf{M}]$, $[\mathbf{C}]$, and $\{\mathbf{R}^{\text{int}}\}$ are constructed by the conceptual expansion of element matrices $[\mathbf{m}]$, $[\mathbf{c}]$, and $\{\mathbf{r}^{\text{int}}\}$ to "structure size" followed by addition of overlapping coefficients, exactly as explained in Sections 2.5 to 2.7. However, as discussed in subsequent sections, the exact manner in which $\{\mathbf{R}^{\text{int}}\}$ is computed is often intimately mated with the dynamic analysis procedure.

When Eqs. 13.2-5 and 13.2-6 are evaluated using the same shape functions $[\mathbf{N}]$ as used in the displacement field interpolation (Eqs. 13.2-2), the results are called *consistent mass* and *consistent damping matrices*. These matrices are symmetric. On the element level, they are generally full, but on the structure level, they have the same sparse topology as the structure stiffness matrix. When ρ and κ_d are nonzero, consistent matrices $[\mathbf{m}]$ and $[\mathbf{c}]$ are positive definite. That is, using the mass matrix for example, the kinetic energy $\frac{1}{2}\{\dot{\mathbf{d}}\}^T[\mathbf{m}]\{\dot{\mathbf{d}}\}$ is positive for any nonzero $\{\dot{\mathbf{d}}\}$.

Consistent damping matrix $[\mathbf{c}]$ is easily evaluated for a Newtonian fluid; its terms are given by Rayleigh [13.3]. In structures we are less interested in viscous damping than in dry friction and hysteresis loss. These energy loss mechanisms are not well understood, and from a practical standpoint Eq. 13.2-6 does not

[1]The term "force" is conventional, even though Eq. 13.2-7 may contain nodal moments if the element has rotational d.o.f.

correctly represent structural damping. In Section 13.4, we present some popular ad hoc damping schemes for structural dynamics.

The internal force vector, Eq. 13.2-7, represents loads at nodes caused by straining of material. Equations 13.2-4 and 13.2-7 are valid for both linear and nonlinear material behavior; that is, in Eq. 13.2-7, $\{\boldsymbol{\sigma}\}$ could be a nonlinear function of strain or strain rate. For linearly elastic material behavior, $\{\boldsymbol{\sigma}\} = [\mathbf{E}][\mathbf{B}]\{\mathbf{d}\}$ and Eq. 13.2-7 becomes

$$\{\mathbf{r}^{\text{int}}\} = [\mathbf{k}]\{\mathbf{d}\} \tag{13.2-9}$$

where the usual definition of the stiffness matrix holds—that is,

$$[\mathbf{k}] = \int_{V_e} [\mathbf{B}]^T[\mathbf{E}][\mathbf{B}]\, dV \tag{13.2-10}$$

When Eq. 13.2-10 is used, Eq. 13.2-4 becomes

$$[\mathbf{m}]\{\ddot{\mathbf{d}}\} + [\mathbf{c}]\{\dot{\mathbf{d}}\} + [\mathbf{k}]\{\mathbf{d}\} = \{\mathbf{r}^{\text{ext}}\} \tag{13.2-11}$$

which can be interpreted as saying that external loads are equilibrated by a combination of inertial, damping, and elastic forces. For the assembled structure, from Eq. 13.2-11,

$$\boxed{[\mathbf{M}]\{\ddot{\mathbf{D}}\} + [\mathbf{C}]\{\dot{\mathbf{D}}\} + [\mathbf{K}]\{\mathbf{D}\} = \{\mathbf{R}^{\text{ext}}\}} \tag{13.2-12}$$

where $\{\mathbf{R}^{\text{ext}}\}$ corresponds to loads $\{\mathbf{R}\}$ of a static problem, but is in general a function of time. Or, returning to Eq. 13.2-4, equations of the assembled structure can be written in the alternative form

$$[\mathbf{M}]\{\ddot{\mathbf{D}}\} + [\mathbf{C}]\{\dot{\mathbf{D}}\} + \{\mathbf{R}^{\text{int}}\} = \{\mathbf{R}^{\text{ext}}\} \tag{13.2-13}$$

which does not require that the material be linearly elastic.

13.3 MASS MATRICES, CONSISTENT AND DIAGONAL

A mass matrix is a discrete representation of a continuous distribution of mass. A *consistent* element mass matrix is defined by Eqs. 13.2-5—that is, by $[\mathbf{m}] = \int \rho[\mathbf{N}]^T[\mathbf{N}]\, dV$. It is termed "consistent" because $[\mathbf{N}]$ represents the same shape functions as are used to generate the element stiffness matrix [13.4]. A simpler and historically earlier formulation is the *lumped* mass matrix, which is obtained by placing particle masses m_i at nodes i of an element, such that $\Sigma\, m_i$ is the total element mass. Particle "lumps" have no rotary inertia unless rotary inertia is arbitrarily assigned, as is sometimes done for the rotational d.o.f. of beams and plates. A lumped mass matrix is diagonal but a consistent mass matrix is not. The two formulations have different merits, and various considerations enter into deciding which one, or what combination of them, is best suited to a particular analysis procedure.

Examples. Consider the uniform bar element shown in Fig. 13.3-1*a*. Shape functions to be used in Eq. 13.2-5 are given by Eq. 3.8-4 with s replaced by x. Mass increment $\rho\, dV$ in Eq. 13.2-5 can be written as $(m/L)\, dx$, where $m = \rho AL$ is the total mass of the element. The consistent and lumped mass matrices are, respectively,

$$[\mathbf{m}] = \frac{m}{6}\begin{bmatrix} 2 & 0 & 1 & 0 \\ 0 & 2 & 0 & 1 \\ 1 & 0 & 2 & 0 \\ 0 & 1 & 0 & 2 \end{bmatrix} \qquad \lceil\mathbf{m}\rfloor = \frac{m}{2}\begin{bmatrix} 1 & 0 & 0 & 0 \\ 0 & 1 & 0 & 0 \\ 0 & 0 & 1 & 0 \\ 0 & 0 & 0 & 1 \end{bmatrix} \tag{13.3-1}$$

The lumped mass is obtained by placing half of the total element mass m as a particle at each node. The particle mass $m/2$ appears four times in $\lceil\mathbf{m}\rfloor$ because four nodal acceleration vectors are resisted by inertia.

For the uniform beam element of Fig. 13.3-1*b*, shape functions are given in Fig. 3.13-2, and Eq. 13.2-5 yields the consistent mass matrix

$$[\mathbf{m}] = \frac{m}{420}\begin{bmatrix} 156 & 22L & 54 & -13L \\ 22L & 4L^2 & 13L & -3L^2 \\ 54 & 13L & 156 & -22L \\ -13L & -3L^2 & -22L & 4L^2 \end{bmatrix} \tag{13.3-2}$$

where $m = \rho AL$ is the total element mass. The (diagonal) lumped mass matrix of the beam is given by

$$\lceil\mathbf{m}\rfloor = \frac{m}{2}\lceil 1 \quad \alpha L^2/210 \quad 1 \quad \alpha L^2/210 \rfloor \tag{13.3-3}$$

where the second and fourth diagonal terms account for rotary inertia. Sometimes rotary inertia is neglected ($\alpha = 0$). If included, it is often selected as the mass moment of inertia I of a uniform slender bar of length $L/2$ and mass $m/2 = \rho AL/2$ spinning about one end; that is, $I = (m/2)(L/2)^2/3$, for which $\alpha = 17.5$. If an element is tapered or has nonuniform density, its mass and stiffness matrices change.

A plane frame element has three d.o.f. per node. Its mass matrix is formed by expanding and then combining the bar and beam mass matrices. An element arbitrarily oriented in xy coordinates requires the coordinate transformation described in Section 7.4.

Further examples appear in Fig. 13.3-2.

Equations 13.3-1 and 13.3-3 illustrate ad hoc mass matrix lumping that is guided by intuition and physical insight. The lumped mass matrices obtained are effective

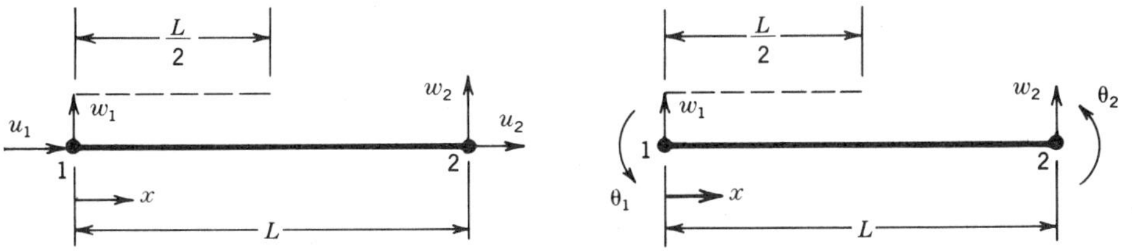

Figure 13.3-1. Uniform bar and beam elements with the respective $\{\mathbf{d}\}$ vectors $\lfloor u_1 \;\; w_1 \;\; u_2 \;\; w_2 \rfloor^T$ and $\lfloor w_1 \;\; \theta_1 \;\; w_2 \;\; \theta_2 \rfloor^T$. Dashed lines show the conceptual shapes and tributary lengths used to obtain the lumped-mass coefficients associated with d.o.f. at node 1.

Constant-strain triangle. With a linear displacement field in each direction, and $\{\mathbf{d}\} = \lfloor u_1 \quad u_2 \quad u_3 \quad v_1 \quad v_2 \quad v_3 \quad w_1 \quad w_2 \quad w_3 \rfloor^T$,

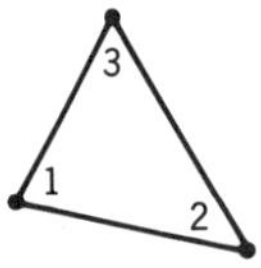

$$\underset{9\times 9}{[\mathbf{m}]} = \lceil \mathbf{Q} \quad \mathbf{Q} \quad \mathbf{Q} \rfloor, \qquad \text{where} \quad [\mathbf{Q}] = \frac{\rho A t}{12}\begin{bmatrix} 2 & 1 & 1 \\ 1 & 2 & 1 \\ 1 & 1 & 2 \end{bmatrix}$$

Bilinear rectangle. With a bilinear displacement field in each direction, and $\{\mathbf{d}\} = \lfloor u_1 \quad u_2 \quad u_3 \quad u_4 \quad v_1 \quad v_2 \quad v_3 \quad v_4 \quad w_1 \quad w_2 \quad w_3 \quad w_4 \rfloor^T$,

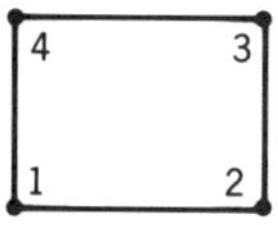

$$\underset{12\times 12}{[\mathbf{m}]} = \lceil \mathbf{Q} \quad \mathbf{Q} \quad \mathbf{Q} \rfloor, \qquad \text{where} \quad [\mathbf{Q}] = \frac{\rho A t}{36}\begin{bmatrix} 4 & 2 & 1 & 2 \\ 2 & 4 & 2 & 1 \\ 1 & 2 & 4 & 2 \\ 2 & 1 & 2 & 4 \end{bmatrix}$$

Figure 13.3-2. Consistent mass matrices for plane elements allowed to move in three dimensions. A = surface area, ρ = uniform mass density, and t = uniform thickness.

and widely used. However, for higher-order elements (e.g., quadratic-displacement plane elements) or elements of irregular shape, intuition can be risky. Accordingly, systematic schemes for lumping are necessary.

HRZ Lumping Scheme. The HRZ scheme [13.5,13.6] is an effective method for producing a diagonal mass matrix. It can be recommended for arbitrary elements. The idea is to use only the diagonal terms of the consistent mass matrix, but to scale them in such a way that the total mass of the element is preserved. Specifically, the procedural steps are as follows.

1. Compute only the diagonal coefficients of the consistent mass matrix.
2. Compute the total mass of the element, m.
3. Compute a number s by adding the diagonal coefficients m_{ii} associated with translational d.o.f. (but not rotational d.o.f., if any) that are mutually parallel and in the same direction.
4. Scale *all* the diagonal coefficients by multiplying them by the ratio m/s, thus preserving the total mass of the element.

As examples, for the bar and beam elements shown in Fig. 13.3-1, the preceding four steps yield, respectively, the diagonal matrices

$$\lceil \mathbf{m} \rfloor = \frac{m}{2} \lceil 1 \quad 1 \quad 1 \quad 1 \rfloor \tag{13.3-4}$$

$$\lceil \mathbf{m} \rfloor = \frac{m}{78} \lceil 39 \quad L^2 \quad 39 \quad L^2 \rfloor \tag{13.3-5}$$

Further examples appear in Fig. 13.3-3.

Test cases to date show that for flexural and low-order finite elements, the

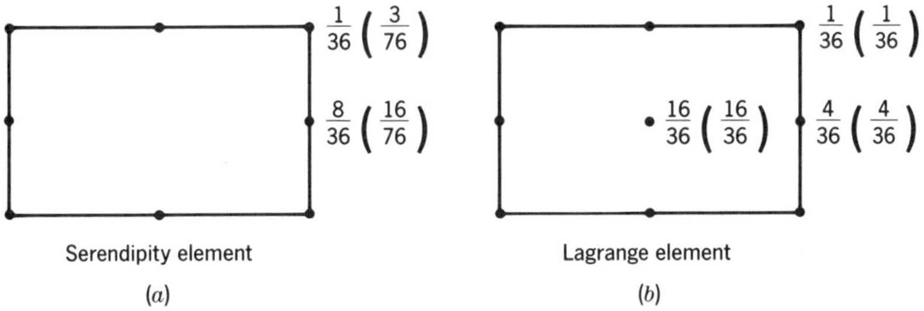

Figure 13.3-3. Diagonal mass matrices of plane rectangular elements obtained using the four-step HRZ scheme [13.5]. Each element has constant thickness. Numbers shown are fractions of the total element mass at each node. First number: based on 2 by 2 Gauss quadrature. Second number (in parentheses): based on 3 by 3 Gauss quadrature. (*a*) Serendipity element. (*b*) Lagrange element.

accuracy of this form of diagonal mass matrix is excellent, often surpassing that of the consistent mass matrix (see Table 13.3-1). For higher-order plane elements, such as the six-node quadratic triangle and the eight-node serendipity quadrilateral, this scheme may be less accurate for transient analysis than an optimally lumped mass matrix [13.7,13.8].

Optimal Lumping. Mass lumping can be thought of as the result of applying an appropriate quadrature rule to evaluate $\int \rho[\mathbf{N}]^T[\mathbf{N}]\, dV$. If the integration points of a quadrature rule coincide with nodal locations of an element having translational d.o.f. only, then no off-diagonal terms are generated and the mass matrix is diagonal. If the element also has rotational d.o.f., then lumping by quadrature produces block-diagonal matrices that are of lesser practical usefulness because they are not diagonal. In the sequel, we shall be concerned with lumping by quadrature for elements with translational d.o.f. only.

Let p be the degree of the highest-degree complete polynomial in [**N**] and m the highest-order derivative in the strain energy expression (e.g., $m = 1$ for elasticity and $m = 2$ for bending). Fix [13.9] has shown that a quadrature rule

TABLE 13.3-1. PERCENTAGE ERRORS OF COMPUTED NATURAL FREQUENCIES OF A SIMPLY SUPPORTED THICK SQUARE PLATE [13.5]. HALF THE PLATE WAS MODELED BY A 4 BY 2 MESH OF EIGHT-NODE 24-D.O.F. ELEMENTS (TABLE 11.3-1). THE AD HOC LUMPED MASS MATRIX HAS EQUAL MASS PARTICLES AT EACH NODE. CONSISTENT MASS MATRIX RESULTS DO NOT GUARANTEE AN UPPER BOUND BECAUSE THE STIFFNESS MATRIX IS BASED ON REDUCED INTEGRATION.

Mode		Type of Mass Matrix Used		
m	n	Consistent (%)	HRZ Lumping (%)	Ad Hoc Lumping (%)
1	1	−0.11	+0.32	+0.32
2	1	−0.40	+0.45	−0.45
2	2	−0.35	−2.75	−4.12
3	1	+5.18	+0.05	−5.75
3	2	+4.68	−2.96	−10.15
3	3	+13.78	−5.18	−19.42
4	2	+16.88	+1.53	+31.70

with degree of precision at least $2(p - m)$ will yield comparable accuracy and cause no loss of convergence rate beyond what is already inherent in the finite element method with consistent mass matrices [13.68]. Diagonal mass matrices integrated in this way are said to be *optimally lumped.*

As an example of optimal lumping, consider the three-node quadratic-displacement bar element shown in Fig. 6.2-1, for which $p = 2$ and $m = 1$. Let the mass density ρ and cross-sectional area A be uniform and the nodes uniformly spaced. The minimum order of integration for optimal mass matrix lumping is $n = 2(2 - 1) = 2$. An integration rule having at least this degree of precision and having integration points at nodal locations is the three-point Newton–Cotes formula [13.12], which is identical to Simpson's rule. (This integration rule will exactly integrate a cubic polynomial.)

$$\int_a^b f(x)\, dx = (b - a)[\tfrac{1}{6}f(x = a) + \tfrac{4}{6}f(x = (a + b)/2) + \tfrac{1}{6}f(x = b)] \qquad (13.3\text{-}6)$$

From Eq. 13.2-5, a single term of the mass matrix is

$$m_{ij} = \rho A \int N_i N_j\, dx = \rho A \int_{-1}^{1} N_i N_j J\, d\xi \qquad (13.3\text{-}7)$$

The Jacobian J is $J = L/2$, so Eq. 13.3-6 yields

$$\begin{aligned} m_{ij} = \rho AL[&\tfrac{1}{6}N_i(\xi = -1)\, N_j(\xi = -1) \\ &+ \tfrac{4}{6}N_i(\xi = 0)\, N_j(\xi = 0) + \tfrac{1}{6}N_i(\xi = 1)\, N_j(\xi = 1)] \end{aligned} \qquad (13.3\text{-}8)$$

Thus $m_{ij} = 0$ for $i \neq j$, and, with node 3 at the middle of the bar,

$$[\mathbf{m}] = \frac{\rho AL}{6}\begin{bmatrix} 1 & 0 & 0 \\ 0 & 1 & 0 \\ 0 & 0 & 4 \end{bmatrix} \qquad (13.3\text{-}9)$$

Equation 13.3-9 also agrees with the mass matrix obtained by HRZ lumping. In addition, the portion of element mass allocated to each d.o.f. is the same as the nodal allocation of a uniform traction load on the edge of a quadratic element as shown by Fig. 4.3-4.

Nodes of Lagrangian elements coincide with integration points of the Lobatto quadrature rule [13.10]. Lobatto integration weights are always positive, hence optimal lumping for Lagrangian elements results in positive definite diagonal mass matrices; that is, each node in the element has a positive mass lump associated with it. Results for the quadratic Lagrange element are shown in Fig. 13.3-4. Additional results appear in [13.7]. In cubic and higher-order Lagrange elements, nodal masses are positive but nodes in the reference element must be at special positions. In general it is not possible to construct an integration rule of a required accuracy that simultaneously permits arbitrary specification of integration point locations and has positive weighting. Thus optimally lumped diagonal mass matrices for triangular and serendipity quadrilateral elements (particularly the quadratic and higher-order elements) frequently have some zero or negative nodal masses as shown in Fig. 13.3-4.

For low-order elements, such as the linear-displacement bar, the constant-strain

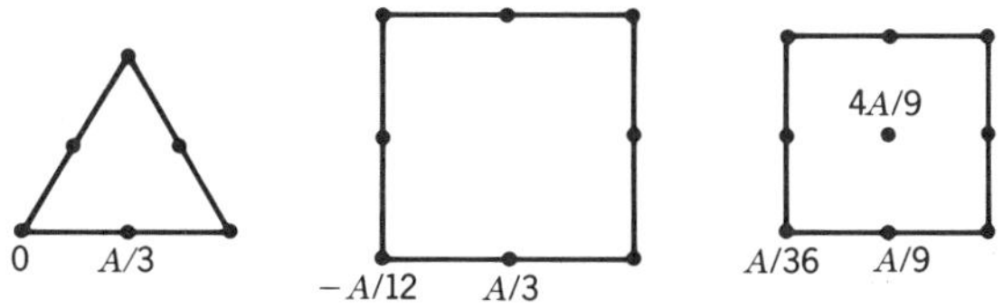

Figure 13.3-4. Optimal mass matrix lumping for some common two-dimensional elements [13.7]. The triangular element is an equilateral reference element in area coordinates. Rectangular elements are isoparametric reference elements. Elements are uniform and have mass proportional to element area A.

triangle, and the bilinear quadrilateral, ad hoc lumping usually gives the same result as optimal lumping. Also, for the quadratic Lagrange element, HRZ lumping and optimal lumping produce the same diagonal mass matrix. For cubic and higher-order Lagrange elements, HRZ lumping and optimal lumping may be different; however, these elements are rarely used in dynamics. For quadratic and higher-order triangular and serendipity quadrilateral elements, HRZ lumping and optimal lumping are markedly different: numerical tests show that the displacement, velocity, and acceleration time-history results for HRZ lumped models can be less accurate than for optimally lumped models in some problems [13.7, 13.8].

Remarks. With any mass matrix, the product $[\mathbf{m}]\{\ddot{\mathbf{d}}\}$ or $\lceil\mathbf{m}\rfloor\{\ddot{\mathbf{d}}\}$ must yield the correct total force on an element according to Newton's law $\mathbf{F} = m\mathbf{a}$ when $\{\ddot{\mathbf{d}}\}$ represents a rigid-body translational acceleration. The rationale is that for convergence to correct results, this kind of motion must be correctly represented because it is the only motion experienced by an element when a mesh is indefinitely refined.

Consistent mass matrices $[\mathbf{m}]$ and $[\mathbf{M}]$ are positive definite (i.e., the kinetic energy $\frac{1}{2}\{\dot{\mathbf{d}}\}^T[\mathbf{m}]\{\dot{\mathbf{d}}\} > 0$ for all $\{\dot{\mathbf{d}}\} \neq \{\mathbf{0}\}$). A lumped mass matrix is positive semidefinite or indefinite if zero or negative masses, respectively, appear on the diagonal. The zeros may or may not make some operations awkward, depending on the algorithm, and negative masses usually (but not always [13.7]) require some special treatment.

If the mesh layout correctly represents the structure volume, elements are compatible and not softened by low-order integration rules, and mass matrices are consistent, then computed natural frequencies are upper bounds to the exact values. If any of these restrictions is violated, such as by the use of any lumping scheme, a bound cannot be guaranteed [13.11]. The upper bound is computed with an error of order $h^{2(q-1)}$, where h and q are defined in Section 18.6 [13.13]. The upper-bound property is illustrated by the numerical example in Section 13.5.

We cannot say that either lumped or consistent mass matrices are best for all problems. Consistent matrices are more accurate for flexural problems, such as beams and shells, but negligibly so if the wavelength of the mode spans more than about four elements [13.14]. Lumped matrices usually yield natural frequencies that are less than the exact values. For a bar element, MacNeal [13.15] finds that $[\mathbf{m}]$ and $\lceil\mathbf{m}\rfloor$ of Eq. 13.3-1 yield natural frequency errors of order h^2 in opposite directions and that an $[\mathbf{m}]$ that is the average of the two yields a natural frequency error of only order h^4. Similar improvement is possible with beam elements [13.46].

In wave propagation problems using linear-displacement field elements, lumped masses give greater accuracy because of fewer spurious oscillations. For higher-order elements, diagonal mass matrices obtained by ad hoc lumping may be deficient in accuracy compared to optimal lumping.

As for efficiency, lumped mass matrices are simpler to form, occupy less storage space, and require less computational effort. Indeed, some methods of dynamic analysis are practicable only with lumped mass matrices. Usually it is more important to have a diagonal mass matrix in time-history analysis than in vibration analysis, as time-history analysis is usually much more expensive.

A practitioner contemplating the use of lumped matrices for higher-order elements having translational d.o.f. only has basically four choices: (1) use lower-order elements instead; (2) use Lagrangian elements with optimal lumping, which always results in positive nodal masses; (3) use non-Lagrangian elements with optimal lumping, which may result in a nonpositive definite mass matrix with diagonal coefficients that can be positive, zero, or negative; or (4) use ad hoc lumping which sacrifices optimal accuracy but maintains positive nodal masses. While option 3 is an accurate and workable computational alternative, the most expedient analyses will result from options 1 and 2. Option 4 may give inaccurate results and should be avoided.

13.4 DAMPING

Damping in structures is not viscous; rather, it is due to mechanisms such as hysteresis in the material and slip in connections. These mechanisms are not well understood. Moreover, they are awkward to incorporate into the equations of structural dynamics, or they make the equations computationally difficult. Therefore, the actual damping mechanism is usually approximated by viscous damping. Comparisons of theory and experiment show that this approach is sufficiently accurate in most cases.

The treatment of damping in computational analyses can be categorized as (1) *phenomenological damping methods,* in which the actual physical dissipative mechanisms such as elastic–plastic hysteresis loss, structural joint friction, or material microcracking are modeled, or (2) *spectral damping methods,* in which viscous damping is introduced by means of specified fractions of critical damping [13.16]. (Critical damping, for which the damping ratio is $\xi = 1$, marks the transition between oscillatory and nonoscillatory response.) Phenomenological methods require detailed models for the dissipative mechanisms and almost always result in nonlinear analyses; hence, they are seldom used. With spectral damping approaches, experimental observations of the vibratory response of structures are used to assign a fraction of critical damping as a function of frequency, or more commonly, a single damping fraction for the entire frequency range of a structure [13.16]. The damping ratio ξ depends on the material and the stress level. In steel piping, ξ ranges from about 0.5% at low stress levels to about 5% at high stress levels. In bolted or riveted steel structures, and in reinforced or prestressed concrete, ξ has the approximate range 2% to 15%.

A popular spectral damping scheme, called *Rayleigh* or *proportional* damping, is to form damping matrix [C] as a linear combination of the stiffness and mass matrices, that is,

$$[\mathbf{C}] = \alpha[\mathbf{K}] + \beta[\mathbf{M}] \tag{13.4-1}$$

where α and β are called, respectively, the stiffness and mass proportional damping constants. Matrix [**C**] given by Eq. 13.4-1 is an *orthogonal* damping matrix because it permits modes to be uncoupled by eigenvectors associated with the undamped eigenproblem (Section 13.6). The relationship between α, β, and the fraction of critical damping ξ at frequency ω is given by the following equation, proof of which is left as an exercise:

$$\xi = \frac{1}{2}\left(\alpha\omega + \frac{\beta}{\omega}\right) \tag{13.4-2}$$

Damping constants α and β are determined by choosing the fractions of critical damping (ξ_1 and ξ_2) at two different frequencies (ω_1 and ω_2) and solving simultaneous equations for α and β. Thus

$$\alpha = 2(\xi_2\omega_2 - \xi_1\omega_1)/(\omega_2^2 - \omega_1^2) \tag{13.4-3a}$$

$$\beta = 2\omega_1\omega_2(\xi_1\omega_2 - \xi_2\omega_1)/(\omega_2^2 - \omega_1^2) \tag{13.4-3b}$$

Shown in Fig. 13.4-1 is the fraction of critical damping versus frequency. Damping attributable to α[**K**] increases with increasing frequency, whereas damping attributable to β[**M**] increases with decreasing frequency. For structures that may have rigid-body motion, it is important that the mass-proportional damping not be excessive. Positive values of β less than about 0.1 per time unit are usually acceptable [13.17].

Usually, ω_1 and ω_2 are chosen to bound the design spectrum. Thus ω_1 is taken as the lowest natural frequency of the structure, and ω_2 is the maximum frequency of interest in the loading or response. For example, in seismic analyses, 30 Hz is often used as the upper frequency because the spectral content of seismic design spectra are insignificant above that frequency.

More general proportional damping schemes are possible in which the fraction

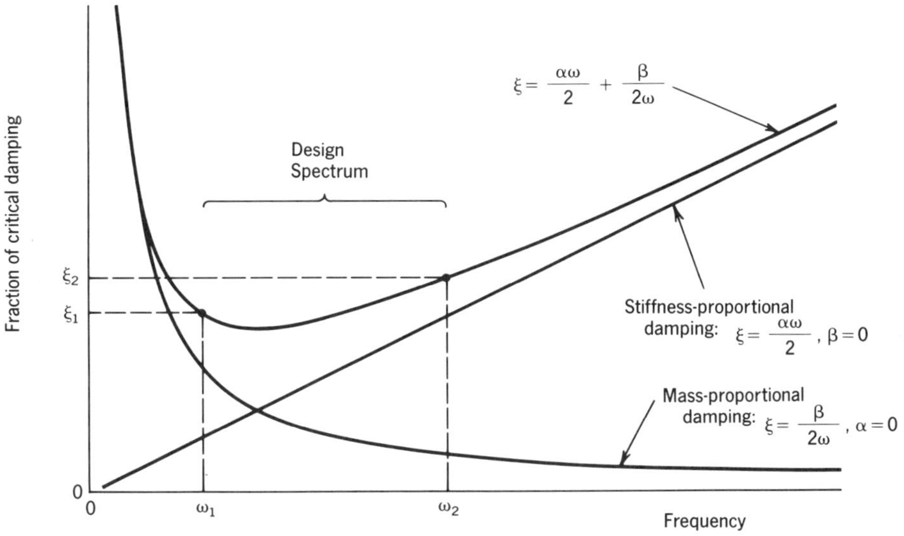

Figure 13.4-1. Fraction of critical damping versus frequency for Rayleigh damping. Contribution of stiffness and mass proportional damping to total damping is also shown.

of critical damping at any desired number of frequencies can be specified [13.18]. However, these schemes usually produce fully populated damping matrices, and hence are rarely used in practice.

13.5 NATURAL FREQUENCIES AND MODE SHAPES

The Eigenvalue Problem. An undamped structure, with no external loads applied to unrestrained d.o.f., undergoes harmonic motion (caused perhaps by initial conditions) in which each d.o.f. moves in phase with all other d.o.f. Thus

$$\{\mathbf{D}\} = \{\overline{\mathbf{D}}\}\sin\omega t \qquad \text{and} \qquad \{\ddot{\mathbf{D}}\} = -\omega^2\{\overline{\mathbf{D}}\}\sin\omega t \tag{13.5-1}$$

where $\{\overline{\mathbf{D}}\}$ = amplitudes of nodal d.o.f. vibration and ω = *circular frequency* (radians per second). The *cyclic frequency* (in Hertz) is $f = \omega/2\pi$ and the *period* is $T = 1/f$ (seconds). Both ω and f are called simply "frequency" and pertain to undamped motion unless stated otherwise.

Combining Eqs. 13.5-1 with Eq. 13.2-12, and with $[\mathbf{C}]$ and $\{\mathbf{R}^{\text{ext}}\}$ both zero, we obtain

$$([\mathbf{K}] - \lambda[\mathbf{M}])\{\overline{\mathbf{D}}\} = \{\mathbf{0}\}, \qquad \text{where} \quad \lambda = \omega^2 \tag{13.5-2}$$

This is the basic statement of the vibration problem. Equation 13.5-2 is called a *generalized eigenproblem* or simply an *eigenproblem*. When the matrix $[\mathbf{K}] - \lambda[\mathbf{M}]$ is nonsingular, Eq. 13.5-2 has only the trivial solution $\{\overline{\mathbf{D}}\} = \{\mathbf{0}\}$. We are interested in nontrivial solutions and hence wish to determine the *eigenvalues* (or *characteristic numbers,* or *latent roots*) λ that satisfy

$$\det([\mathbf{K}] - \lambda[\mathbf{M}]) = 0 \tag{13.5-3}$$

Associated with each eigenvalue λ_i is an *eigenvector* $\{\overline{\mathbf{D}}\}_i$, which is sometimes called a *normal* (or *natural,* or *characteristic,* or *principal*) mode. The lowest nonzero ω_i is called the *fundamental* vibration frequency. Appendix C lists some important properties of eigenvalues and eigenvectors such as orthogonality and linear independence.

If $[\mathbf{K}]$ and $[\mathbf{M}]$ are n_{eq} by n_{eq} matrices, then, under conditions usually satisfied in structural analysis, Eq. 13.5-2 has n_{eq} eigenvalues and n_{eq} eigenvectors (see Appendix C). All eigenvalues are positive if $[\mathbf{K}]$ and $[\mathbf{M}]$ are both positive definite, as is usually the case when $[\mathbf{K}]$ has all rigid-body modes constrained and when $[\mathbf{M}]$ is either a consistent mass matrix or a lumped mass matrix with strictly positive diagonal coefficients. A partly or completely unsupported structure has positive semidefinite $[\mathbf{K}]$ and has one zero eigenvalue associated with each possible rigid-body motion. When $[\mathbf{M}]$ is lumped—that is, a diagonal matrix—some of its coefficients M_{ii} may be zero (which is commonly the case if rotational d.o.f. are present in $\{\overline{\mathbf{D}}\}$ but not associated with rotary inertia) or negative (which is a rare situation that is due to optimal lumping). Typically each zero M_{ii} is associated with an infinite eigenvalue, and each negative M_{ii} is associated with a negative eigenvalue, which gives rise to an imaginary frequency. Some algorithms for eigenvalue extraction require positive definite mass matrices and hence will not work for lumped matrices having some zero or negative diagonal coefficients. When there are negative diagonal coefficients, the reader is referred to algorithms described in Appendix C and in [13.7]. When a diagonal coefficient M_{ii} is zero, the associated d.o.f. can be removed from $\{\bar{\mathbf{D}}\}$ by condensation prior to eigenanalysis.

Rayleigh Quotient. Let $[\mathbf{K}]$ be symmetric and $[\mathbf{M}]$ be positive definite and symmetric. If we premultiply Eq. 13.5-2 by $\{\bar{\mathbf{D}}\}^T$ and solve for λ, we obtain the *Rayleigh quotient*

$$\lambda = \frac{\{\bar{\mathbf{D}}\}^T[\mathbf{K}]\{\bar{\mathbf{D}}\}}{\{\bar{\mathbf{D}}\}^T[\mathbf{M}]\{\bar{\mathbf{D}}\}} \tag{13.5-4}$$

If $\{\bar{\mathbf{D}}\}$ approximates the ith eigenvector with first-order error, then λ approximates the corresponding eigenvalue with second-order error. Thus a casual estimate of $\{\bar{\mathbf{D}}\}$ may result in an accurate estimate of λ. The Rayleigh quotient is, in fact, an extreme value when $\{\bar{\mathbf{D}}\}$ varies in the neighborhood of an exact eigenvector [13.18,13.19]; accordingly, the extraction of an eigenvalue can be approached as an optimization problem. Values of the Rayleigh quotient are bounded by the largest and smallest eigenvalues of the mesh. That is, for any vector $\{\mathbf{v}\}$,

$$\lambda_{\text{min}} \leq \frac{\{\mathbf{v}\}^T[\mathbf{K}]\{\mathbf{v}\}}{\{\mathbf{v}\}^T[\mathbf{M}]\{\mathbf{v}\}} \leq \lambda_{\text{max}} \tag{13.5-5}$$

where λ_{min} and λ_{max} are the smallest and largest eigenvalues of Eq. 13.5-2. Equation 13.5-5 presumes that $[\mathbf{M}]$ is positive definite. If $[\mathbf{M}]$ is indefinite, as may result from optimal lumping or lumping with zero rotary inertia, the Rayleigh quotient may be positive or negative infinity for some choices of $\{\mathbf{v}\}$.

In certain direct integration algorithms, it is necessary to have prior knowledge of the largest eigenvalue of the mesh, λ_{max}. An upper bound to λ_{max} can be obtained by considering the eigenproblem for a single, unsupported element

$$\det([\mathbf{k}] - \lambda'[\mathbf{m}]) = 0 \tag{13.5-6}$$

The largest eigenvalue of Eq. 13.5-6 can often be obtained by hand calculation. If we denote the largest eigenvalue of Eq. 13.5-6 among *all* elements by λ'_{max}, then

$$\lambda_{\text{max}} \leq \lambda'_{\text{max}} \tag{13.5-7}$$

The reason is that the largest eigenvalue is a maximum of the Rayleigh quotient, Eq. 13.5-5. Suppose we consider an alternative Rayleigh quotient instead, in which no constraints of nodal-value sharing are imposed. Then the resulting maximum eigenvalue would be λ'_{max} of Eq. 13.5-6. The actual λ_{max} could be obtained from the alternative Rayleigh quotient by imposing a large number of linear constraints that specify the nodal-value sharing of the original mesh. Optimization theory shows that imposing such constraints can only lower a maximum, and thus $\lambda_{\text{max}} \leq \lambda'_{\text{max}}$. Applying the same argument to λ_{min} shows that $\lambda_{\text{min}} \geq \lambda'_{\text{min}}$, which is consistent with our physical intuition that imposing constraints raises the fundamental frequency.

Example. We consider an elementary example of a matrix eigenproblem and its solution. Consider the uniform one-dimensional unsupported bar with mass density ρ, elastic modulus E, and cross-sectional area A shown in Fig. 13.5-1. With consistent [**m**], Eq. 13.5-2 becomes

$$\left(\frac{AE}{L}\begin{bmatrix} 1 & -1 \\ -1 & 1 \end{bmatrix} - \omega^2 \frac{\rho AL}{6}\begin{bmatrix} 2 & 1 \\ 1 & 2 \end{bmatrix}\right)\begin{Bmatrix} \bar{u}_1 \\ \bar{u}_2 \end{Bmatrix} = \begin{Bmatrix} 0 \\ 0 \end{Bmatrix} \tag{13.5-8}$$

For nontrivial amplitudes $\{\bar{\mathbf{d}}\} = \lfloor \bar{u}_1 \quad \bar{u}_2 \rfloor^T$ to exist, the determinant of the expression in parentheses must vanish. Thus

$$\omega^2(\omega^2 \rho AL - 12AE/L) = 0 \tag{13.5-9}$$

from which $\omega_1 = 0$ and $\omega_2 = (2/L)\sqrt{3E/\rho} = (3.464/L)\sqrt{E/\rho}$. The easiest way to determine the eigenvector associated with any eigenvalue λ_i is to set one d.o.f. in the eigenvector $\{\bar{\mathbf{d}}\}_i$ to an arbitrary nonzero number ($\bar{d}_1 = 1$, for example), substitute the known value of λ_i, and solve for the remaining amplitudes in $\{\bar{\mathbf{d}}\}_i$. (If, by coincidence, the d.o.f. amplitude that was assumed to be nonzero is in fact zero, then the resulting system of equations will be singular and no solution will exist for the remaining amplitudes. Then the eigenvector can usually be found by assuming a nonzero value for one of the other amplitudes.) Thus, from Eq. 13.5-8,

$$\text{for } \omega_1 = 0 \qquad \{\bar{\mathbf{d}}\}_1 = \lfloor 1 \quad 1 \rfloor^T \tag{13.5-10a}$$

$$\text{for } \omega_2 = (3.464/L)\sqrt{E/\rho} \qquad \{\bar{\mathbf{d}}\}_2 = \lfloor 1 \quad -1 \rfloor^T \tag{13.5-10b}$$

The first eigenvector describes a rigid-body translation in the x direction. The second describes an axial straining mode (for which the exact fundamental frequency of a continuous unsupported bar of length L is $(\pi/L)\sqrt{E/\rho}$). Hence, the single-element consistent-mass model overpredicts the exact fundamental frequency by about 10%, thus illustrating the upper-bound property noted in Section 13.3.

If, instead of the consistent mass matrix, the lumped mass matrix $\lceil \mathbf{m} \rfloor = (\rho AL/2) \lceil 1 \quad 1 \rfloor$ is used in Eq. 13.5-4, the computed frequencies are $\omega_1 = 0$ and $\omega_2 = (2/L)\sqrt{E/\rho}$. Thus we see that the upper-bound property is destroyed by lumping. Eigenvectors for the lumped-mass case are the same as those of Eqs. 13.5-10.

In many design situations, we wish to know if severe dynamic excitation of a structure is likely. Therefore we compare the frequency spectrum of the structure with that of the time-dependent loading. If a natural frequency of the structure is

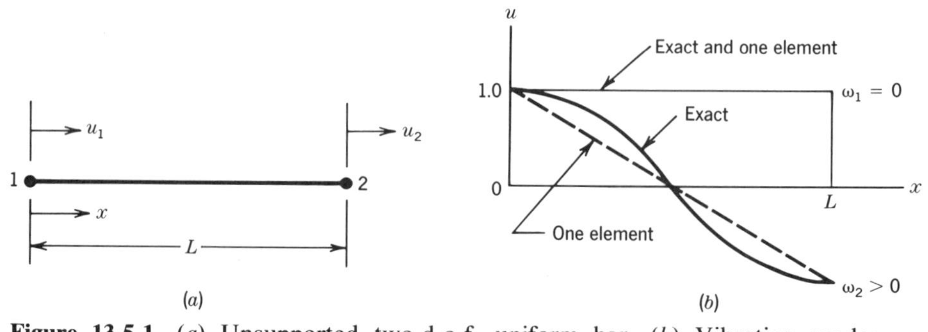

Figure 13.5-1. (*a*) Unsupported two-d.o.f. uniform bar. (*b*) Vibration modes $\omega_1 = 0$ (rigid-body translation) and $\omega_2 > 0$ (axial straining mode).

close to an excitation frequency, then severe vibration and "beating" are likely. This usually necessitates alteration of the structure's natural frequencies by resizing or by adding members or dampers. If frequencies of the structure and the excitation are well separated, the structure still vibrates, but the amplitude of the response is likely to be tolerable.

In static analysis, symmetry can be exploited, for example, by analyzing half of the entire structure. In vibration analysis, symmetry of structure and supports does not imply symmetry of all vibration modes. By imposing symmetry one would exclude all antisymmetric modes, which are probably as important as symmetric modes.

One should be aware that stabilization methods used to suppress mechanisms in underintegrated elements may be associated with relatively low stiffness. The associated nonphysical vibration modes may contaminate that portion of the vibration spectrum of greatest interest [13.70].

13.6 TIME-HISTORY ANALYSIS. MODAL METHODS

In a *time-history* or *dynamic response* problem, we solve Eq. 13.2-12 for $\{\mathbf{D}\}$, $\{\dot{\mathbf{D}}\}$, and $\{\ddot{\mathbf{D}}\}$ as functions of time. When $[\mathbf{M}]$, $[\mathbf{C}]$, and $[\mathbf{K}]$ are known and are time-independent, the problem is linear. When initial values of $\{\mathbf{D}\}$ and $\{\dot{\mathbf{D}}\}$ are prescribed, Eq. 13.2-12 is called an *initial value problem.* If material behavior is nonlinear, then the internal force vector $\{\mathbf{R}^{\text{int}}\}$ replaces $[\mathbf{K}]\{\mathbf{D}\}$ in Eq. 13.2-12, which is then called a *nonlinear initial value problem* (Eq. 13.2-13). For reasons that will become clear, in this section we consider only linear problems, for which $\{\mathbf{R}^{\text{int}}\} = [\mathbf{K}]\{\mathbf{D}\}$.

The *modal* or *mode superposition* method of analysis transforms Eq. 13.2-12 so that $\{\mathbf{D}\}$ and its time derivatives are replaced by $\{\mathbf{Z}\}$ and its time derivatives, where $\{\mathbf{Z}\}$ is a vector of modal amplitudes, or, in other words, a vector of generalized d.o.f. If damping is orthogonal, as Rayleigh damping is, then the transformed equations are uncoupled and each can be solved independently of all others. Solutions of these modal equations are superposed to yield the solution of the original problem [13.1,13.18,13.20,13.21]. Other solution methods for dynamic response are considered in Section 13.9 et seq.

Modal analysis is effective because of special properties that eigenvectors possess. These properties are orthogonality and linear independence (see Appendix C). Eigenvectors $\{\overline{\mathbf{D}}\}$ in Eq. 13.5-2 are orthogonal with respect to the (symmetric) mass and stiffness matrices. That is,

$$\{\overline{\mathbf{D}}\}_i^T[\mathbf{M}]\{\overline{\mathbf{D}}\}_j = 0 \qquad \text{and} \qquad \{\overline{\mathbf{D}}\}_i^T[\mathbf{K}]\{\overline{\mathbf{D}}\}_j = 0, \qquad \text{where} \quad i \neq j \tag{13.6-1}$$

If eigenvectors are normalized with respect to the mass matrix, then the quadratic product of the stiffness matrix yields the undamped natural frequencies. That is, by Eq. 13.5-4 with $\lambda = \omega^2$,

$$\text{if} \quad \{\overline{\mathbf{D}}\}_i^T[\mathbf{M}]\{\overline{\mathbf{D}}\}_i = 1 \qquad \text{then} \qquad \{\overline{\mathbf{D}}\}_i^T[\mathbf{K}]\{\overline{\mathbf{D}}\}_i = \omega_i^2 \tag{13.6-2}$$

(A vector is normalized by dividing each of its terms by the same constant.) If $[\boldsymbol{\phi}]$ is the *modal matrix,* that is, an n_{eq} by n_{eq} matrix whose columns are the normalized eigenvectors, then Eq. 13.6-2 yields

$$[\boldsymbol{\phi}]^T[\mathbf{M}][\boldsymbol{\phi}] = \lceil \mathbf{I} \rfloor \quad \text{and} \quad [\boldsymbol{\phi}]^T[\mathbf{K}][\boldsymbol{\phi}] = \lceil \boldsymbol{\omega}^2 \rfloor \tag{13.6-3}$$

where $\lceil \boldsymbol{\omega}^2 \rfloor$ is a diagonal matrix of the squared natural frequencies. It is called the *spectral matrix*.

Using the property of linear independence of eigenvectors, we can express any vector $\{\mathbf{D}\}$ as a linear combination of eigenvectors. Thus

$$\{\mathbf{D}\} = [\boldsymbol{\phi}]\{\mathbf{Z}\} \tag{13.6-4}$$

where the Z_i in vector $\{\mathbf{Z}\}$ state the proportion of each eigenvector in the transformation. Equation 13.6-4 represents a convenient d.o.f. transformation in which the Z_i are modal amplitudes and, similar to $\{\mathbf{D}\}$, are functions of time. No approximation is introduced by the transformation if all modes of the original system are retained in $[\boldsymbol{\phi}]$.

Mode Displacement Method. In the *mode displacement* method of modal analysis, we premultiply Eq. 13.2-12 by $[\boldsymbol{\phi}]^T$ and combine the result with Eq. 13.6-4. Thus, in view of Eqs. 13.6-3, the coefficients of $\{\ddot{\mathbf{Z}}\}$ and $\{\mathbf{Z}\}$ become the respective diagonal matrices $\lceil \mathbf{I} \rfloor$ and $\lceil \boldsymbol{\omega}^2 \rfloor$. If $[\mathbf{C}]$ is given by Eq. 13.4-1, or by other orthogonal forms [13.1,13.18], then $[\boldsymbol{\xi}] = [\boldsymbol{\phi}]^T[\mathbf{C}][\boldsymbol{\phi}]$ is also a diagonal matrix and the transformed equations are completely uncoupled. Using customary notation, we write the coefficients in a diagonal matrix $[\boldsymbol{\xi}]$ as $2\xi_i\omega_i$, where ξ_i is the fraction of critical damping in mode i. Thus

$$\ddot{Z}_i + 2\xi_i\omega_i\dot{Z}_i + \omega_i^2 Z_i = p_i, \quad \text{where} \quad p_i = \{\boldsymbol{\phi}\}_i^T\{\mathbf{R}^{\text{ext}}\} \tag{13.6-5}$$

in which $\{\boldsymbol{\phi}\}_i$ is the ith column of $[\boldsymbol{\phi}]$. There are as many uncoupled ordinary differential equations as there are d.o.f. In each equation, p_i is a known function of time. Initial values for Eq. 13.6-5 are obtained from known initial values of $\{\mathbf{D}\}$ and $\{\dot{\mathbf{D}}\}$ as follows. Premultiply both sides of Eq. 13.6-4 by $[\boldsymbol{\phi}]^T[\mathbf{M}]$ and take note of Eqs. 13.6-3. Thus

$$\{\mathbf{Z}(t = 0)\} = [\boldsymbol{\phi}]^T[\mathbf{M}]\{\mathbf{D}(t = 0)\} \quad \text{and} \quad \{\dot{\mathbf{Z}}(t = 0)\} = [\boldsymbol{\phi}]^T[\mathbf{M}]\{\dot{\mathbf{D}}(t = 0)\} \tag{13.6-6}$$

When $\{\mathbf{Z}\}$ has been determined as a function of time from Eq. 13.6-5, $\{\mathbf{D}\}$ is obtained from Eq. 13.6-4.

There are many ways to time-integrate Eqs. 13.6-5. In fact, Eq. 13.6-5 can be solved exactly for $Z_i(t)$. If p_i is piecewise linear in time, then exact solutions can be written for Z_i and $\dot{Z}_i$ in terms of $e^{-\xi\omega t}\sin\omega t$ and $e^{-\xi\omega t}\cos\omega t$. However, for more general loading, an exact solution can be tedious and it is more effective to use a direct integration method, as discussed in Section 13.9 et seq.

At first glance, calculating $[\boldsymbol{\phi}]$ in Eq. 13.6-4 seems to be a prohibitive computational expense. But for many problems, the higher-frequency modes participate little in the structural response and therefore *only a small number of low-frequency modes need be used*. Thus, only the first m equations of Eq. 13.6-5, where typically $m \ll n_{\text{eq}}$, are solved and the transformation Eq. 13.6-4 is approximated by

$$\{\mathbf{D}\} \approx \sum_{i=1}^{m} \{\boldsymbol{\phi}\}_i Z_i \quad \text{where} \quad m < n_{\text{eq}} \tag{13.6-7}$$

in which $\{\boldsymbol{\phi}\}_i$ is the ith normalized eigenvector. A measure of the error at time t is denoted by $e(t)$. It can be quantified [13.18] by

$$e(t) \equiv \frac{\|\{\mathbf{R}^{\text{ext}}\} - [\mathbf{M}]\{\ddot{\mathbf{D}}\} - [\mathbf{C}]\{\dot{\mathbf{D}}\} - [\mathbf{K}]\{\mathbf{D}\}\|}{\|\{\mathbf{R}^{\text{ext}}\}\|} \tag{13.6-8}$$

where $\{\mathbf{D}\}$ and its time derivatives are obtained from Eq. 13.6-7 and $\|\ \ \|$ denotes any vector norm. It is assumed in writing Eq. 13.6-8 that $\{\mathbf{R}^{\text{ext}}\} \neq \{\mathbf{0}\}$ at the particular instant $e(t)$ is computed. For an accurate analysis, $e(t)$ should be small (1% or less, as a conjecture) for the entire duration of analysis.

In many structural dynamics problems, more modes participate in the quasistatic response than in the dynamic response. The mode displacement method may have difficulty in reproducing quasistatic deflection shapes of structures when m is small. A modification of the mode displacement method, discussed next, removes this deficiency. Additional comments appear in Section 13.14.

Mode Acceleration Method. In the *mode acceleration method,* we perform modal transformation on only the inertial and viscous terms of Eq. 13.2-12 [13.2, 13.22,13.23,13.24]. This yields

$$[\mathbf{M}][\boldsymbol{\phi}]\{\ddot{\mathbf{Z}}\} + [\mathbf{C}][\boldsymbol{\phi}]\{\dot{\mathbf{Z}}\} + [\mathbf{K}]\{\mathbf{D}\} = \{\mathbf{R}^{\text{ext}}\} \tag{13.6-9}$$

If no rigid-body modes are possible and $[\mathbf{K}]$ is properly formed, then $[\mathbf{K}]^{-1}$ exists and Eq. 13.6-9 can be solved for $\{\mathbf{D}\}$. From Eq. 13.6-9,

$$\{\mathbf{D}\} = [\mathbf{K}]^{-1}\{\mathbf{R}^{\text{ext}}\} - [\mathbf{K}]^{-1}([\mathbf{M}][\boldsymbol{\phi}]\{\ddot{\mathbf{Z}}\} + [\mathbf{C}][\boldsymbol{\phi}]\{\dot{\mathbf{Z}}\}) \tag{13.6-10}$$

Equation 13.6-3 yields $[\mathbf{K}]^{-1}[\boldsymbol{\phi}]^{-T} = [\boldsymbol{\phi}]\lceil \boldsymbol{\omega}^2 \rfloor^{-1}$. Hence, if $[\mathbf{C}]$ is orthogonal, Eq. 13.6-10 can be written as

$$\{\mathbf{D}\} = [\mathbf{K}]^{-1}\{\mathbf{R}^{\text{ext}}\} - [\boldsymbol{\phi}]\lceil \boldsymbol{\omega}^2 \rfloor^{-1}(\{\ddot{\mathbf{Z}}\} + \lceil \boldsymbol{\xi} \rfloor\{\dot{\mathbf{Z}}\}) \tag{13.6-11}$$

where $\lceil \boldsymbol{\xi} \rfloor$ is a diagonal matrix with ith diagonal coefficient $2\xi_i\omega_i$ (proof of Eq. 13.6-11 is left as an exercise). In Eq. 13.6-11, $[\mathbf{K}]$ is the original n_{eq} by n_{eq} stiffness matrix of Eq. 13.2-12. Here $[\boldsymbol{\phi}]$ is n_{eq} by n_{eq}, $\lceil \boldsymbol{\omega}^2 \rfloor$ and $\lceil \boldsymbol{\xi} \rfloor$ are n_{eq} by n_{eq} diagonal matrices, and $\{\mathbf{Z}\}$ is an n_{eq} by 1 vector of modal amplitudes. As with mode displacement analysis, only a small number of modes, m, is usually required for accurate results. Hence, the matrix multiplication indicated in the second term of Eq. 13.6-11 is carried out over only the lowest m eigenvectors and Eq. 13.6-11 is approximated by

$$\{\mathbf{D}\} \approx [\mathbf{K}]^{-1}\{\mathbf{R}^{\text{ext}}\} - \sum_{i=1}^{m} \{\boldsymbol{\phi}\}_i \left(\frac{1}{\omega_i^2}\ddot{Z}_i + \frac{2\xi_i}{\omega_i}\dot{Z}_i\right), \qquad \text{where} \quad m < n_{\text{eq}} \tag{13.6-12}$$

The first term on the right-hand side of Eq. 13.6-12 represents the quasistatic response, and the second term represents the dynamic correction attributable to inertia and viscous effects. To implement the mode acceleration method, we solve Eq. 13.6-5 for $\dot{Z}_i$ and $\ddot{Z}_i$, $i = 1, 2, \ldots, m$ exactly as we would in a mode displacement analysis (note that we do not need Z_i). Then Eq. 13.6-12 is used for superposition (rather than Eq. 13.6-7 as in the mode displacement method) to obtain $\{\mathbf{D}\}$ as a function of time.

Obviously, the mode acceleration algorithm is adept at reproducing the quasistatic structure deflection shape. If the magnitude of the external load (but not its distribution) varies with time, then $\{\mathbf{R}^{ext}\} = s(t)\{\mathbf{S}\}$, where $\{\mathbf{S}\}$ is time-independent and describes the distribution of external load, and $s(t)$ is a time-dependent scalar that describes the amplitude of the external load. We call such loading *proportional loading*. Thus the quasistatic displacement $[\mathbf{K}]^{-1}\{\mathbf{S}\}$ need only be scaled by $s(t)$ at each instant in time. Usually, fewer modes are required in mode acceleration than in a mode displacement analysis of equivalent accuracy [13.24]. On the other hand, for nonproportional loading the mode acceleration method requires the solution of simultaneous equations at each step of the solution (only forward and back substitution after $[\mathbf{K}]$ has been factored). If $\{\mathbf{R}^{ext}\} = \{\mathbf{0}\}$, both methods yield identical results for the same m.

If rigid-body modes are possible, $[\mathbf{K}]$ is singular and the mode acceleration method cannot be employed in the straightforward manner indicated by Eqs. 13.6-11 and 13.6-12. Discussion of this matter appears in [13.2].

Superposition of Ritz Vectors. A disadvantage of the mode displacement and mode acceleration methods is that computation of eigenvectors is expensive. For large structures it is often the most costly part of a dynamic response analysis. Using *Ritz vectors* (which may also be called *basis vectors*), we make a transformation analogous to Eqs. 13.6-4 and 13.6-7. Ritz vectors do not have all of the desirable properties of eigenvectors but are much more economical to compute.

Time-history analysis by superposition of Ritz vectors is a type of *Rayleigh–Ritz* analysis. A *Ritz vector* d.o.f. transformation

$$\{\mathbf{D}\} = [\mathbf{W}]\{\mathbf{y}\} \qquad (13.6\text{-}13)$$

is used where $[\mathbf{W}]$ is an n_{eq} by m matrix whose columns are Ritz vectors $\{\mathbf{w}\}_i$ and $\{\mathbf{y}\}$ is a vector of generalized coordinates. Thus $[\mathbf{W}] = [\mathbf{w}_1 \quad \mathbf{w}_2 \quad \ldots \quad \mathbf{w}_m]$ and $\{\mathbf{y}\}$ is an m by 1 vector whose terms y_i state the proportion of $\{\mathbf{w}\}_i$ in the transformation. The number m of Ritz vectors is chosen by the analyst. There are many ways to obtain the Ritz vectors in Eq. 13.6-13. Some of these procedures are rather arbitrary. If the Ritz vectors are the lowest m eigenvectors of Eq. 13.5-2 and are normalized according to Eq. 13.6-2, then Eq. 13.6-13 reduces to Eq. 13.6-7 of the mode displacement method. It is not necessary that Ritz vectors be close approximations of eigenvectors. Ideally, however, Ritz vectors are linear combinations of the lowest eigenvectors. A reliable method for generating good Ritz vectors (without solving an eigenproblem) is to determine $[\mathbf{W}]$ by solving

$$[\mathbf{K}][\mathbf{W}] = [\mathbf{R}] \qquad (13.6\text{-}14)$$

where $[\mathbf{K}]$ is the n_{eq} by n_{eq} stiffness matrix of the complete structure and $[\mathbf{R}]$ is an n_{eq} by m matrix whose columns are linearly independent load patterns selected by the analyst to excite important (i.e., lower) displacement modes of the structure. In writing Eq. 13.6-14, we assume that $[\mathbf{K}]$ is nonsingular. For treating problems with rigid-body modes, a modified method is required.

The use of Ritz vectors is not limited to problems of dynamics. Ritz vectors constitute a "reduced basis" that serves to reduce the cost of repetitive calculation cycles, which appear in modal methods of dynamics, in nonlinear static problems, and in design optimization.

Ritz Vectors in Time-History Analysis. Often, the load patterns used in Eq. 13.6-14 are obtained from actual load patterns applied to the structure. In dynamic analysis, a question that immediately arises is what load patterns are most effectively used in Eq. 13.6-14 when the loads that shake a structure, $\{\mathbf{R}^{\text{ext}}\}$, are time-dependent. A method called superposition of Ritz vectors effectively addresses this question [13.25]. It is necessary that the external load be representable as a superposition of proportional loads—that is,

$$\{\mathbf{R}^{\text{ext}}\} = \sum_{j=1}^{\ell} s_j(t)\{\mathbf{S}\}_j \qquad (13.6\text{-}15)$$

where $\{\mathbf{S}\}_j$ describes the distribution of the jth load component, $s_j(t)$ is a scalar that describes the time-dependent amplitude of $\{\mathbf{S}\}_j$, and ℓ is the number of $\{\mathbf{S}\}_j$ needed for superposition. For many practical problems ℓ is small.

The dynamic response $\{\mathbf{D}\}_j$ to load component $s_j(t)\{\mathbf{S}\}_j$ is computed independently for each j. Then the total structural response is obtained by superposition

$$\{\mathbf{D}\} = \sum_{j=1}^{\ell} \{\mathbf{D}\}_j \qquad (13.6\text{-}16)$$

In the response analysis for each $\{\mathbf{D}\}_j$, a Ritz vector d.o.f. transformation

$$\{\mathbf{D}\}_j = [\mathbf{W}]_j\{\mathbf{y}\}_j \qquad (13.6\text{-}17)$$

is used where $[\mathbf{W}]_j$ is an n_{eq} by m matrix of Ritz vectors. An algorithm for generating $[\mathbf{M}]$-orthogonal Ritz vectors, when given a load pattern $\{\mathbf{S}\}_j$, is shown in Table 13.6-1. In this algorithm, the first Ritz vector $\{\mathbf{w}\}_1$ is proportional to the

TABLE 13.6-1. Computational procedure for generation of [M]-orthogonal Ritz vectors [13.25].

1. Obtain n_{eq} by n_{eq} mass and stiffness matrices, $[\mathbf{M}]$ and $[\mathbf{K}]$.
2. Factor the stiffness matrix; for example, $[\mathbf{K}] = [\mathbf{L}][\mathbf{L}]^T$.
3. For each proportional load component $\{\mathbf{S}\}_j$, $j = 1, 2, \ldots, \ell$, compute $[\mathbf{W}]_j = [\mathbf{w}_1 \ \ \mathbf{w}_2 \ \ \ldots \ \ \mathbf{w}_m]$ as follows:

   3.1 Solve for the first Ritz vector, $\{\mathbf{w}\}_1$:

   $[\mathbf{K}]\{\mathbf{w}^*\}_1 = \{\mathbf{S}\}_j$ — solve for $\{\mathbf{w}^*\}_1$

   $\{\mathbf{w}\}_1^T[\mathbf{M}]\{\mathbf{w}\}_1 = 1$ — normalize $\{\mathbf{w}^*\}_1$ to yield $\{\mathbf{w}\}_1$

   3.2 Solve for additional Ritz vectors, $\{\mathbf{w}\}_i$; $i = 2, 3, \ldots, m$:

   $[\mathbf{K}]\{\mathbf{w}^*\}_i = [\mathbf{M}]\{\mathbf{w}\}_{i-1}$ — solve for $\{\mathbf{w}^*\}_i$

   $\{\mathbf{w}^{**}\}_i = \{\mathbf{w}^*\}_i - \sum_{k=1}^{i-1} \{\mathbf{w}\}_k^T[\mathbf{M}]\{\mathbf{w}^*\}_i\{\mathbf{w}\}_k$ — $[\mathbf{M}]$-orthogonalize $\{\mathbf{w}^*\}_i$ to yield $\{\mathbf{w}^{**}\}_i$

   $\{\mathbf{w}\}_i^T[\mathbf{M}]\{\mathbf{w}\}_i = 1$ — normalize $\{\mathbf{w}^{**}\}_i$ to yield $\{\mathbf{w}\}_i$

   3.3 Assemble Ritz vectors $\{\mathbf{w}\}_i$ into n_{eq} by m matrix $[\mathbf{W}]_j$.

   3.4 Next load component; $j \leftarrow j + 1$, go to Step 3.1.

quasistatic deflection shape of the structure under loads $\{\mathbf{S}\}_j$. To obtain the second Ritz vector, we impose $\{\mathbf{w}\}_1$ as a nodal acceleration vector; hence, $\{\mathbf{w}\}_2$ is proportional to the quasistatic deflection shape for inertial loads $[\mathbf{M}]\{\mathbf{w}\}_1$. To obtain the third Ritz vector, we impose $\{\mathbf{w}\}_2$ as a nodal acceleration, and so on. If loads $\{\mathbf{S}\}_j$ are zero, as for a structure that moves freely after initial velocities are prescribed, the procedure of Table 13.6-1 must be modified. A possible modification is to simply prescribe $\{\mathbf{w}\}_1$ (e.g., as null except for unity corresponding to a d.o.f. expected to have significant displacement), then go to Step 3.2.

Once a mass-matrix-orthogonal $[\mathbf{W}]_j$ is obtained, Eqs. 13.2-12, 13.6-16, and 13.6-17 are combined and premultiplied by $[\mathbf{W}]_j^T$ to yield

$$\{\ddot{\mathbf{y}}\}_j + [\mathbf{C}]_j\{\dot{\mathbf{y}}\}_j + [\mathbf{K}]_j\{\mathbf{y}\}_j = \{\mathbf{P}\}_j \qquad (13.6\text{-}18)$$

where the transformed stiffness matrix, damping matrix, and load vector are

$$[\mathbf{K}]_j = [\mathbf{W}]_j^T[\mathbf{K}][\mathbf{W}]_j \qquad (13.6\text{-}19a)$$

$$[\mathbf{C}]_j = [\mathbf{W}]_j^T[\mathbf{C}][\mathbf{W}]_j \qquad (13.6\text{-}19b)$$

$$\{\mathbf{P}\}_j = s_j(t)[\mathbf{W}]_j^T\{\mathbf{S}\}_j \qquad (13.6\text{-}19c)$$

After the $\{\mathbf{y}\}_j$ are known (as functions of time) for all j, Eqs. 13.6-16 and 13.6-17 yield $\{\mathbf{D}\} = \{\mathbf{D}(t)\}$. Note that we must generate a $[\mathbf{W}]_j$ and also solve Eqs. 13.6-18 as many times as there are load components in Eq. 13.6-15.

Matrices $[\mathbf{K}]_j$ and $[\mathbf{C}]_j$ have dimension m by m. Therefore, if $m \ll n_{eq}$, Eqs. 13.6-18 represent a much smaller system of simultaneous ordinary differential equations than the original system, Eq. 13.2-12. In general, matrices $[\mathbf{K}]_j$ and $[\mathbf{C}]_j$ are full (but $[\mathbf{M}]_j$ is a unit matrix according to the procedure of Table 13.6-1). Equations 13.6-18 are usually solved for $\{\mathbf{y}\}_j$ and its time derivatives by a direct integration method. Optionally, the Ritz vectors can be made to be also stiffness-matrix-orthogonal using the procedure of [13.25]. This entails greater expense in forming $[\mathbf{W}]_j$, but then $[\mathbf{K}]_j$ is diagonal (as is $[\mathbf{C}]_j$ if damping is orthogonal). Equations 13.6-18 are then completely uncoupled and can be solved independently, either exactly or approximately.

Several features of Ritz mode superposition are apparent from Table 13.6-1. Each Ritz vector is obtained by solving a system of simultaneous algebraic equations. Each solution is relatively inexpensive once $[\mathbf{K}]$ has been factored. As with the mode acceleration method, static structure deflection shapes are accurately modeled. However, since $[\mathbf{K}]$ must be nonsingular, the method cannot treat problems with rigid-body modes unless modified. When the number of loads ℓ necessary for the superposition in Eq. 13.6-13 becomes large, the method loses its attractiveness since an independent set of Ritz vectors $[\mathbf{W}]_j$ must be determined and an independent system of ordinary differential equations, Eqs. 13.6-18, must be solved for each load component. However, for many problems only a few load components are necessary. The reduction in the size of the original problem can be remarkable. For example, for earthquake shaking of structures, the number of Ritz vectors necessary for accurate response analysis is almost always less than 50, whereas equations in the original system may number several thousand [13.25].

Nonlinear Problems. In material-nonlinear problems, [**K**] and [**C**] are time-dependent. Usually [**M**] does not change with time. Modal techniques employ superposition and hence are often assumed to be inapplicable to nonlinear problems. However, nonlinear problems can be accommodated if all nonlinearities are treated as pseudoloads and incorporated with the external load $\{\mathbf{R}^{\text{ext}}\}$. This approach requires that modes be superposed at each time step to obtain {**D**} (and if necessary $\{\dot{\mathbf{D}}\}$) so that the material constitutive law can be evaluated. Pseudoloads are then calculated and transformed back to modal equations, the solution of which is then incremented in time by one step. Strictly speaking, when response is nonlinear $[\boldsymbol{\phi}]$ does not represent normal modes of vibration. Equation 13.6-4 should then be viewed as simply a coordinate transformation [13.26,13.27]. For structural dynamics problems with nonlinearities that are mild and localized in the mesh, mode displacement superposition has been used, sometimes effectively. For severe nonlinearities, convergence of the pseudoload approach is poor. For most nonlinear problems, direct time integration methods are preferable to superposition methods.

13.7 MASS CONDENSATION. GUYAN REDUCTION

In static analyses, problems with 10,000 d.o.f. or more are common. In dynamic analyses in which we ask for natural frequencies and mode shapes, problems with only 1000 d.o.f. can be difficult. The major difficulty is the expense of computing eigenvalues and eigenvectors.

However, *condensation* can be employed to reduce the number of d.o.f., which reduces the expense of computing eigenvalues and eigenvectors and the expense of subsequent calculations. Condensation is detrimental to accuracy, but negligibly so if properly used. However, one may not wish to use condensation if [**M**] has been obtained by optimal lumping, as the use of optimal lumping implies that particular effort is being made to ensure accuracy.

The literature is vast so we present only important ideas [13.1,13.18,13.28]. In Section 13.8, the component mode synthesis method is described, which has significant condensation features.

We present a condensation algorithm known as *Guyan reduction, mass condensation,* or *eigenvalue economization* [13.29]. Equation 13.5-2 is partitioned and written as

$$\left(\begin{bmatrix} \mathbf{K}_{mm} & \mathbf{K}_{ms} \\ \mathbf{K}_{ms}^T & \mathbf{K}_{ss} \end{bmatrix} - \lambda \begin{bmatrix} \mathbf{M}_{mm} & \mathbf{M}_{ms} \\ \mathbf{M}_{ms}^T & \mathbf{M}_{ss} \end{bmatrix}\right) \begin{Bmatrix} \overline{\mathbf{D}}_m \\ \overline{\mathbf{D}}_s \end{Bmatrix} = \begin{Bmatrix} \mathbf{0} \\ \mathbf{0} \end{Bmatrix} \tag{13.7-1}$$

where the m "master" d.o.f. $\{\overline{\mathbf{D}}_m\}$ are to be retained and the s "slave" d.o.f. $\{\overline{\mathbf{D}}_s\}$ are to be removed by condensation. The principal assumption in Guyan reduction is that for *the lowest-frequency modes, inertia forces on slave d.o.f. are much less important than elastic forces transmitted by the master d.o.f.* (subsequently, this assumption will be used as a guideline for selecting whether a d.o.f. should be a master or a slave). In other words, slave d.o.f. are assumed to move quasi-statically in response to the motion of master d.o.f. Thus, in order to obtain a relation between $\{\mathbf{D}_s\}$ and $\{\mathbf{D}_m\}$, we temporarily ignore all mass but $[\mathbf{M}_{mm}]$. From the lower partition of Eq. 13.7-1,

$$\underset{s\times 1}{\{\overline{\mathbf{D}}_s\}} = -\underset{s\times s}{[\mathbf{K}_{ss}]^{-1}}\underset{s\times m}{[\mathbf{K}_{ms}]^T}\underset{m\times 1}{\{\overline{\mathbf{D}}_m\}} \tag{13.7-2}$$

Accordingly, with $n_{eq} = m + s$ and $\lceil\mathbf{I}\rfloor$ an m by m identity matrix,

$$\underset{n_{eq}\times 1}{\begin{Bmatrix}\overline{\mathbf{D}}_m \\ \overline{\mathbf{D}}_s\end{Bmatrix}} = \underset{n_{eq}\times m}{[\mathbf{T}]}\underset{m\times 1}{\{\overline{\mathbf{D}}_m\}}, \qquad \text{where} \quad [\mathbf{T}] = \begin{bmatrix}\mathbf{I} \\ -\mathbf{K}_{ss}^{-1}\mathbf{K}_{ms}^T\end{bmatrix} \tag{13.7-3}$$

Substitution of Eq. 13.7-3 into Eq. 13.7-1 and premultiplication by $[\mathbf{T}]^T$ yields the condensed eigenproblem

$$([\mathbf{K}_r] - \lambda[\mathbf{M}_r])\{\overline{\mathbf{D}}_m\} = \{\mathbf{0}\} \tag{13.7-4}$$

where the reduced matrices are symmetric and are given by

$$\underset{m\times m}{[\mathbf{K}_r]} = [\mathbf{T}]^T[\mathbf{K}][\mathbf{T}] \qquad \text{and} \qquad \underset{m\times m}{[\mathbf{M}_r]} = [\mathbf{T}]^T[\mathbf{M}][\mathbf{T}] \tag{13.7-5}$$

Note that even if $[\mathbf{K}]$ is banded and $[\mathbf{M}]$ is diagonal, $[\mathbf{K}_r]$ and $[\mathbf{M}_r]$ are in general full and $[\mathbf{M}_r]$ is a combination of both mass and stiffness coefficients. If damping matrix $[\mathbf{C}]$ and external loads $\{\mathbf{R}^{ext}\}$ appear in the equation of motion, then condensed damping matrix $[\mathbf{C}_r] = [\mathbf{T}]^T[\mathbf{C}][\mathbf{T}]$ and condensed external loads $\{\mathbf{R}_r^{ext}\} = [\mathbf{T}]^T\{\mathbf{R}^{ext}\}$ appear in the reduced equation of motion

$$[\mathbf{M}_r]\{\ddot{\mathbf{D}}_m\} + [\mathbf{C}_r]\{\dot{\mathbf{D}}_m\} + [\mathbf{K}_r]\{\mathbf{D}_m\} = \{\mathbf{R}_r^{ext}\} \tag{13.7-6}$$

where $\{\mathbf{D}_m\}$ represents the displacements of master d.o.f.

If $[\mathbf{M}]$ is diagonal and slave d.o.f. carry no mass, then $[\mathbf{K}_r]$ is the same matrix as produced by "static condensation," $[\mathbf{M}_r]$ contains the nonzero M_{ii} of $[\mathbf{M}]$, and condensation produces no loss of accuracy. (Static condensation is discussed in Section 8.1.)

In vibration problems, when eigenvalues λ_i and eigenvectors $\{\overline{\mathbf{D}}_m\}_i$ of the reduced system are known, slave modes $\{\overline{\mathbf{D}}_s\}_i$ can be recovered by use of Eq. 13.7-2. However, it is more accurate to recover $\{\overline{\mathbf{D}}_s\}_i$ from the lower partition of Eq. 13.7-1 in which advantage is taken of the previously neglected slave node masses. Thus

$$\{\overline{\mathbf{D}}_s\}_i = -[\mathbf{K}_{ss} - \lambda_i\mathbf{M}_{ss}]^{-1}[\mathbf{K}_{ms}^T - \lambda_i\mathbf{M}_{ms}^T]\{\overline{\mathbf{D}}_m\}_i \tag{13.7-7}$$

Reference 13.34 includes another form of this expression that is more computationally efficient.

Remarks. Why create a detailed finite element model if we subsequently intend to discard many d.o.f. by condensation? Because coarse finite element discretizations usually do not have sufficient detail for accurate stiffness and mass representations. Furthermore, accurate stress computations usually require fine discretizations [13.30].

Because reduction destroys any preexisting band structure and sparsity, m must be considerably smaller than n_{eq} for Guyan reduction to be cost-effective. If original matrices $[\mathbf{M}]$ and $[\mathbf{K}]$ have unusually small bandwidths, it may be prudent to avoid condensation because a gain in efficiency can only be obtained by taking $m \ll n_{eq}$, in which case accuracy may suffer.

If applied only to [**K**], Eq. 13.7-3 becomes the static condensation algorithm, Eq. 8.1-3. There are computational advantages to generating [**T**] in terms of flexibility instead of stiffness [13.28,13.31,13.32].

We see that [**T**] plays the same role as the matrix $[\mathbf{W}]_j$ of Ritz vectors in Eqs. 13.6-19. Indeed, [**T**] can be regarded as a matrix of Ritz vectors and Guyan reduction as a Rayleigh–Ritz method. However, [**T**] and $[\mathbf{W}]_j$ are not the same: [**T**] is obtained more efficiently, without reference to the applied loading, but is not mass-matrix-orthogonal. $[\mathbf{W}]_j$ is more adept at reproducing quasistatic structure deflection shapes, and since it is obtained from the applied loading, it can usually have fewer columns than [**T**] for a given level of accuracy.

If [**M**] is lumped so that $\lceil \mathbf{M}_{mm} \rfloor$ contains *all* the nonzero masses, and $[\mathbf{M}_{ss}]$ is null, then Eq. 13.7-2 follows without the necessity of our principal assumption. This suggests another approach to condensation: the analyst can lump mass at only the d.o.f. to be retained as masters. However, this approach requires experience and is less accurate than condensation with a more populated mass matrix.

In going from Eq. 13.7-1 to Eq. 13.7-4, eigenvalues are raised because constraints are imposed [13.33]. This behavior, and the considerable accuracy possible when $m \ll n_{eq}$, are shown in Fig. 13.7-1. The higher eigenvalues of the original system are absent from the reduced system because several d.o.f. have been discarded. After $\{\overline{\mathbf{D}}_s\}$ has been recovered by use of Eq. 13.7-7, an eigenvalue can be improved by substituting the eigenvector $\{\mathbf{D}\} = \{\overline{\mathbf{D}}_m, \overline{\mathbf{D}}_s\}$ into the Rayleigh quotient of the full system, Eq. 13.5-4.

Master d.o.f. should be those for which inertia is most important. Such d.o.f. have a large mass-to-stiffness ratio. Thus rotational d.o.f. rarely appear as masters. A master d.o.f. should be retained at each node that carries a time-varying applied load or has a time-varying prescribed displacement. Master d.o.f. should not be clustered in one area of the mesh. If they are, some vibration modes may be almost linearly dependent. This is not of major concern, but, if ignored, can lead to the disconcerting appearance of negative eigenvalues in the highest modes of the reduced eigenproblem [13.35].

The selection of master and slave d.o.f. can be automated as follows [13.36]. Diagonal coefficients of [**K**] and [**M**] are scanned, and the d.o.f. i for which K_{ii}/M_{ii}

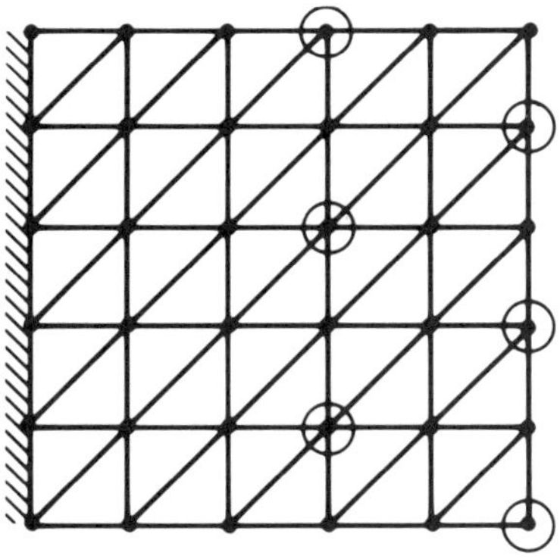

Full system, 90 d.o.f. (one displacement and two rotations at each node)	$\omega_1 = 3.469$ $\omega_2 = 8.535$ $\omega_3 = 21.450$ $\omega_4 = 27.059$
Reduced system, 6 master d.o.f. (lateral displacements at nodes circled)	$\omega_1 = 3.473$ $\omega_2 = 8.604$ $\omega_3 = 22.690$ $\omega_4 = 29.490$

Figure 13.7-1. First four vibration frequencies of a thin, square cantilever plate [13.37]. The analysis uses triangular plate elements and consistent mass matrices.

is largest is selected as the first slave. In case of a tie, the first d.o.f. encountered is taken as a slave. Then [K] and [M] are condensed (by one order). The condensed matrices are now scanned, the largest K_{rii}/M_{rii} is selected as the next slave, and another condensation is performed. This process repeats until a user-specified number of d.o.f. remain. These are the masters, chosen in a near-optimal way.

The number of masters can be chosen automatically by specifying a cut-off frequency, ω_c [13.44]. This frequency is taken to be about three times the highest frequency of interest in the excitation and/or structural response. Modes of the structure having frequencies greater than ω_c are quasistatic with respect to the excitation and can therefore be neglected in the dynamic response, although they may participate in the quasistatic response. Automatic selection of masters and slaves proceeds as described above except that condensation is terminated when the largest ratio K_{rii}/M_{rii} is less than ω_c^2. Another procedure that may be more efficient is given in [13.38].

One may combine manual selection of some master d.o.f. with automatic selection of the rest. A motivation would be to retain as masters those d.o.f. subjected to time-varying loads. It would be very inconvenient to eliminate these d.o.f. as slaves.

The number of master d.o.f., m, may be as low as 1/10 or 1/20 the total number of d.o.f. If there are two or three times as many master d.o.f. as eigenvalues to be computed, the highest computed eigenvalue may err by less than 10%. These estimates are rough and strongly problem-dependent [13.36,13.37].

Mass condensation yields good results if the choice of master d.o.f. is good and if the ratio of master d.o.f. to total d.o.f. (m/n_{eq}) is not too small. As alternatives, the subspace iteration and Lanczos methods of eigenproblem solution employ condensation techniques that may be more reliable than Eq. 13.7-3.

Example: Beam Vibration. Nonzero d.o.f. in the problem of Fig. 13.7-2 are w_1 and θ_2. Accordingly, using the stiffness and mass matrices of Eqs. 4.2-5 and 13.3-2, respectively, we obtain the eigenproblem

$$\left(\frac{EI}{L^3}\begin{bmatrix}12 & 6L\\ 6L & 4L^2\end{bmatrix} - \frac{\omega^2 m}{420}\begin{bmatrix}156 & -13L\\ -13L & 4L^2\end{bmatrix}\right)\begin{Bmatrix}\bar{w}_1\\ \theta_2\end{Bmatrix} = \begin{Bmatrix}0\\ 0\end{Bmatrix} \tag{13.7-8}$$

where $m = \rho AL$ is the mass of the beam. The eigenvalues of this two-d.o.f. system are

$$\omega_1^2 = 6.1362\frac{EI}{mL^3} \quad \text{and} \quad \omega_2^2 = 758.17\frac{EI}{mL^3} \tag{13.7-9}$$

From continuous beam theory, the two lowest frequencies are

$$\omega_1^2 = 6.0881\frac{EI}{mL^3} \quad \text{and} \quad \omega_2^2 = 493.13\frac{EI}{mL^3} \tag{13.7-10}$$

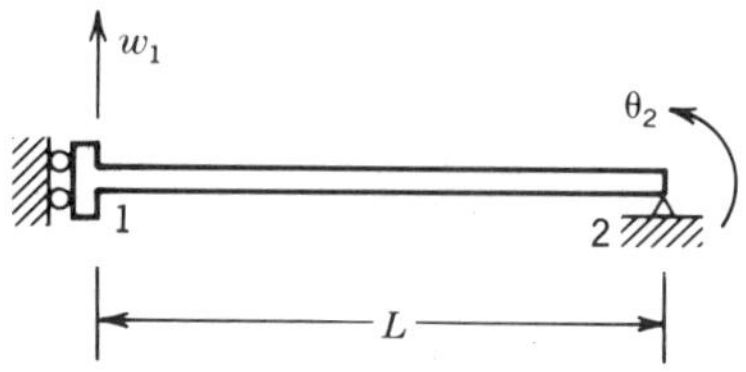

Figure 13.7-2. A uniform beam. The left end is allowed to displace but not to rotate. The right end is simply supported.

and, as expected, are lower than the frequencies of Eqs. 13.7-9. With $\overline{w}_1 = 1$, the amplitude of $\overline{\theta}_2$ in mode 1, from Eq. 13.7-8 and beam theory, respectively, is

$$\overline{\theta}_2 = -1.5704/L \text{ (approximate)} \qquad \overline{\theta}_2 = -1.5708/L \text{ (exact)} \tag{13.7-11}$$

To condense the size of the eigenproblem of Eq. 13.7-8, we employ Guyan reduction and select $\overline{w}_1$ as master and $\overline{\theta}_2$ as slave. Using Eqs. 13.7-3 and 13.7-5, with $K_{ss} = 4EI/L$ and $K_{ms} = 6EI/L^2$, we obtain

$$[\mathbf{T}] = \begin{bmatrix} 1 \\ -3/2L \end{bmatrix} \quad \text{and} \quad \left(\frac{3EI}{L^3} - \omega^2 \frac{204\, m}{420}\right) \overline{w}_1 = 0 \tag{13.7-12}$$

from which $\omega_1^2 = 6.1765EI/mL^3$. This is the only eigenvalue obtainable. As expected, it is higher than ω_1^2 in Eq. 13.7-9. With $\overline{w}_1 = 1$, recovery of $\overline{\theta}_2$ from Eqs. 13.7-2 and 13.7-7, respectively, yields

$$\overline{\theta}_2 = -1.5000/L \quad \text{and} \quad \overline{\theta}_2 = -1.5709/L \tag{13.7-13}$$

An improved estimate of ω_1^2 can be found by substituting $\{\overline{\mathbf{D}}\} = \lfloor 1 \quad -1.5709/L \rfloor^T$ into the Rayleigh quotient (Eq. 13.5-4), with [**K**] and [**M**] taken from Eq. 13.7-8. The resulting eigenvalue is the same as ω_1^2 in Eq. 13.7-9.

A lumped-mass solution for the fundamental vibration frequency is easily obtained for this problem by placing a mass particle $m_1 = m/2$ at node 1. From beam theory, $w_1 = P_1L^3/3EI$, so $K = P_1/w_1 = 3EI/L^3$. Then $\omega^2 = K/m_1 = 6EI/mL^3$. Note that this value is not an upper bound.

13.8 COMPONENT MODE SYNTHESIS

General Remarks. In *component mode synthesis,* or simply *modal synthesis,* a structure is subdivided into *components* or *substructures,* each of which is analyzed independently for natural frequencies and, more importantly, for mode shapes. The component mode shapes are then "assembled" to give displacement shapes or load patterns (either interpretation is possible) of the original structure. These shapes or patterns are not eigenvectors of the original structure, but give rise to Ritz vectors that are subsequently used to transform the original displacement d.o.f. to generalized d.o.f. Thus the size of the system matrices is reduced. Eigenanalysis and/or time-history analysis of the reduced system equations can be performed much more economically and usually with surprising accuracy [13.39].

Component mode synthesis can be regarded as an alternative to Guyan reduction. More importantly, modal synthesis has the managerial advantage of allowing different design groups to work on different parts of a large structure, as discussed in Section 8.14 with respect to static substructuring. There are many methods of component mode synthesis and an extensive literature [13.2,13.40]. In what follows we present only fundamentals.

We will use a "reduced basis" [**W**] to replace d.o.f. {**D**} of the entire (or assembled) structure by a vector of generalized coordinates {**y**} that contains fewer d.o.f. than {**D**}. The relation between {**D**} and {**y**} is stated by Eq. 13.6-13, here repeated:

$$\underset{n_{eq}\times 1}{\{\mathbf{D}\}} = \underset{n_{eq}\times m}{[\mathbf{W}]}\ \underset{m\times 1}{\{\mathbf{y}\}} \tag{13.6-13}$$

in which m is considerably less than n_{eq}. Matrix [**W**] is an array of Ritz vectors. The question of how to establish [**W**] occupies much of our subsequent discussion. Equations 13.2-12 and 13.6-13 yield the reduced dynamic problem

$$[\mathbf{M}_r]\{\ddot{\mathbf{y}}\} + [\mathbf{C}_r]\{\dot{\mathbf{y}}\} + [\mathbf{K}_r]\{\mathbf{y}\} = \{\mathbf{R}_r\} \tag{13.8-1}$$

where $\{\mathbf{R}_r\} = [\mathbf{W}]^T\{\mathbf{R}^{\text{ext}}\}$ and the reduced m by m mass, damping, and stiffness matrices are

$$[\mathbf{M}_r] = [\mathbf{W}]^T[\mathbf{M}][\mathbf{W}] \qquad [\mathbf{C}_r] = [\mathbf{W}]^T[\mathbf{C}][\mathbf{W}] \qquad [\mathbf{K}_r] = [\mathbf{W}]^T[\mathbf{K}][\mathbf{W}] \tag{13.8-2}$$

With $\{\mathbf{y}\} = \{\bar{\mathbf{y}}\} \sin \omega t$ and $\lambda = \omega^2$, the reduced eigenproblem without damping is

$$([\mathbf{K}_r] - \lambda[\mathbf{M}_r])\{\bar{\mathbf{y}}\} = \{\mathbf{0}\} \tag{13.8-3}$$

Of the several methods of establishing an effective [**W**], we will summarize two. In the first, which we label CMS1 for short, [**W**] contains deflection vectors obtained by solving Eq. 13.6-14—that is, $[\mathbf{W}] = [\mathbf{K}]^{-1}[\mathbf{R}]$. Load patterns in [**R**] are assembled eigenvectors of the substructures—that is, the component modes. A modification is needed if [**K**] is singular, as for an unsupported structure; see [13.2]. In the second method we discuss, here labeled CMS2 for short, [**W**] contains the assembled component modes themselves. This method has no difficulty with an unsupported structure. In both of these methods, [**W**] also contains supplementary vectors related to the motion of d.o.f. shared by substructures. The supplementary vectors may be called rigid-body modes, constraint modes, or attachment modes, depending on how the shared d.o.f. are treated. In illustrations that follow we assume that shared d.o.f. are fixed for the substructure analyses that precede synthesis.

Implementation. To illustrate method CMS1, consider a structure that is divided into ℓ substructures. Also assume that substructure k is attached to substructures $k - 1$ and $k + 1$ only; $k = 2, 3, \ldots, \ell - 1$. Denote matrices for substructure k by $[\mathbf{K}_k]$ and $[\mathbf{M}_k]$ where all *attachment d.o.f.* (i.e., d.o.f. that are common to attaching substructures) are *fixed*. Then, for each substructure, the eigenproblem

$$([\mathbf{K}_k] - \lambda[\mathbf{M}_k])\{\bar{\mathbf{D}}_k\} = \{\mathbf{0}\} \tag{13.8-4}$$

is solved for normal modes of vibration $\{\bar{\mathbf{D}}_k\}$. These modes form the columns of the substructure modal matrix $[\boldsymbol{\phi}_k]$, which is n_k by n_ϕ. Here n_k is the number of *interior* d.o.f. of substructure k (i.e., the total number of d.o.f. of the substructure minus the number of attachment d.o.f.) and n_ϕ (usually $n_\phi \ll n_k$) is the number of normal nodes to be determined (the same for each substructure). Ritz vectors [**W**] for the entire structure can be obtained by solving Eq. 13.6-14 with

$$\underset{n_{eq}\times m}{[\mathbf{R}]} = \begin{bmatrix} \boldsymbol{\phi}_1 & \mathbf{0} & \mathbf{0} & \cdots \\ \mathbf{0} & \mathbf{I}_{1,2} & \mathbf{0} & \cdots \\ \boldsymbol{\phi}_2 & \mathbf{0} & \mathbf{0} & \cdots \\ \mathbf{0} & \mathbf{0} & \mathbf{I}_{2,3} & \cdots \\ \vdots & \vdots & \vdots & \\ \boldsymbol{\phi}_\ell & \mathbf{0} & \mathbf{0} & \cdots \end{bmatrix} \tag{13.8-5}$$

where $\lfloor \mathbf{I}_{k,k+1} \rfloor$ is a unit matrix with number of rows equal to the number of attachment d.o.f. between substructures k and $k+1$, and $m = n_\phi + n_a$, where n_a is the total number of attachment d.o.f. in the synthesized model. The first n_ϕ columns of Eq. 13.8-5 represent assembled substructure normal modes (i.e., vertically stacked columns of substructure normal modes $[\boldsymbol{\phi}_k]$) that are imposed as nodal loads in Eq. 13.6-14. However, since attachment d.o.f. for each substructure were fixed, the Ritz modes resulting from these load patterns are not capable of exciting attachment d.o.f. Thus, the first n_ϕ load patterns in Eq. 13.8-5 are supplemented by additional loads $\lfloor \mathbf{I}_{k,k+1} \rfloor$ which physically correspond to successively applying a unit load to each attachment d.o.f. of the assembled structure while keeping all other d.o.f. load-free. The resulting Ritz vectors are called *attachment modes* and correspond to the displacements of *all* d.o.f. of the assembled structure due to the unit applied loads.

Method CMS2 includes what is perhaps the most popular method of component mode synthesis, the Craig–Bampton method [13.43]. Here the assembled substructure normal modes obtained by fixing attachment d.o.f. are used directly as Ritz vectors. These are then supplemented by *constraint modes,* which are deflection shapes of the assembled structure obtained by successively applying a unit displacement to each attachment d.o.f. while keeping all other attachment d.o.f. fixed.

In all methods of component mode synthesis, the reduced eigenproblem, Eq. 13.8-3, is obtained by imposing constraints on the original large system. Thus, frequencies computed from Eq. 13.8-3 are upper bounds to those of the original system equations. Occasionally, it is possible to have all Ritz vectors orthogonal to one of the eigenvectors of the assembled structure. Then that eigenvector will be missing from the reduced eigenproblem, Eq. 13.8-3. In practice, the likelihood of missing lower-spectrum eigenvectors is reduced by using appropriate supplementary Ritz vectors such as attachment modes, constraint modes, and so on.

In the following example problem, method CMS1 is more accurate than method CMS2. Method CMS2 is more prevalent in practice than method CMS1. Method CMS2 has the computational advantage that the constraint modes can be computed by analysis of the separate substructures. In contrast, method CMS1 requires that the stiffness matrix of the entire structure be assembled in order to compute the attachment modes.

Example. Consider axial vibrations of the structure shown in Fig. 13.8-1, whose stiffness and lumped mass matrices are

$$[\mathbf{K}] = \frac{AE}{L}\begin{bmatrix} 1 & -1 & 0 & 0 & 0 \\ -1 & 2 & -1 & 0 & 0 \\ 0 & -1 & 3 & -2 & 0 \\ 0 & 0 & -2 & 4 & -2 \\ 0 & 0 & 0 & -2 & 4 \end{bmatrix} \qquad [\mathbf{M}] = \frac{\rho AL}{2}\begin{bmatrix} 1 & 0 & 0 & 0 & 0 \\ 0 & 2 & 0 & 0 & 0 \\ 0 & 0 & 3 & 0 & 0 \\ 0 & 0 & 0 & 4 & 0 \\ 0 & 0 & 0 & 0 & 4 \end{bmatrix} \tag{13.8-6}$$

Two substructures are created, one consisting of elements 1 and 2 and the other of elements 3, 4, and 5. With node 3 fixed, matrices for substructure 1 are

$$[\mathbf{K}_1] = \frac{AE}{L}\begin{bmatrix} 1 & -1 \\ -1 & 2 \end{bmatrix} \qquad [\mathbf{M}_1] = \frac{\rho AL}{2}\begin{bmatrix} 1 & 0 \\ 0 & 2 \end{bmatrix} \tag{13.8-7}$$

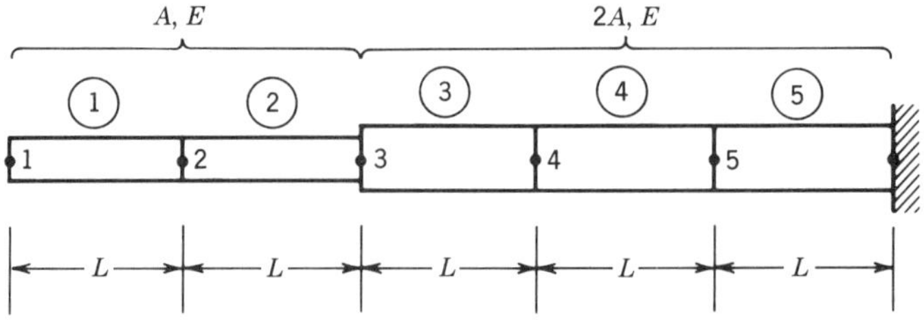

Figure 13.8-1. Bar with variable cross section modeled by five uniform elements of length L.

and for substructure 2

$$[\mathbf{K}_2] = \frac{AE}{L}\begin{bmatrix} 4 & -2 \\ -2 & 4 \end{bmatrix} \qquad [\mathbf{M}_2] = \frac{\rho AL}{2}\begin{bmatrix} 4 & 0 \\ 0 & 4 \end{bmatrix} \tag{13.8-8}$$

In the sequel, we assume that $AE/L = 1$ and $\rho AL/2 = 1$. For substructure 1, the eigenproblem Eq. 13.8-4 is solved with the matrices of Eq. 13.8-7 to yield

$$\lambda_1 = \frac{2-\sqrt{2}}{2} \qquad \lambda_2 = \frac{2+\sqrt{2}}{2} \qquad \{\overline{\mathbf{D}}_1\}_1 = \begin{Bmatrix} 1 \\ \sqrt{2}/2 \end{Bmatrix} \qquad \{\overline{\mathbf{D}}_1\}_2 = \begin{Bmatrix} 1 \\ -\sqrt{2}/2 \end{Bmatrix} \tag{13.8-9}$$

where eigenvectors are normalized so that the first coefficient has unit amplitude. Solving Eq. 13.8-4 using the second substructure's matrices, Eq. 13.8-8, we obtain

$$\lambda_1 = \tfrac{1}{2} \qquad \lambda_2 = \tfrac{3}{2} \qquad \{\overline{\mathbf{D}}_2\}_1 = \begin{Bmatrix} 1 \\ 1 \end{Bmatrix} \qquad \{\overline{\mathbf{D}}_2\}_2 = \begin{Bmatrix} 1 \\ -1 \end{Bmatrix} \tag{13.8-10}$$

For method CMS1, we must evaluate Eq. 13.8-5, which becomes

$$[\mathbf{R}] = \begin{bmatrix} 1 & 1 & 0 \\ \sqrt{2}/2 & -\sqrt{2}/2 & 0 \\ 0 & 0 & 1 \\ 1 & 1 & 0 \\ 1 & -1 & 0 \end{bmatrix} \tag{13.8-11}$$

The first two columns of Eq. 13.8-11 are the assembled component normal modes while the third column has the effect of "releasing" node 3. The resulting Ritz vectors, from Eqs. 13.6-14, 13.8-6, and 13.8-11, are

$$[\mathbf{W}] = \begin{bmatrix} 6.768 & 2.232 & 1.5 \\ 5.768 & 1.232 & 1.5 \\ 4.061 & 0.9393 & 1.5 \\ 3.207 & 0.7929 & 1.0 \\ 1.854 & 0.1464 & 0.5 \end{bmatrix} \quad \text{(method CMS1)} \tag{13.8-12}$$

Reduced stiffness and mass matrices, from Eqs. 13.8-2 and 13.8-12, are

$$[\mathbf{K}_r] = \begin{bmatrix} 15.91 & 4.043 & 4.061 \\ 4.043 & 2.007 & 0.9393 \\ 4.061 & 0.9393 & 1.500 \end{bmatrix} \qquad [\mathbf{M}_r] = \begin{bmatrix} 216.7 & 52.02 & 62.26 \\ 52.02 & 13.27 & 14.74 \\ 62.26 & 14.74 & 18.50 \end{bmatrix} \tag{13.8-13}$$

TABLE 13.8-1. NATURAL FREQUENCIES FOR COMPONENT MODE ANALYSIS OF THE STRUCTURE SHOWN IN FIG. 13.8-1.

Procedure	ω_1	ω_2	ω_3
Original structure (five d.o.f.)	0.2651	0.6156	1.000
Method CMS1 (three d.o.f.)	0.2656	0.8794	1.182
Method CMS2 (three d.o.f.)	0.2719	0.9927	1.268

Natural frequencies of the reduced eigenproblem, Eq. 13.8-3, are reported in Table 13.8-1. Also reported are the first three frequencies of the original structure, whose matrices are given by Eq. 13.8-6.

For method CMS2, we take as Ritz vectors the assembled component normal modes, plus constraint modes [13.43]. Thus

$$[\mathbf{W}] = \begin{bmatrix} 1 & 1 & 1 \\ \sqrt{2}/2 & -\sqrt{2}/2 & 1 \\ 0 & 0 & 1 \\ 1 & 1 & 2/3 \\ 1 & -1 & 1/3 \end{bmatrix} \quad \text{(method CMS2)} \tag{13.8-14}$$

Here the third column of Eq. 13.8-14 (i.e., $\{\mathbf{w}_3\}$) is the deflection shape obtained by solving $[\mathbf{K}]\{\mathbf{w}_3\} = \{\mathbf{p}\}$, where

$$\{\mathbf{w}_3\} = \lfloor u_1 \quad u_2 \quad 1 \quad u_4 \quad u_5 \rfloor^T \tag{13.8-15a}$$

$$\{\mathbf{p}\} = \lfloor 0 \quad 0 \quad p_3 \quad 0 \quad 0 \rfloor^T \tag{13.8-15b}$$

and $[\mathbf{K}]$ is the original structure stiffness matrix of Eq. 13.8-6. As there is but one attachment d.o.f. in this simple problem, the third column of Eq. 13.8-12 and the third column of Eq. 13.8-14 describe the same displacement pattern. Natural frequencies obtained by solving the reduced eigenproblem, Eq. 13.8-3, are given in Table 13.8-1.

As expected, both methods overestimate frequencies obtained by using all d.o.f. of the structure. The discrepancy is greater for higher frequencies. However, higher frequencies are usually of little interest, as structural response is dominated by lower frequencies. The several lowest frequencies are usually estimated accurately by a modest number of component modes.

13.9 TIME-HISTORY ANALYSIS. DIRECT INTEGRATION METHODS

In *direct integration* methods or *step-by-step* methods, a *finite difference approximation* is used to replace the time derivatives appearing in Eq. 13.2-12 or 13.2-13 (i.e., $\{\ddot{\mathbf{D}}\}$ and $\{\dot{\mathbf{D}}\}$) by differences of displacement $\{\mathbf{D}\}$ at various instants of time. Finite difference methods for approximately solving initial value problems have been well studied and have a rich literature [13.47]. Over the past two decades, methods that are particularly effective for transient finite element equations have flourished and continue to be an active area of research [13.48,13.49]. Direct integration is an alternative to modal methods (Section 13.6). For many structural dynamics and wave propagation problems, including those with complicated nonlinearities, direct integration is more expedient. Many methods of direct integration are popular and the choice of method is strongly problem-dependent. In this section explicit and implicit methods are introduced. In Sections 13.10 and 13.11 these methods are discussed in detail.

In direct integration, the approach is to write the equation of motion, Eq. 13.2-12, at a specific instant of time,

$$\boxed{[\mathbf{M}]\{\ddot{\mathbf{D}}\}_n + [\mathbf{C}]\{\dot{\mathbf{D}}\}_n + [\mathbf{K}]\{\mathbf{D}\}_n = \{\mathbf{R}^{\text{ext}}\}_n} \tag{13.9-1}$$

where subscript n denotes time $n\,\Delta t$ and Δt is the size of the time increment or time step. The absence of time step subscripts on matrices $[\mathbf{M}]$, $[\mathbf{C}]$, and $[\mathbf{K}]$ in Eq. 13.9-1 implies linearity. For problems with material nonlinearity, $[\mathbf{K}]$ is a function of displacements and therefore of time as well. Accordingly, from Eq. 13.2-13,

$$\boxed{[\mathbf{M}]\{\ddot{\mathbf{D}}\}_n + [\mathbf{C}]\{\dot{\mathbf{D}}\}_n + \{\mathbf{R}^{\text{int}}\}_n = \{\mathbf{R}^{\text{ext}}\}_n} \tag{13.9-2}$$

$\{\mathbf{R}^{\text{int}}\}_n$ is the internal force vector at time $n\,\Delta t$ due to straining of material. It is obtained by assembling element internal force vectors, $\{\mathbf{r}^{\text{int}}\}_n$, given by Eq. 13.2-7 using $\{\boldsymbol{\sigma}\}_n$. For nonlinear problems, $\{\mathbf{R}^{\text{int}}\}_n$ is a nonlinear function of $\{\mathbf{D}\}_n$ and possibly time derivatives of $\{\mathbf{D}\}_n$. For linear problems, $\{\mathbf{R}^{\text{int}}\}_n = [\mathbf{K}]\{\mathbf{D}\}_n$. In Eq. 13.9-2, $[\mathbf{M}]$ and $[\mathbf{C}]$ are taken as time-independent, although for some problems these may be nonlinear also.

In the following sections, $[\mathbf{M}]$ is assumed to be positive definite. Moreover, unless otherwise stated, $[\mathbf{K}]$ is positive semidefinite; that is, $[\mathbf{K}]$ may permit rigid-body motion.

Difference methods for direct integration of Eqs. 13.9-1 and 13.9-2 can be categorized as *explicit* or *implicit*. Explicit methods have the form

$$\{\mathbf{D}\}_{n+1} = f(\{\mathbf{D}\}_n, \{\dot{\mathbf{D}}\}_n, \{\ddot{\mathbf{D}}\}_n, \{\mathbf{D}\}_{n-1}, \ldots) \tag{13.9-3}$$

and hence permit $\{\mathbf{D}\}_{n+1}$ to be determined in terms of completely historical information consisting of displacements and time derivatives of displacements at time $n\,\Delta t$ and before. Implicit methods have the form

$$\{\mathbf{D}\}_{n+1} = f(\{\dot{\mathbf{D}}\}_{n+1}, \{\ddot{\mathbf{D}}\}_{n+1}, \{\mathbf{D}\}_n, \ldots) \tag{13.9-4}$$

and hence computation of $\{\mathbf{D}\}_{n+1}$ requires knowledge of the time derivatives of $\{\mathbf{D}\}_{n+1}$, which are unknown. Explicit and implicit methods have markedly different properties. This has important practical implications.

Methods that have the general form of Eqs. 13.9-3 and 13.9-4 are called *multistep methods*. When the right-hand sides of Eqs. 13.9-3 and 13.9-4 contain information dating back to time $n\,\Delta t$ only, the methods are called *single-step methods*. When the right-hand sides of Eqs. 13.9-3 and 13.9-4 contain information dating back to time $(n - 1)\,\Delta t$, they are called *two-step methods*. Single-step methods are easy to start from initial conditions. Multistep methods require special starting procedures that may be awkward or may introduce inaccuracies. Poor starting procedures (not described in this book) may degrade the accuracy of the entire analysis [13.69].

13.10 EXPLICIT DIRECT INTEGRATION METHODS

A popular method, which is characteristic of explicit methods in general, is the central-difference method. It approximates velocity and acceleration by

$$\{\dot{\mathbf{D}}\}_n = \frac{1}{2\,\Delta t}(\{\mathbf{D}\}_{n+1} - \{\mathbf{D}\}_{n-1}) \tag{13.10-1}$$

$$\{\ddot{\mathbf{D}}\}_n = \frac{1}{\Delta t^2}(\{\mathbf{D}\}_{n+1} - 2\{\mathbf{D}\}_n + \{\mathbf{D}\}_{n-1}) \tag{13.10-2}$$

Equations 13.10-1 and 13.10-2 are obtained by expanding $\{\mathbf{D}\}_{n+1}$ and $\{\mathbf{D}\}_{n-1}$ in Taylor series about time $n\,\Delta t$:

$$\{\mathbf{D}\}_{n+1} = \{\mathbf{D}\}_n + \Delta t\{\dot{\mathbf{D}}\}_n + \frac{\Delta t^2}{2}\{\ddot{\mathbf{D}}\}_n + \frac{\Delta t^3}{6}\{\dddot{\mathbf{D}}\}_n + \cdots \tag{13.10-3}$$

$$\{\mathbf{D}\}_{n-1} = \{\mathbf{D}\}_n - \Delta t\{\dot{\mathbf{D}}\}_n + \frac{\Delta t^2}{2}\{\ddot{\mathbf{D}}\}_n - \frac{\Delta t^3}{6}\{\dddot{\mathbf{D}}\}_n + \cdots \tag{13.10-4}$$

Subtracting Eq. 13.10-4 from Eq. 13.10-3 yields Eq. 13.10-1 while adding Eqs. 13.10-3 and 13.10-4 yields Eq. 13.10-2. In both cases, terms containing Δt^2 and higher powers are omitted from Eqs. 13.10-1 and 13.10-2. Hence, the central-difference formulas, Eqs. 13.10-1 and 13.10-2, are said to be *second-order accurate*. In other words, the error is $O(\Delta t^2)$ which implies that halving the time step should approximately quarter the error.

Combining Eqs. 13.10-1 and 13.10-2 with Eq. 13.9-1 provides

$$\left[\frac{1}{\Delta t^2}\mathbf{M} + \frac{1}{2\,\Delta t}\mathbf{C}\right]\{\mathbf{D}\}_{n+1} = \{\mathbf{R}^{\text{ext}}\}_n - [\mathbf{K}]\{\mathbf{D}\}_n + \frac{1}{\Delta t^2}[\mathbf{M}](2\{\mathbf{D}\}_n - \{\mathbf{D}\}_{n-1}) + \frac{1}{2\,\Delta t}[\mathbf{C}]\{\mathbf{D}\}_{n-1} \tag{13.10-5}$$

Remarks.

1. Equation 13.10-5 is a system of linear algebraic equations. If [**M**] and [**C**] are diagonal, then the equations are uncoupled and $\{\mathbf{D}\}_{n+1}$ can be obtained without solving simultaneous equations.
2. For small finite element models, [**K**] can be formed and stored in the computer's core memory and the internal force $\{\mathbf{R}^{\text{int}}\}_n = [\mathbf{K}]\{\mathbf{D}\}_n$ at each time step can be obtained by matrix multiplication. However, it is more common, even for linear problems, to compute the internal force vector at each time step by summation of element contributions (i.e., *element-by-element*). Element contributions are given by Eq. 13.2-7. Because the element [**k**] need not be formed or stored, explicit methods can treat large three-dimensional models with comparatively modest computer storage requirements.

3. Starting the method from $n = 0$ requires $\{\mathbf{D}\}_{-1}$, which can be computed from known initial conditions $\{\mathbf{D}\}_0$ and $\{\dot{\mathbf{D}}\}_0$ and Eq. 13.10-4:

$$\{\mathbf{D}\}_{-1} = \{\mathbf{D}\}_0 - \Delta t\{\dot{\mathbf{D}}\}_0 + \frac{\Delta t^2}{2}\{\ddot{\mathbf{D}}\}_0 \tag{13.10-6}$$

where terms with Δt^3 and higher powers are omitted. $\{\ddot{\mathbf{D}}\}_0$ is obtained from the equation of motion, Eq. 13.9-1, at time zero:

$$\{\ddot{\mathbf{D}}\}_0 = [\mathbf{M}]^{-1}(\{\mathbf{R}^{ext}\}_0 - [\mathbf{K}]\{\mathbf{D}\}_0 - [\mathbf{C}]\{\dot{\mathbf{D}}\}_0) \tag{13.10-7}$$

4. To compute $\{\mathbf{D}\}_{n+1}$ requires $\{\mathbf{R}^{int}\}_n$. For nonlinear material constitutive laws that are functions of strain (but not of strain rate), $\{\mathbf{R}^{int}\}_n$ is easy to evaluate because $\{\mathbf{D}\}_n$, and hence the strain at time $n\,\Delta t$, is known. For this reason, explicit methods are well suited to treatment of material nonlinearity.
5. Equation 13.10-5 is *conditionally stable* and requires Δt such that

$$\Delta t \leq 2/\omega_{max} \tag{13.10-8}$$

where ω_{max} is the highest natural frequency of $\det([\mathbf{K}] - \omega^2[\mathbf{M}]) = 0$. If Eq. 13.10-8 is not satisfied, computations will be *unstable*. This is indicated by an obviously erroneous time-history solution that grows unbounded, perhaps by orders of magnitude per time step.[2] (In nonlinear problems, instabilities may be more difficult to detect.)
6. A feature of Eq. 13.10-5 is that stability (i.e., maximum allowable time step size) is not affected by damping. For the central difference method to be economically competitive with implicit methods, both [**M**] and [**C**] must be diagonal. For reasons discussed in Section 13.13, explicit integration is usually more accurate with lumped mass matrices than with consistent mass matrices. However, it is difficult to model damping by spectral methods if [**C**] must be diagonal.
7. It is conceivable that we may know the displacement and velocity of a structure at a given instant (i.e., initial conditions) and wish to integrate backward in time (i.e., use a negative Δt) to determine the configuration of the structure at some instant in the past. The central-difference method is conditionally stable for such applications and requires $-2/\omega_{max} \leq \Delta t \leq 2/\omega_{max}$. Henceforth, we will be concerned with positive Δt only.

Alternative Form for Nondiagonal [C]. A form of the central-difference method that does not require diagonal [**C**] is obtained by approximating the velocity and acceleration by

$$\{\dot{\mathbf{D}}\}_{n-1/2} = \frac{1}{\Delta t}(\{\mathbf{D}\}_n - \{\mathbf{D}\}_{n-1}) \tag{13.10-9}$$

[2]The case $\Delta t = 2/\omega_{max}$ may be called "limiting stability." If $\Delta t = 2/\omega_{max}$, the numerical solution may diverge in certain cases, but only in arithmetic fashion. If $\Delta t > 2/\omega_{max}$, divergence is exponential. Further discussion of these matters appears in Section 13.13.

$$\{\ddot{\mathbf{D}}\}_n = \frac{1}{\Delta t}(\{\dot{\mathbf{D}}\}_{n+1/2} - \{\dot{\mathbf{D}}\}_{n-1/2})$$

$$= \frac{1}{\Delta t^2}(\{\mathbf{D}\}_{n+1} - 2\{\mathbf{D}\}_n + \{\mathbf{D}\}_{n-1}) \tag{13.10-10}$$

The equation of motion, Eq. 13.9-1, is modified by *lagging* the velocity by one-half time step:

$$[\mathbf{M}]\{\ddot{\mathbf{D}}\}_n + [\mathbf{C}]\{\dot{\mathbf{D}}\}_{n-1/2} + [\mathbf{K}]\{\mathbf{D}\}_n = \{\mathbf{R}^{\text{ext}}\}_n \tag{13.10-11}$$

Combination of Eqs. 13.10-9 through 13.10-11 yields

$$\frac{1}{\Delta t^2}[\mathbf{M}]\{\mathbf{D}\}_{n+1} = \{\mathbf{R}^{\text{ext}}\}_n - [\mathbf{K}]\{\mathbf{D}\}_n + \frac{1}{\Delta t^2}[\mathbf{M}](\{\mathbf{D}\}_n + \Delta t\{\dot{\mathbf{D}}\}_{n-1/2}) - [\mathbf{C}]\{\dot{\mathbf{D}}\}_{n-1/2} \tag{13.10-12}$$

If $[\mathbf{M}]$ is lumped, then computation of $\{\mathbf{D}\}_{n+1}$ does not require the solution of simultaneous equations. There are no restrictions on the form of $[\mathbf{C}]$. The method can be started using the initial displacement $\{\mathbf{D}\}_0$ and the approximation $\{\dot{\mathbf{D}}\}_{-1/2} \approx \{\dot{\mathbf{D}}\}_0$. Alternatively, $\{\dot{\mathbf{D}}\}_{-1/2}$ can be approximated by the forward difference formula with negative Δt—that is,

$$\{\dot{\mathbf{D}}\}_{-1/2} = \{\dot{\mathbf{D}}\}_0 - \frac{\Delta t}{2}\{\ddot{\mathbf{D}}\}_0 \tag{13.10-13}$$

where $\{\ddot{\mathbf{D}}\}_0$ is obtained from Eq. 13.10-7. Although the central-difference formulas, Eqs. 13.10-9 and 13.10-10, are second-order accurate, we can only guarantee first-order accuracy in the time integration of Eq. 13.10-11 when $[\mathbf{C}] \neq [\mathbf{0}]$ because viscous forces $[\mathbf{C}]\{\dot{\mathbf{D}}\}_{n-1/2}$ lag by half a time step. However, for practical structures (which are not heavily damped), Eqs. 13.10-5 and 13.10-12 have almost the same accuracy.

The stability condition for Eq. 13.10-12 is

$$\Delta t \leq \frac{2}{\omega_{\max}}(\sqrt{1 + \xi^2} - \xi) \tag{13.10-14}$$

where ξ is the fraction of critical damping at the *highest* undamped natural frequency, $\omega_{\max}$. For proportional damping, ξ at frequency $\omega_{\max}$ can be computed from Eq. 13.4-2. Equation 13.10-14 is more restrictive than Eq. 13.10-8.

Implementation. A computational procedure for central-difference integration of undamped equations of motion with possible material nonlinearity is given in Table 13.10-1 (modification of this scheme to include damping is left as an exercise). In the element internal force evaluation, $\int [\mathbf{B}]^T\{\boldsymbol{\sigma}\}_n\, dV$ requires the same order of quadrature as used for the element stiffness matrix $\int [\mathbf{B}]^T[\mathbf{E}][\mathbf{B}]\, dV$. Therefore, guidelines given in Section 6.11 for stiffness matrix calculation are also applicable to internal force calculation. Note that internal forces must be computed

TABLE 13.10-1. COMPUTATIONAL PROCEDURE FOR DIRECT INTEGRATION BY THE CENTRAL-DIFFERENCE METHOD, USING THE FORM GIVEN IN EQ. 13.10-12 BUT WITHOUT DAMPING.

1. Set initial conditions $\{\mathbf{D}\}_0 = \{\mathbf{D}(t = 0)\}$ and $\{\dot{\mathbf{D}}\}_{-1/2} = \{\dot{\mathbf{D}}(t = 0)\}$, $n = 0$.
2. Assemble $[\mathbf{M}]$.
3. Compute internal force $\{\mathbf{R}^{\text{int}}\}_n$ as follows:
  3.1 Loop over elements $e = 1, 2, \ldots$;
  3.2 Compute element internal force $\{\mathbf{r}_e^{\text{int}}\}_n = \int_{V_e} [\mathbf{B}]^T\{\boldsymbol{\sigma}\}_n \, dV$;
  3.3 Assemble element internal force, $\{\mathbf{r}_e^{\text{int}}\}_n$, into global internal force $\{\mathbf{R}^{\text{int}}\}_n$;
  3.4 Next element; go to Step 3.1.
4. Update displacement by Eq. 13.10-12:

$$\{\mathbf{D}\}_{n+1} = \Delta t^2[\mathbf{M}]^{-1}(\{\mathbf{R}^{\text{ext}}\}_n - \{\mathbf{R}^{\text{int}}\}_n) + \{\mathbf{D}\}_n + \Delta t\{\dot{\mathbf{D}}\}_{n-1/2}$$

5. Update velocity [at time $(n + \frac{1}{2})\,\Delta t$] by Eq. 13.10-9:

$$\{\dot{\mathbf{D}}\}_{n+1/2} = \frac{1}{\Delta t}(\{\mathbf{D}\}_{n+1} - \{\mathbf{D}\}_n)$$

6. Output if desired; $n \leftarrow n + 1$, go to Step 3.

at each time step. This is the most expensive part of the per-time-step cost of an explicit method. Hence, there is considerable motivation to use reduced quadrature to evaluate internal forces. For example, explicit transient analysis with the four-node bilinear quadrilateral element with one-point quadrature will be roughly one-fourth as expensive as analysis using four-point quadrature. In three dimensions, the savings are even more dramatic. However, when reduced integration is used, additional precautions must be taken to prevent mesh instabilities (discussed in Section 6.12). Flanagan and Belytschko [13.52] present an effective scheme using *hourglass control,* or a *stabilization matrix,* that adds artificial stiffness to the element to suppress the zero-energy modes (see also [13.49, 13.53,13.54]).

Stability: Estimation of ω_{max}. The central-difference method, as well as explicit methods in general, is conditionally stable. If Δt is too large, the method fails. If Δt is much smaller than necessary, computations are too expensive. Therefore, it is necessary to determine, or accurately bound, ω_{max} in Eq. 13.10-8 or Eq. 13.10-14. Equation 13.5-7 demonstrates that ω_{max} for the assembled finite element model is bounded by the maximum frequency of the constituent unassembled and unsupported elements. This frequency can often be computed by hand calculation.

Consider the uniform linear-displacement bar element shown in Fig. 13.5-1. With lumped masses, the highest frequency of this element is given in the Example of Section 13.5 as

$$(\omega_{\text{max}})_e = \frac{2c}{L} \tag{13.10-15}$$

where the *dilatational wave speed, acoustic wave speed,* or simply *wave speed,* $c = \sqrt{E/\rho}$ is the speed at which information travels in the bar. From Eq. 13.10-8, we must use $\Delta t \leq 2/\omega_{\text{max}}$. Therefore, if Eq. 13.10-15 represents the maximum element frequency among all elements in a mesh, then stable integration by the central difference method (Eq. 13.10-5 or Eq. 13.10-12 with **[C]** = **[0]**) requires

$$\Delta t \leq \frac{L}{c} \tag{13.10-16}$$

which is called the *CFL condition* after Courant, Friedrichs, and Lewy [13.12,13.50]. The physical interpretation of this condition is that Δt must be small enough that information does not propagate across more than one element per time step. This observation is true only of linear-displacement elements with lumped mass. Note that if a consistent mass matrix is used, the example of Section 13.5 gives $(\omega_{\max})_e = 2\sqrt{3}\,(c/L)$ so that $\Delta t \leq (L/c)/\sqrt{3}$; this is more restrictive than the time step criterion for an element with lumped masses, Eq. 13.10-16. This result is typical. Thus, in addition to providing uncoupled equations and generally more accurate results than consistent mass matrices in explicit integration, lumped mass matrices provide for larger stable time steps.

Higher-order elements yield higher maximum frequencies than lower-order elements. For this reason and for reasons discussed in Section 13.14, one may wish to avoid higher-order elements when doing explicit integration. Similarly, one should avoid penalty constraints, as large penalty numbers make $\omega_{\max}$ very large.

For plane and solid finite elements, it is usually difficult to analytically calculate $(\omega_{\max})_e$. Rather, it is possible to use Eq. 13.10-16 with L replaced by an *effective element diameter, L_e*. Effective element diameters are shown in Fig. 13.10-1 for some low-order displacement field elements. For higher-order displacement field elements, effective diameters are difficult to calculate. As an alternative to calculation, the highest frequency of an element (or of a global system of equations) can be accurately bounded by *Gerschgorin's theorem* [13.12], which, for lumped mass matrices, states that

$$\omega_{\max}^2 \leq \max\left(k_{ii} + \sum_{\substack{j=1\\ j\neq i}}^{n_e} |k_{ij}|\right)\Big/ m_{ii} \qquad \text{where} \quad i = 1, 2, \ldots, n_e \tag{13.10-17}$$

Here n_e is the number of d.o.f. per element. Equation 13.10-17 can be applied to the assembled structure if k_{ii}, k_{ij}, and m_{ii} are replaced by structure coefficients K_{ii}, K_{ij}, and M_{ii}, and n_e becomes n_{eq}, the number of global d.o.f. Equation

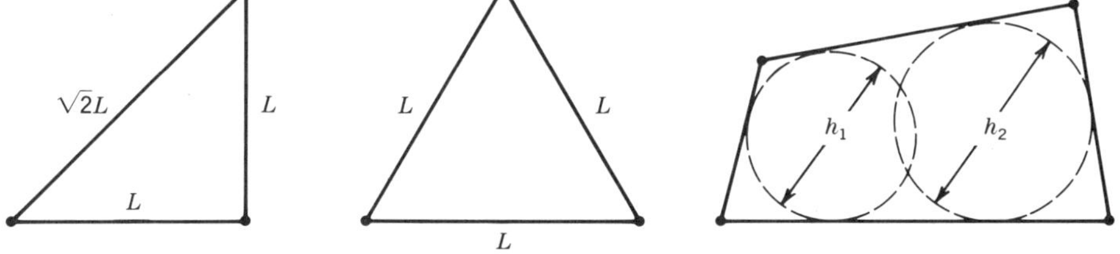

$L_e = \sqrt{2}L$ $L_e = L$ $L_e = \min(h_1, h_2)$

Figure 13.10-1. Effective element diameters for some common two-dimensional elements. Results for three-node elements are exact. Results for the four-node element are exact if the element is rectangular ($h_1 = h_2$) and conservative otherwise. Inscribed circles touch at least three sides of the element.

13.10-17 fails if $m_{ii} = 0$ (or if $M_{ii} = 0$), as may happen with some mass allocation schemes. The difficulty of $m_{ii} = 0$ (but $M_{ii} \neq 0$) is addressed in [13.51].

A useful definition of the relative size of time step used in direct integration is the *Courant number*, C_n, defined as

$$C_n = \frac{\Delta t_{\text{actual}}}{\Delta t_{\text{stable}}} \tag{13.10-18}$$

where Δt_{actual} is the actual time step used and Δt_{stable} is the largest time step permitted for stable explicit integration by a specified explicit method (e.g., Eqs. 13.10-8 or 13.10-14). For reasons discussed in Section 13.13, we recommend a maximum C_n of about 0.95 to 0.98.

Usually Δt_{stable} in Eq. 13.10-18 is not precisely known because $\omega_{\max}$ for the finite element mesh is not precisely known. One can determine a precise value of $\omega_{\max}$ only by a somewhat inconvenient eigenvalue calculation. Instead, practitioners obtain a close upper bound to Δt_{stable} by approximating $\omega_{\max}$ by the largest unconstrained frequency among all elements, $(\omega_{\max})_e$, as obtained perhaps from Eq. 13.10-15 or (as an upper bound) from Eq. 13.10-17.

Example: Wave Propagation. Consider the uniform steel bar and tip loading shown in Fig. 13.10-2. The bar is undamped, initially at rest, and is modeled by 40 equal-length linear-displacement elements, so that $L = L_T/40 = 0.5$ in. The highest element frequency is given by Eq. 13.10-15 as

$$(\omega_{\max})_e = \frac{2}{0.5}\sqrt{\frac{E}{\rho}} = 8.0539(10^5) \text{ rad/sec} \tag{13.10-19}$$

This value is very close to the maximum frequency of the finite element mesh, which is $\omega_{\max} = 8.0523(10^5)$ rad/sec. Therefore, stable integration by the central difference method requires, according to Eq. 13.10-14 with $\xi = 0$, that $\Delta t \leq 2/(\omega_{\max})_e = 2.483(10^{-6})$ sec. The computational procedure of Table 13.10-1 is implemented in the Fortran program shown in Fig. 13.10-3.

The stress time-history response at the midpoint of element 20 (at $x = 9.75$ in.) is shown in Fig. 13.10-4 for $\Delta t = 2.4(10^{-6})$ sec ($C_n = 0.966$) and an analysis duration equivalent to two wave traversals along the bar (83 time steps). The exact solution is also shown. Many important features of numerical solution of wave propagation problems are displayed. One is that stress at $x = 9.75$ in. is zero until sufficient time has elapsed for the stress wave to propagate from the instantaneously loaded tip to the bar midpoint. This is a characteristic feature of a class of partial differential equations called *hyperbolic*, among which the equation of motion is an example. The increase in mean compressive stress from 100 to 200 psi at 0.15 msec is due to wave reflection from the built-in end.

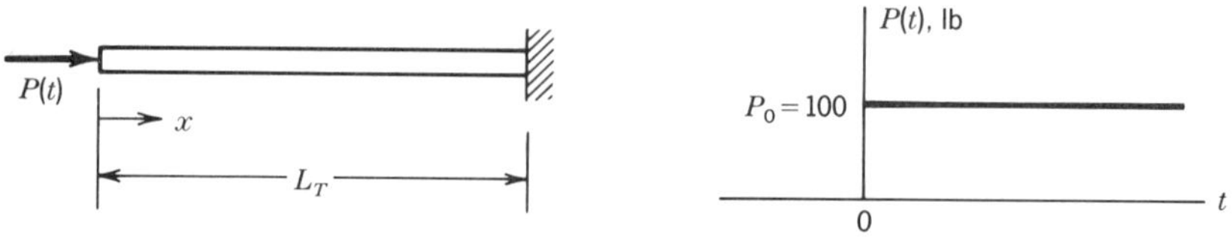

Figure 13.10-2. One-dimensional uniform bar with instantaneous tip loading. The bar is initially at rest. $A = 1.0$ in.^{2}, $E = 30(10^6)$ psi, $\rho = 7.4(10^{-4})$ lb-sec^2/in.^{4}, $L_T = 20$ in. Load $P_0 = 100$ lb is applied at $t = 0$.

```
C---- Program for 1-dimensional wave propagation through NELE equal-
C---- length finite elements using the central difference method
      IMPLICIT DOUBLE PRECISION (A-H,O-Z)
      DIMENSION X(101),RM(101),D(101),V(101),FEXT(101),FINT(101)
      DATA X,RM,FEXT/101*0.,101*0. ,101*0./
C
C     NELE   = number of elements        ELELEN = length of element
C     CSA    = cross-sectional area      DENSTY = mass density
C     E      = elastic modulus           DELT   = time step
C     NSTEP  = number of time steps      IFREQ  = output interval
C
C---- Input mesh data
      READ(5,*)NELE,ELELEN,CSA,DENSTY,E,DELT,NSTEP,IFREQ
      NUMNOD=NELE+1
C---- Generate nodal coordinates and form lumped mass matrix
      DO 10 I=1,NUMNOD
   10 X(I)=FLOAT(I-1)*ELELEN
      DO 20 K=1,NELE
      RM(  K)=RM(  K) + DENSTY*CSA*ELELEN/2.
   20 RM(K+1)=RM(K+1) + DENSTY*CSA*ELELEN/2.
C---- Set at-rest initial conditions and tip-load force
      DO 30 I=1,NUMNOD
      D(I)=0.
   30 V(I)=0.
      FEXT(1)=100.
C---- Integration loop
      DO 100 N=1,NSTEP
C---- Get internal force for last time step
      CALL INTFOR(D,FINT,X,E,CSA,NELE)
C---- Update displacements using the central difference method and
C---- enforce zero-displacement boundary condition at built-in end
      DO 40 I=1,NUMNOD
      DOLD=D(I)
      D(I)=DELT**2*(FEXT(I)-FINT(I))/RM(I) + D(I) + DELT*V(I)
      IF(I.EQ.NUMNOD) D(I)=0.
   40 V(I)=(D(I)-DOLD)/DELT
C---- Output if desired
      SIG20=E*((D(21)-D(20))/ELELEN)
      IF(MOD(N,IFREQ).EQ.0)
     . WRITE(7,1000)N,FLOAT(N)*DELT,D(20),V(20),SIG20
  100 CONTINUE
      STOP
 1000 FORMAT(I5,4E15.4)
      END

      SUBROUTINE INTFOR(D,FINT,X,E,CSA,NELE)
      IMPLICIT DOUBLE PRECISION (A-H,O-Z)
      DIMENSION D(1),FINT(1),X(1)
      NUMNOD=NELE+1
C---- Zero internal force vector
      DO 10 I=1,NUMNOD
   10 FINT(I)=0.
C---- Loop over elements
      DO 20 K=1,NELE
      RL=X(K+1)-X(K)
C---- Compute strain and stress
      STRAIN=(D(K+1)-D(K))/RL
      STRESS=STRAIN*E
C---- Assemble contribution into internal force vector
      F1=-STRESS*CSA
      F2=+STRESS*CSA
      FINT(K   )=FINT(K   )+F1
      FINT(K+1)=FINT(K+1)+F2
   20 CONTINUE
      RETURN
      END
```

Figure 13.10-3. Fortran program for direct integration by the central-difference method using the procedure in Table 13.10-1.

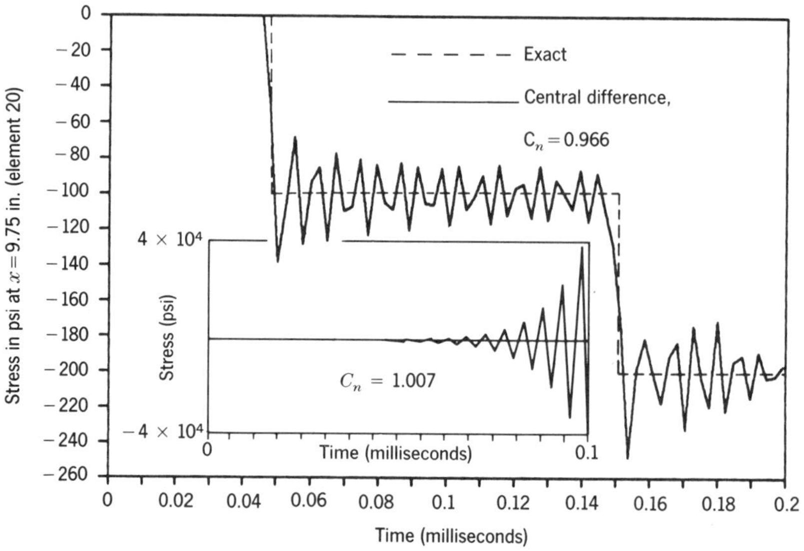

Figure 13.10-4. Stress time history at $x = 9.75$ in. for a 40-element model of the 20-in. bar shown in Fig. 13.10-2 using $\Delta t = 2.4(10^{-6})$ sec ($C_n = 0.966$). Inset shows instability that results from taking Δt too large ($C_n = 1.007$).

Figure 13.10-4 also shows severe *spurious oscillations*, or *noise*, created by the algorithm, as the stress wave passes. Spurious oscillations are high-frequency motions that the mesh cannot accurately resolve and are similar to the Gibbs phenomenon in fitting a finite Fourier series to a piecewise-continuous function. These spurious oscillations are often a nuisance. Stiffness-proportional damping is sometimes employed to attenuate noise although it is usually better practice to employ a different integration method, such as discussed in Section 13.12, which automatically dissipates high-frequency motion.

To demonstrate the instability in explicit integration that results when the maximum stable time step is exceeded, we repeated the analysis using $\Delta t = 2.5(10^{-6})$ sec ($C_n = 1.007$). The results, shown in the inset of Fig. 13.10-4, are typical of instability in linear analysis and demonstrate a solution that increases wildly with each time step until eventually the computer program aborts.

When Δt is slightly less than the stability limit, the numerical solution may display spurious "beating" in which the amplitude of response repeatedly grows and decays. To illustrate beating, we repeated the foregoing analysis using $\Delta t = 2.481(10^{-6})$ sec ($C_n = 0.999$). Results are shown in Fig. 13.10-5. We see that average results are good but spurious oscillations are severe. For some problems, particularly problems having few d.o.f., beating can be more detrimental to accuracy than shown by this example.

Concluding Remarks. When an explicit method is used for a single ordinary differential equation, accuracy is markedly time-step-size-dependent. It may therefore appear that the stability criterion is of only academic interest since Δt must be considerably smaller to achieve satisfactory accuracy. However, this supposition is not true of *systems* of finite element equations (about 20 equations or more) in which it is typically observed that excellent accuracy can be obtained using a time step size just under the stability limit. The reason is that the equations of motion are a *stiff* system of ordinary differential equations, which has a broad frequency spectrum. The stability criterion, Eq. 13.10-8, is based upon the *highest* frequency, or *shortest time scale* phenomenon that the mesh can possibly repro-

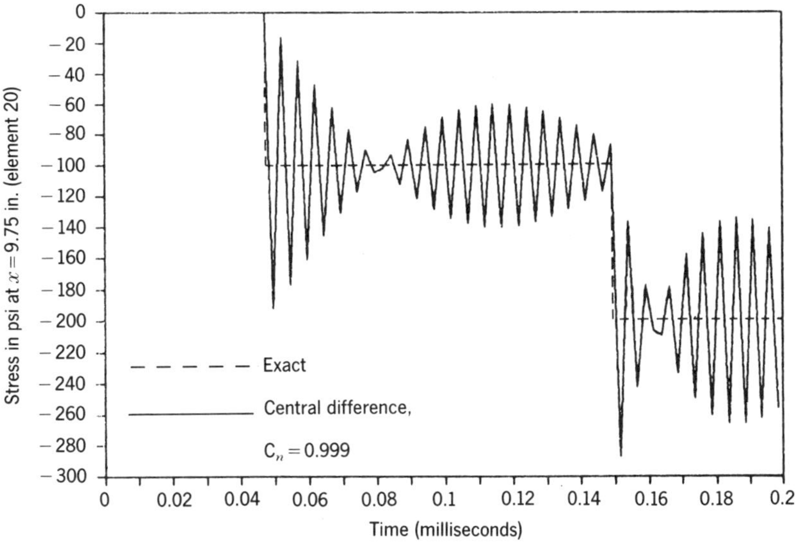

Figure 13.10-5. Stress time history at $x = 9.75$ in. for a 40-element model of the bar shown in Fig. 13.10-2 using $\Delta t = 2.481(10^{-6})$ sec ($C_n = 0.999$). The response demonstrates "beating" from using a time step close to the stability limit.

duce. It is true that motions of the mesh that occur on this time scale are not accurately resolved by taking Δt close to the stability limit. Fortunately, these motions contribute little to the structural response, which is dominated by much lower-frequency, long-time-scale phenomena that *are* accurately resolved. In other words, *we do not require high-frequency phenomena to be accurately resolved; all that we ask is that they be stably integrated.* The implication is that in explicit methods, a Δt satisfying stability criteria is usually satisfactory to guarantee accuracy. Even so, the maximum allowable Δt is often smaller than one would like, because so many time steps may be needed to span the duration of an analysis.

13.11 IMPLICIT DIRECT INTEGRATION METHODS

Most of the useful implicit methods are *unconditionally stable* and have no restriction on the time step size other than as required for accuracy. A popular unconditionally stable implicit method is called the *trapezoidal rule* or the *average acceleration method.* (It is also known as the *Crank–Nicolson method* when applied to *parabolic* partial differential equations, such as the heat conduction equation.) The trapezoidal rule relates displacements, velocities, and accelerations by

$$\{\mathbf{D}\}_{n+1} = \{\mathbf{D}\}_n + \frac{\Delta t}{2}(\{\dot{\mathbf{D}}\}_n + \{\dot{\mathbf{D}}\}_{n+1}) \tag{13.11-1}$$

$$\{\dot{\mathbf{D}}\}_{n+1} = \{\dot{\mathbf{D}}\}_n + \frac{\Delta t}{2}(\{\ddot{\mathbf{D}}\}_n + \{\ddot{\mathbf{D}}\}_{n+1}) \tag{13.11-2}$$

TABLE 13.11-1. COMPUTATIONAL PROCEDURE FOR DIRECT INTEGRATION BY THE TRAPEZOIDAL-RULE METHOD.

1. Form [**K**], [**C**], and [**M**].
2. Set initial conditions $\{\mathbf{D}\}_0 = \{\mathbf{D}(t = 0)\}$ and $\{\dot{\mathbf{D}}\}_0 = \{\dot{\mathbf{D}}(t = 0)\}$; use Eq. 13.9-1 to compute $\{\ddot{\mathbf{D}}\}_0 = [\mathbf{M}]^{-1}(\{\mathbf{R}^{\text{ext}}\}_0 - [\mathbf{C}]\{\dot{\mathbf{D}}\}_0 - [\mathbf{K}]\{\mathbf{D}\}_0)$; $n = 0$.
3. Form effective stiffness matrix, Eq. 13.11-6, and factor it.
4. Form effective load vector, Eq. 13.11-7.
5. Solve $[\mathbf{K}^{\text{eff}}]\{\mathbf{D}\}_{n+1} = \{\mathbf{R}^{\text{eff}}\}_{n+1}$ for $\{\mathbf{D}\}_{n+1}$ by forward- and back-substitution.
6. Update velocity $\{\dot{\mathbf{D}}\}_{n+1}$ and acceleration $\{\ddot{\mathbf{D}}\}_{n+1}$ by Eqs. 13.11-3 and 13.11-4.
7. Output if desired; $n \leftarrow n + 1$, go to Step 4.

Equations 13.11-1 and 13.11-2 can be obtained using Taylor series (this is left as an exercise) in which it is seen that they are second-order accurate. Alternatively, Eqs. 13.11-1 and 13.11-2 can be solved for $\{\dot{\mathbf{D}}\}_{n+1}$ and $\{\ddot{\mathbf{D}}\}_{n+1}$ to provide

$$\{\dot{\mathbf{D}}\}_{n+1} = \frac{2}{\Delta t}(\{\mathbf{D}\}_{n+1} - \{\mathbf{D}\}_n) - \{\dot{\mathbf{D}}\}_n \tag{13.11-3}$$

$$\{\ddot{\mathbf{D}}\}_{n+1} = \frac{4}{\Delta t^2}(\{\mathbf{D}\}_{n+1} - \{\mathbf{D}\}_n) - \frac{4}{\Delta t}\{\dot{\mathbf{D}}\}_n - \{\ddot{\mathbf{D}}\}_n \tag{13.11-4}$$

Combination of Eqs. 13.11-3 and 13.11-4 with the equation of motion, Eq. 13.9-1 at time $(n + 1)\,\Delta t$, yields

$$[\mathbf{K}^{\text{eff}}]\{\mathbf{D}\}_{n+1} = \{\mathbf{R}^{\text{eff}}\}_{n+1} \tag{13.11-5}$$

where the *effective stiffness matrix* and *effective load vector* are, respectively,

$$[\mathbf{K}^{\text{eff}}] = \frac{4}{\Delta t^2}[\mathbf{M}] + \frac{2}{\Delta t}[\mathbf{C}] + [\mathbf{K}] \tag{13.11-6}$$

$$\{\mathbf{R}^{\text{eff}}\}_{n+1} = \{\mathbf{R}^{\text{ext}}\}_{n+1} + [\mathbf{M}]\left(\frac{4}{\Delta t^2}\{\mathbf{D}\}_n + \frac{4}{\Delta t}\{\dot{\mathbf{D}}\}_n + \{\ddot{\mathbf{D}}\}_n\right) + [\mathbf{C}]\left(\frac{2}{\Delta t}\{\mathbf{D}\}_n + \{\dot{\mathbf{D}}\}_n\right) \tag{13.11-7}$$

A flow chart for this algorithm is given in Table 13.11-1.

Remarks.

1. Equation 13.11-5 is a system of coupled linear algebraic equations even if [**M**] and [**C**] are diagonal. For linear problems, $[\mathbf{K}^{\text{eff}}]$ need be formed and factored only once. After the initial expense of factorization, time stepping can be performed for only the cost of forward- and back-substitution.
2. If [**M**] is positive definite, then $[\mathbf{K}^{\text{eff}}]$ is nonsingular even if [**K**] permits rigid-body displacements.
3. The method is easily started from initial conditions $\{\mathbf{D}\}_0$ and $\{\dot{\mathbf{D}}\}_0$ and Eq. 13.10-7.

4. For problems having material nonlinearity, [**K**] and hence [$\mathbf{K}^{\text{eff}}$] are functions of $\{\mathbf{D}\}_{n+1}$, and possibly its time derivative, which are unknowns. Accordingly, [**K**] must be predicted using an estimate for $\{\mathbf{D}\}_{n+1}$. Equation 13.11-5 is then solved for an improved $\{\mathbf{D}\}_{n+1}$; hence, the prediction of [**K**] is improved, and so on. For severe nonlinearity, convergence may be difficult and expensive.
5. In terms of computational efficiency, Eq. 13.11-6 shows that there is no merit to using a lumped mass matrix. In Section 13.13 we show that implicit integration with consistent mass matrices is usually more accurate than with lumped mass matrices.

Choice of Time Step Δt. Since the method is numerically stable for any Δt, time step selection is based on accuracy considerations alone. Compared with explicit methods, the per-time-step cost of an implicit method is high. Thus implicit methods are economically attractive only when Δt can be much larger than would be used in an explicit method. Unconditional stability (which emphatically does *not* imply unconditional accuracy) coupled with the economic need for large Δt tempts many analysts into using time steps that are too large. To select a time step that will provide accurate results, one must identify the highest frequency of interest in the loading or response of a structure. Let this frequency be called ω_u. As an approximation, structure modes with frequency higher than about $3\omega_u$ participate quasistatically in the response while modes with frequency lower than $3\omega_u$ also participate dynamically. With second-order accurate time integration methods (most popular methods are second-order accurate), a minimum of 20 time steps per period of ω_u should provide very good accuracy for modes that participate dynamically in the response; that is, use $\Delta t < (2\pi/\omega_u)/20 \approx 0.3/\omega_u$, unless a smaller Δt is required because of convergence difficulties in nonlinear analysis. As an additional assurance of an accurate solution, analysis should be repeated using a smaller time step than used in the first analysis. In linear problems of a type well suited to implicit integration, a Courant number[3] C_n of about 20 is typical although occasionally it can be as high as 100 and still yield accurate results.

13.12 OTHER IMPLICIT AND EXPLICIT METHODS. MIXED METHODS

The central-difference and trapezoidal-rule methods do not provide automatic dissipation of high-frequency numerical noise, as is sometimes desirable. In what follows, we discuss some methods that often have *dissipation* (also called *artificial viscosity* or *numerical damping*).

The *Houbolt method* [13.55] is obtained by cubic Lagrange interpolation of {**D**} at times $(n - 2)\,\Delta t$ through $(n + 1)\,\Delta t$. Exact differentiation of the interpolant yields

$$\{\dot{\mathbf{D}}\}_{n+1} = \frac{1}{6\,\Delta t}\left(11\,\{\mathbf{D}\}_{n+1} - 18\,\{\mathbf{D}\}_n + 9\,\{\mathbf{D}\}_{n-1} - 2\,\{\mathbf{D}\}_{n-2}\right) \qquad (13.12\text{-}1)$$

[3]The definition of C_n, Eq. 13.10-18, employs Δt_{stable} for *explicit* integration. This definition remains useful in the present context even though most implicit methods are stable for any time step.

$$\{\ddot{\mathbf{D}}\}_{n+1} = \frac{1}{\Delta t^2}(2\{\mathbf{D}\}_{n+1} - 5\{\mathbf{D}\}_n + 4\{\mathbf{D}\}_{n-1} - \{\mathbf{D}\}_{n-2}) \qquad (13.12\text{-}2)$$

This method is implicit and unconditionally stable but provides artificial damping that is too high for low-frequency response. The Houbolt method was once common in general-purpose transient codes but has been supplanted by methods with better algorithmic damping properties and now is more of historical interest.

The *Newmark family of methods* [13.56] is very popular and is given by

$$\{\mathbf{D}\}_{n+1} = \{\mathbf{D}\}_n + \Delta t\{\dot{\mathbf{D}}\}_n + \frac{\Delta t^2}{2}[(1 - 2\beta)\{\ddot{\mathbf{D}}\}_n + 2\beta\{\ddot{\mathbf{D}}\}_{n+1}] \qquad (13.12\text{-}3)$$

$$\{\dot{\mathbf{D}}\}_{n+1} = \{\dot{\mathbf{D}}\}_n + \Delta t[(1 - \gamma)\{\ddot{\mathbf{D}}\}_n + \gamma\{\ddot{\mathbf{D}}\}_{n+1}] \qquad (13.12\text{-}4)$$

where β and γ are chosen by the analyst to control stability and accuracy. Substitution of Eqs. 13.12-3 and 13.12-4 into Eq. 13.9-1 at time $(n + 1)\,\Delta t$ yields equations similar to Eq. 13.10-5 for explicit Newmark methods ($\beta = 0$) and to Eqs. 13.11-5, 13.11-6, and 13.11-7 for implicit Newmark methods ($\beta > 0$). It can be shown that the stability of this algorithm is [13.49]:
unconditional stability when

$$2\beta \geq \gamma \geq \tfrac{1}{2} \qquad (13.12\text{-}5)$$

conditional stability when

$$\gamma \geq \tfrac{1}{2}, \quad \beta < \tfrac{1}{2}, \quad \text{and} \quad \Delta t \leq \frac{\xi(\gamma - \frac{1}{2}) + \sqrt{\gamma/2 - \beta + \xi^2(\gamma - \frac{1}{2})^2}}{\omega_{\max}(\gamma/2 - \beta)} \qquad (13.12\text{-}6)$$

The method is unstable for $\gamma < \frac{1}{2}$. (As a special case, we note that Eq. 13.12-6 yields infinite Δt when $\gamma = \frac{1}{2}$ and $\beta = \frac{1}{4}$—that is, the trapezoidal rule without damping.) According to our definitions of implicit and explicit methods, Newmark's method is implicit unless $\gamma = \beta = 0$, which is unstable for any Δt and therefore cannot be used. However, most analysts refer to Newmark's method with $\beta = 0$ and $\gamma \geq \frac{1}{2}$ as being explicit in which it is noted that, for practical purposes, [**C**] must be null or diagonal to avoid the solution of simultaneous equations. An explicit version of the Newmark method, called a predictor–corrector algorithm, that permits nondiagonal [**C**] is described in [13.61]. Implementation of the explicit and implicit Newmark methods are essentially the same as the computational procedures given in Tables 13.10-1 and 13.11-1, respectively.

A variety of useful techniques obtained from the Newmark family is listed in Table 13.12-1. When $\gamma = \frac{1}{2}$, the methods have no algorithmic damping and are second-order accurate. An exception is the *Fox–Goodwin method* with [**C**] = [**0**], which is fourth-order accurate. Taking $\gamma > \frac{1}{2}$ introduces artificial damping, but also reduces the accuracy of the Newmark methods to first order. For implicit Newmark methods, taking

$$\beta = \tfrac{1}{4}(\gamma + \tfrac{1}{2})^2 \qquad (13.12\text{-}7)$$

maximizes the high-frequency dissipation for a given value of $\gamma > \frac{1}{2}$ [13.46].

TABLE 13.12-1. SUMMARY OF NEWMARK METHODS: u = UNDAMPED, d = DAMPED. $\Omega_{\text{crit}} = \omega_{\max}\,\Delta t_{\max}$. STABILITY REQUIRES $\Delta t \le \Omega_{\text{crit}}/\omega_{\max}$.

	Method	β	γ	Ω_{crit}	Accuracy
Implicit	Artificially damped	$> \gamma/2$	$> 1/2$	$\infty^{(u,d)}$	$O(\Delta t)$
	Average acceleration (trapezoidal rule)	1/4	1/2	$\infty^{(u,d)}$	$O(\Delta t^2)$
	Linear acceleration	1/6	1/2	$2\sqrt{3} \approx 3.464^{(u)}$ Eq. (A)$^{(d)}$	$O(\Delta t^2)$
	Fox–Goodwin (royal road)	1/12	1/2	$\sqrt{6} \approx 2.449^{(u)}$ Eq. (A)$^{(d)}$	$O(\Delta t^4)^{(u)}$ $O(\Delta t^2)^{(d)}$
Explicit	Central difference [**M**], [**C**] diagonal	0	1/2	$2^{(u)}$ Eq. (A)$^{(d)}$	$O(\Delta t^2)$
	Artifically damped [**M**], [**C**] diagonal	0	$> 1/2$	Eq. (A)$^{(u,d)}$	$O(\Delta t)$

$$\Omega_{\text{crit}} = \frac{\xi(\gamma - \frac{1}{2}) + \sqrt{\gamma/2 - \beta + \xi^2(\gamma - \frac{1}{2})^2}}{\gamma/2 - \beta} \qquad \text{(A)}$$

Interestingly, the presence of damping in the explicit Newmark method *raises* the stability limit. This is in contrast to Eqs. 13.10-8 and 13.10-14 for other forms of the central-difference method in which no change and a decrease in stability limit, respectively, are observed. Thus, if the fraction ξ of critical damping at $\omega_{\max}$ is not known, a conservative Δt is obtained by taking $\xi = 0$ in Eq. 13.12-6. However, practical necessity dictates that [**C**] be diagonal in the explicit Newmark method, thus the lagged central-difference algorithm, Eq. 13.10-12 (or a predictor–corrector algorithm [13.61]) is preferable for problems in which physically realistic spectral damping is to be modeled.

The Newmark linear-acceleration method appears to be ideal for problems such as earthquake shaking response analysis in which piecewise linear-acceleration records are typically used as excitation. Unfortunately, this implicit method is only conditionally stable. The *Wilson-θ method* [13.18,13.28,13.46,13.57] is also a linear-acceleration method, but is unconditionally stable.

A disadvantage of the Newmark methods is that algorithmic damping can only be obtained at the expense of reduced accuracy. The *α-method,* proposed by Hilber, Hughes, and Taylor [13.58], does not have this weakness, and with appropriate choice of parameters retains second-order accuracy and provides effective high-frequency dissipation. The method uses the Newmark formulas, Eqs. 13.12-3 and 13.12-4, with the modified equation of motion

$$\begin{aligned}[\mathbf{M}]\{\ddot{\mathbf{D}}\}_{n+1} + (1 + \alpha)[\mathbf{C}]\{\dot{\mathbf{D}}\}_{n+1} - \alpha[\mathbf{C}]\{\dot{\mathbf{D}}\}_n + (1 + \alpha)[\mathbf{K}]\{\mathbf{D}\}_{n+1} \\ - \alpha[\mathbf{K}]\{\mathbf{D}\}_n = (1 + \alpha)\{\mathbf{R}^{\text{ext}}\}_{n+1} - \alpha\{\mathbf{R}^{\text{ext}}\}_n \qquad (13.12\text{-}8)\end{aligned}$$

If the parameters are selected so that $-\frac{1}{3} \le \alpha \le 0$, $\gamma = (1 - 2\alpha)/2$, and $\beta = (1 - \alpha)^2/4$, the method is implicit, unconditionally stable, and second-order accurate [13.46]. When these guidelines are used, with $\alpha = 0$, the method reduces to the trapezoidal rule, which has no dissipation. Decreasing α increases the amount of numerical damping.

Mixed Methods. A current trend in time integration analysis is to create algorithms that combine explicit and implicit methods, so as to capitalize on the strong points of each. Such schemes are called *mixed integration methods*. Belystschko and Mullen [13.59,13.60] developed a method that uses a *nodal partition* to separate nodes into implicit and explicit groups. D.o.f. in the explicit and implicit groups are then integrated by explicit and implicit methods respectively. The motivation for such an approach is that structures often contain spatial subdomains having markedly different time scales. For example, in fluid–structure interaction problems, the time scales associated with the fluid are usually much longer than those associated with the structure. By using an implicit method for the structure and an explicit method for the fluid, we exploit the strong points of each integrator. Hughes and Liu [13.61,13.62] developed an implicit–explicit method similar to Refs. [13.59,13.60] but more implementationally attractive. It uses *element* partitions rather than nodal partitions. *Operator-splitting methods* are mixed methods in which a nonlinear material constitutive law is split into parts that give rise to time-dependent and time-independent terms that are integrated explicitly and implicitly, respectively [13.63,13.64]. *Element-by-element implicit methods* have been developed in which a conventional implicit method is used except that Eqs. 13.11-5 are approximately solved using element-level calculations only [13.65,13.66,13.67]. Thus these methods appear to have the good stability characteristics of implicit methods with the low per-time-step cost of explicit methods. At present, however, these methods are not sufficiently robust and success is very problem-dependent.

13.13 STABILITY ANALYSIS. ACCURACY OF DIRECT INTEGRATION METHODS

In Sections 13.10 and 13.11 many stability and accuracy properties of direct integration methods are stated. In the present section we substantiate some of these properties and offer further suggestions for use of the methods. References [13.18,13.46,13.49] contain extensive discussions.

Stability. When we examine stability, it is sufficient to consider the homogeneous form of the equation of motion obtained by taking $\{\mathbf{R}^{\text{ext}}\} = \{\mathbf{0}\}$. The idea is that if a solution procedure is stable with no external loading, then it will also be stable if $\{\mathbf{R}^{\text{ext}}\}$ is nonzero but bounded. A number of methods for assessing stability are possible. They fall into two broad categories. In the first, called *spectral* or *Fourier stability,* one examines the effects of a time integration method on a single equation of motion obtained by modally uncoupling the original structure equations. In the second, called *energy stability,* one deals with the original structure matrices and establishes the conditions under which a norm of the solution at time $n\ \Delta t$ can be bounded by a norm of the solution at time zero. Spectral stability usually provides more insight and sometimes more precise results. Energy methods sometimes provide results that are slightly more conservative than spectral methods, but can be applied to complicated problems to which spectral methods may be inapplicable. References on energy stability techniques include [13.46,13.49, 13.61,13.63].

Spectral Stability: Central-Difference Method. To illustrate spectral stability, we consider the central-difference method applied to the undamped, uncoupled, homogeneous equation of motion (Eq. 13.6-5 with subscript i dropped for clarity):

$$\ddot{Z} + \omega^2 Z = 0 \qquad (13.13\text{-}1)$$

The solution of Eq. 13.13-1 is purely oscillatory and is given by Eq. 13.13-15. Using the central-difference method, Eq. 13.10-2, we can write a *difference approximation* of Eq. 13.13-1 at time $n\ \Delta t$ as

$$Z_{n+1} + (\omega^2\ \Delta t^2 - 2)Z_n + Z_{n-1} = 0 \qquad (13.13\text{-}2)$$

Because the central-difference method is a two-step method, the exact solution of the difference approximation, Eq. 13.13-2, can be shown to consist of a linear combination of two parts,

$$Z_n = C_1\ \lambda_1^n + C_2\ \lambda_2^n \qquad \text{when } \lambda_1 \neq \lambda_2 \qquad (13.13\text{-}3a)$$

$$Z_n = C_1\ \lambda_1^n + n\ \Delta t\ C_2\ \lambda_1^n \qquad \text{when } \lambda_1 = \lambda_2 \qquad (13.13\text{-}3b)$$

where

$$\lambda_1 = e^{\mu_1\ \Delta t} \qquad \text{and} \qquad \lambda_2 = e^{\mu_2\ \Delta t} \qquad (13.13\text{-}4)$$

in which μ_1 and μ_2 (and hence λ_1 and λ_2) are generally complex. Constants C_1 and C_2 can be determined from initial conditions.

The essence of the stability argument is embodied in Eqs. 13.13-3. Assuming for now that $\lambda_1 \neq \lambda_2$, Eq. 13.13-3a provides an unbounded solution for Z_n if $|\lambda_1| > 1$ or if $|\lambda_2| > 1$. This is instability. If $|\lambda_1| \leq 1$ and $|\lambda_2| \leq 1$, then Z_n will decay or remain steady with time, thus providing stable computation.

If $\lambda_1 = \lambda_2$, then even though $|\lambda_1| = |\lambda_2| = 1$, Z_n will not be bounded and will grow in arithmetic fashion due to the second term of Eq. 13.13-3b. If so, computation will be unstable. Although this situation arises infrequently in practical finite element models, it is well to be aware of its possible occurance.

For methods other than central difference, it is possible to have $\lambda_1 = \lambda_2$, $|\lambda_1| < 1$, and $|\lambda_2| < 1$, in which case Z_n of Eq. 13.3-3b will be bounded.

To determine the criterion for Δt that provides a stable computation, we proceed as follows. We first assume that $\lambda_1 \neq \lambda_2$ and select initial conditions such that C_1 or C_2 is zero. Equation 13.13-3a then becomes

$$Z_n = C\lambda^n, \qquad \text{where} \quad \lambda = e^{\mu\ \Delta t} \qquad (13.13\text{-}5)$$

and where subscripts have been dropped from C and λ. Combining Eq. 13.13-5 with Eq. 13.13-2 and dividing by $C\lambda^n$, we obtain the *characteristic equation*

$$\lambda^2 + (\omega^2\ \Delta t^2 - 2)\lambda + 1 = 0 \qquad (13.13\text{-}6)$$

Solving Eq. 13.13-6 provides the two solutions for λ,

$$\lambda_{1,2} = \tfrac{1}{2}(2 - \omega^2\ \Delta t^2 \pm \omega\ \Delta t\sqrt{\omega^2\ \Delta t^2 - 4}) \qquad (13.13\text{-}7)$$

Although it is possible to determine the $\omega\,\Delta t$ that satisfy $|\lambda_1| \leq 1$ and $|\lambda_2| \leq 1$ directly from Eq. 13.13-7, it is more expedient to note that the solutions of a quadratic equation $a\lambda^2 + b\lambda + c = 0$ satisfy $\lambda_1\lambda_2 = c/a$, where $c/a = 1$ for Eq. 13.13-6. If the radicand of Eq. 13.13-7 is positive, then both λ_1 and λ_2 are real and $\lambda_1\lambda_2 = 1$ indicates that the absolute value of one λ is less than unity while the absolute value of the other λ is greater than unity. Thus instability results since one λ does not satisfy $|\lambda| \leq 1$. If the radicand of Eq. 13.13-7 is negative, then λ_1 and λ_2 are complex conjugates and each is of unit modulus by virtue of $\lambda_1\lambda_2 = 1$. Thus stable computation *always* results if the radicand of Eq. 13.13-7 is negative. If the radicand of Eq. 13.13-7 is zero, then $\lambda_1 = \lambda_2 = -1$. Since the characteristic equation has repeated solutions, Eq. 13.13-3b is the appropriate expression for Z_n in which it is observed that with $|\lambda_1| = 1$, the solution will diverge in arithmetic fashion due to the second term in Eq. 13.13-3b. Hence, stable computation requires

$$\Delta t < \frac{2}{\omega} \tag{13.13-8}$$

This result depends on the form of differential equation and the type of difference approximation. If one of these is changed, say by the addition of damping to Eq. 13.13-1, then in general the stability criterion is also changed.

Note that Eq. 13.13-1 is just one of many uncoupled equations and that our intent is to solve the system of *coupled* equations by direct integration. Thus it is necessary to evaluate Eq. 13.13-6 for each of the n_{eq} frequencies of the model and select the Δt that is most restrictive. This yields

$$\Delta t < \frac{2}{\omega_{\max}} \tag{13.13-9}$$

where $\omega_{\max}$ is the *highest* of the n_{eq} natural frequencies.

Equation 13.13-9 usually appears in the literature as $\Delta t \leq 2/\omega_{\max}$. However, as discussed earlier, using $\Delta t = 2/\omega_{\max}$ in central-difference integration of the undamped equations of motion yields instability and hence should be avoided. In practical problems, $\omega_{\max}$ is rarely known and, as discussed in Section 13.10, is usually bounded by the maximum element frequency among all elements, $(\omega_{\max})_e$. In practical problems, the case $(\omega_{\max})_e = \omega_{\max}$ is rare, so that $(\omega_{\max})_e > \omega_{\max}$ almost always prevails. Accordingly, the time step $\Delta t \leq 2/(\omega_{\max})_e$ almost always provides stable computation.

Spectral Stability: Trapezoidal Rule. To analyze the stability of the undamped, homogeneous equation of motion when integrated by the trapezoidal rule, we begin by summing Eq. 13.13-1 at times $(n - 1)\,\Delta t$ and $(n + 1)\,\Delta t$ with twice Eq. 13.13-1 at time $n\,\Delta t$. This provides

$$\ddot{Z}_{n+1} + 2\ddot{Z}_n + \ddot{Z}_{n-1} + \omega^2(Z_{n+1} + 2Z_n + Z_{n-1}) = 0 \tag{13.13-10}$$

To express the trapezoidal rule in terms of accelerations and displacements only, we subtract Eq. 13.11-1 at time $n\,\Delta t$ from Eq. 13.11-1 at time $(n + 1)\,\Delta t$ and then combine with Eq. 13.11-2 written at times $(n + 1)\,\Delta t$ and $n\,\Delta t$ to eliminate velocities. This provides the trapezoidal formula for second derivatives

$$Z_{n+1} - 2Z_n + Z_{n-1} = \frac{\Delta t^2}{4}(\ddot{Z}_{n+1} + 2\ddot{Z}_n + \ddot{Z}_{n-1}) \tag{13.13-11}$$

Combining Eqs. 13.13-10 and 13.13-11 to eliminate accelerations provides

$$(1 + h)Z_{n+1} + (2h - 2)Z_n + (1 + h)Z_{n-1} = 0, \qquad \text{where} \quad h = \frac{\omega^2 \Delta t^2}{4} \tag{13.13-12}$$

Assuming that $\lambda_1 \neq \lambda_2$, we combine Eqs. 13.13-12 and 13.13-5 and divide by λ^{n-1} to obtain the characteristic equation

$$(1 + h)\lambda^2 + (2h - 2)\lambda + 1 + h = 0 \tag{13.13-13}$$

Solutions of Eq. 13.13-13 for λ are

$$\lambda_{1,2} = \frac{1 - h \pm 2\sqrt{-h}}{1 + h}, \qquad \text{where} \quad h = \frac{\omega^2 \Delta t^2}{4} \tag{13.13-14}$$

In accordance with the discussion following Eq. 13.13-7, $\lambda_1\lambda_2 = 1$ for the trapezoidal rule. In addition, note that the radicand of Eq. 13.13-14 is always negative. Therefore, λ_1 and λ_2 are complex conjugates (hence distinct) and both of unit modulus. Thus $|\lambda| \leq 1$ is satisfied regardless of the value of h. Therefore, trapezoidal-rule integration of the undamped equations of motion is unconditionally stable. The same is true when damping is included, although we have not proved it.

Amplitude and Period Error. To analyze the errors in using the central-difference method and the trapezoidal rule, which are representative of explicit and implicit methods in general, we again focus attention on a single, uncoupled homogeneous equation, Eq. 13.13-1. The exact solution is harmonic and can be written as

$$Z_n^{\text{exact}} = \tilde{C}_1(\cos \omega t + i \sin \omega t) + \tilde{C}_2(\cos \omega t - i \sin \omega t) \tag{13.13-15}$$

where $\tilde{C}_1$ and $\tilde{C}_2$ are determined from initial conditions, and $i = \sqrt{-1}$. The approximate solution obtained by direct integration may display *amplitude error* and *period error*. Amplitude error can be either amplitude increase, which is the same as instability, or amplitude decay, which is more commonly called artificial damping or viscosity. Period error can be either *period elongation* or *period contraction*. These errors are shown in Fig. 13.13-1.

Because λ_1 and λ_2 are complex conjugates (in both the central-difference and trapezoidal-rule methods), μ_1 and μ_2 in Eq. 13.13-4 are also complex conjugates and can be written as

$$\mu_1 = a + ib \qquad \text{and} \qquad \mu_2 = a - ib \tag{13.13-16}$$

where a and b are real. By use of Eq. 13.13-16, Eq. 13.13-4 can be written as

$$\lambda_1^n = e^{an\,\Delta t}e^{ibn\,\Delta t} = e^{an\,\Delta t}(\cos bn\,\Delta t + i \sin bn\,\Delta t) \tag{13.13-17a}$$

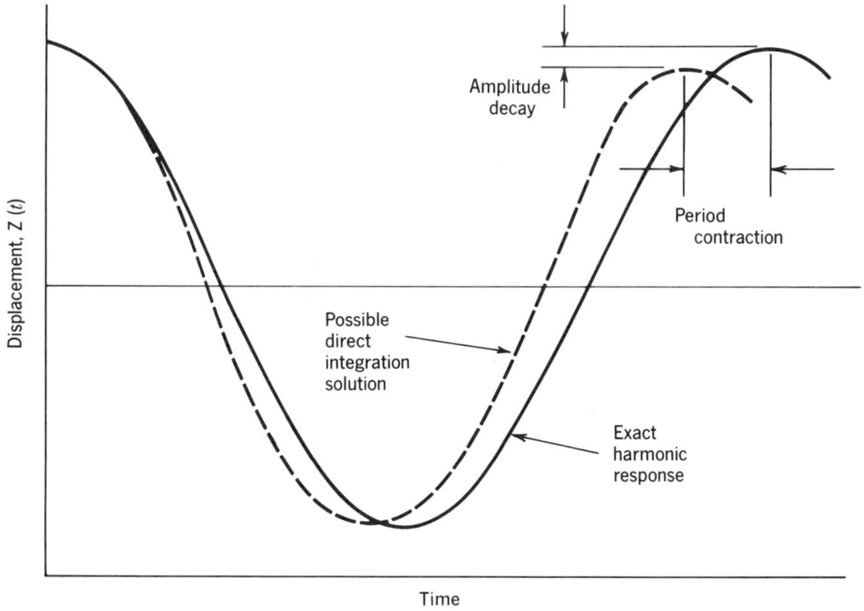

Figure 13.13-1. Possible errors in direct integration.

$$\lambda_2^n = e^{an\,\Delta t}e^{-ibn\,\Delta t} = e^{an\,\Delta t}(\cos bn\,\Delta t - i \sin bn\,\Delta t) \qquad (13.13\text{-}17b)$$

in which the latter forms are obtained by use of DeMoivre's theorem. Combination of Eqs. 13.13-17 and 13.13-3a provides the exact solution to the central-difference approximation:

$$Z_n = C_1\, e^{an\,\Delta t}(\cos bn\,\Delta t + i \sin bn\,\Delta t) + C_2\, e^{an\,\Delta t}(\cos bn\,\Delta t - i \sin bn\,\Delta t) \qquad (13.13\text{-}18)$$

Comparison of Eqs. 13.13-15 and 13.13-18 shows that amplitude error will result unless $a = 0$ and period error will result unless $b = \omega$.

We define period error, P, by

$$P = \frac{2\pi/b}{2\pi/\omega} = \frac{\omega}{b} \qquad (13.13\text{-}19)$$

where $2\pi/\omega$ and $2\pi/b$ are respectively the period of the actual system and the period of the system created by the time integration algorithm. The following types of error are possible:

$P > 1$ period elongation

$P = 1$ no period error

$P < 1$ period contraction

To determine b, we note from Eq. 13.13-17a with $n = 1$ that

$$\frac{\text{Im}(\lambda)}{\text{Re}(\lambda)} = \frac{\sin b\,\Delta t}{\cos b\,\Delta t} = \tan b\,\Delta t \qquad (13.13\text{-}20)$$

For the central-difference method, Eqs. 13.13-7 and 13.13-20 give the period, b, of the numerically integrated solution as

$$b = \frac{1}{\Delta t} \tan^{-1} \frac{\omega \, \Delta t \sqrt{4 - \omega^2 \, \Delta t^2}}{2 - \omega^2 \, \Delta t^2} \tag{13.13-21}$$

Combination of Eqs. 13.13-19 and 13.13-21 yields the period error of the central-difference method as

$$P = \omega \, \Delta t \left[\tan^{-1} \frac{\mathrm{Im}(\lambda)}{\mathrm{Re}(\lambda)} \right]^{-1} = \omega \, \Delta t \left[\tan^{-1} \frac{\omega \, \Delta t \sqrt{4 - \omega^2 \, \Delta t^2}}{2 - \omega^2 \, \Delta t^2} \right]^{-1} \tag{13.13-22}$$

in which the arctangent function is required to yield a positive angle. Equation 13.13-22 represents period contraction and is plotted in Fig. 13.13-2. For the trapezoidal rule, Eqs. 13.13-14, 13.13-19, and 13.13-20 give the period error

$$P = \omega \, \Delta t \left[\tan^{-1} \frac{4\omega \, \Delta t}{4 - \omega^2 \, \Delta t^2} \right]^{-1} \tag{13.13-23}$$

in which the arctangent function is required to yield a positive angle. This represents period elongation and is plotted in Fig. 13.13-2.

Figure 13.13-2 suggests guidelines for mass-matrix selection in direct integration. Natural frequencies obtained by use of consistent mass matrices are overestimated. Hence the periods of modes are underestimated, or contracted. When consistent mass matrices are used with the trapezoidal rule, which has period elongation, period errors in the direct integration of the equations of motion are partially compensatory. This observation is true of implicit methods in general. Use of lumped mass matrices usually underestimates natural frequencies and hence overestimates, or elongates, periods. When lumped mass matrices are used with the central-difference method, period errors are partially compensatory. This observation is also true of explicit methods in general.

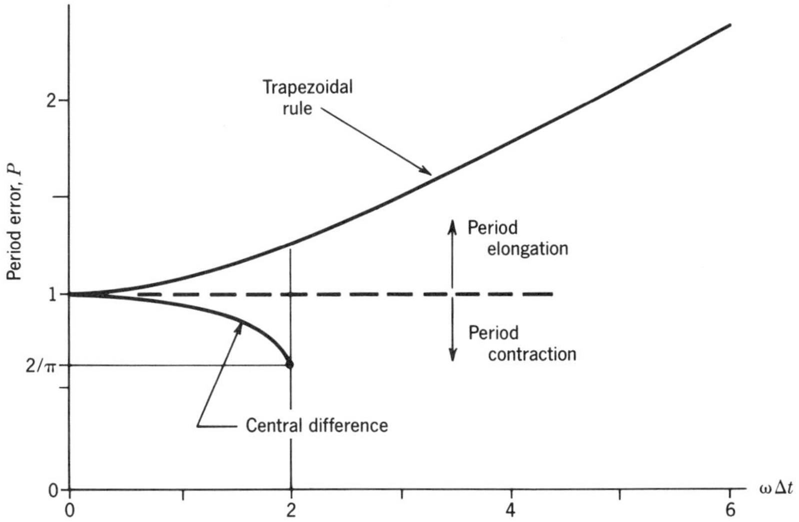

Figure 13.13-2. Period errors for the central-difference and trapezoidal-rule methods.

We emphasize that Fig. 13.13-2 pertains to the mode whose frequency is ω. A multi-d.o.f. structure has many modes. For structural analysis one selects a Δt such that $\omega\, \Delta t$ is small for all modes of practical interest.

The following amplitude errors are possible:

$a > 0$ amplitude growth (instability)

$a = 0$ no amplitude error

$a < 0$ amplitude decay (artificial damping)

Neither the central-difference nor the trapezoidal-rule method has amplitude error. This can be seen immediately by noting that, when integration is stable, Eq. 13.13-3a applies, and $|\lambda_1| = |\lambda_2| = 1$ (although $\lambda_1 \neq \lambda_2$). Hence, Eq. 13.13-3a shows that, although Z_n may be periodic, it will not grow or decay.

To show in a more formal manner that the central-difference method does not have amplitude error, we consider the sum $\lambda_1 + \lambda_2$ as given by Eqs. 13.13-7 and 13.13-17 with $n = 1$. This provides

$$e^{a\,\Delta t} \cos b\,\Delta t = 1 - \tfrac{1}{2}\omega^2\,\Delta t^2 \tag{13.13-24}$$

Combining Eq. 13.13-24 with Eq. 13.13-21 provides

$$a\,\Delta t = \ln \frac{1 - \omega^2\,\Delta t^2/2}{\cos\left[\tan^{-1} \dfrac{\omega\,\Delta t\sqrt{4 - \omega^2\,\Delta t^2}}{2 - \omega^2\,\Delta t^2}\right]} \tag{13.13-25}$$

Evaluation of Eq. 13.13-25 for $0 \leq \omega\,\Delta t < 2$ shows that $a\,\Delta t = 0$, thus verifying that the central-difference method has no amplitude error. However, this does not imply that the amplitude of response as predicted by the central-difference method will agree with the exact response. This is particularly true when using stable time steps that are very close to the stability limit. To show this, consider Eqs. 13.13-15 and 13.13-18. When λ_1 and λ_2 are distinct, C_1 and C_2 in Eq. 13.13-18 are close approximations of $\tilde{C}_1$ and $\tilde{C}_2$ in Eq. 13.13-15. When Δt is very close to the stability limit, λ_1 and λ_2 are almost equal and the expressions within the parentheses of Eq. 13.13-18 are almost linear combinations of one another. In this situation, C_1 and C_2 may differ greatly from $\tilde{C}_1$ and $\tilde{C}_2$. In fact, it is for this reason that the "beating" phenomenon of Fig. 13.10-5 is displayed when Δt is very close to the stability limit. What is implied by the statement of zero-amplitude error is that the envelope of the numerical solution has a mean value that does not grow or decay in comparison with the exact solution.

Following the same procedure used for the central-difference method, we obtain, for the trapezoidal rule,

$$a\,\Delta t = \ln \frac{4 - \omega^2\,\Delta t^2}{(4 + \omega^2\,\Delta t^2)\cos\left[\tan^{-1}\left(\dfrac{4\omega\,\Delta t}{4 - \omega^2\,\Delta t^2}\right)\right]} \tag{13.13-26}$$

which gives $a\,\Delta t = 0$ for all values of $\omega\,\Delta t$, and therefore implies zero-amplitude error.

13.14 CONCLUDING REMARKS ON TIME-HISTORY ANALYSIS

Choice of Method. The choice of method for time-history analysis is strongly problem-dependent. The efficiency of a given method depends on whether the problem is of a wave propagation or a structural dynamics type, the time span for which analysis is required, whether response is linear or nonlinear, and the topology of the finite element mesh.

In wave propagation problems the excitation is usually rich in high-frequency components. Time scales of interest are short and of the order of the acoustic wave traversal time across a structure. Usually we are interested in observing the passage of stress waves through elements and the transients produced. In structural dynamics problems, the excitation and response are characterized by low-frequency, long-time-scale components. Analysis duration is usually long compared to that normally required for a wave propagation problem.

Modal superposition methods are economical when only a small portion of the total number of vibration modes of a model need be used for superposition. Wave propagation problems would require a very large number of modes to be included. Therefore, superposition methods are generally not appropriate for wave propagation problems. In structural dynamics problems, excitation and structural response are dominated by low-frequency components; hence superposition methods can be very effective.

If material response becomes nonlinear or deformations become large, a structure's eigenvalues and eigenvectors change. Because of the expense of solving eigenproblems, it is not prudent to continuously update eigenpairs during nonlinear response. Rather, most modal methods for nonlinear problems treat nonlinearities by *pseudoload* techniques in which loads that account for nonlinearities are transferred to the right-hand side of the equation of motion. These methods are not robust. Convergence is strongly problem-dependent and is often poor. Generally speaking, direct integration methods are preferable for nonlinear problems.

With explicit methods of direct integration, stability typically requires that the time step be small enough that information does not propagate across more than one element per time step (e.g., the CFL condition). Explicit methods are ideal for wave propagation problems in which behavior at the stress wave front is of engineering importance. Here the stability restriction is not a serious disadvantage because a small Δt is necessary for accuracy. Other factors in favor of explicit time integration are easy implementation, accurate treatment of general nonlinearities, and the capability of treating very large problems with only modest computer storage requirements. For structural dynamics problems, time scales and analysis durations are usually long and accuracy considerations alone would permit a Δt much larger than the upper limit of Δt for stable explicit integration. Although explicit methods are often used for structural dynamics problems, they are not as well suited to this class of problems as they are to wave propagation problems.

The only advantage of implicit methods over explicit methods is that they allow a much larger Δt because they are unconditionally stable (conditionally stable implicit methods are not often used). Implicit methods are expensive for wave propagation problems since accuracy requires a small Δt. For long-duration structural dynamics problems, implicit methods are usually more effective than explicit methods, although this depends on mesh topology and severity of nonlinearities.

Compared with explicit methods, implicit methods are more difficult to implement, particularly for nonlinear problems, and they require considerably more computer storage.

Choice of Element and Mesh. When discretizing a structure or a medium, an analyst can choose from simple, low-order elements such as the linear-displacement bar, quadratic beam, and bilinear quadrilateral, or from higher-order elements such as the quadratic Lagrange and serendipity quadrilaterals. In wave propagation problems, discontinuities of strain propagate throughout the model. Lower-order displacement elements are more adept at modeling these discontinuities than are higher-order elements, which tend to produce more numerical noise. Structural dynamics problems tend to have strain fields that vary smoothly with time. Hence, higher-order elements can be used to more advantage than in wave propagation problems. Higher-order elements can also be used effectively in eigenvalue problems.

Guidelines discussed in Chapter 19 for construction of finite element meshes for quasistatic problems are also useful for dynamic problems. Thus, static or dynamic stress analysis requires a finer mesh (particularly near stress raisers) than does quasistatic deflection analysis or calculation of lower vibration frequencies. A dynamic problem may require more elements than the analogous quasistatic problem. For example, in the problem of Fig. 13.10-2, only one linear-displacement bar element is necessary for the quasistatic solution, whereas many more elements are necessary to capture the essential features of the dynamic problem or to calculate accurate natural frequencies and mode shapes.

Element sizes should not change abruptly. If they do, the mass matrix will be a poor discrete representation of the actual continuous mass distribution of the structure. This gives rise to artificial wave reflections and additional numerical noise when waves cross boundaries between elements of markedly different size.

With explicit methods, a lumped mass matrix is preferred for reasons of economy and accuracy. With implicit methods, a consistent mass matrix is preferred for accuracy and is only slightly detrimental to economy.

PROBLEMS

Section 13.1

13.1 A single d.o.f. spring–mass system has natural frequency $\omega_1 = \sqrt{k/m}$. It is excited by a force $P_0 \sin \omega_2 t$. What is the limiting value of the ratio ω_2/ω_1 such that the amplitude of motion differs from the static displacement by less than 10%?

Section 13.2

13.2 Show that under constant acceleration $\{\ddot{\mathbf{d}}\}$, nodal "loads" $[\mathbf{m}]\{\ddot{\mathbf{d}}\}$ with $[\mathbf{m}]$ given by Eq. 13.2-5 are the same as the body force nodal loads given by Eq. 4.1-6.

Section 13.3

13.3 (a) Can a diagonal coefficient in a consistent mass matrix ever be negative? Explain.

(b) Imagine that a beam is vibrated so that nodes of the vibration mode coincide with nodes of the finite element mesh. Would you prefer the consistent mass matrix of Eq. 13.3-2 or the lumped mass matrix of Eq. 13.3-3 with $\alpha = 0$? Is Eq. 13.3-3 with $\alpha \neq 0$ an acceptable alternative?

13.4 (a) Derive the consistent mass matrix given by Eq. 13.3-1.
(b) Imagine that the cross-sectional area of the bar shown in the figure varies linearly from area A_0 at the left end to area γA_0 at the right end, where γ is a constant. Determine the consistent mass matrix associated with d.o.f. u_1 and u_2.
(c) Show that $[\mathbf{m}]$ and $\lceil \mathbf{m} \rfloor$ in Eq. 13.3-1 each yield the correct nodal forces under a rigid-body translational acceleration in the bar's axial direction.

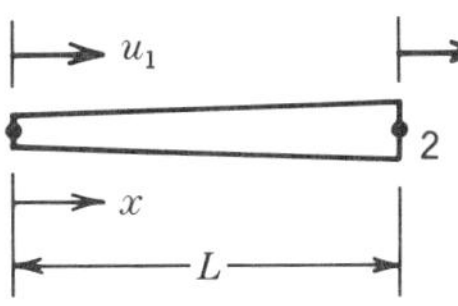

Problem 13.4

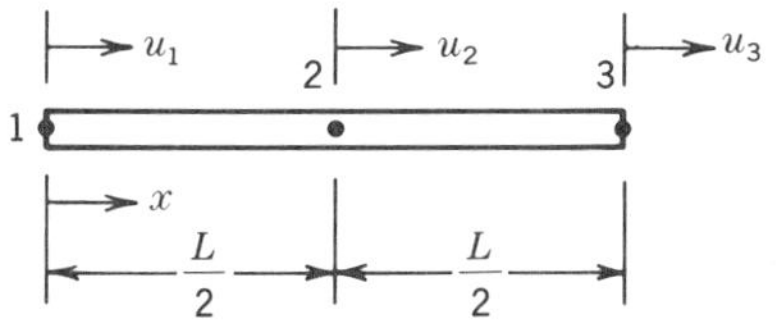

Problem 13.7

13.5 (a) Derive the consistent mass matrix given by Eq. 13.3-2.
(b) Show that Eq. 13.3-2 yields the correct nodal forces and moments under a rigid-body translational acceleration transverse to the axis of the beam.

13.6 Determine the consistent mass matrix of the constant-strain triangle (Fig. 4.2-3). The element has uniform density and thickness. Arrange d.o.f. in the order $\{\mathbf{d}\} = \lfloor u_1 \quad u_2 \quad u_3 \quad v_1 \quad v_2 \quad v_3 \rfloor^T$.

13.7 (a) Derive the consistent mass matrix that operates on d.o.f. u_1, u_2 and u_3 for the uniform quadratic-displacement bar shown.
(b) Using heuristic arguments, derive an ad hoc lumped matrix for this element that agrees with HRZ lumping (Eq. 13.3-9).

13.8 Determine α in Eq. 13.3-3 such that the beam element has the correct kinetic energy $I\omega^2/2$ under rigid-body rotation about its center. Then consider modeling a simply supported beam with one element and computing the natural frequencies of vibration. What conclusions can you draw from this problem?

13.9 For the HRZ mass-lumping procedure:
(a) Verify Eqs. 13.3-4 and 13.3-5.
(b) Verify the nodal masses shown in parentheses for the element shown in Fig. 13.3-3*a*.
(c) Verify the nodal masses shown in parentheses for the element shown in Fig. 13.3-3*b*.
(d) Determine the lumped mass matrix $\lceil \mathbf{m} \rfloor$ for the constant-strain triangle element of Problem 13.6. Consider x- and y-direction accelerations separately.

13.10 For optimal lumping by quadrature verify the nodal masses for the following elements shown in Fig. 13.3-4.
(a) The quadratic-displacement (six-node) triangle.
(b) The quadratic-displacement Lagrange (nine-node) quadrilateral.

13.11 For a uniform beam element having the usual four d.o.f. (see, e.g., Fig. 13.3-1*b*), consider the lateral-displacement field

$$w = \lfloor (1-\xi), \quad (\xi-\xi^2)L/2, \quad \xi, \quad (-\xi+\xi^2)L/2 \rfloor \lfloor w_1 \quad \theta_1 \quad w_2 \quad \theta_2 \rfloor^T$$

where $\xi = x/L$.

(a) Show that w is linear in x if $\theta_1 = \theta_2$. Also show that this field yields the correct displacement w and curvature $w_{,xx}$ under pure bending.

(b) Use Eq. 13.2-5 to evaluate the mass matrix.

(c) Obtain a diagonal mass matrix by applying the HRZ procedure.

Section 13.4

13.12 (a) Determine the Rayleigh proportional damping constants α and β for fractions of critical damping of 3% and 20% at frequencies of 5 and 15 Hz, respectively.

(b) For the values of α and β determined in part (a), draw a graph similar to Fig. 13.4-1. Comment on the fraction of critical damping experienced by frequencies below 5 Hz and above 15 Hz. Is caution called for?

13.13 Consider a particle that is allowed to free-fall from at-rest initial conditions under its own weight due to gravity. If the particle has mass-proportional damping, then the equation governing its velocity v is $\dot{v} + \beta v = g$, where g is the acceleration due to gravity. (*Remark:* The same equation governs the velocity of a particle allowed to sink in a viscous fluid where β is related to a fluid viscosity.)

(a) Determine the analytic solution for v.

(b) Consider the ratio of the damped velocity to the undamped velocity (i.e., $v_{\text{undamped}} = gt$) for values of t of about 1 second. Does this ratio offer guidelines on what values of β are permissible without excessively damping rigid-body modes?

Section 13.5

13.14 (a) Show that the Rayleigh quotient, Eq. 13.5-4, can be regarded as stating an equality between the maximum strain and kinetic energies associated with mode $\{\overline{\mathbf{D}}\}$.

(b) Imagine that redesign produces small changes in $[\mathbf{M}]$ and $[\mathbf{K}]$. Hence, the natural frequency λ_i of each mode $\{\overline{\mathbf{D}}\}_i$ is slightly changed, by an amount $\Delta\lambda_i$. Using the Rayleigh quotient and neglecting terms of higher order, derive an expression for $\Delta\lambda_i$ in terms of λ_i, $\{\overline{\mathbf{D}}\}_i$, $[\mathbf{M}]$, $[\Delta\mathbf{K}]$, and $[\Delta\mathbf{M}]$.

13.15 (a) Prove that the lower inequality of Eq. 13.5-5 is true. *Suggestion:* Express $\{\mathbf{v}\}$ as a linear combination of eigenvectors (each normalized as in Eq. 13.6-2), factor the common term $\lambda_{\min}$ out of the numerator, and argue that when $\{\mathbf{v}\} \neq \{\overline{\mathbf{D}}_{\min}\}$ the numerator is too large and hence overestimates $\lambda_{\min}$.

(b) Prove that the upper inequality of Eq. 13.5-5 is true.

13.16 Consider the following stiffness and mass matrices:

$$[\mathbf{K}] = \begin{bmatrix} 2 & -2 \\ -2 & 5 \end{bmatrix} \qquad [\mathbf{M}] = \begin{bmatrix} 1 & 0 \\ 0 & 1 \end{bmatrix}$$

Exact eigenvalues and eigenvectors for axial vibration are $\lambda_1 = 1$, $\lambda_2 = 6$, $\{\bar{\mathbf{D}}\}_1 = \lfloor 2 \quad 1 \rfloor^T$, and $\{\bar{\mathbf{D}}\}_2 = \lfloor 1 \quad -2 \rfloor^T$. Consider approximate eigenvectors $\lfloor 1.7 \quad 1.0 \rfloor^T$ and $\lfloor 1.2 \quad -2.0 \rfloor^T$ and show that the Rayleigh quotient provides an accurate estimate of λ_1 and λ_2.

13.17 Consider axial vibrations of a uniform bar of length L and mass $m = \rho AL$, free at one end and fixed at the other. Using two-node bar elements, model the bar first by one element, then by two elements of equal length $L/2$. In each case, compute the lowest natural frequency using
(a) the consistent mass matrix $[\mathbf{m}]$.
(b) the lumped mass matrix $\lceil \mathbf{m} \rfloor$.
(c) the average mass matrix $([\mathbf{m}] + \lceil \mathbf{m} \rfloor)/2$.
The exact lowest natural frequency is $\omega_1 = (\pi/2L)\sqrt{E/\rho}$.

13.18 In Problem 13.17 consider the convergence rates of λ_{min} as the mesh is refined. Are they in accord with the rates predicted in Section 13.3?

13.19 The stiffness, consistent mass, and optimally lumped mass matrices for the unsupported, uniform, three-node quadratic bar element shown are

$$\frac{AE}{3L}\begin{bmatrix} 7 & -8 & 1 \\ -8 & 16 & -8 \\ 1 & -8 & 7 \end{bmatrix}, \quad \frac{\rho AL}{30}\begin{bmatrix} 4 & 2 & -1 \\ 2 & 16 & 2 \\ -1 & 2 & 4 \end{bmatrix}, \quad \frac{\rho AL}{6}\begin{bmatrix} 1 & 0 & 0 \\ 0 & 4 & 0 \\ 0 & 0 & 1 \end{bmatrix}$$

(a) For axial vibration, determine the three natural frequencies and mode shapes using the consistent mass matrix.
(b) Repeat part (a) using the optimally lumped mass matrix.
(c) What is the physical significance of the lowest frequency and mode for parts (a) and (b)?
(d) The exact frequencies of an unsupported continuous bar of length L are $\omega_n = (n\pi/L)\sqrt{E/\rho}$; $n = 0, 1, 2, \ldots$. What are the percentage errors of the frequencies computed in parts (a) and (b)?

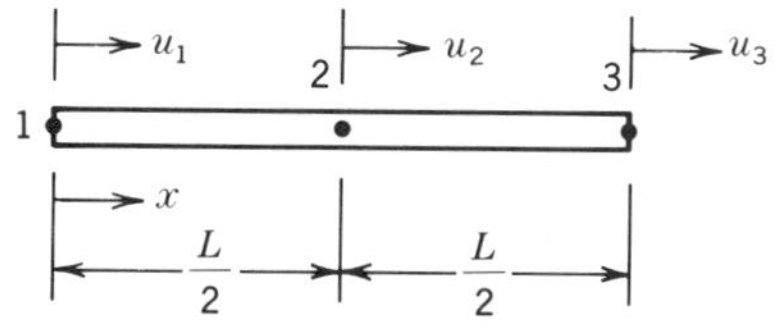

Problem 13.19

13.20 Fix one end of the three-node bar treated in Problem 13.19. Determine the two natural frequencies and mode shapes of axial vibration.
(a) Use the consistent mass matrix.
(b) Use the optimally lumped mass matrix.
(c) Use ad hoc lumping in which particles of mass $\rho AL/3$ are placed at each node.
(d) For parts (a), (b), and (c), estimate the lowest frequency by means of the Rayleigh quotient and the assumed displacement mode $u_1 = 0$, $u_2 = 1$, and $u_3 = 2$.

13.21 Model a simply supported uniform beam of length $2L$ by a single element. Determine the natural frequencies of vibration, where possible, by using the mass matrices cited here. Use the element stiffness matrix given in Eq. 4.2-5. The exact fundamental frequency is $\omega_1 = (\pi^2/4L^2)\sqrt{EI/\rho A}$.

(a) The consistent mass matrix **[m]** for a bar element given by the first of Eqs. 13.3-1 with d.o.f. u_1 and u_2 discarded.
(b) The consistent mass matrix **[m]** for a beam element given by Eq. 13.3-2.
(c) The lumped mass matrix ⌈**m**⌋ given by Eq. 13.3-3 with $\alpha = 0$.
(d) Repeat part (c) with $\alpha = 17.5$.
(e) The lumped mass matrix ⌈**m**⌋ given by Eq. 13.3-5.
(f) The matrix **[m]** given by Problem 13.11(b).

13.22 Consider a uniform cantilever beam of length L, modeled by a single beam element. Repeat parts (a) through (f) of Problem 13.21. The exact fundamental frequency is $\omega = (3.516/L^2)\sqrt{EI/\rho A}$.

13.23 A uniform beam is clamped at both ends. In order to exploit geometric symmetry, only the left half of the beam is modeled for vibration analysis. How would you ensure that analysis of the left half yields all natural frequencies of the original clamped–clamped beam?

Section 13.6

13.24 (a) Verify Eqs. 13.6-1.
(b) Explain a procedure for normalizing a vector (as, for example, $\{\bar{\mathbf{D}}\}_i$ is normalized with respect to **[M]** in Eq. 13.6-2).
(c) Show that Eqs. 13.6-1 are true for the modes stated in Eqs. 13.5-10.
(d) Show that Eqs. 13.6-1 are true for the exact modes stated in Problem 13.16.

13.25 Verify Eq. 13.4-2. *Suggestion:* Compare the uncoupled equation of motion with Rayleigh damping to the single d.o.f. equation $\ddot{Z} + 2\xi\omega\dot{Z} + \omega^2 Z = p$.

13.26 Consider the stiffness and lumped mass matrices given in Problem 13.16. Derive the system of uncoupled ordinary differential equations for a mode displacement response analysis.

13.27 (a) Let $k = m = 1$ in the spring–mass system shown. Let this system be set in axial motion with initial conditions $u_1 = u_2 = \dot{u}_1 = 0$, $\dot{u}_2 = 1$. Compute u_1 and u_2 at times $t = 1, 2, 3, 4,$ and 5 by use of the mode displacement method. Include both modes of the original system in $[\boldsymbol{\phi}]$.
(b) What is the greatest percentage error in u_1 and u_2 if only the lowest mode is used, so that $[\boldsymbol{\phi}]$ becomes a column vector?

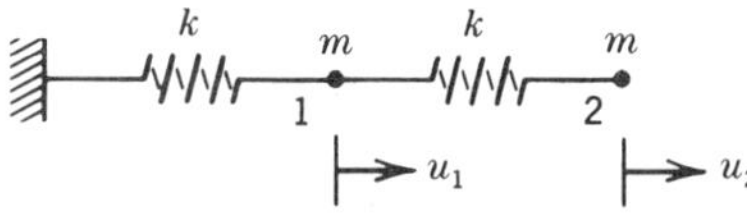

Problem 13.27

13.28 Verify Eq. 13.6-11. *Suggestion:* Insert $[\boldsymbol{\phi}]^{-T}[\boldsymbol{\phi}]^T$ after $[\mathbf{K}]^{-1}$ in the second term of Eq. 13.6-10. Then use orthogonality and the relation $[\boldsymbol{\phi}]^{-1}[\mathbf{K}]^{-1} = \lceil \boldsymbol{\omega}^2 \rfloor^{-1}[\boldsymbol{\phi}]^T$ to obtain Eq. 13.6-11.

13.29 Consider the two-spring, two-mass system described in Problem 13.27. Let this system be undeformed and at rest at time $t = 0$. For each of the following two loadings, use the mode displacement method and then the mode acceleration method to determine $u_1 = u_1(t)$ and $u_2 = u_2(t)$. Retain only the lowest mode $\{\boldsymbol{\phi}\}_1$ in the transformation. Compute numerical values of u_1 and u_2 at times $t = 2, 4, 6, 8$, and 10.

(a) Node 1 is not loaded. A force $F_2 = 1$ is applied to node 2 at $t = 0$.

(b) Forces $F_1 = 1$ and $F_2 = -1$ are applied to nodes 1 and 2 respectively at $t = 0$.

13.30 Show that the mode acceleration method reduces to the mode displacement method if the structure moves freely—that is, with $\{\mathbf{R}^{ext}\} = \{\mathbf{0}\}$.

13.31 Consider applying a Ritz vector analysis to the system of two springs and two masses described in Problem 13.27.

(a) Externally applied loads are zero so arbitrarily assign $\{\mathbf{w}^*\}_1 = \lfloor 1 \quad 0 \rfloor^T$ in Table 13.6-1. Hence, establish a 2 by 2 array $[\mathbf{W}]$ of Ritz vectors and the transformed system of Eq. 13.6-18.

(b) Repeat part (a), now using $\{\mathbf{w}^*\}_1 = \lfloor 0 \quad 1 \rfloor^T$.

(c) If $\{\mathbf{w}^*\}_1$ is arbitrarily taken as $\lfloor 1 \quad 2 \rfloor^T$, and no additional vectors are used, what is the resulting form of Eq. 13.6-18? What fundamental frequency ω does this equation yield? By what other name do you know this method of calculating ω?

Section 13.7

13.32 Mass condensation, starting from Eq. 13.7-1, would be more accurate if the assumption $[\mathbf{M}_{ms}] = [\mathbf{M}_{ss}] = [\mathbf{0}]$ were not made. What is an objection to this approach?

13.33 (a) Show that Eq. 13.7-5 yields $[\mathbf{K}_r] = [\mathbf{K}_{mm}] - [\mathbf{K}_{ms}][\mathbf{K}_{ss}]^{-1}[\mathbf{K}_{ms}]^T$. Where has the relation been seen before?

(b) Derive a similar expression for $[\mathbf{M}_r]$ from Eq. 13.7-5.

13.34 If $\lambda[\mathbf{M}]\{\overline{\mathbf{D}}\}$ in Eq. 13.5-2 is regarded as a vector of inertia loads $\{\mathbf{R}\}$, and $[\mathbf{K}]$ is inverted to become the flexibility matrix $[\mathbf{F}]$, we can write $[\mathbf{F}]\{\mathbf{R}\} = \{\mathbf{D}\}$.

(a) Partition this equation into m master and s slave d.o.f. as in Eq. 13.7-1 and let $\{\mathbf{R}_s\} = \{\mathbf{0}\}$. Derive the transformation

$$\begin{Bmatrix} \overline{\mathbf{D}}_m \\ \overline{\mathbf{D}}_s \end{Bmatrix} = [\mathbf{T}]\{\overline{\mathbf{D}}_m\}, \qquad \text{where} \quad [\mathbf{T}] = \begin{bmatrix} \mathbf{I} \\ \mathbf{F}_{ms}^T\mathbf{F}_{mm}^{-1} \end{bmatrix}$$

(b) Show that this transformation is mathematically the same as that of Eq. 13.7-3.

(c) How can $[\mathbf{F}_{mm}]$ be computed from $[\mathbf{K}]$ and what is its physical meaning?

(d) Why is the transformation of part (a) likely to be more computationally efficient than the form used in Eq. 13.7-3?

13.35 Consider the two-d.o.f. unsupported bar of Fig. 13.5-1. What is the reduced stiffness matrix that results from taking u_2 as a slave d.o.f.? Is this result reasonable?

13.36 Consider application of the procedure for automatic selection of master d.o.f. described in Section 13.7 to the structure shown. Only axial motion is permitted. Recall from Section 2.11 that condensation of a d.o.f. places

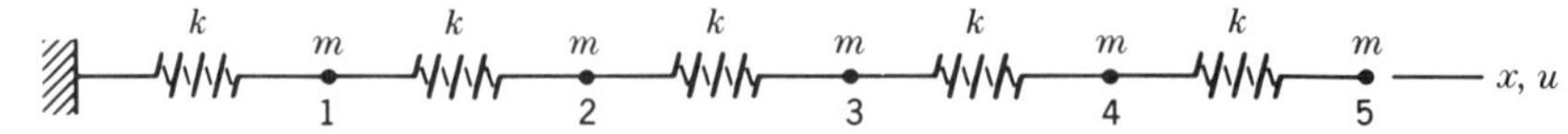

Problem 13.36

the two adjacent springs in series. To simplify this problem, but for no sound theoretical reason, assume that condensation of a mass m effectively adds mass $m/2$ to the two adjacent masses, so that [**M**] remains diagonal. For the structure shown,

(a) choose two masters by making three left-to-right sweeps.
(b) choose two masters by making three right-to-left sweeps.
(c) determine the frequencies ω_1 and ω_2 in parts (a) and (b) and compare results. The exact first two frequencies for the five-d.o.f. structure are $0.2846\sqrt{k/m}$ and $0.8308\sqrt{k/m}$.

13.37 (a) The frequency $\omega_1^2 = 6.1765EI/mL^3$ is computed below Eq. 13.7-12 in the example that closes Section 13.7. Improve this estimate, if possible, by using $\overline{w}_1 = 1$ and the first $\overline{\theta}_2$ of Eq. 13.7-13 in the Rayleigh quotient (Eq. 13.5-4). Use [**K**] and [**M**] from Eq. 13.7-8.
(b) Repeat part (a) using the second $\overline{\theta}_2$ of Eq. 13.7-13.

13.38 (a) In the example problem that closes Section 13.7, is the choice of $\overline{w}_1$ as master and $\overline{\theta}_2$ as slave consistent with the rule of largest M_{ii}/K_{ii}?
(b) Make the other choice, $\overline{\theta}_2$ as master and $\overline{w}_1$ as slave, and compute the frequency and mode shape (analogous to Eqs. 13.7-12 and 13.7-13).
(c) Improve the estimate of ω_1 from part (b) by using its mode shape in the Rayleigh quotient, Eq. 13.5-4, with [**K**] and [**M**] taken from Eq. 13.7-8.

13.39 Apply mass condensation to the system shown. Only axial motion is permitted. Let $k = 1$ and $m = 2$. Determine the fundamental vibration frequency of the reduced system and compare it with the exact value for the original system. Determine the fundamental mode of the reduced system, using first Eq. 13.7-3 and then Eq. 13.7-7. Finally, obtain improved estimates of ω_1 by using each of these modes in the Rayleigh quotient.

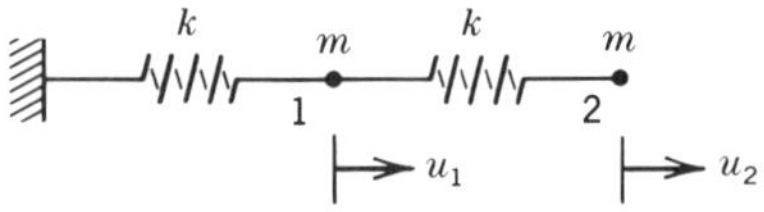

Problem 13.39

13.40 Many methods of solving large eigenproblems require factoring either the stiffness matrix or a combination of the stiffness and mass matrices (e.g., the determinant search and subspace iteration methods). Factoring requires approximately $n_{eq}b^2/2$ operations (i.e., multiplications) where n_{eq} is the number of equations and b is the semibandwidth. For full matrices, the number of operations is about $n_{eq}^3/6$. Consider a system of 5000 equations with $b = 500$. If this system of equations is partitioned into m master and s slave d.o.f., what must m be so that factoring the condensed (full) system is no more expensive than factoring the original (banded) system? What if $b = 100$ instead?

Section 13.8

13.41 Repeat the example that closes Section 13.8 using only the first vector of assembled component normal modes plus the additional mode that accounts for having node 3 fixed in the component mode analyses (i.e., omit the second column of Eq. 13.8-12 and the second column of Eq. 13.8-14).

13.42 Consider axial vibration of the spring–mass system shown with $k = 1$ and $m = 1$. Natural frequencies are $\omega_1 = 0.9246$, $\omega_2 = 1.574$, and $\omega_3 = 2.381$. Create one substructure consisting of the springs to the left of node 2 and another substructure consisting of the springs to the right of node 2. Using component mode synthesis, determine the two natural frequencies of the reduced structure.

(a) Use method CMS1, following the example in the text.

(b) Repeat part (a) except omit the last Ritz vector obtained by applying a unit load to node 2 (only one frequency can be determined).

(c) Use method CMS2, following the example in the text.

(d) Repeat part (c) except omit the last Ritz vector obtained by applying a unit displacement to node 2 (only one frequency can be determined).

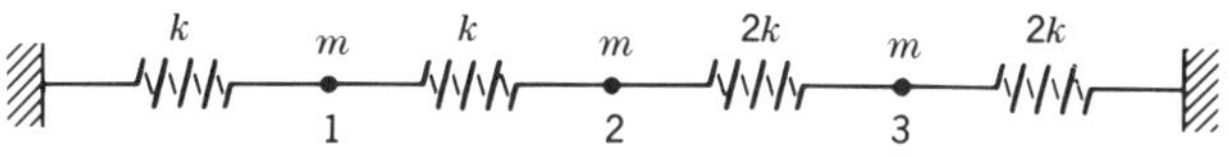

Problem 13.42

13.43 Repeat Problem 13.42, but let each of the four springs have stiffness $k = 1$. Natural frequencies and mode shapes of the original structure are

$$\begin{aligned} \omega_1 &= 0.7654 & \{\bar{\mathbf{D}}\}_1 &= \lfloor 1 \quad \sqrt{2} \quad 1 \rfloor^T \\ \omega_2 &= 1.414 & \{\bar{\mathbf{D}}\}_2 &= \lfloor -1 \quad 0 \quad 1 \rfloor^T \\ \omega_3 &= 1.848 & \{\bar{\mathbf{D}}\}_3 &= \lfloor 1 \quad -\sqrt{2} \quad 1 \rfloor^T \end{aligned}$$

In some cases, a frequency of the original structure may be missing from the reduced structure. Why?

Section 13.9

13.44 The forward and backward Euler direct integration methods are defined by

$$\begin{aligned} \{\mathbf{D}\}_{n+1} &= \{\mathbf{D}\}_n + \Delta t\{\dot{\mathbf{D}}\}_n & &\text{forward Euler} \\ \{\mathbf{D}\}_{n+1} &= \{\mathbf{D}\}_n + \Delta t\{\dot{\mathbf{D}}\}_{n+1} & &\text{backward Euler} \end{aligned}$$

Are these methods explicit or implicit?

Section 13.10

13.45 Using the Taylor series expansion, determine the order of accuracy of the direct integration methods defined in Problem 13.44.

13.46 Verify Eq. 13.10-12.

13.47 Show that Eq. 13.10-12 reduces to Eq. 13.10-5 if damping is zero.

13.48 (a) Apply the Gerschgorin bound, Eq. 13.10-17, with element coefficients replaced by structure coefficients and n_e replaced by n_{eq}, to obtain a bound on the maximum mesh frequency in terms of E, ρ, and L for the model and boundary conditions shown in Fig. 13.10-2. Then divide the bar into 40 elements of equal length and evaluate the bound numerically.

(b) Show that the bound obtained in part (a) agrees precisely with the element bound, Eq. 13.10-15. (*Note:* Usually the Gerschgorin and element bounds do not agree.)

13.49 Consider a model consisting of one linear-displacement bar finite element, with lumped mass, and one end fixed. Do the Gerschgorin bound, Eq. 13.10-17, and the element bound, Eqs. 13.5-7 and 13.10-15, show good agreement with the exact frequency of this model?

13.50 Consider a uniform free–free bar (i.e., an unsupported bar) modeled by equal-length, linear-displacement finite elements. For such a situation, the maximum frequency of the entire model and the maximum unconstrained element frequency are the same. Why?

13.51 In the example of Section 13.10, the element bound was within one-thousandth of a percent of the maximum mesh frequency. Do you expect the agreement to improve or deteriorate as the number of elements in the mesh increases? Why?

13.52 A particle of unit mass is supported by a spring of unit stiffness. There is no damping and no external load. Thus $k = m = \omega = 1$. At time $t = 0$, the particle has zero displacement, zero acceleration, but unit velocity. Use the central-difference method, Eq. 13.10-5, to compute displacement versus time over five time steps. Use a Δt of (a) 1.0, (b) $\sqrt{2.0}$, (c) 2.0, and (d) 3.0. Does there appear to be an amplitude error? Why?

13.53 Repeat the example of Section 13.10 for the bar shown in Fig. 13.10-2 but with the right-hand end unsupported. Modify the Fortran program and compare the displacement, velocity, and stress time histories at the position $x = 9.75$ in. with those for the bar in Fig. 13.10-2. Experiment with time steps of different size. Also examine the time history solutions for position $x = 4.75$ in.

13.54 Repeat the example of Section 13.10 with a mesh of nonuniform length elements. For the first 15 in. of the bar, use 30 elements, each of length 0.5 in. For the remaining 5 in. of the bar, use five elements of length 1 in. Compare the velocity and stress time-histories at the position $x = 9.75$ in. with those of the example of Section 13.10 and comment on any differences.

13.55 For central-difference integration of equations of motion with Rayleigh damping, derive an equation analogous to Eqs. 13.10-5 and 13.10-12, and modify the computational procedure of Table 13.10-1. In deriving this algorithm, write the viscous forces as $\alpha[\mathbf{K}]\{\dot{\mathbf{D}}\}_{n-1/2} + \beta[\mathbf{M}]\{\dot{\mathbf{D}}\}_n$. For the mass-proportional part of the damping, approximate $\{\dot{\mathbf{D}}\}_n$ by Eq. 13.10-1. Note that the stiffness-proportional part of the viscous forces can be obtained element-by-element by summation of the element contributions $\alpha \int [\mathbf{B}]^T\{\dot{\boldsymbol{\sigma}}\}_{n-1/2}\, dV$ (verify this). Based on Eqs. 13.10-8 and 13.10-14, can you suggest what the stability criterion for this scheme will be?

13.56 Using the damping algorithm developed in Problem 13.55, modify the Fortran program of Fig. 13.10-3 to include Rayleigh damping. Use stiffness-proportional damping to give 20% critical damping at the frequency given by Eq. 13.10-15 and zero mass-proportional damping. Repeat the example problem of Section 13.10 and compare results. Experiment with different mass- and stiffness-proportional damping constants.

Section 13.11

13.57 Derive Eq. 13.11-1 using Taylor series and show that $\{\mathbf{D}\}_{n+1}$ is approximated with an error of $O(\Delta t^2)$. *Suggestion:* Write Taylor series for $\{\mathbf{D}\}_{n+1}$ about time $n\ \Delta t$ and $\{\mathbf{D}\}_n$ about time $(n + 1)\ \Delta t$ and then combine them to obtain Eq. 13.11-1 plus higher-order terms.

13.58 Verify Eqs. 13.11-3 through 13.11-7.

13.59 Why do you think the name "trapezoidal rule" is applied to Eqs. 13.11-1 and 13.11-2?

13.60 Repeat Problem 13.52, using the trapezoidal rule (Table 13.11-1). Use four time steps, of magnitude (a) $\Delta t = 2.0$, and (b) $\Delta t = 1.0$.

13.61 In seismic analysis of structures, 30 Hz is usually used as a cutoff frequency; that is, the excitation is composed of components with frequencies lower than 30 Hz. Using this cutoff frequency, what is the largest Δt that should be used for an unconditionally stable implicit method to give accurate results?

Section 13.12

13.62 Show that when $\beta = \frac{1}{4}$ and $\gamma = \frac{1}{2}$, Eqs. 13.12-3 and 13.12-4 can be expressed as Eqs. 13.11-1 and 13.11-2.

13.63 Why is the term "linear acceleration" used when $\beta = \frac{1}{6}$ in Table 13.12-1?

13.64 (a) Use the explicit Newmark method, Eqs. 13.12-3 and 13.12-4 with $\beta = 0$, to derive a computational procedure analogous to Table 13.10-1. State whether [**M**] and [**C**] must be diagonal.
(b) Does this procedure have any advantages in comparison with the central-difference procedure of Eq. 13.10-5?

13.65 (a) Using the results of Problem 13.64, modify the Fortran program of Fig. 13.10-3 to use the explicit Newmark method ($\beta = 0$, $\gamma \geq \frac{1}{2}$).
(b) Repeat the example of Section 13.10 using the algorithm of part (a) with $\gamma = 0.5$ (no artificial damping).
(c) Repeat the example of Section 13.10 using the algorithm of part (a) with $\gamma = 0.6, 0.7, 0.8, 0.9, 1.0$ (increasing artificial damping).

13.66 Use the implicit Newmark method, Eqs. 13.12-3 and 13.12-4 with $\beta > 0$, to derive equations analogous to Eqs. 13.11-5 through 13.11-7.

Section 13.13

13.67 Consider the uncoupled homogeneous equation of motion with damping but without inertia: $2\xi\omega\dot{Z} + \omega^2 Z = 0$. Using spectral stability, determine the stability criterion for direct integration of this equation by:
(a) The central-difference method, Eq. 13.10-1.
(b) The trapezoidal rule. *Suggestion:* Sum the equation of motion at times $n\ \Delta t$ and $(n + 1)\ \Delta t$ and combine with Eq. 13.11-1 to eliminate velocities.
(c) The forward Euler method defined in Problem 13.44.
(d) The backward Euler method defined in Problem 13.44.

13.68 Consider Problem 13.52 again, in which the central-difference method (Eq. 13.10-5) is applied to a spring–mass system for which $k = m = \omega = 1$. Now use $\Delta t = \sqrt{3.96}$ and start the algorithm using $u_0 = 0$ and $u_{-1} = -1$. Follow the motion for at least ten cycles, and observe that the computed amplitude displays "beating" but no net growth.

13.69 (a) Derive Eq. 13.13-23.
(b) Derive Eq. 13.13-26.

13.70 (a) Numerically evaluate Eqs. 13.13-22 and 13.13-25 using $\omega \, \Delta t = 0, 1, \sqrt{2}, 2$ for the period and amplitude errors of the central-difference method.
(b) Numerically evaluate Eq. 13.13-23 and 13.13-26 using $\omega \, \Delta t = 0, 1, 2, 4$ for the period and amplitude errors of the trapezoidal rule.
(c) Analytically show that Eqs. 13.13-25 and 13.13-26 reduce to forms that yield $a \, \Delta t = 0$ for all values of $\omega \, \Delta t$.

13.71 Consider the central-difference solutions obtained in Problem 13.52, parts (a), (b), and (c). What is the period error in each case? Check that the values you obtain agree with Eq. 13.13-22.

13.72 Consider the trapezoidal-rule solutions obtained in Problem 13.60, parts (a) and (b). Using approximations as necessary, determine the period and the period error of each solution. Check that the values you obtain agree with Eq. 13.13-23.

Section 13.14

13.73 The uncoupled equations produced by a modal analysis (Section 13.6) have a lower ω_{max} than ω_{max} of the full system. Hence, in integrating the uncoupled equations, what are the relative merits of explicit and implicit methods? How does the specific choice of modal method affect your answer?

CHAPTER

14

STRESS STIFFENING AND BUCKLING

Bending stiffness is affected by membrane forces. Matrices that account for this effect are formulated and applied to problems such as buckling. The nature of the buckling problem is discussed and warnings given against oversimplification.

14.1 INTRODUCTION

Buckling of bars, frames, plates, and shells may occur as a structural response to membrane forces. Membrane forces act along member axes and tangent to plate and shell midsurfaces. The membrane force in a bar (or column) is the axial load. Membrane forces in a shell are defined by Eqs. 12.1-1.

Buckling occurs when a member or a structure converts membrane strain energy into strain energy of bending with no change in externally applied load. A *critical condition,* at which buckling impends, exists when it is possible that the deformation state may change slightly in a way that makes the loss in membrane strain energy numerically equal to the gain in bending strain energy. In a slender bar of length L, axial stiffness AE/L is much greater than bending stiffness EI/L^3. Similarly, in a thin-walled structure such as a shell, membrane stiffness is typically orders of magnitude greater than bending stiffness. Accordingly, small membrane deformations can store a large amount of strain energy, but comparatively large lateral deflections and cross-section rotations are needed to absorb this energy in bending deformations.

One can also take the view that membrane forces alter the bending stiffness of a structure. Thus buckling occurs when compressive membrane forces are large enough to reduce the bending stiffness to zero for some physically possible deformation mode. If the membrane forces are *reversed*—that is, made tensile rather than compressive—bending stiffness is effectively *increased.* This effect is called *stress stiffening.*

The effects of membrane forces are accounted for by a matrix $[\mathbf{k}_\sigma]$ that augments the conventional stiffness matrix $[\mathbf{k}]$. Matrix $[\mathbf{k}_\sigma]$ has been given various names, as follows: initial stress stiffness matrix, differential stiffness matrix, geometric stiffness matrix, and stability coefficient matrix. In what follows we give $[\mathbf{k}_\sigma]$ the name *stress stiffness matrix.* Matrix $[\mathbf{k}_\sigma]$ is defined by an element's geometry, displacement field, and state of stress. Thus, $[\mathbf{k}_\sigma]$ is independent of elastic properties. (However, by introducing the stress–strain relation, $[\mathbf{k}_\sigma]$ can alternatively be written in terms of elastic properties and strains or deformations.) The structure matrix $[\mathbf{K}_\sigma]$ is built by summing overlapping terms of element matrices $[\mathbf{k}_\sigma]$, in the same way that the conventional $[\mathbf{K}]$ is built by summing overlapping terms of element matrices $[\mathbf{k}]$.

Analysis of a Beam–Column. Consider the simply supported beam shown in Fig.

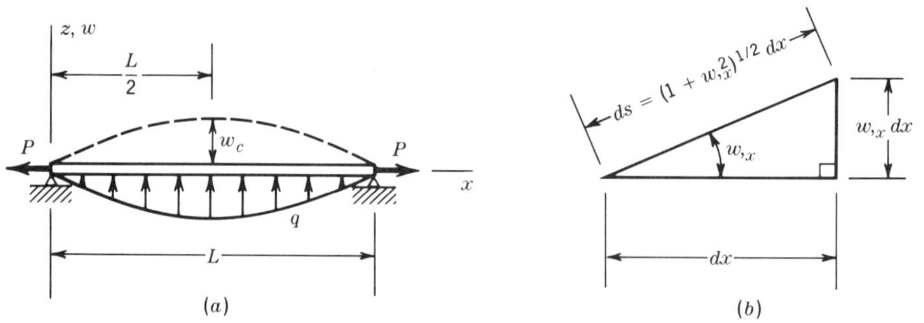

Figure 14.1-1. (*a*) A uniform beam on simple supports. (*b*) Geometric relations for a differential element of length dx.

14.1-1. Axial force P, positive in tension, is regarded as being imposed at the outset, for example, by cooling the bar while not allowing its ends to move toward one another. We will use energy concepts and a single d.o.f. to illustrate the stiffening effect of axial force P and to derive the buckling load $P_{cr} = -\pi^2 EI/L^2$, where the negative sign indicates compression.

Strain energy in bending is given by the standard expression that involves the square of curvature $w_{,xx}$:

$$U_b = \frac{1}{2}\int_0^L EIw_{,xx}^2\, dx \tag{14.1-1}$$

Let a small lateral displacement $w = w(x)$ take place. Thus each differential length dx is changed to a new length ds, where $ds > dx$ because the distance between supports is not allowed to change. From Fig. 14.1-1*b*,

$$ds = (1 + w_{,x}^2)^{1/2}\, dx \approx \left(1 + \frac{w_{,x}^2}{2}\right) dx \tag{14.1-2}$$

where the latter approximation comes from the first two terms of the binomial expansion. The approximation is valid if $w_{,x}^2 << 1$, which restricts this development to small rotations. Axial membrane strain in the bar is therefore

$$\epsilon_m = \frac{ds - dx}{dx} \approx \frac{w_{,x}^2}{2} \tag{14.1-3}$$

In the linear theory of elasticity we ignore terms of order $w_{,x}^2$. But here we seek the consequences of retaining the more important of the higher-order terms that linear theory neglects. We are taking a physical approach to formulating these terms. They may also be obtained by a systematic procedure of linearization, as will be touched upon in Section 14.4.

During a small lateral displacement $w = w(x)$, axial force P in the bar remains practically constant. As each elemental length dx lengthens an amount $\epsilon_m\, dx$, the force P it carries does work (and stores membrane strain energy) in the amount $P\epsilon_m\, dx$. Thus the change in membrane energy is[1]

[1]The same expression would result from the assumptions of a roller support at the right end, constant P, and *constant* length ($\int ds = L$), which would cause the right end to move a distance $u_L = \int \epsilon_m\, dx$, so that force P at $x = L$ gains potential in the amount Pu_L.

$$U_m = P \int_0^L \epsilon_m \, dx = \frac{1}{2} \int_0^L P w_{,x}^2 \, dx \tag{14.1-4}$$

Let us assume that w varies as half a sine wave, which happens to be the exact shape for Euler column buckling. Thus

$$w = w_c \sin \frac{\pi x}{L} \qquad \text{yields} \qquad \begin{cases} U_b = \dfrac{\pi^4 EI}{4L^3} w_c^2 & (14.1\text{-}5a) \\ U_m = \dfrac{\pi^2 P}{4L} w_c^2 & (14.1\text{-}5b) \end{cases}$$

where w_c is the center deflection of the beam.

To do a buckling analysis, we presume that lateral load q is zero. Thus, during buckling, membrane energy is exchanged for bending energy without any input of external work. Therefore

$$U_b + U_m = 0 \qquad \text{yields} \qquad P = -\frac{\pi^2 EI}{L^2} \tag{14.1-6}$$

which is the classical Euler buckling load, independent of w_c so long as w_c is small.

Now consider a deflection problem rather than a buckling problem. Imagine that distributed lateral load q, in the form of a half sine wave with amplitude q_c, is applied to the beam in the positive z direction. With $w = w_c \sin(\pi x/L)$ and $q = q_c \sin(\pi x/L)$, load q has potential

$$\Omega = -\int_0^L q w \, dx = -\frac{q_c L}{2} w_c \tag{14.1-7}$$

The total potential is $\Pi_p = U_b + U_m + \Omega$, and the equilibrium value of w_c is given by $\partial \Pi_p / \partial w_c = 0$. Thus, from Eqs. 14.1-5 and 14.1-7,

$$(k + k_\sigma) w_c = \frac{q_c L}{2} \qquad \text{where} \qquad \begin{cases} k = \dfrac{\pi^4 EI}{2L^3} \\ k_\sigma = \dfrac{\pi^2 P}{2L} \end{cases} \tag{14.1-8}$$

The stiffness coefficient $(k + k_\sigma)$ is the sum of conventional stiffness k and stress stiffness k_σ. If $P = 0$, we obtain $w_c = q_c L^4 / \pi^4 EI$. This result is exact (see Eq. 10.4-11). A tensile load ($P > 0$) decreases the lateral deflection w_c produced by transverse load q. This is the "stress stiffening" effect. If P is compressive ($P < 0$), then w_c is increased, becoming infinite when $P = -\pi^2 EI/L^2$, which again defines the buckling load $P = P_{cr}$. When $P = P_{cr}$, the net stiffness $(k + k_\sigma)$ is zero.

Stress stiffness k_σ can be written in terms of displacement rather than force. Imagine that end $x = L$ of the bar is roller-supported and can have axial displacement u_L. Expressing load P in terms of displacement u_L, we have

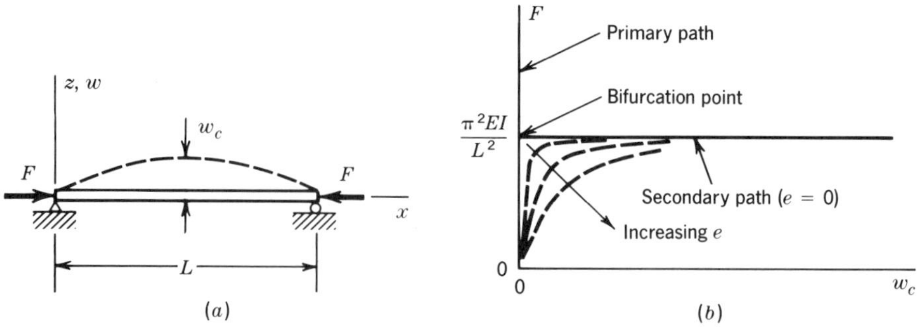

Figure 14.1-2. Bar under compressive load F. In (*b*), e indicates the magnitude of imperfection.

$$P = \frac{AE}{L} u_L \quad \text{and} \quad k_\sigma = \frac{\pi^2}{2L} \frac{AE}{L} u_L = \frac{\pi^2 AE}{2L^2} u_L \tag{14.1-9}$$

Equation 14.1-9 suggests that the following two-stage analysis is possible (although unnecessary in this simple example). In the first stage, one does a conventional static analysis (without k_σ) to determine u_L produced by load P. Hence, from Eq. 14.1-9, k_σ becomes known. One can now use the net stiffness $(k + k_\sigma)$ in Eq. 14.1-8 to determine the lateral displacement w_c produced by lateral load q. Specifically, in this example we obtain $u_L = PL/AE$ and $k_\sigma = \pi^2 P/2L$, exactly as in Eq. 14.1-8. A two-stage analysis is accurate if displacements associated with the first stage are not coupled to displacements associated with the second stage. (If coupling is significant, a multistage analysis is required; see Sections 14.5 and 17.7.) The motivation for using two stages rather than one is that in most structures the distribution of membrane forces is not known a priori and must be determined by the first-stage calculation before the effect of an additional loading can be determined or a buckling analysis performed.

Caution. Buckling theory presumes the existence of a *bifurcation point*. Consider, for example, Fig. 14.1-2. At the bifurcation (buckling) load, two equilibrium configurations are possible: the column could remain straight (primary path) or it could buckle (secondary path). A bifurcation point exists if the column is perfectly straight, perfectly uniform, perfectly free of end moments and lateral loads, and forces F are perfectly centered and perfectly axial. In reality there are always imperfections, whose magnitude we denote by e. If $e \neq 0$, the column displays no bifurcation point and structures in general display "limit points." A computed buckling load is then only an approximation of how much load a structure will carry. The approximation may be quite wrong, and may err on the unconservative (unsafe) side. Most "buckling" problems should be approached as nonlinear problems in which prebuckling deformations are taken into account. These concepts are discussed further in Sections 14.4, 14.5, and 14.7.

These cautionary remarks do not obviate the usefulness of $[\mathbf{k}_\sigma]$ in stress-stiffening and nonlinear analyses.

14.2 STRESS STIFFNESS MATRICES FOR BEAMS AND BARS

In this section, stress stiffness matrices $[\mathbf{k}_\sigma]$ for prismatic members are derived from Eq. 14.1-4 by use of an assumed lateral displacement field $w = w(x)$. Axial force P in the member is presumed known in terms of loads applied to the structure, either a priori or by elastic analysis, depending on whether the structure in which the member resides is statically determinate or not. Attention is restricted to displacements in a plane. Beam and bar elements are placed on a local x axis. Matrices $[\mathbf{k}_\sigma]$ for elements arbitrarily oriented in global coordinates can be obtained by straightforward use of the transformation $[\mathbf{T}]^T[\mathbf{k}_\sigma][\mathbf{T}]$ (see Eq. 7.4-4). Matrices $[\mathbf{k}_\sigma]$ for space frames and space trusses are derivable by use of expressions discussed in Section 14.4.

To begin, we include conventional strain terms as well as $w_{,x}^2/2$ from Eq. 14.1-3 in the strain expression. In this way we show clearly which terms lead to the conventional stiffness matrix $[\mathbf{k}]$ and which to $[\mathbf{k}_\sigma]$.

Plane Beam. The beam in Fig. 14.2-1 can have axial displacement $u = u(x)$ and lateral displacement $w = w(x)$. Membrane strain is $\epsilon_m = u_{,x} + \frac{1}{2} w_{,x}^2$, where the latter term comes from Eq. 14.1-3. At a distance z from the centroidal axis, the contribution of bending to the axial strain is $\epsilon = -zw_{,xx}$, as derived in Eq. 11.1-3. The total axial strain of an arbitrarily located fiber is therefore

$$\epsilon_x = u_{,x} + \tfrac{1}{2} w_{,x}^2 - zw_{,xx} \tag{14.2-1}$$

Each fiber carries uniaxial stress. Strain energy in the element is therefore

$$U = \int_{V_e} \tfrac{1}{2} E\epsilon_x^2 \, dV = \int_0^L \int_A \tfrac{1}{2} E\epsilon_x^2 \, dA \, dx \tag{14.2-2}$$

We substitute Eq. 14.2-1 into Eq. 14.2-2 and note that

$$\int_A dA = A \qquad \int_A z \, dA = 0 \qquad \int_A z^2 \, dA = I \qquad \int_A Eu_{,x} \, dA = P \tag{14.2-3}$$

where P is the axial force, positive in tension. If a term dependent on $w_{,x}^4$ is discarded as negligible in comparison with other terms, we obtain

$$U = \int_0^L \frac{AE}{2} u_{,x}^2 \, dx + \int_0^L \frac{P}{2} w_{,x}^2 \, dx + \int_0^L \frac{EI}{2} w_{,xx}^2 \, dx \tag{14.2-4}$$

The first integral yields $[\mathbf{k}]$ for a bar element; it contains coefficients AE/L and is associated with d.o.f. u_1 and u_2. The third integral yields $[\mathbf{k}]$ for a standard beam element; it contains coefficients such as $12EI/L^3$ and is associated with d.o.f. w_1, θ_1, w_2, and θ_2. The second integral yields $[\mathbf{k}_\sigma]$. This integral was previously seen

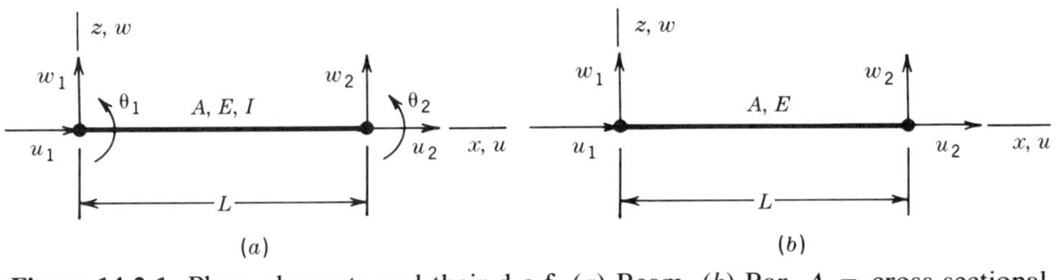

Figure 14.2-1. Plane elements and their d.o.f. (*a*) Beam. (*b*) Bar. A = cross-sectional area, E = elastic modulus, I = moment of inertia of A.

as Eq. 14.1-4. It describes work done, and strain energy stored, when lateral displacement w causes differential elements to stretch an amount $w_{,x}^2\, dx/2$ in the presence of a constant axial force P. The standard $[\mathbf{k}_\sigma]$ for a beam is developed from the integral expression as follows.

With nodal d.o.f. $\{\mathbf{d}\} = \lfloor w_1 \quad \theta_1 \quad w_2 \quad \theta_2 \rfloor^T$ and shape functions $\lfloor \mathbf{N} \rfloor$, lateral displacement w and its first derivative $w_{,x}$ are

$$w = \lfloor \mathbf{N} \rfloor \{\mathbf{d}\} \qquad \text{where} \quad \lfloor \mathbf{N} \rfloor = \lfloor N_1 \quad N_2 \quad N_3 \quad N_4 \rfloor \tag{14.2-5}$$

$$w_{,x} = \lfloor \mathbf{G} \rfloor \{\mathbf{d}\} \qquad \text{where} \quad \lfloor \mathbf{G} \rfloor = \lfloor N_{1,x} \quad N_{2,x} \quad N_{3,x} \quad N_{4,x} \rfloor \tag{14.2-6}$$

The second integral in Eq. 14.2-4 yields

$$\int_0^L \frac{P}{2} w_{,x}^2\, dx = \tfrac{1}{2} \int_0^L w_{,x}^T P w_{,x}\, dx = \tfrac{1}{2} \{\mathbf{d}\}^T [\mathbf{k}_\sigma] \{\mathbf{d}\} \tag{14.2-7}$$

where

$$[\mathbf{k}_\sigma] = \int_0^L \lfloor \mathbf{G} \rfloor^T P \lfloor \mathbf{G} \rfloor\, dx \tag{14.2-8}$$

Force P is constant in this member and can be removed from the integral. Using the standard N_i given in Fig. 3.13-2, we obtain [14.2]

$$[\mathbf{k}_\sigma] = \frac{P}{30L} \begin{bmatrix} 36 & 3L & -36 & 3L \\ 3L & 4L^2 & -3L & -L^2 \\ -36 & -3L & 36 & -3L \\ 3L & -L^2 & -3L & 4L^2 \end{bmatrix} \tag{14.2-9}$$

where P is positive in *tension*. (By inserting rows and columns of zeros, $[\mathbf{k}_\sigma]$ could be written as a 6 by 6 matrix that operates on the d.o.f. $\{\mathbf{d}\} = \lfloor u_1 \quad w_1 \quad \theta_1 \quad u_2 \quad w_2 \quad \theta_2 \rfloor^T$. Coordinate transformation could follow; the resulting $[\mathbf{k}_\sigma]$ could then be used for an arbitrarily oriented member of a plane frame.)

Plane Bar. The foregoing arguments can be repeated, but with curvature $w_{,xx}$ and d.o.f. θ_1 and θ_2 omitted from Eq. 14.2-4. Nonzero terms in $[\mathbf{k}_\sigma]$ are then associated with d.o.f. w_1 and w_2, and matrix $\lfloor \mathbf{G} \rfloor$ describes a rotation of the bar that is independent of x,

$$w_{,x} = \lfloor \mathbf{G} \rfloor \begin{Bmatrix} w_1 \\ w_2 \end{Bmatrix} \qquad \text{where} \quad \lfloor \mathbf{G} \rfloor = \left\lfloor -\frac{1}{L} \quad \frac{1}{L} \right\rfloor \tag{14.2-10}$$

Hence, Eq. 14.2-8 yields

$$[\mathbf{k}_\sigma] = \frac{P}{L} \begin{bmatrix} 1 & -1 \\ -1 & 1 \end{bmatrix} \qquad \text{for} \quad \{\mathbf{d}\} = \lfloor w_1 \quad w_2 \rfloor^T \tag{14.2-11a}$$

or

$$[\mathbf{k}_\sigma] = \frac{P}{L}\begin{bmatrix} 0 & 0 & 0 & 0 \\ 0 & 1 & 0 & -1 \\ 0 & 0 & 0 & 0 \\ 0 & -1 & 0 & 1 \end{bmatrix} \quad \text{for} \quad \{\mathbf{d}\} = \lfloor u_1 \quad w_1 \quad u_2 \quad w_2 \rfloor^T \qquad (14.2\text{-}11\text{b})$$

Remarks. Equations 14.2-11 are exact for small deflections of a bar that may rotate but does not bend. Equation 14.2-9, which allows bending, is approximate because a cubic lateral-displacement field is not exact when a beam carries axial load as well as loads that produce bending. As usual, accuracy is gained by dividing a given beam into two or more elements. A *single* beam–column element can be exact if the element formulation is based on the exact displacement field. Formulation of such an element yields a combined matrix $[\mathbf{k} + \mathbf{k}_\sigma]$, which remains 4 by 4 but has coefficients that are more complicated than coefficients in Eq. 14.2-9.

Note that if $\{\mathbf{d}\}$ represents a small rigid-body rotation, the conventional stiffness matrix $[\mathbf{k}]$ yields zero forces; that is, $[\mathbf{k}]\{\mathbf{d}\} = \{\mathbf{0}\}$. Such is not the case for the stress stiffness matrix; that is, $[\mathbf{k}_\sigma]\{\mathbf{d}\} \neq \{\mathbf{0}\}$. This result does not imply that $[\mathbf{k}_\sigma]$ is in error. Consider, for example, $[\mathbf{k}_\sigma]$ of Eq. 14.2-11. If the element is given a small rotation θ, then axial strain $\epsilon_x = \theta^2/2$ appears and transverse "kickoff" forces of magnitude $P\theta$ appear at the nodes. These forces can be regarded as inseparable from buckling; that is, in the buckled state of a structure, the loading provided by kickoff forces from the various elements is exactly resisted by a deformation state whose associated rotations create the kickoff forces. If *all* higher-order terms were retained in the expression for ϵ_x, the rigid-body rotation of an element would not create nodal forces (see Section 14.4).

When conventional stiffness and stress stiffness are both taken into account, the total or effective stiffness matrix is $[\mathbf{k}] + [\mathbf{k}_\sigma]$ for an element and $[\mathbf{K}] + [\mathbf{K}_\sigma]$ for a structure. Thus, in direct analogy to Eq. 14.1-8, one accounts for the stiffening or weakening effect of axial load on bending stiffness. The inclusion of $[\mathbf{K}_\sigma]$ does not require the inclusion of extra d.o.f. in $\{\mathbf{D}\}$ when the equation $([\mathbf{K}] + [\mathbf{K}_\sigma])\{\mathbf{D}\} = \{\mathbf{R}\}$ is used in place of $[\mathbf{K}]\{\mathbf{D}\} = \{\mathbf{R}\}$. Use of $[\mathbf{K}] + [\mathbf{K}_\sigma]$ to solve buckling problems is discussed in Section 14.5.

14.3 STRESS STIFFNESS MATRIX OF A PLATE ELEMENT

For a flat plate, just as for a bar or a beam, an expression for $[\mathbf{k}_\sigma]$ can be obtained by examination of the work done by constant membrane forces as they act through displacements associated with small lateral deflections. Membrane forces, Fig. 14.3-1, are defined by

$$N_x = \int_{-t/2}^{t/2} \sigma_x \, dz \qquad N_y = \int_{-t/2}^{t/2} \sigma_y \, dz \qquad N_{xy} = \int_{-t/2}^{t/2} \tau_{xy} \, dz \qquad (14.3\text{-}1)$$

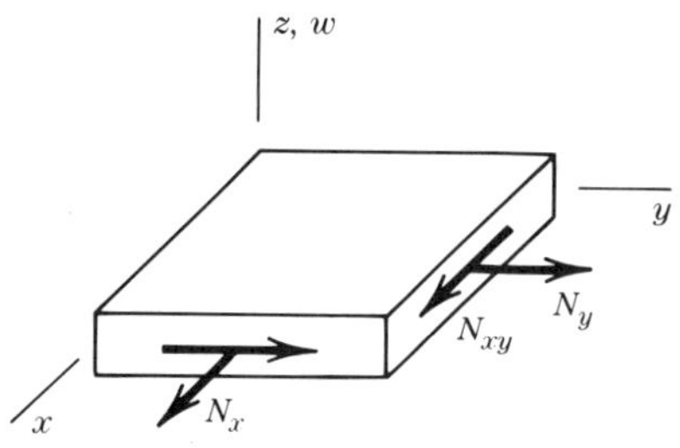

Figure 14.3-1. Differential element of a flat plate, showing membrane forces N_x, N_y, and N_{xy}.

where membrane stresses σ_x, σ_y, and τ_{xy} are either known a priori or calculated by standard static stress analysis, using, for example, plane bilinear isoparametric elements.

Membrane strains associated with small rotations $w_{,x}$ and $w_{,y}$ of the plate mid-surface are [11.1,14.4]

$$\epsilon_x = \tfrac{1}{2}\, w_{,x}^2 \qquad \epsilon_y = \tfrac{1}{2}\, w_{,y}^2 \qquad \gamma_{xy} = w_{,x} w_{,y} \tag{14.3-2}$$

If membrane forces N_x, N_y, and N_{xy} are assumed to be independent of the small lateral deflection $w = w(x,y)$, then the work associated with the membrane forces and the strains of Eqs. 14.3-2 is

$$U_\sigma = \int_A \left(\tfrac{1}{2}\, w_{,x}^2 N_x + \tfrac{1}{2}\, w_{,y}^2 N_y + w_{,x} w_{,y} N_{xy}\right) dA \tag{14.3-3a}$$

$$U_\sigma = \tfrac{1}{2} \iint \begin{Bmatrix} w_{,x} \\ w_{,y} \end{Bmatrix}^T \begin{bmatrix} N_x & N_{xy} \\ N_{xy} & N_y \end{bmatrix} \begin{Bmatrix} w_{,x} \\ w_{,y} \end{Bmatrix} dx\, dy = \tfrac{1}{2}\, \{\mathbf{d}\}^T [\mathbf{k}_\sigma] \{\mathbf{d}\} \tag{14.3-3b}$$

One must choose a displacement field $w = w(x,y)$ whose form is appropriate to the element shape and its d.o.f., for example, Eq. 11.2-5. From the displacement field one obtains rotations, that is,

$$w = \lfloor \mathbf{N} \rfloor \{\mathbf{d}\} \qquad \text{yields} \qquad \begin{Bmatrix} w_{,x} \\ w_{,y} \end{Bmatrix} = \underset{2\times n}{[\mathbf{G}]}\ \underset{n\times 1}{\{\mathbf{d}\}} \tag{14.3-4}$$

where n is the number of d.o.f. per element. Matrix $[\mathbf{G}]$ is in general a function of x and y. Equations 14.3-3b and 14.3-4 yield

$$[\mathbf{k}_\sigma] = \iint [\mathbf{G}]^T \begin{bmatrix} N_x & N_{xy} \\ N_{xy} & N_y \end{bmatrix} [\mathbf{G}]\, dx\, dy \tag{14.3-5}$$

where integration extends over the element area. If the element is of the isoparametric family, shape functions $\lfloor \mathbf{N} \rfloor$ are expressed in terms of dimensionless coordinates ξ and η. Therefore, one must invoke $[\mathbf{J}]$, the Jacobian matrix of Eq. 6.3-11. Thus

$$\begin{Bmatrix} w_{,\xi} \\ w_{,\eta} \end{Bmatrix} = [\mathbf{G}_I]\{\mathbf{d}\} \qquad \text{and} \qquad \begin{Bmatrix} w_{,x} \\ w_{,y} \end{Bmatrix} = [\mathbf{J}]^{-1} \begin{Bmatrix} w_{,\xi} \\ w_{,\eta} \end{Bmatrix} \tag{14.3-6}$$

where $[\mathbf{G}_I]$ contains derivatives of the shape functions with respect to ξ and η. Equations 14.3-3b and 14.3-6 yield

$$[\mathbf{k}_\sigma] = \int_{-1}^{1}\int_{-1}^{1} [\mathbf{G}_I]^T[\mathbf{J}]^{-T} \begin{bmatrix} N_x & N_{xy} \\ N_{xy} & N_y \end{bmatrix} [\mathbf{J}]^{-1}[\mathbf{G}_I]\, J\, d\xi\, d\eta \qquad (14.3\text{-}7)$$

where J is the Jacobian determinant.

Membrane forces N_x, N_y, and N_{xy} may vary over an element. Then, in numerical integration to evaluate $[\mathbf{k}_\sigma]$, different membrane forces would be used at different sampling points.

Note that $[\mathbf{k}_\sigma]$ is determined independently of material properties, except to the extent that material properties may influence computed values of N_x, N_y, and N_{xy}. Accordingly, a given $[\mathbf{k}_\sigma]$ is equally applicable to both isotropic and anisotropic structures.

14.4 A GENERAL FORMULATION FOR $[\mathbf{k}_\sigma]$

In Sections 14.2 and 14.3, each type of element is approached as a special case. It is desirable to also have a general formula for $[\mathbf{k}_\sigma]$, analogous to the formula for the conventional $[\mathbf{k}]$ (Eq. 4.1-5), that may be specialized to particular geometries, and requires only that a specific displacement field be chosen. Such a formula is derived in the present section.[2]

The formula for $[\mathbf{k}_\sigma]$, Eq. 14.4-7, is "linearized" and is limited to small displacements. To demonstrate in a general way that this is so requires comparatively lengthy and complicated arguments. We will omit these arguments [2.1] and instead illustrate the nature of the approximation by means of a simple particular case (Eqs. 14.4-10 to 14.4-16).

Green–Lagrange Strain. Various advanced texts [e.g., 2.1, 3.1, 9.9] discuss the expressions for stress and strain appropriate to problems that involve large deformations. The following equations define a strain measure commonly known as Green–Lagrange strain:

$$\epsilon_x = u_{,x} + \tfrac{1}{2}(u_{,x}^2 + v_{,x}^2 + w_{,x}^2) \qquad (14.4\text{-}1a)$$

$$\epsilon_y = v_{,y} + \tfrac{1}{2}(u_{,y}^2 + v_{,y}^2 + w_{,y}^2) \qquad (14.4\text{-}1b)$$

$$\epsilon_z = w_{,z} + \tfrac{1}{2}(u_{,z}^2 + v_{,z}^2 + w_{,z}^2) \qquad (14.4\text{-}1c)$$

$$\gamma_{xy} = u_{,y} + v_{,x} + (u_{,x}u_{,y} + v_{,x}v_{,y} + w_{,x}w_{,y}) \qquad (14.4\text{-}1d)$$

$$\gamma_{yz} = v_{,z} + w_{,y} + (u_{,y}u_{,z} + v_{,y}v_{,z} + w_{,y}w_{,z}) \qquad (14.4\text{-}1e)$$

$$\gamma_{zx} = w_{,x} + u_{,z} + (u_{,z}u_{,x} + v_{,z}v_{,x} + w_{,z}w_{,x}) \qquad (14.4\text{-}1f)$$

[2]An example application is the buckling analysis of an I beam modeled by flat shell elements. If cross sections of the I beam rotate about the web axis, elements in the flanges move in their own planes. The $[\mathbf{k}_\sigma]$ of Eq. 14.3-5, which presumes that displacements are normal to the element midsurface, is not adequate to model the flange elements. A $[\mathbf{k}_\sigma]$ derived from Eq. 14.4-7, which allows for u, v, and w displacements, *will* be adequate.

The initial terms in Eqs. 14.4-1 are the customary engineering definitions of normal and shear strain ($\epsilon_x = u_{,x}$, etc.). The added terms, in parentheses, become significant if displacement gradients are not small. Green–Lagrange strains are zero for a rigid-body rotation of any magnitude. In Eqs. 14.4-1, all displacement derivatives are computed in the *original* coordinate system, regardless of how large a rigid-body rotation may be superposed on the deformations. This is the "total Lagrangian" approach, in which all displacements are measured in a reference frame that is stationary rather than attached to the deforming structure. The stationary coordinates may also be called "material coordinates" and may be denoted in some papers by uppercase labels X, Y, and Z.

Green–Lagrange normal strains correspond to defining the strain of a line segment by the equation

$$\epsilon = \frac{1}{2}\left[\left(\frac{ds^*}{ds}\right)^2 - 1\right] \tag{14.4-2}$$

where ds and ds^* are respectively the initial and final lengths of the line segment. If $ds \approx ds^*$, Eq. 14.4-2 reduces to the usual small-strain approximation, $\epsilon = (ds^* - ds)/ds$.

Formula for $[\mathbf{k}_\sigma]$. Imagine that initial stresses $\{\boldsymbol{\sigma}_0\}$ prevail. If these stresses are assumed to remain constant as strains $\{\boldsymbol{\epsilon}\}$ occur, the associated work is[3]

$$\int_V \{\boldsymbol{\epsilon}\}^T\{\boldsymbol{\sigma}_0\}\, dV \qquad \text{where} \qquad \begin{cases} \{\boldsymbol{\epsilon}\}^T = \lfloor \epsilon_x \quad \epsilon_y \ . \ . \ . \ \gamma_{zx} \rfloor & \text{(14.4-3a)} \\ \{\boldsymbol{\sigma}_0\} = \lfloor \sigma_{x0} \quad \sigma_{y0} \ . \ . \ . \ \tau_{zx0} \rfloor^T & \text{(14.4-3b)} \end{cases}$$

With $\{\boldsymbol{\epsilon}\}$ given by Eqs. 14.4-1, the integrand $\{\boldsymbol{\epsilon}\}^T\{\boldsymbol{\sigma}_0\}$ first displays the terms $u_{,x}\sigma_{x0} + v_{,y}\sigma_{y0} + \cdots$. These terms lead to nodal loads associated with $\{\boldsymbol{\sigma}_0\}$, as given by Eq. 4.1-6. What remains is

$$U_\sigma = \int_V \left[\tfrac{1}{2}\,(u_{,x}^2 + v_{,x}^2 + w_{,x}^2)\sigma_{x0} + \cdots + (u_{,z}u_{,x} + v_{,z}v_{,x} + w_{,z}w_{,x})\tau_{zx0}\right] dV \tag{14.4-4}$$

If we define

$$\{\boldsymbol{\delta}\} = \lfloor u_{,x} \quad u_{,y} \quad u_{,z} \quad v_{,x} \quad v_{,y} \quad v_{,z} \quad w_{,x} \quad w_{,y} \quad w_{,z} \rfloor^T \tag{14.4-5}$$

then Eq. 14.4-4 can be written in the form

$$U_\sigma = \frac{1}{2}\int_V \{\boldsymbol{\delta}\}^T \begin{bmatrix} \mathbf{s} & \mathbf{0} & \mathbf{0} \\ \mathbf{0} & \mathbf{s} & \mathbf{0} \\ \mathbf{0} & \mathbf{0} & \mathbf{s} \end{bmatrix} \{\boldsymbol{\delta}\}\, dV \qquad \text{where} \quad [\mathbf{s}] = \begin{bmatrix} \sigma_{x0} & \tau_{xy0} & \tau_{zx0} \\ \tau_{xy0} & \sigma_{y0} & \tau_{yz0} \\ \tau_{zx0} & \tau_{yz0} & \sigma_{z0} \end{bmatrix} \tag{14.4-6}$$

[3]This assumption restricts the subsequent development, Eqs. 14.4-3 to 14.4-7, to small strains and small rotations. More advanced arguments [2.1] show that a more elaborate definition of stress than engineering stresses $\{\boldsymbol{\sigma}_0\}$ is required if one is to write a strain energy expression that is meaningful in analyses of large deformations.

This expression is analogous to Eqs. 14.2-7 and 14.3-3b, and yields $[\mathbf{k}_\sigma]$ in an analogous way. Let the element displacement field be given by $\{\mathbf{u}\} = [\mathbf{N}]\{\mathbf{d}\}$, as usual, where $\{\mathbf{u}\} = \lfloor u \quad v \quad w \rfloor^T$ and $\{\mathbf{d}\}$ contains nodal d.o.f. Also let $\{\boldsymbol{\delta}\} = [\mathbf{G}]\{\mathbf{d}\}$, where $[\mathbf{G}]$ is obtained from shape functions $[\mathbf{N}]$ by appropriate differentiation and ordering of terms. Equation 14.4-6 becomes $U_\sigma = \{\mathbf{d}\}^T[\mathbf{k}_\sigma]\{\mathbf{d}\}/2$, where

$$[\mathbf{k}_\sigma] = \int_{V_e} [\mathbf{G}]^T \begin{bmatrix} \mathbf{s} & \mathbf{0} & \mathbf{0} \\ \mathbf{0} & \mathbf{s} & \mathbf{0} \\ \mathbf{0} & \mathbf{0} & \mathbf{s} \end{bmatrix} [\mathbf{G}]\, dV \tag{14.4-7}$$

As an example, consider the bar of Fig. 14.2-1*b*, again with motion restricted to the *xz* plane. For this case all initial stresses are zero except for axial stress σ_{x0}. We assume that u and w are linear in x and require that $v = 0$. Accordingly, with $N_1 = (L - x)/L$ and $N_2 = x/L$, we write

$$\begin{aligned} u &= N_1 u_1 + N_2 u_2 \\ w &= N_1 w_1 + N_2 w_2 \end{aligned} \qquad [\mathbf{G}] = \frac{1}{L}\begin{bmatrix} -1 & 0 & 1 & 0 \\ 0 & -1 & 0 & 1 \end{bmatrix} \tag{14.4-8}$$

Nonzero d.o.f. are $\{\mathbf{d}\} = \lfloor u_1 \quad w_1 \quad u_2 \quad w_2 \rfloor^T$. Also, $\{\boldsymbol{\delta}\} = \lfloor u_{,x} \quad w_{,x} \rfloor^T$. Equation 14.4-7 reduces to

$$[\mathbf{k}_\sigma] = \int_0^L [\mathbf{G}]^T \begin{bmatrix} \sigma_{x0} & 0 \\ 0 & \sigma_{x0} \end{bmatrix} [\mathbf{G}]\, \mathrm{A}\, dx = \frac{P}{L}\begin{bmatrix} 1 & 0 & -1 & 0 \\ 0 & 1 & 0 & -1 \\ -1 & 0 & 1 & 0 \\ 0 & -1 & 0 & 1 \end{bmatrix} \tag{14.4-9}$$

where $P = \sigma_{x0}A$. This $[\mathbf{k}_\sigma]$ is almost the same as that in Eq. 14.2-11, but contains four more nonzero terms. However, note that the additional nonzero terms occupy the same positions as the nonzero terms in the conventional stiffness matrix (see Eq. 2.5-3). Thus, in the net stiffness matrix $[\mathbf{k}] + [\mathbf{k}_\sigma]$, we see coefficients $\pm(AE + P)/L$ (corresponding to d.o.f. u_1 and u_2) and $\pm P/L$ (corresponding to d.o.f. w_1 and w_2). Since $AE >> P$ in any practical problem, the "extra" P/L terms in $[\mathbf{k}_\sigma]$ can be discarded. In this way Eq. 14.4-9 reduces to Eq. 14.2-11.

Full Nonlinearity: An Example. In the foregoing development it is not obvious what approximations are contained in $[\mathbf{k}_\sigma]$. In what follows we allow large rotations, and show by example that use of $[\mathbf{k}_\sigma]$ implies complete linearization of the problem and negligible rotations prior to buckling.

Consider the one-element elastic bar in Fig. 14.4-1. If forces are applied only at the ends, displacements vary linearly.

$$u = \frac{x}{L} u_2 \qquad \text{and} \qquad w = \frac{x}{L} w_2 \tag{14.4-10}$$

Measurements are made in the *original* coordinate system *xz*; that is, x is *not* regarded as an axial coordinate that rotates as the bar rotates. For example, the bar becomes *vertical* if $u_2 = -L$ and $w_2 = \pm L$; nevertheless, one uses the *horizontal* coordinate x in the fields $u = u_2 x/L$ and $w = w_2 x/L$. Thus u_2 is the x-direction component of the displacement of node 2; it is not the stretch of the bar unless $w_2 = 0$. End 2 of the bar is located by coordinates $x = L$ and $z = 0$, regardless of the values of u_2 and w_2.

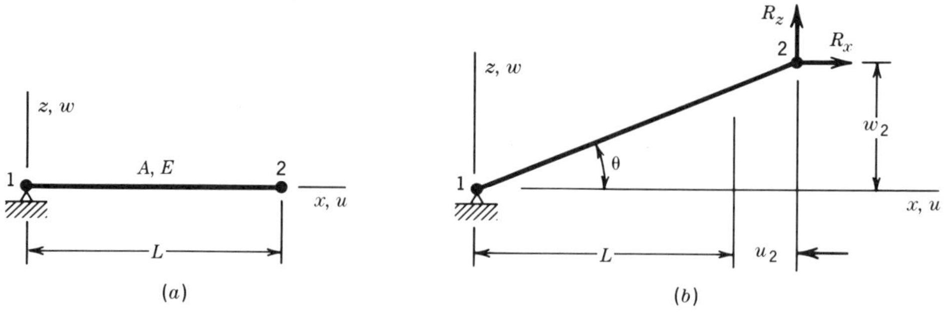

Figure 14.4-1. (*a*) A bar element, hinged at node 1, prior to loading. (*b*) The displaced and deformed bar after forces R_x and R_z are applied to node 2.

The bar moves in the xz plane and carries uniaxial stress. *If strains are small,* the total potential of the bar is

$$\Pi_p = \frac{1}{2}\int_0^L AE\epsilon_x^2\,dx - R_x u_2 - R_z w_2 \tag{14.4-11}$$

where ϵ_x is axial strain, as if the bar occupies its original x-parallel orientation. From Eqs. 14.4-1a and 14.4-10,

$$\epsilon_x = \frac{u_2}{L} + \frac{1}{2}\left(\frac{u_2^2}{L^2} + \frac{w_2^2}{L^2}\right) \tag{14.4-12}$$

The resulting expression for Π_p still allows large rotation of the bar. Static equilibrium prevails when $\partial\Pi_p/\partial u_2 = 0$ and $\partial\Pi_p/\partial w_2 = 0$. Results of these calculations can be written in the form

$$\frac{AE}{L}\left(\begin{bmatrix}1 & 0\\ 0 & 0\end{bmatrix} + \frac{1}{2L}\begin{bmatrix}3u_2 & w_2\\ w_2 & u_2\end{bmatrix} + \frac{1}{2L^2}\begin{bmatrix}u_2^2 & u_2w_2\\ u_2w_2 & w_2^2\end{bmatrix}\right)\begin{Bmatrix}u_2\\ w_2\end{Bmatrix} = \begin{Bmatrix}R_x\\ R_z\end{Bmatrix} \tag{14.4-13}$$

One finds that if u_2 and w_2 represent rigid-body rotation about node 1—that is, if $u_2 = -L(1 - \cos\theta)$ and $w_2 = L\sin\theta$—then $R_x = 0$ and $R_z = 0$ for any rotation θ, no matter how large.

In Eq. 14.4-13, u_2 and w_2 are total displacements. An analogous *incremental* form can be obtained from Eq. 14.4-13 by writing $R_x = R_x(u_2,w_2)$ and $R_z = R_z(u_2,w_2)$, then forming expressions for dR_x and dR_z by differentiation. Thus

$$\left(\underbrace{\frac{AE}{L}\begin{bmatrix}1 & 0\\ 0 & 0\end{bmatrix}}_{[\mathbf{K}]} + \underbrace{\frac{AE}{L^2}\begin{bmatrix}3u_2 & w_2\\ w_2 & u_2\end{bmatrix}}_{[\mathbf{N}_1]} + \underbrace{\frac{AE}{2L^3}\begin{bmatrix}3u_2^2 + w_2^2 & 2u_2w_2\\ 2u_2w_2 & u_2^2 + 3w_2^2\end{bmatrix}}_{[\mathbf{N}_2]}\right)\begin{Bmatrix}du_2\\ dw_2\end{Bmatrix} = \begin{Bmatrix}dR_x\\ dR_z\end{Bmatrix} \tag{14.4-14}$$

This equation describes nodal force increments dR_x and dR_z associated with small nodal displacements du_2 and dw_2, where du_2 and dw_2 are measured from a current reference configuration defined by u_2 and w_2. Thus the reference configuration may display an angle θ far different from its initial value $\theta = 0$.

A correspondence with preceding equations may be recognized as follows. If displacements u_2 and w_2 in Eq. 14.4-14 are sufficiently small, the third matrix in Eq. 14.4-14, which depends quadratically on displacements, may be discarded in comparison with the other two matrices. If in addition we substitute $(AE/L)u_2 = P$, set $w_2 = 0$, and presume that $u_2 << L$, then there is a further simplification to

$$\left(\frac{AE}{L}\begin{bmatrix} 1 & 0 \\ 0 & 0 \end{bmatrix} + \frac{P}{L}\begin{bmatrix} 3 & 0 \\ 0 & 1 \end{bmatrix}\right)\begin{Bmatrix} du_2 \\ dw_2 \end{Bmatrix} = \begin{Bmatrix} dR_x \\ dR_z \end{Bmatrix} \qquad (14.4\text{-}15)$$

The second matrix in Eq. 14.4-15 is analogous to the lower right 2 by 2 submatrix in Eq. 14.4-9, except that the number 3 appears instead of the number 1. Equation 14.4-15 can be reconciled with Eq. 14.2-11 by means of the same arguments that are applied to Eq. 14.4-9; that is, by noting that in practice $AE >> P$. Thus the number 3 is discarded and the second matrix in Eq. 14.4-15 is recognized as $[\mathbf{K}_\sigma]$.

In summary, by discarding the quadratic stiffness terms in Eq. 14.4-14, we obtain two forms of incremental equations:

$$([\mathbf{K}] + [\mathbf{N}_1])\{d\mathbf{D}\} = \{d\mathbf{R}\} \quad \text{and} \quad ([\mathbf{K}] + [\mathbf{K}_\sigma])\{d\mathbf{D}\} = \{d\mathbf{R}\} \qquad (14.4\text{-}16)$$

Equations 14.4-16 are symbolic statements of approximations that can be applied to linearly elastic structures in general, not only to the elastic bar. Both equations are linearized—that is, made to depend on first powers of displacements—by discarding quadratic terms. Either equation may be used to compute a buckling load. Upon buckling, infinitesimal displacements $\{d\mathbf{D}\}$ occur, measured with reference to a prebuckling configuration $\{\mathbf{D}\}$. In computing a buckling load, use of $[\mathbf{K}_\sigma]$ implies that prebuckling rotations in $\{\mathbf{D}\}$ are zero, whereas use of $[\mathbf{N}_1]$ implies only that prebuckling rotations are small. For initially flat plates and initially straight columns, the two formulations yield the same buckling load. In practical problems where prebuckling rotations are not negligible, use of $[\mathbf{N}_1]$ seems always to yield a smaller buckling load than does use of $[\mathbf{K}_\sigma]$, but there is no guarantee that either approach will always err on the high side or the low side of the actual collapse load [14.1].

14.5 BIFURCATION BUCKLING

A bifurcation buckling load is the load for which a reference configuration of the structure and an infinitesimally close (buckled) configuration are both possible equilibrium configurations. As a buckling displacement $\{d\mathbf{D}\}$ takes place from a reference configuration $\{\mathbf{D}\}$, the load does not change. Accordingly, from Eqs. 14.4-16,

$$([\mathbf{K}] + [\mathbf{N}_1])\{d\mathbf{D}\} = \{\mathbf{0}\} \quad \text{or} \quad ([\mathbf{K}] + [\mathbf{K}_\sigma])\{d\mathbf{D}\} = \{\mathbf{0}\} \qquad (14.5\text{-}1)$$

We ask for the level of deformation (symbolized by $[\mathbf{N}_1]$) or the level of stress (symbolized by $[\mathbf{K}_\sigma]$) such that a solution $\{d\mathbf{D}\}$ other than $\{d\mathbf{D}\} = \{\mathbf{0}\}$ is possible.

Use of $[\mathbf{N}_1]$ implies that prebuckling rotations are small but are not to be ignored. Use of $[\mathbf{K}_\sigma]$ implies that prebuckling rotations are either ignored or are zero. The latter is "classical" buckling analysis, as commonly used for straight columns and flat plates.

In what follows we emphasize classical buckling analysis, which uses $[\mathbf{K}_\sigma]$. One begins by applying to the structure a reference level of loading $\{\mathbf{R}\}_{ref}$ and carrying out a standard linear static analysis to obtain membrane stresses in elements (e.g., to determine membrane stresses in a flat plate under thermal load). Hence, we generate a stress stiffness matrix $[\mathbf{K}_\sigma]_{ref}$ appropriate to $\{\mathbf{R}\}_{ref}$. For another load level, with λ a scalar multiplier,

$$[\mathbf{K}_\sigma] = \lambda[\mathbf{K}_\sigma]_{ref} \qquad \text{when} \qquad \{\mathbf{R}\} = \lambda\{\mathbf{R}\}_{ref} \tag{14.5-2}$$

Equations 14.5-2 imply that multiplying all loads R_i in $\{\mathbf{R}\}_{ref}$ by λ also multiplies the intensity of the stress field by λ but does not change the distribution of stresses. Then, since external loads do not change during an infinitesimal buckling displacement $\{d\mathbf{D}\}$,

$$([\mathbf{K}] + \lambda_{cr}[\mathbf{K}_\sigma]_{ref})\{\mathbf{D}\} = ([\mathbf{K}] + \lambda_{cr}[\mathbf{K}_\sigma]_{ref})\{\mathbf{D} + d\mathbf{D}\} = \lambda_{cr}\{\mathbf{R}\}_{ref} \tag{14.5-3}$$

Subtraction of the first equation from the second yields

$$([\mathbf{K}] + \lambda_{cr}[\mathbf{K}_\sigma]_{ref})\{d\mathbf{D}\} = \{\mathbf{0}\} \tag{14.5-4}$$

Equation 14.5-4 defines an eigenvalue problem whose lowest eigenvalue λ_{cr} is associated with buckling. The critical or buckling load is, from Eq. 14.5-2,

$$\{\mathbf{R}\}_{cr} = \lambda_{cr}\{\mathbf{R}\}_{ref} \tag{14.5-5}$$

The eigenvector $\{d\mathbf{D}\}$ associated with λ_{cr} defines the buckling mode. The magnitude of $\{d\mathbf{D}\}$ is indeterminate. Therefore $\{d\mathbf{D}\}$ identifies shape but not amplitude.

A physical interpretation of Eq. 14.5-4 as follows. Terms in parentheses in Eq. 14.5-4 comprise a total or net stiffness matrix $[\mathbf{K}_{net}]$. Since forces $[\mathbf{K}_{net}]\{d\mathbf{D}\}$ are zero, one can say that membrane stresses of critical intensity reduce the stiffness of the structure to zero with respect to buckling mode $\{d\mathbf{D}\}$.

If the foregoing analysis were to use $[\mathbf{N}_1]$ instead of $[\mathbf{K}_\sigma]$, the displacements needed to construct $[\mathbf{N}_1]$ would be those obtained from static analysis under load $\{\mathbf{R}\}_{ref}$. Figure 14.5-1 gives an example of how well these two approaches to buckling analysis compare with the actual collapse load.

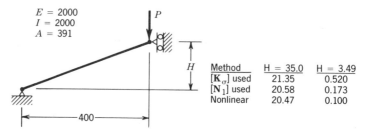

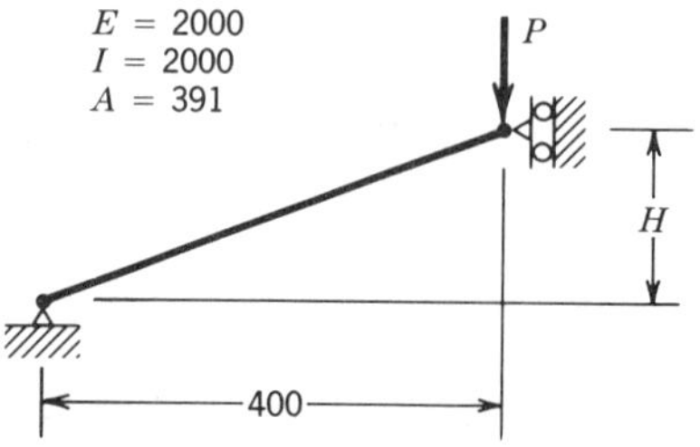

Method	H = 35.0	H = 3.49
$[\mathbf{K}_\sigma]$ used	21.35	0.520
$[\mathbf{N}_1]$ used	20.58	0.173
Nonlinear	20.47	0.100

Figure 14.5-1. An elastic bar, hinged at both ends. The linearized buckling load P_{cr} is compared with the collapse load determined by a more exact nonlinear analysis [14.1].

Computational methods for determining λ_{cr} are numerous (see Appendix C). Eigenvalue extraction methods used to compute natural frequencies and modes of vibration (Section 13.5) can also be applied to buckling problems. An algorithm that requires inversion of $[\mathbf{K}_\sigma]$ may fail, because $[\mathbf{K}_\sigma]$ may not be a positive definite matrix. If only one λ_{cr} is required, it may be wasteful to use a method that automatically extracts several eigenvalues. However, at times one may wish to know the several lowest eigenvalues and their associated buckling modes in order to gain insight into ways of stiffening or supporting the structure so as to make buckling less likely.

By basing $[\mathbf{K}]$ and $[\mathbf{K}_\sigma]$ on the original, undeformed geometry, Eq. 14.5-4 ignores prebuckling nonlinearities that may actually be present; that is, the possible dependence of $[\mathbf{K}]$ and $[\mathbf{K}_\sigma]_{ref}$ on deformation is ignored. One way to account for such nonlinearity is to base $[\mathbf{K}]$ and $[\mathbf{K}_\sigma]_{ref}$ on the configuration just before buckling [14.1]. More specifically, one applies a trial level of load $\{\mathbf{R}\}_{base}$ and performs a *nonlinear* static analysis. A result of this analysis is $[\mathbf{K}_t]$, the "tangent" stiffness matrix of the structure in its current deformed configuration. As compared with $[\mathbf{K}]$ of the structure before loads are applied, $[\mathbf{K}_t]$ is degraded in stiffness because of membrane stresses produced by $\{\mathbf{R}\}_{base}$. A small trial load increment $\{\Delta\mathbf{R}\}$ is applied, and displacements produced by $[\mathbf{K}_t]$ and $\{\Delta\mathbf{R}\}$ are used to compute membrane stresses. Thus $[\mathbf{K}_\sigma]_{ref}$ for the current configuration is established. The linear eigenvalue problem

$$([\mathbf{K}_t] + \Delta\lambda_{cr}[\mathbf{K}_\sigma]_{ref})\{d\mathbf{D}\} = \{\mathbf{0}\} \tag{14.5-6}$$

is solved for $\Delta\lambda_{cr}$. The computed value of $\Delta\lambda_{cr}$ reduces the net stiffness, which is the coefficient of $\{d\mathbf{D}\}$ in Eq. 14.5-6, to zero with respect to the buckling mode. The predicted buckling load is

$$\{\mathbf{R}\}_{cr} = \{\mathbf{R}\}_{base} + \Delta\lambda_{cr}\{\Delta\mathbf{R}\} \tag{14.5-7}$$

Equation 14.5-6 presumes that stresses change in intensity but not in distribution when the load increases an amount $\Delta\lambda_{cr}\{\Delta\mathbf{R}\}$. This assumption becomes more nearly true as $\{\mathbf{R}\}_{base}$ approaches $\{\mathbf{R}\}_{cr}$. By using a sequence of increasing loads $\{\mathbf{R}\}_{base}$, one can approach the correct buckling load arbitrarily closely. At convergence, $\Delta\lambda_{cr} = 0$ and $\{\mathbf{R}\}_{cr} = \{\mathbf{R}\}_{base}$.

Example: Classical Linear Buckling. Consider the uniform column in Fig. 14.5-2. Let the column be modeled by one beam element. We take $[\mathbf{k}]$ from Eqs. 2.4-3 and 4.2-5, and $[\mathbf{k}_\sigma]$ from Eq. 14.2-9. By inspection, we see that the axial load throughout the bar has magnitude P. We arbitrarily choose the reference value of P as -1.0, where the negative sign indicates *compression,* not that the load is directed leftward. Nonzero d.o.f. are u_2, w_2, and θ_2. Thus, for a single element, Eq. 14.5-4 becomes

$$\left(c_2 \begin{bmatrix} c_1/c_2 & 0 & 0 \\ 0 & 12 & -6L \\ 0 & -6L & 4L^2 \end{bmatrix} + \lambda_{cr} \frac{-1}{30L} \begin{bmatrix} 0 & 0 & 0 \\ 0 & 36 & -3L \\ 0 & -3L & 4L^2 \end{bmatrix} \right) \begin{Bmatrix} u_2 \\ w_2 \\ \theta_2 \end{Bmatrix} = \begin{Bmatrix} 0 \\ 0 \\ 0 \end{Bmatrix} \tag{14.5-8}$$

where $c_1 = AE/L$ and $c_2 = EI/L^3$. A solution other than $u_2 = w_2 = \theta_2 = 0$ requires that the expression in parentheses have a zero determinant. Thus we write the char-

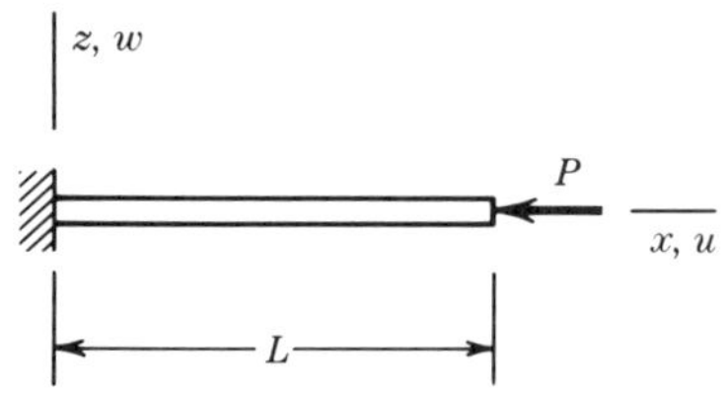

Figure 14.5-2. A uniform elastic bar, fixed at $x = 0$ and free at $x = L$. The exact P_{cr} is $\pi^2 EI/4L^2 = 2.4674EI/L^2$ (in compression).

acteristic polynomial and extract its lowest root (this method is suitable for hand calculation provided there are few d.o.f.). The lowest root in the present case is

$$\lambda_{cr} = 2.4860EI/L^2 \qquad \text{hence} \qquad P_{cr} = \lambda_{cr}(-1.0) = -2.4860EI/L^2 \tag{14.5-9}$$

Matrix $[\mathbf{k}_\sigma]$ in Eq. 14.5-8 comes from Eq. 14.2-9. If instead we use $[\mathbf{k}_\sigma]$ from Eq. 14.2-11, we obtain $P_{cr} = -3EI/L^2$, which is less accurate than P_{cr} in Eq. 14.5-9. Yet both answers are upper bounds to the correct magnitude of P_{cr}, which is $2.4674EI/L^2$. Bounds are discussed further in Section 14.6.

The buckling mode can be computed from Eq. 14.5-8 by setting $\lambda = \lambda_{cr}$, choosing an arbitrary value for one of the d.o.f. (e.g., $\theta_2 = 1$), and solving for the remaining d.o.f. Thus we obtain

$$u_2 = 0 \qquad w_2 = 0.6379L \qquad \theta_2 = 1 \tag{14.5-10}$$

D.o.f. in Eq. 14.5-10 are buckling displacements $\{d\mathbf{D}\}$, measured *relative* to the reference state in which $\lambda_{cr}[\mathbf{K}_\sigma]_{ref}$ is associated with initial membrane stresses. In this reference state, $u_2 = -PL/AE$ and $w_2 = \theta_2 = 0$. We see that u_2 plays no role in Eq. 14.5-8. Indeed, the buckling equation could have been written using w_2 and θ_2 as the only d.o.f.

Example: Condensation of D.O.F. In structural dynamics, one can use condensation to reduce the size of the eigenvalue problem (see Section 13.7). The same can be done in buckling problems, with $[\mathbf{K}_\sigma]$ taking the place of $[\mathbf{M}]$. One might elect to eliminate rotational d.o.f. In the preceding example, to eliminate θ_2 from Eq. 14.5-8, Eq. 13.7-3 becomes

$$[\mathbf{T}] = \begin{bmatrix} 1 & 0 \\ 0 & 1 \\ 0 & 3/2L \end{bmatrix} \qquad \text{where} \quad \frac{3}{2L} = -\left(\frac{L}{4EI}\right)\left(-\frac{6EI}{L^2}\right) \tag{14.5-11}$$

The transformations of Eq. 13.7-5 convert Eq. 14.5-8 to

$$\left(c_2\begin{bmatrix} c_1/c_2 & 0 \\ 0 & 3 \end{bmatrix} + \lambda_{cr}\frac{-1}{30L}\begin{bmatrix} 0 & 0 \\ 0 & 36 \end{bmatrix}\right)\begin{Bmatrix} u_2 \\ w_2 \end{Bmatrix} = \begin{Bmatrix} 0 \\ 0 \end{Bmatrix} \tag{14.5-12}$$

from which $\lambda_{cr} = 2.5EI/L^2$ and $P_{cr} = -2.5EI/L^2$. We see that condensation has slightly increased the magnitude of the computed buckling load.

14.6 REMARKS ON $[\mathbf{K}_\sigma]$ AND ITS USES

Field on Which $[\mathbf{k}_\sigma]$ is Based. A stress stiffness matrix is termed "consistent" if built from the same shape functions used to build the conventional stiffness matrix.

If the structure geometry is well modeled and if elements are compatible and not softened by low-order integration rules, then such a formulation yields an upper bound to the magnitude of the correct buckling load. The "correct" buckling load is the linear bifurcation load of the structure in its reference configuration; it is not necessarily the collapse load of the actual structure. Finite element analysis would yield the correct buckling load if $[\mathbf{k}]$ and $[\mathbf{k}_\sigma]$ were based on fields that include the buckled shape as a possible displacement mode. In the case of buckling of a perfect pin-ended column, this mode is sinusoidal rather than cubic, and the correct buckling load has magnitude $\pi^2 EI/L^2$.

We can base $[\mathbf{k}]$ and $[\mathbf{k}_\sigma]$ on different displacement fields. We recall from the convergence requirements of Section 4.5 that if a strain energy expression involves displacement derivatives of order m, the displacement field must provide inter-element continuity of displacement derivatives of order $m - 1$ as the mesh is refined. Energy integrals that yield $[\mathbf{k}_\sigma]$ involve first derivatives of displacement, so continuity of displacement is all that is required. Thus, for example, we would expect to be able to determine P_{cr} for a pin-ended column by using the conventional beam $[\mathbf{k}]$ but the $[\mathbf{k}_\sigma]$ of Eq. 14.2-11. This is indeed the case but, for a given accuracy, we must divide the column into more elements than when we use the consistent $[\mathbf{k}_\sigma]$ of Eq. 14.2-9.

It is sometimes recommended that $[\mathbf{k}_\sigma]$ for a complicated element be based on a simpler displacement field than that used to construct the conventional stiffness matrix, in order to increase computational efficiency with little loss in accuracy. The "best" $[\mathbf{k}_\sigma]$ is probably intermediate to the consistent $[\mathbf{k}_\sigma]$ and the simplest possible $[\mathbf{k}_\sigma]$. Numerical evidence suggests that computed buckling loads are increased when $[\mathbf{k}_\sigma]$ is simplified. If $[\mathbf{k}_\sigma]$ is generated by numerical integration, a simplified displacement field is effectively employed by adopting a reduced order of quadrature.

A valid $[\mathbf{k}_\sigma]$ must not generate nodal loads during a rigid-body translation. Nodal loads *do* appear when an element rotates. Indeed, from Eq. 14.5-1 one can interpret buckling as a displacement state $\{d\mathbf{D}\}$ in which pseudo-loads $[\mathbf{K}_\sigma]\{d\mathbf{D}\}$ are equal in magnitude to the corresponding resistances $[\mathbf{K}]\{d\mathbf{D}\}$.

Stress stiffness matrices have been devised for many buckling problems, for example, for homogeneous and sandwich plates [14.5-14.7], torsional and torsional–flexural buckling of prismatic members [5.4,14.8], tapered bars and plates [14.6,14.9], and nonconservative problems [14.10].

Applications of $[\mathbf{K}_\sigma]$. A shell of revolution usually has a nonaxisymmetric buckling mode even if geometry, supports, material properties, and loading are all axisymmetric. The buckling mode will probably display many waves in each hoop circle. It is commonly assumed that the buckling mode varies circumferentially as a single Fourier harmonic. Thus buckling analysis is similar to the displacement analysis described in Sections 10.5 and 10.6 [14.11]. First the shell is divided into elements such as those shown in Fig. 12.4-2. Then one selects a specific number n of circumferential waves and computes the corresponding $[\mathbf{K}]_n$ and $[\mathbf{K}_\sigma]_n$. Next one solves the eigenvalue problem to obtain λ_{cr} for n waves. The entire procedure is repeated for $n + 1$ waves, for $n + 2$ waves, and so on. Provided that the initial n is sufficiently small, the lowest of the sequence of λ_{cr} values can be identified as the desired buckling parameter. It is not obvious which mode will govern, and many analyses may be needed: Ref. 14.12 mentions a case where buckling is associated with 39 circumferential waves. If many waves also appear in the *meridional* direction, many elements are needed even if the shell has simple geometry.

Dynamic analysis of undamped structures with membrane forces leads to the equations

$$\text{dynamic response:} \quad [\mathbf{K} + \mathbf{K}_\sigma]\{\mathbf{D}\} + [\mathbf{M}]\{\ddot{\mathbf{D}}\} = \{\mathbf{R}\} \qquad (14.6\text{-}1)$$

$$\text{natural frequencies:} \quad ([\mathbf{K} + \mathbf{K}_\sigma] - \omega^2[\mathbf{M}])\{\overline{\mathbf{D}}\} = \{\mathbf{0}\} \qquad (14.6\text{-}2)$$

where $[\mathbf{M}]$ = mass matrix, $\{\ddot{\mathbf{D}}\}$ = accelerations of nodal d.o.f., ω = circular frequency, and $\{\overline{\mathbf{D}}\}$ = amplitudes of nodal d.o.f. Tensile membrane forces increase the frequencies. Compressive forces decrease them and produce the root $\omega = 0$ if buckling impends.

A structure may have no conventional stiffness $[\mathbf{K}]$. An example is a linkage of pin-connected bars, like a chain, with each link idealized as rigid. Similarly, some elastic structures may have a $[\mathbf{K}]$ that offers no resistance to certain loads. Examples include straight cables and flat membranes, which have no bending stiffness with which to resist lateral loads. Static problems of this type can be analyzed by the equation $[\mathbf{K}_\sigma]\{\mathbf{D}\} = \{\mathbf{R}\}$, where $\{\mathbf{D}\}$ contains d.o.f. associated with small lateral deflection. Analogous dynamic problems, such as a plucked string or a vibrating membrane, can be analyzed by Eqs. 14.6-1 and 14.6-2 with $[\mathbf{K}] = [\mathbf{0}]$.

14.7 REMARKS ON BUCKLING AND BUCKLING ANALYSIS

A real structure may collapse at a load quite different than that predicted by a linear bifurcation buckling analysis. The following remarks, extracted largely from Refs. 14.13 to 14.15, describe types of buckling behavior and caution against oversimplification in analysis. Throughout the discussion it is assumed that the material of the structure remains linearly elastic and that loads are gradually applied.

Figure 14.7-1 illustrates some of the ways a structure may behave. Here P is either the load or is representative of its magnitude, and D is displacement of some d.o.f. of interest. In Fig. 14.7-1*a*, the primary or prebuckling path happens to be linear. At bifurcation, either of two adjacent and infinitesimally close equilibrium positions are possible. Thereafter, for $P > P_{cr}$, a real (imperfect) structure

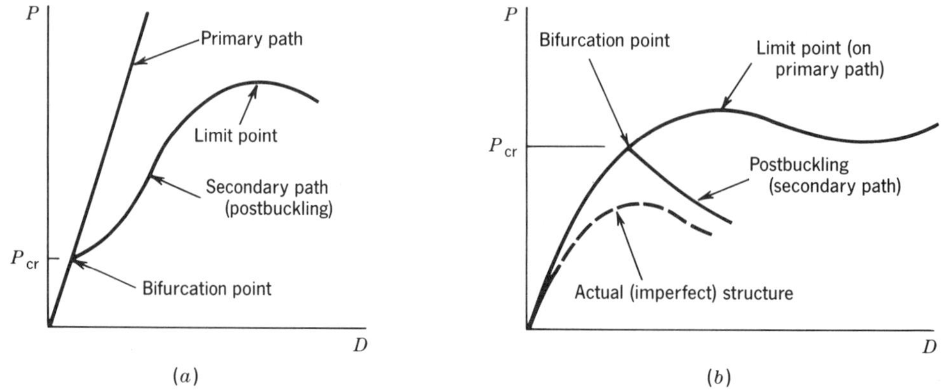

Figure 14.7-1. Possible load versus displacement bahaviors of thin-walled structures.

follows the secondary path. The secondary (postbuckling) path rises, which means that the structure has postbuckling strength. In this case P_{cr} characterizes a local buckling action that has little to do with overall strength. This structure finally collapses at a *limit point,* which is defined as a relative maximum on the P versus D curve for which there is no adjacent equilibrium position. Loose terminology may refer to the limit point load as a buckling load. The action at collapse becomes dynamic, because the slope of the curve becomes negative and the structure releases elastic energy, which is converted into kinetic energy.

A different type of behavior is depicted in Fig. 14.7-1*b*. Here the perfect (idealized) structure has a nonlinear primary path. The postbuckling path falls, so there is no postbuckling strength. If the primary path is close to a falling secondary path, the structure is called *imperfection sensitive,* which means that the collapse load of the actual structure is strongly affected by small changes in direction of loads, manner of support, or changes in geometry. The actual structure, which has imperfections, displays a limit point rather than bifurcation, as shown by the dashed line.

Figure 14.7-2 shows how the response may be affected by overall structure geometry. Figure 14.7-2*c*, for a deep spherical cap, also applies qualitatively to a cylindrical shell under axial compression. The deep cap and the cylindrical shell are imperfection sensitive: if the radius-to-thickness ratio is large, laboratory specimens buckle at roughly one half of the theoretical bifurcation load, even when heroic efforts are made to achieve geometric perfection.

In the absence of prior knowledge about how a structure behaves, one must anticipate that a computed bifurcation buckling load may be far above or far below the actual collapse load, that imperfections may be influential, and that prebuckling nonlinearities may be important. Nonlinearities may arise because pressure loads change direction as the structure deforms, because the deformed shape is more (or less) susceptible to instability than the undeformed shape, or because of the effects of deformation on the membrane stress distribution. Nonlinearities may be accounted for as described in connection with Eqs. 14.5-6 and 14.5-7, or by a

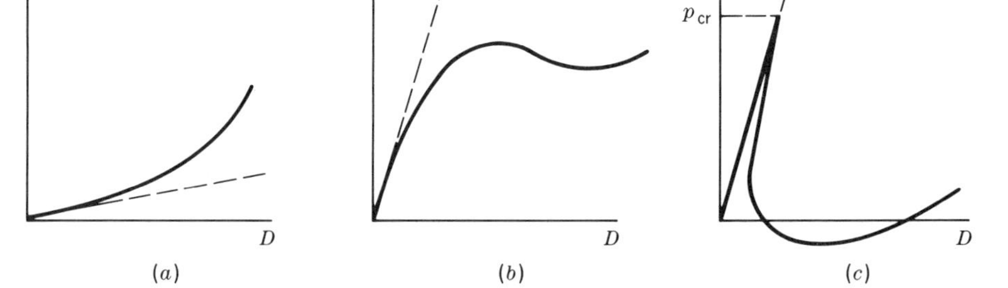

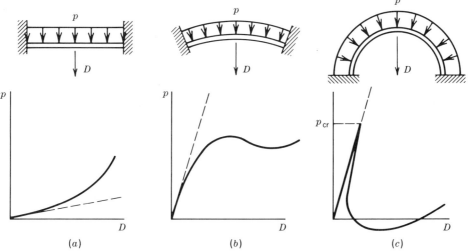

Figure 14.7-2. Pressure p versus center deflection D for thin-walled elastic structures. Dashed line: linear theory. Solid line: nonlinear theory and actual behavior. (*a*) Circular plate. (*b*) Shallow spherical cap. (*c*) Deep spherical cap.

nonlinear analysis that effectively plots load versus displacement and signals collapse when the total stiffness matrix of the structure in its current configuration becomes singular.

Collapse analysis should be approached with caution and with expertise. Novices are cautioned against misuse of computer programs. For example, the axially compressed cylindrical shell is an attractive test case, yet the problem is analytically difficult because several eigenvalues are clustered and correspond to quite different eigenmodes. Even a program that can negotiate the difficulties will not produce the correct result if, misled by the geometric simplicity of the structure, the user has employed so few d.o.f. that the many waves of the actual buckling mode cannot be properly modeled [14.14].

PROBLEMS

Section 14.1

14.1 A straight wooden column has an axial hole that fits closely but without friction around a metal rod, as shown. The rod is tensioned by tightening nuts that bear on the ends of the column. Assume that linear elasticity prevails and that the rod remains precisely centered in the column. Will the column buckle?

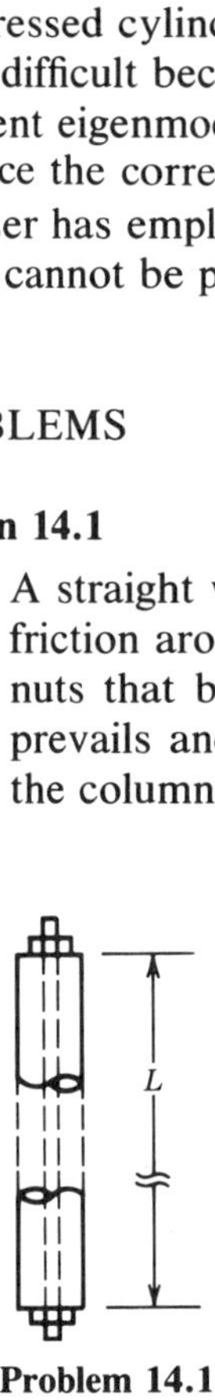

Problem 14.1

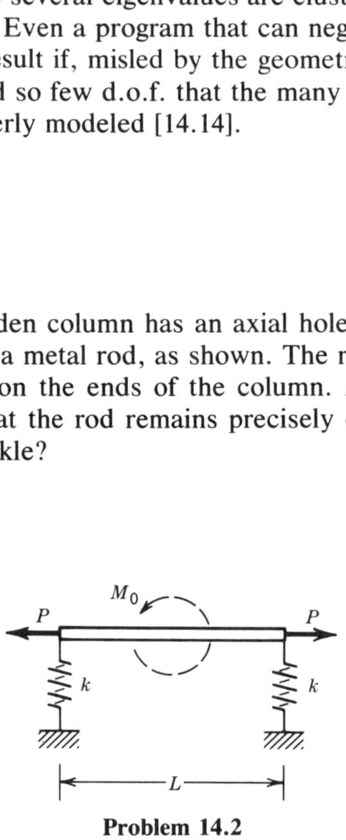

Problem 14.2

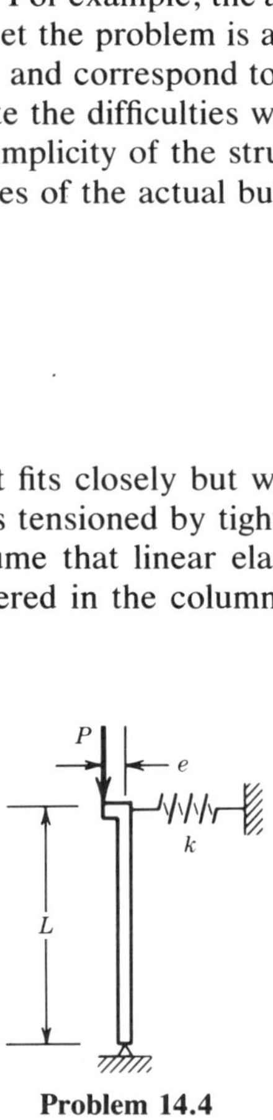

Problem 14.4

14.2 A rigid bar is supported by two springs, each of stiffness k, and loaded by horizontal forces P, as shown. Use the view that loads P do work during a small rotation of the bar (noted in the footnote in Section 14.1), and determine:

(a) the angle of rotation of the rigid bar, in terms of M_0, k, P, and L.

(b) the (compressive) value of P for buckling (when $M_0 = 0$), in terms of k and L.

14.3 For the problem described by Fig. 14.1-1, let $P_{cr} = -\pi^2 EI/L^2$ and w_{c0} represent the value of w_c produced by q alone (when $P = 0$). Plot w_c/w_{c0} versus P/P_{cr} as P goes from $P = P_{cr}$ (compressive) to $P = 5|P_{cr}|$ (tensile).

14.4 A rigid bar is pivoted at the lower end and held by a linear spring at the upper end, as shown. Load P is offset a distance e from the bar axis. Find P_{cr} (for $e = 0$). Also, using small-angle approximations, express lateral deflection Δ of the top in terms of P, e, k, and L. Plot Δ/L versus P/P_{cr} for $e/L = 0$, 0.01, and 0.02.

Section 14.2

14.5 Buckling of a tapered column is to be studied. Each element of the column is tapered. In which element matrices ([**k**] or [$\mathbf{k}_\sigma$]) does the effect of taper appear, and how is it to be included?

14.6 Show that Eq. 14.2-9 is produced by Eq. 14.2-8 and the standard cubic beam shape functions.

14.7 (a) Show that Eq. 14.2-4 results from Eqs. 14.2-1 through 14.2-3.
(b) How must $u_{,x}$ and $w_{,x}$ be related if the term in ϵ_x^2 that contains $w_{,x}^4$ is to be less than 5% of $u_{,x}w_{,x}^2$? Hence, what limiting angle of rotation is indicated if $u_{,x}$ is 0.002?

14.8 Construct a 4 by 4 matrix [$\mathbf{k}_\sigma$] for a uniform beam element, analogous to Eq. 14.2-9, by using the quadratic displacement field $w = (1 - \xi)w_1 + \xi w_2 + (1 - \xi)\xi L(\theta_1 - \theta_2)/2$, where $\xi = x/L$.

14.9 Derive each of the two [$\mathbf{k}_\sigma$] matrices in Eq. 14.2-11 by imposing the appropriate restrictions on [$\mathbf{k}_\sigma$] of Eq. 14.2-9.

14.10 A member of a plane, pin-jointed truss makes an angle β with the global x axis. Determine an expression for [$\mathbf{k}_\sigma$], analogous to Eq. 14.2-11b, that operates on global d.o.f. u_1, w_1, u_2, and w_2. Express your answer in terms of P, L, and β.

14.11 For the system shown, establish a set of two equations that could be solved to determine w_2 and θ_2 in member 1–2 in terms of P, Q, E, I, L, and c. The connection at node 2 transmits no moment.

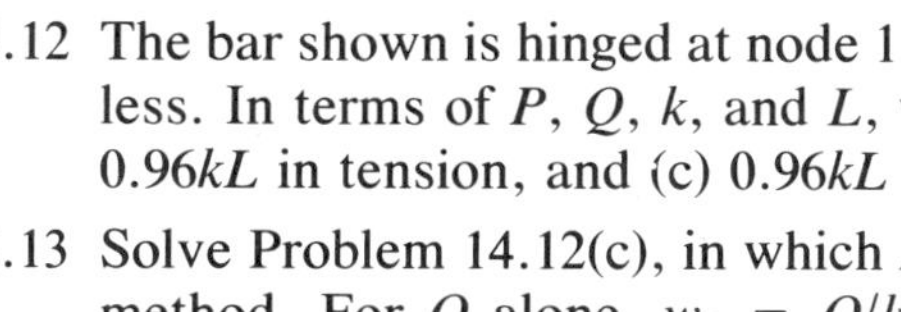

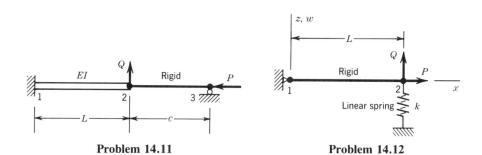

Problem 14.11

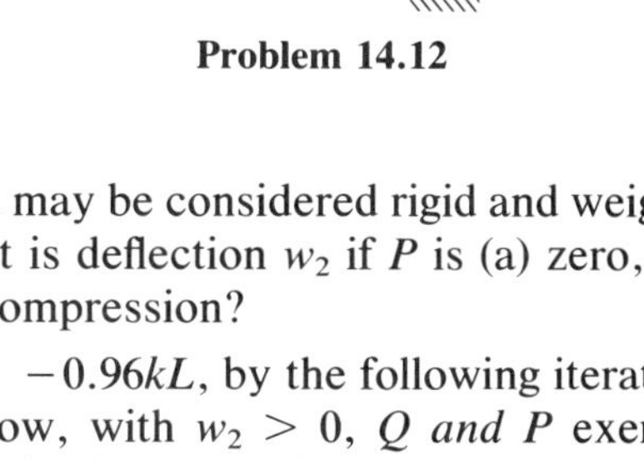

Problem 14.12

14.12 The bar shown is hinged at node 1 and may be considered rigid and weightless. In terms of P, Q, k, and L, what is deflection w_2 if P is (a) zero, (b) $0.96kL$ in tension, and (c) $0.96kL$ in compression?

14.13 Solve Problem 14.12(c), in which $P = -0.96kL$, by the following iterative method. For Q alone, $w_2 = Q/k$. Now, with $w_2 > 0$, Q *and* P exert a moment about node 1. Another analysis therefore yields a w_2 larger than before. The process repeats.

14.14 The column shown is fixed at the left end. Axial load P at the simply supported end has eccentricity e from the centerline. Model the column by one element.
(a) Determine the rotation θ_2 in terms of P, L, E, I, and e.
(b) If $e = 0$, what load P will make $\theta_2 \neq 0$? What is the percentage error of this result?

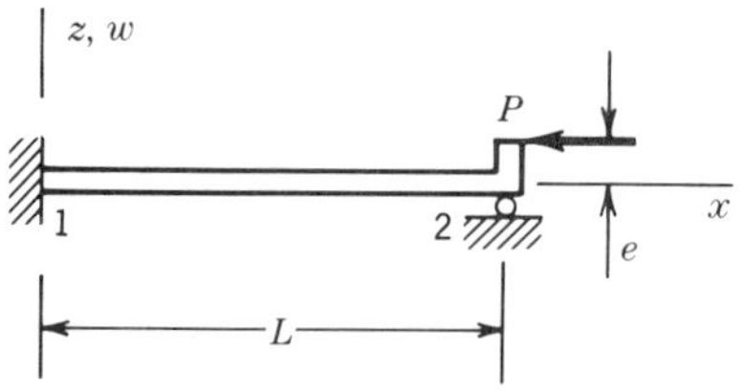

Problem 14.14

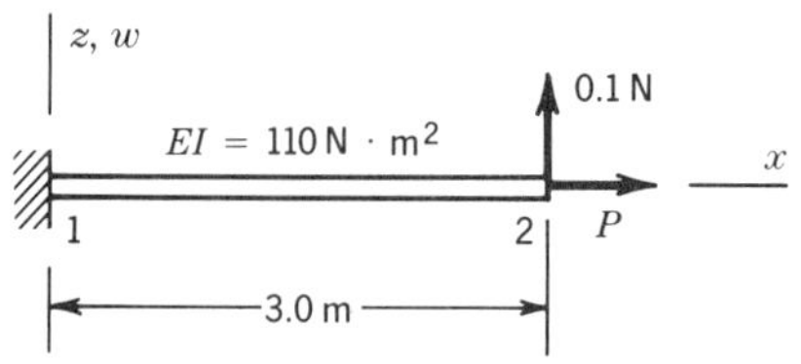

Problem 14.15

14.15 The cantilever beam shown may be subjected to either a tensile or a compressive axial force P. Model the beam by one element. Use w_2 and θ_2 as d.o.f. and take $[\mathbf{k}_\sigma]$ from Eq. 14.2-9.
(a) Determine w_2 if $P = 0$.
(b) Determine w_2 if $P = 30.0$ N in compression.
(c) Determine w_2 if $P = 30.0$ N in tension.

14.16 Repeat Problem 14.15, but use the $[\mathbf{k}_\sigma]$ developed in Problem 14.8.

14.17 In Problem 14.15(b), evaluate the bending moment at $x = 0$ by the calculation $M = EI\lfloor\mathbf{B}\rfloor\{\mathbf{d}\}$, where $\lfloor\mathbf{B}\rfloor$ is based on a cubic field and $\{\mathbf{d}\} = \lfloor 0 \quad 0 \quad w_2 \quad \theta_2\rfloor^T$. Compare this M with that obtained by statics; that is, $M = 0.1L - Pw_2$, where $P = -30.0$ N in this case.

14.18 Let the 4 by 1 displacement vector $\{\mathbf{d}\}$ represent a small rigid-body rotation of a bar or beam about its left end. Determine the nodal forces $[\mathbf{k}_\sigma]\{\mathbf{d}\}$ for a bar (Eq. 14.2-11) and a beam (Eq. 14.2-9). Express your answers in terms of P, L, and w_2.

14.19 Equation 4.1-6 has been used to compute consistent nodal loads on a beam (e.g., see Fig. 4.3-6). If the beam also carries an axial force, are these loads still correct? Explain.

14.20 In Chapter 13 we discussed a diagonal element mass matrix $[\mathbf{m}]$. Why is an analogous diagonal stress stiffness matrix $[\mathbf{k}_\sigma]$ unacceptable if $\{\mathbf{d}\}$ contains only translational d.o.f.?

Section 14.4

14.21 Show that Eq. 14.4-2 yields the small-strain approximation $\epsilon = (ds^* - ds)/ds$ if $ds \approx ds^*$.

14.22 (a) Imagine that the bar element of Fig. 14.2-1*b* is generalized to three dimensions. Thus the bar lies on the x axis of a local coordinate system xyz that is arbitrarily oriented with respect to global coordinates. Nodal d.o.f. consist of three translations at each node (in local directions). What is the 6 by 6 matrix $[\mathbf{k}_\sigma]$ in local coordinates?
(b) How would you establish the $[\mathbf{k}_\sigma]$ that operates on translational d.o.f. in global coordinate directions?

14.23 Show that Eq. 14.3-5 results from appropriate specialization of Eq. 14.4-7 (or of Eq. 14.4-6).

14.24 Consider the eight-node trilinear (solid) isoparametric element. How many rows and columns are there in $[\mathbf{G}]$ of Eq. 14.4-7? Express the G_{ij} in terms of shape function derivatives and coefficients Γ_{ij} of the inverse Jacobian matrix. For convenience, let $\{\mathbf{d}\} = \lfloor u_1 \quad u_2 \quad \ldots \quad u_8 \quad v_1 \quad \ldots \quad w_8\rfloor^T$.

14.25 (a) A three-node plane triangular element is constrained to move only in its plane. Write the formula for $[\mathbf{k}_\sigma]$ in terms of a 2 by 2 matrix $[\mathbf{s}]$ and matrix $[\mathbf{B}]$ of Eq. 5.4-3. Let $\{\mathbf{d}\} = \lfloor u_1 \quad u_2 \quad u_3 \quad v_1 \quad v_2 \quad v_3 \rfloor^T$.
(b) Let the same element be allowed only deflection $w = w(x,y)$ normal to its plane. What then is the formula for $[\mathbf{k}_\sigma]$? Let $\{\mathbf{d}\} = \lfloor w_1 \quad w_2 \quad w_3 \rfloor^T$.

14.26 (a) Show that Eqs. 14.4-11 and 14.4-12 yield Eq. 14.4-13.
(b) Derive Eq. 14.4-14 from Eq. 14.4-13.

14.27 Show that Eq. 14.4-12 yields $\epsilon_x = 0$ for the following rigid-body rotations. In addition, make a sketch that shows the displaced position of the bar.
(a) $u_2 = -1$, $w_2 = 3$, and $L = 5$.
(b) $u_2 = -5$, $w_2 = 5$, and $L = 5$.
(c) $u_2 = -10$, $w_2 = 0$, and $L = 5$.
(d) $u_2 = -L(1 - \cos\theta)$, $w_2 = L \sin\theta$.

14.28 For each of the rigid-body rotations stated in Problem 14.27, show that Eq. 14.4-13 yields $R_x = R_z = 0$.

14.29 If Δ represents an *axial*-displacement increment, the associated force in a uniform bar is $F = AE\Delta/L$. Show that Eq. 14.4-14 yields this result, whether the reference configuration of the bar in Fig. 14.4-1*b* is defined by $\theta = 0$ or by $\theta = \pi/2$. Assume that strains are small.

Section 14.5

14.30 (a) Equation 14.5-8 has two roots, of which the lower, $\lambda = \lambda_{cr}$, is given by Eq. 14.5-9. What is the other root and the corresponding mode shape?
(b) Verify that use of $[\mathbf{k}_\sigma]$ from Eq. 14.2-11 yields $P_{cr} = -3EI/L^2$ for the problem of Fig. 14.5-2. Again use a single element.

14.31 Use the $[\mathbf{k}_\sigma]$ derived in Problem 14.8 to determine the buckling load in Fig. 14.5-2. Use one element. The only d.o.f. needed are w_2 and θ_2.

14.32 The two pin-connected bars shown may be considered rigid and weightless. The linear springs each have stiffness k.
(a) Determine the buckling load P_{cr}.
(b) In the buckling mode, determine w_3 if $w_2 = 1$. Sketch this mode, and show that "kickoff" forces $[\mathbf{K}_\sigma]\lfloor w_2 \quad w_3 \rfloor^T$ are equal in magnitude to forces in the deflected springs.

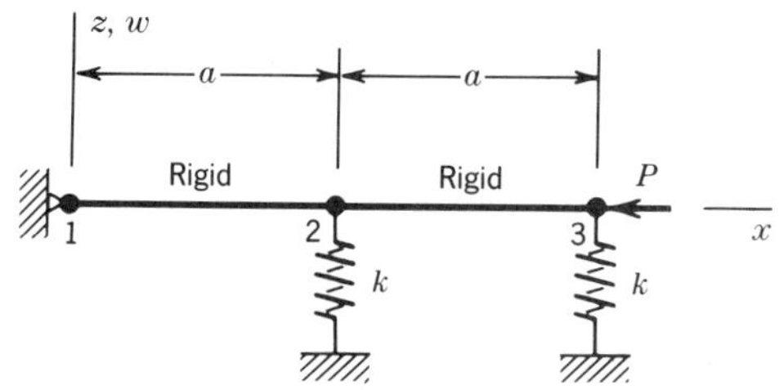

Problem 14.32

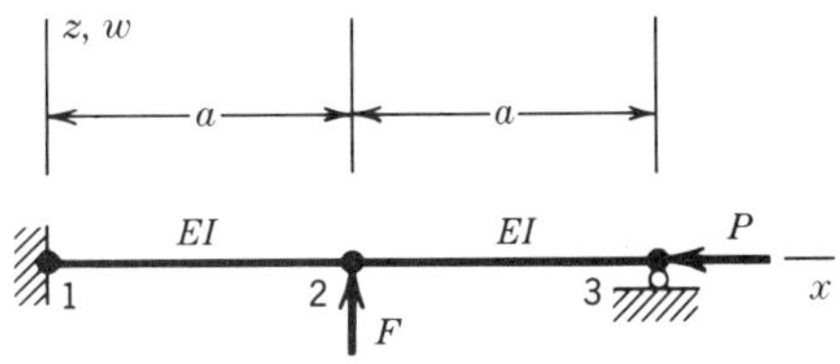

Problem 14.33

14.33 The uniform beam shown is fixed at the left end, simply supported at the right end, and divided into two identical beam elements, each of length a.
(a) Set up a three-equation system $[\mathbf{K} + \mathbf{K}_\sigma]\{\mathbf{D}\} = \{\mathbf{R}\}$, then eliminate d.o.f. θ_2 and θ_3 by condensation, leaving w_2 as the only d.o.f.

(b) Determine P_{cr} (for the case $F = 0$) and compute its percentage error.
(c) Plot w_2/a versus P/P_{cr} if $F = 0.10EI/a^2$.

14.34 Determine buckling loads of the uniform columns shown. Express P_{cr} in terms of E, I, and L. Use the $[\mathbf{k}_\sigma]$ of Eq. 14.2-9.
(a) For Fig. (a), use θ_1 and θ_2 as d.o.f.
(b) For Fig. (b), impose symmetry about the midpoint, so that θ_1 and w_2 are the only d.o.f. needed. Node 2 is not a hinge.
(c) For Fig. (c), let the linear spring have stiffness $k = 2EI/L^3$. The lower end is fully fixed. No axial d.o.f. is needed.
(d) For Fig. (d), let the rotational spring have stiffness $k = EI/L$. The lower end is fully fixed. No axial d.o.f. is needed.
(e) Repeat part (c) letting $k \to \infty$ (top free to rotate but cannot translate).
(f) Repeat part (d) letting $k \to \infty$ (top free to translate but cannot rotate).

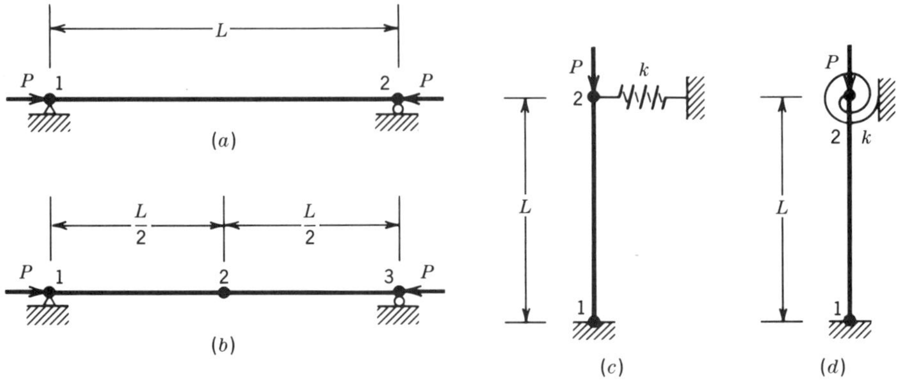

Problem 14.34

14.35 Repeat Problem 14.34, but use the $[\mathbf{k}_\sigma]$ derived in Problem 14.8.

14.36 Repeat Problem 14.34, to the extent possible, if $[\mathbf{k}_\sigma]$ is taken from Eq. 14.2-11.

14.37 (a) Consider use of the condensation technique (described by example in Eqs. 14.5-11 and 14.5-12). How will you decide which d.o.f. to eliminate?
(b) Using condensation, solve Problem 14.15(b).
(c) Using condensation, solve Problem 14.15(c).
(d) Using condensation to eliminate θ_1, solve Problem 14.34(b).
(e) Using condensation to eliminate θ_1, solve Problem 14.35(b).

14.38 The two slender bars shown, which are of lengths L and αL but otherwise identical, are fixed at A and C and welded together at B, where they are simply supported and loaded by axial force P. The possibility of buckling is to be analyzed. Use the two-stage procedure suggested below Eq. 14.1-9: determine axial displacement at B in terms of P, hence compute axial bar forces and $[\mathbf{k}_\sigma]$ for the bars, and finally seek P_{cr} using θ_2 as the only d.o.f.

14.39 The two-element frame shown is fixed at A and at C, and has the same EI throughout. The connection at B is rigid. Assume that d.o.f. that define axial strain are unnecessary because $AE >> EI$. Determine angle β that minimizes the buckling load P_{cr}.

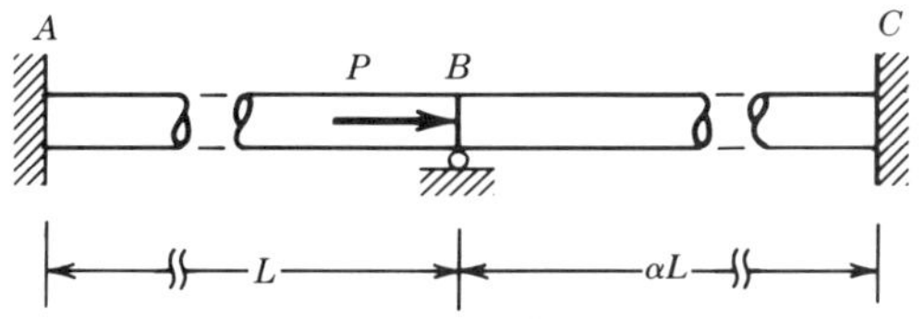

Problem 14.38

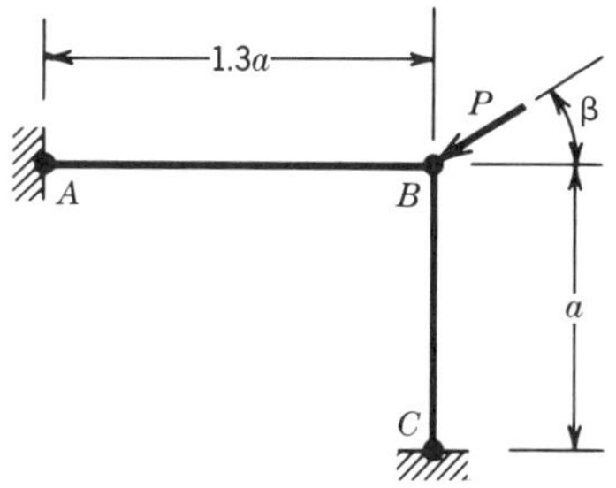

Problem 14.39

14.40 Imagine that a diagonal stress stiffness matrix is proposed for the standard beam element, for which $\{\mathbf{d}\} = \lfloor w_1 \quad \theta_1 \quad w_2 \quad \theta_2 \rfloor^T$. The form of $[\mathbf{k}_\sigma]$ is $\lceil \mathbf{k}_\sigma \rfloor = \lceil 0 \quad c \quad 0 \quad c \rfloor$, where c is a constant.

(a) Determine c by requiring that Eq. 14.2-7 be satisfied when $w_{,x}$ is constant over the element.

(b) Use this $[\mathbf{k}_\sigma]$ to determine P_{cr} for a uniform pin-ended column. Consider a one-element model.

(c) Repeat part (b), but consider a two-element model: impose symmetry about the center of the column, so that nonzero d.o.f. of one-half the column are θ_1 and w_2.

Section 14.6

14.41 The weightless string shown is horizontal, under constant tension T, and carries two particles, each of mass m.

(a) Determine the static deflections of the particles caused by gravity.

(b) Determine the natural frequencies of vibration and the mode shapes.

14.42 Repeat Problem 14.41, but double the mass of the left-hand particle (to mass $2m$).

14.43 A massless and flexible string of length $2L$ hangs from the ceiling. It carries two particles, each of mass m, one at the middle and the other at the lower end.

(a) What is the horizontal deflection of a small horizontal force Q applied to the lower end?

(b) What are the natural frequencies of vibration and the mode shapes?

14.44 The string shown is under tension T and has mass ρ per unit length. Use $[\mathbf{M}]$ and $[\mathbf{K}_\sigma]$ matrices associated with a cubic lateral-displacement field. Omit the conventional stiffness matrix $[\mathbf{K}]$. Solve for the natural frequencies and mode shapes of small-displacement lateral vibrations. (The exact fundamental frequency is $\omega_1^2 = \pi^2 T/4\rho a^2$.)

(a) Use one element. Nonzero d.o.f. are then θ_1 and θ_2.

(b) Use two elements and impose symmetry about the center. Nonzero d.o.f. to be used are then θ_{end} and w_{center}.

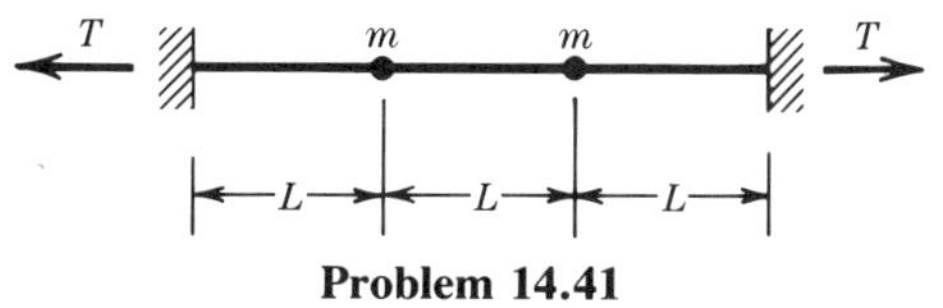

Problem 14.41

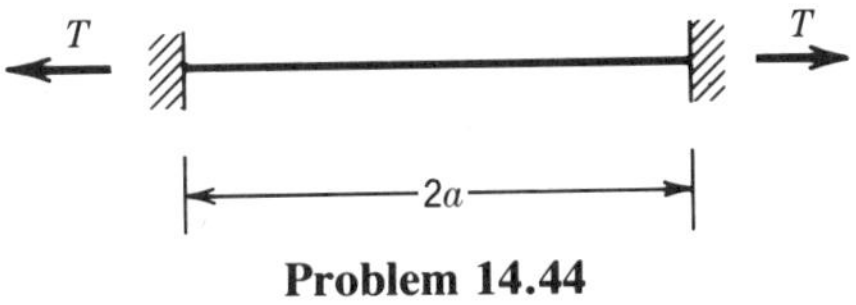

Problem 14.44

14.45 Solve Problem 14.44 using $[\mathbf{k}_\sigma]$ from Eq. 14.2-11, and (a) $[\mathbf{M}]$ based on a cubic lateral-displacement field, and (b) a lumped $[\mathbf{M}]$. (That is, apply each of these mass matrices to each part of Problem 14.44.)

14.46 Model a simply supported beam by a single element. Let $L = 1.0$ m, $A = 0.0002$ m^2, $EI = 300.0$ N·m^2, and $\rho = 2100.0$ kg/m^3. Impose symmetry (and reduce the problem to a single d.o.f.) by setting $\theta_2 = -\theta_1$.
(a) Determine the fundamental frequency ω_1 if there is no axial force.
(b) Determine the axial force that makes the frequency 347 rad/sec.
(c) Determine the frequency if the axial force is 1200 N in compression.

CHAPTER 15

WEIGHTED RESIDUAL METHODS

The construction of approximate solutions of differential equations by means of weighted residual methods is summarized. The Galerkin method, which is the most popular weighted residual method, is used to produce finite element formulations.

15.1 INTRODUCTION

Thus far we have presented the finite element method as a Rayleigh–Ritz method—that is, as an approximation technique that is applied to a variational principle. A variational principle uses an integral expression, called a *functional,* that yields the governing differential equations and nonessential boundary conditions of a problem when operated upon by standard procedures of the calculus of variations. The principle of stationary potential energy is only one of many variational principles.

In an area of physical science other than structural mechanics, a variational principle may be unobtainable. This happens if the differential equation of the problem contains derivatives of odd order. A case in point is fluid mechanics, where, for some types of flow, all that is available are differential equations and boundary conditions. Yet the finite element method can still be applied by means of a *weighted residual method.* Like the Rayleigh–Ritz method, a weighted residual method uses integral expressions that contain the differential equations of a physical problem. Functional and residual formulations are both known as "weak" forms of stating the governing equations of a problem. The differential equations themselves comprise the "strong" form. (The weak form enforces conditions in an average or integral sense, whereas the strong form enforces them at every point.)

The following introductory treatment uses both structural and nonstructural problems to illustrate procedures.

15.2 SOME WEIGHTED RESIDUAL METHODS

This section presents an overview and uses the following notation:

u = dependent variable(s), for example, displacements of a point
x = independent variable(s), for example, coordinates of a point
f,g = functions of x, or constants, or zero
D,B = differential operators

Thus the governing differential equations and nonessential boundary conditions of an arbitrary physical problem are symbolized as

$$Du - f = 0 \qquad \text{in domain } V \tag{15.2-1a}$$

$$Bu - g = 0 \qquad \text{on boundary } S \text{ of } V \tag{15.2-1b}$$

For example, in beam bending Eq. 15.2-1a becomes $EIw_{,xxxx} = q$, where w is lateral deflection and q is distributed lateral load. Thus $D = EId^4/dx^4$, $u = w$, and $f = q$. Equation 15.2-1b symbolizes two equations, namely, $EIw_{,xx} - M_B = 0$ and $EIw_{,xxx} - V_B = 0$, where M_B and V_B are prescribed values of bending moment and transverse shear force at ends of the beam.

In general, the exact solution $u = u(x)$ of Eq. 15.2-1a is unknown and is often difficult to determine. We seek instead an *approximate* solution, $\tilde{u}$. Typically $\tilde{u}$ is a polynomial that satisfies essential boundary conditions and contains undetermined coefficients $a_1, a_2, \ldots, a_n$. Thus $\tilde{u} = \tilde{u}(a,x)$, and $\tilde{u}$ is "admissible" as defined in Section 3.2. To obtain an approximate solution we must determine values of the a_i such that u and $\tilde{u}$ are "close" in some sense.

If $\tilde{u}$ is substituted into Eqs. 15.2-1, equality does not prevail because $\tilde{u}$ is not exact. The discrepancy can be expressed as residuals R_D and R_B, which are functions of x and the a_i:

$$R_D = R_D(a,x) = D\tilde{u} - f \qquad \text{(interior residual)} \tag{15.2-2a}$$

$$R_B = R_B(a,x) = B\tilde{u} - g \qquad \text{(boundary residual)} \tag{15.2-2b}$$

In some physical problems it may happen that all boundary conditions are of the essential class. Then R_B need not enter; only R_D is used in determining the a_i of an approximation $\tilde{u} = \tilde{u}(a,x)$ whose form satisfies essential boundary conditions a priori.

Residuals may vanish for some values of x, but they are not zero for all x unless $\tilde{u}$ is the exact solution, $\tilde{u} \equiv u$. We presume that $\tilde{u}$ is a good approximation of u if residuals are small. Small residuals can be achieved by various schemes, each of which is designed to produce algebraic equations that can be solved for the n coefficients a_i. Some popular schemes are summarized as follows. Their use is illustrated in Section 15.3.

Collocation. For n different values of x, the residuals are set to zero. The method is also called *point collocation*.

$$R_D(a,x_i) = 0 \qquad \text{for} \quad i = 1, 2, \ldots, j - 1 \tag{15.2-3a}$$

$$R_B(a,x_i) = 0 \qquad \text{for} \quad i = j, j + 1, \ldots, n \tag{15.2-3b}$$

Subdomain. Over n different regions of V and S, the integral of the residual is set to zero. The method is also called *subdomain collocation*.

$$\int_{V_i} R_D(a,x)\, dV = 0 \qquad \text{for} \quad i = 1, 2, \ldots, j - 1 \tag{15.2-4a}$$

$$\int_{S_i} R_B(a,x)\, dS = 0 \qquad \text{for} \quad i = j, j + 1, \ldots, n \tag{15.2-4b}$$

Least Squares. The a_i are chosen to minimize a function I:

$$\frac{\partial I}{\partial a_i} = 0 \qquad \text{for} \quad i = 1, 2, \ldots, n \tag{15.2-5}$$

Function I is formed by integrating squares of the residuals,

$$I = \int_V [R_D(a,x)]^2 \, dV + \alpha \int_S [R_B(a,x)]^2 \, dS \tag{15.2-6}$$

where α is an arbitrary scalar multiplier that may be used to achieve dimensional homogeneity and also serves as a penalty number. Larger values of α increase the importance of R_B relative to R_D. The method is also called *continuous least squares*.

Least Squares Collocation. Equation 15.2-5 is still used, but I is redefined. It is now defined in terms of squared residuals at several points i, where i runs from 1 to m and $m \geqq n$:

$$I = \sum_{i=1}^{j-1} [R_D(a,x_i)]^2 + \alpha \sum_{i=j}^{m} [R_B(a,x_i)]^2 \tag{15.2-7}$$

Equation 15.2-5 now yields n equations for the a_i, even when $m > n$. The method is also called *point least squares* and *overdetermined collocation*. If $m = n$, the method becomes simple collocation.

Galerkin. We select "weight functions" $W_i = W_i(x)$ and set the weighted averages of residual R_D to zero. Or, in mathematical terms, we say that R_D is made orthogonal to the weight functions:

$$R_i = \int_V W_i(x) \, R_D(a,x) \, dV = 0 \qquad \text{for} \quad i = 1, 2, \ldots, n \tag{15.2-8}$$

In the Bubnov–Galerkin method, usually called simply the Galerkin method, weight functions W_i are coefficients of the generalized coordinates a_i. Thus $W_i = \partial \tilde{u}/\partial a_i$. In the Petrov–Galerkin method, other forms of W_i are used.

In Galerkin methods, boundary residual R_B is used in combination with integration by parts, so as to introduce nonessential boundary conditions. The procedure is illustrated in subsequent examples.

Remarks. The commonality shared by the foregoing methods is that they all can be loosely symbolized as

$$\int_\Gamma W_i R \, d\Gamma = 0 \tag{15.2-9}$$

where R represents R_D and/or R_B and Γ represents V and/or S. In words, Eq. 15.2-9 says that over the region of interest, the weighted residual has an average value of zero (i.e., $W_i R$ has zero average error). The various weighted residual methods differ in how W_i is defined [15.1]. In the collocation and subdomain methods, the W_i are unit delta or step functions that are nonzero at certain points or over certain regions. In least squares methods, $W_i = \partial R/\partial a_i$. In the Galerkin method, $W_i = \partial \tilde{u}/\partial a_i$.

In solving a problem, whether by the Rayleigh–Ritz method or a weighted residual method, we begin by establishing a trial family of solutions, $\tilde{u} = \tilde{u}(a,x)$. The a_i that define the best form of $\tilde{u}$ are chosen by the stationary–functional conditions $\partial\Pi/\partial a_i = 0$ for the Rayleigh–Ritz method or by Eq. 15.2-9 for a weighted residual method.

Galerkin's method yields a symmetric coefficient matrix if the system of differential equations and boundary conditions is self-adjoint [15.1]. If differential equations *and* a variational principle are both available, then the Galerkin method and the Rayleigh–Ritz method *yield identical solutions when both use the same approximating function* $\tilde{u}$.

Least squares methods *always* produce a symmetric coefficient matrix. Other advantages of least squares include the avoidance of integration in least squares collocation and the "tuning" permitted by adjustment of α in Eqs. 15.2-6 and 15.2-7. However, there are several disadvantages. Despite α, the continuous least squares method may be too strongly influenced by less important residuals. (Separate α_i may be applied to the separate residuals in least squares collocation.) The coefficient matrix tends to be ill conditioned. Because the weights are $W_i = \partial R/\partial a_i$, both R and the W_i contain derivatives of the same order. This means that in continuous least squares, integration by parts cannot reduce the highest-order derivatives of $\tilde{u}$, and elements of higher order are therefore needed so as to achieve the necessary degree of interelement continuity. For example, a two-node, axially loaded bar element would require two d.o.f. at each node—namely, u and $u_{,x}$. This awkwardness could be avoided by reformulating the problem in terms of differential equations of lower order, but again more d.o.f. are required than in a simple displacement formulation, and element d.o.f. are a mixture of force and displacement quantities. Mixed formulations have not been popular in structural mechanics. The least squares collocation method does not require an initial reduction to first-order differential equations [15.4].

15.3 EXAMPLE SOLUTIONS

We illustrate the methods summarized in Section 15.2 by means of the following problem. Let the governing differential equation and nonessential boundary condition be[1]

$$u_{,xx} + cx = 0 \qquad \text{for} \quad 0 < x < L_T \tag{15.3-1a}$$

$$u_{,x} - b = 0 \qquad \text{at} \quad x = L_T \tag{15.3-1b}$$

where u has units of length, c is a constant having units $(\text{length})^{-2}$, and b is a dimensionless constant. The essential boundary condition is $u = 0$ at $x = 0$. A physical interpretation is given in Fig. 15.3-1. Equation 15.3-1a describes this

[1]If derivatives of order $2m$ appear in the differential equation, essential boundary conditions involve derivatives of order zero through $m - 1$, and nonessential boundary conditions involve derivatives of order m through $2m - 1$.

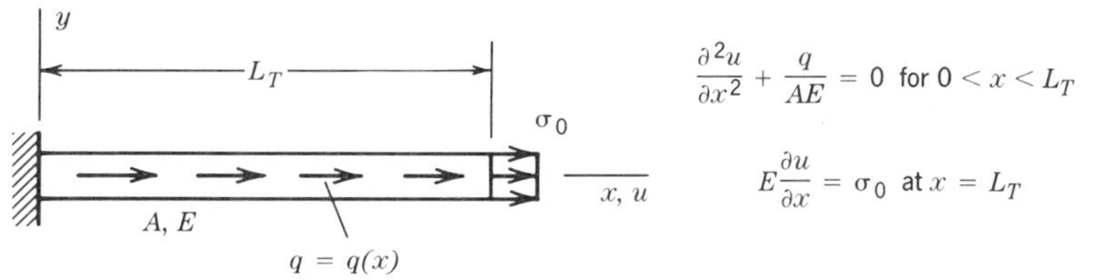

Figure 15.3-1. A physical interpretation of Eq. 15.3-1: axial displacement u of a uniform bar under linearly varying axial load q and end load $\sigma_0 A$, where A is the cross-sectional area. E is the elastic modulus.

problem if q is the linear function $q = q_0 x$ and $c = q_0/AE$. Numerical comparison of exact and approximate solutions appears in Table 15.3-1.

The exact solution of the problem is $u = (3L_T^2 cx - cx^3 + 6bx)/6$, but we pretend that we do not know it. Instead, we seek two-parameter approximate solutions. Let the trial function be

$$\tilde{u} = a_1 x + a_2 x^2 \tag{15.3-2}$$

in which "best" values of a_1 and a_2 are required. Note that $\tilde{u}$ satisfies the essential boundary condition $u = 0$ at $x = 0$. Substitution of Eq. 15.3-2 into Eqs. 15.3-1 yields the interior and boundary residuals

$$R_D = 2a_2 + cx \qquad \text{and} \qquad R_B = (a_1 + 2a_2 L_T) - b \tag{15.3-3}$$

The nonessential boundary condition appears only at $x = L_T$, so R_B is evaluated at $x = L_T$.

As an alternative to Eq. 15.3-2, one can select a trial function $\tilde{u}$ that satisfies Eq. 15.3-1b a priori. Then R_B need not be used in subsequent manipulations. However, the present approach is more in accord with finite element applications of weighted residual methods.

Collocation. We arbitrarily elect to evaluate R_D at $x = L_T/3$. Equations 15.3-3, with $R_D = 0$ and $R_B = 0$, yield

$$a_1 = b + cL_T^2/3 \qquad \text{and} \qquad a_2 = -cL_T/6 \tag{15.3-4}$$

TABLE 15.3-1 RESULTS FOR THE PROBLEM OF EQ. 15.3-1 AND FIG. 15.3-1, FOR THE SPECIAL CASE $b = 0$, $c = L_T = 1$.

Quantity and Location	Exact	Collo-cation	Subdomain and Least Squares	Least Squares Collocation	Galerkin
u at $x = L_T/2$	0.2292	0.1250	0.1875	0.2500	0.2292
u at $x = L_T$	0.3333	0.1667	0.2500	0.3333	0.3333
$u_{,x}$ at $x = 0$	0.5000	0.3333	0.5000	0.6667	0.5833
$u_{,x}$ at $x = L_T/2$	0.3750	0.1667	0.2500	0.3333	0.3333
$u_{,x}$ at $x = L_T$	0	0	0	0	0.0833

Subdomain. We elect to evaluate Eq. 15.2-4a by integrating over the entire span, $x = 0$ to $x = L_T$. Thus Eq. 15.2-4a and $R_B = 0$ yield

$$a_1 = b + cL_T^2/2 \quad \text{and} \quad a_2 = -cL_T/4 \tag{15.3-5}$$

Least Squares. If $\alpha = 1/L_T$ the two terms in Eq. 15.2-6 have the same units, and

$$I = \int_0^{L_T} (2a_2 + cx)^2\, dx + \frac{1}{L_T}[(a_1 + 2a_2L_T) - b]^2 \tag{15.3-6}$$

in which the latter term is already "integrated" over the point $x = L_T$. The equations $\partial I/\partial a_1 = 0$ and $\partial I/\partial a_2 = 0$ are found to yield Eqs. 15.3-5.

Least Squares Collocation. We arbitrarily elect to evaluate R_D at two points, and arbitrarily choose them to be $x = L_T/3$ and $x = L_T$. Residual R_B is as stated in Eq. 15.3-3. Again let $\alpha = 1/L_T$. The three residuals can be written in the form

$$\begin{Bmatrix} R_{D1} \\ R_{D2} \\ R_B \end{Bmatrix} = \begin{bmatrix} 0 & 2 \\ 0 & 2 \\ 1/L_T & 2 \end{bmatrix} \begin{Bmatrix} a_1 \\ a_2 \end{Bmatrix} - \begin{Bmatrix} -cL_T/3 \\ -cL_T \\ b/L_T \end{Bmatrix} \tag{15.3-7}$$

If all three residuals are set to zero, Eqs. 15.3-7 form an overdetermined system (three equations in two unknowns). A least squares solution requires that we form $I = R_{D1}^2 + R_{D2}^2 + R_B^2$ and apply Eq. 15.2-5 for $i = 1$ and $i = 2$. We introduce symbols $[\mathbf{Q}]$, $\{\mathbf{a}\}$, and $\{\mathbf{c}\}$ for arrays on the right-hand side of Eq. 15.3-7 and obtain the least squares solution for a_1 and a_2 as follows:

$$I = \{\mathbf{R}\}^T\{\mathbf{R}\} \quad \text{where} \quad \{\mathbf{R}\} = [\mathbf{Q}]\{\mathbf{a}\} - \{\mathbf{c}\} \tag{15.3-8a}$$

$$I = \{\mathbf{a}\}^T[\mathbf{Q}]^T[\mathbf{Q}]\{\mathbf{a}\} - 2\{\mathbf{a}\}^T[\mathbf{Q}]^T\{\mathbf{c}\} + \{\mathbf{c}\}^T\{\mathbf{c}\} \tag{15.3-8b}$$

$$\left\{\frac{\partial I}{\partial \mathbf{a}}\right\} = \{\mathbf{0}\} \quad \text{yields} \quad [\mathbf{Q}]^T[\mathbf{Q}]\{\mathbf{a}\} = [\mathbf{Q}]^T\{\mathbf{c}\} \tag{15.3-8c}$$

In writing Eq. 15.3-8b we have used the relation $\{\mathbf{c}\}^T[\mathbf{Q}]\{\mathbf{a}\} = \{\mathbf{a}\}^T[\mathbf{Q}]^T\{\mathbf{c}\}$, which is true because each of the two matrix triple products is a scalar. In Eq. 15.3-8c, which is to be solved for $\{\mathbf{a}\}$, the coefficient matrix $[\mathbf{Q}]^T[\mathbf{Q}]$ is symmetric and of the same order as $\{\mathbf{a}\}$. Applying Eq. 15.3-8c to Eq. 15.3-7, we obtain

$$a_1 = b + 2cL_T^2/3 \quad \text{and} \quad a_2 = -cL_T/3 \tag{15.3-9}$$

Galerkin. From Eqs. 15.2-8 and 15.3-1a, the ith residual is

$$R_i = \int_0^{L_T} W_i(\tilde{u}_{,xx} + cx)\, dx \tag{15.3-10}$$

where $i = 1{,}2$ in the present example. It is standard practice in the Galerkin method to begin with integration by parts. A motivation is to reduce the order of differentiation in the integral. If derivatives of order $2m$ appear, the integral is defined if the integrand has continuous derivatives through order $2m - 1$. In the present example, $2m - 1 = 1$, which means that d.o.f. of a finite element model would have to include nodal values of $\tilde{u}$ and $\tilde{u}_{,x}$ in order to achieve the required continuity. Integration by parts reduces $2m - 1$ to $1 - 1 = 0$, so that d.o.f. of a finite element model need only include nodal values of $\tilde{u}$. Integration by parts also serves to introduce nonessential boundary conditions, as follows.

In one dimension, the formula for integration by parts is written in conventional notation as $\int u\, dv = uv - \int v\, du$, where u and v represent functions of x. In the present example, we apply integration by parts to the term $W_i\tilde{u}_{,xx}$ only. Thus $W_i\tilde{u}_{,xx}\, dx$ is regarded as $u\, dv$ in the formula for integration by parts, where $u = W_i$ and $dv = \tilde{u}_{,xx}\, dx = d(\tilde{u}_{,x})$. Equation 15.3-10 becomes

$$R_i = \int_0^{L_T} (-W_{i,x}\tilde{u}_{,x} + W_i cx)\, dx + [W_i\tilde{u}_{,x}]_0^{L_T} \tag{15.3-11}$$

where, from Eq. 15.3-2,

$$W_1 = \frac{\partial \tilde{u}}{\partial a_1} = x \quad \text{and} \quad W_2 = \frac{\partial \tilde{u}}{\partial a_2} = x^2 \tag{15.3-12}$$

Hence, $W_i\tilde{u}_{i,x} = 0$ at $x = 0$. And, from Eq. 15.3-1b, the nonessential boundary condition is $\tilde{u}_{,x} = b$ at $x = L_T$. Therefore, Eq. 15.3-11 becomes

$$R_i = \int_0^{L_T} (-W_{i,x}\tilde{u}_{,x} + W_i cx)\, dx + [W_i b]_{x=L_T} \tag{15.3-13}$$

With Eqs. 15.3-12 and 15.3-13, the conditions $R_1 = 0$ and $R_2 = 0$ yield

$$a_1 = b + 7cL_T^2/12 \quad \text{and} \quad a_2 = -cL_T/4 \tag{15.3-14}$$

Remarks. Table 15.3-1 summarizes results for the particular case $b = 0$, $c = L_T = 1$. The quantity $u_{,x}$ is proportional to stress for the problem of Fig. 15.3-1. In general, different collocation points yield different results, and a large number of collocation points is usually beneficial in least squares collocation. Solutions summarized in Table 15.3-1 are not the best possible.

The example chosen, because of its simplicity, has traits that do not prevail in general, as follows. Collocation and subdomain methods each yield only one equation from R_D (because only a_2 appears in R_D), and α has no effect on the solution in least squares methods (because a_1 happens not to appear in R_D).

15.4 GALERKIN FINITE ELEMENT METHOD

In this section, one-dimensional examples are used to illustrate the finite element form of the Galerkin method. The interpretation of terms that result from integration by parts, and the assembly of elements to form a structure, are explained in the first example.

Uniform Bar, Axial Load. Equilibrium of axial forces in Fig. 15.4-1 requires that

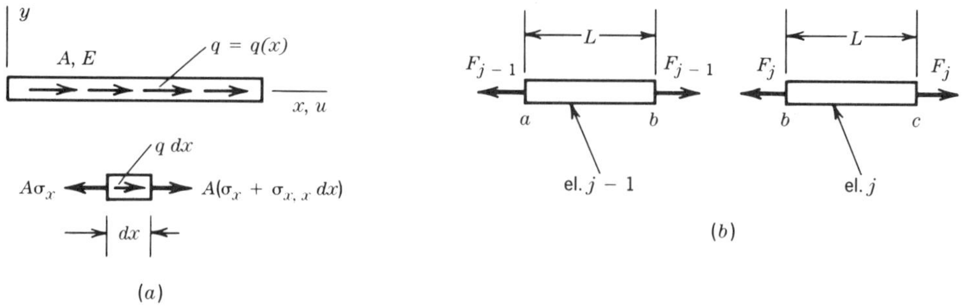

Figure 15.4-1. (a) Uniform elastic bar under distributed axial load q. A = cross-sectional area, E = elastic modulus. (b) Adjacent elements $j - 1$ and j. Node b is shared after assembly of elements.

$A\sigma_{x,x} + q = 0$. Also, $\sigma_x = E\epsilon_x = Eu_{,x}$. Therefore, the governing differential equation in terms of axial displacement u is

$$AEu_{,xx} + q = 0 \tag{15.4-1}$$

At an end where an axial force F is applied, the nonessential boundary condition is

$$AEu_{,x} - F = 0 \tag{15.4-2}$$

At a free end, $F = 0$. Essential boundary conditions consist of prescribed values of u.

Let the bar be divided into *numel* elements of length L. Each element has the assumed displacement field

$$\tilde{u} = \lfloor \mathbf{N} \rfloor \{\mathbf{d}\} \qquad \text{where} \quad \begin{cases} \lfloor \mathbf{N} \rfloor = \left\lfloor \dfrac{L - x}{L} \quad \dfrac{x}{L} \right\rfloor \\ \{\mathbf{d}\} = \lfloor u_1 \quad u_2 \rfloor^T \end{cases} \tag{15.4-3}$$

and $x = 0$ at the left end of the element. D.o.f. u_1 and u_2 are coefficients of modes in the approximating field. Thus u_1 and u_2 play the same role as parameters a_i in Sections 15.2 and 15.3. Weights W_i used in the Galerkin method are therefore

$$W_i = \frac{\partial \tilde{u}}{\partial d_i} = N_i \qquad \text{where} \quad \begin{cases} N_1 = (L - x)/L \\ N_2 = x/L \end{cases} \tag{15.4-4}$$

The Galerkin residual equation, Eq. 15.2-8, becomes

$$\sum_{j=1}^{\text{numel}} \int_0^L N_i(AE\tilde{u}_{,xx} + q)\, dx = 0 \tag{15.4-5}$$

where index i ranges over all shape functions. When elements are assembled, activation of a single d.o.f. activates shape functions in the adjacent elements; that is, linear ramps are activated in elements on either side of the node. Thus, on the structural level, shape functions for all but the first and last nodes are "hat functions," as shown in Fig. 15.4-2. We see that there are as many structural shape functions as there are d.o.f., and therefore as many Galerkin residual equations as there are d.o.f. If AE is constant,[2] integration by parts yields

$$\int_0^L N_i AE\tilde{u}_{,xx}\, dx = [N_i AE\tilde{u}_{,x}]_0^L - \int_0^L N_{i,x} AE\tilde{u}_{,x}\, dx \qquad (15.4\text{-}6)$$

From Eq. 15.4-2, the nonessential boundary condition is $AE\tilde{u}_{,x} = F$ at element ends. Substituting this and Eq. 15.4-6 into Eq. 15.4-5, we obtain

$$\sum_{j=1}^{\text{numel}} \int_0^L (-N_{i,x}AE\tilde{u}_{,x} + N_i q)\, dx + \sum_{j=1}^{\text{numel}} [N_i F]_0^L = 0 \qquad (15.4\text{-}7)$$

We adopt the notation

$$\lfloor \mathbf{B} \rfloor = \lfloor \mathbf{N}_{,x} \rfloor; \qquad \text{then} \qquad \tilde{u}_{,x} = \lfloor \mathbf{B} \rfloor \{\mathbf{d}\} = \left\lfloor -\frac{1}{L} \quad \frac{1}{L} \right\rfloor \begin{Bmatrix} u_1 \\ u_2 \end{Bmatrix} \qquad (15.4\text{-}8)$$

After rearrangement, Eq. 15.4-7 becomes

$$\sum_{j=1}^{\text{numel}} \underbrace{\int_0^L \lfloor \mathbf{B} \rfloor^T AE \lfloor \mathbf{B} \rfloor\, dx}_{[\mathbf{k}]_j} \{\mathbf{d}\} = \sum_{j=1}^{\text{numel}} \underbrace{\int_0^L \lfloor \mathbf{N} \rfloor^T q\, dx}_{\{\mathbf{r}_e\}_j} + \sum_{j=1}^{\text{numel}} \Big[\lfloor \mathbf{N} \rfloor^T F \Big]_0^L \qquad (15.4\text{-}9)$$

Except for the last summation, Eq. 15.4-9 clearly yields the standard formula $[\mathbf{K}]\{\mathbf{D}\} = \{\mathbf{R}\}$, that is,

$$\left(\sum_{j=1}^{\text{numel}} [\mathbf{k}]_j \right) \{\mathbf{D}\} = \sum_{j=1}^{\text{numel}} \{\mathbf{r}_e\}_j + \{\mathbf{P}\} \qquad (15.4\text{-}10)$$

where $\{\mathbf{D}\}$ replaces $\{\mathbf{d}\}$ because of the usual expansion of element matrices to "structure size."

We must explain how the last summation in Eq. 15.4-9 can be regarded as $\{\mathbf{P}\}$,

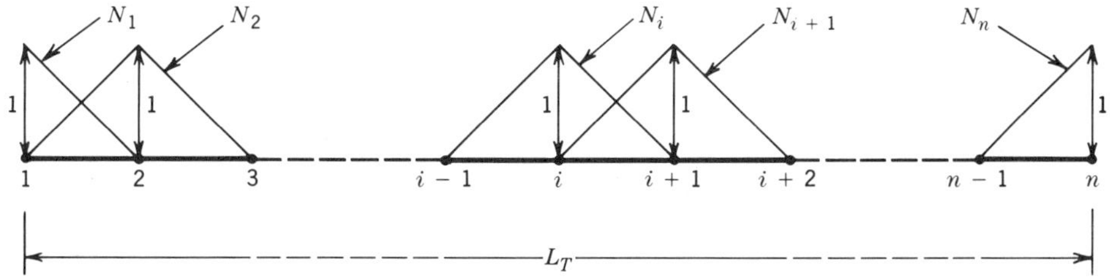

Figure 15.4-2. Shape functions active on the structural level after bar elements have been assembled. The number of d.o.f., and shape functions, is $n = numel + 1$.

[2]If AE is not constant, the differential equation is $d(AEu_{,x})/dx + q = 0$. An equation of this type is treated subsequently (Eq. 15.4-21).

the vector of externally applied concentrated loads. At ends of a typical element, $x = 0$ and $x = L$, respectively,

$$\lfloor \mathbf{N} \rfloor_0 = \lfloor 1 \quad 0 \rfloor \qquad \text{and} \qquad \lfloor \mathbf{N} \rfloor_L = \lfloor 0 \quad 1 \rfloor \tag{15.4-11}$$

For two adjacent elements $j - 1$ and j, Fig. 15.4-1b, the last summation in Eq. 15.4-9 produces the terms

$$\begin{array}{l} \text{node } a \;\text{-----} \\ \text{node } b \;\text{-----} \\ \text{node } c \;\text{---------------} \end{array} \quad - \begin{Bmatrix} 1 \\ 0 \end{Bmatrix} F_{j-1} + \left(\begin{Bmatrix} 0 \\ 1 \end{Bmatrix} F_{j-1} - \begin{Bmatrix} 1 \\ 0 \end{Bmatrix} F_j \right) + \begin{Bmatrix} 0 \\ 1 \end{Bmatrix} F_j \tag{15.4-12}$$

When elements are assembled, as at node b, the resultant axial force $F_{j-1} - F_j$ is produced. This resultant appears in parentheses in Eq. 15.4-12 and is identified as the externally applied load P at node b. Of course, P may be zero; then $F_{j-1} = F_j$. For the problem of Fig. 15.3-1, load vector $\{\mathbf{P}\}$ would contain only one nonzero entry—namely, $F = \sigma_0 A$ associated with the rightmost node. (Expression 15.4-12 is not used in actual computation; it serves only to explain the transition from Eq. 15.4-9 to Eq. 15.4-10.)

Beam Dynamics. In the following example we omit most of the summation signs that indicate assembly of elements and also omit the detailed explanation of nodal loads (as in Eq. 15.4-12). Thus we emphasize the generation of element matrices by the Galerkin method.

With ρ the mass density, the mass per unit length is $\rho_L = \rho A$, where A is the cross-sectional area of the beam. The moment–curvature relation is $EIw_{,xx} = M$. Also, $M_{,x} = V$ and $V_{,x} = q$, where EI = bending stiffness, w = lateral displacement, M = bending moment, V = transverse shear force, and q = transverse load per unit length. The effective inertia load is $q = -\rho_L \ddot{w}$, where $(\dot{\ }) = d(\ )/dt$. Putting all this together, we obtain the governing differential equation

$$EIw_{,xxxx} + \rho_L \ddot{w} = 0 \tag{15.4-13}$$

Nonessential boundary conditions are

$$EIw_{,xx} - M_B = 0 \qquad \text{and} \qquad EIw_{,xxx} - V_B = 0 \tag{15.4-14}$$

where M_B and V_B are prescribed values of bending moment and transverse shear force at ends of the beam. Essential boundary conditions consist of prescribed values of w and $w_{,x}$. The assumed lateral-displacement field $\tilde{w} = \tilde{w}(x)$ and weight functions $W_i = W_i(x)$ are given by

$$\tilde{w} = \lfloor \mathbf{N} \rfloor \{\mathbf{d}\} \qquad \text{and} \qquad W_i = N_i \tag{15.4-15}$$

where $\{\mathbf{d}\} = \lfloor w_1 \quad \theta_1 \quad w_2 \quad \theta_2 \rfloor^T$ and the N_i are the usual cubic shape functions (see Fig. 3.13-2). The Galerkin residual equation for a single element is

$$\int_0^L \lfloor \mathbf{N} \rfloor^T (EI\tilde{w}_{,xxxx} + \rho_L \ddot{\tilde{w}})\, dx = 0 \tag{15.4-16}$$

where L is the element length. With EI constant, two integrations by parts yield

$$\int_0^L \lfloor \mathbf{N} \rfloor^T EI\tilde{w}_{,xxxx}\, dx = \int_0^L \lfloor \mathbf{N}_{,xx} \rfloor^T EI\tilde{w}_{,xx}\, dx + \Big[\lfloor \mathbf{N} \rfloor^T EI\tilde{w}_{,xxx} - \lfloor \mathbf{N}_{,x} \rfloor^T EI\tilde{w}_{,xx} \Big]_0^L \tag{15.4-17}$$

Substitution of Eqs. 15.4-14 and 15.4-17 into Eq. 15.4-16 yields

$$\int_0^L \Big(\lfloor \mathbf{N}_{,xx} \rfloor^T EI\tilde{w}_{,xx} + \rho_L \lfloor \mathbf{N} \rfloor^T \ddot{\tilde{w}} \Big)\, dx + \Big[\lfloor \mathbf{N} \rfloor^T V_B - \lfloor \mathbf{N}_{,x} \rfloor^T M_B \Big]_0^L = 0 \tag{15.4-18}$$

Terms V_B and M_B become part of the load vector $\{\mathbf{R}\}$. The argument is analogous to that used in Eqs. 15.4-11 and 15.4-12. From Eq. 15.4-15,

$$\ddot{\tilde{w}} = \lfloor \mathbf{N} \rfloor \{\ddot{\mathbf{d}}\} \quad \text{and} \quad \tilde{w}_{,xx} = \lfloor \mathbf{B} \rfloor \{\mathbf{d}\}, \quad \text{where} \quad \lfloor \mathbf{B} \rfloor = \lfloor \mathbf{N}_{,xx} \rfloor \tag{15.4-19}$$

Substituting Eq. 15.4-19 into Eq. 15.4-18 and assembling elements, we obtain

$$\sum_{j=1}^{\text{numel}} \Big(\int_0^L \lfloor \mathbf{B} \rfloor^T EI \lfloor \mathbf{B} \rfloor\, dx\, \{\mathbf{d}\} + \int_0^L \rho_L \lfloor \mathbf{N} \rfloor^T \lfloor \mathbf{N} \rfloor\, dx\, \{\ddot{\mathbf{d}}\} \Big) = \{\mathbf{R}\} \tag{15.4-20}$$

which is the standard dynamic equation $[\mathbf{K}]\{\mathbf{D}\} + [\mathbf{M}]\{\ddot{\mathbf{D}}\} = \{\mathbf{R}\}$, where $[\mathbf{K}]$ and $[\mathbf{M}]$ are respectively the structure stiffness and mass matrices and $\{\mathbf{R}\}$ represents time-varying loads.

Heat Flow in a Bar. We consider steady-state heat conduction in a bar with insulated lateral surface, Fig. 15.4-3*a*.[3] Heat flux q is axial and obeys the Fourier heat conduction equation, $q = -kT_{,x}$. Here k is the thermal conductivity of the material. The negative sign indicates that the direction of heat flow is opposite to the direction of temperature increase. In the steady-state condition, the net rate of heat flow out of a differential element is zero. Thus, from Fig. 15.4-3*b*, $d(Aq)/dx = 0$. Combining the two equations $q = -kT_{,x}$ and $d(Aq)/dx = 0$, we obtain

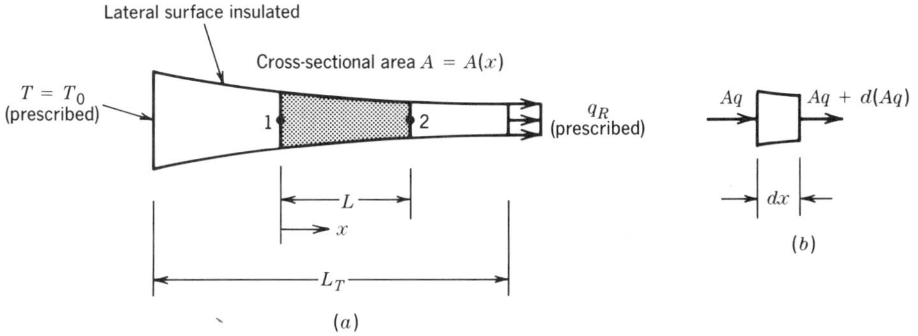

Figure 15.4-3 (a) Heat flow in a tapered bar. A typical element 1–2 is shown shaded. (b) Heat flow through a differential element.

[3]With J = joule (1 N·m), units of the quantities are T, °C; k, J/m·s·°C; q, J/m²·s.

$$\frac{d}{dx}(AkT_{,x}) = 0 \qquad (15.4\text{-}21)$$

as the governing differential equation. We presume that Ak may vary with x. The approximating temperature field is $\tilde{T} = \lfloor \mathbf{N} \rfloor \{\mathbf{T}_e\}$, where, for a two-node element, nodal temperatures are $\{\mathbf{T}_e\} = \lfloor T_1 \quad T_2 \rfloor^T$ and the N_i are given by Eq. 15.4-4. Let q_B indicate a boundary heat flux, that is, a heat flux at either end of the element. We write the Galerkin residual equation, integrate by parts, and substitute the nonessential boundary condition $q_B = -k\tilde{T}_{,x}$. Thus

$$\int_0^L \lfloor \mathbf{N} \rfloor^T (Ak\tilde{T}_{,x})_{,x}\, dx = -\int_0^L \lfloor \mathbf{N}_{,x} \rfloor^T Ak\tilde{T}_{,x}\, dx - \Big[\lfloor \mathbf{N} \rfloor^T A q_B \Big]_0^L = 0 \qquad (15.4\text{-}22)$$

With $\tilde{T}_{,x} = \lfloor \mathbf{N}_{,x} \rfloor \{\mathbf{T}_e\}$, the element equation becomes

$$\int_0^L \lfloor \mathbf{N}_{,x} \rfloor^T Ak \lfloor \mathbf{N}_{,x} \rfloor\, dx\, \{\mathbf{T}_e\} = -\begin{Bmatrix} 0 \\ 1 \end{Bmatrix} A_2 q_{2r} + \begin{Bmatrix} 1 \\ 0 \end{Bmatrix} A_1 q_{1r} = \begin{Bmatrix} A_1 q_{1r} \\ -A_2 q_{2r} \end{Bmatrix} \qquad (15.4\text{-}23)$$

where $A_1 q_{1r}$ and $A_2 q_{2r}$ are respectively heat flow rates at nodes 1 and 2 of the element, each considered positive when heat flows to the right. At the left end of the bar in Fig. 15.4-3*a*, nodal temperature T_0 is prescribed instead of a flux q_0.

It is convenient to adopt the convention that boundary heat flux is considered positive when heat flows into the bar.[4] Using q_1 and q_2 to represent these fluxes, we have $q_1 = q_{1r}$ and $q_2 = -q_{2r}$. Accordingly, for a two-node element in which A and k are constant, Eq. 15.4-23 becomes

$$\frac{Ak}{L}\begin{bmatrix} 1 & -1 \\ -1 & 1 \end{bmatrix} \begin{Bmatrix} T_1 \\ T_2 \end{Bmatrix} = \begin{Bmatrix} Aq_1 \\ Aq_2 \end{Bmatrix} \qquad (15.4\text{-}24)$$

If this uniform element models the entire bar in Fig. 15.4-3*a*, then $T_1 = T_0$ and $q_2 = -q_R$, where the negative sign indicates an outward flux. Thus Eq. 15.4-24 yields $T_2 = T_0 - q_R L/k$ and $q_1 = q_R$. If the boundary conditions were *reversed,* so that $T_2 = T_R$ is prescribed at the right end and inward flux $q_1 = q_0$ is prescribed at the left end, we would obtain $T_1 = T_R + q_0 L/k$ and $q_2 = q_0$.

15.5 INTEGRATION BY PARTS

Integral or "weak" formulations make frequent use of integration by parts. Some useful formulas are now reviewed.

Let **i**, **j**, and **k** be unit vectors in the coordinate directions. Also let F_1, F_2, and F_3 be independent functions of the coordinates, and

$$\mathbf{F} = F_1\mathbf{i} + F_2\mathbf{j} + F_3\mathbf{k} \qquad \text{and} \qquad \boldsymbol{\nu} = \ell\mathbf{i} + m\mathbf{j} + n\mathbf{k} \qquad (15.5\text{-}1)$$

where function $\mathbf{F}$ is defined in a volume V, $\boldsymbol{\nu}$ is a unit outward normal on the

[4]This rule is generalized to multidimensional problems by the convention that boundary flux is considered positive in the direction of the inward surface normal vector.

surface S of V, and ℓ, m, and n are direction cosines of $\boldsymbol{\nu}$. The divergence theorem states that

$$\int_V \nabla\cdot\mathbf{F}\, dV = \int_S \mathbf{F}\cdot\boldsymbol{\nu}\, dS \tag{15.5-2}$$

where $\nabla\cdot\mathbf{F}$ is the divergence of $\mathbf{F}$, for example,

$$\text{rectangular coordinates:}\quad \nabla\cdot\mathbf{F} = \frac{\partial F_1}{\partial x} + \frac{\partial F_2}{\partial y} + \frac{\partial F_3}{\partial z} \tag{15.5-3a}$$

$$\text{cylindrical coordinates:}\quad \nabla\cdot\mathbf{F} = \frac{1}{r}\frac{\partial}{\partial r}(rF_1) + \frac{1}{r}\frac{\partial F_2}{\partial \theta} + \frac{\partial F_3}{\partial z} \tag{15.5-3b}$$

In Eq. 15.5-2, $\mathbf{F}$ and its first partial derivatives must be continuous in V and on S, and integration must proceed over all boundaries, interior as well as exterior.

Let P and Q be functions of the coordinates. Then, for example, $(PQ)_{,x} = P_{,x}Q + PQ_{,x}$. Therefore

$$\int_V PQ_{,x}\, dV = -\int_V P_{,x}Q\, dV + \int_V (PQ)_{,x}\, dV \tag{15.5-4}$$

If we regard PQ as F_1 in Eq. 15.5-3a and let $F_2 = F_3 = 0$, Eq. 15.5-2 allows us to replace the last integral in Eq. 15.5-4 by a surface integral. Thus Eq. 15.5-4 becomes the following formula for integration by parts in rectangular coordinates:

$$\int_V PQ_{,x}\, dV = -\int_V P_{,x}Q\, dV + \int_S PQ\ell\, dS \tag{15.5-5}$$

Analogous formulas for the x and y derivatives are easy to derive.

The same procedure may be applied in cylindrical coordinates. For example,

$$\int_V \frac{1}{r}\frac{\partial}{\partial r}(rPQ)\, dV = \int_V \left[\frac{\partial P}{\partial r}Q + P\frac{1}{r}\frac{\partial}{\partial r}(rQ)\right] dV \tag{15.5-6}$$

We let $F_1 = PQ$ and $F_2 = F_3 = 0$ in Eq. 15.5-3b, solve for the last term in Eq. 15.5-6, and apply Eq. 15.5-2. Thus

$$\int_V P\frac{1}{r}\frac{\partial}{\partial r}(rQ)\, dV = -\int_V \frac{\partial P}{\partial r}Q\, dV + \int_S PQ\ell\, dS \tag{15.5-7}$$

In similar fashion,

$$\int_V \frac{1}{r}P\frac{\partial Q}{\partial \theta}\, dV = -\int_V \frac{1}{r}\frac{\partial P}{\partial \theta}Q\, dV + \int_S PQm\, dS \tag{15.5-8}$$

Formulas for integration by parts in two dimensions can be obtained directly from the preceding formulas by setting $F_3 = 0$ in Eqs. 15.5-3 and presuming that integration with respect to z has already been done across a unit thickness.

15.6 TWO-DIMENSIONAL PROBLEMS

The Quasiharmonic Equation. The "quasiharmonic" equation describes heat conduction and various other physical problems, as explained in more detail in Chapter 16. Here, without specifying the physical problem, we illustrate the formulation of element matrices by the Galerkin method.

Consider a plane region of unit thickness, volume V, and boundary S. The governing equation and the nonessential boundary condition are, respectively,

$$\text{in } V, \quad \frac{\partial}{\partial x}(k_x\phi_{,x}) + \frac{\partial}{\partial y}(k_y\phi_{,y}) + Q = 0 \tag{15.6-1}$$

$$\text{on } S, \quad \ell k_x\phi_{,x} + mk_y\phi_{,y} - q_B = 0 \tag{15.6-2}$$

where $\phi = \phi(x,y)$ is the dependent variable and ℓ and m are direction cosines of an outward normal to S. Known quantities k_x, k_y, and Q may be either constant or functions of x and y. In the nonessential boundary condition, q_B is a prescribed boundary flux, positive when directed into V. Essential boundary conditions, which in general prevail over only a portion of S, consist of prescribed values of ϕ. For the special case $k_x = k_y =$ constant and $Q = 0$, Eq. 15.6-1 becomes Laplace's equation $\nabla^2\phi = 0$. A function ϕ that satisfies $\nabla^2\phi = 0$ is called *harmonic*.

The approximating field $\tilde{\phi}$ is

$$\tilde{\phi} = \lfloor \mathbf{N} \rfloor \{\boldsymbol{\phi}_e\} = \lfloor N_1 \quad N_2 \quad \ldots \quad N_n \rfloor \{\boldsymbol{\phi}_e\} \tag{15.6-3}$$

where n is the number of nodes per element and $\{\boldsymbol{\phi}_e\}$ is the vector of element nodal d.o.f. The Galerkin residual equation is

$$\iint \lfloor \mathbf{N} \rfloor^T \left[\frac{\partial}{\partial x}(k_x\tilde{\phi}_{,x}) + \frac{\partial}{\partial y}(k_y\tilde{\phi}_{,y}) + Q\right] dx\, dy = 0 \tag{15.6-4}$$

Integration by parts (e.g., Eq. 15.5-5) yields

$$\iint \lfloor \mathbf{N} \rfloor^T \frac{\partial}{\partial x}(k_x\tilde{\phi}_{,x})\, dx\, dy = -\iint \lfloor \mathbf{N}_{,x} \rfloor^T k_x\tilde{\phi}_{,x}\, dx\, dy + \int \lfloor \mathbf{N} \rfloor^T k_x\tilde{\phi}_{,x}\ell\, dS \tag{15.6-5a}$$

$$\iint \lfloor \mathbf{N} \rfloor^T \frac{\partial}{\partial y}(k_y\tilde{\phi}_{,y})\, dx\, dy = -\iint \lfloor \mathbf{N}_{,y} \rfloor^T k_y\tilde{\phi}_{,y}\, dx\, dy + \int \lfloor \mathbf{N} \rfloor^T k_y\tilde{\phi}_{,y}m\, dS \tag{15.6-5b}$$

Substitution of Eqs. 15.6-5 and 15.6-2 into Eq. 15.6-4 yields

$$\iint \left(-\lfloor \mathbf{N}_{,x} \rfloor^T k_x\tilde{\phi}_{,x} - \lfloor \mathbf{N}_{,y} \rfloor^T k_y\tilde{\phi}_{,y} + \lfloor \mathbf{N} \rfloor^T Q\right) dx\, dy + \int \lfloor \mathbf{N} \rfloor^T q_B\, dS = 0 \tag{15.6-6}$$

Finally, substitution of $\tilde{\phi}_{,x} = \lfloor \mathbf{N}_{,x} \rfloor \{\boldsymbol{\phi}_e\}$ and $\tilde{\phi}_{,y} = \lfloor \mathbf{N}_{,y} \rfloor \{\boldsymbol{\phi}_e\}$ into Eq. 15.6-6 yields

$$\left[\iint \left(\lfloor \mathbf{N}_{,x}\rfloor^T k_x \lfloor \mathbf{N}_{,x}\rfloor + \lfloor \mathbf{N}_{,y}\rfloor^T k_y \lfloor \mathbf{N}_{,y}\rfloor\right) dx\, dy\right] \{\boldsymbol{\phi}_e\}$$
$$= \iint \lfloor \mathbf{N}\rfloor^T Q\, dx\, dy + \int \lfloor \mathbf{N}\rfloor^T q_B\, dS \quad (15.6\text{-}7)$$

Or, in customary notation, $[\mathbf{k}]\{\boldsymbol{\phi}_e\} = \{\mathbf{r}\}$.

Plane Elasticity. The foregoing manipulations are little changed in application to problems of plane stress or plane strain. We summarize as follows. The governing differential equations are the equilibrium equations, and the nonessential boundary conditions involve boundary tractions Φ_x and Φ_y, that is,

$$\begin{matrix} \sigma_{x,x} + \tau_{xy,y} + F_x = 0 & & \ell\sigma_x + m\tau_{xy} = \Phi_x \\ & \text{and} & \\ \tau_{xy,x} + \sigma_{y,y} + F_y = 0 & & \ell\tau_{xy} + m\sigma_y = \Phi_y \end{matrix} \quad (15.6\text{-}8)$$

where F_x and F_y are body forces per unit volume, and ℓ and m are direction cosines of an outward normal to the boundary. Essential boundary conditions consist of prescribed values of displacements u and v. The element displacement field is

$$\{\tilde{\mathbf{u}}\} = \begin{Bmatrix} \tilde{u} \\ \tilde{v} \end{Bmatrix} = \begin{bmatrix} \lfloor \mathbf{N}\rfloor & \lfloor \mathbf{0}\rfloor \\ \lfloor \mathbf{0}\rfloor & \lfloor \mathbf{N}\rfloor \end{bmatrix} \{\mathbf{d}\} \quad (15.6\text{-}9)$$

where $\{\mathbf{d}\} = \lfloor u_1 \quad u_2 \quad \ldots \quad u_n \quad v_1 \quad v_2 \quad \ldots \quad v_n\rfloor^T$ and $\lfloor \mathbf{N}\rfloor = \lfloor N_1 \quad N_2 \quad \ldots \quad N_n\rfloor$ for an n-node element (e.g., Eqs. 3.12-10 for a four-node rectangle). The arrangement of terms in Eq. 15.6-9 is adopted only for convenience of notation. As there are now two differential equations, there are two Galerkin residual equations, that is,

$$\iint \lfloor \mathbf{N}\rfloor^T(\tilde{\sigma}_{x,x} + \tilde{\tau}_{xy,y} + F_x)\, dx\, dy = 0$$
$$\text{and} \qquad \iint \lfloor \mathbf{N}\rfloor^T(\tilde{\tau}_{xy,x} + \tilde{\sigma}_{y,y} + F_y)\, dx\, dy = 0 \quad (15.6\text{-}10)$$

where $\tilde{\sigma}_x$, $\tilde{\sigma}_y$, and $\tilde{\tau}_{xy}$ are the approximate stress fields produced by Eq. 15.6-9, the strain–displacement relations, and the stress–strain relations. There are four terms in Eqs. 15.6-10 to be integrated by parts. For example, the first such integration yields

$$\iint \lfloor \mathbf{N}\rfloor^T \tilde{\sigma}_{x,x}\, dx\, dy = -\iint \lfloor \mathbf{N}_{,x}\rfloor^T \tilde{\sigma}_x\, dx\, dy + \int \lfloor \mathbf{N}\rfloor^T \tilde{\sigma}_x \ell\, dS \quad (15.6\text{-}11)$$

By this process the nonessential boundary conditions are introduced. Next, we introduce the relations

$$\{\boldsymbol{\sigma}\} = [\mathbf{E}](\{\boldsymbol{\epsilon}\} - \{\boldsymbol{\epsilon}_0\}) + \{\boldsymbol{\sigma}_0\} \quad (15.6\text{-}12)$$
$$\{\boldsymbol{\epsilon}\} = [\partial]\{\tilde{\mathbf{u}}\} = [\mathbf{B}]\{\mathbf{d}\} \quad (15.6\text{-}13)$$

where $[\partial]$ is the differential operator defined in Eq. 1.5-6. The final result given by Eqs. 15.6-10 is the same as given by Eqs. 4.1-5 and 4.1-6.

PROBLEMS

Section 15.3

15.1 Derive the differential equation shown in Fig. 15.3-1.

15.2 Verify that despite collocation at $x = L_T/3$, Eqs. 15.3-2 and 15.3-4 do not yield $\tilde{u} = u$ at $x = L_T/3$. Does this indicate that something is wrong? Explain.

15.3 Show that the Galerkin condition $R_i = 0$ in Eq. 15.3-13 is the same as the stationary-functional condition $\partial\Pi/\partial a_i = 0$ if functional Π is given by

$$\Pi = \int_0^{L_T} \left(\tfrac{1}{2}\, \tilde{u}_{,x}^2 - cx\tilde{u}\right) dx - [b\tilde{u}]_{x=L_T}$$

15.4 Consider the differential equation $u_{,xx} + 4u = 12$, with essential boundary conditions $u = 3$ at $x = 0$ and $u = 1$ at $x = 1$. There are no nonessential boundary conditions. The exact solution is $u = 3 - 2.1995 \sin 2x$. A one-parameter approximating polynomial that meets the essential boundary conditions is $\tilde{u} = 3 - 2x + a(x^2 - x)$. Determine parameter a in the range $0 < x < 1$ by (a) collocation, (b) subdomain, (c) least squares, (d) least squares collocation, and (e) Galerkin methods. Choose points at $x = 0.5$ in part (a) and at $x = \frac{1}{3}$ and $x = \frac{2}{3}$ in part (d). In each case calculate the percentage error of $\tilde{u}$ at $x = 0.5$ and at $x = 0.7$.

15.5 Consider the differential equation $u_{,x} + 2u - 16x = 0$ with the boundary condition $u = 0$ at $x = 0$. The exact solution is $u = 4(e^{-2x} - 1) + 8x$. A two-parameter approximating polynomial that satisfies the boundary condition is $\tilde{u} = a_1 x + a_2 x^2$. Determine a_1 and a_2 in the range $0 < x < 1$ and compute percentage errors of $\tilde{u}$ at $x = 0.5$ and at $x = 0.7$.
(a) Use least squares collocation, with collocation points at $x = 0.25$, $x = 0.50$, and $x = 0.75$.
(b) Use the Galerkin method.

15.6 Solve the problem defined by Eq. 15.3-1 in each of the following ways. Compare your answers with those in Table 15.3-1.
(a) Use collocation, with the sampling point at $x = L_T/2$.
(b) Use subdomain and integrate over the span $0 \le x \le L_T/2$.
(c) Use least squares collocation. Evaluate R_D at $x = 0$, $x = L_T/2$, and $x = L_T$. Use $\alpha = 1/L_T$.
(d) Omit the residual R_{D2} in Eq. 15.3-7.
(e) Show why, in this particular problem, the solution provided by least squares collocation is independent of α.

15.7 In Fig. 15.3-1, let $\sigma_0 = 0$ and let q be the uniform traction $q = q_0$, but acting on the right half $L_T/2 < x < L_T$ only. As in Section 15.3, determine two-parameter approximate solutions by the five methods illustrated. In collocation, choose the point $x = 2L_T/3$; otherwise, proceed as in Section 15.3. Compare exact and approximate results for the case $q_0 = L_T = AE = 1$.

15.8 Consider a uniform, simply supported beam of length L. For each of the following loadings, use the Galerkin method to determine constant a in the approximating lateral displacement field $\tilde{w} = ax(L - x)$. Also compute the percentage error of the predicted midspan deflection.

(a) Sinusoidal distributed lateral load $q = q_0 \sin(\pi x/L)$ over $0 < x < L$.
(b) Uniformly distributed lateral load $q = q_0$ over $0 < x < L$.

Section 15.4

15.9 A cable carries constant axial tension T and contacts an elastic foundation of modulus B (force per unit length per unit of deflection w). The left and right ends of the cable are loaded by the respective forces F_L and F_R, which act perpendicular to the elastic foundation. The sketch shows the cable in its deflected position.
(a) Show that the governing differential equation is $Tw_{,xx} - Bw = 0$.
(b) Use the Galerkin method to establish formulas for element matrices analogous to those in Eq. 15.4-9.

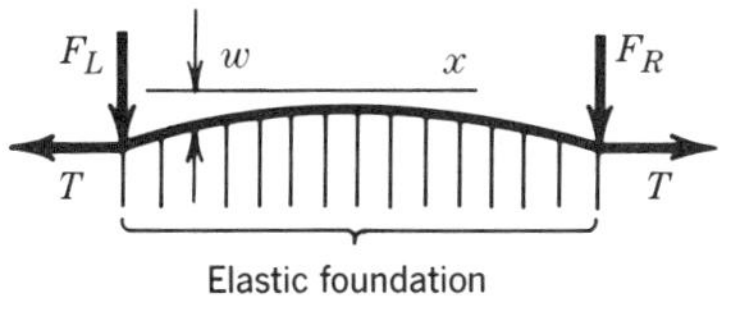

Problem 15.9

15.10 The equation of motion of a string is $Tw_{,xx} - \rho_L \ddot{w} = 0$, where T = constant axial tension, w = small lateral displacement, x = axial coordinate, ρ_L = mass per unit length, and $\ddot{w} = d^2w/dt^2$.
(a) Formulate finite element matrices by the Galerkin method. Let the element have two d.o.f.
(b) Consider a simply supported uniform string of length $2L$. Model it by two elements, each of length L, and of the type formulated in part (a). Solve for the fundamental frequency of vibration. (The exact answer is $\omega^2 = \pi^2 T/4\rho_L L^2$.)
(c) Can the problem of part (b) be solved using a single element? Explain how or why not.

15.11 Let F = constant axial force, positive in tension, and B = elastic foundation modulus (force per unit length per unit of deflection w). With F and B taken into account, the differential equation of a beam becomes $EIw_{,xxxx} - q - Fw_{,xx} + Bw = 0$. Formulate expressions for the element matrices associated with F and B in a form analogous to Eq. 15.4-20.

15.12 For the beam problem, Eqs. 15.4-13 to 15.4-20, demonstrate the treatment of interelement moments and shear forces, in the fashion of Eq. 15.4-12.

15.13 Let the end cross sections of element 1–2 in Fig. 15.4-3 have areas A_1 and A_2, respectively. If A between ends is a linear function of x, what equation replaces Eq. 15.4-24? Assume that k is constant.

15.14 Starting with Eq. 15.4-23, demonstrate the assembly of element thermal "load" vectors. Use Eq. 15.4-12 as a guide.

15.15 In the differential equations cited in (a) and (b) below, $g = g(x)$ and $u = u(x)$. Let the approximating field for an element of length L be $\tilde{u} = \lfloor \mathbf{N} \rfloor \{\mathbf{d}\}$, as usual. In each case, what is a formula for $[\mathbf{k}]$ in the equation $[\mathbf{k}]\{\mathbf{d}\} = \{\mathbf{r}\}$ in terms of g and $\lfloor \mathbf{N} \rfloor$? (You may ignore boundary conditions in this exercise.) (a) $gu_{,x} = 0$. (b) $gu_{,xx} = 0$.

Section 15.6

15.16 Show that Eq. 15.6-7 can also be obtained from Eq. 15.6-3 and the stationary condition of the functional

$$\Pi = \frac{1}{2}\int\int (k_x\phi_{,x}^2 + k_y\phi_{,y}^2 - 2Q\phi)\, dx\, dy - \int q_B\phi\, dS$$

15.17 The Helmholz equation, $p_{,xx} + p_{,yy} + p_{,zz} + (\omega/c)^2p = 0$, governs acoustic modes of vibration in a cavity with rigid walls. Here $p = p(x,y,z)$ represents the amplitude of sinusoidally varying pressure, ω is the circular frequency, and c is the speed of sound in the medium. The boundary condition is $p_{,n} = 0$, where n is a direction normal to the wall. Derive formulas for finite element matrices, using the assumed pressure amplitude field $p = \lfloor \mathbf{N} \rfloor \{\mathbf{p}_e\}$.

15.18 In cylindrical coordinates and with $k_x = k_y = k =$ constant, Eq. 15.6-1 becomes

$$\frac{1}{r}\frac{\partial}{\partial r}\left(r\frac{\partial\phi}{\partial r}\right) + \frac{1}{r^2}\frac{\partial^2\phi}{\partial\theta^2} + \frac{\partial^2\phi}{\partial z^2} + \frac{Q}{k} = 0$$

For a solid of revolution, the nonessential boundary condition is $k(\ell\phi_{,r} + n\phi_{,z}) - q_B = 0$, where ℓ and n are direction cosines of a normal to the surface of the solid. For this problem, formulate equations analogous to Eqs. 15.6-7.

15.19 The differential equation for wind-driven circulation in a shallow lake is

$$\psi_{,xx} + \psi_{,yy} + A\psi_{,x} + B\psi_{,y} + C = 0$$

where ψ is the stream function and A, B, and C are functions of x and y. With h = depth, depthwise average velocities are $u = \psi_{,y}/h$ and $v = -\psi_{,x}/h$. Coordinates x and y are tangent to the lake surface. The nonessential boundary condition is $\psi_{,n} = 0$ on the shoreline, where n is a direction normal to the shoreline. Derive a finite element formulation by the Galerkin method [15.8]. A symbolic result is desired, analogous to Eq. 15.6-7, not details of a particular element.

15.20 Complete the development outlined in Eqs. 15.6-8 to 15.6-13; that is, verify that Eqs. 4.1-5 and 4.1-6 are produced.

15.21 An isotropic, flat disk has unit thickness and inner and outer radii r_i and r_0. The disk is set spinning about its center at constant angular velocity ω. The differential equation of equilibrium is

$$\frac{1}{r}\frac{d}{dr}(r\sigma_r) - \frac{\sigma_\theta}{r} + \rho\omega^2 r = 0$$

where σ_r = radial stress, σ_θ = circumferential stress, and ρ = mass density. Generate formulas for element matrices. Express your results in terms of the shape function matrix and its derivatives, as in Eq. 15.6-7.

15.22 Consider an elastic, axially symmetric solid under axially symmetric loads. Use arguments analogous to those of Eqs. 15.6-8 to 15.6-13 to formulate finite element matrices. For simplicity, omit body forces and initial stresses and strains. Express your results in terms of the shape function matrix and its derivatives, as in Eq. 15.6-7.

16 CHAPTER

HEAT CONDUCTION AND SELECTED FLUID PROBLEMS

Heat conduction equations are reviewed and used to generate finite element formulations. Thermal transients are discussed. Certain problems of acoustics and flow, primarily those whose differential equation is of the same form as the heat conduction equation, are also treated.

16.1 INTRODUCTION TO HEAT CONDUCTION PROBLEMS

Heat conduction analysis may be performed to determine material temperatures and rates of heat flow. The temperature distribution may also be needed in order to perform an analysis for thermally induced stress. Fortunately, it is possible to use a single mesh layout for both problems: a computer program can read a single data file, compute temperatures at nodes, then use these temperatures in stress analysis. Remarks about thermally induced stress, and when a temperature gradient may *not* produce stress, appear in Sections 1.7 and 4.7. Thermally induced nodal loads are accounted for by terms in Eq. 4.1-6.

A finite element formulation of steady-state heat conduction produces equations of the form $[\mathbf{K}]\{\mathbf{T}\} = \{\mathbf{R}\}$, where $\{\mathbf{T}\}$ contains nodal temperatures of the structure. Matrices $[\mathbf{K}]$ and $\{\mathbf{R}\}$ can be generated by the method of making a suitable functional stationary or by a weighted residual method.

Quantities used in our discussion are as follows. The unit of heat or energy is J = 1 joule = 1 N·m.

c = specific heat (J/kg·°C)
h = heat transfer coefficient, or film coefficient (J/m^2·s·°C)
k = thermal conductivity (J/m·s·°C)
Q = rate of internal heat generation per unit volume (J/m^3·s)
q = heat flux per unit area (J/m^2·s)
q_B = prescribed flux normal to a surface (J/m^2·s)
ρ = mass density (kg/m^3)
T = temperature (°C)
T_f = fluid temperature (°C) [used with h]
t = time (s)
$\dot{T}$ = $\partial T/\partial t$ (°C/s)

In the foregoing units, actual fluids and solids display numerical values in the approximate ranges $10^2 < c < 10^4$, $10 < h < 10^5$, and $10^{-2} < k < 500$.

Heat transfer into a solid across a fluid boundary layer is given by $q =$

$h(T_f - T)$, where T is the surface temperature of the solid and T_f is the fluid temperature on the other side of the boundary layer. The heat transfer coefficient h depends on the nature of the fluid, the geometry of the surface, and the dynamics of fluid motion past the surface.

Sources of Q include resistance to electric current and chemical reactions.

In this chapter, the foregoing material properties are taken as independent of temperature unless specifically stated otherwise. Realistically, k is a function of T. Also, $\partial k/\partial T$ may be positive or negative, depending upon the material and sometimes the temperature as well. Similarly, h may depend on temperature. Temperature dependence of k and/or h makes the heat conduction equations nonlinear. This problem is briefly considered in Section 16.5. Radiation is another important source of nonlinearity.

The starting point for heat conduction analysis is the Fourier heat conduction equation, which is

$$q = -k \frac{\partial T}{\partial x} \tag{16.1-1}$$

This equation states that heat flux q in direction x is proportional to the gradient of temperature in direction x. The negative sign indicates that heat flow is opposite to the direction of temperature increase.

16.2 A ONE-DIMENSIONAL EXAMPLE

Consider a straight but tapered bar, Fig. 16.2-1. Heat flows across end surfaces. Due to convection, heat also flows across lateral surfaces with the flux rate $h(T_f - T)$, where T_f is the temperature of surrounding fluid. This flux is directed into the bar if $T_f > T$. Heat is also assumed to be generated internally at rate Q per unit volume. We assume that temperature T in the bar varies only with x, and ask for a finite element formulation that will yield $T = T(x)$ in the steady-state condition.

In the steady-state condition, the net rate of heat flow into any differential element is zero. With volume element $dV = A\,dx$ and surface area increment $dS = p\,dx$, where $p = p(x)$ is the perimeter of the cross section, Fig. 16.2-1b yields

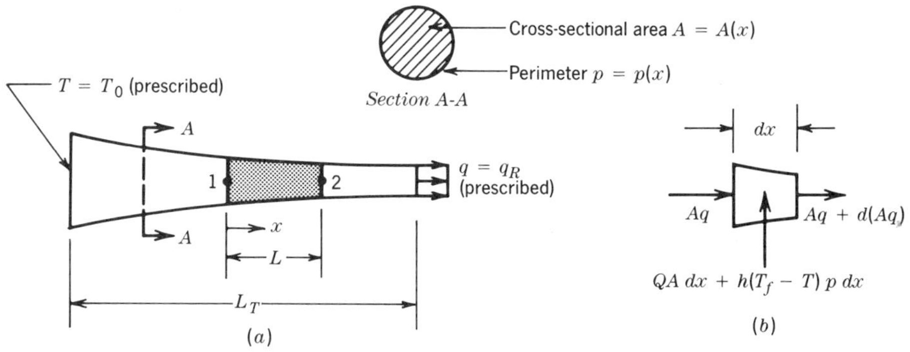

Figure 16.2-1. (a) A bar of varying circular cross section. A typical element 1–2 is shaded. (b) Contributions to heat flow through a differential element.

$$Aq - [Aq + d(Aq)] + QA\,dx + h(T_f - T)p\,dx = 0 \tag{16.2-1}$$

Combining Eqs. 16.1-1 and 16.2-1 and dividing by dx, we obtain the governing differential equation

$$\frac{d}{dx}(AkT_{,x}) + QA + h(T_f - T)p = 0 \tag{16.2-2}$$

Boundary conditions at the ends are

$$T = T_0 \text{ at } x = 0 \qquad \text{and} \qquad T_{,x} = -q_R/k \text{ at } x = L_T \tag{16.2-3}$$

These boundary conditions are respectively essential and nonessential.

A finite element formulation can be developed from the following functional:

$$\Pi = \int [\tfrac{1}{2}AkT_{,x}^2 + \tfrac{1}{2}hpT^2 - (QA + hpT_f)T]\,dx \tag{16.2-4}$$

The standard manipulations of calculus of variations show that Eq. 16.2-2 and the nonessential boundary condition are produced by the stationary condition $d\Pi = 0$ (see Eq. 3.7-6). We interpolate temperature T and temperature gradient $T_{,x}$ along an element from nodal temperatures $\{\mathbf{T}_e\}$:

$$T = \lfloor\mathbf{N}\rfloor\{\mathbf{T}_e\} \qquad \text{and} \qquad T_{,x} = \lfloor\mathbf{N}_{,x}\rfloor\{\mathbf{T}_e\} \tag{16.2-5}$$

For the two-node element shown in Fig. 16.2-1, $\{\mathbf{T}_e\} = \lfloor T_1 \quad T_2\rfloor^T$ and $N_1 = (L - x)/L$, $N_2 = x/L$. Next, because $T = T^T$ and $T^2 = T^TT$, Eqs. 16.2-4 and 16.2-5 yield, for a single element,

$$\Pi_e = \tfrac{1}{2}\{\mathbf{T}_e\}^T \underbrace{\int_0^L \lfloor\mathbf{N}_{,x}\rfloor^T Ak \lfloor\mathbf{N}_{,x}\rfloor\,dx}_{[\mathbf{k}]} \{\mathbf{T}_e\} + \tfrac{1}{2}\{\mathbf{T}_e\}^T \underbrace{\int_0^L \lfloor\mathbf{N}\rfloor^T hp \lfloor\mathbf{N}\rfloor\,dx}_{[\mathbf{h}_{\mathrm{ls}}]} \{\mathbf{T}_e\}$$

$$- \{\mathbf{T}_e\}^T \underbrace{\int_0^L \lfloor\mathbf{N}\rfloor^T QA\,dx}_{\{\mathbf{r}_Q\}} - \{\mathbf{T}_e\}^T \underbrace{\int_0^L \lfloor\mathbf{N}\rfloor^T hpT_f\,dx}_{\{\mathbf{r}_{\mathrm{ls}}\}} \tag{16.2-6}$$

The subscript ls stands for "lateral surface." Quantities such as A, p, and Q are in general functions of x, and may be interpolated from nodal values if so desired.

For the entire finite element structure, Π is given by the summation $\Pi = \Sigma\,\Pi_e$, which implies the expansion of element arrays to "structure size," with the global nodal temperature array $\{\mathbf{T}\}$ replacing the several element arrays $\{\mathbf{T}_e\}$, and with $[\mathbf{K}] = \Sigma\,[\mathbf{k}]$, and so on, following the same matrix assembly procedures described in Section 2.7. Next, Π is made stationary with respect to temperatures $\{\mathbf{T}\}$. Writing the stationary condition for an element rather than for the structure, we have

$$\left\{\frac{\partial\Pi}{\partial\mathbf{T}_e}\right\} = \{\mathbf{0}\} \qquad \text{yields} \qquad ([\mathbf{k}] + [\mathbf{h}_{\mathrm{ls}}])\{\mathbf{T}_e\} = \{\mathbf{r}_Q\} + \{\mathbf{r}_{\mathrm{ls}}\} \tag{16.2-7}$$

After assembly of elements, the corresponding global equations can be written in the form

$$([\mathbf{K}] + [\mathbf{H}_{ls}])\{\mathbf{T}\} = \{\mathbf{R}\} \qquad \text{where} \quad \{\mathbf{R}\} = \{\mathbf{P}\} + \sum \{\mathbf{r}_e\} \tag{16.2-8}$$

where, recalling structural terminology, $[\mathbf{K}] + [\mathbf{H}_{ls}]$ is analogous to a stiffness matrix, $\{\mathbf{T}\}$ is analogous to nodal displacements, $\{\mathbf{r}_e\} = \{\mathbf{r}_Q\} + \{\mathbf{r}_{ls}\}$ is analogous to nodal loads from elements (caused, e.g., by self-weight), and $\{\mathbf{P}\}$ is analogous to externally applied loads. For the bar in Fig. 16.2-1, $\{\mathbf{P}\}$ contains a single nonzero term—namely $-q_R A_R$—which represents the prescribed heat flux at the right end (*out* of the bar, in this case). Our sign convention for boundary values of q is that they are considered positive when heat flows into the body or element. The boundary condition at the left end is imposed by assigning $T = T_0$ at the leftmost node, which is analogous to assigning a known nonzero value to a displacement d.o.f. in a structural problem.

The foregoing formulation is also easy to obtain by the Galerkin method, as described in Section 15.4. Equation 15.4-24 states $[\mathbf{k}]$ for a two-node element.

16.3 HEAT CONDUCTION IN A PLANE

In this section we consider the mathematical formulation of heat flow in a plane of unit thickness. Symbols defined in Section 16.1 are used. Radiation heat transfer is not included.

Governing Equation. For a thermally orthotropic material, Fig. 16.3-1, Eq. 16.1-1 yields the heat fluxes

$$q_r = -k_r T_{,r} \qquad \text{and} \qquad q_s = -k_s T_{,s} \tag{16.3-1}$$

where k_r and k_s are principal thermal conductivities in the principal material directions r and s. Temperature gradients in the various coordinate directions are given by chain rule differentiation, that is, by

$$\begin{Bmatrix} T_{,r} \\ T_{,s} \end{Bmatrix} = [\mathbf{\Lambda}] \begin{Bmatrix} T_{,x} \\ T_{,y} \end{Bmatrix} \qquad \text{where} \quad [\mathbf{\Lambda}] = \begin{bmatrix} x_{,r} & y_{,r} \\ x_{,s} & y_{,s} \end{bmatrix} = \begin{bmatrix} \cos\beta & \sin\beta \\ -\sin\beta & \cos\beta \end{bmatrix} \tag{16.3-2}$$

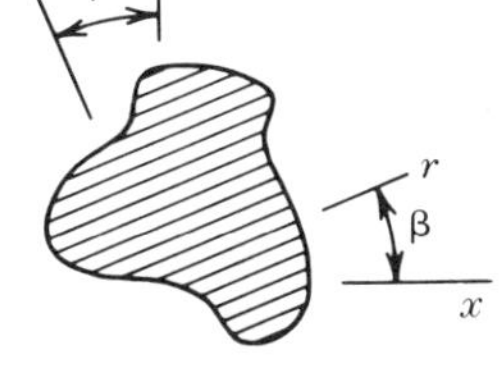

Figure 16.3-1. A layered material with principal directions r and s.

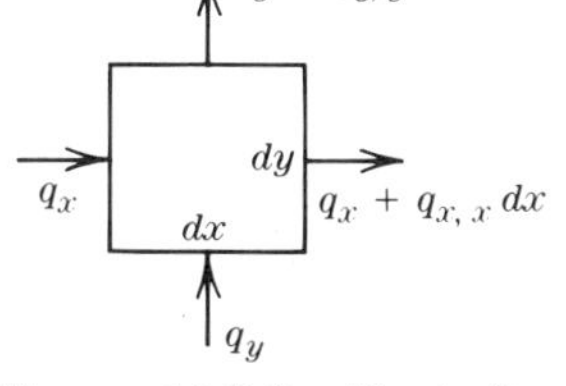

Figure 16.3-2. Heat flux through sides of a differential element.

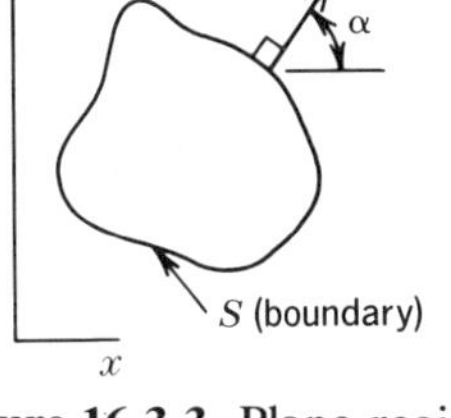

Figure 16.3-3. Plane region with outward normal vector $\boldsymbol{\nu}$ on its boundary S.

Heat flux q is a vector and transforms like displacement, that is, $\lfloor q_x \quad q_y \rfloor^T = [\Lambda]^T \lfloor q_r \quad q_s \rfloor^T$. Combining this with Eqs. 16.3-1 and 16.3-2, we obtain

$$\begin{Bmatrix} q_x \\ q_y \end{Bmatrix} = -[\boldsymbol{\kappa}] \begin{Bmatrix} T_{,x} \\ T_{,y} \end{Bmatrix} \quad \text{where} \quad [\boldsymbol{\kappa}] = \begin{bmatrix} k_x & k_{xy} \\ k_{xy} & k_y \end{bmatrix} = [\Lambda]^T \begin{bmatrix} k_r & 0 \\ 0 & k_s \end{bmatrix} [\Lambda] \tag{16.3-3}$$

For a body of unit thickness, the rate of heat generation in a differential element $dx\,dy$ is $Q\,dx\,dy$. If lateral surfaces of the body are insulated, heat flux across the boundary of a differential element is as depicted in Fig. 16.3-2. The net inward flow of heat is

$$Q\,dx\,dy - (q_{x,x}\,dx)\,dy - (q_{y,y}\,dy)\,dx = (Q - q_{x,x} - q_{y,y})\,dx\,dy \tag{16.3-4}$$

In general, heat flow produces a time rate of change of stored energy—namely, $c\rho\,dx\,dy\,\dot{T}$. Therefore, $Q - q_{x,x} - q_{y,y} = c\rho\dot{T}$. Combining this with Eq. 16.3-3, we obtain

$$\frac{\partial}{\partial x}(k_x T_{,x} + k_{xy} T_{,y}) + \frac{\partial}{\partial y}(k_{xy} T_{,x} + k_y T_{,y}) + Q = c\rho\dot{T} \tag{16.3-5}$$

If a lateral surface z = constant is *not* insulated, so that there is a convective transfer of heat across a lateral surface of the plane body, the heat flux $q = h(T_f - T)$ flows into the body across the lateral surface. If this transfer occurs on *both* lateral surfaces of the body, and if h and T_f are the same on both surfaces, then the term $2h(T_f - T)$ must be added to the left-hand side of Eq. 16.3-5. A portion of a lateral surface may be neither insulated nor subject to convection; instead, an inward flux q_l may be prescribed. Then q_l must be added to the left-hand side of Eq. 16.3-5 (or add q_l/τ for a body of thickness τ).

If the medium is homogeneous and isotropic, then $k_{xy} = 0$ and $k_x = k_y = k$, where k is a constant. Thus Eq. 16.3-5 becomes

$$k(T_{,xx} + T_{,yy}) + Q = c\rho\dot{T} \tag{16.3-6}$$

If, in addition, $Q = 0$ and a steady state prevails ($\dot{T} = 0$), then we obtain Laplace's equation, $T_{,xx} + T_{,yy} = 0$.

Boundary Conditions. One may prescribe temperature on part (or all) of the boundary and heat flux on another part (or all). On any one part, temperature *or* flux is prescribed, not both. In general, the prescribed quantities may be functions of time. The following boundary conditions apply on boundary S, Fig. 16.3-3, not on the lateral surfaces cited below Eq. 16.3-5.

The *essential* boundary condition is a prescription of temperature on part or all of S. Alternative names are *first* or *Dirichlet* boundary condition.

The *nonessential* boundary condition is a prescription of heat flux (possibly zero) on part or all of S. Alternative names are *second* or *Neumann* boundary condition. For a thermally isotropic material, flux in direction $\boldsymbol{\nu}$ of Fig. 16.3-3 is $q_\nu = -kT_{,\nu}$. In addition, by the chain rule,

$$T_{,\nu} = T_{,x}x_{,\nu} + T_{,y}y_{,\nu} = T_{,x}\ell_B + T_{,y}m_B \tag{16.3-7}$$

where ℓ_B and m_B are direction cosines of $\boldsymbol{\nu}$. Adopting the convention that prescribed boundary heat flux q_B is positive when directed into the body, the nonessential boundary condition is therefore $q_B = -q_\nu$, or

$$q_B = k(T_{,x}\ell_B + T_{,y}m_B) \tag{16.3-8}$$

If the body is thermally orthotropic, we write $q_\nu = q_x \cos\alpha + q_y \sin\alpha = q_x\ell_B + q_y m_B$ and obtain q_x and q_y from Eq. 16.3-3. Therefore, with $q_B = -q_\nu$,

$$q_B = (k_x T_{,x} + k_{xy} T_{,y})\ell_B + (k_{xy} T_{,x} + k_y T_{,y})m_B \tag{16.3-9}$$

If there is convection heat transfer across all or part of boundary S, then q_B is replaced by $h(T_f - T)$ for this part of S. Convection heat transfer across a *lateral* surface is not considered part of the boundary condition as it does not appear on S.

Functional. A functional for plane heat conduction is

$$\Pi = \iint \left(\frac{1}{2} \begin{Bmatrix} T_{,x} \\ T_{,y} \end{Bmatrix}^T [\boldsymbol{\kappa}] \begin{Bmatrix} T_{,x} \\ T_{,y} \end{Bmatrix} - QT + \rho c \dot{T} T \right) dx\, dy$$
$$- \int h(T_f T - \tfrac{1}{2}T^2)\, dS - \int q_B T\, dS \tag{16.3-10}$$

in which the surface integrals are each evaluated on the portion of S subject to convection or prescribed flux. If there is also convection across a lateral surface, the term $h(T_f T - T^2/2)$ must be subtracted from the integrand of the double integral, once for each lateral surface involved. With T, $T_{,x}$, and $T_{,y}$ subject to variation, one can show by calculus of variations that the condition $\delta\Pi = 0$ produces Eqs. 16.3-5 and 16.3-9.

16.4 GENERAL SOLIDS AND SOLIDS OF REVOLUTION

The equations of Section 16.3 can be written in a form that also serves for general solids and solids of revolution. The governing equation, Eq. 16.3-5, is

$$\{\partial\}^T([\boldsymbol{\kappa}]\{\mathbf{T}_\partial\}) + Q = c\rho\dot{T} \tag{16.4-1}$$

where $\{\partial\}^T$ is a differential operator and $\{\mathbf{T}_\partial\}$ contains temperature gradients (examples follow). Allowing for either prescribed flux or convection on S, the nonessential boundary condition, Eq. 16.3-9, is

$$q_B = \{\boldsymbol{\mu}\}^T[\boldsymbol{\kappa}]\{\mathbf{T}_\partial\} \qquad \text{or} \qquad h(T_f - T) = \{\boldsymbol{\mu}\}^T[\boldsymbol{\kappa}]\{\mathbf{T}_\partial\} \tag{16.4-2}$$

where $\{\boldsymbol{\mu}\}$ contains direction cosines of a normal to boundary S. The functional, Eq. 16.3-10, is

$$\Pi = \int_V (\tfrac{1}{2}\{\mathbf{T}_\partial\}^T[\boldsymbol{\kappa}]\{\mathbf{T}_\partial\} - QT + \rho c\dot{T}T)\, dV - \int_S (q_B T + hT_f T - \tfrac{1}{2}hT^2)\, dS \tag{16.4-3}$$

where, for the plane problem with unit thickness discussed in Section 16.3, $dV = (1)\,dx\,dy = dx\,dy$, and

$$\{\partial\} = \begin{Bmatrix} \partial/\partial x \\ \partial/\partial y \end{Bmatrix} \qquad \{\mathbf{T}_\partial\} = \begin{Bmatrix} T_{,x} \\ T_{,y} \end{Bmatrix} \qquad \{\boldsymbol{\mu}\} = \begin{Bmatrix} \ell_B \\ m_B \end{Bmatrix} \tag{16.4-4}$$

and $[\boldsymbol{\kappa}]$ is given by Eq. 16.3-3. In structural terms, $\{\mathbf{T}_\partial\}$ is analogous to strains $\{\boldsymbol{\epsilon}\}$ and $[\boldsymbol{\kappa}]$ is analogous to material property matrix $[\mathbf{E}]$.

General Solids. Extension from two dimensions to three requires that we include $T_{,z}$ in $\{\mathbf{T}_\partial\}$, include k_z, k_{yz}, and k_{zx} in $[\boldsymbol{\kappa}]$, and do volume integration over $dV = dx\,dy\,dz$. Equations 16.4-1, 16.4-2, and 16.4-3 apply, with

$$\{\partial\} = \begin{Bmatrix} \partial/\partial x \\ \partial/\partial y \\ \partial/\partial z \end{Bmatrix} \qquad \{\mathbf{T}_\partial\} = \begin{Bmatrix} T_{,x} \\ T_{,y} \\ T_{,z} \end{Bmatrix} \qquad \{\boldsymbol{\mu}\} = \begin{Bmatrix} \ell_B \\ m_B \\ n_B \end{Bmatrix} \tag{16.4-5}$$

If principal thermal conductivities are written as the diagonal matrix $\lceil k_r \quad k_s \quad k_t \rfloor$, then $[\boldsymbol{\kappa}]$ is $[\boldsymbol{\kappa}] = [\boldsymbol{\Lambda}]^T \lceil k_r \quad k_s \quad k_t \rfloor [\boldsymbol{\Lambda}]$, where $[\boldsymbol{\Lambda}]$ is given by Eq. 7.2-1.

Solids of Revolution. The coordinates are r (radial), θ (circumferential), and z (axial). As compared with the general solid, $[\boldsymbol{\kappa}]$ for a solid of revolution is computed in the same way, dr and $r\,d\theta$ replace dx and dy, $T_{,r}$ and $T_{,\theta}/r$ replace $T_{,x}$ and $T_{,y}$, $dV = r\,dr\,d\theta\,dz$, and $dS = r\,d\theta\,db$, where db is an increment of boundary length in an rz plane. Direction cosines in $[\boldsymbol{\Lambda}]$ pertain to angles between principal material axes and coordinate directions r, θ, and z. In Eqs. 16.4-1, 16.4-2, and 16.4-3, we use

$$\{\partial\} = \begin{Bmatrix} (1/r) + \partial/\partial r \\ (1/r)\,\partial/\partial\theta \\ \partial/\partial z \end{Bmatrix} \qquad \{\mathbf{T}_\partial\} = \begin{Bmatrix} T_{,r} \\ T_{,\theta}/r \\ T_{,z} \end{Bmatrix} \qquad \{\boldsymbol{\mu}\} = \begin{Bmatrix} \ell_B \\ 0 \\ n_B \end{Bmatrix} \tag{16.4-6}$$

A boundary-normal $\boldsymbol{\nu}$ has no θ component, so $\ell_B = \cos(\nu,x)$, $m_B = 0$, and $n_B = \cos(\nu,z)$.

If the temperature field and the material properties are axially symmetric, then all derivatives with respect to θ vanish, and the problem is mathematically two-dimensional.

A nonsymmetric temperature field can be treated by Fourier series, as explained for stress analysis in Section 10.5. Thus a three-dimensional problem is replaced by a series of two-dimensional problems. If, for example, θ is a principal material direction and the temperature field is symmetric with respect to the $\theta = 0$ plane, one can use $T = \Sigma\,\overline{T}_n \cos n\theta$, where $\overline{T}_n$ is a function of n, r, and z but is independent of θ. Boundary conditions are written in terms of their Fourier series components, and $\overline{T}$ is determined for $n = 0$, $n = 1$, $n = 2$, and so on. Superposition of these solutions yields the resultant temperature field.

16.5 FINITE ELEMENT FORMULATION

The reader may wish to review Section 3.10, in which element matrices are formulated for a special case of plane heat conduction. The same procedures apply to the more general functional of Eq. 16.4-3 and are summarized as follows.

We write a temperature field T in terms of element nodal temperatures $\{\mathbf{T}_e\}$, and from it compute the required temperature gradients $\{\mathbf{T}_\partial\}$:

$$T = \lfloor \mathbf{N} \rfloor \{\mathbf{T}_e\} \qquad \text{and} \qquad \{\mathbf{T}_\partial\} = [\mathbf{B}]\{\mathbf{T}_e\} \qquad \text{where} \quad [\mathbf{B}] = \{\partial\}\lfloor \mathbf{N} \rfloor \tag{16.5-1}$$

but $(1/r)$ is deleted from row 1 of $\{\partial\}$ if Eq. 16.4-6 is used. Since $T = T^T$, $T^2 = T^T T$, and $\dot{T} = \lfloor \mathbf{N} \rfloor \{\dot{\mathbf{T}}_e\}$, Eq. 16.4-3 can be written, for one element,

$$\Pi_e = \tfrac{1}{2}\{\mathbf{T}_e\}^T([\mathbf{k}] + [\mathbf{h}])\{\mathbf{T}_e\} + \{\mathbf{T}_e\}^T\big([\mathbf{c}]\{\dot{\mathbf{T}}_e\} - \{\mathbf{r}_q\} - \{\mathbf{r}_Q\} - \{\mathbf{r}_h\}\big) \tag{16.5-2}$$

where

$$\begin{aligned}
[\mathbf{k}] &= \int_{V_e} [\mathbf{B}]^T[\boldsymbol{\kappa}][\mathbf{B}]\, dV & \{\mathbf{r}_q\} &= \int_{S_e} \lfloor \mathbf{N} \rfloor^T q_B \, dS \\
[\mathbf{h}] &= \int_{S_e} \lfloor \mathbf{N} \rfloor^T h \lfloor \mathbf{N} \rfloor \, dS & \{\mathbf{r}_Q\} &= \int_{V_e} \lfloor \mathbf{N} \rfloor^T Q \, dV \\
[\mathbf{c}] &= \int_{V_e} \lfloor \mathbf{N} \rfloor^T \rho c \lfloor \mathbf{N} \rfloor \, dV & \{\mathbf{r}_h\} &= \int_{S_e} \lfloor \mathbf{N} \rfloor^T h T_f \, dS
\end{aligned} \tag{16.5-3}$$

By adding the Π_e contributions of the elements, we obtain Π of the assembled system. Assembly implies the usual expansion of element arrays to "structure size," so that the global nodal temperature array $\{\mathbf{T}\}$ replaces $\{\mathbf{T}_e\}$ of each element, $[\mathbf{K}] = \Sigma\, [\mathbf{k}]$, and so on. Equations that make Π stationary are $\{\partial\Pi/\partial\mathbf{T}\} = \{\mathbf{0}\}$, that is,

$$([\mathbf{K}] + [\mathbf{H}])\{\mathbf{T}\} + [\mathbf{C}]\{\dot{\mathbf{T}}\} = \{\mathbf{R}_q\} + \{\mathbf{R}_Q\} + \{\mathbf{R}_h\} \tag{16.5-4}$$

In a plane problem with convection heat transfer across a lateral surface, additional terms appear, as noted below Eq. 16.3-10. For convection across *one* lateral surface, we define

$$[\mathbf{h}_{\text{ls}}] = \iint \lfloor \mathbf{N} \rfloor^T h \lfloor \mathbf{N} \rfloor \, dx\, dy \qquad \text{and} \qquad \{\mathbf{r}_{\text{ls}}\} = \iint \lfloor \mathbf{N} \rfloor^T h T_f \, dx\, dy \tag{16.5-5}$$

The terms $[\mathbf{H}_{\text{ls}}]\{\mathbf{T}\}$ and $\{\mathbf{R}_{\text{ls}}\}$ must be added to the left- and right-hand sides, respectively, of Eq. 16.5-4.

Remarks. If the medium is isotropic, $[\boldsymbol{\kappa}]$ becomes the diagonal matrix $k\lceil 1 \quad 1 \quad 1 \rfloor$ in general solids and solids of revolution, or $k\lceil 1 \quad 1 \rfloor$ in plane problems and axisymmetric solids with axisymmetric temperature distribution.

The simplest special form of Eq. 16.5-4 is $[\mathbf{K}]\{\mathbf{T}\} = \{\mathbf{0}\}$, which represents steady-state conditions without internal sources or sinks, some of the nodal temperatures T_i in $\{\mathbf{T}\}$ prescribed, and no heat flow across the boundary other than that implicitly associated with the prescribed T_i.

Prescribed nodal temperatures—and their spatial derivatives, if also used as nodal d.o.f. in $\{\mathbf{T}\}$—can be treated like prescribed nodal displacements in structural mechanics (Section 2.10). Thus, as in Fig. 2.10-7, the right-hand side of Eq. 16.5-4 is modified, and ones and zeros appear in the coefficient matrix. An alternative procedure, which corresponds to Fig. 2.10-6, is to add a large conductivity K_D to the appropriate diagonal coefficient in **[K]** and augment the corresponding coefficient on the right-hand side by $K_D\overline{T}$, where $\overline{T}$ is the prescribed nodal temperature. This treatment greatly increases the maximum eigenvalues of the system and may therefore cause trouble in a direct integration analysis of how temperature varies with time [16.1].

Matrices defined by Eqs. 16.5-3 and matrices used in structural mechanics have the following analogies of form:

[**k**] is analogous to a conventional stiffness matrix.
[**h**] is analogous to an elastic foundation stiffness matrix.
[**c**] is analogous to a mass matrix.
$\{\mathbf{r}_Q\}$ is analogous to nodal loads from body force.
$\{\mathbf{r}_q\}$, $\{\mathbf{r}_h\}$ are analogous to nodal loads from surface traction.

Matrix [**c**] is analogous to mass matrix [**m**] in that both multiply time derivatives of nodal d.o.f. and both provide resistance to time rates of change. But [**c**] multiplies *first* derivatives and [**m**] multiplies *second* derivatives.

Like [**m**], [**c**] may be formulated as a consistent matrix or as a lumped matrix. The simplification provided by lumping may be accompanied by loss of accuracy. For an element with n nodal temperatures in $\{\mathbf{T}_e\}$, a lumped form of [**c**] results from multiplying ρc by the element volume V_e and assigning $\rho c V_e/n$ to each element node. Thus $\rho c V_e/n$ appears n times in a diagonal matrix [**c**]. Similar ad hoc lumping can be used to simplify calculation of [**h**] or $\{\mathbf{r}_Q\}$. An $\{\mathbf{r}_Q\}$ or $\{\mathbf{R}_Q\}$ vector that contains a single nonzero term represents a point source or a point sink at a node. Proportional and optimal capacity-lumping schemes can be employed, in direct analogy to mass-lumping schemes described in Section 13.3, and having the same trade-offs of advantage and disadvantage.

If the element is formulated in isoparametric fashion, then $dV = J\, d\xi\, d\eta\, d\zeta$ and the usual coordinate transformations are invoked. If the problem is essentially plane but the body is of varying thickness, then $dV = \tau\, dx\, dy$, where thickness τ is a function of x and y. If the body is a solid of revolution, then $dV = r\, dr\, d\theta\, dz$ or $dV = rJ\, d\xi\, d\eta\, d\theta$, in which radius r should not be taken as constant over an element unless the element is far from the axis of revolution.

Matrices [**h**], $\{\mathbf{r}_q\}$, and $\{\mathbf{r}_h\}$ are zero unless the element has an edge or face on boundary S and either convection or prescribed flux is associated with that part of S. Even then, these matrices receive nonzero contributions only from the element edge or face that forms part of S.

It often happens that [$\boldsymbol{\kappa}$] must be regarded as a function of temperature. For steady-state conditions, a simple but perhaps inefficient way to solve this nonlinear problem is as follows. Estimate numerical values of conductivities, generate [**K**], and solve for {**T**}. Use these temperatures to obtain improved estimates of conductivities, generate a new [**K**], and solve for the new {**T**}. Repeat until convergence. Refinements of this iterative process are of course possible [16.2]. Techniques discussed in Chapter 17 can also be used.

Example. For the four-node plane isoparametric element depicted in Fig. 16.5-1, the formulation of certain matrices is outlined as follows. The element temperature field is

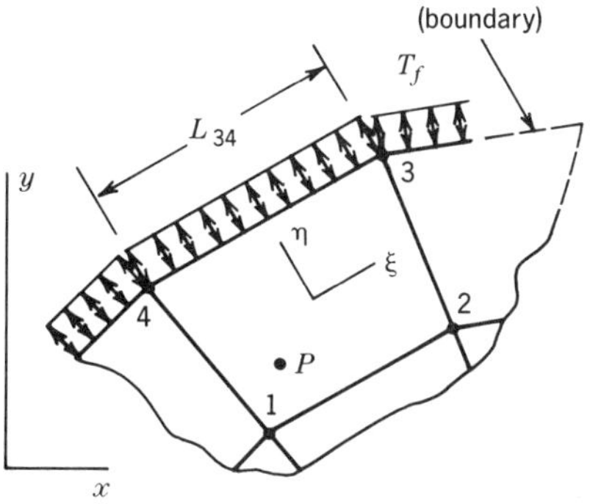

Figure 16.5-1. Bilinear isoparametric element at the boundary of a plane structure where convection occurs.

$T = \lfloor \mathbf{N} \rfloor \{\mathbf{T}_e\} = N_1 T_1 + N_2 T_2 + N_3 T_3 + N_4 T_4$, where the N_i are given by Eq. 6.3-2. Temperature gradients $\{\mathbf{T}_\partial\}$ are

$$\begin{Bmatrix} T_{,x} \\ T_{,y} \end{Bmatrix} = [\mathbf{J}]^{-1} \begin{Bmatrix} T_{,\xi} \\ T_{,\eta} \end{Bmatrix} = \underbrace{[\mathbf{J}]^{-1} \begin{bmatrix} N_{1,\xi} & N_{2,\xi} & N_{3,\xi} & N_{4,\xi} \\ N_{1,\eta} & N_{2,\eta} & N_{3,\eta} & N_{4,\eta} \end{bmatrix}}_{[\mathbf{B}]} \begin{Bmatrix} T_1 \\ T_2 \\ T_3 \\ T_4 \end{Bmatrix} \qquad (16.5\text{-}6)$$

where Jacobian matrix $[\mathbf{J}]$ is defined by Eq. 6.3-11. The conductivity matrix $[\mathbf{k}]$ for an element of thickness τ becomes

$$[\mathbf{k}] = \int_{-1}^{1} \int_{-1}^{1} [\mathbf{B}]^T [\boldsymbol{\kappa}] [\mathbf{B}] \tau J \, d\xi \, d\eta \qquad (16.5\text{-}7)$$

where τ may be a function of ξ and η.

Convection matrix $[\mathbf{h}]$ receives a contribution from side 3–4 only. Along side 3–4, $N_1 = N_2 = 0$, $N_3 = (1 + \xi)/2$, $N_4 = (1 - \xi)/2$, and $J = L_{34}/2$. Therefore, if τ and h are independent of ξ,

$$[\mathbf{h}] = \int_{-1}^{1} \lfloor \mathbf{N} \rfloor^T h \lfloor \mathbf{N} \rfloor \tau J \, d\xi = \frac{h \tau L_{34}}{6} \begin{bmatrix} 0 & 0 & 0 & 0 \\ 0 & 0 & 0 & 0 \\ 0 & 0 & 2 & 1 \\ 0 & 0 & 1 & 2 \end{bmatrix} \qquad (16.5\text{-}8)$$

In similar fashion, for $\{\mathbf{r}_h\}$ we obtain

$$\{\mathbf{r}_h\} = \int_{-1}^{1} \lfloor \mathbf{N} \rfloor^T h T_f \tau J \, d\xi = T_f \frac{h \tau L_{34}}{2} \begin{Bmatrix} 0 \\ 0 \\ 1 \\ 1 \end{Bmatrix} \qquad (16.5\text{-}9)$$

If a heat input Q_P (units J/s) is prescribed at point P, we obtain the resulting $\{\mathbf{r}_Q\}$ vector from Eq. 16.5-3 by saying that $Q = 0$ except at point P,

$$\{\mathbf{r}_Q\} = \int_{V_e} \lfloor \mathbf{N} \rfloor^T Q \, dV = \lfloor \mathbf{N} \rfloor_P^T \int_{V_e} Q \, dV = \lfloor \mathbf{N} \rfloor_P^T Q_P \qquad (16.5\text{-}10)$$

where $\lfloor \mathbf{N} \rfloor_P$ is the value of $\lfloor \mathbf{N} \rfloor$ at point P.

16.6 THERMAL TRANSIENTS

In thermal problems, as in structural mechanics, a time-varying solution may be obtained by the modal method or by direct temporal integration. If material properties are not temperature-dependent, and the solution is dominated by a few of the lowest eigenmodes and is needed over a long time span, then the modal method is favored. If the problem is nonlinear, or the solution displays sharp transients (which require many eigenmodes for an accurate description) and is needed over a short time span, then direct integration is favored.

By either method, the equations to be solved have the form

$$[\mathbf{K}_T]\{\mathbf{T}\} + [\mathbf{C}]\{\dot{\mathbf{T}}\} = \{\mathbf{R}\} \tag{16.6-1}$$

This equation is the same as Eq. 16.5-4, in which $[\mathbf{K}_T]$ contains all matrices that premultiply $\{\mathbf{T}\}$, and $\{\mathbf{R}\}$ contains all vectors on the right-hand side. One seeks to determine $\{\mathbf{T}\}$ as a function of time when $[\mathbf{K}_T]$ and $[\mathbf{C}]$ are known, $\{\mathbf{R}\}$ is a known function of time, and the initial temperatures at time $t = 0$ are known. Unless stated otherwise, we presume that $[\mathbf{K}_T]$ and $[\mathbf{C}]$ are independent of $\{\mathbf{T}\}$.

Modal Method. The procedure is very similar to that used for structural dynamics (Section 13.6). It is outlined as follows. One first considers the eigenproblem

$$([\mathbf{K}_T] - \lambda[\mathbf{C}])\{\overline{\mathbf{T}}\} = \{\mathbf{0}\} \tag{16.6-2}$$

If each eigenvector $\{\overline{\mathbf{T}}\}_i$ is normalized with respect to $[\mathbf{C}]$, that is, if $\{\overline{\mathbf{T}}\}_i^T[\mathbf{C}]\{\overline{\mathbf{T}}\}_i = 1$, then

$$[\boldsymbol{\phi}]^T[\mathbf{C}][\boldsymbol{\phi}] = \lceil \mathbf{I} \rfloor \qquad \text{and} \qquad [\boldsymbol{\phi}]^T[\mathbf{K}_T][\boldsymbol{\phi}] = \lceil \boldsymbol{\lambda} \rfloor \tag{16.6-3}$$

where $[\boldsymbol{\phi}]$ is the *modal matrix;* that is, a matrix whose columns are the normalized eigenvectors $\{\overline{\mathbf{T}}\}_1$, $\{\overline{\mathbf{T}}\}_2$, and so on, $\lceil \mathbf{I} \rfloor$ is a unit matrix, and $\lceil \boldsymbol{\lambda} \rfloor$ is the (diagonal) *spectral matrix* $\lceil \boldsymbol{\lambda} \rfloor = \lceil \lambda_1 \;\; \lambda_2 \;\; . \; . \; . \;\; \lambda_n \rfloor$. Nodal temperatures are transformed to generalized temperatures $\{\mathbf{Z}\}$ by

$$\{\mathbf{T}\} = [\boldsymbol{\phi}]\{\mathbf{Z}\} \tag{16.6-4}$$

where the Z_i in $\{\mathbf{Z}\}$ state the proportion of each eigenvector in the transformation. We substitute Eq. 16.6-4 into Eq. 16.6-1, premultiply by $[\boldsymbol{\phi}]^T$, and take note of Eqs. 16.6-3. Thus we obtain uncoupled equations, each having the form

$$\dot{Z}_i + \lambda_i Z_i = p_i \qquad \text{where} \quad p_i = \{\boldsymbol{\phi}\}_i^T\{\mathbf{R}\} \tag{16.6-5}$$

Here $\{\boldsymbol{\phi}\}_i$ is the ith column of $[\boldsymbol{\phi}]$, and i runs from 1 to m, where m is typically much less than the total number of d.o.f.; that is, only the first few columns of $[\boldsymbol{\phi}]$ are used. After Eqs. 16.6-5 are integrated, $\{\mathbf{Z}\} = \{\mathbf{Z}(t)\}$ is known, and Eq. 16.6-4 yields $\{\mathbf{T}\} = \{\mathbf{T}(t)\}$.

Variants of modal methods used in structural mechanics can also be applied to thermal transient analysis [16.3].

Direct Integration. Consider two temperature states, separated by time increment

Δt, and denoted by $\{\mathbf{T}\}_n$ and $\{\mathbf{T}\}_{n+1}$. A temporal integration scheme, known as the generalized trapezoidal rule, is based on the assumption that the two temperature states have the relation

$$\{\mathbf{T}\}_{n+1} = \{\mathbf{T}\}_n + \{(1 - \beta)\dot{\mathbf{T}}_n + \beta\dot{\mathbf{T}}_{n+1}\}(\Delta t) \tag{16.6-6}$$

Like Newmark's method for the second-order equations of structural dynamics, Eq. 16.6-6 contains a factor β that the analyst may select. Next we write Eq. 16.6-1 for time t and again for time $t + \Delta t$, then multiply the first equation by $1 - \beta$ and the second by β. Thus

$$(1 - \beta)\big([\mathbf{K}_T]\{\mathbf{T}\}_n + [\mathbf{C}]\{\dot{\mathbf{T}}\}_n\big) = (1 - \beta)\{\mathbf{R}\}_n \tag{16.6-7a}$$

$$\beta\big([\mathbf{K}_T]\{\mathbf{T}\}_{n+1} + [\mathbf{C}]\{\dot{\mathbf{T}}\}_{n+1}\big) = \beta\{\mathbf{R}\}_{n+1} \tag{16.6-7b}$$

Equations 16.6-7 are added, then Eq. 16.6-6 is used to eliminate time derivatives of temperature. This step requires that [**C**] not change with time. The result is

$$\left(\frac{1}{\Delta t}[\mathbf{C}] + \beta[\mathbf{K}_T]\right)\{\mathbf{T}\}_{n+1} = \left(\frac{1}{\Delta t}[\mathbf{C}] - (1 - \beta)[\mathbf{K}_T]\right)\{\mathbf{T}\}_n + (1 - \beta)\{\mathbf{R}\}_n + \beta\{\mathbf{R}\}_{n+1} \tag{16.6-8}$$

From a known $\{\mathbf{T}\}_0$ at $t = 0$, Eq. 16.6-8 yields $\{\mathbf{T}\}_1$ at $t = \Delta t$. Then, using $\{\mathbf{T}\}_1$, we determine $\{\mathbf{T}\}_2$ at $t = 2(\Delta t)$, and so on. If Δt is not changed, the coefficient of $\{\mathbf{T}\}_{n+1}$ need be generated and forward-reduced only once; the equations are then repeatedly solved for a sequence of right-hand sides.

Depending on β, time step Δt in Eq. 16.6-8 may have an upper limit if the algorithm is to be numerically stable. If $\beta < \frac{1}{2}$ the largest Δt for stability is [16.4]

$$\Delta t_{cr} = \frac{2}{(1 - 2\beta)\lambda_{max}} \tag{16.6-9}$$

where λ_{max} is the largest eigenvalue of Eq. 16.6-2. If $\beta \geq \frac{1}{2}$ the algorithm is unconditionally stable; that is, stability (but not accuracy) is guaranteed as Δt becomes indefinitely large. Names associated with various schemes are as follows:

$\beta = 0$ forward difference or Euler (conditionally stable)
$\beta = \frac{1}{2}$ Crank–Nicolson or trapezoidal rule (unconditionally stable)
$\beta = \frac{2}{3}$ Galerkin (unconditionally stable)
$\beta = 1$ backward difference (unconditionally stable)

If $\beta = 0$, the algorithm is termed *explicit*. If $\beta > 0$, it is termed *implicit*. If [**C**] is a diagonal matrix and $\beta = 0$, the computational effort per time step is small but so is Δt_{cr}. Among implicit methods, the choice $\beta = \frac{1}{2}$ is popular, but sharp transients may excite annoying oscillations in the solution. Oscillations can be reduced by using a smaller value of Δt or numerically damped by using a value of β somewhat greater than $\frac{1}{2}$. If the problem is *nonlinear* the only unconditionally stable form of Eq. 16.6-8 is $\beta = 1$; however, it is not particularly accurate [16.4].

Various other direct integration algorithms are available [16.5]. Detailed discussion of selected time-stepping algorithms may be found in the latter sections of Chapter 13.

16.7 RELATED PROBLEMS. FLUID FLOW

Several physical phenomena are described by the same form of differential equation that describes heat conduction, Eq. 16.3-5 or 16.4-1. For steady-state conditions and a homogeneous and isotropic material, this equation has the form

$$k\nabla^2\phi + Q = 0 \tag{16.7-1}$$

where ∇^2 is the Laplacian operator (defined in Eq. 16.7-4). Phenomena described by Eq. 16.7-1, or by its less specialized form, include the following:

Heat conduction (ϕ = temperature)
Viscous flow in a pipe (ϕ = axial velocity)
Groundwater flow (ϕ = hydraulic head)
Pressurized membrane (ϕ = membrane deflection)
Elastic torsion (ϕ = stress function or warping function)
Electric conduction (ϕ = voltage)
Electrostatics (ϕ = field potential)
Magnetostatics (ϕ = magnetic potential)
Potential flow (ϕ = velocity potential or stream function)

For the first four phenomena cited, k represents conductivity, viscosity, permeability, and surface tension, and Q represents internal heat generated, pressure gradient, flow associated with a source or a sink, and pressure, respectively.

With appropriate definition of variables, material properties, and boundary conditions, problems in any of these areas can be addressed by use of the computational procedures already established for heat conduction analysis. Indeed, several of these problems can be solved by appropriate use of a computer program for *structural* analysis [16.6]. For example, if a viscous incompressible fluid flows so slowly that inertia effects are negligible, the two-dimensional problem is described by the equation $\nabla^4\psi = 0$, where ψ is the stream function defined in Eq. 16.7-3b [16.7]. This equation has the same form as the equation $\nabla^4 w = q/D$ that describes bending of a flat plate. It is highly recommended that use of any of these analogies be preceded by study of the specific problem area in question.

Potential Flow. A particularly simple example of Eq. 16.7-1 is provided by plane *potential flow*—that is, irrotational flow of an incompressible and inviscid fluid. Let u and v represent flow velocities in the x and y directions, respectively. The conditions of irrotationality and continuity are, respectively,

$$u_{,y} - v_{,x} = 0 \qquad \text{and} \qquad u_{,x} + v_{,y} = 0 \tag{16.7-2}$$

For analysis, we can use either a potential function $\phi = \phi(x,y)$ or a stream function $\psi = \psi(x,y)$, which are defined such that

$$\text{Potential function } \phi\text{:} \quad u = \phi_{,x} \qquad \text{and} \qquad v = \phi_{,y} \tag{16.7-3a}$$

$$\text{Stream function } \psi: \qquad u = \psi_{,y} \qquad \text{and} \qquad v = -\psi_{,x} \tag{16.7-3b}$$

If n and s are arbitrarily oriented, right-handed, orthogonal coordinates in the xy plane, then $\phi_{,n}$ represents flow velocity in the positive n direction and $\psi_{,n}$ represents flow velocity in the negative s direction (see Fig. 16.7-1*b*).

Substitution of Eqs. 16.7-3 into Eqs. 16.7-2 shows that one of the two equations becomes $0 \equiv 0$ while the other becomes Laplace's equation. Thus the problem is described by

$$\nabla^2\phi = 0 \qquad \text{or by} \qquad \nabla^2\psi = 0 \qquad \text{where} \quad \nabla^2 = \frac{\partial^2}{\partial x^2} + \frac{\partial^2}{\partial y^2} \tag{16.7-4}$$

A finite element formulation produces element matrices of the form

$$[\mathbf{k}]\{\boldsymbol{\phi}_e\} = \{\mathbf{r}\} \qquad \text{or} \qquad [\mathbf{k}]\{\boldsymbol{\psi}_e\} = \{\mathbf{r}\} \tag{16.7-5}$$

where $\{\boldsymbol{\phi}_e\}$ and $\{\boldsymbol{\psi}_e\}$ contain nodal values of ϕ and ψ, respectively, $[\mathbf{k}]$ is obtained from Eqs. 16.5-3 with $[\boldsymbol{\kappa}]$ a unit matrix and $dV = (1)\, dx\, dy$, and $\{\mathbf{r}\}$ comes from $\{\mathbf{r}_q\}$ of Eqs. 16.5-3 with q_B representing the prescribed flow rate $\phi_{,n}$ or $\psi_{,s}$ in a direction n normal to boundary S.

An example application is depicted in Fig. 16.7-1. Uniform flow at velocity u_0 enters at the left. Flow velocities in the neighborhood of the cylindrical obstacle are desired. Because of symmetry about horizontal and vertical centerlines, only one quadrant need be modeled. Other known symmetries of the flow pattern, when associated with Eqs. 16.7-3, dictate boundary conditions shown in Fig. 16.7-1. For example, $v = 0$ along AB, CD, and DE; therefore, $\phi_{,y} = 0$ along these

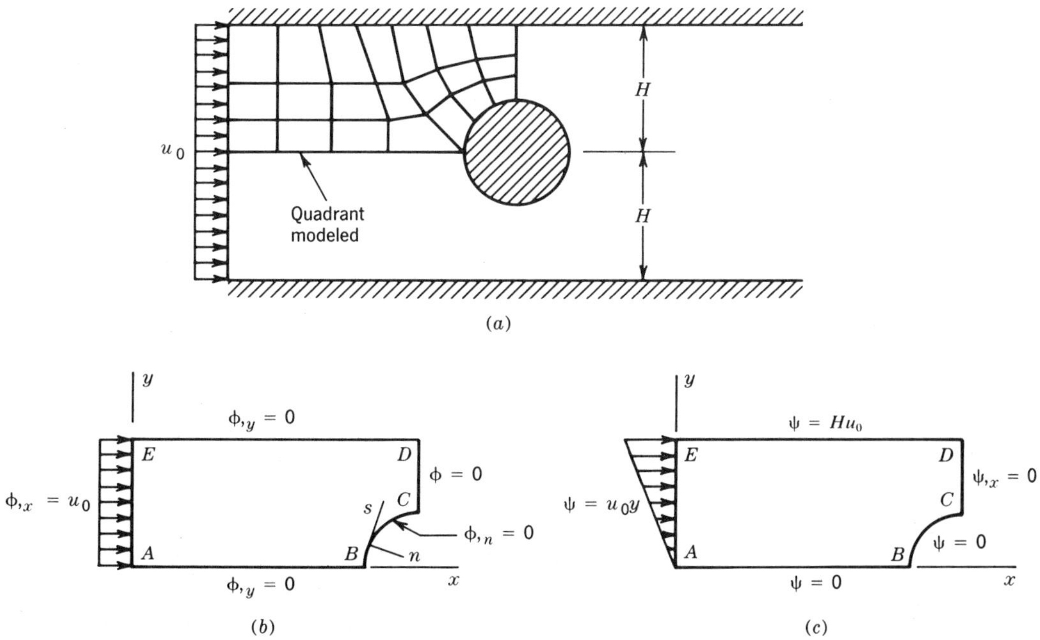

Figure 16.7-1. (*a*) Flow around a cylindrical obstacle in a rectangular channel. (*b,c*) Boundary conditions associated with potential function and stream function solutions, respectively, in the quadrant modeled.

lines. The fluid has no velocity normal to the cylinder; therefore $\phi_{,n} = 0$ along BC. A nodal value of ϕ must be assigned in order to make the coefficient matrix nonsingular, but the numerical value is arbitrary because only derivatives of ϕ are of interest. If $\phi = 0$ is prescribed at C, the value $\phi = 0$ is dictated all along CD because of the condition $\phi_{,y} = 0$ along CD. Analogous remarks apply to use of ψ rather than ϕ. At points such as D, both essential and nonessential boundary conditions appear. Here one may prefer to use $\phi_D = 0$ or $\psi_D = Hu_0$ rather than assigning the load terms $r_D = 0$ and letting ϕ_D or ψ_D remain an unknown to be calculated by solving the global equivalent of Eqs. 16.7-5. If the formulation uses ϕ (or ψ) *and* its x and y derivatives as nodal d.o.f., values of ϕ and a derivative of ϕ may be applied at a single node, although not to a single d.o.f. at that node. When all nodal d.o.f. are known, gradients are calculated (e.g., by Eq. 16.5-6 with T replaced by ϕ or ψ), and Eq. 16.7-3 yields the required flow velocities.

16.8 FLUID VIBRATION AND WAVES, PRESSURE FORMULATION

The governing differential equation, which is Eq. 16.8-4 in the discussion that follows, is called the *wave equation*. It describes phenomena in which energy is propagated by waves and has applications in problems of sound propagation, the sloshing of liquid in a container, and fluid–structure interaction. Its finite element expression has nodal pressures as d.o.f.

Formulation. We consider a fluid without viscosity. Part of the total pressure at any point may be a hydrostatic pressure. The remaining pressure, denoted by p, is associated with motion of the fluid. Pressure gradients $p_{,x}$, $p_{,y}$, and $p_{,z}$ may exist in the respective coordinate directions. Hence, if Newton's law $\mathbf{F} = m\mathbf{a}$ is applied to a differential element of volume $dx\,dy\,dz$, we obtain

$$p_{,x} = -\rho\ddot{u} \qquad p_{,y} = -\rho\ddot{v} \qquad p_{,z} = -\rho\ddot{w} \tag{16.8-1}$$

where u, v, and w are displacements in the x, y, and z directions, and ρ is the mass density, which is assumed to be essentially constant despite compressibility of the fluid. We differentiate each of Eqs. 16.8-1 with respect to its own spatial coordinate, then add. Thus

$$p_{,xx} + p_{,yy} + p_{,zz} = -\rho(\ddot{u}_{,x} + \ddot{v}_{,y} + \ddot{w}_{,z}) \tag{16.8-2a}$$

or

$$\nabla^2 p = -\rho \frac{d^2}{dt^2}(\epsilon_x + \epsilon_y + \epsilon_z) \tag{16.8-2b}$$

where $\epsilon_x = u_{,x}$, and so on. Bulk modulus B is defined as

$$B = -\frac{p}{dV/V} = -\frac{p}{\epsilon_x + \epsilon_y + \epsilon_z} \tag{16.8-3}$$

The negative sign appears because volume decreases under increasing pressure. Equations 16.8-2b and 16.8-3 yield

$$\nabla^2 p = \frac{\rho}{B}\ddot{p} \qquad \text{or} \qquad \nabla^2 p = \frac{1}{c^2}\ddot{p} \tag{16.8-4}$$

where c is the speed of sound in the medium, $c = \sqrt{B/\rho}$. Equation 16.8-4 is the *wave equation*.

Equation 16.8-4 must be solved in a volume V, subject to boundary conditions on its surface S. The essential boundary condition is $p = 0$, which prevails on a free fluid surface if waves are negligible. The nonessential boundary condition, which prevails on a solid boundary, is

$$\frac{\partial p}{\partial n} = -\rho\ddot{u}_n \tag{16.8-5}$$

where n = outward normal direction and $\ddot{u}_n$ = acceleration of the boundary in direction n. For a rigid boundary, $\ddot{u}_n = 0$. If the effect of small-amplitude waves on a fluid surface is to be included, we write $p = \rho g w_s$, where g = acceleration of gravity and w_s = surface elevation relative to the mean surface level. Combining $p = \rho g w_s$ with $p_{,z} = -\rho\ddot{w}_s$ from Eqs. 16.8-1, we obtain

$$\frac{\partial p}{\partial z} = -\frac{1}{g}\ddot{p} \tag{16.8-6}$$

as the boundary condition on a fluid surface with small-amplitude waves.

A functional for this problem is

$$\Pi = \int_V \left(\frac{p_{,x}^2 + p_{,y}^2 + p_{,z}^2}{2} + \frac{\rho}{B}p\ddot{p}\right) dV + \int_{S_s} \rho\ddot{u}_n p \, dS + \int_{S_f} \frac{1}{g} p\ddot{p} \, dS \tag{16.8-7}$$

where S_s = solid boundary and S_f = fluid boundary with wave action. As in Eqs. 16.8-4, ρ/B may be replaced by $1/c^2$. The stationary condition $\delta\Pi = 0$ yields Eqs. 16.8-4, 16.8-5, and 16.8-6.

A finite element formulation follows the familiar pattern. Pressure p within an element is interpolated from nodal d.o.f. $\{\mathbf{P}_e\}$, where $\{\mathbf{P}_e\}$ may contain nodal pressures only or may also contain spatial derivatives of nodal pressures. Thus

$$p = \lfloor\mathbf{N}\rfloor\{\mathbf{P}_e\} \qquad \text{and} \qquad \ddot{p} = \lfloor\mathbf{N}\rfloor\{\ddot{\mathbf{P}}_e\} \tag{16.8-8}$$

We define the following global matrices, where summation signs indicate assembly of element matrices. The notation is analogous to that in Eqs. 16.5-3.

$$[\mathbf{K}] = \sum \int_{V_e} (\lfloor\mathbf{N}_{,x}\rfloor^T\lfloor\mathbf{N}_{,x}\rfloor + \lfloor\mathbf{N}_{,y}\rfloor^T\lfloor\mathbf{N}_{,y}\rfloor + \lfloor\mathbf{N}_{,z}\rfloor^T\lfloor\mathbf{N}_{,z}\rfloor) \, dV \tag{16.8-9a}$$

$$[\mathbf{C}] = \sum \int_{V_e} \frac{\rho}{B}\lfloor\mathbf{N}\rfloor^T\lfloor\mathbf{N}\rfloor \, dV \tag{16.8-9b}$$

$$[\mathbf{H}] = \sum \int_{S_f} \lfloor\mathbf{N}\rfloor^T\lfloor\mathbf{N}\rfloor \frac{1}{g} \, dS \qquad \{\mathbf{R}_s\} = \sum \int_{S_s} \lfloor\mathbf{N}\rfloor^T\rho\ddot{u}_n \, dS \tag{16.8-9c}$$

Equation 16.8-7 becomes

$$\Pi = \tfrac{1}{2}\{\mathbf{P}\}^T[\mathbf{K}]\{\mathbf{P}\} + \{\mathbf{P}\}^T([\mathbf{C}] + [\mathbf{H}])\{\ddot{\mathbf{P}}\} + \{\mathbf{P}\}^T\{\mathbf{R}_s\} \tag{16.8-10}$$

where $\{\mathbf{P}\}$ is the *global* array of nodal d.o.f. The stationary condition $\{\partial\Pi/\partial\mathbf{P}\} = \{\mathbf{0}\}$ yields the finite element formulation

$$[\mathbf{K}]\{\mathbf{P}\} + ([\mathbf{C}] + [\mathbf{H}])\{\ddot{\mathbf{P}}\} = -\{\mathbf{R}_s\} \tag{16.8-11}$$

One must often deal with a boundary at infinity, that is, a boundary so distant that reflected waves do not appear for the duration of the event of interest. Such a condition can be modeled by "infinite elements" or by other analytical procedures [16.9–16.11].

Sloshing and Acoustic Modes. Let walls that support or contain a fluid be rigid, so that $\ddot{u}_n = 0$. Thus the forcing function becomes zero, and the fluid vibrates in one of its natural modes. The pressure becomes $p = \bar{p} \sin \omega t$, where ω is a natural frequency and amplitude $\bar{p}$ is a function of the spatial coordinates but is independent of time t. Equation 16.8-4 becomes

$$\nabla^2\bar{p} + \omega^2 \frac{\rho}{B}\bar{p} = 0 \qquad \text{or} \qquad \nabla^2\bar{p} + \omega^2 \frac{1}{c^2}\bar{p} = 0 \tag{16.8-12}$$

Similarly, a finite element formulation appears from Eq. 16.8-11 if we set $\{\mathbf{R}_s\} = \{\mathbf{0}\}$ and $\{\mathbf{P}\} = \{\overline{\mathbf{P}}\} \sin \omega t$:

$$[[\mathbf{K}] - \omega^2([\mathbf{C}] + [\mathbf{H}])]\{\overline{\mathbf{P}}\} = \{\mathbf{0}\} \tag{16.8-13}$$

This eigenvalue problem can be solved to obtain natural frequencies ω_i and the corresponding pressure modes $\{\overline{\mathbf{P}}\}_i$. If $[\mathbf{C}] = [\mathbf{0}]$, Eq. 16.8-13 yields the slosh frequencies of an incompressible liquid in a rigid container. If $[\mathbf{H}] = [\mathbf{0}]$, Eq. 16.8-13 yields the acoustic modes of a fluid in a cavity with rigid walls.

As a special case, consider the one-dimensional problem of acoustic modes in a pipe with rigid walls that lies along an x axis (Fig. 16.8-1). Thus, in Eq. 16.8-7, $p_{,y} = p_{,z} = 0$ and $\ddot{u}_n = 0$. Volume dV becomes $A\,dx$, where $A = A(x)$ for a pipe of variable cross section. The integral over S_f is discarded because there are no surface waves, and the integral over S_s is zero because $\ddot{u}_n = 0$. Thus, only special forms of Eqs. 16.8-9a and 16.8-9b are used in Eq. 16.8-13. As an alternative to specialization, one may begin with a functional for this particular problem. It is

$$\Pi = \int \left(\bar{p}_{,x}^2 - \frac{\omega^2}{c^2}\bar{p}^2\right) A\,dx \tag{16.8-14}$$

where $\bar{p}$ is the pressure amplitude. By making the usual definitions and substitutions (analogous to Eqs. 16.8-8 and 16.8-9), we may develop a finite element formulation directly from Eq. 16.8-14 [16.12]. As for boundary conditions: at a closed end, $\bar{p}_{,x} = 0$ and $\bar{p}$ is unknown; at an open end, $\bar{p} = 0$ and $\bar{p}_{,x}$ is unknown.

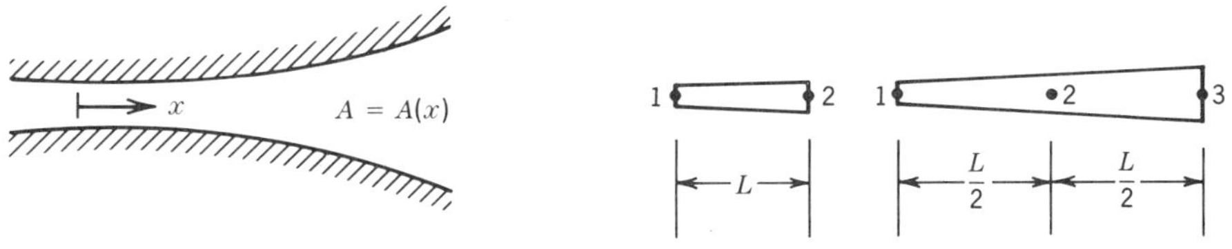

Figure 16.8-1. (*a*) A pipe of variable cross-sectional area A. (*b*) Possible finite elements for acoustic mode analysis.

16.9 FLUID–STRUCTURE INTERACTION

Remarks. A structure may be surrounded by fluid or may contain fluid. A phenomenon of fluid–structure interaction exists when fluid and structure both move and exert forces upon one another associated with the motion. Different categories of the phenomenon may be identified [16.10].

1. Problems having large relative motion; for example, aerodynamic flutter. Behavior is dominated by flow characteristics of the fluid.
2. Problems having limited fluid displacement; for example, a reservoir behind a dam under earthquake load, or wave action on an offshore structure.

In this section we are concerned with the second category.

One analytical approach is to model the structure by finite elements as usual, then load its wetted surface by the pressures described by Eq. 16.8-11. This approach results in a system of equations that uses displacement d.o.f. of the structure and pressure d.o.f. at nodes of the wetted surface. The coefficient matrices are unsymmetric. With the simplifying assumptions of no surface waves and incompressibility, the pressure d.o.f. may be eliminated, leaving a smaller system of equations that includes only the structural d.o.f. and has symmetric coefficient matrices. In this smaller system the fluid is represented by a "virtual" mass matrix that is added to the mass matrix of the structure. However, the simplifying assumption of incompressibility may be crude, as seems to be the case for a relatively stiff submerged structure [16.9].

Another analytical approach is to model the fluid by displacement-based elements, as if the fluid were a structure with special material properties. This "mock-fluid" approach is described in some detail in the following discussion. As compared with the pressure-load approach, the mock-fluid approach typically requires more d.o.f. but produces symmetric matrices with smaller bandwidth or wave front.

Other approaches are also available, for example, the boundary element method.

Mock-Fluid Elements. We will model the fluid by elements that have displacement d.o.f., as do solid elements. Displacements are assumed to be small. Our needs are for a compressibility stiffness matrix $[\mathbf{k}_B]$, a wave or slosh stiffness matrix $[\mathbf{k}_S]$, and a mass matrix $[\mathbf{m}]$. This formulation appears to be reliable for static analysis and for calculation of natural frequencies of fluid in rigid or flexible containers, but does not always accurately predict natural frequencies of elastic structures surrounded by fluid [16.15].

The strain energy per unit volume associated with compressibility is $B\epsilon_V^2/2$, where B is the bulk modulus, and ϵ_V is the volumetric strain, $\epsilon_V = dV/V$. The expression $B\epsilon_V^2/2$ can be obtained from the standard expression $\{\boldsymbol{\epsilon}\}^T[\mathbf{E}]\{\boldsymbol{\epsilon}\}/2$ by recognizing that $\epsilon_V = \epsilon_x + \epsilon_y + \epsilon_z$ and using [E] from Eq. 9.4-7 with $G = 0$. If displacements $\{\mathbf{u}\} = \lfloor u \quad v \quad w \rfloor^T$ are interpolated over an element in standard fashion—that is, $\{\mathbf{u}\} = [\mathbf{N}]\{\mathbf{d}\}$—then

$$\epsilon_V = \lfloor \mathbf{B}_V \rfloor \{\mathbf{d}\} \qquad \text{where} \quad \lfloor \mathbf{B}_V \rfloor = \left\lfloor \frac{\partial}{\partial x} \quad \frac{\partial}{\partial y} \quad \frac{\partial}{\partial z} \right\rfloor [\mathbf{N}] \tag{16.9-1}$$

Let us now restrict attention to *linear* elements, such as the four-node plane element shown in Fig. 16.9-1. We elect to evaluate ϵ_V at only the *center* of the element and regard its value there as representative of the element as a whole. Thus, with $\epsilon_{V0} = \lfloor \mathbf{B}_{V0} \rfloor \{\mathbf{d}\}$ the center value of ϵ_V, the strain energy of compressibility in an element of volume V_e is

$$U = \tfrac{1}{2}\{\mathbf{d}\}^T[\mathbf{k}_B]\{\mathbf{d}\} \qquad \text{where} \quad [\mathbf{k}_B] = BV_e\lfloor \mathbf{B}_{V0} \rfloor^T \lfloor \mathbf{B}_{V0} \rfloor \tag{16.9-2}$$

It is not necessary to use Eq. 16.9-2 to generate $[\mathbf{k}_B]$. One can use instead the formula for [k] of a structural element (Eq. 4.1-5), integrate using a single Gauss point at the element center, and take [E] from Eq. 9.4-7 but with $G = 0$. Thus, with reduced integration, bulk modulus B provides the element with resistance only to volume change of the element as a whole, and $[\mathbf{k}_B]$ of Eq. 16.9-2 is produced.

Note that the constraint of near-incompressibility is enforced at each Gauss point used to evaluate $[\mathbf{k}_B]$. Accordingly, if $[\mathbf{k}_B]$ were formed by full integration of $B\lfloor \mathbf{B}_V \rfloor^T \lfloor \mathbf{B}_V \rfloor$, $[\mathbf{k}_B]$ would exhibit undesirable stiffnesses. For example, the plane four-node element of Fig. 16.9-1*a* would yield zero strain energy only when $\{\mathbf{d}\}$ represents rigid-body motion or a pure shear deformation. Thus the slosh mode in Fig. 16.9-2*b* would be resisted by bulk modulus B, which is clearly unreasonable.

One may contemplate the use of elements of higher order than the linear elements discussed thus far. One would then probably use more than one Gauss point per element, yet underintegrate so as to avoid locking due to too many penalty constraints, as discussed in Sections 9.4 and 9.5. However, higher-order elements seem not to be used in practice.

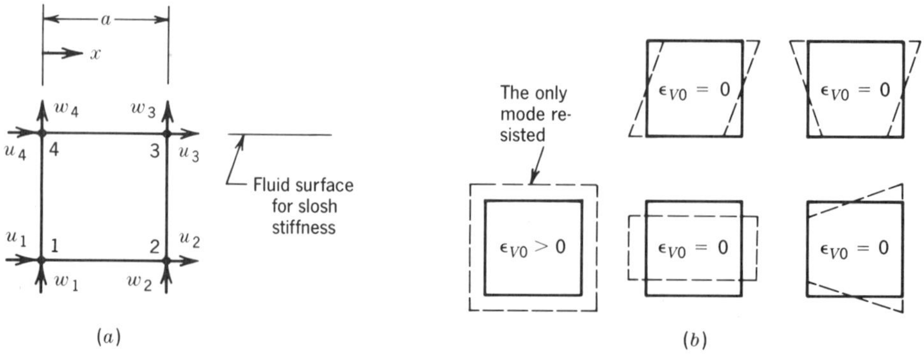

Figure 16.9-1. (*a*) Plane rectangular element. (*b*) Deformation modes of a plane element, showing volumetric strain at the element center. Rigid-body modes are not shown.

A slosh stiffness matrix $[\mathbf{k}_S]$ is needed only if there is a free surface with waves or an oscillating boundary between fluids of different density. The energy associated with small-amplitude waves on a surface S is

$$U = \int_S \tfrac{1}{2}\rho g w_s^2 \, dS \qquad \text{where} \quad w_s = \lfloor \mathbf{N}_S \rfloor \{\mathbf{d}\} \tag{16.9-3}$$

and ρ = mass density of the fluid, g = acceleration of gravity, w_s = surface displacement in the upward direction, and $\lfloor \mathbf{N}_S \rfloor$ operates only on element d.o.f. in surface S. For example, for the plane element of Fig. 16.9-1a, $w_s = \lfloor \mathbf{N}_S \rfloor \{\mathbf{d}\} = w_3 x/a + w_4(a - x)/a$. Equation 16.9-3 becomes

$$U = \tfrac{1}{2}\{\mathbf{d}\}^T[\mathbf{k}_S]\{\mathbf{d}\} \qquad \text{where} \quad [\mathbf{k}_S] = \int_S \lfloor \mathbf{N}_S \rfloor^T \lfloor \mathbf{N}_S \rfloor \rho g \, dS \tag{16.9-4}$$

The element mass matrix $[\mathbf{m}]$ is formulated in standard fashion, as though the element were elastic (e.g., see Eq. 13.2-5).

If desired, a viscous damping matrix $[\mathbf{c}]$ can be included [16.13]

$$[\mathbf{c}] = \int_{V_e} [\mathbf{B}]^T[\mathbf{E}_G][\mathbf{B}]\mu \, dV \tag{16.9-5}$$

where $[\mathbf{B}]$ is the standard strain–displacement matrix of Eq. 4.1-3, $[\mathbf{E}_G]$ is the first rectangular matrix on the right-hand side of Eq. 9.4-7 (exclusive of G), and μ is the dynamic coefficient of viscosity [13.3].

As is evident from Eq. 16.9-2, $[\mathbf{k}_B]$ is a matrix of rank one. Accordingly, the assembled structure may have several zero-energy modes, called *circulation modes*. They involve no volume change of the fluid and no surface waves. One such mode is shown in Fig. 16.9-2a. A way to slightly restrain circulation modes is to make G a small positive number in Eq. 9.4-7 and use the structural formulation to obtain $[\mathbf{k}_B]$, as explained in the paragraph that follows Eq. 16.9-2. (This is not the same as attributing viscosity to the fluid.) Another way to inhibit circulation is to augment $[\mathbf{k}_B]$ by the penalty matrix

$$[\mathbf{k}_c] = \alpha \int_{V_e} [\mathbf{B}_c]^T[\mathbf{B}_c] \, dV \tag{16.9-6}$$

where α is a penalty number and $\{\mathbf{d}\}^T[\mathbf{k}_c]\{\mathbf{d}\}$ is the integral over the element of the square of the curl of the displacement field [16.14]. For a plane element in

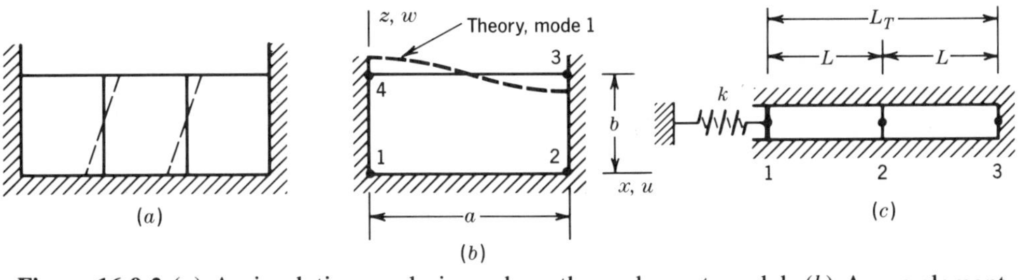

Figure 16.9-2 (a) A circulation mode in a plane three-element model. (b) A one-element model for slosh in a plane rectangular tank. (c) A two-element model for acoustic analysis of a pipe with an elastic end.

the xz plane, $(\text{curl } \mathbf{u})^2 = (u_{,z} - w_{,x})^2$ and $[\mathbf{B}_c] = \left\lfloor \frac{\partial}{\partial z} \quad -\frac{\partial}{\partial x} \right\rfloor [\mathbf{N}]$. If $\alpha = B$, circulation modes disappear and triangular elements yield good results [16.14].

If circulation modes are unrestrained, they may appear with frequencies indistinguishable from other low frequencies of greater interest [16.14]. Note also that $[\mathbf{k}_B] >> [\mathbf{k}_S]$ (Eqs. 16.9-2 and 16.9-4). Accordingly, if both are used in an analysis, there is a risk that $[\mathbf{k}_S]$ will "fall off the end" of computer words, as discussed in Section 18.2.

Example: Plane Slosh Analysis. Consider the one-element model shown in Fig. 16.9-2*b*. Displacements in the y (out of plane) direction are assumed to be zero. The element has eight d.o.f., but because of the rigid walls, only w_3 and w_4 are unrestrained. Rather than invoke all eight d.o.f. and then eliminate six, we elect to use only w_3 and w_4 at the outset. Now $u = 0$, and

$$w = \frac{xz}{ab} w_3 + \frac{(a - x)z}{ab} w_4 \tag{16.9-7}$$

Using $[\mathbf{k}_B]$, $[\mathbf{k}_S]$, and the consistent mass matrix $[\mathbf{m}]$, we obtain, for a unit thickness,

$$\left(\frac{Ba}{4b}\begin{bmatrix} 1 & 1 \\ 1 & 1 \end{bmatrix} + \frac{\rho g a}{6}\begin{bmatrix} 2 & 1 \\ 1 & 2 \end{bmatrix}\right)\begin{Bmatrix} w_3 \\ w_4 \end{Bmatrix} + \frac{\rho ab}{18}\begin{bmatrix} 2 & 1 \\ 1 & 2 \end{bmatrix}\begin{Bmatrix} \ddot{w}_3 \\ \ddot{w}_4 \end{Bmatrix} = \begin{Bmatrix} R_3 \\ R_4 \end{Bmatrix} \tag{16.9-8}$$

where R_3 and R_4 are time-varying loads. Typically $(Ba/4b) >> (\rho ga/6)$, which implies the condition $w_3 = -w_4$. This is a constraint of zero volume change. Imposing this constraint, and setting $R_3 = R_4 = 0$ and $\lfloor w_3 \quad w_4 \rfloor = \lfloor \overline{w}_3 \quad \overline{w}_4 \rfloor \sin \omega t$, we obtain an eigenvalue problem that yields $\omega^2 = 3g/b$. The theoretical result is $\omega^2 = (\pi g/a) \tanh(\pi b/a)$, which yields $\omega^2 = 3.13g/b$ for $a = b$.

Example: Acoustic Modes. Air in a uniform pipe is modeled by two two-node elements (Fig. 16.9-2*c*). At the left end there is a massless plug connected to an elastic spring of stiffness k. For this problem, there is no slosh stiffness, $\epsilon_V = \epsilon_x$, and $[\mathbf{k}_B]$ has the same form as the stiffness matrix of a two-node elastic bar. To $[\mathbf{k}_B]$ we add the structure (spring) stiffness k. Thus, with $[\mathbf{m}]$ the consistent mass matrix and the boundary condition $u_3 = 0$ already imposed, the structure equations are

$$\left(\frac{BA}{L}\begin{bmatrix} 1 & -1 \\ -1 & 2 \end{bmatrix} + \begin{bmatrix} k & 0 \\ 0 & 0 \end{bmatrix}\right)\begin{Bmatrix} u_1 \\ u_2 \end{Bmatrix} + \frac{\rho AL}{6}\begin{bmatrix} 2 & 1 \\ 1 & 4 \end{bmatrix}\begin{Bmatrix} \ddot{u}_1 \\ \ddot{u}_2 \end{Bmatrix} = \begin{Bmatrix} R_1 \\ R_2 \end{Bmatrix} \tag{16.9-9}$$

where A is the constant cross-sectional area. The fundamental acoustic frequency in a pipe with *rigid* ends is obtained by setting $u_1 = 0$, $R_2 = 0$, and $u_2 = \overline{u}_2 \sin \omega t$. Thus

$$\omega^2 = \frac{3B}{\rho L^2} = \frac{3c^2}{L^2} = \frac{12c^2}{L_T^2} \quad \text{and} \quad \omega = 3.46\frac{c}{L_T} \tag{16.9-10}$$

The exact result (for infinite k) is $\omega = \pi c/L_T$.

PROBLEMS

Section 16.2

16.1 Obtain Eq. 16.2-7, with matrices defined as in Eq. 16.2-6, by use of the Galerkin method.

16.2 The three-node bar shown is uniform. Temperature T_1 is prescribed, $Q = 0$, node 3 is insulated ($q_3 = 0$), and heat is transferred across the lateral surface by convection. In the units used in Section 16.1, let $k = 180$, $h = 12$, $A = 0.1$, $p = 1$, and $L = 2$.

(a) Use two two-node elements to determine T_2 and T_3 in terms of T_1 and T_f.

(b) Repeat part (a), but use a single three-node element of length $2L$. Note that $[\mathbf{h}_{ls}]$ has the form of a mass matrix, and use an "optimally lumped" form (see Eq. 13.3-9).

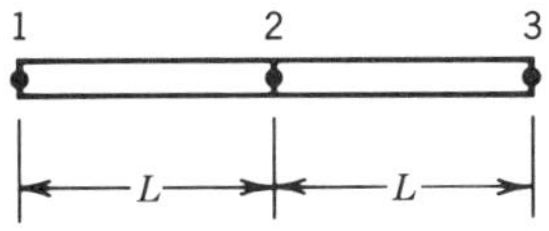

Problem 16.2

Section 16.3

16.3 Verify that $[\boldsymbol{\kappa}]$ is as defined by Eq. 16.3-3 (resolve q_x and q_y into components q_r and q_s).

16.4 Imagine that an essentially two-dimensional body has a gradual variation in thickness τ—that is, $\tau = \tau(x,y)$. What changes must be made in the equations developed in Section 16.3?

16.5 Verify that Eq. 16.3-10 yields the correct governing equation and nonessential boundary conditions from the condition $\delta\Pi = 0$. Include the term $2h(T_fT - T^2/2)$ in Π, so as to account for convection on two identical lateral surfaces.

Section 16.4

16.6 Devise an example that shows why θ must be a principal material direction if T in a solid of revolution is to be symmetric with respect to the $\theta = 0$ plane.

16.7 Using conventional scalar notation like that in Eq. 16.3-5, write Eqs. 16.4-1 and 16.4-2 for the following special cases.

(a) Solid of revolution with an axially symmetric temperature field and isotropic material.

(b) The plane problem in polar coordinates ($T_{,z} = 0$). Let $[\boldsymbol{\kappa}]$ be full, as for an anisotropic material.

16.8 Derive the governing equation of Problem 16.7(b) from first principles, analogous to the Cartesian coordinate derivation in Eqs. 16.3-4 and 16.3-5.

Section 16.5

16.9 The three-node triangle shown is to be used for heat conduction analysis. The body in question is homogeneous, plane, isotropic, and of unit thickness.

(a) Evaluate [**k**] in terms of k and nodal coordinates.

(b) Evaluate [**h**] in terms of h and nodal coordinates if only side 1–3 transfers heat by convection.

(c) Write the "lumped" forms of [**h**] and [**c**].

(d) Write $\{\mathbf{r}_Q\}$ in terms of Q and nodal coordinates if Q is constant over the element.

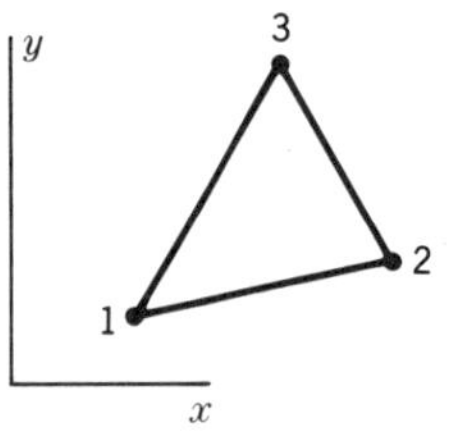

Problem 16.9

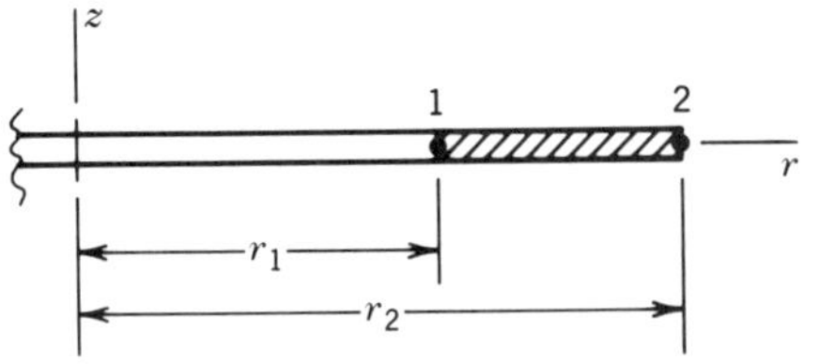

Problem 16.11

16.10 Repeat Problem 16.9, but regard the body as a solid of revolution, so that r replaces x and axis of revolution z replaces y.

16.11 The solid of revolution element shown is homogeneous, isotropic, flat, and of unit thickness. Surfaces z = constant are insulated so that heat flows only radially. Assume that T varies linearly with r in the element and let Q be constant. Evaluate the matrices defined by Eqs. 16.5-3 in terms of r_1, r_2, and material properties k, c, and so on.

16.12 The element shown is a quadratic triangle with straight sides and midside nodes. Evaluate $\{\mathbf{r}_Q\}$ due to heat input Q_P (units J/s), if Q_P is

(a) Concentrated at node 1.

(b) Concentrated at node 4.

(c) Concentrated at the centroid of the triangle.

(d) Uniformly distributed along a line between nodes 5 and 6. Use a simple approximation.

(e) Improve your answer to part (d) by devising a better approximation.

16.13 The uniform bar element shown is described by Eqs. 16.2-6 and 16.2-7. Conductivity k is a function of temperature. Let T_a be the average temperature in the element. In the units used in Section 16.1, let $k = 100 - 0.4T_a$, $h = 30$, $A = 0.1$, $p = 1$, $L = 2$, and $T_f = 200$. Heat flows across the lateral surface. If $T = 400°\text{C}$ at node 1 and the right end is insulated ($q_R = 0$ in Fig. 16.2-1), what is the steady-state value of T_2? Use an iterative solution method.

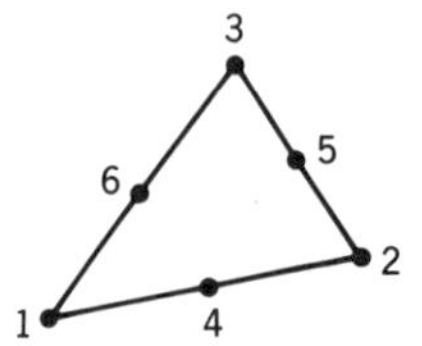

Problem 16.12

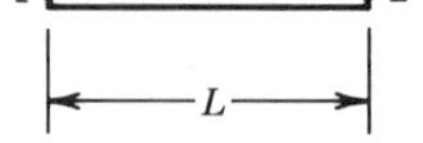

Problem 16.13

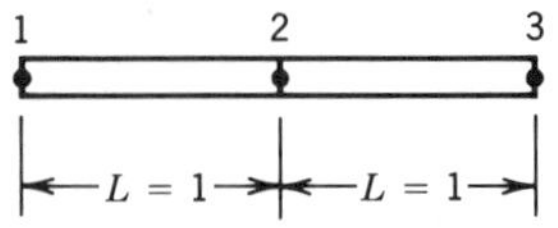

Problem 16.14

16.14 The uniform bar shown is modeled by two linear elements. Lateral surfaces are insulated. Heat flows into the bar at node 3 at the prescribed rate $20A$, where A is the cross-sectional area of the bar and the units are those used in Section 16.1. The temperature at node 1 is kept at zero. Let conductivity in an element be given by $k = 2 + 0.04T_a$, where T_a is the average temperature in an element. Perform two steps of an iterative solution for T_2 and T_3, starting with $T_a = 0$ in each element.

Section 16.6

16.15 Prove Eqs. 16.6-3.

16.16 The sketch shows the actual variation of temperature with time at a certain node. Imagine that we start at point A and use Eq. 16.6-6 to predict T at time t_{n+1}. Consider $\beta = 0$, $\beta = 0.5$, and $\beta = 1.0$. For each of these three values make a sketch that shows how the predicted temperature compares with T_B.

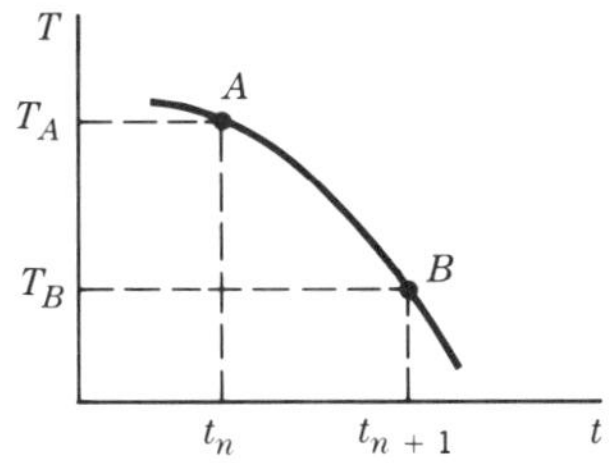

Problem 16.16

16.17 Show that Eq. 16.6-8 follows from Eqs. 16.6-6 and 16.6-7.

16.18 Imagine that the bar of Fig. 16.2-1 is initially at zero temperature and then is subjected to a heat flux at the right end. The flux appears at time $t = 0$ and thereafter remains constant. The left end is kept at zero temperature. For a one-element model, physical constants are such that Eq. 16.6-1 becomes $6T + 2\dot{T} = 3$, where T is the temperature of the node where heat flux is imposed. We are to compute $T = T(t)$ by use of Eq. 16.6-8.
(a) To what T should $T(t)$ converge as t becomes large?
(b) What is the exact solution for $T = T(t)$?
(c) What is Δt_{cr} for Euler's method?
In what follows, use Eq. 16.6-8 with the values of β and Δt given. Take five time steps in each case. Compare results with exact values and with the other direct integration results.
(d–g) With $\Delta t = 0.1$, take β as (d) 0, (e) $\frac{1}{2}$, (f) $\frac{2}{3}$, (g) 1.0.
(h–k) With $\Delta t = 1.0$, take β as (h) 0, (i) $\frac{1}{2}$, (j) $\frac{2}{3}$, (k) 1.0.

Section 16.7

16.19 Equations 16.5-3 apply to the problem of groundwater flow if T = hydraulic head (meters of water), $[\boldsymbol{\kappa}]$ = permeability coefficients, q = seepage velocity, and Q = flow rate per unit volume. The sketch represents the plan view of a square array of four wells in a homogeneous, horizontal aquifer of infinite extent and constant thickness τ between impermeable strata above and below it. The hydraulic head at a large distance from the group of wells is h_0. The hydraulic head in the neighborhood of a typical well is

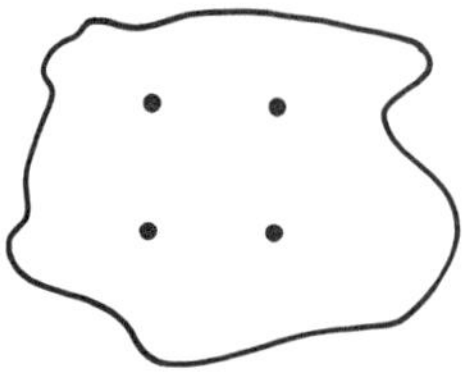

Problem 16.19

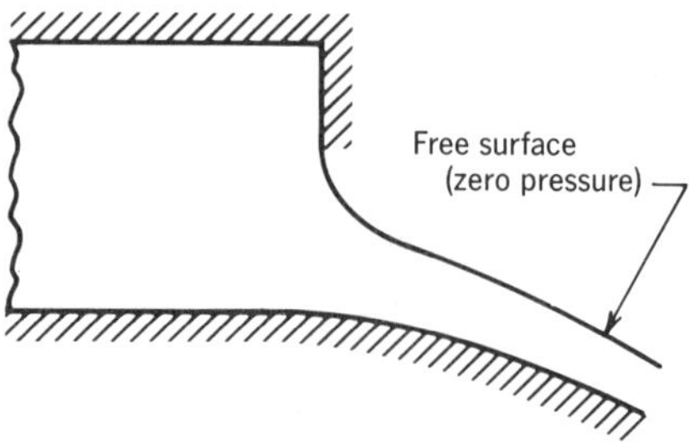

Problem 16.21

desired. Each well pumps at constant flow rate Q_P m^3/s. Outline how to set up a finite element solution for steady-state $T = T(x,y)$ in the aquifer, with attention to mesh layout, prescribed d.o.f., and load terms.

16.20 Derive Eqs. 16.7-2.

16.21 The sketch depicts a steady flow that emerges from an enclosure, thereafter to flow with a "free surface" (open to the air). Bernoulli's equation is $\frac{1}{2}(u^2 + v^2) + (p/\rho) + gy =$ constant, where $p =$ pressure, $\rho =$ mass density, $g =$ acceleration of gravity, and $y =$ elevation above a datum. Locating the free surface requires an iterative solution process. Outline the major steps of this process [16.8].

Section 16.8

16.22 (a) Derive Eqs. 16.8-1.
(b) Show that $dV/V = \epsilon_x + \epsilon_y + \epsilon_z$ in Eq. 16.8-3.
(c) Show that with Π defined by Eq. 16.8-7, the condition $\delta\Pi = 0$ yields Eqs. 16.8-4, 16.8-5, and 16.8-6.

16.23 The formulation of Eqs. 16.8-9 is to be applied with a plane rectangular element of unit thickness (see sketch).
(a) Generate element matrices **[k]**, **[c]**, and **[h]**, in terms of a, b, ρ, B, and g.
(b) Imagine that the sketch represents a one-element model of an incompressible liquid in a rectangular tank that is open at the top. Write the appropriate form of Eq. 16.8-13 for the case $a = b = h$.
(c) To reduce the number of equations, set $p_1 = -p_2$ and $p_4 = -p_3$, then solve for the frequency of the slosh mode, again with $a = b = h$. (Do not expect high accuracy. The theoretical result is stated below Eq. 16.9-8.)

16.24 (a) Show that the functional $\Pi = \int [\bar{p}_{,x}^2 + \bar{p}_{,y}^2 + \bar{p}_{,z}^2 - (\omega\bar{p}/c)^2]\, dV$ yields the latter form of Eq. 16.8-12 from the stationary condition $\delta\Pi = 0$.
(b) Derive Eq. 16.8-13 (with **[H]** = **[0]**) from the functional stated in part (a).

16.25 (a) Specialize Eq. 16.8-7 to the one-dimensional acoustic mode problem, Fig. 16.8-1. Use this result to show that the condition $\delta\Pi = 0$ yields the governing differential equation $d(Ap_{,x})/dx - (A/c^2)\ddot{p} = 0$.
(b) Using Eq. 16.8-14, obtain the equation given by the condition $\delta\Pi = 0$. Check your result by specializing the differential equation given in part (a).

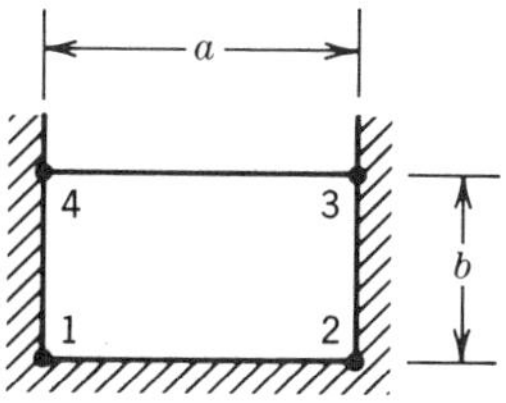

Problem 16.23

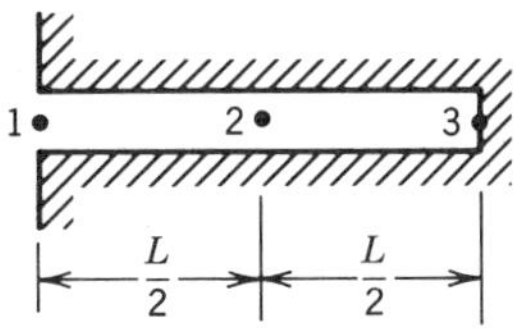

Problem 16.27

16.26 Consider a uniform pipe of length L and having closed ends. Model the pipe by a single element. Determine the frequency of the fundamental acoustic mode by use of the following element formulations. The exact result is $\omega = \pi c/L$.

(a) Use an element with d.o.f. p_1 and p_2 at ends 1 and 2.

(b) Use a four d.o.f. element. The d.o.f. at each end are p and $p_{,x}$.

16.27 The sketch represents a three-node model of air in a uniform pipe. The end at the left is open. Determine ω for the acoustic mode of lowest frequency. The exact result is $\omega = \pi c/2L$.

(a) Use a single quadratic element.

(b) Use two two-node (linear) elements.

Section 16.9

16.28 The sketch represents a rectangular tank of liquid. Beam EFGH is fixed at E and H and is in contact with the liquid surface. Imagine that motion is confined to the xz plane.

(a) If the liquid were *absent,* what would be the appearance of the first two vibration modes of the beam? Sketch them. Which has the higher frequency?

(b) If the liquid is restored, what now are the appearances of the first two vibration modes that involve bending of the beam? Sketch them.

(c) If beam and liquid are each divided into three elements, with displacement d.o.f. and nodes at points A–H, how many d.o.f. are nonzero after boundary conditions are imposed? Identify these d.o.f.

(d) Are beam and liquid elements compatible? If no, is this acceptable? If yes, how is it accomplished?

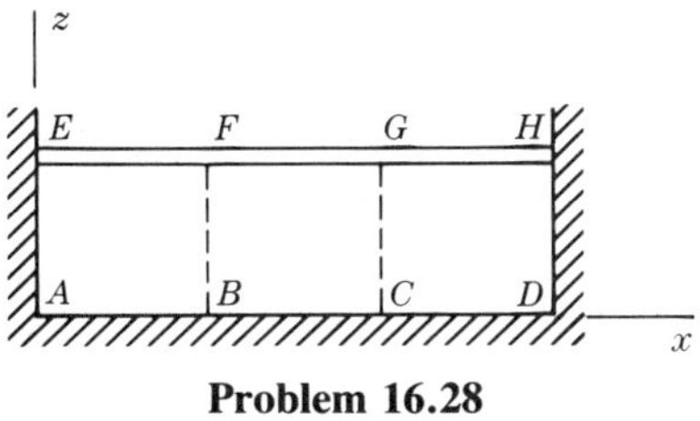

Problem 16.28

16.29 Imagine that a very small amount of liquid is to be analyzed, and that the effects of surface tension σ (force per unit length) are to be included. Determine an expression for the appropriate stiffness matrix that operates on displacement d.o.f. $\{\mathbf{d}\}$.

16.30 For the three-element fluid model shown in Fig. 16.9-2*a*, sketch all possible circulation modes. Show the direction of the circulation in each case. *Note:* A simple reversal of the direction of all fluid displacements is not considered a different mode.

16.31 (a) Write an expression for $[\mathbf{B}_c]$ in Eq. 16.9-6, in the form of an operator matrix times shape function matrix $[\mathbf{N}]$. Assume that the problem is three-dimensional.
(b) Generate $[\mathbf{k}_c]$ of Eq. 16.9-6 for the two-d.o.f. element described by Eq. 16.9-7.

16.32 Verify the correctness of the rectangular matrices in Eq. 16.9-8.

16.33 Use a three-node, mock-fluid model to solve for the lowest acoustic frequency in a uniform pipe with one open end (see Problem 16.27).
(a) Use a single quadratic element.
(b) Use two two-node (linear) elements.

CHAPTER **17**

AN INTRODUCTION TO SOME NONLINEAR PROBLEMS

Solution methods for nonlinear equations are discussed. Selected problems are described and given a finite element formulation. Emphasis is given to nonlinearities arising from material properties and changes in geometry.

17.1 INTRODUCTION

In structural mechanics, a problem is nonlinear if the stiffness matrix or the load vector depends on the displacements. Nonlinearity in structures can be classed as *material nonlinearity* (associated with changes in material properties, as in plasticity) or as *geometric nonlinearity* (associated with changes in configuration, as in large deflections of a slender elastic beam). In heat transfer, nonlinearity may arise from temperature-dependent conductivity (which makes the coefficient matrix depend on temperature) and from radiation (which makes the radiative heat flux a nonlinear function of temperature). In general, for a time-independent problem symbolized as $[\mathbf{K}]\{\mathbf{D}\} = \{\mathbf{R}\}$, in *linear* analysis both $[\mathbf{K}]$ and $\{\mathbf{R}\}$ are regarded as independent of $\{\mathbf{D}\}$, whereas in *non*linear analysis $[\mathbf{K}]$ and/or $\{\mathbf{R}\}$ are regarded as functions of $\{\mathbf{D}\}$.

The classifications "linear" and "nonlinear" are artificial in that physical reality presents various problems, some of which can be satisfactorily *approximated* by linear equations. We are fortunate that linear approximations are quite good for many problems of stress analysis and heat conduction. Nonlinear approximations are more difficult to formulate, and solving the resulting equations may cost 10 to 100 times as much as a linear approximation having the same number of d.o.f.

Many physical situations present nonlinearities too large to be ignored. Stress–strain relations may be nonlinear in either a time-dependent or a time-independent way. A change in configuration may cause loads to alter their distribution and magnitude or cause gaps to open or close. Mating parts may stick or slip. Welding and casting processes cause the material to change in conductivity, modulus, and phase. The generation and shedding of vortices in fluid flow past a structure produces oscillatory loads on the structure. Pre-buckling rotations alter the effective stiffness of a shell and change its buckling load. Thus we see that nonlinear effects may vary in type and may be mild or severe.

An analyst must understand the physical problem and must be acquainted with various solution strategies. A single strategy will not always work well, and may not work at all for some problems. Several attempts may be needed in order to obtain a satisfactory result.

Nevertheless, nonlinear analyses are undertaken more often than in the past. In part, this is because computing costs have declined and capable software has become available. In addition, more demands are placed on structures: they must function at higher temperatures and pressures, offer earthquake resistance, and provide crashworthiness. Forming and extrusion processes must be analyzed in an attempt to reduce production costs. Plastics, elastomers, and composites are used with increasing frequency as structural materials; they display material nonlinearity well below the limits of their useful strengths [17.1].

A comprehensive discussion of nonlinearity, even in structural mechanics alone, would require at least one complete volume. The following introductory treatment contains a sampling of nonlinear problems and presents some of the basic procedures for solving the associated equations.

17.2 SOME SOLUTION METHODS

A representative time-independent nonlinear problem can be stated as $[\mathbf{K}]\{\mathbf{D}\} = \{\mathbf{R}\}$, where $\{\mathbf{R}\}$ is known and $[\mathbf{K}]$ is a function of $\{\mathbf{D}\}$ that can be computed for a given $\{\mathbf{D}\}$. We are required to compute $\{\mathbf{D}\}$—for example, to compute the displacement state associated with known loads. In what follows we introduce some of the available computational methods. For simplicity, a one-dimensional problem is chosen as the principal example.

Consider a nonlinear spring, Fig. 17.2-1. The source of the nonlinearity is unimportant in the present discussion. We imagine that the spring stiffness k is composed of a constant term k_0 and a term k_N that depends on deformation. Displacement u is caused by load P and is given by the equation

$$(k_0 + k_N)u = P \qquad \text{where} \quad k_N = f(u) \tag{17.2-1}$$

We ask for the value of u when P is given. In order to mimic a realistic problem, we assume that k_N is known in terms of u, and therefore that P can be calculated in terms of u, but that an explicit solution for u in terms of P is not available. Instead, iterative methods are needed to determine u, as follows.

Direct Substitution. Let a load P_A be applied to a softening spring (for which $k_N < 0$). For the first iteration, assume that $k_N = 0$. Therefore, as the first approximation of displacement u_A produced by P_A, we compute $u_1 = P_A/k_0$. Using u_1 we compute the new stiffness approximation $k_0 + k_{N1} = k_0 + f(u_1)$, and then the new displacement approximation u_2. Thus we generate the sequence of approximations

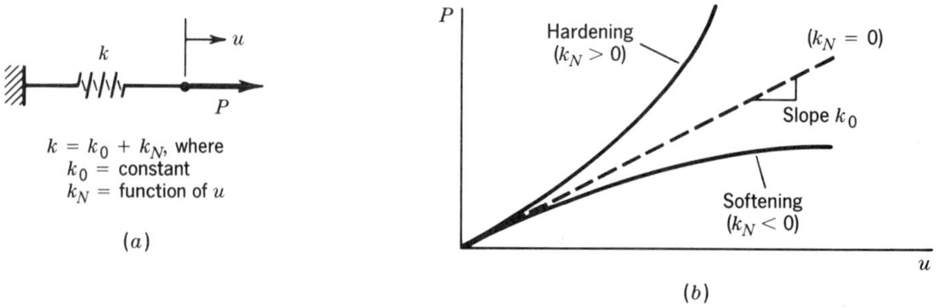

Figure 17.2-1. (*a*) A nonlinear spring. (*b*) When $u > 0$, there is hardening if $k_N > 0$ and softening if $k_N < 0$. When $u = 0$, we assume that $k_N = 0$.

$$u_1 = k_0^{-1}P_A,\ u_2 = (k_0 + k_{N1})^{-1}P_A,\ \ldots,\ u_{i+1} = (k_0 + k_{Ni})^{-1}P_A \tag{17.2-2}$$

These calculations are interpreted graphically in Fig. 17.2-2*a*. We see that the approximate stiffnesses $k_0 + k_{Ni}$ can be regarded as secants of the actual curve, each emanating from $P = u = 0$. After several iterations, the secant stiffness is $k_0 + k_N \approx P_A/u_A$, and the correct solution $u = u_A$ is closely approximated.

In an alternative form of direct substitution, nonlinear terms $k_N u$ are taken to the right-hand side. Thus, instead of Eq. 17.2-2, we have the sequence

$$u_1 = k_0^{-1}P_A,\ u_2 = k_0^{-1}(P_A - k_{N1}u_1),\ \ldots,\ u_{i+1} = k_0^{-1}(P_A - k_{Ni}u_i) \tag{17.2-3}$$

(Equations 17.2-2 and 17.2-3 will not yield the same values of u_2, u_3, etc., but upon convergence both will yield the result $u_\infty = u_A$.) Equation 17.2-3 is interpreted graphically in Fig. 17.2-2*b*. The effective loads applied in the second and third iterations in Fig. 17.2-2*b* are

$$P_A - k_{N1}u_1 = P_A + (k_0u_1 - [k_0 + k_{N1}]\,u_1) = P_A + (P_a - P_1) \tag{17.2-4a}$$

$$P_A - k_{N2}u_2 = P_A + (k_0u_2 - [k_0 + k_{N2}]\,u_2) = P_A + (P_I - P_2) \tag{17.2-4b}$$

It may be helpful to note that $P_a - P_1 = P_I - P_A$ and $P_I - P_2 = P_{II} - P_A$. The sequence of pseudoloads $P_a - P_1, P_I - P_2, \ldots$, must converge if Eq. 17.2-3 is to converge to $u = u_A$. Failure to converge is more likely with hardening structures than with softening structures.

If convergence difficulties arise, underrelaxation may help. Thus, rather than updating a calculated value u_{i+1} to its full value, we update instead to

$$u_{i+1} = u_i + \beta(\Delta u_{i+1}) \tag{17.2-5a}$$

or, changing the form by the substitution $\Delta u_{i+1} = u_{i+1} - u_i$, we have

$$u_{i+1} = \beta u_{i+1} + (1 - \beta)u_i \tag{17.2-5b}$$

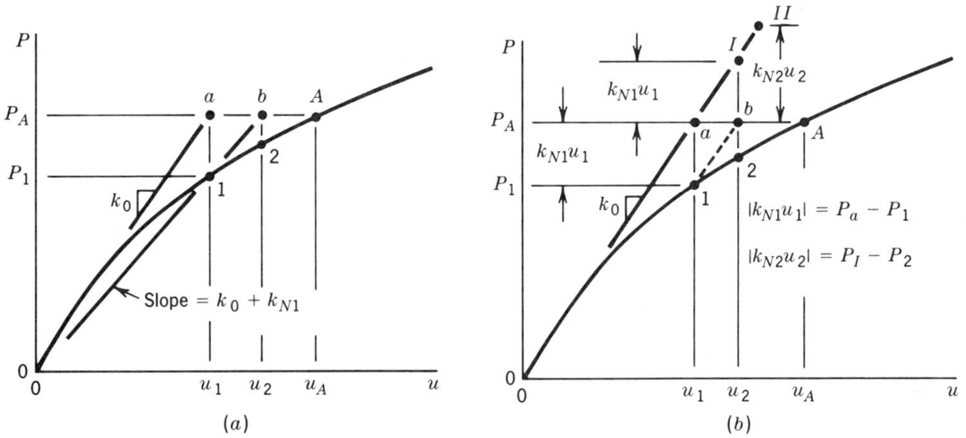

Figure 17.2-2. Graphical interpretations of direct substitution. (*a*) According to Eq. 17.2-2. (*b*) According to Eq. 17.2-3. Lines *aI* and 1*b* are parallel.

where "=" means "is replaced by," as in Fortran, and β is a number in the range $0 < \beta < 1$.

Note that for a multi-d.o.f. structure, k is a stiffness matrix rather than a scalar. Thus Eq. 17.2-2 requires that a new matrix $[\mathbf{K}_0 + \mathbf{K}_N]$ be formed and reduced in each iteration, whereas Eq. 17.2-3 requires but one formation and reduction of $[\mathbf{K}_0]$. However, Eq. 17.2-3 will require more iterative cycles than Eq. 17.2-2 in order to reach a prescribed accuracy.

Newton–Raphson (N–R). Imagine that we have applied load P_A and somehow determined the corresponding displacement u_A. That is, from Eq. 17.2-1,

$$(k_0 + k_{NA})u_A = P_A \qquad \text{where} \quad k_{NA} = f(u_A) \tag{17.2-6}$$

The load is now increased to a value P_B and the corresponding displacement u_B is sought. A truncated Taylor series expansion of $P = f(u)$ about u_A is

$$f(u_A + \Delta u_1) = f(u_A) + \left(\frac{dP}{du}\right)_A \Delta u_1 \tag{17.2-7}$$

where

$$\frac{dP}{du} = \frac{d}{du}(k_0 u + k_N u) = k_0 + \frac{d}{du}(k_N u) = k_t \tag{17.2-8}$$

and k_t is called the *tangent stiffness*. We seek Δu_1 for which $f(u_A + \Delta u_1) = P_B$. Thus, with $f(u_A) = P_A$ and k_t evaluated at A, Eq. 17.2-7 becomes

$$P_B = P_A + (k_t)_A \,\Delta u_1 \qquad \text{or} \qquad (k_t)_A \,\Delta u_1 = P_B - P_A \tag{17.2-9}$$

where $P_B - P_A$ can be interpreted as a load imbalance—that is, as the difference between the applied load P_B and the force $P_A = (k_0 + k_{NA})u_A$ in the spring when its stretch is u_A. The solution process is depicted in Fig. 17.2-3. After computing

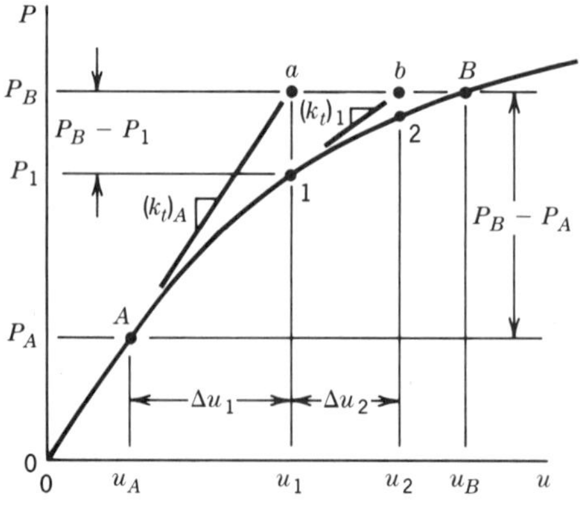

Figure 17.2-3. N–R solution for u_B caused by P_B, starting from point A.

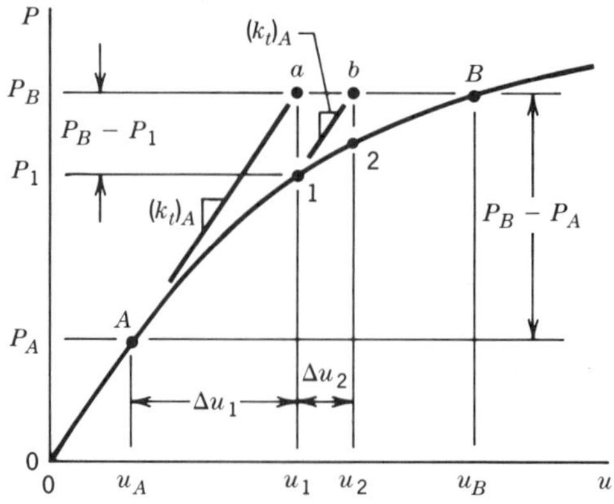

Figure 17.2-4. Modified N–R solution for u_B caused by P_B, starting from point A.

Δu_1, we update the displacement estimate to $u_1 = u_A + \Delta u_1$. For the next iteration, we obtain a new tangent stiffness $(k_t)_1$ by use of Eq. 17.2-8 with $u = u_1$, and obtain a new load imbalance $P_B - P_1$, where P_1 comes from Eq. 17.2-1 with $u = u_1$. The updated displacement estimate is $u_2 = u_1 + \Delta u_2$, where Δu_2 is obtained by solving $(k_t)_1 \, \Delta u_2 = P_B - P_1$.

Remarks. Methods discussed in this section extend directly to multiple d.o.f., where $k = k_0 + k_N$ becomes $[\mathbf{K}] = [\mathbf{K}_0 + \mathbf{K}_N]$, P becomes $\{\mathbf{R}\}$, and u becomes $\{\mathbf{D}\}$. In one dimension, if the stiffness can be stated as $k = k_0 + k_N$, the tangent stiffness k_t is easily obtained (Eq. 17.2-8). Such a simple expression is not available if there are multiple d.o.f. However, in practice, the physics of the problem usually allows us to calculate the tangent-stiffness matrix $[\mathbf{K}_t]$. Neither $[\mathbf{K}]$ nor $[\mathbf{K}_t]$ need be symmetric in a nonlinear problem, but in some situations symmetry prevails or can be achieved by manipulation.

In a multi-d.o.f. context, N–R iteration involves repeated solution of the equations $[\mathbf{K}_t]_i\{\Delta\mathbf{D}\}_{i+1} = \{\Delta\mathbf{R}\}_{i+1}$, where tangent-stiffness matrix $[\mathbf{K}_t]$ and load imbalance $\{\Delta\mathbf{R}\}$ are updated after each cycle. The solution process seeks to reduce the load imbalance, and consequently $\{\Delta\mathbf{D}\}$, to zero.

Modified Newton–Raphson. This method differs from the N–R method only in that the tangent stiffness either is not updated or is updated infrequently. Thus, in multi-d.o.f. problems, we avoid the expensive repetitions of forming and reducing the tangent-stiffness matrix $[\mathbf{K}_t]$. However, more iterative cycles are needed in order to reach a prescribed accuracy. The process is depicted one-dimensionally in Fig. 17.2-4.

If $[\mathbf{K}_t]$ is referred to the initial configuration, the modified N–R method becomes almost identical to the direct substitution method of Eq. 17.2-3. The only difference is that modified N–R computes u_{i+1} by adding Δu_{i+1} to u_i, and Eq. 17.2-3 computes u_{i+1} directly.

Incremental Methods. The foregoing discussion is concerned with locating a single point on the curve of P versus u. If the entire curve is required, one can approximate it as a series of points by applying an iterative process repeatedly: for example, in Fig. 17.2-3, after convergence under load P_B, increase the load to P_C and again iterate until convergence, then increase the load to P_D, and so on. If instead the load is increased in *each* computational cycle, the solution method may be called *incremental* rather than iterative.

The simplest incremental method is Euler's method of solving a first-order differential equation. To explain Euler's method we write Eq. 17.2-1 as $P = f(u)$, define $k_t = dP/du$, and consider load increments ΔP. Starting from $P = 0$ at $u = 0$, we compute successively

$$u_1 = 0 + (k_t)_0^{-1} \, \Delta P_1 \qquad \text{where} \quad (k_t)_0 = k_t \text{ at } u = 0 \qquad (17.2\text{-}10a)$$

$$u_2 = u_1 + (k_t)_1^{-1} \, \Delta P_2 \qquad \text{where} \quad (k_t)_1 = k_t \text{ at } u = u_1 \qquad (17.2\text{-}10b)$$

$$u_3 = u_2 + (k_t)_2^{-1} \, \Delta P_3 \qquad \text{where} \quad (k_t)_2 = k_t \text{ at } u = u_2 \qquad (17.2\text{-}10c)$$

and in general $u_{i+1} = u_i + (k_t)_i^{-1} \, \Delta P_{i+1}$. The process is depicted in Fig. 17.2-5.

A disadvantage of the foregoing method is apparent in Fig. 17.2-5: the approximate solution drifts further from the exact solution with every step. Progressive

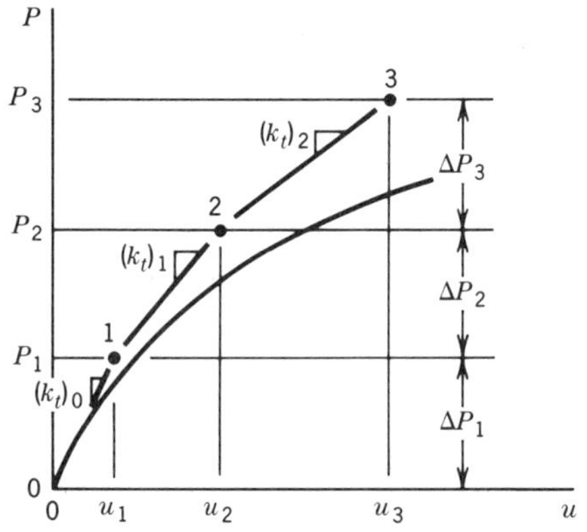

Figure 17.2-5. Purely incremental solution of the equation $P = f(u)$.

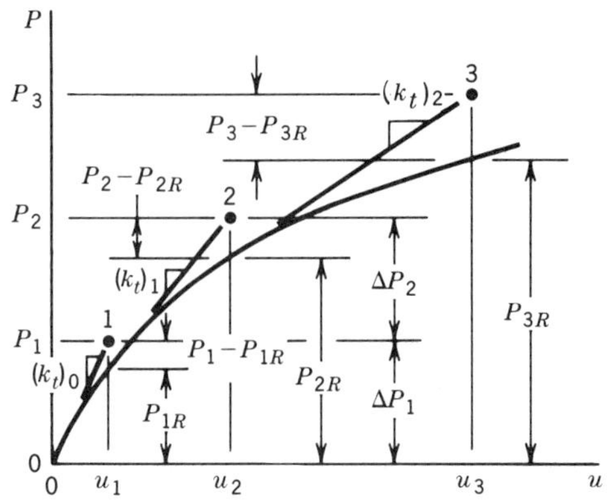

Figure 17.2-6. Incremental solution of $P = f(u)$ with load corrections $(P_i - P_{iR})$.

drift can be eliminated by introducing the load imbalance as a corrective term. Load imbalance has the same meaning in the present context as in the N–R method. With this corrective term we obtain

$$u_{i+1} = u_i + (k_t)_i^{-1} [\Delta P_{i+1} + (P_i - P_{iR})] \tag{17.2-11}$$

where P_i is the externally applied load at step i ($P_i = \Sigma\, \Delta P_i$ summed through step i), and P_{iR} is the resisting load of the spring, $P_{iR} = (k_0 + k_{Ni})u_i$ from Eq. 17.2-1. This method has been called "incremental with one-step N–R correction." It is depicted in Fig. 17.2-6. Computed points 1, 2, . . . , do not lie on the curve, but they do not progressively drift away from the curve as in Fig. 17.2-5.

Quasi-Newton Methods. Inverse-Broyden. In Fig. 17.2-7, displacements u_1 and u_2 are computed by two cycles of modified N–R iteration. Then a secant to the curve is established through points 1 and 2, and a step is taken along the secant. The next step, not shown, would be along a secant through points 2 and 3. With more iterations, leading to convergence, the secant stiffness approaches the exact tangent stiffness at A. Steps in secant directions are not quite as profitable as steps in tangent directions, as in the N–R method, but secant-stiffness steps are much cheaper and are more stable than tangent-stiffness steps.

One expression of the quasi-Newton concept is the inverse-Broyden method.

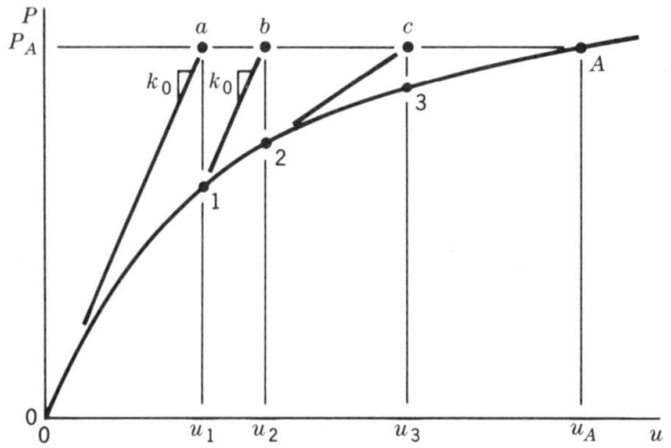

Figure 17.2-7. Two modified N–R iterations, followed by a secant step along a line through points 1 and 2.

The label "inverse" means that the *inverse* of the stiffness matrix is updated, not the stiffness matrix itself. A complete explanation is beyond the scope of this text. References include [17.2–17.6]. In outline, the method operates as follows. Imagine that a given load $\{\mathbf{R}\}_A$ has been applied; we now wish to determine the corresponding $\{\mathbf{D}\}$ by iteration. We write

$$\{\mathbf{D}\}_{i+1} = \{\mathbf{D}\}_i + \{\Delta\mathbf{D}\}_{i+1} \qquad \text{where} \quad \{\Delta\mathbf{D}\}_{i+1} = [\mathbf{K}]_i^{-1}\{\Delta\mathbf{R}\}_{i+1} \tag{17.2-12}$$

where $\{\Delta\mathbf{R}\}_{i+1}$ is the load imbalance (as in Eq. 17.2-9) and $[\mathbf{K}]_i$ can be regarded as a secant-stiffness matrix. After several iterations i, both $\{\Delta\mathbf{R}\}_{i+1}$ and $\{\Delta\mathbf{D}\}_{i+1}$ become small, and $\{\mathbf{D}\}_{i+1}$ is a good approximation of $\{\mathbf{D}\}$ under loads $\{\mathbf{R}\}_A$. The second of Eqs. 17.2-12 is expanded as follows,

$$[\mathbf{K}]_i^{-1}\{\Delta\mathbf{R}\}_{i+1} = [\mathbf{K}]_I^{-1}\{\Delta\mathbf{R}\}_{i+1} + \sum_{k=1}^{i} \lfloor\mathbf{p}\rfloor_k^T \lfloor\mathbf{v}\rfloor_k \{\Delta\mathbf{R}\}_{i+1} \tag{17.2-13}$$

where $[\mathbf{K}]_I^{-1}$ is an estimate of $[\mathbf{K}]^{-1}$ at the outset of iteration at a given load level. Thus, in each iteration, $[\mathbf{K}]_I^{-1}$ is improved by the addition of one more rank 1 matrix $\lfloor\mathbf{p}\rfloor^T\lfloor\mathbf{v}\rfloor$. Update matrices $\lfloor\mathbf{p}\rfloor^T\lfloor\mathbf{v}\rfloor$ are in general unsymmetric, but with continued updating and eventual convergence, a symmetric $[\mathbf{K}]_i^{-1}$ may be approached.

The computational efficiency of the procedure, which may make it the method of choice, arises as follows. Neither $[\mathbf{K}]_i$ nor $[\mathbf{K}]_i^{-1}$ is ever written out as a square matrix. Matrix $[\mathbf{K}]_I$ is forward-reduced, to act as $[\mathbf{K}]_I^{-1}$, only once. Thereafter, as i increases, only imbalances $\{\Delta\mathbf{R}\}_{i+1}$ are treated, by forward-reduction and back-substitution. Also, each product $\lfloor\mathbf{p}\rfloor_k^T\lfloor\mathbf{v}\rfloor_k\{\Delta\mathbf{R}\}_{i+1}$ requires one vector-times-vector multiplication and one vector-times-scalar multiplication. Thus the procedure is "vectorizable" on vector-processing computers.

In nonstructural problems, the negative of $[\mathbf{K}]$ may be called the *Jacobian* and $\{\Delta\mathbf{R}\}$ called the *residual*.[1] In some problems, nothing may be known about $[\mathbf{K}]_I$. Then a unit matrix might be assumed, $[\mathbf{K}]_I = [\mathbf{I}]$. However, the better the estimate of $[\mathbf{K}]_I$, the faster the convergence. Similarly, starting with a good estimate of the solution $\{\mathbf{D}\}$ will reduce the residual and speed convergence. If more than roughly 30 to 60 iterations are used, depending on the machine precision, the method may fail because successive updates $\lfloor\mathbf{p}\rfloor^T\lfloor\mathbf{v}\rfloor$ become linearly dependent. Then one can discard the updates and start afresh with a new $[\mathbf{K}]_I$, even if it is again $[\mathbf{K}]_I = [\mathbf{I}]$. A fresh start is also recommended when starting to iterate with a new $\{\mathbf{R}\}$. As a termination criterion, Eq. 17.2-16 below is recommended.

A Fortran version of the algorithm appears in Fig. 17.2-8.

If $[\mathbf{K}_I]$ is known to be symmetric and positive definite, one may apply the "BFGS method," which is a powerful quasi-Newton method related to the inverse-Broyden method [17.4, 17.5].

Termination. The efficiency of a nonlinear solution method can be measured by its "order of termination." Let e_i represent a measure of the error after the ith iterative cycle. If e_i is sufficiently small, it is often possible to bound e_{i+1}. Practical possibilities include the following, illustrated here for a single d.o.f.

[1]The Jacobian is the negative of the tangent stiffness because of our choice of sign in writing the residual as external force minus internal force and our desire that tangent stiffness reduce to conventional stiffness in the linear case. Other conventions are possible.

```
      SUBROUTINE INVBDN (K,A,Q,U,DU,NEQ,MBAND,V,DV,NBRY,ROI,IER)
C   Subroutine to carry out ONE inverse-Broyden iteration.
C   K = current iteration number, A = initial estimate of Jacobian (e.g.
C   stiffness) matrix, Q = current residual, U = current solution, DU =
C   increment to update U, NEQ = number of unknowns, MBAND = semiband-
C   width of A, V and DV = saved Broyden vectors, NBRY = number of Broy-
C   den vectors allowed = number of iterations allowed, ROI = saved con-
C   stant, IER = 132 if error detected in current iteration.
      IMPLICIT DOUBLE PRECISION (A-H,O-Z)
C---  In DIMENSION line, Q(1), U(1), etc. act as Q(NEQ), U(NEQ), etc.
      DIMENSION A(NEQ,1),Q(1),U(1),DU(1),V(NBRY,1),DV(NBRY,1),ROI(1)
      DATA EPS /1.D-12/
      IER = 0
C---  Calling program has computed residual Q and REDUCED form of A.
C---  Call SOLVER to do only reduction and back-substitution of Q.
      CALL SOLVER (A,Q,NEQ,MBAND,2)
      IF (K .GT. 1)  GO TO 150
      DO 100 I=1,NEQ
  100 DU(I) = Q(I)
      GO TO 500
C---  Update U using the K-1 Broyden vectors already established.
  150 KM1 = K - 1
      DO 400 I=1,KM1
      CK = 0.D0
      DO 200 J=1,NEQ
  200 CK = CK + DV(I,J)*Q(J)
      CONS = CK*ROI(I)
      DO 300 J=1,NEQ
  300 Q(J) = Q(J) + CONS*(DV(I,J)-V(I,J))
  400 CONTINUE
C---  Establish the Kth Broyden vectors. Use them to further update U.
  500 DENO = 0.D0
      DO 600 J=1,NEQ
      V(K,J) = Q(J) + DU(J)
      DV(K,J) = DU(J)
  600 DENO = DENO + DU(J)*V(K,J)
C---  If DENO < EPS, we may have a singular Jacobian, or may have con-
C---  verged but not recognized it due to faulty termination criterion.
      IF (DABS(DENO) .GE. EPS)  GO TO 650
      IER = 132
      RETURN
  650 CK = 0.D0
      ROI(K) = 1.D0/DENO
      DO 700 J=1,NEQ
  700 CK = CK + DU(J)*Q(J)
      CONS = ROI(K)*CK - 1.D0
      DO 800 J=1,NEQ
      DU(J) = Q(J)*CONS
  800 U(J) = U(J) + DU(J)
      RETURN
      END
```

Figure 17.2-8. Fortran coding of the inverse-Broyden algorithm. Subroutine SOLVER appears in Appendix B as Fig. B.2-3 (note that SOLVER is for a band-symmetric matrix). The best equation solver available should be used and can easily be substituted.

$$\text{linear:} \quad |e_{i+1}| \leq C_1|e_i| \tag{17.2-14a}$$

$$\text{superlinear:} \quad |e_{i+1}| \leq C_2|e_i||e_{i-1}| = C_3|e_i|^{1.6} \tag{17.2-14b}$$

$$\text{quadratic:} \quad |e_{i+1}| \leq C_4 e_i^2 \tag{17.2-14c}$$

where C_1 through C_4 are constants. Of the methods we have discussed, only N–R exhibits quadratic termination. The inverse-Broyden method exhibits a generalized superlinear termination. Note that both N–R and inverse-Broyden methods will fail if $P = f(u)$ exhibits zero slope at the intended solution u_A (i.e., if $f'(u_A) = 0$, which is a "limit point").

A termination criterion for an iterative process can be of various forms. One that is usually good is as follows. We define [17.4]

$$\text{CNORM} = \left(\sum \Delta D_j^2\right)^{1/2} \left(\sum D_j^2\right)^{-1/2} \tag{17.2-15a}$$

$$\text{RNORM} = \left(\sum \Delta R_j^2\right)^{1/2} \left(\sum R_j^2\right)^{-1/2} \tag{17.2-15b}$$

where, in structural problems, ΔD_j, D_j, ΔR_j, and R_j mean respectively displacement increment, displacement, load increment, and load. Summations span all terms (NEQ in Fig. 17.2-8). Thus CNORM and RNORM are ratios of Euclidean norms. In a nonstructural problem, $(\Sigma\ R_j^2)^{1/2}$ might become the product of a modulus and the square of a characteristic length, or the product of viscosity, the reciprocal of a characteristic time, and the square of a characteristic length. We terminate when

$$\max(\text{CNORM},\text{RNORM}) \leq tol \tag{17.2-16}$$

where *tol* is chosen to balance accuracy requirements against machine precision. Possible choices are $tol = 10^{-5}$ in 64-bit arithmetic and $tol = 10^{-3}$ in 32-bit arithmetic. For the inverse-Broyden method at least, it is recommended that *both* CNORM and RNORM be less than *tol*, as Eq. 17.2-16 requires. In the rare case that $\{\mathbf{D}\} = \{\mathbf{0}\}$ is a possible solution or iterate, the denominator of CNORM becomes very small. Then one can safely set the denominator to unity.

Concluding Remarks. Hardening structures are usually more difficult to analyze than softening structures. Iterative processes are more likely to converge slowly or fail to converge (Fig. 17.2-9).

There is no need to maintain strict separation between solution methods. For example, we could adopt a modified N–R strategy, but occasionally update the tangent-stiffness matrix. Often such an update is most effective when done immediately *after* one iteration at a new load level using the old stiffness. Apparently this approach would work well in Fig. 17.2-9*b*. Underrelaxation would also help in this example.

Two-dimensional sketches in the present section, for *P* versus *u*, are representative of multi-d.o.f. problems if the structure carries a single load $R_i = P$ and

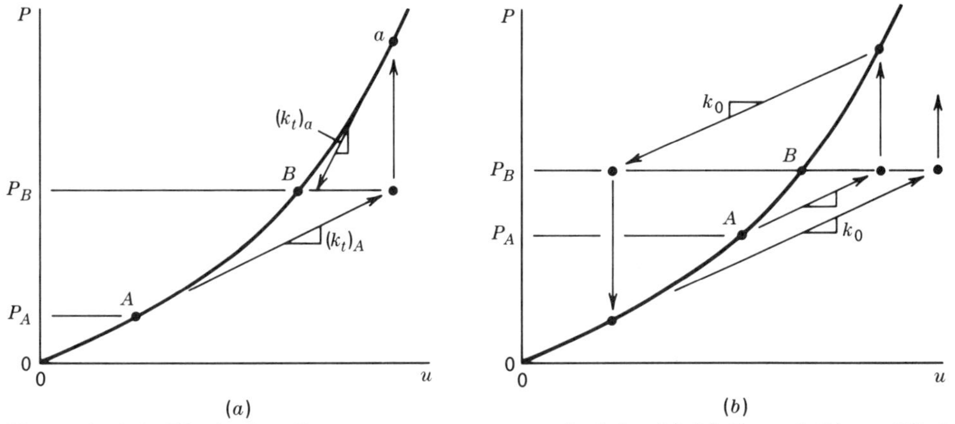

Figure 17.2-9. Hardening *P* versus *u* curves, attacked by (*a*) N–R, and (*b*) modified N–R methods.

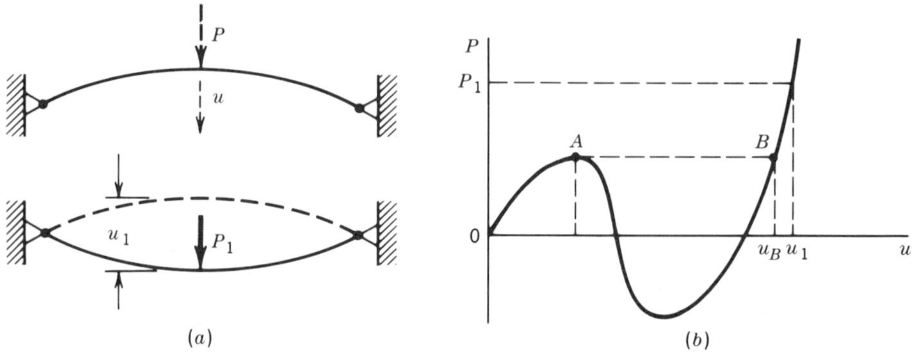

Figure 17.2-10. (*a*) Shallow arch under load P. (*b*) Load versus displacement plot, showing limit point at A.

$D_i = u$ is its displacement. In effect, the structure then acts as a single nonlinear spring in resisting R_i. If there is more than one d.o.f., a two-dimensional sketch provides an incomplete and possibly misleading representation.

One can provide displacement increments rather than load increments—that is, use "displacement control" rather than "load control." However, in a typical problem one does not know in advance how the D_i in $\{\mathbf{D}\}$ will be related, so the appropriate increments ΔD_i are unknown.

In Fig. 17.2-10, displacement control yields the entire curve of P versus u. Under load control, the physical structure will experience a sudden "snap" from A to B when the limit point at A is reached. Computational methods of traversing limit points have been devised [17.7, 17.8].

Another way to traverse limit points is called *viscous relaxation*. In Fig. 17.2-10, imagine that displacement u_1 corresponding to load P_1 is required, but that the entire curve for $0 < u < u_1$ is not required. One can imagine that the structure is immersed in a viscous fluid. When load P_1 is applied, the structure moves slowly and without snapping. When motion ceases, the static displacement u_1 is achieved. Computationally, we solve a dynamic problem using step-by-step integration in time, including the nonlinear stiffness matrix and a damping matrix but omitting the mass matrix [17.9,17.10]. The method is most useful when there are strong geometric nonlinearities.

17.3 ONE-DIMENSIONAL ELASTIC–PLASTIC ANALYSIS

Plastic Action. A material is called nonlinear if stresses $\{\sigma\}$ and strains $\{\epsilon\}$ are related by a strain-dependent matrix rather than a matrix of constants. Thus the computational difficulty is that equilibrium equations must be written using material properties that depend on strains, but strains are not known in advance. Plastic flow is often a cause of material nonlinearity. In the present section we use the case of uniaxial stress to introduce the formulation and solution of elastic–plastic problems.

Imagine that yielding has already occurred; then a strain increment $d\epsilon$ takes place (Fig. 17.3-1*a*). This strain increment can be regarded as composed of an elastic contribution $d\epsilon^e$ and a plastic contribution $d\epsilon^p$, so that $d\epsilon = d\epsilon^e + d\epsilon^p$. The corresponding stress increment $d\sigma$ can be written in various ways,

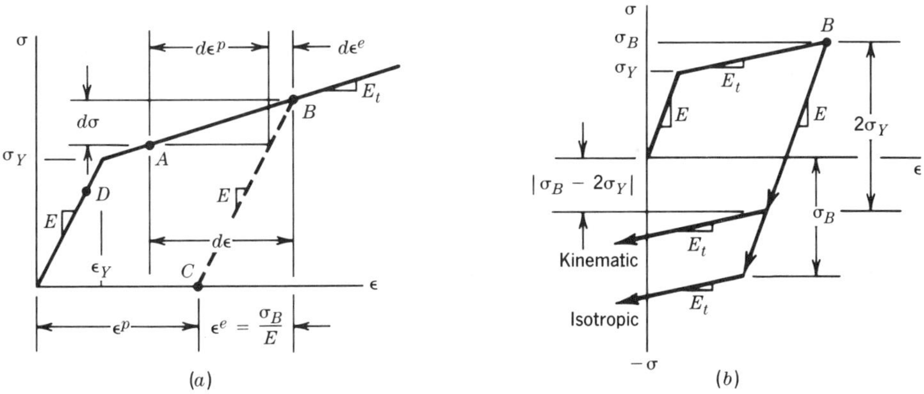

Figure 17.3-1. (*a*) Stress–strain plot in uniaxial stress, idealized as two straight lines, where σ_Y is the stress at first onset of yielding. (*b*) Kinematic and isotropic hardening rules.

$$d\sigma = E(d\epsilon - d\epsilon^p) \qquad d\sigma = E_t\, d\epsilon \qquad \text{and} \qquad d\sigma = H\, d\epsilon^p \tag{17.3-1}$$

where H is called the strain-hardening parameter. Substitution of the first and third of Eqs. 17.3-1 into the second yields

$$H = \frac{E_t}{1 - (E_t/E)} \qquad \text{or} \qquad E_t = E\left(1 - \frac{E}{E + H}\right) \tag{17.3-2}$$

where E_t is the tangent modulus. When written in this form, the expression for E_t is similar to a more general expression used for multiaxial states of stress. If E is finite and $E_t = 0$, then $H = 0$, and the material is called "elastic–perfectly plastic."

A summary of elastic–plastic action in uniaxial stress is as follows. The *yield criterion* states that yielding begins when $|\sigma|$ reaches σ_Y, where in practice σ_Y is usually taken as the tensile yield strength. Subsequent plastic deformation may alter the stress needed to produce renewed or continued yielding; this stress exceeds the initial yield strength σ_Y if $E_t > 0$. A *flow rule* can be written in multidimensional problems. It leads to a relation between stress increments $\{d\boldsymbol{\sigma}\}$ and strain increments $\{d\boldsymbol{\epsilon}\}$. In uniaxial stress this relation is simply $d\sigma = E_t\, d\epsilon$, which describes the increment of stress produced by an increment of strain. Note, however, that if the material has yet to yield or is unloading, then $d\sigma = E\, d\epsilon$ (e.g., in Fig. 17.3-1*a*, complete unloading from point B leads to point C and a permanent strain ϵ^p). Finally, there is a *hardening rule,* which describes how the yield criterion is changed by the history of plastic flow. For example, imagine that unloading occurs from point B in Fig. 17.3-1*a*. With reloading from point C, response will be elastic until $\sigma > \sigma_B$, when renewed yielding occurs. If we assume that yielding reappears when $|\sigma| > \sigma_B$, whether σ is tensile or compressive, we have adopted the "isotropic hardening" rule (Fig. 17.3-1*b*). However, for common metals, such a rule is in conflict with the observed behavior that yielding reappears at a stress of approximate magnitude $\sigma_B - 2\sigma_Y$ when loading is reversed. Accordingly, a better match to observed behavior is provided by the "kinematic hardening" rule, which (for uniaxial stress) says that a total elastic range of $2\sigma_Y$ is preserved.

The discussion in the foregoing paragraph does not require that postelastic response be idealized as a straight line. In other words, E_t need not be constant.

As a simple application of one-dimensional plasticity, imagine that a tapered bar is to be loaded by an axial force P (Fig. 17.3-2). Material properties are those depicted in Fig. 17.3-1. The bar is modeled by two-d.o.f. bar elements, each of constant cross section. For elastic conditions, the element stiffness matrix is given by Eq. 2.4-5, where $E = d\sigma/d\epsilon$ when $|\sigma| < \sigma_Y$. Upon yielding, the stress–strain relation becomes $E_t = d\sigma/d\epsilon$. Accordingly, letting E_{ep} represent the "elastic–plastic" stiffness, we write the element tangent-stiffness matrix as

$$[\mathbf{k}_t] = \frac{AE_{\text{ep}}}{L}\begin{bmatrix} 1 & -1 \\ -1 & 1 \end{bmatrix} \tag{17.3-3}$$

where $E_{\text{ep}} = E$ if the yield criterion is not exceeded or if unloading is taking place, and $E_{\text{ep}} = E_t$ if plastic flow is involved.

In numerical solutions, material may make the transition from elastic to plastic within an iterative cycle of the solution process. For example, imagine that $d\epsilon$ spans ϵ_D to ϵ_A in Fig. 17.3-1a. The problem of "rounding the corner" can be addressed by combining E and E_t according to the fraction m of the total step $d\epsilon$ that is elastic. Thus let

$$E_{\text{ep}} = mE + (1 - m)\,E_t \qquad \text{where} \quad m = \frac{\epsilon_Y - \epsilon_D}{\epsilon_A - \epsilon_D} \tag{17.3-4}$$

Alternatively, by substituting stresses for strains and using the fictitious stress $\sigma^* = E\epsilon_A$, we can write m in terms of stresses, as $m = (\sigma_Y - \sigma_D)/(\sigma^* - \sigma_D)$. Refinements of this scheme are possible [17.11].

The present discussion excludes thermal strains and creep strains. In general, these effects may appear in combination with elastic–plastic action. Then the total strain increment $d\epsilon^{\text{tot}}$ is a combination of elastic, plastic, thermal, and creep strain components. The strain increment $d\epsilon = d\epsilon^e + d\epsilon^p$ used in elastic–plastic analysis excludes thermal and creep strains. Thus

$$d\epsilon^e + d\epsilon^p = d\epsilon^{\text{tot}} - d\epsilon^T - d\epsilon^C \tag{17.3-5}$$

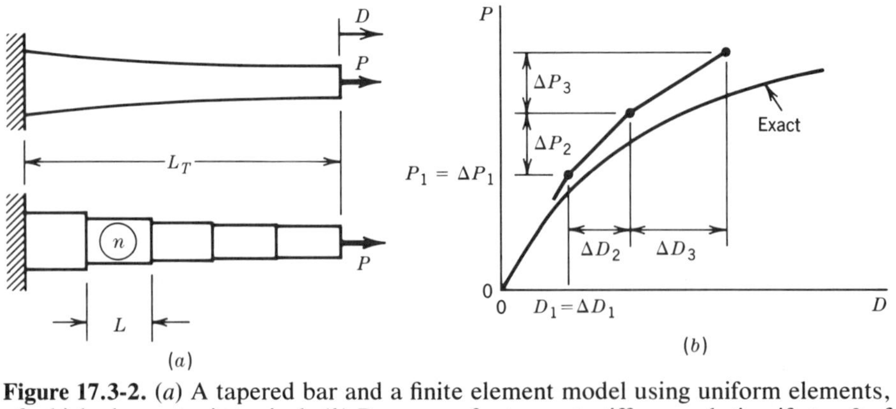

Figure 17.3-2. (a) A tapered bar and a finite element model using uniform elements, of which element n is typical. (b) Progress of a tangent-stiffness solution if step 3 of the algorithm is omitted. D = displacement of load P.

Tangent-Stiffness Method. Consider the tapered bar depicted in Fig. 17.3-2*a*. It is desired to trace the quasistatic load versus displacement curve and determine element stresses by means of a finite element model and load increments ΔP. Increments are small but not infinitesimal, so that $d\epsilon$ becomes $\Delta\epsilon$, and the numerical solution is not exact. A numerical representation of the stress–strain relation must be stored, so that σ, E, and E_t can be obtained for any ϵ. The algorithm outlined below requires that we also store, and update after each computational cycle, the nodal displacements $\{\mathbf{D}\}$, element strains ϵ, and element stresses σ. With two-d.o.f. bar elements (Eq. 17.3-3), σ and ϵ are constant over each element length L.

1. For the first computational cycle ($i = 1$), assume $E_{ep} = E$ for all elements. Apply the first load increment, $\{\Delta\mathbf{R}\}_1$.
2. Using the current strains, determine the current E_{ep} in each element. Use Eq. 17.3-3 to obtain $[\mathbf{k}_t]_n$ for each element n. Obtain the current structure tangent stiffness $[\mathbf{K}_t]_{i-1} = \Sigma\ [\mathbf{k}_t]_n$. Solve $[\mathbf{K}_t]_{i-1}\{\Delta\mathbf{D}\}_i = \{\Delta\mathbf{R}\}_i$ for $\{\Delta\mathbf{D}\}_i$. (For the bar of Fig. 17.3-2*a*, ΔP at the right end is the only nonzero entry in $\{\Delta\mathbf{R}\}_i$.) From $\{\Delta\mathbf{D}\}_i$, obtain current strain increments $\Delta\epsilon_i$ for each element.
3. *Optional.* If any elements make the elastic-to-plastic transition, use Eq. 17.3-4 to revise E_{ep} for each such element, and go back to step 2. Without changing the applied load $\{\Delta\mathbf{R}\}_i$, repeat steps 2 and 3 until convergence, which may be defined as $\Delta\epsilon$ being less than a prescribed fraction of the accumulated total ϵ in every element. These operations represent secant-stiffness iterations (see Fig. 17.2-2*a*) within one of the load steps of the tangent-stiffness procedure.
4. Update: $\{\mathbf{D}\}_i = \{\mathbf{D}\}_{i-1} + \{\Delta\mathbf{D}\}_i$, and for each element, $\epsilon_i = \epsilon_{i-1} + \Delta\epsilon_i$ and $\sigma_i = \sigma_{i-1} + \Delta\sigma_i$, where $\Delta\sigma_i = (E_{ep})_i\ \Delta\epsilon_i$. For the first cycle ($i = 1$), initial values (subscript $i - 1$) of displacement, strain, and stress are typically all zero if one starts from the unloaded configuration, but are nonzero if one starts from a state in which plastic action impends.
5. Apply the next load increment and return to step 2.
6. Stop when $\Sigma\ \{\Delta\mathbf{R}\}_i$ reaches the total applied load.

Three cycles of the foregoing algorithm are depicted in Fig. 17.3-2*b*. Each cycle produces a line segment whose slope corresponds to the current stiffness. Drift from the exact path can be reduced by using smaller load increments, by exercising step 3 previously discussed, and by using "corrective loads," which are discussed in Section 17.5. Step 3 can be avoided by using load increments $\{\Delta\mathbf{R}\}_i$ that bring a single element to the verge of yielding as each load increment is added. This is easily accomplished by scaling the incremental tangent-stiffness solutions.

The foregoing incremental procedure is essentially a Newton–Raphson method; that is, a new tangent-stiffness matrix is used in each computational cycle.

Initial-Stiffness Method. Again we seek displacements and stresses in a structure in which plastic action occurs. One can apply the iterative method described by Eq. 17.2-3 and Fig. 17.2-2*b*. Thus the original *elastic* stiffness matrix is used at all times. The effects of plastic action are regarded as initial stresses that produce fictitious loads, which are combined with the load actually applied (accordingly, this procedure is often called the *initial-stress method*). This procedure avoids the expense of repeatedly forming and factoring a tangent-stiffness matrix, but may converge slowly if plastic strains are large or widespread [17.12].

In Fig. 17.3-3*a*, imagine that we seek the strain ϵ_B associated with stress σ_B. We can obtain ϵ_B using only the *elastic* modulus E by writing $\epsilon_B = \sigma_C/E$. Here σ_C is the fictitious stress $\sigma_C = \sigma_B + E\,\Delta\epsilon^p$.

In computation, ϵ^p can be obtained by accumulating the plastic strain increments $\Delta\epsilon^p$ produced in the iterative cycles. From Eqs. 17.3-1,

$$\Delta\epsilon^p = \frac{1}{H}\Delta\sigma = \frac{1}{H}E_t\,\Delta\epsilon = \frac{E}{E+H}\Delta\epsilon = \left(1 - \frac{E_t}{E}\right)\Delta\epsilon \qquad (17.3\text{-}6)$$

In computation, the progression to ϵ_B is made in a series of steps, as shown in Fig. 17.3-3*b*, using "supplementary loads" defined in Eq. 17.3-7. It is not necessary that the stress–strain relation be piecewise linear.

The calculation procedure for a structure such as that in Fig. 17.3-2*a* is as follows.

1. Compute the elastic stiffness matrix $[\mathbf{K}]$ (which is identically the initial tangent-stiffness matrix). Solve $[\mathbf{K}]\{\mathbf{D}\} = \{\mathbf{R}\}$ for $\{\mathbf{D}\}$, where $\{\mathbf{R}\}$ is proportional to the actual load but of arbitrary level. From this solution, scale $\{\mathbf{R}\}$ so that it becomes $\{\mathbf{R}_Y\}$, which causes yielding to impend. Scale $\{\mathbf{D}\}$ similarly and call the result $\{\mathbf{D}\}_{\text{old}}$. Subsequent load increments may be chosen as $\{\Delta\mathbf{R}\} = 0.05\,\{\mathbf{R}_Y\}$ or as $\{\Delta\mathbf{R}\} = (E_T/E)\{\mathbf{R}_Y\}$, whichever is greater [17.13]. Initialize supplementary loads $\{\Delta\mathbf{R}_s\}$ to zero.
2. Solve the equations $[\mathbf{K}]\{\Delta\mathbf{D}\} = \{\Delta\mathbf{R}\} + \{\Delta\mathbf{R}_s\}$ for $\{\Delta\mathbf{D}\}$.
3. Update displacements: $\{\mathbf{D}\}_{\text{new}} = \{\mathbf{D}\}_{\text{old}} + \{\Delta\mathbf{D}\}$.
4. In each element, calculate the strain increment $\Delta\epsilon$ associated with $\{\Delta\mathbf{D}\}$. Update element stress by adding $\Delta\sigma$ to the existing stress σ, using $\Delta\sigma = E\,\Delta\epsilon$ if $\sigma < \sigma_Y$ and $\Delta\sigma = E_t\,\Delta\epsilon$ if $\sigma > \sigma_Y$. For elements that make the elastic-to-plastic transition by the addition of $\Delta\sigma$, evaluate m by Eq. 17.3-4 and recompute $\Delta\sigma$ as $\Delta\sigma = Em\,\Delta\epsilon$.
5. For all elements that display plastic strains ($|\sigma| > \sigma_Y$ in Fig. 17.3-3), calculate plastic strain increments according to Eq. 17.3-6. In this calculation, use $(1 - m)\,\Delta\epsilon$ rather than $\Delta\epsilon$ for elements that make the elastic-to-plastic

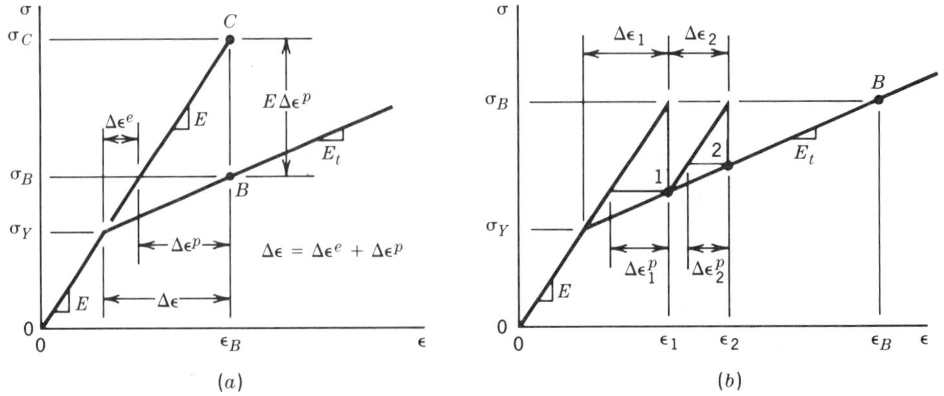

Figure 17.3-3. (*a*) $E\,\Delta\epsilon^p$ is regarded as an initial stress. (*b*) Iterative approach to the solution point B.

transition (see Eq. 17.3-4). Generate the supplementary loads by summing element contributions:

$$\{\Delta \mathbf{R}_s\} = \sum \{\Delta \mathbf{r}_s\} \qquad \text{where} \quad \{\Delta \mathbf{r}_s\} = \int_0^L \lfloor \mathbf{B} \rfloor^T E \, \Delta\epsilon^p A \, dx \qquad (17.3\text{-}7)$$

Solve the equations $[\mathbf{K}]\{\Delta \mathbf{D}\} = \{\Delta \mathbf{R}_s\}$ for $\{\Delta \mathbf{D}\}$. Return to step 3.

6. Repeat steps 3 through 5 until convergence. Then apply another load increment $\{\Delta \mathbf{R}\}$ and return to step 2.

7. Stop when $\{\mathbf{R}_Y\} + \Sigma\{\Delta \mathbf{R}\}$ reaches the total applied load.

17.4 SMALL-STRAIN PLASTICITY RELATIONS

Multiaxial states of stress can be analyzed if the theory in Section 17.3 is generalized. The following is a summary. In our discussion we use the engineering definition of shear strain (e.g., $\gamma_{xy} = u_{,y} + v_{,x}$), not the tensor definition (e.g., $\epsilon_{xy} = (u_{,y} + v_{,x})/2$).

General. Plasticity theory has three parts: a yield criterion, a flow rule, and a hardening rule. The general theory and its various special forms are contrived to fit experimental data.

Yield Criterion. We define a *yield function* F, which is a function of stresses $\{\boldsymbol{\sigma}\}$ and quantities $\{\boldsymbol{\alpha}\}$ and W_p associated with the hardening rule. Yielding occurs when

$$F(\boldsymbol{\sigma},\boldsymbol{\alpha},W_p) = 0 \qquad (17.4\text{-}1)$$

where $\{\boldsymbol{\alpha}\}$ and W_p are defined by Eq. 17.4-3. Specifically, if we evaluate F using given values of $\{\boldsymbol{\sigma}\}$, $\{\boldsymbol{\alpha}\}$, and W_p, then the possible results are $F < 0$ and $F = 0$. Respectively, these results mean that the material is in the elastic range or is yielding. The result $F > 0$ is not physically possible, as it indicates a state of stress that does not satisfy the constitutive law (e.g., σ_C in Fig. 17.3-3 is not physically possible). Similarly, the respective results $dF < 0$ and $dF = 0$ imply elastic unloading and continued yielding. The result $dF > 0$ is not possible in the plastic regime.

Flow Rule. We define a *plastic potential* Q, which has units of stress and is a function of the stresses, $Q = Q(\boldsymbol{\sigma},\boldsymbol{\alpha},W_p)$. With $d\lambda$ a scalar that may be called a "plastic multiplier," plastic strain increments are given by

$$\{d\boldsymbol{\epsilon}^p\} = \left\{\frac{\partial Q}{\partial \boldsymbol{\sigma}}\right\} d\lambda \qquad (17.4\text{-}2)$$

Thus $d\epsilon_x^p = (\partial Q/\partial \sigma_x)\, d\lambda$, and so on. The flow rule is called "associated" if $Q = F$ and "nonassociated" otherwise. Associated flow rules are commonly used for ductile metals, but nonassociated rules are better suited to soil and granular materials.

Hardening Rule. In Eq. 17.4-1, $\{\boldsymbol{\alpha}\}$ locates the center of the yield surface in stress space. Initially, before any plastic strains appear, $\{\boldsymbol{\alpha}\} = \{\mathbf{0}\}$. In "kinematic hardening," the center moves in the direction of plastic straining, so that $\{\boldsymbol{\alpha}\}$ becomes nonzero. Parameter W_p describes how the yield surface grows. In "isotropic hardening," W_p is nonzero but $\{\boldsymbol{\alpha}\}$ is zero. Quantities $\{\boldsymbol{\alpha}\}$ and W_p are defined as

$$\{\boldsymbol{\alpha}\} = \int C\,\{d\boldsymbol{\epsilon}^p\} \qquad \text{and} \qquad W_p = \int \{\boldsymbol{\sigma}\}^T\{d\boldsymbol{\epsilon}^p\} \tag{17.4-3}$$

where C can be assumed to be a material constant [17.13,17.14]. For purely kinematic hardening, $C = H$ (Fig. 17.4-1). W_p can be identified as plastic work per unit volume. (Use of W_p in F implies a "work-hardening" model. Alternatively, W_p can be replaced by an effective plastic strain ϵ^p_{ef}, which implies a "strain-hardening" model. Either model can be used to represent isotropic hardening.)

An incremental stress–strain relation, analogous to the relation $\{\boldsymbol{\sigma}\} = [\mathbf{E}]\{\boldsymbol{\epsilon}\}$ of elasticity but valid into the elastic–plastic regime, can be derived as follows. First, we differentiate Eq. 17.4-1:

$$dF = 0 = \left\{\frac{\partial F}{\partial \boldsymbol{\sigma}}\right\}^T\{d\boldsymbol{\sigma}\} + \left\{\frac{\partial F}{\partial \boldsymbol{\alpha}}\right\}^T\{d\boldsymbol{\alpha}\} + \frac{\partial F}{\partial W_p}\,dW_p \tag{17.4-4}$$

From Eqs. 17.4-3 we obtain $\{d\boldsymbol{\alpha}\} = C\{d\boldsymbol{\epsilon}^p\}$ and $dW_p = \{\boldsymbol{\sigma}\}^T\{d\boldsymbol{\epsilon}^p\}$. In addition, in multidimensional analogy to Eq. 17.3-1, we have

$$\{d\boldsymbol{\sigma}\} = [\mathbf{E}]\{d\boldsymbol{\epsilon}^e\} = [\mathbf{E}](\{d\boldsymbol{\epsilon}\} - \{d\boldsymbol{\epsilon}^p\}) \tag{17.4-5}$$

where $\{d\boldsymbol{\epsilon}\}$ is assumed to contain no creep or thermal strains (see Eq. 17.3-5). Making these substitutions into Eq. 17.4-4, using Eq. 17.4-2 to eliminate $\{d\boldsymbol{\epsilon}^p\}$, and solving for the plastic multiplier $d\lambda$, we obtain

$$d\lambda = \{\mathbf{C}_\lambda\}^T\{d\boldsymbol{\epsilon}\} \tag{17.4-6}$$

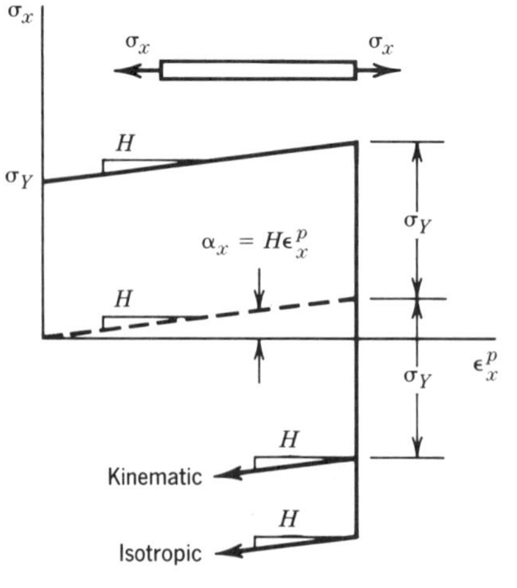

Figure 17.4-1. Stress versus plastic strain in uniaxial stress.

where

$$\{\mathbf{C}_\lambda\}^T = \frac{\left\{\dfrac{\partial F}{\partial \boldsymbol{\sigma}}\right\}^T [\mathbf{E}]}{\left\{\dfrac{\partial F}{\partial \boldsymbol{\sigma}}\right\}^T [\mathbf{E}] \left\{\dfrac{\partial Q}{\partial \boldsymbol{\sigma}}\right\} - C \left\{\dfrac{\partial F}{\partial \boldsymbol{\alpha}}\right\}^T \left\{\dfrac{\partial Q}{\partial \boldsymbol{\sigma}}\right\} - \dfrac{\partial F}{\partial W_p} \{\boldsymbol{\sigma}\}^T \left\{\dfrac{\partial Q}{\partial \boldsymbol{\sigma}}\right\}} \tag{17.4-7}$$

Finally, substituting Eq. 17.4-2 into Eq. 17.4-5, we obtain

$$\{d\boldsymbol{\sigma}\} = [\mathbf{E}]\left(\{d\boldsymbol{\epsilon}\} - \left\{\frac{\partial Q}{\partial \boldsymbol{\sigma}}\right\} d\lambda\right) \quad \text{or} \quad \{d\boldsymbol{\sigma}\} = [\mathbf{E}_{\text{ep}}]\{d\boldsymbol{\epsilon}\} \tag{17.4-8}$$

where

$$[\mathbf{E}_{\text{ep}}] = [\mathbf{E}] - [\mathbf{E}] \left\{\frac{\partial Q}{\partial \boldsymbol{\sigma}}\right\} \{\mathbf{C}_\lambda\}^T \tag{17.4-9}$$

Equation 17.4-9 can be regarded as a generalized form of tangent modulus E_t (see Eq. 17.3-2). Matrix $[\mathbf{E}_{\text{ep}}]$ is symmetric if $F = Q$. It is valid even if the material is elastic–perfectly plastic. It can be used to generate a tangent-stiffness matrix $[\mathbf{k}_t]$, which expresses the relation between increments of nodal displacement and the resulting increments of nodal load,

$$[\mathbf{k}_t] = \int_{V_e} [\mathbf{B}]^T [\mathbf{E}_{\text{ep}}][\mathbf{B}]\, dV \tag{17.4-10}$$

where $[\mathbf{E}_{\text{ep}}]$ is given by Eq. 17.4-9 if $F = 0$ and $dF = 0$, but is replaced by elastic coefficients $[\mathbf{E}]$ if $F < 0$ or if $dF < 0$.

The von Mises Criterion, Kinematic Hardening. Equation 17.4-9 does not presuppose particular forms of F and Q. Commonly used forms are those of the von Mises yield criterion and its associated flow rule. These forms are popular for analysis of isotropic ductile metals.

To begin, we must introduce *deviatoric stresses* $\{\mathbf{s}\}$, which are associated with distortion of shape but produce no volume change. By definition,

$$\{\mathbf{s}\} = \{\boldsymbol{\sigma}\} - \sigma_m \lfloor 1 \quad 1 \quad 1 \quad 0 \quad 0 \quad 0 \rfloor^T \quad \text{where} \quad \sigma_m = \tfrac{1}{3}(\sigma_x + \sigma_y + \sigma_z) \tag{17.4-11}$$

Stress σ_m is the mean or average normal stress. Thus $s_x = \sigma_x - \sigma_m, \ldots,$ $s_{zx} = \tau_{zx}$. For convenience, we define portions $\{\mathbf{s}_\sigma\}$ and $\{\mathbf{s}_\tau\}$ of $\{\mathbf{s}\}$ as follows:

$$\{\mathbf{s}\} = \begin{Bmatrix} \mathbf{s}_\sigma \\ \mathbf{s}_\tau \end{Bmatrix} \quad \text{where} \quad \{\mathbf{s}_\sigma\} = \begin{Bmatrix} s_x \\ s_y \\ s_z \end{Bmatrix} \quad \text{and} \quad \{\mathbf{s}_\tau\} = \begin{Bmatrix} \tau_{xy} \\ \tau_{yz} \\ \tau_{zx} \end{Bmatrix} \tag{17.4-12}$$

Similarly, $\{\boldsymbol{\alpha}\}$ of Eq. 17.4-3 is split into portions $\{\boldsymbol{\alpha}_\sigma\}$ and $\{\boldsymbol{\alpha}_\tau\}$. With σ_Y the yield strength in a uniaxial tensile test, the yield function is

$$F = [\tfrac{3}{2}(\{s_\sigma\} - \{\boldsymbol{\alpha}_\sigma\})^T(\{s_\sigma\} - \{\boldsymbol{\alpha}_\sigma\}) + 3(\{s_\tau\} - \{\boldsymbol{\alpha}_\tau\})^T(\{s_\tau\} - \{\boldsymbol{\alpha}_\tau\})]^{1/2} - \sigma_Y \tag{17.4-13}$$

in which the positive root of the bracketed expression is intended. As before, σ_Y is taken as the *initial* yield strength (unchanged by subsequent plastic strains). For uniaxial stress σ_x, with $\{\boldsymbol{\alpha}\}$ initially zero, Eq. 17.4-13 reduces to $F = |\sigma_x| - \sigma_Y$, so that $|\sigma_x| = \sigma_Y$ defines the onset of yielding.

To obtain an "associated" theory, we take $Q = F$. Thus, after some manipulation,

$$\{d\boldsymbol{\epsilon}^p\} = \left\{\frac{\partial F}{\partial \boldsymbol{\sigma}}\right\} d\lambda = \left(\frac{3}{2\sigma_Y}\begin{Bmatrix} s_\sigma - \boldsymbol{\alpha}_\sigma \\ \mathbf{0} \end{Bmatrix} + \frac{3}{\sigma_Y}\begin{Bmatrix} \mathbf{0} \\ s_\tau - \boldsymbol{\alpha}_\tau \end{Bmatrix}\right) d\lambda \tag{17.4-14}$$

which is known as the Prandtl–Reuss relation. Similarly, one concludes that $\{\partial Q/\partial \boldsymbol{\alpha}\} = -\{\partial F/\partial \boldsymbol{\sigma}\}$. Because isotropic hardening is omitted in this example, F does not contain W_p, so $\partial F/\partial W_p = 0$ in Eq. 17.4-7.

All quantities necessary for the construction of an elastic–plastic solution algorithm are now at hand. An algorithm is outlined in Section 17.5.

Similar but specialized relations may be written for elastic–plastic problems of plates, in which the material may carry in-plane loads as well as bending loads [17.15]. Without such specialized relations, a thickness-direction numerical integration is required in each computational cycle, which is quite expensive.

If the postyield portion of the stress–strain relation is not to be idealized as a straight line, one must store the following data for an isotropic material: E, ν, σ_Y, and a functional or tabular representation of H or E_t versus ϵ^p_{ef}, where ϵ^p_{ef} is an effective plastic strain defined by

$$\epsilon^p_{\text{ef}} = \frac{\sqrt{2}}{3}\Big[(\epsilon^p_x - \epsilon^p_y)^2 + (\epsilon^p_y - \epsilon^p_z)^2 + (\epsilon^p_z - \epsilon^p_x)^2 + \frac{3}{2}\{(\gamma^p_{xy})^2 + (\gamma^p_{yz})^2 + (\gamma^p_{zx})^2\}\Big]^{1/2} \tag{17.4-15}$$

in which the positive root of the bracketed expression is intended. In the plastic range where Poisson's ratio is 0.5, uniaxial stress σ_x produces $\epsilon^p_{\text{ef}} = |\epsilon^p_x|$, so that data from a tension test are easily plotted and converted to a numerical representation. In computations with multiaxial states of stress and strain, all terms in Eq. 17.4-15 may be needed to compute ϵ^p_{ef}.

Specialization to Uniaxial Stress. Let σ_x be the only nonzero stress in $\{\sigma\}$. For kinematic hardening, with σ_Y the initial yield strength,

$$F = Q = [(\sigma_x - \alpha_x)^2]^{1/2} - \sigma_Y = |\sigma_x - \alpha_x| - \sigma_Y \tag{17.4-16}$$

Hence, with "sgn" denoting "the sign of,"

$$\frac{\partial F}{\partial \sigma_x} = \frac{\partial Q}{\partial \sigma_x} = \operatorname{sgn}(\sigma_x - \alpha_x) \quad \text{and} \quad \frac{\partial F}{\partial \alpha_x} = -\operatorname{sgn}(\sigma_x - \alpha_x) \tag{17.4-17}$$

In addition, from Fig. 17.4-1, $\alpha_x = H\epsilon_x^p$; that is, $C = H$. Accordingly, with the term containing W_p in Eq. 17.4-7 set to zero, we obtain from Eqs. 17.4-2 and 17.4-6

$$d\epsilon_x^p = \frac{\partial Q}{\partial \sigma_x} d\lambda = \frac{\partial Q}{\partial \sigma_x} C_\lambda \, d\epsilon_x = \frac{E}{E + H} d\epsilon_x \qquad (17.4\text{-}18)$$

which agrees with Eq. 17.3-6. From Eq. 17.4-9 we obtain

$$E_{\text{ep}} = E - E\frac{E}{E + H} = E\left(1 - \frac{E}{E + H}\right) \qquad (17.4\text{-}19)$$

which agrees with Eq. 17.3-2.

17.5 ELASTIC–PLASTIC ANALYSIS PROCEDURES

In the present section we summarize the tangent-stiffness method and the initial-stiffness method. The same two algorithms are discussed in a one-dimensional context in Section 17.3. The loading history and the geometry, support conditions, and material properties are assumed to be known. We seek the deformations and stresses in the body as a function of load. With either solution method, the load is incremented in several steps. The tangent-stiffness method allows large but expensive steps, while the initial-stiffness method uses small but inexpensive steps. It is not always clear which method will be better in particular problems. Detailed discussion of algorithms may be found in [17.11–17.23].

When the material behavior is nonlinear, material properties in an element are dictated by material properties at a finite number of sampling points in each element. Typically these points are quadrature stations of a numerical integration rule. At each point one must keep a record of strains and update the record in each computational cycle. The number of points must be small to reduce computational expense. Accordingly, some analysts prefer simple elements, which may require only one sampling point per element. A contrary argument is that many sampling points are needed to accurately capture the spread of yielding in individual elements. In simple terms, the choice is between many simple elements and a smaller number of more sophisticated elements.

In what follows we will assume that strain increments $\{d\epsilon\}$ include elastic components $\{d\epsilon^e\}$ and plastic components $\{d\epsilon^p\}$, but that thermal strains $\{d\epsilon^T\}$ and creep strains $\{d\epsilon^C\}$ have already been subtracted out.

We presume that a tensile test of the material has been performed, and a numerical representation of its stress–strain curve is stored. We also presume that specific choices of yield criterion, flow rule, and hardening rule have been made. If we choose the von Mises yield criterion, the Prandtl–Reuss flow relations, either kinematic or isotropic hardening, and a bilinear stress–strain relation, then we need store only E, ν, σ_Y, and either E_t or H for an isotropic material. Alternatively, to represent a more general stress–strain relation, either E_t or H may be defined as a function of ϵ_{ef}^p (Eq. 17.4-15). Then, in computation, we must record and update the value of ϵ_{ef}^{p} at each sampling point, and use it to obtain the current value of E_t or H.

Tangent-Stiffness Method. Loads $\{\mathbf{R}\}$ on the structure are applied in increments $\{\Delta\mathbf{R}\}_1$, $\{\Delta\mathbf{R}\}_2$, and so on, so that $\{\mathbf{R}\} = \Sigma\,\{\Delta\mathbf{R}\}_i$. The first load increment might be contrived to place only the most highly stressed sampling point on the verge of yield, but we will not make this assumption. Procedural steps are as follows.

1. At the outset, $\{\boldsymbol{\epsilon}\} = \{\boldsymbol{\sigma}\} = \{\boldsymbol{\alpha}\} = \{\mathbf{0}\}$, $W_p = 0$, and $[\mathbf{E}_{\text{ep}}] = [\mathbf{E}]$ for all sampling points. These values prevail in the first computational cycle ($i = 1$). Apply the first load increment, $\{\Delta\mathbf{R}\}_1$.

2. Use the current conditions $\{\boldsymbol{\sigma}\}_{i-1}$, $\{\boldsymbol{\alpha}\}_{i-1}$, and $W_{p(i-1)}$ to evaluate $[\mathbf{E}_{\text{ep}}]_{i-1}$ for each sampling point. Note that $[\mathbf{E}_{\text{ep}}]_{i-1} = [\mathbf{E}]$ for sampling points that have yet to yield ($F < 0$ for the current $\{\boldsymbol{\sigma}\}_{i-1}$, $\{\boldsymbol{\alpha}\}_{i-1}$, and $W_{p(i-1)}$) or are unloading ($dF < 0$ for the most recent *changes* in $\{\boldsymbol{\sigma}\}$, $\{\boldsymbol{\alpha}\}$, and W_p). Evaluate $[\mathbf{k}_t]$ for each element n. The structure tangent-stiffness matrix is formed by the usual assembly, $[\mathbf{K}_t]_{i-1} = \Sigma\,[\mathbf{k}_t]_n$. Solve for structure displacement increments $\{\Delta\mathbf{D}\}_i$ and strain increments $\{\Delta\boldsymbol{\epsilon}\}_i$ at element sampling points from the equations

$$[\mathbf{K}_t]_{i-1}\{\Delta\mathbf{D}\}_i = \{\Delta\mathbf{R}\}_{i-1} \qquad \text{and} \qquad \{\Delta\boldsymbol{\epsilon}\}_i = [\mathbf{B}]\{\Delta\mathbf{d}\}_i \tag{17.5-1}$$

For sampling points in the plastic range, compute increments as follows.

From Eq. 17.4-6:
$$\Delta\lambda_i = \int \{\mathbf{C}_\lambda\}^T\{d\boldsymbol{\epsilon}\} \approx \{\mathbf{C}_\lambda\}_{i-1}^T\{\Delta\boldsymbol{\epsilon}\}_i \tag{17.5-2}$$

From Eq. 17.4-2:
$$\{\Delta\boldsymbol{\epsilon}^p\}_i = \int \left\{\frac{\partial Q}{\partial\boldsymbol{\sigma}}\right\} d\lambda \approx \left\{\frac{\partial Q}{\partial\boldsymbol{\sigma}}\right\}_{i-1} \Delta\lambda_i \tag{17.5-3}$$

From Eq. 17.4-5:
$$\{\Delta\boldsymbol{\sigma}\}_i = [\mathbf{E}](\{\Delta\boldsymbol{\epsilon}\}_i - \{\Delta\boldsymbol{\epsilon}^p\}_i) \tag{17.5-4}$$

From Eq. 17.4-3:
$$\{\Delta\boldsymbol{\alpha}\}_i = \int C\{d\boldsymbol{\epsilon}^p\} \approx C\{\Delta\boldsymbol{\epsilon}^p\}_i \tag{17.5-5}$$

From Eq. 17.4-3:
$$\Delta W_{pi} = \int \{\boldsymbol{\sigma}\}^T\{d\boldsymbol{\epsilon}^p\} \approx \{\boldsymbol{\sigma}\}_i^T\{\Delta\boldsymbol{\epsilon}^p\}_i \tag{17.5-6}$$

Typically, a solution will use $\{\boldsymbol{\alpha}\}$ or W_p but not both. For sampling points in the elastic range, Eqs. 17.5-2 through 17.5-6 are not used. Instead, one computes only $\{\Delta\boldsymbol{\sigma}\} = [\mathbf{E}]\{\Delta\boldsymbol{\epsilon}\}$.

3a. *Optional.* For sampling points that make the elastic-to-plastic transition, compute the fraction m of the current increment that is elastic. For example, writing Eq. 17.4-13 in the form $F = Y - \sigma_Y$,

$$m = \frac{\sigma_Y - Y_{i-1}}{Y_i - Y_{i-1}} \tag{17.5-7}$$

where Y_i is based on current stresses, computed by use of elastic coefficients, and temporarily updated for this step only. The revised $[\mathbf{E}_{ep}]_{i-1}$ is $m[\mathbf{E}]$ plus $(1 - m)$ times the right-hand side of Eq. 17.4-9; this reduces to

$$[\mathbf{E}_{ep}]_{i-1} = [\mathbf{E}] - (1 - m)[\mathbf{E}] \left\{ \frac{\partial Q}{\partial \boldsymbol{\sigma}} \right\}_{i-1} \{\mathbf{C}_\lambda\}_{i-1}^T \tag{17.5-8}$$

Repeat steps 2 and 3a until convergence but without making the updates final (step 4 below) until convergence. In applying Eq. 17.5-2, use $(1 - m)\{\Delta\boldsymbol{\epsilon}\}_i$ rather than $\{\Delta\boldsymbol{\epsilon}\}_i$, as $\{\mathbf{C}_\lambda\}_{i-1}$ is zero for the elastic portion of the increment.

3b. *Optional.* Without changing the load or recalculating $\{\Delta\mathbf{D}\}_i$, one can evaluate Eqs. 17.5-2 through 17.5-6 more accurately by dividing the increment $\{\Delta\boldsymbol{\epsilon}\}_i$ into subincrements. After each such subincremental cycle, one updates $\{\boldsymbol{\sigma}\}_i$, $\{\boldsymbol{\alpha}\}_i$, and so on (Eqs. 17.5-10). Note that $\{\mathbf{C}_\lambda\}_{i-1}$ is zero in elastic subincrements if the sampling point makes the elastic-to-plastic transition within the current load step.

3c. *Optional (but recommended).* "Corrective" loads are introduced to prevent progressive drift, as discussed in connection with Figs. 17.2-5 and 17.2-6. Thus the *next* load increment is not simply $\{\Delta\mathbf{R}\}_{i+1}$; rather, it is

$$\{\Delta\mathbf{R}\}_{i+1} + \{\Delta\mathbf{R}_c\}_i \quad \text{where} \quad \{\Delta\mathbf{R}_c\} = \{\mathbf{R}\}_i - \sum \int_{V_e} [\mathbf{B}]^T\{\boldsymbol{\sigma}\}_i \, dV \tag{17.5-9}$$

and $\{\mathbf{R}\}_i$ is the *total* externally applied load in cycle i. The summation spans all elements of the structure and expresses the loads that elements apply to nodes because they have stresses $\{\boldsymbol{\sigma}\}_i$. Stresses $\{\boldsymbol{\sigma}\}_i$ are updated values (Eqs. 17.5-10).

4. Update the solution:

$$\begin{aligned} \{\mathbf{D}\}_i &= \{\mathbf{D}\}_{i-1} + \{\Delta\mathbf{D}\}_i \qquad & \{\boldsymbol{\sigma}\}_i &= \{\boldsymbol{\sigma}\}_{i-1} + \{\Delta\boldsymbol{\sigma}\}_i \\ \{\boldsymbol{\alpha}\}_i &= \{\boldsymbol{\alpha}\}_{i-1} + \{\Delta\boldsymbol{\alpha}\}_i \qquad & (W_p)_i &= (W_p)_{i-1} + (\Delta W_p)_i \end{aligned} \tag{17.5-10}$$

5. Apply the next load increment and return to step 2.

6. Stop when $\Sigma\ \{\Delta\mathbf{R}\}_i$ reaches the total applied load.

Exercising steps 3a, 3b, and/or 3c permits load increments to be larger without increasing the error. Step 3c can be exercised repeatedly *within* a given load increment, either with or without updates in the structure matrix $[\mathbf{K}_t]$; the effect is depicted in Figs. 17.2-3 and 17.2-4. If step 3c is used in this way there should be no test for unloading ($dF < 0$) until these cycles are complete because intermediate cycles can give false indications. Use of step 3b within this cycling gives a three-level process: subincrements within iterations within load increments.

Flow rules often allow little or no volume change. Thus, if plastic strains become extremely large, the response becomes nearly incompressible, and fully integrated elements may encounter numerical difficulties associated with locking of the mesh. Use of selective reduced integration is recommended.

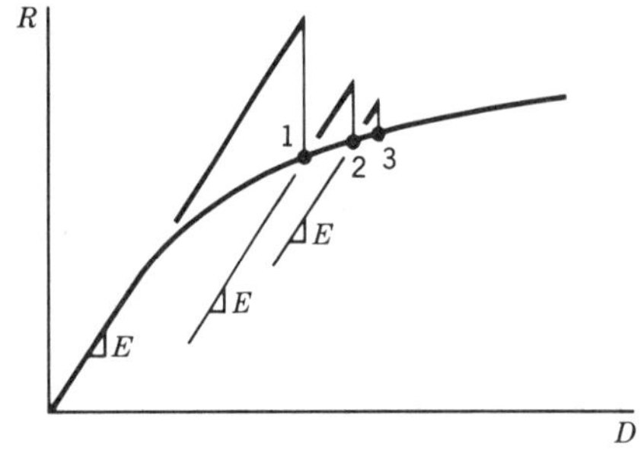

Figure 17.5-1. Load versus displacement plot for a representative d.o.f. D in a multi-d.o.f. model, showing convergence of the initial stiffness method.

Initial-Stiffness Method. The step-by-step procedure remains as outlined in Section 17.3. We need make only a few modifications in step 5 to allow for the multidimensionality of stress. Specifically, in place of Eq. 17.3-6, we use Eqs. 17.5-1 through 17.5-6 to obtain plastic strain increments and update the yield criterion and the flow rule. Fraction m, if used, is given by Eq. 17.5-7. These calculations and updates are performed at each sampling point that sustains plastic strains. Equation 17.3-7 is replaced by

$$\{\Delta \mathbf{R}_s\} = \sum \{\Delta \mathbf{r}_s\} \qquad \text{where} \quad \{\Delta \mathbf{r}_s\} = \int_{V_e} [\mathbf{B}]^T[\mathbf{E}]\{\Delta \boldsymbol{\epsilon}^p\}\, dV \qquad (17.5\text{-}11)$$

Note that sampling points that unload in a computational cycle ($\Delta F < 0$) make no contribution to $\{\Delta \mathbf{R}_s\}$.

In the one-dimensional problem of Fig. 17.3-2a, the applied load must be carried by every element. The initial-stiffness method then converges slowly (Fig. 17.3-3b). In conditions of multiaxial stress, the yielding of one element is accompanied by a transfer of load to other elements, and the initial-stiffness method converges in fewer iterations (Fig. 17.5-1).

17.6 NONLINEAR DYNAMIC PROBLEMS

When the frequency of excitation exceeds about one-third the structure's lowest natural frequency of vibration, inertia becomes important and the problem is dynamic rather than quasistatic. Of the methods of response history analysis discussed in Chapter 13, direct time integration methods are usually the most effective for nonlinear problems. In this section, explicit and implicit methods are briefly discussed with particular emphasis on nonlinearities due to plasticity. Most remarks, however, are true of material nonlinearities in general. Direct integration methods that also include geometric nonlinearities are discussed [17.24]. It is assumed that the reader is familiar with Sections 13.9 through 13.14.

Explicit Methods. Treatment of nonlinearities by explicit methods is usually straightforward, accurate, and effective. All remarks of Section 13.10 are valid, but elaboration is necessary for nonlinear problems, as follows.

Explicit methods require that the internal force of each element, $\{\mathbf{r}^{\text{int}}\}_n$, be calculated before the new displacement $\{\mathbf{D}\}_{n+1}$ can be computed. Element-by-element calculation of $\{\mathbf{r}^{\text{int}}\}_n$ using Eq. 13.2-7 requires that element stresses $\{\boldsymbol{\sigma}\}_n$ be known. For linear problems, $\{\boldsymbol{\sigma}\}_n = [\mathbf{E}][\mathbf{B}]\{\mathbf{D}\}_n$ in which $\{\mathbf{D}\}_n$ is known. For plasticity, stress increment $\{d\boldsymbol{\sigma}\} \approx \{\Delta \boldsymbol{\sigma}\}$ can be computed from the strain increment, $\{\Delta\epsilon\} = [\mathbf{B}](\{\mathbf{D}\}_n - \{\mathbf{D}\}_{n-1})$, and the constitutive law, Eq. 17.4-8. Hence, the stress at time $n\ \Delta t$ is given by $\{\sigma\}_n = \{\sigma\}_{n-1} + \{\Delta\sigma\}$ and $\{\mathbf{r}^{\text{int}}\}_n$ can be obtained from Eq. 13.2-7.

As with linear problems, the accuracy of an explicit solution is usually assured when the time-step stability criterion is satisfied. In Chapter 13, stability criteria are cited for several explicit methods as applied to *linear* problems, for example, Eqs. 13.10-8, 13.10-14, and 13.12-6 with $\beta = 0$. Extensive computational experience suggests that these criteria are also valid for nonlinear problems provided that one uses the instantaneous value of ω_{max}, which is a function of material properties, element geometry, and mesh geometry.

Many materials display "softening" behavior in which the tangent modulus decreases with increasing stress or strain level. Examples of such materials are ductile and brittle solids. For these materials, it can often be shown that the instantaneous value of ω_{max} will not exceed the ω_{max} for linearly elastic response [17.25]. Thus, a time step that is stable for purely elastic response will also be stable for nonlinear response. Alternatively, it is possible to change Δt during a problem solution by continuously monitoring ω_{max} through the use of an element frequency bound. However, for most problems the possible increase in Δt is small, as ω_{max} often continues to be governed by elements of the mesh that experience little or no plastic deformation.

A material that displays "stiffening" has a tangent modulus that exceeds its initial modulus. In such a material ω_{max} may increase, making the stability criterion more restrictive. In problems with geometric nonlinearity, ω_{max} may increase or decrease. In these situations it is usually necessary to monitor ω_{max} through the use of element bounds during the course of a computational solution [17.24].

Numerical instability is usually easy to detect in linear problems because the solution grows without limit. In nonlinear problems, with elastic–plastic or other energy-dissipating materials, extra energy introduced into the system by the numerical instability may be dissipated by plastic work or some other irreversible mechanism so that it is possible for the instability to be arrested [17.26]. An "arrested instability" is often difficult to detect because the solution, although in error by 10% to 100% or more, may appear to be reasonable.

Energy Balance Check. In the analysis of nonlinear dynamic problems by explicit methods, it is usually advisable to perform an energy balance check to help assure stable and accurate computation. Ideally, the energy at time $(n + 1)\ \Delta t$ in a system started from at-rest initial conditions should satisfy the equation

$$W_{n+1}^{\text{int}} + T_{n+1} = W_{n+1}^{\text{ext}} \tag{17.6-1}$$

where W represents work and T represents kinetic energy. Physically, Eq. 17.6-1 states that the work of external loads is converted to kinetic energy and to energy either stored elastically or dissipated by plastic deformations. The separate terms in Eq. 17.6-1 are explained as follows.

The internal work, W_{n+1}^{int}, represents the work done by nodal loads that are developed from straining of material and is given by

$$W_{n+1}^{\text{int}} = W_n^{\text{int}} + \int_{n\,\Delta t}^{(n+1)\,\Delta t} \dot{W}^{\text{int}}\,dt \tag{17.6-2}$$

Noting that $\dot{W}_n^{\text{int}} = \{\dot{\mathbf{D}}\}_n^T\{\mathbf{R}^{\text{int}}\}_n$ and approximating the integral in Eq. 17.6-2 by the trapezoidal rule, we obtain

$$W_{n+1}^{\text{int}} = W_n^{\text{int}} + \frac{\Delta t}{2}\left(\{\dot{\mathbf{D}}\}_n^T\{\mathbf{R}^{\text{int}}\}_n + \{\dot{\mathbf{D}}\}_{n+1}^T\{\mathbf{R}^{\text{int}}\}_{n+1}\right) \tag{17.6-3}$$

Equation 17.6-3 is appropriate for use with explicit methods that compute velocities at whole time steps (e.g., the Newmark method with $\beta = 0$). When velocities are known at half time steps, as in the central-difference method, then Eq. 17.6-3 can be written as

$$W_{n+1}^{\text{int}} = W_n^{\text{int}} + \frac{\Delta t}{2}\{\dot{\mathbf{D}}\}_{n+1/2}^T\left(\{\mathbf{R}^{\text{int}}\}_n + \{\mathbf{R}^{\text{int}}\}_{n+1}\right) \tag{17.6-4}$$

The external work, W_{n+1}^{ext}, represents the work of the externally applied loads and is given by

$$W_{n+1}^{\text{ext}} = W_n^{\text{ext}} + \int_{n\,\Delta t}^{(n+1)\,\Delta t} \{\dot{\mathbf{D}}\}^T\{\mathbf{R}^{\text{ext}}\}\,dt \tag{17.6-5}$$

Difference expressions for W_{n+1}^{ext} can be obtained from Eqs. 17.6-3 and 17.6-4 by replacing superscript "int" by "ext." The kinetic energy, T_n, is given by

$$T_n = \tfrac{1}{2}\{\dot{\mathbf{D}}\}_n^T[\mathbf{M}]\{\dot{\mathbf{D}}\}_n \tag{17.6-6}$$

or, if half-time-step velocities are known, by

$$T_n = \tfrac{1}{2}(T_{n-1/2} + T_{n+1/2}) \tag{17.6-7}$$

To construct an energy balance, we note that, in general, Eq. 17.6-1 is not satisfied exactly. To measure the quality of a solution, we can use

$$W_n^{\text{int}} + T_n - |W_n^{\text{ext}}| \le e(W_n^{\text{int}} + T_n + |W_n^{\text{ext}}|) \tag{17.6-8}$$

where e is a tolerance and absolute magnitude bars are a precaution to avoid small negative values of W^{ext} due to numerical errors. Terms within parentheses on the right-hand side of Eq. 17.6-8 represent the total energy in the system. The left-hand side is the energy error. A stable explicit computation should satisfy Eq. 17.6-8 with $e \le 0.02$ [17.24]. If satisfaction of Eq. 17.6-8 requires $e \ge 0.05$, even for models with hundreds of elements and using thousands of time steps, then instability should be suspected.

Example: Dynamic Plasticity, Explicit Algorithm. Consider the example of Section 13.10 but with the elastic–plastic isotropic hardening model of steel depicted in Fig. 17.3-1b, for which $E = 30(10^6)$ psi, $E_t = E/4$, and $\sigma_Y = 40(10^3)$ psi. The 20-in. bar, shown in Fig. 13.10-2, is initially at rest and is modeled by 40 equal-length linear-displacement elements, so that $L = L_T/40 = 0.5$ in. The tip loading is as shown in Fig. 13.10-2, except that $P_0 = 2A\sigma_Y = 80(10^3)$ lbs. Since $E_t < E$, a time step that is stable for linear elastic response will also be stable for nonlinear response. Hence, with $\rho = 7.4(10^{-4})$ lb-sec^2/in.^{4}, the highest element frequency is $(\omega_{\max})_e = 2\sqrt{E/\rho}/L = 8.054(10^5)$ rad/sec. Thus stable integration by the central-difference method requires $\Delta t \leq 2/(\omega_{\max})_e = 2.483(10^{-6})$ sec according to Eq. 13.10-14 with $\xi = 0$.

The computational procedure in Table 13.10-1 and the Fortran program in Fig. 13.10-3 were used except that Subroutine INTFOR in Fig. 13.10-3 was replaced by the code in Fig. 17.6-1 (which does not include the energy balance check of Eq. 17.6-8). Additional alterations to the main program in Fig. 13.10-3 include adding SIGMA(101) and SIGYLD(101) to the DIMENSION statement, adding the tangent modulus (ET) and initial yield strength (σ_Y in Fig. 17.3-1) to the READ statement, initializing the stress in each element to zero and the yield strength in each element to σ_Y, modifying the tip load, the CALL INTFOR statement, and the WRITE statement.

The Fortran code in Fig. 17.6-1 generates the internal force vector for a model with elastic–plastic behavior. The yield function, called YLDFUN in Fig. 17.6-1, is obtained from the discussion following Eq. 17.4-13 as $F = |\sigma_x| - \sigma_Y$, where σ_x in element I is called SIGMA(I) in Fig. 17.6-1. Because of isotropic hardening, the stress required for renewed or continued yielding exceeds σ_Y; this stress is called σ_B in Fig. 17.3-1*b* and SIGYLD(I) in Fig. 17.6-1. The code uses modulus E when operating on the linearly elastic part of the stress–strain diagram shown in Fig. 17.3-1 and modulus ET when operating on the plastic part of the diagram. When states of stress make the transition from elastic

```
      SUBROUTINE INTFOR(D,DOLD,FINT,SIGMA,X,SIGYLD,E,ET,CSA,NELE)
      IMPLICIT DOUBLE PRECISION (A-H,O-Z)
      DIMENSION D(1),DOLD(1),FINT(1),SIGMA(1),X(1),SIGYLD(1)
      NUMNOD=NELE+1
C---- Zero internal force vector.
      DO 10 I=1,NUMNOD
   10 FINT(I)=0.
C---- Loop over elements.
      DO 30 K=1,NELE
      RL=X(K+1)-X(K)
C---- Compute strain increment.
      EPSOLD=(DOLD(K+1)-DOLD(K))/RL
      EPSNEW=(   D(K+1)-   D(K))/RL
      EPSINC=EPSNEW-EPSOLD
C---- Compute stress increment (DSIGMA) assuming elastic response.
      DSIGMA=E*EPSINC
C---- Check if DSIGMA satisfies yield function  .LT. 0.
      YLDFUN=DABS(SIGMA(K)+DSIGMA)-SIGYLD(K)
      IF (YLDFUN.LT.0.) GO TO 20
C---- If material was plastic before strain increment, and
C---- YLDFUN .GE. 0., then entire strain increment is plastic.
      IF (DABS(SIGMA(K)).EQ.SIGYLD(K)) THEN
      DSIGMA=ET*EPSINC
      GO TO 20
      ELSE
C---- Compute the portion of the strain increment that is elastic.
      RATIO=(SIGYLD(K)-DABS(SIGMA(K)))/DABS(DSIGMA)
      DSIGMA=RATIO*EPSINC*E + (1.-RATIO)*EPSINC*ET
      ENDIF
C---- Update stress and account for isotropic hardening.
   20 CONTINUE
      SIGMA(K)=SIGMA(K)+DSIGMA
      IF (DABS(SIGMA(K)).GT.SIGYLD(K)) SIGYLD(K)=DABS(SIGMA(K))
C---- Assemble contribution into internal force vector.
      F1=-SIGMA(K)*CSA
      F2=+SIGMA(K)*CSA
      FINT(K)=FINT(K)+F1
      FINT(K+1)=FINT(K+1)+F2
   30 CONTINUE
      RETURN
      END
```

Figure 17.6-1. Fortran program to compute the internal force vector for a mesh of linear displacement bar elements with elastic–plastic, isotropic-hardening material.

to plastic, the fraction of the strain increment that is elastic is computed according to the procedure described in step 3a of Section 17.5. In the Fortran code, the elastic fraction of the strain increment, m in Eq. 17.5-7, is called RATIO.

The stress time-history results for the midpoint of element 20 (at $x = 9.75$ in.) are shown in Fig. 17.6-2 for $\Delta t = 2.4(10^{-6})$ sec ($C_n = 0.966$) and 83 time steps. Two separate stress waves arrive at different times. These correspond to an elastic wave traveling at speed $c_e = \sqrt{E/\rho}$, which arrives first, followed by a plastic wave traveling at speed $c_p = \sqrt{E_t/\rho} = \frac{1}{2}c_e$. Thus, at $x = 9.75$ in., an elastic wave arrives at $t = 0.0484$ msec and a plastic wave arrives at $t = 0.0968$ msec.

The solution is devoid of noise until after the passage of the plastic wave. The reason is that during the initial passage of the elastic stress wave, numerical noise causes stress excursions above the initial yield stress. These excursions are transmitted at the plastic wave speed, and hence do not arrive at element 20 until approximately 0.11 msec has elapsed.

Implicit Methods. Compared with explicit methods, implicit methods for nonlinear problems are less attractive in every respect save one—that is, the possibility of using large time steps permitted by the excellent stability properties of popular implicit methods. Nonlinearities present the same difficulty to both static and implicit dynamic solution algorithms: stiffness is a function of displacements, which are not known in advance. For example, in Eq. 13.11-5, $[\mathbf{K}^{\text{eff}}]$ is a function of $\{\mathbf{D}\}_{n+1}$, which is unknown. Methods for addressing this difficulty are analogous to the tangent-stiffness and initial-stiffness methods described in Section 17.3 for nonlinear quasistatic problems.

Tangent-Stiffness (Implicit) Method. In the tangent-stiffness method, the internal force in the equations of motion, Eq. 13.9-2, is written as

$$\{\mathbf{R}^{\text{int}}\}_{n+1} = \{\mathbf{R}^{\text{int}}\}_n + [\mathbf{K}_t]\{\Delta\mathbf{D}\} \tag{17.6-9}$$

where

$$\{\Delta\mathbf{D}\} = \{\mathbf{D}\}_{n+1} - \{\mathbf{D}\}_n \tag{17.6-10}$$

Combining Eqs. 17.6-9 and 17.6-10 with the equations of motion, Eq. 13.9-2, and the trapezoidal rule equations, Eqs. 13.11-3 and 13.11-4, we obtain

$$[\mathbf{K}^{\text{eff}}]\{\Delta\mathbf{D}\} = \{\mathbf{R}^{\text{eff}}\}_{n+1} \tag{17.6-11}$$

where

$$[\mathbf{K}^{\text{eff}}] = \frac{4}{\Delta t^2}[\mathbf{M}] + \frac{2}{\Delta t}[\mathbf{C}] + [\mathbf{K}_t] \tag{17.6-12}$$

and

$$\{\mathbf{R}^{\text{eff}}\}_{n+1} = \{\mathbf{R}^{\text{ext}}\}_{n+1} - \{\mathbf{R}^{\text{int}}\}_n + [\mathbf{M}]\left(\frac{4}{\Delta t}\{\dot{\mathbf{D}}\}_n + \{\ddot{\mathbf{D}}\}_n\right) + [\mathbf{C}]\{\dot{\mathbf{D}}\}_n \tag{17.6-13}$$

Note that $[\mathbf{K}_t]$ must be predicted using $\{\mathbf{D}\}_n$ (and possibly $\{\dot{\mathbf{D}}\}_n$ if strain rate effects are important) and must be factored at least once each time step during nonlinear

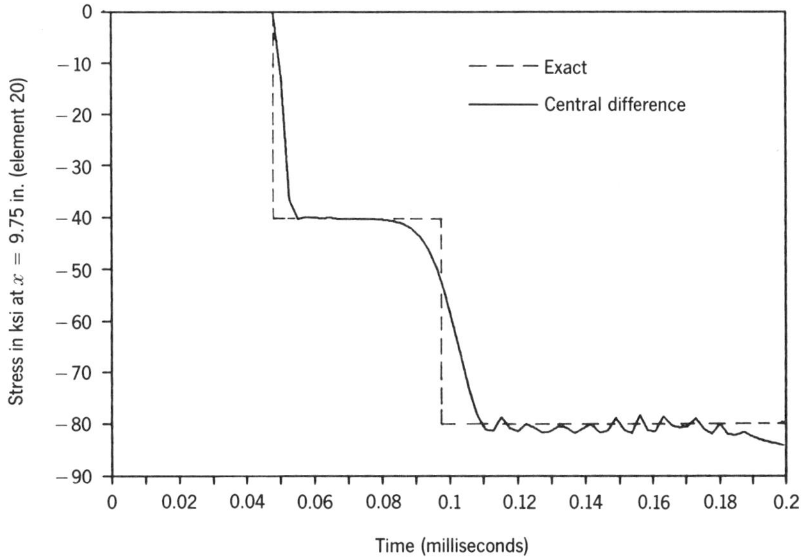

Figure 17.6-2. Stress versus time at $x = 9.75$ in. for a 40-element model of a 20-in. bar with elastic–plastic material using $\Delta t = 2.4(10^{-6})$ sec ($C_n = 0.966$). Figure 13.10-2 applies, except that $P_0 = 80{,}000$ lb.

response. If $[\mathbf{K}_t]$ is not an accurate prediction of the true tangent-stiffness matrix from time $n\ \Delta t$ to time $(n + 1)\ \Delta t$, then the solution of Eq. 17.6-11 for $\{\Delta\mathbf{D}\}$ will be in error. The error in nodal forces—that is, the residual—is given by the imbalance in the equation of motion (Eq. 13.9-2) as

$$\{\mathbf{R}^{\text{err}}\} = \{\mathbf{R}^{\text{ext}}\}_{n+1} - [\mathbf{M}]\{\ddot{\mathbf{D}}\}_{n+1} - [\mathbf{C}]\{\dot{\mathbf{D}}\}_{n+1} - \{\mathbf{R}^{\text{int}}\}_{n+1} \tag{17.6-14}$$

where $\{\mathbf{R}^{\text{int}}\}_{n+1}$ is computed using Eq. 17.6-9 and an improved tangent-stiffness matrix (i.e., $[\mathbf{K}_t]$ obtained using $\{\mathbf{D}\}_{n+1} = \{\mathbf{D}\}_n + \{\Delta\mathbf{D}\}$, which is obtained by solving Eq. 17.6-11). Alternatively, $\{\mathbf{R}^{\text{int}}\}_{n+1}$ can be computed element-by-element in the same way as for explicit direct integration. If measures are not taken to control the growth of $\{\mathbf{R}^{\text{err}}\}$, the solution will diverge in a manner similar to the instability displayed by explicit methods when the stability criterion is violated [17.26]. A computational procedure for the trapezoidal rule with error control is given in Table 17.6-1. Note that the procedure for error control is essentially the pseudo-load approach, described later in this section, applied within a time step. If enough iterations are performed within each time step to guarantee that W^{err} of Table 17.6-1 is bounded for the entire solution, then the trapezoidal rule algorithm is unconditionally stable, although not necessarily accurate [17.27]. To assure accuracy, W^{err} should be small.

Initial-Stiffness (Implicit) Method. In the initial-stiffness method, the initial tangent-stiffness matrix of the structure is used throughout the analysis and corrective loads due to nonlinearities are transferred to the right-hand side of the equations solved at each iteration. With this method, internal forces are given by

$$\{\mathbf{R}^{\text{int}}\}_{n+1} = [\mathbf{K}]\{\mathbf{D}\}_{n+1} - \{\Delta\mathbf{R}_s\}_{n+1} \tag{17.6-15}$$

TABLE 17.6-1. COMPUTATIONAL PROCEDURE FOR DIRECT INTEGRATION OF MATERIAL NONLINEAR PROBLEMS BY THE TRAPEZOIDAL RULE METHOD WITH TANGENTIAL STIFFNESS. SUPERPOSED TILDE ($\tilde{\ }$) DENOTES QUANTITIES OBTAINED FROM AN ESTIMATED $[\mathbf{K}_t]$.

1. Form $[\mathbf{C}]$ and $[\mathbf{M}]$ (also form $[\mathbf{K}_t]$ if $\{\mathbf{D}(t = 0)\} \neq \{\mathbf{0}\}$).
2. Set initial conditions $\{\mathbf{D}\}_0 = \{\mathbf{D}(t = 0)\}$ and $\{\dot{\mathbf{D}}\}_0 = \{\dot{\mathbf{D}}(t = 0)\}$; use Eq. 13.9-1 to compute $\{\ddot{\mathbf{D}}\}_0 = [\mathbf{M}]^{-1}(\{\mathbf{R}^{\text{ext}}\}_0 - [\mathbf{C}]\{\dot{\mathbf{D}}\}_0 - [\mathbf{K}]\{\mathbf{D}\}_0)$; use Eq. 13.2-7 to compute $\{\mathbf{R}^{\text{int}}\}_0$; $n = 0$.
3. Form $[\mathbf{K}_t]$ and compute $[\mathbf{K}^{\text{eff}}] = \dfrac{4}{\Delta t^2}[\mathbf{M}] + \dfrac{2}{\Delta t}[\mathbf{C}] + [\mathbf{K}_t]$.
4. Form $\{\mathbf{R}^{\text{eff}}\}_{n+1} = \{\mathbf{R}^{\text{ext}}\}_{n+1} - \{\mathbf{R}^{\text{int}}\}_n + [\mathbf{M}]\left(\dfrac{4}{\Delta t}\{\dot{\mathbf{D}}\}_n + \{\ddot{\mathbf{D}}\}_n\right) + [\mathbf{C}]\{\dot{\mathbf{D}}\}_n$.
5. Solve $[\mathbf{K}^{\text{eff}}]\{\Delta\tilde{\mathbf{D}}\} = \{\mathbf{R}^{\text{eff}}\}_{n+1}$ for $\{\Delta\tilde{\mathbf{D}}\}$.
6. Compute $\{\tilde{\dot{\mathbf{D}}}\}_{n+1} = \dfrac{2}{\Delta t}\{\Delta\tilde{\mathbf{D}}\} - \{\dot{\mathbf{D}}\}_n$ and $\{\tilde{\ddot{\mathbf{D}}}\}_{n+1} = \dfrac{4}{\Delta t^2}\{\Delta\tilde{\mathbf{D}}\} - \dfrac{4}{\Delta t}\{\dot{\mathbf{D}}\}_n - \{\ddot{\mathbf{D}}\}_n$.
7. Compute $\{\tilde{\mathbf{R}}^{\text{int}}\}_{n+1} = \sum \{\tilde{\mathbf{r}}^{\text{int}}\}_{n+1} = \sum \int [\mathbf{B}]^T\{\tilde{\sigma}\}_{n+1}\, dV$ using $\{\tilde{\mathbf{D}}\}_{n+1} = \{\mathbf{D}\}_n + \{\Delta\tilde{\mathbf{D}}\}$.
8. Compute $\{\mathbf{R}^{\text{err}}\} = \{\mathbf{R}^{\text{ext}}\}_{n+1} - [\mathbf{M}]\{\tilde{\ddot{\mathbf{D}}}\}_{n+1} - [\mathbf{C}]\{\tilde{\dot{\mathbf{D}}}\}_{n+1} - \{\tilde{\mathbf{R}}^{\text{int}}\}_{n+1}$.
9. If $W^{\text{err}} = \Delta t\{\tilde{\dot{\mathbf{D}}}\}_{n+1}^T\{\mathbf{R}^{\text{err}}\} >$ tolerance, add $\{\mathbf{R}^{\text{err}}\}$ to $\{\mathbf{R}^{\text{ext}}\}_{n+1}$ and go to Step 4.
10. If tolerance is satisfied, update histories $\{\mathbf{D}\}_{n+1} = \{\mathbf{D}\}_n + \{\Delta\tilde{\mathbf{D}}\}$, $\{\dot{\mathbf{D}}\}_{n+1} = \{\tilde{\dot{\mathbf{D}}}\}_{n+1}$, $\{\ddot{\mathbf{D}}\}_{n+1} = \{\tilde{\ddot{\mathbf{D}}}\}_{n+1}$, $\{\mathbf{R}^{\text{int}}\}_{n+1} = \{\tilde{\mathbf{R}}^{\text{int}}\}_{n+1}$; $n \leftarrow n + 1$, go to Step 3.

where $[\mathbf{K}]$ is the *initial* tangent-stiffness matrix and $\{\Delta\mathbf{R}_s\}$ is a vector of nodal loads computed so that the material constitutive law is satisfied. Combining Eq. 17.6-15 with the equations of motion, Eq. 13.9-2, and the trapezoidal rule equations, Eqs. 13.11-3 and 13.11-4, we obtain

$$[\mathbf{K}^{\text{eff}}]\{\mathbf{D}\}_{n+1} = \{\mathbf{R}^{\text{eff}}\}_{n+1} \tag{17.6-16}$$

where

$$[\mathbf{K}^{\text{eff}}] = \frac{4}{\Delta t^2}[\mathbf{M}] + \frac{2}{\Delta t}[\mathbf{C}] + [\mathbf{K}] \tag{17.6-17}$$

and

$$\begin{aligned}\{\mathbf{R}^{\text{eff}}\}_{n+1} = \{\mathbf{R}^{\text{ext}}\}_{n+1} &+ [\mathbf{M}]\left(\frac{4}{\Delta t^2}\{\mathbf{D}\}_n + \frac{4}{\Delta t}\{\dot{\mathbf{D}}\}_n + \{\ddot{\mathbf{D}}\}_n\right) \\ &+ [\mathbf{C}]\left(\frac{2}{\Delta t}\{\mathbf{D}\}_n + \{\dot{\mathbf{D}}\}_n\right) + \{\Delta\mathbf{R}_s\}_{n+1}\end{aligned} \tag{17.6-18}$$

From Eq. 17.5-11, the corrective loads can be written as

$$\{\Delta\mathbf{R}_s\}_{n+1} = \sum \int_{V_e} [\mathbf{B}]^T\{\boldsymbol{\sigma}^{\text{err}}\}\, dV \tag{17.6-19}$$

where $\{\boldsymbol{\sigma}^{\text{err}}\}$ is the linear-stress prediction less the actual stresses.

Before Eq. 17.6-16 can be solved for $\{\mathbf{D}\}_{n+1}$, it is necessary to use information at time $n\ \Delta t$, for example, from an estimate of $\{\mathbf{D}\}_{n+1}$, so that $\{\Delta\mathbf{R}_s\}_{n+1}$ can be computed. If solution vector $\{\mathbf{D}\}_{n+1}$ is not in close agreement with the prior estimate (and usually it is not), then $\{\Delta\mathbf{R}_s\}_{n+1}$ must be recomputed using the improved $\{\mathbf{D}\}_{n+1}$ and the solution of Eq. 17.6-16 repeated. This process continues until convergence of $\{\mathbf{D}\}_{n+1}$ and $\{\Delta\mathbf{R}_s\}_{n+1}$. As an example, consider a uniform linear-displacement bar element with the material shown in Fig. 17.3-3*a*. If ϵ_B is the strain obtained from the predicted $\{\mathbf{D}\}_{n+1}$, then $\sigma_C = E\epsilon_B$ is the stress that would result if material behavior were linearly elastic. Stress σ_C, which violates the material's constitutive law, is implied by the term $[\mathbf{K}]\{\mathbf{D}\}_{n+1}$ in Eq. 17.6-15. To satisfy the constitutive law, we compute a "corrective" stress $\sigma_C - \sigma_B$. The corresponding corrective load in the element is then $A(\sigma_C - \sigma_B)$ and the corrective internal load vector for the element is $\{\Delta\mathbf{r}_s\}_{n+1} = A(\sigma_C - \sigma_B)\lfloor -1 \quad 1\rfloor^T$.

The principal advantage of this method is that $[\mathbf{K}^{\text{eff}}]$ need be formed and factored only once. Corrective loads $\{\Delta\mathbf{R}_s\}_{n+1}$ are estimated using only information available at time $n\ \Delta t$. Since this estimate is rarely satisfactory, several iterations must usually be performed within a time step to improve $\{\Delta\mathbf{R}_s\}_{n+1}$ so that the material constitutive law is satisfied. For problems with mild nonlinearities, this method can be more economical than the tangent-stiffness method. When nonlinearities are severe, widespread, or both, this method often demonstrates very poor convergence or divergence and may require a time step that scarcely exceeds a Courant number of unity.

Remarks on Implicit Methods. Solution of nonlinear problems by implicit methods is not easy and there are many pitfalls. Convergence is usually the major difficulty. It is usually good practice to repeat a solution using a smaller time step. Obviously the results of analyses using a different time step should show good agreement if one is to have faith in them.

Of the tangent-stiffness and initial-stiffness methods, the tangent-stiffness method has better convergence properties, but is often much more expensive. However, the initial-stiffness approach sometimes does not converge. A hybrid solution strategy consisting of the initial-stiffness approach with occasional stiffness matrix reformation can be effective and is often used in practice. This strategy can be particularly effective when combined with the inverse-Broyden method of Fig. 17.2-8 to update the initial and revised stiffness matrices.

17.7 A PROBLEM HAVING GEOMETRIC NONLINEARITY

Consider the plane cantilever beam shown in Fig. 17.7-1*a*. We seek the quasistatic deflections produced by loads P and M_L. We assume that the beam is slender and that its material is linearly elastic at all times. For small deflections, linear theory is adequate; for example, the root moment is $M_0 = PL_T + M_L$ because moment arm L_T is almost independent of load. For larger deflections, the moment arm H of force P is less than L_T, and $M_0 = PH + M_L$, where H depends on P and M_L. In such a problem, the nonlinearity is called *geometric nonlinearity*. The name implies that deformations significantly alter the location or distribution of loads, so that equilibrium equations must be written with respect to the deformed geometry, which is not known in advance.

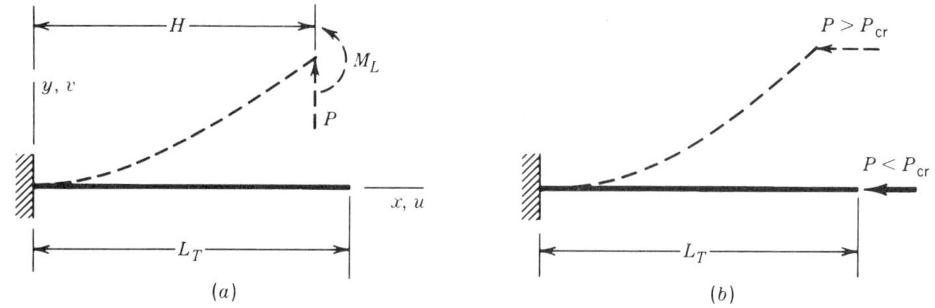

Figure 17.7-1. (*a*) Cantilever beam under tip loading. (*b*) Column under axial load.

To solve this problem we can use plane frame elements—that is, straight elements that have two nodes, six d.o.f., axial stiffness AE, and bending stiffness EI. The equilibrium configuration under a given loading will be obtained by the Newton–Raphson method of Fig. 17.2-3. The particulars given in what follows allow us to solve geometrically nonlinear problems of various plane structures that can be modeled by plane frame elements. Other elements [17.31] and other solution algorithms can also be used, but the same physical concept applies to all: we seek a displacement state in which the *deformed* structure is in equilibrium with loads applied to it [17.28, 17.29].

A Computational Algorithm. A structure—for example, one of those in Fig. 17.7-1—is divided into finite elements in the usual way. We can begin with zero initial displacements, $\{\mathbf{D}\}_0 = \{\mathbf{0}\}$, or with a better estimate for $\{\mathbf{D}\}_0$ if one is available. A load $\{\mathbf{R}\}$ is applied and the corresponding $\{\mathbf{D}\}$ is sought by iteration as follows. Computational details are explained subsequently.

1. Form the tangent-stiffness matrix $[\mathbf{K}_t]_i$ of the structure in the current configuration $\{\mathbf{D}\}_i$ (which is $\{\mathbf{D}\}_0$ initially, $\{\mathbf{D}\}_1$ after the first computational cycle, etc.).
2. Use displacements $\{\mathbf{D}\}_i$ to form resisting loads $\{\mathbf{R}_R\}_i$, which are loads applied to structure nodes by the deformed elements. (In the first computational cycle, these loads are zero if $\{\mathbf{D}\}_0 = \{\mathbf{0}\}$.)
3. Solve for displacement increments $\{\Delta\mathbf{D}\}_{i+1}$ and update the configuration to $\{\mathbf{D}\}_{i+1}$:

$$\{\mathbf{D}\}_{i+1} = \{\mathbf{D}\}_i + \{\Delta\mathbf{D}\}_{i+1} \qquad \text{where} \quad \{\Delta\mathbf{D}\}_{i+1} = [\mathbf{K}_t]_i^{-1}(\{\mathbf{R}\} + \{\mathbf{R}_R\}_i) \tag{17.7-1}$$

   Net loads $\{\mathbf{R}\} + \{\mathbf{R}_R\}_i$ are an imbalance between externally applied and internally generated nodal loads. The load imbalance drives the structure toward a configuration that reduces the imbalance. In an equilibrium configuration, the imbalance is zero.

4. Check for convergence; for example, see if $\|\Delta\mathbf{D}_{i+1}\| < e\|\mathbf{D}_{i+1}\|$, where e is a small number chosen by the analyst. If not converged, return to step 1.

Clearly, many alternatives are possible within this overall strategy. Modified Newton–Raphson cycles can be invoked by updating $[\mathbf{K}_t]$ only occasionally. Or, the inverse-Broyden method can be used. A final load level $\{\mathbf{R}\}$ can be approached in several stages, with convergence required in each stage before going on to the next. Indeed, the latter alternative, or underrelaxation, may be mandatory in some problems that are reluctant to converge (or tend to converge to the wrong result, which is possible if there is more than one equilibrium state that is mathematically possible).

In view of the many physical and computational alternatives that can be tested, coding the foregoing algorithm is a good educational device. Numerous test cases are available [17.32, 17.33]. A particularly simple one is that of Fig. 17.7-1*a* under tip moment M_L alone. The *y*-direction deflection of the tip is

$$v_{\text{tip}} = \frac{EI}{M_L}\left(1 - \cos\frac{M_L L_T}{EI}\right) \tag{17.7-2}$$

which is valid for all values of M_L provided that the material remains linearly elastic.

Some Computational Details. Figure 17.7-2 shows a typical element. For the structures of Fig. 17.7-1, initial angle α_0 is zero for all elements. Global coordinates xy are fixed in space. A local system $x'y'$ is attached to each element and moves with it: we attach the origin $x' = y' = 0$ to node 1 and direct the x' axis through node 2. Thus, in a *local* system $x'y'$, three nodal d.o.f. are always zero: $u_1 = v_1 = v_2 = 0$. *All* element d.o.f. in *global* directions, D_1 through D_6, are in general nonzero. We presume that elements are small enough that *local* rotations are small—that is, that $|\theta_1| << 1$ and $|\theta_2| << 1$.

In the deformed and displaced configuration, element length projections x_L and y_L on global axes xy and the orientation α of the local x' axis are

$$x_L = x_0 + D_{41} \qquad y_L = y_0 + D_{52} \qquad \alpha = \arctan(y_L/x_L) \tag{17.7-3a}$$

where

$$D_{41} = D_4 - D_1 \qquad \text{and} \qquad D_{52} = D_5 - D_2 \tag{17.7-3b}$$

If $\alpha_0 = 0$ and α is computed in Fortran as DATAN2(YL,XL), then α can reach $\pm\pi$ before it becomes ambiguously defined.

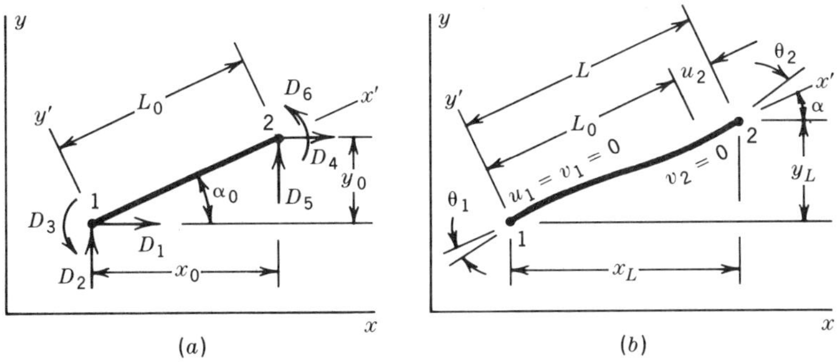

Figure 17.7-2. (*a*) Plane frame element, shown before any deformation or motion, but identifying global d.o.f. D_1 through D_6. (*b*) The same element after deformation and motion, showing d.o.f. in local coordinates $x'y'$.

Element d.o.f. in the local system $x'y'$ are $\{\mathbf{d}'\} = \lfloor 0 \quad 0 \quad \theta_1 \quad u_2 \quad 0 \quad \theta_2 \rfloor^T$, where

$$\theta_1 = D_3 - (\alpha - \alpha_0) \qquad \theta_2 = D_6 - (\alpha - \alpha_0) \tag{17.7-4a}$$

$$u_2 = \frac{1}{L + L_0} [(2x_0 + D_{41}) D_{41} + (2y_0 + D_{52}) D_{52}] \tag{17.7-4b}$$

where again $D_{41} = D_4 - D_1$ and $D_{52} = D_5 - D_2$. The expression for u_2 in Eq. 17.7-4b is more accurate than the expression $u_2 = L - L_0$ (in which $L^2 = x_L^2 + y_L^2$ and $L_0^2 = x_0^2 + y_0^2$) because Eq. 17.7-4b avoids finding the small difference between large numbers. Equation 17.7-4b is obtained by writing $L^2 - L_0^2$, substituting for x_L^2 and y_L^2 from Eq. 17.7-3, factoring $L^2 - L_0^2$, and solving for $u_2 = L - L_0$ [17.30]. In the denominator, $L + L_0 \approx 2L_0$, as we presume that strains are small.

In the local system $x'y'$, loads $\{\mathbf{r}'\}$ applied to nodes by the distorted element can be obtained from Eq. 4.1-6, using for $\{\boldsymbol{\sigma}_0\}$ the stresses produced by local d.o.f. $\{\mathbf{d}'\}$, that is, $\{\boldsymbol{\sigma}_0\} = [\mathbf{E}][\mathbf{B}]\{\mathbf{d}'\}$. An alternative calculation is

$$\{\mathbf{r}'\} = -[\mathbf{k}']\{\mathbf{d}'\} \tag{17.7-5}$$

in which the element stiffness matrix $[\mathbf{k}']$ in the local system does not change as the local system moves and the element deforms. Here $[\mathbf{k}']$ is given by Fig. 7.5-2*a*. If axial forces are significant, particularly in compression, one should add to this matrix a stress stiffness matrix $[\mathbf{k}_\sigma]$, for example, Eq. 14.2-11b (with zeros added to expand the matrix to size 6 by 6). In $[\mathbf{k}_\sigma]$, the element axial force P is given by $P = (AE/L)u_2$. Thus P is deformation-dependent, and may be considered unknown at the outset of the iterative process.

Referred to global coordinates xy, the element stiffness matrix and element nodal load vector are

$$[\mathbf{k}] = [\mathbf{T}]^T[\mathbf{k}'][\mathbf{T}] \qquad \text{and} \qquad \{\mathbf{r}\} = [\mathbf{T}]^T\{\mathbf{r}'\} \tag{17.7-6}$$

where, for the frame element, transformation matrix $[\mathbf{T}]$ is given by Eq. 7.5-7 (with β replaced by α). In general, each element has a different α and therefore requires a different $[\mathbf{T}]$. The structure tangent-stiffness matrix and resisting load vector in global coordinates are formed by the usual assembly process, which is symbolized by

$$[\mathbf{K}_t] = \sum [\mathbf{k}] \qquad \text{and} \qquad \{\mathbf{R}_R\} = \sum \{\mathbf{r}\} \tag{17.7-7}$$

Note that coordinate transformation is not the essence of an analysis procedure for geometric nonlinearity, but only the vehicle adopted here to establish what *is* essential: the properties of the system in its current configuration.

17.8 OTHER NONLINEAR PROBLEMS

Nonlinear computational mechanics continues to be an active area of research. Some of the nonlinear structural problems we have not discussed in this book are as follows.

Materials that creep are analyzed by use of constitutive relations and computational algorithms very similar to those used for analysis of plastic deformations. Indeed, creep and plasticity analyses are often combined [17.12, 17.34].

Forming processes such as extrusion and rolling involve large plastic deformations. Elastic response may be considered negligible, and the material analyzed as if it were a viscous fluid. "Springback" after the forming process is complete can be modeled if a viscoelastic material is invoked. Casting processes involve considerable heat transfer calculations and changes of phase [17.35].

Problems of moving contact fields, such as the rolling contact of a tire on pavement, have been addressed by special algorithms [17.36]. Other contact problems, either static or dynamic, include bearings, joints in rock, and gaps that may open or close. Special elements for such problems have been devised [17.13, 17.37, 17.38].

Membranes may be deflected by pressure, as when a balloon is inflated. Such a problem typically involves large strains, large deflections, and large rotations. Pressure loads change in direction, as they continue to act normal to the membrane. Element nodal loads $\{\mathbf{r}_e\}$ also change in magnitude as pressure increases and as the element surface area subjected to pressure increases [17.39]. If a membrane is initially flat and initially unstressed, it has no resistance to lateral load. To avoid a singular structure stiffness matrix in the first iterative cycle, one can add a fictitious initial stress for the first cycle only, thereafter to be replaced by the computed stress and corresponding stress stiffness matrix.

Cable problems are somewhat similar to membrane problems. One can follow the standard approach of dividing each cable of a cable network into many elements (e.g., two-node bar elements). However, this approach is inefficient if deflections are large. A method that treats an entire cable as a single element appears to be much more economical [17.40, 17.41].

A nonlinear vibration problem leads to an eigenvalue problem of the standard symbolic form—that is, $([\mathbf{K}] - \omega^2[\mathbf{M}])\{\overline{\mathbf{D}}\} = \{\mathbf{0}\}$; however, stiffness matrix $[\mathbf{K}]$ is a function of $\{\overline{\mathbf{D}}\}$, the nodal amplitudes [17.42].

PROBLEMS

Section 17.2

17.1 Sketch the progress of the two forms of direct substitution, as in Fig. 17.2-2, but let the curves be concave up, as for a hardening structure.

17.2 Load P acts on a nonlinear spring, as shown. Let $k = 0.2 - u$ and $P = 0.006$. Apply three cycles of direct substitution according to Eq. 17.2-3. Also, apply three cycles of modified Newton–Raphson iteration, using k at $u = 0$. Thus, show that both methods yield the same results.

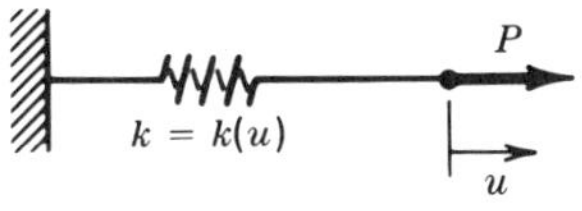

Problem 17.2

17.3 In Problem 17.2, what is the expression for the tangent stiffness k_t in terms of displacement u?

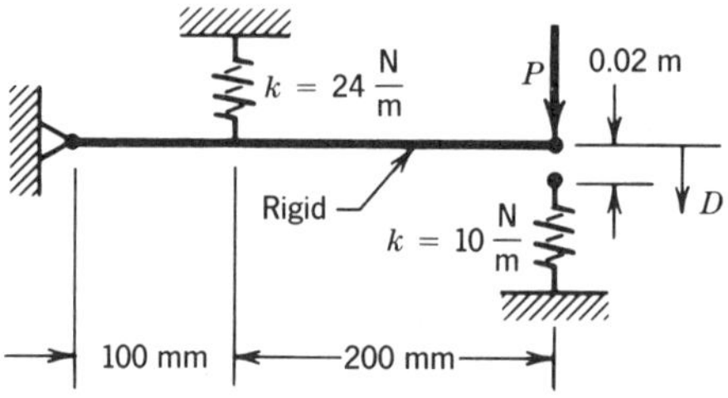

Problem 17.4

17.4 The bar shown in the sketch is rigid. Contact with the right-hand spring is made only when the 0.02-m gap closes. Set up an equation for displacement D of load P, then apply three cycles of a direct-substitution solution for D when $P = 0.24$ N, as follows:

(a) Use the procedure in which stiffness K is repeatedly updated but the right-hand side (i.e., load P) is unchanged.

(b) Use the procedure in which K is unchanged from its initial value, but the effective load is repeatedly updated.

17.5 The bar shown is of length L when unstressed, where $L^2 = a^2 + c^2$. When load P is zero, displacement D is also zero. The bar has axial stiffness AE/L, rolls without friction at B, and does not buckle as a column. Assume that the roller is constrained to remain in contact with the wall.

(a) For $a >> c$, show that $\Pi_p = U - PD = (AE/8a^3)(-2cD + D^2)^2 - PD$.

(b) Show that the equilibrium values of D are given by roots of the equation $P = (AE/2a^3)(2cD - D^2)(c - D)$.

(c) Determine expressions for the secant stiffness and the tangent stiffness.

(d) Show that limit points are at $D = c(1 \pm \sqrt{3}/3)$.

(e) Let constants of this system be such that points A and F on the P versus D curve are at $P_A = 241$, $D_A = 0.211$, $P_F = 250$, and $D_F = 1.080$. After convergence at $P = 200$, P is increased to 250, and the following sequence of displacements D is generated by a Newton–Raphson algorithm: 0.173, 0.219, 0.071, 0.143, 0.190, 0.249, 0.199, 0.294, 0.235, 0.175, 0.222, 0.108, 0.166, 0.210, 1.178, 1.096, 1.080, 1.080. Explain this path to convergence by sketching it on the P versus D plot.

(f) Similarly, sketch the path that would be taken by a *modified* Newton–Raphson algorithm.

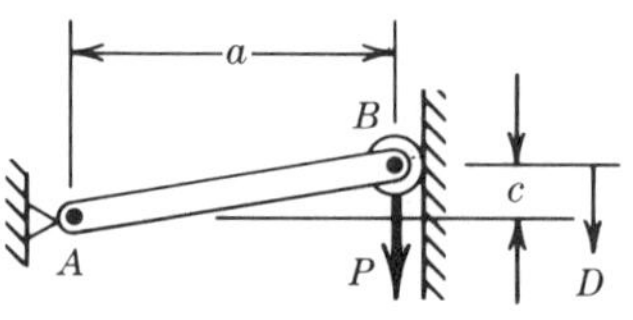

(a)

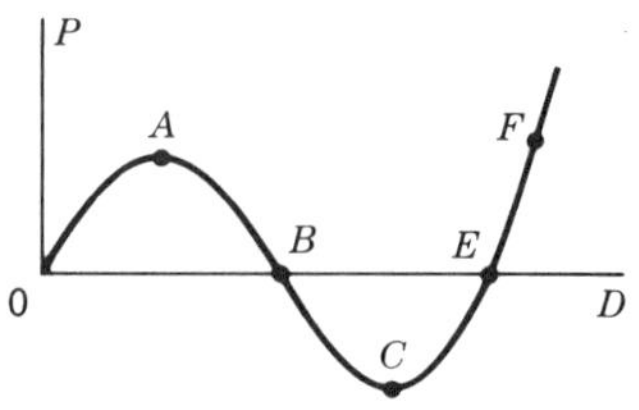

(b)

Problem 17.5

17.6 Let loads P_1 and P_2 be functions of displacements D_1 and D_2, that is, $P_1 = f_1(D_1,D_2)$ and $P_2 = f_2(D_1,D_2)$. Let D_A and D_B be exact values of D_1

and D_2 produced by loads P_A and P_B. Let D_A^* and D_B^* be approximations of D_A and D_B. Assume that $D_A = D_A^* + \Delta D_A$ and $D_B = D_B^* + \Delta D_B$. Derive the following equations (analogous to Eq. 17.2-9):

$$\begin{bmatrix} \partial P_1/\partial D_1 & \partial P_1/\partial D_2 \\ \partial P_2/\partial D_1 & \partial P_2/\partial D_2 \end{bmatrix}_{D_A^*, D_B^*} \begin{Bmatrix} \Delta D_A \\ \Delta D_B \end{Bmatrix} = \begin{Bmatrix} P_A - f_1(D_A^*, D_B^*) \\ P_B - f_2(D_A^*, D_B^*) \end{Bmatrix}$$

17.7 The sketch shows a nonlinear load versus deflection curve. In the exercises of this problem we pretend that P and dP/dD can be found when D is known, but that an explicit expression for D in terms of P is not available. Solve for D, using the situations and methods indicated. Sketch the progress of each solution on a plot of P versus D.

(a) What D is predicted by five cycles of Newton–Raphson iteration if $P = 8$, starting from $P = D = 0$?

(b) What D is predicted by five cycles of modified Newton–Raphson iteration if $P = 8$, starting from $P = 7.5$, $D = 3$?

(c) Repeat part (b) but use three cycles and update the tangent stiffness after the first cycle.

(d) What D is predicted by four purely incremental (Euler's method) steps of $\Delta P = 2$, starting from $P = 1$ and going to $P = 9$? *Given:* $D = 0.11111$ at $P = 1$.

(e) Repeat part (d) but include a force imbalance correction at every step.

(f) What D is predicted by five cycles of the direct-substitution algorithm of Eq. 17.2-2 if $P = 8$, starting from $P = D = 0$?

(g) What D is predicted by five cycles of the secant method depicted in Fig. 17.2-7 if $P = 8$, starting from $P = D = 0$?

17.8 The introductory remarks of Problem 17.7 again apply, but now to the hardening curve sketched.

(a) What D is predicted by five cycles of Newton–Raphson iteration if $P = 0.8$, starting from $P = D = 0$?

(b) What D is predicted by five cycles of modified Newton–Raphson iteration if $P = 3$, starting from $P = 1.5$, $D = 6$?

(c) Repeat part (b) but apply an underrelaxation factor $\beta = 0.6$ to the increments ΔD.

(d) Repeat part (b) but use four cycles and update the tangent stiffness after the first cycle.

(e) What D is predicted by three purely incremental (Euler's method) steps

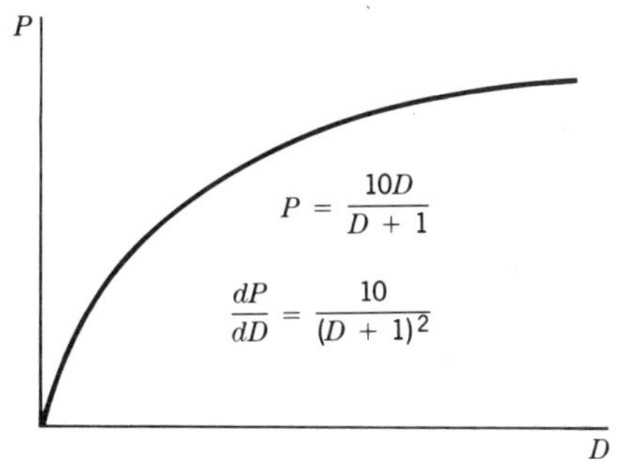

Problem 17.7

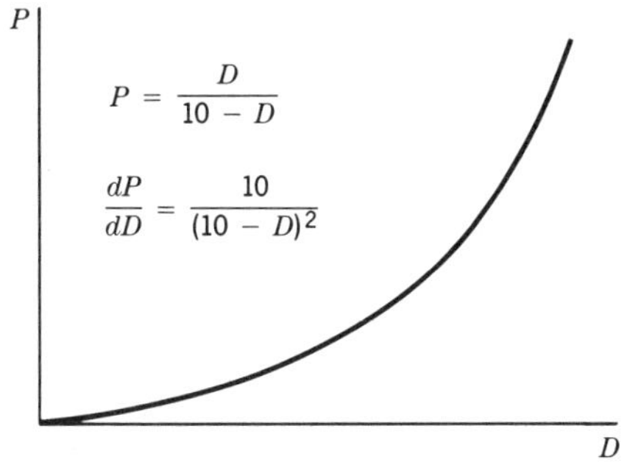

Problem 17.8

of $\Delta P = 1$, starting from $P = 1.5$ and going to $P = 4.5$? *Given:* $D = 6$ at $P = 1.5$.

(f) Repeat part (e) but include a force imbalance correction at every step.

(g) What D is predicted by four cycles of the direct-substitution algorithm of Eq. 17.2-2 if $P = 4$, starting from $P = D = 0$?

(h) Repeat part (g) but apply an underrelaxation factor $\beta = 0.3$ in Eq. 17.2-5b.

(i) What D is predicted by five cycles of the secant method in Fig. 17.2-7 if $P = 0.8$, starting from $P = D = 0$?

(j) Repeat part (i), but apply an underrelaxation factor $\beta = 0.3$ to the increments ΔD.

17.9 Use four cycles of the secant method depicted in Fig. 17.2-7 to calculate iterates u_i for the spring problem posed in Problem 17.2.

17.10 The Newton–Raphson method to solve $f(x) = 0$ can be formulated by defining an iteration function

$$g(x) = x - \frac{f(x)}{f'(x)}$$

and seeking a solution x^* that satisfies $x^* = g(x^*)$ by taking a given x_0 and iterating: $x_{i+1} = g(x_i)$.

(a) Verify that x^* satisfies $f(x^*) = 0$ provided that $f'(x^*) \neq 0$.

(b) Use a three-term exact Taylor series for $g(x)$, that is,

$$g(x) = g(x^*) + g'(x^*)(x - x^*) + \tfrac{1}{2}\, g''(\bar{x})(x - x^*)^2$$

where $\bar{x}$ lies between x and x^*, to verify that the Newton–Raphson method terminates quadratically.

17.11 Use Eq. 17.2-1, $(k_0 + k_N)u = P$, to show that the principle of superposition does not apply to a nonlinear problem.

Section 17.3

17.12 The bar shown is to be modeled by a single element. Thus there is one d.o.f.—namely, the displacement u_2 of load P. Apply the tangent-stiffness method to determine u_2 for $P = 3.0$ kN and for $P = 6.0$ kN. Use two load increments of $\Delta P = 3.0$ kN. Start from $P = 0$. Use two cycles of step 3 of the algorithm. Plot u_2 versus P, showing the exact solution and the progress of the incremental solution.

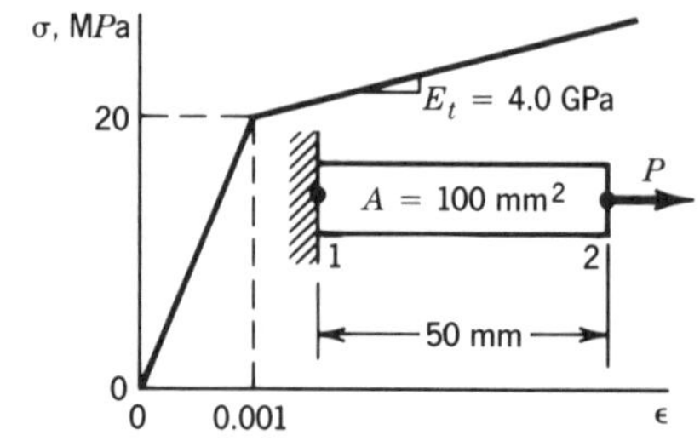

Problem 17.12

17.13 Repeat Problem 17.12 but apply only the 3.0-kN load. Start from P = 2.0 kN, apply ΔP = 1.0 kN, and carry out five cycles of the initial-stiffness method. Compute the percentage error in the resulting u_2.

17.14 The two bars shown are fixed to rigid walls at their outer ends and are welded together where load P is applied. The material behavior is shown in the sketch. Use the tangent-stiffness method and the successive load increments $\Delta P = 20$, $\Delta P = 10$, and $\Delta P = -30$. Determine the corresponding values of displacement D. Show results on a plot of P versus D.

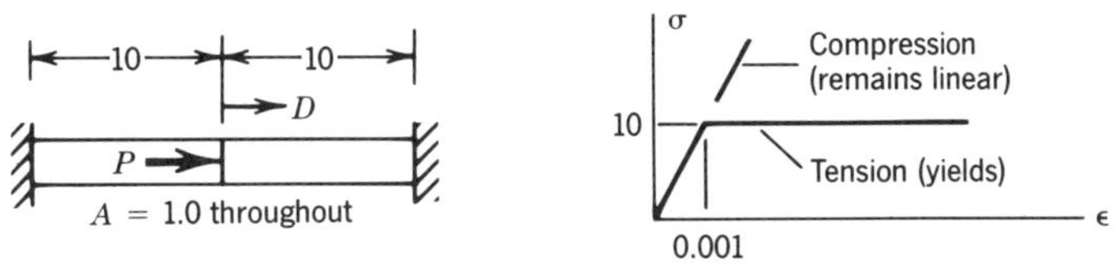

Problem 17.14

17.15 For the bar sketched in Problem 17.14, apply the single load increment ΔP = 10 after the yield point value of load P is reached. Determine the value of displacement D predicted by five cycles of the initial-stiffness algorithm.

17.16 The horizontal bar in the sketch is perfectly rigid and is constrained to remain horizontal as load P and displacement v increase. The three vertical bars are elastic–perfectly plastic with $A = 1$, $E = 1$ and $L = 2$. These bars have the respective yield point loads $F_1 = 2$, $F_2 = 4$, and $F_3 = 6$. Use the tangent-stiffness method to generate the P versus v relation. Use three steps. Scale each step so that one bar begins to yield at the end of the step.

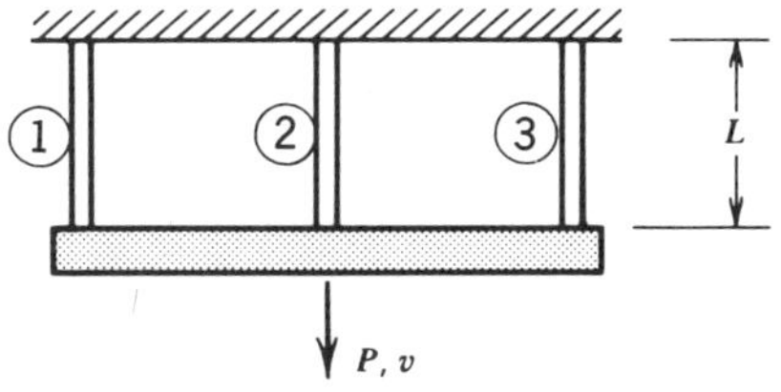

Problem 17.16

17.17 A weightless and rigid block B is pushed down in a frictionless guide by force P. The force versus deflection plot for each of two supporting bars is given in the sketch. Solve for displacement D of block B under a force P = 24 N, as follows.

(a) Determine the exact solution.

(b) Use the tangent-stiffness method. Let the first load increment be ΔP = 19 N.

(c) Use the initial-stiffness method. Apply one load increment ΔP = 5 N starting from the exact solution at P = 19 N.

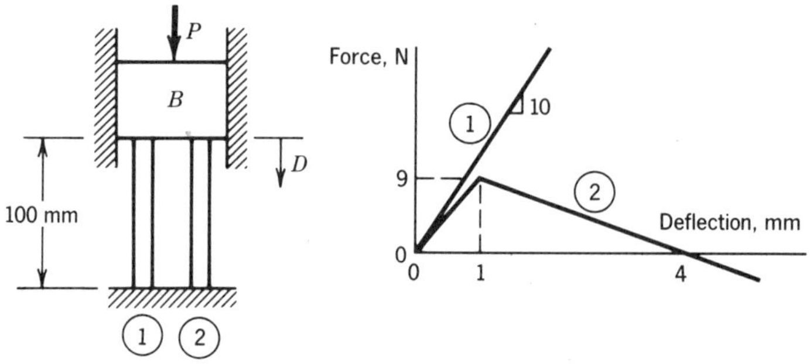

Problem 17.17

17.18 Assume that members of a truss carry only uniaxial stress, and that members in compression will buckle elastically at their critical loads without yielding. Tensile members are elastic–perfectly plastic. Outline a tangent-stiffness algorithm for computation of the displacements produced by applied loads. For simplicity, assume that load reversal does not occur in any member.

17.19 In both the initial-stiffness algorithm and the tangent-stiffness algorithm, factor m of Eq. 17.3-4 is used only to correct a trial solution after it is computed. Can the correction be *anticipated* instead? Explain.

Section 17.4

17.20 Verify Eq. 17.4-7.

17.21 Show that Eq. 17.4-13 becomes $F = \sigma_a - \sigma_Y$ when a state of uniaxial stress σ_a is defined in the following ways. Let $\{\boldsymbol{\alpha}\} = \{\mathbf{0}\}$.
(a) $\sigma_x = \sigma_a$, $\sigma_y = \sigma_z = \tau_{xy} = \tau_{yz} = \tau_{zx} = 0$.
(b) $\sigma_x = \sigma_y = \tau_{xy} = \sigma_a/2$, $\sigma_z = \tau_{yz} = \tau_{zx} = 0$.
(c) $\sigma_x = 0.8\sigma_a$, $\sigma_y = 0.2\sigma_a$, $\tau_{xy} = 0.4\sigma_a$, $\sigma_z = \tau_{yz} = \tau_{zx} = 0$.

17.22 Verify that Eq. 17.4-14 follows from Eq. 17.4-13.

17.23 Imagine that plastic action in bending is to be modeled and that several sampling points are used in the thickness direction (d in the sketch).
(a) Why should the sampling points pertain to a trapezoidal or Simpson quadrature rule rather than to a Gauss–Legendre quadrature rule?
(b) Imagine that the stress distribution shown prevails across the depth of a beam of rectangular cross section. What is the percentage error of the computed bending moment M_c, if M_c is integrated from the stress distribution using a two-point Gauss rule? Repeat the calculation using a three-point Gauss rule.
(c) Repeat part (b), but use trapezoidal rules instead of Gauss rules. Try three, five, seven, and then nine sampling points.

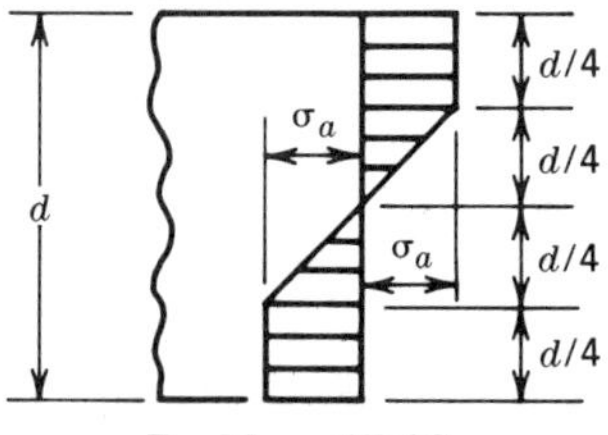

Problem 17.23

17.24 Imagine that a state of uniaxial stress causes plastic strains. Show that Eq. 17.4-15 yields the correct value of ϵ_{ef}^{p} if
(a) The stress is parallel to the x axis.
(b) The stress acts at 45 degrees to the x and y axes.

Section 17.5

17.25 Describe the steps of a tangent-stiffness solution algorithm in which each load increment causes a single sampling point to be brought to the initiation of yielding.

17.26 Consider a plane structure modeled by finite elements. The material is isotropic but brittle: it cracks when the tensile stress in any direction exceeds a value σ_t. Outline a tangent-stiffness algorithm for predicting deformations caused by increasing load. How will the collapse load be detected by this algorithm?

17.27 Imagine that corrective loads $\{\Delta\mathbf{R}_c\}$ are to be computed for a mesh of elements having internal d.o.f. Should internal d.o.f. carry loads that result from $\{\boldsymbol{\sigma}\}$, or should these loads be omitted from internal d.o.f.? If carried, should they be distributed to remaining d.o.f. (that is, condensed) by means of elastic element stiffness equations?

17.28 Imagine that a plane beam of rectangular cross section is modeled by plane finite elements. The material is linearly elastic, but elastic moduli in tension and compression are different. Outline an algorithm that will calculate the stresses produced by a pure bending load. What are the comparative merits of tangent-stiffness and initial-stiffness solutions?

Section 17.6

17.29 Consider an elastic–perfectly plastic material with an associated flow rule. The constitutive law for such a material is given by Eqs. 17.4-7 through 17.4-9 with $C = 0$, $\partial F/\partial W_p = 0$, and $F = Q$. Equation 17.4-9 can be written as $[\mathbf{E}_{ep}] = [\mathbf{E}] + [\mathbf{E}_p]$, where $[\mathbf{E}_p] = -[\mathbf{E}]\{\partial F/\partial\boldsymbol{\sigma}\}\{\mathbf{C}_\lambda\}^T$. Using Eq. 17.4-7, show that $[\mathbf{E}_p]$ is negative semidefinite.

17.30 Starting with the basic definition of the rate of internal work for an element e as

$$\dot{W}_e^{\text{int}} = \int_{V_e} \{\dot{\boldsymbol{\epsilon}}\}^T\{\boldsymbol{\sigma}\}\, dV$$

show that the rate of internal work for the entire structure is $\dot{W}^{\text{int}} = \{\dot{\mathbf{D}}\}^T\{\mathbf{R}^{\text{int}}\}$.

17.31 Starting with $\dot{W}_n^{\text{int}} = \{\dot{\mathbf{D}}\}_n^T\,\{\mathbf{R}^{\text{int}}\}_n$, show that for linearly elastic material behavior, $W_n^{\text{int}} = \frac{1}{2}\{\mathbf{D}\}_n^T[\mathbf{K}]\{\mathbf{D}\}_n$. Note: $\frac{d}{dt}(\{\mathbf{D}\}^T[\mathbf{K}]\{\mathbf{D}\}) = 2\{\dot{\mathbf{D}}\}^T[\mathbf{K}]\{\mathbf{D}\}$ if $[\mathbf{K}]$ is symmetric.

17.32 Verify that Eq. 17.6-3 results from Eq. 17.6-2.

17.33 Of the energies W^{int}, W^{ext}, and T, and the energy rates $\dot{W}^{\text{int}}$ and $\dot{W}^{\text{ext}}$, which are always nonnegative and which can be positive, zero, or negative? Assume that the material is linearly elastic.

17.34 Repeat the example of Section 17.6 using different time steps and longer analysis duration.

17.35 Repeat the example of Section 17.6 using the nonlinearly elastic stiffening material shown (in which loading and unloading are both on the same path). It will be necessary to write a subroutine similar to Fig. 17.6-1 for this constitutive law. Note that the transition between slopes E and E_t may occur during both loading and unloading. Analyze response for approximately one wave traversal along the bar. Carefully address: (a) the largest stable time step; (b) the correctness of the stress wave shape, and (c) the wave arrival time(s).

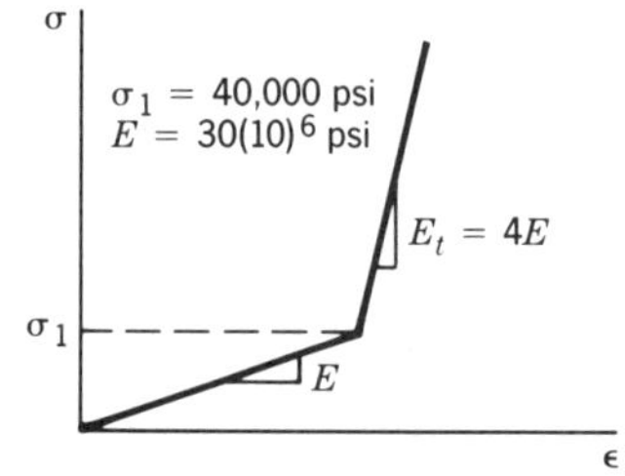

Problem 17.35

17.36 Derive Eqs. 17.6-11 through 17.6-13.

Section 17.7

17.37 (a) If $P = 1.884P_{cr}$ in Fig. 17.7-1b, the tip of the column rotates 120° from its unloaded position. To what other equilibrium configuration might the numerical process converge? Answer qualitatively, without calculation.

(b) Repeat part (a) if force P is supplemented by a vertically directed tip force $Q = 0.001P_{cr}$. Sketch your answer.

17.38 Verify Eq. 17.7-2, and show that it agrees with the formula $M_L L_T^2/2EI$ of elementary beam theory if M_L is small.

17.39 (a) Derive the expression for u_2, Eq. 17.7-4b.

(b) Show that $u_2 = 0$ for a rigid-body rotation α about node 1, starting from $\alpha_0 = 0$.

17.40 Imagine that $M_L = 0$ in Fig. 17.7-1a and that the right end of the cantilever beam has rotated about 45° under the action of force P. Let the beam be divided into several elements. For the rightmost element, qualitatively sketch loads $[\mathbf{k}']\{\mathbf{d}'\}$ of Eq. 17.7-5 in the local system $x'y'$ and loads $\{\mathbf{r}\}$ of Eq. 17.7-6 in the global system xy. Show these loads in the directions they actually act. Which of these loads will be zero after the iterative numerical process has converged?

17.41 A long uniform beam has bending stiffness EI, weight q per unit length, and rests on a flat horizontal surface, as shown. A vertical force F, where $F < qL_T/2$, is applied to one end.

(a) Analytically determine the length L_s that lifts off the surface ($L_s < L_T$).

(b) Imagine that the beam is modeled by several elements. Outline a numerical algorithm for calculating length L_s.

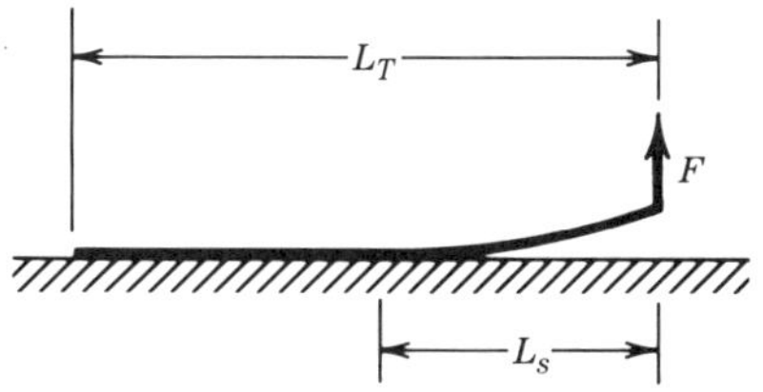

Problem 17.41

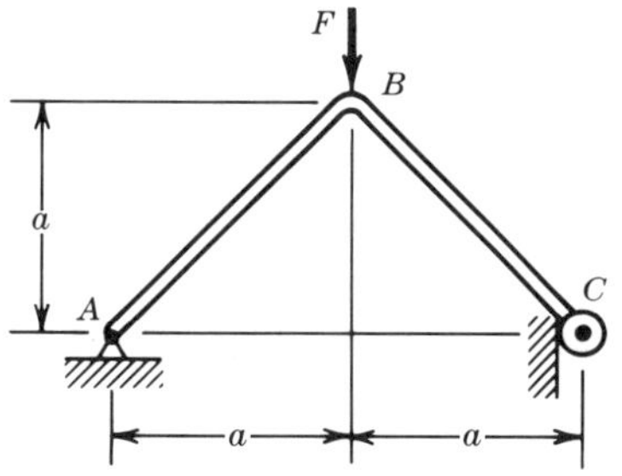

Problem 17.42

17.42 The angle frame shown has uniform bending stiffness EI. At C, a small frictionless roller contacts a vertical surface.

(a) Solve for the vertical deflection at C caused by force F, using mechanics of materials methods. Assume that this deflection is small.

(b) Solve by a computer program, for example, based on the algorithm described in Section 17.7. What difficulty appears in the first iterative cycle, and how will you overcome it?

17.43 The algorithm described in Section 17.7 can be applied to a plane body modeled by four-node quadrilaterals. Describe in detail how one might track the rigid-body motion of a quadrilateral and compute its d.o.f. $\{\mathbf{d}'\}$ in a local coordinate system attached to the element. Assume that all displacements lie in the xy plane.

17.44 When P and its displacement D are zero, the springs are unstressed (see sketch). Spring stiffness k_0 is constant.

(a) Determine the secant stiffness (the ratio of P to D) for small values of D ($D << L$).

(b) What is the tangent stiffness for small values of D?

(c) Determine an expression for the secant stiffness if D may be moderate to large.

(d) Describe an iterative way to calculate D for a given P, with particular emphasis on the first step.

17.45 The frame shown is built of slender members that may be assumed to remain linearly elastic at all times. Connections at A and B are frictionless pins.

(a) Qualitatively sketch the anticipated relation between load P and its vertical displacement D.

(b) How would you generate the P versus D relation numerically? That is, what difficulties do you anticipate, and how might you avoid or overcome them?

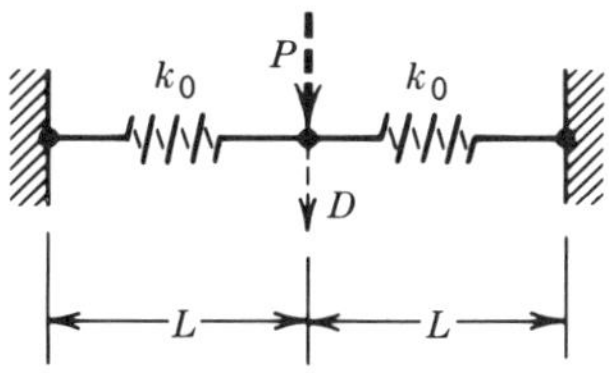

Problem 17.44

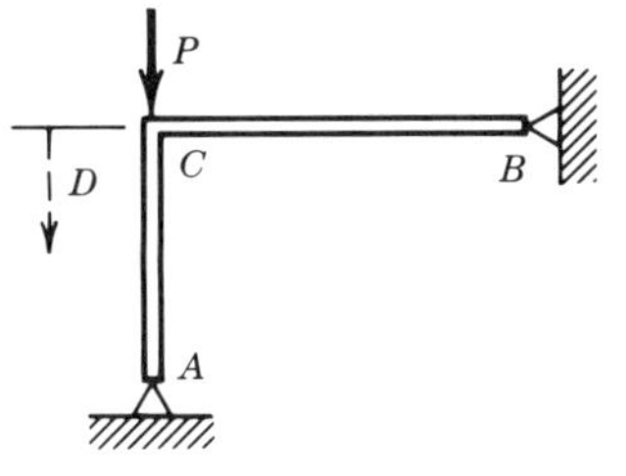

Problem 17.45

18 CHAPTER

NUMERICAL ERRORS AND CONVERGENCE

Sources of computational error are categorized. Methods of detecting and avoiding such errors are presented. The relation between element size and solution error is discussed. Tests of element quality are reviewed.

18.1 INTRODUCTION. ERROR CLASSIFICATION

Computed results are rarely exact. Some of the many reasons are as follows. We divide a structure into elements whose displacement fields exclude many of the physically possible deformation modes. The type, number, and shapes of elements may be chosen within a broad range of possibilities, and some choices are better than others. The computer represents numbers by a finite number of bits or digits.

Numerical difficulties may arise even when the analyst makes no outright blunder in using a computer program. In this chapter we assume that there are no outright blunders, that is, that the program used is appropriate to the task at hand, that the program is free of bugs, that the choices of element types, shapes, and quadrature orders are suitable, and that elements pass patch tests and do not lock. Some choices remain, such as the specific type and number of elements, their arrangement, and the way in which relatively stiff regions are treated. How these choices are made can increase or decrease *numerical error,* which is caused by the inability of a computer to store and process numbers in infinite precision.

There is no single definitive test of solution accuracy short of knowing the correct result by other means. A calculation that survives one error test may fail another. A usually reliable error test may fail in particular cases. For example, given a single-precision matrix $[\mathbf{H}]$, one may calculate $[\mathbf{X}] = [\mathbf{H}]^{-1}$, then $[\mathbf{Y}] = [\mathbf{X}]^{-1}$. If $[\mathbf{Y}]$ agrees with $[\mathbf{H}]$, one expects $[\mathbf{X}]$ to be correct. In fact, when $[\mathbf{H}]$ is a tenth-order Hilbert matrix ($H_{ij} = (i + j - 1)^{-1}$, which is notoriously ill conditioned), one study found that $[\mathbf{Y}] = [\mathbf{H}]$ to seven-digit accuracy, yet $[\mathbf{X}]$ had coefficients in error by three orders of magnitude [18.1].

Granted that some loss of precision is possible or even likely in finite element calculations, one should not introduce gratuitous errors: numerical constants, such as π and Gauss point coordinates and weights, should be written with as many accurate digits as the machine can accommodate.

Terminology. The following terms are useful in subsequent discussion [18.2]. This abbreviated list shows at once that there are several aspects to the study of numerical errors.

Modeling error refers to the difference between a physical system and its mathematical model. For example, an actual plate may be mathematically modeled by Kirchhoff plate theory. Numerical analysis is performed on the mathematical model. It is this analysis that is subject to the following errors.

Discretization error refers to the error caused by representing the infinitely many d.o.f. of a continuous mathematical model by a finite number of d.o.f. in its discretized form. For example, the aforementioned Kirchhoff plate is divided into finite elements, thus introducing discretization error.

Round-off error is caused by use of a finite number of bits or digits to represent real numbers. The last digit retained may be rounded or may be obtained by simple truncation (chopping). The *round-off limit* is the smallest floating point number ϵ such that, in the computer, $1.0 + \epsilon > 1.0$. For example, one may find $\epsilon = 7(10^{-15})$ on some machine in double-precision arithmetic.

Inherited error at any stage of calculation is the sum of previous discretization and round-off errors.

Manipulation error refers to round-off error introduced by an algorithm. For example, an equation solver performs numerical operations such as $A_{22} - (A_{21}/A_{11})A_{12}$, where the A_{ij} are matrix coefficients. The division and multiplication are each followed by rounding of the result to computer-word length, and the subtraction may lose several significant digits if A_{22} and $(A_{21}/A_{11})A_{12}$ are almost equal. For example, if $R = D_1 - D_2$, where $D_1 = 1.23456$ and $D_2 = 1.23455$, then $R = 1.00000(10^{-5})$, which contains but one significant digit in its six-digit mantissa and is therefore far less accurate than either D_1 or D_2.

In a finite element context, inherited error is present after disparate element stiffnesses are added to form structural stiffness coefficients, and manipulation error is produced by solving the equations. Alternative terms for these errors are *truncation* and *rounding*. Inherited error and manipulation error have the same source: the round-off limit is not zero.

18.2 ILL-CONDITIONING

Concepts from Algebra. Consider the set of equations

$$\begin{bmatrix} 1.00 & -1.00 \\ -1.00 & 1.02 \end{bmatrix} \begin{Bmatrix} x \\ y \end{Bmatrix} = \begin{Bmatrix} 4.00 \\ -2.00 \end{Bmatrix} \quad \text{for which} \quad \begin{Bmatrix} x \\ y \end{Bmatrix} = \begin{Bmatrix} 104 \\ 100 \end{Bmatrix} \tag{18.2-1}$$

and the very similar set of equations

$$\begin{bmatrix} 1.00 & -1.00 \\ -1.00 & 1.01 \end{bmatrix} \begin{Bmatrix} x \\ y \end{Bmatrix} = \begin{Bmatrix} 4.00 \\ -2.00 \end{Bmatrix} \quad \text{for which} \quad \begin{Bmatrix} x \\ y \end{Bmatrix} = \begin{Bmatrix} 204 \\ 200 \end{Bmatrix} \tag{18.2-2}$$

A 1% change in one coefficient has changed the results by a factor of two. These equation sets are both *ill conditioned,* which means that their solutions are sensitive to small changes in either the coefficient matrix or the vector of constants.

The solution of Eq. 18.2-1 can be regarded as the intersection point in xy space of two straight lines, one representing the equation $x - y = 4$ and the other representing the equation $-x + 1.02y = -2$. When plotted, the two lines are seen to be almost parallel. Accordingly, when the second equation is changed to

$-x + 1.01y = -2$ (Eq. 18.2-2), the second line is rotated slightly, and the intersection point changes markedly.

In matrix terminology, the rows of an ill-conditioned matrix are almost linearly dependent. For 2 by 2 systems such as Eqs. 18.2-1 and 18.2-2, this means that the second row of the coefficient matrix is almost a scalar multiple of the first row.

A Gauss elimination solution of these equations changes the second diagonal coefficient in Eq. 18.2-1 to $1.02 - 1.00 = 0.02$. If only two digits were retained, all coefficients would be represented as 1.0 and the matrix would be singular. Gauss elimination would produce $1.0 - 1.0 = 0.0$ as the second diagonal coefficient and the computation $y = 2.0/0.0$ would be attempted.

Comparatively Flexible Support. In Fig. 18.2-1, the circumstance $k_1 >> k_2$ promotes ill-conditioning but the circumstance $k_2 >> k_1$ does not. This is easy to see by examining the structure equations $[\mathbf{K}]\{\mathbf{D}\} = \{\mathbf{R}\}$, which are

$$\begin{bmatrix} k_1 & -k_1 \\ -k_1 & k_1 + k_2 \end{bmatrix} \begin{Bmatrix} u_1 \\ u_2 \end{Bmatrix} = \begin{Bmatrix} P \\ 0 \end{Bmatrix} \tag{18.2-3}$$

The rows of $[\mathbf{K}]$ are almost linearly dependent if $k_1 >> k_2$, but not if $k_2 >> k_1$. In a Gauss elimination solution, we calculate the reduced coefficient $K_{22} = (k_1 + k_2) - k_1$, which yields an inaccurate result if $k_1 >> k_2$ and k_2/k_1 is close to the round-off limit.

To make the point numerically, imagine that $k_1 = 40$ and $k_2 = 0.0014$. If the last digit of k_2 is to appear in the number $k_1 + k_2$, the computer word mantissa must store at least six digits and k_1 must be represented as 40.0000. (That k_1 is not physically known to six-digit accuracy does not matter.) If five digits were stored, Gauss elimination would produce a reduced K_{22} of 0.0010 rather than the correct reduced value $K_{22} = 0.0014$. If only four digits were stored, $k_1 + k_2$ would be represented as 40.00, and $[\mathbf{K}]$ would be singular. That is, $[\mathbf{K}]$ would represent only the single spring k_1, unsupported and free to translate as a rigid body.

In the foregoing example, the modeling and discretization errors are zero, but round-off produces an inherited error in $[\mathbf{K}]$ that becomes obvious during subsequent manipulations.

In summary, and in structural terminology, a major cause of ill-conditioning in practical finite element models is a large difference in stiffnesses, with the stiffer region being supported by the more flexible region. This circumstance shifts essential numerical information to the latter digits of stiffness coefficients K_{ij}. These latter digits may be so few in number that the solution is worthless. Physically,

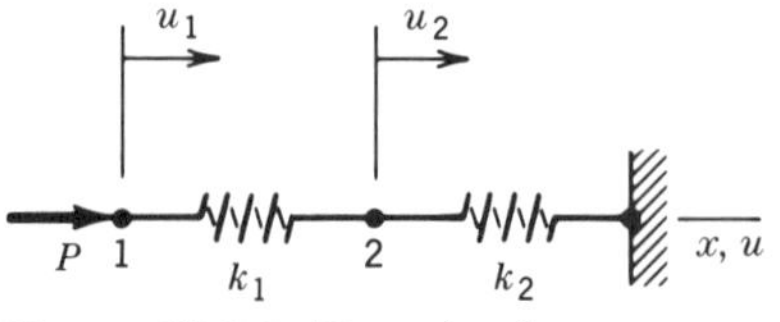

Figure 18.2-1. Two-d.o.f. structure with linear springs of stiffness k_1 and k_2.

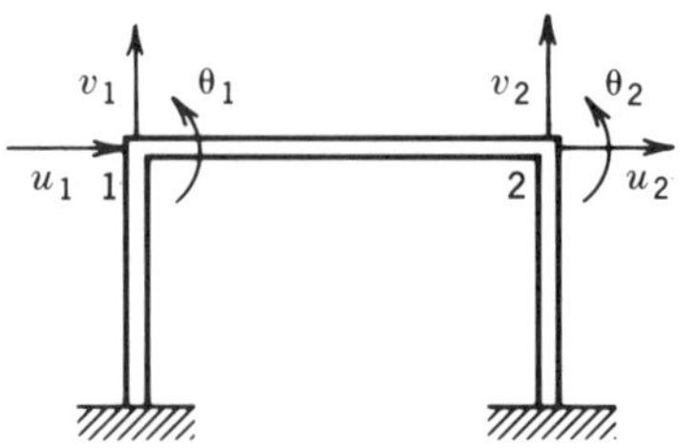

Figure 18.2-2. A plane frame with six nonzero d.o.f.

the stiffer region has one or more displacement states that are almost rigid-body motions within a more flexible supporting structure. The limiting case is a structure without *any* supports: it has *only* rigid-body motion in static analysis, and its stiffness matrix is singular.

In nonstructural problems, similar difficulties may arise. For example, in heat conduction analysis one might encounter a region of high conductivity imbedded in a region of low conductivity.

One way to avoid difficulty is to use longer computer words—that is, double precision rather than single precision. But if equations are generated by use of single-precision arithmetic and are ill conditioned, will it help to solve them by use of double-precision arithmetic? From our discussion, the answer is *no*. Information already discarded cannot be recovered by subsequent manipulation, however accurately done.

Structures susceptible to ill-conditioning include thin shells, for which membrane stiffness is much greater than bending stiffness. For the same reason, the frame of Fig. 18.2-2 may be troublesome: the axial stiffness of horizontal member 1–2 greatly exceeds its bending stiffness, so horizontal load will produce $u_1 \approx u_2$. For such a problem it is usually best to *enforce* $u_1 = u_2$ by application of a constraint. Thus, a source of ill-conditioning is removed.

An Important Special Case. The three-member structure in Fig. 18.2-3*a* has springs of stiffness αk and k. If α is large, a stiff structure (the spring of stiffness αk) is supported against rigid-body motion by a flexible structure (the springs of stiffness k). Structural equations of this system are

$$\begin{bmatrix} k(1 + \alpha c^2) & k\alpha cs \\ k\alpha cs & k(1 + \alpha s^2) \end{bmatrix} \begin{Bmatrix} u_1 \\ v_1 \end{Bmatrix} = \begin{Bmatrix} P \\ 0 \end{Bmatrix} \qquad \text{where} \qquad \begin{matrix} c = \cos\beta \\ s = \sin\beta \end{matrix} \tag{18.2-4}$$

In u_1v_1 space, the straight lines corresponding to these two equations have the respective slopes

$$-\frac{1 + \alpha c^2}{\alpha cs} \qquad \text{and} \qquad -\frac{\alpha cs}{1 + \alpha s^2} \tag{18.2-5}$$

If $\alpha >> 1$, the two slopes are nearly the same, and ill-conditioning is therefore

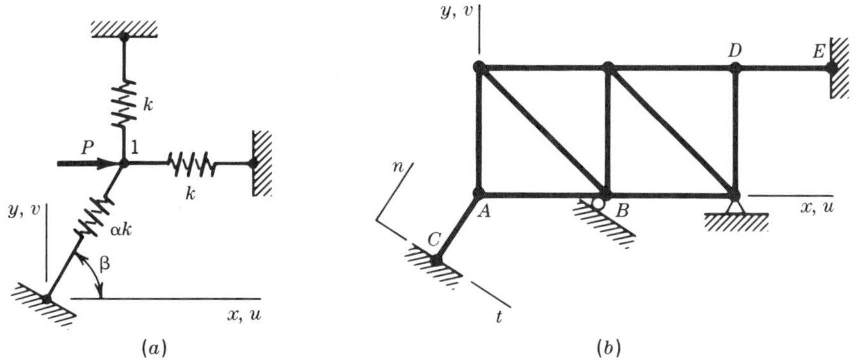

Figure 18.2-3. (*a*) Three-spring structure, having d.o.f. u_1 and v_1 at node 1. (*b*) Skew supports at *A* and *B* in a plane truss.

indicated. However, there are two exceptions: for $c = 0$ ($\beta = \pi/2$) and for $s = 0$ ($\beta = 0$), spring αk contributes no off-diagonal terms to the structure stiffness matrix, and the equations are well conditioned for any value of α.

The practical implication of the foregoing analysis is as follows. In Fig. 18.2-3*b*, let each node i have the usual d.o.f. u_i and v_i. A very stiff bar *DE* approximates the constraint $u_D = 0$ but does not produce ill-conditioning because $\beta = 0$ for bar *DE*. At node *A*, imagine that the constraint of zero n-direction motion is desired. One can approximate this constraint condition by inserting a very stiff bar *AC*. This invites ill-conditioning if d.o.f. at *A* are x- and y-direction displacements, but not if d.o.f. at *A* are n- and t-direction displacements (introduced by coordinate transformation prior to assembly of elements). However, if coordinate transformation has been invoked, one may as well simply set the n-direction displacement at *A* to zero (as is implied at node B).

Avoiding Trouble. We have seen that large stiffness differences may be troublesome. The ideal, not always attainable in practice, is a model without large discrepancies in stiffness.

Arbitrary adjustments in modeling may be appropriate. A comparatively stiff region may be modeled as perfectly rigid by use of a constraint transformation, as has been noted in connection with Fig. 18.2-2. On the other hand, perhaps a stiff region can be made more flexible. For example, in Fig. 18.2-2 one may be able to decrease the axial stiffness AE/L of bar 1–2 by a factor of (say) 100: if the members are slender, bending will still dominate the solution, and u_1 and u_2 will still greatly exceed the relative motion $u_2 - u_1$.

Another possibility in modeling is to use relative motions, rather than absolute motions, for troublesome d.o.f. For example, in Fig. 18.2-1 one could introduce the d.o.f. $u_r = u_1 - u_2$. Instead of Eqs. 18.2-3 we now have

$$\begin{bmatrix} k_1 & 0 \\ 0 & k_2 \end{bmatrix} \begin{Bmatrix} u_r \\ u_2 \end{Bmatrix} = \begin{Bmatrix} P \\ P \end{Bmatrix} \tag{18.2-6}$$

Equations 18.2-6 are well conditioned for all values of k_1 and k_2. Equations 18.2-6 must be obtained by using u_r as a d.o.f. in element formulation and assembly. No purpose would be served by introducing u_r by coordinate transformation of the assembled equations, Eqs. 18.2-3, as the coefficient $k_1 + k_2$ already contains the error we seek to avoid.

After a good finite element model is prepared, subsequent manipulation errors may be negligible. However, it is recommended that double-precision arithmetic be used for all phases of a finite element analysis, unless the round-off limit is about 10^{-14} in single precision or the problem has very few d.o.f.

18.3 THE CONDITION NUMBER

A numerical measure of the ill-conditioning in a coefficient matrix [**K**] is the *condition number*, denoted here by $C(\mathbf{K})$. A large condition number warns that a finite element solution *may* contain appreciable error: the anticipated error may be realized for some systems and some loadings but not for others.

Definition and Interpretation. The *spectral condition number* of a matrix [**K**], which we will call simply the condition number, is defined as

$$C(\mathbf{K}) = \frac{\lambda_{\text{max}}}{\lambda_{\text{min}}} \tag{18.3-1}$$

where λ_{max} and λ_{min} are largest and smallest eigenvalues of [**K**]. (It is best to scale [**K**] before calculating $C(\mathbf{K})$, as described in connection with Eqs. 18.3-3 to 18.3-6.)

It can be shown [18.1] that for each power of ten in the ratio $\lambda_{\text{max}}/\lambda_{\text{min}}$, the operations of equation solving lose about one digit of accuracy in the displacement mode associated with λ_{min}. In a direct method of solving equations, such as Gauss elimination, the difficulty may materialize near the end of the forward-reduction phase, when the difference is computed between two numbers that are almost equal. Error contamination spreads during the back-substitution phase. The estimated accuracy loss is

$$\text{accurate digits lost} \approx \log_{10} \frac{\lambda_{\text{max}}}{\lambda_{\text{min}}} = \log_{10} C(\mathbf{K}) \tag{18.3-2}$$

For example, imagine that $C(\mathbf{K}) = 10^5$. If computer words have seven-digit capacity, only two reliable digits may remain in the computed displacements. With fourteen-digit capacity, nine accurate digits may remain.

The estimate given by Eq. 18.3-2 is based on the inherited error, caused by round-off, that exists at the *outset* of equation solving. The estimate does not include the manipulation error of the solution algorithm.

For stress analysis purposes, the estimate in Eq. 18.3-2 may be pessimistic because it ignores the load vector. Consider Fig. 18.3-1. For both load cases shown, the percentage errors in computed values of u_1 and u_2 are larger than the percentage error in the computed relative displacement $u_1 - u_2$. In stress analysis one seeks the largest strains. The strain of largest magnitude is u_2/L in Fig. 18.3-1*a* but is $(u_1 - u_2)/L$ in Fig. 18.3-1*b*. Accordingly, ill-conditioning may be of little consequence in Fig. 18.3-1*b*. In general, for stress analysis one should estimate the number accurate digits lost as $\log_{10}(\lambda_{\text{max}}/\lambda_k)$, where λ_k is the lowest eigenvalue of [**K**] whose associated eigenmode is not approximately orthogonal to the load vector. Usually $\lambda_k = \lambda_{\text{min}}$, but not necessarily. Thus, in Fig. 18.3-1*b*, the loss of digits in the strain $(u_1 - u_2)/L$ is more accurately predicted by using

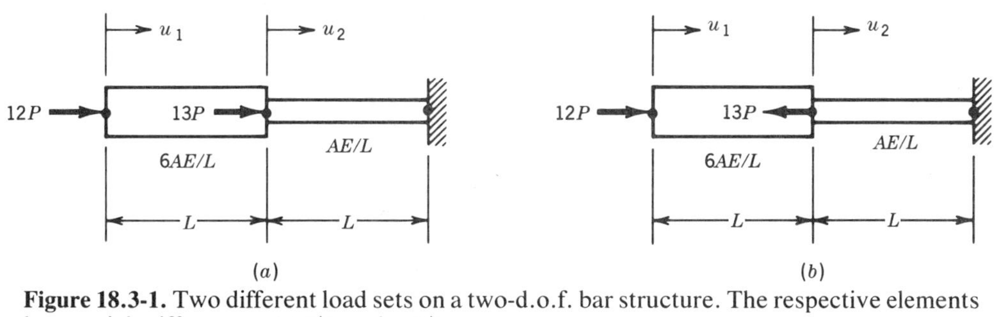

Figure 18.3-1. Two different load sets on a two-d.o.f. bar structure. The respective elements have axial stiffnessess $6AE/L$ and AE/L.

the λ_k associated with the eigenmode in which u_1 and u_2 are of opposite sign than by using $\lambda_k = \lambda_{min}$ (in whose eigenmode u_1 and u_2 are of the same sign).

Calculation and Scaling. Besides ignoring the load vector, Eq. 18.3-2 may overestimate error because $C(\mathbf{K})$ is "artificially" high. In Fig. 18.2-1 for example, $C(\mathbf{K})$ is large when $k_1 >> k_2$ and when $k_2 >> k_1$, yet the case $k_2 >> k_1$ is well conditioned. One can arrange for $C(\mathbf{K})$ to be large only when [**K**] is truly ill conditioned by scaling [**K**] before calculating $C(\mathbf{K})$, as follows. One constructs a diagonal scaling matrix $\lceil \mathbf{S} \rfloor$ from diagonal coefficients in [**K**], then transforms [**K**] to the scaled matrix [$\mathbf{K}_s$]:

$$[\mathbf{K}_s] = \lceil \mathbf{S} \rfloor [\mathbf{K}] \lceil \mathbf{S} \rfloor \qquad \text{where} \quad S_{ii} = \frac{1}{\sqrt{K_{ii}}} \tag{18.3-3}$$

Diagonal coefficients of [$\mathbf{K}_s$] are unity. The extreme eigenvalues λ_{max} and λ_{min} of [$\mathbf{K}_s$] are to be used in Eq. 18.3-2.

The eigenvalue problem that yields λ_{max} and λ_{min} of the scaled matrix [$\mathbf{K}_s$] is

$$([\mathbf{K}_s] - \lambda \lceil \mathbf{I} \rfloor)\{\mathbf{D}\} = \{\mathbf{0}\} \tag{18.3-4}$$

which may be regarded as a vibration problem with a unit mass matrix and natural frequencies $\omega_i^2 = \lambda_i$. An alternative form, which yields the same eigenvalues as Eq. 18.3-4, is produced by substituting $\{\mathbf{D}\} = \lceil \mathbf{S} \rfloor^{-1}\{\mathbf{D}_1\}$ and premultiplying by $\lceil \mathbf{S} \rfloor^{-1}$. Thus [18.3]

$$(\lceil \mathbf{S} \rfloor^{-1}[\mathbf{K}_s]\lceil \mathbf{S} \rfloor^{-1} - \lambda \lceil \mathbf{S} \rfloor^{-1}\lceil \mathbf{S} \rfloor^{-1})\{\mathbf{D}_1\} = \{\mathbf{0}\} \tag{18.3-5}$$

which is the same as

$$([\mathbf{K}] - \lambda \lceil K_{11} \quad K_{22} \quad \ldots \quad K_{nn} \rfloor)\{\mathbf{D}_1\} = \{\mathbf{0}\} \tag{18.3-6}$$

Therefore, λ_{max} and λ_{min} of the scaled matrix [$\mathbf{K}_s$] can be calculated using the unscaled [**K**] and a diagonal "mass" matrix that is simply the principal diagonal of [**K**]. Equation 18.3-6 shows why an isolated stiff region raises $C(\mathbf{K})$: an isolated large "mass" K_{ii}, if not held by supports, reduces the lowest "frequency" but has little effect on the highest "frequency."

Scaling is used only to avoid obtaining an unrealistically pessimistic result from Eq. 18.3-2 because of "artificial" ill-conditioning. Scaling need not be applied to the [**K**] used in solving equations. Scaling of [**K**] has no effect on the accuracy of a direct-solution algorithm such as Gauss elimination, assuming that the scaling process itself introduces no manipulation error, and provided that the choice of pivots and the sequence of eliminations are unchanged [18.4].

Because Eq. 18.3-2 is only an approximation, λ_{max} and λ_{min} need not be computed accurately. A close upper bound on λ_{max} of [$\mathbf{K}_s$] can be obtained from the Gerschgorin bound, Eq. 13.10-17. Thus we add the magnitudes of coefficients in each row of [$\mathbf{K}_s$], then choose the largest such row sum, that is,

$$\lambda_{max} \approx \max Q_i \qquad \text{where} \quad Q_i = \sum_{j=1}^{n} |K_{sij}| \tag{18.3-7}$$

where n is the order of $[\mathbf{K}_s]$. Unfortunately, there is no corresponding simple estimate of λ_{min}. The expense of computing λ_{min} is comparable to the expense of solving equations, which means that Eq. 18.3-2 does not provide an inexpensive a priori estimate of solution accuracy.

Causes of Ill-Conditioning. A finite element model tends to produce ill-conditioned equations if an element or a patch of elements can respond to loads with large rigid-body motion but little deformation. Examples include (a) high-modulus inclusions, (b) plate elements that allow transverse shear strain but have large transverse shear stiffness because they are thin, (c) elements of severe shape distortion or large aspect ratio, and (d) stiff supports such as spring αk in Fig. 18.2-3. (Analogous difficulties can appear in nonstructural problems.) In the foregoing cases one can reduce or eliminate the trouble by changing the model. In the respective examples, one can (a) make the inclusion rigid by applying constraints, (b) use thin-plate elements or arbitrarily decrease the shear stiffness, (c) remodel using more regular and compact element shapes, and (d) use differently directly d.o.f. at the offending node.

Ill-conditioning of a stiffness matrix may also be caused by mixing elements of different size and by using a fine mesh. It has been found that [18.5]

$$C(\mathbf{K}) = b\left(\frac{h_{max}}{h_{min}}\right)^{2m-1} N^{2m/n} \tag{18.3-8}$$

where b = a positive constant independent of h_{max} and h_{min},
h_{min} = smallest node spacing in any element of the mesh,
h_{max} = greatest node spacing in any element of the mesh,
N = number of elements,
$2m$ = differential equation order,
n = dimensionality.

To elaborate, the differential equation that describes the physical problem uses displacements as dependent variables, has $2m$ as the highest derivative of the dependent variable(s), and requires n independent variables. As examples, for an axially loaded bar, $2m/n = 2/1$; for a beam, $2m/n = 4/1$; in plane stress, $2m/n = 2/2$; in thin-plate bending, $2m/n = 4/2$; in three-dimensional solids, $2m/n = 2/3$; for a thin shell, $2m/n = 4/3$. Thus, in a beam problem, if the length ratio h_{max}/h_{min} of elements is changed from 1/1 to 10/1, $C(\mathbf{K})$ increases by a factor of 1000. If the number of elements is doubled, $C(\mathbf{K})$ increases by a factor of 16. If *both* of these changes are made, $C(\mathbf{K})$ increases by a factor of 16,000.

As a practical matter, we would like to know the dependencies of the condition number on important mesh and problem parameters. For the *unscaled* stiffness matrix, Fried [18.5] has shown that

$$\left(\frac{1}{\lambda_1 c_{max}}\right)\frac{\max\limits_{1\le\ell\le N}(\Lambda_\ell^k)}{\max\limits_{1\le\ell\le N}(\Lambda_\ell^m)} \le C(\mathbf{K}) \le \left(\frac{c_{max}}{\lambda_1}\right)\frac{\max\limits_{1\le\ell\le N}(\Lambda_\ell^k)}{\min\limits_{1\le\ell\le N}(\lambda_\ell^m)} \tag{18.3-9}$$

where, using *unscaled* element stiffness and mass matrices,

Λ_ℓ^k, Λ_ℓ^m = maximum eigenvalue of the stiffness and mass matrices, respectively, of element ℓ,

λ_ℓ^m = minimum eigenvalue of the mass matrix of element ℓ,

λ_1 = minimum eigenvalue of the continuous problem,

N = number of elements in the structure,

c_{max} = maximum number of elements meeting at a single node.

The mass matrix cited need not be the consistent mass matrix; it may be lumped provided that no λ_ℓ^m is zero because a d.o.f. is assigned zero mass. Alternative notation for λ_1 is ω_1^2, where ω_1 is the fundamental vibration frequency of the actual structure. Mass density cancels in the denominators and so may be taken as unity. One can use estimates of Λ_ℓ^k and Λ_ℓ^m obtained from the Gershgorin bound, Eq. 13.10-17. Hence, if λ_1 is known, Eq. 18.3-9 provides estimated numerical bounds on $C(\mathbf{K})$. We see from Eq. 18.3-9 that the bound on $C(\mathbf{K})$ becomes less certain as more elements are connected to a node. However, c_{max} is typically 4, 6, or even 8, which indicates a probable uncertainty of about one digit in $\log_{10} C(\mathbf{K})$.

The condition number may be strongly affected by Poisson's ratio ν. Fried [18.18] has shown that b in Eq. 18.3-8 is given by

$$b = \frac{b_1}{1 - 2\nu} \tag{18.3-10}$$

where b_1 is a constant that is independent of E and ν (provided that E and ν are constant throughout the mesh). Equation 9.4-12 shows that b is a penalty number α, that is, $\alpha = b/3b_1$. Near the incompressibility limit of $\nu = 0.5$, Eq. 18.3-10 can increase $C(\mathbf{K})$ by orders of magnitude. If computer words carry about p digits each, and if the value of α or b approaches $10^{p/2}$ as suggested at the end of Section 9.4, then we might expect an accuracy loss of almost $p/2$ digits in the solution process. Such a loss may be of concern if pressures in nearly incompressible media are computed by the penalty method (see Eqs. 9.5-1 and 9.6-2), for the following reason. Strain ϵ_V in Eq. 9.5-1 is small because strains $u_{,x}$, $v_{,y}$, and $w_{,z}$ almost cancel one another when added to produce ϵ_V, even when using computer words of p-digit accuracy. We now propose to calculate ϵ_V by adding three strains that are each accurate in only about the leading $p/2$ digits. If ϵ_V is to be computed with useful accuracy, a more rigorous analysis suggests that α should be approximately $10^{p/3}$. In practice the estimate $\alpha \approx 10^{p/3}$ appears to be sufficiently pessimistic that a value of α approaching $10^{p/2}$ can usually be used.

18.4 DIAGONAL DECAY ERROR TESTS

Here we describe a simple and inexpensive test for round-off errors that appear during a direct method of solving equations, such as the Gauss elimination method. The test can warn of possible trouble and can be used to terminate execution if serious trouble is indicated [18.6].

Assume that the coefficient matrix $[\mathbf{K}]$ is symmetric and positive definite. Then, as each equation is processed, that is, as each unknown is eliminated, a subtraction operation reduces the magnitude of diagonal coefficients K_{ii} that correspond to d.o.f. i yet to be eliminated (however, each K_{ii} remains positive). Thus, each K_{ii}

tends to accumulate round-off errors. A simple example of the decay of diagonal coefficients appears in Fig. 2.11-1. Another example appears in connection with Eq. 18.2-3, where Gauss elimination of u_1 produces the reduced coefficient $K_{22} = (k_1 + k_2) - k_1$. There it is seen that the reduced K_{22} may have no accurate digits if $k_1 >> k_2$ and computer words are too short. The reduced K_{22}, now perhaps greatly in error, is used as a pivot in eliminating u_2, causing error to propagate throughout the solution.

A simple test of accuracy is to check the amount of decay of each K_{ii} before it acts as a pivot. Let P_{ii} be the reduced value of K_{ii} just before it is used as a pivot in eliminating the ith d.o.f. The diagonal decay ratio

$$r_i = \frac{K_{ii}}{P_{ii}} \tag{18.4-1}$$

where K_{ii} is the original diagonal coefficient, is an approximate measure of the number of digits of accuracy lost; for example, roughly six accurate digits have been lost if any r_i is 10^6, leaving only $p - 6$ accurate digits if computer words contain p digits.

Table 18.4-1 shows numerical results obtained from a beam problem on a computer that carries slightly more than eleven digits in double precision arithmetic. We see that it is best if node numbers progress from the free end to the fixed end. (Unfortunately, this arrangement is not practical for all finite element structures.) We also see that trouble may not be detected until reduction is almost complete: in the case where node N is free, trouble appears only in the next-to-last equation, when the equation solver detects that node N is unsupported.

The example in Table 18.4-1 uses elements of equal length. Consider the case $N = 1000$, now with elements mixed in length but varying in length only between 0.99999 and 1.00001. Then the tip-to-root numbering gives $w = 1.0269$ while diagonal decay ratios are essentially unchanged from their values of 8.0 and 2.0. This example shows the possibility of substantial accuracy loss that is not detected by the diagonal decay ratio.

If a large coefficient appears only on the diagonal of [**K**], without corresponding

TABLE 18.4-1. EFFECT OF NUMBER OF BEAM ELEMENTS AND NODE NUMBERING ON ACCURACY, WHERE DIAGONAL DECAY RATIO r_i IS DEFINED BY EQ. 18.4-1 AND w IS THE RATIO OF THE COMPUTED DEFLECTION OF LOAD P TO ITS DEFLECTION ACCORDING TO BEAM THEORY.

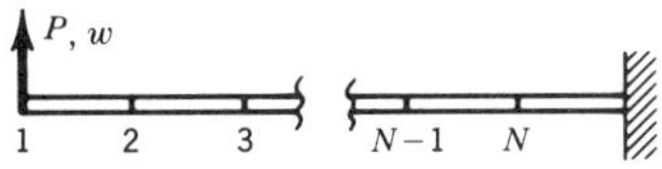

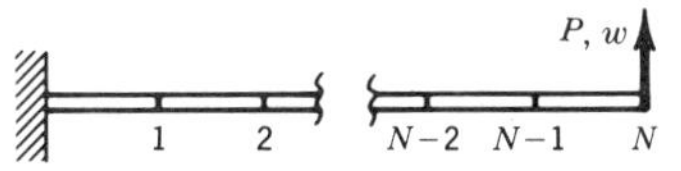

$EI = 8(10)^6$, Length = 1000	N Elements[a] 2N Nonzero D.O.F.				N Elements[a] 2N Nonzero D.O.F.			
Number of elements	w	r_{2N-2}	r_{2N-1}	r_{2N}	w	r_{2N-2}	r_{2N-1}	r_{2N}
$N = 10$	1.0000	8.0	2.0	8.0	1.0000	5.3	$1(10)^3$	$4(10)^1$
$N = 100$	1.0000	8.0	2.0	8.0	0.9993	7.7	$1(10)^6$	$4(10)^2$
$N = 1000$	1.0000	8.0	2.0	8.0	0.1197	8.0	$2(10)^8$	$2(10)^3$

[a]Elements are of equal length (see text).

large off-diagonal coefficients, it does not lead to large diagonal decay. Thus, $k_2 >> k_1$ is acceptable in Eq. 18.2-3, and $\beta = 0$ or $\beta = \pi/2$ is acceptable in Fig. 18.2-3*a*.

It is the *decay* of diagonals that is significant, not their smallness. Small pivots *per se* do not provoke large error. The causes of a large decay ratio r_i are the causes of ill-conditioning: inadequate supports, mechanisms, isolated stiff regions, and so on. It is interesting that the condition number of [**K**] is the same for both beams in Table 18.4-1, roughly 10^{12} when $N = 1000$, yet the severe loss of accuracy predicted by this large $C(\mathbf{K})$ materializes for only one of the two beams.

18.5 RESIDUALS

Consider the equation

$$\{\Delta\mathbf{D}\} = [\mathbf{K}]^{-1}\{\Delta\mathbf{R}\} \qquad \text{where} \quad \{\Delta\mathbf{R}\} = \{\mathbf{R}\} - [\mathbf{K}]\{\mathbf{D}\} \tag{18.5-1}$$

where $\{\Delta\mathbf{R}\}$ is called the *residual*. If the equations $[\mathbf{K}]\{\mathbf{D}\} = \{\mathbf{R}\}$ could be solved exactly, that is, so that the solution algorithm does not contaminate $\{\mathbf{D}\}$ with any round-off error, then $\{\Delta\mathbf{R}\} = \{\mathbf{0}\}$. Actually, there is some contamination, and $\{\Delta\mathbf{D}\}$ can be regarded as a measure of the probable round-off error in $\{\mathbf{D}\}$. A similar scalar error measure is

$$e = \frac{\{\mathbf{D}\}^T\{\Delta\mathbf{R}\}}{\{\mathbf{D}\}^T\{\mathbf{R}\}} \tag{18.5-2}$$

Physically, e is the ratio of work done by residual loads to work done by actual loads when both act through displacements $\{\mathbf{D}\}$. If $|e|$ is (say) 10^{-8} or less, round-off contamination is probably negligible [18.7]. However, if the equations $[\mathbf{K}]\{\mathbf{D}\} = \{\mathbf{R}\}$ are ill conditioned, it is *possible* that inaccurate solutions will yield small residuals and a small e.

A measure of error in vibration analysis, similar to Eq. 18.5-2, is

$$e = \frac{\{\overline{\mathbf{D}}\}^T([\mathbf{K}]\,\{\overline{\mathbf{D}}\} - \omega^2[\mathbf{M}]\{\overline{\mathbf{D}}\})}{\{\overline{\mathbf{D}}\}^T[\mathbf{K}]\{\overline{\mathbf{D}}\}} \tag{18.5-3}$$

in which the quantity in parentheses would be zero if frequency ω and mode shape $\{\overline{\mathbf{D}}\}$ were uncontaminated by round-off error.

One can use Eq. 18.5-1 as an iterative improvement scheme. Thus the computed $\{\Delta\mathbf{D}\}$ is added to the existing $\{\mathbf{D}\}$, a new $\{\Delta\mathbf{R}\}$ is formed, a new $\{\Delta\mathbf{D}\}$ is computed and added to the current $\{\mathbf{D}\}$, and so on. If this method is to succeed, $|e|$ must be less than unity, and the $\{\mathbf{R}\}$ and $[\mathbf{K}]$ used to compute $\{\Delta\mathbf{R}\}$ must be represented with greater precision than the $[\mathbf{K}]^{-1}$ used to compute $\{\Delta\mathbf{D}\}$. With $[\mathbf{K}]$ represented accurately and $[\mathbf{K}]^{-1}$ represented inaccurately, convergence is toward the exact solution of $[\mathbf{K}]\{\mathbf{D}\} = \{\mathbf{R}\}$, as is desired, rather than toward the exact solution of $\{\mathbf{D}\} = [\mathbf{K}]^{-1}\{\mathbf{R}\}$. Iterative improvement reduces the rounding error introduced during equation solving but does not reduce the inherited error present at the outset of equation solving.

If a set of equations is seriously ill conditioned, it is usually better to rework the finite element model so as to improve its condition than to make heroic attempts to improve a poor solution by iteration. "If a thing is not worth doing, it is not worth doing well" [18.8].

Interpretation of {ΔR}. A small residual does not guarantee that equations have been solved accurately. Of two approximate solutions, it is possible that the solution of lesser accuracy will yield the smaller residuals [18.8].

In structural mechanics, a small $\{\Delta\mathbf{R}\}$ indicates that applied loads $\{\mathbf{R}\}$ are balanced by resisting loads $[\mathbf{K}]\{\mathbf{D}\}$ arising from the deformed structure. Checking that $\{\Delta\mathbf{R}\} \approx \{\mathbf{0}\}$ is sometimes called a "statics check" or an "equilibrium check." Passing an equilibrium check is no guarantee that results are accurate. For example, one may compute deflections and stresses that are physically unrealistic because the mesh is too coarse, yet find $\{\Delta\mathbf{R}\} \approx \{\mathbf{0}\}$ (merely because the equation solver is working properly).

In summary, small residuals are a necessary but not sufficient condition for accuracy. Reference 18.21 suggests a residual norm test: that $\|\Delta\mathbf{R}\| << \|\mathbf{R}\|$ if the solution is sufficiently accurate. Reference 18.21 also suggests that the solution is probably acceptable if either the condition number $C(\mathbf{K})$ alone or the residual norm test alone indicates trouble, but probably unacceptable if both $C(\mathbf{K})$ and the residual norm test indicate trouble.

18.6 DISCRETIZATION ERROR: ANALYSIS

In representing a mathematical continuum by finite elements, we select the number, type, and shape of elements, the grading of the mesh, allocate distributed loads to nodes, and represent support conditions by fixing certain d.o.f. The approximation inherent in this process is called *discretization error*. Unfavorable discretization can provoke subsequent numerical difficulty, for example, as described in Section 18.2. However, in the present section we are concerned only with discretization error itself, that is, with the discrepancy between the discretized model and the mathematical model, with the latter being taken as correct.

A simple rule is that the volume of the discretized structure should be correct. Thus, if a circular region is modeled by a polygon of straight-sided elements, the polygon should neither inscribe nor circumscribe the circle; rather, element sides should intersect the circle so that the area of the polygon equals the area of the circle. Similarly, straight elements that model an arch should not be chords, but longer, so that the sum of element lengths equals the arch length.

Error Analysis. Discretization error can sometimes be determined by an order of error analysis. Consider, for example, the axially loaded bar of Fig. 18.6-1*a*. From the differential element, the equation of axial equilibrium is $A\sigma_{x,x} + q = 0$. Or, substituting the stress–strain relation $\sigma_x = Eu_{,x}$, the equilibrium equation is

$$u_{,xx} + \frac{q}{AE} = 0 \qquad (18.6\text{-}1)$$

Let the discretized model consist of standard two-d.o.f. bar elements. It is not hard to show that *if the load integral is evaluated consistently* (i.e., by Eq. 4.1-6), and if A and E are constant, then finite element displacements are exact at the nodes (see Problems 18.28 and 18.29). This does not mean that finite element displacements are exact everywhere: the exact solution is in general not piecewise linear, so there is error *between* the nodes.

The behavior of the error in the bar model can be understood by doing a comparatively simple error analysis. The error of the bar model is in many ways either characteristic of, or very relevant to, the behavior of the error in more complicated and realistic problems. We cannot hope to obtain discretization error bounds in all practical problems. But many model problems can be analyzed, each reflecting salient features of a more practical problem. What we touch on here is most directly applicable to linear elasticity problems with smooth (but possibly varying) elastic coefficients, with distributed (not point) loading, in bodies whose geometry does not induce strong singularities (no cracks). The arguments and conclusions rest on a fundamental fact of finite element error analysis: *for a sufficiently refined mesh,* the error in the finite element solution can be bounded by the error in approximating the exact solution by shape function interpolation that is exact at nodes [18.9]. Thus, whether or not the finite element solution is exact at nodes (and in general it is not), the error in nodal interpolation tells the story. The proviso of "a sufficiently refined mesh" is intended to exclude models having large discretization error, such as a beam modeled by a few constant-strain triangles, whose computed nodal displacements are greatly in error.

Accordingly, in the bar problem we analyze the error of linear interpolation between exact nodal displacements u_i and u_{i+1}. In the ith element, displacement error $e = e(x)$ is

$$e(x) = u(x) - u_i\left(1 - \frac{x - x_i}{h_i}\right) - u_{i+1}\left(\frac{x - x_i}{h_i}\right) \tag{18.6-2}$$

where $u(x)$ is the exact solution. Also, $u_i = u(x_i)$ and $u_{i+1} = u(x_{i+1})$, where x_i and x_{i+1} are the nodal locations. Element length is $h_i = x_{i+1} - x_i$. There is no need to require all element lengths to be equal. In the ith element, error $e(x)$ is a smooth function. At nodes, there are discontinuities in the first derivative[1] $e'(x)$, as shown by Fig. 18.6-1c. We note that $e(x_i) = e(x_{i+1}) = 0$.

One can determine an upper bound on $e(x)$ by the following argument [18.9]. Let z be an axial coordinate $x_i \leq z \leq x_{i+1}$ such that $e'(z) = 0$. That there is such a point is obvious from Fig. 18.6-1b and follows rigorously from Rolle's theorem of elementary calculus. In what follows we assume that $u''(x)$ is continuous in the interval $x_i \leq x \leq x_{i+1}$. The change in e' can be computed by integrating e''. Thus, with s a dummy variable, $e'(z) = 0$, and using Eq. 18.6-2,

$$e'(x) - e'(z) = e'(x) = \int_z^x e''(s)\, ds = \int_z^x u''(s)\, ds \tag{18.6-3}$$

because the linear terms in Eq. 18.6-2 do not contribute to $e''(x)$. In addition,

$$\left|\int_z^x u''(s)\, ds\right| \leq \int_z^x |u''(s)|\, ds \leq \int_{x_i}^{x_{i+1}} |u''(s)|\, ds \leq h_i\left(\max_{x_i \leq x \leq x_{i+1}} |u''(x)|\right) \tag{18.6-4}$$

[1]In the remainder of this section we use the notation $e_{,x} = e'$, $e_{,xx} = e''$, and so on.

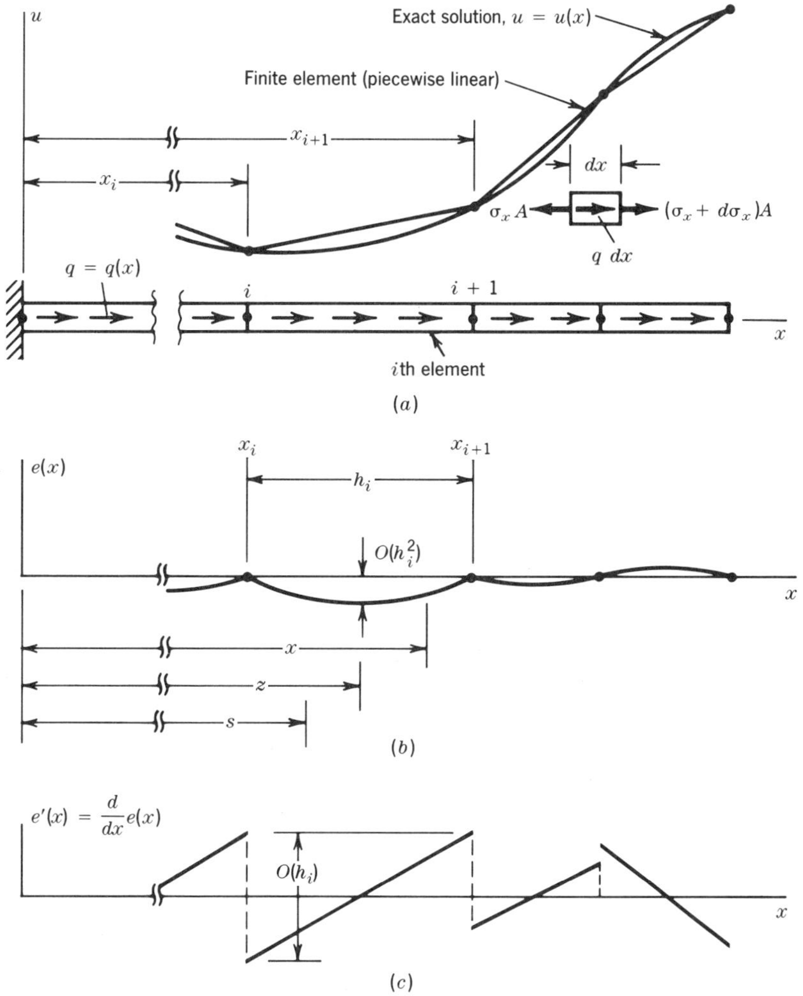

Figure 18.6-1. The behavior of the error in a uniform bar under distributed axial load q. The exact axial displacement is $u = u(x)$ and the error is $e = e(x)$.

Therefore, on the ith element the error in strains $e'(x)$ is bounded by

$$|e'(x)| \leq h_i \left(\max_{x_i \leq x \leq x_{i+1}} |u''(x)| \right) \tag{18.6-5}$$

We can also bound the displacement error $e(x)$ by observing that it must have greatest magnitude at $x = z$, where $e'(z) = 0$. Now z must be closer to x_i or to x_{i+1}. Assume that z is closer to x_i, and compute $e(x_i)$ by a three-term Taylor series, with exact remainder, expanded about $x = z$,

$$e(x_i) = e(z) + (x_i - z)e'(z) + \tfrac{1}{2}(x_i - z)^2 e''(s) \tag{18.6-6}$$

where s is the remainder evaluation point on element i. But $e(x_i) = 0$, $e'(z) = 0$, and $e''(s) = u''(s)$ from Eq. 18.6-2, so

$$e(z) = -\tfrac{1}{2}(x_i - z)^2 u''(s) \tag{18.6-7}$$

According to the assumption that z is closer to x_i than to x_{i+1}, we conclude that $|z - x_i| \leq h_i/2$. Therefore, the error in displacements, $e(x)$, is bounded by

$$e(x) \leq \tfrac{1}{8}h_i^2 \left(\max_{x_i \leq x \leq x_{i+1}} |u''(x)| \right) \tag{18.6-8}$$

One may verify that the same result is obtained when z is assumed to be closer to x_{i+1} than to x_i.

We note that the existence of a z for which the error in strain is zero is the rationale for the existence of the optimal stress calculation points discussed in Section 6.13.

Important features of the foregoing discussion are as follows.

1. The strain error is proportional to element size, and the displacement error is proportional to the square of element size.
2. The error estimates are proportional to derivatives one order higher than the degree of the shape functions (for the bar, second derivatives and linear shape functions are involved).
3. Displacements are most accurate at or near element nodes. Strains are most accurate in element interiors, that is, at or near Gauss points.

Using the notation "O" for "order," we say that Eq. 18.6-5 displays a discretization error $O(h)$ in strain and Eq. 18.6-8 displays a discretization error $O(h^2)$ in displacement. Thus, if $h = \max(h_i)$ is cut in half to produce two elements, h is halved; the error in strain is approximately halved and the error in displacement is approximately quartered.

Remarks. We would like to generalize the foregoing conclusions to other elements and to strain energy error as well [18.10]. For this purpose we define symbols as follows:

h = approximate "characteristic length" of an element: length of a linear element; length of the longest line segment connecting two points in a plane or solid element

$q - 1$ = degree of highest *complete* polynomial in the element displacement field

$2m$ = order of highest derivative in the governing equilibrium equation expressed in terms of displacements

For the bar of Fig. 18.6-1, $q - 1 = 1$ and $2m = 2$. For the standard four-d.o.f. beam element, $q - 1 = 3$ and $2m = 4$. In plane and solid problems with bilinear and trilinear elements discussed in Chapter 6, $q - 1 = 1$ and $2m = 2$. For bending of thin plates with twelve d.o.f. elements based on Eq. 11.2-5, $q - 1 = 3$ and $2m = 4$.

Because an element can fit exactly a displacement field of degree $q - 1$, it therefore has error $O(h^q)$ in representing the polynomial fields of degree q and higher that in general are present in the exact solution. The error in strains (and therefore stresses) is proportional to the error in the rth derivative of the displacement field; that is, the stress error is $O(h^{q-r})$. For plane and solid elements, $r = 1$. For beam and plate elements, for which stresses are dictated by curvatures, $r = 2$. Thus the standard four-d.o.f. beam element has displacement error $O(h^4)$ and stress error $O(h^2)$.

In plane problems it is sometimes possible to use optimal points for stress calculation, as noted in this section and in Section 6.13. Then the stress error is $O(h^q)$ rather than $O(h^{q-1})$.

A strain energy expression contains squares of the mth displacement derivatives. Therefore, the strain energy error is $O(h^{2q-2m})$.

Again we remark that the foregoing error estimates are based on the assumption that nodal loads are evaluated consistently, that is, by Eq. 4.1-6. If ad hoc load lumping is used, or if Eq. 4.1-6 is evaluated by too low a quadrature rule, the estimates *may not apply*. For example, the beam deflection data seen at the end of Section 4.3 use ad hoc load lumping and display displacement error $O(h^2)$ rather than the expected $O(h^4)$.

Singularities. It is a general principle in finite element error estimation that constants involved in an error bound are proportional to the qth partial derivative of the exact solution. If this derivative is infinite, the error bound does not rigorously apply. For example, in problems with cracks, displacement derivatives through the second are singular at the crack tip; therefore, even with linear elements (for which $q - 1 = 1$), our estimates do not apply. Other problems may have singularities in higher derivatives. We often do not know the order of the lowest singular derivative, and therefore we cannot estimate the factor by which error will be reduced by subdividing the mesh.

However, although singularities may slow convergence, the presence of singularities in derivatives higher than the second does not rule out convergence. A polynomial of degree $q - 1$ necessarily contains also a polynomial of degree $q - p$ for $1 < p \leq q$. If the $(q - p + 1)$-order derivative of the exact solution is the highest-order derivative that is nonsingular, then we can apply our previous estimates with $q - 1$ replaced by $q - p$ to deduce that the error in displacements is $O(h^{q-p+1})$, the error in strains (calculated from first derivatives) is $O(h^{q-p})$, and so on. The implication of this is that *for each derivative of order q or less that is singular in the exact solution, we risk losing a power of h accuracy in the finite element solution*. Such an estimate may be pessimistic, and often is. But *when it holds, elements of degree $q - p$ will have the same order of error as elements of degree $q - 1$*, and the lower-degree elements will have less computational expense.

As an example, consider again the axially loaded bar, with a step change in loading *within an element*. Specifically, for the element that spans $x = x_i$ to $x = x_{i+1}$, let $x_i < x_q < x_{i+1}$ and

$$q = 0 \text{ for } x \leq x_q \qquad \text{and} \qquad q = 1 \text{ for } x > x_q \tag{18.6-9}$$

Then, from Eq. 18.6-1 $u_{,xx}$ is defined for all x, $\max|u''(x)| = 1/AE$, and Eqs. 18.6-5 and 18.6-8 make sense. However, $u'''(x)$ is a delta function at $x = x_q$, so $\max|u'''(x)|$ is undefined. Accordingly, the error estimates remain as stated in Eqs. 18.6-5 and 18.6-8, whether the element is linear, quadratic, or of yet higher degree: the $O(h^3)$ displacement accuracy one would normally expect from a quadratic element is not available. Full accuracy of a quadratic element could be recovered by placing a node at the step change in load, for example, set $x_i = x_q$ or set $x_{i+1} = x_q$.

For C^0 elements in general, we do not expect to lose much accuracy when properties such as A, E, ν, thickness, and so on, display sudden jumps at inter-element boundaries. But if properties jump *within* elements or if the exact solution displays "blow-up" singularities in higher derivatives, considerable accuracy may be lost.

18.7 DISCRETIZATION ERROR: ESTIMATION AND EXTRAPOLATION

In the latter part of Section 18.6 we argued that error e in a computed quantity could be represented as $e = O(h^{q-r})$, where $r = 0$ for displacement error, $r = 1$ for stress or strain error in elasticity, and so on. A restatement of this result is

$$e \approx Ch^{q-r} \tag{18.7-1}$$

where h is again the "characteristic length" and C is a problem-dependent constant that is influenced by element aspect ratio (ratio of longest side to shortest side), size ratio of the largest and smallest elements in the mesh, element type, quadrature rule, the $(q - p + 1)$-order derivative of the exact solution (as in Section 18.6), and other factors [18.19]. Error e in a computed quantity ϕ may indeed refer to the error at a point, but more generally is

$$e = \|\boldsymbol{\phi} - \boldsymbol{\phi}^h\| \tag{18.7-2}$$

where $\boldsymbol{\phi}$ is the exact value, $\boldsymbol{\phi}^h$ is the computed value, and the norm symbol $(\|\cdot\|)$ indicates an averaging process; for example, e may represent the root-mean-square of several pointwise values, or an integrated average over a portion of the body, or the square root of strain energy in the entire body, and so on. If a single value of $\boldsymbol{\phi}$ is intended, then $e = |\phi - \phi^h|$.

The *relative* error is $e_r = \|\boldsymbol{\phi} - \boldsymbol{\phi}^h\|/\|\boldsymbol{\phi}\|$. A *rough* indicator of this error is

$$e_r \approx \rho_1\rho_2 h^{q-r} \qquad \text{where} \quad h = \frac{1}{N^{1/n}} \tag{18.7-3}$$

Here q = one plus the degree of the highest complete polynomial in the element displacement field, r is defined above Eq. 18.7-1, ρ_1 = largest element aspect ratio, ρ_2 = ratio of characteristic length of the largest element to characteristic length of the smallest element, h = "dimensionless length," N = number of elements in the mesh, and n = spatial dimension (n = 1, 2, or 3 for line, plane, and solid problems, respectively). Thus h is the characteristic length of an element in a domain that has been scaled so that its length, area, or volume is unity. The analyst has some control over all quantities in Eq. 18.7-3 except n. Our interpretation of e_r is as follows. If e_r is about 0.1 times the acceptable percentage error, the results are probably reliable (e.g., e_r = 1% if 10% error is acceptable). If e_r is considerably smaller than this, the mesh may already be finer than necessary. If e_r approaches or exceeds unity, further study is required. "Further study" may

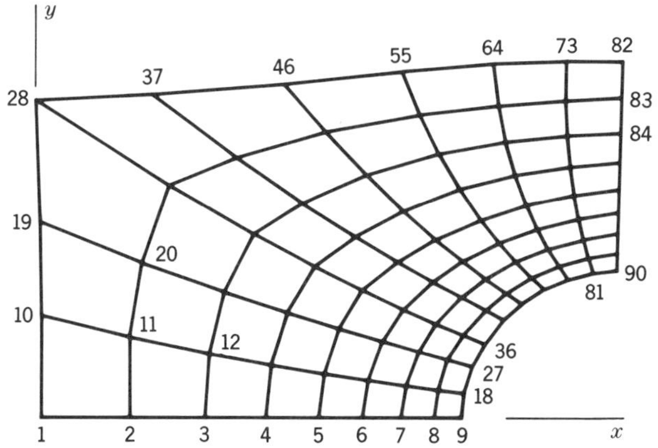

Figure 18.7-1. A "Laplacian" plane mesh. Coordinates x_i and y_i of each interior node are equal to the average of the coordinates of the four adjacent nodes, for example, $x_{11} = (x_2 + x_{10} + x_{12} + x_{20})/4$.

mean mesh refinement, alteration of the mesh without changing the number of elements, or assessment of the physical reasonability of the results if computational limitations do not permit mesh refinement or rearrangement.

However, Eq. 18.7-3 may be very pessimistic. One can easily devise a patch test in which results are exact yet $e_r > 1000$. Clearly, the role of stress gradients is not addressed by Eq. 18.7-3. In preparing or refining a mesh, an analyst must draw upon experience and intuition about gradients present and how well they can (or should) be modeled: if a test quantity such as stress or strain energy changes appreciably across interelement boundaries or within individual elements, mesh refinement may be needed.

Consider, for example, Fig. 18.7-1. Here $N = 72$, $n = 2$, and we estimate $\rho_1 = 2$ and $\rho_2 = 8$. Therefore, if elements are bilinear ($q = 2$), we estimate $e_0 \approx 0.22$ for displacements ($r = 0$) and $e_1 \approx 1.9$ for stresses ($r = 1$). These estimates indicate that results may not be reliable. This may indeed be the case if we seek the effect of a local disturbance, say a concentrated force at node 36. On the other hand, if nodes along the left edge are fixed and nodes along the right edge are loaded by a uniform traction, displacements and stresses near node 81 would probably be reliable because the mesh is smoothly graded, finer where stress gradients are likely to be larger, and fine enough at the location of interest. A uniform mesh ($\rho_1 = \rho_2 \doteq 1$) with elements the size of the smallest element in Fig. 18.7-1 might improve accuracy slightly, but would display $e_0 = O(10^{-3})$ and $e_1 = O(10^{-2})$, and would be much more costly. We conclude that by mesh grading and suiting the mesh to the problem, one can "beat the estimate."

Multimesh Extrapolation. It is possible to use Eq. 18.7-1 and the computed results from two or more different meshes to extrapolate to an improved result. Ideally, this result has zero error. What rigor there is in this process depends on *completely regular mesh refinement*. That is, in each refinement, nodes and interelement boundaries of the coarser mesh are preserved, both in number and location, while adding new nodes and new boundaries; ρ_1 and ρ_2 of Eq. 18.7-3 remain constant; element types and quadrature rules are not changed; corner nodes stay corner nodes and side nodes stay side nodes (Fig. 18.7-2). In addition, the quantity to

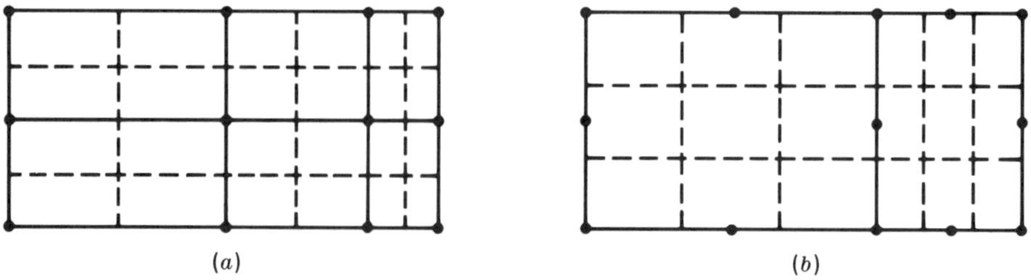

Figure 18.7-2. Dashed lines indicate new element boundaries introduced by regular mesh subdivision. New nodes introduced by subdivision are not shown. (*a*) Bilinear elements; divisions by 2 shown. (*b*) Quadratic elements; divisions by 3 shown.

be extrapolated must be calculated at a location fixed in space and fixed in position relative to an element (e.g., at a corner node) in all refinements. If such regularity is lacking, extrapolation can still yield a useful estimate, but without a sound basis in theory.

If only two meshes are used, and the convergence rate is unknown, we have little choice but to assume that convergence is linear ($e = O(h)$; see Fig. 18.7-3*a*). With data from three meshes, we may conclude that convergence is indeed linear, or that quadratic (or higher) convergence is a possibility. A too-coarse mesh may fail to display a definite trend. Incompatible elements often display nonmonotonic convergence (curve *AD* in Fig. 18.7-3*b*). With nonmonotonic convergence, two meshes yield no reliable prediction: clearly, two data points can be extrapolated to a result that is better, worse, or no different, depending on the points chosen. If convergence is monotonic and p is the order of error, $e = O(h^p)$, then an improved result ϕ^0 is obtainable by Richardson extrapolation [18.11],

$$\phi^0 = \frac{\phi_1 h_2^p - \phi_2 h_1^p}{h_2^p - h_1^p} \tag{18.7-4}$$

where, as examples, $p = 1$ for $e = O(h)$ in Fig. 18.7-3*a* and $p = 2$ for $e = O(h^2)$ in Fig. 18.7-3*b*. Equation 18.7-4 usually gives an inexact ϕ^0 because conditions

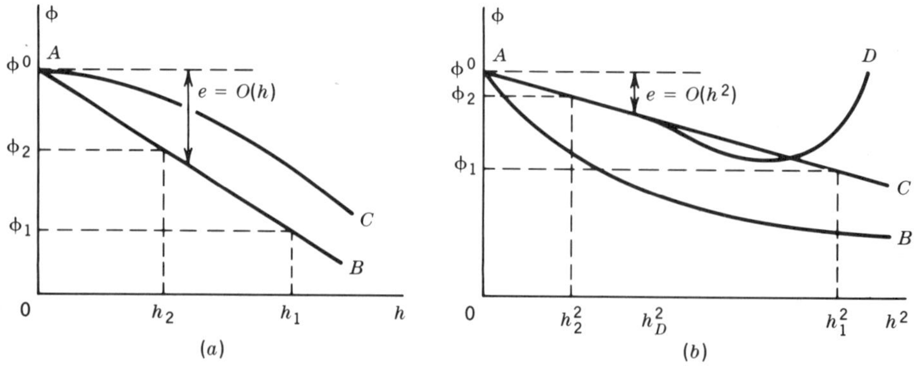

Figure 18.7-3. Convergence of a quantity ϕ with mesh refinement. *AB*, linear convergence. *AC*, quadratic convergence, *AD*, nonmonotonic convergence (but quadratic convergence when $h < h_D$).

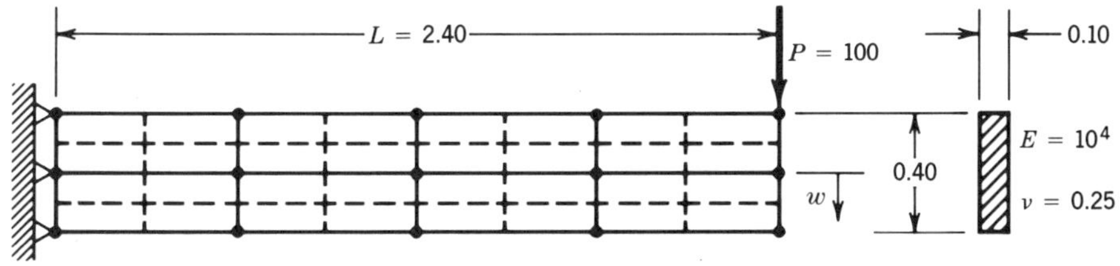

Figure 18.7-4. End-loaded cantilever beam. Mesh $N = 8$ is shown. Mesh $N = 32$ is suggested by dashed lines. All elements are bilinear.

necessary for exactness do not prevail. Rather than regarding ϕ^0 as the exact value, it seems preferable to regard

$$e = \frac{\phi_2 - \phi^0}{\phi^0} 100\% \tag{18.7-5}$$

as an estimate of percentage errors in mesh h_2.

Example: Regular Refinement. The cantilever beam problem shown in Fig. 18.7-4 was solved using three different meshes and three different formulations for each mesh. Results are presented in Fig. 18.7-5 and in Table 18.7-1 using the following notation. Element stiffness matrices were integrated using 2 by 2 quadrature ("Full") or one-point quadrature ("Reduced" and "H–G"), where "H–G" refers to hourglass control (see Eq. 6.12-4). No mechanism is possible in the reduced-integration case because of complete fixity at the left end. The problem was solved on a microcomputer whose capacity did not permit the use of a finer mesh than $N = 128$.

Figure 18.7-5 suggests that $e = O(h)$ for the full-integration case and $e = O(h^2)$ for the reduced-integration case. Nevertheless, all extrapolated values in Table 18.7-1 were computed using $p = 2$ in Eq. 18.7-4.

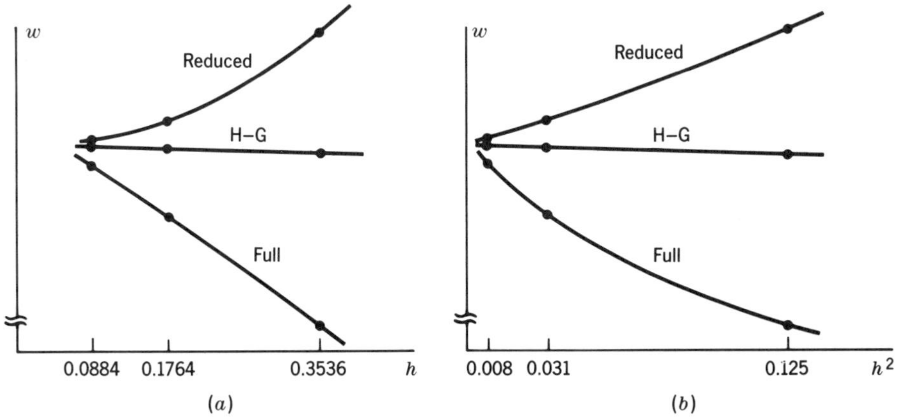

Figure 18.7-5. Plots of data from Table 18.7-1, for the problem of Fig. 18.7-4: (*a*) w versus h, (*b*) w versus h^2.

TABLE 18.7-1. COMPUTED DEFLECTION w IN FIG. 18.7-4. BEAM THEORY, INCLUDING TRANSVERSE SHEAR DEFORMATION, GIVES $w = (PL^3/3EI) + (6PL/5AG) = 8.640 + 0.180 = 8.820$. EXTRAPOLATION IS BASED ON $p = 2$ IN EQ. 18.7-4.

Mesh Data		Computed Results (w)			Extrapolated Results (w)		
N	$h = 1/N^{1/2}$	Full	Reduced	H–G	Full	Reduced	H–G
8	0.3536	4.562	11.440	8.572			
					7.978	8.617	8.768
32	0.1768	7.124	9.323	8.719			
					8.695	8.788	8.788
128	0.0884	8.302	8.922	8.771			

Example: Irregular Refinement. The beam in Fig. 18.7-6 contains a large central hole. We seek the largest principal stress. Each mesh shown is composed of bilinear elements, for which a 2 by 2 integration rule was used. Load P is uniformly distributed across the right end. The upper half of the beam was modeled, with all nodes on $x = 0$ fixed and all nodes on $y = 0$ allowed only vertical displacement. Even so, the microcomputer could not accommodate regular mesh refinement. The meshes shown violate most rules of regular refinement. Indeed, the meshes could be substantially improved because they are undesirably coarse close to the hole and unnecessarily fine far from the hole. Stresses were evaluated at element centers. What is plotted as σ_{max} is the largest element center stress from each mesh, regardless of the element in which it appears.

For pure bending load, photoelastic data indicate that σ_{max} appears at the top edge of the hole and is $\sigma_{max} = 98.4$ [18.20]. One can perhaps assume that the error in σ_{max} is $O(h)$. Hence, a least squares fit of a straight line to the three data points yields

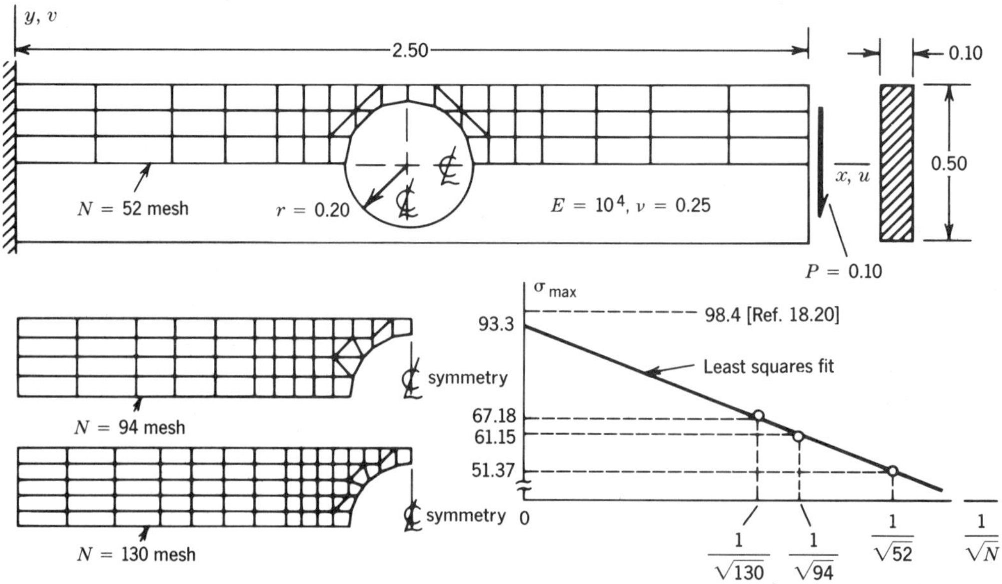

Figure 18.7-6. Cantilever beam with a large central hole. Each mesh is symmetric about both centerlines. Results of irregular mesh refinement are shown. (The authors are grateful to S-C. Liang and D. Rusche for doing the computations.)

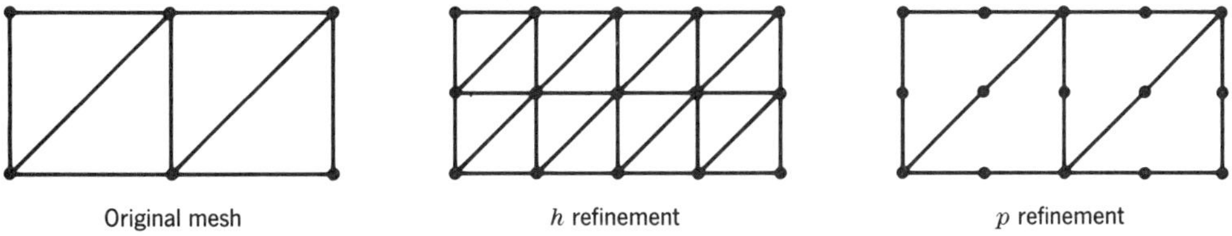

Figure 18.7-7. The h and p versions of refinement of a plane mesh.

$\sigma_{max} = 93.3$, as shown. This extrapolated result is remarkably good, and perhaps somewhat fortuitous. At the very least, the plotted results show that none of the three meshes by itself yields a reliable result.

The h version and the p version. The h and p versions of the finite element method are different ways of adding d.o.f. to the model, so as to reduce discretization error in a subsequent analysis. The h version refers to decreasing the characteristic length (h) of elements, by dividing each existing element into two or more elements, but without changing the types of elements used. The p version refers to increasing the degree of the highest complete polynomial (p) in elements, by adding nodes to elements, adding d.o.f. (e.g., derivative d.o.f.) to nodes, or both, but without changing the number of elements used. In Fig. 18.7-7, the next stage of p refinement might be to add derivative d.o.f. without changing the number of nodes.

A sequence of successively refined meshes produces convergence toward correct results. The process is known as h convergence or p convergence, depending on the method of adding d.o.f. A computer program is termed "adaptive" if addition of d.o.f. and reanalysis can be accomplished with a minimum of direction from the analyst. The program is called "self-adaptive" if it can automatically decide where additional d.o.f. are most needed in the model, prepare a suitable new model, reanalyze, and keep repeating the process until a preselected convergence tolerance is achieved [18.12,18.13]. Adaptive h refinement can continue until limits imposed by computer capacity and numerical noise are reached. Adaptive p refinement can continue until the highest-order polynomial coded in the program is used.

Error estimates and monotonic convergence of successive solutions are made possible by completely regular mesh refinement, which is known in the present context as a "hierarchial" procedure. Hierarchial refinement produces a mesh that allows, as special cases, all displacement modes that were possible in the unrefined mesh. Both refinements in Fig. 18.7-7 are hierarchial. With either the h version or the p version, clever programming can incorporate computations done in the preceding mesh rather than repeating them entire for the current mesh.

18.8 TESTS OF ELEMENT QUALITY

Eigenvalue Test. The eigenvalue test is one of several tests of element quality. The test can detect zero-energy deformation modes, lack of invariance, and absence of rigid-body motion capability. It can also be used to estimate the relative quality of competing elements [18.14,18.15]. We will describe the calculations first and then comment on how to interpret results.

Let loads $\{\bar{\mathbf{r}}\}$ applied to element nodes be proportional to element nodal displacements $\{\mathbf{d}\}$ through a factor λ:

$$[\mathbf{k}]\{\mathbf{d}\} = \{\bar{\mathbf{r}}\} = \lambda\{\mathbf{d}\} \qquad \text{or} \qquad ([\mathbf{k}] - \lambda[\mathbf{I}])\{\mathbf{d}\} = \{\mathbf{0}\} \tag{18.8-1}$$

This is an eigenproblem. Eigenvalues λ_i are called eigenvalues of $[\mathbf{k}]$. There are as many λ_i as there are d.o.f. in $\{\mathbf{d}\}$. Not all λ_i need be different. To each λ_i there corresponds an eigenvector $\{\mathbf{d}\}_i$. If each $\{\mathbf{d}\}_i$ is normalized so that $\{\mathbf{d}\}_i^T\{\mathbf{d}\}_i = 1$, premultiplication of Eq. 18.8-1 by $\{\mathbf{d}\}_i^T$ yields

$$\{\mathbf{d}\}_i^T[\mathbf{k}]\{\mathbf{d}\}_i = \lambda_i \qquad \text{or} \qquad 2U_i = \lambda_i \tag{18.8-2}$$

where U_i is strain energy in the element when its nodal d.o.f. are the normalized displacements $\{\mathbf{d}\}_i$ (see Eq. 3.3-9). Usually the element is unrestrained for the eigenvalue test so that $[\mathbf{k}]$ is the complete element stiffness matrix. In an existing computer program, one can conveniently compute the λ_i as the squared natural vibration frequencies of an unsupported element that has unit mass attached to each d.o.f. in $\{\mathbf{d}\}$. Note that λ_i is unchanged if the deformation is reversed—that is, if $\{\mathbf{d}\}_i$ is replaced by $-\{\mathbf{d}\}_i$.

Equation 18.8-2 shows that $[\mathbf{k}]$ should yield $\lambda_i = 0$ when $\{\mathbf{d}\}_i$ represents any rigid-body motion. There are three linearly independent rigid-body motions possible in the plane. Therefore three of the λ_i should be zero for a plane element. Six should be zero for a general solid or shell element, but only one for a solid or shell of revolution element (if only axially symmetric states are permitted).

Zero-energy modes (mechanisms) also yield zero eigenvalues. The associated $\{\mathbf{d}\}_i$ may appear in combination with the $\{\mathbf{d}\}_i$ of a rigid-body motion.

In testing an element, we first check that $[\mathbf{k}]$ has as many $\lambda_i = 0$ values as expected. Too few suggests that the element lacks a desired capability for rigid-body motion without strain. Too many suggests the presence of one or more mechanisms.

Nonzero eigenvalues are real and positive if $[\mathbf{k}]$ is symmetric and positive semidefinite. If eigenvalues change when the element is reoriented in global coordinates, the element is not geometrically isotropic. Similar modes, such as the flexural modes of Fig. 18.8-1, should be associated with equal eigenvalues if the material is isotropic.

Eigenvalues of $[\mathbf{k}]$ can sometimes be used to compare different formulations of a given element type, for example, isoparametric versus hybrid formulations of

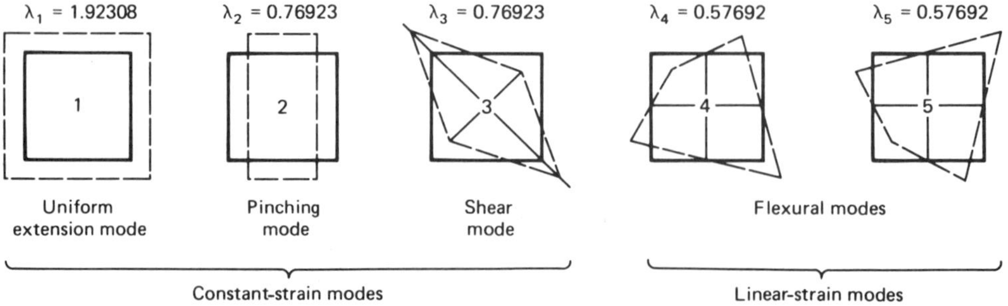

Figure 18.8-1. Nonzero eigenvalues and corresponding eigenvectors (deformation modes) of a square bilinear element in plane strain [18.16]. $E = 1.0$, $\nu = 0.3$, side length $= 1.0$.

a four-node plane element. Properly formulated elements of the same size, shape, and material properties should be equally stiff in their constant strain modes. Their stiffness matrices should therefore have eigenvalues in common. Stiffnesses, and therefore eigenvalues, may differ in other modes. For compatible elements based on assumed displacement fields, the λ_i are either exact or are upper bounds on the correct strain energy. Therefore, when comparing elements of the same size, shape, [**E**], node placement, and number and type of d.o.f., the element with the lowest strain energy is best. The stiffness matrix of this element has the lowest trace (tr[**k**] equals the sum of the eigenvalues of [**k**]). This argument fails if any λ_i is not an upper bound, which happens if the element contains a mechanism.

Other Tests. Remarks. In the *single-element test* [18.17], the response of a single element to a certain loading is examined as one changes the element aspect ratio, skewness, or taper. For example, in Fig. 18.8-2, a tip-loaded cantilever beam can be repeatedly analyzed as L/H is varied from a small value to a large value. It may happen that of two competing element formulations A and B, both work well when $L/H = 1$ but A is much more accurate than B when $L/H >> 1$. Then A has proved superior in this particular test. The test is very easy to perform. When used with but one element formulation, rather than in comparing different formulations, the test provides information about element behavior that is useful in modeling.

Another type of element test compares strain energies [18.15]. For a prescribed deformation state, strain energy in the mathematical continuum is computed over the volume spanned by the element. Strain energy in the element is also computed by imposing nodal d.o.f. consistent with the prescribed deformation state. The two energies are compared. The test is repeated, using other deformation states.

Other tests have already been noted, including the patch test (Section 4.6) and applying a finite element model to various problems whose solutions are already known. The latter test is neither definitive nor general, but one would be foolish not to use it.

In testing a new element, or in becoming acquainted with an unfamiliar element, it is appropriate to use a variety of tests. No single test is likely to be decisive, except perhaps in discovering a fatally flawed element. Element behavior tends to be case-dependent, so that the ranking of competing elements is likely to be different in different test cases and is likely to depend on whether stress or displacement is taken as the indicator of quality. The ideal element, probably never to be discovered, passes patch tests, yields good results in a coarse mesh, converges rapidly with mesh refinement, is almost insensitive to shape distortion, is geometrically isotropic, has no zero- or low-energy deformation modes when assembled with other elements, can be joined to elements of different type, rests on simple theory, and is economical to formulate.

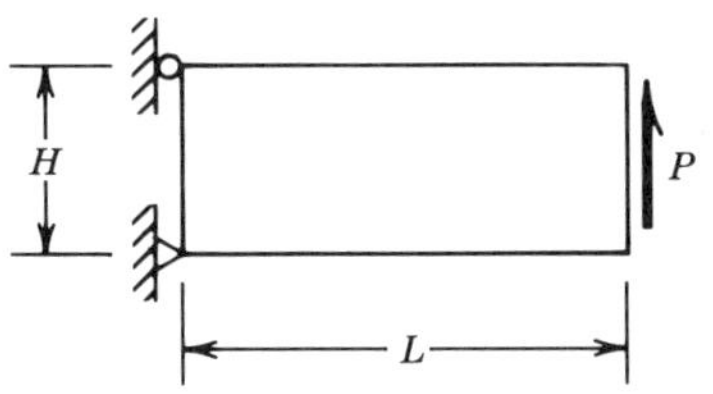

Figure 18.8-2. Single-element model of a cantilever beam under transverse tip load.

18.9 CONCLUDING REMARKS

In this chapter we have discussed tests that can be applied to elements, to an assembly of elements, or to a set of equations. If an error or a potential difficulty is present, no single test is certain to detect it. Even if several tests are applied and passed, one should be aware that tests can detect errors but cannot prove their absence, and that a trouble-free finite element structure may not be a good model of physical reality.

Physical situations in structural mechanics that make numerical error more likely include elements with great shape distortion or large aspect ratio, an element whose shear or membrane stiffness is much larger than its bending stiffness, stiff elements used to approximate a rigid region, and a very fine mesh. This is a list of guidelines rather than firm rules, as in each situation one can identify one or more special cases in which the anticipated trouble does not materialize.

On most digital computers, one should use double-precision data for all constants and double-precision arithmetic for all phases of generating structural equations and solving them. Partial double precision in equation solving and double-precision solution of equations generated in single precision are dangerous practices. It is risky to generalize from a single example, yet it is interesting that practical problems with $1.5(10^6)$ d.o.f. have been solved with good engineering accuracy.

PROBLEMS

Section 18.2

18.1 (a) Solve Eqs. 18.2-1 graphically by plotting the slope of each line and determining the intersection point.
(b) Repeat part (a), using Eqs. 18.2-2 instead.

18.2 Consider the equations $x + y = 2$ and $x + 1.01y = 2.01$. Show that the solution is sensitive in small changes in both the coefficient matrix and the vector of constants.

18.3 Show that Eq. 18.2-3 can be written in the form of Eq. 9.3-5, with $k = k_2$ and penalty number $\alpha = (k_1/k_2) - 1$. What happens when α becomes very large?

18.4 In Eq. 18.2-4, determine the reduced coefficient K_{22} produced by Gauss elimination. Show that this result predicts trouble if α is large.

18.5 The two-element beam shown is uniform, fixed at the left end, and simply supported at the right end. Show that the structural equations become ill conditioned if the scalar multiplier α is very small.

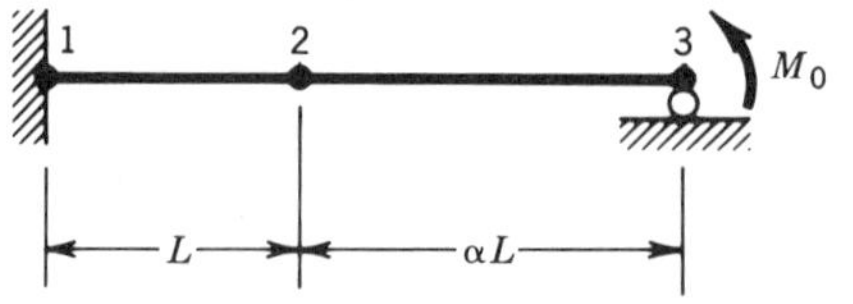

Problem 18.5

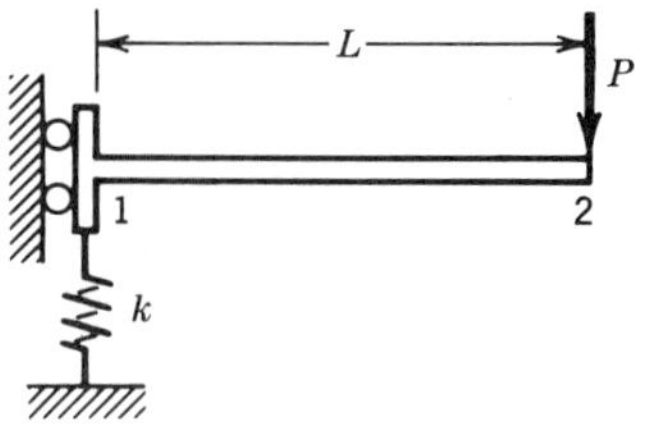

Problem 18.7

18.6 (a) Obtain Eqs. 18.2-6 by coordinate transformation of Eqs. 18.2-3. Show that large cancellation error may be present.
(b) Obtain Eqs. 18.2-6 by applying coordinate transformation to [**k**] of the left element in Fig. 18.2-1, then assembling the two elements.

18.7 The left end of the one-element cantilever beam shown rests on a soft spring of stiffness k. Rotation is prevented at the left end.
(a) Solve for the deflection of load P using w_1, w_2, and θ_2 as nonzero d.o.f. Show that the equations are ill conditioned if $k << EI/L^3$.
(b) Form an element stiffness matrix that operates on d.o.f. w_1, θ_1, w_{21}, and θ_{21}, where w_{21} and θ_{21} are the "relative" d.o.f. $w_{21} = w_2 - (w_1 + L\theta_1)$ and $\theta_{21} = \theta_2 - \theta_1$. Impose one boundary condition, include the soft spring, and again solve for the deflection of load P. Show that the equations are not ill conditioned.

18.8 For a general structure, how many d.o.f. can be "relative" and how many must be "absolute"? (See Problem 18.7(b) for an illustration of these terms.)

Section 18.3

18.9 (a) How small can $C(\mathbf{K})$ be?
(b) What is $C(\mathbf{K})$ if [**K**] is diagonal?
(c) Physically, what is implied if $C(\mathbf{K})$ is infinite?

18.10 (a) Let $k_1 = 10k_2$ in Fig. 18.2-1. Compute the condition numbers of the unscaled matrix [**K**] and of the scaled matrix [$\mathbf{K}_s$].
(b) Repeat part (a), now with $k_2 = 10k_1$.

18.11 (a) The two-spring system shown has d.o.f. u_1 and u_2. For what value of scalar c is the condition number of the unscaled stiffness matrix a minimum, and what is its minimum value?
(b) Repeat part (a), but use the scaled stiffness matrix.

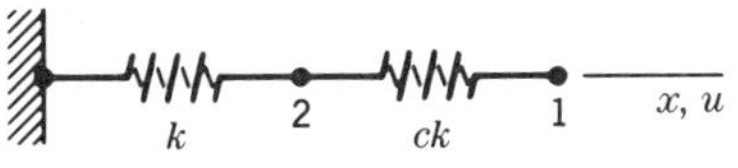

Problem 18.11

18.12 Let $\beta = 45°$ in Fig. 18.2-3*a*. Compute $C(\mathbf{K})$ in terms of α using (a) the unscaled matrix [**K**], and (b) the scaled matrix [$\mathbf{K}_s$].

18.13 (a) Evaluate Eq. 18.3-2 for the structure shown in Fig. 18.3-1.
(b) Let $AE/L = 174$ in Fig. 18.3-1*a*. Let a hypothetical computer retain only three significant digits per word, so that coefficients in [**K**] are represented as $K_{11} = -K_{12} = -K_{21} = 1040$ and $K_{22} = 1220$. If subsequent manipulations are exact, what are the percentage errors in the computed values of u_1, u_2, and $(u_1 - u_2)/L$? Does Eq. 18.3-2 seem to apply?
(c) Repeat part (b) for the problem of Fig. 18.3-1*b*.

18.14 Interchange the bars in Fig. 18.3-1; for example, let the left bar have stiffness 174 and the right bar have stiffness 6(174) = 1044.
(a) Compute $C(\mathbf{K})$, using first the unscaled [**K**] and then the scaled matrix [$\mathbf{K}_s$].
(b) Repeat Problem 18.13(b), where now $K_{11} = -K_{12} = -K_{21} = 174$ and $K_{22} = 1220$.

18.15 A uniform beam is modeled by a single standard beam element. What is the condition number of the scaled stiffness matrix if the beam is (a) simply supported, and (b) cantilevered? (There are two nonzero d.o.f. in each case.)

18.16 Why, in Eq. 18.3-8, can $N^{2m/n}$ be replaced by h^{-2m}? (Here it is easiest to imagine that h, the span of a typical element, is uniform throughout the mesh.)

18.17 The bar shown is uniform and is built of standard two-node elements. In axial vibration, its fundamental frequency ω_1 is given by $\omega_1^2 = \pi^2 E/4L_T^2\rho$, where ρ is the mass density.

(a) Let all elements have the same length L. Use Eq. 18.3-9 to bound $C(\mathbf{K})$ in terms of L_T and L.

(b) Evaluate this bound numerically, using two elements ($L = L_T/2$). Also evaluate the exact $C(\mathbf{K})$, and compare.

(c) Repeat part (b), but use two unequal elements, of lengths L and $2L$, respectively.

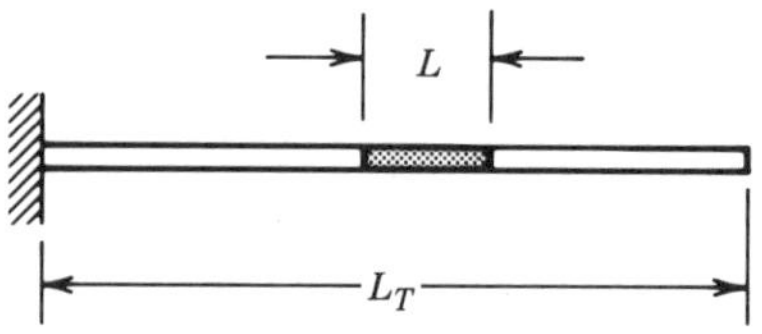

Problem 18.17

Section 18.4

18.18 (a) Write [**K**] for the structure shown, in which each spring has stiffness $k = 100$. By examination of the first few steps of Gauss elimination, deduce an expression for the reduced diagonal coefficient K_{ii} in terms of k and i. Hence, what are the diagonal decay ratios after the 99th and 100th eliminations?

(b) Repeat part (a), but sequence the node numbers from right to left, so that node 1 carries load P.

18.19 In each part of this problem, imagine that the analyst has forgotten to specify any displacement boundary conditions, so that the structure is unsupported. In what equation (first, second, . . ., last) of the system [**K**]{**D**} = {**R**} will the diagonal decay test detect this trouble?

(a) The train of springs in Problem 18.18.

(b) The beams of Table 18.4-1.

(c) A plane frame having three d.o.f. per node.

(d) A solid of revolution having two d.o.f. (radial and axial) per node.

(e) A plane structure having two d.o.f. per node.

18.20 Compute the diagonal decay ratio in terms of α for the problem described by Fig. 18.2-3a. Show that it is not large if $\beta = 0$ or if $\beta = \pi/2$.

18.21 With unit axial loads at nodes 2 and 3 in the structure shown, the exact

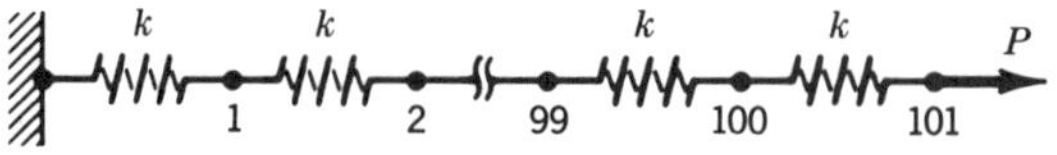

Problem 18.18

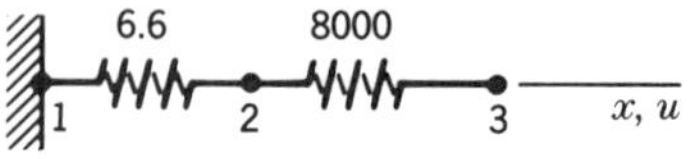

Problem 18.21

structure equations are $8006.6u_2 - 8000u_3 = 1$ and $-8000u_2 + 8000u_3 = 1$. Compute the exact values of u_2 and u_3. Also compute approximate values under the assumption that the computer rounds numbers to only four digits after each operation of Gauss elimination. Does the accuracy loss agree with that predicted by (a) the diagonal decay ratio and (b) the condition number of **[K]**?

18.22 What Fortran statements should be added, and where, if a test for diagonal decay is to be incorporated in the equation solver provided in Fig. B.2-3?

Section 18.5

18.23 Consider the two equations $1.78u_1 + 1.06u_2 = 2.88$ and $0.94u_1 + 0.56u_2 = 1.52$. What residual vector $\{\Delta\mathbf{R}\}$ is given by the approximate solution $u_1 = 1.88$, $u_2 = -0.44$, and what is e of Eq. 18.5-2? What is the exact solution? Are the equations ill conditioned?

18.24 Consider the ill-conditioned equations $u_1 + u_2 = 2$, $u_1 + 1.0001u_2 = 2.0001$. What residuals and what e are given by the approximate solution $u_1 = 2.0$, $u_2 = 0.0$, and by the approximate solution $u_1 = u_2 = 1.1$? Which of the two approximate solutions is most nearly correct?

18.25 A linear spring of stiffness 28 N/m is loaded by a 0.5-N force. Using the approximate value $k^{-1} \approx 0.040$ m/N, we compute the approximate displacement $u \approx 0.020$ m. Improve this result by the iterative method associated with Eq. 18.5-1, using the correct k and the approximate k^{-1}.

18.26 Consider the two equations $(u_1/3) - (u_2/3) = 1$ and $-(u_1/3) + (7u_2/12) = 0$. Let the stiffness matrix and its inverse be approximated as

$$[\mathbf{K}] \approx \begin{bmatrix} 0.3 & -0.3 \\ -0.3 & 0.5 \end{bmatrix} \quad \text{and} \quad [\mathbf{K}]^{-1} \approx \begin{bmatrix} 8.0 & 5.0 \\ 5.0 & 5.0 \end{bmatrix}$$

(a) Use the foregoing approximate matrices in 18.5-1. Apply three cycles of iterative improvement. Do displacements u_1 and u_2 appear to be converging toward correct values?

(b) Use the approximate $[\mathbf{K}]^{-1}$ to compute $\{\Delta\mathbf{D}\} = [\mathbf{K}]^{-1}\{\Delta\mathbf{R}\}$ but the *correct* **[K]** to compute $\{\Delta\mathbf{R}\}$. Again apply three cycles of iterative improvement and assess the results.

18.27 The cantilever beam shown is built of four constant-strain triangles (described in Chapter 5). For what loadings at nodes A and B will computed results be exact? For which of these loadings will residuals be essentially zero? For each loading plot the qualitative variation of σ_x along the x axis, according to your expectation of the finite element results.

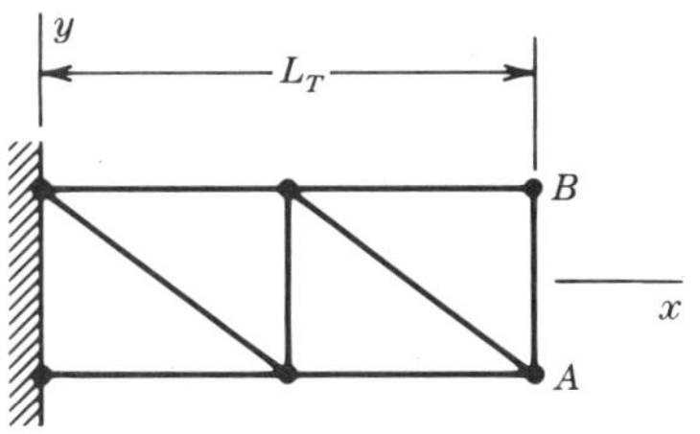

Problem 18.27

Section 18.6

18.28 For the bar problem of Fig. 18.6-1, show that the finite element solution yields exact displacements at the nodes when A and E are constant and the load integral, $\int \lfloor \mathbf{N} \rfloor^T q \, dx$, is evaluated consistently over length h_i of each element. Assume that $q = q(x)$ is a continuous function for $x_i \leq x \leq x_{i+1}$. *Suggestion:* First show that the exact solution is

$$AEu(x) = -\int_0^x \int_0^s q(\ell)\, d\ell \, ds + x \int_0^{L_T} q(\ell)\, d\ell \tag{a}$$

Then evaluate the nodal load vector consistently. Integrate each entry by parts to show that the ith entry of the nodal load vector is

$$\int_0^{h_{i-1}} q(x + x_{i-1})\, dx - \frac{1}{h_{i-1}} \int_0^{h_{i-1}} \int_0^x q(s + x_{i-1})\, ds\, dx + \frac{1}{h_i} \int_0^{h_i} \int_0^x q(s + x_i)\, ds\, dx \tag{b}$$

The last integral is absent in the load at L_T. Since the response at the nodes, x_i, to given nodal loads is exact for bar elements, deduce the response to these nodal loads exactly, and show that the result coincides with Eq. (a) at the nodes.

18.29 From Problem 18.28 we may deduce that the error function $e(x)$ defined in Eq. 18.6-2 has the expansion given in Eq. 18.6-3 on element i. Show that $e'(x_i) = (h_i/2)u''(\bar{x})$ for an $\bar{x}$ in the ith element. *Suggestion:* From Eq. (a) and Eq. 18.6-2, we may easily deduce that

$$e'(x_i) = \frac{1}{AE} \int_{x_i}^{L_T} q(x)\, dx - \frac{u_{i+1} - u_i}{h_i} \tag{c}$$

Express u_{i+1} and u_i using Eq. (a), since these values have been shown to be exact in problem 18.28. Apply the mean value theorem for integrals to the resulting expression.

18.30 If a uniform beam carries a smooth and continuous distributed lateral load, the standard beam element yields exact values of nodal displacements and rotations. Therefore, over one element, error $e(x)$ in lateral displacement has the appearance shown. A Taylor series analysis shows that the leading term in an expansion like Eq. 18.6-3 for $e(x)$ is $\frac{1}{2}(x - x_i)^2 e''(\bar{x})$ for an $\bar{x}$ in the element. Show that $e(x) = O(h^4)$. *Suggestion:* Use the N_i of Fig. 3.13-2, and show that if $u(0)$, $u'(0)$, $u(h)$, and $u'(h)$ are exact, then $e(h/2) = O(h^4)$ and $e'(h/2) = O(h^3)$ on the element. Then

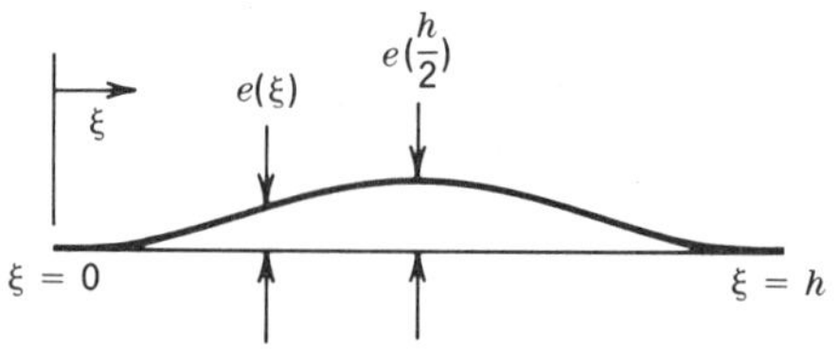

Problem 18.30

$$e(x) = e(h/2) + \xi e'(h/2) + (\xi^2/2)e''(h/2) + \cdots \quad \text{(d)}$$

and $e(h) = e'(h) = 0$ implies that $e''(h/2) = O(h^2)$, from which the result follows.

18.31 Assume that the loops in Fig. 18.6-1*b* are portions of parabolas. Hence, construct a geometric argument that shows that error e_{max} is quartered if element length h is halved. Similarly, show that error e'_{max} is halved if h is halved.

Section 18.7

18.32 Imagine that stresses are to be calculated using the plane meshes shown in Fig. 18.7-2. Including meshes suggested by dashed lines, a total of four different meshes are indicated. For each of the four, what relative stress error is estimated by Eq. 18.7-3?

18.33 Derive Eq. 18.7-4.

18.34 Why would it be inadvisable to apply Eq. 18.7-4 to results obtained by use of the DKT plate element described in Section 11.4?

18.35 For meshes $N = 1$, 2, and 4, the respective computed values of displacement at a certain point in a plane mesh are 4.16, 4.64, and 4.76 units. What is the deflection predicted by extrapolation?

18.36 Recompute the six "extrapolated results" in Table 18.7-1, this time using $p = 1$ in Eq. 18.7-4.

18.37 What value of σ_{max} is predicted by linear extrapolation in Fig. 18.7-6? Use all three possible combinations of two data points. What is the average of these three values?

18.38 Verify that $\sigma_{max} = 93.3$ is the value predicted at $N = \infty$ in Fig. 18.7-6 by linear regression (least squares fit of a straight line).

18.39 Use Eq. 18.7-4, with the appropriate value of p, to predict the converged value of stress at point A in Fig. 4.7-3*a* and in Fig. 4.7-3*b*.
(a) Use the "standard" results in Table 4.7-1.
(b) Use the iterated results in Table 4.7-1.

18.40 Use Eq. 18.7-4, with a value of p appropriate to the result to be extrapolated, to predict the converged values of v_C, σ_A, and σ_B in the following example problems.
(a) "QM6" results in Fig. 8.3-3.
(b) "Bilinear" results in Fig. 8.3-3.
(c) "Equation 8.4-5" results in Fig. 8.6-2.
(d) "Figure 8.6-1" results in Fig. 8.6-2.

18.41 In Problem 13.17, axial vibrations of a bar were analyzed using three different mass matrices and meshes $N = 1$ and $N = 2$ for each mass matrix. The squared fundamental frequency is $\omega^2 = cAE/mL$, where $c = \pi^2/4$ for the exact result. For the respective mass matrices, the computations yield the following approximate values of c:

Mesh $N = 1$	3.000	2.000	2.400
Mesh $N = 2$	2.597	2.343	2.463

Use Eq. 18.7-4 to extrapolate to $N = \infty$ for each of the three sequences of two values.

Section 18.9

18.42 Various situations that may promote numerical error are noted in the second paragraph of Section 18.9. For each of the four, cite an exception, for which the difficulty does not materialize.

18.43 Imagine that you must model a very stiff beam on a very soft elastic foundation. What will you do to assure that deflections and bending moments of the beam are computed accurately?

18.44 With mesh refinement, accuracy may improve, decline, or not change. For each of these three possibilities, give examples of problems (or situations, or meshes) that would behave in this way.

CHAPTER 19

MODELING, PROGRAMS, AND PROGRAMMING

Guidelines are discussed for producing a finite element model of an actual structure. Advice is given regarding the writing and acquisition of computer programs.

19.1 MODELING

General. Modeling is an art based on the ability to visualize physical interactions. All basic and applied knowledge of physical problems, finite elements, and solution algorithms contributes to modeling expertise. Little is published regarding modeling. Practitioners tend to learn by doing and by talking with others. The documentation of a large general-purpose program usually contains some modeling advice in connection with specific example problems [e.g., 19.1].

The documentation cannot be ignored: *before* all else fails, study the directions! Quite possibly, this study will not resolve all questions regarding terminology, symbols, assumptions, default conditions, and so on. Small test problems, sometimes involving a single element, can be run to clarify matters. Small test problems can help resolve questions about the sensitivity of an element type to aspect ratio and other shape distortions, whether beam and plate elements allow for transverse shear deformation, how distributed loads are treated, sensitivity to ill-conditioning, whether an eigensolver has trouble with zero or repeated eigenvalues (for the latter test one could use a three-dimensional beam of circular cross section [19.16]). These exercises will also increase confidence in use of the program and improve understanding of how various elements behave.

In modeling, the principal difficulty faced by a typical user of a computer program is not understanding the physical action and boundary conditions of the actual structure, and the limitations of applicable theory, well enough to prepare a satisfactory model. Another difficulty is not understanding the behaviors of various elements, and the program's options and limitations, well enough to make an intelligent choice among them. The result may be a poor specification of the problem to be solved, a model that fails to reflect important features of the physical problem, fine detail irrelevant to the problem, a solution based on inappropriate loading or support conditions, and a surplus of computed results which are not properly examined and questioned [19.2]. Automatic mesh generators make it easy to use too much fine detail. Powerful graphic postprocessors may smooth stress discontinuities that should warn of a need for local refinement or may hide questionable results by attractive display. It is possible that most finite element analyses are so flawed that they are worthless, and many experts feel that the situation is not improving. The reader is advised to review Section 1.8 of this book, "Warning: The Computed Answer May Be Wrong."

Advice contained in the following discussion should not be regarded as a set of inflexible rules. An experienced and competent analyst may find exceptions that can sometimes be exploited to advantage.

Cost and Dimensionality. Occasionally one may elect to analyze a one- or two-dimensional model rather than a two- or three-dimensional model, particularly when doing an initial simplified analysis. A rough guide to relative costs can be obtained by comparing the equation-solving costs of one-, two-, and three-dimensional meshes of comparable elements. For example, let us fill a square with N^2 elements and a cube with N^3 elements. Thus each mesh has N elements per side. If N is large, elements have corner nodes only, and each node has a single d.o.f., then the plane mesh produces rought N^2 equations having a semibandwidth of roughly N, for which the equation-solving expense is roughly proportional to $N^2(N)^2 = N^4$. The corresponding numbers for the solid mesh are N^3, N^2, and $N^3(N^2)^2 = N^7$. Thus the rough measure of cost increases by the factor $N^7/N^4 = N^3$ in going from two dimensions to three. Even for a mesh only ten elements on a side, this is a thousandfold increase.

Which Element is Best? The appropriate answer is another question: Best for what? Element performance is problem-dependent. An element or mesh that works well in one situation may work badly in another. The analyst must understand how various elements behave in various situations, and must understand the physics of the problem well enough to make an intelligent choice of elements and mesh.

A rough guideline, which falls well short of being a rule, is that elements of intermediate complexity work well for many problems. Thus, one would usually avoid using a great many of the simplest elements or a very few of the most complicated elements.

Start Simply. A problem of moderate or large complexity should not be swallowed whole. One might begin with rough approximations from "back of the envelope" calculations. Numerically, a good beginning that often provides considerable insight is a "stick model," which is a model built of a few bar and beam elements. A stick model is simple to prepare, cheap to run, and gives approximate results. If a more refined model gives greatly different results, the analyst should seek the reason for the discrepancy.

A stick model, or a coarse-mesh model, can be used to guide subsequent refinement. If symmetry is to be exploited, a simple model serves to check anticipated symmetries and perhaps discover additional symmetries. A two-dimensional model may serve as an early analysis step in a three-dimensional problem, and may sometimes make a three-dimensional analysis unnecessary.

If a dynamic or nonlinear analysis is contemplated, a linear static analysis of the proposed final mesh might be done beforehand as a relatively cheap test that may disclose flaws in the model. Loads applied in the static analysis should be contrived to produce strain distributions and gradients comparable to those anticipated in the subsequent dynamic or nonlinear analysis.

Structure to Model. Modeling is more than just laying out a mesh. A focus on mesh layout or the minutiae of modeling may be at the expense of grasping important physical aspects that strongly affect the actual behavior.

In attempting to simulate reality with a mathematical model, the analyst must come to grips with the physics of the problem. What are the loads? What are the boundary conditions? Which actions are important and which are unimportant? Is the problem quasistatic or dynamic? If dynamic, is damping important? If so, how should it be represented? Is buckling a possibility? Is the material isotropic? Do properties depend on temperature or strain rate? Is plastic flow involved? If so, is it localized or widespread? Are there other nonlinearities that demand attention? These and other questions suggested by the problem at hand must be addressed before one can decide what element types and what arrangement of specific elements will produce a good model, and, if the problem is dynamic or nonlinear, what solution algorithm will produce reliable results at acceptable cost.

Broad guidelines include the following. Include all real structure in the model; do not omit parts on the untested assumption that they carry little load or little stress. If a curved boundary is modeled as a polygon or a faceted surface, do so in a way that preserves the correct volume of the structure. In the analysis of thermally induced stresses, arrange element sizes and types so that the complexity of the temperature field can be approximately matched by the complexity of the strain field. If the temperature field is discontinuous across an interelement boundary, make sure the program does not "smear" the temperature change across elements by interpolation from nodal temperatures. Use consistent nodal loads rather than ad hoc lumping of loads. Use a relatively coarse mesh where gradients are known to be low and a relatively fine mesh where gradients are known to be high.

In a coarse mesh, different element arrangements can produce a different model than intended. Consider, for example, a rectangular plate with clamped edges (Fig. 19.1-1). Unshaded elements are completely inactive because all their d.o.f. are set to zero. Thus the plate is modeled by only the shaded elements. The model differs from the structure in size and shape, and symmetries of behavior may be lost.

Anticipate the Results and Know the Goal. If results were known in advance, it would be comparatively easy to prepare an adequate model. Similarly, if the probable results can be *anticipated,* a cheaper and better model will result. If locations of high stress are known in advance, it may be possible to model remote locations rather crudely. If the severity of stress gradients is anticipated, one can estimate the proper element size. If the goal of analysis is only to assess deflections, not to compute stresses, than a comparatively coarse mesh may suffice.

Expect to Revise. Ideally, the original model is adequate and only a single analysis is performed. Far more often, the first analysis discloses inadequacies of the

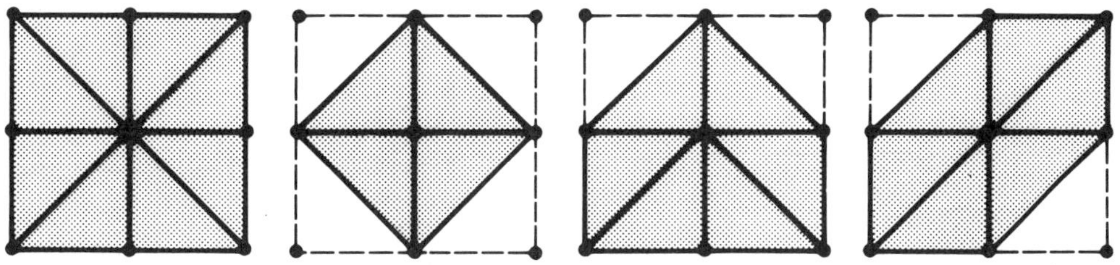

Figure 19.1-1. Four arrangements of eight triangular elements to model a rectangular plate with clamped edges.

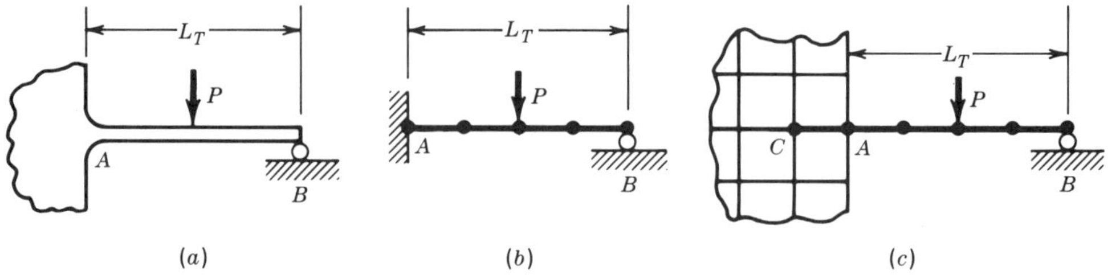

Figure 19.1-2. Beam carrying load P, with elastic support at A and simple support at B. (a) Actual structure. (b) Simple model. (c) Model with elastic support at A.

model, and one or more revisions are needed. Rather than regard the revisions as attempts to correct previous failures, it is better to regard them as expected steps in an investigation that proceeds from the overly simple to the adequate [19.14].

For example, at the outset one may not know enough about stress gradients or the behavior of available elements to immediately generate an adequate model. Then, rather than attempting to overwhelm ignorance by a very refined initial model, it is usually easier and cheaper to start with a crude model and refine it in successive analyses until it is adequate. We may also consider gaining the necessary modeling insight by analysis of a different problem, related but simpler, and preferably one for which analytical or experimental results are known.

Supports. Typically, in expositions of theory, structural supports are idealized as completely rigid or as ideally hinged (i.e., a simple support). Actual supports lie somewhere between completely fixed and ideally hinged. Consider, for example, Fig. 19.1-2. The left end of the beam is elastically supported. The simple model, Fig. 19.1-2b, might be analyzed twice, first with fixity at A and then with a hinge at A, in an attempt to bound the correct response. In Fig. 19.1-2c the actual support elasticity is modeled by elements. One might terminate the beam model at A and use constraint relations to couple the beam rotation θ_A to d.o.f. of the elastic support elements, as discussed in connection with Fig. 7.7-1a. Alternatively, the beam model might be extended into the support, as shown, with translational d.o.f. of the leftmost beam element coupled to translational d.o.f. of the elastic support elements. The associated incompatibilities of displacement along AC are ignored.

Superficially innocuous changes in support conditions can substantially affect results. Consider Fig. 19.1-3. In Fig. 19.1-3a, strains ϵ_y associated with the Poisson effect are prohibited at the left end. In Fig. 19.1-3b, they are permitted, as is

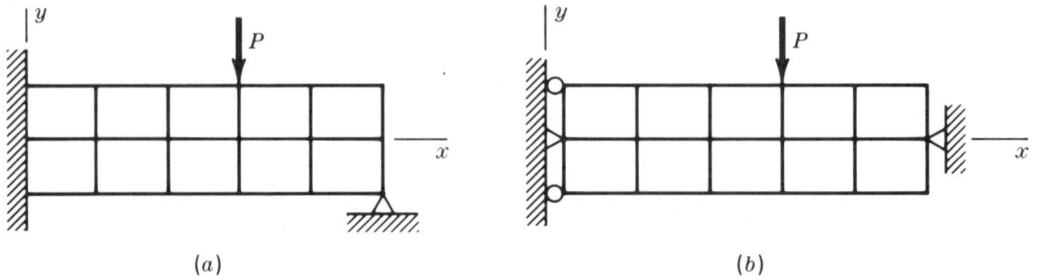

Figure 19.1-3. Two different models of a propped cantilever beam.

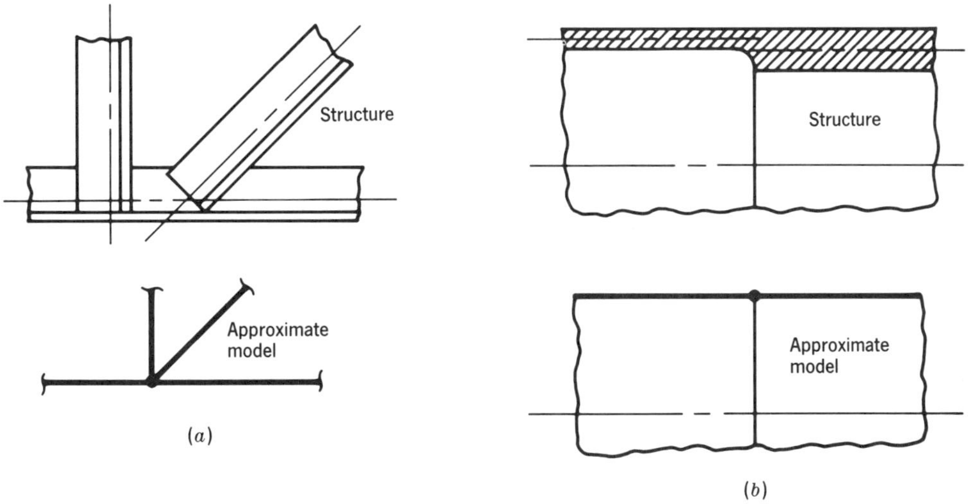

Figure 19.1-4. Models that somewhat misrepresent the structure. (*a*) Welded connection of three angles whose axes are not concurrent. (*b*) Cylindrical vessel with step change in thickness and midsurfaces offset.

presumed by elementary beam theory. This change in support conditions may produce large changes in σ_y near the fixed end if Poisson's ratio is nonzero. In Fig. 19.1-3*a* the hinge support acts to resist rotation of the right end; in Fig. 19.1-3*b* it does not because it lies on the neutral surface of bending.

Joints and Other Modifiers of Stiffness. Junctions between members may not have as much stiffness as a simple model attributes to them. For example, in Fig. 19.1-4, offsets produce significant local bending action in the actual structures, but this action is lacking in the approximate models that do not represent the offsets.

The effect of joints is often underestimated but may have appreciable effect on global behavior. Stiffeners, swages, corrugations, and perforations also have an effect upon stiffness that cannot be ignored [19.2]. However, detailed modeling—for example, of individual weld lines or of individual spot welds—is usually not appropriate unless the joint itself is the object of study.

Some bodies have geometric irregularities that can be "smeared." An example is a boiler tube sheet, which is a flat plate pierced by a regular pattern of identical holes. Effective elastic moduli and flexural rigidities for tube sheets have been established by analytical and experimental methods [19.15]. Thus overall response can be analyzed as though the tube sheet were homogeneous. However, stress analysis must acknowledge the presence of individual holes.

A related structure is a plate with a corrugated core, Fig. 19.1-5. If effective stiffnesses are known, the plate may be analyzed as if it were a homogeneous orthotropic plate. One way to obtain the effective stiffnesses is to model a small

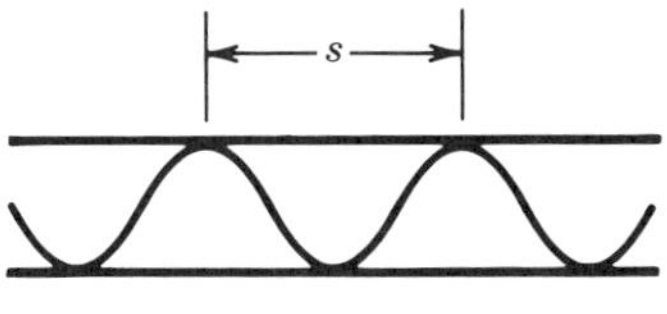

Figure 19.1-5. Cross section of a plate with thin facings and a corrugated core, viewed parallel to axes of the corrugations.

part of the structure in suitable fine detail (e.g., span s in Fig. 19.1-5), apply boundary displacements to this mesh consistent with a state of constant strain or constant curvature, compute the resulting boundary force or boundary moment, and finally obtain the required stiffness as the ratio of force to displacement (or of moment to rotation). If necessary, the same process can be applied to the aforementioned tube sheet to obtain its effective stiffness coefficients.

Element Shapes, Connection, and Grading. An element performs best if its shape is compact and regular. An element tends to stiffen and lose accuracy as its aspect ratio increases, as its corner angles become markedly different from one another, as sides become curved, or as side nodes (if present) become nonuniformly spaced. Figure 19.1-6 shows element shapes that are usually undesirable.

Different elements have different sensitivities to shape distortion. Accordingly, an all-purpose guideline must be vague: keep aspect ratios near unity, corner angles of quadrilaterals near 90°, side nodes at midsides, and sides straight. Elements derived as planar may behave badly if warped to fit a curved surface [19.4,19.5].

There are, of course, exceptions. Elements of large aspect ratio may be used in areas where the strain gradient is almost zero. Side nodes may be moved to quarter-points to produce crack-tip elements. Sides may be curved to fit a curved boundary (but sides of the element interior to the mesh should be straight).

Poor elements (Fig. 19.1-6) and poor element connections (Fig. 19.1-7) may produce only locally poor results. Usually, if the surrounding mesh is satisfactory, spurious gradients caused by a local mesh error die out rather than propagate, in accord with Saint-Venant's principle. The same is usually true of errors caused by ad hoc (but statically equivalent) nodal lumping of distributed loads.

If a mesh is graded rather than uniform, as is usually the case, grading should be done in a way that produces no great discrepancy in size between adjacent elements. Figure 19.1-8 shows three examples of mesh grading that use quadrilaterals. If triangles are also permitted, the range of possibilities increases. In general, adjacent elements should not differ greatly in stiffness. As a working rule, if E and V_e represent elastic modulus and element volume, the ratio E/V_e should not change by more than a factor of roughly 3 in going from one element to the next.

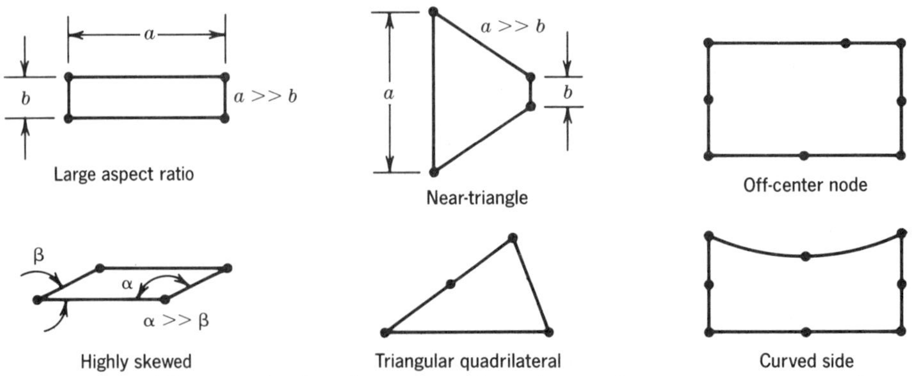

Figure 19.1-6. Elements having shape distortions that tend to promote poor results.

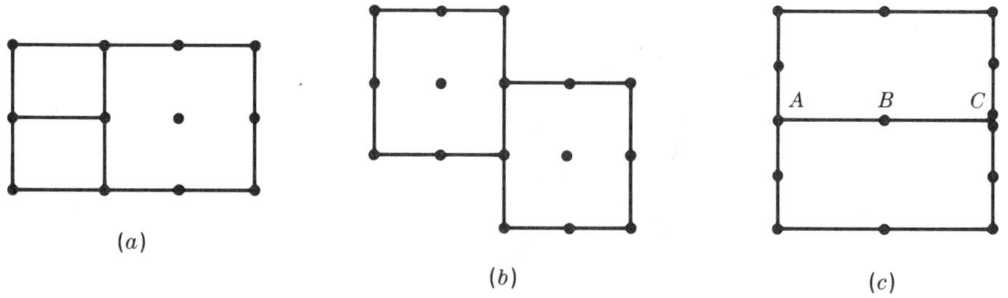

Figure 19.1-7. Poor element connections. (*a*) Two bilinear elements and one quadratic element. (*b*) Two quadratic elements. (*c*) Two quadratic elements, connected at *A* and *B* but not at *C* (as if to model a crack from *B* to *C*).

Checking the Model. A model should be checked *before* results are computed; afterward, there will be even greater reluctance to do the job. Ideally, a model is checked by an analyst who was not directly involved in its preparation and is therefore more likely to be objective.

Graphical display makes it comparatively easy to detect gross errors, such as a misplaced node or a missing element. Preprocessors offer color, shrink plots, rotation, sectioning, exploded views, and removal of hidden lines as aids in the checking process (Figs. 19.1-9 and 19.1-10).

Tests and warnings should already be coded into the program and should be exercised by a "check run" that precedes actual solution. Such tests may include examination for overly distorted element shapes, checking that adequate supports are provided, and searching for poor element connections like those in Fig. 19.1-7. The check run may also estimate the time required for the actual solution. All error messages and warnings produced by the program should be investigated, whether they appear during the check run or later.

Built-in error tests cannot be given all responsibility for the success of an analysis. The program cannot know whether the element type is appropriate, if supports are properly located, whether data have been supplied using consistent units, and so on. Responsibility resides with the user.

Ill-Conditioning, Locking, and Instability. Great stiffness discrepancies between elements, poor choice of quadrature rule, and a Poisson ratio near 0.5 in plane strain and solid problems may provoke ill-conditioning, locking, or instability. These are dangerous difficulties because they can seriously degrade results rather than making results so peculiar that it becomes obvious that something is wrong.

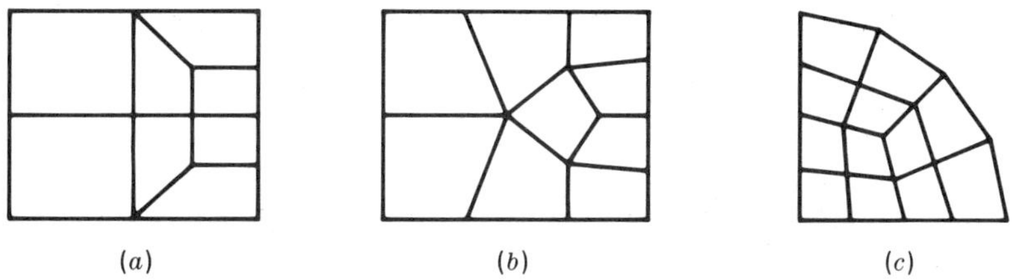

Figure 19.1-8. (*a*, *b*) Transitions from coarse to finer mesh that avoid abrupt size changes. (*c*) Possible mesh of quadrilaterals on one quadrant of a circular plate.

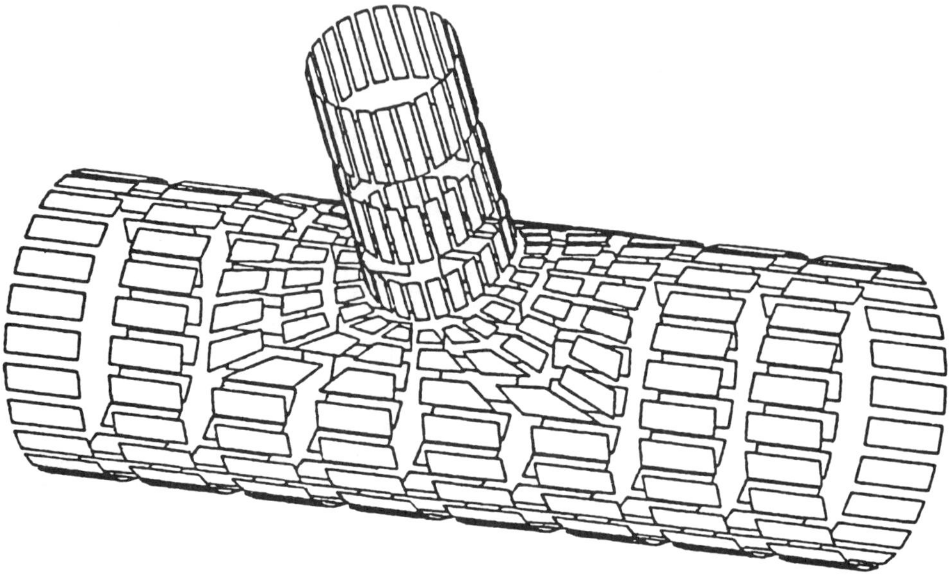

Figure 19.1-9. The intersection of two cylindrical shells. The "shrink plot" shows elements at about 75% of their actual sizes.

Some important "don'ts" are as follows. Do not support a stiff element by flexible elements; instead, impose rigid-body constraints on the stiff element. Do not "fake" a skew support (Fig. 18.2-3). Do not let Poisson's ratio approach 0.5 in plane strain and solid problems unless a special formulation is used. Do not let three-dimensional elements or Mindlin plate and shell elements become extremely thin. Do not use a minimal integration rule without being aware of possible mechanisms.

Some of these difficulties can be detected by error tests in the coding, such as a test for the condition number of the structure stiffness matrix or a test for diagonal decay during equation solving. Such tests are usually a posteriori rather than a priori and may be optimistic or pessimistic.

Stresses. At optimal stress points, computed stresses may be as accurate as displacements. Usually, however, stresses are less accurate than displacements. Accordingly, a finer mesh is needed for stress analysis than for displacement

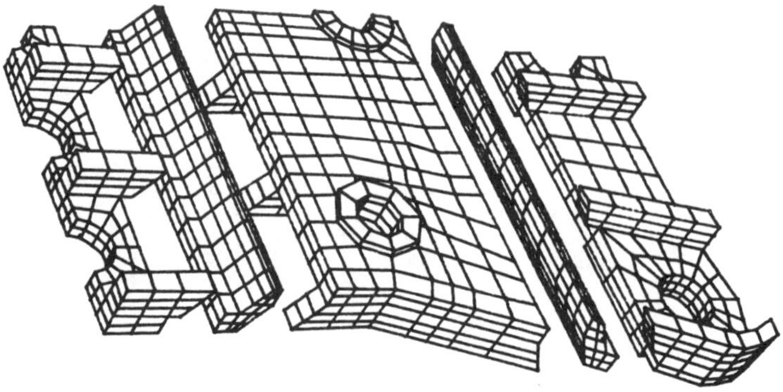

Figure 19.1-10. An exploded view of a machine part. (*Courtesy of Algor Interactive Systems Inc., Pittsburgh, Pennsylvania.*)

analysis, and stresses are not considered reliable if displacements are suspect. Analogously, in vibration analysis, mode shapes are not considered reliable if natural frequencies are suspect.

A postprocessor can usually display stress contours, which typically are smoothed, perhaps by working from nodal average stresses. Too much smoothing can make stresses appear more accurate than they really are. It may be preferable to avoid averaging between elements. For example, one may plot contours of the von Mises effective stress, element by element and without any averaging of stresses at nodes shared by elements. These contours may be plotted as "stress bands" [19.3] by designating equally spaced stress intervals, locating the areas of each element that fall into each interval, and using a different color to plot each interval. Typically, in an adequately refined mesh, the bands are slightly discontinuous across interelement boundaries but a global contour pattern is evident upon visual inspection. If no such global pattern is apparent, the mesh is too coarse. If the bands appear perfectly continuous, the mesh is finer than necessary. In Fig. 19.1-11*a*, discontinuities are probably too pronounced for the solution to be considered acceptable. In Fig. 19.1-11*b*, discontinuities are sufficiently small that the solution may be considered acceptable.

Stresses must not be averaged across a step change in modulus or a step change in thickness. Specifically, if ϵ_t is the mechanical strain tangent to a boundary between two different materials, then, in the absence of initial strains, the corresponding stress σ_t is proportional to $E_1\epsilon_t$ on one side and to $E_2\epsilon_t$ on the other. And, if a bar has a step change in cross-sectional area from A_1 to A_2 and no load is applied at the step, axial stresses on either side of the step are in the ratio A_1/A_2.

Mesh Refinement. A need for refinement of all or part of the mesh may be indicated by visual inspection of discontinuities in the stress bands just cited. Analogous numerical indices may be coded. As an example, consider a four-node plane element that is not very good at modeling bending. Strain energy U_0 per unit volume might be selected as a reference quantity, and the ratio of the greatest change in U_0 across the element to the value of U_0 at the element center taken as an index. If the value of this index exceeds a prescribed tolerance, either for individual elements or for a patch of elements, a need for refinement is indicated. A similar numerical test, not limited to low-order elements, could be based on changes in a reference quantity between adjacent elements.

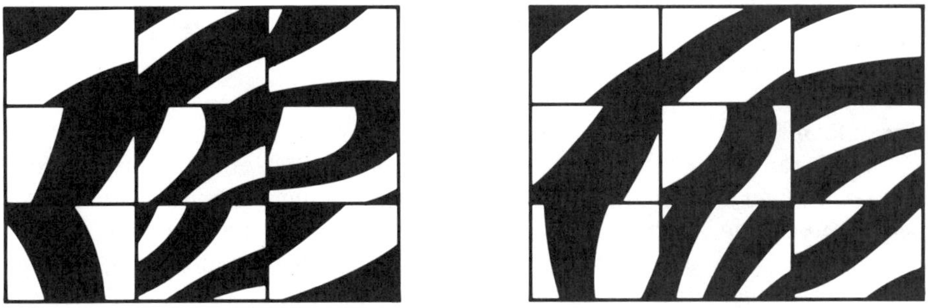

(*a*) (*b*)

Figure 19.1-11. Hypothetical stress bands in 3 by 3 patches of plane rectangular elements, representing solutions that are (*a*) probably inadequate, and (*b*) probably adequate.

If a model tends to be ill conditioned, further refinement may make results worse rather than better (see Table 18.4-1).

If refinement does little to change results, one has evidence (but not proof) that results are satisfactory. Similarly, if the mesh is already rather fine, one may try a coarser mesh, again to discover if there are significant changes. Analogously, in a response history analysis, one might see whether results are significantly changed by a change in the number of modes used in a modal method or by a change in the time step Δt of direct integration.

Local Analysis. If localized mesh refinement is necessary, we need not reanalyze the entire structure with the local refinement imbedded in it. The portion of the structure that contains the refined mesh can be analyzed separately. It is loaded by whatever prescribed loads may be present and, along the boundary where it has been cut free of the rest of the structure, loaded by the displacements computed in the preceding analysis of the entire structure. If refinement adds nodes along this boundary, interpolation is needed to obtain the prescribed displacements of the added nodes.

In such an analysis, a correction may be advisable. Typically the original mesh is stiffer than the refined mesh. Therefore boundary displacements to be imposed on the local refinement may be underestimated, which results in underestimation of stresses in the local refinement. An ad hoc correction scheme is as follows. For the original local mesh and the refined local mesh, respectively, compute nodal loads $[\mathbf{K}]\{\mathbf{D}\}$ produced by prescribed boundary displacements $\{\mathbf{D}\}$. These loads, $\{\mathbf{R}\}_0$ and $\{\mathbf{R}\}_r$, will differ because $[\mathbf{K}]$ is changed by refinement. The ratio of their norms, $\|\mathbf{R}_0\|/\|\mathbf{R}_r\|$, usually exceeds unity and can be used as an approximate corrective multiplier to stresses computed in the local refinement.

Vibrations and Dynamics. If the dynamic load includes frequencies of interest up to ω_u, then the mesh should be able to accurately represent modes associated with frequencies up to about $3\omega_u$, and a mode superposition analysis should include frequencies up to about $3\omega_u$. The time step Δt in a direct integration analysis should be approximately $0.3/\omega_u$ or less, and must provide numerical stability if the integration method is conditionally stable. If a reduced basis is used for eigenvalue computations, there should be roughly four times as many master d.o.f. as eigenvalues to be accurately computed.

In direct integration there should be a match between the type of algorithm and the mass matrix; for example, lumped masses are best for an explicit algorithm. Abrupt changes in element size should be avoided, as such changes tend to produce spurious wave reflections and numerical noise.

Nonlinear Problems. Typically one must make many more trial runs in order to solve a nonlinear problem than to solve a linear problem. Not only must blunders be discovered and removed, but solution strategy must be guided by what is learned in preceding attempts. Here, much more than in linear problems, it is wise to start simply and not attempt the complete solution all at once. One might elect to solve a linear form of the problem first, then add nonlinearities one by one. Thus blunders are more easily discovered, the effect of each nonlinearity is more apparent, useful information is gained from each trial, and the risk of failure with a large, complicated, and expensive model is reduced [2.1].

Nonlinear analyses tend to be very expensive. It is therefore necessary that no

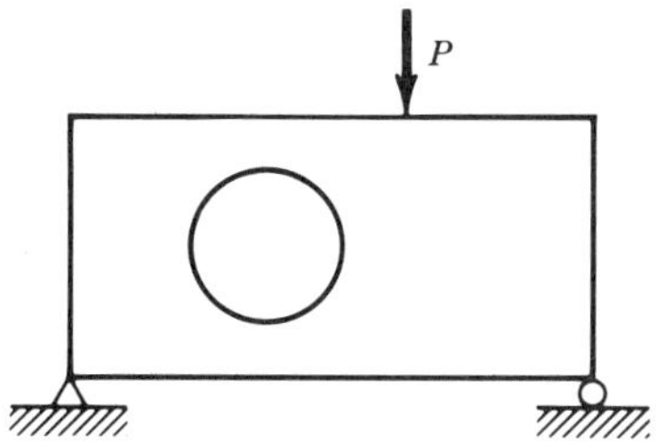

Figure 19.1-12. Flat plate containing a large hole.

superfluous nonlinearities be introduced. For example, imagine that yielding near the edge of the hole in Fig. 19.1-12 is to be analyzed. It is likely that yielding will also appear at the point of load application and at the support points. Yielding at these points will be treated with due respect (and due expense) by the algorithm [19.14]. Perhaps these concentrated forces result from oversimplification of loads and supports. If so, or if yielding in these locations is indeed ignorable, one might prevent undesired yielding by assigning a high yield strength to elements adjacent to the concentrated loads.

Miscellaneous Perils. Carelessness or lack of adequate understanding can lead to puzzling or misleading results. If the problem involves vibration, buckling, or nonlinear behavior, then axisymmetric geometry and axisymmetric loads do not guarantee axisymmetric response: unless symmetry is known to prevail, it should not be imposed by choice of boundary conditions. A quarter-point element for crack analysis can be too large or too small: thus, mesh refinement may make results worse. Incompatible and underintegrated elements may display a dependence on Poisson's ratio in problems that should be independent of Poisson's ratio. Anisotropy adversely affects accuracy [8.34]. If plane elements are warped so that element nodes are not all coplanar, results may be erratic and very sensitive to changes in the mesh [19.4,19.5]. If convergence with mesh refinement is not monotonic, extrapolation of results from two different meshes may give a worse result than is given by either mesh. Imperfections of load, geometry, supports, and mesh may be far more important in a buckling problem than in a static problem. Buckling, collapse, and nonlinear analyses demand more expertise than static stress analysis.

Check the Results. Computed results should be checked for "self-consistency," for example, by checking that intended supports do indeed have zero displacement and that any symmetries of the finite element model are represented in stress and displacement results. Computed results should be compared with whatever else is available that can be used for comparison. Examples include "back of the envelope" calculations, approximate analytical models, experimental data, textbook and handbook cases, preceding numerical analyses of similar problems, numerical analysis of a related but simpler problem, and results for the same problem predicted by a different program (which ideally should be based on a different numerical method). All these results should be regarded with some skepticism: analytical models incorporate idealizations, mistakes may be made in mathematics, textbooks and handbooks may contain errors, numerical solutions are subject to errors in coding and in data preparation, and experiments may be improperly performed and the results misinterpreted. When the inevitable disagreements appear, the reason for the discrepancy should be sought, and the amount of disagreement satisfactorily explained.

The time spent in processing and checking output should equal or exceed the time previously spent in data preparation. As with model preparation, an objective critique of the work should be obtained from a very competent analyst who is not directly involved with the project [19.14].

19.2 PROGRAMMING AND PROGRAMS

Programming. The writing of finite element programs is done by researchers and software vendors, who do so of necessity, and by students, who do so as an aid to learning. Those who use finite elements as a tool should buy or lease a program rather than write one. The few who write programs should do so in a way that makes the code easy to maintain and improve. These points are discussed in more detail later in this section.

Fortran is the language of all major finite element programs and is likely to remain so for the foreseeable future because of the large investment already made in Fortran software. (However, critical parts of a Fortran program may be coded in assembly language for the sake of efficiency.) Ideally, coding is guided by a previously prepared user's manual, in order to impose discipline on the developers and to produce a user-oriented product. The code should be built on a data base structure, and should be modular, that is, divided into logical subsets, each composed of one or more subroutines. The code should have mnemonic names for variables, monotonically increasing statement numbers, and adequate comment statements so that personnel other than the original programmer can read and maintain it. Much additional advice about good programming practice is available, especially in the computer science literature.

Dynamic Storage Allocation. Different problems require different amounts of computer memory, so it is inefficient to use fixed dimensions for arrays. Moreover, in different problems, the fractions of the total memory used by different phases of a single analysis run may differ. These difficulties are overcome by dynamic storage allocation, in which the dimensions of arrays are set at the time of execution. This procedure is common in finite element programs, but is often unfamiliar to the student. It is explained as follows.

In Fig. 19.2-1, array A contains most of the storage space to be used by the program. Various subroutines will use this same space but call it by other names. Imagine that Subroutine INPUT is to use one-dimensional arrays X and Y, which must each contain NUMNP entries, and the two-dimensional array ID, which must contain NDOF rows and NUMNP columns. The "pointers" N1, N2, and N3 identify the starting addresses in array A for arrays X, Y, and ID, respectively.[1] In Subroutine INPUT we find the statements

```
SUBROUTINE INPUT (X,Y,ID,NUMNP,NDOF)
DIMENSION X(1),Y(1),ID(NDOF,1)
```

[1]When both real and integer quantities appear in blank common, the compiler must assign the same word length to reals and integers if the addressing is to work properly.

```
      COMMON A(15000)
      LIM = 15000
C ---------(STATEMENTS NOT ESSENTIAL TO THIS EXAMPLE OMITTED HERE)
      N1 = 1
      N2 = N1 + NUMNP
      N3 = N2 + NUMNP
      N4 = N3 + NUMNP*NDOF - 1
      IF (N4 .GT. LIM)  CALL ERROR (N4)
      CALL INPUT (A(N1),A(N2),A(N3),NUMNP,NDOF)
C ---------(STATEMENTS NOT ESSENTIAL TO THIS EXAMPLE OMITTED HERE)
      N1 = 1
      N2 = N1 + NEQ*MBAND
      N3 = N2 + NEQ - 1
      IF (N3 .GT. LIM)  CALL ERROR (N3)
      CALL BUILD (A(N1),A(N2),NEQ,MBAND)
C ---------(STATEMENTS NOT ESSENTIAL TO THIS EXAMPLE OMITTED HERE)
      END
```

Figure 19.2-1. Hypothetical main routine that uses dynamic dimensioning. Subroutine ERROR (not shown) terminates execution if the problem is too large.

Arrays X, Y, and ID are stored in consecutively addressed cells of blank common. Since starting addresses have already been defined, it is not necessary to state the actual array size; this explains the 1's used in the dimension statement. Array ID is stored by columns, so that the address in blank common of ID(M,N) is N3 + NDOF∗(N-1) + M-1. Thus the dimension statement must say ID(NDOF,1) rather than ID(1,1), so that the background bookkeeping can locate the proper address.[2] (For a three-dimensional array, the first two dimensions must be provided; only the third can be 1.)

Later in the program, imagine that arrays X, Y, and ID have been stored elsewhere—for example, on a disk file. The memory space they occupied is now available for other use. In Fig. 19.2-1 we imagine that the space is now to be used to store the structure stiffness matrix S and load vector R. Accordingly, new pointers are computed. In Subroutine BUILD we find the statements

```
SUBROUTINE BUILD (S,R,NEQ,MBAND)
DIMENSION S(NEQ,1),R(1)
```

To change the capacity of the entire program, we change only the first two statements in Fig. 19.2-1, which necessitates recompiling only one subroutine.

Documentation. The quality of documentation is of major concern, whether we must prepare it or must read it in order to use a program. Unfortunately, documentation is often difficult to use, perhaps because it is often written by people so familiar with the program that they are unable to take the user's viewpoint. It is safest to assume when preparing documentation that most users know nothing about the program and that the remainder will often overlook what is obvious to the programmer. Good documentation is expensive to write and to maintain, but it is important: a good program may fail in the marketplace if its documentation is unreadable, and a useful program may be abandoned if its documentation does not keep up with program changes and enhancements [19.6,19.7].

A general-purpose commercial finite element program is accompanied by thou-

[2]Some compilers have an option that checks subscripts in executable statements against array bounds in DIMENSION statements. A logic error will then be signaled. To avoid this signal one can disable the checking option or replace each 1 in the foregoing DIMENSION statement by NUMNP.

sands of pages of documentation, typically divided as follows: theoretical manual (describes the analytical basis and limitations of algorithms), user's manual (describes available elements and data preparation), example problems manual (describes test cases, showing input data, output data, and comparison with theory), and programming or systems manual (describes computer science topics, incomprehensible to most engineers). More recent developments are introductory handbooks, which describe a simple and often-used subset of the program; online interactive help, which provides instant explanation of a command or an error message; and videotapes, which offer training lectures on aspects of program use [19.7].

Desirable features in documentation, not always present, include the following: index, nomenclature, table of elements, summary description of element behavior, estimates of timing and cost, modeling suggestions, glossary of errors and error messages, limitations of major algorithms, and lists of element quirks and frequently made errors [19.7]. At worst, a manual tersely explains acronyms in terms of other acronyms, so that making use of its information resembles trying to determine the function of an unfamiliar machine by reading its parts list.

Before buying an expensive program, it is wise to read some of its manuals. Documentation can usually be purchased separately from software, and far more cheaply.

Costs. The computational expense of a finite element analysis varies widely. At one extreme, when the computation is done on a personal computer that would otherwise be idle, the expense is almost zero. Toward the other extreme, when a large nonlinear problem occupies most of a day's running time on a supercomputer, the expense approaches an engineer's annual salary. Overall, hundreds of millions of dollars are spent each year on finite element modeling and computer costs.

Similarly, the cost of writing a program varies widely (but is usually high). In one study of programming efficiency, each of several programmers with from 2 to 11 years of experience coded a logic problem. The ratio of best to worst was 25/1 for coding time, 5/1 for code size, and 13/1 for running time. There was no correlation between productivity and experience [19.8]. Another study covered more than 400 military software projects that lasted from 1 month to 8 years in length and involved from 2 to 200 people at a time. Time expended on design, coding, testing, and documentation was included in the study. Productivity was measured in lines of code per person per month, which we abbreviate here as "lines." Average productivity over the life of each project ranged from 5 to 5000 lines. Average productivity over all projects was 200 lines, with two-thirds of the projects having average productivities between 75 and 550 lines [19.9]. From these data, we conclude that it rarely makes sense to write a program if one can be purchased or leased instead.

After release of a software system, it must be maintained. Bugs must be corrected and enhancements added to keep the program alive in the marketplace. Over the lifetime of a significant software system, maintenance costs far exceed development costs.

Programs for engineering are far outsold by programs for business. Yet the programs for engineering cost much more to develop, maintain, and support [19.10]. Programs for personal computers are proliferating, however no new major software system has entered the market for many years. This is no surprise when

TABLE 19.2-1. CHARACTERISTICS OF STRUCTURAL MECHANICS SOFTWARE [19.11]. K = MULTIPLIER OF 1000.

Type of Program	Single Element Static	Multiple Element Static	Large General Purpose
Number of statements	500–2000	2K–10K	100K–600K
Development cost[a]	\$5K–\$50K	\$25K–\$200K	\$2000K–\$10,000K
Pages of documentation	20–100	50–500	2000–7000
Machine words needed	10K–30K	20K–50K	50K–150K
Diagnostics	few–moderate	few–moderate	extensive
Cost per run[a]	\$1–\$100	\$1–\$500	\$10–\$10K

[a]When this table was written, a graduating engineer started work at roughly \$1000 per month.

one considers the high start-up costs (Table 19.2-1). Furthermore, even with a quality product, the entrepreneur must persevere for years, and incur additional costs that may exceed the development costs, in order to penetrate a reluctant market: users develop loyalties to software with which they are familiar and tend to be oblivious to other options. User suspiciousness is not without foundation. Unless there is good reason to believe that software will be supported for many years, one cannot justify the expense of acquiring it or the much higher expense of developing competence in its use.

Some programs of quality are available cheaply, usually because they are of limited scope or because no commitment of maintenance or user support is supplied with the program. Examples include certain versions of the BOSOR programs for thin shells of revolution and certain versions of the SAP program for linear static and dynamic structural analysis.

Commercial Programs. Hundreds of finite element programs are available, from small to large. Large general-purpose analysis systems share the following traits [19.12].

Generality. Many element types are provided, so that almost any conceivable structure, supports, and boundary conditions can be treated. Linear problems of statics are certainly included; linear dynamics and heat transfer are almost certain to be included. Certain nonlinear capabilities, magnetics, or other special features are probably included.

Large Size. The source code (not available to the user) comprises 100,000 to 600,000 statements and is the result of a development effort of from fifty to several hundred man years.

Worldwide Distribution. The system is installed at tens or hundreds of locations in various countries.

Large User Community. There are thousands of users in industry, data centers, consulting firms, research establishments, and universities.

User Support. In additon to documentation, the vendor offers hotline support, consulting, training courses, user conferences, and newsletters.

Portability. The program is available on a variety of machines, from supercomputers to minicomputers. Some versions of the program may be available on personal computers.

Maintenance. Updated versions of the program are released every year or two. Each new release is thoroughly tested, so that few bugs remain.

In choosing a software vendor, companies may compile a list of candidates by study of industry journals and by talking with associates at other companies. The list is shortened by discarding products that do not meet the company's needs, seem either overpriced or suspiciously inexpensive, or are new and untried. Programs that remain on the list are tested to see if the promotional claims are met and if the product will solve the company's everyday problems. The first program tested may stand out: if it is complex, others will not seem comprehensive enough; if it is comparatively easy to use, others will seem too difficult [19.10]. Probably none of the systems will be as easy to use as one might wish. For this reason service to the user is very important. A software vendor must support the product in order to survive amidst rising user expectations.

APPENDIX **A**

MATRICES: SELECTED DEFINITIONS AND MANIPULATIONS

This appendix summarizes portions of matrix theory that are frequently used in finite element analysis. Symbols used are arbitrary and imply no particular physical meaning. Further explanation may be found in any of several references, for example [A.1].

Multiplication. Let [**C**] be the product [**A**][**B**] ([**A**] premultiplies [**B**]; [**B**] postmultiplies [**A**]). Then [**C**] is

$$\underset{m\times q}{[\mathbf{C}]} = \underset{m\times n}{[\mathbf{A}]}\ \underset{n\times q}{[\mathbf{B}]} \qquad \text{where} \quad C_{ij} = \sum_{k=1}^{n} A_{ik}B_{kj} \tag{A.1}$$

where i ranges from 1 to m and j ranges from 1 to q. Matrices [**A**] and [**B**] must be *conformable* for multiplication. That is, the number of columns in [**A**] must equal the number of rows in [**B**]. Fortran statements that accomplish the multiplication in Eq. A.1 are

```
      DO 20 I=1,M
      DO 20 J=1,Q
      SUM = 0.
      DO 10 K=1,N
   10 SUM = SUM + A(I,K)*B(K,J)
   20 C(I,J) = SUM
```

To accomplish the multiplication $[\mathbf{A}]^T[\mathbf{B}]$ (for which $[\mathbf{A}]^T$ and [**B**] must be conformable), one replaces A(I,K) by A(K,I) in statement 10.

The transpose of a product is the product of the transposes *in reverse order,* that is

$$([\mathbf{A}][\mathbf{B}])^T = [\mathbf{B}]^T[\mathbf{A}]^T \tag{A.2}$$

Linear Dependence. A set of vectors $\{\mathbf{v}\}_1, \{\mathbf{v}\}_2, \ldots, \{\mathbf{v}\}_n$ is linearly dependent if, for some j,

$$\sum_{i\neq j} \alpha_i\{\mathbf{v}\}_i = \{\mathbf{v}\}_j \tag{A.3}$$

where the α_i are scalars. For example, if $3\{\mathbf{v}\}_2 + 4\{\mathbf{v}\}_4 = \{\mathbf{v}\}_5$, then vector 5 is linearly dependent on vectors 2 and 4, regardless of n. Columns (or rows) of a matrix may be regarded as vectors; thus Eq. A.3 pertains to a matrix that has one or more linearly dependent columns (or rows).

Rank. Singularity. The rank of a matrix [**A**] is defined as the order of the largest nonzero determinant in [**A**]. An equivalent definition of rank is the maximum number of linearly independent rows (or columns) in [**A**]. As examples, the following matrices have ranks 2, 2, and 1 respectively.

$$\begin{bmatrix} 1 & 3 & -3 \\ 2 & 4 & -2 \\ 3 & 5 & -1 \end{bmatrix} \quad \begin{bmatrix} 1 & -1 & 0 & 2 \\ -1 & 2 & -1 & -1 \\ 0 & -1 & 1 & -1 \end{bmatrix} \quad \begin{bmatrix} 2 & 4 & -2 \\ 1 & 2 & -1 \\ 3 & 6 & -3 \end{bmatrix}$$

A matrix whose rank is less than its order is said to be *rank-deficient*. The rank of a null matrix is zero. A square matrix $[\mathbf{A}] = \{\mathbf{a}\}\{\mathbf{b}\}^T$ has rank 1 or is null, regardless of order. A square matrix whose rank is less than its order is called *singular*. The rank of a matrix product is

$$\text{rank}([\mathbf{A}][\mathbf{B}]) \leq \min(\text{rank}\,[\mathbf{A}],\ \text{rank}\,[\mathbf{B}]) \tag{A.4}$$

Quadratic Forms. If [**A**] is a real square matrix and {**x**} is a real vector of the same order, then scalar F is called a *quadratic form*, where

$$F = \{\mathbf{x}\}^T[\mathbf{A}]\{\mathbf{x}\} \tag{A.5}$$

Now imagine that we calculate all possible values of F as follows. Let coefficients x_i in {**x**} be allowed to assume independently any and all real values except for all x_i simultaneously zero. Then matrix [**A**] is called

positive definite if $F > 0$ for all {**x**}
positive semidefinite if $F \geq 0$ for all {**x**}
negative semidefinite if $F \leq 0$ for all {**x**}
negative definite if $F < 0$ for all {**x**}

and simply *indefinite* if F can be either positive or negative. For example, the following two matrices are respectively positive definite and positive semidefinite:

$$\begin{bmatrix} 1 & -1 \\ -1 & 2 \end{bmatrix} \quad \begin{bmatrix} 1 & -1 \\ -1 & 1 \end{bmatrix}$$

If a square matrix is positive definite or negative definite, it is also nonsingular.

Differentiation. Differentiation of a matrix is accomplished by differentiating each of its terms. For example, if $\lfloor \mathbf{x} \rfloor = \lfloor 1 \quad y \rfloor$, then $d\lfloor \mathbf{x} \rfloor/dy = \lfloor 0 \quad 1 \rfloor$.

Let $\{\mathbf{x}\} = \lfloor x_1 \quad x_2 \quad \dots \quad x_n \rfloor^T$ and let [**A**] be an arbitrary n by n square matrix that does not depend on the x_i. Suppose that the quadratic form

$$\phi = \tfrac{1}{2}\{\mathbf{x}\}^T[\mathbf{A}]\{\mathbf{x}\} \tag{A.6}$$

is to be differentiated with respect to each of the x_i. The result is conveniently stated as a vector,

$$\left\{\frac{\partial \phi}{\partial \mathbf{x}}\right\} = \left\lfloor \frac{\partial \phi}{\partial x_1} \quad \frac{\partial \phi}{\partial x_2} \quad \cdots \quad \frac{\partial \phi}{\partial x_n} \right\rfloor^T = \tfrac{1}{2}([\mathbf{A}] + [\mathbf{A}]^T)\{\mathbf{x}\} \tag{A.7}$$

as may be verified by writing ϕ in terms of the x_i and the A_{ij}, taking the derivatives, and gathering terms. If [**A**] is *symmetric*, then

$$\left\{\frac{\partial \phi}{\partial \mathbf{x}}\right\} = [\mathbf{A}]\{\mathbf{x}\} \qquad \text{and} \qquad \frac{\partial^2 \phi}{\partial x_i \partial x_j} = A_{ij} = A_{ji} \tag{A.8}$$

As a special case, if [**A**] is a unit matrix, then $\{\partial\phi/\partial\mathbf{x}\} = \{\mathbf{x}\}$.

Let $\{\mathbf{x}\} = \lfloor x_1 \quad x_2 \quad \ldots \quad x_n \rfloor^T$, $\{\mathbf{y}\} = \lfloor y_1 \quad y_2 \quad \ldots \quad y_m \rfloor^T$, and [**A**] be an arbitrary n by m matrix that does not depend on the x_i. Suppose that the scalar $\psi = \{\mathbf{x}\}^T[\mathbf{A}]\{\mathbf{y}\}$ is to be differentiated with respect to each of the x_i. The result is conveniently stated as a vector:

$$\left\{\frac{\partial \psi}{\partial \mathbf{x}}\right\} = \left\lfloor \frac{\partial \psi}{\partial x_1} \quad \frac{\partial \psi}{\partial x_2} \quad \cdots \quad \frac{\partial \psi}{\partial x_n} \right\rfloor^T = [\mathbf{A}]\{\mathbf{y}\} \tag{A.9}$$

As for differentiation with respect to the y_i, we note that since ψ is scalar,

$$\psi = \psi^T = \{\mathbf{y}\}^T[\mathbf{A}]^T\{\mathbf{x}\} \tag{A.10}$$

therefore, if [**A**] does not depend on the y_i,

$$\left\{\frac{\partial \psi}{\partial \mathbf{y}}\right\} = \left\lfloor \frac{\partial \psi}{\partial y_1} \quad \frac{\partial \psi}{\partial y_2} \quad \cdots \quad \frac{\partial \psi}{\partial y_m} \right\rfloor^T = [\mathbf{A}]^T\{\mathbf{x}\} \tag{A.11}$$

As a special case, if [**A**] is a unit matrix, then

$$\psi = \{\mathbf{x}\}^T\{\mathbf{y}\} \qquad \left\{\frac{\partial \psi}{\partial \mathbf{x}}\right\} = \{\mathbf{y}\} \qquad \left\{\frac{\partial \psi}{\partial \mathbf{y}}\right\} = \{\mathbf{x}\} \tag{A.12}$$

B APPENDIX

SIMULTANEOUS ALGEBRAIC EQUATIONS

Selected algorithms for equation solving are described. Fortran coding is provided for some of these algorithms.

B.1 INTRODUCTION

Methods of computational mechanics usually produce large systems of simultaneous algebraic equations. Solution algorithms for these problems can be categorized as either *direct* or *iterative*.

Direct methods provide solutions within a fixed number of steps. The number of steps can be calculated a priori from knowledge of the size of the problem and the specific procedure elected. The solution obtained would be exact if infinitely precise arithmetic were possible. Storage requirements can be large, especially for three-dimensional finite element problems for which matrices are not narrowly banded and "fill-in" terms are created during processing.

Iterative (or *indirect*) methods provide approximate solutions that improve with continued iteration. The number of iterations needed is not known a priori: it depends on the size of the problem, the specific algorithm elected, the convergence criterion, and the numerical conditioning of the problem. (For ill-conditioned problems, iterative methods converge slowly and sometimes diverge.) Storage requirements are less than for direct methods, especially when bandwidths are large, because no "fill-in" terms are created. In this circumstance an iterative method may also be faster than a direct method. Convergence takes fewer iterations if a good initial guess is available. However, iterative methods are usually not competitive with direct methods except in specific areas such as reanalysis, optimization, and large three-dimensional problems.

In linear static analysis, equation solving typically accounts for roughly one-quarter of the "number-crunching" cost. In nonlinear analysis the fraction may be much higher.

Equation solving (and eigenvalue extraction) requires many multiplications, and roughly an equal number of additions or subtractions. Addition and subtraction are done much faster than multiplication. Accordingly, one can estimate the relative speed of competing algorithms by counting the number of multiplications required for each.

Some computers have vector processors that compute the product of two vectors almost as quickly as the product of two scalars. For such a machine we seek an algorithm that can be "vectorized."

Often a problem is too large for all data to be stored in primary (core) memory. Many algorithms use out-of-core storage. In a virtual memory machine, the swapping of data in and out of core need not be explicitly coded; it is handled automatically by the operating system. However, if data are badly structured, a virtual memory machine may *thrash,* that is, spend an inordinate amount of time swapping information in and out of core.

Small problems, and well-conditioned problems, can be analyzed in single precision arithmetic (i.e., with approximately 32 bits per word). However, double-precision arithmetic is usually recommended for *all* calculations in a finite element analysis.

There are many ways to store sparse matrices, preserve their sparsity, avoid multiplications by 0's and 1's, and so on [B.1]. No one selection of options is best in all cases, as different problems yield matrices of different topology.

Actual listings of many algorithms are published [B.2–B.6, for example]. Others, already compiled and callable as subroutines, appear in widely available software packages such as IMSL, NAg, and LINPAC. Further information on related software and procedures appears in Appendix C.

B.2 SOLUTION OF SIMULTANEOUS LINEAR ALGEBRAIC EQUATIONS BY GAUSS ELIMINATION

We wish to solve the system of equations[1]

$$[\mathbf{A}]\{\mathbf{x}\} = \{\mathbf{c}\} \tag{B.2-1}$$

where $[\mathbf{A}]$ is an n_{eq} by n_{eq} matrix and $\{\mathbf{x}\}$ and $\{\mathbf{c}\}$ are n_{eq} by 1 vectors. It is assumed that $[\mathbf{A}]$ is nonsingular and has constant coefficients, $\{\mathbf{c}\}$ is given, and $\{\mathbf{x}\}$ is to be determined. The concept of Gauss elimination is to combine the rows of Eq. B.2-1 in such a way that coefficient matrix $[\mathbf{A}]$ is transformed into upper triangular form. This is the *forward-reduction* phase. The equations are then sufficiently uncoupled to enable $\{\mathbf{x}\}$ to be determined by *back-substitution* [A.1, pp. 12–13].

The foregoing technique is detailed in Fig. B.2-1, and Fortran coding for it is given in Fig. B.2-2. During elimination, the ith equation, $1 \le i < n_{eq}$, is used to reduce to zero all entries below the diagonal in the ith column. Thus, in the ith elimination, one divides by the ith diagonal coefficient A_{ii}. This simple strategy would fail if A_{ii} were very small or zero. However, if $[\mathbf{A}]$ is a stiffness matrix, A_{ii} is sufficiently large unless the structure is nearly unstable or badly modeled.

When Fig. B.2-2 is studied, it may help to compare the coding with the following equation, which shows the result of one elimination (K = 2 in Fig. B.2-2, with NEQ = 3).

$$\begin{bmatrix} A_{11} & A_{12} & A_{13} \\ 0 & A_{22}\text{-}(A_{21}/A_{11})A_{12} & A_{23}\text{-}(A_{21}/A_{11})A_{13} \\ 0 & A_{32}\text{-}(A_{31}/A_{11})A_{12} & A_{33}\text{-}(A_{31}/A_{11})A_{13} \end{bmatrix} \begin{Bmatrix} x_1 \\ x_2 \\ x_3 \end{Bmatrix} = \begin{Bmatrix} c_1 \\ c_2\text{-}(A_{21}/A_{11})c_1 \\ c_3\text{-}(A_{31}/A_{11})c_1 \end{Bmatrix} \tag{B.2-2}$$

An operation count shows that, for a large system of n_{eq} equations, forward-reduction in Fig. B.2-2 requires about $n_{eq}^3/3$ multiplications [A.1, pp. 14–16]. If $[\mathbf{A}]$ were symmetric, and Fig. B.2-2 were altered accordingly, about $n_{eq}^3/6$ multiplications would be needed. In either case, back-substitution requires about $n_{eq}^2/2$ multiplications. Clearly, for a large system of equations, most of the solution time is spent in doing forward-reduction.

If needed, the determinant of $[\mathbf{A}]$ can be calculated as the product of reduced diagonal coefficients in $[\mathbf{A}]$. Thus, after statement 10 in Fig. B.2-2, insert the statements

```
      DET = 1.0D0
      DO 15 I=1,NEQ                    (B.2-3)
   15 DET = DET*A(I,I)
```

However, DET is usually large, and may overflow. In some cases DET may be very small and may underflow. Overflow (or underflow) is less likely if statements B.2-3 are replaced by the statements

[1]In some technical papers the symbolism $\{\mathbf{x}\} = [\mathbf{A}]^{-1}\{\mathbf{c}\}$ merely implies solution for unknowns $\{\mathbf{x}\}$, not that matrix inversion is the numerical procedure actually used.

Forward-Reduction Phase

For $k = 2, 3, \ldots, n_{eq}$
  For $i = k, k+1, \ldots, n_{eq}$

$$r_{i,k-1} = A_{i,k-1}/A_{k-1,k-1}$$
$$c_i = c_i - r_{i,k-1}\, c_{k-1}$$

  For $j = k, k+1, \ldots, n_{eq}$

$$A_{ij} = A_{ij} - r_{i,k-1}\, A_{k-1,j}$$

Back-Substitution Phase

$$x_{n_{eq}} = c_{n_{eq}}/A_{n_{eq}n_{eq}}$$

For $k = n_{eq} - 1, n_{eq} - 2, \ldots, 1$

$$x_k = \frac{1}{A_{kk}}\left(c_k - \sum_{j=k+1}^{n_{eq}} A_{kj}\, x_j\right)$$

Figure B.2-1. Algorithm for Gauss elimination solution of the system of n_{eq} simultaneous linear algebraic equations [**A**]{**x**} = {**c**}, where {**c**} is specified; [**A**] = A_{ij}, {**c**} = c_i, {**x**} = x_i. In this figure, "=" means "is replaced by," as in Fortran.

```
   DETLN = 0.D0
   DO 15 I=1,NEQ
15 DETLN = DETLN + DLOG(A(I,I))
```
(B.2-4)

which calculate the natural logarithm of the determinant.

Band-Symmetric Matrix [A]. Let [**A**] be symmetric and banded [A.1, pp. 52–56]. With b the semibandwidth, coefficients in [**A**] to the right of its diagonal can be stored in a rectangular array of size n_{eq} by b, with all diagonal matrix coefficients A_{ii} in column 1 of the array (see Fig. 2.8-3b). A Gauss elimination equation solver for this storage format appears in Fig. B.2-3. It requires about $n_{eq}b^2/2$ multiplications for forward-reduction of [**A**] and about $n_{eq}b$ multiplications each for forward-reduction of {**c**} and for back-substitution.

Figure B.2-3 exploits the fact that the "uneliminated" portion of [**A**] (i.e., the southeast corner of [**A**], below the eliminated equations) remains symmetric at all stages of reduction (see Eq. B.2-2). Thus the equation solver may use A_{ij} where one expects to see A_{ji}.

During elimination of a row of [**A**], coefficients in this row "cast a shadow" of width b below the row in which reside all coefficients that are modified by that elimination (Fig. B.2-4a). Accordingly, in Fig. B.2-3, the DO 780 and DO 750 statements carry calculations

```
      SUBROUTINE SOL (A,C,X,NEQ)
      IMPLICIT DOUBLE PRECISION (A-H,O-Z)
      DIMENSION A(NEQ,NEQ),C(NEQ),X(NEQ)
C---  Forward reduction phase.
      DO 10 K=2,NEQ
      DO 10 I=K,NEQ
      R = A(I,K-1)/A(K-1,K-1)
      C(I) = C(I) - R*C(K-1)
      DO 10 J=K,NEQ
   10 A(I,J) = A(I,J) - R*A(K-1,J)
C---  Back substitution phase (results stored in X).
      X(NEQ) = C(NEQ)/A(NEQ,NEQ)
      DO 30 K=NEQ-1,1,-1
      X(K) = C(K)
      DO 20 J=K+1,NEQ
   20 X(K) = X(K) - A(K,J)*X(J)
   30 X(K) = X(K)/A(K,K)
      RETURN
      END
```

Figure B.2-2. Fortran statements for the Gauss elimination algorithm given in Fig. B.2-1.

```
      SUBROUTINE SOLVER (A,C,NEQ,MBAND,IFLAG)
      IMPLICIT DOUBLE PRECISION (A-H,O-Z)
      DIMENSION A(NEQ,1),C(1)
C---  Treat the case of one or more independent equations.
      IF (MBAND .GT. 1)  GO TO 690
      DO 680 N=1,NEQ
  680 C(N) = C(N)/A(N,1)
      RETURN
  690 NEQP = NEQ + 1
      NEQM = NEQ - 1
      GO TO (700,800), IFLAG
C---  Forward reduction of the coefficient matrix [A].
  700 DO 790 N=1,NEQM
      LIM = MIN(MBAND,NEQP-N)
      DO 780 L=2,LIM
      DUM = A(N,L)/A(N,1)
      I = N + L - 1
      J = 0
      DO 750 K=L,LIM
      J = J + 1
  750 A(I,J) = A(I,J) - DUM*A(N,K)
      A(N,L) = DUM
  780 CONTINUE
  790 CONTINUE
C---  Forward reduction of the constant vector {c}.
  800 DO 830 N=1,NEQM
      LIM = MIN(MBAND,NEQP-N)
      DO 820 L=2,LIM
      I = N + L - 1
      C(I) = C(I) - A(N,L)*C(N)
  820 CONTINUE
  830 C(N) = C(N)/A(N,1)
      C(NEQ) = C(NEQ)/A(NEQ,1)
C---  Back substitution. Former unknowns {x} overwrite {c}
      DO 860 N=NEQM,1,-1
      LIM = MIN(MBAND,NEQP-N)
      DO 850 L=2,LIM
      K = N + L - 1
      C(N) = C(N) - A(N,L)*C(K)
  850 CONTINUE
  860 CONTINUE
      RETURN
      END
```

Figure B.2-3. Subroutine for Gauss elimination solution of [**A**]{**x**} = {**c**} for band-symmetric matrix [**A**] (see Fig. 2.8-3*b*). NEQ = number of equations, MBAND = semibandwidth. Except for the case MBAND = 1, start at statement 700 (IFLAG = 1) for the first {**c**} and at statement 800 (IFLAG = 2) for any additional vectors {**c**}.

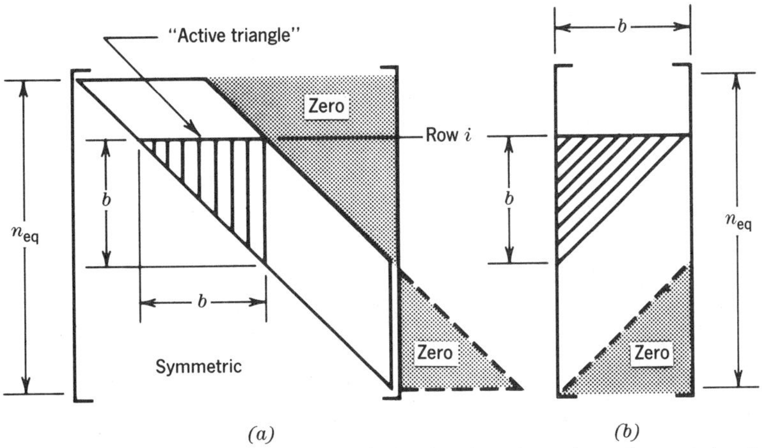

Figure B.2-4. The "active triangle"; that is, the triangular patch of coefficients in the stored band affected by elimination of row i during forward reduction, in (*a*) full matrix format, and (*b*) band storage format.

to the right and to the bottom, respectively, of the active triangle. The MIN function selects LIM = MBAND except when this choice would cause processing to extend into the "zero triangle" in Fig. B.2-4*b*. Then LIM = NEQ + 1 − N is chosen.

The determinant of [**A**] can be calculated from Fig. B.2-3 according to statements B.2-3 or B.2-4, but with A(I,I) replaced by A(I,1).

If a new {**c**} is to be processed for the same [**A**], one can enter Fig. B.2-3 at statement 800. Alternatively, with modest reprogramming, all constant vectors can be placed in a rectangular array and processed in a single pass through the subroutine.

A recommended modification of Fig. B.2-3 is to test for *decay of diagonal coefficients* (Section 18.4).

A possible modification of Fig. B.2-3 would avoid "do-nothing" computations that occur if DUM happens to be zero. One simply tests DUM in Fig. B.2-3 by an IF statement as soon as it is computed, and transfers to statement 780 if DUM = 0. For example, the triangular block of zeros in Fig. B.2-5 remains zero throughout reduction of the matrix, so no purpose is served by exercising the innermost DO loop of Fig. B.2-3 on these coefficients. However, Fig. B.2-3 can be vectorized by a sophisticated compiler. Then the IF test would be of little help, or even detrimental, as it may inhibit vectorization on many compilers.

Another possible modification in Fig. B.2-3 would be to use a semibandwidth appropriate to the row being eliminated; that is, use MBAND(N) rather than taking MBAND as constant throughout the process. Then one must count semibandwidths of rows and be aware that each may change from its initial value when fill-in terms are created by forward-reduction.

Remarks. The Gauss elimination procedure of Fig. B.2-3 is row-oriented and uses a constant semibandwidth b. Some other forms of Gauss elimination are column-oriented and exploit the differing heights above the diagonal exhibited by various columns (see Fig. 2.8-1*b*). These procedures are known as *profile*, *skyline*, or *active column* solvers [B.2–B.6]. Some topologies of coefficient matrix [**A**] are treated more efficiently by an active column solver than by a band solver. Figure B.2-5 is a case in point. An active column solver can avoid storage and processing of both blocks of zero coefficients. A band solver (Fig. B.2-3) stores and processes the triangular block of zeros, all to no useful purpose.

The *wave front* or *frontal* method is an arrangement of Gauss elimination in which assembly of structural equations alternates with their solution. The sequence in which equations are processed is driven by element numbering rather than by node numbering. The first equations to be eliminated are those associated with element 1 only. Then the adjacent element, element 2, makes its contribution of stiffness coefficients to the system of equations. If any additional equations are fully summed—that is, if any additional d.o.f. are shared by elements 1 and 2 only—these equations are eliminated. The next elimination awaits contributions from one or more additional elements. This repetitive alternation between assembly and solution can be viewed as a "wave" that sweeps over the structure

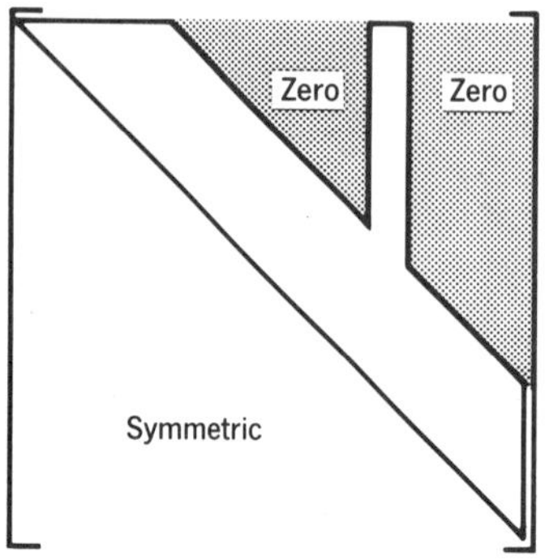

Figure B.2-5. Matrix topology better suited to an active column solver than to a band solver.

in a pattern dictated by the element numbering. For efficiency, consecutive element numbers should run "across" the structure—that is, in the direction that spans the smallest number of nodes.

If efficiently coded, and if presented with a structure whose nodes (or elements) are numbered for efficient processing, active column solvers and wavefront solvers appear to be equally efficient. Active column solvers appear to be easier to program and understand.

C APPENDIX

EIGENVALUES AND EIGENVECTORS

This appendix discusses the nature of the eigenproblem and suggests solution methods appropriate to particular forms that may be encountered.

C.1 THE EIGENPROBLEM

Let $[\mathbf{A}]$ and $[\mathbf{B}]$ be n by n square matrices. In an eigenvalue problem one seeks values of a scalar λ such that the matrix equation

$$([\mathbf{A}] - \lambda[\mathbf{B}])\{\mathbf{x}\} = \{\mathbf{0}\} \tag{C.1-1}$$

has solutions other than the trivial solution $\{\mathbf{x}\} = \{\mathbf{0}\}$. There are at most n solutions λ_i, not necessarily all distinct. The λ_i are called *eigenvalues* (other names include *characteristic values, latent roots, proper values,* and *principal values*). Corresponding to each λ_i there is an $\{\mathbf{x}\}_i$, called an *eigenvector* (other names include *characteristic vector, proper vector, principal vector, normal mode, natural mode,* and *principal mode*). Equation C.1-1 is called a *generalized eigenproblem* or simply an *eigenproblem*. If $[\mathbf{B}]$ happens to be the identity matrix, Eq. C.1-1 is called a *standard eigenproblem* [A.1] and the associated λ_i are called eigenvalues of $[\mathbf{A}]$.

Eigenproblems arise in vibration analysis (where $\sqrt{\lambda_i} = \omega_i$ is a vibration frequency and $\{\mathbf{x}\}_i$ is the vibration mode) and in buckling analysis (where λ_i indicates the critical load and $\{\mathbf{x}\}_i$ is the buckling mode). In other physical problems the eigenproblem must be solved to obtain information needed for the modal method of time-history analysis. Section 13.5 discusses eigenproblems to the extent needed to solve simple vibration problems.

C.2 THE STANDARD EIGENPROBLEM

Consider an eigenproblem of the form

$$([\mathbf{A}'] - \lambda\lceil\mathbf{B}\rfloor)\{\mathbf{x}'\} = \{\mathbf{0}\} \tag{C.2-1}$$

where $[\mathbf{A}']$ is an arbitrary n by n square matrix and $\lceil\mathbf{B}\rfloor$ is diagonal and nonsingular. There are always n eigenvalues of Eq. C.2-1. One may define a diagonal matrix $\lceil\mathbf{T}\rfloor$ and a transformation of $\{\mathbf{x}'\}$ by

$$T_{ii} = \frac{1}{\sqrt{B_{ii}}} \qquad \text{and} \qquad \{\mathbf{x}'\} = \lceil\mathbf{T}\rfloor\{\mathbf{x}\} \tag{C.2-2}$$

Premultiplication of Eq. C.2-1 by $\lceil\mathbf{T}\rfloor$ and substitution from Eq. C.2-2 yields the *standard eigenproblem*

$$([\mathbf{A}] - \lambda\lceil\mathbf{I}\rfloor)\{\mathbf{x}\} = \{\mathbf{0}\} \tag{C.2-3}$$

where $[\mathbf{A}] = \lceil\mathbf{T}\rfloor[\mathbf{A}']\lceil\mathbf{T}\rfloor$. Equations C.2-1 and C.2-3 have the same eigenvalues λ_i. An eigenvector $\{\mathbf{x}'\}_i$ of the original system is recovered from the corresponding eigenvector $\{\mathbf{x}\}_i$ by the operation $\{\mathbf{x}'\}_i = \lceil\mathbf{T}\rfloor\{\mathbf{x}\}_i$. (The foregoing transformation is convenient if $\lceil\mathbf{B}\rfloor$ is the diagonal mass matrix of a structural dynamics problem.)

Properties of Eq. C.2-3 that may be useful in engineering applications are as follows. Most of these results can be deduced from [A.1, pp. 243–360].

1. If $[\mathbf{A}]$ is real and symmetric, the λ_i are real.
2. If $[\mathbf{A}]$ is real, symmetric, and positive semidefinite, there are no negative λ_i. The number of nonzero λ_i equals the rank of $[\mathbf{A}]$.
3. If $[\mathbf{A}]$ is real, symmetric, and positive definite, all λ_i are positive.
4. If $[\mathbf{A}]$ is real and positive definite but unsymmetric, the matrix $([\mathbf{A}] + [\mathbf{A}]^T)$ has positive eigenvalues.
5. If $\{\mathbf{x}\}_i$ is an eigenvector, so is $c\{\mathbf{x}\}_i$, where c is an arbitrary nonzero scalar.
6. If all λ_i are distinct, all eigenvectors are distinct and linearly independent.
7. A λ_i repeated k times may or may not have k independent associated eigenvectors.
8. Let $[\mathbf{G}]$ be a square matrix, nonsingular and the same order as $[\mathbf{A}]$ but otherwise arbitrary. A matrix $[\mathbf{C}]$, obtained by the "similarity transformation" $[\mathbf{C}] = [\mathbf{G}]^{-1}[\mathbf{A}][\mathbf{G}]$, has the same eigenvalues as $[\mathbf{A}]$. If eigenvectors of $[\mathbf{C}]$ are $\{\mathbf{x}_c\}$, eigenvectors of $[\mathbf{A}]$ are $[\mathbf{G}]\{\mathbf{x}_c\}$.
9. If $[\mathbf{A}]$ is real and $[\mathbf{A}]^T[\mathbf{A}] = [\mathbf{A}][\mathbf{A}]^T$ (e.g., $[\mathbf{A}]$ is real and symmetric), then eigenvectors are orthogonal, that is, $\{\mathbf{x}\}_i^T\{\mathbf{x}\}_j = 0$ if $i \neq j$.
10. If eigenvectors are scaled so that $\{\mathbf{x}\}_i^T\{\mathbf{x}\}_i = 1$, then $\{\mathbf{x}\}_i^T[\mathbf{A}]\{\mathbf{x}\}_i = \lambda_i$ (see the Rayleigh quotient for the real symmetric case, Eq. 13.5-4 or Eq. C.3-9).
11. If eigenvectors are scaled so that $\{\mathbf{x}\}_i^T\{\mathbf{x}\}_i = 1$ and $[\mathbf{A}]$ is symmetric and nonsingular, then

$$[\mathbf{A}] = \sum_{i=1}^{n} \lambda_i\{\mathbf{x}\}_i\{\mathbf{x}\}_i^T \quad \text{and} \quad [\mathbf{A}]^{-1} = \sum_{i=1}^{n} \frac{1}{\lambda_i}\{\mathbf{x}\}_i\{\mathbf{x}\}_i^T \tag{C.2-4}$$

    where n is the order of $[\mathbf{A}]$.

C.3 THE GENERAL EIGENPROBLEM

Lumping and condensation procedures can result in eigenproblems of the form of Eq. C.1-1 in which $[\mathbf{A}]$ and particularly $[\mathbf{B}]$ may be of a more general nature than considered thus far. However, *we will assume that* $[\mathbf{A}]$ *and* $[\mathbf{B}]$ *are symmetric*. The following terminology and observations are useful in understanding these problems.

1. The matrix of linear polynomials $A_{ij} - \lambda B_{ij}$, that is,

$$[\mathbf{A}] - \lambda[\mathbf{B}] \tag{C.3-1}$$

    is called a *pencil* (of $[\mathbf{A}]$ and $[\mathbf{B}]$).
2. Eigenvalues of the pencil arise from nontrivial solutions of Eq. C.1-1 and thus are values of λ that make the determinant of the pencil vanish,

$$\det([\mathbf{A}] - \lambda[\mathbf{B}]) = 0 \tag{C.3-2}$$

    (as in Eq. 13.5-3, for example).

3. The polynomial of degree n or less

$$q(\lambda) = \det([\mathbf{A}] - \lambda[\mathbf{B}]) \tag{C.3-3}$$

is called the *characteristic polynomial* of the pencil.

4. In the case described by Eq. C.2-1, and consequently in the standard eigenproblem, $q(\lambda)$ is of degree n and has n eigenvalues as roots (counting multiplicity).

5. In general, $q(\lambda)$ may be of degree less than n. Matrix $[\mathbf{B}]$ must be singular for this to occur. In the most extreme case it is possible (when $[\mathbf{A}]$ also is singular) to have $q(\lambda) = 0$ for all λ. Fortunately this rarely happens in practical problems [C.1–C.3]. The roots of $q(\lambda)$ are called the *finite eigenvalues*. Infinite eigenvalues, if any, do not correspond to roots of $q(\lambda)$.

6. Let $[\mathbf{R}]$ and $[\mathbf{S}]$ be *nonsingular* matrices of numbers (not involving λ). The transformed pencil

$$[\mathbf{R}][\mathbf{A}][\mathbf{S}] - \lambda[\mathbf{R}][\mathbf{B}][\mathbf{S}] \tag{C.3-4a}$$

has the same characteristic polynomial as Eq. C.3-1, that is,

$$\det([\mathbf{R}][\mathbf{A}][\mathbf{S}] - \lambda[\mathbf{R}][\mathbf{B}][\mathbf{S}]) = \det([\mathbf{A}] - \lambda[\mathbf{B}]) = q(\lambda) \tag{C.3-4b}$$

7. The pencil of Eq. C.3-1 in general may have fewer than n independent eigenvectors. The case of Eq. C.2-1 ($[\mathbf{B}]$ diagonal, $B_{ii} > 0$) is one in which there is a *complete* set of *independent* eigenvectors $[\mathbf{Q}]$, whose n columns are the independent solutions of Eq. C.1-1.

8. Whenever there is a complete set of eigenvectors $[\mathbf{Q}]$, the pencil of Eq. C.3-1 is called *diagonalizable*. The special case of $[\mathbf{B}]$ positive definite is discussed in [A.1, pp. 343–346].

9. In the most common cases of diagonalizable pencils (when $[\mathbf{B}]$ is positive definite) the eigenvectors are *orthogonal with respect to* $[\mathbf{B}]$, and may be normalized so that $[\mathbf{Q}]$ *simultaneously diagonalizes* $[\mathbf{A}]$ and $[\mathbf{B}]$,

$$[\mathbf{Q}]^T[\mathbf{A}][\mathbf{Q}] = \lceil \mathbf{\Lambda} \rfloor \tag{C.3-5a}$$

$$[\mathbf{Q}]^T[\mathbf{B}][\mathbf{Q}] = \lceil \mathbf{I} \rfloor \tag{C.3-5b}$$

where $\lceil \mathbf{\Lambda} \rfloor$ is a diagonal matrix with $\Lambda_{ii} = \lambda_i$. A set of $[\mathbf{B}]$-orthogonal vectors, $[\mathbf{Q}]$, that satisfies Eq. C.3-5b is called *orthonormal with respect to* $[\mathbf{B}]$. As a special case, if $[\mathbf{B}]$ is a unit matrix $\lceil \mathbf{I} \rfloor$, then $[\mathbf{Q}]$ contains orthogonal unit vectors.

When $[\mathbf{B}]$ is positive definite, case 9 applies. This can be seen by transforming the pencil to standard form by using the Cholesky decomposition [A.1, p. 195],

$$[\mathbf{B}] = [\mathbf{L}][\mathbf{L}]^T \tag{C.3-6}$$

and applying Eqs. C.3-4 with $[\mathbf{R}] = [\mathbf{L}]^{-1}$ and $[\mathbf{S}] = [\mathbf{L}]^{-T}$ to obtain the pencil

$$[\mathbf{L}]^{-1}[\mathbf{A}][\mathbf{L}]^{-T} - \lambda\lceil \mathbf{I} \rfloor \tag{C.3-7}$$

Note that $[\mathbf{L}]^{-1}[\mathbf{A}][\mathbf{L}]^{-T}$ is symmetric and *there are n independent and orthonormal eigenvectors*. If $[\mathbf{X}]$ is a matrix whose columns are these eigenvectors, then the matrix of eigenvectors of the original pencil, Eq. C.3-1, is

$$[\mathbf{Q}] = [\mathbf{L}]^{-T}[\mathbf{X}] \tag{C.3-8}$$

which are orthonormal with respect to [**B**] in the sense of Eq. C.3-5b. Note that [**L**] generalizes the diagonal transformation matrix ⌈**T**⌋ of Eq. C.2-2 to the case in which [**B**] need not be diagonal. When [**A**] is positive definite, all $\lambda_i > 0$. When [**A**] is positive semidefinite, $\lambda_i \geq 0$. In structural problems, the $\lambda_i = 0$ are associated with rigid-body motions or mechanisms.

Either ad hoc or optimal mass lumping may lead to a problem in which [**A**] is positive semidefinite and [**B**] is diagonal with some $B_{ii} = 0$ (i.e., some massless d.o.f.). Consideration of the generalized *Rayleigh quotient* (see Eqs. 13.5-4 and 13.5-5),

$$\lambda_{\text{min}} \leq \frac{\{\mathbf{x}\}^T[\mathbf{A}]\{\mathbf{x}\}}{\{\mathbf{x}\}^T[\mathbf{B}]\{\mathbf{x}\}} \leq \lambda_{\text{max}} \tag{C.3-9}$$

usually indicates that there are as many infinite eigenvalues, $\lambda_{\text{max}} = \infty$, as there are $B_{ii} = 0$. It is possible to construct pathological forms of [**A**] and [**B**] that violate this rule. Pathological forms will not arise in finite element analysis if rigid-body motions (or mechanisms) are associated with positive kinetic energy [C.2]. It is up to the analyst to ensure that this condition prevails. If there are no pathological forms, [**A**] is positive semidefinite, and there are m massless d.o.f. ($B_{ii} = 0$), then there are m infinite eigenvalues and $n - m$ finite eigenvalues associated with $n - m$ independent eigenvectors {**x**}. Thus $q(\lambda)$ has degree $n - m$. The infinite eigenvalues are associated with m independent eigenvectors {**x**} such that [**B**]{**x**} = {**0**}.

Infinite eigenvalues can be observed in a more general context by considering the finite eigenvalues of the *inverse eigenproblem,* whose pencil is

$$[\mathbf{B}] - \mu[\mathbf{A}] \tag{C.3-10}$$

Many of the following statements are also true in a more general context, but if we limit our attention to the *diagonalizable case,* it is easy to see that the correspondence to the pencil of Eq. C.3-1 is that

$$\frac{1}{\mu_i} = \lambda_i \tag{C.3-11}$$

where $\mu_i = 0$ corresponds to $\lambda_i = \infty$. The complete set of eigenvectors is the same for both pencils. Equation C.3-11 defines an infinite eigenvalue even when [**B**] is nondiagonal. The pencils of Eqs. C.3-1 and C.3-10 *may not be diagonalizable in general.* When one is, the other clearly will be also, and this will be true whenever

(a) Either [**A**] or [**B**] is positive definite.

(b) [**A**] and [**B**] are positive semidefinite and there are no ill-disposed vectors (i.e. vectors such that [**A**]{**x**} = {**0**} and [**B**]{**x**} = {**0**}).

(c) [**A**] is semidefinite and [**B**] is an indefinite matrix arising from an optimal lumping formula, which leads to no indeterminate vectors [C.2] (i.e. vectors such that $\{\mathbf{x}\}^T[\mathbf{A}]\{\mathbf{x}\} = 0$ and $\{\mathbf{x}\}^T[\mathbf{B}]\{\mathbf{x}\} = 0$).

Case (c) arises in optimal lumping in which some masses (quadrature weights) are negative, as explained in Section 13.3. No known optimal lumping scheme leads to indeterminate vectors. It should be noted that ill-disposed vectors are a special case of indeterminate vectors, and that in pathological cases indeterminate vectors are possible even without massless d.o.f., because an indefinite [**B**] always has vectors {**x**} such that $\{\mathbf{x}\}^T[\mathbf{B}]\{\mathbf{x}\} = 0$. Pencils that fall outside of restrictions (a) through (c) can be highly pathological but are almost never observed in finite element practice.

In the most general diagonalizable cases—that is, (a) through (c) with [**B**] singular—we must revise our notion of orthogonality of [**Q**]. Simultaneous diagonalization of Eq. C.3-1 is possible, and

$$[\mathbf{Q}]^T[\mathbf{A}][\mathbf{Q}] = \lceil\boldsymbol{\Lambda}^+\rfloor \qquad \text{(C.3-12a)}$$

$$[\mathbf{Q}]^T[\mathbf{B}][\mathbf{Q}] = \lceil\boldsymbol{\Omega}\rfloor \qquad \text{(C.3-12b)}$$

where $\Lambda_{ii}^+ = \lambda_i$ for $0 \leq \lambda_i < \infty$, $\Lambda_{ii}^+ = 1$ for $\lambda_i = \infty$. In case (c), $\lambda_i \leq 0$ with $|\lambda_i|$ "very large" are introduced for each negative mass [C.1] and $\Lambda_{ii}^+ = |\lambda_i|$ for $\lambda_i < 0$. $\lceil\boldsymbol{\Omega}\rfloor$ is also diagonal, and $\Omega_{ii} = 1$ for $0 \leq \lambda_i < \infty$, $\Omega_{ii} = 0$ for $\lambda_i = \infty$, and $\Omega_{ii} = -1$ for $\lambda_i < 0$. In case (c), the negative signs can only be transferred from $\lceil\boldsymbol{\Omega}\rfloor$ to $\lceil\boldsymbol{\Lambda}^+\rfloor$ by a transformation of the pencil of the form of Eqs. C.3-4 in which $[\mathbf{R}] \neq [\mathbf{Q}]^T$. The zeros on the diagonal of $\lceil\boldsymbol{\Omega}\rfloor$ cannot be avoided. Thus *the eigenvectors with* $0 \leq \lambda_i < \infty$ *are orthonormal* in the usual sense. The other eigenvectors can have "negative lengths" or be "orthogonal to themselves." Their physical meaning has been sacrificed in such cases.

C.4 REMARKS ON SPECIAL FORMS

We consider the general eigenproblem,

$$([\mathbf{A}] - \lambda[\mathbf{B}])\{\mathbf{x}\} = \{\mathbf{0}\} \quad \text{or} \quad ([\mathbf{K}] - \omega^2[\mathbf{M}])\{\mathbf{D}\} = \{\mathbf{0}\} \qquad \text{(C.4-1)}$$

where the latter form uses the symbolism adopted for vibration problems in Section 13.5. In what follows we assume that [**A**] and [**B**] are symmetric, and consider how to proceed under circumstances that may arise in finite element modeling.

If [**A**] and [**B**] are positive definite, there is no special difficulty. One may proceed directly to solve the eigenproblem (see Section C.5).

[**B**] will be positive semidefinite and diagonal if "lumping" is used with $B_{ii} = 0$ for some d.o.f. x_i; for example, if mass particles have no rotary inertia or are not attached to all nodes (this is not a recommended practice). One can eliminate the x_i for which $B_{ii} = 0$ in a diagonal [**B**]. The procedure is discussed in Section 8.1. In the condensed system, matrices are of lower order than initially and [**B**] is positive definite. In the case where $B_{ii} = 0$ because of optimal lumping, condensation is not recommended, since it will lose the accuracy the scheme is designed to retain. Several algorithms discussed in the next section and in Ref. C.1 are either not affected if [**B**] fails to be positive definite or can be easily modified to work in this case.

In the structural context, a positive semidefinite [**A**] implies an unsupported structure such as a spacecraft. This circumstance presents no obstacle to some algorithms for solving the eigenproblem. Other algorithms may require that [**A**] be nonsingular. This restriction can be met by using an *eigenvalue shift*. Thus we select a scalar c and substitute $\lambda = \lambda' - c$ into Eq. C.1-1, to obtain

$$([\mathbf{A} + c\mathbf{B}] - \lambda'[\mathbf{B}])\{\mathbf{x}\} = \{\mathbf{0}\} \qquad \text{(C.4-2)}$$

in which $[\mathbf{A} + c\mathbf{B}]$ is positive definite if $\{\mathbf{x}\}_i^T[\mathbf{B}]\{\mathbf{x}\}_i > 0$ for every $\{\mathbf{x}\}_i$ for which $\{\mathbf{x}\}_i^T[\mathbf{A}]\{\mathbf{x}\}_i = 0$. We solve for shifted eigenvalues λ_i', then obtain actual eigenvalues $\lambda_i = \lambda_i' - c$. Eigenvectors are not changed by shifting. If $|c|$ is too small, $[\mathbf{A} + c\mathbf{B}]$ is almost singular. If $|c|$ is too large, convergence of the eigensolver may be slow. A possible choice is $c = 0.01r$, where r is the ratio of the trace of [**A**] to the trace of [**B**]. Surprisingly, perhaps, shifting usually yields a positive definite $[\mathbf{A} + c\mathbf{B}]$ even when [**B**] is indefinite, as when [**B**] arises from an optimal lumping scheme [C.1]. If the problem is structural and [**B**] is diagonal, one can interpret Eq. C.4-2 physically by saying that a spring of stiffness cB_{ii} has been added between each d.o.f. x_i and ground, thus preventing any rigid-body motion.

C.5 SOLUTION ALGORITHMS

Algorithms for eigenvalue extraction are plentiful. The best choice for a particular eigenproblem depends on the order n of the matrices, their sparsity, the number of eigenvalues to be extracted, and where in the eigenspectrum the eigenvalues of interest are located. In what follows we indicate some choices commonly made in practice, but make little or no attempt to explain the workings of various algorithms. These details may be found in references such as [C.4–C.12]. At the outset we note that the various methods are often used in combination rather than as stand-alone algorithms.

Writing out the characteristic polynomial, $q(\lambda) = \det[\mathbf{A} - \lambda\mathbf{B}] = 0$, and calculating its roots is a method suitable only for hand calculation, for example, when there are only one or two roots. If n is large, the method is costly and inaccurate.

If all n eigenvalues are required and n is relatively small (roughly $n < 200$), the *Jacobi method* is a good choice.

If [**A**] and [**B**] are narrowly banded and only a few eigenvalues are required (e.g., only λ_1 and λ_2), *determinant search* may be appropriate. Effectively, one seeks zeros of the characteristic polynomial by repeated factoring of $[\mathbf{A} - \lambda\mathbf{B}]$ for trial values of λ. To ensure that no root is skipped, one can use the *Sturm sequence property*.[1]

The Sturm sequence property is also useful in calculating the eigenvalues that appear in a prescribed range. The associated eigenvectors must be calculated by a different method. A slightly modified Sturm sequence property holds for indefinite [**B**] resulting from optimal lumping [C.1].

The *inverse iteration* or *inverse power* method computes the lowest eigenvalue. Or, when used with an eigenvalue shift, it computes the eigenvalue closest to the "shift point." The associated eigenvector is automatically computed as part of the process. The inverse power method does not require a positive definite [**B**].

The *subspace iteration* method uses k trial vectors that are iteratively improved as calculation proceeds. If m is the number of eigenvalues required, k may be taken as the smallest of the three numbers $2m$, $m + 8$, and n. Typically, $k << n$. There is a similarity between subspace iteration and mass condensation (Section 13.7).

Householder reduction and *Givens reduction* transform a standard or general eigenproblem into one in which the symmetric [**A**] and [**B**] are tridiagonal, which then can be easily solved by one of the other methods cited here. The transformation involved can be chosen to respect band storage format.

The *Lanczos method* is probably the most efficient algorithm available. Early difficulties with the method have been overcome. It is usually not available as a library package at computer installations, but appears destined to supplant other eigenproblem algorithms among the finite element community [C.5]. Like the Householder and Givens methods, the Lanczos method is a tridiagonal transformation method. It can be arranged so that it respects band or skyline/profile storage format. The resulting tridiagonal system is easily solved by other means.

The *QR* and *QZ methods* can solve a general eigenproblem but are particularly suited to extracting all eigenvalues and eigenvectors of a tridiagonal system resulting from Givens, Householder, or Lanczos transformations. The QZ method is complicated, but is the only known method capable of handling cases where [**A**] and [**B**] are singular and do not satisfy restrictions (a) through (c) of Section C.3. Thus the QZ method is useful for testing small systems whose properties are unknown a priori.

[1]We can use Subroutine SOLVER, Fig. B.2-3, to exploit the Sturm sequence property. Thus we select a numerical value of λ, place $[\mathbf{A}] - \lambda[\mathbf{B}]$ in array A of Fig. B.2-3, and complete the loop on statement 790. At this time the number of eigenvalues *exceeded* by the selected λ is equal to the number of negative diagonal coefficients (which appear in column 1 of array A in Fig. B.2-3).

Standard library packages (e.g., EISPACK, IMSL, NAg; Refs. C.10–C.12) provide several eigensolvers, usually tailored to exploit matrix properties such as symmetry or bandedness. The storage formats used, particularly in the more specialized algorithms, often disagree with storage formats for [**A**] and [**B**] in the program that generates these arrays.

An explanation of many of the methods described here, aimed at an engineering audience, can be found in Chapter 5 of [C.7]. A guide to obtaining software from EISPACK, IMSL, NAg, and other packages is given in the Appendix of [C.7].

REFERENCES

CHAPTER 1

1.1 J. Ergatoudis, B. M. Irons, and O. C. Zienkiewicz, "Three-Dimensional Analysis of Arch Dams and Their Foundations," *Research Report No. C/R/74/67,* University of Wales, Swansea, 1968.

1.2 K. Wieghardt, "Über einen Grenzübergang der Elastizitätslehre und seine Andwendung auf die Statik hochgradig statisch unbestimmter Fachwerke," *Verhandlungen des Vereins z. Beförderung des Gewerbefleisses, Abhandlungen,* Vol. 85, 1906, pp. 139–176.

1.3 W. Riedel, "Beiträge zur Lösung des ebenen Problems eines elastichen Körpers mittels der Airyschen Spannungsfunktion," *Zeitschrift für Angewandte Mathematik und Mechanik,* Vol. 7, No. 3, 1927, pp. 169–188.

1.4 A. Hrennikoff, "Solution of Problems in Elasticity by the Framework Method," *J. Appl. Mech.,* Vol. 8, No. 4, 1941, pp. A169–A175.

1.5 R. Courant, "Variational Methods for the Solution of Problems of Equilibrium and Vibrations," *Bulletin of the American Mathematical Society,* Vol. 49, 1943, pp. 1–23.

1.6 S. Levy, "Structural Analysis and Influence Coefficients for Delta Wings," *J. Aero. Sci.,* Vol. 20, No. 7, 1953, pp. 449–454.

1.7 R. W. Clough, "The Finite Element Method After Twenty-Five Years: A Personal View," *Computers & Structures,* Vol. 12, No. 4, 1980, pp. 361–370.

1.8 M. J. Turner, R. W. Clough, H. C. Martin, and L. J. Topp, "Stiffness and Deflection Analysis of Complex Structures," *J. Aero. Sci.,* Vol. 23, No. 9, 1956, pp. 805–823.

1.9 J. H. Argyris and S. Kelsey, *Energy Theorems and Structural Analysis,* Butterworths, London, 1960 (collection of papers published in *Aircraft Engineering* in 1954 and 1955).

1.10 J. Robinson, *Early FEM Pioneers,* Robinson & Associates, Dorset, England, 1985.

CHAPTER 2

2.1 K. J. Bathe, *Finite Element Procedures in Engineering Analysis,* Prentice-Hall, Englewood Cliffs, NJ, 1982.

2.2 G. C. Everstine, "A Comparison of Three Resequencing Algorithms for the Reduction of Matrix Profile and Wavefront," *Int. J. Num. Meth. Engng.,* Vol. 14, No. 6, 1979, pp. 837–853.

2.3 S. W. Sloan, "An Algorithm for Profile and Wavefront Reduction of Sparse Matrices," *Int. J. Num. Meth. Engng.,* Vol. 23, No. 2, 1986, pp. 239–251.

CHAPTER 3

3.1 H. L. Langhaar, *Energy Methods in Applied Mechanics,* John Wiley & Sons, New York, 1962.

3.2 K. Washizu, *Variational Methods in Elasticity and Plasticity,* 3rd Ed., Pergamon Press, Oxford, England, 1982.

3.3 P. Tong, "Exact Solution of Certain Problems by the Finite Element Method," *AIAA Jnl.,* Vol. 7, No. 1, 1969, pp. 178–180.

3.4 A. A. Ball, "The Interpolation Function of a General Serendipity Rectangular Element," *Int. J. Num. Meth. Engng.,* Vol. 15, No. 5, 1980, pp. 773–778.

CHAPTER 4

4.1 R. J. Melosh, "Basis for Derivation of Matrices for the Direct Stiffness Method," *AIAA Jnl.,* Vol. 1, No. 7, 1963, pp. 1631–1637.

4.2 R. Narayanaswami and H. M. Adelman, "Inclusion of Transverse Shear Deformation in Finite Element Displacement Formulations," *AIAA Jnl.,* Vol. 12, No. 11, 1974, pp. 1613–1614 (discussion: Vol. 13, No. 9, pp. 1253–1254; application to plates: Vol. 12, No. 12, pp. 1761–1763).

4.3 R. J. Melosh and D. W. Lobitz, "On a Numerical Sufficiency Test for Monotonic Convergence of Finite Element Models," *AIAA Jnl.,* Vol. 13, No. 5, 1975, pp. 675–678.

4.4 W. E. Haisler and J. A. Stricklin, "Rigid-Body Displacements of Curved Elements in the Analysis of Shells by the Matrix Displacement Method," *AIAA Jnl.,* Vol. 5, No. 8, 1967, pp. 1525–1527.

4.5 G. P. Bazeley, Y. K. Cheung, B. M. Irons, and O. C. Zienkiewicz, "Triangular Elements in Plate Bending—Conforming and Nonconforming Solutions," *Proc. First Conf. on Matrix Methods in Structural Mechanics,* Wright-Patterson Air Force Base, Ohio, 1965 (AFFDL-TR-66-80, Nov. 1966; AD-646-300, N.T.I.S.), pp. 547–576.

4.6 R. L. Taylor, J. C. Simo, O. C. Zienkiewicz, and A. C. H. Chan, "The Patch Test—A Condition for Assessing FEM Convergence," *Int. J. Num. Meth. Engng.,* Vol. 22, No. 1, 1986, pp. 39–62.

4.7 I. U. Ojalvo, "Improved Thermal Stress Determination by Finite Element Methods," *AIAA Jnl.,* Vol. 12, No. 8, 1974, pp. 1131–1132.

4.8 J. Pittr and H. Hartl, "Improved Stress Evaluation Under Thermal Load for Simple Finite Elements," *Int. J. Num. Meth. Engng.,* Vol. 15, No. 10, 1980, pp. 1507–1515.

4.9 G. Loubignac, G. Cantin, and G. Touzot, "Continuous Stress Fields in Finite Element Analysis," *AIAA Jnl.,* Vol. 15, No. 11, 1977, pp. 1645–1647.

4.10 R. D. Cook and X. Huang, "Continuous Stress Fields by the Finite Element-Difference Method," *Int. J. Num. Meth. Engng.,* Vol. 22, No. 1, 1986, pp. 229–240.

4.11 R. T. Severn, "Inclusion of Shear Deflection in the Stiffness Matrix for a Beam Element," *J. of Strain Analysis,* Vol. 5, No. 4, 1970, pp. 239–241.

4.12 R. E. Cornwell and R. D. Cook, "Improvement in Peak Stress Estimates Through Post-Processing," *Finite Elements in Analysis and Design* (to appear).

4.13 I. C. Taig, "Finite Element Analysis in Industry—Expertise or Proficiency?," in *Accuracy, Reliability, and Training in FEM Technology,* J. Robinson, ed., Pitman Press, England, 1984.

CHAPTER 5

5.1 K. Bell, "A Refined Triangular Plate Bending Finite Element," *Int. J. Num. Meth. Engrg.,* Vol. 1, No. 1, 1969, pp. 101–122.

5.2 N. M. Ferrers, *An Elementary Treatise on Trilinear Coordinates, the Method of Reciprocal Polars and the Theory of Projections,* Macmillan, London, 1861.

5.3 J. B. Mertie, "Transformation of Trilinear and Quadriplanar Coordinates to and from Cartesian Coordinates," *The American Mineralogist,* Vol. 49, Nos. 7/8, 1964, pp. 926–936.

5.4 R. H. Ghallagher, *Finite Element Analysis: Fundamentals,* Prentice-Hall, Englewood Cliffs, NJ, 1975.

5.5 C. A. Felippa, "Refined Finite Element Analysis of Linear and Nonlinear Two-Dimensional Structures," Ph.D. dissertation, University of California, Berkeley, 1966 (also available as PB-178-418 and PB-178-419, N.T.I.S.; computer programs in the latter).

5.6 G. Subramanian and C. Jayachandrabose, "Convenient Generation of Stiffness Matrices for the Family of Plane Triangular Elements," *Computers & Structures,* Vol. 15, No. 1, 1982, pp. 85–89.

CHAPTER 6

6.1 B. M. Irons, "Engineering Applications of Numerical Integration in Stiffness Methods," *AIAA Jnl.,* Vol. 4, No. 11, 1966, pp. 2035–2037.

6.2 A. H. Stroud and D. Secrest, *Gaussian Quadrature Formulas,* Prentice-Hall, Englewood Cliffs, NJ, 1966.

6.3 S. W. Sloan, "A Fast Stiffness Formulation for Finite Element Analysis of Two-Dimensional Solids," *Int. J. Num. Meth. Engng.,* Vol. 17, No. 9, 1981, pp. 1313–1323.

6.4 A. K. Gupta, "Efficient Numerical Integration of Element Stiffness Matrices," *Int. J. Num. Meth. Engng.,* Vol. 19, No. 9, 1983, pp. 1410–1413.

6.5 D. A. Dunavant, "High Degree Efficient Symmetrical Gaussian Quadrature Rules for the Triangle," *Int. J. Num. Meth. Engng.,* Vol. 21, No. 6, 1985, pp. 1129–1148.

6.6 A. K. Noor and C. M. Anderson, "Computerized Symbolic Manipulation in Nonlinear Finite Element Analysis," *Computers & Structures,* Vol. 13, Nos. 1–3, 1981, pp. 379–403.

6.7 I. Ergatoudis, B. M. Irons, and O. C. Zienkiewicz, "Curved Isoparametric, 'Quadrilateral' Elements for Finite Element Analysis," *Int. J. Solids Structures,* Vol. 4, No. 1, 1968, pp. 31–42.

6.8 P. C. Hammer and A. H. Stroud, "Numerical Evaluation of Multiple Integrals II," *Math. Tables and Other Aids to Comp.,* Vol. 12, No. 64, 1958, pp. 272–280.

6.9 B. M. Irons, "Quadrature Rules for Brick Based Finite Elements," *Int. J. Num. Meth. Engng.,* Vol. 3, No. 2, 1971, pp. 293–294.

6.10 T. K. Hellen, "Effective Quadrature Rules for Quadratic Solid Isoparametric Finite Elements," *Int. J. Num. Meth. Engng.,* Vol. 4, No. 4, 1972, pp. 597–599.

6.11 B. Verhegghe and G. H. Powell, "Control of Zero-Energy Modes in 9-Node Plane Element," *Int. J. Num. Meth. Engng.,* Vol. 23, No. 5, 1986, pp. 863–869.

6.12 R. D. Cook and Z. H. Feng, "Control of Spurious Modes in the Nine-Node Quadrilateral Element," *Int. J. Num. Meth. Engng.,* Vol. 18, No. 10, 1982, pp. 1576–1580.

6.13 W. K. Liu, J. S. J. Ong, and R. A. Uras, "Finite Element Stabilization Matrices: A Unification Approach," *Comp. Meth. Appl. Mech. Engng.,* Vol. 53, No. 1, 1985, pp. 13–46.

6.14 E. Hinton and J. S. Campbell, "Local and Global Smoothing of Discontinuous Finite Element Functions Using a Least Squares Method," *Int. J. Num. Meth. Engng.,* Vol. 8, No. 3, 1974, pp. 461–480.

6.15 J. Barlow, "Optimal Stress Locations in Finite Element Models," *Int. J. Num. Meth. Engng.,* Vol. 10, No. 2, 1976, pp. 243–251 (discussion: Vol. 11, No. 3, p. 604).

6.16 A. Peano, "Inadmissible Distortion of Solid Elements and Patch Test Results," *Comm. in Appl. Num. Meth.,* Vol. 3, No. 2, 1987, pp. 97–101.

CHAPTER 7

7.1 A. K. Gupta and P. S. Ma, "Error in Eccentric Beam Formulation," *Int. J. Num. Meth. Engng.,* Vol. 11, No. 9, 1977, pp. 1473–1477.

7.2 R. E. Miller, "Reduction of the Error in Eccentric Beam Modelling," *Int. J. Num. Meth. Engng.,* Vol. 15, No. 4, 1980, pp. 575–582.

CHAPTER 8

8.1 E. L. Wilson, "The Static Condensation Algorithm," *Int. J. Num. Meth. Engng.,* Vol. 8, No. 1, 1974, pp. 198–203.

8.2 R. D. Cook and V. N. Shah, "A Cost Comparison of Two Static Condensation–Stress Recovery Algorithms," *Int. J. Num. Meth. Engng.,* Vol. 12, No. 4, 1978, pp. 581–588.

8.3 R. L. Taylor, P. J. Beresford, and E. L. Wilson, "A Non-Conforming Element for Stress Analysis," *Int. J. Num. Meth. Engng.,* Vol. 10, No. 6, 1976, pp. 1211–1219.

8.4 M. Fröier, L. Nilsson, and A. Samuelsson, "The Rectangular Plane Stress Element by Turner, Pian and Wilson," *Int. J. Num. Meth. Engng.,* Vol. 8, No. 2, 1974, pp. 433–437.

8.5 D. J. Allman, "A Compatible Triangular Element Including Vertex Rotations for Plane Elasticity Analysis," *Computers & Structures,* Vol. 19, No. 1–2, 1984, pp. 1–8.

8.6 R. D. Cook, "On the Allman Triangle and a Related Quadrilateral Element," *Computers & Structures,* Vol. 22, No. 6, 1986, pp. 1065–1067.

8.7 P. Tong and T. H. H. Pian, "A Variational Principle and the Convergence of a Finite-Element Method Based on Assumed Stress Distribution," *Int. J. Solids Structures,* Vol. 5, No. 5, 1969, pp. 463–472.

8.8 T. H. H. Pian, "Derivation of Element Stiffness Matrices by Assumed Stress Functions," *AIAA Jnl.,* Vol. 2, No. 7, 1964, pp. 1333–1336 (discussion: Vol. 3, No. 1, 1965, pp. 186–187).

8.9 J. P. Wolf, "Alternate Hybrid Stress Finite Element Models," *Int. J. Num. Meth. Engng.,* Vol. 9, No. 3, 1975, pp. 601–615.

8.10 T. H. H. Pian and K. Sumihara, "Rational Approach for Assumed Stress Finite Elements," *Int. J. Num. Meth. Engng.,* Vol. 20, No. 9, 1984, pp. 1685–1695.

8.11 R. D. Cook, "A Plane Hybrid Element with Rotational D.O.F. and Adjustable Stiffness," *Int. J. Num. Meth. Engng.,* Vol. 24, No. 8, 1987, pp. 1499–1508.

8.12 M. F. Kanninen and C. H. Popelar, *Advanced Fracture Mechanics,* Oxford University Press, New York, 1985.

8.13 D. P. Rooke and D. J. Cartwright, *Compendium of Stress Intensity Factors,* Her Majesty's Stationary Office, London, 1976.

8.14 R. S. Barsoum, "On the Use of Isoparametric Finite Elements in Linear Fracture Mechanics," *Int. J. Num. Meth. Engng.,* Vol. 10, No. 1, 1976, pp. 25–37.

8.15 R. S. Barsoum, "Letter to the Editor," *Int. J. Num. Meth. Engng.,* Vol. 18, No. 9, 1982, pp. 1420–1422.

8.16 L. P. Harrop, "The Optimum Size of Quarter-Point Crack Tip Elements," *Int. J. Num. Meth. Engng.,* Vol. 18, No. 7, 1982, pp. 1101–1103.

8.17 N. A. B. Yahia and M. S. Shephard, "On the Effect of Quarter-Point Element Size on Fracture Criteria," *Int. J. Num. Meth. Engng.,* Vol. 21, No. 10, 1985, pp. 1911–1924.

8.18 V. E. Saouma and D. Schwemmer, "Numerical Evaluation of the Quarter-Point Crack Tip Element," *Int. J. Num. Meth. Engng.,* Vol. 20, No. 9, 1984, pp. 1629–1641.

8.19 D. M. Parks and E. M. Kamenetzky, "Weight Functions From Virtual Crack Extension," *Int. J. Num. Meth. Engng.,* Vol. 14, No. 11, 1979, pp. 1693–1706.

8.20 A. D. Kerr, "Elastic and Viscoelastic Foundation Models," *J. Appl. Mech.,* Vol. 31, No. 3, 1964, pp. 491–498.
8.21 M. S. Cheung, "A Simplified Finite Element Solution for the Plates on Elastic Foundation," *Computers & Structures,* Vol. 8, No. 1, 1978, pp. 139–145.
8.22 Z. Feng and R. D. Cook, "Beam Elements on Two-Parameter Elastic Foundations," *J. of Engng. Mech.,* Vol. 109, No. 6, 1983, pp. 1390–1402.
8.23 O. C. Zienkiewicz, P. Bettess, T. C. Chaim, and C. Emson, "Numerical Methods for Unbounded Field Problems and a New Infinite Element Formulation," in *Computational Methods for Infinite Domain Media Structure Interaction,* A. J. Kalinowski, ed., Am. Soc. Mech. Engrs., New York, 1981.
8.24 P. Bettess and J. A. Bettess, "Infinite Elements for Static Problems," *Engineering Computations,* Vol. 1, No. 1, 1984, pp. 4–16.
8.25 J. M. M. C. Marques and D. R. J. Owen, "Infinite Elements in Quasi-Static Materially Nonlinear Problems," *Computers & Structures,* Vol. 18, No. 4, 1984, pp. 739–751.
8.26 R. T. Fenner, "The Boundary Integral Equation (Boundary Element) Method in Engineering Stress Analysis," *J. of Strain Analysis,* Vol. 18, No. 4, 1983, pp. 199–205.
8.27 J. Mackerle and T. Andersson, "Boundary Element Software in Engineering," *Advances in Engng. Software,* Vol. 6, No. 2, 1984, pp. 66–102 (lists 656 references).
8.28 S. J. Fenves et al., eds., *Numerical and Computer Methods in Structural Mechanics,* Academic Press, New York, 1973 (see papers by D. Bushnell, pp. 291–336, and by S. W. Key and R. D. Krieg, pp. 337–352).
8.29 J. S. Arora, "Survey of Reanalysis Techniques," *J. Struct. Div., Proc. ASCE,* Vol. 102, No. ST4, 1976, pp. 783–802 (lists 89 references).
8.30 U. Kirsch, *Optimum Structural Design,* McGraw-Hill, New York, 1981.
8.31 J. S. Przemienicki, "Matrix Structural Analysis of Substructures," *AIAA Jnl.,* Vol. 1, No. 1, 1963, pp. 138–147.
8.32 A. K. Noor, H. A. Kamel and R. E. Fulton, "Substructuring Techniques—Status and Projections," *Computers & Structures,* Vol. 8, No. 5, 1978, pp. 621–632.
8.33 P. G. Glockner, "Symmetry in Structural Mechanics," *J. Struct. Div., Proc. ASCE,* Vol. 99, No. ST1, 1973, pp. 71–89.
8.34 A. K. Noor and R. A. Camin, "Symmetry Considerations for Anisotropic Shells," *Comp. Meth. Appl. Mech. Engng.,* Vol. 9, No. 3, 1976, pp. 317–335.
8.35 P. D. Mangalgiri, B. Dattaguru, and T. S. Ramamurthy, "Specification of Skew Conditions in Finite Element Formulation," *Int. J. Num. Meth. Engng.,* Vol. 12, No. 6, 1978, pp. 1037–1041.
8.36 O. C. Zienkiewicz and F. C. Scott, "On the Principle of Repeatability and Its Application in Analysis of Turbine and Pump Impellers," *Int. J. Num. Meth. Engng.,* Vol. 4, No. 3, 1972, pp. 445–448.
8.37 R. H. MacNeal and R. L. Harder, "A Refined Four-Noded Membrane Element with Rotational Degrees of Freedom," *Computers & Structures,* Vol. 28, No. 1, 1988, pp. 75–84.

CHAPTER 9

9.1 J. F. Abel and M. S. Shephard, "An Algorithm for Multipoint Constraints in Finite Element Analysis," *Int. J. Num. Meth. Engng.,* Vol. 14, No. 3, 1979, pp. 464–467.
9.2 O. C. Zienkiewicz, *The Finite Element Method,* 3rd Ed., McGraw-Hill, London, 1977.
9.3 C. A. Felippa, "Iterative Procedures for Improving Penalty Function Solutions of Algebraic Systems," *Int. J. Num. Meth. Engng.,* Vol. 12, No. 5, 1978, pp. 821–836.
9.4 T. J. R. Hughes, R. L. Taylor, and W. Kanoknukulchai, "A Simple and Efficient

Finite Element for Plate Bending," *Int. J. Num. Meth. Engng.*, Vol. 11, No. 10, 1977, pp. 1529–1543.

9.5 D. S. Malkus, *Finite Element Analysis of Incompressible Solids*, Ph.D. Dissertation, Boston University, Boston, 1976.

9.6 D. S. Malkus and T. J. R. Hughes, "Mixed Finite Element Methods-Reduced and Selective Integration Techniques: A Unification of Concepts," *Comp. Meth. Appl. Mech. Engng.*, Vol. 15, No. 1, 1978, pp. 68–81.

9.7 T. J. R. Hughes, *A Course in the Finite Element Method: Linear Static and Dynamic Finite Element Analysis*, Prentice-Hall, Englewood Cliffs, NJ, 1987.

9.8 R. L. Spilker and N. I. Munir, "A Hybrid Stress Quadratic Serendipity Displacement Mindlin Plate Element," *Computers & Structures*, Vol. 12, No. 1, 1980, pp. 11–21.

9.9 Y. C. Fung, *Foundations of Solid Mechanics*, Prentice-Hall, Englewood Cliffs, NJ, 1965.

9.10 M. Engelman, R. L. Sani, P. M. Gresho, and M. Bercovier, "Consistent vs. Reduced Integration Penalty Methods for Incompressible Media Using Several Old and New Elements," *Int. J. for Num. Meth. in Fluids*, Vol. 2, No. 1, 1982, pp. 25–42.

9.11 C. Johnson and J. Pitkäranta, "Analysis of Some Mixed Finite Element Methods Related to Reduced Integration," *Mathematics of Computation*, Vol. 38, No. 158, 1982, pp. 375–400.

9.12 E. T. Olsen, *Stable Finite Elements for Non-Newtonian Flows; First Order Elements Which Fail the LBB Condition*, Ph.D. Dissertation, Illinois Institute of Technology, Chicago, 1983.

9.13 J. Pitkäranta and R. Stenberg, "Error Bounds for the Approximation of the Stokes Problem Using Bilinear/Constant Elements on Irregular Quadrilateral Meshes," Report MAT-A222, Helsinki University of Technology, Espoo, Finland, 1984.

9.14 T. J. R. Hughes, W. K. Liu, and A. Brooks, "Finite Element Analysis of Incompressible Viscous Flows by the Penalty Function Formulation," *J. of Comp. Phys.*, Vol. 30, No. 1, 1979, pp. 1–60.

CHAPTER 10

10.1 E. L. Wilson, "Structural Analysis of Axisymmetric Solids," *AIAA Jnl.*, Vol. 3, No. 12, 1965, pp. 2269–2274.

10.2 J. G. Crose and R. M. Jones, *SAAS III: Finite Element Analysis of Axisymmetric and Plane Solids with Different Orthotropic, Temperature-Dependent Material Properties in Tension and Compression*, Aerospace Corp., San Bernardino, CA, 1971 (AD-729-188, N.T.I.S.).

10.3 O. C. Zienkiewicz and Y. K. Cheung, "Stresses in Shafts," *The Engineer* (London), Vol. 224, No. 5835, 1967, pp. 696–697.

10.4 T. Belytschko, "Finite Elements for Axisymmetric Solids Under Arbitrary Loadings with Nodes on Origin," *AIAA Jnl.*, Vol. 10, No. 11, 1972, pp. 1532–1533 (discussion and closure: Vol. 11, No. 9, pp. 1357–1358).

10.5 J. Padovan, "Quasi-Analytical Finite Element Procedures for Axisymmetric Anisotropic Shells and Solids," *Computers & Structures*, Vol. 4, No. 3, 1974, pp. 467–483.

10.6 J. G. Crose, "Stress Analysis of Axisymmetric Solids with Asymmetric Properties," *AIAA Jnl.*, Vol. 10, No. 7, 1972, pp. 866–871.

10.7 M. Sedaghat and L. R. Herrmann, "A Nonlinear, Semi-Analytical Finite Element Analysis for Nearly Axisymmetric Solids," *Computers & Structures*, Vol. 17, No. 3, 1983, pp. 389–401.

10.8 E. L. Wilson and P. C. Pretorius, "A Computer Program for the Analysis of

Prismatic Solids,'' *Report UC-SESM-70-21,* Civil Engineering Department, University of California, Berkeley, 1970 (PB-196-462, N.T.I.S.).

10.9 O. C. Zienkiewicz and J. M. Too, ''The Finite Prism in Analysis of Thick Simply Supported Bridge Boxes,'' *Proc. Inst. Civil Engrs.,* Vol. 53, Part 2, 1972, pp. 147–172.

10.10 G. A. Greenbaum, L. D. Hofmeister, and D. A. Evenson, ''Pure Moment Loading of Axisymmetric Finite Element Models,'' *Int. J. Num. Meth. Engng.,* Vol. 5, No. 4, 1973, pp. 459–463.

10.11 K. J. Bathe and C. A. Almeida, ''A Simple and Effective Pipe Elbow Element-Linear Analysis,'' *J. Appl. Mech.,* Vol. 47, No. 1, 1980, pp. 93–100.

10.12 K. J. Han and P. L. Gould, ''Line Node and Transitional Shell Element for Rotational Shells,'' *Int. J. Num. Meth. Engng.,* Vol. 18, No. 6, 1982, pp. 879–895.

10.13 W. M. Chen and P. Tsai, ''The Combined Use of Axisymmetric and Solid Elements for Three Dimensional Stress Analysis,'' *Computers & Structures,* Vol. 18, No. 4, 1984, pp. 689–694.

CHAPTER 11

11.1 S. Timoshenko and S. Woinowsky-Krieger, *Theory of Plates and Shells,* 2nd Ed., McGraw-Hill, New York, 1959.

11.2 F. J. Plantema, *Sandwich Construction,* John Wiley & Sons, New York, 1966.

11.3 J. M. Whitney and A. W. Leissa, ''Analysis of Heterogeneous Anisotropic Plates,'' *J. Appl. Mech.,* Vol. 36, No. 2, 1969, pp. 261–266.

11.4 M. M. Hrabok and T. M. Hrudey, ''A Review and Catalog of Plate Bending Finite Elements,'' *Computers & Structures,* Vol. 19, No. 3, 1984, pp. 479–495.

11.5 R. H. Gallagher, *Finite Element Analysis: Fundamentals,* Prentice-Hall, Englewood Cliffs, NJ, 1975.

11.6 R. J. Melosh, ''Basis for Derivation of Matrices for the Direct Stiffness Method,'' *AIAA Jnl.,* Vol. 1, No. 7, 1963, pp. 1631–1637 (discussion: Vol. 2, No. 2, 1964, p. 403; Vol. 2, No. 6, 1964, p. 1161; Vol. 3, No. 6, 1965, pp. 1215–1216).

11.7 Y. K. Cheung, *Finite Strip Method in Structural Analysis,* Pergamon Press, Oxford, England, 1976.

11.8 T. J. R. Hughes and M. Cohen, ''The 'Heterosis' Finite Element for Plate Bending,'' *Computers & Structures,* Vol. 9, No. 5, 1978, pp. 445–450.

11.9 E. Hinton and H. C. Huang, ''A Family of Quadrilateral Mindlin Plate Elements with Substitute Shear Strain Fields,'' *Computers & Structures,* Vol. 23, No. 3, 1986, pp. 409–431.

11.10 J. A. Stricklin et al., ''A Rapidly Converging Triangular Plate Element,'' *AIAA Jnl.,* Vol. 7, No. 1, 1969, pp. 180–181.

11.11 J. L. Batoz, K. J. Bathe, and L. W. Ho, ''A Study of Three-Node Triangular Plate Bending Elements,'' *Int. J. Num. Meth. Engng.,* Vol. 15, No. 12, 1980, pp. 1771–1812.

11.12 J. L. Batoz, ''An Explicit Formulation for an Efficient Triangular Plate Bending Element,'' *Int. J. Num. Meth. Engng.,* Vol. 18, No. 7, 1982, pp. 1077–1089.

11.13 C. Jeyachandrabose and J. Kirkhope, ''An Alternative Formulation for the DKT Plate Bending Element,'' *Int. J. Num. Meth. Engng.,* Vol. 21, No. 7, 1985, pp. 1289–1293.

11.14 T. J. R. Hughes, M. Cohen, and M. Haroun, ''Reduced and Selective Integration Techniques in the Finite Element Analysis of Plates,'' *Nucl. Engng. Design,* Vol. 46, No. 1, 1978, pp. 203–222.

11.15 M. P. Rossow, ''Efficient C^0 Finite-Element Solutions of Simply Supported Plates of Polygonal Shape,'' *J. Appl. Mech.,* Vol. 44, No. 2, 1977, pp. 347–349.

11.16 T. J. R. Hughes and T. E. Tezduyar, ''Finite Elements Based on Mindlin Plate

Theory with Particular Reference to the Four-Node Bilinear Isoparametric Element," *J. Appl. Mech.*, Vol. 48, No. 3, 1981, pp. 587–596.

11.17 J. Donea and L. G. Lamain, "A Modified Representation of Transverse Shear in C^0 Quadrilateral Plate Elements," *Comp. Meth. Appl. Mech. Engng.*, Vol. 63, No. 2, 1987, pp. 183–207.

CHAPTER 12

12.1 F. Kikuchi and K. Tanizawa, "Accuracy and Locking-Free Property of the Beam Element Approximation for Arch Problems," *Computers & Structures*, Vol. 19, No. 1–2, 1984, pp. 103–110.

12.2 G. Prathap, "The Curved Beam/Deep Arch/Finite Ring Element Revisited," *Int. J. Num. Meth. Engng.*, Vol. 21, No. 3, 1985, pp. 389–407.

12.3 H. R. Meck, "An Accurate Polynomial Displacement Function for Finite Ring Elements," *Computers & Structures*, Vol. 11, No. 4, 1980, pp. 265–269.

12.4 P. M. Mebane and J. A. Stricklin, "Implicit Rigid Body Motion in Curved Elements," *AIAA Jnl.*, Vol. 9, No. 2, 1971, pp. 344–345.

12.5 G. Prathap and C. R. Babu, "An Isoparametric Quadratic Thick Curved Beam Element," *Int. J. Num. Meth. Engng.*, Vol. 23, No. 9, 1986, pp. 1583–1600.

12.6 S. F. Pawsey and R. W. Clough, "Improved Numerical Integration of Thick Shell Finite Elements," *Int. J. Num. Meth. Engng.*, Vol. 3, No. 4, 1971, pp. 575–586.

12.7 M. A. Crisfield, "Explicit Integration and the Isoparametric Arch and Shell Elements," *Comm. in Appl. Num. Meth.*, Vol. 2, No. 2, 1986, pp. 181–187.

12.8 N. Carpenter, H. Stolarski, and T. Belytschko, "Improvements in 3-Node Triangular Shell Elements," *Int. J. Num. Meth. Engng.*, Vol. 23, No. 9, 1986, pp. 1643–1667.

12.9 O. C. Zienkiewicz, *The Finite Element Method*, 3rd Ed., McGraw-Hill, London, 1977.

12.10 C. R. Babu and G. Prathap, "A Field-Consistent Two-Noded Axisymmetric Shell Element," *Int. J. Num. Meth. Engng.*, Vol. 23, No. 7, 1986, pp. 1245–1261.

12.11 A. Tessler, "An Efficient, Conforming Axisymmetric Shell Element Including Transverse Shear and Rotary Inertia," *Computers & Structures*, Vol. 15, No. 5, 1982, pp. 567–574.

12.12 B. M. Irons and A. Razzaque, "Further Modification to Ahmad's Shell Element," *Int. J. Num. Meth. Engng.*, Vol. 5, No. 4, 1973, pp. 588–589.

12.13 T. Belytschko et al., "Implementation and Application of a 9-Node Lagrange Shell Element With Spurious Mode Control," *Computers & Structures*, Vol. 20, No. 1–3, 1985, pp. 121–128.

12.14 V. T. Nicholas and E. Citipitioglu, "A General Isoparametric Finite Element Program SDRC SUPERB," *Computers & Structures*, Vol. 7, No. 2, 1977, pp. 303–313.

12.15 R. H. MacNeal and R. L. Harder, "A Proposed Standard Set of Problems to Test Finite Element Accuracy," *Finite Elements in Analysis and Design*, Vol. 1, No. 1, 1985, pp. 3–20.

12.16 S. S. Murthy and R. H. Gallagher, "Patch Test Verification of a Triangular Thin-Shell Element Based on Discrete Kirchhoff Theory," *Comm. in Appl. Num. Meth.*, Vol. 3, No. 2, 1987, pp. 83–88.

12.17 H. Stolarski and T. Belytschko, "Membrane Locking and Reduced Integration for Curved Elements," *J. Appl. Mech.*, Vol. 49, No. 1, 1982, pp. 172–176.

CHAPTER 13

13.1 R. W. Clough and J. Penzien, *Dynamics of Structures*, McGraw-Hill, New York, 1975.

13.2 R. R. Craig, Jr., *Structural Dynamics,* John Wiley & Sons, New York, 1981.
13.3 J. W. S. Rayleigh, *Theory of Sound,* 2nd Ed., Vols. I and II, Dover Publications, New York, 1945 (originally published in 1894).
13.4 J. S. Archer, "Consistent Matrix Formulations for Structural Analysis Using Finite Element Techniques," *AIAA Jnl.,* Vol. 3, No. 10, 1965, pp. 1910–1918.
13.5 E. Hinton, T. Rock, and O. C. Zienkiewicz, "A Note on Mass Lumping and Related Processes in the Finite Element Method," *Earthquake Engng. Struct. Dynamics,* Vol. 4, No. 3, 1976, pp. 245–249.
13.6 K. S. Surana, "Lumped Mass Matrices with Non-Zero Inertia for General Shell and Axisymmetric Shell Elements," *Int. J. Num. Meth. Engng.,* Vol. 12, No. 11, 1978, pp. 1635–1650.
13.7 D. S. Malkus and M. E. Plesha, "Zero and Negative Masses in Finite Element Vibration and Transient Analysis," *Comp. Meth. Appl. Mech. Engng.,* Vol. 59, No. 3, 1986, pp. 281–306.
13.8 D. S. Malkus, M. E. Plesha, and M. R. Liu, "Reversed Stability Conditions in Transient Finite Element Analysis," *Comp. Meth. Appl. Mech. Engng.,* Vol. 68, No. 1, 1988, pp. 97–114.
13.9 G. J. Fix, "Effects of Quadrature Errors in Finite Element Approximation of Steady State Eigenvalue and Parabolic Problems," in *Mathematical Foundations of the Finite Element Method,* I. Babuska and A. K. Aziz, eds., Academic Press, New York, 1972, pp. 525–556.
13.10 Z. Kopal, *Numerical Analysis,* John Wiley & Sons, New York, 1955.
13.11 S. H. Crandall, *Engineering Analysis,* McGraw-Hill, New York, 1956.
13.12 E. Isaacson and H. B. Keller, *Analysis of Numerical Methods,* John Wiley & Sons, New York, 1966.
13.13 E. Dokumaci, "A Critical Examination of Discrete Models in Vibration Problems of Continuous Systems," *J. Sound Vibration,* Vol. 53, No. 2, 1977, pp. 153–164.
13.14 T. Belytschko, "A Survey of Numerical Methods and Computer Programs for Dynamic Structural Analysis, *Nucl. Engng. Design,* Vol. 37, No. 1, 1976, pp. 23–34.
13.15 R. H. MacNeal, ed., *The NASTRAN Theoretical Manual* (*Level 16.0*), NASA-SP-221(03), march 1976 (N79-27531, N.T.I.S.).
13.16 T. Belytschko and W. L. Mindle, "The Treatment of Damping in Transient Computations," in *Damping Applications for Vibration Control,* P. J. Torvik, ed., ASME AMD, Vol. 38, 1980, pp. 123–132.
13.17 M. E. Plesha, "Mixed Time Integration for the Transient Analysis of Jointed Media," *Int. J. Num. An. Meth. Geomech.,* Vol. 10, No. 1, 1986, pp. 91–110.
13.18 K. J. Bathe, *Finite Element Procedures in Engineering Analysis,* Prentice-Hall, Englewood Cliffs, NJ, 1982.
13.19 L. Fox, *An Introduction to Numerical Linear Algebra,* Oxford University Press, New York, 1965.
13.20 W. C. Hurty and M. F. Rubinstein, *Dynamics of Structures,* Prentice-Hall, Englewood Cliffs, NJ, 1964.
13.21 O. C. Zienkiewicz, R. W. Lewis, and K. G. Stagg, *Numerical Methods in Offshore Engineering,* John Wiley & Sons, Chichester, England, 1978.
13.22 N. R. Maddox, "On the Number of Modes Necessary for Accurate Response and Resulting Forces in Dynamic Analysis," *J. Appl. Mech.,* Vol. 42, No. 2, 1975, pp. 516–517.
13.23 O. E. Hansteen and K. Bell, "On the Accuracy of Mode Superposition Analysis in Structural Dynamics," *Earthquake Engng. Struct. Dynamics,* Vol. 7, No. 5, 1979, pp. 405–411.
13.24 R. E. Cornwell, R. R. Craig, Jr., and C. P. Johnson, "On the Application of the Mode-Acceleration Method to Structural Engineering Problems," *Earthquake Engng. Struct. Dynamics,* Vol. 11, No. 5, 1983, pp. 679–688.
13.25 E. L. Wilson, M. Y. Yuan, and J. M. Dickens, "Dynamic Analysis by Direct Superposition of Ritz Vectors," *Earthquake Engng. Struct. Dynamics,* Vol. 10, No. 6, 1982, pp. 813–821.

13.26 V. N. Shah, G. J. Bohm, and A. N. Nahavandi, "Modal Superposition Method for Computationally Economical Nonlinear Structural Analysis," *J. Pressure Vessel Tech.*, ASME, Vol. 101, No. 2, 1979, pp. 134–141.

13.27 J. A. Stricklin and W. E. Haisler, "Formulations and Solution Procedures for Nonlinear Structural Analysis," *Computers & Structures,* Vol. 7, No. 1, 1977, pp. 125–136.

13.28 K. J. Bathe and E. L. Wilson, *Numerical Methods in Finite Element Analysis,* Prentice-Hall, Englewood Cliffs, NJ, 1976.

13.29 R. J. Guyan, "Reduction of Stiffness and Mass Matrices," *AIAA Jnl.,* Vol. 3, No. 2, 1965, p. 380.

13.30 I. U. Ojalvo, "Computer Methods for Determining Vibration Modes of Complex Structures," *Shock and Vibration Digest,* Vol. 5, No. 5, May, 1983, pp. 3–10.

13.31 C. A. Felippa, "Refined Finite Element Analysis of Linear and Nonlinear Two-Dimensional Structures," Ph.D. dissertation, University of California, Berkeley, 1966 (also available as PB-178-418 and PB-178-419, N.T.I.S.; computer programs in the latter).

13.32 J. D. Sowers, "Condensation of Free Body Mass Matrices Using Flexibility Coefficients," *AIAA Jnl.,* Vol. 16, No. 3, 1978, pp. 272–273.

13.33 M. Geradin, "Error Bounds for Eigenvalue Analysis by Elimination of Variables," *J. Sound Vibration,* Vol. 19, No. 2, 1971, pp. 111–132.

13.34 R. L. Kidder, "Reduction of Structural Frequency Equations," *AIAA Jnl.,* Vol. 11, No. 6, 1973, p. 892 (discussion and closure: Vol. 13, No. 5, 1975, pp. 701–703).

13.35 V. B. Watwood, T. Y. Chow, Z. Zudans, and W. H. Miller, "Combined Analysis and Evaluation of Piping Systems Using the Computer," *Nucl. Engng. Design,* Vol. 27, No. 3, 1974, pp. 334–342.

13.36 R. D. Henshell and J. H. Ong, "Automatic Masters for Eigenvalue Economization," *Earthquate Engng. Struct. Dynamics,* Vol. 3, No. 4, 1975, pp. 375–383.

13.37 R. G. Anderson, B. M. Irons, and O. C. Zienkiewicz, "Vibration and Stability of Plates Using Finite Elements," *Int. J. Solids Structures,* Vol. 4, No. 10, 1968, pp. 1031–1055.

13.38 K. W. Matta, "Selection of Degrees of Freedom for Dynamic Analysis," *J. Pressure Vessel Tech.,* Vol. 109, No. 1, 1987, pp. 65–69.

13.39 W. C. Hurty, "Dynamic Analysis of Structural Systems Using Component Modes," *AIAA Jnl.,* Vol. 3, No. 4, 1965, pp. 678–685.

13.40 R. R. Craig, Jr., "A Review of Time-Domain and Frequency Domain Component Mode Synthesis Methods," *Int. J. Analytical and Experimental Modal Analysis,* Vol. 2, No. 2, 1987, pp. 59–72.

13.41 R. H. MacNeal, "A Hybrid Method of Component Mode Synthesis," *Computers & Structures,* Vol. 1, No. 4, 1971, pp. 581–601.

13.42 S. Rubin, "An Improved Component-Mode Representation," *AIAA/ASME/SAE 15th Structures, Structural Dynamics and Materials Conference,* Las Vegas, Nevada, 1974, pp. 17–19.

13.43 R. R. Craig, Jr. and M. C. C. Bampton, "Coupling of Substructures for Dynamic Analysis," *AIAA Jnl.,* Vol. 6, No. 7, 1968, pp. 1313–1319.

13.44 V. N. Shah and M. Raymund, "Analytical Selection of Masters for the Reduced Eigenvalue Problem," *Int. J. Num. Meth. Engng.,* Vol. 18, No. 1, pp. 89–98, 1982.

13.45 M. Paz, *Structural Dynamics: Theory and Computation,* 2nd ed., Van Nostrand Reinhold, New York, 1984.

13.46 T. J. R. Hughes, *The Finite Element Method: Linear Static and Dynamic Finite Element Analysis,* Prentice-Hall, Englewood Cliffs, NJ, 1987.

13.47 R. D. Richtmyer and K. W. Morton, *Difference Methods for Initial Value Problems,* Interscience, New York, 1967.

13.48 T. J. R. Hughes and T. Belytschko, "A Précis of Developments in Computational Methods for Transient Analysis," *J. Appl. Mech.,* Vol. 50, No. 4b, 1983, pp. 1033–1041.

13.49 T. Belytschko and T. J. R. Hughes, eds., *Computational Methods for Transient Analysis,* North Holland, Amsterdam, 1983.
13.50 R. Courant, K. O. Friedrichs, and H. Lewy, "Uber die partiellen Differenzengleichungen der mathematischen Physik," *Math. Ann.*, Vol. 100, 1928, pp. 32–74. (Translated by P. Fox: "On the Partial Differential Equations of Mathematical Physics," New York University, Courant Institute of Mathematical Sciences, Report NYO-7689, 1956.)
13.51 M. E. Plesha, "Eigenvalue Estimation for Dynamic Contact Problems," *J. of Engng. Mech.,* Vol. 113, No. 3, 1987, pp. 457–462.
13.52 D. D. Flanagan and T. Belytschko, "A Uniform Strain Hexahedron and Quadrilateral with Orthogonal Hourglass Control," *Int. J. Num. Meth. Engng.,* Vol. 17, No. 5, 1981, pp. 679–706. (Also see Errata, Vol. 19, No. 3, 1983, pp. 467–468.)
13.53 T. Belytschko, J. S. J. Ong, W. K. Liu, and J. M. Kennedy, "Hourglass Control in Linear and Nonlinear Problems," *Comp. Meth. Appl. Mech. Engng.,* Vol. 43, No. 3, 1984, pp. 251–276.
13.54 W. K. Liu and T. Belytschko, "Efficient Linear and Nonlinear Heat Conduction with a Quadrilateral Element," *Int. J. Num. Meth. Engng.,* Vol. 20, No. 5, 1984, pp. 931–948.
13.55 J. C. Houbolt, "A Recurrence Matrix Solution for the Dynamic Response of Elastic Aircraft," *J. Aero. Sci.,* Vol. 17, No. 9, 1950, pp. 540–550.
13.56 N. M. Newmark, "A Method of Computation for Structural Dynamics," *J. Engng. Mech. Div., Proc. ASCE,* Vol. 85, No. EM3, 1959, pp. 67–94.
13.57 E. L. Wilson, "A Computer Program for the Dynamic Stress Analysis of Underground Structures," *SESM Report No. 68-1,* Division of Structural Engineering and Structural Mechanics, University of California, Berkeley, 1968.
13.58 H. M. Hilber, T. J. R. Hughes, and R. L. Taylor, "Improved Numerical Dissipation for Time Integration Algorithms in Structural Dynamics," *Earthquake Engng. Struct. Dynamics,* Vol. 5, No. 3, 1977, pp. 283–292.
13.59 T. Belytschko and R. Mullen, "Mesh Partitions of Explicit-Implicit Time Integration," in *Formulation and Computational Algorithms in Finite Element Analysis,* K. J. Bathe, J. T. Oden, and W. Wunderlich, eds., MIT Press, Cambridge, MA, 1977, pp. 673–690.
13.60 T. Belytschko and R. Mullen, "Stability of Explicit-Implicit Mesh Partitions in Time Integration," *Int. J. Num. Meth. Engng.,* Vol. 12, No. 10, 1978, pp. 1575–1586.
13.61 T. J. R. Hughes and W. K. Liu, "Implicit-Explicit Finite Elements in Transient Analysis: Stability Theory," *J. Appl. Mech.,* Vol. 45, No. 2, 1978, pp. 371–374.
13.62 T. J. R. Hughes and W. K. Liu, "Implicit-Explicit Finite Elements in Transient Analysis: Implementation and Numerical Examples," *J. Appl. Mech.,* Vol. 45, No. 2, 1978, pp. 375–378.
13.63 T. J. R. Hughes, K. S. Pister, and R. L. Taylor, "Implicit-Explicit Finite Elements in Nonlinear Transient Analysis," *Comp. Meth. Appl. Mech. Engng.,* Vol. 17/18, Part I, 1979, pp. 159–182.
13.64 T. J. R. Hughes, W. K. Liu, and A. Brooks, "Finite Element Analysis of Incompressible Viscous Flows by the Penalty Function Formulation," *J. of Comp. Phys.,* Vol. 30, No. 1, 1979, pp. 1–60.
13.65 M. Ortiz, P. M. Pinsky, and R. L. Taylor, "Unconditionally Stable Element-by-Element Algorithms for Dynamic Problems," *Comp. Meth. Appl. Mech. Engng.,* Vol. 36, No. 2, 1983, pp. 223–239.
13.66 T. J. R. Hughes, I. Levit, and J. Winget, "Unconditionally Stable Element-by-Element Implicit Algorithm for Heat Conduction Analysis," *J. of Engng. Mech.,* Vol. 109, No. 2, 1983, pp. 576–585.
13.67 T. J. R. Hughes, I. Levit, and J. Winget, "An Element-by-Element Solution Algorithm for Problems of Structural and Solid Mechanics," *Comp. Meth. Appl. Mech. Engng.,* Vol. 36, No. 2, 1983, pp. 241–254.
13.68 I. Fried and D. S. Malkus, "Finite Element Mass Matrix Lumping by Numerical

Integration With No Convergence Rate Loss," *Int. J. Solids Structures,* Vol. 11, No. 4, 1975, pp. 461–466.
13.69 G. M. Hulbert and T. J. R. Hughes, "An Error Analysis of Truncated Starting Conditions in Step-by-Step Time Integration: Consequences for Structural Dynamics," *Earthquake Engng. Struct. Dynamics,* Vol. 15, No. 7, 1987, pp. 901–910.
13.70 R. A. Brockman, "Dynamics of the Bilinear Mindlin Plate Element," *Int. J. Num. Meth. Engng.,* Vol. 24, No. 12, 1987, pp. 2343–2356.

CHAPTER 14

14.1 S. C. Chang and J. J. Chen, "Effectiveness of Linear Bifurcation Analysis for Predicting the Nonlinear Stability Limits of Structures," *Int. J. Num. Meth. Engng.,* Vol. 23, No. 5, 1986, pp. 831–846.
14.2 R. H. Gallagher and J. Padlog, "Discrete Element Approach to Structural Stability Analysis," *AIAA Jnl.,* Vol. 1, No. 6, 1963, pp. 1437–1439.
14.3 E. Chwalla, "Second Order Theory," in *Handbook of Engineering Mechanics,* W. Flügge, ed., McGraw-Hill, New York, 1962.
14.4 S. P. Timoshenko and J. M. Gere, *Theory of Elastic Stability,* 2nd Ed., McGraw-Hill, New York, 1961.
14.5 J. S. Przemieniecki, "Discrete-Element Methods for Stability Analysis," *Aeronautical J.,* Vol. 72, No. 12, 1968, pp. 1077–1086.
14.6 R. A. Tinawi, "Anisotropic Tapered Elements Using Displacement Models," *Int. J. Num. Meth. Engng.,* Vol. 4, No. 4, 1972, pp. 475–489.
14.7 R. D. Cook, "Finite Element Buckling Analysis of Homogeneous and Sandwich Plates," *Int. J. Num. Meth. Engng.,* Vol. 9, No. 1, 1975, pp. 39–50.
14.8 R. S. Barsoum and R. H. Gallagher, "Finite Element Analysis of Torsional and Torsional-Flexural Stability Problems," *Int. J. Num. Meth. Engng.,* Vol. 2, No. 3, 1970, pp. 335–352.
14.9 R. H. Gallagher and C. H. Lee, "Matrix Dynamic and Instability Analysis with Nonuniform Elements," *Int. J. Num. Meth. Engng.,* Vol. 2, No. 2, 1970, pp. 265–275.
14.10 R. S. Barsoum, "Finite Element Method Applied to the Problem of Stability of a Non-Conservative System," *Int. J. Num. Meth. Engng.,* Vol. 3, No. 1, 1971, pp. 63–87.
14.11 D. Bushnell, "Analysis of Ring-Stiffened Shells of Revolution Under Combined Thermal and Mechanical Loadings," *AIAA Jnl.,* Vol. 9, No. 3, 1971, pp. 401–410.
14.12 D. R. Navaratna, T. H. H. Pian, and E. A. Witmer, "Analysis of Elastic Stability of Shells of Revolution by the Finite Element Method," *AIAA Jnl.,* Vol. 6, No. 2, 1968, pp. 355–361.
14.13 A. D. Kerr and M. T. Soifer, "The Linearization of the Prebuckling State and Its Effect on the Determined Stability Loads," *J. Appl. Mech.,* Vol. 36, No. 4, 1969, pp. 775–783.
14.14 D. Bushnell, "Buckling of Shells—Pitfall for Designers," *AIAA Jnl.,* Vol. 19, No. 9, 1981, pp. 1183–1226.
14.15 D. Bushnell, "Computerized Analysis of Shells—Governing Equations," *Computers & Structures,* Vol. 18, No. 3, 1984, pp. 471–536.

CHAPTER 15

15.1 S. H. Crandall, *Engineering Analysis,* McGraw-Hill, New York, 1956.
15.2 B. A. Finlayson and L. E. Scriven, "The Method of Weighted Residuals—A Review," *Applied Mechanics Reviews,* Vol. 19, No. 9, 1966, pp. 735–748.

15.3 E. D. Eason, "A Review of Least-Squares Methods for Solving Partial Differential Equations," *Int. J. Num. Meth. Engng.*, Vol. 10, No. 5, 1976, pp. 1021–1046.

15.4 W. L. Kwok, Y. K. Cheung, and C. Delcourt, "Application of Least Squares Collocation Technique in Finite Element and Finite Strip Formulation," *Int. J. Num. Meth. Engng.*, Vol. 11, No. 9, 1977, pp. 1391–1404.

15.5 M. F. N. Mohsen, "Some Details of the Galerkin Finite Element Method," *Appl. Math. Modelling,* Vol. 6, No. 3, 1982, pp. 165–170.

15.6 B. A. Finlayson, *The Method of Weighted Residuals and Variational Principles,* Academic Press, New York, 1972.

15.7 P. C. M. Lau and C. A. Brebbia, "The Cell Collocation Method in Continuum Mechanics," *Int. J. Mech. Sci.,* Vol. 20, No. 2, 1978, pp. 83–95.

15.8 R. H. Gallagher, J. A. Liggett, and S. T. K. Chan, "Finite Element Shallow Lake Circulation Analysis," *J. Hydraulics Div., Proc. ASCE,* Vol. 99, No. HY7, 1973, pp. 1083–1096.

CHAPTER 16

16.1 E. L. Wilson, K. J. Bathe and F. E. Peterson, "Finite Element Analysis of Linear and Nonlinear Heat Transfer," *Nucl. Engng. Design,* Vol. 29, No. 1, 1974, pp. 110–124.

16.2 J. F. Lyness, D. R. J. Owen, and O. C. Zienkiewicz, "The Finite Element Analysis of Engineering Systems Governed by a Non-Linear Quasi-Harmonic Equation," *Computers & Structures,* Vol. 5, No. 1, 1975, pp. 65–79.

16.3 B. Nour-Omid, "Lanczos Method for Heat Conduction Analysis," *Int. J. Num. Meth. Engng.,* Vol. 24, No. 1, 1987, pp. 251–262.

16.4 T. J. R. Hughes, "Unconditionally Stable Algorithms for Nonlinear Heat Conduction," *Comp. Meth. Appl. Mech. Engng.,* Vol. 10, No. 2, 1977, pp. 135–139.

16.5 T. J. R. Hughes, I. Levit, and J. Winget, "Element-by-Element Implicit Algorithms for Heat Conduction," *J. of Engng. Mech.,* Vol. 109, No. 2, 1983, pp. 576–585.

16.6 G. C. Everstine, "Structural Analogies for Scalar Field Problems," *Int. J. Num. Meth. Engng.,* Vol. 17, No. 3, 1981, pp. 471–476.

16.7 P. Tong and Y. C. Fung, "Slow Particulate Flow in Channels and Tubes—Application to Biomechanics," *J. Appl. Mech.,* Vol. 38, No. 4, 1971, pp. 721–728.

16.8 S. T. K. Chan, B. E. Larock, and L. R. Herrmann, "Free-Surface Ideal Fluid Flows by Finite Elements," *J. Hydraulics Div., Proc. ASCE,* Vol. 99, No. HY6, 1973, pp. 959–974.

16.9 O. C. Zienkiewicz, R. W. Lewis, and K. G. Stagg, *Numerical Methods in Offshore Engineering,* John Wiley & Sons, Chichester, England, 1978.

16.10 O. C. Zienkiewicz and P. Bettess, "Fluid-Structure Dynamic Interaction and Wave Forces. An Introduction to Numerical Treatment," *Int. J. Num. Meth. Engng.,* Vol. 13, No. 1, 1978, pp. 1–16.

16.11 R. J Astley and W. Eversman, "Finite Element Formulations for Acoustical Radiation," *J. Sound Vibration,* Vol. 88, No. 1, 1983, pp. 47–64.

16.12 A. Craggs, "A Note on the Theory and Application of a Simple Pipe Acoustic Element," *J. Sound Vibration,* Vol. 85, No. 2, 1982, pp. 292–295.

16.13 R. D. Cook, "Comment on 'Discrete Element Idealization of an Incompressible Liquid for Vibration Analysis' and 'Discrete Element Structural Theory of Fluids,' " *AIAA Jnl.,* Vol. 11, No. 5, 1973, pp. 766–767.

16.14 M. A. Hamdi, Y. Ousset, and G. Verchery, "A Displacement Method for the Analysis of Vibrations of Coupled Fluid-Structure Systems," *Int. J. Num. Meth. Engng.,* Vol. 13, No. 1, 1978, pp. 139–150.

16.15 L. G. Olson and K. J. Bathe, "A Study of Displacement-Based Fluid Finite Elements for Calculating Frequencies of Fluid and Fluid-Structure Systems," *Nucl. Engng. Design,* Vol. 76, No. 2, 1983, pp. 137–151.

CHAPTER 17

17.1 *ANSYS News,* Swanson Analysis Systems Inc., Houston, PA, Second Issue, 1982.

17.2 A. Ralston and P. Rabinowitz, *A First course in Numerical Analysis,* 2nd ed., McGraw-Hill, New York, 1978.

17.3 D. M. Gay, "Some Convergence Properties of Broyden's Method," *SIAM J. on Num. Anal.,* Vol. 16, No. 4, 1979, pp. 623–630.

17.4 *FIDAP Theoretical Manual,* Fluid Dynamics International, P.O. Box 1855, Evanston, IL 60201, 1986, pp. 7-1, 7-34.

17.5 J. E. Dennis and R. Schnabel, *Numerical Methods for Nonlinear Equations and Unconstrained Optimization,* Prentice-Hall, Englewood Cliffs, NJ, 1983.

17.6 R. E. White, *An Introduction to the Finite Element Method with Applications to Nonlinear Problems,* John Wiley & Sons, New York, 1985.

17.7 M. A. Crisfield, "An Arc-Length Method Including Line Searches and Accelerations," *Int. J. Num. Meth. Engng.,* Vol. 19, No. 9, 1983, pp. 1269–1289.

17.8 E. Riks, "Progress in Collapse Analysis," *J. Pressure Vessel Tech.,* Vol. 109, No. 1, 1987, pp. 33–41.

17.9 R. L. Webster, "On the Static Analysis of Structures with Strong Geometric Nonlinearity," *Computers & Structures,* Vol. 11, Nos. 1/2, 1980, pp. 137–145.

17.10 O. C. Zienkiewicz and R. Löhner, "Acclerated 'Relaxation' or Direct Solution? Future Prospects for FEM," *Int. J. Num. Meth. Engng.,* Vol. 21, No. 1, 1985, pp. 1–11.

17.11 S. W. Sloan, "Substepping Schemes for the Numerical Integration of Elastoplastic Stress-Strain Relations," *Int. J. Num. Meth. Engng.,* Vol. 24, No. 5, 1987, pp. 893–911.

17.12 A. Levy and B. Pifko, "On Computational Strategies for Problems Involving Plasticity and Creep," *Int. J. Num. Meth. Engng.,* Vol. 17, No. 5, 1981, pp. 747–771.

17.13 P. C. Kohnke, *ANSYS Engineering Analysis System Theoretical Manual,* Swanson Analysis Systems, Inc., Houston, PA, March 1986.

17.14 D. R. J. Owen and E. Hinton, *Finite Elements in Plasticity,* Pineridge Press Ltd., Swansea, U.K., 1980.

17.15 H. R. Evans, D. O. Peska, and A. R. Taherian, "The Analysis of the Nonlinear and Ultimate Load Behavior of Plated Structures by the Finite Element Method," *Engineering Computations,* Vol. 2, No. 4, 1985, pp. 271–284.

17.16 P. V. Marcal, "Large Deflection Analysis of Elastic-Plastic Shells of Revolution," *AIAA Jnl.,* Vol. 8, No. 9, 1970, pp. 1627–1633.

17.17 D. Bushnell, "Large Deflection Elastic-Plastic Creep Analysis of Axisymmetric Shells," in *Numerical Solution of Nonlinear Structural Problems,* R. F. Hartung, ed., ASME AMD, Vol. 38, 1973, pp. 103–138.

17.18 J. A. Stricklin, W. E. Haisler, and W. A. Von Riesemann, "Computation and Solution Procedures for Nonlinear Analysis by Combined Finite Element-Finite Difference Methods," *Computers & Structures,* Vol. 2, Nos. 5/6, 1972, pp. 955–974.

17.19 J. A. Stricklin, W. E. Haisler, and W. A. Von Riesemann, "Formulation, Computation and Solution Procedures for Material and/or Geometric Nonlinear Structural Analysis by the Finite Element Method," *Report SC-CR-72-3102,* Sandia Laboratories, Albuquerque, NM, July 1972.

17.20 L. D. Hofmeister, G. A. Greenbaum, and D. A. Evenson, "Large Strain, Elasto-Plastic Finite Element Analysis," *AIAA Jnl.,* Vol. 9, No. 7, 1971, pp. 1248–1254.

17.21 D. Bushnell, "BOSOR5—Program for Buckling of Elastic-Plastic Complex Shells of Revolution Including Large Deflections and Creep," *Computers & Structures,* Vol. 6, No. 3, 1976, pp. 221–239.

17.22 G. C. Nayak and O. C. Zienkiewicz, "Elasto-Plastic Stress Analysis. A Generalization for Various Constitutive Relations Including Strain Softening," *Int. J. Num. Meth. Engng.,* Vol. 5, No. 1, 1972, pp. 113–135.

17.23 A. Grill and K. Sorimachi, "The Thermal Loads in the Finite Element Analysis of Elasto-Plastic Stresses," *Int. J. Num. Meth. Engng.*, Vol. 14, No. 4, 1979, pp. 499–505.

17.24 T. Belytschko and T. J. R. Hughes, eds., *Computational Methods for Transient Analysis,* North Holland, Amsterdam, 1983.

17.25 M. E. Plesha, "Eigenvalue Estimation for Dynamic Contact Problems," *J. of Engng. Mech.,* Vol. 113, No. 3, 1987, pp. 457–462.

17.26 T. Belytschko, "A Survey of Numerical Methods and Computer Programs for Dynamic Structural Analysis," *Nucl. Engng. Design,* Vol. 37, No. 1, 1976, pp. 23–34.

17.27 T. Belytschko and D. F. Schoeberle, "On the Unconditional Stability of an Implicit Algorithm for Nonlinear Structural Dynamics," *J. Appl. Mech.,* Vol. 42, No. 4, 1975, pp. 865–869.

17.28 M. S. Gadala, M. A. Dokainish, and G. A. Oravas, "Formulation Methods of Geometric and Material Nonlinearity Problems," *Int. J. Num. Meth. Engng.,* Vol. 20, No. 5, 1984, pp. 887–914.

17.29 K. Mattiasson, A. Bengtsson, and A. Samuelsson, "On the Accuracy and Efficiency of Numerical Algorithms for Geometrically Nonlinear Structural Analysis," in *Finite Element Methods for Nonlinear Problems,* P. G. Bergan, K. J. Bathe, and W. Wunderlich, eds., Springer, Berlin, 1986.

17.30 T. Belytschko and B. J. Hsieh, "Non-Linear Transient Finite Element Analysis with Convected Coordinates," *Int. J. Num. Meth. Engng.,* Vol. 7, No. 3, 1973, pp. 255–271.

17.31 D. W. Murray and E. L. Wilson, "Finite-Element Large Deflection Analysis of Plates," *J. Engr. Mech. Div., Proc. ASCE,* Vol. 95, No. EM1, 1969, pp. 143–165.

17.32 R. Schmidt and D. A. DaDeppo, "A Survey of Literature on Large Deflection of Nonshallow Arches. Bibliography of Finite Deflections of Straight and Curved Beams, Rings, and Shallow Arches," *J. Industrial Math. Soc.,* Vol. 21, Part 2, 1971, pp. 91–114.

17.33 J. H. Lau, "Large Deflections of Beams with Combined Loads," *J. Engr. Mech. Div., Proc. ASCE,* Vol. 108, No. EM1, 1982, pp. 180–185.

17.34 M. D. Snyder and K. J. Bathe, "A Solution Procedure for Thermo-Elastic-Plastic-Creep Problems," *Nucl. Engng. Design,* Vol. 64, No. 1, 1981, pp. 49–80.

17.35 J. F. T. Pittman, O. C. Zienkiewicz, R. D. Wood, and J. M. Alexander, *Numerical Analysis of Forming Processes,* John Wiley & Sons, Chichester, England, 1984.

17.36 J. Padovan, S. Tovichakchaikul, and I. Zeid, "Finite Element Analysis of Steadily Moving Contact Fields," *Computers & Structures,* Vol. 18, No. 2, 1984, pp. 191–200.

17.37 C. S. Desai, M. M. Zaman, J. G. Lightner, and H. J. Siriwardane, "Thin-Layer Element For Interfaces and Joints," *Int. J. Num. An. Meth. Geomech.,* Vol. 8, No. 1, 1984, pp. 19–43.

17.38 E. L. Wilson, "Finite Elements for Foundations, Joints and Fluids," in *Finite Elements in Geomechanics,* G. Gudehus, ed., John Wiley & Sons, London, 1977, pp. 319–350.

17.39 K. Schweizerhof and E. Ramm, "Displacement Dependent Pressure Loads in Nonlinear Finite Element Analysis," *Computers & Structures,* Vol. 18, No. 6, 1984, pp. 1099–1114.

17.40 A. H. Peyrot and A. M. Goulois, "Analysis of Cable Structures," *Computers & Structures,* Vol. 10, No. 5, 1979, pp. 805–813.

17.41 H. B. Jayaraman and W. C. Knudson, "A Curved Element for the Analysis of Cable Structures," *Computers & Structures,* Vol. 14, Nos. 3–4, 1981, pp. 325–333.

17.42 M. Sathyamoorthy, "Nonlinear Vibrations of Plates—A Review," *Shock and Vibration Digest,* Vol. 15, No. 6, 1983, pp. 3–16.

CHAPTER 18

18.1 R. A. Rosanoff, J. F. Gloudeman, and S. Levy, "Numerical Conditioning of Stiffness Matrix Formulations for Frame Structures," *Proc. Second Conf. on Matrix Meth. in Struct. Mech.*, Wright-Patterson AFB, Ohio, 1968, (AFFDL-TR-68-150, Dec. 1969; AD-703-685, N.T.I.S.), pp. 1029–1060.

18.2 S. Utku and R. J. Melosh, "Solution Errors in Finite Element Analysis," *Computers & Structures,* Vol. 18, No. 3, 1984, pp. 379–393.

18.3 S. Kelsey, K. N. Lee, and C. K. Mak, "The Condition of Some Finite Element Coefficient Matrices," in *Computer Aided Engineering,* G. M. L. Gladwell, ed., University of Waterloo Press, Waterloo, Ontario, 1971.

18.4 G. Forsythe and C. B. Moler, *Computer Solution of Linear Algebraic Systems,* Prentice-Hall, Englewood Cliffs, NJ, 1967.

18.5 I. Fried, "Condition of Finite Element Matrices Generated from Nonuniform Meshes," *AIAA Jnl.,* Vol. 10, No. 2, 1972, pp. 219–221.

18.6 B. M. Irons, "Roundoff Criteria in Direct Stiffness Solutions," *AIAA Jnl.,* Vol. 6, No. 7, 1968, pp. 1308–1312.

18.7 R. H. MacNeal, ed., *The NASTRAN Theoretical Manual* (*Level 16.0*), NASA-SP-221(03), March 1976 (N79-27531, N.T.I.S.).

18.8 B. Noble, *Applied Linear Algebra,* Prentice-Hall, Englewood Cliffs, NJ, 1969.

18.9 G. Strang and G. J. Fix, *An Analysis of the Finite Element Method,* Prentice-Hall, Englewood Cliffs, NJ, 1973 (now copyrighted by Wellesley–Cambridge Press, Wellesley, MA, 1988).

18.10 I. Fried, "Accuracy and Condition of Curved (Isoparametric) Finite Elements," *J. Sound Vibration,* Vol. 31, No. 3, 1973, pp. 345–355.

18.11 S. H. Crandall, *Engineering Analysis,* McGraw-Hill, New York, 1956.

18.12 J. Robinson, "An Introduction to Hierarchial Displacement Elements and the Adaptive Technique," *Finite Elements in Analysis and Design,* Vol. 2, No. 4, 1986, pp. 377–388.

18.13 N. Kikuchi, "Adaptive Grid-Design Methods for Finite Element Analysis," *Comp. Meth. Appl. Mech. Engng.,* Vol. 55, Nos. 1–2, 1986, pp. 129–160.

18.14 G. L. Rigby and G. M. McNeice, "A Strain Energy Basis for Studies of Element Stiffness Matrices," *AIAA Jnl.,* Vol. 10, No. 11, 1972, pp. 1490–1493.

18.15 J. O. Dow, T. H. Ho, and H. D. Cabiness, "Generalized Finite Element Evaluation Procedure," *J. of Struct. Engng.,* Vol. 111, No. 6, 1984, pp. 435–452.

18.16 W. P. Doherty, E. L. Wilson, and R. L. Taylor, "Stress Analysis of Axisymmetric Solids Utilizing Higher-Order Quadrilateral Finite Elements," *Report UC-SESM-69-3,* Civil Engineering Department, University of California, Berkeley, 1969, (PB-190-321, NTIS).

18.17 J. Robinson, "A Single Element Test," *Comp. Meth. Appl. Mech. Engng.,* Vol. 7, No. 2, 1976, pp. 191–200.

18.18 I. Fried, "Influence of Poisson's Ratio on the Condition of the Finite Element Stiffness Matrix," *Int. J. Solids Structures,* Vol. 9, No. 3, 1973, pp. 323–329.

18.19 P. G. Ciarlet, *The Finite Element Method for Elliptic Problems,* North Holland, New York, 1978.

18.20 R. J. Roark and W. C. Young, *Formulas for Stress and Strain,* 5th ed., McGraw-Hill, New York, 1975.

18.21 Anon., *A Finite Element Primer,* Dept. of Trade and Industry, National Engineering Laboratory, Glasgow G75 OQU, U.K., 1986.

CHAPTER 19

19.1 G. J. DeSalvo, *ANSYS Engineering Analysis System Verification Manual,* Swanson Analysis Systems Inc., Houston, PA, 1985.

19.2 I. C. Taig, "Finite Element Analysis in Industry—Expertise or Proficiency?," in *Accuracy, Reliability, and Training in FEM Technology* (Proc. of Fourth World Congress), J. Robinson, ed., Robinson and Associates, Wimborne, England, 1984.

19.3 T. Sussman and K. J. Bathe, "Studies of Finite Element Procedures—Stress Band Plots and the Evaluation of Finite Element Meshes," *Engineering Computations,* Vol. 3, No. 3, 1986, pp. 178–191.

19.4 R. T. Haftka and J. C. Robinson, "Effect of Out-of-Planeness of Membrane Quadrilateral Finite Elements," *AIAA Jnl.,* Vol. 11, No. 5, 1973, pp. 742–744.

19.5 R. H. MacNeal, ed., *The NASTRAN Theoretical Manual* (*Level 16.0*), NASA-SP-221(03), March 1976 (N79-27531, N.T.I.S.).

19.6 K. Christensen, "Writing Easy-to-Use Programs for Computers," *Mechanical Engineering,* Sept. 1983, pp. 66–69.

19.7 H. H. Fong, "An Evaluation of Eight U.S. General Purpose Finite-Element Computer Programs," *23rd AIAA/ASME/ASCE/AHS Structures, Structural Dynamics, and Materials Conference,* 1982, pp. 145–160.

19.8 P. Naur, B. Randell, and J. N. Buxton, *Software Engineering: Concepts and Techniques,* Petrocelli-Charter, New York, 1976.

19.9 J. R. Rice, *Numerical Methods, Software, and Analysis,* McGraw-Hill, New York, 1983.

19.10 R. Evans, "Guidelines for the Selection of Analysis Software," *Mechanical Engineering,* March 1987, pp. 42–43.

19.11 J. A. Swanson, "The Development of General Purpose Software, or What is a Software Supplier?," in *Structural Mechanics Computer Programs: Surveys, Assessments, and Availability,* W. Pilkey et al., eds., University Press of Virginia, Charlottesville, 1974, pp. 687–702.

19.12 E. Schrem, "Status and Trends in Finite Element Software," in *State-of-the-Art Surveys on Finite Element Technology,* A. K. Noor and W. Pilkey, eds., ASME, New York, 1983, pp. 325–340.

19.13 K. Bell, "Some Thoughts on Design, Development and Maintenance of Engineering Software," *Advances in Engng. Software,* Vol. 8, No. 2, 1986, pp. 66–72.

19.14 Anon., *A Finite Element Primer,* Dept of Trade and Industry, National Engineering Laboratory, Glasgow G75 OQU, U.K., 1986.

19.15 T. Slot and W. J. O'Donnell, "Effective Elastic Constants for Thick Perforated Plates with Square and Triangular Penetration Patterns," *J. Eng. Industry,* Vol. 93, No. 4, 1971, pp. 935–942.

19.16 C. Meyer, ed., *Finite Element Idealization,* Am. Soc. of Civil Engrs., New York, 1987.

APPENDIX A

A.1 G. Strang, *Linear Albegra and Its Applications,* 3rd Ed., Harcourt-Brace-Jovanovich, San Diego, 1988.

APPENDIX B

B.1 I. S. Duff, "A Survey of Sparse Matrix Research," *Proceedings of the IEEE,* Vol. 65, No. 4, 1977, pp. 500–535 (cites 604 references).

B.2 W. F. Tinney and J. W. Walker, "Direct Solutions of Sparse Network Equations by Optimally Ordered Triangular Factorization," *Proceedings of the IEEE,* Vol. 55, No. 11, 1967, pp. 1801–1809.

B.3 D. P. Mondkar and G. H. Powell, "Towards Optimal In-Core Equation Solving," *Computers & Structures,* Vol. 4, No. 3, 1974, pp. 531–548.

B.4 E. L. Wilson and H. H. Dovey, "Solution or Reduction of Equilibrium Equations for Large Complex Structural Systems," *Advances in Engng. Software,* Vol. 1, No. 1, 1978, pp. 19–25.

B.5 C. A. Felippa, "Solution of Linear Equations with Skyline-Stored Symmetric Matrix," *Computers & Structures,* Vol. 5, No. 1, 1975, pp. 13–29.

B.6 E. Mendelssohn and M. Baruch, "Solution of Linear Equations with a Symmetrically Skyline-Stored Nonsymmetric Matrix," *Computers & Structures,* Vol. 18, No. 2, 1984, pp. 215–246.

APPENDIX C

C.1 D. S. Malkus and M. E. Plesha, "Zero and Negative Masses in Finite Element Vibration and Transient Analysis," *Comp. Meth. Appl. Mech. Engng.,* Vol. 59, No. 3, 1986, pp. 281–306.

C.2 D. S. Malkus and X. Qiu, "Divisor Structure of Finite Element Eigenproblems Arising from Negative and Zero Masses," *Comp. Meth. Appl. Mech. Engng.,* Vol. 66, No. 3, 1988, pp. 365–368.

C.3 D. S. Malkus, M. E. Plesha, and M-R. Liu, "Reversed Stability Conditions in Transient Finite Element Analysis," *Comp. Meth. Appl. Mech. Engng.,* Vol. 68, No. 1, 1988, pp. 97–114.

C.4 K. J. Bathe, *Finite Element Procedures in Engineering Analysis,* Prentice-Hall, Englewood Cliffs, NJ, 1982.

C.5 T. J. R. Hughes, *The Finite Element Method: Linear Static and Dynamic Finite Element Analysis,* Prentice-Hall, Englewood Cliffs, NJ, 1987.

C.6 H. Kardestuncer and D. H. Norrie, eds., *The Finite Element Handbook,* McGraw-Hill, New York, 1987.

C.7 G. Strang, *Introduction to Applied Mathematics,* Wellesley-Cambridge Press, Wellesley, MA, 1986.

C.8 J. H. Wilkinson, *The Algebraic Eigenvalue Problem,* Clarendon Press, Oxford, England, 1965.

C.9 J. R. Rice. *Numerical Methods, Software, and Analysis, IMSL Reference Edition,* McGraw-Hill, New York, 1983.

C.10 B. T. Smith et al., *Matrix Eigensystem Routines—EISPACK Guide,* Lecture Notes in Computer Science No. 6, 2nd Ed., Springer-Verlag, New York, 1976.

C.11 NAg Library Manual, Numerical Algorithms Group, Downers Grove, IL 60515.

C.12 IMSL Library Manual, International Mathematical and Statistical Library, Houston, TX 77036-5085.

INDEX